PESTICIDE MANUFACTURING

AND TOXIC MATERIALS CONTROL ENCYCLOPEDIA

Pesticide Manufacturing and Toxic Materials Control Encyclopedia

Edited by Marshall Sittig

NOYES DATA CORPORATION
Park Ridge, New Jersey, U.S.A.
1980

Published in the United States of America by
Noyes Data Corporation
Noyes Building, Park Ridge, New Jersey 07656

Library of Congress Cataloging in Publication Data

Sittig, Marshall.
Pesticide manufacturing and toxic materials control
encyclopedia.

(Chemical technology review ; 168) (Environmental
health review ; 3) (Pollution technology review ; 69)
"A thorough revision of our previous Pesticide
process encyclopedia published in 1977."
Bibliography: p.
Includes index.
1. Pesticides--Dictionaries. I. Title.
II. Series. III. Series: Environmental health
review ; no. 3. IV. Series: Pollution technology
review ; 69.
TP248.P47S52 1980   668'.65   80-19373
ISBN 0-8155-0814-X

This book is dedicated to Dr. Morton H. Seelenfreund
without whose technical expertise this book could not
have been completed.

FOREWORD

Agricultural pesticides, properly used, are essential in supplying the food requirements of the world's ever growing population. The economic and social benefits arising from the use of agricultural chemicals become obvious to any student of rural economy. Modern agricultural sufficiency is maintained only by the judicious use of pesticides. The benefits consist not only in increased yields of produce, but also in increased quality. In many cases the improvement in quality has been such, that a high percentage of the crop would not have been marketable if pesticidal chemicals had not been used.

It is appreciated that some of these materials are subject to restrictions or even bans on their use in some countries. In many less developed countries, however, concern for the prevention of spread of pest-borne disease and concern for adequate food supplies outweigh some environmental and ecological considerations.

Current attacks on the toxicity of today's pesticides notwithstanding, it has been estimated that throughout the United States about five dollars are saved for every dollar spent on the war against harmful pests.

The data on manufacturing processes in this volume is drawn primarily from the U.S. and British patent literature. The data on product toxicity, process emissions and restrictions on product use is drawn primarily from published and unpublished data obtained from the Environmental Protection Agency.

The manufacture of modern pesticides presents a technology that requires high levels of chemical and microbial sophistication. For each chemical pesticide manufacturing process the raw materials descriptions and synthetic procedures for the intermediates are ample and to the point. Reaction conditions are depicted minutely and in great detail with weights of the ingredients, catalysts, temperatures and special precautions to be observed. The actual product isolation by optimum separation techniques is described. Often this is followed by a waste and effluent diagram indicating the recommended disposal, such as landfill or waste treatment plant.

The book contains a total of 514 subject entries arranged in an alphabetical and encyclopedic fashion by common or generic name.

This is designed to be a "book with a conscience." The chemical industry is under attack in many quarters because of incidents such as the Love Canal in New York which apparently contains many pesticide manufacturing wastes.

A true perspective on the pesticide industry is hard to achieve. On one hand we have environmental activists who would have us renounce all chemicals. On the other hand we have chemical manufacturers who tell us that dioxins are naturally occurring chemicals in the environment and who, apparently, have knowingly dumped toxic chemicals in areas of public exposure.

This volume attempts to gather all available information on commercially important pesticide products with attention to:

- manufacturing processes
- exposures and allowable exposures in the factory workplace
- effluent discharges of concern to factory neighbors
- toxicity of each product to mammals
- allowable residues of these pesticides on agricultural products which is of concern to the general public whether they are 10, 100 or 1,000 miles away from the manufacturing plant.

The predecessor volume, *Pesticide Process Encyclopedia,* published in 1977 was widely accepted and is now out of print. It contained 558 subject entries and gave "recipes" for the manufacture of all 558 commercial pesticide materials.

In the preparation of this volume, over 100 out-of-date pesticides have been deleted and 72 new ones added so that 514 timely entries are now available. It is important to note that not one single entry retained from the old edition is unchanged. All have been modernized and expanded.

Beyond that, basic production process information has been expanded, wherever possible by information on pollutants and pollution control, product toxicity and residue tolerances for raw agricultural commodities, animal feeds and foodstuffs.

This book should be of interest to:

- pesticide manufacturers
- chemical raw material suppliers
- formulators
- applicators
- growers
- food processors

It should be useful to:

- environmentalists
- legislators
- lawyers
- chemists
- engineers
- industrial hygienists
- public health officials

CONTENTS

Contents

Contents

INTRODUCTION

This book is designed to present in handy, useful form available information on manufacture of pesticides and toxic materials control problems in manufacture and use.

The process information is drawn primarily from U.S. and British patents and is the result of extensive searching and cross-checking extending into early 1980.

Further all references are cited here, even to expired patents. It may indeed be that technical knowledge in an older, expired patent is still valid and is even more valuable if protection has expired and the process is now in the public domain. With more recent materials, patents are identified as a guide to possible licensing. Also disclosure of the patented process may suggest nonpatented process alternatives to the skilled chemist or chemical engineer.

The information on toxic materials control is drawn from dozens of published and unpublished documents obtained from the U.S. Environmental Protection Agency. Under the Federal Insecticide, Fungicide, and Rodenticide Act (FIFRA), EPA is responsible for the registration of all pesticide products which are distributed or sold in the United States. EPA must register a pesticide if (among other things) it performs its intended function (i.e., it is effective) without unreasonable adverse effects on the environment (i.e., it is safe). This is in addition to the traditional role of EPA in setting forth procedures for pollutant identification, control and regulation in effluents to air, water and solid wastes.

Particular acknowledgment should be made to Dr. Charles R. Worthing, editor of the latest (1979) edition of the *Pesticide Manual* published by the British Crop Protection Council. It is the most timely and authoritative volume on many aspects of pesticides and the reader is referred to that volume (referenced in almost every section of this volume) for information on:

>—physical properties
>—chemical properties
>—biological properties
>—toxicity
>—product analysis
>—product formulations

It is hoped that this volume and the *Pesticide Manual* will complement one another for a wide public of pesticide manufacturers, suppliers to the pesticide industry, regulatory agencies in countries around the world and commercial appliers of pesticide materials.

1

The volume is arranged alphabetically by the common name used in the United States. The chemical name is that used by *Chemical Abstracts*. For the convenience of the reader, the volume is indexed in a variety of ways:

—by trade name
—by raw material

This volume then covers everything from some of the older arsenicals to a variety of pesticides which are being substituted for materials like DDT. It includes several of the new pheromones or insect sex attractants which may act as pesticides themselves (by causing confusion and reducing mating efficiency) or simply as attractants (directing the selective application of conventional pesticides).

WHAT IS A PESTICIDE?

A pesticide is defined (B-15) as (1) any substance or mixture of substances intended for preventing, destroying, repelling, or mitigating any pest, or (2) any substance intended for use as a plant regulator, defoliant, or desiccant.

PESTICIDE MANUFACTURE—THE NATURE OF THE INDUSTRY

In 1976, there were 139 pesticide manufacturing plants (excluding industrial chemicals with minor pesticide uses, which are primarily products of other industries) distributed throughout 34 states in the United States (B-15). These plants generally employ unit operations and equipment similar to those used by the chemical processing industry (reaction kettles, driers, filters, etc.). The raw material common to the most pesticides is elemental chlorine, which is used directly on site in the production of chlordane, toxaphene, 2,4-D, 2,4,5-T, atrazine, captan, carbaryl, and mercuric chloride. Chlorine is also used to prepare raw materials brought in for production of DDT, aldrin, and perhaps trifluralin and alachlor. Raw materials of an unusually hazardous nature include hydrogen cyanide, carbon disulfide, various amines, and concentrated acids and caustics. The phosphorus pentasulfide (P_2S_5) used in manufacturing organophosphorus pesticides, the hexachlorocyclopentadiene (C_5Cl_6) used for cyclodiene pesticides, the phosgene used to make carbaryl, and numerous other raw materials, as well as solvents such as xylene, toluene, and similar materials, present potential health hazards.

In terms of worldwide pesticide production, there are three major producing nations—the U.S., Japan and Germany. The U.S. is first in importance with Japan a close second. German production is about one-fourth of U.S. production, so there is approximately a 50-40-10 relationship between production volumes in the three countries.

Due to increasingly stringent regulations, rising costs, insect resistance to pesticide chemicals, and greatly increased reliance on herbicides for cultivation, changes are occurring in pest control strategies; however, chemical pesticides should continue to play a significant role in pest management. Pesticide production beyond the next few years is difficult to estimate because of diverse changes in government regulations, the influence of research on new products and on application rates of products, and a variety of economic factors. Production of synthetic organic pesticides is estimated to increase by an average of 1% annually until 1985 with average annual herbicide growth of approximately 2%, average annual insecticide growth unchanged from the present level, and average annual fungicide growth of approximately 1.8%. At this predicted growth rate, total synthetic organic pesticide production in 1985 will be 8.06 x 10^5 metric tons in the U.S. alone. World markets are about three times U.S. sales and are expected to reach $10 billion by 1984 (B-28).

Opportunities for new pesticide chemicals will, of course, continue. The average life expectancy of a pesticide is estimated to be about 10 years, during which time cheaper, more effective and safer compounds are developed. Several organic pesticides, however, are exhibiting life-spans far in excess of 10 years (B-15).

The fastest growing pesticide market areas are Africa, Central and South America, Asia and the Far East and the Middle East. The rate of growth projected for Africa (B-28) is by far the highest, about 10 times the rate projected for the rest of the world.

In 1966 U.S. farmers used approximately 300 million pounds of pesticides for crop protection; by 1976 pesticide use had doubled to more than 600 million pounds. This escalation reflects the dramatic increase in herbicide use over the 10-year period, while insecticide and fungicide use has increased only slightly. In contrast, the number of new pesticides introduced each year has declined steadily from a high of about 30 in 1967 to less than 10 in 1975. Although there are more than 1,200 chemicals labeled for pesticide use and thousands of registered pesticide formulations, farmers currently use a relatively small number of major pesticides: 17 herbicides, 20 insecticides, and 6 fungicides account for more than 80% of all pesticides used (B-19).

Table 1 shows production and use statistics for 50 of the most common pesticide products (B-20).

Table 1: Fifty of the Most Common U.S. Insecticides, Herbicides, and Fungicides

Pesticide	1976 Production (10^3 t/yr AI)*	Use (%)			
		Agricultural	Home and Garden	Industrial/ Commercial	Government
. . . . Insecticides and Rodenticides					
Aldicarb	0.5–2	99	<1	<1	<1
Carbaryl	22–45	76	14	4	6
Carbofuran	2–6	99	<1	<1	<1
Chlordane	6–13	20	33	43	4
Chloropicrin	2–6	–	–	–	–
Diazinon	2–6	43	28	17	12
Dichlorvos	0.5–2	<1	15	70	15
Disulfoton	2–6	98	<1	<1	1
Endosulfan	0.5–2	85	5	5	5
Ethion	0.5–2	99	<1	<1	<1
Fensulfothion	0.5–2	99	<1	<1	<1
Heptachlor	2–6	75	<1	25	<1
Lindane and BHC	0.5–2	70	15	15	<1
Malathion	13–22	31	31	25	13
Methoxychlor	2–6	10	40	40	10
Methyl parathion	13–22	99	<1	<1	<1
Monocrotophos	2–6	99	<1	<1	<1
Naled	0.5–2	25	25	25	25
Parathion	6–13	99	<1	<1	<1
Phorate	2–6	99	<1	<1	<1
Ronnel	0.5–6	20	<1	60	20
Toxaphene	13–22	98	<1	1	<1
Weighted average		74	10	12	4
. . . . Herbicides					
Alachlor	6–13	99	<1	<1	<1
Atrazine	>45	96	1	2	1
Bromacil	2–6	13	<1	77	10
Chloramben	6–13	99	<1	<1	<1
2,4-D	6–13	76	6	12	6
Dalapon	0.5–2	15	5	15	65
Dicamba	0.5–2	40	10	40	10
Diuron	2–6	37	<1	57	6

(continued)

Table 1: (continued)

Pesticide	1976 Production (10³ t/yr AI)*	Agricultural	Home and Garden	Industrial/ Commercial	Government
. . . . .Use (%)					

. . . . .Herbicides . . . .

Pesticide	1976 Production (10³ t/yr AI)*	Agricultural	Home and Garden	Industrial/ Commercial	Government
Methanearsonics (MSMA, DSMA)	22–45	66	8	21	5
Picloram	0.5–2	35	<1	65	<1
Propachlor	6–13	99	<1	<1	<1
Propanil	2–6	99	<1	<1	<1
Silvex	0.5–2	99	<1	<1	<1
Simazine	2–6	20	<1	80	<1
2,4,5-T	2–6	60	<1	40	<1
Thiocarbamates (butylate, EPTC)	13–22	75	25	<1	<1
Trifluralin	6–13	98	1	1	<1
Weighted average		74	4	18	4

. . . . Fungicides . . . .

Pesticide	1976 Production	Agricultural	Home and Garden	Industrial/ Commercial	Government
Benomyl	2–6	70	30	<1	<1
Captafol	0.5–2	70	30	<1	<1
Captan	6–13	62	37	<1	<1
Dodine	0.5–2	70	30	<1	<1
Ferbam	0.5–2	50	50	<1	<1
Folpet	0.5–2	65	35	<1	<1
Maneb	2–6	79	21	<1	<1
Metham	2–6	75	25	<1	<1
Pentachlorophenol	13–22	<1	3	97	<1
Trichlorophenol	6–13	<1	<1	99	<1
Zineb	0.5–2	85	15	<1	<1
Weighted average		46	21	33	<1
Overall average		69	10	18	3

*Metric tons per year active ingredient ranges.

Source: Reference (B-20).

Commercial pesticide products are produced in two sequential operations: manufacturing and formulating. The manufacturing operation produces the active pesticide ingredients by chemical synthetic procedures. The active ingredients are then transformed into formulated products by diluting them with solvents, by spraying them onto clay, or by mixing them with other carriers. By their nature, formulating operations are primarily batch mixing and blending operations (B-18).

Pesticide manufacturers usually operate capital-intensive, integrated chemical synthesis plants and, with a few exceptions, produce many other chemical products in addition to pesticides. In 1972, the average pesticide manufacturing plant employed approximately 185 workers, 100 of whom were employed directly in production operations. The employees normally include chemists, engineers, managers, skilled chemical operators, pipefitters, electricians, and laborers.

Manufacturing includes pretreatment of reactants (change of size, temperature, and state), reaction, purification, and posttreatment of products (change of size and state). Raw materials for manufacture are delivered in bulk by pipeline, railroad car, barge, etc., or may be produced in-plant, often as by-products of other reactions. Manufacturers may produce several pesticide active ingredients at a single plant, and such a plant usually consists of several separate but interconnectable production areas or subplants.

A subplant contains the equipment necessary for carrying out all the unit processes and operations, such as reaction, distillation, filtration, and mixing, which are necessary to synthesize a product from raw materials. Subplant hardware may include mills, screens, hoppers, tanks, reactors, absorption columns, cooling towers, stills, filters, centrifuges, dryers, etc. Process monitoring, sampling, and analysis are performed to determine temperatures, pressures, flowrates, densities, and composition changes in order to control the chemical reactions (B-18).

POLLUTION PROBLEMS IN PESTICIDE MANUFACTURE

The pesticide manufacturing industry with its associated sources of emissions is represented schematically in Figure 1.

PESTICIDE FORMULATION

Successful use of pesticides depends to a large extent on the formulation of the product into a preparation which can be used directly or can be reformulated and brought most effectively into contact with the target pests in a safe and environmentally acceptable manner. The most important types of formulations are powders, dusts, wettable powders, emulsifiable concentrates, granules, and aerosols. The type of formulation most suitable for a specific application is dependent on a large number of factors, including physicochemical properties and biological efficiency of the active ingredient, host-pest relationship, characteristics of the available production/application equipment, and economic and environmental considerations (B-1).

Most pesticides are formulated in mixing equipment that is used only for pesticide formulations. The most important unit operations involved are dry mixing and grinding of solids, dissolving solids, and blending. Formulation systems are virtually all batch mixing operations. Formulation units may be completely enclosed within a building or may be in the open, depending primarily on the geographical location of the plant.

Individual formulation units are normally not highly sophisticated systems. Rather, they are comparatively uncomplicated batch-blending systems that are designed to meet the requirements of a given company, location, rate of production, and available equipment (B-4).

The scale on which pesticides are produced covers a broad range: Undoubtedly, many of the small firms, having only one product registration, produce only a few hundred pounds of formulated pesticides each year. At least one plant that operated in the range of 100,000,000 pounds of formulated product per year has been identified. The bulk of pesticide formulations, however, is apparently produced by independent formulators operating in the 20,000,000 to 40,000,000 pounds per year range (B-4).

Wastes from formulation of the pesticides originate from spills, "offspec" batches, equipment clean-up and mixing and grinding operations. The exact quantity of the waste is affected to a large extent by the in-plant management practices of good housekeeping. The quantity of land-destined wastes generated in the pesticide formulation industry is estimated at 0.0033 kg of waste per kilogram of production, with the waste containing about 40% active ingredients (B-1).

Pesticide formulating establishments are generally smaller than manufacturing establishments. The average formulator employs 32 workers, 18 of whom are employed in production. Employees may include engineers, chemists, operators, and laborers. There are major variations among formulators both in size and in operating practices. Seventy-one percent of all formulating establishments employ less than 20 employees. Only 6% of all establishments employ more than 100 employees, but these larger plants dominate total production. Formulators with 100 or more employees account for 56% of production, whereas formulators with less than 20 employees account for only 12.5% of all production (B-18).

Figure 1: Schematic Representation of Pesticide Manufacturing Plant Emissions

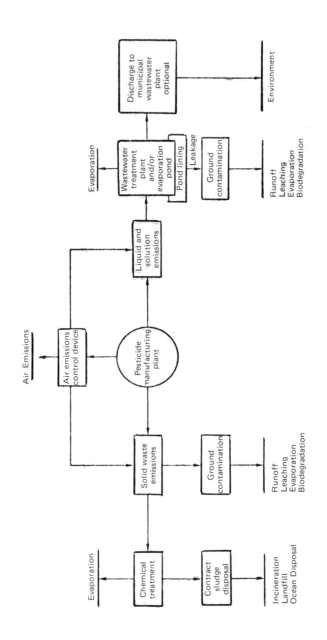

Dusts and Wettable Powders

Dusting is frequently the most inexpensive and simplest method for applying pesticides. Pesticidal dusts consist of a physical mixture of active ingredients with fine particles (3-30μ) of an inert carrier/diluent. The concentration of active ingredient may range from 0.1 to 20%. The carriers most widely used are organic flour, sulfur, silicon oxides, lime, gypsum, talc, pyrophyllite and bentonite. In making dusts, the ingredients are usually ground in a hammer, impact, vertical roller, or fluid-energy mill (B-1).

Wettable powders are solid formulations which can be dispersed in water to produce stable suspensions for more effective application by spraying. A surface active agent (1-2%) is usually used to impart wettability and suspendibility of the powder. Wettable powders are usually formulated in a manner similar to that for dust production. Wettable powders, however, are generally more concentrated in the active ingredient (15-95%). Emulsifiable concentrates are pesticide formulations which upon dilution with water yield stable emulsions suitable for spraying plants and surfaces. The concentrates are prepared by dissolving the active ingredients (15-80%) and a surface active agent (less than 5%) in a water-emulsifiable organic solvent. The surface active emulsifiers commonly used include calcium sulfonates, polyethylene and polypropylene glycols, various soaps, and salts of naphthenic acids (B-1).

Dusts and powders are manufactured by mixing the technical material with the appropriate inert carrier, and grinding this mixture to obtain the correct particle size. Mixing can be effected by a number of rotary or ribbon blender type mixers. See Figure 2.

Particulate emissions from grinding and blending processes can be most efficiently controlled by baghouse systems. Vents from feed hoppers, crushers, pulverizers, blenders, mills, and cyclones are typically routed to baghouses for product recovery. This method is preferable to the use of wet scrubbers, however even scrubber effluent can be largely eliminated by recirculation (B-4).

Granules

Granulated pesticide formulations are prepared by impregnating granular (0.2-1 mm particle size) inert carriers with liquid pesticides or their solutions. The inert carriers most commonly used include clay, vermiculite, bentonite and diatomaceous earth. The impregnation involves mixing or spraying of the inert granules with the active ingredient in a liquid form (B-1).

Granules are formulated in systems similar to the mixing sections of dust plants. The active ingredient is adsorbed onto a sized, granular carrier such as clay or a botanical material. This is accomplished in various capacity mixers that generally resemble cement mixers.

If the technical material is a liquid, it can be sprayed directly onto the granules. Solid technical material is usually melted or dissolved in a solvent in order to provide adequate dispersion on the granules. The last step in the formulation process, prior to intermediate storage before packaging, is screening to remove fines (B-4).

Liquid Formulations

For applications as aerosols, the active ingredient is commonly dissolved in water or a suitable organic solvent and then packaged with a propellant into various types of pressurized containers. The pesticides solution usually contains less than 2% active ingredient and may also contain a number of special purpose additives (e.g., for odor-masking or for enhancing the effectiveness of the active ingredient) (B-1).

A typical liquid unit is depicted in Figure 3. Technical grade pesticide is usually stored in its original shipping container in the warehouse section of the plant until it is needed. When technical material is received in bulk, however, it is transferred to holding tanks for storage.

Figure 2: Dry Formulation Unit

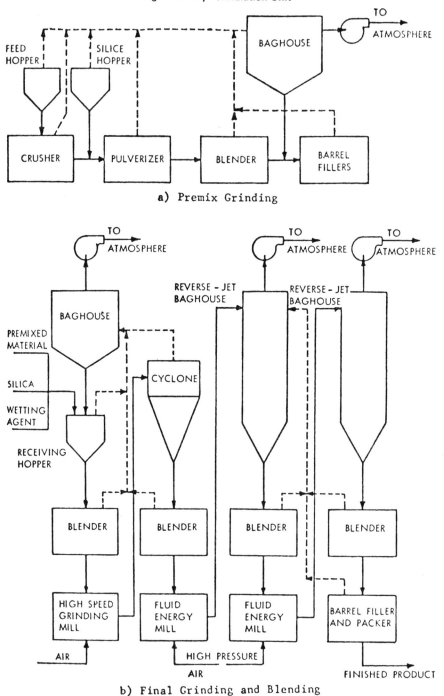

a) Premix Grinding

b) Final Grinding and Blending

Source: Reference (B-4)

Figure 3: Liquid Formulation Unit

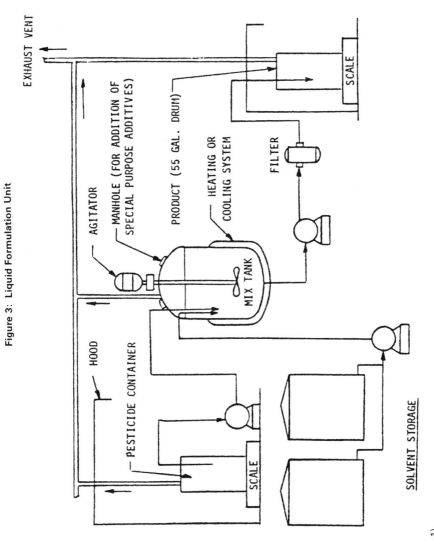

Source: Reference (B-1)

Batch-mixing tanks are frequently open-top vessels with a standard agitator. The mix tank may or may not be equipped with a heating/cooling system. When solid technical material is to be used, a melt tank is required before this material is added to the mix tank. Solvents are normally stored in bulk tanks. The quantity of appropriate solvent is either metered into the mix tank, or determined by measuring the tank level. Necessary blending agents (emulsifiers, synergists, etc.) are added directly from the mix tank. The formulated material is frequently pumped to a holding tank before being put into containers for shipment. Before being packaged, many liquid formulations must be filtered by conventional cartridge filters or equivalent polishing filters.

Air pollution control equipment used on liquid formulation units typically involves an exhaust system at all potential sources of emission. Storage and holding tanks, mix tanks, and container-filling lines are normally provided with an exhaust connection or hood to remove any vapors. The exhaust from the system normally discharges to a scrubber system or to the atmosphere (B-4).

Packaging and Storage

The last operation conducted at the formulation plant is packaging the finished pesticide into a marketable container. This is usually done in conventional filling and packaging units. Frequently, the same liquid filling line is used to fill products from several formulation units; the filling and packaging line is simply moved from one formulation unit to another. Packages of almost every size and type are used, including 1-, 2-, and 5-gallon cans, 30- and 55-gallon drums, glass bottles, bags, cartons and plastic jugs.

On-site storage, as a general rule, is minimized. The storage facility is very often a building completely separate from the actual formulation and filling operation. In almost all cases, the storage area is at least located in a part of the building separate from the formulation units in order to avoid contamination and other problems. Technical material, except for bulk shipments, is usually stored in a special section of the product storage area (B-4).

PESTICIDE USE

Table 1 gave some information as to areas of use of some important pesticides.

In the overall consideration of the toxic materials problems posed by pesticides, an important aspect is the regulatory situation regarding allowable product use and allowable residue tolerances on agricultural commodities. These aspects are noted, wherever information is available, for each of the products given detailed coverage later in this volume.

Details of pesticide application are obviously beyond the scope of this volume. The reader is referred to the British *Pest and Disease Control Handbook* (B-29) for details of application to particular crops.

TOXIC MATERIALS CONTROL

As pointed out by the author in a paper presented in New Zealand (B-30), a consideration of toxic chemicals should be holistic in nature and should consider toxic hazards:

- In the workplace
- For factory neighbors
- For consumers

ENGINEERING CONTROLS IN MANUFACTURE AND FORMULATION

Details of recommended engineering controls have been published by the National Institute for Occupational Safety and Health (B-18). Space does not permit a lengthy discussion of these matters in this volume.

SAFE WORK PRACTICES IN MANUFACTURING AND FORMULATION

The National Institute for Occupational Safety and Health (B-18) has discussed:

- Monitoring
- Personal protective clothing and equipment
- Housekeeping, hygiene and sanitation
- Emergency procedures
- Maintenance
- Support of work practices (training and supervision)

in the manufacture and formulation of pesticides. Again, space does not permit detailed discussion of such matters in this volume.

A health and safety guide for pesticide formulators has been published by the U.S. Public Health Service (B-6).

POLLUTION CONTROL IN PESTICIDE MANUFACTURE

Some information on air emissions and air pollution control in pesticide manufacture is available as a result of a study by Monsanto Research Corp. (B-15). Additional information on air, water and solid waste problems encountered in pesticide manufacture has been developed by Midwest Research Institute (B-16).

A comprehensive review of water pollution control problems and solutions in pesticide manufacture has been presented by the US EPA (B-4). This treatment was subsequently updated by the US EPA (B-13).

One option for the disposal of process wastes, product wastes and waste pesticide containers is the proper landfill. A study of selected landfills designated as pesticide disposal sites has been published (B-2). A state-of-the-art report on pesticide disposal research has been prepared by Midwest Research Institute (B-11).

RESTRICTIONS ON PESTICIDE EXPOSURE AND USE

Restrictions on pesticide exposure and use to date have taken the following forms:

- Recommendations for maximum concentrations in air—usually prepared by ACGIH and by NIOSH and finally implemented by OSHA (as described in the paragraphs which follow immediately). These are primarily applicable to the workplace.

- Criteria for allowable concentrations in water proposed by US EPA (again described in the pages which follow). These are primarily applicable to factory effluents and to watercourses which receive these effluents and which may in turn be sources of public drinking water.

- Restrictions on use set by EPA which take two forms (prohibitions on certain uses and allowable tolerances on certain crops). These primarily affect the consumer.

Concentrations in Air

The question of allowable concentrations of toxic substances in workroom air has been the traditional domain of the American Conference of Governmental Industrial Hygienists who publish annually a pocket guide to their recommendations (B-23). This pocket guide is backed up by a volume entitled *Documentation of the Threshold Limit Values* which gives the rationale for the determination of the suggested values along with literature references. The ACGIH values are usually then adopted by the U.S. Government, through the chain of the National Institute for Occupational Safety and Health (NIOSH) who recommends and the Occupational Safety and Health Administration (OSHA) who regulates such limits.

In each case where such threshold limit values are available for pesticide products, they are cited under the appropriate product in this volume.

The U.S. Environmental Protection Agency has not yet addressed itself to allowable environmental concentrations of pesticides. However, it is in the process of setting limits on airborne carcinogens and this may result in limitations for specific pesticides.

Concentrations in Water

The allowable limits on toxic substances in water is the domain of the U.S. Environmental Protection Agency, although this area has also been addressed by the National Research Council (B-22).

In this volume in the section on each product, recommended limits are cited where available from:

- A 1976 U.S. EPA publication on *Quality Criteria for Water* (B-12).

- The 1977 National Research Council publication on *Drinking Water and Health* (B-22).

- The 65 reports issued by the U.S. Environmental Protection Agency in 1979 on individual priority toxic pollutants, many of them pesticides, as summarized in a single volume (B-26).

Registration and Rebuttable Presumptions Against Registration

As indicated in the Introduction to this volume, the U.S. Environmental Protection Agency is responsible under the Federal Insecticide, Fungicide and Rodenticide Act (FIFRA) for the registration of all pesticide products which are distributed or sold in the U.S.

The EPA may register a pesticide if (among other things) it performs its intended function (i.e., it is effective) without unreasonable adverse effects on the environment (i.e., it is safe).

Now, either with a new and unknown product or with an older product already in use and about which questions as to safety have arisen, the agency has set up (as of 1975) a process of rebuttable presumption against registration (RPAR) as described by Wells (B-31) which is intended to accomplish the following:

- To weigh the risks of the use of suspect pesticides against the benefits, and

- To do this risk/benefit analysis in an open, informal process with public participation.

The RPAR process consists of the following steps as outlined by Wells (B-31):

- Investigation of risk. This involves an intense review of data which have triggered the process by qualified scientists. The registrant is notified and requested to supply additional information and a world-wide literature search is made as well as a gathering of all available information on exposure to the pesticide in question. This results in a position document which is published in the *Federal Register* along with a Notice of Presumption Against Registration or Reregistration when the data so indicate.

- The rubuttal period. A 45-day period and a 60-day allowable extension (usually requested and granted) is allowed for the registrant to reply.

  How are risks rebutted? This can be done in two ways: (1) prove that the study or studies upon which the presumption is based are not scientifically valid or (2) prove that actual exposure to the compound will not cause the effects described (B-31).

- The development of the risk/benefit analysis. The first step of the risk/benefit analysis phase is to assess the information on risk—Is the original position on risk rebutted? Information received during the rebuttal period can, of course, cause the agency to change its position on the risk of the pesticide. In the event that all the triggers have been successfully rebutted, the pesticide is returned to the registration process and the RPAR is terminated.

  Information may also be received which indicates greater hazard than originally thought to exist. New information on risk may cause an acceleration of the regulatory process if the concern is serious enough to indicate an imminent hazard. The agency is mandated to take action to protect man and the environment in any instance where information becomes available which demonstrates the existence of imminent hazard. The existence of an ongoing RPAR does not relieve the agency of this mandate. A rebuttal comment in response to the 2,4,5-T RPAR led to new information which resulted in an acceleration of the regulatory process on 2,4,5-T.

The full range of regulatory options which are available under the law range from full registration to full cancellation. Between the two ends of the regulatory spectrum are the options that deal with labeling changes, classification, and use pattern changes. The impact of each of these options on the benefits and risk of each use of the pesticide must be considered. The potential risk of alternative pesticides must be a consideration. As one moves through the regulatory spectrum from registration to cancellation the expense to the industry and user usually increases and the potential risk decreases. The final decision is reached only after examining the consequences on benefits of each option—not just a justification on the basis of risk reduction. The final decision is that which presents the best balance of risks and benefits (B-31).

In the section on each individual pesticide an indication is given as to whether an RPAR has been issued by EPA and also indications are given of whether this process (or other determinations) has resulted in cancellation, voluntary cancellation, labelling restrictions, or restrictions on use.

Table 2 gives a listing of those pesticides for which RPARs have been issued or were under consideration by EPA as of the end of 1979.

Table 2: RPAR's Issued or Under Consideration by EPA

1. Amitraz
2. Arsenic acid
3. Benomyl
4. BHC
5. Cacodylic acid
6. Calcium arsenate
7. Captan
8. Carbaryl
9. Carbon tetrachloride
10. Chlordecone
11. Chlorobenzilate
12. Diallate
13. Dibromochloropropane
14. 1,1-Dichloro-2,2-bis(4-ethylphenyl)ethane (Ethylan)
15. Dichlorvos
16. Diflubenzuron
17. Dimethoate
18. DSMA
19. Endrin
20. EPN
21. Erbon
22. Ethylene dibromide
23. Ethylene oxide
24. Fluoroacetamide
25. Lead arsenate
26. Lindane
27. Maleic hydrazide
28. Mancozeb
29. Maneb
30. Metiram
31. Monuron
32. MSMA
33. Nabam
34. Paraquat
35. Pentachlorophenol
36. Piperonyl butoxide
37. Pronamide
38. Quintozene
39. Ronnel
40. Rotenone
41. Silvex
42. Sodium arsenite
43. Sodium fluoroacetate
44. Sodium trichlorophenate
45. Strychnine
46. 2,4,5-T
47. Thiophanate-methyl
48. Toxaphene
49. Triallate
50. S,S,S-Tributyl phosphorotrithioate
51. Tributyl phosphorotrithioite (Merphos)
52. Trichlorfon
53. Trifluralin
54. Zineb

Residue Tolerances

As part of the registration of a pesticide, the U.S. Environmental Protection Agency must establish a tolerance for a pesticide residue on food, if the proposed labeling of the pesticide product bears directions for use on food or if the intended use of the pesticide results or may reasonably be expected to result, directly or indirectly, in a residue of the pesticide becoming a component of a food.

Tolerances are set for pesticides by EPA upon the submission of a tolerance petition by a person who wants the tolerance established. The establishment of a tolerance does not of itself authorize use of the particular pesticide involved on a food crop for which it has not been registered by the EPA under FIFRA.

A tolerance is the legal maximum residue of a pesticide chemical allowed to remain in or on a food. Under the Federal Food, Drug and Cosmetic Act (FFDCA), if the residue exceeds the tolerance level then the food is "adulterated" and is subject to seizure by the Food and Drug Administration (FDA). The tolerance requirements of the FFDCA are also applicable to imported food, whether or not the pesticide is registered for use in the United States.

The EPA has been charged with the responsibility for establishing tolerances for residue of pesticide chemicals in or on raw agricultural commodities, food, and feed under the Federal Food, Drug and Cosmetic Act. This responsibility was transferred from the Food and Drug Administration, Department of Health, Education, and Welfare, to the Administrator of the EPA by the President of the United States in Reorganization Plan no. 3 with the creation of the agency in 1970.

Regulations pertaining to pesticides are found in titles 40 and 21 of the *Code of Federal Regulations (CFR)*. Within title 40 of the *CFR*, tolerance regulations can be found in part 180, "Tolerances and Exemptions for Pesticide Chemicals in or on Raw Agriculture Commodities." Title 21 of the *CFR* contains two parts which are the EPA's responsibility: part 193, "Tolerances for Pesticides in Food Administered by the Environmental Protection Agency," and part 561, "Tolerances for Pesticides in Animal Feeds Administered by the Environmental Protection Agency."

The *Code of Federal Regulations* is a codification of permanent regulations published in the *Federal Register,* and it is kept up to date by the individual issues of the *Federal Register.* These two publications are used together to determine the latest version of any given rule. Because of the many regulations established with regard to pesticide tolerances, however, the substantive and numerous documents which appear quite regularly throughout the year often prove cumbersome to a constant user of this material.

The Office of Pesticide Programs has prepared a document (B-27) to serve as an index to tolerance regulations. As such, this document is considered neither legally sufficient nor binding to a pesticide user, registrant, manufacturer, distributor, or formulator.

The pertinent sections of this document have been extracted and appended to the sections of this volume dealing with individual pesticides.

Tolerances and exemptions established for pesticide chemicals in or on the general catagory of raw agricultural commodities listed in column A apply to the corresponding specific raw agricultural commodities listed in column B. However, a tolerance or exemption for a specific commodity in column B does not apply to the general category in column A.

A	B
Bananas	Bananas, plaintains
Beans	Green beans, lima beans, navy beans, red kidney beans, snap beans, cowpeas, black-eyed peas
Celery	Anise (fresh leaves and stalks only), celery
Cherries	Sour cherries, sweet cherries

(continued)

A	B
Citrus fruits	Grapefruit, lemons, limes, oranges, tangelos, tangerines, citrons, kumquats and hybrids of these
Corn grain	Includes field corn, sweet corn, and popcorn
Cucurbits	Cantaloupes, casabas, cranshaws, cucumbers, honey balls, honeydew melons, melon hydrids, muskmelons, Persian melons, pumpkins, summer squash, watermelons and hybrids, winter squash
Forage grasses	Any grasses (either green or cured) that will be fed to or grazed by livestock, all pasture and range grasses, all grasses grown for hay or silage, corn grown for fodder or silage, sorghum grown for hay or silage, small grains grown for hay grazing or silage
Forage legumes	Any crop belonging to the family leguminosae that is grown for forage (hay, grazing, silage, etc.) alfalfa, beans (for forage), clovers, cowpeas (for forage), cowpea hay, lespedeza, lupines, peanuts (for forage), pea hay, trefoil, velvet beans (for forage) soybeans (for forage), soybean hay, peas (for forage), pea vine hay.
Fruiting vegetables	Eggplants, peppers, pimentos, tomatoes
Grain crops	Any crop belonging to the family graminae that produces mature seed is used for food and feed, barley, wheat, corn (field corn, sweet corn, popcorn), millet, milo, oats, rice, rye, sorghum (grain), wheat
Leafy vegetable	Anise (fresh leaf and stock only), beet greens (tops), broccoli, brussels sprouts, cabbage, cauliflower, celery, Chinese cabbage, collards, dandelion, endive, escarole, fennel, kale, kohlrabi, lettuce, mustard greens, parsely, rhubarb, salsify tops, spinach, sugar beet tops, Swiss chard, turnip greens (tops), watercress.
Melons	Cantaloupes, casabas, cranshaws, honeyballs, honeydew melons, muskmelons, Persian melons and hybrids of these, watermelons and their hybrids
Nuts	Almonds, Brazil nuts, bush nuts, butternuts, cashews, chestnuts, filberts, hazelnuts, hickory nuts, macadamia nuts, pecans, walnuts
Onions	Dry bulb onions, green onions, garlic, leeks, shallots, spring onions
Onions (dry bulb only)	Garlic, onions (dry bulb only)
Peppers	All varieties of peppers including pimentos and bell, hot, and sweet peppers
Pome fruits	Apples, crabapples, pears, quinces
Poultry	Chickens, ducks, geese, guinea, pheasant, pigeons, quail, turkeys
Root crops-vegetables	Beets, carrots, chicory, garlic, green onions, parsnips, potatoes, radishes, rutabagas, salsify, shallots, spring onions, sugar beets, sweet potatoes, turnips, yams
Seed and pod vegetables (dry or succulent)	Black-eyed peas, cowpeas, dill, edible soybeans, field beans, field peas, garden peas, green beens, kidney beans, lima beans, navy beans, okra, peas, pole beans, snap beans, string beans, wax beans, other beans and peas, lentils
Small fruits	Blackberries, blueberries, boysenberries, cranberries, currants, dewberries, elderberries, gooseberries, grapes, huckleberries, loganberries, raspberries, strawberries, youngberries
Stone fruit	Apricots, cherries (sour and sweet) damsons, nectarines, pawpaws, peaches, plums, prunes
Tangerines	Tangelos, tangerines
Turnip tops or greens	Broccoli raab (raab, raab salad), Hanover salad, turnip tops (turnip greens)

ENVIRONMENTALLY ACCEPTABLE ALTERNATIVES TO CONVENTIONAL PESTICIDE USE

Some ways of avoiding toxic chemicals problems in manufacture, formulation, application and use of conventional pesticides are to use alternative materials or alternative application techniques which minimize toxic chemicals impact.

BIODEGRADABLE PESTICIDES

It was early recognized that the organochlorine pesticides were, in many cases, both toxic to mammals and persistent in the environment.

Thus, an early effort was to provide substitute pesticides. One tabulation of the results of an EPA program to develop such substitutes is shown in Table 3.

Table 3: Proposed Substitute Insecticides
(Registrations Cancelled)

Proposed Substitutes	DDT	Aldrin	Dieldrin	Chlordane	Heptachlor
Phorate	X	X	—	X	X
Demeton	X	—	—	—	—
Methyl parathion	X	—	—	—	—
Parathion	X	X	—	X	X
Malathion	X	—	—	—	—
Guthion	X	—	—	—	—
Aldicarb	X	—	—	—	—
Azodrin	X	—	—	—	—
Diazinon	X	X	—	X	X
Dimethoate	X	—	—	—	—
Fenthion	X	—	—	—	—
Methomyl	X	—	—	—	—
Crotoxyphos	X	—	—	—	—
Chlorpyrifos	X	X	—	X	X
Buxten	—	X	—	X	X
Carbonfuran	X	X	—	X	X
Counter	—	X	—	X	X
Dasanite	X	X	—	X	X

(continued)

Table 3: (continued)

Proposed Substitutes	DDT	Aldrin	Dieldrin	Chlordane	Heptachlor
Disulfoton	X	X	—	X	X
Dyfonate	X	X	—	X	X
Landrin	—	X	—	X	X
Trichlorfon	X	X	—	X	X
Dacthal	—	—	—	X	—
Aspon	X	X	X	X	—
Siduron	—	—	—	X	—
Ethion	X	X	X	X	—
Propoxur	X	X	X	X	X
Acephate	X	—	X	X	—
Methoxychlor	X	—	X	X	—

Source: Reference (B-15)

The initial broad trend was from the persistent organochlorine compounds to the less persistent organophosphorus compounds.

As toxicity information develops it is necessary to constantly reasses the substitute program.

Instead of quite different chemical compounds as substitutes for toxic and persistent pesticides like DDT, another approach that has been suggested is to produce biodegradable analogs of DDT.

Thus, *R.L. Metcalf, I. Kapoor and A. Hirwe; U.S. Patent 3,787,505; January 22, 1974; assigned to University of Illinois Foundation* have found from metabolic studies in insects, mice, and in a model ecosystem with several food chains that certain asymmetrical DDT analogues with substituent groups readily attacked by multifunction oxidase (MFO) enzymes are substantially biodegradable and do not appear to be stored readily in animal tissues or concentrated in food chains. Insecticidal activity studies involving flies and mosquitoes have further indicated that these DDT analogues are persistent, biodegradable insecticides, of very low mammalian toxicity. These compositions are effective and persistent insecticides in inanimate situations; yet, when such compositions are absorbed into living organisms they contain one or more points readily susceptible for attack by the MFO enzymes, promoting rapid detoxication of the insecticide. Such compositions have many advantages as safe, relatively stable, and potentially inexpensive residual insecticides.

The present asymmetrical DDT analogues contain at least one p-substituent group such as methoxy ($-OCH_3$), methio ($-SCH_3$) or methyl ($-CH_3$) which in the presence of MFO enzymes acts as a substrate for MFO enzymes and thereby biodegrades or metabolically converts such analogues into environmentally acceptable products. The compositions also contain a second, different p-substituent group which at least in part contributes to the insecticidal activity and/or the relatively low mammalian toxicity exhibited by the compositions and may also contribute to their biodegradability.

The asymmetrical DDT analogues are highly insecticidal, yet biodegradable when contacted by MFO enzymes, and are thus of low toxicity to mammals. They have the following formula:

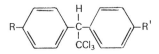

where R and R' are different and R is selected from the group consisting of CH_3, $-OCH_3$, $-OC_2H_5$, $-OC_3H_7$ and R' is selected from the group consisting of $-SCH_3$ and $-CH_3$.

The preparation or synthesis of the asymmetrical DDT analogues can be accomplished by Bayer or Friedel-Crafts condensation reactions between chloral and the appropriate substituted benzenes, for example, toluene, thioanisole, alkoxybenzenes, etc.

A later development in diphenyl methane biodegradable pesticides has been reported by *R.L. Metcalf and J.R. Coats; U.S. Patent 4,003,950; January 18, 1977; assigned to University of Illinois Foundation.*

These are biodegradable insecticides having the formula:

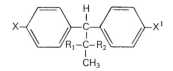

where R_1 is H or Cl, R_2 is CH_3 or Cl, X and X^1 are the same or different and each is selected from the group consisting of C_1 to C_3 alkyl groups and C_1 to C_1 alkoxy groups.

MECHANICAL CONTROL OF TOXIC PESTICIDES

A variety of techniques have been proposed and developed for ameliorating the effect of high concentrations of what may be fairly toxic pesticides. These techniques may involve:

1. Controlled release of the active material from a bound form over a period of time.

2. Ultra-low volume application in which special spraying compositions and techniques are employed to minimize exposure to toxic chemicals.

3. Spraying of binary agents (akin to binary nerve gases) which combine to give the toxic pesticide at the point where it meets the pests.

Controlled Release Pesticides

Controlled release pesticide technology can be defined as the release of a pest control agent from a polymeric carrier, binder, absorbent, or encapsulant at a slow continuous effective rate into the environment. Output is prolonged (from weeks to perhaps years) in contrast to the short-time effective interval common to many conventional pesticide formulations.

Controlled release pesticide formulations offer the possibility of achieving acceptable or improved pest control by releasing low levels of pesticide (which may already be EPA-registered in conventional formulations) at a fairly uniform rate ovar a prolonged time interval. In this way, satisfactory protection from pests is attained for longer periods of time with fewer applications and overall smaller quantities of the slow-release pesticides (compared to conventional ones). Thus, the slow-release pesticides are utilized more efficiently and with greater safety (considering that lower levels and smaller quantities of pesticide are needed) to nontarget organisms, including man, and the environment (B-24).

To illustrate this point, the potent insecticide methyl parathion is lethal to humans at a relatively low level of exposure, so that workers cannot enter sprayed areas for at least one day and then only with elaborate protective garmentation. The same insecticide encapsulated, known as Pencap M, can be handled with less precautions, and workers can enter sprayed areas 24 hours after treatment without protective clothing. In addition, Pencap M requires fewer applications (compared to conventional formulations of methyl parathion), smaller amounts of insecticide, and is less harmful to wildlife.

Aside from the economic advantages of reduced amounts of pesticide required for pest control and the lesser number of treatments, some (but not all) controlled-release formulations have demonstrated another unexpected advantage in trials. Extremely low concentrations of pesticides continuously applied over weeks or months have destroyed target pests (generally organisms living in water) without harm to existing biota.

Ultra-Low Volume Application

In endosulfan application, for example, as described by *H. Frensch, K. Albrecht and N. Taubel; U.S. Patent 3,996,375; December 7, 1976; assigned to Hoechst AG, Germany*, the active agent is used in the form of emulsifiable concentrates and wettable powders which, immediately prior to application, are diluted with water and then applied in the form of the spray liquors so obtained by means of spraying apparatus. Generally, from 400 to 600 liters of spray liquor per hectare are employed. However, this requires large amounts of water which, especially in the tropics, often are not available. Furthermore, at elevated temperature, above all in the case of application by plane, the water content of the spray mist quickly evaporates and the active substance is frequently driven off thus resulting in uneven coverage of the treated area. It was therefore desirable to develop a formulation that is essentially free from water, which is in the form of an ultra low volume (ULV) concentrate.

ULV application techniques are already known. In these processes liquid active substances or solutions of active substances in application amounts of from 1 to 5 liters per hectare are spread by means of special nozzles. In order to attain a good and homogeneous spreading at such small amounts, the liquids have to be applied in very small droplets of from 75 to 120 microns in diameter, that is, in a far finer distribution than in the conventional spraying of emulsifiable concentrates or wettable powders. ULV formulations therefore have to contain high-boiling solvents in order to prevent quick evaporation and drift-off of the active substance and to prevent crystallization or agglomeration already in the atomizer nozzle. The flash point (closed) of these solvents should therefore be above 55°C.

On the other hand, the formulation should not be too viscous, in order to ensure a good and homogeneous atomization of the composition. Although it is possible to spread products having a viscosity of up to 49 centipoises (cp), a more uniform distribution of the droplets is achieved using formulations the viscosity of which is below 20 cp. For the preparation of ULV formulations of solid substances, solvents are therefore required which possess solubility, low volatility, low viscosity and, above all, a good plant compatibility.

However, the usual paraffinic hydrocarbons and vegetable oils which answer these requirements are unsuitable because of the low solubility (5 to 8%) of endosulfan in these solvents. On the other hand the aromatic solvents generally used in emulsifiable concentrates and which are relatively well tolerated by plants, such as xylene, methylethylketone or cyclohexanone have excessive volatility. Finally, high-boiling aromatic hydrocarbon fractions and high-boiling ketones such as isophorone are more or less phytotoxic; the toxicity degree rising with increasing boiling point.

Furthermore, when testing solutions containing such high-boiling aromatic solvents of for instance, N-methyl-pyrrolidone, and an endosulfan content of about 25 weight percent, it turned out that the active substance separated rapidly in the form of coarse crystals from the atomized droplets, thus reducing the insecticidal effect. It is therefore apparent that usual technical solvents commonly used in the preparations of the pesticide formulations do not produce useful ULV formulations of endosulfan.

It has now been found that these drawbacks can be overcome and a stable, technically applicable ULV formulation of endosulfan is obtained by combining (a) 15 to 35 weight percent of endosulfan; (b) 60 to 84.5 weight percent of a solvent mixture consisting of 15 to 85 weight percent of a liquid ester of (C_1 to C_{12}) monoalcohols with (C_2 to C_{10}) carboxylic acids, the esters containing at least 8 and, in the case of esters of a monovalent acid a maximum of 12, in the case of esters of a bivalent acid a maximum of 32 carbon atoms, and of 85 to 15 weight percent of aromatic hydrocarbons having boiling ranges of from 168° to 250°C; and (c) 0.5 to 5 weight percent of an epoxide as stabilizer.

The ULV formulations described impart an excellent insecticidal effect to endosulfan as well as a long-duration activity, so that considerable amounts of the active substance are saved as compared to the known emulsion concentrates and wettable powder formulations.

Binary Agents as Pesticides

A technique described by *R.T. Kemp, Jr.; U.S. Patent 4,004,000; January 18, 1977; assigned to U.S. Secretary of the Navy* involves combining at least two relatively stable reactants to form a nonstable toxic principle, the reactants being combined just prior to actual use of the principle.

With chemicals, there is a limit to the degree of nonpersistence that can be achieved, because a substantial shelf-life is required for the practical distribution and use of chemical pesticides. A pesticidal system in which the product is synthesized at the site of application, however, has no shelf-life constraint and if nontoxic reactants are employed in such systems, they are capable of being both safe and nonpolluting.

One such combination proposed is trimethyl phosphite plus chloral (which yield DDVP). Various methods of combining the active reactants to form the toxic insecticidal principle may be employed. For example, a spray device may be used which has two spray nozzles and is constructed so that the two streams from the nozzles intersect. Or two separate spray devices may be used in time series (i.e., one after another) to spray the same area. Or the active reactants can be placed in the same container and mixed just prior to the spraying.

The greater effectiveness of the binary spray system over the conventional pesticide may be attributable to the highly reactive character of the nascent product, i.e., it is to be expected that the insecticidal principle will have more high-energy molecules just after its formation. The binary systems are safer because nontoxic ingredients are used to form toxic products which are intrinsically unstable and therefore self-limiting with respect to toxicity.

The binary reaction can be limited to liquid-phase reactions employing spray droplets too heavy to drift, thus preventing contamination of areas adjacent to the target. A further advantage of the self-limiting character of the binary product is that it provides the means to achieve the ultimate degree of nonpersistence and ecological compatibility that can be obtained with chemicals.

BIOLOGICAL CONTROLS

Increasing interest has been expressed in the use of nonchemical pest control methods and, in particular, in biological methods. Advocates of these methods note that biological controls have long existed in nature. They point to the successful development of rust-resistant wheat varieties in this century which greatly reduced the damage from such pests; the introduction in 1888 of the vedalia (a small lady beetle) to control the cottony-cushion scale in California citrus crops; the commercialization of a pathogenic agent, *Bacillus thuringiensis*, for use on larvae that chew on certain plants; the release of radiation-sterilized screwworms to suppress natural populations; the recent identification of substances emitted by some insects to attract mates; and the strong desire by many persons to eat "organically grown" foods. These developments have led many people to speak optimistically of a "new generation" of pesticides to replace the chemicals now in use (B-8).

Efforts by entomologists to spur government interest in the use of parasites and predators for control of agricultural pests began over 100 years ago in the United States. The development of methods of biological control—the term was first used in 1920—has been accompanied by an undercurrent of conflicts between federal, state, academic, and private sector parties over allocation of research funds, policy issues, and jurisdictional authorities and responsibilities. The introduction of the lady beetle in 1888 involved a most interesting clash of professional and personal conflicts, as well as the first comparison of the relative merits of biological methods and chemical methods, and contributed to a degree to a polarization that continues to the present.

The rate and extent to which the biological methods might substitute for chemicals are, however, quite uncertain at present (B-8). The overall benefits and costs of the widespread adoption of biological pest control methods are unknown. In addition, the risks of unexpected adverse consequences of the use of biologicals have not been systematically evaluated. Hence, "preliminary comprehensive technology assessment" of the adoption of biological substitutes, including the forecasting of the possible courses of development of each of several biological control strategies and the probable consequences or impacts of those developments was undertaken by Midwest Research Institute in 1977 (B-8).

As noted by *L. Jurd; U.S. Patent 4,049,722; September 20, 1977; assigned to the U.S. Secretary of Agriculture,* effective means of biologically controlling insect populations encompasses two distinct concepts. First, a chemical may be administered to the insects, which then become sexually sterile. The sexually sterilized insects mate with fertile insects, but the eggs laid do not yield any progeny. The result is a decrease in population of the insects.

Another method of biological control involves administering a chemical to the insects, with the result that the female species do not posit (lay) any eggs. Consequently, no progeny are produced and a decrease in insect population is thus attained. Although the above methods of biological control encompass two distinct ideas, the chemical compounds which produce the above effects may be termed generally as antiprocreants, that is, compounds which act either as chemosterilants or as oviposition inhibitors and prevent procreation of the species.

The biological method of insect control offers many advantages over the usual method of applying an insecticide to insects or their habitat. For example, it avoids harm to humans, animals, and useful insects (bees, for instance).

In controlling insects by sterilization or oviposition inhibition, a suitable compound is administered to a group of insects and these are then released in a locus where insects of the same species are present. As noted above, the treated insects mate with fertile ones but without producing progeny so that the overall population is decreased.

More specifically, L. Jurd states that some polybutyl-2-cinnamylphenols, particularly 4,6-di-tert-butyl-2-cinnamylphenol, are useful for insect control and especially as insect chemosterilants and oviposition inhibitors.

Pheromones

Pheromones are chemical substances secreted by one insect and received by another insect in which they produce a specific reaction. They are agents of a chemical communication system and as such they may serve to attract and excite members of the opposite sex or simply to perform other functions such as marking a trail leading to a food source.

The pheromones may be used in various ways:

- To lure insects into traps.

- To overpower the natural odor emitted by females, preventing the confused males from locating and mating with the females, which then lay infertile eggs and die.

The manufacture of several of these pheromones are described in this volume. They include:

Codlelure
Methoprene
Trimedlure

INTEGRATED PEST MANAGEMENT

Many agricultural experts, as well as industry or government spokesmen, have contended that the chemicals are economically indispensible at present, and that alternatives are not now available. Consequently, some authorities speak of "integrated pest management" programs in which the best of the biological methods would be combined with cultural and genetic approaches and backed up by chemical and physical methods as needed to control the pests. The anticipated result would be a net benefit to the environment and human health. [IPM methods are still far short of the objectives of some environmentalist groups and organic gardening enthusiasts (B-8).]

The concept of integrated pest management dates back to the 1950s and has come to mean the optimum application of all techniques to realize economical control with minimum ill effects on nontarget species, the food chain, and the environment. Integrated pest management techniques, many of which may grow in importance, are listed in Table 4 (B-15).

Table 4: Integrated Pest Management Options

Traditional chemical manipulation
  Herbicides
  Insecticides
  Fungicides

Behavioral manipulations
  Juvenile hormones .
  Pheromones

Environmental manipulations
  Plant spacing
  Crop rotation
  Water management
  Soil preparation (plowing, cultivation, destruction of crop residues)
  Sanitation
  Fertilization
  Mixed plant culture
  Planting time
  Land use management
  Regional crop planting
  Timing of flooding
  Vegetation clearings
  Metal guards
  Isolation, quarantine

Genetic manipulations
  Resistant varieties (genetic resistance)
  Lethal genes
  Male producing genes
  Generation control genes
  Cytoplasmic incompatibility
  Chromosome translocations
  Gene replacement
  Sterile male

(continued)

Table 4: (continued)

Ecology manipulations
    Natural enemies
        Parasite
        Predator
        Pathogen
    Competition
    Eradication of alternate hosts

Source: Reference (B-15)

Behavioral manipulation refers to a technique designed to affect the communication systems of pests by sending special signals or altering existing signals. Insects are very sensitive to odors, and very small quantities of insect attractants can be used to lure them to traps. The attractants, which are usually food- or sex-based, may be effective for distances up to 1.6 km. They are usually highly specific, attracting only a few closely related species, and then often only males. Attractants may be used either to monitor the presence of a pest or in bait traps, to which large numbers of males would be drawn and destroyed.

Another promising technique of pest management is the use of juvenile hormones. Also called insect growth regulators, these compounds do not kill, but instead interfere with the insect's normal development. Altosid SR-10, a juvenile hormone recently approved by the EPA for commercial use, is imbedded in a slow-release matrix of 1-μm-diameter polymer spheres. The water suspension has shown low toxicity to nontarget species; it degrades rapidly, and small volumes are required.

Genetic manipulation is another technique of integrated pest control that is receiving increased attention. It takes two basic forms: development of crop varieties genetically resistant to pests, and use of genetically modified members of pest species (such as sterile males) to interfere with reproduction or other functions (as noted above under "Biological Controls"). Chromosome modification, usually leading to pests that produce no offspring or offspring with lethal genes, is under intensive study. Genetic techniques will probably be used only on specialized groups, in limited areas, and for special problems. Massive use of these techniques is not believed to be close at hand (B-15).

THE A TO Z OF INDIVIDUAL PESTICIDES

In the pages which follow, individual pesticides are listed alphabetically by generic name (the generic name used in the U.S.A. where multiple generic names are used—with an indication of the alternative generic names used in other countries).

The function of the compound is indicated briefly. The *Chemical Abstracts (CA)* version of the chemical name is given. The formula of the compound is then shown. Trade names and compound identification numbers used by the manufacturers are then given. The paragraphs on manufacture are drawn in most cases from the referenced U.S. or British patents.

Under toxic materials control, the following outline is used to present the information that is available.

1. Process Wastes and Their Control
   a. Air
   b. Water
   c. Solid waste
   d. Product wastes

2. Toxicity. Information is quoted where extensive investigations have been made on the toxicity of various materials. In most cases, however the LD_{50} value for rats is quoted as a simple index of mammalian toxicity.

Classification	Oral LD_{50}-rats (mg/kg)
Insignificantly toxic	>5,000
Slightly toxic	500-5,000
Moderately toxic	50-499
Highly toxic	5-49
Extremely toxic	<5

   The rating scale used for the acute mammalian toxicity of pesticides was one that is recognized by various authors on the subject (B-16).

3. Allowable Limits on Exposure and Use
   a. Air. This usually involves the recommendation by

the American Conference of Governmental Industrial Hygienists (ACGIH) on threshold limit values in the factory workplace.

b. Water. These involve criteria set by the U.S. Environmental Protection Agency.

c. Product Use. These involve two types of actions in general:

    (1) Qualitative restrictions on use—rebuttable presumptions against registration, label requirements or cancellation.

    (2) Quantitative restrictions on use—allowable tolerance levels on various individual raw agricultural commodities, animal feeds and human foods.

The following glossary discusses the terms used in tolerances quoted in the sections on individual pesticides.

GLOSSARY

appli = application
c-i met = cholinesterase-inhibiting metabolites
carb = carbamates
conc = concentrated
ct.f = cattle feed
epwrr = edible portion with rind removed
exc = except
gt.f = goat feed
hg.f = hog feed
hr.f = horse feed
i (in ppm column) = interim tolerance
inc = including
k+cwhr = kernel plus cob with husk removed
ls.f = livestock feed
mbyp = meat by-products
min = minimum
N (in ppm column) = negligible residues
nmt = not more than
non-per bag/pkgd rac = non-perishable packaged or bagged raw agricultural commodity
nonmed = nonmedicated
ppm = part(s) per million
post-h = postharvest application
pre-h = preharvest application
pre-s = preslaughter application
prods = products

"40 *CFR* Reference" defines the section in the *U.S. Code of Federal Regulations* under Title 40 entitled "Protection of Environment" when tolerances in or on raw agricultural commodities are listed.

Title 21 of the *Code of Federal Regulations* entitled "Food and Drugs" contains the regulations on pesticides in foods and in animal feeds.

A

ACEPHATE

Function: Insecticide (effective against aphids and thrips) (1)(2)

Chemical Name: O,S-dimethyl acetyl phosphoroamidothioate

Formula:

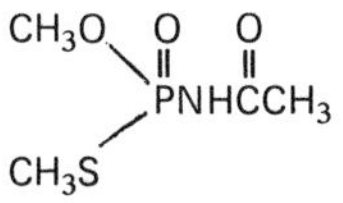

$$CH_3O \diagdown \quad \underset{\parallel}{O} \quad \underset{\parallel}{O}$$
$$\diagup PNHCCH_3$$
$$CH_3S$$

Trade Names: Ortho 12420 (Chevron Chemical)
Orthene® (Chevron Chemical)
Ortran® (Hokko Chemical)
Ortran® (Takeda Chemical Industries)

Manufacture (3)(4)(5)

Acephate may be produced by the acetylation of O,O-dimethylphosphoroamidothioate, (which is prepared from O,O-dimethylphosphorothioic acid chloride and ammonia). 14.1 g (0.1 mol) of O-methyl-S-methylphosphoroamidothioate was dissolved in 100 ml benzene in a flask. 7.85 (0.1 mol) acetyl chloride was added to this solution. This mixture was brought to reflux—HCl being evolved at that point. This mixture was then stirred overnight at ambient temperature. Supernatant liquid was decanted and the solvent was stripped off at 30° to 40°C, 12 mm Hg. An oil remained which solidified on standing. This solid was filtered and washed with ether to yield 7 g of impure O-methyl-S-methyl-N-acetylphosphoroamidothioate. This material melted at 64° to 68°C (3).

Toxicity

Acephate is slightly toxic with an LD_{50} value of 900 mg/kg for rats.

Allowable Limits on Exposure and Use

Product Use: The tolerances set by the US EPA for acephate in or on raw agricultural commodities are as follows:

	40 *CFR* Reference	Parts per Million
Beans, dry; nmt 1 ppm c-i met	180.108	3.0
Beans, succulent; nmt 1 ppm c-i met	180.108	3.0
Cattle, fat	180.108	0.1
Cattle, mbyp	180.108	0.1
Cattle, meat	180.108	0.1
Celery	180.108	10.0
Cotton, seed	180.108	2.0
Eggs	180.108	0.1
Goats, fat	180.108	0.1
Goats, mbyp	180.108	0.1
Goats, meat	180.108	0.1
Hogs, fat	180.108	0.1
Hogs, mbyp	180.108	0.1
Hogs, meat	180.108	0.1
Horses, fat	180.108	0.1
Horses, mbyp	180.108	0.1
Horses, meat	180.108	0.1
Lettuce, head	180.108	10.0
Milk	180.108	0.1
Peppers, bell; ≤ 1 ppm c-i met	180.108	4.0
Poultry, fat	180.108	0.1
Poultry, mbyp	180.108	0.1
Poultry, meat	180.108	0.1
Sheep, fat	180.108	0.1
Sheep, mbyp	180.108	0.1
Sheep, meat	180.108	0.1
Soybeans	180.108	1.0

The tolerances set by the US EPA for acephate in animal feeds are as follows (the *CFR* reference is to Title 21):

	CFR Reference	Parts per Million
Cottonseed hulls	561.20	4.0
Cottonseed meal	561.20	8.0
Soybean meal	561.20	4.0

References

(1) Worthing, C.R., *Pesticide Manual,* 6th ed., p. 1, British Crop Protection Council (1979).
(2) Spencer, E.Y., *Guide to the Chemicals Used in Crop Protection,* 6th ed., p. 1, London, Ontario, Agriculture Canada (January 1973).
(3) Magee, P.L., U.S. Patent 3,716,600, February 13, 1973, assigned to Chevron Research Co.
(4) Platt, J.L., Jr., U.S. Patent 3,732,344, May 8, 1973, assigned to Chevron Research Co.
(5) Magee, P.S., U.S. Patent 3,845,172, October 29, 1974, assigned to Chevron Research Co.

ACROLEIN

Function: Aquatic herbicide (1)(2)(8) and bactericide (9)

Chemical Name: 2-propenal

Formula: $CH_2=CH-CHO$

Trade Names: Aqualin® (Shell)
Magnacide H®

Manufacture

    (A) By oxidation of acetaldehyde (3)
    (B) By oxidation of propylene in liquid phase (4)
    (C) By oxidation of propylene in vapor phase (5)(6)
    (D) By oxidation of allyl alcohol (7)

Process Wastes and Their Control

The high toxicity of waste quantities of this water-soluble liquid can be greatly decreased by treatment with water and an excess of 10% sodium bisulfite. All users of this hazardous product should be familiar with manufacturers' recommendations. (B-3).

Toxicity

Acrolein is highly toxic with an LD_{50} of 46 mg/kg for rats. Acrolein is highly volatile and apparently does not persist for extended periods in an aqueous environment. Only limited acute- and subchronic-toxicity data are available. In view of the relative paucity of data on the mutagenicity, carcinogenicity, teratogenicity, and long-term oral toxicity of acrolein, estimates of the effects of chronic oral exposure at low levels cannot be made with any confidence. It is recommended that studies to produce such information be conducted before limits in drinking water can be established (B-22).

Acrolein has been shown to produce a great variety of disorders in mammalian animals and man. However, it has not been shown to be a teratogen and only a mild to weak mutagen, if one at all, depending on the test system employed. Though it has been suspected as a carcinogen or cytotoxigen, information does not definitively produce evidence of confirmation (B-26).

Freshwater acute toxicity values as low as 61 μg/l have been reported. A chronic fish value of 21.8 μg/l has been demonstrated. Acrolein has been found to bioconcentrate 344 times in a freshwater fish. Saltwater acute toxicity in one fish species was found to be 240 μg/l. No bioconcentration or chronic data are available for marine species (B-26).

Allowable Limits on Exposure and Use

Air: The threshold limit value for acrolein in air is 0.25 mg/m³ or 0.10 ppm as of 1979. The tentative value for the short term exposure limit (STEL) is 0.8 mg/m³ or 0.3 ppm (B-23).

In the Soviet Union, the maximum permissible daily concentration of acrolein in the atmosphere is 0.1 mg/m³. This study did not specify whether this level is intended as an occupational or ambient air quality standard (B-26).

Water: In water, EPA has suggested for acrolein a criterion to protect freshwater aquatic life of 1.2 μg/l as a 24-hour average; the concentration would not exceed 2.7 μg/l at any time. For acrolein the criterion to protect saltwater aquatic life has been suggested as 0.88 μg/l as a 24-hour average and the concentration should not exceed 2.0 μg/l at any time. For the protection of human health from the adverse effects of acrolein ingested through the consumption of water and contaminated aquatic organisms a criterion of 6.5 μg/l is suggested (B-26).

References

(1) Worthing, C.R., *Pesticide Manual,* 6th ed., p. 2, British Crop Protection Council (1979).
(2) Spencer, E.Y., *Guide to the Chemicals Used in Crop Protection,* 6th ed., p. 2, London, Ontario, Agriculture Canada (January 1973).
(3) MacLean, A.F., U.S. Patent 3,517,006, August 1, 1950, assigned to Celanese Corp. of America.

(4) Fujiwara, Y., et al, U.S. Patent 3,172,914, March 9, 1965, assigned to Nippon Oil Co.
(5) Johnson, A.J., U.S. Patent 3,102,147, August 27, 1963, assigned to Shell Oil Co.
(6) Sennewald, K., et al, U.S. Patent 3,359,325, December 19, 1967, assigned to Knapsack AG.
(7) Groll, H.P.A. and De Jong, H.W., U.S. Patent 2,042,220, May 26, 1936, assigned to Shell Development Co.
(8) Van Overbeek, J., U.S. Patent 2,959,476, November 8, 1960, assigned to Shell Oil Co.
(9) Legator, M., U.S. Patent 3,987,475, June 6, 1961, assigned to Shell Oil Co.

ACRYLONITRILE

Function: Insecticide (for fumigation of cereals) (1)(2)

Chemical Name: Propenenitrile

Formula: $CH_2=CH-CN$

Trade Names: Acrylon®
Acritet®
Carbacryl®
Ventox® (Degesch AG)

Manufacture

    (A) From acetylene and hydrogen cyanide (3)(4)
    (B) From acrolein, ammonia and air (5)
    (C) From acrolein cyanohydrin (a by-product from acrylonitrile manufacture), ammonia and air (6)
    (D) From lactonitrile by pyrolysis (7)
    (E) From propylene and ammonia (8)
    (F) From propylene and NO (9)

Process Wastes and Their Control

Acrylonitrile is a very toxic liquid fumigant. It is formulated as mixtures with carbon tetrachloride. A procedure which would safely and effectively cause the combustion of carbon tetrachloride would also provide a disposal method for the acrylonitrile mixture (see carbon tetrachloride) (B-3).

Toxicity

Acrylonitrile is moderately toxic with an acute oral toxicity, LD_{50}, value of 93 mg/kg for rats. At present the body of evidence produced in both toxicity studies on laboratory animals and occupational epidemiologic studies on man suggests that acrylonitrile may be a human carcinogen. Thus, NIOSH has recently voiced its opinion that "acrylonitrile must be handled in the workplace as a suspect human carcinogen." This judgment of NIOSH is based primarily on (1) A preliminary epidemiologic study of E.I. duPont de Nemours and Co., Inc. on acrylonitrile polymerization workers from one particular textile fibers plant (Camden, S.C.); in this study, it was ascertained that a substantial excess risk (doubling over expected) of lung and colon cancers occurred between 1969 and 1975 in a cohort exposed between 1950 and 1955. (2) Interim results from ongoing 2-year studies on laboratory rats performed by Dow Chemical Co., and reported by the Manufacturing Chemists Association in which, by either drinking water or inhalation routes of acrylonitrile exposure, laboratory rats developed CNS tumors and zymbal gland carcinomas, not evident in control animals. Mammary region masses were also in excess upon exposure to 80 ppm.

Aside from suggestive evidence of carcinogenicity in man and animals, numerous workers have reported on the other genotoxic characteristics of acrylonitrile (embryotoxicity, mutagenicity and teratogenicity) in laboratory animals. Even though there is some controversy over the chronic effects of acrylonitrile, the acute toxicity of acrylonitrile is well known and the compound appears to exert part of its toxic effect through the release of inorganic cyanide (B-26).

Acrylonitrile has been reported as acutely toxic to fish at concentrations as low as 10,100 μg/l and to the invertebrate, *Daphnia magna*, at 7,550 μg/l. Chronic toxic effects were not seen in *Daphnia magna* at concentrations up to and including 3,600 μg/l. The bluegill, *Lepomis macrochirus*, concentrated acrylonitrile in its tissues by a factor of 48. The only available datum for a saltwater organism is a 96-hour LC_{50} of 24,500 μg/l for the pinfish, *Lagodon rhomboides*.

Allowable Limits on Exposure and Use

Air: The existing standards for acrylonitrile in various countries and various years appear in Table 5.

Table 5: Standards for Acrylonitrile Air Exposure Levels in Various Countries (between 1970–1974)

Year	Country	Air Standard ppm	Air Standard mg/m³
1970	U.S.S.R.	0.2	0.435
1970	Federal Republic of Germany	20.0	43.5
1970	England	20.0	43.5
1974	U.S.A.	20.0	43.5

Source: Reference (B-26)

It is evident that at this time, the Russian standard is substantially less (two orders of magnitude) than the American and West European standards. However, the standard may be exceeded significantly. A study of a Yugoslavian acrylic fiber plant indicated that their in-plant concentrations of acrylonitrile begin to approach the TLV in the U.S.A. Other investigators have noted that the U.S.S.R. air standards are often exceeded, although it is unlikely that higher concentrations occur throughout the day (B-26).

The threshold limit value for acrylonitrile in air is still conditionally set at 45 mg/m (20 ppm) as of 1979. The proposed short term exposure limit is 65 mg/m³ (30 ppm). Both limits bear the notation "skin" indicating that cutaneous absorption should be prevented so that the threshold limit value is not invalidated. Further, acrylonitrile is designated as a human carcinogen so that no exposure (respiratory, skin or oral) should be permitted (B-23).

By an emergency temporary standard (ETS), however, the Occupational Safety and Health Administration (OSHA) amended in early 1978 (10) its present standard concerning employee exposure to acrylonitrile (AN) and reduced the permissible exposure level from 20 parts acrylonitrile per million parts of air (20 ppm), as an 8-hour time-weighted average concentration, to 2 ppm with a ceiling level of 10 ppm for any 15-minute period during the 8-hour day (B-25).

In addition, the standard includes an action level of 1 ppm as an 8-hour time-weighted average. Provision is also made for specific exemptions in the standard for certain operations involving the processing, use and handling of products fabricated from polyacrylonitrile (PAN). The basis for the ETS is OSHA's determination that laboratory and epidemiological data indicate that continued exposure to acrylonitrile presents a cancer hazard to workers. A grave danger therefore exists for workers exposed to AN, necessitating the issuance of an emergency standard to protect them.

In addition, the ETS requires the measurement and control of employee exposure, personal protective equipment and clothing, employee training, medical surveillance, work practices, and record keeping. It should be noted that as recently as September 29, 1977 (11), the permissible exposure level was 4 ppm in air, so this limit is under continuing reevaluation. The economic impact of such changes is also being considered (12). The ETS was made permanent in late 1978 (13).

Water: For water, EPA has subsequently suggested a limit to protect freshwater aquatic life of 130 μg/l as a 24-hour average and the concentration should not exceed 300 μg/l at any time.

For acrylonitrile the criterion to protect saltwater aquatic life is 130 μg/l as a 24-hour average and the concentration should not exceed 290 μg/l at any time.

For the maximum protection of human health from the potential carcinogenic effects of exposure to acrylonitrile through ingestion of water and contaminated aquatic organisms, the ambient water concentration is zero. Concentrations of acrylonitrile estimated to result in additional lifetime cancer risks ranging from no additional risk to an additional risk of 1 in 100,000 have been suggested by EPA. EPA is considering setting criteria at an interim target risk level in the range of 10^{-5}, 10^{-6} or 10^{-7} with corresponding criteria of 0.08 μg/l, 0.008 μg/l and 0.0008 μg/l respectively. If water alone is consumed, the water concentration should be less than 0.16 μg/l to keep the lifetime cancer risk below 10^{-5} (B-26).

Product Use: Some acrylonitrile pesticide products have been subject to voluntary cancellation due to possible oncogenicity, teratogenicity and neurotoxicity, according to US EPA notices dated June 19, June 30 and August 5, 1978 (B-17).

References

(1) Martin, H. and Worthing, C.R., *Pesticide Manual,* 4th ed., p. 3, British Crop Protection Council (November 1974).
(2) Migrdichian, V., U.S. Patent 2,356,075, August 15, 1944, assigned to American Cyanamid Co.
(3) Pursell, H.P., Jr., U.S. Patent 3,267,128, August 16, 1966, assigned to American Cyanamid Co.
(4) Kremer, V.W., et al., U.S. Patent 3,053,881, September 11, 1962, assigned to DuPont.
(5) Bewley, T., U.S. Patent 2,836,614, May 27, 1958, assigned to Distillers Co., Ltd.
(6) Sennewald, K., et al, U.S. Patent 3,372,986, March 12, 1968, assigned to Knapsack-Griesheim AG.
(7) Sennewald, K., et al, U.S. Patent 3,079,420, February 26, 1963, assigned to Knapsack-Griesheim AG.
(8) Sennewald, K., et al, U.S. Patent 3,226,422, December 28, 1965, assigned to Knapsack-Griesheim AG.
(9) England, D.C., et al, U.S. Patent 3,023,226, February 27, 1962, assigned to DuPont.
(10) *Federal Register* 43, No. 11, 2586-2621 (January 17, 1978).
(11) National Institute for Occupational Safety and Health, *Criteria for a Recommended Standard: Occupational Exposure to Acrylonitrile,* NIOSH Doc. No. 78-116, Wash., DC (1978).
(12) Department of Labor, *Economic Impact Assessment for Acrylonitrile,* Wash., DC, Occupational Safety and Health Administration (February 21, 1978).
(13) *Federal Register* 43, No. 192, 45762-45819 (October 3, 1978).

ALACHLOR

Function: Preemergence herbicide (1)(2)(3)(4)

Chemical Name: 2-chloro-2',6'-diethyl-N-methoxymethylacetanilide

Formula:

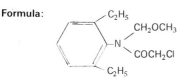

Trade Names: CP 50144
Lasso® (Monsanto)

Manufacture (5)

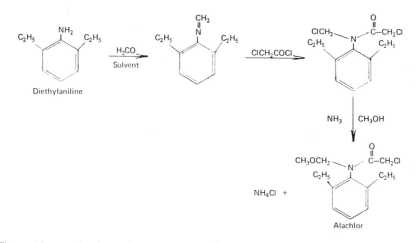

Figure 4 is a production and waste schematic for this process.

Process Wastes and Their Control

Air: Air emissions of 1.5 kg of hydrocarbons per metric ton of alachlor have been reported for the manufacturing process (B-15).

Water: The waste NH_4Cl from alachlor production is combined with wastewaters and discharged with liquid effluent. The solvent is recovered and used as fuel (B-10).

Solid Waste Disposal: This compound is hydrolyzed under strongly acid or alkaline conditions, to chloroacetic acid, methanol, formaldehyde and 2,6-diethylaniline. Incineration is recommended as a disposal procedure (B-3).

Toxicity

Alachlor is slightly toxic with an oral LD_{50} value for rats of 1,200 mg/kg. Alachlor and its relatives, butachlor and propachlor appear to be fairly well tolerated by mammals (B-22). Both alachlor and butachlor are apparently tolerated by rats at up to 100 mg/kg/day in the diet, except for increased liver weight in female rats fed butachlor.

The existing toxicity data for these compounds are largely those produced by the manufacturer for registration purposes. Based on the available data, ADI's were calculated at 0.1, 0.1 and 0.01 mg/kg/day for alachlor, propachlor and butachlor, respectively.

Figure 4: Production and Waste Schematic for Alachlor

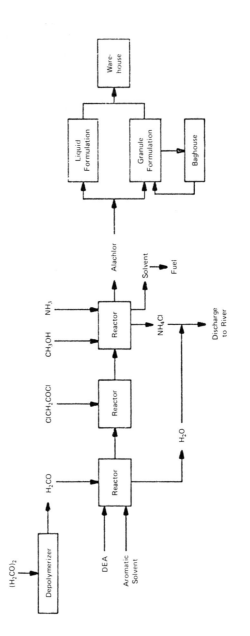

Apparently, no long-term toxicity studies have been completed that would contribute information on reproductive effects or carcinogenic potential of these acetanilides or their degradation products, which include aniline derivatives. These studies are needed (B-22).

Allowable Limits on Exposure and Use

Product Use: The tolerances set by the US EPA for alachlor in or on raw agricultural commodities are as follows:

	40 *CFR* Reference	Parts per Million
Beans, field, dry	180.249	0.1 N
Beans, forage	180.249	0.2 N
Beans, hay	180.249	0.2 N
Beans, lima, green	180.249	0.1 N
Cattle, fat	180.249	0.02 N
Cattle, mbyp	180.249	0.02 N
Cattle, meat	180.249	0.02 N
Corn, fodder	180.249	0.2 N
Corn, forage	180.249	0.2 N
Corn, fresh (inc. sweet) (k+cwhr)	180.249	0.05 N
Corn, grain	180.249	0.2 N
Cotton, forage	180.249	0.2 N
Cotton, seed	180.249	0.05 N
Eggs	180.249	0.02 N
Goats, fat	180.249	0.02 N
Goats, mbyp	180.249	0.02 N
Goats, meat	180.249	0.02 N
Hogs, fat	180.249	0.02 N
Hogs, mbyp	180.249	0.02 N
Hogs, meat	180.249	0.02 N
Horses, fat	180.249	0.02 N
Horses, mbyp	180.249	0.02 N
Horses, meat	180.249	0.02 N
Milk	180.249	0.02 N
Peanuts	180.249	0.05 N
Peanuts, forage	180.249	3.0
Peanuts, hay	180.249	3.0
Peanuts, hulls	180.249	1.5
Peas (pods removed)	180.249	0.1 N
Peas, forage	180.249	0.2 N
Peas, hay	180.249	0.2 N
Potatoes	180.249	0.1 N
Poultry, fat	180.249	0.02 N
Poultry, mbyp	180.249	0.02 N
Poultry, meat	180.249	0.02 N
Sheep, fat	180.249	0.02 N
Sheep, mbyp	180.249	0.02 N
Sheep, meat	180.249	0.02 N
Soybean, hay	180.249	0.2 N
Soybeans	180.249	0.2 N
Soybeans, forage	180.249	0.2 N

References

(1) Worthing, C.R., *Pesticide Manual,* 6th ed., p. 3, British Crop Protection Council (1979).
(2) Spencer, E.Y., *Guide to the Chemicals Used in Crop Protection,* 6th ed., p. 3, London, Ontario, Agriculture Canada (January 1973).
(3) Olin, J.F., U.S. Patent 3,442,945, May 6, 1969, assigned to Monsanto Co.
(4) Olin, J.F., U.S. Patent 3,547,620, December 15, 1970, assigned to Monsanto Co.
(5) Midwest Research Institute, *The Pollution Potential in Pesticide Manufacturing,* Wash., DC, U.S. Environmental Protection Agency (June 1972).

ALDICARB

Function: Systemic insecticide (1)(2)(3)

Chemical Name: 2-methyl-2-(methylthio)-propanal O-[(methylamino)carbonyl)]oxime

Formula:

$$CH_3S\text{—}\underset{\underset{CH_3}{|}}{\overset{\overset{CH_3}{|}}{C}}\text{—}CH=NO\overset{\overset{O}{\|}}{C}NHCH_3$$

Trade Names: UC 21149
                  Temik® (Union Carbide)

Manufacture (4)

$$2(CH_3)_2C=CH_2 + 2NaNO_2 + 4HCl \longrightarrow 2NaCl + 2H_2O + [ClC(CH_3)_2CH_2NO]_2$$

                                                                        2-chloro-2-methyl-
                                                                        1-nitrosopropane dimer

$$[ClC(CH_3)_2CH_2NO]_2 + 2CH_3SNa \longrightarrow 2NaCl + 2CH_3SC(CH_3)_2CH=NOH$$

                                                                2-methyl-2-(methylthio)-
                                                                propionaldehyde oxime

$$CH_3SC(CH_3)_2CH=NOH + CH_3NCO \longrightarrow CH_3SC(CH_3)_2CH=NO\text{—}\overset{\overset{}{}}{\underset{\underset{O}{\|}}{C}}\text{—}NHCH_3$$

                                                                        Aldicarb

Figure 5 is a production and waste schematic for aldicarb manufacture.

Figure 5: Production and Waste Schematic for Aldicarb

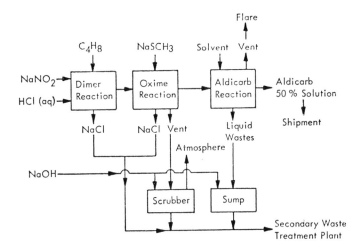

Source: Reference (4)

Process Wastes and Their Control

Air: Air emissions of 205 kg SO_2, 1.5 kg of hydrocarbons and 0.5 kg of aldicarb per metric ton of pesticide have been reported (B-15) for the aldicarb production process.

Toxicity

Aldicarb is extremely toxic with an oral LD_{50} value for rats of 0.93 mg/kg. Aldicarb and methomyl are highly toxic, oxime-carbamate insecticides with increasing uses on food crops, cotton, and ornamentals. Both compounds act systemically and are readily metabolized and degraded in organisms and in the environment. It is not likely that either compound will appear as a major contaminant of drinking water. The mode of action of these materials is inhibition of acetylcholinesterase. Acute toxicity is quite high, but because of the rapid breakdown of the compounds in organisms and the environment, chronic toxicity is not a major problem.

The chronic toxicity data reported in the literature have not been sufficient as yet for WHO/FAO to establish an acceptable daily intake for either aldicarb or methomyl. An ADI at 0.001 mg/kg/day for aldicarb was calculated based on the available data (B-22).

Allowable Limits on Exposure and Use

Product Use: The tolerances set by the EPA for aldicarb in or on raw agricultural commodities are as follows:

	40 *CFR* Reference	Parts per Million
Bananas	180.269	0.3
Beans, dry	180.269	0.1
Beets, sugar	180.269	0.05
Beets, sugar, tops	180.269	1.0
Cattle, fat	180.269	0.01
Cattle, mbyp	180.269	0.01
Cattle, meat	180.269	0.01
Coffee, beans	180.269	0.1
Cotton, seed	180.269	0.1
Goats, fat	180.269	0.01
Goats, mbyp	180.269	0.01
Goats, meat	180.269	0.01
Hogs, fat	180.269	0.01
Hogs, mbyp	180.269	0.01
Hogs, meat	180.269	0.01
Horses, fat	180.269	0.01
Horses, mbyp	180.269	0.01
Horses, meat	180.269	0.01
Milk	180.269	0.002
Oranges	180.269	0.3
Peanuts	180.269	0.05
Peanuts, hulls	180.269	0.5
Pecans	180.269	0.5
Potatoes	180.269	0.1
Sheep, fat	180.269	0.01
Sheep, mbyp	180.269	0.01
Sheep, meat	180.269	0.01
Sugarcane	180.269	0.02
Sugarcane, fodder	180.269	0.1
Sugarcane, forage	180.269	0.1
Sweet potatoes	180.269	0.1

The tolerances set by the US EPA for aldicarb in animal feeds are as follows (the *CFR* Reference is to Title 21):

	CFR Reference	Parts per Million
Citrus pulp, dried	561.30	0.6
Cottonseed hulls	561.30	0.3

References

(1) Worthing, C.R., *Pesticide Manual,* 6th ed., p. 4, British Crop Protection Council (1979).
(2) Spencer, E.Y., *Guide to the Chemicals Used in Crop Protection,* 6th ed., p. 4, London, Ontario, Agriculture Canada (January 1973).
(3) Payne, L.K., Jr. and Weiden, H.J., U.S. Patent 3,217,037, November 9, 1965, assigned to Union Carbide Corp.
(4) Midwest Research Institute, *The Pollution Potential in Pesticide Manufacturing,* Wash., DC, U.S. Environmental Protection Agency (June 1972).

ALDOXYCARB

Function: Insecticide, miticide and nematocide (1)(2)

Chemical Name: 2-methyl-2-(methylsulfonyl)propanal-O-[(methylamino)carbonyl]oxime

Formula:

$$CH_3-\overset{\overset{O}{\uparrow}}{\underset{\underset{O}{\downarrow}}{S}}-\overset{\overset{CH_3}{|}}{\underset{\underset{CH_3}{|}}{C}}-CH=N-O-\overset{\overset{O}{\|}}{C}-NHCH_3$$

Trade Names: UC 21865 (Union Carbide Corp.)
Standak® (Union Carbide Corp.)

Manufacture (2)

Aldoxycarb is made by the oxidation of aldicarb as follows: To 9.5 grams of 2-methyl-2-methylthiopropionaldehyde N-methylcarbamoyloxime (0.05 mol) in 50 ml of ethyl acetate was added 35.6 grams of a 21.2% solution of peracetic acid in ethyl acetate (0.1 mol of peracetic acid). The temperature of the well-stirred reaction medium was maintained at 25°-30°C with cooling while the peracetic acid solution was added over a forty-five minute period. The reaction mixture was then allowed to stand for sixteen hours. The resulting solid was collected by filtration and recrystallized from ethyl acetate. There was obtained 7 grams of a white crystalline solid which had a melting point of 132°-133°C.

Toxicity

The acute oral LD_{50} value for rats is about 27 mg/kg, which is highly toxic.

References

(1) Worthing, C.R., *Pesticide Manual,* 6th ed., p. 5, British Crop Protection Council (1979).
(2) Payne, L.K., Jr. and Weiden, M.H.J., U.S. Patent 3,217,037, November 9, 1965, assigned to Union Carbide Corp.

ALDRIN

Function: Insecticide (effective against soil insects) (1)(2)(3)

Chemical Name: 1,2,3,4,10,10-hexachloro-1α,4α,4aβ,5α,8α,8aβ-hexahydro-1,4:5,8-dimethanonaphthalene

Formula:

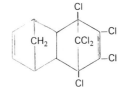

Trade Names: Compound 118
Octalene® (Julius Hyman and Co.)
Aldrec®
Algran®
Soilgrin®

Manufacture (3)(4)(5)(6)

Aldrin is made by the reaction of hexachlorocyclopentadiene with bicyclo[2.2.1]-2,5-heptadiene, according to the equation:

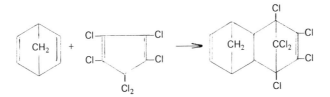

Hexachlorocyclopentadiene may be made by the reaction of n-pentane and chlorine as described by J.T. Patton et al (7) and by E.T. McBee et al (8) or by the reaction of cyclopentadiene and chlorine as described by Lidov (9). Bicyclo[2.2.1]heptadiene may be made by the condensation of cyclopentadiene and acetylene as described by F. Reicheneder et al (10) and by J. Hyman et al (11).

The overall reaction scheme for aldrin manufacture is shown in Figure 6.

The final condensation reaction may be carried out at temperatures of 80° to 150°C. The preferred temperature is approximately 120°C (4). This reaction is conducted at atmospheric pressure or at pressures slightly above atmospheric. Actually it is the vapor pressure of the lowest boiling reactant which determines the operating pressure and when this is above 80°C, the reaction can be carried out at atmospheric pressure (3). Operating pressures of as much as 4 kg/cm² (about 60 psi) may be used, however (4).

Reaction times of as long as 18 hours may be used in batch operations. However, in continuous operation, a residence time of 2 hours with 2 reactors and 7 hours with 3 reactors may be adequate (4). This reaction is carried out in the liquid phase in the absence of added reaction solvents. Solvents such as toluene may be used; the presence of such solvents increases the reaction time required but makes temperature control easier (3).

This reaction may be carried out continuously in an apparatus furnished with a tangential supply line and having a greatest dimension no more than four times as large as the smallest (4). The batch version of this reaction may be conducted in a closed kettle equipped with a reflux condenser.

At the end of the reaction, excess bicycloheptadiene is removed by distillation. The crude

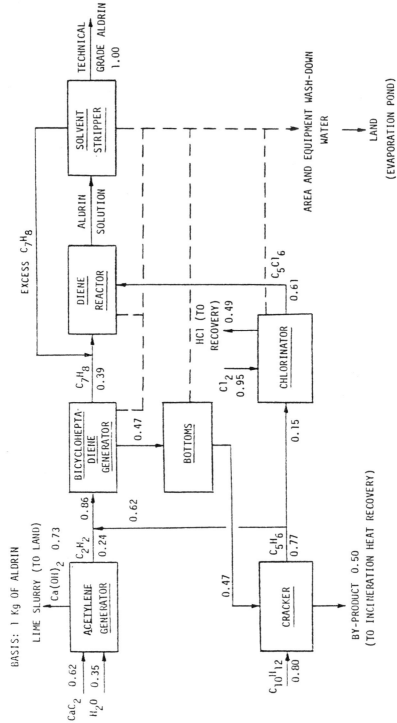

Figure 6: Aldrin Manufacture

Source: Reference (B-1)

product residue may then be distilled at 1 mm of mercury pressure and a foreshot discarded before the bulk of the product is collected. This material distilled over may be further purified by crystallization if desired.

Process Wastes and Their Control

Aldrin and dieldrin were manufactured in the United States by Shell Chemical Company until the EPA prohibited their manufacture in 1974 under the Federal Insecticide, Fungicide and Rodenticide Act. They are currently manufactured by Shell Chemical Company in Holland (B-26).

Water: Liquid wastes containing aldrin to the level of solubility are sent to an asphalt-lined evaporation basin in a typical plant (B-10).

Product Wastes: Aldrin is very stable thermally with no decomposition noted at 250°C. Aldrin (along with the structurally related compounds dieldrin and isodrin) is remarkably stable to alkali (in contrast to chlordane and heptachlor) and refluxing with aqueous or alcoholic caustic has no effect.

Incineration methods for aldrin disposal have been recommended by the MCA and the combustion of aldrin in polyethylene on a small scale gave more than 99% decomposition. Aldrin can be degraded by active metals such as sodium in alcohol (a reaction which forms the basis of the analytical method for total chlorine), but this method is not suitable for the layman.

A disposal method suggested for materials contaminated with aldrin, dieldrin or endrin consists of burying 8 to 12 feet underground in an isolated area away from water supplies, with a layer of clay, a layer of lye and a second layer of clay beneath the wastes (B-3).

Toxicity

Aldrin is a moderately toxic material with an oral LD_{50} value for rats of 67 mg/kg. During the past decade, considerable information has been generated concerning the toxicity and potential carcinogenicity of the two organochlorine pesticides aldrin and dieldrin. These two pesticides are usually considered together since aldrin is readily epoxidized to dieldrin in the environment. Both are acutely toxic to most forms of life including arthropods, mollusks, invertebrates, amphibians, reptiles, fish, birds and mammals. Dieldrin is extremely persistent in the environment. By means of bioaccumulation it is concentrated manyfold as it moves up the food chain (B-26).

Early work by Treon and Cleveland in 1955 suggested that aldrin and dieldrin may have tumor-inducing potential, especially in the liver. Since that time, several conflicting reports of the hepatocarcinogenicity in mice, rats and dogs have appeared in literature. Studies have been carried out mainly by the U.S. Food and Drug Administration, the National Cancer Institute, and by the manufacturer, Shell Chemical Company. There has been much debate over the type and significance of hepatic damage caused by aldrin and dieldrin. In order to ascertain the human risks associated with aldrin and dieldrin, evaluations of the toxic effects of these pesticides have been carried out on workers in the Shell Chemical Company. The evaluations include epidemiological studies in addition to the more routine toxicity studies. However, it is felt that the number of workers with high exposures was too small and the time interval too short to determine whether or not aldrin and dieldrin represent a cancer threat to humans (B-26).

Allowable Limits on Exposure and Use

Air: The threshold limit value for aldrin in air has been set at 0.25 mg/m^3 as of 1979. The tentative short term exposure limit is 0.75 mg/m^3. These limits carry the notation "skin," indicating that cutaneous absorption should be prevented so the threshold limit value is not invalidated (B-23).

NIOSH has recommended, as of October 1978, that aldrin be held to the lowest reliably detectable level in air which is 0.15 mg/m^3 on a time weighted average basis.

Water: In water, EPA has set (B-12) a limit of 0.003 μg/l for the protection of freshwater and marine aquatic life. They note that the persistence, bioaccumulation potential and carcinogenicity of aldrin cautions minimum human exposure.

In water, EPA has subsequently suggested a limit for aldrin/dieldrin to protect freshwater aquatic life of 0.0019 μg/l as a 24-hour average and the concentration should not exceed 1.2 μg/l at any time (B-26). For aldrin/dieldrin the criterion to protect saltwater aquatic life is 0.0069 μg/l as a 24-hour average and the concentration should not exceed 0.16 μg/l at any time.

For the maximum protection of human health from the potential carcinogenic effects of exposure to aldrin through ingestion of water and contaminated aquatic organisms, the ambient water concentration is zero. Concentrations of aldrin estimated to result in additional lifetime cancer risks ranging from no additional risk to an additional risk of 1 in 100,000 have been suggested by EPA. EPA is considering setting criteria at an interim target risk level in the range of 10^{-5}, 10^{-6} or 10^{-7} with corresponding criteria of 4.6 x 10^{-2} ng/l, 4.6 x 10^{-3} ng/l and 4.6 x 10^{-4} ng/l, respectively (B-26).

Product Use: Aldrin and dieldrin have been the subject of litigation bearing upon the contention that these substances cause severe aquatic environmental change and are potential carcinogens. In 1970, the U.S. Department of Agriculture cancelled all registrations of these pesticides based upon a concern to limit dispersal in or on aquatic areas. In 1972, under the authority of the Fungicide, Insecticide, Rodenticide Act as amended by the Federal Pesticide Control Act of 1972, USCS Section 135, et. seq., an EPA order lifted cancellation of all registered aldrin and dieldrin for use in deep ground insertions for termite control, nursery clipping of roots and tops of nonfood plants, and mothproofing of woolen textiles and carpets where there is no effluent discharge.

In 1974, cancellation proceedings disclosed the severe hazard to human health and suspension of registration of aldrin and dieldrin use was ordered; production was restricted for all pesticide products containing aldrin or dieldrin. However, formulated products containing aldrin and dieldrin are imported from Europe each year solely for subsurface soil injection for termite control. Therefore, limits that protect all receiving water uses must be placed on aldrin and dieldrin. The litigation has produced the evidentiary basis for the Administrator's conclusions that aldrin/dieldrin are carcinogenic in mice and rats, approved the EPA's extrapolation to humans of data derived from tests on animals, and affirmed the conclusions that aldrin and dieldrin pose a substantial risk of cancer to humans, which constitutes an "imminent hazard" to man (B-26).

The US EPA cancelled all uses of aldrin as of March 18, 1971 (B-17) except those in the following list:

1. Subsurface ground insertion for termite control.
2. Dipping of nonfood roots and tops.
3. Moth-proofing by manufacturing processes in a closed system.

The tolerances set by the EPA for aldrin in or on raw agricultural commodities are as follows:

	40 *CFR* Reference	Parts per Million
Alfalfa	180.135	0.0
Apples	180.135	0.0

(continued)

	40 *CFR* Reference	Parts per Million
Apricots	180.135	0.0
Asparagus	180.135	0.1
Barley, grain	180.135	0.02 i
Barley, straw	180.135	0.1 i
Beans	180.135	0.0
Beets, garden	180.135	0.0
Beets, garden, tops	180.135	0.0
Beets, sugar	180.135	0.0
Beets, sugar, tops	180.135	0.0
Broccoli	180.135	0.1
Brussels sprouts	180.135	0.1
Cabbage	180.135	0.1
Cantaloupe	180.135	0.1
Carrots	180.135	0.0
Cauliflower	180.135	0.1
Celery	180.135	0.1
Cherries	180.135	0.1
Clover	180.135	0.0
Collards	180.135	0.0
Corn, forage	180.135	0.0
Corn, grain	180.135	0.0
Corn, pop	180.135	0.0
Cowpeas	180.135	0.0
Cowpeas, hay	180.135	0.0
Cranberries	180.135	0.1
Cucumbers	180.135	0.1
Eggplant	180.135	0.1
Endive (escarole)	180.135	0.0
Garlic	180.135	0.0
Grapefruit	180.135	0.05 i
Grapes	180.135	0.1
Horseradish	180.135	0.0
Kale	180.135	0.0
Kohlrabi	180.135	0.0
Leeks	180.135	0.0
Lemons	180.135	0.05 i
Lespedeza	180.135	0.0
Lettuce	180.135	0.1
Limes	180.135	0.05 i
Mangoes	180.135	0.1
Muskmelons	180.135	0.1
Mustard greens	180.135	0.0
Nectarines	180.135	0.1
Oats, grain	180.135	0.02 i
Oats, straw	180.135	0.1 i
Onions	180.135	0.0
Oranges	180.135	0.05 i
Parsnips	180.135	0.0
Peaches	180.135	0.1
Peanuts	180.135	0.0
Peanuts, hay	180.135	0.0
Pears	180.135	0.0
Peas	180.135	0.0
Peas, black-eyed	180.135	0.0
Peas, cowpeas	180.135	0.0
Peas, hay	180.135	0.0
Peppers	180.135	0.1
Pimentos	180.135	0.1
Pineapples	180.135	0.1
Plums (fresh prunes)	180.135	0.1

(continued)

	40 *CFR* Reference	Parts per Million
Potatoes	180.135	0.1
Pumpkins	180.135	0.1
Quinces	180.135	0.0
Radishes	180.135	0.0
Rice, grain	180.135	0.05 i
Rice, straw	180.135	0.1 i
Rutabagas	180.135	0.0
Rye, grain	180.135	0.02 i
Rye, straw	180.135	0.1 i
Salsify, roots	180.135	0.0
Salsify, tops	180.135	0.0
Shallots	180.135	0.0
Sorghum, forage	180.135	0.0
Sorghum, grain	180.135	0.0
Soybeans	180.135	0.0
Soybeans, hay	180.135	0.0
Spinach	180.135	0.0
Squash, summer	180.135	0.1
Squash, winter	180.135	0.1
Strawberries	180.135	0.1
Sweet potatoes	180.135	0.1
Swiss chard	180.135	0.0
Tangerines	180.135	0.05 i
Tomatoes	180.135	0.1
Turnips	180.135	0.0
Turnips, tops	180.135	0.0
Watermelons	180.135	0.1
Wheat, grain	180.135	0.02 i
Wheat, straw	180.135	0.1 i

References

(1) Worthing, C.R., *Pesticide Manual,* 6th ed., p. 6, British Crop Protection Council (1979).
(2) Spencer, E.Y., *Guide to the Chemicals Used in Crop Protection,* 6th ed., p. 5, London, Ontario, Agriculture Canada (January 1973).
(3) Lidov, R.E., U.S. Patent 2,635,977, April 21, 1953, assigned to Shell Development Co.
(4) Luijckx, W.L.L.M. et al, U.S. Patent 2,917,552, December 15, 1959, assigned to Shell Development Co.
(5) Midwest Research Institute, *The Pollution Potential in Pesticide Manufacturing* Wash., DC, U.S. Environmental Protection Agency (June 1972).
(6) Schmerling, L., U.S. Patent 2,911,447, November 3, 1959, assigned to Universal Oil Products Co.
(7) Patton, J.T. et al, U.S. Patent 2,960,543, November 15, 1960, assigned to Wyandotte Chemical Corp.
(8) McBee, E.T. et al, U.S. Patent 2,509,160, May 23, 1950, assigned to Purdue Research Foundation.
(9) Lidov, R.E., U.S. Patent 2,900,420, August 18, 1959, assigned to Shell Development Co.
(10) Reicheneder, F. et al, U.S. Patent 3,073,872, January 15, 1963, assigned to Badische Anilin- und Soda-Fabrik.
(11) Hyman, J., Freireich, E. and Lidov, R.E., U.S. Patent 2,875,256, February 24, 1959, assigned to Shell Development Co.

ALLETHRIN

Function: Contact insecticide (effective against house flies) (1)(2)

Chemical Name: 2-methyl-4-oxo-3-(2-propenyl)-2-cyclopenten-1-yl 2,2-dimethyl-3-(2-methyl-1-propenyl) cyclopropane carboxylate

Formula:

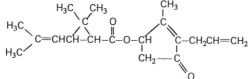

Trade Name: Pynamin® (Sumitomo Chemical Co.)

Manufacture (3)(4)(5)(6)

Allethrin of high purity and biological acitivity not obtainable by prior methods of manufacture can be produced by the reaction of 2-allyl-3-methyl-2-cyclopenten-4-ol-1-one (allethrolone) with chrysanthemum monocarboxylic anhydride (5). The reaction can be illustrated by the following equation:

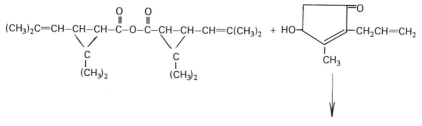

allethrin + chrysanthemum monocarboxylic acid

The chrysanthemum monocarboxylic acid which is liberated in the reaction can be separated and recovered with good efficiency for conversion to the anhydride again, as in a cyclic process.

The reaction can be readily carried out simply by heating the chrysanthemum monocarboxylic anhydride and the cyclopentenolone dissolved in a suitable solvent. In accord with well-established principles, the time required for the reaction to near completion is dependent upon the temperature. In general, the reaction temperature should not exceed 200°C in order to avoid decomposition. At temperatures below 100°C, the reaction rate tends to proceed too slowly for the procedure to be practicable. A temperature of about 150° to 175°C is preferred.

As a medium for carrying out the reaction, any material which is capable of dissolving the reactants and which at the same time is not reactive with them or with the allethrin product at the temperatures employed can be used. If desired, the reaction can be carried out under reflux by choosing a solvent having a boiling temperature in the range of the reaction temperature. Representative of materials that can be employed as solvents are the following: diisopropyl ether, benzene, toluene, xylene, dibutyl ether, butyl ethyl ether, dihexyl ether. Dibutyl ether is preferred.

Usually a heating period of about 3 to 6 hours will be required for the reaction. At the end of that time the reaction mixture is diluted with additional solvent to diminish saponification of allethrin during the washing of the product with alkali. The solution is washed with alkali to remove the chrysanthemum monocarboxylic acid and then with water to remove the alkali.

The wash liquids are then extracted with more solvent to minimize the mechanical loss of allethrin. The washed oil and the extracts are then combined and stripped of low boiling material by conventional procedures, as by reduced pressures, elevated temperatures, sparging with a nonreactive gas and the like. The allethrin is obtained as a residue product.

Process Wastes and Their Control

Solid Waste Disposal: Allethrin is detoxified by hydrogenation of the double bonds in the molecule. It is more stable than natural pyrethrin to UV radiation and heat (B-3).

Allowable Limits on Exposure and Use

Product Use: The tolerances set by the EPA for allethrin in or on raw agricultural commodities are as follows:

	40 *CFR* Reference	Parts per Million
Apples (post-h)	180.113	4.0
Apples (pre-h)	180.100	exempt
Barley, grain (post-h)	180.113	2.0
Beans (pre-h)	180.100	exempt
Blackberries (post-h)	180.113	4.0
Blueberries (huckleberries) (post-h)	180.113	4.0
Boysenberries (post-h)	180.113	4.0
Broccoli (pre-h)	180.100	exempt
Brussel sprouts (pre-h)	180.100	exempt
Cabbage (pre-h)	180.100	exempt
Cauliflower (pre-h)	180.100	exempt
Cherries (post-h)	180.113	4.0
Citrus (pre-h)	180.100	exempt
Collards (pre-h)	180.100	exempt
Corn, grain (post-h)	180.113	2.0
Crabapples (post-h)	180.113	4.0
Currants (post-h)	180.113	4.0
Dewberries (post-h)	180.113	4.0
Figs (post-h)	180.113	4.0
Gooseberries (post-h)	180.113	4.0
Grapes (post-h)	180.113	4.0
Guavas (post-h)	180.113	4.0
Horseradish (pre-h)	180.100	exempt
Kale (pre-h)	180.100	exempt
Kohlrabi (pre-h)	180.100	exempt
Lettuce (pre-h)	180.100	exempt
Loganberries (post-h)	180.113	4.0
Mangoes (post-h)	180.113	4.0
Milo, grain (post-h)	180.113	2.0
Mushrooms (pre-h)	180.100	exempt
Muskmelons (post-h)	180.113	4.0
Mustard greens (pre-h)	180.100	exempt
Oats, grain (post-h)	180.113	2.0
Oranges (post-h)	180.113	4.0
Peaches (post-h)	180.113	4.0
Peaches (pre-h)	180.100	exempt
Pears (post-h)	180.113	4.0
Pears (pre-h)	180.100	exempt
Peppers (pre-h)	180.100	exempt
Pineapples (post-h)	180.113	4.0
Plums (fresh prunes) (post-h)	180.113	4.0
Radishes (pre-h)	180.100	exempt
Raspberries (post-h)	180.113	4.0
Rutabagas (pre-h)	180.100	exempt

(continued)

	40 CFR Reference	Parts per Million
Rye, grain (post-h)	180.113	2.0
Sorghum, grain (post-h)	180.113	2.0
Tomatoes (post-h)	180.113	4.0
Tomatoes (pre-h)	180.100	exempt
Turnips (pre-h)	180.100	exempt
Wheat, grain (post-h)	180.113	2.0

References

(1) Worthing, C.R., *Pesticide Manual*, 6th ed., p. 7, British Crop Protection Council (1979).
(2) Spencer, E.Y., *Guide to the Chemicals Used in Crop Protection*, 6th ed., p. 7, London, Ontario, Agriculture Canada (January 1973).
(3) Schechter, M.S. and LaForge, F.B., U.S. Patent 2,574,500, November 13, 1951, dedicated to the free use of the People in the territory of the United States.
(4) Sanders, H.J. and Taff, A.W., *Ind. Eng. Chem.* 46, 414 (1954).
(5) Stansbury, H.A., Jr. and Guest, H.R., U.S. Patent 2,768,965, October 30, 1956, assigned to Union Carbide and Carbon Corp.
(6) LaForge, F.B. et al, *J. Org. Chem.* 12, 199-202 (1947).

ALLIDOCHLOR

Function: Selective preemergence herbicide (effective against certain grasses) (1)(2)

Chemical Name: 2-chloro-N,N-di-2-propenylacetamide

Formula: $ClCH_2CON(CH_2CH=CH_2)_2$

Trade Names: CP 6343
Randox® (Monsanto)
CDAA

Manufacture (3)(4)

Allidochlor is made by the reaction of diallylamine and chloroacetyl chloride. The following is a specific example (3) of the conduct of the process.

A one liter flask provided with a reflux condenser and a stirring mechanism was charged with 500 cc of ethylene dichloride and 97.2 g of diallylamine. The flask and its contents was cooled to −10°C by immersing the flask in a freezing mixture. While maintaining the temperature at −10°C with the vigorous agitation of the reagents, 66.5 g of chloroacetyl chloride was added gradually through a one-half hour period. After the reagents were combined, the mixture was stirred until the temperature rose to 15° to 20°C at which time 50 cc of water was added.

The solvent layer was separated and washed with 50 cc of a saturated sodium chloride solution. The solvent was evaporated in a vacuum and the resultant oil distilled at 92°C at 2 mm total pressure. The product so prepared was identified as alpha-chloro-N,N-diallyl acetamide.

Process Wastes and Their Control

Product Waste Disposal: Concentrated HCl at reflux temperature hydrolyzes the compound to chloroacetic acid and diallyl amine. The latter compound, a secondary amine, may be potentially hazardous: it may react with nitrites in the environment to form a nitrosoamine, a suspected carcinogen (B-3).

Toxicity

Allidochlor shows an acute oral LD_{50} value for rats of 700 mg/kg. Hence, it is considered to be slightly toxic.

Allowable Limits on Exposure and Use

Product Use: The tolerances set by the US EPA for allidochlor (N,N-diallyl-2-chloroacetamide) in or on raw agricultural commodities are as follows:

	40 *CFR* Reference	Parts per Million
Beans, castor	180.282	0.05 N
Beans, dry	180.282	0.05 N
Beans, lima	180.282	0.05 N
Beans, lima, forage	180.282	0.05 N
Beans, snap	180.282	0.05 N
Beans, snap, forage	180.282	0.05 N
Cabbage	180.282	0.05 N
Celery	180.282	0.05 N
Corn, field, fodder	180.282	0.05 N
Corn, field, forage	180.282	0.05 N
Corn, fresh (inc. sweet) (k+cwhr)	180.282	0.05 N
Corn, grain	180.282	0.05 N
Corn, pop, fodder	180.282	0.05 N
Corn, pop, forage	180.282	0.05 N
Corn, sweet, fodder	180.282	0.05 N
Corn, sweet, forage	180.282	0.05 N
Onions	180.282	0.05 N
Peas	180.282	0.05 N
Peas, forage	180.282	0.05 N
Potatoes	180.282	0.05 N
Sorghum, forage	180.282	0.05 N
Sorghum, grain	180.282	0.05 N
Soybeans	180.282	0.05 N
Soybeans, forage	180.282	0.05 N
Sugarcane	180.282	0.05 N
Sweet potatoes	180.282	0.05 N
Tomatoes	180.282	0.05 N

References

(1) Martin, H. and Worthing, C.R., *Pesticide Manual,* 4th ed, p. 8, British Crop Protection Council (November 1974).
(2) Spencer, E.Y., *Guide to the Chemicals Used in Crop Protection,* 6th ed., p. 8, London, Ontario, Agriculture Canada (January 1973).
(3) Hamm, P.C. and Speziale, A.J., U.S. Patent 2,864,683, December 16, 1958, assigned to Monsanto Chemical Co.
(4) Speziale, A.J. and Hamm, P.C., *J. Am. Chem. Soc.* 78, 2556 (1956).

ALLYL ALCOHOL

Function: Herbicide (1)(5)

Chemical Name: 2-propen-1-ol

Formula: $CH_2{=}CH{-}CH_2OH$

Trade Names: None

Manufacture

Allyl alcohol is produced by the high temperature chlorination of propylene to give allyl chloride (2) followed by the hydrolysis of the allyl chloride to allyl alcohol (3). Alternatively it may be made by the isomerization of propylene oxide (4).

Process Wastes and Their Control

Product Waste Disposal: Prior to incineration, dilution of this highly toxic liquid with a flammable solvent is recommended. It is miscible with water and most organic solvents. The compound is hazardous to wildlife and is toxic to plants and seeds. It has no lasting effect on soil although temporary sterilization occurs (B-3).

Toxicity

Allyl alcohol has an acute oral LD_{50} value for rats of 64 mg/kg, hence it is considered moderately toxic. The harmful effects of allyl alcohol have been noted as follows (B-25).

Local: Liquid and vapor are highly irritating to eyes and upper respiratory tract. Skin irritation and burns have occurred from contact with liquid but are usually delayed in onset and may be prolonged.

Systemic: Local muscle spasms occur at sites of percutaneous absorption. Pulmonary edema, liver and kidney damage, diarrhea, delirium, convulsions, and death have been observed in laboratory animals, but have not been reported in man.

Allowable Limits on Exposure and Use

Air: The threshold limit value of allyl alcohol is 5 mg/m^3 (2 ppm) as of 1979. The tentative short term exposure limit is 10 mg/m^3 (4 ppm). Both limits bear the notation "skin" indicating that cutaneous absorption should be prevented so that the threshold limit value is not invalidated (B-23).

References

(1) Worthing, C.R., *Pesticide Manual,* 6th ed., p. 9, British Crop Protection Council (1979).
(2) Brown, D., U.S. Patent 3,120,568, February 4, 1964, assigned to Halcon International.
(3) Goldstein, R.F. and Wadams, A.L., *The Petroleum Chemicals Industry,* p. 183, London, E. and F.N. Spon, Ltd. (1967).
(4) Rowton, R.L., U.S. Patent 3,238,264, March 1, 1966, assigned to Jefferson Chemical Co.
(5) Hughes, W.J., U.S. Patent 2,773,331, December 11, 1956, assigned to Shell Development Co.

ALUMINUM PHOSPHIDE

Function: Insecticide (effective in grain fumigation) (1)(2)

Chemical Name: Aluminum phosphide

Formula: AlP

Trade Names: Detia GAS-EX-B (Dr. W. Freyberg Chem. Fabrik)
Detia GAS-EX-T (Dr. W. Freyberg Chem. Fabrik)

Detia Phosphine Pellets (Dr. W. Freyberg Chem. Fabrik)
Delicia Gastoxin (DIA-Chemie, East Germany)
Phostoxin (Degesch AG)

Manufacture

Aluminum phosphide can be manufactured in a high degree of purity and with a phosphide content of about 90 to 98% without any special difficulty, by heating aluminum and phosphorus (3).

It is frequently desirable that during the production of the phosphine, it should be easily detectable, even when highly dilute. This is easily accomplished by the addition of sulfur during manufacture. This leads to the formation of aluminum sulfide which decomposes with the evolution of H_2S, which has a much more easily detected odor than the phosphine, thus resulting in an early warning of the presence of the phosphine (2).

Process Wastes and Their Control

Product Wastes: A technique has been developed (4) for detoxifying phosphide-containing pesticides by treatment with aqueous alkaline reagents optionally containing oxidants and/or surfactants.

Toxicity

Phosphine is a potent mammalian poison. Aluminum phosphide itself is said to be hazardous at a concentration of 1.0 ppm in air.

Allowable Limits on Exposure and Use

Product Use: The tolerance set by the US EPA for aluminum phosphide in or on raw agricultural commodities are as follows:

	CFR Reference	Parts per Million
Almonds (post-h)	180.225	0.1
Barley (post-h)	180.225	0.1
Brazil nuts (post-h)	180.225	0.1
Cashews (post-h)	180.225	0.1
Cocoa beans (post-h)	180.225	0.1
Coffee beans (post-h)	180.225	0.1
Corn (post-h)	180.225	0.1
Corn, pop (post-h)	180.225	0.1
Cotton, seed (post-h)	180.225	0.1
Dates (post-h)	180.225	0.1
Filberts (post-h)	180.225	0.1
Millet (post-h)	180.225	0.1
Oats (post-h)	180.225	0.1
Peanuts (post-h)	180.225	0.1
Pecans (post-h)	180.225	0.1
Pistachio nuts (post-h)	180.225	0.1
Rice (post-h)	180.225	0.1
Rye (post-h)	180.225	0.1
Safflower seed	180.225	0.1
Sorghum (post-h)	180.225	0.1
Soybeans (post-h)	180.225	0.1
Sunflower, seed (post-h)	180.225	0.1
Vegetables, seed and pod (exc. soybean)	180.225	0.01
Walnuts (post-h)	180.225	0.1
Wheat (post-h)	180.225	0.1

The tolerances set by the US EPA for aluminum phosphide in animal feeds are as follows (the *CFR* reference is to Title 21):

	CFR **Reference**	**Parts per Million**
Feeds, animal	561.40	0.1

The tolerances set by the EPA for aluminum phosphide in animal feeds are as follows (the *CFR* Reference is to Title 21):

	CFR **Reference**	**Parts per Million**
Processed foods	193.20	0.01

References

(1) Worthing, C.R., *Pesticide Manual,* 6th ed., p. 10, British Crop Protection Council (1979).
(2) Freyberg, W. and Haupt, W., U.S. Patent 2,117,158, May 10, 1938, assigned to Ernst Freyberg Chem. Fabrik Delitia.
(3) White, W.E. and Bushey, A.H., *J. Am. Chem. Soc.* 66, 1666 (1944).
(4) Praxl, W. and Ehret, R., U.S. Patent 4,180,557, December 25, 1979, assigned to D. Werner Freyberg, Germany.

AMETRYNE

Function: Herbicide (1)(2)(3)

Chemical Name: N-ethyl-N'-(1-methylethyl)-6-(methylthiol)-1,3,5-triazine-2,4-diamine

Formula:

Trade Names: G-34162 (Ciba-Geigy)
Gesapax® (Ciba-Geigy)
Evik® (Ciba-Geigy)

Manufacture (4)

Ametryne is made by reacting atrazine (which see) with methyl mercaptan in the presence of NaOH.

Toxicity

The oral acute toxicity of ametryne to rats is 1,150 mg/kg, which puts it in the slightly toxic category.

Allowable Limits on Exposure and Use

Product Use: The tolerances set by the US EPA for ametryne in or on raw agricultural commodities are as follows:

	40 *CFR* Reference	Parts per Million
Bananas	180.258	0.25
Corn, fodder	180.258	0.5
Corn, forage	180.258	0.5
Corn, fresh (inc. sweet) (k+cwhr)	180.258	0.25
Corn, grain	180.258	0.25
Grapefruit	180.258	0.1 N
Melons	180.158	10.0
Oranges	180.258	0.1 N
Pineapples	180.258	0.25
Pineapples, fodder	180.258	0.25
Pineapples, forage	180.258	0.25
Potatoes	180.258	0.25
Sugarcane	180.258	0.25
Sugarcane, fodder	180.258	0.25
Sugarcane, forage	180.258	0.25

References

(1) Worthing, C.R., *Pesticide Manual,* 6th ed., p. 12, British Crop Protection Council (1979).
(2) Spencer, E.Y., *Guide to the Chemicals Used in Crop Protection,* 6th ed., p. 16, London, Ontario, Agriculture Canada (January 1973).
(3) Gysin, H. and Knusli, E., U.S. Patent 2,909,420, October 20, 1959, assigned to J.R. Geigy AG.
(4) Rufener, W., Berger, R. and Riethmann, J., U.S. Patent 3,558,622, January 26, 1971, assigned to Geigy Chemical Corp. and J.R. Geigy AG.

AMINOCARB

Function: Nonsystemic insecticide (1)(2)

Chemical Name: 4-dimethylamino-3-methylphenyl methylcarbamate

Formula:

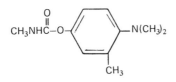

Trade Names: Bayer 44646 (Bayer)
Matacil® (Bayer)

Manufacture (3)

30.2 grams (0.2 mol) of 4-dimethylamino-3-methylphenol are treated with a solution of 12.5 grams (0.22 mol) of methylisocyanate in anhydrous dioxane. After the addition of 3 drops of triethylamine, the reaction mixture warms up very slightly. In order to complete the reaction, the mixture is further heated in a closed flask to 60°C for an hour. The solvent is distilled off under vacuum, the residue digested with water, filtered off and dried in a desiccator. After recrystallization from ligroin a product of MP 82° to 83°C is obtained.

Toxicity

The acute oral LD_{50} for rats shown by aminocarb is 50 mg/kg; this is moderately toxic.

References

(1) Worthing, C.R., *Pesticide Manual*, 6th ed., p. 13, British Crop Protection Council (1979). Council (1979).
(2) Spencer, E.Y., *Guide to the Chemicals Used in Crop Protection*, 6th ed., p. 13, London, Ontario, Agriculture Canada (January 1973).
(3) Heiss, R., Bocker, E. and Unterstenhofer, G., U.S. Patent 3,134,806, May 26, 1964, assigned to Farbenfabriken Bayer AG.

AMITRAZ

Function: Acaricide (1)(2)

Chemical Name: N'-(2,4-dimethylphenyl)-N-[[2,4-dimethylphenyl)imino] methyl] N-methylmethaniminamide

Formula:

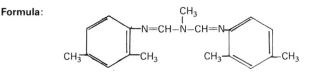

Trade Names: BTS 27,419 (Boots Co., Ltd.)
Taktic® (Boots Co., Ltd.)
Mitac® (Boots Co., Ltd.)
Triatox® (Coopers)

Manufacture (2)(3)

A solution of 19.4 g N-2,4-dimethylphenyl-N'-methylformamidine and 0.3 g p-toluenesulfonic acid in 195 ml dry xylene was refluxed under anhydrous conditions for 48 hours, causing the evolution of methylamine. The xylene was distilled off under reduced pressure and the solid residue was crystallized twice from isopropyl alcohol to yield 1,5-di(2,4-dimethylphenyl)-3-methyl-1,3,5-triazapenta-1,4-diene, MP 88° to 89°C.

The formamidine used in the above preparation is obtained in the following manner. A mixture of 55.1 g 2,4-dimethylaniline hydrochloride, 83.7 g p-toluenesulfonyl chloride and 150 ml N-methylformamide was stirred with occasional cooling to maintain the temperature at 20° to 35°C. When the exothermic reaction had subsided, the mixture was stirred at room temperature for 4 hours, poured into a mixture of ice and water, and basified with 10N sodium hydroxide solution, keeping the temperature of the mixture below 10°C.

The precipitated solid was filtered, washed with water until free from alkali, dried at room temperature, and recrystallized from cyclohexane to give N-2,4-dimethylphenyl-N'-methyl-formamidine, MP 75° to 76°C.

Toxicity

The acute oral LD_{50} value for rats is 800 mg/kg. This puts amitraz in the sightly toxic category.

Allowable Limits on Exposure and Use

Product Use: A rebuttable presumption against registration for amitraz has been issued on April 6, 1977 by US EPA on the basis of oncogenicity (B-17).

References

(1) Worthing, C.R., *Pesticide Manual,* 6th ed., p. 16, British Crop Protection Council (1979).
(2) Harrison, I.R., McCarthy, J.F., Palmer, B.H. and Kozlik, A., British Patent 1,327,935, August 22, 1973, assigned to The Boots Co., Ltd.
(3) Harrison, I.R., McCarthy, J.F. and Palmer, B.H., U.S. Patent 3,781,355, December 25, 1973, assigned to Boots Pure Drug Co., Ltd.

AMITROLE (AMINOTRIAZOLE IN U.K.)

Function: Herbicide (1)(2)(3)

Chemical Name: 1-H-1,2,4-triazol-3-amine

Formula:

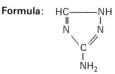

Trade Names: Weedazol® (Amchem Products)
Cytrol® (Amchem Products)
Azolan® (Amchem Products)
Amitrol-T®
Amizol®

Manufacture (3)

Amitrole is made by the condensation of formic acid with aminoguanidine. To 68 g (0.5 mol) of aminoguanidine bicarbonate in a 500 ml round-bottomed flask is added carefully the cold dilute sulfuric acid (0.24 mol), made from 24.5 g of concentrated acid (SG 1.84) and 50 ml of water. After the gas evolution has subsided, the solution is heated for 1 hour on a steam bath and then evaporated to dryness under a pressure of about 15 millimeters.

To the white residue are added 25 g of 98 to 100% formic acid and 2 to 3 drops of concentrated nitric acid. The mixture is heated for 24 hours on a steam bath. The resulting syrup is dissolved in 100 ml of water and, with the temperature at 50°C is carefully treated with 25 g of anhydrous sodium carbonate. The solution is then placed in an evaporating dish and evaporated to dryness on the steam bath. The residue is extracted twice by boiling it with 200 ml portions of absolute ethanol, and the alcohol solutions are filtered. The alcohol is removed by evaporation and the residue triturated with 100 ml of a mixture of equal parts of dry ethanol and dry ether and collected on a filter. The yield of crude 3-amino-1,2,4-triazole is 33 to 36 g (79 to 86%). The melting point varies.

The crude product is purified by dissolving it in 140 ml of boiling absolute ethanol, treating it with 1 g of Norite, and filtering. To the filtrate is added 50 ml of ether, and the solution is placed in a refrigerator for 48 hours. The aminotriazole crystallizes and is collected by filtration. It weighs 20 to 25 g (60 to 73% recovery) and melts at 153°C after softening at 148°C.

Process Wastes and Their Control

Product Waste Disposal: Amitrole is resistant to hydrolysis and the action of oxidizing agents. Burning the compound with polyethylene is reported to result in >99% decomposition (B-3).

Toxicity

This white solid herbicide is insignificantly toxic to mammals ($LD_{50} = 14,700$), but is a goitrogen at high concentrations. Its toxicity to fish is also low (B-3).

References

(1) Worthing, C.R., *Pesticide Manual*, 6th ed., p. 14, British Crop Protection Council (1979).
(2) Spencer, E.Y., *Guide to the Chemicals Used in Crop Protection*, 6th ed., p. 14, London, Ontario, Agriculture Canada (January 1973).
(3) Allen, W.W., U.S. Patent 2,670,282, February 23, 1954, assigned to American Chemical Paint Co.

AMS

Function: Herbicide (1)(2)(3)

Chemical Name: Monoammonium sulfamate

Formula: $NH_4SO_3NH_2$

Trade Names: Ammate® (DuPont)
Amcide® (Albright and Wilson)

Manufacture

(A) By converting sulfamic acid, made by the action of fuming sulfuric acid on urea, to the ammonium salt (4)(5).
(B) By the action of nongaseous sulfur trioxide on liquid ammonia (6)(7)(8).

Toxicity

The acute oral toxicity (LD_{50}) for rats of this product is 3,900 mg/kg. This means it is only slightly toxic.

Allowable Limits on Exposure and Use

Air: The threshold limit value for ammonium sulfamate in air is 10 mg/m^3 as of 1979. The tentative short term exposure limit is 20 mg/m^3 (B-23). The NIOSH/OSHA limit on ammonium sulfamate in air is 15 mg/m^3 (B-21).

Product Use: The tolerances set by the EPA for AMS in or on raw agricultural commodities are as follows:

	40 *CFR* Reference	Parts per Million
Apples	180.188	5.0
Pears	180.188	5.0
Turnips, greens	180.187	7.0

References

(1) Worthing, C.R., *Pesticide Manual*, 6th ed., p. 16, British Crop Protection Council (1979).
(2) Spencer, E.Y., *Guide to the Chemicals Used in Crop Protection*, 6th ed., p. 15, London, Ontario, Agriculture Canada (January 1973).

(3) Cupery, M.E. and Tanberg, A.P., U.S. Patent 2,277,744, March 31, 1942, assigned to DuPont.
(4) Baumgarten, P., U.S. Patent 2,102,350, December 14, 1937, assigned to DuPont.
(5) Rohrmann, C.A., U.S. Patent 2,487,480, November 8, 1949, assigned to DuPont.
(6) Tauch, E.G., U.S. Patent 2,426,420, August 26, 1947, assigned to DuPont.
(7) Ito, Y., U.S. Patent 3,404,949, October 8, 1968, assigned to Agency of Industrial Science and Technology.
(8) Sasaki, N., Sugahara, H., Maehara, T., Tsutsumi, K. and Egami, Y., U.S. Patent 3,404,950, October 8, 1968, assigned to Toyo Koatsu Industries, Inc.

ANCYMIDOL

Function: Plant growth regulator (1)(2)

Chemical Name: α-cyclopropyl-α-(4-methoxyphenyl)-5-pyrimidinemethanol

Formula:

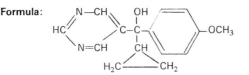

Trade Names: EL-531 (Eli Lilly and Co.)
A-Rest® (Eli Lilly and Co.)
Reducymol® (Eli Lilly and Co.)

Manufacture

To a solution of 0.1 mol of 4-methoxy-benzoylcyclopropane in 250 ml of a mixture of equal volumes of tetrahydrofuran and ether and cooled to −125°C was added a solution of 0.1 mol of 5-bromopyrimidine in the same mixed solvent. The mixture was stirred and maintained at about −125°C in a cooling bath composed of liquid nitrogen and ethanol, and to the cooled solution were added 60 ml of a 15% solution of n-butyl lithium in n-hexane, and the reaction mixture was stirred overnight.

The reaction product mixture was washed successively with 10% aqueous ammonium chloride solution and water and dried over anhydrous potassium carbonate. The dried organic solution was evaporated to dryness to yield a solid product. The solid was extracted with ether and the undissolved solid washed twice with ether. The ether-insoluble material was ancymidol.

Toxicity

The acute oral LD_{50} value for rats is 4,500 mg/kg, which puts it in the slightly toxic category.

References

(1) Worthing, C.R., *Pesticide Manual,* 6th ed., p. 17, British Crop Protection Council (1979).
(2) Davenport, J.D., Hackler, R.E. and Taylor, H.M., British Patent 1,218,623, January 6, 1971, assigned to Eli Lilly and Co.

ANILAZINE

Function: Fungicide (1)(2)(3)

Chemical Name: 4,6-dichloro-N-(2-chlorophenyl)-1,3,5-triazin-2-amine

Formula:

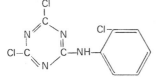

Trade Names: B-622®
Dyrene® (Chemagro Div., Baychem Corp.)
Kemate®

Manufacture

Anilazine is made by the reaction of cyanuric chloride with o-chloroaniline (4). By reacting cyanuric chloride with o-chloroaniline white crystals of 2,4-dichloro-6-(o-chloroanilino)-s-triazine were obtained in 98.7% yield. This material had a melting point of 155° to 157°C after recrystallization from a mixture of benzene and isohexanes and a chlorine content of 38.1, the theoretical content being 38.6%.

Process Wastes and Their Control

Product Waste Disposal: While anilazine is readily hydrolyzed by treatment with sodium hydroxide at elevated temperatures, o-chloroaniline may be formed as a product, and therefore this method is not recommended for disposal (B-3).

Toxicity

The acute oral LD_{50} for rats is 2,700 mg/kg (slightly toxic).

Allowable Limits on Exposure and Use

Product Use: The tolerances set by US EPA for anilazine in or on raw agricultural commodities are as follows:

	40 CFR Reference	Parts per Million
Blackberries	180.158	10.0
Blueberries (huckleberries)	180.158	10.0
Celery	180.158	10.0
Cranberries	180.158	10.0
Cucumbers	180.158	10.0
Dewberries	180.158	10.0
Garlic	180.158	1.0
Loganberries	180.158	10.0
Onions, dry bulb	180.158	1.0
Onions, green	180.158	10.0
Potatoes	180.158	1.0
Pumpkins	180.158	10.0
Raspberries	180.158	10.0
Shallots	180.158	10.0
Squash, summer	180.158	10.0
Squash, winter	180.158	10.0
Strawberries	180.158	10.0
Tomatoes	180.158	10.0

References

(1) Martin, H. and Worthing, C.R., *Pesticide Manual*, 4th ed., p. 19, British Crop Protection Council (November 1974).
(2) Spencer, E.Y., *Guide to the Chemicals Used in Crop Protection*, 6th ed., p. 17, London, Ontario, Agriculture Canada (January 1973).
(3) Wolf, C.N. et al, *Science* 121, 61 (1955).
(4) Wolf, C.N., U.S. Patent 2,720,480, October 11, 1955, assigned to Ethyl Corp.

ANTHRAQUINONE

Function: Bird repellent (used in seed treatment) (1)(2)

Chemical Name: 9,10-anthracenedione

Formula:

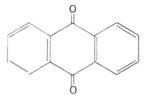

Trade Name: Morkit® (Bayer)

Manufacture

    (A) By the oxidation of anthracene with air (3).
    (B) By the oxidation of anthracene with nitric acid (4).
    (C) By the condensation of naphthoquinone with butadiene to give tetrahydroanthraquinone, followed by air oxidation (5).
    (D) By the condensation of phthalic anhydride with benzene (6).

Toxicity

The acute oral LD_{50} for rats is 5,000 (very slightly toxic).

References

(1) Worthing, C.R., *Pesticide Manual*, 6th ed., p. 18, British Crop Protection Council (1979). Council (1979).
(2) Spencer, E.Y., *Guide to the Chemicals Used in Crop Protection*, 6th ed., p. 18, London, Ontario, Agriculture Canada (January 1973).
(3) Gross, G., U.S. Patent 3,062,842, November 6, 1962, assigned to Ciba, Ltd.
(4) Alexander, W.N., U.S. Patent 2,821,534, January 28, 1958, assigned to General Aniline and Film Corp.
(5) Weyker, R.G. et al, U.S. Patent 2,938,913, May 31, 1960, assigned to American Cyanamid Co.
(6) Caesar, P.D. et al, U.S. Patent 2,401,225, May 28, 1946, assigned to Socony-Vacuum Oil Co.

ANTU

Function: Rodenticide (1)(2)(3)

Chemical Name: 1-naphthalenylthiourea

Formula:

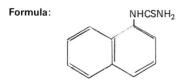

NHCSNH₂

Trade Name: Krysid®

Manufacture (4)

ANTU is made by the reaction of 1-naphthylamine with ammonium thiocyanate.

Process Wastes and Their Control

Product Waste Disposal: ANTU, a solid rodenticide, is stable in light and air (B-3).

Toxicity

The acute oral toxicity LD_{50} for rats is 6-8 mg/kg (highly toxic).

Allowable Limits on Exposure and Use

Air: The threshold limit value for ANTU in air is 0.3 mg/m³ as of 1979. The tentative short term exposure limit has been set at 0.9 mg/m³ (B-23).

References

(1) Worthing, C.R., *Pesticide Manual,* 6th ed., p. 19, British Crop Protection Council (1979).
(2) Spencer, E.Y., *Guide to the Chemicals Used in Crop Protection,* 6th ed., p. 19, London, Ontario, Agriculture Canada (January 1973).
(3) Richter, C.P., U.S. Patent 2,390,848, December 11, 1945, assigned to the U.S. Secretary of War.
(4) Beilstein, 4th ed, 12, 1241 (1929).

ARSENIC ACID

Function: Cotton desiccant (1)(2)

Chemical Name: Arsenic acid

Formula: H_3AsO_4

Trade Names: Desiccant L-10®
Scorch®

Manufacture (3)

Arsenic acid is produced on a batch basis, to be used as such, or as an intermediate in the manufacture of arsenate pesticides. A schematic of the arsenic acid manufacturing process is shown in Figure 7. Nitric acid is mixed with arsenic trioxide to form arsenic acid. Unspent

Figure 7: Arsenic Acid Production

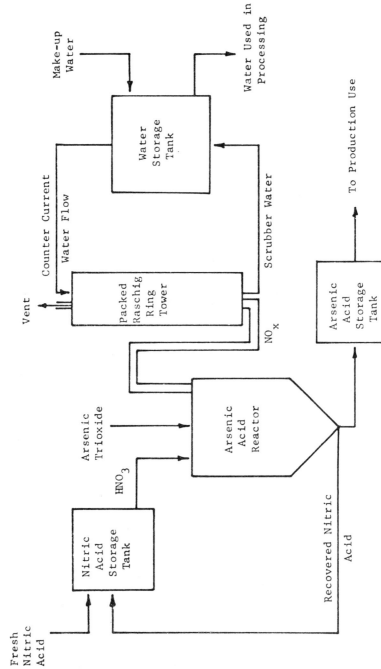

nitric acid is recovered. Arsenic acid is stored, to be utilized as such, or in the manufacture of calcium and lead arsenate.

The arsenic acid reaction produces nitrous oxide fumes, which are scrubbed in a Raschig ring packed tower, with countercurrent (recycled) water flow. The water is returned to a storage tank, where a portion is employed for processing and vat cleanup. The remainder is reused in the NO_x scrubber tower. The portions used for processing and vat cleanup are evaporated to dryness in rotary gas-fired driers, to recover product. There is thus no wastewater associated with this process.

Toxicity

Arsenic acid has an acute oral LD_{50} for rats of 100 mg/kg (moderately toxic).

Allowable Limits on Exposure and Use

Air: The threshold limit value for arsenic and its compounds in air has been conditionally set at 0.5 mg/m^3 as of 1979. An intended change to 0.2 mg/m^3 has been indicated, however (B-23).

Water: In water, EPA has set (B-12) limits of 50 μg/l of arsenic for domestic water supplies based on health considerations and 100 μg/l for water used for irrigation of crops.

In water, EPA has subsequently suggested (B-26) a limit for arsenic of 57 μg/l as a 24-hour average to protect freshwater aquatic life. The concentration should never exceed 130 μg/l at any time.

The criterion to protect saltwater aquatic life is 29 μg/l as a 24-hour average, never to exceed 67 μg/l.

For human health protection from potential carcinogenic effects through ingestion of water and contaminated aquatic organisms, the ambient water concentration should be zero. At a risk level of 10^{-5}, arsenic concentration should be 0.02 μg/l.

Product Use: A rebuttable presumption against registration was issued on October 18, 1978 by US EPA on the basis of oncogenicity, teratogenicity and mutagenicity.

A tolerance set by the US EPA for arsenic acid in or on raw agricultural commodities is as follows:

	40 *CRF* Reference	Parts per Million
Cotton, seed	180.180	4.0

References

(1) Spencer, E.Y., *Guide to the Chemicals Used in Crop Protection,* 6th ed., p. 21, London, Ontario, Agriculture Canada (January 1973).
(2) Culver, W.H., U.S. Patent 3,130,035, April 21, 1964, assigned to Pennsalt Chemicals Corp.
(3) Paterson, J.W., *State of the Art of the Inorganic Chemicals Industry: Inorganic Pesticides,* Report EPA-600/2-74/0092, Washington, D.C., Environmental Protection Agency (March 1975).

ASULAM

Function: Herbicide (1)(2)(3)

Chemical Name: Methyl [(4-aminophenyl) sulfonyl]-carbamate

Formula:

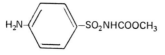

$$H_2N-\text{C}_6H_4-SO_2NHCOOCH_3$$

Trade Names: M & B 9057 (May and Baker, Ltd.)
Asulox® (May and Baker, Ltd.)

Manufacture (3)

Asulam is made by the reaction of 4-aminobenzenesulfonamide (sulfanilamide) (after suitable protection of the amino group) with methyl chloroformate. The reaction is carried out in an aqueous or organic medium in the presence of a basic condensing agent, preferably at a temperature from 10° to 20°C in aqueous media and at the reflux temperature in organic media. Preferably the reaction is carried out in water in the presence of sodium hydroxide or in acetone in the presence of potassium carbonate.

Toxicity

The acute oral LD_{50} value for rats is greater than 5,000 mg/kg; hence this product is insignificantly toxic.

Allowable Limits on Exposure and Use

Product Use: The tolerance set by the US EPA for asulam in or on raw agricultural commodities is as follows:

	40 *CFR* Reference	Parts per Million
Sugarcane	180.360	0.1

References

(1) Worthing, C.R., *Pesticide Manual,* 6th ed., p. 21, British Crop Protection Council (1979).
(2) Spencer, E.Y., *Guide to the Chemicals Used in Crop Protection,* 6th ed., p. 22, London, Ontario, Agriculture (January 1973).
(3) Carpenter, K., Heywood, B.J., Parnell, E.W., Metivier, J. and Boesch, R., British Patent 1,040,541, September 1, 1966, assigned to May and Baker, Ltd.

ATRAZINE

Function: Herbicide (1)(2)(3)

Chemical Name: 6-chloro-N-ethyl-N'-(1-methylethyl)-1,3,5-triazine-2,4-diamine

Formula:

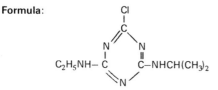

Trade Names: G-30027 (Ciba-Geigy)
Gesaprim® (Ciba-Geigy)
In Primatol® (Ciba-Geigy)

Manufacture

Atrazine is prepared by the reaction of cyanuric chloride with one equivalent of ethylamine followed by one equivalent of isopropylamine (in the presence of an acid-binding agent).

Figure 8 is a production and waste schematic for atrazine manufacture.

The following is the production chemistry involved.

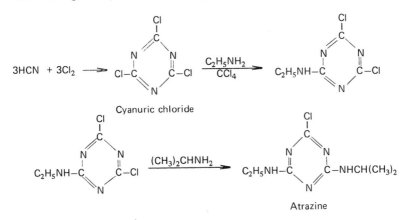

Atrazine

Process Wastes and Their Control

Air: Air emissions of 1.9 kg of hydrocarbons per metric ton of pesticide produced have been reported (B-15).

Waste gases from the cyanuric chloride manufacturing unit pass through the alkali scrubber to prevent the emission of hydrogen cyanide, HCl, and cyanuric chloride into the atmosphere (B-7).

Dusts are emitted in the formulation and packaging areas. Losses are estimated to be less than 1%. Fabric filters or wet scrubbers are employed to control dust emissions (B-10).

Water: Scrubber water as well as the liquid wastes from the filtration and formulation are treated and then discharged to a river or to a deep well.

Sanitary wastes from the plant are chlorinated before they are discharged. The BOD of the waste going to the river is 500 lb/day at the 100 million lb/yr production rate (B-10).

The present method of disposal of the alkali scrubbing wastes from cyanuric acid manufacture is deep well disposal (B-7). This is subsequent to filtration (to remove the insoluble residues) and neutralization. Deep well injection is not an acceptable means of disposal for these scrubber wastes, however, because there is:

1. Possible infiltration of brine into water table.
2. Presence of residues such as sodium cyanide and cyamelide.
3. Nondetoxification of cyanuric acid component.

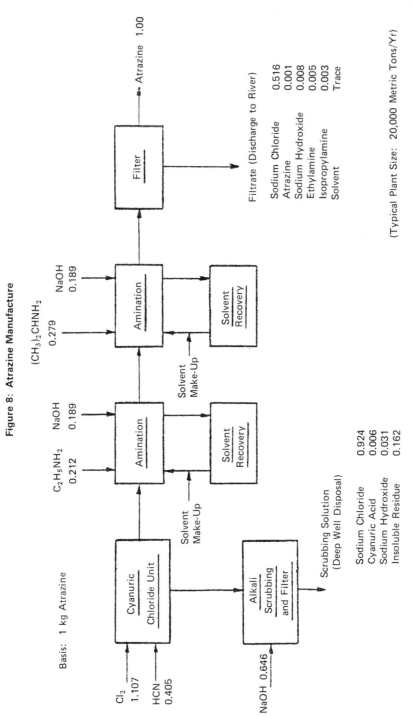

Figure 8: Atrazine Manufacture

Basis: 1 kg Atrazine

(Typical Plant Size: 20,000 Metric Tons/Yr)

Source: Reference (B-1)

To successfully treat the alkali scrubber wastes from cyanuric chloride manufacture it is necessary to detoxify the cyanuric acid and insoluble residues, and then treat the detoxified wastes which are contained within an 8.2% saline solution. A process (see Figure 9) that meets these needs would comprise: pH adjustment followed by ozonation, then biotreatment, plus evaporation and/or sale.

Product Waste Disposal

Atrazine is hydrolyzed by either acid or base as shown:

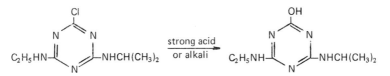

The hydroxy compounds are generally herbicidally inactive, but their complete environmental effects are uncertain. However, the method appears suitable for limited use and quantities of triazine (B-3).

Atrazine underwent >99% decomposition when burned in a polyethylene bag, and combustion with a hydrocarbon fuel would appear to be a generally suitable method for small quantities. Combustion of larger quantities would probably require the use of a caustic wet scrubber to remove nitrogen oxides and HCl from the product gases (B-3).

Toxicity

Atrazine has an acute oral LD_{50} for rats of about 3,000, hence this product is only sightly toxic.

Atrazine, propazine and simazine all appear to have low chronic toxicity. The only good carcinogenicity feeding study done on these compounds did not reveal a significant increase in cancer incidence over controls. On the basis of these chronic studies, an ADI was calculated for each of these compounds. The ADI for atrazine is 0.0215 mg/kg/day, for propazine 0.0464 mg/kg/day, and for simazine 0.215 mg/kg/day (B-22).

Allowable Limits on Exposure and Use

Air: The threshold limit value for atrazine in air has been set at 10 mg/m³ as of 1979 (B-23).

Product Use: The tolerances set by the US EPA for atrazine in or on raw agricultural commodities are as follows:

	40 CFR Reference	Parts per Million
Cattle, fat	180.220	0.02 N
Cattly, mbyp	180.220	0.02 N
Cattle, meat	180.220	0.02 N
Corn, field, fodder	180.220	15.0
Corn, field, forage	180.220	15.0
Corn, fresh (inc. sweet) (k+cwhr)	180.220	0.25
Corn, grain	180.220	0.25
Corn, pop, fodder	180.220	15.0
Corn, pop, forage	180.220	15.0
Corn, sweet, fodder	180.220	15.0
Corn, sweet, forage	180.220	15.0
Eggs	180.220	0.02 N
Goats, fat	180.220	0.02 N
Goats, mbyp	180.220	0.02 N

(continued)

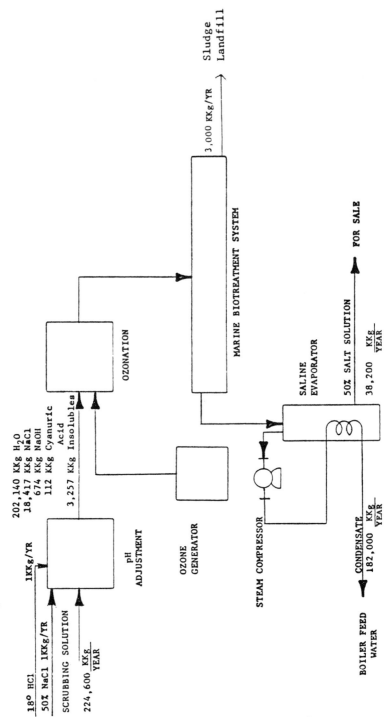

Figure 9: Improved Process for Treatment of Alakli Scrubber Wastes from Cyanuric Chloride Manufacture

Source: Reference (B-7)

	40 *CFR* Reference	Parts per Million
Goats, meat	180.220	0.02 N
Grasses, range	180.220	4.0
Grasses, rye, perennial	180.220	15.0
Hogs, fat	180.220	0.02 N
Hogs, mbyp	180.220	0.02 N
Hogs, meat	180.220	0.02 N
Horses, fat	180.220	0.02 N
Horses, mbyp	180.220	0.02 N
Horses, meat	180.220	0.02 N
Macadamia nuts	180.220	0.25
Milk	180.220	0.02 N
Pineapples	180.220	0.25
Pineapples, fodder	180.220	10.0
Pineapples, forage	180.220	10.0
Poultry, fat	180.220	0.02 N
Poultry, mbyp	180.220	0.02 N
Poultry, meat	180.220	0.02 N
Sheep, fat	180.220	0.02 N
Sheep, mbyp	180.220	0.02 N
Sheep, meat	180.220	0.02 N
Sorghum, fodder	180.220	15.0
Sorghum, forage	180.220	15.0
Sorghum, grain	180.220	0.25
Sugarcane	180.220	0.25
Sugarcane, fodder	180.220	0.25
Sugarcane, forage	180.220	0.25
Wheat, fodder	180.220	5.0
Wheat, grain	180.220	0.25
Wheat, straw	180.220	5.0

References

(1) Worthing, C.R., *Pesticide Manual,* 6th ed., p. 22, British Crop Protection Council (1979).
(2) Spencer, E.Y., *Guide to the Chemicals Used in Crop Protection,* 6th ed., p. 23, London, Ontario, Agriculture Canada (January 1973).
(3) Gysin, H. and Knusli, E., U.S. Patent 2,891,855, June 23, 1959, assigned to J.R. Geigy AG.
(4) Midwest Research Institute, *The Pollution Potential in Pesticide Manufacturing,* Wash. DC, U.S. Environmental Protection Agency (June 1972).

AZINPHOS-ETHYL

Function: Insecticide and acaricide (1)(2)

Chemical Name: O,O-diethyl S-[(4-oxo-1,2,3-benzotriazin-3(4H)-yl)methyl] phosphorodithioate

Formula:

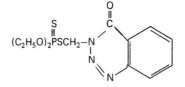

Trade Names: Bayer 16,259
R 1513 (Bayer)
Gusathion A® (Bayer)
Ethyl Guthion®

Manufacture (3)

This type of compound can be obtained by reacting N-halogenomethyl-benzazimides, which may be substituted in the benzene nucleus with the salts of dialkyldithio-phosphoric acids. The reaction is preferably carried out in an inert diluent. Suitable diluents are especially ketones, but alcohols or other solvents such as benzene or toluene may also be employed.

Due to the high reactivity of N-halogenomethyl-benzazimides, the reaction starts readily at room temperature; it is advantageous, however, to complete the reaction at slightly elevated temperatures. In the case of N-chloromethylbenzazimide (see azinphos-methyl for preparation) and the sodium salt of diethyldithiophosphoric acid, the reaction proceeds according to the following equation:

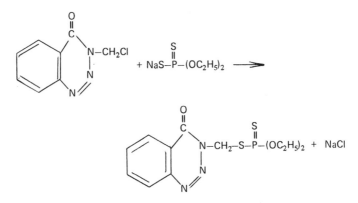

The N-halogenomethyl-benzazimides employed as starting products for this process are obtainable from the corresponding methylol compounds with the aid of halogenating agents in known manner. The compounds obtained are either solid, crystalline substances having a low melting point or nondistillable, water-insoluble oils.

Process Wastes and Their Control

Air: Air emissions of 0.16 kg of SO_2, 0.13 kg of NO_2, 0.75 kg of hydrocarbons and 0.5 kg of ethion per metric ton of pesticide produced have been reported (B-15).

Product Waste: This compound is thermally stable but "readily" hydrolyzed by alkali. Analysis is accomplished by alkaline hydrolysis to give dialkyl phosphorodithioic acid. Other chemical properties are similar to azinphos-methyl (B-3).

Toxicity

The acute oral LD_{50} for rats is 17.5 mg/kg (highly toxic).

References

(1) Worthing, C.R., *Pesticide Manual,* 6th ed., p. 23, British Crop Protection Council (1979).
(2) Spencer, E.Y., *Guide to the Chemicals Used in Crop Protection,* 6th ed., p. 24, London, Ontario, Agriculture Canada (January 1973).
(3) Lorenz, W., U.S. Patent 2,758,115, August 7, 1965, assigned to Farbenfabriken Bayer AG.

AZINPHOS-METHYL

Function: Insecticide and acaricide (1)(2)

Chemical Name: O,O-dimethyl-S-[(4-oxo-1,2,3-benzotriazin-3(4H)-yl)methyl] phosphorodithioate

Formula:

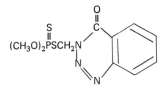

Trade Names: Bayer 17,147 (Bayer)
R-1582 (Bayer)
Guthion® (Bayer)
Gusathion M® (Bayer)

Manufacture (3)

In a first step, N-hydroxymethyl benzazimide may be synthesized as follows: 250 g of benzazimide are slightly heated with 1.6 liters of 30% formaldehyde and 300 ml of water while stirring on the water bath for 15 minutes. The mixture solidifies to a viscous paste. By filtering the paste with suction, the methylol compound is obtained in almost colorless needles. MP, 134° to 135°C; yield, 277 to 300 g, i.e, 92 to 100% of the theoretical.

Alternatively, the methylol compound can be obtained by dissolving the benzazimide, which is still wet from its manufacture, with twice the quantity of a 30% formaldehyde solution and eight times the quantity of water on the water bath, filtering the solution with some charcoal, and allowing the solution to cool. The yield of N-hydroxymethyl-benzazimide is almost quantitative.

Then, this product is converted to N-chloromethyl benzazimide in a second step as follows: 700 g (4 mols) of dry N-hydroxymethyl-benzazimide are suspended in 1.5 to 2 liters of chloroform. 1.8 mols (about 130 ml) of thionylchloride are added rapidly, the temperature rising to 35°C. Another quantity of thionylchloride, 3.8 mols (about 270 ml), is run in so as to obtain a temperature of 40° to 50°C. The mixture is stirred at 60° to 65°C for one hour, the residue separated by filtering and the solvent distilled off. The solidified residue is dissolved in 1 liter of acetone and the solution is poured into 4 liters of water with vigorous stirring to prevent the chloride from precipitating in clc⁺s. The mixture is filtered with suction, the filter cake washed until free from acid, and the resulting N-chloromethyl-benzazimide dried in air. MP, 124°C; yield, 590 to 700 g, i.e., 75 to 89% of the theoretical. By recrystallizing from ten times the quantity of isopropyl alcohol, the product is obtained in an entirely pure condition; MP, 125°C.

The following is then a specific example of the conduct of the overall process for the manufacture of azinphos-methyl. 38 g of O,O-dimethyldithiophosphoric acid are neutralized in 100 cc of acetone and 5 cc of water with 28 g of sodium bicarbonate. At 25° to 30°C, 40 g of N-chloromethyl-benzazimide (MP 125°C) dissolved in 400 cc of acetone are added. After stirring at 50° to 60°C for one hour the precipitated sodium chloride is separated by filtering and the solvent distilled.

The remaining oil is taken up in benzene, washed first with sodium bicarbonate solution and then with water. After drying over sodium sulfate the solvent is distilled off. The product is left in the form of an oil and solidifies. After recrystallizing from methanol, colorless crystals with a 72°C MP are obtained. The yield amounts to 12 g.

Process Wastes and Their Control

Air: Air emissions of 0.16 kg of SO_2, 0.13 kg of NO_2, 0.75 kg of hydrocarbons and 0.5 kg of guthion per metric ton of pesticide produced have been reported (B-15).

Water: The wastewater treatment system for a plant making azinphos-methyl and disulfoton is shown in Figure 10.

This system treats disulfoton and azinphos-methyl manufacturing wastewater and waters from nonorganophosphorus (nitrogen-containing) agricultural chemical production. The wastewater treatment system consists of: solvent removal; gross pH adjustment; polymer reaction tank of unspecified function; an equalization tank which has a 3-day hold to adjust for variations in flow and concentration fluctuation; dilution to 60% with well water as the wastes leave the equalization tank; a splitter tank which directs wastewater to either one or both biotrains; biotrains which are aerobic liquor mass digestors; and a final clarifier before discharge.

The production site personnel would not permit sampling of the raw production wastewater. A midtreatment sample was taken from the site composite sampler after the splitter tank (2302). This sample was at pH 7.6 and the average flow rate during the sampling period was 1,430 gal/min. The posttreatment sample was taken from the site's composite sampler after the final clarifier (2303). This sample was at pH 7.3 and the average flow rate during the sampling period was 1,260 gal/min.

The wastewater treatment system at the site is designed to remove organic compounds by the following mechanism: solvent stripping; adsorption onto sludge and polymer; neutral hydrolysis, aerobic biodegradation, oxidation, and adsorption to secondary sludge. Because a pretreatment sample could not be taken, an overall evaluation of the treatment system could not be made. Analyses of midtreatment and posttreatment samples allow an evaluation of the aerobic biodegradation and secondary clarifier (B-14).

Product Waste Disposal: Although this compound is chemically stable in storage, it is decomposed at elevated temperatures with evolution of gas, and rapidly decomposed in cold alkali to form anthranilic acid and other decomposition products. Fifty percent hydrolysis at pH 9 and 70°C requires 0.6 hour; 8.9 hours at pH 5 and 70°C, 240 days at pH 5 and 20°C (B-3).

Toxicity

Azinphos-methyl exhibits an acute oral LD_{50} for rats of 16.4 mg/kg (highly toxic).

Azinphos-methyl is an organophosphorus insecticide whose principal application is in agriculture. Its mode of action is inhibition of the enzyme acetylcholinesterase. It has high acute toxicity, and its chronic toxicity is moderate. Based on 2-year feeding studies in rats and dogs, an ADI at 0.0125 mg/kg/day was calculated.

There is a pressing need for studies on the metabolism of azinphos-methyl in mammalian systems. It is difficult to understand how a compound could have come to be so extensively used when so little is known of its fate in mammalian systems, as well as in soil and the environment.

Studies on the potential of azinphos-methyl for mutagenicity, teratogenicity, and carcinogenicity need to be conducted. There is almost nothing in the literature on the behavior of this compound in these respects. Data on the behavior of azinphos-methyl in water and the likelihood of its appearing in drinking water are needed. Studies on its environmental transport would also be useful in this respect (B-22).

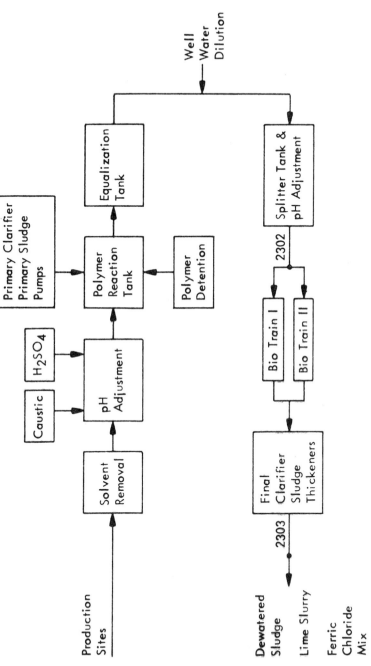

Figure 10: Wastewater Treatment System for Disulfoton and Azinphos-Methyl Production

Source: Reference (B-14)

Allowable Limits on Exposure and Use

Air: The threshold limit value for azinphos-methyl in air has been set at 0.2 mg/m^3 as of 1979. The tentative short term exposure limit value is 0.6 mg/m^3 (B-23).

Water: In water, EPA set criteria (B-12) of 0.01 μg/l azinphos-methyl for the protection of freshwater and marine aquatic life.

Product Use: The tolerances set by the US EPA for azinphos-methyl in or on raw agriculture commodities are as follows:

	40 *CFR* Reference	Parts per Million
Alfalfa	180.154	2.0
Alfalfa, hay	180.154	5.0
Almonds	180.154	0.3
Almonds, hulls	180.154	10.0
Apples	180.154	2.0
Apricots	180.154	2.0
Artichokes	180.154	2.0
Barley, grain	180.154	0.2
Barley, straw	180.154	2.0
Beans, dry	180.154	0.3
Beans, snap	180.154	2.0
Blackberries	180.154	2.0
Blueberries	180.154	5.0
Boysenberries	180.154	2.0
Broccoli	180.154	2.0
Brussels sprouts	180.154	2.0
Cabbage	180.154	2.0
Cattle, fat	180.154	0.1
Cattle, mbyp	180.154	0.1
Cattle, meat	180.154	0.1
Cauliflower	180.154	2.0
Celery	180.154	2.0
Cherries	180.154	2.0
Citrus fruits	180.154	2.0
Clover	180.154	2.0
Clover, hay	180.154	5.0
Cotton, seed	180.154	0.5
Crabapples	180.154	2.0
Cranberries	180.154	2.0
Cucumbers	180.154	2.0
Eggplant	180.154	0.3
Filberts	180.154	0.3
Goats, fat	180.154	0.1
Goats, mbyp	180.154	0.1
Goats, meat	180.154	0.1
Gooseberries	180.154	5.0
Grapes	180.154	5.0
Grass, pasture, hay	180.154	5.0
Grasses, pasture (green)	180.154	2.0
Horses, fat	180.154	0.1
Horses, mbyp	180.154	0.1
Horses, meat	180.154	0.1
Kiwi fruit	180.154	10.0
Loganberries	180.154	2.0
Melons	180.154	2.0
Milk	180.154	0.04 N
Nectarines	180.154	2.0

(continued)

	40 *CFR* Reference	Parts per Million
Nuts, pistachio	180.154	0.3
Oats, grain	180.154	0.2
Oats, straw	180.154	2.0
Onions	180.154	2.0
Peaches	180.154	2.0
Pears	180.154	2.0
Peas, black-eyed	180.154	0.3
Pecans	180.154	0.3
Peppers	180.154	0.3
Plums (fresh prunes)	180.154	2.0
Potatoes	180.154	0.3
Quinces	180.154	2.0
Raspberries	180.154	2.0
Rye, grain	180.154	0.2
Rye, straw	180.154	2.0
Sheep, fat	180.154	0.1
Sheep, mbyp	180.154	0.1
Sheep, meat	180.154	0.1
Soybeans	180.154	0.2
Spinach	180.154	2.0
Strawberries	180.154	2.0
Sugarcane	180.154	0.3
Tomatoes (pre- & post-h)	180.154	2.0
Walnuts	180.154	0.3
Wheat, grain	180.154	0.2
Wheat, straw	180.154	2.0

The tolerances set by the US EPA for azinphos-methyl in animal feeds are as follows (the *CFR* reference is to Title 21):

	CFR Reference	Parts per Million
Citrus pulp, dried (ct, gt, sh.f)	561.180	5.0
Sugarcane bagasse (ct, gt, sh.f)	561.180	1.5

The tolerances set by the US EPA for azinphos-methyl in food are as follows (the *CFR* reference is to Title 21):

	CFR Reference	Parts per Million
Soybean oil	193.150	1.0

References

(1) Worthing, C.R., *Pesticide Manual,* 6th ed., p. 24, British Crop Protection Council (1979).
(2) Spencer, E.Y., *Guide to the Chemicals Used in Crop Protection,* 6th ed., p. 25, London, Ontario, Agriculture Canada (January 1973).
(3) Lorenz, W., U.S. Patent 2,758,115, August 7, 1965, assigned to Farbenfabriken Bayer AG.

AZIPROTRYN

Function: Preemergence herbicide (1)(2)

Chemical Name: 4-azido-N-(1-methylethyl)-6-(methylthio)-1,3,5-triazin-2-amine

Formula:

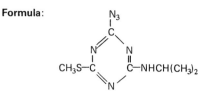

Trade Names: C-7019 (Ciba-Geigy)
Mesoranil® (Ciba-Geigy)
Brasoran® (Ciba-Geigy)

Manufacture (3)

Aziprotryn is made by the reaction of cyanuric chloride with one equivalent of isopropyl amine (in the presence of NaOH), followed by the reaction of one equivalent of sodium methyl mercaptide. The final step involves reaction with one equivalent of sodium azide.

Toxicity

The acute oral LD_{50} for rats is 3,600 mg/kg (slightly toxic).

References

(1) Worthing, C.R., *Pesticide Manual,* 6th ed., p. 25, British Crop Protection Council (1979).
(2) Spencer, E.Y., *Guide to the Chemicals Used in Crop Protection,* 6th ed., p. 26, London, Ontario, Agriculture Canada (January 1973).
(3) Ciba, Ltd., British Patent 1,093,376, November 29, 1967.

B

BACILLUS THURINGIENSIS

Function: Insecticide and larvicide (1)

Chemical Name: Structure not known

Formula: Protein of unknown structure

Trade Names: Agritol®
Biotrol® (Nutrilite Products, Inc.)
Bakthane® (Rohm and Haas)
Dipel® (Abbott Laboratories)
Larvatrol®
Thuricide® (Int. Minerals and Chemical Corporation)
Tribactur®
Sporeine®

Manufacture (2)(3)(4)(5)

The manufacturing process consists of subjecting a nutrient mixture feed to the successive process steps of:

[1] Sterilization	[3] Fermentation	[5] Drying
[2] Inoculation	[4] Separation	

More specifically, the process (3) provides a microbial insecticide of high spore content and potency produced by inoculating the surface of a nutrient medium with *Bacillus thuringiensis,* passing first stage air into the inoculated medium at a pressure of not more than about five pounds per square inch and at a rate of from about 0.2 to about 1.2, and preferably about 1.0 volume of air per volume of medium per hour until substantial sporulation occurs, the first stage air being introduced into the medium at a temperature of from about 25° to about 35°C and at a relative humidity of at least 95%, maintaining the medium at a temperature of from 25° to 35°C substantially throughout the period the first stage air is passed therethrough and thereafter passing second stage air having a relative humidity substantially lower and a temperature not in excess of 60°C but substantially higher than the first stage air through the sporulated nutrient medium to reduce the moisture content thereof to not more than about 10% by weight.

During this latter step most of the moisture is taken off by evaporation with only a negligible loss of the water soluble products of the fermentation. The art is familiar with nutrient media useful for this propagation of *Bacillus thuringiensis.* Such media appropriately contain protein and carbohydrate materials.

Alternatively media consisting essentially of essential mineral, vitamin, and nitrogen supplying materials such as are present in ordinary nutrient agar may be utilized. A nutrient medium found to be particularly useful in the production of the microbial insecticides comprises in weight percent, 30 to 40 soy bean meal; 15 to 20, sugar, e.g., dextrose; 15 to 20 fish meal; and 15 to 20 dried milk.

Bacillus thuringiensis, in common with Bacillus species generally, is an amylase producer. Accordingly, in lieu of carbohydrates other than starches, an appropriate nutrient medium may comprise a mixture of bran and expanded perlite. *Bacillus thuringiensis,* in utilization of such a medium produces amylase which, in turn, breaks down the starch content of the bran to assimilable carbohydrates such as sugars.

In a preferred embodiment, the nutrient medium is adsorbed onto a particulate inorganic carrier and nutrient substrate. Appropriate organic substrates include bran, wheat middlings, red dog flour, alfalfa meal, corn meal, peanut meal, oat hulls, rice hulls, oatmeal, corn stalks, corn cobs, kudzu vines, sorghum stalks, beet pulp, soybean vines, sweet potato vines, sweet potatoes, Irish potatoes, cottonseed meal and the like.

Vegetable materials utilized as a carrier for or as a part of the nutrient medium are preferably comminuted to provide a high ratio of surface area to volume and hence encourage vigorous bacterial growth. Experimental evidence indicates that organic carrier materials such as bran afford appreciable quantities of nutrients to *Bacillus thuringiensis.*

Utilization of inorganic carriers affords a microbial insecticide culture which may be comminuted to a selected mesh size more readily than comparable cultures propagated on media composed entirely from organic materials. Preferred inorganic carriers include expanded volcanic glasses such as perlite, obsidian, and the like; exfoliated vermiculite, pumice, volcanic ash, calcined diatomaceous earth and similar materials preferably characterized by a substantial degree of friability requisite to facilitate comminution of microbial insecticide culture.

Optimum results are obtained with mixtures of inorganic and organic carrier materials. Preferred mixed carrier media comprise from about 20% to about 80% by volume of inorganic material and about 80% to 20% by volume of organic material. A particularly appropriate carrier medium comprises from about 20% to about 80% by volume of expanded perlite and about 80% to about 20% by volume of an organic material such as bran.

Figure 11 is a block flow diagram of the overall process with particular attention to waste generation and waste treatment.

A process developed in Switzerland (6) is one in which a *Bacillus thuringiensis* microorganism is cultivated in a nutrient medium under formation of spores in such a way that the early lysis of the cells caused in the course of the submerged cultivation by the ventilation prior to the end of the formation of spores is prevented.

Process Wastes and Their Control

Air: Air emissions of 0.5 kg of particulates per metric ton of pesticide produced have been reported (B-15).

Water: The liquid wastes from the fermentation tank are either passed to an evaporation pond or are sterilized with heat, treated and discharged to a lake (B-10).

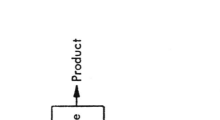

Figure 11: Production and Waste Schematic for *Bacillus thuringiensis*

Source: Reference (5)

Solid Wastes: This material degrades readily in the environment and poses little or no disposal problem (B-3).

Toxicity

There is no evidence of acute or chronic toxicity in rats.

Allowable Limits on Exposure and Use

Product Use: According to the list of tolerances set by US EPA for pesticides in or on raw agricultural commodities, *Bacillus thuringiensis* is exempt from use tolerance restrictions (see 40 *CFR* 180.101).

References

(1) Worthing, C.R., *Pesticide Manual*, 6th ed., p. 26, British Crop Protection Council (1979).
(2) Megna, J.C., U.S. Patent 3,073,749, January 15, 1963, assigned to Bioferm Corporation.
(3) Mechalas, B.J., U.S. Patent 3,086,922, April 23, 1963, assigned to Nutrilite Products, Inc.
(4) Drake, B.B. and Smythe, C.V., U.S. Patent 3,087,865, April 30, 1963, assigned to Rohm and Haas Company.
(5) Midwest Research Institute, *The Pollution Potential in Pesticide Manufacturing*, Wash. DC, U.S. Environmental Protection Agency (June 1972).
(6) Zamola, B. and Kaufez, F., U.S. Patent 4,133,716, January 9, 1979, assigned to CRC, Compagnia Ricerca Chimca, SA Switzerland.

BARBAN

Function: Selective postemergence herbicide (1)(2)

Chemical Name: 4-chloro-2-butynyl 3-chlorophenyl carbamate

Formula:

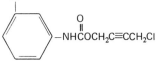

Trade Names: S-847 (Spencer Chemical Company)
Carbyne® (Spencer Chemical Company)
Oatax®

Manufacture

Barban may be prepared by the condensation of 4-chloro-2-butyn-1-ol with chlorophenyl isocyanate according to the equation (3):

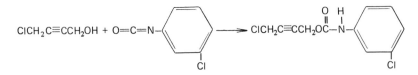

An alternative possibility is to react 4-chloro-2-butynyl chloroformate with 3-chloroaniline according to the equation:

$$\text{ClCH}_2\text{C} \equiv \text{CCH}_2\overset{\overset{\text{O}}{\|}}{\text{OC}}\text{Cl} + \text{H}_2\text{N} - \langle \bigcirc \rangle \longrightarrow \text{ClCH}_2\text{C} \equiv \text{CCH}_2\overset{\overset{\text{O}}{\|}}{\text{OC}} - \overset{\overset{\text{H}}{|}}{\text{N}} - \langle \bigcirc \rangle + \text{HCl}$$

This reaction may be carried out at room temperature or at somewhat higher and lower temperatures. This reaction is carried out at essentially atmospheric pressure. This reaction proceeds fairly rapidly, 3 hours being adequate for completion. This reaction is carried out in the liquid phase in the presence of an inert solvent. Solvents such as benzene, ether, carbon tetrachloride and chloroform may be used for this purpose.

Essentially anhydrous conditions must be maintained if highest yields are to be attained. The presence in the reaction mixture of a basic substance such as pyridine is desirable to catalyze the reaction. A jacketed kettle equipped with a reflux condenser is suitable for the conduct of this reaction. The product is fairly insoluble in a number of solvents. Hence, it is precipitated from solution and recovered by filtration.

Toxicity

The acute oral toxicity (LD_{50}) for rats is 1,350 mg/kg (slightly toxic).

Allowable Limits on Exposure and Use

Product Use: The tolerances set by the EPA for barban in or on raw agricultural commodities are as follows:

	40 *CFR* Reference	Parts per Million
Barley	180.268	0.1 N
Beets, sugar	180.268	0.1 N
Beets, sugar, tops	180.268	0.1 N
Flax, seed	180.268	0.1 N
Lentils	180.268	0.1 N
Mustard, seed	180.268	0.1 N
Peas	180.268	0.1 N
Safflower, seed	180.268	0.1 N
Soybeans	180.268	0.1 N
Sunflower, seed	180.268	0.1 N
Wheat	180.268	0.1 N

References

(1) Worthing, C.R., *Pesticide Manual,* 6th ed., p. 27, British Crop Protection Council (1979).
(2) Spencer, E.Y., *Guide to the Chemicals Used in Crop Protection,* 6th ed., p. 29, London, Ontario, Agriculture Canada (January 1973).
(3) Hopkins, T.R., U.S. Patent 3,906,614, September 29, 1959, assigned to Spencer Chemical Company.

BENAZOLIN

Function: Postemergence herbicide (1)(2)(3)

Chemical Name: 4-chloro-2-oxo-3-benzothiazoline-acetic acid

Formula:

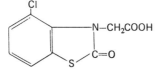

Trade Names: RD-7693 (Boots Co., Ltd.)
Cornox CWK® (Boots Co., Ltd.)
Ley Cornox® (Boots Co., Ltd.)
Tricornox® (Boots Co., Ltd.)
Legumex-Extra® (Fisons Pest Control, Ltd.)

Manufacture

Benazolin is made by the reaction of ethyl chloroacetate with 4-chloro-2-hydroxy-benzothi-azole, followed by hydrolysis to remove the ethyl group. The 4-chloro-2-hydroxy-benzothi-azole is, in turn, made starting with N-(2-chlorophenyl) thiourea which is subjected to ring closure to give 2-amino-4-chlorobenzothiazole. The 2-amino-4-chlorobenzothiazole is then converted to 4-chloro-2-hydroxy-benzothiazole.

Toxicity

The acute oral LD$_{50}$ for rats is more than 4,800 mg/kg (very slightly toxic).

References

(1) Worthing, C.R., *Pesticide Manual,* 6th ed., p. 28, British Crop Protection Council (1979).
(2) Spencer, E.Y., *Guide to the Chemicals Used in Crop Protection,* 6th ed., p. 31, London, Ontario, Agriculture Canada (January 1973).
(3) Jayne, D.W., Jr., Day, H.M. and Nolan, K.G., U.S. Patent 2,468,075, April 26, 1949, assigned to American Cyanamid Company.

BENDIOCARB

Function: Insect contact and stomach poison (1)(2)

Chemical Name: 2,2-dimethylbenzo-1,3-dioxol-4-yl-N-methyl carbamate

Formula:

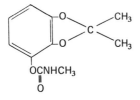

Trade Names: NC 6897 (Fisons, Ltd.)
Ficam® (Fisons, Ltd.)

Manufacture (3)(4)

Bendiocarb is made by the reaction of methyl isocyanate with 2,2-dimethyl-1,3-benzodioxol-4-ol. The following is a specific example of this process. 2,2-dimethyl-4-hydroxy-1,3-benzodioxal (14 parts) in benzene (25 parts) was treated with methyl isocyanate (6 parts) and a few drops of triethylamine with cooling. After standing for 30 minutes, the crystals of 2,2-di-methyl-1,3-benzodioxol-4-yl-N-methylcarbamate which formed were filtered and washed with benzene then with petroleum ether (BP below 40°C) to yield the pure compound as a white solid, MP 129° to 130°C (16 parts, 85% yield).

Toxicity

The acute oral LD_{50} for rats is 180 mg/kg (B-5), which indicates moderate toxicity.

References

(1) Worthing, C.R., *Pesticide Manual,* 6th ed., p. 29, British Crop Protection Council (1979).
(2) Spencer, E.Y., *Guide to the Chemicals Used in Crop Protection,* 6th ed., p. 312, London, Ontario, Agriculture Canada (January 1973).
(3) Gates, P.S. and Gillon, J., British Patent 1,220,056, January 20, 1971, assigned to Fisons Pest Control, Ltd.
(4) Gates, P.S. and Gillon, J., U.S. Patent 3,736,338, May 29, 1973, assigned to Fisons Ltd.

BENFLURALIN

Function: Preemergence herbicide (1)(2)(3)

Chemical Name: N-butyl-N-ethyl-2,6-dinitro-4-(trifluoromethyl)benzeneamine

Formula: $CH_3CH_2-N-CH_2CH_2CH_2CH_3$

Trade Names: EL-110 (Eli Lilly and Co.)
Balan® (Eli Lilly and Co.)
Bonalan® (Eli Lilly and Co.)
Quilan® (Eli Lilly and Co.)
Benefin®
Bethrodine®
Binnell®

Manufacture

Benfluralin is made from 1-chloro-4-trifluoromethylbenzene which in turn may be made by the reaction of 4-chloro-3,5-dinitrobenzoic acid with sulfur tetrafluoride (3). The product of the first step is then reacted with butylethylamine to give benfluralin.

Process Wastes and Their Control

Air: Air emissions for benfluralin manufacture have been reported as follows (B-15):

Constituent	Kilograms per Metric Ton Pesticide Produced
Particulates	0.94
SO_2	1.20

(continued)

Constituent	Kilograms per Metric Ton Pesticide Produced
NO_x	0.54
Hydrogen chloride	3.2
HF	0.12

Water: The wastewaters from benfluralin manufacture contain NaCl (B-10).

Product Wastes: This liquid herbicide is reported to be susceptible to decomposition by ultraviolet radiation (B-3).

Toxicity

Benfluralin is relatively nontoxic (LD_{50} = 10,000) (B-3).

Allowable Limits on Exposure and Use

Product Use: The tolerances set by the EPA for benfluralin in or on raw agricultural commodities are as follows:

	40 *CFR* Reference	Parts per Million
Alfalfa	180.208	0.05 N
Clover	180.208	0.05 N
Lettuce	180.208	0.05 N
Peanuts	180.208	0.05 N
Trefoil, birdsfoot	180.208	0.05 N

References

(1) Worthing, C.R., *Pesticide Manual,* 6th ed., p. 30, British Crop Protection Council (1979).
(2) Spencer, E.Y., *Guide to the Chemicals Used in Crop Protection,* 6th ed., p. 32, London, Ontario, Agriculture Canada (January 1973).
(3) Soper, Q.F., U.S. Patent 3,257,190, June 21, 1966, assigned to Eli Lilly and Co.

BENODANIL

Function: Systemic fungicide (1)

Chemical Name: 2-iodo-N-phenylbenzamide

Formula:

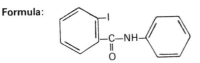

Trade Name: BAS 3170F (BASF)

Manufacture

Benodanil is made by the reaction of iodobenzoic acid, phosgene and aniline.

Toxicity

The acute oral LD_{50} value for rats is over 6,000 mg/kg showing that this product is insignificantly toxic.

Reference

(1) Worthing, C.R., *Pesticide Manual,* 6th ed., p. 31, British Crop Protection Council (1979).

BENOMYL

Function: Fungicide (1)(2)(3) and acaricide (4)

Chemical Name: Methyl 1-[(butylamino)carbonyl]-1H-benzimidazol-2-ylcarbamate

Formula:

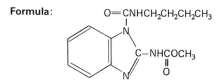

Trade Names: DuPont 1991 (DuPont)
Benlate® (DuPont)

Manufacture

Compounds of this type can be prepared (3)(5) by reacting 2-benzimidazolecarbamates with isocyanates in accordance with the following reaction:

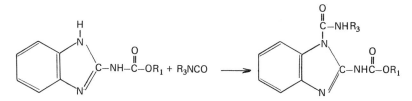

The reaction as set forth above can be carried out in different inert solvents such as chloro-form, carbon tetrachloride, methylene chloride, benzene, or cyclohexane. Mixtures of these solvents can also be used. The reaction can also be carried out without solvent by combin-ing the two reactants in a closed system and subjecting them to shear or impact force, e.g., by use of a mix muller.

The reaction temperature, in general, is not critical and can be anywhere in between the freezing point and the boiling point of the reaction mixture, provided this boiling point is be-low the temperature at which reactants and products decompose. Ambient temperature is preferred.

An alternate method for preparing the compounds involves reacting 2-benzimidazolecar-bamates with a base such as sodium hydroxide to form the sodium derivative and then re-acting the sodium derivative with a carbamyl chloride to form the desired product. This method is illustrated by the following equations:

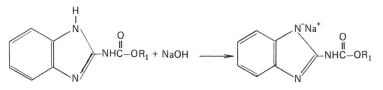

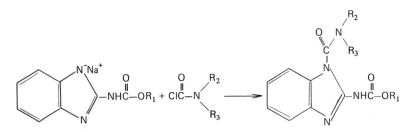

The following is a specific example of the conduct of benomyl manufacture (3). To a slurry of 19.1 parts of methyl 2-benzimidazolecarbamate and 600 parts of chloroform are added 9.9 parts of n-butylisocyanate. The reaction is stirred at room temperature until a clear solution is present or until only a small amount of solid is present.

Any solids present are removed by filtration. The solvent is removed from the filtrate under reduced pressure and essentially pure methyl 1-(butylcarbamoyl)-2-benzimidazolecarbamate is obtained by triturating the remaining white solid with hexane and collecting the product by filtration.

Toxicity

The acute oral LD_{50} for rats is in excess of 9,500 mg/kg. This product is thus insignificantly toxic.

Allowable Limits on Exposure and Use

Air: The threshold limit value for benomyl in air has been set at 10 mg/m^3 (0.8 ppm) as of 1979. The tentative short term exposure limit value is 15 mg/m^3 (1.3 ppm) (B-23).

Product Use: A rebuttable presumption against registration for benomyl has been issued on December 6, 1978 by US EPA on the basis of reduction in nontarget species, mutagenicity, teratogenicity, reproductive effects, and hazard to wildlife.

The tolerances set by the EPA for benomyl in or on raw agricultural commodities are as follows:

	40 *CFR* Reference	Parts per Million
Almonds, hulls	180.294	1.0
Apples (pre- & post-h)	180.294	7.0
Apricots (pre- & post-h)	180.294	15.0
Avocados	180.294	1.0
Bananas (pre- & post-h)	180.294	1.0
Bananas, pulp	180.294	0.2 N
Beans	180.294	2.0
Beans, vines, forage	180.294	50.0
Beets, sugar, roots	180.294	0.2
Beets, sugar, tops	180.294	15.0
Blackberries	180.294	7.0
Blueberries	180.294	7.0
Boysenberries	180.294	7.0
Cattle, fat	180.294	0.1
Cattle, mbyp	180.294	0.1
Cattle, meat	180.294	0.1
Celery	180.294	3.0
Cherries (pre- & post-h)	180.294	15.0

(continued)

	40 *CFR* Reference	Parts per Million
Citrus fruits (pre- & post-h)	180.294	10.0
Cucumbers	180.294	1.0
Dewberries	180.294	7.0
Eggs	180.294	0.1
Goats, fat	180.294	0.1
Goats, mbyp	180.294	0.1
Goats, meat	180.294	0.1
Grapes	180.294	10.0
Hogs, fat	180.294	0.1
Hogs, mbyp	180.294	0.1
Hogs, meat	180.294	0.1
Horses, fat	180.294	0.1
Horses, mbyp	180.294	0.1
Horses, meat	180.294	0.1
Loganberries	180.294	7.0
Mangoes	180.294	3.0
Melons	180.294	1.0
Milk	180.294	0.1
Mushrooms	180.294	10.0
Nectarines (pre- & post-h)	180.294	15.0
Nuts	180.294	0.2 N
Peaches (pre- & post-h)	180.294	15.0
Peanuts	180.294	0.2
Peanuts, forage	180.294	15.0
Peanuts, hay	180.294	15.0
Peanuts, hulls	180.294	2.0
Pears (pre- & post-h)	180.294	7.0
Pineapples (post-h)	180.294	35.0
Plums (fresh prunes) (pre- & post-h)	180.294	15.0
Poultry, fat	180.294	0.1
Poultry, liver	180.294	0.2
Poultry, mbyp (exc. liver)	180.294	0.1
Poultry, meat	180.294	0.1
Pumpkins	180.294	1.0
Raspberries	180.294	7.0
Rice	180.294	5.0
Rice, straw	180.294	15.0
Sheep, fat	180.294	0.1
Sheep, mbyp	180.294	0.1
Sheep, meat	180.294	0.1
Soybeans	180.294	0.2
Squash, summer	180.294	1.0
Squash, winter	180.294	1.0
Strawberries	180.294	5.0
Tomatoes	180.294	5.0

The tolerances set by the US EPA for benomyl in animal feeds are as follows (the *CFR* reference is to Title 21):

	CFR Reference	Parts per Million
Apple pomace, dried (post- & pre-h)	561.50	70.0
Citrus pulp, dried (post- & pre-h)	561.50	50.0
Grape pomace, dried	561.50	125.0
Raisin waste	561.50	125.0
Rice hulls (pre-h)	561.50	20.0

The tolerances set by the US EPA for benomyl in food are as follows (the *CFR* reference is to Title 21):

	40 *CFR* Reference	Parts per Million
Raisins	193.30	50.0

References

(1) Worthing, C.R., *Pesticide Manual,* 6th ed., p. 32, British Crop Protection Council (1979).
(2) Spencer, E.Y., *Guide to the Chemicals Used in Crop Protection,* 6th ed., p. 33, London, Ontario, Agriculture Canada (January 1973).
(3) Klopping, H.L., U.S. Patent 3,631,176, December 28, 1971, assigned to DuPont.
(4) Klopping, H.L., U.S. Patent 3,541,213, November 17, 1970, assigned to DuPont.
(5) E.I. DuPont de Nemours and Company, British Patent 1,238,977 (July 14, 1971).

BENSULIDE

Function: Herbicide (1)(2)

Chemical Name: O,O-bis(1-methylethyl)-S-[2-[(phenylsulfonyl)amino] ethyl] phosphoro-dithioate

Formula:

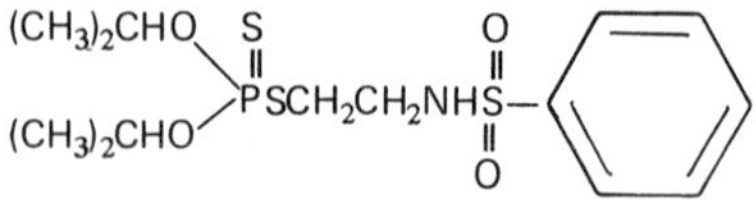

Trade Names: R-4461 (Stauffer Chemical Co.)
Betasan® (Stauffer Chemical Co.)
Prefar® (Stauffer Chemical Co.)

Manufacture

These compounds are prepared by the following reaction:

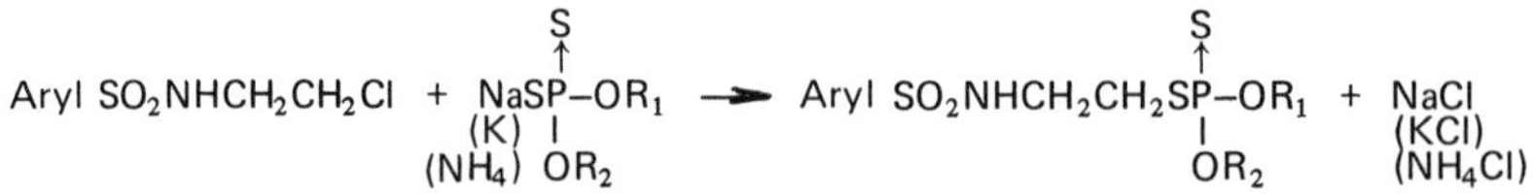

$$\text{Aryl } SO_2NHCH_2CH_2Cl + \underset{\substack{(K)\\(NH_4)}}{NaS}\overset{\overset{S}{\uparrow}}{P}\text{—}OR_1 \longrightarrow \text{Aryl } SO_2NHCH_2CH_2S\overset{\overset{S}{\uparrow}}{P}\text{—}OR_1 + \underset{\substack{(KCl)\\(NH_4Cl)}}{NaCl}$$

The beta-chloro intermediates only have been indicated in the above reaction merely for economic reasons. However, the beta-bromo and beta-iodo intermediates will also yield the equivalent phosphate derivatives. The completeness of the above reaction is dependent upon the type of solvent used. In general, polar solvents are better than nonpolar solvents and certain catalysts such as tertiary amines and dimethyl formamide aid in bringing the reaction to completion. The following examples illustrate the preparation of this class of compounds.

Example 1: The preparation of N-(beta-O,O-diethyldithiophosphorylethyl)-benzene sulfonamide is as follows. A mixture of 116 grams (0.53 mol) of N-(beta-chloroethyl)-benzene sulfonamide, 162.4 grams (0.8 mol) of ammonium-diethyldithiophosphate and 250 ml of methyl ethyl ketone was stirred and refluxed for 4 hours. The solvent was removed on a steam bath with an air jet, and the residue taken up in ethyl ether and washed three times with dilute aqueous sodium chloride.

The ether solution was dried over anhydrous magnesium sulfate, filtered and the ether removed on the steam bath with an air jet. The viscous product weighed 170 grams (90% of theory).

Example 2: The preparation of N-(beta-O,O-diisopropyldithiophosphorylethyl)-benzene sulfonamide (bensulfide) is as follows. By an analogous procedure as for Example 1, a mixture of 13.2 grams (0.06 mol) of N-(beta-chloroethyl)-benzene sulfonamide, 25.2 grams (0.1 mol) of potassium-diisopropyl dithiophosphate and 100 ml of methyl ethyl ketone yielded after a 4 hour reflux period, 20.8 grams (88% theory) of viscous product.

Process Wastes and Their Control

Product Wastes: Bensulide is stable at 80°C for 50 hours, but decomposes at 200°C in 18 to 40 hours. The manufacturer states that 20% hydrochloric acid is used for disposal. Alkaline hydrolysis may be effective also (B-3).

Toxicity

The acute oral LD_{50} value for rats is 770 mg/kg (slightly toxic).

Allowable Limits on Exposure and Use

Product Use: The tolerances set by the EPA for bensulide in or on raw agricultural commodities are as follows:

	40 *CFR* Reference	Parts per Million
Carrots	180.241	0.1 N
Cotton, seed	180.241	0.1 N
Cucurbits	180.241	0.1 N
Onions, dry bulb	180.241	0.1 N
Vegetables, fruiting	180.241	0.1 N
Vegetables, leafy	180.241	0.1 N

References

(1) Worthing, C.R., *Pesticide Manual,* 6th ed., p. 33, British Crop Protection Council (1979).
(2) Spencer, E.Y., *Guide to the Chemicals Used in Crop Protection,* 6th ed., p. 35, London, Ontario, Agriculture Canada (January 1973).
(3) Fancher, L.W. and Dewald, C.L., U.S. Patent 3,205,253, September 7, 1965, assigned to Stauffer Chemical Company.

BENTAZON

Function: Herbicide (1)

Chemical Name: 3-(1-methylethyl)-(1H)-2,1,3-benzothiadiazin-4(3H)-one 2,2-dioxide

Formula:

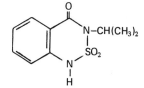

Trade Name: BAS 351H (BASF)

Manufacture (1)

Bentazon is made by the reaction of anthranilic acid with isopropylsulfamoyl chloride to give N-(isopropylsulfamoyl) anthranilic acid which is then cyclized with phosgene to give bentazon.

Toxicity

The acute oral LD_{50} for rats is 1,100 mg/kg (slightly toxic).

Allowable Limits on Exposure and Use

Product Use: The tolerances set by the US EPA for bentazon in or on raw agricultural commodities are as follows:

	CFR Reference	Parts per Million
Beans (exc. soybeans), dry	180.355	0.05
Beans (exc. soybeans), vinehay	180.355	3.0
Beans, lima	180.355	0.05
Beans, succulent	180.355	0.5
Cattle, fat	180.355	0.05
Cattle, mbyp	180.355	0.05
Cattle, meat	180.355	0.05
Corn, fodder	180.355	3.0
Corn, forage	180.355	3.0
Corn, fresh (inc. sweet) (k+cwhr)	180.355	0.05
Corn, grain	180.355	0.05
Eggs	180.355	0.05
Goats, fat	180.355	0.05
Goats, mbyp	180.355	0.05
Goats, meat	180.355	0.05
Hogs, fat	180.355	0.05
Hogs, mbyp	180.355	0.05
Hogs, meat	180.355	0.05
Horses, fat	180.355	0.05
Horses, mbyp	180.355	0.05
Horses, meat	180.355	0.05
Milk	180.355	0.02
Mint	180.355	1.0
Peanuts	180.355	0.05
Peanuts, hay	180.355	3.0
Peanuts, hulls	180.355	0.3
Peas, dry	180.355	0.05
peas, dry, vinehay	180.355	3.0
Peas, succulent	180.355	0.5
Poultry, fat	180.355	0.05
Poultry, mbyp	180.355	0.05
Poultry, meat	180.355	0.05
Rice	180.355	0.05
Rice, straw	180.355	3.0
Sheep, fat	180.355	0.05
Sheep, mbyp	180.355	0.05
Sheep, meat	180.355	0.05
Soybeans	180.355	0.05
Soybeans, hay	180.355	0.3

The tolerances set by the EPA for bentazon in animal feeds are as follows (the CFR Reference is to Title 21):

	CFR Reference	**Parts per Million**
Mint hay	561.51	4.0

Reference

(1) Worthing, C.R., *Pesticide Manual,* 6th ed., p. 34, British Crop Protection Council (1979).

BENZENE HEXACHLORIDE

Function: Insecticide (1)(2)

Chemical Name: 1,2,3,4,5,6-hexachlorocyclohexane

Formula:

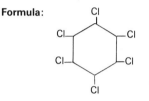

Trade Names: Agrocide®
Ambiocide®
Benzanex®
Benzex®
Gammacide®
Gammacoid®
Gammexane®
Gamaspra®
Gamtox®
Gyben®
Hexdow®
Isatox®
Lintox-Lexone®
Trivex-T®

Manufacture

The addition chlorination of benzene produces benzene hexachloride, more properly known as hexachlorocyclohexane by the reaction:

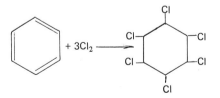

Benzene hexachloride, as it is commonly known, exists in the form of various stereoisomers. One of these, the gamma isomer, is the only one which possesses insecticidal activity. Materials containing at least 99% of the gamma isomer are called "lindane," and will be discussed in more detail in a subsequent section.

It generally is desired to control the feed ratio of chlorine to benzene so as to produce a product essentially consisting of a saturated solution of benzene hexachloride in benzene, approximately 20 to 60 grams of benzene hexachloride per 100 grams of benzene, depending upon the temperature, and preferably about 22 grams of benzene at a temperature of 30°C.

The amount of excess of benzene is preferably the minimum quantity required to maintain the hexachlorocyclohexane product in solution until it reaches the distillation apparatus. When the reaction is carried out at approximately 50°C, the molar ratio of chlorine to benzene employed is approximately 0.25 to 1, the actual combined ratio in hexachlorocyclohexane being 3 to 1. At 80°C a sufficient excess is provided if the ratio is as high as 2 to 1, whereas at 10°C the ratio cannot be much higher than 0.1 to 1 (3).

It is advantageous to operate at as low a temperature as possible (4). However, the reaction of chlorine and benzene to form benzene hexachloride is an exothermic reaction, and thus throughput capacity of a reaction apparatus and the cost of refrigerating equipment must be considered as balancing factors in determining the most economical operating conditions for any process of this type.

The reaction temperature can be varied from −60°C or lower to about 70°C. In the preferred embodiment, the temperature must be maintained below 5°C, with temperatures of between 0° and −30°C generally being most suitable. It is possible to conduct the reaction at lower temperatures, but the additional problem of adequate cooling frequently mitigates against the use of lower temperatures.

The reaction may be performed at subatmospheric and superatmospheric pressures, as well as atmospheric pressure. Cooling of the reaction may be accomplished by conducting the reaction at subatmospheric pressure and refluxing the most volatile component of the reaction mixture.

As a result of operating with a low concentration of chlorine in the large excess of benzene in the circulating stream entering the reaction zone, a short residence time of the circulating stream in the reaction zone, i.e., about 2 to 4 minutes, is sufficient to complete the addition of the chlorine present to benzene. These two factors, low chlorine concentration and short residence time in the reaction zone, seem to favor the addition chlorination of the benzene in preference to any substitution chlorination of the benzene hexachloride (5).

Solvents which are essentially inert at reaction condition and will depress the freezing point of the reaction mixture are satisfactory for operations in which gaseous or liquid chlorine are fed into a mixture of benzene and an anhydrous solvent. Even those solvents which only chlorinate slightly or which upon chlorination yield commercially useful materials may be used.

Typically, various chlorinated hydrocarbons may be used, such as methylene chloride (6). A process using methylene chloride is shown in Figure 12. The presence of both acetic anhydride and carbon tetrachloride in the reaction mixture, each being present in substantial excess of the volume of unreacted benzene present is indicated to be very desirable (7).

It is well known that benzene hexachloride may be prepared by the addition reaction of benzene and chlorine in the absence of a chlorination substitution catalyst such as ferric or aluminum chloride. This addition chlorination reaction is facilitated by resort to various catalytic means such as actinic light, that is, light usually about 2500 to 4500 A in wave length or organic peroxides.

Even the presence of minor amounts of a substitution catalyst such as ferric chloride in the reaction mixture has resulted in the preferential formation of chlorine substitution products, e.g., chlorobenzenes, rather than benzene hexachloride. For example, substitution chlorination occurs when the reactants have been in contact with iron or iron-containing alloys such as steel. Consequently, one of the conventional precautions that has been practiced prior to this process involved performing the reaction in nonferrous equipment, notably glass or nickel. This materially increases the cost of equipment used in this process.

However, the addition chlorination of benzene may be readily practiced in ferrous or other metallic equipment without encountering undue substitution chlorination by performing the reaction under substantially anhydrous conditions, e.g., by maintaining the water content in the reaction mixture such that it does not exceed 50 ppm by weight of the mixture (6).

The presence of compounds having the general formula $(RO)_2SO_2$, and methyl sulfate in particular, is claimed to enhance the gamma isomer content of the product (8). Acetic anhydride and acetyl chloride are claimed to perform similar functions (9).

Clarke et al (4) have described a continuous process for the manufacture of benzene hexachloride comprising the steps of continuously reacting benzene and chlorine in the presence of actinic light in an excess of benzene in a first reaction zone to react between about 70 and 95% of the chlorine, recycling a major portion of the reaction solution containing chlorine and benzene hexachloride in benzene solvent through the first reaction zone in a weight ratio above 4:1 relative to the weight of fresh feed reactants, and introducing the remaining portion of the reaction solution into a second zone to substantially complete the reaction of chlorine with benzene. An integrated process for the manufacture of BHC and trichlorobenzene from chlorine and benzene has been described (10).

The manufacture of tri- and tetra-chlorobenzenes as products of BHC manufacture has been described by Bareis (11).

Figure 12: Process Flow Chart for BHC Manufacture by Process Using Methylene Chloride Solvent

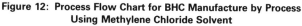
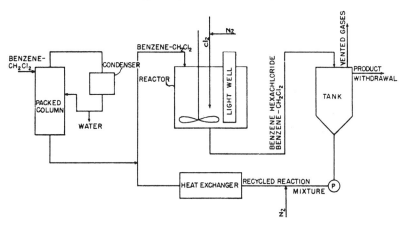

Source: Reference (6)

Toxicity

The acute oral LD_{50} value for rats is 125-200 mg/kg for mixed isomers of BHC. This is in the moderately toxic category.

The chronic toxicity of the BHC isomers is clearly related to the tumorigenic effects so far observed only in rodents. The α-isomer is the most strongly implicated; its activity is sufficient to account for the degree of hepatoma formation observed with technical BHC administration in mice. Lindane is a weaker tumorigen in mice, and is so far a questionable tumorigen in rats.

As of 1972, the FAO/WHO ADI for lindane was set at 0.0125 mg/kg/day. Later, that value was reduced to 0.001 mg/kg/day and held under temporary status because of the newer data concerning carcinogenicity. A full-scale reevaluation of the chronic toxicity of lindane was made in 1977 by the FAO/WHO. The EPA has recently announced plans to issue a presumptive notice that lindane is too hazardous for continued registered use, with the intention of reevaluating its administrative position on this insecticide.

In light of the above and taking into account the carcinogenic risk projections it is suggested that very strict criteria be applied when limits for BHC isomers are established (B-22).

Allowable Limits on Exposure and Use

Product Use: A rebuttable presumption against registration for BHC was issued on October 19, 1976 on the basis of oncogenicity, fetotoxicity and reproductive effects. Technical grade BHC was subject to voluntary cancellation on this basis.

In an action dated July 21, 1978, the US EPA eliminated all registered products containing BHC. In some instances these products were voluntarily cancelled, in others lindane was substituted for BHC nongamma isomers. These nongamma isomers may not be sold, manufactured or distributed for use in the United States.

The tolerances set by the US EPA for benzene hexachloride in or on raw agricultural commodities are as follows:

	40 *CFR* Reference	Parts per Million
Apples	180.140	1.0
Apricots	180.140	1.0
Asparagus	180.140	1.0
Avocados	180.140	1.0
Broccoli	180.140	1.0
Brussels sprouts	180.140	1.0
Cabbage	180.140	1.0
Cauliflower	180.140	1.0
Celery	180.140	1.0
Cherries	180.140	1.0
Collards	180.140	1.0
Cucumbers	180.140	1.0
Eggplant	180.140	1.0
Grapes	180.140	1.0
Kale	180.140	1.0
Kohlrabi	180.140	1.0
Lettuce	180.140	1.0
Melons	180.140	1.0
Mustard, greens	180.140	1.0
Nectarines	180.140	1.0
Okra	180.140	1.0
Onions, dry bulb only	180.140	1.0
Peaches	180.140	1.0
Pears	180.140	1.0
Pecans	180.140	0.01 N
Peppers	180.140	1.0
Plums (fresh prunes)	180.140	1.0
Pumpkins	180.140	1.0

(continued)

	40 *CFR* Reference	Parts per Million
Spinach	180.140	1.0
Squash, summer	180.140	1.0
Squash, winter	180.140	1.0
Strawberries	180.140	1.0
Swiss chard	180.140	1.0
Tomatoes	180.140	1.0

The tolerances set by the US EPA for benzene hexachloride in food are as follows (the *CFR* reference is to Title 21):

	CFR Reference	Parts per Million
Peppers, dehydrated (paprika)	193.35	5.0

References

(1) Worthing, C.R., *Pesticide Manual,* 6th ed., p. 42, British Crop Protection Council (1979).
(2) Spencer, E.Y., *Guide to the Chemicals Used in Crop Protection,* 6th ed., p. 40, London, Ontario, Agriculture Canada (January 1973).
(3) Pianfetti, J.A., et al, U.S. Patent 2,759,982, August 21, 1956, assigned to Food Machinery and Chemical Corp.
(4) Clarke, J.T., et al, U.S. Patent 2,729,603, January 3, 1956, assigned to Ethyl Corp.
(5) Nicolaisen, B.H., U.S. Patent 2,744,862, May 8, 1956, assigned to Olin Mathieson Chemical Corp.
(6) Twiehaus, H.C., U.S. Patent 2,744,145, May 1, 1956, assigned to Columbia Southern Chemical Corp.
(7) Pitt, H.M., et al, U.S. Patent 2,857,437, October 21, 1958, assigned to Stauffer Chemical Co.
(8) Neubauer, J.A., et al, U.S. Patent 2,858,260, October 28, 1958, assigned to Columbia-Southern Chemical Corp.
(9) Neubauer, J.A., et al, U.S. Patent 2,797,195, June 25, 1957, assigned to Columbia-Southern Chemical Corp.
(10) Calingaert, G., U.S. Patent 2,773,103, December 4, 1956, assigned to Ethyl Corp.
(11) Bareis, C.W., U.S. Patent 2,938,929, May 31, 1960, assigned to Olin Mathieson Chemical Corp.

BENZOXIMATE

Function: Nonsystemic acaricide (1)(2).

Chemical Name: Benzoic acid anhydride with 3-chloro-N-ethoxy-2,6-dimethoxy benzenecarboximidic acid

Formula:

Trade Name: Citrazon® (Nippon Soda Co.)

Manufacture (2)

To a solution of 88.8 grams (3.42 mols) ethyl 3-chloro-2,6-dimethoxy benzohydroxamate and 4.4 grams (0.0439 mol) triethylamine in one liter of chloroform and 34.2 grams (0.410 mol) of 40% sodium hydroxide aqueous solution, 4.81 grams (3.42 mols) benzoyl chloride is added dropwise at −10°C for about a half hour and the mixture is stirred for two more hours at the same temperature.

After the termination of the reaction, the temperature is gradually raised to room temperature. The water layer is separated and removed. To the organic solvent layer, 7 grams of 48% of an aqueous solution of sodium hydroxide is added and shaken to form a product in the water layer. A chloroform layer is dried and evaporated to obtain 115.6 grams of crystals. It was recrystallized from xylene to obtain 100 grams of pure crystals having a melting point of 69° to 71°C.

Toxicity

The acute oral LD_{50} value for rats is more than 15,000 mg/kg, which rates this product as insignificantly toxic.

References

(1) Worthing, C.R., *Pesticide Manual,* 6th ed., p. 35, British Crop Protection Council (1979).
(2) Noguchi, T., Asada, M., Sakimoto, R. and Hashimoto, K., U.S. Patent 3,821,402, June 28, 1974.

BENZOYLPROP-ETHYL

Function: Herbicide (1)(2)(3)

Chemical Name: N-benzoyl-N-(3,4-dichlorophenyl)alanine ethyl ester

Formula:

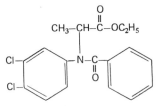

Trade Names: WL-17,731 (Shell-U.K. and Europe)
Suffix® (Shell-U.K. and Europe)
Enavene® (Shell-U.S. and Canada)

Manufacture

The preferred synthetic route (3) comprises: [1] reaction of 3,4-dichloroaniline and 2-chloropropionic acid in the presence of sodium bicarbonate; [2] esterification of product from [1] with ethanol; and [3] benzoylation with benzoyl chloride. The steps are described in more detail as follows.

To a solution of 3,4-dichloroaniline (2,686 grams, i.e., 16.6 M) in isopropanol (8,400 ml) were added water (500 ml) and 2-chloropropionic acid (3,600 grams, i.e., 33.2 M). This mixture was warmed to 40°C and sodium bicarbonate (5,600 grams, i.e., 66.4 M) was added in successive portions before heating under reflux for 113 hours.

After cooling the reaction mixture was poured into water (100 liters) and the unreacted 3,4-dichloroaniline removed by filtration. The filtrate was acidified to pH 3 to 4 with concentrated hydrochloric acid and the resultant precipitate filtered, washed and dried to yield N-(3,4-dichlorophenyl)alanine (2,455 grams, i.e., 10.6 M), mp 148° to 149°C.

Hydrochloric acid gas was passed into a solution of N-(3,4-dichlorophenyl)alanine (2,475 grams, i.e., 10.6 M) in absolute ethanol (10 liters) while heating under reflux for 6 hours. After allowing the mixture to stand overnight the bulk of the ethanol (ca 8 liters) was removed under reduced pressure and the remaining solution poured into water (10 liters).

This aqueous solution was neutralized with sodium bicarbonate and extracted with methylene chloride (3 x 1 liter). The extract was washed, dried over anhydrous magnesium sulfate, and the solvent removed under reduced pressure to leave N-(3,4-dichlorophenyl)alanine ethyl ester as a red/brown oil. The product obtained in this manner was purified by triturating with hexane over an ice-bath to give a final yield of 2,176 grams, i.e., 8.3 M of the intermediate compound, mp 37° to 38°C.

A mixture of N-(3,4-dichlorophenyl)alanine ethyl ester (2,176 grams, i.e., 8.3 M), benzoyl chloride (1,450 grams, i.e., 10.4 M) and dry benzene (5 liters) was heated under reflux. After 4 hours a further portion of benzoyl chloride (290 grams, i.e., 2.1 M) was added and heating under reflux continued for an additional 20 hours.

After cooling the solvent was removed under reduced pressure to leave N-benzoyl-N-(3,4-dichlorophenyl)alanine ethyl ester as a dark brown oil. The product obtained in this manner was purified by triturating with hexane over an ice-bath to give a final yield of 2,200 grams, i.e., 6.1 M of the required benzoylprop-ethyl, mp 50° to 52°C.

Toxicity

The acute oral LD_{50} value for rats is 1,555 mg/kg (slightly toxic).

References

(1) Worthing, C.R., *Pesticide Manual*, 6th ed., p. 36, British Crop Protection Council (1979).
(2) Spencer, E.Y., *Guide to the Chemicals Used in Crop Protection*, 6th ed., p. 256, London, Ontario, Agriculture Canada (January 1973).
(3) Yates, J. and Payne, D.H., British Patent 1,164,160; September 17, 1969, assigned to Shell International Research.

BENZTHIAZURON

Function: Preemergence herbicide (1)

Chemical Name: N-2-benzothiazolyl-N'-methylurea

Formula:

Trade Names: Bayer 60618
Gatnon® (Bayer)

Manufacture

57 parts of methylisocyanate dissolved in 1,000 parts of benzene is mixed with 150 parts of 2-aminobenzothiazole. The reacting mass is agitated and heated at reflux temperature for 2 hours. The reaction mass is then filtered while hot to obtain 155 parts of white 3-(2-benzo-thiazolyl)-1-methylurea, mp 265°C with decomposition.

Toxicity

The acute oral LD_{50} for rats is 1,280 mg/kg which puts this product in the slightly toxic category.

References

(1) Worthing, C.R., *Pesticide Manual,* 6th ed., p. 37, British Crop Protection Council (1979).
(2) Searle, N.E., U.S. Patent 2,756,135, July 24, 1956, assigned to DuPont.

S-BENZYL-DI-sec-BUTYLTHIOCARBAMATE

Function: Rice field herbicide and rice growth stimulant (1)(2)

Chemical Name: S-phenylmethyl bis(1-methylpropyl) carbamothioate

Formula:

$(sec\text{-}C_4H_9)_2-N-CO-S-CH_2-$

Trade Name: Drepamon® (Montedison SpA)

Manufacture (2)

Di-sec-butylcarbamoyl chloride has been prepared by reacting phosgene with di-sec-butyl-amine as follows: 250 grams (2.5 mols) of phosgene were absorbed, at about 5°C, under thoroughly anhydrous conditions, in 1,000 ml of anhydrous benzene. 645 grams of anhydrous di-sec-butylamine (5 mols) were subsequently added under stirring, while keeping the temperature at 0° to 5°C. A voluminous white precipitate consisting of amine hydrochloride thereupon formed.

Stirring was continued for about 3 hours, and then 500 ml of ice water introduced in order to dissolve the precipitate. The two liquid phases thus formed were separated. The organic phase was washed with 500 ml of ice water, then dried over $CaCl_2$ and concentrated under reduced pressure (15 mm Hg) at 50° to 60°C.

The resulting liquid (450 grams) was distilled under high vacuum. 425 grams of di-sec-butyl-carbamoyl chloride were thus obtained, having distilled at 85° to 90°C at 0.1 mm Hg. The yield was 89% of the theoretical value.

2.3 grams of sodium metal (0.1 mol) cut into small pieces were suspended in 25 ml of anhydrous benzene. 18.6 grams (0.15 mol) of benzyl mercaptan were added and the mixture heated under countercurrent stirring until gas generation ceased. 19.1 grams (0.1 mol) of di-sec-butylcarbamoyl chloride were then added, and the heating continued under reflux for 2 hours. After cooling to 20°C, the sodium chloride which formed was removed by filtration, and the benzene solvent was evaporated at 50° to 60°C under a reduced pressure of about 15 mm Hg. The residue was then distilled under a pressure of 0.1 mm Hg and the fraction distilling at 130° to 132°C was recovered. 20 grams of S-benzyl-N,N-di-sec-butylthiocarbamate were thus obtained.

Toxicity

The acute oral LD_{50} value for rats is more than 10,000 mg/kg which makes this product insignificantly toxic.

References

(1) Worthing, C.R., *Pesticide Manual*, 6th ed., p. 38, British Crop Protection Council (1979).
(2) Pellegrini, G., Losco, G., Quattrini, A. and Arsura, E., U.S. Patent 3,930,838, January 6, 1976, assigned to Montecatini Edison SpA.

BIFENOX

Function: Herbicide (1)(2)(3)

Chemical Name: Methyl 5-(2,4-dichlorophenoxy)-2-nitrobenzoate

Formula:

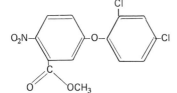

Trade Names: MC-4379 (Mobil)
Modown® (Mobil)

Manufacture (2)(3)

Compounds of this type are readily prepared by the Ullman ether synthesis reaction between the alkali metal (Na, K) salt of a halophenol and a 5-halo(Cl, Br)-2-nitrobenzoic acid or an ester, amide, or salt thereof. The 5-halo-2-nitrobenzoic acid is readily prepared by nitrating a m-halotoluene, followed by oxidation of the methyl group by well-known procedures.

For example, a stirred solution of methyl-5-chloro-2-nitrobenzoate and the potassium salt of 2,4-dichlorophenol in dimethyl sulfoxide was heated at 90°C for 17 hours. The cooled reaction mixture was diluted with water and then extracted with ether. The combined ether fractions were washed with 10% sodium hydroxide solution and then with a saturated aqueous sodium chloride solution. The ether solution was dried (Na_2SO_4) and the solvent evaporated to give a dark oily bifenox product.

Toxicity

The acute oral LD_{50} for rats is more than 6,400 mg/kg which is in the insignificantly toxic category.

Allowable Limits on Exposure and Use

Product Use: The tolerances set by the EPA for bifenox in or on raw agricultural commodities are as follows:

	40 *CFR* Reference	Parts per Million
Barley, grain	180.351	0.05
Barley, straw	180.351	0.05
Corn, field, fodder	180.351	0.05
Corn, field, forage	180.351	0.05
Corn, field, grain	180.351	0.05
Oats, grain	180.351	0.05
Oats, straw	180.351	0.05
Rice, grain	180.351	0.05
Rice, straw	180.351	0.05
Sorghum, forage	180.351	0.05
Sorghum, grain	180.351	0.05
Soybeans	180.351	0.05
Soybeans, forage	180.351	0.05
Soybeans, hay	180.351	0.05
Wheat, grain	180.351	0.05
Wheat, straw	180.351	0.05

References

(1) Worthing, C.R., *Pesticide Manual*, 6th ed., p. 40, British Crop Protection Council (1979).
(2) Thiessen, R.J., U.S. Patent 3,652,645, March 28, 1972, assigned to Mobil Oil Corp.
(3) Thiessen, R.J., U.S. Patent 3,776,715, December 4, 1973, assigned to Mobil Oil Corp.

BINAPACRYL

Function: Acaricide (1)(2)(3)

Chemical Name: 2-(1-methylpropyl)-4,6-dinitrophenyl-3-methyl-2-butenoate

Formula:

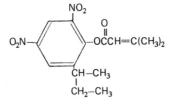

Trade Names: HOE 2784 (Farbwerke Hoechst)
Acricid® (Farbwerke Hoechst)
Endosan® (Farbwerke Hoechst)
Morocide® (Farbwerke Hoechst)
Ambox®
NIA 9044 (FMC)

Manufacture (3)(4)

The reaction is suitably carried out by dissolving 4,6-dinitro-2-sec-butylphenol and dimethyl-acrylic acid chloride at molar proportion in an inert organic solvent, for example, benzene or carbon tetrachloride and adding in the course of about 15 to 30 minutes a quantity slightly exceeding the calculated quantity of a tertiary amine, for example, pyridine or dimethylaniline, and so adjusting the rate of feed of the latter that, due to the exothermic reaction, the boiling temperature is reached.

Boiling under reflux is continued for one to two hours and the whole is then cooled, and, after introduction into a separating vessel, there is added water and additional hydrochloric acid; the mixture is then so often washed with this acid solution until all of the tertiary amine is dissolved out, and subsequently, it is so often washed with solutions that are alkaline with soda until all the acid components are removed, and finally it is washed to neutral with water. The solvent is then completely removed under reduced pressure. There are obtained in this manner esters that are technically pure and can be applied in formulated form without ado. The binapacryl product melts at 67° to 69°C.

Process Wastes and Their Control

Product Wastes: Binapacryl, a toxic miticide, is decomposed by strong acids or bases to the very toxic 2-sec-butyl-4,6-dinitrophenol (dinoseb, see below) and 3-methyl-2-butenoic acid. The product is slowly decomposed by UV radiation. It is not subject to oxidation (B-3).

References

(1) Worthing, C.R., *Pesticide Manual,* 6th ed., p. 41, British Crop Protection Council (1979).
(2) Spencer, E.Y., *Guide to the Chemicals Used in Crop Protection,* 6th ed., p. 42, London, Ontario, Agriculture Canada (January 1973).
(3) Scherer, O., Reichner, K. and Emmel, L.F., U.S. Patent 3,123,522, March 3, 1964, assigned to Farbwerke Hoechst.
(4) Scherer, O., Reichner, K. and Habicht, H., U.S. Patent 3,370,085, February 20, 1968, assigned to Farbwerke Hoechst AG.

BIS(TRIBUTYL TIN) OXIDE

Function: Fungicide and bactericide (1)

Chemical Name: Hexabutyldistannoxane

Formula: $(C_4H_9)_3SnOSn(C_4H_9)_3$

Trade Name: TBTO® (M & T Chemicals)

Manufacture

Bis(tributyl tin) oxide is made by the reaction of butyl magnesium chloride Grignard reagent on stannic oxide to give tributyl tin chloride. Hydrolysis of tributyl tin chloride gives bis(tributyl tin) oxide.

Toxicity

The acute oral LD_{50} value for rats is 200 mg/kg (moderately toxic).

Reference

(1) Worthing, C.R., *Pesticide Manual,* 6th ed., p. 45, British Crop Protection Council (1979).

BROMACIL

Function: Herbicide (1)(2)(3)(4)

Chemical Name: 5-bromo-6-methyl-3-(1-methylpropyl)-2,4(1H,3H)-pyrimidinedione

Formula:

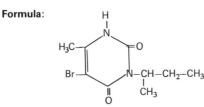

Trade Names: DuPont Herbicide 976 (DuPont)
Hyvar X® (DuPont)
Istemul®
Urox B®

Manufacture (3)(4)(5)

The reaction chemistry is believed to be approximately as follows:

$$sec-C_4H_9NH_2 + COCl_2 + NH_3 \longrightarrow sec-C_4H_9NHCONH_2 + 2HCl$$

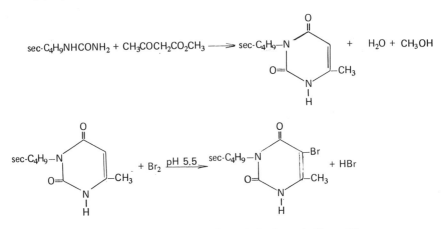

A tentative production and waste control schematic is shown in Figure 13.

Process Wastes and Their Control

Air: Air emissions of 0.4 kg of hydrocarbons, 1.0 kg of HBr and 0.5 kg of bromine per metric ton of pesticide produced have been reported (B-15).

Water: The NaBr waste stream is passed to a bromine recovery process.

Product Wastes: Bromacil is stable in water and aqueous bases. It is also stable at temperatures up to the melting point. It decomposes slowly in strong acid (B-3).

Toxicity

The acute oral LD_{50} value for rats is 5,200 mg/kg which rates as insignificantly toxic.

Bromacil is low in both acute and chronic toxicity. It appears that 1,250 ppm is a no-adverse-

Figure 13: Production and Waste Schematic for Bromacil

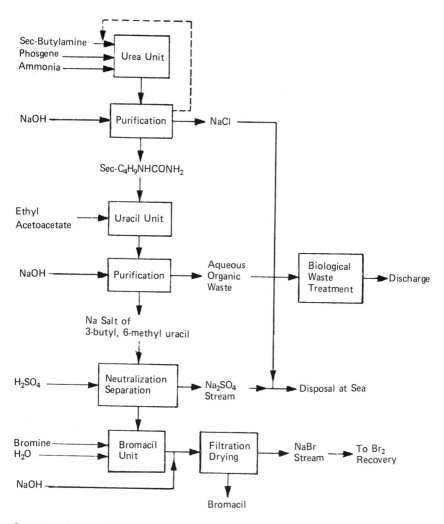

Source: Reference (5)

effect dietary concentration of bromacil in dogs. However, rats fed this concentration of bromacil in the diet exhibited abnormal thyroid pathology. In a 2-year feeding study the no-adverse-effect dose for rats was 12.5 mg/kg/day. Based on these data an ADI was calculated at 0.0125 mg/kg/day (B-22).

Allowable Limits on Exposure and Use

Air: The threshold limit value for bromacil in air is 10 mg/m^3 (1 ppm) as of 1979. The tentative short term exposure limit is 20 mg/m^3 (2 ppm) (B-23).

Product Use: The tolerances set by the US EPA for bromacil in or on raw agricultural commodities are as follows:

	40 CFR Reference	Parts per Million
Citrus fruits	180.210	0.1
Pineapples	180.210	0.1

References

(1) Worthing, C.R., Pesticide Manual, 6th ed., p. 50, British Crop Protection Council (1979).
(2) Spencer, E.Y., Guide to the Chemicals Used in Crop Protection, 6th ed., p. 51, London, Ontario, Agriculture Canada (January 1973).
(3) Loux, H.M., U.S. Patent 3,235,357, February 15, 1966, assigned to DuPont.
(4) Loux, H.M., U.S. Patent 3,352,862, November 14, 1967, assigned to DuPont.
(5) Midwest Research Institute and RvR Consultants, Production, Distribution, Use and Environmental Impact Potential of Selected Pesticides, Washington, DC, Council on Environmental Quality (August 1, 1974).

BROMOFENOXIM

Function: Herbicide (1)(2)(3)

Chemical Name: 3,5-dibromo-4-hydroxybenzaldehyde-O-(2,4-dinitrophenyl) oxime

Formula:

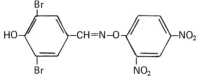

Trade Names: C-9122 (Ciba-Geigy)
Faneron® (Ciba-Geigy)
Bromfenim®

Manufacture (3)

Bromofenoxim is made by the reaction of the sodium salt of 3,5-dibromo-4-hydroxybenzaldoxime with 1-chloro-2,4-dinitrobenzene. The reaction may be carried out in a solvent, for example in ethanol, methanol, acetonitrile, or dioxane, as a rule at room temperature; in many cases it is accompanied by a spontaneous rise in temperature. The oxime ethers obtained in this manner are very easy to isolate by diluting the reaction solution with water. The ethers precipitate and may, if desired, be recrystallized.

Toxicity

The acute oral LD_{50} for rats is about 1,200 mg/kg (slightly toxic).

References

(1) Worthing, C.R., Pesticide Manual, 6th ed., p. 51, British Crop Protection Council (1979).
(2) Spencer, E.Y., Guide to the Chemicals Used in Crop Protection, 6th ed., p. 53, London, Ontario, Agriculture Canada (January 1973).
(3) Ciba, Ltd., British Patent 1,096,037, December 20, 1967.

BROMOPHOS

Function: Insecticide (1)(2)(3)

Chemical Name: O-(4-bromo-2,5-dichlorophenyl)-O,O-dimethyl phosphorothioate

Formula:

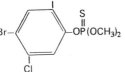

Trade Names: S-1942 (C.H. Boeringer Sohn)
Nexion® (C.H. Boeringer Sohn)
Brofene®
Brophene®

Manufacture (3)

272.9 grams of 2,5-dichloro-4-bromophenyl sodium (1.0 mol) were suspended in 700 cc of chlorobenzene and the resulting solution was admixed with 176.6 grams of O,O-dimethyl-thiophosphoric acid chloride (1.1 mols). The resulting mixture was heated for 8 hours at 100°C accompanied by stirring after 5 grams of potassium bromide were added to the reaction mixture as a catalyst.

After termination of the reaction, the sodium chloride which had precipitated out was filtered off. The filtrate was washed with dilute sodium hydroxide and was then dried over sodium sulfate. After distilling off the chlorobenzene, a colorless oil was obtained which crystallized after a short period of time to give a 95% yield of raw O,O-dimethyl-O-(2,5-dichloro-4-bromo-phenyl)-thiophosphate.

Upon purification by distillation or by recrystallization from methanol, 320 grams (85.6% of theory) of the pure product having a melting point of 51°C and a boiling point of 140° to 142°C at 0.01 mm Hg were obtained.

Process Wastes and Their Control

Product Waste Disposal: This compound is 50% hydrolyzed at pH 13 and 22°C in 3.5 hours. Bromophos is stable at pH 9. Dichlorobromophenol is the product of hydrolysis (B-3).

Toxicity

The acute oral LD_{50} value for rats is 3,750 mg/kg (slightly toxic).

References

(1) Worthing, C.R., *Pesticide Manual,* 6th ed., p. 52, British Crop Protection Council (1979).
(2) Spencer, E.Y., *Guide to the Chemicals Used in Crop Protection,* 6th ed., p. 54, London, Ontario, Agriculture Canada (January 1973).
(3) Sehring, R. and Zeile, K., U.S. Patent 3,275,718, September 27, 1966, assigned to C.H. Boehringer Sohn.

BROMOPHOS-ETHYL

Function: Insecticide (1)(2)

Chemical Name: O-(4-bromo-2,5-dichlorophenyl)-O,O-diethyl phosphorothioate

Formula:

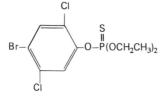

Trade Names: S-2225 (C.H. Boehringer Sohn)
Nexagan® (C.H. Boehringer Sohn)

Manufacture (2)

Using the procedure described for bromophos above, 272.9 grams of 2,5-dichloro-4-bromo-phenyl sodium were reacted with 186.5 grams of O,O-diethyl-thiophosphoric acid chloride to form 380 grams (96% of theory) of O,O-diethyl-O-(2,5-dichloro-4-bromo-phenyl)-thio-phosphate.

Process Wastes and Their Control

Product Waste Disposal: Bromophos-ethyl is slowly hydrolyzed via deethylation at pH 9 in aqueous alcohol solution. At higher pH, the phenol is removed (B-3).

Toxicity

The acute oral LD_{50} value for rats is about 100 mg/kg (moderately toxic).

References

(1) Worthing, C.R., *Pesticide Manaul,* 6th ed., p. 53, British Crop Protection Council (1979).
(2) Sehring, R. and Zeile, K., U.S. Patent 3,275,718, September 27, 1966, assigned to C.H. Boehringer Sohn.

BROMOPROPYLATE

Function: Acaricide

Chemical Name: 1-methylethyl 4-bromo-α-(4-bromophenyl)-α-hydroxybenzeneacetate

Formula:

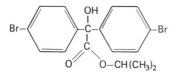

Trade Names: GS 19851 (Ciba Geigy)
Neoron (Ciba Geigy)
Acarol (Ciba Geigy)
Phenisobromolate

Manufacture

Bromopropylate is made by the reaction of 4,4'-dibromobenzilic acid with 2-propanol. To a solution of 123.5 parts of the acid in 300 parts of benzene is added 1.8 parts H_2SO_4 and 47 parts isopropanol. The reaction is conducted under reflux for 25 hours. The solvent is then distilled off and the residue recrystallized in petroleum ether.

Toxicity

The acute oral LD_{50} value for rats is over 5,000 mg/kg which puts this product in the insignificantly toxic category.

References

(1) Worthing, C.R., *Pesticide Manual,* 6th ed., p. 54, British Crop Protection Council (1979).
(2) Spencer, E.Y., *Guide to the Chemicals Used in Crop Protection,* 6th ed., p. 399, London, Ontario, Agriculture Canada (January 1973).
(2) Agripat, S.A., Belgian Patent 691,105, December 13, 1965.

BROMOXYNIL

Function: Herbicide (1)(2)(3)

Chemical Name: 3,5-dibromo-4-hydroxyphenyl cyanide

Formula:

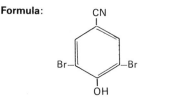

Trade Names: MB 10,064 (May and Baker, Ltd.)
Buctril® (May and Baker, Ltd.)
Brominil® (Amchem Products, Inc.)

Manufacture

Bromoxynil may be made by: [1] bromination of 4-hydroxybenzaldehyde followed by reaction with hydroxylamine to give the oxime, followed by dehydration of the oxime to the nitrile, or; [2] the action of sodium hypobromite as the brominating agent on 4-hydroxybenzonitrile.

Process Wastes and Their Control

Product Wastes: This selective herbicide is a 4-hydroxydibromo analogue of dichlobenil (although the latter is somewhat less toxic), and is marketed as the water-soluble alkali salts or as the oil-soluble octanoate ester. It is less persistent than dichlobenil, and disposal on the ground surface appears to be practical for small amounts. Incineration of bromoxynil would produce HBr and nitrogen oxides which should be removed in a caustic scrubber (B-3).

Toxicity

The acute oral LD_{50} value for rats is about 250 mg/kg (moderately toxic).

Allowable Limits on Exposure and Use

Product Use: The tolerances set by the EPA for bromoxynil in or on raw agricultural commodities are as follows:

	40 *CFR* Reference	Parts per Million
Barley, forage, green	180.324	0.1 N
Barley, grain	180.324	0.1 N
Barley, straw	180.324	0.1 N
Cattle, fat	180.324	0.1 N
Cattle, mbyp	180.324	0.1 N
Cattle, meat	180.324	0.1 N
Flax, seed	180.324	0.1 N
Flax, straw	180.324	0.1 N
Goats, fat	180.324	0.1 N
Goats, mbyp	180.324	0.1 N
Goats, meat	180.324	0.1 N
Hogs, fat	180.324	0.1 N
Hogs, mbyp	180.324	0.1 N
Hogs, meat	180.324	0.1 N
Horses, fat	180.324	0.1 N
Horses, mbyp	180.324	0.1 N
Horses, meat	180.324	0.1 N
Oat, forage, green	180.324	0.1 N
Oats, grain	180.324	0.1 N
Oats, straw	180.324	0.1 N
Rye, forage, green	180.324	0.1 N
Rye, grain	180.324	0.1 N
Rye, straw	180.324	0.1 N
Sheep, fat	180.324	0.1 N
Sheep, mbyp	180.324	0.1 N
Sheep, meat	180.324	0.1 N
Wheat, forage, green	180.324	0.1 N
Wheat, grain	180.324	0.1 N
Wheat, straw	180.324	0.1 N

References

(1) Worthing, C.R., *Pesticide Manual,* 6th ed., p. 55, British Crop Protection Council (1979).
(2) Spencer, E.Y., *Guide to the Chemicals Used in Crop Protection,* 6th ed., p. 55, London, Ontario, Agriculture Canada (January 1973).
(3) Hart, R.D. and Harris, H.E., U.S. Patent 3,397,054, August 13, 1968, assigned to Schering Corp.

BROMOXYNIL OCTANOATE

Function: Herbicide (1)(2)

Chemical Name: 2,6-dibromo-4-cyanophenyl octanoate

Formula:

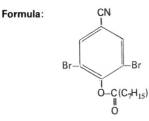

Trade Names: MB 10731 (May and Baker, Ltd.)
16272RP (May and Baker, Ltd.)
Buctril® (May and Baker, Ltd.)
Bronate® (Rhodia, Inc.)
Brominal (Amchem Products, Inc.)

Manufacture

Bromoxynil octanoate is made from bromoxynil by reacting it with octanoic acid or with octanoyl chloride in the presence of pyridine.

Toxicity

The acute oral LD_{50} value for rats is 250 mg/kg (moderately toxic).

References

(1) Worthing, C.R., *Pesticide Manual,* 6th ed., p. 56, British Crop Protection Council (1979).
(2) Spencer, E.Y., *Guide to the Chemicals Used in Crop Protection,* 6th ed., p. 56, London, Ontario, Agriculture Canada (January 1973).
(3) Heywood, B.J. and Leeds, W.G., British Patent 1,067,033, April 26, 1967, assigned to May and Baker, Ltd.

BRONOPOL

Function: Bacteriostat (1)(2)

Chemical Name: 2-bromo-2-nitro-1,3-propanediol

Formula:

$$HOCH_2-\overset{\displaystyle Br}{\underset{\displaystyle NO_2}{C}}-CH_2OH$$

Trade Name: Onyxide 500®

Manufacture

Bronopol is made by the bromination of the sodium salt of 2-nitropropane-1,3-diol (3).

Toxicity

The acute oral LD_{50} value for rats is 180 mg/kg (B-5). This makes bronopol a moderately toxic material.

References

(1) Worthing, C.R., *Pesticide Manual,* 6th ed., p. 57, British Crop Protection Council (1979).
(2) Clark, N.G., Croshaw, B. and Spooner, D.F., British Patent 1,193,954, June 3, 1970, assigned to Boots Pure Drug Company, Ltd.
(3) Darzens, E., *Compt. Rend.* 225, 943 (1947).

BUFENCARB

Function: Insecticide (1)(2)(3)4)

Chemical Name: 3-(1-methylbutyl)phenyl methylcarbamate and 3-(1-ethylpropyl)phenyl methylcarbamate (3:1)

Formula:

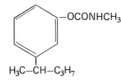   and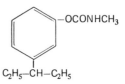

Trade Names: Ortho 5353 (Chevron Chemical)
Bux® (Chevron Chemical)

Manufacture (3)(4)

Bufencarb is made by the reaction of a mixture of alkylated phenols with methyl isocyanate. The preparation of the individual alkylphenols and their conversion to the corresponding carbamates will be described.

Preparation of m-(1-Methylbutyl)Phenol (3)—A total of 448 grams of chlorobenzene and 26.0 grams of $AlCl_3$ were charged to a liter flask equipped with an agitator, thermometer, condenser and dropping funnel. 213.2 grams of n-amyl-chloride were then added dropwise at 20° to 24°C over a 2 hour period.

The resultant deep-red solution was quenched in cold dilute HCl and the organic phase separated. This was washed three times with water, dried over Na_2SO_4 and the excess chlorobenzene stripped at atmospheric pressure. The residue was then distilled in vacuo. The desired 1-methylbutyl chlorobenzene was collected at 116° to 119°C at 23 mm. A total yield of 198 grams was obtained.

A high pressure bomb was charged with 150 ml of the above 1-methylbutyl chlorobenzene, 90 grams NaOH, 6.0 grams Cu_2Cl_2 and 1,050 ml of water. This mixture was then heated at 325°C for 6 hours. After cooling, the aqueous phase was extracted with ether to remove caustic insoluble materials. It was then acidified, extracted with ether, dried and the ether removed. The residual oil was distilled under a mechanical vacuum. The desired (1-methylbutyl)phenol was collected at 85° to 87°C at 0.7 mm.

Preparation of m-(1-Methylbutyl)Phenol N-Methylcarbamate (3)—37.0 grams of (1-methylbutyl)phenol were dissolved in 50 ml of water containing 9.0 grams NaOH, and this solution was added all at once to a solution of 26.0 grams of phosgene in 150 ml of toluene at −20°C. The temperature rose immediately to 30°C. After stirring for 15 minutes, the toluene phase was separated, washed, dried and stripped. The resultant oil was distilled and yielded 15 grams of the desired (1-methylbutyl)phenyl chloroformate boiling at 76° to 81°C at 0.5 mm.

15 grams of the chloroformate was dissolved in 50 ml of benzene and 20 ml of 33% aqueous methylamine were added at icebath temperature with good agitation. Upon completion of the reaction, the benzene phase was separated, washed with water, dried and stripped. The remaining oil was distilled in vacuo and 10 grams of m-(1-methylbutyl)phenyl N-methyl-carbamate was collected at 136° to 139°C and 0.5 mm pressure. The nitrogen analysis was calculated at 6.34%; found 6.14, 6.07%. The infrared spectrum of the compound indicated a very high meta content with only a negligible amount of para-isomer.

Preparation of m-(1-Ethylpropyl)Phenol (4)—18.3 grams of Mg were suspended in 50 ml of ether and treated with 85.0 grams of bromoethane in 100 ml of ether. After all the Mg had dissolved, 50.0 grams of methyl-3-methoxy benzoate were added, keeping the temperature below 20°C. Toward the end of the reaction a thick precipitate formed. The complex was destroyed by adding 80 grams of H_2SO_4 in 200 ml of water. The ether phase was separated, washed with water, dried over Na_2SO_4 and stripped of ether. A drop of H_2SO_4 was added to the residual oil and then distilled. A yield of 47 grams of the m-pentenyl-anisole was obtained, boiling at 87° to 91°C and 0.5 mm. This oil was reduced by dissolving in 100 ml of 95% ethanol, adding 100 mg PtO_2 and hydrogenating at an initial pressure of 50 lb/in². The pressure dropped from 50 lb/in² seven times during a 4 hour period at room temperature. When hydrogen was no longer absorbed, the catalyst was filtered and the solvent stripped. The oil remaining was the crude m-(1-ethylpropyl)anisole.

This oil was then dissolved in 250 ml of acetic acid containing 60 grams of 48% aqueous HBr. The solution was heated at 110°C for 16 hours. The acids were then stripped and the residual oil extracted in aqueous NaOH. After acidification the oil was extracted with ether, washed with water, dried over Na_2SO_4 and stripped. Distillation of the oil yielded 25 grams of the desired m-(1-ethylpropyl)phenol boiling at 80° to 85°C at 0.4 mm.

Preparation of m-(1-Ethylpropyl)Phenyl N-Methylcarbamate (4)—16 grams of the phenol, 6.0 grams of methyl isocyanate, and a drop of pyridine were sealed in a tube and heated for 16 hours at 100°C. After cooling, the oil was removed and distilled. The fraction boiling at 140°C at 0.4 mm was collected and crystallized from mixed hexanes to yield 10.5 grams of white crystals melting at 70° to 72°C.

Toxicity

The acute oral LD_{50} for rats is 87 mg/kg which is in the moderately toxic category.

Allowable Limits on Exposure and Use

Product Use: The tolerances set by the US EPA for bufencarb in or on raw agricultural commodities are as follows:

	40 *CFR* Reference	Parts per Million
Corn, fodder	180.255	0.05 N
Corn, forage	180.255	0.05 N
Corn, fresh (inc. sweet) (k+cwhr)	180.255	0.05 N
Corn, grain	180.255	0.05 N
Rice	180.255	0.05 N
Rice, straw	180.255	0.05 N

References

(1) Worthing, C.R., *Pesticide Manual,* 6th ed., p. 58, British Crop Protection Council (1979).
(2) Spencer, E.Y., *Guide to the Chemicals Used in Crop Protection,* 6th ed., p. 271, London, Ontario, Agriculture Canada (January 1973).
(3) Ospenson, J.N., Kohn, G.K. and Moore, J.E., U.S. Patent 3,062,864, November 6, 1962, assigned to California Research Corp.
(4) Kohn, G.K., Moore, J.E. and Ospenson, J.N., U.S. Patent 3,062,867, November 6, 1962, assigned to California Research Corp.

BUPIRIMATE

Function: Systemic fungicide (1)(2) (especially active against powdery mildews).

Chemical Name: 5-butyl-2-(ethylamino)-6-methyl-4-pyrimidinyl dimethylsulfamate

Formula:

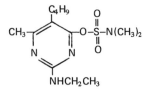

Trade Names: PP-588 (I.C.I.)
Nimrod® (I.C.I.)

Manufacture (2)

2-ethylamino-4-methyl-5-n-butyl-6-hydroxypyrimidine (ethirimol, which see) (32.1 grams, 0.11 mol) and sodium hydroxide (4.0 grams, 0.1 mol) were mixed in toluene (300 ml) and the mixture heated under reflux to remove the water formed during the preparation of the sodium salt of the 6-hydroxy pyrimidine. Dimethylsulfamoyl chloride (14.35 grams, 0.1 mol) in toluene (40 ml) was added and the whole mixture was heated under reflux for 3½ hours, and then cooled. The solution was washed three times with sodium hydroxide solution (200 ml of a 10% solution) and then twice with water (200 ml). The solution was then dried with anhydrous magnesium sulfate and the toluene distilled off to yield a dark brown oil, yield 22.15 grams (70%).

Toxicity

The acute oral LD_{50} value for rats is about 4,000 mg/kg (slightly toxic).

References

(1) Worthing, C.R., *Pesticide Manual,* 6th ed., p. 59, British Crop Protection Council (1979).
(2) Cole, A.M., Turner, J.A.W. and Snell, B.K., British Patent 1,400,710, July 23, 1975, assigned to Imperial Chemical Industries, Ltd.

BUTACARB

Function: Insecticide (1)(2) (especially against sheep blowfly larvae).

Chemical Name: 3,5-bis(1,1-dimethylethyl)phenyl methylcarbamate

Formula:

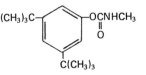

Trade Names: RD-14639 (Boots Co., Ltd.)
BTS-14639 (Boots Co., Ltd.)

Manufacture (2)

N,N-dimethylaniline (23 cc) was added dropwise to a mechanically stirred mixture of 3,5-di-tert-butylphenol (30.9 grams, 0.15 mol) and a solution of phosgene (18 grams, 0.18 mol) in dry toluene (180 cc) keeping the temperature at 20°C by external cooling. Stirring was continued for 4 hours at room temperature and then the flask was immersed in ice-water and gaseous methylamine bubbled into the stirred reaction mixture at such a rate that the temperature did not exceed 25°C. When the mixture became alkaline the flow of methylamine was interrupted, and stirring was continued for a further 30 minutes. The alkaline suspension was poured into a mixture of crushed ice and an excess of 2 N HCl, and the toluene layer was separated and washed with 2 N HCl, until free from dimethylaniline. After being washed with water followed by N NaOH to remove any unchanged dibutylphenol, the organic layer was finally washed with water, dried (Na_2SO_4), and distilled under reduced pressure to remove toluene. Crystallization of the residue from aqueous alcohol and subsequently from light petroleum (BP 60° to 80°C) gave 3,5-di-tert-butylphenyl N-methylcarbamate (20.4 grams MP 100° to 102°C).

Toxicity

The acute oral LD_{50} for rats is about 1,800 mg/kg which rates as slightly toxic.

References

(1) Martin, H. and Worthing, C.R., *Pesticide Manual,* 5th ed., p. 59, British Crop Protection Council (January 1977).
(2) Fraser, J. and Harrison, I.R., British Patent 987,254, March 24, 1965, assigned to Boots Pure Drug Co., Ltd.

BUTACHLOR

Function: Preemergence herbicide (1)(2)(3)

Chemical Name: N-(butoxymethyl)-2-chloro-2',6'-diethylacetanilide

Formula:

Trade Name: Machete® (Monsanto)

Manufacture (2)(3)

Butachlor is made by a process sequence very similar to that for alachlor (which see), except that butanol is used in place of methanol in the final reaction step. Thus, the raw materials are: diethylaniline, formaldehyde, chloroacetyl chloride, and n-butanol.

Process Wastes and Their Control

Air: Air emissions of 1.5 kg of hydrocarbons per metric ton of pesticide produced have been reported (B-15).

Product Wastes: Butachlor is an analog of alachlor and should be disposed of in the same manner (B-3).

Toxicity

The acute oral LD_{50} for rats is 3,120 mg/kg (B-5) (slightly toxic).

Butachlor and its relatives, alachlor and propachlor, appear to be fairly well tolerated by mammals (B-22). Both alachlor and butachlor are apparently tolerated by rats at up to 100 mg/kg/day in the diet, except for increased liver weight in female rats fed butachlor.

The existing toxicity data for these compounds are largely those produced by the manufacturer for registration purposes. Based on the available data, ADIs were calculated at 0.1, 0.1, and 0.01 mg/kg/day for alachlor, propachlor and butachlor, respectively.

Apparently, no long-term toxicity studies have been completed that would contribute information on reproductive effects or carcinogenic potential of these acetanilides or their degradation products, which include aniline derivatives. These studies are needed (B-22).

Allowable Limits on Exposure and Use

Product Use: The tolerances set by the US EPA for butachlor in animal feeds are as follows (the *CFR* reference is to Title 21):

	CFR Reference	Parts per Million
Rice bran	561.55	0.5
Rice hulls	561.55	1.0

References

(1) Worthing, C.R., *Pesticide Manual,* 6th ed., p. 60, British Crop Protection Council (1979).
(2) Olin, J.F., U.S. Patent 3,442,945, May 6, 1969, assigned to Monsanto Company.
(3) Olin, J.F., U.S. Patent 3,547,620, December 15, 1970, assigned to Monsanto Company.

BUTAM

Function: Preemergent herbicide (1)(2)(3)

Chemical Name: 2,2-Dimethyl-N-(1-methylethyl)-N-phenylmethyl) propanamide

Formula:

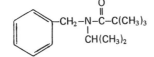

Trade Name: GPC-5544 (Gulf Oil Chemical Co.)

Manufacture (2)(3)

A one-liter reaction flask fitted with a power stirrer, heating mantle, dropping funnel, thermometer, water-cooled condenser and drying tube was charged with 149.2 g (1.0 mol) N-benzyl-N-isopropylamine, 350 ml of benzene and 111.1 g (1.1 mols) of triethylamine. The dropping funnel contained 126.6 g (1.05 mols) of pivalyl chloride which was added dropwise to the stirred reaction mixture. After the addition was completed, the reaction mixture was stirred and heated at 75° to 80°C for 18 hours. The mixture was cooled and the amine salt was collected on a vacuum filter. The filtrate was transferred to a separatory funnel and extracted with water which was followed by dilute aqueous hydrochloric acid. The organic phase was dried over sodium sulfate. After removing the drying agent, the solvent was evaporated. The liquid residue was transferred to a pot which was appropriately fitted for a simple vacuum distillation. The forecut material proved to be mainly trimethylacetyl chloride and trimethylacetic acid. The product fraction weighed 188.6 g, BP 86-87/0.07 mm.

Toxicity

The acute oral LD_{50} value for rats is 6,210 mg/kg which is insignificantly toxic.

References

(1) Worthing, C.R., *Pesticide Manual,* 6th ed., p. 61, British Crop Protection Council (1979).
(2) Cahoy, R.P., U.S. Patent 3,974,218, August 10, 1976, assigned to Gulf Research and Development Co.
(3) Cahoy, R.P., U.S. Patent 3,707,366, December 26, 1972, assigned to Gulf Research and Development Co.

BUTAMIPHOS

Function: Herbicide (1)(2)

Chemical Name: O-ethyl O-(5-methyl-2-nitrophenyl)-1-methylpropylphos-phoramidothioate

Formula:

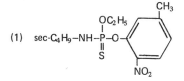

(1)

Trade Names: S-2846 (Sumitomo Chemical Co.)
Cremart® (Sumitomo Chemical Co.)

Manufacture (2)

The compound may be synthesized by reacting a thionophosphoric chloride of the formula,

$$\text{(2)} \quad sec\text{-}C_4H_9\text{-}NH\text{-}\underset{\underset{S}{\|}}{\overset{\overset{OR}{|}}{P}}\text{-}Cl$$

wherein R is ethyl with 3-methyl-6-nitrophenol, in an organic solvent in the presence of an acid binding agent.

Examples of the solvents used include aromatic solvents such as benzene and toluene, ketones such as acetone and methyl isobutyl ketone, and acetonitrile, etc.

Examples of the acid binding agents include inorganic bases such as potassium carbonate and sodium hydroxide, tertiary amines such as pyridine and triethylamine, and a mixture thereof.

The reaction temperature varies depending on the kind of solvent or acid binding agent to be used. The reaction is preferably carried out at a temperature of from room temperature to about 120°C for two to several hours.

Toxicity

The acute oral LD_{50} value for rats is 630 to 790 mg/kg which is slightly toxic.

References

(1) Worthing, C.R., *Pesticide Manual,* 6th ed., p. 62, British Crop Protection Council (1979).
(2) Satomi, T., Mukai, K., Mine, A., Hino, N., Tateishi, K. and Hirano, M., U.S. Patent 3,936,433, February 3, 1976, assigned to Sumitomo Chemical Co., Ltd.

BUTHIDAZOLE

Function: Herbicide (1)(2)

Chemical Name: 3-[5-(1,1-dimethylethyl)-1,3,4-thiadiazol-2-yl]-4-hydroxy-1-methyl-2-imidazolidinone

Formula:

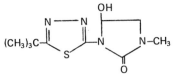

Trade Names: VEL-5026 (Velsicol Chemical Co.)
Ravage® (Velsicol Chemical Co.)

Manufacture (2)

(A) Preparation of 5-tert-Butyl-1,3,4-Thiadiazol-2-yl Isocyanate Dimer—A saturated solution of phosgene in ethyl acetate (100 ml) was charged into a glass reaction vessel equipped with a mechanical stirrer. A slurry of 5-tert-butyl-2-amino-1,3,4-thiadiazole (10 grams) in ethyl acetate (300 ml) was added to the reaction vessel and the resulting mixture was stirred for a period of about 16 hours resulting in the formation of a precipitate. The reaction mixture was then purged with nitrogen gas to remove unreacted phosgene. The purged mixture was then filtered to recover the desired product 5-tert-butyl-1,3,4-thiadiazol-2-yl isocyanate dimer as a solid having a MP of 261° to 263°C.

(B) Preparation of the Dimethyl Acetal of 2-[1-Methyl-3-(5-tert-Butyl-1,3,4-Thiadiazol-2-yl)-Ureido]Acetaldehyde—A mixture of 5-tert-butyl-1,3,4-thiadiazol-2-yl isocyanate dimer (6 grams), the dimethyl acetal of 2-methylaminoacetaldehyde (3.9 grams) and benzene (50 ml) was charged into a glass reaction flask equipped with a mechanical stirrer and reflux condenser. The reaction mixture was heated at reflux, with stirring for a period of about 5 minutes. After this time the reaction mixture was stripped of benzene to yield an oil which solidified upon standing. The resulting solid was then recrystallized from pentane to yield the desired product the dimethyl acetal of 2-[1-methyl-3-(5-tert-butyl-1,3,4-thiadiazol-2-yl)ureido]acetaldehyde having a MP of 80° to 82°C.

(C) Preparation of 1-(5-tert-Butyl-1,3,4-Thiadiazol-2-yl)-3-Methyl-5-Hydroxy-1,3-Imidazolidin-2-one— The dimethyl acetal of 2-[1-methyl-3-(5-tert-butyl-1,3,4-thiadiazol-2-yl)ureido]acetaldehyde (16 grams), concentrated hydrochloric acid (10 ml) and water (500 ml) were charged into a glass reaction vessel equipped with a mechanical stirrer, thermometer and reflux condenser. The reaction mixture was heated at reflux for a period of about 15 minutes. The reaction mixture was filtered while hot and the filtrate was then cooled, resulting in the formation of a precipitate. The precipitate was recovered by filtration, dried and was recrystallized from a benzene-hexane mixture to yield the desired product 1-(5-tert-butyl-1,3,4-thiadiazol-2-yl)-3-methyl-5-hydroxy-1,3-imidazolidin-2-one having a MP of 133° to 134°C.

Toxicity

The acute oral LD_{50} for rats is about 1,500 mg/kg (slightly toxic).

References

(1) Worthing, C.R., *Pesticide Manual,* 6th ed., p. 63, British Crop Protection Council (1979).
(2) Krenzer, J., U.S. Patent 3,904,640, September 9, 1975, assigned to Velsicol Chemical Corp.

BUTHIOBATE

Function: Fungicide (1)

Chemical Name: Butyl [4-(1,1-dimethylethyl)phenyl]-methyl 3-pyridinylcarbonimido-dithioate

Formula:

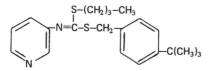

Trade Names: S-1358 (Sumitomo Chemical Co.)
Denmert® (Sumitomo Chemical Co.)

Manufacture (2)

In 100 ml of methanol was dissolved 2.3 grams (0.1 mol) of metallic sodium and 31.6 grams (0.1 mol) of p-tert-butylbenzyl-N-3-pyridyl dithiocarbamate was then added to the resulting solution at 15°C to dissolve it in the solution. Into the thus obtained solution was dropped at 15° to 20°C 15.1 grams (0.11 mol) of n-butyl bromide (n-C_4H_9Br). The resulting mixture was maintained at the temperature for 2 hours and then at 40°C for 1 hour. The resulting reaction mixture was poured into 400 ml of ice water, and the separated oil material was extracted with 200 ml of ethyl acetate. The thus obtained ethyl acetate layer was washed once with 200 ml of water, dried with anhydrous sodium sulfate and then subjected to distillation under reduced pressure to obtain 33.1 grams (yield: 89%) of a pale yellow, oily material.

Toxicity

The acute oral LD_{50} value for rats is 2,700 to 4,900 mg/kg which is slightly toxic.

References

(1) Worthing, C.R., *Pesticide Manual,* 6th ed., p. 64, British Crop Protection Council (1979).
(2) Tanaka, S., Ozaki, T., Mine, A., Tanaka, K., Yamamoto, S., Ooishi, T., Hino, N. and Satomi, T., U.S. Patent 3,832,351, August 27, 1974, assigned to Sumitomo Chemical Co.

BUTOCARBOXIM

Function: Insecticide

Chemical Name: 3-Methylthio-2-butanone O-[(methylamino)carbonyl]oxime

Formula:

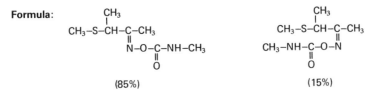

(85%)   (15%)

Trade Names: Co 755 (Wacker-Chemie)
Drawin 755® (Wacker-Chemie)

Manufacture

(A) Production of the Oxime—Into a solution of 2 mols NaOH in 310 grams of water, 1 mol of methyl mercaptan is piped in gaseous form. After that, under vigorous stirring and external cooling with ice, 1 mol of 2-chlorbutanone-3 is added drop by drop. The dripping speed is set in such a manner that the temperature remains between 30° and 50°C. After completed addition 1 mol of hydroxylamine hydrochloride is immediately added and stirred for 7 hours at 20°C. After this the aqueous layer is separated. The proton resonance spectrum of the slightly yellowish organic phase shows only 2 isomeric oximes which contain about 4% water. From analysis, the isomeric proportion of 85:15 is determined. The total yield amounts to 99% of theoretical. After distilling in a vacuum, BP 65°C/0.05 mm Hg, one obtains 97.5% of theoretical 2-thiomethylbutanone-3-oxime. The isomeric proportion remains unchanged at 85:15.

(B) Conversion of the Purified Oxime with Methylisocyanate—The oxime from (A), after addition of 0.05 ml triethylamine with the molar quantity of methylisocyanate, is stirred and externally cooled with ice to maintain the temperature at 35°C. After letting the mixture stand for 8 hours at room temperature 2-methylthiobutanone-N-methylcarbamyloxime is obtained as a colorless oily liquid. Conversion and yield are quantitative.

Toxicity

The acute oral LD_{50} value for rats is 158 to 240 mg/kg which is moderately toxic.

References

(1) Worthing, C.R., *Pesticide Manual,* 6th ed., p. 65, British Crop Protection Council (1979).
(2) Muller, F., Lohringer, W., Milles, K., Braunling, H. and Prigge, H., U.S. Patent 3,816,532, June 11, 1974, assigned to Consortium for Elektrochemische Industrie GmbH, Germany.

BUTOPYRONOXYL

Function: Insect repellent (1)(2)(3)

Chemical Name: Butyl-3,4-dihydro-2,2-dimethyl-4-oxo-2H-pyran-6-carboxylate

Formula:

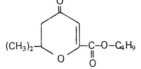

Trade Name: Indalone® (U.S. Industrial Chemicals Inc.)

Manufacture (3)

Eleven and one-half grams of sodium was dissolved in 200 cc of dry n-butyl alcohol to yield the condensing agent. A reactant mixture consisting of 54 grams of mesityl oxide and 101 grams of n-butyl oxalate was slowly added while the temperature of the reaction mixture was maintained within the range of from 40° to 45°C by external cooling. In this manner a pasty suspension of the sodium salt of the enol form was obtained.

This was diluted with 300 cc of toluene and the mixture neutralized by agitation with cold 10% sulfuric acid. The washed toluene solution of the product was fractionally distilled under reduced pressure. 96 grams of material boiling at 115° to 123°C at 3 mm was obtained. This product was an equilibrium mixture consisting predominantly of butopyronoxyl.

Toxicity

The acute oral LD_{50} for rats is 7,840 mg/kg (insignificantly toxic).

References

(1) Worthing, C.R., *Pesticide Manual,* 6th ed., p. 66, British Crop Protection Council (1979).
(2) Spencer, E.Y., *Guide to the Chemicals Used in Crop Protection,* 6th ed., p. 57, London, Ontario, Agriculture Canada (January 1973).
(3) Ford, J.H., U.S. Patent 2,138,540, November 29, 1938, assigned to Kilgore Development Corp.

BUTOXYCARBOXIM

Function: Insecticide (1)(2)

Chemical Name: 3-Methylsulfonyl-2-butanone O-[(methylamino)carbonyl]oxime

Formula:

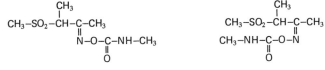

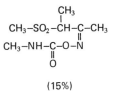

(85%)   (15%)

Trade Names: Co 859 (Wacker-Chemie)
Plant Pin® (Wacker-Chemie)

Manufacture (2)

Butoxycarboxim is made by the oxidation of butocarboxim as follows: 1 mol of butocarboxim is dissolved in 250 cm³ of glacial acetic acid and at 50°C is slowly compounded with stirring and cooling with 2 mol of a 30% peracetic acid in acetic acid ethyl ester. After evaporating the acetic acid ethyl ester in vacuum butoxycarboxim is obtained in practically quantitative yield.

Toxicity

The acute oral LD_{50} value for rats is 458 mg/kg which is moderately toxic.

References

(1) Worthing, C.R., *Pesticide Manual,* 6th ed., p. 67, British Crop Protection Council (1979).
(2) Muller, F., Lohringer, W., Milles, K., Braunling, H. and Prigge, H., U.S. Patent 3,816,532, June 11, 1974, assigned to Consortium für Elektrochemische Industrie GmbH, Germany.

BUTRALIN

Function: Herbicide (1)(2)

Chemical Name: 4-(1,1-dimethylethyl)-N-(1-methylpropyl)-2,6-dinitrobenzamine

Formula:

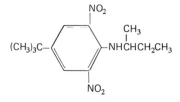

Trade Names: A-820
              Amex 820®

Manufacture (2)

In the first step of the process para-tert-butylphenol is nitrated by treating the phenol with nitric acid in acetic acid solution to give the corresponding 2,6-dinitrophenol in good yield. The chlorination of the 2,6-dinitro-4-butylphenol can be carried out by employing as the chlorinating agent thionyl chloride, phosphorus oxychloride or phosphorus pentachloride in the presence of a complexing agent such as formamide, acetamide, propionamide or their alkylated derivatives, e.g, dimethyl formamide (DMF), etc.

The chlorinating agent and acetamide are used in about equimolar amounts or if desired the chlorinating agent can be used in excess though the acetamide and chlorinating agent should ordinarily both be present in amounts of equal to or greater than the 2,6-dinitro-4-butylphenol intermediate on a molar basis.

The reaction is conveniently carried out in an inert organic solvent such as toluene, xylene and the like. A preferred reaction temperature is the reflux temperature of the reaction mixture (about 110° to 120°C) though higher or lower temperatures can also be employed with correspondingly shorter or longer reaction times. The reaction is ordinarily completed in about 12 to 16 hours at which time the excess chlorinating agent can be readily stripped from the reaction mixture by evaporating under reduced (about 30 mm) pressure.

The desired chlorinated compound remains in the supernatant liquid, while the chlorinating agent-DMF complex separates as a heavy liquid and is easily removed from the liquid supernatant. The product obtained in this way can be precipitated from an inert organic solvent such as n-hexane to give pure 2,6-dinitro-4-tert-butyl-chlorobenzene in good yield.

One gram of 2,6-dinitro-4-tert-butylchlorobenzene was allowed to react with 1 gram of sec-butylamine by adding the amine dropwise to a refluxing mixture of 50 ml dry toluene, and the 2,6-dinitro-4-tert-butylchlorobenzene. After complete addition, the mixture was refluxed eight hours, cooled to room temperature, the amine hydrochloride filtered off, and toluene and unreacted amine were removed under reduced pressure. The thick material that resulted was dissolved in hot ethanol and the product was crystallized upon cooling to yield N-sec-butyl-4-tert-butyl-2,6-dinitroaniline.

Allowable Limits on Exposure and Use

Product Use: The tolerances set by the US EPA for butralin in or on raw agricultural commodities are as follows:

	40 *CFR* Reference	Parts per Million
Beans, lima	180.358	0.1
Cotton, seed	180.358	0.1
Peas, southern	180.358	0.1
Soybeans	180.358	0.1
Soybeans, forage	180.358	0.1
Watermelons	180.358	0.1

References

(1) Worthing, C.R., *Pesticide Manual,* 6th ed., p. 68, British Crop Protection Council (1979).
(2) Damiano, J.J., U.S. Patent 3,672,866, June 27, 1972, assigned to Amchem Products, Inc.

BUTURON

Function: Herbicide (1)

Chemical Name: N'-(4-chlorophenyl)-N-methyl-N-(1-methyl-2 propynyl) urea

Formula:

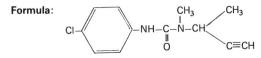

Trade Names: H-95 (BASF)
Eptapur® (BASF)
Arisan®
Butyron®

Manufacture

Buturon is made by the reaction of 4-chloro-phenyl isocyanate and methyl butynyl amine, $CH_3NHCH(CH_3)C{\equiv}CH$.

Toxicity

The acute oral LD_{50} value for rats is 3,000 mg/kg (slightly toxic).

Reference

(1) Worthing, C.R., *Pesticide Manual,* 6th ed., p. 69, British Crop Protection Council (1979).

BUTYLAMINE

Function: Fungicidal fumigant

Chemical Name: 2-butaneamine

Formula: $CH_3-CH_2-CH-CH_3$
                                $|$
                               NH_2

Trade Names: Tutane® (Eli Lilly & Co.)
                 Butafume® (BASF AG)

Manufacture

sec-Butylamine may be prepared by the reaction of 2-butanol with ammonia in the liquid phase over a nickel catalyst at 200° to 220°C and 500 to 1,000 pounds pressure.

Toxicity

The acute oral LD_{50} for rats is 380 mg/kg (moderately toxic).

Allowable Limits on Exposure and Use

Product Use: The tolerances set by the EPA for butylamine in or on raw agricultural commodities are as follows:

	40 *CFR* Reference	Parts per Million
Cattle, fat	180.321	0.75
Cattle, kidney	180.321	3.0
Cattle, mbyp (exc. kidney)	180.321	0.75
Cattle, meat	180.321	0.75
Citrus fruits (post-h)	180.321	30.0
Milk	180.321	0.75

The tolerances set by the US EPA for sec-butylamine in animal feeds are as follows (the *CFR* reference is to Title 21):

	CFR Reference	Parts per Million
Citrus molasses (ct. f.) (post-h)	561.60	90.0
Citrus pulp, dried (ct. f.) (post-h)	561.60	90.0

Reference

(1) Worthing, C.R., *Pesticide Manual,* 6th ed., p. 70, British Crop Protection Council (1979).

BUTYLATE

Function: Herbicide (1)(2)

Chemical Name: S-ethyl bis(2-methylpropyl) carbamothioate

Formula:

$$C_2H_5SCN[CH_2CH(CH_3)_2]_2$$

with O double-bonded to the carbon (carbamoyl group):

O
‖
$C_2H_5SCN[CH_2CH(CH_3)_2]_2$

Trade Names: R-1910 (Stauffer Chemical Co.)
Sutan (Stauffer Chemical Co.)

Manufacture (3)

Butylate is made by the reaction of di-isobutylamine with phosgene to give the carbamoyl chloride which is reacted with ethanethiol to give butylate. Sodium is dispersed in xylene using oleic acid as the stabilizing agent until a particle size of 5 to 200 μ in diameter is obtained. The dispersion of sodium is then transferred to a reactor which has been previously flushed out with argon (or other inert gas such as nitrogen). A solution of ethanethiol dissolved in xylene is then gradually added to the sodium dispersion over an interval of 30 minutes. The temperature is maintained at 25° to 36°C by cooling.

The sodium ethylmercaptide forms as finely divided crystals which make an easily stirrable slurry. This suspension is heated to reflux, the heat is turned off, and diisobutyl-carbamyl chloride is added over an interval to the refluxing slurry. The heat of reaction is sufficient to keep the xylene refluxing. After all of the diisobutyl-carbamyl chloride has been added, the mixture is refluxed for an additional 3 hours. It is then cooled, filtered from sodium chloride which has formed during the reaction, and the solvent is removed under reduced pressure. The residual liquid is then distilled under vacuum to give butylate.

Process Wastes and Their Control

Air: Air emissions of 205 kg of SO_2, 1.5 kg of hydrocarbons and 0.5 kg of butylate per metric ton of pesticide produced have been reported (B-15).

Toxicity

The acute oral LD_{50} value for rats is about 4,600 mg/kg (very slightly toxic).

Allowable Limits on Exposure and Use

Product Use: The tolerances set by the US EPA for butylate in or on raw agricultural commodities are as follows:

	40 *CFR* Reference	Parts per Million
Corn, field, fodder	180.232	0.1 N
Corn, field, forage	180.232	0.1 N
Corn, fresh (inc. sweet) (k+cwhr)	180.232	0.1 N
Corn, grain	180.232	0.1 N
Corn, pop, fodder	180.232	0.1 N
Corn, pop, forage	180.232	0.1 N
Corn, sweet, fodder	180.232	0.1 N
Corn, sweet, forage	180.232	0.1 N

References

(1) Worthing, C.R., *Pesticide Manual,* 6th ed., p. 71, British Crop Protection Council (1979).
(2) Spencer, E.Y., *Guide to the Chemicals Used in Crop Protection,* 6th ed., p. 63, London, Ontario, Agriculture Canada (January 1973).
(3) Tilles, H. and Antognini, J., U.S. Patent 2,913,327, November 17, 1959, assigned to Stauffer Chemical Company.

2-(4-tert-BUTYLPHENOXY)-1-METHYLETHYL 2-CHLOROETHYL SULFITE

Function: Acaricide (1)

Chemical Name: 2-(p-tert-butylphenoxy)isopropyl 2-chloroethyl sulfite

Formula:

$$(CH_3)_3C-\underset{}{\bigcirc}-OCH_2\underset{CH_3}{\overset{}{C}}HO\overset{O}{\overset{\|}{S}}OCH_2CH_2Cl$$

Trade Names: Aramite® (Uniroyal, Inc.)
Aracide®
Aratron®
Niagaramite®
Ortho-Mite®

Manufacture (2)

In a first step, 2-chloroethyl chlorosulfinate was prepared as follows. Thionyl chloride (70 grams) was added to ethylene chlorohydrin (40 grams) at such a rate that the temperature did not rise above 35°C. Hydrogen chloride was evolved rapidly. The mixture was allowed to stand for 2 days protected from the moisture of the air by a calcium chloride tube. The reaction mixture was fractionally distilled, yielding 72 grams of 2-chloroethyl chlorosulfinate, a water-white liquid which boiled at 90° to 94°C at 30 mm Hg.

In a separate reaction, propylene glycol mono-p-tert-butylphenyl ether was prepared as follows. Propylene oxide (116 grams) was added through a dropping funnel to a solution of p-tert-butylphenol (300 grams) and sodium (4 grams) in 500 cc of anhydrous alcohol. The resulting solution was refluxed for 18 hours.

Most of the alcohol was removed by distillation and the resulting concentrate was washed with water. Fractionation through a 6 inch helix packed column under reduced pressure yielded a fore-run (31 grams) which boiled below 140°C at 8 mm Hg. The main fraction (353 grams) was water-white liquid which distilled almost constantly at 140°C at 8 mm Hg and solidified at about 35° to 40°C.

Then, in the preparation of the final product, to a rapidly stirred solution of the propylene glycol mono-p-tert-butylphenyl ether (61.2 grams) and pyridine (29.7 grams) in 250 cc of dry benzene, was added 2-chloroethyl chlorosulfinate (61.2 grams). The reaction flask was cooled in a cold water bath throughout the addition. Stirring was continued for a few minutes after the addition was complete.

The pyridine hydrochloride was removed by vacuum filtration. The filtrate was washed in cold water and dried over calcium chloride. After topping under reduced pressure, the product was distilled. The main fraction (yield 75.5 grams) was a yellow liquid which boiled at 200° to 210°C at 7 mm Hg.

Process Wastes and Their Control

Solid Wastes: This material is readily hydrolyzed by alkalis or mineral acids (acids result in the liberation of sulfur dioxide). In sunlight it breaks down relatively rapidly with the evolution of sulfur dioxide (B-3).

Toxicity

The acute oral LD$_{50}$ value for rats is 3,900 mg/kg (slightly toxic) (B-5).

Allowable Limits on Exposure and Use

Product Use: This product was voluntarily cancelled on the basis of oncogenicity according to US EPA in a notice dated April 12, 1977.

The tolerances set by the US EPA for this material in or on raw agricultural commodities are as follows:

	40 *CFR* Reference	Parts per Million
Alfalfa	180.107	0.0
Apples	180.107	0.0
Beans, green	180.107	0.0
Blueberries	180.107	0.0
Cantaloups	180.107	0.0
Celery	180.107	0.0
Corn, sweet, forage	180.107	0.0
Corn, sweet, kernel	180.107	0.0
Cucumbers	180.107	0.0
Grapefruit	180.107	0.0
Grapes	180.107	0.0
Lemons	180.107	0.0
Muskmelons	180.107	0.0
Oranges	180.107	0.0
Peaches	180.107	0.0
Pears	180.107	0.0
Plums	180.107	0.0
Raspberries	180.107	0.0
Soybeans (whole plant)	180.107	0.0
Strawberries	180.107	0.0
Tomatoes	180.107	0.0
Watermelons	180.107	0.0

References

(1) Spencer, E.Y., *Guide to the Chemicals Used in Crop Protection,* 6th ed., p. 20, London, Ontario, Agriculture Canada (January 1973).
(2) Harris, N.D., Tate, H.D. and Zukel, J.W., U.S. Patent 2,529,494, November 14, 1950, assigned to U.S. Rubber Co.

2-sec-BUTYLPHENYL METHYLCARBAMATE

Function: Insecticide (1)

Chemical Name: 2-(1-methylpropyl)phenyl methylcarbamate

Formula:

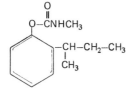

Trade Name: Osbac® (Sumitomo Chemical Company, Ltd.)

Manufacture

2-sec-butylphenyl methylcarbamate is made by reacting 2-sec-butylphenyl chloroformate with methyl amine. The 2-sec-butylphenyl chloroformate is, in turn, made by reacting 2-sec-butylphenol with phosgene.

Toxicity

The acute oral LD_{50} value for rats is 410 mg/kg which rates as moderately toxic.

Reference

(1) Worthing, C.R., *Pesticide Manual,* 6th ed., p. 72, British Crop Protection Council (1979).

C

CACODYLIC ACID

Function: Herbicide (1)(2)(3)

Chemical Name: Dimethylarsinic acid

Formula:

$$(CH_3)_2 \overset{\overset{\displaystyle O}{\displaystyle \|}}{As} OH$$

Trade Names: Ansar® (Ansul Chemical Co.)
Phytar® (Ansul Chemical Co.)
Arsan®
Silvisar®

Manufacture (4)(5)(6)

A solution of disodium methane-arsonate was prepared from arsenic trioxide, sodium hydroxide and methyl chloride (6). The disodium methane-arsonate produced was then reduced by the well-known process involving calcium chloride and sulfur dioxide, and the calcium sulfate filtered off leaving a solution of arsenosomethane ($CH_3As{=}O$) at a pH of less than 7. Thirty-four hundred grams of the solution was charged to an agitated pressure reactor which had an internal reactor volume of 5,480 ml. The charge volume amounted to 2,700 ml so that the volume of the space above the charge was 2,780 ml. The charge had the following approximate analysis:

	Gram Mols
Arsenosomethane	3.54
Arsenic trioxide	0.164
Calcium chloride	0.221
Sulfur dioxide	0.117

As long as the pH of the charge solution was less than 10, no decomposition or disproportionation was noted (4). The pressure reactor was then sealed and evacuated to 15 mm mercury absolute pressure followed by introduction of methyl chloride to a pressure of about 5 psi gauge. The evacuation-purging step was repeated three times to reduce the oxygen content in the vessel to a very low level. The calculated amount of oxygen in the reactor following the evacuation and pressurizing cycle was calculated as 5×10^{-9} mol. Methyl chloride was then introduced to the reactor and for the remainder of the process a positive pressure of methyl chloride was maintained to eliminate any possible leaks of air into the reactor.

Eight hundred and seven grams of a 50% aqueous solution of sodium hydroxide (10.07 mols) was pumped into the reactor, the agitator was started, and additional methyl chloride added to maintain the positive pressure. The temperature was raised to about 80°C and additional methyl chloride supplied until the excess sodium hydroxide was neutralized, i.e., an end point pH of about 5.5 to 6.5 was reached. The reaction solution containing the product was worked up in the usual manner well known to the art. Based on arsenic, about 98% of the arsenosomethane was converted into cacodylic acid.

In a slightly different variation of the process, as shown in Figure 14 (5), 258 g sodium hydroxide, 163.2 g arsenic trioxide and 960 ml water were mixed and allowed to react for 5 minutes. The temperature rose to 176°F. The resulting solution was then pumped to a pressure reactor and held at 175°F. The closed system was then pressurized with methyl chloride by means of a compressor; and the system was maintained at 175 psig for 45 minutes, utilizing an appropriate by-pass type pressure control system.

Approximately 317 g methyl chloride were used while the reaction was proceeding. The reactor was equipped with an agitator which was operated continually to enhance heat and mass transfer. The unreacted methyl chloride gas was then bled off, reducing the reactor pressure to atmospheric. Sulfur dioxide was then bubbled into the closed system for 15 minutes. Approximately 121 g sulfur dioxide were introduced into the reactor. The pressure increased to 35 psig and the temperature to 205°F. The solution was then purged by bubbling nitrogen at 20 psig until all remaining unreacted sulfur dioxide was removed.

150 g sodium hydroxide were dissolved in 200 ml water and the solution added to the reactor and reacted at about 175°F. The closed system was again pressurized with methyl chloride by means of a compressor and the system maintained at 175°F and 175 psig for 30 minutes. Approximately 250 g methyl chloride were added.

The reactant solution was removed from the reactor and centrifuged in a perforated basket-type centrifuge. 1,420 ml filtrate and 172.7 g solids were collected. The filtrate was treated with 77 ml of 0.8 M sodium hypochlorite at ambient conditions. This solution, which had a pH of 8.15, was acidified with 147 ml of 20°Be hydrochloric acid. After acidification, the pH was 4.2 and the volume of liquid was 1,630 ml.

The acidified solution was then split in half. An 815-ml aliquot of the solution was evaporated at 230°F for 2 hours 13 minutes. 546 ml distillate were collected. The remaining slurry was centrifuged and 173 ml solution and 184.7 g solid collected. The filtrate was approximately 3.7 molar in cacodylic acid. 50 ml of this filtrate was additionally dried, yielding 37.8 g of 67% pure cacodylic acid. The other 815 ml aliquot was evaporated at 230°F for 1 hour and 22 minutes. 500 ml distillate were collected. The remaining slurry was centrifuged and 216 ml filtrate and 171.3 g solids were collected. The filtrate was approximately 3.3 molar in cacodylic acid.

Process Wastes and Their Control

Air: Air emissions from cacodylic acid manufacture have been reported (B-15) as follows:

Constituent	Kilograms per Metric Ton Pesticide Produced
Arsenic trioxide	3×10^{-6}
Methyl chloride	0.05
Methanol	0.05

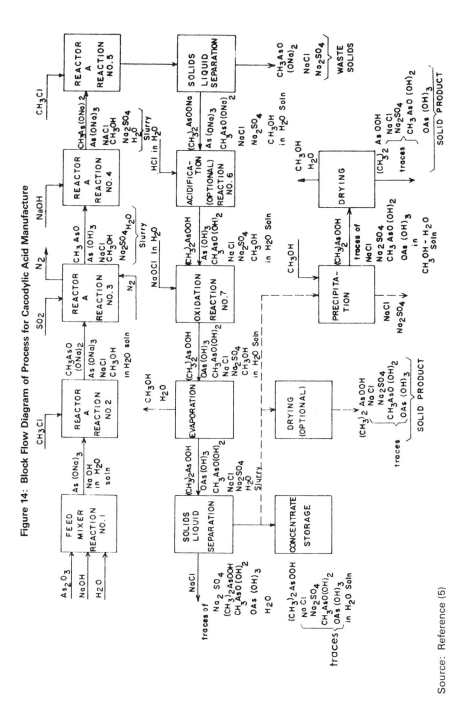

Figure 14: Block Flow Diagram of Process for Cacodylic Acid Manufacture

Source: Reference (5)

Water: At one plant all the liquid streams are recycled for reuse. A number of multiple effect evaporators are employed in the solution recycling systems.

Solid Waste: The process does, however, create a solid waste which is a mixture of sodium chloride and sodium sulfate containing 1 to 1½% cacodylate contaminants (B-10).

Toxicity

The acute oral LD_{50} for rats is 700 mg/kg (slightly toxic) (B-5).

Allowable Limits on Exposure and Use

Product Use: Issuance of a rebuttable presumption against registration was being considered by US EPA as of September 1979 on the basis of possible oncogenicity, mutagenicity, teratogenicity, fetotoxicity and reproductive effects.

The tolerances set by the US EPA for cacodylic acid in or on raw agricultural commodities are as follows:

	40 *CFR* Reference	Parts per Million
Cattle, fat	180.311	0.7
Cattle, kidney	180.311	1.4
Cattle, liver	180.311	1.4
Cattle, mbyp (exc. kidney, liver)	180.311	0.7
Cattle, meat	180.311	0.7
Cotton, seed	180.311	2.8

References

(1) Worthing, C.R., *Pesticide Manual,* 6th ed., p. 202, British Crop Protection Council (1979).
(2) Spencer, E.Y., *Guide to the Chemicals Used in Crop Protection,* 6th ed., p. 70, London, Ontario, Agriculture Canada (January 1973).
(3) Sprague, M.A., U.S. Patent 3,056,668, October 2, 1962, assigned to Ansul Chemical Co.
(4) Moyerman, R.M., and Ehman, P.J., U.S. Patent 3,173,937, March 16, 1965; assigned to Ansul Chemical Co.
(5) Schanhals, L.R., U.S. Patent 3,322,805, May 30, 1967, assigned to O.M. Scott and Sons Co.
(6) Miller, G.E. and Seaton, S.G., U.S. Patent 2,442,372, June 1, 1948, assigned to the Secretary of War.

CALCIUM ARSENATE

Function: Insecticide (1)(2)

Chemical Name: Calcium arsenate

Formula: $Ca_3(AsO_4)_2$

Trade Name: Pencal®

Manufacture

Both calcium and lead arsenate are produced by similar reaction processes. Lime or lead oxide is mixed with arsenic acid to produce calcium or lead arsenate, respectively (Figure 15). The products are piped to drum driers. Water vapor from the driers is collected and vented to stacks. Cleanup water from the batch mix vats is either saved to mix with the next batch, or evaporated in the drum driers to recover product. All product spills are caught in spill-pans below each processing unit. Spilled liquids are recycled back to the processing line. There is thus no wastewater associated with the manufacture of either calcium or lead arsenate.

Figure 15: Production of Lead and Calcium Arsenate

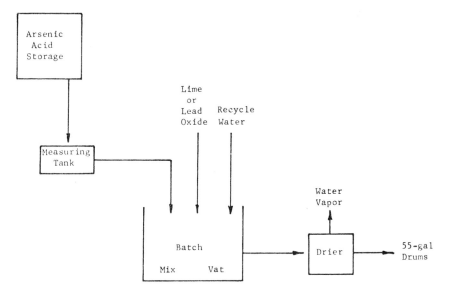

Source: Reference (3)

An improved technique has been described in some detail (4). It produces a product of high bulk density which adheres to foliage even in heavy rain.

Process Wastes and Their Control

Air: Air emissions of 3×10^{-6} kg of arsenic trioxide per metric ton of pesticide produced have been reported (B-15).

Toxicity

The acute oral LD_{50} value for rats is 35-100 mg/kg (highly toxic).

Allowable Limits on Exposure and Use

Air: The threshold limit value for calcium arsenate in air is 1.0 mg/m^3 as of 1979 (B-23). NIOSH has recommended a value of 2 μg/m^3 on a 15-minute ceiling basis (B-21).

Product Use: A rebuttable presumption against registration was issued on October 18, 1978 by US EPA on the basis of oncogenicity, teratogenicity and mutagenicity.

The tolerances set by the US EPA for calcium arsenate in or on raw agricultural commodities are as follows:

	40 *CFR* Reference	Parts per Million
Asparagus	180.192	3.5
Beans	180.192	3.5
Blackberries	180.192	3.5
Blueberries (huckleberries)	180.192	3.5
Boysenberries	180.192	3.5
Broccoli	180.192	3.5
Brussels sprouts	180.192	3.5
Cabbage	180.192	3.5
Carrots	180.192	3.5
Cauliflower	180.192	3.5
Celery	180.192	3.5
Collards	180.192	3.5
Corn	180.192	3.5
Cucumbers	180.192	3.5
Dewberries	180.192	3.5
Eggplant	180.192	3.5
Kale	180.192	3.5
Kohlrabi	180.192	3.5
Loganberries	180.192	3.5
Melons	180.192	3.5
Peppers	180.192	3.5
Pumpkins	180.192	3.5
Raspberries	180.192	3.5
Rutabagas, tops	180.192	3.5
Rutabagas, with tops	180.192	3.5
Rutabagas, without tops	180.192	3.5
Spinach	180.192	3.5
Squash	180.192	3.5
Squash, summer	180.192	3.5
Strawberries	180.192	3.5
Tomatoes	180.192	3.5
Turnips, tops	180.192	3.5
Turnips, with tops	180.192	3.5
Turnips, without tops	180.192	3.5
Youngberries	180.192	3.5

References

(1) Worthing, C.R., *Pesticide Manual,* 6th ed., p. 74, British Crop Protection Council (1979).
(2) Spencer, E.Y., *Guide to the Chemicals Used in Crop Protection,* 6th ed., p. 71, London, Ontario, Agriculture Canada (January 1973).
(3) Patterson, J.W., *State of the Art for the Inorganic Chemical Industry: Inorganic Pesticides,* Report EPA-600/2-74-0093, Wash., DC, Environmental Protection Agency (March 1975).
(4) Les Veaux, J.F., U.S. Patent 3,715,562, August 16, 1955, assigned to Food Machinery and Chemical Corp.

CALCIUM CYANAMIDE

Function: Herbicide and fertilizer (1)(2)

Chemical Name: Calcium cyanamide

Formula:

Trade Name: Cyanamid®

Manufacture

Calcium cyanamide may be made by [a] reaction of calcium carbide and nitrogen (3), or [b] reaction of calcium carbonate and ammonia (4), both reactions being carried out at high temperatures.

Toxicity

The acute oral LD_{50} value for rats is 1,400 mg/kg (slightly toxic). The harmful effects of calcium cyanamide exposure have been summarized as follows (B-25):

Local: Calcium cyanamide is a primary irritant of the mucous membranes of the respiratory tract, eyes, and skin. Inhalation may result in rhinitis, pharyngitis, laryngitis, and bronchitis. Conjunctivitis, keratitis, and corneal ulceration may occur. An itchy erythematous dermatitis has been reported and continued skin contact leads to the formation of slowly healing ulcerations on the palms and between the fingers. Sensitization occasionally develops. Chronic rhinitis and perforation of the nasal septum have been reported after long exposures. All local effects appear to be due to the caustic nature of cyanamide.

Systemic: Calcium cyanamide causes a characteristic vasomotor reaction. There is erythema of the upper portions of the body, face, and arms, accompanied by nausea, fatigue, headache, dyspnea, vomiting, oppression in the chest, and shivering. Circulatory collapse may follow in the more serious cases. The vasomotor response may be triggered or intensified by alcohol ingestion. Pneumonia or lung edema may develop. Cyanide ion is not released in the body, and the mechanism of toxic action is unknown.

Allowable Limits on Exposure and Use

Air: The threshold limit value for calcium cyanamide in air is 0.5 mg/m^3 as of 1979. A tentative short term exposure limit value has been set at 1.0 mg/m^3 (B-23).

References

(1) Martin, H. and Worthing, C.R., *Pesticide Manual,* 5th ed., p. 72, British Crop Protection Council (January 1977).
(2) Spencer, E.Y., *Guide to the Chemicals Used in Crop Protection,* 6th ed., p. 73, London, Ontario, Agriculture Canada (January 1973).
(3) Fischer, T. et al, U.S. Patent 2,917,371, December 15, 1959, assigned to Suddeutsche Kalkstickstoffwerke AG, Germany.
(4) Schaus, O.O., U.S. Patent 3,039,848, June 19, 1962, assigned to American Cyanamid Co.

CALCIUM CYANIDE

Function: Insecticidal fumigant (1)(2)

Chemical Name: Calcium cyanide

Formula: Ca(CN)$_2$

Trade Name: Cyanogas® (American Cyanamid)

Manufacture

Calcium cyanide may be produced by [a] fusion of sodium chloride with calcium cyanamide or [b] reaction of liquid hydrogen cyanide with calcium carbide. The product of this second reaction is more nearly $CaH_2(CN)_4$, calcium acid cyanide.

Toxicity

The acute oral LD_{50} value for rats is 39 mg/kg (B-5) which makes this a highly toxic substance.

Allowable Limits on Exposure and Use

Product Use: The tolerances set by the US EPA for calcium cyanide in or on raw agricultural commodities are as follows:

	40 *CFR* Reference	Parts per Million
Barley, grain (post-h)	180.125	25.0
Buckwheat, grain (post-h)	180.125	25.0
Corn, grain (post-h)	180.125	25.0
Cucumbers	180.125	5.0
Lettuce	180.125	5.0
Oats, grain (post-h)	180.125	25.0
Radishes	180.125	5.0
Rice, grain (post-h)	180.125	25.0
Rye, grain (post-h)	180.125	25.0
Sorghum, grain (post-h)	180.125	25.0
Tomatoes	180.125	5.0
Wheat, grain (post-h)	180.125	25.0

References

(1) Worthing, C.R., *Pesticide Manual,* 6th ed., p. 75, British Crop Protection Council (1979).
(2) Spencer, E.Y., *Guide to the Chemicals Used in Crop Protection,* 6th ed., p. 74, London, Ontario, Agriculture Canada (January 1973).

CAPTAFOL

Function: Fungicide (1)(2)

Chemical Name: 3a,4,7,7a-tetrahydro-2-(1,1,2,2-tetrachloroethyl)thio-1H-isoindole-1,3(2H)-dione

Formula:

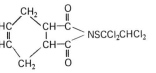

Trade Names: Difolatan® (Chevron Chemical Co.)
Sanspor®
Sulfenimide®
Folcid®

Manufacture (3)

Tetrahydrophthalic anhydride may first be produced by the reaction of butadiene with maleic anhydride. The tetrahydrophthalic anhydride is then reacted with ammonia to give tetrahydrophthalimide.

The solution or suspension of the alkali metal salt of the imide is vigorously stirred, preferably at low temperature, while the desired sulfenyl halide is added. The sulfenyl halide may be contained in a suitable inert solvent, such as petroleum ether or mixed hexanes, if desired. After a sufficient reaction period, the product is isolated by filtration or by removing the solvent by distillation. The product then may, if desired, be recrystallized from a suitable solvent, such as methanol or aromatic solvent. Specific examples of the conduct of this process are as follows (3):

Example 1—A mixture of 9.0 g cis-Δ^1-tetrahydrophthalimide and 200 ml ice water was vigorously stirred while 4.8 g of 50% sodium hydroxide was added. Then 14.0 g of 1,2,2,2-tetrachloroethylsulfenyl chloride were added and the agitation continued for about 5 minutes. The crude product was removed by filtration, recrystallized from methanol, and analyzed to be N-(1,2,2,2-tetrachloroethyl)-cis-Δ^1-cyclohexene-1,2-dicarboximide.

Example 2—A mixture of 5.4 g sodium methoxide, 60 ml of methanol, 12.8 g of cis-Δ^1-tetrahydrophthalimide, and 60 ml of methanol was agitated while 20.0 g of 1,1,2,2-tetrachloroethylsulfenyl chloride were added, causing a solid to separate. The solid was filtered, water-washed, recrystallized from methanol, and analyzed to be N-(1,1,2,2-tetrachloroethylthio)-cis-Δ^1-cyclohexene-1,2-dicarboximide.

Process Wastes and Their Control

Air: Air emissions for captafol manufacture have been reported as follows (B-15).

Component	Kilograms per Metric Ton Pesticide Produced
Captafol	0.05
Butadiene	1.0
Carbon disulfide	0.5

Product Wastes: Captafol is slowly hydrolyzed by water; rapidly hydrolyzed in alkaline solution. It decomposes slowly when heated at its melting point (B-3).

Toxicity

The acute oral LD_{50} value for rats is 2,500 mg/kg (B-5) which is classified as slightly toxic.

Allowable Limits on Exposure and Use

Air: The threshold limit value for captafol in air is 0.1 mg/m^3 as of 1979. The limit bears the notation "skin" indicating that cutaneous absorption should be prevented so that the threshold limit value is not invalidated.

Product Use: The tolerances set by the US EPA for captafol in or on raw agricultural commodities are as follows:

	40 *CFR* Reference	Parts per Million
Apples	180.267	0.25
Apricots	180.267	30.0
Blueberries	180.267	35.0
Cherries, sour	180.267	50.0

(continued)

	40 *CFR* Reference	Parts per Million
Cherries, sweet	180.267	2.0
Citrus fruits	180.267	0.5
Corn, fresh (inc. sweet) (k+cwhr)	180.267	0.1 N
Cranberries	180.267	8.0
Cucumbers	180.267	2.0
Macadamia nuts	180.267	0.1 N
Melons	180.267	5.0
Nectarines	180.267	2.0
Onions	180.267	0.1 N
Peaches	180.267	30.0
Peanuts, hulls	180.267	2.0
Peanuts, meat (hulls removed)	180.267	0.05
Pineapples	180.267	0.1 N
Plums (fresh prunes)	180.267	2.0
Potatoes	180.267	0.5
Taro (corm)	180.267	0.02
Tomatoes	180.267	15.0

References

(1) Worthing, C.R., *Pesticide Manual,* 6th ed., p. 77, British Crop Protection Council (1979).
(2) Spencer, E.Y., *Guide to the Chemicals Used in Crop Protection,* 6th ed., p. 75, London, Ontario, Agriculture Canada (January 1973).
(3) Kohn, G.K., U.S. Patent 3,178,447, April 13, 1965, assigned to California Research Corp.

CAPTAN

Function: Fungicide (1)(2)

Chemical Name: 3a,4,7,7a-tetrahydro-2-[(trichloromethyl)thio]-1H-isoindole-1,3(2H)-dione

Formula:

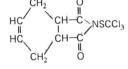

Trade Names: SR406 (Chevron Chemical Co.)
Orthocide 406® (Chevron Chemical Co.)

Manufacture

Perchloromethyl mercaptan is a first intermediate (4)(5).

$$CS_2 + 3Cl_2 \xrightarrow{I_2} CCl_3SCl + SCl_2$$

Then, as in captafol manufacture, butadiene is reacted with maleic anhydride and that product is reacted with ammonia. Then the imide is reacted with CCl_3SCl to give captan (6)(7)(8)(9)(10)(11) (see Figure 16).

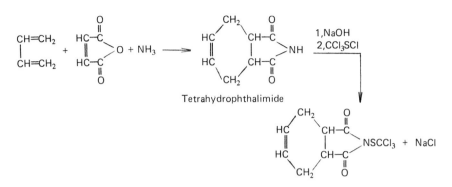

Tetrahydrophthalimide

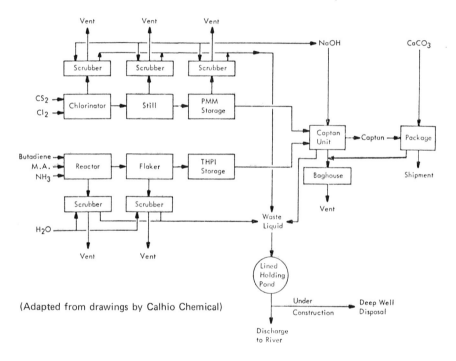

Figure 16: Production and Waste Schematic for Captan

(Adapted from drawings by Calhio Chemical)

Source: Reference (3)

Process Wastes and Their Control

Air: Air emissions from captan manufacture have been reported as follows (B-15):

Component	Kilograms per Metric Ton Pesticide Produced
Captan	0.05
Butadiene	1.0
Carbon disulfide	0.5

Water and Solid Wastes: Overall process wastes and losses at one captan plant have been summarized as follows (B-10):

Material	Form	Amount Produced (lb/lb AI)	Disposition
Active ingredient	Particulates	~4 lb/day	
Solvents	Liquid	10 tons/year	Asphalt-lined settling pond; discharge
Solid paper		10 tons/year	Local collector
Metal		25 tons/year	Local scrap dealers
Miscellaneous chemicals		1,200 lb/year	Buried on plant property

Product Wastes: Captan decomposes fairly readily in alkaline media (pH>8). It is hydrolytically stable at neutral or acid pH but decomposes when heated alone at its melting point (B-3).

Toxicity

The acute oral LD_{50} value for rats is 9,000 mg/kg (insignificantly toxic). Most of the chronic-oral-toxicity data on captan and folpet suggest that the no-adverse-effect or toxicologically safe dosage of these agents is about 1,000 ppm (50 mg/kg/day). However, on the basis of fetal mortality observed in monkeys exposed to captan (12.5 mg/kg/day), the acceptable daily intake of captan and folpet has been established at 0.1 mg/kg of body weight by the FAO/WHO. Based on long-term feeding studies results in rats and dogs, ADI's were calculated at 0.05 mg/kg/day for captan and 0.16 mg/kg/day for folpet (B-22).

Allowable Limits on Exposure and Use

Air: The threshold limit value for captan in air is 5 mg/m³ as of 1979. A tentative short term exposure limit is 15 mg/m³ (B-23).

Product Use: Issuance of a rebuttable presumption against registration for captan was being considered by US EPA as of September 1979 on the basis of possible oncogenicity, mutagenicity and teratogenicity.

The tolerances set by the US EPA for captan in or on raw agricultrual commodities are as follows:

	40 *CFR* Reference	Parts per Million
Almond, hulls	180.103	100.0
Almonds	180.103	2.0 i
Apples	180.103	25.0
Apricots	180.103	50.0
Avocados	180.103	25.0 i
Beans, dry	180.103	25.0 i
Beans, succulent	180.103	25.0 l
Beets, greens	180.103	100.0
Beets, roots	180.103	2.0
Blackberries	180.103	25.0
Blueberries (huckleberries)	180.103	25.0
Broccoli	180.103	2.0
Brussels sprouts	180.103	2.0
Cabbage	180.103	2.0
Cantaloups	180.103	25.0
Carrots	180.103	2.0
Cauliflower	180.103	2.0
Celery	180.103	50.0

(continued)

	40 *CFR* Reference	Parts per Million
Cherries	180.103	100.0
Collards	180.103	2.0
Corn, sweet (k+cwhr)	180.103	2.0
Cotton, seed	180.103	2.0
Crabapples	180.103	25.0
Cranberries	180.103	25.0
Cucumbers	180.103	25.0
Dewberries	180.103	25.0
Eggplant	180.103	25.0
Garlic	180.103	25.0
Grapefruit	180.103	25.0 i
Grapes	180.103	50.0
Honeydew	180.103	25.0
Kale	180.103	2.0
Leeks	180.103	50.0
Lemons	180.103	25.0 i
Lettuce	180.103	100.0
Limes	180.103	25.0 i
Mangoes	180.103	50.0
Muskmelons	180.103	25.0
Mustard, greens	180.103	2.0
Nectarines	180.103	50.0
Onions, dry bulb	180.103	25.0
Onions, green	180.103	50.0
Oranges	180.103	25.0 i
Peaches	180.103	50.0
Pears	180.103	25.0
Peas, dry	180.103	2.0
Peas, succulent	180.103	2.0
Peppers	180.103	25.0
Pimentos	180.103	25.0
Pineapples	180.103	25.0 i
Plums (fresh prunes)	180.103	50.0
Potatoes	180.103	25.0 i
Pumpkins	180.103	25.0
Quinces	180.103	25.0
Raspberries	180.103	25.0
Rhubarb	180.103	25.0
Rutabagas, roots	180.103	2.0
Shallots	180.103	50.0
Soybeans, dry	180.103	2.0
Soybeans, succulent	180.103	2.0
Spinach	180.103	100.0
Squash, summer	180.103	25.0
Squash, winter	180.103	25.0
Strawberries	180.103	25.0
Tangerines	180.103	25.0 i
Taro (corm)	180.103	0.25 i
Tomatoes	180.103	25.0
Turnips, greens	180.103	2.0
Turnips, roots	180.103	2.0
Watermelons	180.103	25.0

The tolerances set by the EPA for captan in food are as follows (the *CFR* Reference is to Title 21):

	CFR Reference	Parts per Million
Raisins, washed (pre- & post-h)	193.40	50.0

References

(1) Worthing, C.R., *Pesticide Manual,* 6th ed., p. 78, British Crop Protection Council (1979).
(2) Spencer, E.Y., *Guide to the Chemicals Used in Crop Protection,* 6th ed., p. 76, London, Ontario, Agriculture Canada (January 1973).
(3) Midwest Research Institute, *The Pollution Potential in Pesticide Manufacturing,* Wash., DC, Environmental Protection Agency (June 1972).
(4) Ohsol, E.O. et al., U.S. Patent 2,575,290, November 13, 1951, assigned to Standard Oil Development Co.
(5) Churchill, J.W., U.S. Patent 2,666,081, January 12, 1954, assigned to Mathieson Chemical Corp.
(6) Kittleson, A.R., U.S. Patent 2,553,770, May 22, 1951, assigned to Standard Oil Development Co.
(7) Kittleson, A.R. et al, U.S. Patent 2,553,771, May 22, 1951, assigned to Standard Oil Development Co.
(8) Kittleson, A.R., U.S. Patent 2,553,776, May 22, 1951, assigned to Standard Oil Development Co.
(9) Kittleson, A.R., U.S. Patent 2,653,155, September 22, 1953, assigned to Standard Oil Development Co.
(10) Kittleson, A.R. et al, U.S. Patent 2,856,410, October 14, 1958, assigned to Esso Research and Engineering Co.
(11) Kittleson, A.R., U.S. Patent 2,713,058, July 12, 1955, assigned to Esso Research and Engineering Co.

CARBARYL

Function: Insecticide (1)(2)

Chemical Name: 1-naphthalenyl methylcarbamate

Formula:

OCONHCH$_3$

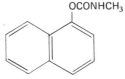

Trade Name: Sevin® (Union Carbide)

Manufacture (3)

In a first step, sodium 1-naphthoxide is reacted with phosgene according to the equation:

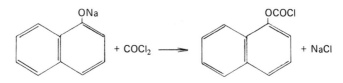

and in a second step that intermediate is reacted with methylamine to give 1-naphthyl-N-methyl carbamate (4)(5) according to the equation:

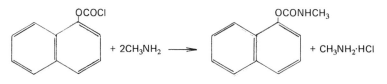

A process flow diagram is shown in Figure 17.

Figure 17: Production and Waste Schematic for Carbaryl

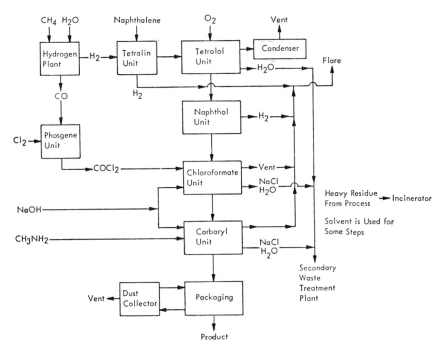

Source: Reference (3)

Process Wastes and Their Control

Air: Air emissions from carbaryl manufacture have been reported to consist of 1.5 kg of hydrocarbons and 0.5 kg of carbaryl per metric ton of pesticide produced. Considerable care, through the use of special scrubbers and by enforcing safety precautions, must be taken to ensure that phosgene does not escape to the atmosphere (B-10). Toxic vents are either flared or go to NaOH scrubbers. Nontoxic vents go to a condenser and are then vented to the atmosphere. Standard hoods are used in the packaging area and the recovered material is recycled (B-10).

Water: Wastewaters contain high total solids but almost no suspended solids and little pesticidal compounds. A typical waste stream from carbamate manufacture will contain:

COD	10,000 mg/l
BOD_5	Nil

(continued)

Total solids	40,000 mg/l
pH	7-10
Suspended solids	Nil
Sodium	8,000 mg/l
Chlorides	100 mg/l
Phosphates	Nil
Organic nitrogen	500 mg/l
Sulfates	20,000 mg/l
Product	Nil
Toxicity	Low
Flow per pound of product	Not available

At one plant, all pesticide-containing wastewater goes to the plant's secondary waste treatment system and then to a river. This effluent contains only 0.01 to 1 ppm carbaryl (B-10).

Solid Wastes: The heavy residue solid wastes are burned. One shutdown for cleaning is made per year, but numerous maintenance clean-ups are made and the washings go to the process waste treatment system (B-10).

Toxicity

The acute oral LD_{50} value for rats is 400 mg/kg (moderately toxic) (B-5). Carbaryl was the first of the carbamate insecticides to be introduced and is still the most heavily used. The mode of action of carbaryl is inhibition of the acetylcholinesterase, although there is evidence that the inhibition is reversible under some conditions, in contrast with that caused by the organophosphorus insecticides. Hence, chronic studies have been concerned with measuring factors in addition to decrease in cholinesterase activity, to determine a no-adverse-effect dosage.

Long-term studies have been conducted in rats and dogs and have led to the establishment of no-adverse-effect concentrations for these species. In addition, corroborating studies have been done in rats and man for shorter periods. Carbaryl has a known mode of action, adequate chronic-toxicity studies have been done, and there is no evidence of teratogenicity, mutagenicity, or carcinogenicity. An ADI was calculated at 0.082 mg/kg/day based on these data. There are no pressing research needs with respect to carbaryl. Continued monitoring of the presence and amounts of carbaryl in food and water will be necessary (B-25).

The signs and symptoms observed as a consequence of exposure to carbaryl in the workplace environment (6) are manifestations of excessive cholinergic stimulation, e.g., nausea, vomiting, mild abdominal cramping, dimness of vision, dizziness, headache, difficulty in breathing, and weakness (B-25).

Allowable Limits on Exposure and Use

Air: The threshold limit value for carbaryl in air is 5 mg/m^3 as of 1979. A tentative short term exposure limit is 10 mg/m^3 (B-23).

Product Use: Issuance of a rebuttable presumption against registration was being considered by US EPA as of September 1979 on the basis of possible teratogenicity and fetotoxicity.

The tolerances set by the EPA for carbaryl in or on raw agricultural commodities are as follows:

	40 *CFR* Reference	Parts per Million
Alfalfa	180.169	100.0
Alfalfa, hay	180.169	100.0

(continued)

	40 *CFR* Reference	Parts per Million
Almonds	180.169	1.0
Almonds, hulls	180.169	40.0
Apples	180.169	10.0
Apricots	180.169	10.0
Asparagus	180.169	10.0
Bananas	180.169	10.0
Barley, fodder, green	180.169	100.0
Barley, grain	180.169	0.0
Barley, straw	180.169	100.0
Beans	180.169	10.0
Beans, forage	180.169	100.0
Beans, hay	180.169	100.0
Beets, garden, roots	180.169	5.0
Beets, garden, tops	180.169	12.0
Beets, sugar, tops	180.169	100.0
Blackberries	180.169	12.0
Blueberries	180.169	10.0
Boysenberries	180.169	12.0
Broccoli	180.169	10.0
Brussels sprouts	180.169	10.0
Cabbage	180.169	10.0
Cabbage, Chinese	180.169	10.0
Carrots	180.169	10.0
Cauliflower	180.169	10.0
Celery	180.169	10.0
Cherries	180.169	10.0
Chestnuts	180.169	1.0
Citrus fruits	180.169	10.0
Clover	180.169	100.0
Clover, hay	180.169	100.0
Collards	180.169	12.0
Corn (inc sweet) (k+cwhr)	180.169	5.0
Corn, fodder	180.169	100.0
Corn, forage	180.169	100.0
Cotton, forage	180.169	100.0
Cotton, seed	180.169	5.0
Cranberries	180.169	10.0
Cucumbers	180.169	10.0
Dandelion	180.169	12.0
Dewberries	180.169	12.0
Eggplant	180.169	10.0
Endive (escarole)	180.169	10.0
Filberts (hazelnuts)	180.169	1.0
Grapes	180.169	10.0
Grasses	180.169	100.0
Grasses, hay	180.169	100.0
Horseradish	180.169	5.0
Kale	180.169	12.0
Kohlrabi	180.169	10.0
Lettuce	180.169	10.0
Loganberries	180.169	12.0
Maple sap	180.169	0.5
Melons	180.169	10.0
Mustard, greens	180.169	12.0
Nectarines	180.169	10.0
Oats, fodder, green	180.169	100.0
Oats, grain	180.169	0.0
Oats, straw	180.169	100.0
Okra	180.169	10.0
Olives	180.169	10.0

(continued)

	40 *CFR* Reference	Parts per Million
Parsley	180.169	12.0
Parsnips	180.169	5.0
Peaches	180.169	10.0
Peanuts	180.169	5.0
Peanuts, hay	180.169	100.0
Pears	180.169	10.0
Peas, cowpeas	180.169	5.0
Peas, cowpeas, forage	180.169	100.0
Peas, cowpeas, hay	180.169	100.0
Peas, vines	180.169	100.0
Peas, with pods	180.169	10.0
Pecans	180.169	1.0
Peppers	180.169	10.0
Plums (fresh prunes)	180.169	10.0
Potatoes	180.169	0.2 N
Poultry, fat	180.169	5.0
Poultry, meat	180.169	5.0
Pumpkins	180.169	10.0
Radishes	180.169	5.0
Raspberries	180.169	12.0
Rice	180.169	5.0
Rice, straw	180.169	100.0
Rutabagas	180.169	5.0
Rye, fodder, green	180.169	100.0
Rye, grain	180.169	0.0
Rye, straw	180.169	100.0
Salsify, roots	180.169	5.0
Salsify, tops	180.169	10.0
Sorghum, forage	180.169	100.0
Sorghum, grain	180.169	10.0
Soybeans	180.169	5.0
Soybeans, forage	180.169	100.0
Soybeans, hay	180.169	100.0
Spinach	180.169	12.0
Squash, summer	180.169	10.0
Squash, winter	180.169	10.0
Strawberries	180.169	10.0
Sweet potatoes	180.169	0.2
Swiss chard	180.169	12.0
Tomatoes	180.169	10.0
Turnips, roots	180.169	5.0
Turnips, tops	180.169	12.0
Walnuts	180.169	1.0
Wheat, fodder, green	180.169	100.0
Wheat, grain	180.169	0.0
Wheat, straw	180.169	100.0

References

(1) Worthing, C.R., *Pesticide Manual,* 6th ed., p. 79, British Crop Protection Council (1979).

(2) Spencer, E.Y., *Guide to the Chemicals Used in Crop Protection,* 6th ed., p. 79, London, Ontario, Agriculture Canada (January 1973).

(3) Midwest Research Institute, *The Pollution Potential in Pesticide Manufacturing,* Wash., DC, Environmental Protection Agency (June 1972).

(4) Lambrech, J.A., U.S. Patent 2,903,478, September 8, 1959, assigned to Union Carbide Corp.

(5) Lambrech, J.A., U.S. Patent 3,009,855, November 21, 1961, assigned to Union Carbide Corp.

(6) National Institute for Occupational Safety and Health, *Criteria for a Recommended Standard: Occupational Exposure to Carbaryl,* NIOSH Doc. No. 77-107 (1977).

CARBENDAZIM

Function: Fungicide (1)(2)

Chemical Name: Methyl 1H-benzimidazol-2-yl-carbamate

Formula:

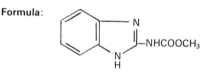

Trade Names: Hoe 17411 OF (Hoechst AG)
Derosal® (Hoechst AG)
BAS 346F (BASF)
Bavistin® (BASF)
Delsene® (DuPont)

Manufacture

The compound 2-benzimidazolecarbamic acid, methyl ester is prepared by the following method. A mixture of 228 parts of thiourea and 110 parts of water is treated over a five minute period with 244 parts of dimethyl sulfate. Rapid agitation is used throughout the whole procedure. The temperature of the reaction mixture rises to 95°C, then begins to subside. The material is brought to reflux by application of heat and held at reflux for 30 minutes, then cooled to −30°C, diluted with 1,800 parts of water, and treated with 535 parts of methyl chloroformate in one portion. A 25% solution of aqueous sodium hydroxide is added at such a rate as to keep the pH of the reaction mixture between 6 and 7 and the temperature below 25°C.

When the pH of the mixture reaches 6.9 and the rate of change of pH has become negligible, the addition of base is stopped. The amount of base required is 1,085 parts by volume of 25% solution. The temperature at the end of this addition is 23°C. Immediately after completion of the base addition, 360 parts of glacial acetic acid is added over a 20 minute period, followed by 324 parts of o-phenylenediamine in one portion. The resulting mixture is slowly warmed to 80°C and held there for 30 minutes, then cooled to 27°C and the light tan solid product isolated by filtration, washed well with water and acetone, and air-dried.

Toxicity

The acute oral LD_{50} value for rats is more than 15,000 mg/kg which is insignificantly toxic.

References

(1) Worthing, C.R., *Pesticide Manual,* 6th ed., p. 80, British Crop Protection Council (1979).
(2) Klopping, H.L., U.S. Patent 3,657,443, April 18, 1972, assigned to E.I. DuPont de Nemours and Co.

CARBETAMIDE

Function: Herbicide (1)(2)(3)

Chemical Name: N-ethyl-2-[(phenylamino)carbonyl]oxypropanamide

Formula:

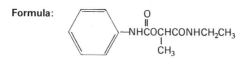

$$-NHCOCHCONHCH_2CH_3$$
$$\overset{|}{CH_3}$$

Trade Names: RP-11,561 (Rhone-Poulenc)
Legurame® (Rhone-Poulenc)
Carbetamex®

Manufacture (3)

Ethylamine and methyl lactate are first reacted to give N-ethyl lactamide (4). The lactamide is then reacted with phenyl isocyanate (3). To a solution of phenyl isocyanate (23.8 g) in benzene (75 cc) is added N-ethyl-lactamide (23.4 g). After standing for 20 minutes at room temperature, the mixture is refluxed for 4 hours. After cooling, the benzene is driven off and the residual oil is taken up in water (150 cc). The product is filtered off and dried in vacuo over sulfuric acid. On recrystallization from a mixture of ethyl acetate and petroleum ether (1 vol:3 vol), there is obtained carbetamide.

Process Wastes and Their Control

Product Waste Disposal: This product, which contains both an amide and a carbamate function, is stable under normal storage conditions. It is hydrolyzed in strong acid at elevated temperatures. The compound can also be hydrolyzed under strong basic conditions, but both acidic and basic hydrolysis yields aniline which is much more toxic than the starting material (B-3).

Toxicity

The acute oral LD_{50} for rats is 11,000 mg/kg which means that this product is insignificantly toxic.

References

(1) Worthing, C.R., *Pesticide Manual,* 6th ed., p. 81, British Crop Protection Council (1979).
(2) Spencer, E.Y., *Guide to the Chemicals Used in Crop Protection,* 6th ed., p. 80, London, Ontario, Agriculture Canada (January 1973).
(3) Metivier, J., U.S. Patent 3,177,061, April 6, 1965, assigned to Rhone-Poulenc SA.
(4) Ratchford, *J. Org. Chem.* 15, 326 (1950).

CARBOFURAN

Function: Systemic insecticide, acaricide (1)(2)(4)(5)

Chemical Name: 2,3-dihydro-2,2-dimethyl-7-benzofuranyl methyl carbamate

Formula:

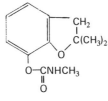

Trade Names: FMC 10242 (FMC Corp.)
Furadan® (FMC Corp.)
Bay 70143 (Bayer)
Curaterr (Bayer)

Manufacture (6)(7)

The reaction sequence is as follows (3):

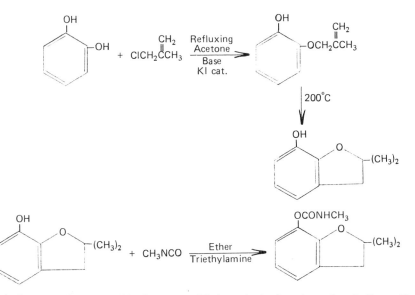

Carbofuran may be prepared by the process (5) shown in the flow sheet given in Figure 18, whereby 2-methallyloxyphenol was cyclized and rearranged to form a 7-hydroxybenzofuran, followed by esterification to form the carbamate. The starting material 2-methylallyloxyphenol was prepared as follows. To a stirred mixture of 322 parts of catechol in 300 parts of dry acetone was slowly added, under nitrogen atmosphere, 401 parts of potassium carbonate and 481 parts of potassium iodide.

The mass was heated to reflux temperature, and 262 parts of methallyl chloride was added slowly. The mixture was refluxed for 30 hours, allowed to cool and stand for 18 hours, filtered, and the filtrate concentrated under reduced pressure. The residual oil was extracted with chloroform, and the chloroform solution was washed with water, dried, and concentrated. The residual oil was distilled to give 213 parts of 2-methallyloxyphenol, BP 78.5°C to 83.0°C/0.55 mm, n_D^{25} 1.5300.

7-Hydroxy-2,2-dimethyl-2,3-dihydrobenzofuran was prepared as follows. A round-bottom flask containing 131 parts 2-methallyloxyphenol was heated slowly with stirring. At 200°C, an exothermic reaction occurred, and the temperature of the mixture in the flask increased rapidly to 275°C. The temperature was controlled at 275°C by external cooling. The thick syrup was distilled under reduced pressure to yield colorless liquid 7-hydroxy-2,2-dimethyl-2,3-dihydrobenzofuran, BP 78° to 80°C/0.35 to 40 mm, n_D^{25} 1.5401.

The carbofuran product was prepared as follows. A cold solution of 16.4 parts 7-hydroxy-2,2-dimethyl-2,3-dihydrobenzofuran in 14 parts of ether was treated with 5.8 parts methyl isocyanate and 0.1 part triethylamine. The mixture was stirred at room temperature, and a white crystalline product precipitated. Separation of the solid yielded 17.5 parts of product, MP 151° to 152°C. Recrystallization from methylcyclohexane gave a purer product.

Figure 18: Production and Waste Schematic for Carbofuran

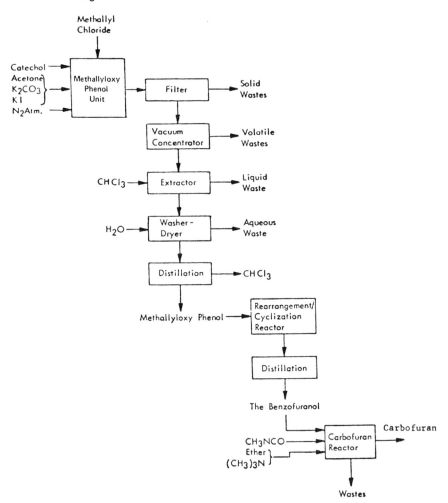

Source: Reference (3)

Process Wastes and Their Control

Air: Air emissions from carbofuran manufacture have been reported (B-15) to consist of 1.5 kg of hydrocarbons and 0.5 kg of carbofuran per metric ton of pesticide produced.

Toxicity

The acute oral LD_{50} value for rats is only 8-14 mg/kg which rates as highly toxic.

Allowable Limits on Exposure and Use

Air: The threshold limit value for carbofuran in air is 0.1 mg/m³ as of 1979 (B-23).

Product Use: The tolerances set by the US EPA for carbofuran in or on raw agricultural commodities are as follows:

	40 *CFR* Reference	Parts per Million
Alfalfa, fresh; nmt 5 ppm carb*	180.254	10.0
Alfalfa, hay; nmt 20 ppm carb	180.254	40.0
Bananas	180.254	0.1
Beets, sugar	180.254	0.1
Beets, sugar, tops; nmt 1 carb	180.254	2.0
Beets, sugar, tops	180.254	0.1 N
Cattle, fat; nmt 0.02 carb	180.254	0.05
Cattle, mbyp; nmt 0.02 carb	180.254	0.05
Cattle, meat; nmt 0.02 carb	180.254	0.05
Coffee beans	180.254	0.1
Corn, fodder, nmt 5 ppm carb	180.254	25.0
Corn, forage; nmt 5 ppm carb	180.254	25.0
Corn, grain nmt 0.1 carb	180.254	0.2
Goats, fat; nmt 0.02 carb	180.254	0.05
Goats, mbyp; nmt 0.02 carb	180.254	0.05
Goats, meat; nmt 0.02 carb	180.254	0.05
Hogs, fat; nmt 0.02 carb	180.254	0.05
Hogs, mbyp; nmt 0.02 carb	180.254	0.05
Hogs, meat; nmt 0.02 carb	180.254	0.05
Horses, fat; nmt 0.02 carb	180.254	0.05
Horses, mbyp; nmt 0.02 carb	180.254	0.05
Horses, meat; nmt 0.02 carb	180.254	0.05
Milk; nmt 0.02 carb	180.254	0.1
Peanuts; nmt 0.1 carb	180.254	0.2
Peanuts, hulls; nmt 1 carb	180.254	5.0
Peppers; nmt 0.2 carb	180.254	1.0
Potatoes; nmt 1 carb	180.254	2.0
Rice	180.254	0.2
Rice, straw; nmt 0.2 carb	180.254	1.0
Sheep, fat; nmt 0.02 carb	180.254	0.05
Sheep, mbyp; nmt 0.02 carb	180.254	0.05
Sheep, meat; nmt 0.02 carb	180.254	0.05
Sorghum, fodder; nmt 0.5 carb	180.254	3.0
Sorghum, grain	180.254	0.1
Sorghum, forage nmt 0.5 carb	180.254	3.0
Strawberries; nmt 0.2 carb	180.254	0.5
Sugarcane	180.254	0.1

*carb = carbamate metabolites

References

(1) Worthing, C.R., *Pesticide Manual,* 6th ed., p. 82, British Crop Protection Council (1979).
(2) Spencer, E.Y., *Guide to the Chemicals Used in Crop Protection,* 6th ed., p. 81, London, Ontario, Agriculture Canada (January 1973).
(3) Midwest Research Institute and RvR Consultants, *Production, Distribution, Use and Environmental Impact Potential of Selected Pesticides,* Wash. DC, Council on Environmental Quality (March 15, 1974).
(4) Scharpf, W.G., U.S. Patent 3,474,170, October 21, 1969, assigned to FMC Corp.
(5) Scharpf, W.G., U.S. Patent 3,474,171, October 21, 1969, assigned to FMC Corp.
(6) Orwoll, E.F., U.S. Patent 3,356,690, December 5, 1967, assigned to FMC Corp.
(7) Heiss, R., Bocker, E., Behrenz, W. and Unterstenhofer, G., U.S. Patent 3,470,299, September 30, 1969, assigned to Farbenfabriken Bayer AG, Germany.

CARBON DISULFIDE

Function: Fumigant (1)(2)

Chemical Name: Carbon disulfide

Formula: CS_2

Trade Name: None

Manufacture

Carbon disulfide may be made by:

(a) the reaction of carbon monoxide and sulfur (3)

$$2CO + 2S \longrightarrow \tfrac{1}{2}CO_2 + COS + \tfrac{1}{2}CS_2$$

(b) the reaction of coke and sulfur (4)

$$C + 2S \longrightarrow CS_2$$

(c) the reaction of methane and hydrogen sulfide (5)

$$CH_4 + 2H_2S \longrightarrow CS_2 + 4H_2$$

Process Wastes and Their Control

Product Waste Disposal: This compound is a very flammable liquid which evaporates rapidly. It burns with a blue flame to carbon dixoide (harmless) and sulfur dioxide. Sulfur dioxide has a strong suffocating odor; 1,000 ppm in air is lethal to rats. The pure liquid presents an acute fire and explosion hazard. The Manufacturing Chemists Association suggests the following disposal procedure (B-3):

> "All equipment or contact surfaces should be grounded to avoid ignition by static charges. Absorb on vermiculite, sand, or ashes and cover with water. Transfer underwater in buckets to an open area. Ignite from a distance with an excelsior train. If quantity is large, carbon disulfide may be recovered by distillation and repackaged for use."

Toxicity

The acute oral LD_{50} value for rats is not available. The lowest published lethal dose for humans is 14 mg/kg. Carbon disulfide has been demonstrated to produce disturbances in reproduction as well as teratogenic effects in animals when inhaled. There is no data available on teratogenicity following oral exposure.

In view of the relative paucity of data on the mutagenicity, carcinogenicity, and long-term oral toxicity of carbon disulfide, estimates of the effects of chronic oral exposure at low levels cannot be made with any confidence. It is recommended that studies to produce such information be conducted before limits in drinking water are established.

The harmful effects of carbon disulfide have been cited as follows (B-25):

Local: Carbon disulfide vapor in sufficient quantities is severely irritating to eyes, skin, and mucous membranes. Contact with liquid may cause blistering with second and third degree burns. Skin sensitization may occur. Skin absorption may result in localized degeneration of peripheral nerves which is most often noted in the hands. Respiratory irritation may result in bronchitis and emphysema, though these effects may be overshadowed by systemic effects.

Systemic: Intoxication from carbon disulfide is primarily manifested by psychological, neurological, and cardiovascular disorders. Recent evidence indicates that once biochemical alterations are initiated they may remain latent; clinical signs and symptoms then occur following subsequent exposure.

Following repeated carbon disulfide exposure, subjective psychological as well as behavioral disorders have been observed. Acute exposures may result in extreme irritability, uncontrollable anger, suicidal tendencies, and a toxic manic depressive psychosis. Chronic exposures have resulted in insomnia, nightmares, defective memory, and impotency. Less dramatic changes include headache, dizziness, and diminished mental and motor ability, with staggering gait and loss of coordination.

Neurological changes result in polyneuritis. Animal experimentation has revealed pyramidal and extrapyramidal tract lesions and generalized degeneration of the myelin sheaths of peripheral nerves. Chronic exposure signs and symptoms include retrobulbar and optic neuritis, loss of sense of smell, tremors, paresthesias, weakness, and, most typically, loss of lower extremity reflexes.

Atherosclerosis and coronary heart disease have been significantly linked to exposure to carbon disulfide. Atherosclerosis develops most notably in the blood vessels of the brain, glomeruli, and myocardium. Abnormal electroencephalograms and retinal hypertension typically occur before renal involvement is noted. Any of the above three areas may be affected by chronic exposure, but most often only one aspect can be observed. A significant increase in coronary heart disease mortality has been observed in carbon disulfide workers. Studies also reveal higher frequency of angina pectoris and hypertension. Abnormal electrocardiograms may also occur and are also suggestive of carbon disulfide's role in the etiology of coronary disease.

Other specific effects include chronic gastritis with the possible development of gastric and duodenal ulcers; impairment of endocrine activity, specifically adrenal and testicular; abnormal erythrocytic development with hypochromic anemia; and possible liver dysfunction with abnormal serum cholesterol. Also in women, chronic menstrual disorders may occur. These effects usually occur following chronic exposure and are subordinate to the other symptoms.

Recently human experience and animal experimentation have indicated several possible biochemical changes. Carbon disulfide and its metabolites (i.e., dithiocarbamic acids and isothiocyanates) show amino acid interference, cerebral monoamine oxidase inhibition, endocrine disorders, lipoprotein metabolism interference, blood protein, and zinc level abnormalities, and inorganic metabolism interference due to chelating of polyvalent ions. The direct relationship between these biochemical changes and clinical manifestations is only suggestive.

Allowable Limits on Exposure and Use

Air: The threshold limit value for carbon disulfide in air is 60 mg/m^3 (20 ppm) as of 1979 but an intended change will lower these values to 30 mg/m^3 (10 ppm). The tentative short term exposure limit is 90 mg/m^3 (30 ppm) (B-23).

NIOSH has recommended a 3 mg/m^3 (1 ppm) value on a time weighted average basis and a 30 mg/m^3 (10 ppm) value for 15-minute exposure (B-21).

Product Use: The tolerances set by the US EPA for CS_2 in or on raw agricultural commodities are as follows:

	40 *CFR* Reference	Parts per Million
Barley, grain (post-h)	180.100	exempt
Corn, grain (post-h)	180.100	exempt
Oats, grain (post-h)	180.100	exempt
Rice, grain (post-h)	180.100	exempt
Rye, grain (post-h)	180.100	exempt
Sorghum, grain, milo (post-h)	180.100	exempt
Wheat, grain (post-h)	180.100	exempt

The tolerances set by the US EPA for carbon disulfide as a fumigant in food are as follows (*CFR* Reference is to Title 21):

	CFR Reference	Parts per Million
Cereal grains, milled fractions	193.225	125.0
Corn grits, malted beverages	193.230	125.0
Rice, cracked, malted beverages	193.230	125.0

References

(1) Worthing, C.R., *Pesticide Manual,* 6th ed., p. 83, British Crop Protection Council (1979).
(2) Spencer, E.Y., *Guide to the Chemicals Used in Crop Protection,* 6th ed., p. 82, London, Ontario, Agriculture Canada (January 1973).
(3) Adcock, W.A. et al, U.S. Patent 2,935,380, May 3, 1960, assigned to Pan American Petroleum Corp.
(4) Kimberlin, C.N. et al, U.S. Patent 2,789,037, April 16, 1957, assigned to Esso Research and Engineering Co.
(5) Reid, J. et al, U.S. Patent 3,170,763, February 23, 1965, assigned to Shawinigan Chemicals, Ltd.

CARBON TETRACHLORIDE

Function: Fumigant (1)(2)

Chemical Name: Tetrachloromethane

Formula: CCl_4

Trade Name: None

Manufacture

Carbon tetrachloride may be made by:

(a) the chlorination of carbon disulfide (3)

$$CS_2 + 3Cl_2 \rightarrow CCl_4 + S_2Cl_2$$

(b) the chlorination of methane (4)

$$CH_4 + 4Cl_2 \rightarrow CCl_4 + 4HCl$$

Process Wastes and Their Control

Product Disposal: The MCA Manual recommends evaporation in a fume hood as a disposal method. The rate of degradation in the atmosphere is uncertain. Carbon tetrachloride has been used as a fire extinguishing agent, but it can give dangerous amounts of the much more toxic compound phosgene ($COCl_2$) and users should not remain in unventilated areas. Complete combustion of CCl_4 at flame temperatures produces CO_2 and corrosive hydrogen chloride (B-3).

Carbon tetrachloride may be quite stable under certain environmental conditions and the hydrolytic breakdown of carbon tetrachloride in water is estimated to require 70,000 years for 50% decomposition. This decomposition is considerably accelerated in the presence of metals such as iron. Hydrolytic decomposition as a means of removal from water appears to be insignificant as compared to evaporation. An evaporative half-life of 29 minutes of CCl_4 in water has been determined at ambient temperatures (B-26).

Toxicity

The acute oral LD_{50} value for rats is 5,730 mg/kg (insignificantly toxic). This compound has a relatively high oral LD_{50} compared to most insecticides, but is a toxic inhalant and is also absorbed through the skin (B-3).

The acute-toxicity effects of carbon tetrachloride are best characterized as hepatic nodular hyperplasia and cirrhosis and renal dysfunction in both experimental animals and man. It had no mutagenic potential in in vitro and in vivo test systems. Its teratogenic potential has not been firmly established. Carcinogenic bioassays have produced hepatomas in mice, rats, and hamsters, associated in most cases with regenerative nodular hyperplasia or postnecrotic cirrhosis.

In light of the above and taking into account the rise projections it is suggested that very strict criteria be applied when limits for carbon tetrachloride in drinking water are established (B-22).

The harmful effects of CCl_4 have been summarized as follows (B-25):

Local: Carbon tetrachloride solvent removes the natural lipid cover of the skin. Repeated contact may lead to a dry, scaly, fissured dermatitis. Eye contact is slightly irritating, but this condition is transient.

Systemic: Excessive exposure may result in central nervous system depression, and gastrointestinal symptoms may also occur. Following acute exposure, signs and symptoms of liver and kidney damage may develop. Nausea, vomiting, abdominal pain, diarrhea, enlarged and tender liver, and jaundice result from toxic hepatitis. Diminished urinary volume, red and white blood cells in the urine, albuminuria, coma, and death may be consequences of acute renal failure. The hazard of systemic effects is increased when carbon tetrachloride is used in conjunction with ingested alcohol.

Occupational exposure to CCl_4 and its hazards have been reviewed (5).

Allowable Limits on Exposure and Use

Air: The threshold limit value for CCl_4 in air is 65 mg/m³ (10 ppm) as of 1979. However, an intended change has been indicated to 30 mg/m³ (5 ppm) with the notation that CCl_4 is "an industrial substance suspect of carcinogenic potential for man." The short term exposure limit for CCl_4 is tentatively set at 130 mg/m³ (20 ppm). The limits for CCl_4 all bear the notation "skin" indicating that cutaneous absorption should be prevented so that the threshold limit value is not invalidated.

NIOSH has recommended a 12.6 mg/m³ (2 ppm) ceiling as a one hour ceiling value (B-21).

Water: In water, EPA has suggested (B-26) a criterion to protect freshwater aquatic life of 620 µg/l as a 24-hour average and the concentration should never exceed 1,400 µg/l at any time.

For carbon tetrachloride, the criterion to protect saltwater aquatic life is 2,000 µg/l as a 24-hour average and the concentration should never exceed 4,600 µg/l at any time.

For the maximum protection of human health from the potential carcinogenic effects of exposure to carbon tetrachloride through ingestion of water and contaminated aquatic organisms, the recommended ambient water concentration is zero. Concentrations of carbon tetrachloride estimated to result in additional lifetime cancer risks ranging from no additional risk to an additional risk of 1 in 100,000 have been set by US EPA. They are considering setting criteria at an interim target risk level in the range of 10^{-5}, 10^{-6}, or 10^{-7} with corresponding criteria of 2.6 $\mu g/l$, 0.26 $\mu g/l$, and 0.026 $\mu g/l$, respectively (B-26).

Product Use: Issuance of a rebuttable presumption against registration was being considered by US EPA as of September 1979 on the basis of possible oncogenicity, nephrotoxicity and hepatotoxicity.

The tolerances set by the US EPA for CCl_4 in or on raw agricultural commodities are as follows:

	40 *CFR* Reference	Parts per Million
Barley, grain (post-h)	180.100	exempt
Corn, grain (post-h)	180.100	exempt
Oats, grain (post-h)	180.100	exempt
Rice, grain (post-h)	180.100	exempt
Rye, grain (post-h)	180.100	exempt
Sorghum, grain, milo (post-h)	180.100	exempt
Wheat, grain (post-h)	180.100	exempt

The tolerances set by the US EPA for carbon tetrachloride as a fumigant in food are as follows (*CFR* Reference is to Title 21):

	CFR Reference	Parts per Million
Cereal grains, milled fractions	193.225	125.0
Corn grits, malted beverages	193.230	125.0
Rice, cracked, malted beverages	193.230	125.0

References

(1) Worthing, C.R., *Pesticide Manual,* 6th ed., p. 84, British Crop Protection Council (1979).
(2) Spencer, E.Y., *Guide to the Chemicals Used in Crop Protection,* 6th ed., p. 83, London, Ontario, Agriculture Canada (January 1973).
(3) Dehn, F.C. et al, U.S. Patent 3,234,293, February 8, 1966, assigned to Pittsburgh Plate Glass Co.
(4) Burks, W.M., et al, U.S. Patent 3,126,419, March 24, 1964, assigned to Stauffer Chemical Co.
(5) National Institute for Occupational Safety and Health, *Criteria for a Recommended Standard: Occupational Exposure to Carbon Tetrachloride,* NIOSH Doc. No. 76-133 (1976).

CARBOPHENOTHION

Function: Insecticide and acaricide (1)(2)

Chemical Name: S-[[(4-chlorophenyl)thio]methyl] O,O-diethyl phosphorodithioate

Formula:

Trade Names: R-1313 (Stauffer Chemical Co.)
Trithion® (Stauffer Chemical Co.)
Garrathion® (Stauffer Chemical Co.)
Acarithion®

Manufacture (3)

Thiophenol is reacted with formaldehyde and hydrochloric acid to give chloromethyl chlorophenyl sulfide. About 15.0 g (0.08 mol) of p-chlorophenyl-chloromethyl sulfide, 25.8 g (0.12 mol) of sodium diethyl dithiophosphate and 150 ml of 97% isopropanol were refluxed two hours. The bulk of the alcohol was distilled off and the residue treated with 100 ml of cold water and 100 ml of 30° to 60°C petroleum ether and the mixture transferred to a separatory funnel and shaken thoroughly.

The lower aqueous layer was discarded and the petroleum ether layer washed several times with cold water, dried over anhydrous sodium carbonate, filtered and the petroleum ether distilled off on the steam bath. The product, a light yellow colored liquid, weighed 15.8 g (59.0% based on the p-chlorophenyl-chloromethyl sulfide).

Process Wastes and Their Control

Product Waste Disposal: This compound is reported to be "rapidly" decomposed by hypochlorite. Hydrolysis rates are 37% in 2 hours in NaOH (pH 13.1) at 20°C and no decomposition in HCl (pH 1.0) in 24 hours at 60°C (B-3).

Toxicity

The acute oral LD_{50} value for rats is about 32 mg/kg (highly toxic).

Allowable Limits on Exposure and Use

Product Use: The tolerances set by the EPA for carbophenothion in or on raw agricultural commodities are as follows:

	40 *CFR* Reference	Parts per Million
Alfalfa, fresh	180.156	5.0
Alfalfa, hay	180.156	5.0
Almonds, hulls	180.156	10.0
Apples	180.156	0.8
Apricots	180.156	0.8
Beans, dry	180.156	0.1 N
Beans, lima, succulent	180.156	0.8
Beans, snap, succulent	180.156	0.8
Beans, straw	180.156	5.0
Beets, garden, roots	180.156	0.8
Beets, garden, tops	180.156	0.8
Beets, sugar, roots	180.156	5.0
Beets, sugar, tops	180.156	5.0
Blueberries	180.156	4.0
Cantaloups	180.156	0.8
Cattle, fat of meat	180.156	0.1
Cherries	180.156	0.8
Clover, fresh	180.156	5.0
Clover, hay	180.156	5.0
Corn (k+cwhr)	180.156	0.2
Corn, forage	180.156	5.0
Cotton, seed (undelinted)	180.156	0.2

(continued)

	40 *CFR* Reference	Parts per Million
Crabapples	180.156	0.8
Cucumbers	180.156	0.8
Eggplant	180.156	0.8
Figs	180.156	0.8
Goats, fat of meat	180.156	0.1
Grapefruit	180.156	2.0
Grapes	180.156	0.8
Hogs, fat of meat	180.156	0.1
Lemons	180.156	2.0
Limes	180.156	2.0
Milk	180.156	0.0
Nectarines	180.156	0.8
Olives	180.156	0.8
Onions, dry bulb	180.156	0.8
Onions, green	180.156	0.8
Oranges	180.156	2.0
Peaches	180.156	0.8
Pears	180.156	0.8
Peas, succulent	180.156	0.8
Pecans	180.156	0.1 N
Peppers	180.156	0.8
Pimentos	180.156	0.8
Plums (fresh prunes)	180.156	0.8
Quinces	180.156	0.8
Sheep, fat of meat	180.156	0.1
Sorghum, forage	180.156	5.0
Sorghum, grain	180.156	2.0
Soybeans, succulent	180.156	0.8
Spinach	180.156	0.8
Squash, summer	180.156	0.8
Strawberries	180.156	0.8
Tangerines	180.156	2.0
Tomatoes	180.156	0.8
Walnuts	180.156	0.1 N
Watermelons	180.156	0.8

The tolerances set by the US EPA for carbophenothion in animal feeds are as follows (the *CFR* Reference is to Title 21):

	CFR Reference	Parts per Million
Citrus meal (ct.f)	561.70	10.0
Citrus pulp (ct.f) dehydrated	561.70	10.0

The tolerances set by the US EPA for carbophenothion in food are as follows (the *CFR* Reference is to Title 21):

	CFR Reference	Parts per Million
Tea, dried	193.50	20.0

References

(1) Worthing, C.R., *Pesticide Manual,* 6th ed., p. 85, British Crop Protection Council (1979).
(2) Spencer, E.Y., *Guide to the Chemicals Used in Crop Protection,* 6th ed., p. 84, London, Ontario, Agriculture Canada (January 1973).
(3) Fancher, L.W., U.S. Patent 2,793,224, May 21, 1957, assigned to Stauffer Chemical Co.

CARBOXIN

Function: Fungicide (1)(2)(3)(4)(5)

Chemical Name: 5,6-dihydro-2-methyl-N-phenyl-1,4-oxathiin-3-carboxamide

Formula:

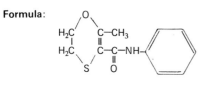

Trade Names: D-735 (Uniroyal, Inc.)
Vitavax® (Uniroyal, Inc.)
DMOC
Vitawax®

Manufacture (3)(4)(5)

Carboxin may be made by one of 2 routes (4):

    (a) starting with acetoacetanilide
    (b) starting with ethyl acetoacetate

Route (a) is as follows. To a stirred suspension of acetoacetanilide (150 g, 0.845 mol) and dry benzene (1 liter) was added sulfuryl chloride (72 ml or 120 g, 0.890 mol) dropwise over a period of 1½ hours. The stirring was continued for ½ hour more. The product was filtered (the filtrate used in a second run in place of dry benzene gave a higher yield of alpha-chloroacetoacetanilide); washed with water and benzene and dried. The yield was 131 g (73.5%), the MP was 136° to 138°C.

To a stirred suspension of alpha-chloroacetoacetanilide (63.5 g or 0.3 mol) and dry benzene (300 ml) was added a solution of KOH (20.4 g), 2-mercaptoethanol (22.2 ml or 22.5 g, 0.3 mol) and methanol (40 ml) dropwise over a period of two hours, keeping the temperature below 30°C. The mixture was stirred for one hour more. The potassium chloride which precipitated was filtered.

The solvents were removed from the filtrate by distillation. Benzene was added to the residue and then washed with water until neutral. The solution was acidified with p-toluenesulfonic acid (0.8 g) and heated under reflux using a Dean-Stark trap to collect water. The water collected was 5 ml (theory 5.3 ml). The solution was washed with water and the benzene removed. The residue solidified and was crystallized from 95% ethanol. The yield was 45.8 g (65%); the MP was 93° to 95°C.

Route (b) is, on the other hand, as follows. To a stirred and cooled solution of ethyl acetoacetate (260 g or 2 mols) was added sulfuryl chloride (270 g or 2 mols) over 3 hours, keeping the temperature between 0° and 5°C. The reaction mixture was left overnight. The SO_2 and HCl were removed on a water pump. The residual dark liquid was distilled at reduced pressure. After a small fore-run the liquid distilling between 88° and 90°C (at 16 mm) was collected. The yield was 300 g (91%).

To a cooled and stirred solution of ethyl alpha-chloroacetoacetate (33 g or 0.2 mol) and dry benzene (200 ml) was added a solution of potassium hydroxide (13.6 g), 2-mercaptoethanol (15.0 ml or 15.6 g) and methanol (30 ml) over a period of 1½ hours keeping the temperature below 30°C. The reaction mixture was stirred for ½ hour more. The potassium chloride formed was filtered.

The solvents were removed from the filtrate. Benzene was added to the residue and then washed with water. The benzene solution was acidified with p-toluenesulfonic acid and the water (3.4 ml; theory 3.6 ml) was collected by azeotropic distillation using the Dean-Stark trap. The reaction mixture was cooled, washed with water and then the benzene removed. The residue was distilled under high vacuum; BP (1 mm) 107° to 110°; yield 23 g (61.2%).

To a solution of ethyl 2,3-dihydro-6-methyl-1,4-oxathiin-5-carboxylate (188 g) in 95% ethanol (50 ml) was added a solution of NaOH (60 g) in water (400 ml). The reaction mixture was heated under reflux until the two layers became homogeneous (about ½ hour). The solution was cooled, diluted with water and acidified with dilute HCl. The white solid which precipitated was filtered at once, washed with water and dried in air. The yield was 134 g (84%); the MP was 178° to 180°C. Recrystallized material from ethanol melts at 180° to 181°C.

To a suspension of 2,3-dihydro-5-carboxy-6-methyl-1,4-oxathiin (32 g or 0.2 mol) in chloroform (200 ml) was added thionychloride (16 ml) and the solution was heated under reflux. Hydrogen chloride and sulfur dioxide were evolved and all the solids went into solution in 2 hours. The excess thionylchloride and solvent were removed in vacuo.

To the residue dissolved in chloroform (or benzene) was added a solution of aniline (37.2 g) in chloroform (or benzene), portionwise. The aniline hydrochloride which formed was filtered. The filtrate was washed with very dilute HCl solution and then with water. The chloroform (or benzene) was removed and the residue solidified at once. It was recrystallized from 95% ethanol. The yield was 38 g (80%); the MP was 93° to 94°C.

Process Wastes and Their Control

Product Waste Disposal: Carboxin is resistant to mild oxidative and hydrolytic conditions. Alkaline hydrolysis yields the more toxic aniline. Incineration appears to be the method of choice for disposal (B-3).

Toxicity

The acute oral LD_{50} value for rats is 3,200 mg/kg which rates as slightly toxic.

Allowable Limits on Exposure and Use

Product Use: The tolerances set by the US EPA for carboxin in or on raw agricultural commodities are as follows:

	40 *CFR* Reference	Parts per Million
Alfalfa, forage	180.301	5.0
Barley, forage	180.301	0.5
Barley, grain	180.301	0.2
Barley, straw	180.301	0.2
Cattle, fat	180.301	0.1
Cattle, mbyp	180.301	0.1
Cattle, meat	180.301	0.1
Corn, fodder	180.301	0.2
Corn forage	180.301	0.2
Corn, fresh (inc. sweet) (k+cwhr)	180.301	0.2
Corn, grain	180.301	0.2
Cotton, seed	180.301	0.2 N
Eggs	180.301	0.01
Goats, fat	180.301	0.1
Goats, mbyp	180.301	0.1
Goats, meat	180.301	0.1
Hogs, fat	180.301	0.1

(continued)

	40 *CFR* Reference	Parts per Million
Hogs, mbyp	180.301	0.1
Hogs, meat	180.301	0.1
Horses, fat	180.301	0.1
Horses, mbyp	180.301	0.1
Horses, meat	180.301	0.1
Milk	180.301	0.01
Oats, forage	180.301	0.5
Oats, seed	180.301	0.2
Oats, straw	180.301	0.2
Peanuts	180.301	0.2 N
Peanuts, hay	180.301	0.2 N
Peanuts, hulls	180.301	0.2 N
Poultry, fat	180.301	0.1
Poultry, mbyp	180.301	0.1
Poultry, meat	180.301	0.1
Rice	180.301	0.2
Rice, straw	180.301	0.2
Sheep, fat	180.301	0.1
Sheep, mbyp	180.301	0.1
Sheep, meat	180.301	0.1
Sorghum, fodder	180.301	0.2
Sorghum, forage	180.301	0.2
Sorghum, grain	180.301	0.2
Soybeans	180.301	0.2
Walnuts	180.300	0.5
Wheat, forage	180.301	0.5
Wheat, grain	180.301	0.2
Wheat, straw	180.301	0.2

References

(1) Worthing, C.R., *Pesticide Manual,* 6th ed., p. 86, British Crop Protection Council (1979).
(2) Spencer, E.Y., *Guide to the Chemicals Used in Crop Protection,* 6th ed., p. 85, London, Ontario, Agriculture Canada (January 1973).
(3) Von Schmeling, B., Kulka, M., Thiara, D.S. and Harrison, W.A., U.S. Patent 3,249,499, May 3, 1966, assigned to United States Rubber Co.
(4) Kulka, M., Thiara, D.S. and Harrison, W.A., U.S. Patent 3,393,202, July 16, 1968, assigned to Uniroyal, Inc.
(5) Von Schmeling, B. and Kulka, M., U.S. Patent 3,454,391, July 8, 1969, assigned to Uniroyal, Inc.

CARTAP

Function: Insecticides (1)(2)

Chemical Name: S,S'-[2-(dimethylamino)-1,3-propanediyl] carbamothioate

Formula: $(CH_3)_2NCH(CH_2SCNH_2)_2$
$$\underset{O}{\overset{\parallel}{}}$$

Trade Names: T 1258 (Takeda Chemical Industries, Ltd.)
Padan® (Takeda Chemical Industries, Ltd.)

Manufacture

1,3-dichloropropyl dimethylamine may be reacted with sodium thiocyanate to give a dithio-cyanato compound which is then hydrated to cartap using HCl in methanol. For example, hydrogen chloride gas is introduced for 2 hours into 12 parts by weight of 1,3-dithiocyanato-2-dimethyl-aminopropane hydrochloride suspended in 30 parts by volume of methanol in order to cause a reaction to take place. Upon completion of the ensuing reaction, the methanol is distilled off under reduced pressure from the reaction mixture to obtain a residue which is then dried. The residue is recrystallized from methanol to give 13.7 parts by weight of colorless needle-like crystals of cartap.

Toxicity

The acute oral LD_{50} value for rats is 325 mg/kg which rates as moderately toxic.

References

(1) Worthing, C.R., *Pesticide Manual,* 6th ed., p. 87, British Crop Protection Council (1979).
(2) Spencer, E.Y., *Guide to the Chemicals Used in Crop Protection,* 6th ed., p. 86, London, Ontario, Agriculture Canada (January 1973).
(3) Konishi, I., Okutan, T. and Soma, T., U.S. Patent 3,332,943, July 25, 1967, assigned to Takeda Chemical Industries, Ltd.

CHLORALOSE

Function: Rodenticide and narcotic for birds

Chemical Name: α-chloralose

Formula:

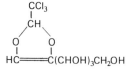

Trade Names: Somio[R]
Alphakil[R]

Manufacture

Chloralose is made by the reaction of chloral on glucose.

Toxicity

The acute oral LD_{50} value for rats is 400 mg/kg (moderately toxic).

References

(1) Worthing, C.R., *Pesticide Manual,* 6th ed., p. 89, British Crop Protection Council (1979).
(2) Spencer, E.Y., *Guide to the Chemicals Used in Crop Protection,* 6th ed., p. 89, London, Ontario, Agriculture Canada (January 1973).

CHLORAMBEN

Function: Herbicide (1)(2)(3)(4)

Chemical Name: 3-amino-2,5-dichlorobenzoic acid

Formula:

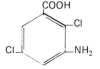

Trade Names: ACP-M-728 (Amchem Products)
Amiben® (Amchem Products)
Amoben®
Vegiben®

Manufacture

Preparation of 3-amino-2,5-dichlorobenzoic acid is indicated by way of example as follows. 23.6 g (0.1 mol) of pure 3-nitro-2,5-dichlorobenzoic acid (MP 220° to 221°C) and 20 g (0.169 mol) of granular tin are admixed in a flask to which is then added 100 ml of water and 100 ml of concentrated hydrochloric acid. The flask is then heated to 90° to 100°C, this temperature being thereafter maintained until complete solution takes place (usually about 6 hours). After complete solution has occurred, the entire solution is poured into a mixture of 1,000 ml of water and ice, whereby a crude precipitate is formed. This precipitate is collected and re-slurried in 1,000 ml of water to remove any inorganic salt contaminants. The crude precipitate is then recrystallized from boiling water (solubility 6 g/100 ml) and dried in an oven at 100°C. A white crystalline product is recovered in about an 80% yield based upon the 3-nitro-2,5-dichlorobenzoic acid, this product being 3-amino-2,5-dichlorobenzoic acid.

Process Wastes and their Control

Air: Air emissions from chloramben manufacture have been reported (B-15) to consist of 1.0 kg hydrogen chloride and 0.5 kg chlorine per metric ton of pesticide produced.

Product Waste Disposal: Chloramben is stable to heat, oxidation, and hydrolysis in acidic or basic media. The stability is comparable to that of benzoic acid. Chloramben is decomposed by sodium hypochlorite solution (B-3).

Toxicity

The acute oral LD_{50} value for rats is 3,500 mg/kg (B-5) which rates as slightly toxic.

The available data on chloramben are very sparse. Much additional information is needed regarding its chronic toxicity, teratogenicity, and carcinogenicity before limits can be confidently set. It is possible that many pertinent studies have been conducted by the manufacturer and could be made available for evaluation.

No-observed-adverse-effect doses for chloramben were at 250 mg/kg/day and 500 mg/kg/day in dogs and rats, respectively, in feeding studies. Based on these data an ADI was calculated at 0.25 mg/kg/day (B-22).

Allowable Limits on Exposure and Use

Product Use: The tolerances set by the US EPA for chloramben in or on raw agricultural commodities are as follows:

	40 *CFR* Reference	Parts per Million
Beans, dry	180.266	0.1 N
Beans, lima	180.266	0.1 N
Beans, snap	180.266	0.1 N
Beans, vines	180.266	0.1 N
Cantaloups	180.266	0.1 N
Corn, field, fodder	180.266	0.1 N
Corn, field, forage	180.266	0.1 N
Corn, field, grain	180.266	0.1 N
Cucumbers	180.266	0.1 N
Peanuts	180.266	0.1 N
Peanuts, forage	180.266	0.1 N
Peppers	180.266	0.1 N
Pumpkins	180.266	0.1 N
Soybeans	180.266	0.1 N
Soybeans, forage	180.266	0.1 N
Squash, summer	180.266	0.1 N
Squash, winter	180.266	0.1 N
Sunflower, seed	180.266	0.1 N
Sweet potatoes	180.266	0.1 N
Tomatoes	180.266	0.1 N

References

(1) Worthing, C.R., *Pesticide Manual,* 6th ed., p. 90, British Crop Protection Council (1979).
(2) Spencer, E.Y., *Guide to the Chemicals Used in Crop Protection,* 6th ed., p. 90, London, Ontario, Agriculture Canada (January 1973).
(3) McLane, S.R., Bishop, J.R. and Raman, H.P., U.S. Patent 3,014,063, December 19, 1961, assigned to Amchem Products, Inc.
(4) McLane, S.R., Bishop, J.R. and Raman, H.P., U.S. Patent 3,174,842, March 23, 1965, assigned to Amchem Products, Inc.

CHLORANIL

Function: Fungicide (1)(2)(3)

Chemical Name: 2,3,5,6-tetrachloro-2,5-cyclohexadiene-1,4-dione

Formula:

Trade Name: Spergon® (Uniroyal, Inc.)

Manufacture

A number of methods may be found in the literature for preparation of this compound, the more common of which are:

1. Chlorination of nitroaniline followed by reduction to dichloro-p-phenylene diamine and then simultaneous

chlorination and oxidation by potassium chlorate and hydrochloric acid;

2. Chlorination of aniline as described by H.H. Fletcher (4);

3. The reaction of concentrated hydrochloric acid on benzoquinone in the presence of hydrogen peroxide (30%); and

4. The oxidation of trichlorophenol with chromic acid, or passage of chlorine into 2,4,6-trichlorophenol in sulfuric acid monohydrate and chlorosulfonic acid at 85° to 90°C as described, for example, by F.N. Alquist et al (5).

Most of these methods, however, are costly or not too well suited to commercial production on a large scale. In some cases, too, the product as produced is in such an impure state as to require additional purification techniques to render it suitable to meet commercial specifications. Thus, an improved process has been suggested by J.E. Fox (6). It involves reaction of cyclohexane with hydrochloric acid and oxygen according to the equation:

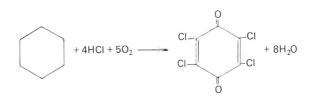

The reaction may be carried out by mixing cyclohexane and hydrogen chloride with oxygen or air and passing the mixture over the catalyst in a heated reaction chamber. If desired, the three reactants may be introduced in separate streams, or the air or oxygen may be introduced into the mixture of cyclohexane and hydrogen chloride. However, care should be taken not to permit the hydrocarbon and oxygen or air to reach reaction temperature in the absence of the hydrogen chloride.

Generally, stoichiometric quantities of the reactants are employed. Thus, as indicated in the equation above, 4 mols of hydrogen chloride and 5 mols of oxygen are fed for every mol of cyclohexane. A slight excess of oxygen or air is desirable. Excesses of either or both of the other reactants, however, do not adversely affect the reaction, but no advantages are to be gained in the use of either of these reactants in excess. Unreacted cyclohexane may be recycled in the process and thus excellent yields may be obtained.

Either aqueous hydrochloric acid or anhydrous hydrogen chloride may be employed. Because of the exothermic nature of the reaction, aqueous HCl is preferred since the heat capacity of the accompanying water facilitates maintenance of the desired temperature in the reactor. The oxidative chlorination of cyclohexane may be effected at a temperature within the range from about 180° to 250°C. Preferably, however, reaction temperature is maintained in the range from about 220° to 240°C at atmospheric pressure under vapor phase conditions.

The use of an oxidation catalyst is essential in order to obtain good yields of chloranil. Generally, all of the oxidation catalysts known in the art are effective, to some degree, for producing the reaction. To be preferred, however, are the oxides or salts of copper, iron, cobalt, aluminum and bismuth and mixtures thereof. In general, best results are obtained by the use of a copper catalyst which may be in the metallic state or as copper oxide or a copper salt such as the chloride, for example.

Preferably, the catalysts are deposited on refractory supports such as pumice, silica gel, alumina gel, porcelain, or the like. An eminently suitable catalyst for the reaction and one with which excellent results have been secured is a mixture of copper chloride, cobalt chloride, and ferric chloride deposited on alumina. Another excellent catalyst is prepared by coprecipitating copper and aluminum hydroxides from solutions of copper and aluminum chlorides, drying the precipitate, breaking up the dried filter cake, and screening the particles to secure preferably a 4 to 8 mesh size.

A preheater reactor of vertical tubular design may be used. It should be enclosed in a jacket for a heat transfer agent such as Aroclor. The lower portion of the tube may be packed with catalyst pellets and the upper or preheater section may be packed with glass helices. The product of the reaction may be readily recovered by condensing the reaction gases downstream of the reactor by means of suitable cooling. The crude product which is thus recovered as a solid material may be readily separated from unreacted cyclohexane by filtration and purified by crystallization.

Process Wastes and Their Control

Product Waste Disposal: Chloranil, a chlorinated quinone, is partially dechlorinated in alkaline solutions to form alkali salts of chloranilic acid, $C_6Cl_2O_2(OH)_2$ and is quantitatively reduced by potassim iodide to give tetrachlorohydroquinone $C_6Cl_4(OH)_2$ (B-3).

Toxicity

The acute oral LD_{50} value for rats is 4,000 mg/kg (very slightly toxic).

Allowable Limits on Exposure and Use

Product Use:

Chloranil was the subject of voluntary cancellation of all products as of January 19, 1977 on the basis of oncogenicity (B-17).

References

(1) Martin, H. and Worthing, C.R., *Pesticide Manual,* 5th ed., p. 90, British Crop Protection Council (January 1977).
(2) Spencer, E.Y., *Guide to the Chemicals Used in Crop Protection,* 6th ed., p. 91, London, Ontario, Agriculture Canada (January 1973).
(3) Ter Horst, W.P., U.S. Patent 2,349,771, May 23, 1944, assigned to United States Rubber Co.
(4) Fletcher, H.H., U.S. Patent 2,422,089, June 10, 1947, assigned to United States Rubber Co.
(5) Alquist, F.N. et al, U.S. Patent 2,414,008, January 7, 1947, assigned to Dow Chemical Co.
(6) Fox, J.E., U.S. Patent 2,722,537, November 1, 1955, assigned to Monsanto Chemical Co.

CHLORBENSIDE

Function: Acaricide (1)

Chemical Name: 4-chlorobenzyl-4-chlorophenyl sulfide

Formula:

Trade Names: Chlorparacide® (Boots Pure Drug)
Chlorsulphacide® (Boots Pure Drug)
Mitox®
Orthocide®
Chlorocide®
Elimite®

Manufacture

Chlorbenside is made by the reaction of p-chlorophenyl mercaptan and p-chlorobenzyl chloride. The preparation of the analogous compound, p-chlorophenyl-p-nitrobenzyl sulfide from p-chlorophenyl mercaptan and p-nitrobenzyl chloride has been described (2).

Process Wastes and Their Control

Product Wastes Disposal: Chlorbenside is oxidized by air or other oxidizing agents to the corresponding sulfoxide and sulfone, both of which have approximately the same acaricidal effects as the sulfide. The relatively low toxicity of chlorbenside (rats tolerated a diet of 10,000 mg/kg/day for 3 weeks) and its low persistence in soil suggests that burial or careful burning would be acceptable (with care to prevent exposure to combustion products, HCl and SO_2) (B-3).

Toxicity

The acute oral LD_{50} value for rats is greater than 10,000 mg/kg which rates as insignificantly toxic.

Allowable Limits on Exposure and Use

Product Use: The tolerances set by the US EPA for chlorbenside in or on raw agriculture commodities are as follows:

	40 *CFR* Reference	Parts per Million
Apples	180.168	3.0
Crabapples	180.168	3.0
Eggplant	180.168	3.0
Nectarines	180.168	3.0
Peaches	180.168	3.0
Pears	180.168	3.0
Plums (fresh prunes)	180.168	3.0
Quinces	180.168	3.0

Reference

(1) Spencer, E.Y., *Guide to the Chemicals Used in Crop Protection,* 6th ed., p. 92, London, Ontario, Agriculture Canada (January 1973).
(2) Stevenson, H.A., Brookes, R.F., Higgons, D.J. and Cranham, J.E., British Patent 738,170, October 12, 1955, assigned to Boots Pure Drug Co., Ltd., England.

CHLORBROMURON

Function: Herbicide (1)(2)(3)(4)

Chemical Name: N-(4-bromo-3-chlorophenyl)-N'-methyl-N'-methoxyurea

Formula:

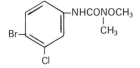

Trade Names: C 6313 (Ciba AG)
Maloran® (Ciba AG)

Manufacture (2)(3)(4)

107 g of N-(3-chlorophenyl)-N'-methoxy-N'-methylurea are dissolved in 500 cc of glacial acetic acid, 41 g of anhydrous sodium acetate are added, and the bromination is carried out at 80°C with 80 g Br_2 dissolved in glacial acetic acid. After 1 hour at 80°C, the mixture is poured into ice.

The yield of crude N-(4-bromo-3-chlorophenyl)-N'-methyl-N'-methoxyurea is 121 g. The product melts at 94° to 96°C and after recrystallization from acetonitrile, melts at 94.5° to 95.5°C.

Toxicity

The acute oral LD_{50} value for rats is 4,287 mg/kg (B-5) which rates as very slightly toxic.

Allowable Limits on Exposure and Use

Product Use: The tolerances set by the US EPA for chlorbromuron in or on raw agricultural commodities are as follows:

	40 *CFR* Reference	Parts per Million
Cattle, fat	180.279	0.1 N
Cattle, mbyp	180.279	0.1 N
Cattle, meat	180.279	0.1 N
Corn, fodder	180.279	0.2 N
Corn, forage	180.279	0.2 N
Corn, fresh (inc. sweet) (k+cwhr)	180.279	0.2 N
Corn, grain	180.279	0.2 N
Goats, fat	180.279	0.1 N
Goats, mbyp	180.279	0.1 N
Goats, meat	180.279	0.1 N
Hogs, fat	180.279	0.1 N
Hogs, mbyp	180.279	0.1 N
Hogs, meat	180.279	0.1 N
Horses, fat	180.279	0.1 N
Horses, mbyp	180.279	0.1 N
Horses, meat	180.279	0.1 N
Potatoes	180.279	0.2 N
Poultry, fat	180.279	0.1 N
Poultry, mbyp	180.279	0.1 N
Poultry, meat	180.279	0.1 N
Sheep, fat	180.279	0.1 N
Sheep, mbyp	180.279	0.1 N
Sheep, meat	180.279	0.1 N
Soybeans	180.279	0.2 N
Soybeans, forage	180.279	0.2 N
Wheat, grain	180.279	0.2 N
Wheat, straw	180.279	0.2 N

References

(1) Worthing, C.R., *Pesticide Manual,* 6th ed., p. 91, British Crop Protection Council (1979).
(2) Ciba, Ltd., British Patent 965,313, July 29, 1964.
(3) Martin, H., Aebi, H. and Ebner, L., U.S. Patent 3,497,541, February 24, 1970, assigned to Ciba, Ltd.
(4) Martin, H., Aebi, H. and Ebner, L., U.S. Patent 3,617,249, November 2, 1971, assigned to Ciba, Ltd.

CHLORBUFAM

Function: Preemergence herbicide

Chemical Name: 1-methyl-2-propynyl-3-chlorophenyl-carbamate

Formula:

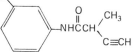

Trade Name: BiPC (BASF)

Manufacture

Chlorbufam is made by the reaction of 3-chlorophenyl isocyanate and methyl propynol, $HC\equiv CCH(CH_3)OH$.

Toxicity

The acute oral LD_{50} value for rats is 2,500 mg/kg (slightly toxic).

Reference

(1) Worthing, C.R., *Pesticide Manual,* 6th ed., p. 92, British Crop Protection Council (1979).

CHLORDANE

Function: Insecticide (1)(2)

Chemical   Name: 1,2,4,5,6,7,8,8-octachloro-2,3,3a,4,7,7a-hexahydro-4,7-methano-1H-indene

Formula:

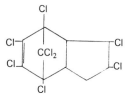

Trade Names: Velsicol 168 (Velsicol)
Octachlor® (Velsicol)
Chlordan®
Octa-Klor®
Chlorogran®
Chlor-Kil®
Prentox®
Penticklor®
Corodane®
Synklor®

Manufacture (3)(4)(5)(6)(7)

The feed materials for chlordane manufacture are hexachlorocyclopentadiene and cyclopentadiene which react according to the Diels-Alder condensation:

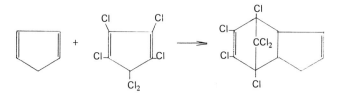

which produces a material known as chlordene which is subjected to addition chlorination
by one mol of chlorine to give chlordane as follows:

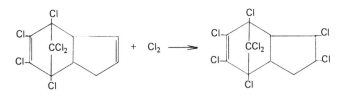

The reactants in the initial condensation step may be utilized in an equimolar ratio; however, an excess of either reactant can be present.

Hexachlorocyclopentadiene may be made by (a) chlorination of pentanes (4) and (b) chlorination of cyclopentadiene. Cyclopentadiene may be produced as a raw material for chlordane manufacture by the vapor phase cracking of naphtha. The cyclopentadiene is frequently
separated in the form of its dimer, dicyclopentadiene, which may be cracked to give the
monomer.

Sulfuryl chloride is the preferred agent for effecting the chlorination step. A stoichiometric
amount of sulfuryl chloride or an excess thereof should be employed. The use of sulfuryl
chloride is superior to the use of elemental chlorine; the latter gives substitution chlorination products in addition to the desired simple addition product. Such a complex product
mixture is a viscous liquid, hard to separate and hard to use in insecticide formulation, in
contrast to the crystalline product produced using sulfuryl chloride.

The Diels-Alder reaction temperature may be maintained at 70° to 85°C according to M.
Kleiman (5). The exothermic reaction is usually controlled below 100°C in any event at atmospheric pressure. The chlorination step proceeds satisfactorily at any temperature between room temperature and 120°C at atmospheric pressure. Reaction periods of from 1 to
10 hours have been cited for the Diels-Alder condensation step by S.H. Herzfeld et al (6). A
reaction time of less than about one hour at the reflux temperature of sulfuryl chloride (69°C)

is generally sufficient in the chlorination step, according to M. Kleiman (5). The adduct of hexachlorocyclopentadiene and cyclopentadiene can be prepared by simply mixing the reactants in the presence or absence of a solvent, but preferably in the absence of a solvent, according to M. Kleiman (5).

A solvent may be used for the chlorination; the ideal solvent is merely excess sulfuryl chloride which can be recovered when the reaction is completed. No catalyst is needed for the Diels-Alder condensation. A metal halide catalyst of the Friedel-Crafts type such as aluminum chloride, stannic chloride, ferric chloride, arsenic trichloride or antimony pentachloride or a mixture thereof is used in the chlorination.

The initial condensation step may be carried out in a conventional stirred, jacketed reaction kettle equipped with a reflux condenser. The same or a similar vessel is suitable for the subsequent chlorination step. The product of the condensation reaction crystallizes upon cooling the reaction mass. It may be purified by recrystallization from methanol if desired. When the chlorination step is complete, any solvent or excess sulfuryl chloride can be removed under vacuum. The residue, containing the desired chlordane product, can be washed with a hydrocarbon solvent, washed with caustic and water, and directly crystallized to give a relatively pure crystalline chlordane. Figure 19 is a flow diagram of the process.

Figure 19: Production and Waste Schematic for Chlordane

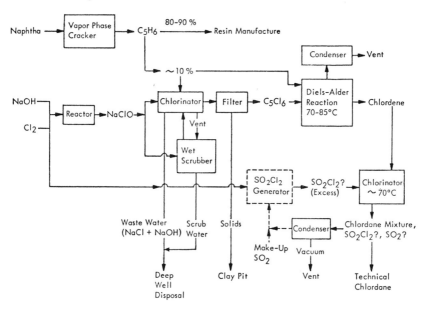

Source: Reference (3)

Process Wastes and Their Control

Air: Air emissions from chlordane manufacture have been reported (B-15) to consist of 1.0 kg of hydrocarbons and 0.5 kg of hydrogen chloride per metric ton of pesticide produced.

Water: Spent hypochlorite wastewater contains 2% NaOH and 400 ppm hexachlorocyclopentadiene. This wastewater is deep-well injected at one plant (B-10).

Product Wastes: Chlordane is readily dehydrochlorinated in alkali to form "nontoxic" products, a reaction catalyzed by traces of iron. The environmental hazards of the products are uncertain. Chlordane is completely dechlorinated by sodium in isopropyl alcohol. The MCA recommends incineration methods for disposal of chlordane (B-3).

Toxicity

The acute oral LD_{50} value for rats is 475 mg/kg which is moderately toxic.

Chlordane has been demonstrated to be highly toxic to aquatic organisms, to bioconcentrate in many aquatic species, and to persist for prolonged periods in the environment. It is toxic to avian and mammalian species and exhibits carcinogenic activity in mice. Thus, chlordane in water is a hazard to both aquatic and terrestrial life (B-26).

Allowable Limits on Exposure and Use

Air: The threshold limit value for chlordane in air is set at 0.5 mg/m^3 as of 1979. The tentative short term exposure limit is 2.0 mg/m^3 (B-23).

Water: In water, EPA has set (B-12) a limit of 0.01 μg/l of chlordane for the protection of freshwater aquatic life and a limit of 0.005 μg/l for the protection of marine aquatic life. It was noted that persistence, bioaccumulation potential and carcinogenicity of chlordane caution minimum human exposure.

In water, EPA has subsequently suggested (B-26) a limit to protect freshwater aquatic life of 0.024 μg/l as a 24-hour average and the concentration should not exceed 0.36 μg/l at any time.

For chlordane the criterion to protect saltwater aquatic life is 0.0091 μg/l as a 24-hour average and the concentration should not exceed 0.18 μg/l at any time.

For the maximum protection of human health from the potential carcinogenic effects of exposure to chlordane through ingestion of water and contaminated aquatic organisms, the ambient water concentration is zero. Concentrations of chlordane estimated to result in additional lifetime cancer risks ranging from no additional risk to an additional risk of 1 in 100,000 have been calculated by US EPA (B-26). The US EPA is considering setting criteria at an interim target risk level in the range of 10^{-5}, 10^{-6}, or 10^{-7} with corresponding criteria of 1.2 ng/l, 0.12 ng/l, and 0.012 ng/l, respectively.

Canadian Drinking Water Standards list a tentative maximum permissible limit for chlordane of 3 μg/l, which is applicable to raw water supplies in Canada (B-26).

Product Use: The tolerances set by the EPA for chlordane in or on raw agricultural commodities are as follows:

	40 *CFR* Reference	Parts per Million
Apples	180.122	0.3
Apricots	180.122	0.3
Beans	180.122	0.3
Beets, greens (alone)	180.122	0.3
Beets, with tops	180.122	0.3
Beets, without tops	180.122	0.3
Blackberries	180.122	0.3
Blueberries (huckleberries)	180.122	0.3
Boysenberries	180.122	0.3
Broccoli	180.122	0.3

(continued)

	40 *CFR* Reference	Parts per Million
Brussels sprouts	180.122	0.3
Cabbage	180.122	0.3
Carrots	180.122	0.3
Cauliflower	180.122	0.3
Celery	180.122	0.3
Cherries	180.122	0.3
Citrus fruits	180.122	0.3
Collards	180.122	0.3
Corn	180.122	0.3
Cucumbers	180.122	0.3
Dewberries	180.122	0.3
Eggplant	180.122	0.3
Grapes	180.122	0.3
Kale	180.122	0.3
Kohlrabi	180.122	0.3
Lettuce	180.122	0.3
Loganberries	180.122	0.3
Melons	180.122	0.3
Nectarines	180.122	0.3
Okra	180.122	0.3
Onions	180.122	0.3
Papayas	180.122	0.3
Peaches	180.122	0.3
Peanuts	180.122	0.3
Pears	180.122	0.3
Peas	180.122	0.3
Peppers	180.122	0.3
Pineapples	180.122	0.3
Plums (fresh prunes)	180.122	0.3
Potatoes	180.122	0.3
Quinces	180.122	0.3
Radishes, tops	180.122	0.3
Radishes, with tops	180.122	0.3
Radishes, without tops	180.122	0.3
Raspberries	180.122	0.3
Rutabagas	180.122	0.3
Rutabagas, tops	180.122	0.3
Rutabagas, with tops	180.122	0.3
Rutabagas, without tops	180.122	0.3
Squash	180.122	0.3
Squash, summer	180.122	0.3
Strawberries	180.122	0.3
Sweet potatoes	180.122	0.3
Tomatoes	180.122	0.3
Turnips, greens	180.122	0.3
Turnips, with tops	180.122	0.3
Turnips, without tops	180.122	0.3
Youngberries	180.122	0.3

Under the provisions of the US EPA Administrator's acceptance of the settlement plan to phase out certain uses of the pesticides chlordane and heptachlor, most registered products containing chlordane will be effectively cancelled or their applications for registration denied by December 31, 1980 (B-17). A summary of those uses not affected by this settlement, as well as a summary of those uses affected (Phase Out Uses) by this settlement, including the pest to the controlled, the site of application, the end-use dates and use restirctions, follows:

    1. Uses not affected

        (a) Subsurface ground insertion for termite control
            (clarified by 40 *FR* 30522, July 21, 1975, to apply to

the use of emulsifiable or oil concentrate formulations for controlling subterranean termites on structural sites such as buildings, houses, barns, and sheds, using current control practices).
   (b) Dipping of roots or tops of nonfood plants.
2. Uses affected (Phase Out Uses)
   (a) Registrations for control of ants on citrus in the States of California and Texas will be effectively cancelled or denied by December 31, 1979. Restricted to use on acreage under an Integrated Pest Management program involving the release of beneficial insects (parasites and predators) to maintain populations of harmful insects (scales, mites, mealybugs) at an acceptable level. In California, this use is permitted only pursuant to the provisions of the appropriate State of California permit and prescription use programs.
   (b) Registrations for control of imported fire ants on lands not presently used or to be used for food or feed production or grazing for a period of two years following treatment are effectively cancelled or denied by December 31, 1980. Distribution and use is restricted to nine states (AL, AR, FL, GA, LA, MS, NC, SC and TX). Use shall be restricted to mound treatment; broadcast or aerial application is prohibited.
   (c) Registrations for control of cutworms on grapes in the State of California will be effectively cancelled or denied by July 1, 1980. This use shall be restricted to application by ground equipment only. Use shall be permitted only pursuant to the provisions of the appropriate State of California permit and prescription use programs.
   (d) Registrations for control of grasshoppers on flax will be effectively cancelled or denied by October 1, 1978.
   (e) Registrations for control of white grubs, strawberry rootworm, strawberry root weevil or crown girdler, strawberry crown borer and black vine weevil on strawberries will be effectively cancelled or denied by August 1, 1979. This use shall be permitted in California only pursuant to the provisions of the appropriate State of California permit and prescription use programs.
   (f) Registrations for control of imported fire ants and Japanese beetle larvae on nursery stock, for compliance with Federal or State Quarantines, and for control of the black vine weevil on nursery stock for compliance with State Nursery Certification Regulations will be effectively cancelled or denied by December 31, 1979. Restricted to use on land with nursery stock grown for balled and burlapped, bare root or container stock. Use on turf is prohibited.

> Note: Phase Out Uses listed above may be applied by certified applicators only. The end-use dates for Phase Out Uses should be those dates listed above, unless the production limitations imposed by FIFRA Docket No. 326 et al

is exceeded earlier. Pesticide products in existence 90 days before the effective date of cancellation or denial of a Phase Out Use may: (1) be distributed, sold or otherwise moved in commerce, and used; provided that the pesticide shall not be used inconsistent with its labeling, and (2) may be relabeled by or under the authority of a registrant for another Phase Out Use not already cancelled or denied, and any pesticide product so produced shall not count against the production limitation for the other Phase Out Use.

References

(1) Worthing, C.R., *Pesticide Manual,* 6th ed., p. 93, British Crop Protection Council (1979).
(2) Spencer, E.Y., *Guide to the Chemicals Used in Crop Protection,* 6th ed., p. 94, London, Ontario, Agriculture Canada (January 1973).
(3) Midwest Research Institute, *The Pollution Potential in Pesticide Manufacturing,* Wash., DC, Environment Protection Agency (June 1972).
(4) McBee, E.T. and Baranauckas, C.F., U.S. Patent 2,509,160, assigned to Purdue Research Foundation.
(5) Kleiman, M., U.S. Patent 2,598,561, May 27, 1952, assigned to Velsicol Corp.
(6) Herzfeld, S.H. et al, U.S. Patent 2,606,910, August 12, 1952, assigned to Velsicol Corp.
(7) Hyman, J., U.S. Patent 2,519,190, August 15, 1960, assigned to Velsicol Corp.

CHLORDECONE

Function: Insecticide (1)(2)(3)(4)

Chemical Name: 1,1a,3,3a,4,5,5,5a,5b,6-decachloro-octahydro-1,3,4-metheno-2H-cyclo-buta[cd]pentalen-2-one

Formula:

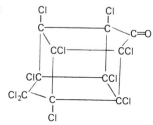

Trade Names: GC-1189 (Allied Chemical)
Kepone® (Allied Chemical)

Manufacture

Chlordecone is made by condensation of 2 mols of hexachlorocyclopentadiene in the presence of SO_3, followed by hydrolysis. In a typical example (3), a charge of 188 parts (0.69 mol) of hexachlorocyclopentadiene was cooled to 5° to 10°C, and to the agitated charge was added gradually 940 parts of 60% oleum [containing 565 parts (7.1 mols) of free SO_3]. After addition of all the oleum, which required about one hour, the mixture whose temperature had risen progressively to about 70°C, was added slowly to a large volume (5,000 parts) of water to dilute the acid.

The crude product precipitated immediately, upon contact of the charge with the water, as a white solid. The product was filtered from the spent acid, stirred three times with fresh water, and filtered after each water wash to remove most of the sulfuric acid. The product was further purified by dissolving it in 500 parts 95% ethanol, reprecipitating by the addition of 500 parts water, filtering and drying. One-hundred twenty-six parts of purified chlordecone were obtained representing a yield of 72% of theoretical.

Process Wastes and Their Control

Product Wastes: A process has been developed (5) which effects chlordecone degradation by treatment of aqueous wastes with UV radiation in the presence of hydrogen in aqueous sodium hydroxide solution. Up to 95% decomposition was effected by this process.

Chlordecone previously presented serious disposal problems because of its great resistance to bio- and photodegradation in the environment. It is highly toxic to normally-occurring degrading microorganisms. Although it can undergo some photodecomposition when exposed to sunlight to the dihydro compound (leaving a compound with 8 chloro substituents) that degradation product does not significantly reduce toxicity.

Toxicity

The acute oral LD_{50} value for rats is 95 mg/kg which rates as moderately toxic.

Chlordecone is persistent in the environment. Test results clearly suggest that liver lesions, including cancer, were induced in both sexes of rats and mice fed chlordecone. In addition, the time to detection of the first hepatocellular carcinoma observed at death was shorter for treated than for control mice and, in both sexes and both species, it appeared inversely related to the dose.

In light of the above and taking into account the carcinogenic risk projections it is suggested that very strict criteria be applied when limits for chlordecone in drinking water are established.

Apparently, little is known about the pharmacokinetics of chlordecone and its mechanisms of toxicity. There is a pressing need for systematic investigation of absorption, distribution, biotransformation, and excretion in humans and experimental animals, to gain an understanding of its toxicity and to provide a basis for rational therapy.

There is also very little information on the environmental transport mechanisms of chlordecone and its degradation products, its persistence, and its degradation in soil. Chlordecone residues have been found in food crops grown in rotation with chlordecone-treated tobacco (B-22).

Allowable Limits on Exposure and Use

Air: NIOSH recommended an environmental exposure limit for chlordecone of 1.0 $\mu g/m^3$ as a ceiling value on a 15-minute exposure basis in 1976 (6).

Product Use: A rebuttable presumption against registration of chlordecone was issued by US EPA on March 25, 1976 on the basis of oncogenicity.

The trademarked Kepone and products of six formulations were the subject of voluntary cancellation according to a US EPA notice dated July 27, 1977.

In a series of decisions, the first of which was issued on June 17, 1976, the EPA effectively cancelled all registered products containing Kepone as of May 1, 1978. A summary of Kepone products, their registration numbers, effective cancellation dates, and disposition of uses of existing stocks follows:

1. Inaccessible Products
    (a) Antrol Ant Trap, Reg. No. 475-11; Black Flag Ant Trap, Reg. No. 475-82; Grant's Ant Trap, Reg. No. 1663-21; Grant's Roach Trap, Reg. No. 1663-22; Grant's Ant Control, Reg. No. 1663-24; and Dead Shot Ant Killer, Reg. No. 274-23 were cancelled as of May 11, 1977. Distribution, sale, and use of existing stocks formulated prior to May 11, 1977, was permitted until such stocks were exhausted.
    (b) Black Leaf Ant Trap, Reg. No. 5887-63; Hide Roach and Ant Trap, Reg. No. 3325-4; Lilly's Ant Trap With Kepone, Reg. No. 460-17; T.N.T. Roach and Ant Killer, Reg. No. 2095-2; Johnston's No-Roach Traps, Reg. No. 2019-19; Mysterious Ant Trap With Kepone, Reg. No. 395-19; Magikil Ant Trap With Kepone, Reg. No. 395-21; Magikil Roach Trap With Kepone, Reg. No. 395-25; Ant-Not Ant Trap, Reg. No. 358-20; Nott Roach Trapp, Reg. No. 358-129; E-Z Ant Trap Contains Kepone, Reg. No. 506-109; Tat Ant Trap, Reg. No. 506-126; and Ant Check Ant Trap, Reg. No. 506-129 will be effectively cancelled on May 1, 1978. Kepone already in the formulation process may be formulated into inaccessible products between now and May 1, 1978. Distribution, sale and use of Kepone products formulated prior to May 1, 1978 will be permitted until such stocks are exhausted.

2. Accessible Products: all of these products were cancelled as of December 13, 1977. Distribution, sale, and use of these products is now unlawful.

> Note: The following definitions are included in order to distinguish between the two categories of Kepone products. Inaccessible products: Include those enclosed Kepone traps made from metal or plastic as well as metal stakes containing enclosed Kepone bait which are hammered into the ground. Accessible products: Includes those which, in normal use, would be removed from their containers, as well as foil or cardboard covered traps.

References

(1) Martin, H. and Worthing, C.R., *Pesticide Manual,* 5th ed., p. 94, British Crop Protection Council (January 1977).
(2) Spencer, E.Y., *Guide to the Chemicals Used in Crop Protection,* 6th ed., p. 96, London, Ontario, Agriculture Canada (January 1973).
(3) Gilbert, E.E. and Gioloto, S.L., U.S. Patent 2,616,928, November 4, 1952, Reissue 24,435, February 25, 1958, assigned to Allied Chemical and Dye Corp.
(4) Gilbert, E.E. and Gioloto, S.L., U.S. Patent 2,616,825, November 17, 1950, Reissue 24,749, December 15, 1959, assigned to Allied Chemical Corp.
(5) Kitchens, J.A.F., U.S. Patent 4,144,152, March 13, 1979, assigned to Atlantic Research Corp.
(6) National Institute for Occupational Safety and Health, *Recommended Standard for Occupational Exposure to Kepone,* Wash. DC, (January 27, 1976).

CHLORDIMEFORM

Function: Acaricide (1)(2)(3)(4)

Chemical Name: N'-(4-chloro-2-methylphenyl)-N,N-dimethylmethaninimide

Formula:

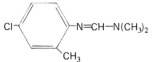

Trade Names: C-8514 (Ciba-Geigy)
Galecron® (Ciba-Geigy)
Schering 36,268 (Schering)
Fundal® (Schering)
Chlorfenamidine®

Manufacture (3)(4)

15.4 parts by weight of phosphorus oxychloride were slowly added with stirring to 7.4 parts by weight of dimethylformamide dissolved in 50 parts by weight of ethylene chloride. The addition period was about 30 to 40 minutes and the temperature was allowed to rise to 45°C. To the solution then were subsequently added 14.2 parts by weight of 2-amino-5-chlorotoluene while stirring was continually maintained and the temperature of the reaction mixture was allowed to rise to 65°C. After completion of the 2-amino-5-chlorotoluene addition, the mixture was stirred for two additional hours at about 65° to 70°C. Then the temperature of the solution was adjusted to about 5°C by external cooling, and then a strong base such as sodium hydroxide solution or triethylamine was added with vigorous stirring to maintain the pH at 11 to 11.5. The ethylene chloride layer was separated, washed with water, and dried over sodium sulfate, followed by evaporation of the solvent under diminished pressure which yielded as residue 16.7 parts by weight of an oily liquid, nearly 85% of the theoretical amount; BP 156° to 157°C/0.4 mm.

Process Wastes and Their Control

Product Wastes: This miticide is hydrolyzed in alkaline, neutral or acidic media, first to N-formylchlorotoluidine and then to 4-chlorotoluidine (a less toxic product). The hydrolysis proceeds slowly in acidic media. The hydrochloride is stable at 50°C for 70 hours, but decomposes at its melting point of 225° to 227°C. It is reported to decompose in fire and may give off poisonous fumes (B-3).

Toxicity

The acute oral LD_{50} value for rats is 250 mg/kg (moderately toxic).

Allowable Limits on Exposure and Use

Product Use: The tolerances set by the US EPA for chlordimeform in or on raw agricultural commodities are as follows:

	40 *CFR* Reference	Parts per Million
Apples	180.285	3.0
Broccoli	180.285	2.0
Brussels sprouts	180.285	2.0

(continued)

	40 *CFR* Reference	Parts per Million
Cabbage	180.285	2.0
Cattle, fat	180.285	0.25
Cattle, mbyp	180.285	0.25
Cattle, meat	180.285	0.25
Cauliflower	180.285	2.0
Cherries	180.285	5.0
Cotton, seed	180.285	5.0
Eggs	180.285	0.05
Goats, fat	180.285	0.25
Goats, mbyp	180.285	0.25
Goats, meat	180.285	0.25
Hogs, fat	180.285	0.25
Hogs, mbyp	180.285	0.25
Hogs, meat	180.285	0.25
Horses, fat	180.285	0.25
Horses, mbyp	180.285	0.25
Horses, meat	180.285	0.25
Milk	180.285	0.05
Nectarines	180.285	5.0
Peaches	180.285	5.0
Pears	180.285	5.0
Plums (fresh prunes)	180.285	4.0
Poultry, fat	180.285	0.25
Poultry, mbyp	180.285	0.25
Poultry, meat	180.285	0.25
Sheep, fat	180.285	0.25
Sheep, mbyp	180.285	0.25
Sheep, meat	180.285	0.25
Tomatoes	180.285	1.0
Walnuts	180.285	0.1

The tolerances set by the US EPA for chlordimeform in food are as follows (the *CFR* Reference is to Title 21):

	CFR Reference	Parts per Million
Prunes, dried	193.60	15.0

References

(1) Worthing, C.R., *Pesticide Manual,* 6th ed., p. 95, British Crop Protection Council (1979).
(2) Spencer, E.Y., *Guide to the Chemicals Used in Crop Protection,* 6th ed., p. 97, London, Ontario, Agriculture Canada (January 1973).
(3) Arndt, H., U.S. Patent 3,378,437, April 16, 1968, assigned to Schering AG, Germany.
(4) Arndt, H. and Steinhausen, W., U.S. Patent 3,502,720, March 24, 1970, assigned to Schering AG, Germany.

CHLORFENETHOL

Function: Acaricide (1)(2)

Chemical Name: 4-chloro-α-(4-chlorophenyl)-α-methylbenzenemethanol

Formula:

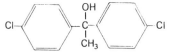

Trade Name: Dimite® (Sherwin-Williams Co.)

Manufacture

Chlorfenethol may be made by the condensation of 4,4-dichlorobenzophenone and methyl magnesium bromide (3) according to the equation:

$$ClC_6H_4COC_6H_4Cl + CH_3MgBr \longrightarrow ClC_6H_4 \overset{\overset{\displaystyle OMgBr}{|}}{\underset{\underset{\displaystyle CH_3}{|}}{C}} C_6H_4Cl$$

followed by hydrolysis of the intermediate:

$$ClC_6H_4 \overset{\overset{\displaystyle OMgBr}{|}}{\underset{\underset{\displaystyle CH_3}{|}}{C}} C_6H_4Cl + H_2O \longrightarrow ClC_6H_4 \overset{\overset{\displaystyle OH}{|}}{\underset{\underset{\displaystyle CH_3}{|}}{C}} C_6H_4Cl + MgOHBr$$

The coupling with the Grignard reagent may be carried out at reflux temperature (35° to 80°C). The hydrolysis step may be carried out at 0° to 5°C. Both steps are carried out at atmospheric pressure. The reaction time for the initial condensation of the ketone with the Grignard reagent is about 2 hours. The Grignard condensation step is carried out in the liquid phase in ether/benzene solution. The hydrolysis step is carried out in the liquid phase using water and ice.

No catalyst is used in either the Grignard condensation or the hydrolysis steps. An agitated reaction vessel equipped with a reflux condenser is suitable for the Grignard condensation. A tank or vat may be used for the hydrolysis step. The product of the Grignard addition precipitates from the initial reaction mixture. It is not separated but the mixture is simply poured onto ice. The organic layer is distilled to give an 80% yield of a product melting at 68°C.

An alternative route (4) involves p-dichlorobenzene as a starting material. It is converted to a Grignard reagent, then reacted with ethyl acetate and finally hydrolyzed to give the bis(p-chlorophenyl)ethanol product.

Process Wastes and Their Control

Product Wastes: Chlorfenethol can be dehydrated with difficulty to the ethylenic product 4,4-dichlorodiphenyl ethylene:

$$(ClC_6H_4)_2C(OH)-CH_3 \rightarrow (ClC_6H_4)_2CH=CH_2 + H_2O$$

Only 61% was reacted in 24 hours at 195°C, but in the presence of 0.1 N H_2SO_4 in alcohol, 80% was dehydrated in 5 hours at reflux (B-3).

Toxicity

The acute oral LD_{50} value for rats is 500 mg/kg (B-5) which rates as slightly to moderately toxic.

References

(1) Worthing, C.R., *Pesticide Manual,* 6th ed., p. 97, British Crop Protection Council (1979).
(2) Spencer, E.Y., *Guide to the Chemicals Used in Crop Protection,* 6th ed., p. 98, London, Ontario, Agriculture Canada (January 1973).
(3) Rutruff, R.F. et al, U.S. Patent 2,430,586, November 11, 1947, assigned to Sherwin-Williams Co.
(4) Metal and Thermit Corp., British Patent 831,421 (March 30, 1960).

CHLORFENPROP-METHYL

Function: Herbicide (1)(2)(3)

Chemical Name: Methyl α-4-dichlorobenzene propanoate

Formula:

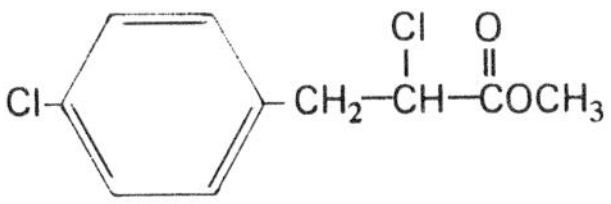

Trade Names: Bayer 70533 (Bayer)
Bisidin® (Bayer)

Manufacture (3)

In a first step p-chloroaniline is diazotized. The diazo compound is reacted with methyl acrylate in the presence of $CuCl_2$ in a second step which is conducted in an acetone solvent at 10° to 60°C. The product of the second step is the chloro ester, $ClC_6H_4CH_2CHClCOOCH_3$.

Toxicity

The acute oral LD_{50} value for rats is about 1,190 mg/kg (slightly toxic).

References

(1) Worthing, C.R., *Pesticide Manual,* 6th ed., p. 98, British Crop Protection Council (1979).
(2) Spencer, E.Y., *Guide to the Chemicals Used in Crop Protection,* 6th ed., p. 107, London, Ontario, Agriculture Canada (January 1973).
(3) Eue, L., Hack, H., Westphal, K. and Wegler, R., British Patent 1,077,194, July 26, 1967, assigned to Farbenfabriken Bayer AG, Germany.

CHLORFENVINPHOS

Function: Insecticide (1)(2)

Chemical Name: 2-chloro-1-(2,4-dichlorophenyl)-ethenyl diethyl phosphate

Formula:

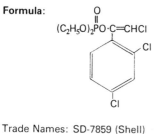

Trade Names: SD-7859 (Shell)
Birlane® (Shell)
Sapecron® (Ciba-Geigy)

Manufacture (3)(4)

113 g of 2,4-dichloracetophenone was added over a period of one-half hour to 100 g of tri-ethyl phosphite. During the addition of the dichloroacetophenone the temperature was maintained at 10° to 30°C by cooling as required. The reaction mixture then was stripped by heating on the steam bath under about 2 mm mercury pressure leaving a crude product consisting of 179 g of a clear amber liquid. The liquid was molecularly distilled at 80°C under about 6 x 10^{-3} mm mercury pressure to give 163 g of a clear liquid product.

Toxicity

The acute oral LD$_{50}$ for rats is 25 mg/kg which rates as highly toxic.

Allowable Limits on Exposure and Use

Product Use: The tolerances set by the US EPA for chlorfenvinphos in or on raw agriculture commodities are as follows:

	40 *CFR* Reference	Parts per Million
Cattle, fat (pre-s min 3 days)	180.322	0.15
Eggs	180.322	0.005
Goats, fat	180.322	0.005
Hogs, fat	180.322	0.005
Horses, fat	180.322	0.005
Milk, fat (= N in whole milk)	180.322	0.1 N
Poultry, fat	180.322	0.005
Sheep, fat (pre-s min 3 days)	180.322	0.15

References

(1) Worthing, C.R., *Pesticide Manual,* 6th ed., p. 100, British Crop Protection Council (1979).
(2) Spencer, E.Y., *Guide to the Chemicals Used in Crop Protection,* 6th ed., p. 100, London, Ontario, Agriculture Canada (January 1973).
(3) Whetstone, R. and Harman, D., U.S. Patent 3,116,201, December 31, 1963, assigned to Shell Oil Co.
(4) Whetstone, R.R. and Harman, D., U.S. Patent 2,956,073, October 11, 1960, assigned to Shell Oil Co.

CHLORFLURECOL-METHYL

Function: Herbicide (1)(2)(3)(4)

Chemical Name: 2-Chloro-9-hydroxy-9H-fluorene-9H-carboxylic acid methyl ester

Formula:

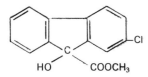

Trade Name: IT-3456 (E. Merck AG)

Manufacture

9-Hydroxyfluorene-9-carboxylic acid is first made from anthraquinone. It is then esterfied with methanol.

Toxicity

The acute oral LD$_{50}$ for rats is more than 12,800 mg/kg which is in the insignificantly toxic category.

References

(1) Worthing, C.R., *Pesticide Manual*, 6th ed., p. 101, British Crop Protection Council (1979).
(2) E. Merck AG, British Patent 1,051,652, December 14, 1966.
(3) E. Merck AG, British Patent 1,051,653, December 14, 1966.
(4) E. Merck AG, British Patent 1,051,654, December 14, 1966.

CHLORIDAZON
(formerly Pyrazon)

Function: Herbicide (1)(2)

Chemical Name: 5-amino-4-chloro-2-phenyl-3(2H)-pyridazinone

Formula:

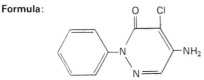

Trade Names: H-119 (BASF)
Pyramin® (BASF)
Alicep®
Pyrazone®
Pyrazonyl®

Manufacture

Pyrazon may be made by reacting mucochloric acid with phenylhydrazine followed by treatment of the intermediate with ammonia. The following is a specific example (3) of the ammonia reaction step.

1,000 parts by weight of 1-phenyl-4,5-dichloropyridazone-(6) are mixed in a pressure vessel holding 8 liters with 4,000 parts of water and 600 parts of anhydrous ammonia are forced into the mixture under pressure. Then the vessel is heated for 6 hours at 120°C, the pressure rising to 40 atmospheres. The reaction product is then allowed to cool and decompressed and the pressure vessel flushed repeatedly with nitrogen. The water is then separated from the solid substance and the latter washed repeatedly with dilute hydrochloric acid and then with water. By crystallization from methanol, 588 parts of 1-phenyl-4-amino-5-chloropyrida-zone-(6) with the melting point 202° to 204°C are obtained.

Process Wastes and Their Control

Product Wastes: This crystalline herbicide is stable in acid media. It persists in sandy loam about 20 weeks and is decomposed to a compound (4-amino-5-chloropyridaz-6-one) which is nonphytotoxic (B-3).

Toxicity

The acute oral LD_{50} value for rats is 2,424 mg/kg which is slightly toxic.

Allowable Limits on Exposure and Use

Product Use: The tolerances set by the EPA for chloridazon in or on raw agricultural commodities are as follows:

	40 *CFR* Reference	Parts per Million
Beets	180.316	0.1 N
Beets, sugar	180.316	0.1 N
Beets, sugar, tops	180.316	1.0
Beets, tops	180.316	1.0
Milk	180.316	0.01 N

References

(1) Worthing, C.R., *Pesticide Manual,* 6th ed., p. 102, British Crop Protection Council (1979).
(2) Spencer, E.Y., *Guide to the Chemicals Used in Crop Protection,* 6th ed., p. 441, London, Ontario, Agriculture Canada (January 1973).
(3) Reicheneder, F., Dury, K., Dury, J.M. and Fischer, A., U.S. Patent 3,210,353, October 5, 1965, assigned to Badische Anilin- & Soda-Fabrik AG.

CHLORMEPHOS

Function: Insecticide (1)(2)(3)

Chemical Name: S-chloromethyl O,O-diethylphosphorodithioate

Formula:

$$(C_2H_5O)_2\overset{\displaystyle S}{\overset{\displaystyle \|}{P}}SCH_2Cl$$

Trade Name: MC 2188 (Murphy Chemical, Ltd.)

Manufacture (3)

Chlormephos is made by the reaction of sodium diethyl dithiophosphate and bromochloro-methane.

Toxicity

The acute oral LD_{50} for rats is only 7 mg/kg which puts the material in the highly to extremely toxic category.

References

(1) Worthing, C.R., *Pesticide Manual,* 6th ed., p. 103, British Crop Protection Council (1979).
(2) Spencer, E.Y., *Guide to the Chemicals Used in Crop Protection,* 6th ed., p. 103, London, Ontario, Agriculture Canada (January 1973).
(3) Pianka, M., British Patent 1,258,922, December 30, 1971, assigned to Murphy Chemical Co., Ltd.

CHLORMEQUAT

Function: Plant Growth regulant (1)(2)(3)(4)

Chemical Name: 2-chloro-N,N,N-trimethylethanammonium ion

Formula: $ClCH_2CH_2N^+(CH_3)_3$ Br

Trade Names: AC-38,555 (American Cyanamid)
Cycocel® (American Cyanamid)

Manufacture

Chlormequat is made by reacting ethylene dibromide and trimethylamine. The following illustrates the process (3) of making the plant growth regulators. One equivalent of a dihalide aliphatic compound, e.g., 1,2-dibromoethane, is put in a pressure flask containing 100 ml toluene per mol of expected product. The flask is cooled in an ice bath and one equivalent of cold amine, i.e., trimethylamine, added. The flask is sealed and allowed to stand overnight at a temperature sufficient for the equation to yield a solid cake of reaction product.

$$(CH_3)_3N + CH_2Br-CH_2Br \rightarrow CH_2Br-CH_2N^+(CH_3)_3 \cdot Br^-$$

The solid cake of product is broken up and removed from the flask with the aid of more toluene and petroleum ether. The precipitate is filtered and washed with petroleum ether and recrystallized from either 100 or 95% ethanol, methanol or toluene. The bromo or chloro compounds are stable solids and can be stored indefinitely. They are exceedingly water-soluble and are usable directly in controlling plant growth.

Process Wastes and Their Control

Product Waste: This crystalline, water-soluble solid, melts at 245°C with decomposition. Incineration would be a highly effective disposal method (B-3). Heating the product with strong aqueous alkali would result in decomposition with the evolution of trimethylamine and other gaseous products.

Toxicity

The acute oral LD_{50} for rats is 670 mg/kg for chlormequat chloride which puts this material in the slightly to moderately toxic category.

Allowable Limits On Exposure and Use

Product Use: The tolerances set by the US EPA for chlormequat in food are as follows (the *CFR* Reference is to Title 21):

	CFR Reference	Parts per Million
Sugarcane molasses	193.70	6.0

References

(1) Worthing, C.R., *Pesticide Manual,* 6th ed., p. 104, British Crop Protection Council (1979).
(2) Spencer, E.Y., *Guide to the Chemicals Used in Crop Protection,* 6th ed., p. 104, London, Ontario, Agriculture Canada (January 1973).
(3) Tolbert, N.E., U.S. Patent 3,156,554, November 10, 1964, assigned to Research Corp.
(4) British Patent 944,807, December 18, 1963, assigned to Research Corp.

CHLOROACETIC ACID

Function: Herbicide

Chemical Name: Chloroacetic acid

Formula: $ClCH_2COOH$

Trade Name: Monoxone® (Plant Protection, Ltd.)

Manufacture

(a) By hydrolysis of trichloroethylene using sulfuric acid

$$CHCl{=}CCl_2 + 2H_2O \rightarrow ClCH_2COOH + 2HCl$$

(b) By chlorination of acetic acid
(c) From the reaction of acetic acid and dichloroacetic acid (2)

Toxicity

The acute oral LD_{50} value for rats is 650 mg/kg expressed as sodium chloroacetate which puts this material in the slightly to moderately toxic class.

References

(1) Worthing, C.R., *Pesticide Manual,* 6th ed., p. 105, British Crop Protection Council (1979).
(2) Opitz, W. et al, U.S. Patent 3,071,615, January 1, 1963, assigned to Knapsack-Griesheim AG.

CHLOROBENZILATE

Function: Acaricide (1)(2)(3)

Chemical Name: Ethyl-4-chloro-α-(4-chlorophenyl)-α-hydroxybenzeneacetate

Formula:

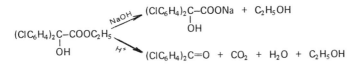

Trade Names: G-23992 (Ciba-Geigy)
Akar® (Ciba-Geigy)
Folbex® (Ciba-Geigy)
Acaraben®
Acarben®

Manufacture (3)

Chlorobenzilate is made by the reaction of 4,4'-dichlorobenzilic acid with diethyl sulfate.

Process Wastes and Their Control

Product Wastes: Chlorobenzilate is easily hydrolyzed in strong alkali or acid. The dichloro-benzilic acid is unstable and readily decarboxylates.

$$(ClC_6H_4)_2C-COOC_2H_5 \xrightarrow[H^+]{NaOH} \begin{cases} (ClC_6H_4)_2C-COONa + C_2H_5OH \\ \quad\quad\quad OH \\ (ClC_6H_4)_2C=O + CO_2 + H_2O + C_2H_5OH \end{cases}$$

Chlorobenzilate is dehalogenated by sodium in isopropyl alcohol (B-3).

Toxicity

The acute oral LD_{50} value for rats is 700 mg/kg (B-5) (slightly to moderately toxic).

Allowable Limits on Exposure and Use

Product Use: A rebuttable presumption against registration for chlorobenzilate was issued by US EPA on May 26, 1976 on the basis of oncogenicity and testicular effects.

In an action dated February 13, 1979, the US EPA issued notice of cancellation and denial of registrations of chlorobenzilate products for uses other than citrus uses in Florida, Texas, California and Arizona. Notwithstanding the above, registration of chlorobenzilate products for citrus use in these four states will also be cancelled or denied unless registrants or applicants for new registrations modify the terms or conditions of registration as follows:

1. Classification of chlorobenzilate products for these citrus uses for restricted use, for use only by or under the supervision of certified applicators.

2. Modification of the labeling of chlorobenzilate products for these citrus uses to include the following:

   a. Restricted Use Pesticide—For retail sale and to use only by certified applicators or persons under their direct supervision and only for those uses covered by the certified applicator's certification.

   b. General Precautions—

      (1) Take special care to avoid getting chlorobenzilate in eyes, on skin, or on clothing.

      (2) Avoid breathing vapors or spray mist.

      (3) In case of contact with skin, wash as soon as possible with soap and plenty of water.

      (4) If chlorobenzilate gets on clothing, remove contaminated clothing and wash affected parts of body with soap and water. If the extent of contamination is unknown, bathe entire body thoroughly. Change to clean clothing.

      (5) Wash hands with soap and water each time before eating, drinking or smoking.

      (6) At the end of the work day, bathe entire body with soap and plenty of water.

      (7) Wear clean clothes each day and launder before reusing.

  c. Required Clothing and Equipment for Application—

      (1) One-piece overalls which have long sleeves and long pants constructed of finely-woven fabric as specified in the USDA/EPA *Guide for Commercial Applicators*.

      (2) Wide-brimmed hat.

      (3) Heavy-duty fabric work gloves.

      (4) Any article which has been worn while applying chlorobenzilate must be cleaned before reusing. Clothing which has been drenched or has otherwise absorbed concentrated pesticide must be buried or burned.

      (5) Facepiece respirator of the type approved for pesticide spray applications by the National Institute for Occupational Safety and Health.

      (6) Instead of the clothing and equipment specified above, the applicator can use an enclosed tractor cab which provides a filtered air supply. Aerial application may be conducted without the specified clothing and equipment.

  d. Handling Precautions—Heavy duty rubber or neoprene gloves and apron must be worn during loading, unloading, and equipment clean-up.

The tolerances set by the US EPA for chlorobenzilate in or on raw agricultural commodities are as follows:

	40 *CFR* Reference	Parts per Million
Almond, hulls	180.109	15.0
Almonds	180.109	0.2
Apples	180.109	5.0
Cattle, fat	180.109	0.5
Cattle, mbyp	180.109	0.5
Cattle, meat	180.109	0.5
Citrus fruits	180.109	5.0
Cotton, seed	180.109	0.5
Melons	180.109	5.0
Pears	180.109	5.0
Sheep, fat	180.109	0.5
Sheep, mbyp	180.109	0.5
Sheep, meat	180.109	0.5
Walnuts	180.109	0.2

References

(1) Worthing, C.R., *Pesticide Manual,* 6th ed., p. 106, British Crop Protection Council (1979).
(2) Spencer, E.Y., *Guide to the Chemicals Used in Crop Protection,* 6th ed., p. 106, London, Ontario, Agriculture Canada (January 1973).
(3) Hafliger, E., U.S. Patent 2,745,780, May 15, 1956, assigned to J.R. Geigy, AG.

CHLOROMETHIURON

Function: Acaricide (1)(2)

Chemical Name: N'-(4-chloro-2-methylphenyl)-N,N-dimethylthiourea

Formula:

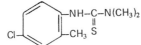

Trade Names: CGA 13444 (Ciba-Geigy AG)
Dipofene® (Ciba-Geigy AG)

Manufacture (2)

110 g of 4-chloro-2-methylphenyl isothiocyanate are dissolved in 300 ml of acetonitrile, and 75 ml of aqueous dimethylamine solution of 40% strength are added. The temperature of the solution rises; 30 minutes later the product formed is precipitated by adding 1 liter of water. Yield: 131 g MP 173° to 175°C.

Toxicity

The acute oral LD_{50} value for rats is 2,500 mg/kg which is slightly toxic.

References

(1) Worthing, C.R., *Pesticide Manual*, 6th ed., p. 107, British Crop Protection Council (1979).
(2) Ciba, Ltd., British Patent 1,138,714, January 1, 1969.

CHLORONEB

Function: Systemic fungicide (1)(2)(3)

Chemical Name: 1,4-dichloro-2,5-dimethoxybenzene

Formula:

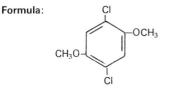

Trade Names: Soil Fungicide 1823 (DuPont)
Demosan® (DuPont)
Tersan SP®

Manufacture

Chloroneb may be made by (a) chlorination of p-dimethoxybenzene or (b) the reaction of quinone with HCl to give monochlorohydroquinone. Oxidation of the hydroquinone gives monochloroquinone. Addition of HCl again gives dichlorohydroquinone. Diazomethane reacts with the dichlorohydroquinone to give chloroneb (3).

Process Wastes and Their Control

Product Wastes: Chloroneb is stable at its boiling point (268°C), in water in the presence of dilute acids or alkalis, and in the common organic solvents (B-3).

Toxicity

The acute oral LD_{50} value for rats is more than 11,000 mg/kg which is insignificantly toxic.

Allowable Limits on Exposure and Use

Product Use: The tolerances set by the US EPA for chloroneb in or on raw agricultural commodities are as follows:

	40 *CFR* Reference	Parts per Million
Beans	180.257	0.1 N
Beans, vines, forage	180.257	2.0
Beets, sugar, roots	180.257	0.1 N
Cattle, fat	180.257	0.2
Cattle, mbyp	180.257	0.2
Cattle, meat	180.257	0.2
Cotton, forage	180.257	2.0
Cotton, seed	180.257	0.1 N
Goats, fat	180.257	0.2
Goats, mbyp	180.257	0.2
Goats, meat	180.257	0.2
Hogs, fat	180.257	0.2
Hogs, mbyp	180.257	0.2
Hogs, meat	180.257	0.2
Horses, fat	180.257	0.2
Horses, mbyp	180.257	0.2
Horses, meat	180.257	0.2
Milk	180.257	0.05 N
Sheep, fat	180.257	0.2
Sheep, mbyp	180.257	0.2
Sheep, meat	180.257	0.2
Soybeans	180.257	0.1 N
Soybeans, vines, forage	180.257	2.0

References

(1) Worthing, C.R., *Pesticide Manual,* 6th ed., p. 108, British Crop Protection Council (1979).
(2) Spencer, E.Y., *Guide to the Chemicals Used in Crop Protection,* 6th ed., p. 110, London, Ontario, Agriculture Canada (January 1973).
(3) Scribner, R.M. and Soboczenski, U.S. Patent 3,265,564, August 9, 1966, assigned to DuPont.

CHLORONITROPROPANE

Function: Fungicide (1)(2)

Chemical Name: 1-chloro-2-nitropropane

Formula: $ClCH_2CH(NO_2)CH_3$

Trade Names: NIA 5961 (FMC)
Lanstan® (FMC)
Korax®

Manufacture

The active fungicidal 1-chloro-2-nitropropane may be prepared by known methods, such as by the reaction of 2-nitro-1-propanol with a chlorinating agent such as sulfuryl chloride or thionyl chloride in the presence of a pyridine base, or by the vapor phase reaction of chlorine and nitrogen dioxide with propene. Still other methods have been described in the literature. The following example illustrates one of the methods known to be useful for the preparation of 1-chloro-2-nitropropane.

Substantially anhydrous chlorine was bubbled through 750 g of substantially anhydrous 2-nitropropane for approximately 44 hours at 55° in a glass tube exposed to three 275-watt lights. During this period, 318 g of chlorine was consumed. The reaction mixture was fractionally distilled to recover unreacted 2-nitropropane and 1-chloro-2-nitropropane, which boiled at 78° to 80°C at 25 mm pressure. The yield of 1-chloro-2-nitropropane was 306 g.

Process Wastes and Their Control

Products Wastes: This soil fungicide, is chemically similar to chloropicrin and disposal would be similiar (C-3).

Toxicity

The acute oral LD_{50} for rats is 197 mg/kg which rates as moderately toxic.

Allowable Limits on Exposure and Use

Air: The threshold limit value for chloronitropropane in air is 100 mg/m³ (20 ppm) as of 1979. An intended change may lower the value to 10 mg/m³ (2 ppm), however (B-23).

References

(1) Spencer, E.Y., *Guide to the Chemicals Used in Crop Protection,* 6th ed., p. 111, London, Ontario, Agriculture Canada (January 1973).
(2) Willard, J.R. and Maitlen, E.G., U.S. Patent 3,078,209, February 19, 1963, assigned to FMC Corp.

CHLOROPHACINONE

Function: Anticoagulant rodenticide (1)(2)

Chemical Name: 2-[(4-chlorophenyl)phenylacetyl]-1H-indene-1,3(2H)-dione

Formula:

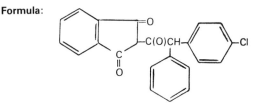

Trade Names: L.M.91 (Lipha SA)
Caid®
Liphadione®
Raviac®
Drat® (May & Baker Ltd.)
Quick® (Rhone-Poulenc)
Saviac® (Aulagne-Chimiotechnic)

Manufacture (2)

(a) Preparation of 1-Bromophenyl-Acetone—67 g (0.5 mol) of phenyl acetone dissolved in 300 cc of carbon disulfide, are introduced into a 1-liter flask equipped with a stirrer device, a condenser and a dropping funnel, and 80 g (0.5 mol) of bromine are added dropwise while stirring. After completing the addition, stirring is continued in the cold for 2 hours; when the release of hydrobromic acid has ceased, the flask is heated on a water bath with a slight reflux until the release has terminated and then a strong current of carbon dioxide is introduced into the flask. The solution which is obtained is deep green in color.

(b) Preparation of 1'-p-Chlorophenyl-1'-Phenyl-2-Propanone—The same flask as before is used and there are successively introduced thereinto:

> 133.5 g of aluminum chloride (1 mol)
> 250 cc of carbon disulfide
> 67.5 g of chlorobenzene (0.6 mol)

After having started the stirrer device, the solution of 1-bromophenyl acetone previously obtained is gradually introduced by way of the dropping funnel over a period of 1 hour. The flask is then heated with slight reflux, while continuing the stirring until the release of hydrobromic acid ceases, this taking 2 to 3 hours.

The carbon disulfide is distilled on the water bath, and then the blackish and oily liquid which remains is carefully poured on to 1,000 g of ice to which are added 200 cc of concentrated hydrochloric acid. The oily layer which forms is taken up in ether and washed with water and then with a concentrated solution of sodium bicarbonate. The solution in the ether is then dried over calcium chloride. The ether is distilled and the product is rectified in a flask with a Vigreux column. The clear yellow liquid distills at 150°C/0.5 mm Hg. The yield is 76% with respect to the initial phenyl acetone.

(c) Preparation of (1'-p-Chlorophenyl-1'-Phenyl)-2-Acetyl-Indane-1-3-Dione—A flask similar to the foregoing is used, the flask comprising a reflux condenser above which is an ordinary condenser. Sodium methylate is prepared by the action of 3.45 g of sodium (0.15 mol) on 30 cc of dry methanol. When the solution has completely reacted, 100 cc of benzene are added while stirring and the solution is slowly distilled to dryness.

150 cc of benzene and 24.25 g (0.125 mol) of methyl phthalate are then introduced and the substance is boiled under reflux while stirring. A solution of 12.2 g of 1'-p-chlorophenyl-1'-phenyl-2-propanone (0.05 mol) in 50 cc of benzene is added dropwise over 1 hour, while distilling 50 cc of a mixture of methanol and benzene. A fresh solution of 12.2 g (0.05 mol) of this same ketone in 50 cc of benzene, to which has been added 9.7 g (0.05 mol) of methyl

phthalate, is added dropwise over a period of 1 hour, while again distilling 50 cc of a mixture of methanol and benzene.

The benzene is distilled off and the temperature is brought to 130° to 140°C for 1¼ hours while stirring. The mass of the product becomes thick and deep red in color. 100 cc of ethanol (96° GL) are then added and the mass is stirred under reflux until the product dissolves. The ethanol solution is concentrated to about 50 cc, then acidified in an ice bath with concentrated hydrochloric acid. The product solidifies immediately or after adding a small quantity of acetone while stirring.

The crude product is centrifuged, washed with water, and recrystallized from ethanol or acetone, as light yellow silky needles with a melting point of 138°C. The yield is 58%.

Toxicity

The acute oral LD_{50} value for rats is only 2.1 mg/kg (B-5) which classifies it as highly to extremely toxic.

References

(1) Worthing, C.R., *Pesticide Manual,* 6th ed., p. 109, British Crop Protection Council (1979).
(2) Molho, D., Boschetti, E. and Fontaine, L., U.S. Patent 3,153,612, October 20, 1964, assigned to Lipha (Lyonnaise Industrielle Pharmaceutique).

4-CHLOROPHENYL PHENYL SULFONE

Function: Acaricide (1)

Chemical Name: 4-chlorodiphenyl sulfone

Formula:

Trade Names: R-242 (Stauffer)
Sulphenone® (Stauffer)

Manufacture (2)

The feed materials are benzene sulfonic acid and monochlorobenzene which react according to the following equation:

$$C_6H_5SO_3H + C_6H_5Cl \rightarrow C_6H_5SO_2C_6H_4Cl + H_2O$$

The benzene sulfonic acid mixture fed should contain at least about 25% by weight of monochlorobenzene, to insure its being a liquid at the temperature at which it is introduced into the reaction zone; the monochlorobenzene content of the benzene sulfonic acid can be increased to as much as 20 to 40% and even more by weight of the benzene sulfonic acid.

The monochlorobenzene vapor fed should be at least twice that required stoichiometrically to react with the benzene sulfonic acid and produce the desired p-monodichlorophenyl sulfone; one can use as much as ten times the quantity of monochlorobenzene vapor to sweep out the water. However, one reaches a point whereat the added vapor cannot be circulated economically due to the heat requirements.

The process is one which is best carried on under a pressure which is relatively elevated with respect to atmospheric, and pressures of the order of 6 to 15 pounds gauge are preferred, although one can utilize successfully pressures of as much as 35 to 40 pounds gauge. The reaction is carried out at 250° to 270°C in the liquid phase. A reaction time of 3 to 4 minutes is used and no catalyst is employed.

Process Wastes and Their Control

Product Wastes: The monochlorodiphenyl sulfone is very resistant to hydrolysis of oxidation. Incineration under proper conditions to prevent air pollution would be effective (scrubbers to absorb HCl and SO_2) (B-3).

Toxicity

The acute oral LD_{50} value for rats is 1,400 mg/kg (slightly toxic) (B-5).

Allowable Limits on Exposure and Use

Product Use: The tolerances set by the US EPA for chlorophenyl phenyl sulfone in or on raw agricultural commodities are as follows:

	40 *CFR* Reference	Parts per Million
Apples	180.112	8.0
Peaches	180.112	8.0
Pears	180.112	8.0

References

(1) Spencer, E.Y., *Guide to the Chemicals Used in Crop Protection,* 6th ed., p. 117, London, Ontario, Agriculture Canada (January 1973).
(2) Bender, H. et al, U.S. Patent 2,593,001, April 15, 1952, assigned to Stauffer Chemical Co.

CHLOROPICRIN

Function: Fumigant insecticide (1)(2)

Chemical Name: Trichloronitromethane

Formula: CCl_3NO_2

Trade Names: Acquinite®
Larvacide®
Nemax®
Picfume®
Chlor-o-Pic®

Manufacture

Chloropicrin may be made (a) from bleaching powder and picric acid (hence the derivation of the name); or (b) by the chlorination of nitromethane (3) in the presence of caustic. In a typical example, 550 gallons of water were added to 500 gallons of 16% sodium hypochlorite by weight. To the resulting mixture was added 25 gallons of 61% nitromethane by volume over a period of 8 minutes with agitation. The amount of nitromethane was 2% in excess of the theoretical amount. The yield of chloropicrin was 43.5 gallons, or 94.5% of the theoretical amount possible.

Process Wastes and Their Control

Product Wastes: Chloropicrin (B-3) reacts readily with alcoholic sodium sulfite solutions to produce methanetrisulfonic acid (which is relatively nonvolatile and less harmful). This reaction has been recommended for treating spills and cleaning equipment. Although not specifically suggested as a decontamination procedure, the rapid reaction of chloropicrin with ammonia to produce guanidine (LD_{50} = 500) could be used for detoxification (B-3).

The Manufacturing Chemists Association suggests two procedures for disposal of chloropicrin:

1. Pour or sift over soda ash. Mix and wash slowly into large tank. Neutralize and pass to sewer with excess water.

2. Absorb on vermiculite. Mix and shovel into paper boxes. Drop into incinerator with afterburner and scrubber.

Toxicity

This intensely irritating liquid fumigant is lethal (LD_{50} = 0.8) to most wildlife at very low concentration when confined. It is also toxic to plants when injected into soil and in low concentrations in air it may decrease germination (B-3).

Allowable Limits on Exposure and Use

Air: The threshold limit value for chloropicrin in air is 0.7 mg/m^3 (0.1 ppm) as of 1979. The tentative short term exposure limit is 2.0 mg/m^3 (0.3 ppm) (B-23).

Product Use: The tolerances set by the US EPA for chloropicrin in or on raw agricultural commodities are as follows:

	40 *CFR* Reference	Parts per Million
Barley, grain (post-h)	180.100	exempt
Buckwheat, grain (post-h)	180.100	exempt
Corn, grain (post-h)	180.100	exempt
Oats, grain (post-h)	180.100	exempt
Rice, grain (post-h)	180.100	exempt
Rye, grain (post-h)	180.100	exempt
Sorghum, grain (post-h)	180.100	exempt
Wheat, grain (post-h)	180.100	exempt

References

(1) Worthing, C.R., *Pesticide Manual,* 6th ed., p. 112, British Crop Protection Council (1979).
(2) Spencer, E.Y., *Guide to the Chemicals Used in Crop Protection,* 6th ed., p. 118, London, Ontario, Agriculture Canada (January 1973).
(3) Wilhelm, J.M., U.S. Patent 3,106,588, October 8, 1963.

CHLOROPROPYLATE

Function: Acaricide (1)(2)

Chemical Name: 1-methylethyl-4-chloro-α-(4-chlorophenyl)-α-hydroxybenzene acetate

Formula:

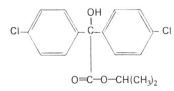

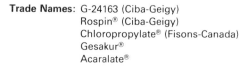

Trade Names: G-24163 (Ciba-Geigy)
Rospin® (Ciba-Geigy)
Chloropropylate® (Fisons-Canada)
Gesakur®
Acaralate®

Manufacture (3)

Chloropropylate is made by reacting 4,4'-dichlorobenzilic acid with isopropanol.

Toxictity

The acute oral LD$_{50}$ value for rats is more than 5,000 mg/kg which rates as insignificantly toxic.

Allowable Limits on Exposure and Use

Product Use: The tolerances set by the US EPA for chloropropylate in or on raw agricultural commodities are as follows:

	40 *CFR* Reference	Parts per Million
Apples	180.218	5.0
Pears	180.218	5.0

References

(1) Worthing, C.R., *Pesticide Manual,* 6th ed., p. 113, British Crop Protection Council (1979).
(2) Spencer, E.Y., *Guide to the Chemicals Used in Crop Protection,* 6th ed., p. 119, London, Ontario, Agriculture Canada (January 1973).
(3) Hafliger, E., U.S. Patent 2,745,780, May 15, 1966, assigned to J.R. Geigy, AG.

CHLOROTHALONIL

Function: Fungicide (1)(2)(4)

Chemical Name: 2,4,5,6-tetrachloro-1,3-benzene-dicarbonitrile

Formula:

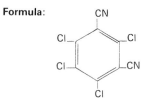

Trade Names: Daconil 2787® (Diamond Shamrock)
Termil® (Diamond Shamrock)
Exotherm Termil® (Diamond Shamrock)
Forturf®
Bravo®

Manufacture

Chlorothalonil is made from tetrachloroisophthalic acid chloride (3). The acid chloride is converted to the corresponding amide by dissolving the acid chloride in a suitable organic solvent and introducing ammonia. A wide range of solvents may be employed in this reaction with the preferred solvents being xylene or dioxane. Either anhydrous or aqueous ammonia may be employed, depending upon whether the solvent used is miscible with water.

The reaction between the acid chloride and ammonia is exothermic and the reactor may be cooled to maintain a maximum temperature of about 40° to 60°C. The aromatic amide which is formed is insoluble in most organic solvents and this material precipitates from the solution. Ammonium chloride is formed as a by-product and is separated from the desired amide.

If a water-miscible solvent such as dioxane is used, and aqueous ammonia is employed as the source of ammonia, sufficient water may be added to the reaction mixture to dissolve the ammonium chloride, and the solid precipitate is primarily the desired halogenated aromatic amide. If a solvent such as xylene, which is not miscible with water, is used or if anhydrous ammonia is employed, the precipitate comprising the amide and the ammonium chloride is filtered and then washed with water to remove the ammonium chloride and again filtered to recover the desired amide.

The ring-chlorinated aromatic amide is dried to remove all traces of water and is reacted with a dehydrating agent such as phosphorus pentoxide or phosphorus oxychloride with the preferred dehydrating agent being phosphorus oxychloride. The reaction time and temperature will be governed by the reactivity of the dehydrating agent employed.

Typical conditions when phosphorus oxychloride is used are a temperature of about 75° to 110°C and a reaction time of about 1 to 6 hours. The product of the dehydration reaction is an aromatic nitrile, having chlorine atoms occupying the same positions on the aromatic ring as they did in the acid chloride starting material. Alternatively the product may be made by the chlorination of isophthalodinitrile. The isophthalodinitrile is made by the ammonoxidation of m-xylene (4).

Process Wastes and Their Control

Product Wastes: Chlorothalonil is stable in aqueous acid and alkali and to ultraviolet light and is noncorrosive (B-3).

Toxicity

The acute LD_{50} value for rats is over 10,000 mg/kg which is insignificantly toxic.

Allowable Limits on Exposure and Use

Product Use: The tolerances set by the US EPA for chlorothalonil in or on raw agricultural commodities are as follows:

	40 *CFR* Reference	Parts per Million
Beans, snap	180.275	5.0
Broccoli	180.275	5.0

(continued)

	40 *CFR* Reference	Parts per Million
Brussels sprouts	180.275	5.0
Cabbage	180.275	5.0
Carrots	180.275	1.0
Cauliflower	180.275	5.0
Celery	180.275	15.0
Corn, sweet (k+cwhr)	180.275	1.0
Cucumbers	180.275	5.0
Fruits, passion	180.275	3.0
Melons	180.275	5.0
Onions, dry bulb	180.275	0.5
Onions, green	180.275	5.0
Peanuts	180.275	0.3
Potatoes	180.275	0.1 N
Pumpkins	180.275	5.0
Squash, summer	180.275	5.0
Squash, winter	180.275	5.0
Tomatoes	180.275	5.0

References

(1) Worthing, C.R., *Pesticide Manual,* 6th ed., p. 114, British Crop Protection Council (1979).
(2) Spencer, E.Y., *Guide to the Chemicals Used in Crop Protection,* 6th ed., p. 120, London, Ontario, Agriculture Canada (January 1973).
(3) Battershell, R.D. and Bluestone, H., U.S. Patent 3,290,353, December 6, 1966, assigned to Diamond Alkali Co.
(4) Battershell, R.D. and Bluestone, H., U.S. Patent 3,331,735, July 18, 1967, assigned to Diamond Alkali Co.

CHLOROXURON

Function: Herbicide (1)(2)(3)

Chemical Name: N'-[4-(4-chlorophenoxy)phenyl]-N,N-dimethylurea

Formula:

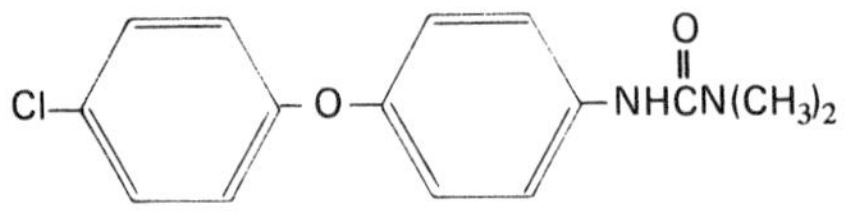

Trade Names: C-1933 (Ciba-Geigy)
Tenoran® (Ciba-Geigy)
Norex®

Manufacture (3)(4)

290 g of 4-p-chlorophenoxyphenyl isocyanate (boiling point at 0.2 mm: 136° to 138°C) are dissolved in 290 cc of acetone and stirred vigorously into 220 cc of dimethylamine of 40% strength in 1,500 cc of water, N-4-(p-chlorophenoxy)-phenyl-N',N'-dimethyl urea being precipitated. After some hours, this is collected, washed with water and dried in vacuo at 65°C. The yield of crude product is 332 g. The melting point is 143° to 147°C. Recrystallized from alcohol, the product melts at 151° to 152°C.

Process Wastes and Their Control

Product Wastes: Acid or alkaline hydrolysis yields 4-(4-chlorophenoxy)aniline and dimethyl-amine. Continued heating in boiling water yields the symmetrically substituted urea, N,N'-(4-chlorophenoxy)phenylurea (B-3).

Toxicity

The acute oral LD_{50} value for rats is more than 3,000 mg/kg which rates as slightly toxic.

Allowable Limits on Exposure and Use

Product Use: The tolerances set by the US EPA for chloroxuron in or on raw agricultural commodities are as follows:

	40 *CFR* Reference	Parts per Million
Carrots	180.216	0.1
Celery	180.216	0.1
Onions, dry bulb	180.216	0.1
Soybeans	180.216	0.15
Soybeans, forage	180.216	0.15
Strawberries	180.216	0.1

References

(1) Worthing, C.R., *Pesticide Manual,* 6th ed., p. 115, British Crop Protection Council (1979).
(2) Spencer, E.Y., *Guide to the Chemicals Used in Crop Protection,* 6th ed., p. 123, London, Ontario, Agriculture Canada (January 1973).
(3) Martin, H. and Aebi, H., U.S. Patent 3,119,682, January 28, 1964, assigned to Ciba Ltd.
(4) Martin, H. and Aebi, H., U.S. Patent 3,060,235, October 23, 1962, assigned to Ciba Ltd.

CHLORPHONIUM

Function: Plant growth regulator (1)(2)

Chemical Name: Tributyl-2,4-dichlorobenzyl-phosphonium ion

Formula:

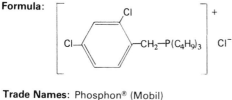

Trade Names: Phosphon® (Mobil)
Fosfon®
Phosphon®
Phosphone®

Manufacture

The phosphonium halide compounds may be prepared by methods, such as the method disclosed by Dye (3). One method by which the compounds have been prepared is to add a slight (ca 5%) excess of the appropriate organic halide to a cold solution of the phosphine in a solvent, and following a brief heating period (reflux) to complete the reaction, removing the solvent and excess halide. The product is usually a white solid which requires no further purification for use as a plant-growth regulant.

The following is a specific example of such a process. 3,4-dichlorobenzyltri-n-butylphosphonium chloride was prepared by refluxing a solution of 10.1 g (0.05 mol) tri-n-butyl-phosphine, 15.6 g (0.08 mol) 3,4-dichlorobenzyl chloride, and 100 cc sodium-dried dioxane, in a stream of dry nitrogen. White needles began to separate after 35 minutes, and the charge quickly set to a thick slurry. Although reaction probably was nearly complete at this point, another 100 cc dioxane was added; and refluxing was continued for another 3½ hours.

The mixture was cooled and the solid filtered with suction, rinsed first with dioxane and then with ether, and dried over sulfuric acid. The product, tiny elongated plates, weighed 19.1 g (96% of theory).

Toxicity

The acute oral LD_{50} value for rats is 178 mg/kg (moderately toxic).

References

(1) Worthing, C.R., *Pesticide Manual,* 6th ed., p. 116, British Crop Protection Council (1979).
(2) Goyette, L.E., U.S. Patent 3,268,323, August 23, 1966, assigned to Mobil Oil Corp.
(3) Dye, W.T., Jr., U.S. Patent 2,703,814, March 8, 1955, assigned to Monsanto Chemical Co.

CHLORPHOXIM

Function: Insecticide (1)(2)(3)

Chemical Name: 7-(2-chlorophenyl)-4-ethoxy-3,5-dioxa-6-aza-4-phosphaoct-6-ene-8-nitrile-4-sulfide

Formula:

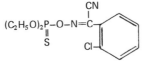

Trade Names: Bay SRA 7747 (Bayer AG)
Baythion C® (Bayer AG)

Manufacture (2)(3)

57 g (0.3 mol) O,O-diethyl-thionophosphoric acid ester chloride are added dropwise at 20° to 30°C, while cooling, to a suspension of 85 g (0.36 mol) of the sodium salt of α-oximino-2-chlorophenyl-acetic acid nitrile [prepared according to M.R. Zimmermann, *Journal für praktische Chemie* (2), 66, 377/1902] in 250 cc acetone, and, after stirring for one hour, the mixture is poured into water. The precipitated oil rapidly solidifies in crystalline form. The crystallisate is filtered off with suction and washed with water. By recrystallization from a mixture of ether and petroleum ether, O,O-diethyl-thionophosphoryl-α-oximino-2-chlorophenyl-acetic acid nitrile is obtained in the form of heavy, colorless crystals of melting point 64°C. The yield is 90 g (90% of theory).

Toxicity

The acute oral LD$_{50}$ value for rats is more than 2,500 mg/kg (slightly toxic).

References

(1) Worthing, C.R., *Pesticide Manual,* 6th ed., p. 117, British Crop Protection Council (1979).
(2) Lorenz, W., Fest, C., Hammann, I., Federmann, M., Flucke, W. and Stendel, W., U.S. Patent 3,591,662, July 6, 1971, assigned to Farbenfabriken Bayer AG.
(3) Lorenz, W., Fest, C., Hammann, I., Federmann, M., Flucke, W. and Stendel, W., U.S. Patent 3,689,648, September 5, 1972, assigned to Farbenfabriken Bayer AG.

CHLORPROPHAM

Function: Herbicide (1)(2)

Chemical Name: 1-methylethyl-3-chlorophenyl carbamate

Formula:

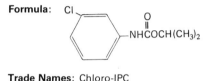

Trade Names: Chloro-IPC
Mirvale®

Manufacture (3)

Chloropropham may be prepared by the reaction of isopropyl chloroformate and meta-chloroaniline according to the equation:

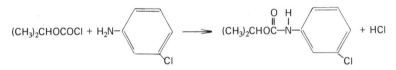

A preferred method of procedure is to add isopropyl chloroformate to a mixture of meta-chloroaniline and the base, employing substantially one mol of each reactant. Equally good results are obtained, however, when an excess, for example, up to 100% or more by molar ratio, of meta-chloroaniline is used or when any organic tertiary amine is used as the HCl acceptor. When an excess of meta-chloroaniline is used, a correspondingly smaller amount of base may be used. The reaction may also be performed by adding the base to a mixture of meta-chloroaniline and isopropyl chloroformate.

Various basic compounds may be employed in the reaction. Inorganic bases such as the oxides, hydroxides, carbonates and bicarbonates of sodium, potassium, calcium, barium, strontium, and magnesium or other alkaline earth metal or alkali metal; organic bases such as pyridine, dimethyl aniline, and quaternary ammonium bases such as trimethyl phenyl ammonium hydroxide are included among those bases which may be employed in the reaction.

The reaction is carried out at 0° to 10°C at atmospheric pressure in an aqueous alkaline medium. The chloroformate is added to the chloroaniline over a 3-hour period in a typical example and the mixture is stirred for one additional hour after reactant addition is completed. A stirred jacketed reaction vessel of corrosion-resistant construction is employed for the conduct of this reaction.

Upon completion of the reaction, benzene is added to the reaction mixture to effect separation of an organic layer. The resulting benzene solution containing the product is washed with dilute hydrochloric acid and then with water. After drying over a solid desiccant, the solution may be filtered and distilled under reduced pressure to give the product in the form of a colorless, viscous liquid which crystallizes slowly on standing at temperatures slightly below room temperature.

The following is an example of the conduct of a commercial scale process (4). A reaction mixture of 800 pounds of water, 600 pounds of meta-chloroaniline and 1,000 pounds of perchloroethylene was established in a jacketed reaction kettle. This mixture was then agitated by use of a turbo mixer, while brine was circulated in the jacket until the temperature of the reaction mixture was about 4° or 5°C (40°F). Thereafter, a 50% by weight aqueous sodium hydroxide solution and isopropyl chloroformate were simultaneously added from separate weight tanks to the reaction mixture in the ratio of 0.6 pound of caustic to 1 pound of isopropyl chloroformate.

Addition of these two reactants is done at a rate such that the reaction temperature may be maintained at between 4° and 5°C. After 710 pounds of isopropyl chloroformate and 425 pounds of caustic have been added, the addition is stopped and the mixture is agitated for 15 minutes.

If standard aniline tests reveal the presence of free chloroaniline in the reaction mixture, batchwise addition of 10 pounds of a 50% aqueous sodium hydroxide solution and 15 pounds of isopropyl chloroformate are made until no chloraniline is present. Thereafter, the organic phase was separated from the inorganic phase by simple phase separation. The organic phase was then washed with two gallons of water and two gallons of hydrochloric acid. This washing was followed with a washing with 200 gallons of water and sufficient 50% aqueous sodium hydroxide solution to raise the pH of the mixture to 8.0. The washed organic phase was then topped at 140°C and 10 ml mercury to remove the perchloroethylene, and leaving the product as residue in a molten state.

Toxicity

The acute oral LD_{50} value for rats is 1,200 mg/kg (B-5) which rates as slightly toxic.

Allowable Limits on Exposure and Use

Product Use: The tolerances set by the US EPA for chlorpropham in or on raw agricultural commodities are as follows:

	40 *CFR* Reference	Parts per Million
Potatoes (post-h)	180.181	50.0

References

(1) Worthing, C.R., *Pesticide Manual*, 6th ed., p. 118, British Crop Protection Council (1979).
(2) Spencer, E.Y., *Guide to the Chemicals Used in Crop Protection*, 6th ed., p. 124, London, Ontario, Agriculture Canada (January 1973).
(3) Witman, E.D., U.S. Patent 2,695,225, November 23, 1954, assigned to Columbia-Southern Chemicals Corp.
(4) Strain, F., U.S. Patent 2,734,911, February 14, 1956, assigned to Columbia-Southern Chemicals Corp.

CHLORPYRIFOS

Function: Insecticide (1)(2)(3)

Chemical Name: O,O-diethyl O-3,5,6-trichloro-2-pyridyl phosphorothioate

Formula:

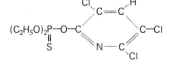

Trade Names: Dowco 179 (Dow)
Dursban® (Dow)
Lorsban® (Dow)

Manufacture (3)

The compound is prepared by several methods. In a preferred method, the compound is prepared by reacting a phosphorochloridate or phosphorochloridothioate having the formula:

with an alkali metal or tertiary amine salt of a halopyridinol having the formula:

$$R—O\text{-alkali metal} \quad \text{or} \quad R—OH\text{·tertiary amine}$$

The reaction conveniently is carried out in an inert organic liquid such as acetone, dimethyl-formamide, carbon tetrachloride, chloroform, benzene, toluene, isobutyl methyl ketone, or methylene dichloride. The amounts of the reagents to be employed are not critical, some of the desired product being obtained when employing any proportion of the reactants. In the preferred method of operation, good results are obtained when employing substantially equimolecular proportions of the pyridinol salt and phosphorochloridate or phosphorochloridothioate. The reaction takes place smoothly at the temperature range from 0° to 100°C with the production of the desired product and chloride by-product.

In carrying out the reaction, the reactants are mixed and contacted together in any convenient fashion, and the resulting mixture maintained for a period of time in the reaction temperature range to complete the reaction. Following the completion of the reaction, the reaction mixture is washed with water and any organic reaction medium removed by fractional distillation under reduced pressure to obtain the desired product as a residue. This product can be further purified by conventional procedures, such as washing with water and dilute aqueous alkali metal hydroxide, solvent extraction and recrystallization.

In an alternative procedure, the phosphoroamidates or phosphoroamidothioates can be prepared by reacting a phosphorodichloridate or phosphorodichloridothioate having the formula:

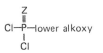

with an alkali metal salt or tertiaryamine salt of a halopyridinol, to form an intermediate halopyridyl phosphorochloridate or phosphorochloridothioate having the structure:

$$R-O-\overset{\overset{\displaystyle Z}{\|}}{\underset{\underset{\displaystyle Cl}{|}}{P}}-\text{lower alkoxy}$$

This intermediate is thereafter reacted with an amino compound such as ammonia or a lower alkyl amine to produce the desired phosphoramidate or phosphoramidothioate. The reaction conveniently is carried out in a reaction medium such as benzene, acetone, dimethylformamide, chloroform, carbon tetrachloride, or methylene chloride. Good results are obtained when employing substantially equimolecular proportions of the lower alkoxy phosphorodichloridate or phosphorodichloridothioate and halopyridinol salt and at least two molecular proportions of amino compound.

The reaction with the halopyridinol salt is somewhat exothermic and is carried out at temperatures of from −50° to 25°C. The reaction between the intermediate halopyridyl phosphorochloridate or phosphorochloridothioate and the amino compound is also exothermic and takes place at the temperature range of from −10° to 60°C. The temperature can be controlled by regulating the rate of mixing and contacting the reactants together and by external cooling.

The by-product in both steps of the reaction is chloride. In the first step, the chloride appears as alkali metal or tertiaryamine chloride. In the second step, the chloride is removed as the hydrochloride salt of the amine or ammonia reactant. Following the reaction, the desired product can be separated in accordance with the conventional procedures as previously described.

In an alternative procedure, the compounds can be prepared by reacting phosphorus oxychloride or phosphorus thiochloride with a halopyridinol salt as previously defined to form an intermediate halopyridyl phosphorodichloridate or phosphorodichloridothioate. Good results are obtained when employing substantially equimolecular proportions of the reagents. The reaction takes place readily at temperatures of from −50° to 80°C with the production of the desired product and halide of reaction.

The intermediate is thereafter reacted with an alkali metal alcoholate or an amino compound such as ammonia or a lower-alkylamine, or successively with two or three reagents to produce the desired compound either as a monoester, diester or triester product. The reaction takes place at temperatures at which chloride of reaction is formed.

This chloride appears in the reaction mixture as alkali metal chloride or amine chloride depending upon whether an alcoholate or amine product is employed as a reactant. Good results are obtained when operating at temperatures of from −10° to 60°C and employing substantially stoichiometric amounts of the reactants. Upon completion of the reaction, the desired product is separated by the conventional procedures as previously described. Chloropyrifos is made by the reaction of O,O-diethylphosphorochloridothioate (O,O-diethyl thiophosphoric acid chloride) and 3,5,6-trichloro-2-pyridinol.

Process Wastes and Their Control

Air: Air emissions from chlorpyrifos manufacture have been reported (B-15) to consist of 205 kg of sulfur dioxide and 0.5 kg of chlorpyrifos per metric ton of pesticide produced.

Product Wastes: This compound is 50% hydrolyzed in aqueous MeOH solution at pH 6 in 1,930 days, and in 7.2 days at pH 9.96. Spray mixture of <1% concentration are destroyed with an excess of 5.25% sodium hypochlorite in <½ hour at 100°C, and in 24 hours at 30°C.

Concentrated (61.5%) mixtures are essentially destroyed by treatment with 100:1 volumes of the above sodium hypochlorite solution and steam in 10 minutes (B-3).

Toxicity

The acute oral LD_{50} value for rats is 135-163 mg/kg (moderately toxic).

Allowable Limits on Exposure and Use

Air: The threshold limit value for chlorpyrifos in air is 0.2 mg/m^3 as of 1979. The tentative short term exposure limit is 0.6 mg/m^3. Both limits bear the notation "skin" indicating that cutaneous absorption should be prevented so the threshold limit value is not invalidated.

Product Use: The tolerances set by the US EPA for chlorpyrifos in or on raw agricultural commodities are as follows:

	40 *CFR* Reference	Parts per Million
Almond hulls	180.342	0.05
Almonds	180.342	0.05
Apples	180.342	0.05
Bananas	180.342	0.25
Bananas, pulp (no peel)	180.342	0.05
Beans, lima	180.342	0.05
Beans, lima, forage	180.342	1.0
Beans, snap	180.342	0.05
Beans, snap, forage	180.342	1.0
Beets, sugar, roots	180.342	0.2
Beets, sugar, tops	180.342	0.05
Cattle, fat	180.342	1.5
Cattle, mbyp	180.342	1.5
Cattle, meat	180.342	1.5
Corn, field, grain	180.342	0.1
Corn, fodder	180.342	0.1
Corn, forage	180.342	0.1
Corn, fresh (inc. sweet) (k+cwhr)	180.342	0.1
Cottonseed	180.342	0.5
Eggs	180.342	0.01
Goats, fat	180.342	0.1
Goats, mbyp	180.342	0.1
Goats, meat	180.342	0.1
Hogs, fat	180.342	0.1
Hogs, mbyp ·	180.342	0.1
Hogs, meat	180.342	0.1
Horses, fat	180.342	0.1
Horses, mbyp	180.342	0.1
Horses, meat	180.342	0.1
Milk, fat (0.01 ppm-N in whole milk)	180.342	0.25
Peaches	180.342	0.05
Pears	180.342	0.05
Plums (fresh prunes)	180.342	0.05
Poultry, fat (exc. turkeys)	180.342	0.01
Poultry, mbyp (exc. turkeys)	180.342	0.01
Poultry, meat (exc. turkeys)	180.342	0.01
Sheep, fat	180.342	0.1
Sheep, mbyp	180.342	0.1
Sheep, meat	180.342	0.1 N
Sorghum, fodder	180.342	6.0
Sorghum, forage	180.342	1.5
Sorghum, grain	180.342	0.75

(continued)

	40 *CFR* Reference	Parts per Million
Sweet potatoes	180.342	0.1
Tomatoes	180.342	0.5
Turkeys, fat	180.342	0.2
Turkeys, mbyp	180.342	0.2
Turkeys, meat	180.342	0.2

The tolerances set by the US EPA for chlorpyrifos in animal feeds are as follows (the *CFR* Reference is to Title 21):

	CFR Reference	Parts per Million
Beets, sugar, molasses	561.98	3.0
Beets, sugar, pulp (dried)	561.98	1.0

References

(1) Worthing, C.R., *Pesticide Manual,* 6th ed., p. 119, British Crop Protection Council (1979).
(2) Spencer, E.Y., *Guide to the Chemicals Used in Crop Protection,* 6th ed., p. 203, London, Ontario, Agriculture Canada (January 1973).
(3) Rigterink, R.H., U.S. Patent 3,244,586, April 5, 1966, assigned to Dow Chemical Co.

CHLORPYRIFOS-METHYL

Function: Insecticide (1)

Chemical Name: O,O-dimethyl O-3,5,6-trichloro-2-pyridyl phosphorothioate

Formula:

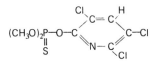

Trade Names: Dowco 214 (Dow)
Reldan® (Dow)

Manufacture (2)

Chlorpyrifos-methyl is made by the reaction of O,O-dimethyl phosphorochloridothioate on 3,5,6-trichloro-2-pyridinol.

Toxicity

The acute oral LD_{50} value for rats is 941 mg/kg (B-5) which rates as slightly toxic.

References

(1) Worthing, C.R., *Pesticide Manual,* 6th ed., p. 120, British Crop Protection Council (1979).
(2) Rigterink, R.H., U.S. Patent 3,244,586, April 5, 1966, assigned to Dow Chemical Co.

CHLORTHAL-DIMETHYL

Function: Preemergence herbicide (1)(2)

Chemical Name: Dimethyl-2,3,5,6-tetrachloro-1,4-benzenedicarboxylate

Formula:

Trade Names: DAC893 (Diamond-Shamrock)
Dacthal® (Diamond-Shamrock)

Manufacture

Dimethyl 2,3,5,6-tetrahaloterephthalate may be prepared by halogenating terephthalyl dichloride, e.g., by chlorinating terephthalyl dichloride as described by N. Rabjohn (5), and reacting the resultant tetrahaloterephthalyl dichloride with methanol, preferably at reflux. The process has also been described by R.F. Lindemann (3).

An improved route involving the reaction of hexachloro-p-xylene with terephthalic acid, followed by chlorination of the crude reaction product and, finally, esterification with methanol of the crude chlorination product has been described by E. Zinn et al (4).

The first (fusion condensation) reaction in this improved process is as follows:

$$C_6H_4(COOH)_2 + C_6H_4(CCl_3)_2 \rightarrow 2C_6H_4(COCl)_2 + 2HCl$$

and the second (chlorination) reaction is as follows:

$$C_6H_4(COCl)_2 + 4Cl_2 \rightarrow C_6Cl_4(COCl)_2 + 4HCl$$

and the final (esterification) reaction is as follows:

$$C_6Cl_4(COCl)_2 + 2CH_3ONa \rightarrow C_6Cl_4(COOCH_3)_2 + 2NaCl$$

The fusion step is carried out at a temperature of from 100° to 200°C and preferably from 120° to 160°C. The chlorination step is carried out at a temperature of 150° to 210°C and preferably from 170° to 190°C. The esterification step is carried out at a temperature within the range of 50° to 100°C. This may conveniently be the reflux temperature of the methanol employed at the operating pressure.

The fusion step is carried out at atmospheric pressure. The chlorination step is desirably carried out at superatmospheric pressures up to 200 psig and preferably at pressures of 90 to 120 psig. The esterification step may be carried out at pressures between atmospheric and 50 psig.

The fusion reaction may be conducted in a reaction period of about 3.5 hours. The chlorination reaction may be carried out in a period ranging from 8 to 12 hours. When sufficient pressure is employed to maintain the esterification reaction temperature in the range of 90° to 100°C, esterification usually can be carried out in about two hours or less.

Generally, it is preferred to regulate the pressure and temperature so as to effect esterification over a somewhat longer time of about 5 to 7 hours, although there may well be instances where an extremely short reaction time would be desired, e.g., as in a continuous operation in which case the use of superatmospheric pressure and the resulting higher temperature is recommended.

The initial fusion step is of course carried out in the liquid melt phase. No added solvents are used. A small amount of solvent is desirable for the conduct of the chlorination step. A non-chlorinatable solvent such as CCl_i in an amount of from 0.5 to 5.0% and preferably from 1 to 2% of the total reaction mixture is used. Such a solvent addition serves to prevent clogging of the equipment due to sublimation of the reactants during chlorination.

The esterification step is carried out in the liquid phase in methanol solution. The fusion step is conducted in the presence of 0.01 to 1.0% of ferric chloride based on the total weight of charge. In the chlorination step, ferric chloride is used as a catalyst in an amount up to about 1.0% by weight of the total chlorination reaction charge.

No catalyst is used in the esterification step. The fusion step may be carried out in a glass-lined or steel reactor. The chlorination may be carried out in the same reactor as the fusion, thereby avoiding any handling of the intermediate terephthalyl dichloride. The esterification step is carried out in a glass-lined or steel reactor equipped with a reflux condenser. The overall reaction scheme for chlorthal-dimethyl manufacture is shown in Figure 20 (4).

Figure 20: Flow Scheme for Chlorthal-Dimethyl Manufacture

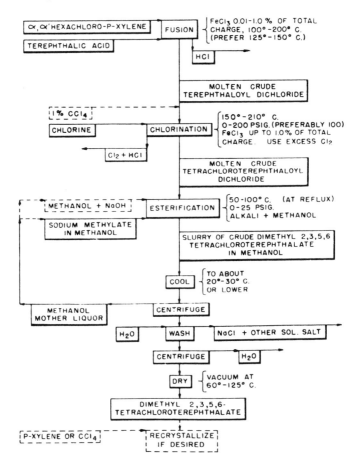

Source: Reference (4)

From the esterification step, there is obtained a slurry of crude dimethyl 2,3,5,6-tetrachloro-terephthalate in methanol, typically at a concentration of about 1.5 to 2.5 lb/gal, and at an esterification temperature of about 65°C. It desirably is cooled to about 20° to 30°C or lower, depending upon the capacity of the cooling equipment employed and the thus-cooled slurry is centrifuged or filtered to separate methanol which can be recycled, as indicated in the drawing, to the esterification operation.

The solid product is then washed with hot water while retained in the centrifuge, typically at a temperature of about 70° to 80°C, or reslurried in water to dissolve out NaCl and any other sodium salts or water-soluble materials which may be present. The thus-washed material is again centrifuged or filtered to remove water and is dried, typically in a vacuum dryer or other dryer operating at a temperature of about 70° to 125°C to obtain dimethyl 2,3,5,6-tetra-chloroterephthalate, typically about 80 to 92% pure. In those instances where an extremely pure product is desired, the dimethyl 2,3,5,6-tetrachloroterephthalate may be subjected to a conventional recrystallization from a solvent such as p-xylene or carbon tetrachloride as in-dicated in the drawing.

Process Wastes and Their Control

Product Wastes: DCPA, the dimethyl ester of tetrachloroterephthalic acid, has a half-life in soil of about 100 days and is only moderately toxic. It is very insoluble in water, but soluble in organic solvents. Incineration of these solutions could be used for disposal with proper precautions for the HCl formed, i.e., caustic scrubbers would be required for any large amounts (B-3).

Toxicity

The acute oral LD_{50} value for rats is greater than 3,000 mg/kg (slightly toxic).

Allowable Limits on Exposure and Use

Product Use: The tolerances set by the US EPA for chlorthal-dimethyl in or on raw agricul-tural commodities are as follows:

	40 *CFR* Reference	Parts per Million
Beans, field, dry	180.185	2.0
Beans, mung, dry	180.185	2.0
Beans, snap, succulent	180.185	2.0
Broccoli	180.185	1.0
Brussels sprouts	180.185	1.0
Cabbage	180.185	1.0
Cantaloups	180.185	1.0
Cauliflower	180.185	1.0
Collards	180.185	2.0
Corn, field, fodder	180.185	0.4
Corn, field, forage	180.185	0.4
Corn, grain (inc. field & pop)	180.185	0.05
Corn, pop, fodder	180.185	0.4
Corn, pop, forage	180.185	0.4
Corn, sweet (k+cwhr)	180.185	0.05
Corn, sweet, fodder	180.185	0.4
Corn, sweet, forage	180.185	0.4
Cotton, seed	180.185	0.2
Cress, upland	180.185	5.0
Cucumbers	180.185	1.0
Eggplant	180.185	1.0
Garlic	180.185	1.0

(continued)

	40 *CFR* Reference	Parts per Million
Honeydew, melons	180.185	1.0
Horseradish	180.185	2.0
Kale	180.185	2.0
Lettuce	180.185	2.0
Mustard, greens	180.185	5.0
Onions	180.185	1.0
Peas, southern, black-eyed	180.185	2.0
Peppers	180.185	2.0
Pimentos	180.185	2.0
Potatoes	180.185	2.0
Rutabagas	180.185	2.0
Soybeans	180.185	2.0
Squash, summer	180.185	1.0
Squash, winter	180.185	1.0
Strawberries	180.185	2.0
Sweet potatoes	180.185	2.0
Tomatoes	180.185	1.0
Turnips	180.185	2.0
Turnips, greens	180.185	5.0
Watermelons	180.185	1.0
Yams	180.185	2.0

References

(1) Worthing, C.R., *Pesticide Manual,* 6th ed., p. 120, British Crop Protection Council (1979).
(2) Spencer, E.Y., *Guide to the Chemicals Used in Crop Protection,* 6th ed., p. 125, London, Ontario, Agriculture Canada (January 1973).
(3) Lindemann, R.F., U.S. Patent 2,923,634, February 2, 1960, assigned to Diamond Alkali Co.
(4) Zinn, E. et al, U.S. Patent 3,052,712, September 4, 1962, assigned to Diamond Alkali Co.
(5) Rabjohn, N., *J. Am. Chem. Soc.* 70, 3518 (1948).

CHLORTHIAMID

Function: Herbicide and seedicide (1)(2)

Chemical Name: 2,6-dichlorobenzenecarbothioamide

Formula:

Trade Names: WL 5792 (Shell Research, Ltd.)
Prefix®

Manufacture (2)

Triethylamine (25 g, 0.25 mol) was added to a solution of 2,6-dichlorobenzonitrile (34.4 g, 0.2 mol) in dry pyridine (150 ml). Dry hydrogen sulfide was bubbled through the stirred mixture. The reaction temperature rose to about 35°C and then commenced to fall, and the reaction mixture changed in color from yellow to dark red. After hydrogen sulfide had been passed for two hours, the mixture was poured into water (1,000 ml) giving a red oil which solidified on stirring. The solid was collected, washed well with water and air-dried. During drying, the melting point rose from about 60° to 144° to 148°C giving a pale yellow solid (40.5 g, theory 41.5 g). It is believed that the initial precipitate may be amine salt of the thiol form

with either pyridine and/or triethylamine and that on drying the base evaporates.

Crystallization from a mixture of benzene and light petroleum gave 2,6-dichlorothiobenz-amide as stout, white prisms, MP 151° to 152°C, in 70% yield.

Toxicity

The acute oral LD_{50} value for rats is 757 mg/kg which is slightly to moderately toxic.

References

(1) Worthing, C.R., *Pesticide Manual,* 6th ed., p. 121, British Crop Protection Council (1979).
(2) Yates, J., British Patent 987,253, March 24, 1965, assigned to Shell Research Ltd.

CHLORTHIOPHOS

Function: Insecticide and acaricide (1)(2)

Chemical Name: O-2,5-dichloro-4-(methylthio)-phenyl O,O-diethyl phosphorothioate

Formula:

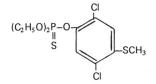

Trade Names: S-2957 (C.H. Boehringer Sohn)
Celathion® (Celamerck GmbH)

Manufacture (2)

(a) Preparation of 2,5-dichloro-4-methylmercapto-phenol—A mixture of 455 g of 2,4,5-tri-chlorophenyl methyl sulfide, 370 g of an aqueous 48% sodium hydroxide solution and one liter of methanol was heated for six hours at 160°C in an autoclave. Thereafter, the reaction mixture was vacuum filtered, the methanol was distilled out of the filtrate, and the distillation residue was diluted with water. The dilute aqueous solution was filtered through charcoal, the filtrate was acidified with concentrated hydrochloric acid, and the acid solution was extracted with methylene chloride. The extract solution was dried over sodium sulfate, and the methylene chloride was distilled off, leaving as a residue 355 g (85% of theory) of 2,5-dichloro-4-methylmercapto-phenol, BP 110°C at 0.05 mm Hg. After recrystallization from benzene the product had a MP of 110° to 112°C.

(b) Preparation of O,O-diethyl-O-(2,5-dichloro-4-methylmercapto-phenyl)-thionophosphate—
188 g (1.1 mols) of O,O-diethyl-thionophosphoric acid chloride were slowly added dropwise at
60°C to a mixture of 209 g (1 mol) of 2,5-dichloro-4-methylmercapto-phenol, 40 g of sodium
hydroxide and 150 cc of water. The resulting mixture was stirred for three hours at 60°C, and
then 100 cc of 2 N sodium hydroxide were added thereto, and the mixture was allowed to
cool. Thereafter, the aqueous phase was separated and extracted several times with toluene.
The organic extract solutions were combined, dried over sodium sulfate, and the toluene
was distilled off. 322 g (92% of theory) of O,O-diethyl-O-(2,5-dichloro-4-methylmercapto-
phenyl)-thionophosphate were obtained, BP 150° to 151°C at 0.001 mm Hg.

Toxicity

The acute oral LD_{50} value for rats is only 9.1 mg/kg which is highly toxic.

References

(1) Worthing, C.R., *Pesticide Manual,* 6th ed., p. 122, British Crop Protection Council (1979).
(2) Sehring, R. and Buck, W., U.S. Patent 3,600,472, August 17, 1971, assigned to C.H.
    Boehringer Sohn.

CODLELURE

Function: Insect sex attractant (for codling moth)

Chemical Name: (8E,10E)-8,10-dodecadien-1-ol

Formula: $CH_3CH=CHCH=CH(CH_2)_7OH$

Trade Name: Codlemone® (Zoecon Corp.)

Manufacture

A route is disclosed (1) for the synthesis of trans-8-trans-10-dodecadien-1-ol. There are also
disclosed methods of preparing the relatively inactive cis-trans, trans-cis, and cis-cis isomers
thereof. It should be noted however, that the active isomer is thermodynamically the more
stable isomer, and therefore the other isomers are convertible thereto. While methods of
preparing the nonpreferred isomers are disclosed, it is preferred to prepare the active desired isomer directly rather than via the inactive compounds.

In the preparation of trans-8-trans-10-dodecadien-1-ol, a monoalkyl ester of azelaic acid,
suitably the methyl ester is prepared by esterifying azelaic acid with the appropriate alkanol,
suitably a lower alkanol such as methanol. The ester is converted to an alkyl-8-bromoocta-
noate, by a Hunsdiecker reaction, or, preferably by the action of bromine in the presence of
red mercuric oxide in the absence of light. The action of an appropriate oxidizing agent in
the presence of a mild base upon the alkyl-8-bromooctanoate, suitably pyridine-N-oxide in
the presence of sodium bicarbonate yields the corresponding alkyl-8-oxooctanoate.

This compound may then be treated with one of two Wittig reagents. Reaction with 1-tri-
phenylphosphonium trans-2-butene bromide, prepared from trans-crotyl bromide and tri-
phenylphosphine, with the alkyl-8-oxooctanoate yields cis-trans-8-trans-10-dodecadienoic
acid. The acid is reduced to the corresponding alcohol, suitably by means of an organometal-
lic reducing agent such as lithium aluminum hydride or sodium dihydro-bis-2-(methoxy
ethoxy) aluminate, yielding, cis-trans-8-trans-10-dodecadien-1-ol which is converted to the
desired trans-8-trans-10-dodecadien-1-ol by treatment with a mild free radical source such
as ultraviolet light or iodine in the presence of sunlight.

Toxicity

An LD$_{50}$ value for rats was not available but toxicity is believed to be insignificant.

Reference

(1) Roelofs, W., Comeau, A. and Hill, A., U.S. Patent 3,852,419, December 3, 1974.

COPPER CARBONATE

Function: Fungicide (1)(2)

Chemical Name: [Carbonato(2)-dihydroxy]dicopper

Formula: $Cu(OH)_2 \cdot CuCO_3$

Trade Name: None

Manufacture (3)

Copper shot is reacted on a batch basis with steam-heated sulfuric acid, to form a concentrated copper sulfate solution. To manufacture copper carbonate, soda-ash is mixed with the copper sulfate concentrate in a batch reactor. Lime is then added to neutralize the mix to pH 6.3 to 6.5, thereby precipitating copper carbonate. The precipitate is settled, vacuum filtered and dried. The supernatant from the precipitation tank, and the filtrate are discharged as wastewater. Average wastewater flow is approximately 11,000 gallons per ton of copper carbonate product. Of this, approximately 7,000 gallons is vacuum filter wash and water pump flow.

Currently, discharge is to a·holding pond and thence into a nearby waterway. Plans are being formulated, in cooperation with the appropriate state regulatory agency, to design and construct a precipitation treatment plant to remove copper and other heavy metals from the wastewater. This will also likely include provision for reductions in suspended solids.

Toxicity

The acute oral LD$_{50}$ value for rabbits is 159 mg/kg which indicate a fair degree of toxicity.

Allowable Limits on Exposure and Use

Product Use: The tolerances set by the US EPA for copper carbonate in or on raw agricultural commodities are as follows:

	40 *CFR* Reference	Parts per Million
Pears (post-h)	180.136	3.0

References

(1) Martin, H. and Worthing, C.R., *Pesticide Manual,* 5th ed., p. 124, British Crop Protection Council (January 1977).
(2) Spencer, E.Y., *Guide to the Chemicals Used in Crop Protection,* 6th ed., p. 128, London, Ontario, Agriculture Canada (January 1973).
(3) Patterson, J.W., *State of the Art of the Inorganic Chemicals Industry: Inorganic Pesticides,* Report EPA-600/2-74-0092, Wash., DC, EPA (March 1975).

COPPER OXYCHLORIDE

Function: Fungicide (1)(2)

Chemical Name: Copper chloride hydroxide

Formula: $3Cu(OH)_2 \cdot CuCl_2$

Trade Names: Recop® (Sandoz)
Cupravit® ICI
Fernacot® ICI
Pere-col® ICI

Manufacture

Copper oxychloride is made by the action of air on scrap copper in cupric chloride-sodium chloride solution.

Toxicity

The acute oral LD_{50} value for rats is 1,440 mg/kg (slightly toxic).

References

(1) Worthing, C.R., *Pesticide Manual,* 6th ed., p. 126, British Crop Protection Council (1979).
(2) Spencer, E.Y., *Guide to the Chemicals Used in Crop Protection,* 6th ed., p. 131, London, Ontario, Agriculture Canada (January 1973).

COPPER SULFATE

Function: Fungicide (1)(2)

Chemical Name: Copper sulfate

Formula: $CuSO_4 \cdot 5H_2O$

Trade Names: Blue Vitriol
Blue Copperas

Manufacture (3)

Copper shot is reacted on a batch basis with steam-heated sulfuric acid to form a supersaturated copper sulfate solution. When the solution is allowed to cool, copper sulfate crystals form. The supernatant mother liquor is decanted and either reused for the next batch of copper sulfate crystals, or used in the production of tri-basic copper sulfate or copper carbonate.

Any washwater used in cleanup is retained in the crystallization vats, and used in the copper sulfate production process. Steam to heat the solution is produced on site, from chemically softened creek water. Sludge from the softening process, plus boiler blowdown, constitute the only wastewaters associated with the copper sulfate process. These are discharged directly to a holding pond, where the sludge settles. Pond overflow is to an adjacent waterway.

Toxicity

The acute oral LD_{50} value for rats is 300 mg/kg which is moderately toxic.

Allowable Limits on Exposure and Use

Product Use: The tolerances set by the US EPA for copper sulfate in or on raw agricultural commodities are as follows:

	40 *CFR* Reference	Parts per Million
Eggs	180.102	exempt
Fish	180.102	exempt
Irrigated crops	180.102	exempt
Meat	180.102	exempt
Milk	180.102	exempt
Shellfish	180.102	exempt

References

(1) Martin, H. and Worthing, C.R., *Pesticide Manual,* 5th ed., p. 127, British Crop Protection Council (January 1977).
(2) Spencer, E.Y., *Guide to the Chemicals Used in Crop Protection,* 6th ed., p. 133, London, Ontario, Agriculture Canada (January 1973).
(3) Patterson, J.W., *State of the Art of the Inorganic Chemicals Industry: Inorganic Pesticides,* Report EPA-600/2-74-0092, Wash., DC, Environmental Protection Agency (March 1975).

COUMACHLOR

Function: Rodenticide (1)(2)

Chemical Name: 3-[1-(4-chlorophenyl)-3-oxobutyl]-4-hydroxy-2H-1-benzopyran-2-one

Formula:

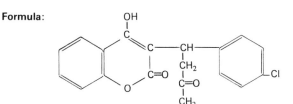

Trade Names: Geigy Rodenticide Exp. 332 (Ciba-Geigy)
Tomorin® (Ciba-Geigy)
Ratilan® Ciba-Geigy)

Manufacture (3)

Coumachlor is made by the condensation of 3-carbethoxy-4-hydroxycoumarin and p-chlorobenzalacetone.

Toxicity

The acute oral LD_{50} for rats is 900 mg/kg (slightly toxic).

References

(1) Worthing, C.R., *Pesticide Manual*, 6th ed., p. 127, British Crop Protection Council (1979).
(2) Spencer, E.Y., *Guide to the Chemicals Used in Crop Protection*, 6th ed., p. 135, London, Ontario, Agriculture Canada (January 1973).
(3) Starr, D.F. and DiSanto, C.C., U.S. Patent 2,752,360, June 26, 1956, assigned to S.B. Penick and Co., Inc.

COUMAPHOS

Function: Insecticide (1)

Chemical Name: O-3-chloro-4-methylcoumarin-7-yl O,O-diethylphosphorothioate

Formula:

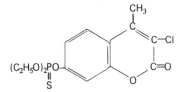

Trade Names: Bayer 21/199 (Bayer)
Resitox® (Bayer)
Asuntol®
Co-Ral®
Muscatox®

Manufacture (2)

The feed materials for coumaphos manufacture are diethyl phosphorothionochloridate and 3-chloro-4-methyl-7-hydroxy coumarin which react according to the following equation:

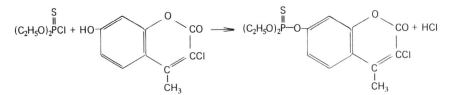

The reaction is usually carried out in the presence of an acid binding agent such as potassium carbonate. This process is carried out at a reaction temperature in the range of 50° to 100°C. This reaction is carried out at atmospheric pressure. Reaction periods of from 15 minutes to several hours are used. This reaction is preferably carried out in the presence of a diluent in which the coumarin is soluble. Particularly suitable solvents of this kind are ketones such as acetone and methyl ethyl ketone.

A small quantity of finely divided copper is used as catalyst, and a stirred jacketed kettle of conventional design may be used. After reaction, the mixture is cooled, salts are filtered off and the filtrate is freed of solvent by evaporation. The residue may be ethanol recrystallized.

Process Wastes and Their Control

Product Wastes: This compound is completely decomposed on heating with concentrated alkali. It is oxidized with HNO_3 or other oxidizing agents to the phosphate analogue, Coroxon. Dilute alkali (pH 8-12) causes an opening of the pyrone ring, which can be closed again by acidification to yield the original compound (B-3).

Toxicity

The acute oral LD_{50} for rats is only 16 mg/kg which is highly toxic.

Allowable Limits on Exposure and Use

Product Use: The tolerances set by the US EPA for coumaphos in or on raw agricultural commodities are as follows:

	40 *CFR* Reference	Parts per Million
Cattle, fat	180.189	1.0
Cattle, mbyp	180.189	1.0
Cattle, meat	180.189	1.0
Eggs	180.189	0.1
Goats, fat	180.189	1.0
Goats, mbyp	180.189	1.0
Goats, meat	180.189	1.0
Hogs, fat	180.189	1.0
Hogs, mbyp	180.189	1.0
Hogs, meat	180.189	1.0
Horses, fat	180.189	1.0
Horses, mbyp	180.189	1.0
Horses, meat	180.189	1.0
Milk, fat (= N in whole milk)	180.189	0.5
Poultry, fat	180.189	1.0
Poultry, mbyp	180.189	1.0
Polutry, meat	180.189	1.0
Sheep, fat	180.189	1.0
Sheep, mbyp	180.189	1.0
Sheep, meat	180.189	1.0

References

(1) Spencer, E.Y., *Guide to the Chemicals Used in Crop Protection,* 6th ed., p. 136, London, Ontario, Agriculture Canada (January 1973).
(2) Schrader, G., U.S. Patent 2,748,146, May 29, 1956, assigned to Farbenfabriken Bayer AG.

COUMATETRALYL

Function: Rodenticide (1)(2)

Chemical Name: 4-hydroxy-3-(1,2,3,4-tetrahydro-naphthalenyl)-2H-1-benzopyran-2-one

Formula:

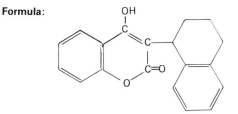

Trade Names: B-25634 (Bayer)
Racumin 57® (Bayer)

Manufacture (3)

Eight parts by weight 4-hydroxy-coumarin and 11 parts by weight α-tetralol are dissolved in 5 parts by volume glacial acetic acid at a temperature of about 100°C. Then 1 part by volume sulfuric acid is slowly added (60°Be) so that a clear solution is formed. The reaction mixture is heated for another hour at a temperature of 110° to 120°C, then it is poured into water, and the reaction product is taken up in ether.

The ethereal layer is extracted with diluted soda lye, the extract is filtered over carbon black and acidified with diluted acetic acid. The precipitated reaction product is filtered off and dried, yield: 10 parts by weight of 3-(α-tetrahydronaphthyl)-4-hydroxy-coumarin (MP 186° to 187°C, recrystallized from alcohol).

Toxicity

The acute oral LD_{50} value for rats is only 17 mg/kg which is highly toxic.

References

(1) Worthing, C.R., *Pesticide Manual,* 6th ed., p. 128, British Crop Protection Council (1979).
(2) Spencer, E.Y., *Guide to the Chemicals Used in Crop Protection,* 6th ed., p. 137, London, Ontario, Agriculture Canada (January 1973).
(3) Enders, E. and Muller, A., U.S. Patent 2,952,689, September 13, 1960, assigned to Farbenfabriken Bayer AG.

CRIMIDINE

Function: Rodenticide (1)(2)

Chemical Name: 2-chloro-N,N,6-trimethyl-4-pyrimidinamine

Formula:

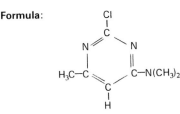

Trade Names: W-491 (Bayer)
Castrix® (Bayer)
Crimidin®

Manufacture (3)

Crimidine is made by the condensation of urea and ethyl acetoacetate followed by chlorination and reaction with dimethylamine.

Toxicity

The acute oral LD_{50} value for rats is only 1.25 mg/kg which is extremely toxic.

References

(1) Worthing, C.R., *Pesticide Manual,* 6th ed., p. 129, British Crop Protection Council (1979).
(2) Spencer, E.Y., *Guide to the Chemicals Used in Crop Protection,* 6th ed., p. 139, London, Ontario, Agriculture Canada (January 1973).
(3) Westphal, K., U.S. Patent 2,219,858, October 29, 1940, assigned to Winthrop Chemical Co., Inc.

CROTOXYPHOS

Function: Insecticide (1)(2)

Chemical Name: 1-phenylethyl (E)-3-[(dimethoxyphosphinyl)oxy]-2-butenoate

Formula:

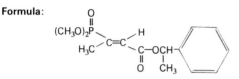

Trade Names: SD-4294 (Shell)
Ciodrin® (Shell)

Manufacture

Crotoxyphos is made by the reaction of trimethyl phosphite with alpha-methylbenzyl-2-chloroacetoacetate (3) as shown by the reaction:

$$(CH_3O)_3P + CH_3COCHClCOOCHC_6H_5 \longrightarrow (CH_3O)_2\overset{\overset{O}{\|}}{P}OC{=}CHCOOCHC_6H_5 + CH_3Cl$$

(with CH₃ groups shown on the chloroacetoacetate and on the product)

In a typical example (4), a solution of 20 g of alpha-methylbenzyl alpha-2-chloroacetoacetate, 3 ml of glacial acetic acid and 13.65 g of trimethyl phosphite was allowed to warm spontaneously to 50°C, then heated for 4.25 hours at 95° to 102°C. The resulting mixture then was stripped to a kettle temperature of 110°C, at 0.35 mm mercury pressure to yield 26 g of product.

Process Wastes and Their Control

Product Wastes: Hydrolysis of this product occurs readily; at 38°C it is 50% hydrolyzed after 35 hours at pH 9, or after 87 hours at pH 1. All formulations are unstable on most solid carriers (B-3).

Toxicity

The acute oral LD_{50} for rats is about 50 mg/kg which is moderately to highly toxic.

Allowable Limits on Exposure and Use

Product Use: The tolerances set by the US EPA for crotoxyphos in or on raw agriculture products are as follows:

	40 *CFR* Reference	Parts per Million
Cattle, fat	180.280	0.02 N
Cattle, mbyp	180.280	0.02 N
Cattle, meat	180.280	0.02 N
Goats, fat	180.280	0.02 N
Goats, mbyp	180.280	0.02 N
Goats, meat	180.280	0.02 N
Hogs, fat	180.280	0.02 N
Hogs, mbyp	180.280	0.02 N
Hogs, meat	180.280	0.02 N
Milk	180.280	0.02 N
Sheep, fat	180.280	0.02 N
Sheep, mbyp	180.280	0.02 N
Sheep, meat	180.280	0.02 N

References

(1) Worthing, C.R., *Pesticide Manual,* 6th ed., p. 130, British Crop Protection Council (1979).
(2) Spencer, E.Y., *Guide to the Chemicals Used in Crop Protection,* 6th ed., p. 140, London, Ontario, Agriculture Canada (January 1973).
(3) Whetstone, R.R. et al, U.S. Patent 2,982,686, May 2, 1961, assigned to Shell Oil Co.
(4) Tieman, C.H. and Stiles, A.R., U.S. Patent 3,068,268, December 11, 1962, assigned to Shell Oil co.

CRUFOMATE

Function: Systemic insecticide (1)(2)(3)

Chemical Name: 2-chloro-4-(1,1-dimethylethyl)phenyl methyl methylphosphoramidate

Formula:

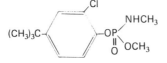

Trade Names: Dowco 132 (Dow)
Ruelene® (Dow)

Manufacture (3)

4-tert-butyl-2-chlorophenol was reacted with phosphorus oxychloride, POCl₃, to give 4-tert-butyl-2-chlorophenyl phosphorodichloridate in an initial step. 16 g (0.5 mol) of methanol in 50 ml of methylene chloride was added dropwise with stirring to a solution of 151 g (0.5 mol) of 4-tertiary-butyl-2-chlorophenyl phosphorodichloridate in 400 ml of methylene chloride. The temperature of the reaction mixture was maintained below 30°C during the addition.

After completion of the addition, the reaction mixture was maintained at 25°C while the mixture was agitated and nitrogen gas was bubbled therethrough to complete the reaction and to remove the hydrogen chloride by-product and to obtain the intermediate 4-tertiary-butyl-2-chlorophenyl methyl phosphorochloridate.

337 g (one-half of the reaction mixture above prepared containing 0.25 mol of 4-tertiary-butyl-2-chlorophenyl methyl phosphorochloridate) was cooled to 3°C and a solution of 16 g

(0.5 mol) of methylamine in 50 ml of methylene chloride was added thereto portionwise and with stirring over a period of about one-half hour. The temperature was maintained below 15°C during the addition. After completion of the addition the mixture was warmed to 30°C to complete the reaction and obtain 4-tertiary-butyl-2-chlorophenyl methyl methylphosphoramidate product and methylamine-hydrochloride by-product.

The reaction mixture was then washed several times with water to remove the methylamine-hydrochloride by-product, warmed to remove the methylene chloride solvent by vaporization and to recover the product as a viscous, colorless oil. The latter was crystallized from petroleum ether (boiling range 30° to 60°C) to obtain a purified product melting at 60° to 60.5°C. The product had a nitrogen content of 4.90% and a phosphorus content of 11.93%. The theoretical values are 4.64% and 11.78%, respectively.

Process Wastes and Their Control

Product Wastes: Crufomate decomposes above pH 7.0 in alkaline media (B-3).

Toxicity

The acute oral LD_{50} for rats is 460 mg/kg (B-5) which is slightly to moderately toxic.

Allowable Limit on Exposure and Use

Air: The threshold limit value for crufomate in air is 5 mg/m^3 as of 1979. The tentative short term exposure limit is 20 mg/m^3 (B-23).

Product Use: The tolerances set by the US EPA for crufomate in or on raw agricultural commodities are as follows:

	40 *CFR* Reference	Parts per Million
Cattle, fat	180.295	1.0
Cattle, mbyp	180.295	1.0
Cattle, meat	180.295	1.0
Goats, fat	180.295	1.0
Goats, mbyp	180.295	1.0
Goats, meat	180.295	1.0
Sheep, fat	180.295	1.0
Sheep, mbyp	180.295	1.0
Sheep, meat	180.295	1.0

References

(1) Worthing, C.R., *Pesticide Manual,* 6th ed., p. 131, British Crop Protection Council (1979).
(2) Spencer, E.Y., *Guide to the Chemicals Used in Crop Protection,* 6th ed., p. 141, London, Ontario, Agriculture Canada (January 1973).
(3) Wasco, J.L., Wade, L.L. and Landram, J.F., U.S. Patent 2,929,762, March 22, 1960, assigned to Dow Chemical Co.

CRYOLITE

Function: Insecticide (1)(2)

Chemical Name: Sodium aluminofluoride

Formula: Na_3AlF_6

Trade Name: Koyoside® (Pennsalt)

Manufacture

Cryolite may be obtained by (a) mining natural mineral cryolite or (b) synthesis by the reaction of aluminum oxide, sodium chloride and hydrogen fluoride (3).

Toxicity

The acute oral LD_{50} value for rats is 200 mg/kg (B-5) which is moderately toxic.

Allowable Limits on Exposure and Use

Product Use: The tolerances set by the US EPA for cryolite in or on raw agricultural commodities are as follows:

	40 *CFR* Reference	Parts per Million
Apples	180.145	7.0
Apricots	180.145	7.0
Beans	180.145	7.0
Beets, greens (alone)	180.145	7.0
Beets, with tops	180.145	7.0
Beets, without tops	180.145	7.0
Blackberries	180.145	7.0
Blueberries (huckleberries)	180.145	7.0
Boysenberries	180.145	7.0
Broccoli	180.145	7.0
Brussels sprouts	180.145	7.0
Cabbage	180.145	7.0
Carrots	180.145	7.0
Cauliflower	180.145	7.0
Citrus fruits	180.145	7.0
Collards	180.145	7.0
Corn	180.145	7.0
Cranberries	180.145	7.0
Cucumbers	180.145	7.0
Dewberries	180.145	7.0
Eggplant	180.145	7.0
Grapes	180.145	7.0
Kale	180.145	7.0
Kohlrabi	180.145	7.0
Lettuce	180.145	7.0
Loganberries	180.145	7.0
Melons	180.145	7.0
Mustard, greens	180.145	7.0
Nectarines	180.145	7.0
Okra	180.145	7.0
Peaches	180.145	7.0
Peanuts	180.145	7.0
Pears	180.145	7.0
Peas	180.145	7.0
Peppers	180.145	7.0
Plums (fresh prunes)	180.145	7.0
Pumpkins	180.145	7.0
Quinces	180.145	7.0
Radishes, tops	180.145	7.0
Radishes, with tops	180.145	7.0

(continued)

	40 *CFR* Reference	Parts per Million
Radishes, without tops	180.145	7.0
Raspberries	180.145	7.0
Rutabagas, tops	180.145	7.0
Rutabagas, with tops	180.145	7.0
Rutabagas, without tops	180.145	7.0
Squash	180.145	7.0
Squash, summer	180.145	7.0
Strawberries	180.145	7.0
Tomatoes	180.145	7.0
Turnips, greens	180.145	7.0
Turnips, with tops	180.145	7.0
Turnips, without tops	180.145	7.0
Youngberries	180.145	7.0

References

(1) Worthing, C.R., *Pesticide Manual,* 6th ed., p. 132, British Crop Protection Council (1979).
(2) Spencer, E.Y., *Guide to the Chemicals Used in Crop Protection,* 6th ed., p. 142, London, Ontario, Agriculture Canada (January 1973).
(3) Howard, H., U.S. Patent 1,475,155, November 20, 1923, assigned to Grasselli Chemical Co.

CUPROUS OXIDE

Function: Fungicide

Chemical Name: Copper oxide

Formula: Cu_2O

Trade Names: Perenox® (Plant Protection, Ltd.)
Copper-Sardez® (Sardez)
Yellow Cuprocide® (Rohm & Haas)

Manufacture

Cuprous oxide may be made by (a) precipitation of copper salts by alkali in the presence of reducing agents (3) or (b) electrolytic oxidation of metallic copper.

Toxicity

The acute oral LD_{50} value for rats is 470 mg/kg which is slightly to moderately toxic.

References

(1) Worthing, C.R., *Pesticide Manual,* 6th ed., p. 133, British Crop Protection Council (1979).
(2) Spencer, E.Y., *Guide to the Chemials Used in Crop Protection,* 6th ed., p. 143, London, Ontario, Agriculture Canada (January 1973).
(3) Rowe, P.J., U.S. Patent 2,474,497, June 28, 1949, assigned to Lake Chemical Co.

CYANAZINE

Function: Herbicide (1)(2)(3)(4)(5)

Chemical Name: 2-[[4-chloro-6-(ethylamino)-1,3,5-triazin-2-yl]amino]-2-methylpropanenitrile

Formula:

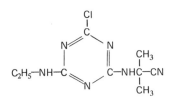

Trade Names: Fortrol®
Bladex®
Payze®

Manufacture (3)(4)(5)

Cyanuric chloride is suspended in acetone and aminoisobutyronitrile, cooled to 0°C, and then 50% NaOH solution is added with stirring. When the mixture becomes neutral, ethylamine and NaOH are added to give the final product.

Toxicity

The acute oral LD_{50} value for rats is 182 mg/kg. This is moderately toxic.

Allowable Limits on Exposure and Use

Product Use: The tolerances set by the US EPA for cyanazine in or on raw agricultural commodities are as follows:

	40 *CFR* Reference	Parts per Million
Corn, fodder	180.307	0.2
Corn, forage	180.307	0.2
Corn, fresh (inc. sweet) (k+cwhr)	180.307	0.05
Corn, grain	180.307	0.05
Cotton, seed	180.307	0.05
Sorghum, fodder	180.307	0.05
Sorghum, forage	180.307	0.05
Sorghum, grain	180.307	0.05
Wheat, forage, green	180.307	0.1
Wheat, grain	180.307	0.1
Wheat, straw	180.307	0.1

References

(1) Worthing, C.R., *Pesticide Manual,* 6th ed., p. 134, British Crop Protection Council (1979).
(2) Spencer, E.Y., *Guide to the Chemicals Used in Crop Protection,* 6th ed., p. 47, London, Ontario, Agriculture Canada (January 1973).
(3) Deutsche Gold- und Silber-Scheideanstalt, British Patent 1,132,306, October 30, 1968.
(4) Schwarze, W., U.S. Patent 3,505,325, April 7, 1970, assigned to Deutsche Gold- und Silber-Scheideanstalt.
(5) Schwarze, W., U.S. Patent 3,620,710, November 16, 1971, assigned to Deutsche Gold- und Silber-Scheideanstalt.

CYANOFENPHOS

Function: Insecticide (1)(2)

Chemical Name: O-4-cyanophenyl O-ethyl phenylphosphonothioate

Formula:

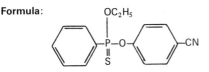

Trade Names: S-4087 (Sumitomo Chemical)
Surecide® (Sumitomo Chemical)

Manufacture

Cyanofenphos is made by the reaction of 4-cyanophenol with O-ethyl phenyl phosphono-chloridothionate. The following is a specific example (3) of the process. A mixture of 11.9 g (0.1 mol) of 4-cyanophenol and 7.2 g of potassium carbonate in methyl isobutyl ketone is heated to 120°C, whereby the generating carbon dioxide and an azeotropic mixture of water and a part of methyl isobutyl ketone are removed therefrom, to leave a suspension of potassium 4-cyanophenolate. 20 g (0.1 mol) of O-ethyl chlorothionobenzenephosphonate is added dropwise to the suspension at approximately 30°C, and, thereafter, heated to 80°C for 4 hours. After cooling, the organic solvent layer is washed with water, with an aqueous sodium carbonate solution, and again with water, followed by drying on anhydrous sodium sulfate.

The methyl isobutyl ketone solvent is distilled off in vacuo, to leave a light brown oily substance which weighs 25.3 g. The oily substance is dissolved in toluene, and the solution is subjected to an active alumina-column chromatography. The purified O-ethyl-O-(4-cyanophenyl) thionobenzenephosphonate is a pale yellow oil, which crystallizes on standing, MP 93° to 96°C.

Toxicity

The acute oral LD_{50} value for rats is 89 mg/kg which is moderate toxicity.

References

(1) Worthing, C.R., *Pesticide Manual,* 6th ed., p. 135, British Crop Protection Council (1979).
(2) Spencer, E.Y., *Guide to the Chemicals Used in Crop Protection,* 6th ed., p. 257, London, Ontario, Agriculture Canada (January 1973).
(3) Kuramoto, S., Nishizawa, Y., Sakamoto, H. and Mizutan, T., U.S. Patent 3,308,016, March 7, 1967, assigned to Sumitomo Chemical Co., Ltd.
(4) Sumitomo Chemical Co., British Patent 929,738, June 26, 1963.

1-(2-CYANO-2-METHOXYIMINOACETYL)-3-ETHYLUREA

Function: Fungicide (1)(2)

Chemical Name: 2-cyano-N-[(ethylamino)-carbonyl]-2-(methoxyimino)acetamide

Formula:

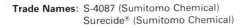

Trade Names: DPX-3217 (DuPont)
Curzate® (DuPont)

Manufacture (2)

2-Cyano-N-ethylcarbamoyl-2-methoxyiminoacetamide, melting point 160° to 161°C, may be prepared from 1-cyanoacetyl-3-ethylurea and sodium nitrite in aqueous acetic acid or other suitable acid such as hydrochloric acid followed by methylation of the oxime with diazomethane.

This methylation may also be carried out by dissolving the free oxime in dimethylformamide, adding a molar equivalent of a base such as sodium methoxide to convert the free oxime to a salt such as the sodium salt, adding methyl iodide (preferably in excess) and stirring the mixture at room temperature. Methyl bromide may be used instead of methyl iodide.

Toxicity

The acute oral LD_{50} value for rats is about 1,200 mg/kg which is slightly toxic.

References

(1) Worthing, C.R., *Pesticide Manual,* 6th ed., p. 136, British Crop Protection Council (1979).
(2) Davidson, S.H., U.S. Patent 3,957,847, May 18, 1976, assigned to E.I. DuPont de Nemours and Co.

CYANOPHOS

Function: Insecticide (1)(2)(3)

Chemical Name: O-4-cyanophenyl O,O-dimethyl phosphorothioate

Formula:

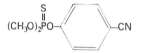

Trade Names: S-4084 (Sumitomo Chemical)
Cyanox® (Sumitomo Chemical)

Manufacture (3)(4)

Cyanophos is made by the reaction of 4-cyanophenol with $(CH_3O)_2PSCl$. The following is a specific example (3) of the conduct of this process. To a mixture of 23.8 g of 4-cyanophenol and 27.6 g of anhydrous potassium carbonate in 200 cc of methyl isobutyl ketone, 32.2 g of O,O-dimethyl chlorothionophosphate was added drop by drop at 60°C under stirring. After the addition of the phosphate was completed, stirring of the mixture was further continued for 8 hours at 60° to 80°C to complete the reaction.

Water was added to dissolve the precipitated inorganic compound, and the organic layer was separated, washed with water and dried over anhydrous sodium sulfate. After distilling off the methyl isobutyl ketone in vacuo, 45.2 g of reddish brown oil product was obtained. For further purification, the crude product was subjected to column-chromatography using active carbon and active alumina, obtaining a pale yellow oily product having a refractive index $n_D^{21.2}$ 1.5457.

Toxicity

The acute oral LD_{50} value for rats is 610 mg/kg which is slightly to moderately toxic.

References

(1) Worthing, C.R., *Pesticide Manual,* 6th ed., p. 138, British Crop Protection Council (1979).
(2) Spencer, E.Y., *Guide to the Chemicals Used in Crop Protection,* 6th ed., p. 144, London, Ontario, Agriculture Canada (January 1973).
(3) Kuramoto, S., Nishizawa, Y., Sakamoto, H. and Mizutani, T., U.S. Patent 3,150,040, September 22, 1964, assigned to Sumitomo Chemical Co., Ltd.
(4) Bernhart, D.N., U.S. Patent 3,792,132, February 12, 1974, assigned to Stauffer Chemical Co.

CYCLOATE

Function: Herbicide (1)(2)

Chemical Name: S-ethyl cyclohexylethyl carbamothioate

Formula:

$$C_2H_5SCN \begin{matrix} O \\ \| \\ \end{matrix} \text{(cyclohexyl)} \\ | \\ C_2H_5$$

Trade Names: R-2063 (Stauffer)
Ro-Neet® (Stauffer)

Manufacture (3)(4)

About 21 g (0.165 mol) of N-ethyl cyclohexylamine was dissolved in 125 cc of ethyl ether and the solution was cooled in an ice bath to 5°C. To this mixture was then slowly added 10 g (0.080 mol) of ethyl chlorothiolformate. The precipitated amine hydrochloride was then removed by filtration and the filter cake was washed with a little ether.

The washings were combined with the original filtrate and the ether solution of product was concentrated on a steam bath. The residual liquid was transferred to a 50 cc round-bottom flask and the solvent was completely removed at 25°C and 25 mm Hg pressure by means of a Rinco rotating film evaporator. There was obtained 15.9 g (91.8% yield) of ethyl ethylcyclohexylthiocarbamate.

Toxicity

The acute oral LD_{50} value for rats is 3,160 mg/kg (slightly toxic).

Allowable Limits on Exposure and Use

Product Use: The tolerances set by the US EPA for cycloate in or on raw agricultural commodities are as follows:

	40 *CFR* Reference	Parts per Million
Beets, garden, roots	180.212	0.05 N
Beets, garden, tops	180.212	0.05 N
Beets, sugar, roots	180.212	0.05 N
Beets, sugar, tops	180.212	0.05 N
Spinach	180.212	0.05 N

References

(1) Worthing, C.R., *Pesticide Manual,* 6th ed., p. 139, British Crop Protection Council (1979).
(2) Spencer, E.Y., *Guide to the Chemicals Used in Crop Protection,* 6th ed., p. 145, London, Ontario, Agriculture Canada (January 1973).
(3) Tilles, H. and Antognini, J., U.S. Patent 3,175,897, March 30, 1965, assigned to Stauffer Chemical Co.
(4) Tilles, H. and Antognini, J., U.S. Patent 3,185,720, May 25, 1965, assigned to Stauffer Chemical Co.

CYCLOHEXIMIDE

Function: Fungicide (1)(2)

Chemical Name: 3-[2-(3,5-dimethyl-2-oxocyclohexyl)-2-hydroxyethyl]- glutarimide

Formula:

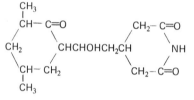

Trade Name: Acti-Dione® (Upjohn)

Manufacture (3)

Cycloheximide is produced in fermentation by *Streptomyces griseus* and recovered as a by-product of streptomycin manufacture.

Process Wastes and Their Control

Product Wastes: Cycloheximide is stable in neutral or acidic solutions but decomposes rapidly in basic solution at room temperature. Admixture with chlordane for some unknown reason causes rapid loss of activity (B-3).

Toxicity

The acute oral LD_{50} value for rats is only 2 mg/kg which rates as extremely toxic.

Allowable Limits on Exposure and Use

Product Use: The tolerances set by the US EPA for cycloheximide in or on raw agricultural commodities are as follows:

	40 *CFR* Reference	Parts per Million
Citrus fruits	180.336	0.1

References

(1) Worthing, C.R., *Pesticide Manual,* 6th ed., p. 140, British Crop Protection Council (1979).

(2) Spencer, E.Y., *Guide to the Chemicals Used in Crop Protection,* 6th ed., p. 146, London, Ontario, Agriculture Canada (January 1973).
(3) Whiffen, A.J., Emerson, R.L. and Bohonos, N., U.S. Patent 2,574,519, November 13, 1951, assigned to The Uphohn Co.

CYCLOPRATE

Function: Acaricide

Chemical Name: Hexadecyl cyclopropane carboxylate

Formula:

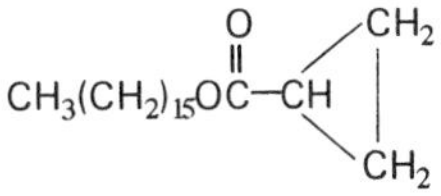

$$CH_3(CH_2)_{15}OC-CH\begin{matrix}CH_2\\ \ \\CH_2\end{matrix}$$

Trade Name: Zardex® (Zoecon Corp.)

Manufacture

Cycloprate may be made by the reaction of cyclopropane carboxylic acid chloride and hexadecylalcohol.

Toxicity

The acute oral LD_{50} value for rats is 12,200 mg/kg (insignificantly toxic) (B-5)

Reference

(1) Henrick, C.A. and Staal, G.B., U.S. Patent 3,925,461, December 9, 1975, assigned to Zoecon Corp.

CYCLURON

Function: Herbicide (1)(2)

Chemical Name: N'-cyclooctyl-N,N-dimethylurea

Formula:

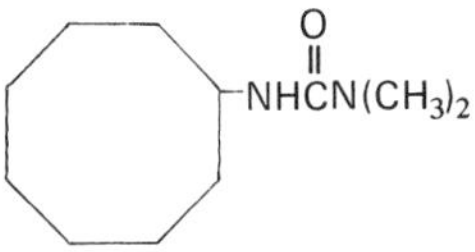

$$-NHCN(CH_3)_2$$

Trade Names: A component of Alipur® (BASF)
Cyclouron®

Manufacture (2)

Dry hydrogen chloride is led into a solution of 127 parts of cyclooctylamine in 410 parts of dry dioxane at room temperature until the whole of the amine has been converted into the hydrochloride. Then dry phosgene is led into the solution kept at 90°C until the color of the solution had changed to yellow-brown, which requires about 2 to 3 hours. After distilling off about 100 parts of dioxane (whereby the excess of phosgene and hydrogen chloride are removed), dimethylamine is led into the solution of the cyclooctyl isocyanate formed at room temperature until it is saturated and the reaction mixture kept at 50°C for 1 hour after the end of the introduction; it is then allowed to cool and the precipitated 1-cyclooctyl-3,3-dimethylurea is filtered off by suction, suspended in water and after repeated filtration by suction dried in a vacuum drying cabinet at 60°C. If necessary the compound can be recrystallized from cyclohexane. 168 parts of 1-cyclooctyl-3,3-dimethylurea of the MP 137° to 137.5°C are obtained.

Toxicity

The acute oral LD_{50} value is 2,600 mg/kg (slightly toxic).

References

(1) Worthing, C.R., *Pesticide Manual,* 6th ed., p. 141, British Crop Protection Council (1979).
(2) Badische Anilin- & Soda-Fabrik AG, British Patent 812,120, April 15, 1959.

CYHEXATIN

Function: Acaricide (1)(2)(3)

Chemical Name: Tricyclohexylhydroxystannane

Formula: $(C_6H_{11})_3SnOH$

Trade Names: Dowco 213 (Dow Chemical Co.)
Plictran® (Dow Chemical Co.)

Manufacture

Cyclohexyl chloride may be reacted with $SnCl_4$ and metallic sodium to give tricyclohexyl tin chloride. The tricyclohexyl tin chloride may then be reacted with alkali to give the tricyclohexyl tin hydroxide.

Toxicity

The acute oral LD$_{50}$ value for rats is 540 mg/kg which is slightly to moderately toxic.

Allowable Limits on Exposure and Use

Air: The threshold limit value for cyhexatin in air is 5 mg/m^3 as of 1979. The tentative short term exposure limit is 10 mg/m^3 (B-23).

Product Use: The tolerances set by the EPA for cyhexatin in or on raw agricultural commodities are as follows:

	40 *CFR* Reference	Parts per Million
Almonds	180.144	0.5
Almonds, hulls	180.144	60.0

(continued)

	40 *CFR* Reference	Parts per Million
Apples	180.144	2.0
Cattle, fat	180.144	0.2
Cattle, kidney	180.144	0.5
Cattle, liver	180.144	0.5
Cattle, mbyp (exc. kidney, liver)	180.144	0.2
Cattle, meat	180.144	0.2
Citrus fruits	180.144	2.0
Goats, fat	180.144	0.2
Goats, kidney	180.144	0.5
Goats, liver	180.144	0.5
Goats, mbyp (exc. kidney, liver)	180.144	0.2
Goats, meat	180.144	0.2
Hogs, fat	180.144	0.2
Hogs, kidney	180.144	0.5
Hogs, liver	180.144	0.5
Hogs, mbyp (exc. kidney, liver)	180.144	0.2
Hogs, meat	180.144	0.2
Hops	180.144	30.0
Horses, fat	180.144	0.2
Horses, kidney	180.144	0.5
Horses, liver	180.144	0.5
Horses, mbyp (exc. kidney, liver)	180.144	0.2
Horses, meat	180.144	0.2
Macadamia nuts	180.144	0.5
Milk, fat (= N in whole milk)	180.144	0.05
Nectarines	180.144	4.0
Peaches	180.144	4.0
Pears	180.144	2.0
Plums (fresh prunes)	180.144	1.0
Sheep, fat	180.144	0.2
Sheep, kidney	180.144	0.5
Sheep, liver	180.144	0.5
Sheep, mbyp (exc. kidney, liver)	180.144	0.2
Sheep, meat	180.144	0.2
Strawberries	180.144	3.0
Walnuts	180.144	0.5

The tolerances set by the US EPA for cyhexatin in animal feeds are as follows (the *CFR* Reference is to Title 21):

	CFR Reference	Parts per Million
Apple pomace, dried	561.400	8.0
Citrus, pulp, dried	561.400	8.0

The tolerances set by the US EPA for cyhexatin in food are as follows (the *CFR* Reference is to Title 21):

	CFR Reference	Parts per Million
Hops, dried	193.430	90.0
Prunes, dried	193.430	4.0

References

(1) Worthing, C.R., *Pesticide Manual,* 6th ed., p. 142, British Crop Protection Council (1979).

(2) Kenaga, E.E., U.S. Patent 3,264,177, August 2, 1966, assigned to Dow Chemical Co.
(3) Kenaga, E.E., U.S. Patent 3,389,048, June 18, 1968, assigned to Dow Chemical Co.

CYPRAZINE

Function: Herbicide (1)(2)(3)(4)

Chemical Name: 6-chloro-N-cyclopropyl-N'-(1-methylethyl)-1,3,5-triazine-2,4-diamine

Formula:

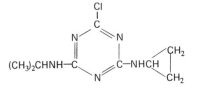

Trade Names: Outfox® (Gulf Oil)
S-6115

Manufacture

Cyprazine is made from cyanuric chloride as a starting material. It is reacted in a first step with cyclopropylamine and in a second step with isopropylamine.

Toxicity

The acute oral LD_{50} value for rats is about 1,200 mg/kg which is slightly toxic.

Allowable Limits on Exposure and Use

Product Use: The tolerances set by the US EPA for cyprazine in or on raw agricultural commodities are as follows:

	40 *CFR* Reference	Parts per Million
Corn, fodder	180.306	0.1 N
Corn, forage	180.306	0.1 N
Corn, fresh (inc. sweet) (k+cwhr)	180.306	0.1 N
Corn, grain	180.306	0.1 N

References

(1) Martin, H. and Worthing, C.R., *Pesticide Manual,* 4th ed., p. 146, British Crop Protection Council (November 1974).
(2) Spencer, E.Y., *Guide to the Chemicals Used in Crop Protection,* 6th ed., p. 147, London, Ontario, Agriculture Canada (January 1973).
(3) Neighbors, R.P. and Phillips, L.V., U.S. Patent 3,451,802, June 24, 1969, assigned to Gulf Research and Development Co.
(4) Neighbors, R.P. and Phillips, L.V., U.S. Patent 3,503,971, March 31, 1970, assigned to Gulf Research and Development Co.

D

2,4-D

Function: Systemic herbicide (1)(2)(4)

Chemical Name: 2,4-dichlorophenoxyacetic acid

Formula:

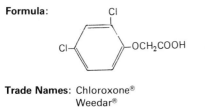

Trade Names: Chloroxone®
Weedar®
Weedone®
Salvo®

Manufacture (3)

2,4-D is made by the following reactions and as shown in the flow diagram in Figure 21.

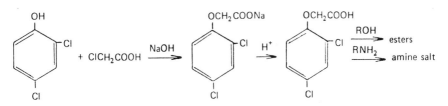

An improved process for 2,4-D manufacture is claimed by Manske (5). His process is based on the discovery that phenol gives a better yield of phenoxyacetic acid than 2,4-dichloro-phenol gives of 2,4-dichlorophenoxyacetic acid when they are reacted under comparable conditions with sodium hydroxide and chloroacetic acid. For example, the yield of phenoxy-acetic acid from sodium hydroxide, phenol and chloracetic acid under favorable conditions reached 94%.

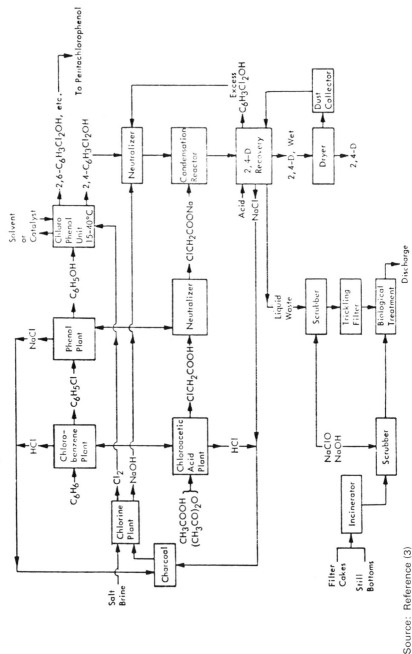

Figure 21: Production and Waste Schematic for 2,4-D (Dow Chemical)

Source: Reference (3)

Under similar conditions which were most favorable, the yield of 2,4-dichlorophenoxyacetic acid from 2,4-dichlorophenol, sodium hydroxide and chloroacetic acid was seldom more than 84%. Furthermore, the purity of the phenoxyacetic acid was slightly higher than that of the 2,4-dichlorophenoxyacetic acid.

The following is a specific example of the conduct of Manske's process. Phenoxyacetic acid having a melting point of 99°C was employed. Three hundred and four parts by weight of this material were heated to about 110°C in a vessel which itself was resistant to the corrosive action of chlorine and hydrogen chloride. A stream of chlorine was introduced into the molten phenoxyacetic acid. As chlorination proceeded, the 4-chlorophenoxyacetic acid which was formed tended to crystallize from the reaction mixture. The temperature was gradually raised to maintain the latter in solution so that by the time 2 atoms of chlorine had been introduced for every molecule of phenoxyacetic acid originally present, the temperature was approximately 150°C.

The dissolved chlorine and hydrogen chloride were then removed by a stream of air and the liquid product was withdrawn and allowed to solidify by cooling to a pale yellow crystalline mass. The yield of crude, 2,4-dichlorophenoxyacetic acid thus obtained was 442 parts by weight, that is 100%. It melted nonsharply at around 130°C. It was virtually free of chlorinated phenols. When it was recrystallized from benzene it melted at 137° to 138°C and the yield of once recrystallized product was about 375 parts by weight or 85% of theory. If desired, the 15% of impure 2,4-dichlorophenoxyacetic acid remaining in the mother liquor from the benzene recrystallization can be worked up to give more pure 2,4-dichlorophenoxyacetic acid.

Process Wastes and Their Control

Air: Air emissions from the manufacture of 2,4-D acid and its salts and esters have been reported (B-15) to consist of 1.0 kg of 2,4-dichlorophenol, 1.0 kg of chloroacetic acid and 1.5 kg of ammonia per metric ton of pesticide produced.

Water: There are three waste streams from the 2,4-D recovery step. At one of the plants manufacturing 2,4-D, dusts are collected and recycled. Liquid wastes are chemically treated, passed through a special trickling filter, and then to a biological waste treatment plant. Solid wastes, still bottoms, etc., are incinerated; the combustion gases are scrubbed; the water is chemically treated, and then sent to a biowaste plant (B-10).

Waste streams from chlorinated hydrocarbon herbicides such as 2,4-D include large amounts of sodium chloride, hydrochloric acid, some caustic, and organics including solvents, phenols, chlorophenols and chlorophenoxy acids. These wastes arise from acidification, washing steps, phase separation steps, incomplete yields and chlorination of the phenolic compounds. A typical waste stream can be characterized by:

COD	8,300 mg/l
BOD$_3$	6,300 mg/l
Total solids	104,000 mg/l
Suspended solids	2,500 mg/l
pH	0.5
Chlorides	52,000 mg/l
Chlorophenols	112 mg/l
Chlorophenoxy acids	235 mg/l
Nitrogen	low
Phosphorus	low
Flow	30 lb COD/lb product

The chlorinated herbicide wastes can vary considerably from plant to plant and even in the same plant (B-10).

Product Disposal: This compound, 2,4-Cl$_2$C$_6$H$_3$OCH$_2$COOH, is considered to be nonpersistent, although it may last more than one year in dry areas. Studies at 100 lb/acre have shown it is degraded by spreading on the land. Since the ester forms are somewhat volatile, they should probably be hydrolyzed first (see above) to avoid drift, if this method of disposal is used.

Incineration of phenoxys is effective in one second at 1800°F using a straight combustion process or at 900°F using catalytic combustion. Over 99% decomposition was reported when small amounts of 2,4-D were burned in a polyethylene bag.

Chlorination at pH 3 and 85°F for 10 minutes (using an excess of either gaseous chlorine or a sodium hypochlorite solution) renders the phenoxys nonherbicidal, but the products were unstated and are possibly objectionable. Reduction of 2,4-D by sodium or lithium in liquid ammonia has been reported as has oxidation with boric oxide in a Parr bomb, but neither method is suitable as a general disposal technique. Addition of soluble calcium (or magnesium) salts to solutions of 2,4-D acid or salts (but not the ester forms) produces the insoluble (solid) calcium 2,4-D salt. This reaction can aid in collecting the 2,4-D but the product is still herbicidally active.

Degradation of 2,4-D is achieved in biological treatment facilities (B-3).

Toxicity

The acute toxicity of 2,4-D is moderate with an LD_{50} value for rats of 375 mg/kg. No-adverse-effect doses for 2,4-D were up to 62.5 mg/kg/day and 10 mg/kg/day in rats and dogs, respectively. Based on these data, an ADI (acceptable daily intake) was calculated at 0.0125 mg/kg/day.

The acceptable daily intake of 2,4-D has been established at 0.3 mg/kg by FAO/WHO. On the basis of electron-capture gas chromatography, the detection limit for 2,4-D in water is 1 ppb.

There are substantial disagreements in the results of subchronic and chronic toxicity studies with 2,4-D, perhaps reflecting the use of different formulations or preparations. In view of these deficiencies and the variability of the results, additional, properly consititued toxicity studies should be undertaken (B-22).

Allowable Limits on Exposure and Use

Air: The threshold limit value for 2,4-D in air is 10 mg/m^3 as of 1979. The tentative short term exposure limit is 20 mg/m^3 (B-23).

Water: In water, EPA has set (B-12) a criterion of 100 μg/l of 2,4-D for domestic water supplies on a health protection basis.

Product Use: The US EPA as of October 12, 1967 (B-17) issued a restriction on 2,4-D to the effect that products bearing directions for use on small grains (barley, oats, rye, or wheat) must bear the following label precaution: "Do not forage or graze treated grain fields within 2 weeks after treatment with 2,4-D."

The tolerances set by the US EPA for 2,4-D in or on raw agricultural commodities are as follows:

	40 *CFR* Reference	Parts per Million
Apples	180.142	5.0
Asparagus	180.142	5.0
Avocados	180.142	0.1 N

(continued)

	40 *CFR* Reference	Parts per Million
Barley, forage	180.142	20.0
Barley, grain	180.142	0.5
Blueberries	180.142	0.1
Cattle, fat	180.142	0.2
Cattle, kidney	180.142	2.0
Cattle, mbyp (exc. kidney)	180.142	0.2
Cattle, meat	180.142	0.2
Citrus fruits	180.142	0.1 N
Citrus fruits (inc. pre-h & post-h)	180.142	5.0
Corn, fodder	180.142	20.0
Corn, forage	180.142	20.0
Corn, fresh (inc. sweet) (k+chwhr)	180.142	0.5
Cotton, seed	180.142	0.1 N
Cranberries	180.142	0.5
Cucurbits	180.142	0.1 N
Eggs	180.142	0.05
Fish	180.142	1.0
Fruits, pome	180.142	0.1 N
Fruits, small	180.142	0.1 N
Fruits, stone	180.142	0.1 N
Goats, fat	180.142	0.2
Goats, kidney	180.142	2.0
Goats, mbyp (exc. kidney)	180.142	0-2.0
Goats, meat	180.142	0.2
Grain crops	180.142	0.1 N
Grapes	180.142	0.5
Grasses, forage	180.142	0.1 N
Grasses, hay	180.142	300.0
Grasses, pasture	180.142	1000.0
Grasses, rangeland	180.142	1000.0
Hogs, fat	180.142	0.2
Hogs, kidney	180.142	2.0
Hogs, mbyp (exc. kidney)	180.142	0.2
Hogs, meat	180.142	0.2
Hops	180.142	0.1 N
Horses, fat	180.142	0.2
Horses, kidney	180.142	2.0
Horses, mbyp (exc. kidney)	180.142	0.2
Horses, meat	180.142	0.2
Legumes, forage	180.142	0.1 N
Lemons (post-h)	180.142	5.0
Milk	180.142	0.1
Nuts	180.142	0.1 N
Oats, forage	180.142	20.0
Oats, grain	180.142	0.5
Pears	180.142	5.0
Potatoes	180.142	0.2
Poultry	180.142	0.05
Quinces	180.142	5.0
Rice, straw	180.142	20.0
Rye, forage	180.142	20.0
Rye, grain	180.142	0.5
Sheep, fat	180.142	0.2
Sheep, kidney	180.142	2.0
Sheep, mbyp (exc. kidney)	180.142	0.2
Sheep, meat	180.142	0.2
Shellfish	180.142	1.0
Sorghum	180.142	0.5
Sorghum, fodder	180.142	20.0

(continued)

	40 *CFR* Reference	Parts per Million
Sorghum, forage	180.142	20.0
Strawberries	180.142	0.1 N
Sugarcane	180.142	2.0
Vegetables, fruiting	180.142	0.1 N
Vegetables, leafy	180.142	0.1 N
Vegetables, root crop	180.142	0.1 N
Vegetables, seed & pod	180.142	0.1 N
Wheat, forage	180.142	20.0
Wheat, grain	180.142	0.5

The tolerances set by the US EPA for 2,4-D in animal feeds are as follows (the *CFR* Reference is to Title 21):

	CFR Reference	Parts per Million
Barley, milled fractions	561.100	2.0
Oats, milled fractions	561.100	2.0
Rye, milled fractions	561.100	2.0
Sugarcane bagasse	561.100	5.0
Sugarcane molasses	561.100	5.0
Wheat, milled fractions	561.100	2.0

The tolerances set by the US EPA for 2,4-D in food are as follows (the *CFR* Reference is to Title 21):

	CFR Reference	Parts per Million
Barley, milled fractions (exc. flour)	193.100	2.0
Oats, milled fractions (exc. flour)	193.100	2.0
Rye, milled fractions (exc. flour)	193.100	2.0
Sugarcane bagasse	193.100	5.0
Sugarcane molasses	193.100	5.0
Water, potable	193.100	0.1 N
Wheat, milled fractions (exc. flour)	193.100	2.0

References

(1) Worthing, C.R., *Pesticide Manual*, 6th ed., p. 145, British Crop Protection Council (1979).
(2) Spencer, E.Y., *Guide to the Chemicals Used in Crop Protection*, 6th ed., p. 148, London, Ontario, Agriculture Canada (January 1973).
(3) Midwest Research Institute, *The Pollution Potential in Pesticide Manufacturing*, Wash., DC, Environmental Protection Agency (June 1972).
(4) Jones, F.D., U.S. Patent 2,390,941, December 11, 1945, assigned to American Chemical Co.
(5) Manske, R.H.F., U.S. Patent 2,471,575, May 31, 1949, assigned to United States Rubber Co.

DALAPON

Function: Contact herbicide (1)(2)

Chemical Name: Sodium 2,2-dichloropropionate

Formula: CH_3CCl_2COONa

Trade Names: Dowpon® (Dow)
              Radapon® (Dow)
              Basfapon® (BASF)

Manufacture (3)

Dichloropropionic acid is produced by the direct chlorination of propionic acid in the presence of catalysts. Alkali metal salts of 2,2-dichloropropionic acid are well known and widely used as selective herbicides. They have not found application in the control of grass weeds. However, commercial production of these salts has been handicapped by the fact that neutralization of 2,2-dichloropropionic acid, when carried out in water is accompanied by decomposition in greater or less degree of the 2,2-dichloropropionic acid.

For example, when the 2,2-dichloropropionic acid is neutralized with an alkali material, such as sodium hydroxide, in water at or near the boiling temperature under standard conditions, the acid and the resulting salts are unstable and break down or decompose more or less promptly with the formation of herbicidally useless or nearly useless degradation products.

When the same procedure is carried out at lower temperatures the breakdown of the acid and salts proceeds at a lower rate of speed. However, in any case the neutralization in water results in the formation of a water solution of at least the products of such neutralization, and all common and commercially economical means for removing such water to obtain a dry product cause breakdown of the desired product to a greater or less extent.

According to an improved process developed by Maylott and Meyer (4), alkali salts of 2,2-dichloropropionic acid are prepared by carrying out the procedural steps of contacting 2,2-dichloropropionic acid with an essentially dry alkali compound of an alkali metal in a liquid chlorinated hydrocarbon as reaction medium, and thereafter separating in its solid phase an alkali salt of 2,2-dichloropropionic acid.

Process Wastes and Their Control

Air: Air emissions from dalapon manufacture have been reported (B-15) as follows:

Component	Kilograms per Metric Ton Pesticide Produced
Hydrogen chloride	0.5
Propionic acid	0.5
Dalapon	0.1

Product Wastes: This product is dehydrochlorinated by alkali above 120°C.

$$CH_3CCl_2COONa \xrightarrow[120^\circ C]{Base} CH_2=CClCOONa$$

It undergoes hydrolysis of both chlorine atoms in aqueous solution at or above 50°C (and slowly even at 25°C) to give herbicidally inactive pyruvic acid:

$$CH_3CCl_2COONa \xrightarrow{H_2O} CH_3COCOOH$$

Dalapon reacts with sodium or lithium in liquid ammonia. Dalapon underwent charring when heated alone, underwent exothermic decomposition when heated with the oxidants KNO_3 and $KClO_3$ (above 187° and 140°C, respectively), and was >99% decomposed when burned in a polyethylene bag, but toxic phosgene gas was detected in the products (B-3).

Toxicity

The acute oral LD_{50} value for rats is 3,860 mg/kg (B-5) which is slightly toxic.

Allowable Limits on Exposure and Use

Product Use: The tolerances set by the US EPA for dalapon in or on raw agricultural commodities are as follows:

	40 *CFR* Reference	Parts per Million
Almonds	180.150	10.0
Almonds, hulls	180.150	50.0
Apples	180.150	3.0
Apricots	180.150	1.0
Asparagus	180.150	30.0
Bananas	180.150	5.0
Beans	180.150	1.0
Beans, straw	180.150	1.0
Beets, sugar, roots	180.150	5.0
Beets, sugar, tops	180.150	5.0
Cattle, mbyp	180.150	0.2
Cattle, meat	180.150	0.2
Coffee beans	180.150	2.0
Corn, ear, dried (k+c)	180.150	10.0
Corn, fodder	180.150	5.0
Corn, forage	180.150	5.0
Corn, fresh (inc. sweet) (k+cwhr)	180.150	5.0
Corn, grain	180.150	10.0
Cotton, seed	180.150	35.0
Cranberries	180.150	5.0
Eggs	180.150	0.3
Flax, seed	180.150	75.0
Goats, mbyp	180.150	0.2
Goats, meat	180.150	0.2
Grapefruit	180.150	5.0
Grapes	180.150	3.0
Grasses, pasture	180.150	10.0
Grasses, range	180.150	10.0
Hogs, mbyp	180.150	0.2
Hogs, meat	180.150	0.2
Lemons	180.150	5.0
Limes	180.150	5.0
Macadamia nuts	180.150	1.0
Milk	180.150	0.1 N
Oranges	180.150	5.0
Peaches	180.150	15.0
Pears	180.150	3.0
Peas, shelled	180.150	15.0
Peas, unshelled	180.150	15.0
Peas, vines, with pod	180.150	15.0
Peas, vines, without pod	180.150	15.0
Pecans	180.150	0.1 N
Pineapples	180.150	3.0
Plums	180.150	1.0
Potatoes	180.150	10.0
Poultry (exc. kidney)	180.150	3.0
Poultry, kidney	180.150	9.0
Sheep, mbyp	180.150	0.2
Sheep, meat	180.150	0.2

(continued)

	40 *CFR* Reference	Parts per Million
Sorghum	180.150	1.0
Sorghum, forage	180.150	5.0
Soybeans	180.150	1.0
Soybeans, straw	180.150	1.0
Sugarcane	180.150	0.1 N
Tangerines	180.150	5.0
Walnuts	180.150	5.0

The tolerances set by the US EPA for dalapon in animal feeds are as follows (the *CFR* Reference is to Title 21):

	CFR Reference	Parts per Million
Citrus pulp, dehydrated (ct.f)	561.110	20.0

The tolerances set by the US EPA for dalapon in food are as follows (the *CFR* Reference is to Title 21):

	CFR Reference	Parts per Million
Water, potable	193.105	0.2

References

(1) Worthing, C.R., *Pesticide Manual,* 6th ed., p. 146, British Crop Protection Council (1979).
(2) Spencer, E.Y., *Guide to the Chemicals Used in Crop Protection,* 6th ed., p. 151, London, Ontario, Agriculture Canada (January 1973).
(3) Barrons, K.C., U.S. Patent 2,642,354, June 16, 1953, assigned to Dow Chemical Co.
(4) Maylott, A.O. and Meyer, R.H., U.S. Patent 3,007,964, November 7, 1961, assigned to Dow Chemical Co.

DAMINOZIDE

Function: Plant growth regulant (1)(2)

Chemical Name: Butanedioic acid mono(2,2-dimethylhydrazide)

Formula:

$$CH_2-\overset{\overset{\textstyle O}{\|}}{C}NHN(CH_3)_2$$
$$CH_2-\underset{\underset{\textstyle O}{\|}}{C}-OH$$

Trade Names: B-995 (Uniroyal, Inc.)
Alar® (Uniroyal, Inc.)
B-Nine® (Uniroyal, Inc.)

Manufacture (3)(4)(5)

A specific example of the preparation of succinic acid dimethylhydrazide is as follows. To a solution of 50 g (0.50 mol) of succinic anhydride in 150 g of acetonitrile was added a solution of 30 g (0.50 mol) of 1,1-dimethylhydrazine in 50 g of acetonitrile. After the exothermic reaction had subsided the mixture was cooled in an ice bath and the precipitate collected by filtration. A yield of 74 g (92%) of N-dimethylaminosuccinamic acid as a colorless, crystalline solid of MP 154° to 156°C was obtained. A sample of this material after crystallization from ethanol gave colorless crystals of MP 154° to 155°C.

Process Wastes and Their Control

Product Wastes: This product is stable at 50°F for 5 months, stable in water for over 2 months, and stable in solution at 80°F for 21 days. It is rapidly hydrolyzed in boiling dilute HCl and boiling 50% sodium hydroxide solutions (B-3).

Toxicity

The acute oral LD_{50} value for rats is 8,400 mg/kg (insignificantly toxic).

Allowable Limits on Exposure and Use

Product Waste: The tolerances set by the US EPA for daminozide in or on raw agricultural commodities are as follows:

	40 *CFR* Reference	Parts per Million
Apples	180.246	30.0
Brussels sprouts	180.246	20.0
Cattle, fat	180.246	0.2
Cattle, mbyp	180.246	0.2
Cattle, meat	180.246	0.2
Cherries, sour	180.246	55.0
Cherries, sweet	180.246	30.0
Eggs	180.246	0.2
Goats, fat	180.246	0.2
Goats, mbyp	180.246	0.2
Goats, meat	180.246	0.2
Grapes	180.246	10.0
Hogs, fat	180.246	0.2
Hogs, mbyp	180.246	0.2
Hogs, meat	180.246	0.2
Horses, fat	180.246	0.2
Horses, mbyp	180.246	0.2
Horses, meat	180.246	0.2
Melons	180.246	3.0
Milk	180.246	0.02 N
Nectarines	180.246	30.0
Peaches	180.246	30.0
Peanuts	180.246	30.0
Peanuts, hay	180.246	20.0
Peanuts, hulls	180.246	10.0
Pears	180.246	20.0
Peppers	180.246	1.0
Plums (fresh prunes)	180.246	50.0
Poultry, fat	180.246	0.2
Poultry, kidney	180.246	2.0
Poultry, mbyp (exc. kidney)	180.246	0.2
Poultry, meat	180.246	0.2
Sheep, fat	180.246	0.2
Sheep, mbyp	180.246	0.2
Sheep, meat	180.246	0.2
Tomatoes	180.246	40.0

The tolerances set by the US EPA for daminozide in animal feeds are as follows (the *CFR* Reference is to Title 21):

	CFR Reference	Parts per Million
Peanut meal	561.360	90.0

The tolerances set by the US EPA for daminozide in food are as follows (the *CFR* Reference is to Title 21):

	CFR Reference	Parts per Million
Prunes, dried	193.410	135.0

References

(1) Worthing, C.R., *Pesticide Manual,* 6th ed., p. 147, British Crop Protection Council (1979).
(2) Spencer, E.Y., *Guide to the Chemicals Used in Crop Protection,* 6th ed., p. 469, London, Ontario, Agriculture Canada (January 1973).
(3) Hageman, H.A. and Hubbard, W.L., U.S. Patent 3,240,799, May 15, 1966, assigned to United States Rubber Company.
(4) Hageman, H.A. and Hubbard, W.L., U.S. Patent 3,334,991, August 8, 1967, assigned to Uniroyal, Inc.
(5) Hageman, H.A. and Hubbard, W.L., U.S. Patent 3,257,414, June 21, 1966, assigned to United States Rubber Company.

DAZOMET

Function: Fungicide and nematocide (1)(2)

Chemical Name: Tetrahydro-3,5-dimethyl-2H-1,3,5-thiadiazine-2-thione

Formula:

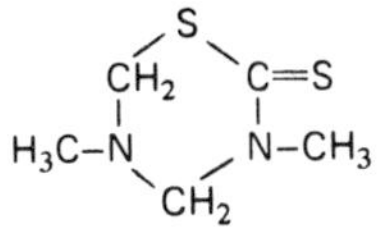

Trade Names: N-521 (Stauffer)
Crag Fungicide 974® (Union Carbide)
Mylone® (Union Carbide)
Crag Nemacide®
Mico-Fume®

Manufacture

Dazomet may be produced by:

(A) The interaction of carbon disulfide, methylamine and formaldehyde (3); or
(B) The action of formaldehyde on methylammonium methyldithiocarbamate (4).

Process Wastes and Their Control

Product Wastes: Dazomet, a crystalline fungicide, structurally related to the dithiocarbamates, decomposes upon heating at 100°C (its MP) to form methylisothiocyanate and dimethylthiourea. The decomposition rate is accelerated by moisture. Acid hydrolysis results in the formation of carbon disulfide (see Carbon Disulfide) (B-3).

Toxicity

The acute oral LD_{50} value for rats is 640 mg/kg (slightly toxic).

References

(1) Worthing, C.R., *Pesticide Manual,* 6th ed., p. 148, British Crop Protection Council (1979).
(2) Spencer, E.Y., *Guide to the Chemicals Used in Crop Protection,* 6th ed., p. 152, London, Ontario, Agriculture Canada (January 1973).
(3) Delepine, M., *Bull. Soc. Chim.* 15, 891 (1897).
(4) Bodendorf, K., *J. Prakt. Chem.* 126, 233 (1930).

2,4-DB

Function: Herbicide (1)(2)(4)

Chemical Name: 4-(2,4-dichlorophenoxy)butanoic acid

Formula: $OCH_2CH_2CH_2COOH$

Trade Names: MB 2878 (May and Baker, Ltd.)
Embutox (May and Baker, Ltd.)
Butyrac®
Butoxone SB®

Manufacture (3)
The feed materials for 2,4-DB manufacture are 2,4-dichlorophenol and butyrolactone which react according to the following equations:

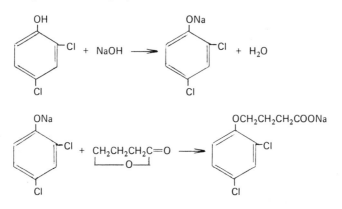

From 0.5 to 2.0 mols of sodium hydroxide per mol of phenol is the preferred feed ratio for the first reaction. In the second step, the lactone should be present in a mol ratio of from 0.5:1 to 2:1 based on the phenol. The conversion of phenol to alkali phenolate is carried out at a temperature of 50° to 200°C. The reaction of the phenolate with butyrolactone is carried out at 135° to 210°C and preferably at 160° to 165°C. The reaction of phenol with alkali is carried out at atmospheric pressure, as is the subsequent reaction of the phenolate with butyrolactone.

Reaction periods of from 0.5 to 24.0 hours may be used in the conversion of the phenol to the alkali phenolate. One hour was the period used in one typical example. The reaction of alkali phenoxide with butyrolactone may be conducted over a period of 1 to 24 hours. A more typical period is from 1 to 6 hours.

The conversion of phenol to phenoxide is preferably conducted in the presence of from 1.0 to 9.0 parts by weight, based on the phenol, of a solvent comprising an alcohol and a hydrocarbon. Suitable alcohols may contain 1 to 10 carbon atoms but butanol is the preferred alcohol; when it is used, higher product yields are obtained. Suitable hydrocarbons may contain from 5 to 14 carbon atoms and should preferably form a constant boiling mixture with the alcohol employed. A particularly desirable hydrocarbon solvent is commercial nonane (a C_7 to C_{10} fraction). Volume ratios of alcohol to hydrocarbon in the solvent mixture may vary from 1:2 to 1:1.

The second step is conducted in essentially the same liquid medium except that water formed in the first step should preferably be removed by distillation prior to the second step. No catalyst is employed in the conversion of phenol to alkali phenoxide, or in the subsequent reaction of the phenoxide with butyrolactone. An agitated reactor having a means for heating and cooling the reaction mixture and fitted with a distillation column is well suited to the conduct of both steps of this reaction. The scheme of operations is shown In Figure 22.

Figure 22: Process Flow Scheme for 2,4-DB Manufacture

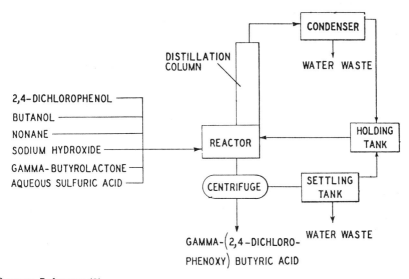

Source: Reference (3)

The initial alkoxide product from the first reaction step is not separated but is used directly in the second step with butyrolactone. The alkali metal salt of the phenoxyalkanoic acid produced in the second step is neutralized with a mineral acid such as sulfuric, hydrochloric or phosphoric acid.

The final product can be recovered by centrifugation, filtration or other means for liquid-solid separation. The effluent from the solids-liquid separation may be introduced to a settling tank or other quiescent zone where two layers, a water layer and an organic layer separate out. The water layer, containing water-soluble salts such as Na_2SO_4 and the like formed by the reaction, is drawn off and discarded. The organic layer, containing substantially all of the unreacted phenol and lactone in addition to the alcohol-hydrocarbon solvent, is recycled to the reactor.

If it is desired, traces of unreacted phenol remaining in the product may be removed by any of several well-known means such as by steam distillation or by solvent extraction with suitable solvents, such as petroleum ether, hexane, heptane or nonane. The final product may then be dried under reduced pressure. It can be further purified by slurrying in petroleum ether, diethyl ether, nonane or other hydrocarbon solvent, centrifuging or filtering the resulting slurry, and drying the precipitate thus produced. Such purification is generally not necessary, however, since the product is substantially free of impurities.

Process Wastes and Their Control

The 2,4-DB is removed and cycled through a centrifuge process. The wastes from the reaction and the centrifuge are cycled through a recovery process to regain excess reactants. These are then recycled to the system (B-10).

Toxicity

The acute oral LD_{50} for rats is 700 mg/kg (slightly toxic).

Allowable Limits on Exposure and Use

Product Use: The tolerances set by the US EPA for 2,4-DB in or on raw agricultural commodities are as follows:

	40 *CFR* Reference	Parts per Million
Alfalfa	180.331	0.2 N
Clover	180.331	0.2 N
Peanuts	180.331	0.2 N
Soybeans	180.331	0.2 N
Soybeans, hay	180.331	0.2 N
Trefoil, birdsfoot	180.331	0.2 N

References

(1) Worthing, C.R., *Pesticide Manual,* 6th ed., p. 149, British Crop Protection Council (1979).
(2) Spencer, E.Y., *Guide to the Chemicals Used in Crop Protection,* 6th ed., p. 153, London, Ontario, Agriculture Canada (January 1973).
(3) Hogsett, J.N., U.S. Patent 3,076,025, January 29, 1963, assigned to Union Carbide Corp.
(4) Wain, R.L., U.S. Patent 2,863,754, December 9, 1958, assigned to National Research Development Corp.

DDT

Function: Insecticide (1)(2)

Chemical Name: 1,1'-(2,2,2-trichloroethylidene)bis(4-chlorobenzene)

Formula:

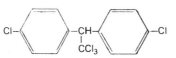

Trade Names: Gesarol® (Ciba-Geigy)
Buesarol® (Ciba-Geigy)
Necoid® (Ciba-Geigy)
Anofox®
Dinocide®
Genitox®
Gesapon®
Gesarex®
Neocidol®
Zerdane®

Manufacture (3)

The basic reaction for the synthesis of DDT is the Bayer condensation of chlorobenzene with chloral (7) according to the equation:

$$CCl_3CHO + 2C_6H_5Cl \rightarrow (p\text{-}C_6H_4Cl)_2CHCCl_3 + H_2O$$

Excess chlorobenzene is used to obtain good yields of DDT and the molar ratio of chlorobenzene to chloral may vary from 2:1 up to 4.5:1. A ratio of from 2.36:1 to 3.2:1 is recommended, with from 2.9:1 to 3.0:1 being a particularly preferred ratio (4).

The formation of DDT is exothermic, releasing about 210 calories per gram when oleum is used as the condensing agent. Thus, the reactants are initially cooled to a temperature in the range of 0° to 30°C and cooling is continued to maintain the reaction temperature in this range. The production of DDT is carried out at atmospheric pressure.

The time required for oleum addition when oleum is used as the condensing agent in DDT manufacture may vary from 2 to 8 hours and in practice is more usually from 4 to 7 hours. The time varies, of course, with batch size, with oleum strength, and with the heat transfer characteristics of the system. An addition time of from 1 to 5 hours, and preferably 4 hours is specified by Miller and Lasco. After oleum addition, the reaction mass may be further agitated from 0.5 to 4.0 hours to effect completion of the reaction.

This process is carried out in the liquid phase with no added reaction solvents present; the mass consists simply of chloral, chlorobenzene, acid and product. Oleum may be used as the condensing agent as noted above. It is not really a catalyst, of course, but a condensing agent or aid to the reaction since typically 3.75 to 4.50 mols of oleum are used per mol of chloral. Chlorosulfonic acid and fluorosulfonic acid have been proposed as replacements for oleum in this process.

The batch preparation of DDT may be carried out in a closed reaction vessel equipped with an agitator and a jacket and/or internal cooling coils. An external heat exchanger circuit may also be used to provide effective heat transfer. Early DDT production reactions were of glass-lined construction but it has been found satisfactory to use steel vessels which are of course more economical.

A continuous system for DDT manufacture has been proposed as described by J.R. Callaham (5).

For economic DDT manufacture, the chloral feed may be dehydrated with spent acid from the process as described by Stange (6).

Figure 23 shows a production and waste schematic for DDT manufacture. Much of DDT manufacture is in fact product recovery. The actual condensation reaction may be considered as merely the first of five steps, the other four being separation, neutralization and purification, solidification, and acid recovery (4).

Process Wastes and Their Control

Air: Air emissions from DDT manufacture have been reported (B-15) to consist of 0.5 kg of DDT, 0.5 kg of chlorobenzene and 0.5 kg of chloral per metric ton of pesticide produced.

Water: The DDT process contains three waste recovery steps. One step removes the C_6H_5Cl from the separating and processing streams and recycles it to the reaction step. The second step first removes the acid from the separator stream. This acid is cycled through an acid recovery plant from which the recovered acid is recycled to the initial reaction step. The waste acid from the recovery plant is then combined with the dilute caustic in a neutralizer. The final step is the recovery of waste from clean-up operations (B-10).

Wastes resulting from the process include spent hydrochloric and sulfuric acids, monochloral benzene, sodium monochloral benzene sulfonate, chloral, NaOH caustic washwaters, chlorobenzene, sulfonic acid, and some product. The waste streams may contain DDT in the 1-5 ppm range with DDE and other related compounds present in amounts up to four times the DDT level. The pH of the waste is low and the salt content is high. At one plant, the primary waste stream containing waste acid and dilute caustic is neutralized and sent to a Class I dump (B-10).

A report by one manufacturer (8) gives results of a study both to test the feasibility of detoxification of DDT manufacturing wastes, using solvent extraction, and to develop a practical process, if possible. Three different liquid-liquid contacting devices were tested: all provided reasonably good extraction of DDT and homologs from the caustic aqueous phase. Unfortunately, major phase separation problems resulted in excessive losses of monochlorobenzene solvent to the aqueous phase. Efforts to improve coalescence/phase separation were unsuccessful. Further development of a solvent extraction process for detoxification of DDT manufacturing waste cannot be justified.

Product Wastes: DDT is chemically and biologically quite stable and its degradation poses a serious challenge. Specific methods which have been studied are the following (B-3):

> 1. *Reaction with Base*—At the time of its discovery in 1874 DDT was reported to undergo a dehydrochlorination reaction in strongly basic solution, i.e., caustic alkali in alcohol, to give a product now known as DDE.

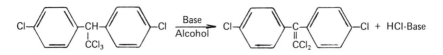

> In the late 1940s the dehydrochlorination was shown to be complete in 30 to 60 minutes at room temperature in 0.1N alkali solution and to occur also with organic bases such as amines. The dehydrochlorination reaction (frequently referred to incorrectly as a hydrolysis) continues to be mentioned as a degradation method, but it is not suitable as a disposal method: although

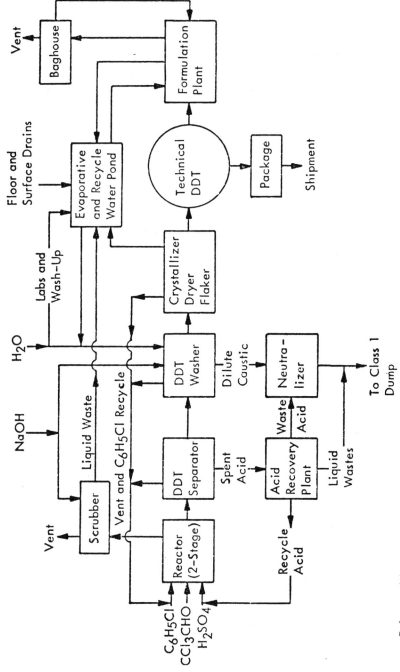

Figure 23: Production and Waste Schematic for DDT

Source: Reference (3)

the DDE is insecticidally nearly inert, it is persistent in the environment and·is suspected of being biomagnified and causing undesirable effects in fish and fowl.

2. *Oxidation*—DDT is quite resistant to oxidation and is not oxidized by chromic oxide in glacial acetic acid or by nitric acid, although the latter nitrates the aromatic rings.

3. *Reduction*—The reduction of DDT by active metals has been known for several years, but does not constitute a practical disposal method.

4. *Pyrolysis-Combustion-Incineration*—Incineration has been successfully used on a large scale for several years and huge incinerator equipment with scrubbers to catch HCl, a combustion product, are in use at several facilities such as Hooker Chemical, Dow Chemical and other producers of chlorinated hydrocarbon products. One incinerator operates at 900° to 1400°C with air and steam added which precludes formation of Cl_2. A few companies also construct incinerator-scrubber combinations of smaller size, e.g., a system built by Garver-Davis, Inc., of Cleveland, Ohio, for the Canadian government can handle 200 to 500 lb DDT/day as a kerosene solution.

5. *Biological Methods*—The biological degradation of DDT has received considerable interest, including numerous studies of the microbiological degradation under anaerobic and aerobic conditions. The consensus appears to be that anaerobic degradation is faster, but the most common product DDD, $(ClC_6H_1)_2CHCHCl_2$, is resistant to further anaerobic breakdown. Aerobic degradation, on the other hand, may cleave the aromatic rings and it is possible that a combination is effective in nature. The method is inappropriate as a disposal technique because of the low concentrations of DDT and carefully controlled conditions that appear to be required and the slow degradation rates.

Toxicity

The acute oral LD_{50} value for rats is about 115 mg/kg (moderately toxic).

DDT is also of moderate acute toxicity to man and most other organisms. However, its extremely low solubility in water (0.0012 ppm) and high solubility in fat (100,000 ppm) result in great bioconcentration. Its principal breakdown product, DDE, has very similar properties. Both compounds are also highly persistent in living organisms, so the major concern about DDT toxicity is related to its chronic effects.

In light of the above and taking into account the carcinogenic risk projections, it was suggested (B-22) that very strict criteria be applied when limits for DDT and DDE in drinking-water are established.

Allowable Limits on Exposure and Use

Air: The threshold limit value for DDT in air is 1 mg/m^3 as of 1979. The tentative short term exposure limit is 3 mg/m^3 (B-23).

NIOSH has recommended as of October 1978 that DDT be held to the lowest reliably detectable level which is 0.5 mg/m^3 on a time weighted average basis. They further noted that skin contact was to be avoided.

Water: In water, EPA set (B-12) a criterion of 0.001 μg/l for the protection of both freshwater and marine aquatic life. They go on to caution that persistence, bioaccumulation and carcinogenicity require that human exposure be held to a minimum.

In water, EPA has subsequently suggested (B-26) a limit to protect freshwater aquatic life of 0.00023 μg/l as a 24-hour average and that the concentration should not exceed 0.41 μg/l at any time.

For DDT and metabolites the criterion to protect saltwater aquatic life is 0.0067 μg/l as a 24-hour average and the concentration should not exceed 0.021 μg/l at any time.

For the maximum protection of human health from the potential carcinogenic effects of exposure to DDT through ingestion of water and contaminated aquatic organisms, the ambient water concentration is zero. Concentrations of DDT estimated to result in additional lifetime cancer risks ranging from no additional risk to an additional risk of 1 in 100,000 have been developed by US EPA. The agency is considering setting criteria at an interim target risk level in the range of 10^{-5}, 10^{-6}, or 10^{-7} with corresponding criteria of 0.98 ng/l, 0.098 ng/l, and 0.0098 ng/l, respectively. If water alone is consumed, the water concentration should be less than 0.36 μg/l to keep the lifetime cancer risk below 10^{-5} (B-26).

Product Use: In actions dated January 15, 1971 and July 7, 1972, the US EPA (B-17) cancelled all uses of DDT products, except the following lists of uses:

1. The U.S. Public Health Service and other Health Service Officials for control of vector diseases.

2. The USDA or military for health quarantine.

3. In drugs, for controlling body lice. (To be dispensed only by a physician.)

4. In the formulation for prescription drugs for controlling body lice.

The tolerances set by the US EPA for DDT in or on raw agricultural commodities are as follows:

	40 *CFR* Reference	Parts per Million
Apples	180.147	0.5
Apricots	180.147	0.5
Artichokes	180.147	1.0
Asparagus	180.147	1.0
Avocados	180.147	3.5
Beans (exc. dry)	180.147	7.0
Beans, dry	180.147	0.5
Beets (roots and tops)	180.147	1.0
Blackberries	180.147	0.5
Blueberries (huckleberries)	180.147	0.5
Boysenberries	180.147	0.5
Broccoli	180.147	1.0
Brussels sprouts	180.147	1.0
Cabbage	180.147	1.0
Carrots	180.147	3.5
Cattle, fat of meat	180.147	5.0

(continued)

	40 *CFR* Reference	Parts per Million
Cauliflower	180.147	1.0
Celery	180.147	1.0
Cherries	180.147	0.5
Citrus fruits	180.147	3.5
Collards	180.147	1.0
Corn, sweet (k+cwhr) (not for feed)	180.147	3.5
Cotton, seed	180.147	4.0
Cranberries	180.147	7.0
Cucumbers	180.147	0.5
Currants	180.147	0.5
Dewberries	180.147	0.5
Eggplant	180.147	0.5
Endive (escarole)	180.147	1.0
Goats, fat of meat	180.147	5.0
Gooseberries	180.147	0.5
Grapes	180.147	7.0
Guavas	180.147	0.5
Hogs, fat of meat	180.147	5.0
Hops, fresh	180.147	20.0
Horses, fat of meat	180.147	5.0
Kale	180.147	1.0
Kohlrabi	180.147	1.0
Lettuce	180.147	7.0
Loganberries	180.147	0.5
Mangoes	180.147	0.5
Melons	180.147	0.5
Milk	180.147	0.05
Mushrooms	180.147	1.0
Mustard, greens	180.147	1.0
Nectarines	180.147	0.5
Okra	180.147	1.0
Onion, dry bulb only	180.147	1.0
Papayas	180.147	3.5
Parsnips (roots and tops)	180.147	1.0
Peaches	180.147	0.5
Peanuts	180.147	0.5
Pears	180.147	0.5
Peas	180.147	0.5
Peppermint, hay	180.147	50.0
Peppers	180.147	7.0
Pineapples	180.147	7.0
Plums (fresh prunes)	180.147	0.5
Potatoes (soilless)	180.147	1.0
Pumpkins	180.147	0.5
Quinces	180.147	0.5
Radishes (roots and tops)	180.147	1.0
Raspberries	180.147	0.5
Rutabagas (roots and tops)	180.147	1.0
Sheep, fat of meat	180.147	5.0
Soybeans (dry form)	180.147	1.5
Spearmint, hay	180.147	50.0
Spinach	180.147	1.0
Squash	180.147	0.5
Squash, summer	180.147	0.5
Strawberries	180.147	0.5
Sweet potatoes	180.147	1.0
Swiss chard	180.147	1.0
Tomatoes	180.147	7.0
Turnips (roots and tops)	180.147	1.0
Youngberries	180.147	0.5

The tolerances set by the US EPA for DDT in animal feeds are as follows (the *CFR* Reference is to Title 21):

	CFR Reference	Parts per Million
Tomato pomace, dried (dog & cat food)	561.120	100.0

The tolerances set by the US EPA for DDT in food are as follows (the *CFR* Reference is to Title 21):

	CFR Reference	Parts per Million
Dairy prods. mfgd	193.120	1.25
Hops, dried	193.110	80.0
Peppermint oil	193.110	100.0
Soybean, crude oil	193.110	6.0
Spearmint oil	193.110	100.0

References

(1) Worthing, C.R., *Pesticide Manual,* 6th ed., p. 150, British Crop Protection Council (1979).
(2) Spencer, E.Y., *Guide to the Chemicals Used in Crop Protection,* 6th ed., p. 156, London, Ontario, Agriculture Canada (January 1973).
(3) Midwest Research Institute, *The Pollution Potential in Pesticide Manufacturing,* Wash., DC, Environmental Protection Agency (June 1972).
(4) Miller, G.A. and Lasco, R.H., U.S. Patent 2,932,672, April 12, 1960, assigned to Diamond Alkali Company.
(5) Callaham, J.R., *Chem. Met. Eng.* 51, No. 10, 109-114 (1944).
(6) Stange, H., U.S. Patent 2,763,698, September 18, 1956, assigned to Olin Mathieson Chemical Corporation.
(7) Muller, P., U.S. Patent 2,329,074, September 7, 1943, assigned to J.R. Geigy AG.
(8) Sobleman, M., Sweeney, K.H. and Calimag, E.D. (Montrose Chemical Corp. of Calif.), *Development of a DDT Manufacturing and Processing Plant Waste Treatment System,* Report EPA-600/2-78-125, Wash., DC, U.S. Environmental Protection Agency (June 1978).

DEHYDROACETIC ACID

Function: Fungicide (1)(2)

Chemical Name: 3-acetyl-6-methyl-2H-pyran-2,4(3H)-dione

Formula:

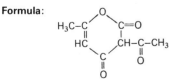

Trade Name: DHA

Manufacture

Dehydroacetic acid is made by refluxing ethyl acetoacetate with sodium bicarbonate. It can also be made by the polymerization of diketene.

Toxicity

The acute oral LD_{50} for rats is 1,000 mg/kg (slightly toxic).

References

(1) Worthing, C.R., *Pesticide Manual,* 6th ed., p. 152, British Crop Protection Council (1979).
(2) Spencer, E.Y., *Guide to the Chemicals Used in Crop Protection,* 6th ed., p. 160, London, Ontario, Agriculture Canada (January 1973).

DEMEPHION

Function: Systemic insecticide and acaricide (1)

Chemical Name: O,O-dimethyl O-2-methylthioethyl phosphorothioate and
O,O-dimethyl S-2-methylthioethyl phosphorothioate

Formula: $(CH_3O)_2POCH_2CH_2SCH_3$ and $(CH_3O)_2PSCH_2CH_2SCH_3$
         $\overset{\|}{S}$ $\overset{\|}{O}$

Trade Names: Cymetox® (Cyanamid)
        Atlasetox® (Atlas Products & Services, Ltd.)
        Pyracide® (BASF)
        Tinox® also known as Methyl-Demeton-Methyl

Manufacture (2)

Demephion is made by the stepwise esterification and transesterification of thiophosphoryl chloride; first with methanol to give $(CH_3O)_2P(S)OCl$, then with beta-hydroxyethyl methyl sulfide (made from methyl mercaptan and ethylene oxide) to give demephion.

Toxicity

The acute oral LD_{50} for rats is 15 mg/kg (B-15) which is highly toxic.

References

(1) Worthing, C.R., *Pesticide Manual,* 6th ed., p. 153, British Crop Protection Council (1979).
(2) Fabrenfabriken Bayer AG, British Patent 814,332 (June 3, 1959).

DEMETON

Function: Systemic insecticide and acaricide (1)(2)

Chemical Name: O,O-diethyl O-2-ethylthioethyl phosphorothioate and
O,O-diethyl S-2-ethylthioethyl phosphorothioate

Formula: $(C_2H_5O)_2POCH_2CH_2SC_2H_5$ and $(C_2H_5O)_2PSCH_2CH_2SC_2H_5$
         $\overset{\|}{S}$ $\overset{\|}{O}$

Trade Name: Systox® (Bayer)

Manufacture (3)

Beta-hydroxyethyl ethyl sulfide may be made as the first reactant by reaction of ethyl mercaptan and ethylene oxide according to the following reaction.

$$C_2H_5SH \ + \ \overset{\displaystyle O}{\overset{\diagdown}{CH_2\text{--}CH_2}} \ \longrightarrow \ C_2H_5SCH_2CH_2OH$$

The feed materials for demeton production are O,O-diethyl phosphorochloridothionate and beta-hydroxyethyl ethyl sulfide which has been reacted with sodium hydroxide.

$$\overset{\displaystyle S}{\overset{\|}{(C_2H_5O)_2PCl}} \ + \ NaOCH_2CH_2SC_2H_5 \ \longrightarrow \ \overset{\displaystyle S}{\overset{\|}{(C_2H_5O)_2POCH_2CH_2SC_2H_5}} \ + \ NaCl$$

The production of demeton may be carried out at room temperature or at temperatures up to 90° or 100°C. A typical operating temperature might be 40°C, and 40° to 60°C might be a preferred operating range. The reaction is carried out at essentially atmospheric pressure. A typical reaction period is 2 hours for this process.

The reaction is preferably carried out in an inert diluent. Suitable diluents include benzene, chlorobenzene, acetone and methyl ethyl ketone. The reaction can be accelerated by the addition of metallic copper (3). A stirred, jacketed kettle of conventional design may be used for the conduct of this reaction.

After completion of the reaction, water is added to dissolve the salt formed. The aqueous layer is removed and the organic (product plus reaction solvent) layer is fractionated under reduced pressure to give a 65 to 75% yield of product. An alternative route (4) involves the reaction of alkyl thiocyanates with alkali salts of dialkyl phosphites.

A specific example of the process relevant to the production of demeton is as follows. Seven grams of sodium are suspended in 150 cc of toluene. Forty-two grams of diethyl phosphite are added and the mixture is stirred at 40° to 50°C until all sodium is dissolved. With stirring, 44 grams of β-ethylmercapto-ethyl thiocyanate (boiling at 90°C at 1 mm pressure) are added drop by drop at a temperature of 40°C. The mixture is kept for one hour at 50°C and is then worked up in the usual manner. Thus, 28 grams of

$$(C_2H_5O)_2PO\text{--}S\text{--}C_2H_4\text{--}S\text{--}C_2H_5$$

boiling at 137° to 141°C at 1 mm pressure are obtained. The product is a light yellow oil which is insoluble in water.

Process Wastes and Their Control

Product Wastes: The thiono-and thiolo-isomers of this mixture are 50% hydrolyzed in 75 minutes and 0.85 minute, respectively at 20°C and pH 13. At pH 9 and 70°C, the half life of demeton is 1.25 hours, but at pH 1-5 it is over 11 hours (B-3).

Toxicity

The acute oral LD_{50} value for rats is 1.7 mg/kg (B-5) which is extremely toxic.

Allowable Limits on Exposure and Use

Air: The threshold limit value for demeton in air is 0.1 mg/m^3 (0.01 ppm) as of 1979. The tentative short term exposure limit is 0.3 mg/m³ (0.03 ppm) (B-23). These limits bear the notation "skin," indicating that cutaneous absorption should be prevented so that the threshold limit value is not invalidated.

Water: In water, EPA has set (B-12) a criterion of 0.1 μg/l for the protection of freshwater and marine aquatic life.

Product Use: The tolerances set by the US EPA for demeton in or on raw agricultural commodities are as follows:

	40 *CFR* Reference	Parts per Million
Alfalfa, fresh	180.105	5.0
Alfalfa, hay	180.105	12.0
Almonds	180.105	0.75
Almonds, hulls	180.105	5.0
Apples	180.105	0.75
Apricots	180.105	0.75
Barley, fodder, green	180.105	5.0
Barley, grain	180.105	0.75
Barley, Straw	180.105	5.0
Beans	180.105	0.3
Beets, sugar	180.105	0.5
Beets, sugar, tops	180.105	5.0
Broccoli	180.105	0.75
Brussels sprouts	180.105	0.75
Cabbage	180.105	0.75
Cauliflower	180.105	0.75
Celery	180.105	0.75
Clover, fresh	180.105	5.0
Clover, hay	180.105	12.0
Cotton, seed	180.105	0.75
Eggplant	180.105	0.3
Filberts	180.105	0.75
Grapefruit	180.105	0.75
Grapes	180.105	1.25
Hops	180.105	1.25
Lemons	180.105	0.75
Lettuce	180.105	0.75
Muskmelons	180.105	0.75
Nectarines	180.105	0.75
Oats, fodder, green	180.105	5.0
Oats, grain	180.105	0.75
Oats, straw	180.105	5.0
Oranges	180.105	0.75
Peaches	180.105	0.75
Pears	180.105	0.75
Peas	180.105	0.75
Pecans	180.105	0.75
Peppers	180.105	0.75
Plums (fresh prunes)	180.105	0.75
Potatoes	180.105	0.75
Sorghum, forage	180.105	0.2
Sorghum, grain	180.105	0.2
Strawberries	180.105	0.75
Tomatoes	180.105	0.75
Walnuts	180.105	0.75
Wheat, fodder, green	180.105	5.0
Wheat, grain	180.105	0.75
Wheat, straw	180.105	5.0

The tolerances set by the US EPA for demeton in animal feeds are as follows (the *CFR* Reference is to Title 21):

	CFR Reference	Parts per Million
Sugar beet pulp, dehydrated (ct.f)	561.130	5.0

References

(1) Worthing, C.R., *Pesticide Manual,* 6th ed., p. 154, British Crop Protection Council (1979).
(2) Spencer, E.Y., *Guide to the Chemicals Used in Crop Protection,* 6th ed., p. 161, London, Ontario, Agriculture Canada (January 1973).
(3) Schrader, G., U.S. Patent 2,571,989, October 16, 1951, assigned to Farbenfabriken Bayer AG.
(4) Schrader, G., U.S. Patent 2,597,534, May 20, 1952, assigned to Farbenfabriken Bayer AG.

DEMETON-S-METHYL

Function: Systemic insecticide and acaricide

Chemical Name: O,O-dimethyl S-2-ethylthioethyl phosphorothioate

Formula: $(CH_3O)_2 \overset{\text{O}}{\underset{\|}{P}} SCH_2CH_2SC_2H_5$

Trade Names: Bayer 18436
Bayer 25/154
Meta-Systox® (Bayer)

Manufacture

Demeton-S-methyl is made by the reaction of beta-hydroxyethyl ethyl sulfide (made in turn from ethyl mercaptan and ethylene oxide). It is reacted with O,O-dimethyl hydrogen phosphorothioate to give demeton-S-methyl. See demeton for similar synthesis details.

Process Wastes and Their Control

Product Wastes: Demeton-S-methyl is less stable than demeton. Fifty percent hydrolysis at 70°C requires 1.25 hours at pH 9 and 4.9 hours at pH 3 (B-3).

Toxicity

The acute oral LD_{50} for rats is 60 mg/kg (moderately toxic).

Allowable Limits on Exposure and Use

Air: The threshold limit value for demeton-S-methyl is 0.5 mg/m³ as of 1979. The tentative short term exposure limit is 1.5 mg/m³. Further, it is noted that cutaneous absorption should be prevented so that the TLV is not invalidated.

Product Use: The tolerances set by the US EPA for demeton-s-methyl in animal feeds are as follows (the *CFR* Reference is to Title 21):

	CFR Reference	Parts per Million
Apple pomace, dried	561.80	25.0
Cotton, seed, hulls	561.80	10.0

References

(1) Worthing, C.R., *Pesticide Manual,* 6th ed., p. 155, British Crop Protection Council (1979).
(2) Spencer, E.Y., *Guide to the Chemicals Used in Crop Protection,* 6th ed., pp. 163-64, London, Ontario, Agriculture Canada (January 1973).

DEMETON-S-METHYL SULFONE

Function: Systemic insecticide

Chemical Name: S-2-ethylsulfonylethyl O,O-dimethyl phosphorothioate

Formula:

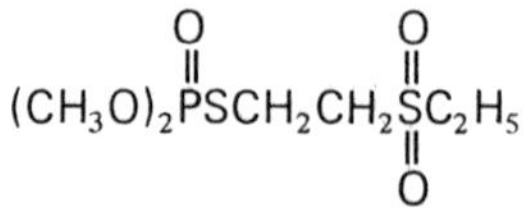

$$(CH_3O)_2\overset{\overset{O}{\|}}{P}SCH_2CH_2\overset{\overset{O}{\|}}{\underset{\underset{O}{\|}}{S}}C_2H_5$$

Trade Names: Bayer 20315 (Bayer)
Bayer E158 (Bayer)
Bayer M3/158 (Bayer)
Metaisosystox-Sulfon® (Bayer)

Manufacture

Demeton-S-methyl sulfone is made by the oxidation of demeton-S-methyl with $KMnO_4$.

Toxicity

The acute oral LD_{50} value for rats is 37.5 mg/kg (highly toxic).

References

(1) Worthing, C.R., *Pesticide Manual,* 6th ed., p. 156, British Crop Protection Council (1979).
(2) Spencer, E.Y., *Guide to the Chemicals Used in Crop Protection,* 6th ed., p. 165, London, Ontario, Agriculture Canada (January 1973).

DESMEDIPHAM

Function: Postemergence herbicide (1)(2)

Chemical Name: Ethyl (3-[[(phenylamino)carbonyl]oxy])phenyl) carbamate

Formula:

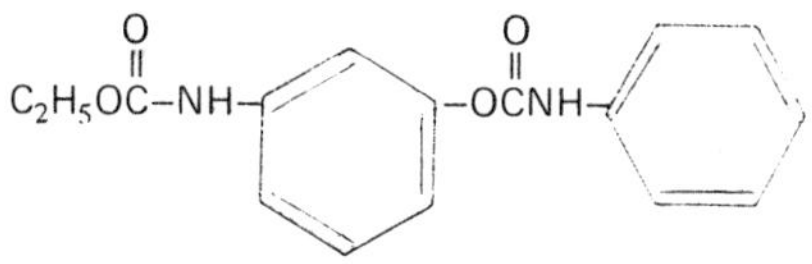

Trade Names: Schering 38107 (Schering AG)
Betanal AM® (Schering AG)

Manufacture (2)

Desmedipham is made by the reaction of phenyl isocyanate with 3-ethoxycarbonylamino-phenol. m-Aminophenol is the initial starting material which is reacted with ethyl chloro-formate to give the ethoxycarbonylaminophenol.

The preparation of various related m-(carbamoyloxy) carbanilates has been described in some detail (3).

Toxicity

The acute oral LD_{50} for rats is nearly 10,000 mg/kg which is insignificantly toxic.

Allowable Limits on Exposure and Use

Product Use: The tolerances set by the US EPA for desmedipham in or on raw agricultural commodities are as follows:

	40 *CFR* Reference	**Parts per Million**
Beets, sugar, tops	180.353	0.2 N

References

(1) Worthing, C.R., *Pesticide Manual,* 6th ed., p. 157, British Crop Protection Council (1979).
(2) Schering AG, British Patent 1,127,050, September 11, 1968.
(3) Wilson, K.R. and Hill, K.L., U.S. Patent 3,404,975, October 8, 1968, assigned to FMC Corp.

DESMETRYNE

Function: Postemergence herbicide (1)(2)

Chemical Name: N-methyl-N'-(1-methylethyl)-6-(methylthio)-1,3,5-triazine-2,4-diamine

Formula:

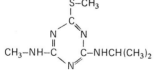

Trade Names: G-34360 (Ciba-Geigy)
Semeron® (Ciba-Geigy)

Manufacture

Desmetryne is made by the reaction of cyanuric chloride in turn with equivalents of iso-pro-pylamine, methylamine and methyl mercaptan. The preparation and more particularly the formulation of this material have been described (3).

Toxicity

The acute oral LD_{50} value for rats is about 1,390 mg/kg (slightly toxic).

References

(1) Worthing, C.R., *Pesticide Manual,* 6th ed., p. 158, British Crop Protection Council (1979).
(2) Spencer, E.Y., *Guide to the Chemicals Used in Crop Protection,* 6th ed., p. 168, London, Ontario, Agriculture Canada (January 1973).
(3) J.R. Geigy AG, British Patent 814,948 (June 17, 1959).

DIALIFOS (DIALIFOR)

Function: Insecticide and acaricide (1)(2)

Chemical Name: S-[2-chloro-1-(1,3-dihydro-1,3-dioxy-2H-isoindol-2-yl)ethyl] O,O-diethyl phosphorodithioate

Formula:

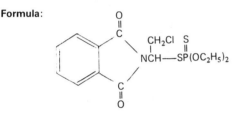

Trade Names: Hercules 14503® (Hercules)
Torak® (Hercules)

Manufacture (3)

N-vinyl phthalimide is reacted with chlorine in an initial step to give N-(1,2-dichloroethyl) phthalimide. To a solution of 61 parts of N-(1,2-dichloroethyl) phthalimide in 390 parts acetonitrile was added 56 parts ammonium diethyldithiophosphate while stirring at about 25°C. After the initial reaction appeared to be complete, the mixture was heated to 50°C for 2 hours, and it was then cooled and filtered.

The acetonitrile was removed, and benzene was used in its place. The benzene solution was washed neutral and free of water-soluble materials. The benzene was evaporated to obtain 82 parts of O,O-diethyl S-[2-chloro-1-(N-phthalimido)ethyl] phosphorodithioate as a viscous oil, a portion of which, after crystallization from a toluene-hexane mixture, melted at 62° to 64°C.

Process Wastes and Their Control

Product Wastes: Dialifos is reported to be readily hydrolyzed by strong alkali (B-3).

Toxicity

The acute oral LD_{50} value for rats is 5 mg/kg (highly toxic).

Allowable Limits on Exposure and Use

Product Use: The tolerances set by the US EPA for dialifos in or on raw agricultural commodities are as follows:

	40 CFR Reference	Parts per Million
Apples	180.326	1.5
Beans, forage	180.327	0.05 N
Cattle, fat	180.326	0.15 N
Cattle, mbyp	180.326	0.15 N
Cattle, meat	180.326	0.15 N
Citrus fruits	180.326	3.0
Eggs	180.326	0.01 N
Goats, fat	180.326	0.15 N
Goats, mbyp	180.326	0.15 N
Goats, meat	180.326	0.15 N
Grapes	180.326	1.0
Milk, fat (= N in whole milk)	180.326	0.15 N
Nuts, pecans	180.326	0.01 N
Poultry, fat	180.326	0.05 N
Poultry, mbyp	180.326	0.05 N
Poultry, meat	180.326	0.05 N
Sheep, fat	180.326	0.15 N
Sheep, mbyp	180.326	0.15 N
Sheep, meat	180.326	0.15 N

The tolerances set by the US EPA for dialifos in animal feeds are as follows (the CFR Reference is to Title 21):

	CFR Reference	Parts per Million
Apple pomace, dried	561.140	40.0
Citrus, pulp, dried	561.140	15.0
Grape pomace, dried	561.140	20.0
Raisin waste	561.140	10.0

The tolerances set by the US EPA for dialifos in food are as follows (the CFR Reference is to Title 21):

	CFR Reference	Parts per Million
Raisins	193.130	2.0

References

(1) Worthing, C.R., *Pesticide Manual,* 6th ed., p. 160, British Crop Protection Council (1979).
(2) Spencer, E.Y., *Guide to the Chemicals Used in Crop Protection,* 6th ed., p. 169, London, Ontario, Agriculture Canada (January 1973).
(3) Jamison, J.D., U.S. Patent 3,355,353, November 23, 1967, assigned to Hercules, Inc.

DIALLATE

Function: Preemergence herbicide (1)(2)

Chemical Name: S-(2,3-dichloro-2-propenyl) bis(1-methylethyl)carbamothioate

Formula:

$$[(CH_3)_2CH]_2 N\overset{\displaystyle O}{\overset{\displaystyle \|}{C}}SCH_2CCl\!=\!CHCl$$

Trade Names: CP 15336 (Monsanto)
Avadex® (Monsanto)
DATC
DDTC
2,3-DCDT

Manufacture (3)(4)(5)

To a stirred solution of 202.4 grams (2.0 mols) of diisopropylamine in 1,000 ml of dry ethyl ether at −10° to 0°C there was bubbled in carbonoxysulfide until the gain in weight was 120 grams. This addition required 30 minutes and the mixture was then stirred at −10° to 0°C for an additional 90 minutes. Thereupon 145.4 grams (1.0 mol) of 1,2,3-trichloropropene was added in one portion and the mixture stirred at 25° to 30°C for 24 hours. The by-product salt was removed by filtration and the excess ether removed in vacuo. The residue was distilled in vacuo and the fraction boiling at 145° to 159°C at 9 mm collected. The cis- and trans-2,3-dichloroallyl diisopropylthiocarbamate was an amber liquid analyzing 11.83% sulfur as compared to 11.87% calculated for $C_{10}H_{17}Cl_2NOS$.

The same product was readily prepared in aqueous medium by bubbling 76.5 grams (1.1 mols) of 85% carbonoxysulfide into a mixture of 105 grams of 97% diisopropylamine, 200 ml of water and 160 grams of 25% sodium hydroxide at 2° to 6°C. There was then added 145.5 grams of 1,2,3-trichloropropene in one portion. The temperature was kept below 12°C for about an hour by cooling with an ice bath. The ice bath was then removed. The temperature rose slowly to 30°C in 4 hours, 400 ml of water was added, the organic layer separated, washed with water and stripped at 95° to 100°C at 14 mm. The residue was then distilled, collecting the fraction boiling at 118° to 119°C at 1 mm.

Toxicity

The acute oral LD_{50} value for rats is 395 mg/kg (moderately toxic).

Allowable Limits on Exposure and Use

Product Use: A rebuttable presumption against registration was issued on May 31, 1977 by US EPA on the basis of oncogenicity.

The tolerances set by the US EPA for diallate in or on raw agricultural commodities are as follows:

	40 *CFR* Reference	Parts per Million
Alfalfa, fresh	180.277	0.05 N
Alfalfa, hay	180.277	0.05 N
Barley, forage	180.277	0.05 N
Barley, grain	180.277	0.05 N
Barley, straw	180.277	0.05 N
Beets, sugar, roots	180.277	0.05 N
Beets, sugar, tops	180.277	0.05 N
Clover, fresh	180.277	0.05 N
Clover, hay	180.277	0.05 N
Corn, field, fodder	180.277	0.05 N
Corn, field, forage	180.277	0.05 N
Corn, field, grain	180.277	0.05 N
Flax, seed	180.277	0.05 N
Lentils	180.277	0.05 N
Peas	180.277	0.05 N
Peas, forage	180.277	0.05 N

(continued)

	40 CFR Reference	Parts per Million
Peas, hay	180.277	0.05 N
Potatoes	180.277	0.05 N
Safflower, seed	180.277	0.05 N
Soybeans	180.277	0.05 N
Soybeans, forage	180.277	0.05 N
Soybeans, hay	180.277	0.05 N

References

(1) Worthing, C.R., *Pesticide Manual,* 6th ed., p. 161, British Crop Protection Council (1979).
(2) Spencer, E.Y., *Guide to the Chemicals Used in Crop Protection,* 6th ed., p. 170, London, Ontario, Agriculture Canada (January 1973).
(3) Harman, M.W. and D'Amico, J.J., U.S. Patent 3,330,821, July 11, 1967, assigned to Monsanto Company.
(4) Monsanto Chemical Co., British Patent 882,111 (November 15, 1961).
(5) Harman, M.W. and D'Amico, J.J., U.S. Patent 3,330,643, July 11, 1967, assigned to Monsanto Company.

N,N-DIALLYLDICHLOROACETAMIDE

Function: Herbicide enhancer (1)

Chemical Name: 2,2-dichloro-N,N-di-2-propenyl acetamide

Formula:

$$Cl_2CHCN(CH_2CH{=}CH_2)_2$$
$$\overset{O}{\overset{\|}{}}$$

Trade Name: R25788 (Stauffer)

Manufacture

N-diallyldichloroacetamide is made by the reaction of dichloroacetylchloride with diallyl-amine in dichloromethane as a solvent.

Toxicity

The acute oral LD_{50} value for rats is 2,000 mg/kg (slightly toxic).

Reference

(1) Worthing, C.R., *Pesticide Manual,* 6th ed., p. 162, British Crop Protection Council (1979).

DIAMIDAFOS

Function: Nematocide (1)(2)

Chemical Name: Phenyl N,N'-dimethylphosphorodiamidate

Formula:

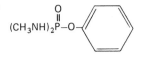

$$(CH_3NH)_2P-O-$$

Trade Names: Dowco-169 (Dow Chemical Company)
Nellite® (Dow Chemical Company)

Manufacture (2)

This compound may be prepared by reacting one mol of O-phenylphosphorochloridate with at least 4 mols of methylamine. The reaction is carried out in an inert reaction medium such as benzene and takes place at temperatures of from 0° to 40°C with the production of the desired product and amine hydrochloride of reaction. Upon completion of the reaction, the reaction mixture is filtered to separate the hydrochloride and the solvent removed from the filtrate by distillation to obtain the desired product as a residue.

Toxicity

The acute oral LD_{50} value for rats is 200 mg/kg which is moderately toxic.

References

(1) Martin, H. and Worthing, C.R., *Pesticide Manual,* 5th ed., p. 413, British Crop Protection Council (January 1977).
(2) Youngson, C.R., U.S. Patent 3,005,749, October 24, 1961, assigned to Dow Chemical Co.

DIAZINON

Function: Insecticide (1)(2)

Chemical Name: O,O-diethyl O-[6-methyl-2-(1-methylethyl)-4-pyrimidinyl] phosphorothioate

Formula:

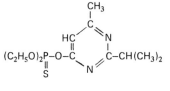

Trade Names: G-24480 (Ciba-Geigy)
Basudin® (Ciba-Geigy)
Neocidoc® (Ciba-Geigy)
Nucidol® (Ciba-Geigy)
Diazitol®
Sarolex®
Spectracide®

Manufacture (3)(4)

The reaction chemistry for the production of diazinon is believed to be as follows (3).

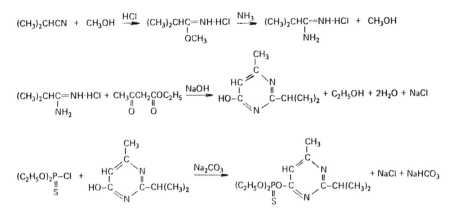

Figure 24 is a production and waste schematic flow diagram for diazinon manufacture. In the first step of the production process, isobutyronitrile is reacted with methanol and HCl to produce the corresponding imido ester hydrochloride. There are no by-products to this reaction except excess hydrochloric acid which is recycled. In this reaction, the alcohol acts as a shielding agent.

In the second step, the ester is treated with ammonia to form the corresponding amidine. There are also no by-products to this step; the methyl alcohol is recovered by distillation and recycled.

The next step is the most critical and most difficult step in the production of diazinon. The amidine is condensed with acetoacetic ester in a reaction which is very sensitive to control. It is relatively easy to obtain 70% yields, but, in order for the process to be economically successful, 95% yields must be obtained. The resulting pyrimidine is insoluble in the reaction mixture (the sodium salt, however, is very soluble) and it is separated along with small amounts of other insoluble inorganic compounds (this separation does not generate filter solids which would require disposal).

The aqueous stream plus alcohols and other organic products (including acetone) are subjected to a stripping process in which pure materials are separated and recycled. The residual aqueous solution is transferred to a biological sewage treatment plant and the residual materials from this process are incinerated. The pyrimidine product is subjected to a drying step.

The next step is the reaction of the pyrimidine with diethylthiophosphoryl chloride. The manufacture of this ester causes the most notorious waste treatment problems in the production of phosphorus-sulfur pesticides of this kind. The reaction is reported to proceed in 85% yield. There are toxic by-products formed during this reaction. These are the combined esters related to TEPP. Some of the TEPP by-products contain sulfur. These materials are removed from the reaction mixture, evidently by a filtration step, and are decontaminated by a treatment with acid. The reaction mixture is treated with sodium carbonate. The resulting product is sodium bicarbonate, which, along with sodium chloride, is also removed in the filtration step. The filtration results in the removal of a haze which is formed in the brown reaction mixture. It is believed that the brown material in the diazinon solution is a sulfur polymer. Diazinon itself is water white and almost odorless.

There is no air pollution problem in the production of diazinon. However, carbon dioxide is evolved during the pyrimidine-phosphoryl chloride operation (caused by the addition of sodium carbonate to the acidic reaction mixture). The equipment can be run either continuously or batchwise.

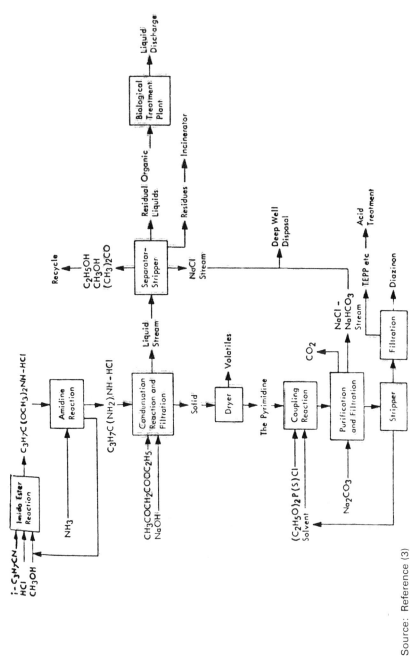

Figure 24: Production and Waste Schematic for Diazinon

Process Wastes and Their Control

Air: Air emissions from diazinon manufacture have been reported (B-15) to consist of 205 kg of SO_2 and 0.5 kg of diazinon per metric ton of pesticide produced.

Water: The two major waste streams from this process are at the filtration steps. At one plant, the first waste stream containing NaCl and $NaHCO_3$ is deep well injected. The waste stream from final filtration contains other organophosphates and is further processed (B-10).

Figure 25 shows the wastewater treatment system for diazinon production. The wastewater from the diazinon production unit enters the acid destruct holding pond. This wastewater stream does not include cooling water from the diazinon production and therefore is not diluted. The total flow for the 24-hour sampling period at the pretreatment sample point (01 on Figure 25) was 91,560 gallons at a pH of 10.6. The acid destruct pond was at pH 1.0. As indicated in Figure 25, an additional wastewater stream labeled "Acid Waste" (cyanuric chloride) also enters the acid destruct holding pond. The pretreatment sample was taken from an in-line flow dependent composite sampler used by the production site for their own samples.

The diazinon wastewater has a several day hold-up in the acid destruct pond. Diazinon is relatively unstable (T_{50} at pH 3.1 is 12 hours) to acid hydrolysis when compared to other dialkyl phosphorothioates. The acid treatment is intended to be the primary destruction method for diazinon. The acid hydrolysis of diazinon minimizes the formation of diazoxon.

The effluent from the acid destruct holding pond was sampled at Location 02 (Figure 25). The total flow for the sampling period was 250,000 gallons. This resulted in about a threefold dilution of the toxic waste (diazinon) by the acid waste stream.

Cyanide, when present, is removed by caustic chlorination, the pH is adjusted to 6.5, and wastewater from other production units is mixed before the combined waste stream enters an aerobic biological treatment system. The water is then clarified and pH adjusted before river discharge. The posttreatment sample (03) was taken at this location. The 24-hour composite sample was taken from an in-line flow proportional sampler. The flow volume for the final effluent during 24-hour sampling was 1.4 million gallons. This implies a 15-fold dilution of the toxic waste (diazinon) stream.

The waste treatment system for diazinon wastewater is designed to remove material by the following method: acid hydrolysis, neutral hydrolysis, biological degradation, adsorption onto biomass, and vaporization. Additionally, compounds detected at low levels in the pretreatment (01) sample may be below the detection limits in the final effluent due to the dilution resulting from additional wastewater inputs into the treatment system.

Product Wastes: Diazinon is hydrolyzed in acid media about 12 times as rapidly as parathion, and at about the same rate as parathion in alkaline media. In excess water this compound yields diethylthiophosphoric acid and 2-isopropyl-4-methyl-6-hydroxypyrimidine. With insufficient water, highly toxic tetraethyl monothiopyrophosphate is formed (B-3).

Toxicity

The acute oral LD_{50} value for rats is 300-850 mg/kg (1) which is moderately toxic.

Diazinon is a widely used organophosphorus insecticide with applications in agriculture, homes and gardens, and structural pest control. It may well be used in situations that would lead to contamination of drinking-water. The mode of action of diazinon, as with other organophosphorus insecticides, is inhibition of the enzyme cholinesterase. Its acute toxicity, however, in comparison with other organophosphates, is only moderate. Its metabolism is straightforward and leads to metabolites that have little toxic potential. Subchronic- and

Figure 25: Wastewater Treatment Sampling Points for Diazinon Production

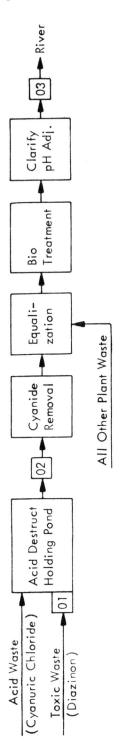

chronic-feeding studies are sufficiently complete and indicate little problem with the use of diazinon. An ADI was calculated at 0.002 mg/kg/day based on these data.

The data needed for the toxicologic evaluation of diazinon are fairly complete, and there is no pressing need for research to evaluate its safety. There is little information available on the actual presence or absence of diazinon in drinking-water or in sources of drinking-water. Studies on the environmental transport and persistence of diazinon would be useful in this respect (B-22).

Allowable Limits on Exposure and Use

Air: The threshold limit value for diazinon in air is 0.1 mg/m^3 as of 1979. The tentative short term exposure limit is 0.3 mg/m^3. Both limits bear the notation "skin," indicating that cutaneous absorption should be prevented so that the threshold limit value is not invalidated.

Product Use: The tolerances set by the US EPA for diazinon in or on raw agricultural commodities are as follows:

	40 *CFR* Reference	Parts per Million
Alfalfa, fresh	180.153	40.0
Alfalfa, hay	180.153	10.0
Almonds	180.153	0.5
Almonds, hulls	180.153	3.0
Apples	180.153	0.75
Apricots	180.153	0.75
Bananas	180.153	0.2
Bananas, pulp	180.153	0.1
Beans, forage	180.153	25.0
Beans, guar	180.153	0.1
Beans, guar, forage	180.153	0.1
Beans, hay	180.153	10.0
Beans, lima	180.153	0.75
Beans, snap	180.153	0.75
Beets, roots	180.153	0.75
Beets, sugar, roots	180.153	0.75
Beets, sugar, tops	180.153	10.0
Beets, tops	180.153	0.75
Blackberries	180.153	0.75
Blueberries	180.153	0.75
Boysenberries	180.153	0.75
Broccoli	180.153	0.75
Brussels sprouts	180.153	0.75
Cabbage	180.153	0.75
Carrots	180.153	0.75
Cattle, fat (pre-s)	180.153	0.75
Cattle, mbyp (pre-s)	180.153	0.75
Cattle, meat (pre-s)	180.153	0.75
Cauliflower	180.153	0.75
Celery	180.153	0.75
Cherries	180.153	0.75
Citrus fruits	180.153	0.75
Clover, fresh	180.153	40.0
Clover, hay	180.153	10.0
Coffee beans	180.153	0.2
Collards	180.153	0.75
Corn, (inc. sweet) (k+cwhr)	180.153	0.75
Corn, forage	180.153	40.0

(continued)

	40 *CFR* Reference	Parts per Million
Cotton, seed	180.153	0.2
Cowpeas	180.153	0.1
Cowpeas, forage	180.153	0.1
Cranberries	180.153	0.75
Cucumbers	180.153	0.75
Dandelion	180.153	0.75
Dewberries	180.153	0.75
Endive (escarole)	180.153	0.75
Figs	180.153	0.75
Filberts	180.153	0.5
Grapes	180.153	0.75
Grasses (nmt 40 ppm 24 hrs after appli)	180.153	60.0
Grasses, hay	180.153	10.0
Hops	180.153	0.75
Kale	180.153	0.75
Kiwi fruit	180.153	0.75
Lespedeza	180.153	1.0
Lettuce	180.153	0.75
Loganberries	180.153	0.75
Melons	180.153	0.75
Mushrooms	180.153	0.75
Mustard, greens	180.153	0.75
Nectarines	180.153	0.75
Olives	180.153	1.0
Onions	180.153	0.75
Parsley	180.153	0.75
Parsnips	180.153	0.75
Peaches	180.153	0.75
Peanuts	180.153	0.75
Peanuts, forage	180.153	40.0
Peanuts, hay	180.153	10.0
Peanuts, hulls	180.153	10.0
Pears	180.153	0.75
Peas, vines	180.153	25.0
Peas, vines, hay	180.153	10.0
Peas, with pods	180.153	0.75
Pecans (shell removed)	180.153	0.5
Peppers	180.153	0.75
Pineapples	180.153	0.75
Pineapples, forage	180.153	40.0
Plums (fresh prunes)	180.153	0.75
Potatoes	180.153	0.1
Radishes	180.153	0.75
Raspberries	180.153	0.75
Rutabagas	180.153	0.75
Sheep, fat (pre-s)	180.153	0.75
Sheep, mbyp (pre-s)	180.153	0.75
Sheep, meat (pre-s)	180.153	0.75
Sorghum, forage	180.153	10.0
Sorghum, grain	180.153	0.75
Soybeans	180.153	0.1
Soybeans, forage	180.153	0.1
Spinach	180.153	0.75
Squash, summer	180.153	0.75
Squash, winter	180.153	0.75
Strawberries	180.153	0.75
Sugarcane	180.153	0.75
Sweet potatoes	180.153	0.1
Swiss chard	180.153	0.75

(continued)

	40 *CFR* Reference	Parts per Million
Tomatoes	180.153	0.75
Trefoil, birdsfoot	180.153	40.0
Trefoil, birdsfoot, hay	180.153	10.0
Turnips, roots	180.153	0.75
Turnips, tops	180.153	0.75
Walnuts	180.153	0.5
Watercress	180.153	0.75

References

(1) Worthing, C.R., *Pesticide Manual,* 6th ed., p. 163, British Crop Protection Council (1979).
(2) Spencer, E.Y., *Guide to the Chemicals Used in Crop Protection,* 6th ed., p. 171, London, Ontario, Agriculture Canada (January 1973).
(3) Midwest Research Institute and RvR Consultants, *Production, Distribution, Use and Environmental Impact Potential of Selected Pesticides,* Washington, DC, Council on Environmental Quality (March 15, 1974).
(4) Gysin, H. and Margot, A., U.S. Patent 2,754,243, July 10, 1956, assigned to J.R. Geigy AG.

DIBROMOCHLOROPROPANE

Function: Nematocide (1)(2)(3)(4)

Chemical Name: 1,2-dibromo-3-chloropropane

Formula: $CH_2BrCHBrCH_2Cl$

Trade Names: OS1897 (Shell)
Nemagon® (Shell)
Fumazone® (Dow)

Manufacture

Dibromochloropropane is made by the liquid phase addition of bromine to allyl chloride.

Process Wastes and Their Control

Air: Air emissions from dibromochloropropane manufacture have been reported (B-15) to consist of 1.0 kg hydrocarbons and 0.5 kg of hydrogen chloride per metric ton of pesticide produced.

Product Wastes: Dibromochloropropane is reported to be stable to neutral and acid media. It is hydrolyzed by alkali to 2-bromoallyl alcohol. For recommended disposal procedure see ethylene dibromide (B-3).

Toxicity

The acute oral LD_{50} value for rats is 170-300 (moderately toxic).

The possible effects on the health of employees chronically exposed to DBCP may include sterility, diminished renal function, and degeneration and cirrhosis of the liver. In addition, ingestion of daily doses of DBCP by mice and rats has been found to result in the appearance of gastric cancers in both sexes of both species and in mammary cancers in female rats. Although an increased risk for cancer has not been seen with inhalation exposures, these results are not definitive, therefore the risk of cancer due to occupational exposure to DBCP remains a continuing concern.

There are indications from in vitro experiments that mutagenic effects may occur also, but there has been no study yet of this possibility with mammalian subjects. Employees should be told of these possible effects and informed that some 20-25 years of experience in the manufacture and formulation of DBCP has not yet called such effects in employees of the pesticide industry to the notice of physicians and epidemiologists (B-25).

Allowable Limits on Exposure and Use

Air: Although ACGIH (B-23) does not list this compound, OSHA has a standard of 1 ppb on an 8-hour time weighted average basis with the notation that eye and skin contact are to be avoided. Further, NIOSH has recommended a 10 ppb (0.1 mg/m^3) ceiling on a 30-minute exposure basis.

Product Use: A rebuttable presumption against registration was issued by US EPA on September 22, 1977 on the basis of oncogenicity and reproductive effects.

Then, as of November 3, 1977, EPA in a further action (B-17) suspended all registrations of end use products, subject to the following:

1. *Specific Food Use Suspension*—The uses of DBCP products on broccoli, brussel sprouts, cabbage, carrots, cauliflower, celery, cucumber, eggplant, endive, lettuce, melons, parsnips, peanuts, peppers, radishes, squash, strawberries, tomatoes, and turnips are prohibited and existing stocks already in the hands of users may not be used on any of these crops under any circumstances. Sale, distribution or other movement in commerce is also prohibited.

2. *Conditional Suspension*—All other end uses are also suspended. The sale and distribution, and the use of all pesticide products registered for these additional uses is prohibited unless registrants have fulfilled the classification and labeling requirements listed below or, those users (not to include users who also distribute) who already have existing stocks of products in their possession comply with the classification and labeling requirements listed below.

   a. Turf and ornamental uses:

      1) All such pesticide products may only be used (this includes the application, mixing, and loading of the pesticide) by certified commercial applicators or persons under their direct supervision, and only for those uses covered by the certified commercial applicator's certification.

      2) People and animals must be kept off treated areas until the material has been washed into the soil.

      3) Persons preparing the pesticide for application or involved in application must:

         A. Wear a respirator jointly approved by the Mining Enforcement and Safety Administration and the National Institute of Occupational Safety and Health; and

         B. Wear impermeable protective fullbody clothing to avoid skin or eye contact.

      4) The material must not be allowed to remain on the skin.

      5) If clothing or shoes become contaminated they must be removed promptly and not worn again until completely free of the material.

b. All other uses:

1) All such pesticide products may only be used (this includes the application, mixing, and loading of the pesticide) by certified applicators or persons under their direct supervision, and only for those uses covered by the certified applicators certification.

2) Persons preparing the pesticide for application or involved in application must:

A. Same as 2a, 3) and A above.

B. Wear impermeable protective fullbody clothing to avoid any skin or eye contact, except that drivers of subsurface soil injection equipment are only required to wear impermeable protective fullbody clothing during mixing, loading and transfer operations, and during the calibration of and performance of maintenance upon the soil injection equipment.

3) Same as 2a and 4) above.

4) Same as 2a and 5) above.

The tolerances set by the US EPA for DBCP in or on raw agricultural commodities are as follows:

	40 *CFR* Reference	Parts per Million
Almonds	180.197	50.0
Almonds, hulls	180.197	75.0
Apricots	180.197	5.0
Bananas, pulp	180.197	75.0
Bananas, whole	180.197	125.0
Beans, lima	180.197	75.0
Beans, snap	180.197	75.0
Blackberries	180.197	25.0
Boysenberries	180.197	25.0
Broccoli	180.197	50.0
Brussels sprouts	180.197	50.0
Cabbage	180.197	50.0
Carrots	180.197	75.0
Cauliflower	180.197	50.0
Celery	180.197	75.0
Cherries	180.197	15.0
Citrus fruits	180.197	20.0
Cotton, seed	180.197	25.0
Cucumbers	180.197	25.0
Dewberries	180.197	25.0
Eggplant	180.197	50.0
Endive (escarole)	180.197	130.0
Figs	180.197	75.0
Grapes	180.197	25.0
Lettuce	180.197	130.0
Loganberries	180.197	25.0
Melons	180.197	50.0
Nectarines	180.197	5.0
Okra	180.197	75.0
Parsnips	180.197	75.0
Peaches	180.197	5.0
Peanuts	180.197	50.0
Peppers	180.197	50.0
Pineapples	180.197	50.0
Plums (fresh prunes)	180.197	15.0

(continued)

	40 *CFR* Reference	Parts per Million
Radishes	180.197	75.0
Raspberries	180.197	25.0
Soybeans	180.197	125.0
Squash, summer	180.197	25.0
Strawberries	180.197	10.0
Tomatoes	180.197	50.0
Turnips	180.197	75.0
Walnuts, English	180.197	10.0

The tolerances set by the US EPA for inorganic bromides in animal feed resulting from fumigation with 1,2-dibromo-3-chloropropane are as follows (*CFR* Reference is to Title 21):

	CFR Reference	Parts per Million
Barley, milled fractions	561.260	125.0
Citrus pulp, dehydrated	561.260	90.0
Corn, milled fractions	561.260	125.0
Dog food	561.260	400.0
Oats, milled fractions	561.260	125.0
Rice, milled fractions	561.260	125.0
Rye, milled fractions	561.260	125.0
Sorghum, grain, milo, milled fractions	561.260	125.0
Wheat, milled fractions	561.260	125.0

The tolerances set by the US EPA for inorganic bromides in food resulting from fumigation with 1,2-dibromo-3-chloropropane are as follows (*CFR* Reference is to Title 21):

	CFR Reference	Parts per Million
Barley, flours	193.250	125.0
Biscuit mixes	193.250	125.0
Bread Mixes	193.250	125.0
Breading	193.250	125.0
Cake mixes	193.250	125.0
Cereal flours & related prods.	193.250	125.0
Cheese, Parmesan	193.250	325.0
Cheese, Roquefort	193.250	325.0
Cookie mixes	193.250	125.0
Eggs, dried	193.250	400.0
Figs, dried	193.250	250.0
Foods, processed except below	103.250	125.0
Herbs, processed	193.250	400.0
Macaroni products	193.250	125.0
Malt beverages	193.250	25.0
Milo (sorghum), flours	193.250	125.0
Noodle products	193.250	125.0
Oats, flours	193.250	125.0
Pie mixes	193.250	125.0
Rice, cracked	193.250	125.0
Rice, flours	193.250	125.0
Rye, flours	193.250	125.0
Soya, flour	193.250	125.0
Spices, processed	193.250	400.0
Tomato products, conc.	193.250	250.0
Vegetables, dried	193.250	125.0

References

(1) Worthing, C.R., *Pesticide Manual,* 6th ed., p. 164, British Crop Protection Council (1979).

(2) Spencer, E.Y., *Guide to the Chemicals Used in Crop Protection,* 6th ed., p. 172, London, Ontario, Agriculture Canada (January 1973).

(3) Schmidt, C.T., U.S. Patent 2,937,936, May 24, 1960, assigned to Pineapple Research Institute of Hawaii.

(4) Swezey, A.W., U.S. Patent 3,049,472, August 14, 1962, assigned to Dow Chemical Co.

DIBUTYL PHTHALATE

Function: Insect repellent (1)(2)

Chemical Name: Dibutyl 1,2-benzenedicarboxylate

Formula:

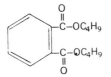

Trade Name: DBP

Manufacture (3)

Dibutyl phthalate is made by the esterification of 1-butanol with phthalic anhydride. The reaction is conducted in the liquid phase at atmospheric pressure and at a temperature of 150° to 200°C in a stirred, jacketed steel kettle. A sulfuric acid catalyst is used and the reaction time is 1 to 3 hours. Yields of 90% are obtained.

Toxicity

The acute oral LD_{50} value for rats is more than 20,000 mg/kg (insignificantly toxic).

The major effect of di-n-butylphthalate in animals involves disturbances in reproduction and teratogenicity. There is some data available on mutagenicity, and the chronic feeding studies did not show any evidence of carcinogenicity.

An ADI was calculated, on the basis of the chronic toxicity data, to be 0.11 mg/kg/day (B-22).

Allowable Limits on Exposure and Use

Air: The threshold limit value for dibutyl phthalate in air is 5 mg/m^3 as of 1979. The tentative short term exposure limit is 10 mg/m^3 (B-23).

Water: In water, a criterion was set by EPA (B-12) at a level of 3 μg/l of phthalate esters for the protection of freshwater aquatic life.

In water, EPA has subsequently suggested (B-26) a limit for the protection of human health from the toxic properties of phthalate esters ingested through water and through contaminated aquatic organisms. The ambient water criteria for dibutyl phthalate was determined to be 5 mg/l. They felt data were insufficient to develop criteria for freshwater or saltwater aquatic life.

References

(1) Martin, H. and Worthing, C.R., *Pesticide Manual,* 5th ed., p. 163, British Crop Protection Council (January 1977).

(2) Spencer, E.Y., *Guide to the Chemicals Used in Crop Protection,* 6th ed., p. 173, London, Ontario, Agriculture Canada (January 1973).

(3) Aldridge, C.L. et al, U.S. Patent 2,834,801, May 13, 1958, assigned to Esso Research and Engineering Company.

DICAMBA

Function: Herbicide (1)(2)

Chemical Name: 3,6-dichloro-2-methoxybenzoic acid

Formula:

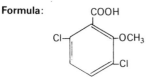

Trade Names: Velsicol 58-CS-11 (Velsicol)
Banvel® (Velsicol)
Mediben® (Velsicol)

Manufacture (3)(4)

Step 1: Preparation of 2,5-Dichlorophenol—1,2,4-trichlorobenzene (250 grams, 1.4 mols) and sodium hydroxide (250 grams, 6.3 mols) were dissolved in 1,100 cc of methanol, and the solution was charged into a rocking bomb of 4 liter capacity. The solution was heated in the sealed bomb at 190°C for 4 hours, during which time the pressure in the bomb rose to 600 psi. The reaction mixture was removed from the cooled bomb, and the residual solid sodium salt of the phenol was dissolved in hot water, and the solution was filtered.

The combined aqueous and methanolic solutions were then acidified with hydrochloric acid, whereupon an oil separated which was taken up in ether. Drying of the ether solution over magnesium sulfate, filtration, and removal of the ether in vacuo produced an oily residue which on distillation under 2 mm pressure gave 200 grams (90% of theory) of a yellow oil boiling at 70°C. The oil solidified on standing to yellowish-white, solid 2,5-dichlorophenol having a melting point of 57°C.

Step 2: Preparation of 3,6-Dichlorosalicylic Acid—2,5-dichlorophenol (200 grams, 1.2 mols) prepared as described in Step 1 was dissolved in a solution of potassium hydroxide (73 grams, 1.2 mols) in 50 ml water. The solution was added to 1.5 liters of xylene, and the mixture was heated to remove the water azeotropically. When the last of the water was removed, the salt of the phenol went into solution.

The solution was then placed in a 1 gallon capacity autoclave fitted with stirring apparatus, and the autoclave was pressured to 500 psi with carbon dioxide. The mixture was then heated and stirred at 130° to 140°C for 8 hours. On cooling and opening of the autoclave, the potassium salt of the product was present as a solid admixed with xylene solution of unreacted phenol. The salt was dissolved in hot water, and the solution was filtered and acidified with hydrochloric acid to give a white precipitate, which was filtered and pressed dry to give 73 g (42% theory based on phenol utilized) of 3,6-dichlorosalicylic acid having a MP of 183°C.

Step 3: Preparation of 2-Methoxy-3,6-Dichlorobenzoic Acid—3,6-dichlorosalicylic acid (210 g, 0.87 mol) prepared as described in Step 2 was dissolved in a solution of sodium hydroxide

(139 g, 3.48 mols) in 900 ml water. The solution was cooled to 20°C, and dimethyl sulfate (219 g, 1.74 mols) was added to the vigorously stirred solution. The mixture was stirred for 20 minutes while the temperature was maintained below 35°C by ice-cooling. Another portion of dimethyl sulfate (139 g) was added, and the mixture was stirred for 10 minutes while the temperature was maintained below 45°C.

The mixture was then refluxed for 2 hours, treated with a solution of 69.6 g (1.74 mols) of sodium hydroxide in 250 ml water, and refluxed for an additional 2 hours. The cooled reaction mixture was acidified to Congo red with hydrochloric acid. the precipitated solid was filtered, dissolved in ether, dried over magnesium sulfate, and filtered. Removal of the ether in vacuo gave a viscous oil, which when dried to a solid at room temperature in a vacuum oven, washed with cold pentane and again dried gave 125 g (65% of theory) of a pale yellow solid melting at 113° to 115°C. Crystallization of the solid from pentane gave white crystals of 2-methoxy-3,6-dichlorobenzoic acid melting at 114° to 116°C.

Process Wastes and Their Control

Air: Air emissions from dicamba manufacture have been reported (B-15) to consist of 1.0 kg of hydrocarbons and 0.5 kg of hydrogen chloride per metric ton of pesticide produced.

Product Wastes: Dicamba is stable to oxidation and is resistant to acid and strong alkali. It is degraded by sodium or lithium in liquid ammonia (B-3).

Toxicity

The acute oral LD_{50} value for rats is 1,040 mg/kg (B-5) which is slightly toxic.

Dicamba produced no adverse effect when fed to rats at up to 19.3 mg/kg/day and 25 mg/kg/day in subchronic and chronic studies. The no-adverse-effect dose in dogs was 1.25 mg/kg/day in a 2-year feeding study. Based on these data an ADI was calculated at 0.00125 mg/kg/day.

A detection limit of 1 ppb for dicamba by electron-capture gas chromatography has been reported. Additional studies are needed to clarify the finding of toxicity in subchronic experiments on various strains of rats in the absence of adverse effects in rats fed higher dicamba concentrations over a 2-year period. Because toxicity was not observed in chronic toxicity studies in dogs, additional chronic studies should be conducted at higher dosages to establish a minimal-toxic-effect dosage (B-22).

Allowable Limits on Exposure and Use

Product Use: The tolerances set by the US EPA for dicamba in or on raw agricultural commodities are as follows:

	40 *CFR* Reference	Parts per Million
Asparagus	180.227	3.0
Barley, grain	180.227	0.5
Barley, straw	180.227	0.5
Corn, fodder	180.227	0.5
Corn, forage	180.227	0.5
Corn, grain	180.227	0.5
Grasses, hay	180.227	40.0
Grasses, pasture	180.227	40.0
Grasses, rangeland	180.227	40.0
Milk	180.227	0.05 N
Oats, grain	180.227	0.5

(continued)

	40 *CFR* Reference	Parts per Million
Oats, straw	180.227	0.5
Sorghum, fodder	180.227	3.0
Sorghum, forage	180.227	3.0
Sorghum, grain	180.227	3.0
Wheat, grain	180.227	0.5
Wheat, straw	180.227	0.5

References

(1) Worthing, C.R., *Pesticide Manual,* 6th ed., p. 165, British Crop Protection Council (1979).
(2) Spencer, E.Y., *Guide to the Chemicals Used in Crop Protection,* 6th ed., p. 175, London, Ontario, Agriculture Canada (January 1973).
(3) Richter, S.B., U.S. Patent 3,013,054, December 12, 1961, assigned to Velsicol Chemical Corporation.
(4) Richter, S.B., U.S. Patent 3,297,427, January 10, 1967, assigned to Velsicol Chemical Corporation.

DICHLOBENIL

Function: Herbicide (1)(2)(6)

Chemical Name: 2-6-dichlorobenzonitrile

Formula:

Trade Names: 2,6-DBN
Code 133®
Casoron® (Philips-Duphar B.V.)
H-133 (Philips-Duphar B.V.)

Manufacture

Dichlobenil may be made starting with dichlorobenzoyl chloride which is reacted with ammonium chloride to give dichlorobenzamide. The benzamide is heated with aluminum chloride to effect dehydration to the nitrile (3).

Alternatively 2-chloro-6-nitrobenzonitrile or 2,6-dinitrobenzonitrile may be chlorinated to give 2,6-dichlorobenzonitrile (4).

Dichloroaniline is another suitable starting material (5). A solution of nitrosyl sulfuric acid was prepared by adding sodium nitrite (240 g) with stirring to concentrated (98%) sulfuric acid (2,840 g, 1,540 ml) cooled at 0°C. When the addition was complete the mixture was warmed to 55°C to complete solution and cooled to 20°C. This temperature was maintained by suitable cooling during the dropwise addition with stirring of a solution of 2,6-dichloro-aniline (500 g, 3.08 mols) in glacial acetic acid (1,230 ml). After 1 hour at room temperature (about 20°C) the diazonium solution was added portionwise to a stirred solution of potassium cyanide (1,026 g, 15.8 mols), cuprous cyanide (342 g, 3.82 mols), and sodium carbonate monohydrate (3,000 g, 29 mols) in water (3,500 ml). Cooling of the reaction vessel and the rate of addition were regulated so that the reaction temperature did not exceed 15°C. When the addition was complete, stirring was continued for 30 minutes. The mixture was then heated at 80°C with benzene (3,000 ml), and the benzene layer siphoned off and evaporated to dryness. A brown solid residue of 2,6-dichlorobenzonitrile remained. The crude product (356 g, 2.07 mols, 67% yield), MP 139° to 141°C, could be purified by crystallization from petroleum ether to give white needles, MP 142°C.

Process Wastes and Their Control

Product Wastes: This herbicide is stable to heat and to acids, and is moderately persistent in soils. It is hydrolyzed by alkali to the benzamide, but this does not appear to be a sufficiently complete degradation for disposal purposes. Incineration would be preferable as a disposal method (B-3).

Toxicity

The acute oral LD_{50} value for rats is 3,160 mg/kg which is slightly toxic.

Allowable Limits on Exposure and Use

Product Use: The tolerances set by the US EPA for dichlobenil in or on raw agricultural commodities are as follows:

	40 *CFR* Reference	Parts per Million
Almonds, hulls	180.231	0.15 N
Apples	180.231	0.15 N
Avocados	180.231	0.15 N
Blackberries	180.231	0.15 N
Blueberries	180.231	0.15 N
Citrus fruits	180.231	0.15 N
Cranberries	180.231	0.15 N
Figs	180.231	0.15 N
Fruits, stone	180.231	0.15 N
Grapes	180.231	0.15 N
Mangoes	180.231	0.15 N
Nuts	180.231	0.15 N
Pears	180.231	0.15 N
Raspberries	180.231	0.15 N

References

(1) Spencer, E.Y., *Guide to the Chemicals Used in Crop Protection,* 6th ed., p. 177, London, Ontario, Agriculture Canada (January 1973).
(2) Koopman, H. and Daams, J., U.S. Patent 3,027,248, March 27, 1962, assigned to North American Philips Co., Inc.
(3) Norris, J.F. and Klemka, A.J., *Jour. Am. Chem. Soc.* 62, 1432-5 (1940).
(4) Hackmann, J.T. and Ten Haken, P., British Patent 861,899, March 1, 1961, assigned to Shell Research Ltd.
(5) Higson, H.M., British Patent 862,937, March 15, 1961, assigned to Shell Research Ltd.
(6) Worthing, C.R., *Pesticide Manual,* 6th ed., p. 166, British Crop Protection Council (1979).

DICHLOFENTHION

Function: Nematocide (1)(2)

Chemical Name: O-2,4-dichlorophenyl O,O-diethyl phosphorothioate

Formula:

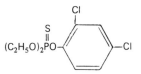

Trade Names: VC-13® (Virginia-Carolina Chem. Corp.)
VC-13 Nemacide® (Virginia-Carolina Chem. Corp.)
Mobilawn® (Mobil Oil Corp.)

Manufacture (3)

A mixture of 1 mol of diethyl phosphorochloridothionate, 1 mol of anhydrous sodium carbonate and 1 mol of 2,4-dichlorophenol and about 500 ml of methyl ethyl ketone is refluxed for about 6 to 9 hours. The mixture is then cooled and filtered and the filtrate stripped of the methyl ethyl ketone by distillation under slightly reduced pressure. The residue of the distillation in ethyl ether and the solution is washed first with water and then with saturated sodium chloride and then dried over anhydrous sodium sulfate. The ether is then removed under reduced pressure to produce technical products of sufficient purity for commercial use. The technical products of the diethyl ester have been found to contain more than 90% of the ester. By fractionally distilling the technical products under reduced pressure the ester contents may be raised to above 90% and in some instances to analytical purity.

The relatively pure products obtained by fractional distillation are all colorless liquids while the technical products vary from light yellow to colorless and all of the products are somewhat soluble in many organic solvents but substantially insoluble in water.

An improved process (4) comprises the addition, in the absence of a catalyst, of an aqueous solution of an alkali metal hydroxide, the dry alkali metal hydroxide constituting preferably at least 45 to 50% of the total solution weight, to a solution of an O,O-dialkyl phosphorohalidothioate and a phenol. The process may be carried out at any temperature within the range of about 50° to 110°C. The reaction proceeds smoothly and the heat evolved in the early stages supplies most of the energy required to sustain the reaction.

The following is a specific example of this improved process. A reactor was charged with 840 parts of 2,4-dichlorophenol. 923 parts of O,O-diethyl phosphorochloridothioate was added with stirring and enough heat to dissolve the two raw materials. 412 parts of 50% aqueous sodium hydroxide was fed to the phosphorochloridothioate-phenol mixture at a temperature of 50°C, using refrigeration as necessary to maintain the addition temperature.

After the aqueous NaOH addition was complete, the reaction mixture, while stirring, was raised to a temperature of 110°C and maintained at 110°C for 2 hours. 124 parts of 25% sodium hydroxide, 1,098 parts of water and 2,410 parts of hexane were loaded into a wash tank. Stirring of the wash mixture was begun and the reactor charge was dropped into the wash tank and the wash product mixture was stirred for 15 minutes. Stripping was stopped, the aqueous layer was allowed to settle and was drawn off.

The product hexane layer was stirred an additional 15 minutes with 488 parts of water and 124 parts of 25% NaOH. The aqueous layer again was allowed to settle and was drawn off. The hexane product layer was transferred to a still and the hexane was removed; first, to a pot temperature of 85°C at atmospheric pressure, and then to a final pot temperature of 93° to 100°C at 50 mm Hg. There was obtained 1,487 parts of product which represented a 96.3% yield based on the O,O-diethyl phosphorochloridothioate used.

Toxicity

The acute oral LD_{50} value for rats is 270 mg/kg (moderately toxic).

References

(1) Worthing, C.R., *Pesticide Manual,* 6th ed., p. 167, British Crop Protection Council (1979).
(2) Spencer, E.Y., *Guide to the Chemicals Used in Crop Protection,* 6th ed., p. 178, London, Ontario, Agriculture Canada (January 1973).
(3) Boyer, W.P., U.S. Patent 2,761,806, September 4, 1956, assigned to Virginia-Carolina Chemical Corp.
(4) Smithey, W.R., Jr., U.S. Patent 3,004,054, October 10, 1961, assigned to Virginia Carolina Chemical Corp.

DICHLOFLUANID

Function: Fungicide (1)(2)

Chemical Name: 1,1-dichloro-N-[(dimethylamino)sulfonyl]-1-fluoro-N-phenylmethane-sulfenamide

Formula:

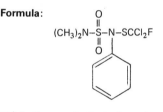

Trade Names: B-47,531 (Bayer)
Euparen® (Bayer)
Elvaron® (Bayer)

Manufacture

Dichlofluanid is made by the reaction of sulfuryl chloride with, in turn, dimethylamine and aniline. The dimethylphenylsulfonamide thus produced is reacted with fluorodichloromethyl sulfenyl chloride to give dichlofluanid.

Toxicity

The acute oral LD_{50} value for rats is about 500 mg/kg which is slightly to moderately toxic.

References

(1) Worthing, C.R., *Pesticide Manual*, 6th ed., p. 168, British Crop Protection Council (1979).
(2) Spencer, E.Y., *Guide to the Chemicals Used in Crop Protection,* 6th ed., p. 179, London, Ontario, Agriculture Canada (January 1973).

DICHLONE

Function: Fungicide (1)(2)(3)(4)

Chemical Name: 2,3-dichloro-1,4-naphthalenedione

Formula:

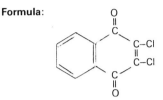

Trade Names: USR604 (Uniroyal, Inc.)
Phygon® (Uniroyal, Inc.)

Manufacture

Dichlone may be made by a process (5) in which 1,4-naphthoquinone is chlorinated in a single step with molecular chlorine in the presence of a catalyst consisting of an N,N-dialkyl acylamide (especially N,N-dimethylformamide or N,N-dimethyl acetamide), the chlorination reaction being carried out at a temperature of 80° to 120°C and in the presence of 0.1 to 2% by weight of the catalyst based upon the quantity of 1,4-naphthoquinone.

The reaction medium is an organic solvent in the presence of the catalyst and chlorine gas, especially chlorinated aromatic and catenary hydrocarbons such as tetrachloroethane and o-dichlorobenzene. The crystals of 2,3-dichloronaphthoquinone-(1,4) are filtered from the mother liquor which is treated with an additional quantity of the N,N-dialkyl acylamide to produce further quantitites of the 2,3-dichloronaphthoquinone-(1,4). Alternatively, naphthalene may be the starting material (6). The light catalyzed chlorination of naphthalene in an inert halogenated solvent at a temperature of 20° to 60°C produces a substantially quantitative yield of 1,2,3,4-tetrachloro-1,2,3,4-tetrahydronaphthalene. Hydrolysis of the 1,2,3,4-tetrachloro-1,2,3,4-tetrahydronaphthalene by refluxing with water to produce 1,4-dihydroxy-2,3-dichloro-1,2,3,4-tetrahydronaphthalene followed by nitric acid oxidation results in the production of 2-chloro-1,4-naphthaquinone which is chlorinated to produce 2,3-dichloro-1,4-naphthoquinone. Figure 26 illustrates the essentials of this process.

The preparation of related aromatic halohydroquinones has been described (7).

Toxicity

The acute oral LD_{50} value for rats is 1,300 mg/kg (slightly toxic).

Allowable Limits on Exposure and Use

Product Use: The tolerances set by the US EPA for dichlone in or on raw agricultural commodities are as follows:

	40 *CFR* Reference	Parts per Million
Apples	180.118	3.0
Beans	180.118	3.0
Celery	180.118	3.0
Cherries	180.118	3.0
Peaches	180.118	3.0
Plums (fresh prunes)	180.118	3.0
Strawberries	180.118	15.0
Tomatoes	180.118	3.0

Figure 26: Block Flow Diagram for the Manufacture of Dichlone from Naphthalene

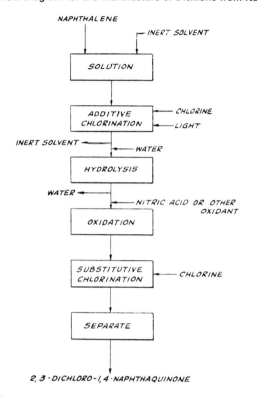

Source: Reference (6)

References

(1) Martin, H. and Worthing, C.R., *Pesticide Manual,* 5th ed., p. 169, British Crop Protection Council (January 1977).
(2) Spencer, E.Y., *Guide to the Chemicals Used in Crop Protection,* 6th ed., p. 180, London, Ontario, Agriculture Canada (January 1973).
(3) Ter Horst, W.P., U.S. Patent 2,302,384, November 17, 1942, assigned to United States Rubber Company.
(4) Ter Horst, W.P., U.S. Patent 2,349,772, May 23, 1944, assigned to United States Rubber Company.
(5) Dannhauser, K., Wiedemann, O. and Leitsmann, R., U.S. Patent 3,484,461, December 16, 1969, assigned to Chemische Fabrik von Heyden AG, Germany.
(6) Buzbee, L.R. and Ecke, G.G., U.S. Patent 3,433,812, March 18, 1969, assigned to Koppers Co., Inc.
(7) Gaertner, V.R., U.S. Patent 2,750,427, June 12, 1956, assigned to Monsanto Chemical Co.

para-DICHLOROBENZENE

Function: Insecticidal fumigant (1)(2)

Chemical Name: 1,4-dichlorobenzene

Formula:

Trade Name: Paracide

Manufacture (3)

p-Dichlorobenzene and o-dichlorobenzene are both produced almost entirely as by-products of the production of monochlorobenzene. The o- and p-dichlorobenzenes are obtained in approximately equal amounts, along with smaller amounts of tri- and tetrachlorobenzenes and are separated by distillation. The proportion of by-product in monochlorobenzene production can easily be increased by longer chlorination. The reaction chemistry is as follows.

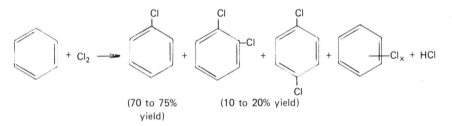

(70 to 75%   (10 to 20% yield)
yield)

At Dow Chemical (3), the HCl by-product is apparently recycled to chlorine production while trichlorobenzenes are recovered. Dow says they have developed a proprietary process to reduce the so-called "tetra-tar" (which can run from 5 to 10% of the monochlorobenzene yield) back to dichlorobenzenes.

At Monsanto (3), the HCl is recovered as muriatic acid with only small amounts escaping through vents or going to a waste treatment plant. Polychlorobenzenes are not recovered and go to an approved landfill with other liquid plant wastes. The landfill is on Monsanto-owned land and is surrounded by wells that are periodically tested. Monsanto says that plans are being developed to incinerate their chlorobenzene wastes. The entire dichlorobenzene work area is ventilated and the air is monitored for p-dichlorobenzene levels. Ventilation air goes to a wet scrubber; the wash water collects little dichlorobenzene and is eventually discharged.

A process for the production of dichlorobenzene from monochlorobenzene has been described by White (4). It is carried out in a fluidized bed of ferric chloride catalyst at 150° to 190°C and atmospheric pressure. A reaction time of 7 to 70 minutes is employed and a product yield of 50 to 70% is obtained.

Process Wastes and Their Control

Air: The basic waste product is the HCl by-product of the chlorination process. This HCl is scrubbed with benzene or chlorobenzene and then is absorbed in water to give hydrochloric acid. The benzene or chlorobenzene from the scrubbing process is circulated into the chlorination step of the process (B-10).

Water: At one plant, liquid plant wastes containing polychlorobenzenes go to an approved landfill (B-10).

Toxicity

The acute oral LD$_{50}$ value for rats is 500 mg/kg (B-5) (slightly to moderately toxic).

PDB is used in such a manner that large amounts of it could enter surface water and humans could obtain substantial exposure by inhalation. In spite of its tremendous use in the United States (68 million pounds) and its suspected involvement in human blood dyscrasias, there is very little adequate information on its toxicity. An ADI was calculated at 0.0134 mg/kg/day based on the available data.

Gas-chromatographic methods have been developed for PDB with a sensitivity of 380 pg/cm peak height, and PDB concentrations as low as 1.0 ppb in water have been analyzed.

Apparently, no chronic-toxicity studies have been performed with PDB. There is no information on the reproductive effects, teratogenicity, mutagenicity, or carcinogenicity of PDB. This lack of information is disturbing, in view of the suspected role of PDB in human leukemia and its apparent ability to undergo metabolic activation and covalent binding to tissue constituents. Particularly disturbing is the very high degree of toxicity in rats that received o-dichlorobenzene at 0.1 or 0.01 mg/kg/day. The no-adverse-effect dosage in that study (0.001 mg/kg/day) was 1/13,400 of that found in other similar rat studies. The reason for this marked difference should be established (B-22).

Human exposure to dichlorobenzene is reported to cause hemolytic anemia and liver necrosis, and 1,4-dichlorobenzene has been found in human adipose tissue. In addition, the dichlorobenzenes are toxic to nonhuman mammals, birds, and aquatic organisms and impart an offensive taste and odor to water. The dichlorobenzenes are metabolized by mammals, including humans, to various dichlorophenols, some of which are as toxic as the dichlorobenzenes (B-26).

Allowable Limits on Exposure and Use

Air: The threshold limit value for p-DCB in air is 450 mg/m^3 (75 ppm) as of 1979. The tentative short term exposure limit is 675 mg/m^3 (110 ppm) (B-23).

The Russian maximal allowable concentration (MAC) value for 1,4-DCB is 20 mg/m^3, much lower than U.S. standards. The US EPA (1977b) has published multimedia environmental goals (MEGs) for health related estimated permissible concentrations in air for 1,4-DCB of 1.07 mg/m^3 (0.18 ppm). The Russian and MEG Limits appear to recognize sensory perception levels more closely than the OSHA and ACGIH values (B-26).

Water: In water, EPA has suggested (B-26) a criterion to protect freshwater aquatic life of 190 μg/l as a 24-hour average; the concentration should not exceed 440 μg/l at any time.

For 1,4-dichlorobenzene the criterion to protect saltwater aquatic life is 15 μg/l as a 24-hour average and the concentrations should not exceed 34 μg/l at any time.

For the protection of human health from the toxic properties of dichlorobenzene ingested through water and through contaminated aquatic organisms, the ambient water criterion is determined to be 230 μg/l total dichlorobenzene (all isomers combined).

References

(1) Worthing, C.R., *Pesticide Manual*, 6th ed., p. 169, British Crop Protection Council (1979).
(2) Spencer, E.Y., *Guide to the Chemicals Used in Crop Protection*, 6th ed., p. 183, London, Ontario, Agriculture Canada (January 1973).
(3) Midwest Research Institute and RvR Consultants, *Production, Distribution, Use and Environmental Impact Potential of Selected Pesticides*, Washington, DC, Council on Environmental Quality (March 15, 1974).

(4) White, W.A., U.S. Patent 3,029,296, April 10, 1962, assigned to Monsanto Chemical Co.

1,1-DICHLORO-2,2-BIS(4-ETHYLPHENYL) ETHANE

Function: Insecticide (1)(2)(4)

Chemical Name: 1,1'-(2,2-dichloroethylidene)bis(4-ethylbenzene)

Formula:

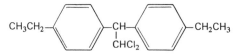

Trade Names: Q-137 (Rohm & Haas Co.)
Perthane® (Rohm & Haas Co.)
Ethylan

Manufacture

The feed materials for ethylan manufacture are 2,2-dichlorovinyl ethyl ether and ethylbenzene (6) which react according to the equation:

$$CCl_2{=}CHOC_2H_5 + 2C_6H_5C_2H_5 \rightarrow CHCl_2CH(C_6H_4C_2H_5)_2 + C_2H_5OH$$

The temperature during the condensation reaction may be maintained at 5° to 75°C, and more preferably at 10° to 25°C. This reaction is preferably conducted under reduced pressure below 300 mm of mercury and preferably at a pressure between 20 and 100 mm according to J.W. Nemec et al (5).

The acid condensing agent may be added to the well-stirred reactants over a period of 1.0 to 2.25 hours. The entire mixture is then agitated at reaction temperature for from 1 to 3 hours and then allowed to settle for ½ hour prior to separation.

If desired, the aromatic compound may be used in excess and thus also serve as a solvent, according to E.F. Meitzner et al (3). While an organic solvent is not essential, it is often convenient to use one during the mixing of the reactants or the working up of the reaction products. For such purposes, such solvents as naphtha or ethylene dichloride may be used. As a condensing agent for the desired reaction between the halogenated product and the aromatic compound, there may be used any strongly acidic condensing agent, such as sulfuric acid, oleum, tetraphosphoric acid, toluene, or benzene sulfonic acids, aluminum chloride, zinc chloride (particularly with some free hydrogen chloride), boron trifluoride and its coordination complexes, and the like.

Sulfuric acid is the condensing agent of choice for use in this process. It may commonly be the usual 95 to 99% acid of commerce or acid containing up to 20% of sulfur trioxide. Sulfuric acid may in general have a strength between 90 and 120%. In place of sulfuric acid, it is possible to use alkane-sulfonic acids or arenesulfonic acids or mixtures of such acids with sulfuric acid.

The amount of condensing agent used may vary from about 0.5 mol up to about 2 mols per mol of aromatic compound to be condensed. The preferred proportion is 0.7 to 1.5 mols per mol of the compound.

A reactor equipped with an agitator, a cooling jacket and a condenser leading to a source of vacuum is the conventional equipment for the conduct of this reaction. The reaction products in solution may be readily washed with water, neutralized and separated from the condensing agent. The solvent may then be stripped off, unreacted starting materials removed as by distillation, and the condensation products obtained as a residue which may, if desired, be purified by extraction, recrystallization, or by treatment with activated carbon. Reportedly, dichloracetaldehyde can also be condensed with ethylbenzene to give ethylan.

Process Wastes and Their Control

Product Wastes: Ethylan is readily dehydrochlorinated like DDT. It undergoes some thermal decomposition above 125°F and the ethyl groups are readily oxidized to carboxylic acid groups (B-3).

Toxicity

The acute oral LD_{50} value for rats is 8,170 mg/kg which is insignificantly toxic.

Allowable Limits on Exposure and Use

Product Use: Issuance of a rebuttable presumption against registration was being considered by EPA as of September 1971 on the basis of possible oncogenicity and reproductive effects.

The tolerances set by the US EPA for this product in or on raw agricultural commodities are as follows:

	40 *CFR* Reference	Parts per Million
Apples	180.139	15.0
Broccoli	180.139	15.0
Brussels sprouts	180.139	15.0
Cabbage	180.139	15.0
Cauliflower	180.139	15.0
Cherries	180.139	15.0
Kohlrabi	180.139	15.0
Lettuce	180.139	15.0
Meat	180.139	0.0
Milk	180.139	0.0
Pears	180.139	15.0
Spinach	180.139	15.0

References

(1) Worthing, C.R., *Pesticide Manual,* 6th ed., p. 170, British Crop Protection Council (1979).
(2) Spencer, E.Y., *Guide to the Chemicals Used in Crop Protection,* 6th ed., p. 258, London, Ontario, Agriculture Canada (January 1973).
(3) Meitzner, E.F. et al, U.S. Patent 2,464,600, March 15, 1949, assigned to Rohm & Haas Co.
(4) Craig, W.E., Shropshire, E.Y. and Wilson, H.F., U.S. Patent 2,881,111, April 7, 1959, assigned to Rohm & Haas Co.
(5) Nemec, J.W., et al, U.S. Patent 2,883,428, April 21, 1959, assigned to Rohm & Haas Co.
(6) McKeever, C.H. and Nemec, J.W., U.S. Patent 2,917,553, December 15, 1959, assigned to Rohm & Haas Co.

DI-(2-CHLOROETHYL) ETHER

Function: Soil fumigant (1)(2)

Chemical Name: 1,1'-oxybis(2-chloroethane)

Formula: $(ClCH_2CH_2)_2O$

Trade Name: Chlorex®

Manufacture

Di-(2-chloroethyl) ether may be made by:

    (A) Passage of chlorine and ethylene into an ethylene chlorohydrin solution at 80°C;
    (B) Direct chlorination of ethyl ether;
    (C) Treatment of ethylene chlorohydrin with sulfuric acid.

Process Wastes and Their Control

Product Wastes: The MCA recommends two disposal procedures: pouring on the ground and then either allowing to evaporate or igniting from a safe distance, and; dissolving in a flammable solvent followed by incineration (B-3).

Toxicity

The acute oral LD_{50} value for rats is 75 mg/kg (B-5) which is moderately toxic.

This soil fumigant has a high acute mammalian toxicity and air containing more than 15 ppm should not be breathed. (It does not, however, appear to have the carcinogenic properties present in α-chloro ethers.) It is also phytotoxic and should be disposed with care (B-3).

Allowable Limits on Exposure and Use

Air: The threshold limit value for this material in air is 30 mg/m^3 (5 ppm) as of 1979. The tentative short term exposure limit is 60 mg/m^3 (10 ppm). These limits bear the notation "skin" indicating that cutaneous absorption should be prevented so that the threshold limit value will not be invalidated (B-23).

The current U.S. Federal standard is 15 ppm (90 mg/m^3) on a 15-minute ceiling basis (B-21).

Water: In water, EPA has suggested (B-26) a limit for the maximum protection of human health from the potential carcinogenic effects of exposure to bis(2-chloroethyl) ether through ingestion of water and contaminated aquatic organisms of zero concentration. A lifetime cancer risk of 1 in 100,000 would permit a concentration of 0.42 μg/l.

References

(1) Martin, H. and Worthing, C.R., *Pesticide Manual,* 4th ed., p. 177, British Crop Protection Council (November 1974).
(2) Spencer, E.Y., *Guide to the Chemicals Used in Crop Protection,* 6th ed., p. 185, London, Ontario, Agriculture Canada (January 1973).

1,1-DICHLORO-1-NITROETHANE

Function: Fumigant (1)

Chemical Name: 1,1-dichloro-1-nitroethane

Formula: CCl$_2$(NO$_2$)CH$_3$

Trade Name: Ethide® (Commercial Solvents)

Manufacture

1,1-Dichloro-1-nitroethane is made by the chlorination of the sodium salt of nitroethane.

Toxicity

The acute oral LD$_{50}$ value for rats is 410 mg/kg (moderately toxic) (B-5).

Allowable Limits on Exposure and Use

Air: The threshold limit value for dichloronitroethane in air is 60 mg/m^3 (10 ppm) as of 1979. However, intended changes will result in a threshold limit value of 10 mg/m^3 (2 ppm) and a short term exposure limit of 60 mg/m^3 (10 ppm) (B-23).

Reference

(1) Spencer, E.Y., *Guide to the Chemicals Used in Crop Protection,* 6th ed., p. 187, London, Ontario, Agriculture Canada (January 1973).

DICHLOROPHEN

Function: Fungicide (1)(2)

Chemical Name: 2,2'-methylenebis(4-chlorophenol)

Formula:

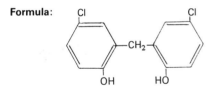

Trade Names: G-4® (Sindar Corp.)
Preventol GD®
Dicestal®
Didroxane®
Di-phenthane 70®
Hyosan®
Parabis®
Plath-Lyse®
Teniathane®
Teniatol®

Manufacture (3)

2,520 grams of sulfuric acid (93% H$_2$SO$_4$ content) is stirred and cooled to 0°C. A solution of 552 grams of p-chlorophenol in 305 grams of methyl alcohol is run into the acid, the temperature being kept below 10°C. The mixture is cooled to −5°C and a solution of 170 grams of aqueous formaldehyde solution (37% CH$_2$O in water) in 332 grams of methyl alcohol is introduced at a more or less uniform rate over a period of 4 hours. The temperature of the reaction mixture is not allowed to rise above 0°C. After all of the formaldehyde-containing solution has been added, the batch is stirred for 3 hours longer at a temperature of −5° to 0°C.

Enough ice is then added to the contents of the reaction chamber in order to reduce the sulfuric acid concentration to 70%. 2,2'-dihydroxy-5,5'-dichlorodiphenylmethane is extracted from the resulting mixture with a mixture of 1,069 grams of isopropyl ether and 1,575 grams of toluene. Ice is added until the acid concentration is about 30%. The acid layer is removed and the solvent layer is washed acid-free.

Most of the isopropyl ether is removed therefrom by atmospheric distillation with a fractionating column, the temperature of the escaping vapors not being permitted to exceed 90°C. From the residue, about 280 grams of pure 2,2'-dihydroxy-5,5'-dichlorodiphenylmethane, MP of 177° to 178°C, crystallize. The product is filtered, washed with toluene and dried at about 100°C.

By concentrating the mother liquor remaining after the foregoing crystallization and filtration, another 225 grams of substantially pure 2,2'-dihydroxy-5,5'-dichlorodiphenylmethane are obtained. This latter crop may be crystallized from toluene in order to convert it into 2,2'-dihydroxy-5,5'-dichlorodiphenylmethane of MP of 177° to 178°C.

Toxicity

The acute oral LD$_{50}$ value for rats is 2,690 mg/kg (B-5) which is slightly toxic.

References

(1) Worthing, C.R., *Pesticide Manual,* 6th ed., p. 172, British Crop Protection Council (1979).
(2) Spencer, E.Y., *Guide to the Chemicals Used in Crop Protection,* 6th ed., p. 188, London, Ontario, Agriculture Canada (January 1973).
(3) Gump, W.S. and Luthy, M., U.S. Patent 2,334,408, November 16, 1943, assigned to Burton T. Bush, Inc.

O-2,4-DICHLOROPHENYL-O-ETHYL PHENYLPHOSPHONOTHIOATE

Function: Insecticide (1)(2)

Chemical Name: See common name

Formula:

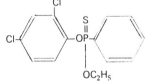

Trade Names: S-7 (Nissan Chemical Industries, Ltd.)
S-Seven® (Nissan Chemical Industries, Ltd.)

Manufacture

This material is made by the condensation of sodium-2,4-dichlorophenolate with O-ethyl phenylphosphonochloridothioate.

Toxicity

An acute oral LD$_{50}$ value for rats is not available. The value for mice is 275 mg/kg.

References

(1) Worthing, C.R., *Pesticide Manual*, 6th ed., p. 173, British Crop Protection Council (1979).
(2) Otsubo, I., Ura, Y., Sato, S., Hayakawa, M. and Sakata, K., U.S. Patent 3,318,764, May 9, 1967, assigned to Nissan Chemical Industries, Ltd.

3-(3,5-DICHLOROPHENYL)-1-ISOPROPYLCARBAMOYLHYDANTOIN

Function: Fungicide (1)(2)

Chemical Name: 3-(3,5-dichlorophenyl)-N-(1-methylethyl)-2,4-dioxo-1-imidazoline carboxamide

Formula:

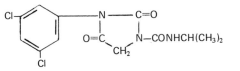

Trade Names: 26,019 R.P. (Rhone-Poulenc)
Rovral® (Rhone-Poulenc)

Manufacture (2)

3-(3,5-dichlorophenyl)hydantoin, MP 199°C, employed as starting material can be prepared according to the method described by Dhar, *J. Soc. Ind. Research,* 20c, 145 (1961). It involves the cyclization of 3-(3,5-dichlorophenyl)ureidoacetic acid, prepared from glycine and 3,5-dichlorophenyl isocyanate.

Then, isopropyl isocyanate (4.6 g) and triethylamine (5.5 g) are added to a solution of 3-(3,5-dichlorophenyl)hydantoin (11 g) in acetone (150 cc). After 30 minutes heating under reflux followed by cooling, the acetone is evaporated under reduced pressure. The residue obtained is washed with petroleum ether (BP 50° to 70°C; 250 cc) and recrystallized from diisopropyl ether to yield 1-isopropylcarbamoyl-3-(3,5-dichlorophenyl)hydantoin (11 g) melting at about 136°C.

Toxicity

The acute oral LD$_{50}$ value for rats is 3,500 mg/kg (slightly toxic).

References

(1) Martin, H. and Worthing C.R., *Pesticide Manual,* 5th ed., p. 174, British Crop Protection Council (January 1977).
(2) Sauli, M., U.S. Patent 3,755,350, August 28, 1973, assigned to Rhone-Poulenc SA.

1,1-DI-(4-CHLOROPHENYL)-2-NITROPROPANE
AND -BUTANE MIXTURE

Function: Insecticide (1)(2)

Chemical Name: 1,1'-(2-nitropropylidene)bis(4-chlorobenzene) and 1,1'-(2-nitrobutylidene)-
bis(4-chlorobenzene)

Formula:

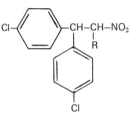

$$R = CH_3 \text{ or } C_2H_5$$

Trade Names: Prolan® (R = CH₃)
Butan® (R = C₂H₅)
Dilan® = mixture

Manufacture (3)

One thousand four hundred and twenty milliliters of sulfuric acid (2 to 3% SO_3) was mixed with 9.35 mols (1,050 g) of chlorobenzene and cooled to 10°C in an ice-salt bath. Three and eight-tenths mols (820 g) of 1-p-chlorophenyl-2-nitro-1-propanol was then added dropwise to the thoroughly stirred mixture at such a rate that the temperature remained between 10° and 15°C. After the addition of the nitro alcohol, the mixture was stirred at room temperature for 10 hours, then poured over cracked ice and allowed to stand several hours.

The mixture was extracted with ether and the ether solution washed with sodium bicarbonate solution, saturated sodium bisulfite solution, and water. The ether was then evaporated and the residue steam distilled. The residue from the steam distillation was dissolved in ether, the ether solution dried with anhydrous sodium sulfate, and the ether evaporated. The residue was a viscous oil which gradually crystallized. After several recrystallizations from absolute ethanol, a white solid remained which melted at 80.5° to 81.5°C. The yield of 1,1-bis-p-chlorophenyl-2-nitropropane was 437 g, 37%.

The 1-p-chlorophenyl-2-nitro-1-propanol used in the above condensation was obtained by condensing p-chlorobenzaldehyde with nitroethane. A convenient method of effecting this condensation is as follows. Four mols (562 g) of p-chlorobenzaldehyde was added with stirring to a solution of 440 g of sodium bisulfite in 2,000 ml of distilled water and stirred at room temperature for 2 hours. Simultaneously, 4.4 mols (330 g) of nitroethane was dissolved slowly in a solution of 180 g of sodium hydroxide in 800 ml of water, which was cooled in an ice-salt bath.

These solutions were mixed and stirred at room temperature for 12 hours, then poured into a separatory funnel. The two layers were separated, the aqueous layer was extracted with ether and the ether portion was added to the organic layer. The ether solution was extracted with saturated sodium bisulfite solution until the unreacted aldehyde was removed. The ether solution was then dried with anhydrous sodium sulfate and the ether evaporated. The conversion to 1-p-chlorophenyl-2-nitro-1-propanol was 820 g, 95%.

The 2-nitro-1,1-bis(p-chlorophenyl)butane may be prepared by condensing 1-p-chlorophenyl-2-nitro-1-butanol with chlorobenzene, following the same procedure as that given above for condensing 1-p-chlorophenyl-2-nitro-1-propanol with chlorobenzene. Also, the 1-p-chlorophenyl-2-nitro-1-butanol may be prepared by a condensation of p-chlorobenzaldehyde with

1-nitropropane, substantially as described above for the preparation of 1-p-chlorophenyl-2-nitro-1-propanol.

Process Wastes and Their Control

Product Wastes: Dilan is a mixture of 21.3% Prolan (R = CH_3) and 42.7% Bulan (R = C_2H_5), both of which are degraded when heated:

$$2(ClC_6H_4)_2CH-CH(NO_2)R \rightarrow 2(ClC_6H_4)_2CH-CO-R + N_2O + H_2O$$

The reaction is aided by light and acid and also by alkali. Oxidizing agents produce a mixture of the corresponding ketones and acids which are described as nontoxic (B-3).

Toxicity

The acute oral LD_{50} value for rats is 475 mg/kg (B-5) which is moderately toxic.

References

(1) Martin, H. and Worthing, C.R., *Pesticide Manual,* 5th ed., p. 176, British Crop Protection Council (January 1977).
(2) Spencer, E.Y., *Guide to the Chemicals Used in Crop Protection,* 6th ed., p. 375, London, Ontario, Agriculture Canada (January 1973).
(3) Hass, H.B. and Blickenstaff, R.T., U.S. Patent 2,516,186, July 25, 1960, assigned to Purdue Research Foundation.

3,6-DICHLOROPICOLINIC ACID

Function: Herbicide (1)(2)

Chemical Name: 3,6-dichloro-2-pyridinecarboxylic acid

Formula:

Trade Names: Dowco 290 (Dow Chemical Co.)
Lontrel® (Dow Chemical Co.)

Manufacture (2)

3.0 g (0.011 mol) of 3,6-dichloro-2-(trichloromethyl)pyridine and 10 ml of concentrated nitric acid were mixed together and the resulting mixture heated at reflux temperature for 70 minutes. During the heating, a reaction took place with the formation of the desired 3,6-dichloropicolinic acid product. After completion of the heating, the reaction mixture was cooled and thereafter poured over ice to obtain 3,6-dichloropicolinic acid product as a precipitate. The latter was recovered by filtration and recrystallized from benzene to obtain a purified product melting at 152° to 153°C.

The 3,6-dichloro-2-(trichloromethyl)pyridine starting material may be prepared by photochlorinating 3-chloro-2-(trichloromethyl)pyridine while the temperature is maintained from about 120° to about 130°C for about six hours, and thereafter cooling and recrystallizing from a hydrocarbon solvent such as hexane. The 3-chloro-2-(trichloromethyl)pyridine employed in such preparation may be prepared by photochlorinating picoline at temperature of from about 50° to 150°C in the presence of a small amount of water, followed by fractional distillation of the reaction mixture, recovering the portion boiling at about 100° to 104°C at 2 mm of mercury pressure and recrystallizing the distillate from hexane.

Toxicity

The acute oral LD_{50} value for rats is nearly 5,000 mg/kg which is only slightly toxic.

References

(1) Worthing, C.R., *Pesticide Manual*, 6th ed., p. 175, British Crop Protection Council (1979).
(2) Johnston, H., U.S. Patent 3,317,549, May 2, 1967, assigned to Dow Chemical Co.

1,2-DICHLOROPROPANE

Function: Insecticidal fumigant (1)(2)

Chemical Name: 1,2-dichloropropane

Formula: $CH_2CICHCICH_3$

Trade Name: Dowfume®

Manufacture (3)

Propylene dichloride may be made by reacting propylene and chlorine in the liquid phase in a tower-type steel reactor in the presence of an iron oxide catalyst. Temperatures of 45°C and pressures of 25 to 30 psi are typical. Product yields are 90%.

Toxicity

The acute oral LD_{50} value for rats is 1,900 mg/kg (B-5) which is slightly toxic.

Allowable Limits on Exposure and Use

Air: The threshold limit value for 1,2-dichloropropane in air is 350 mg/m³ (75 ppm) as of 1979. The tentative short term exposure limit is 510 mg/m³ (110 ppm) (B-23).

Water: For the protection of human health from adverse effects through ingestion of contaminated fish and water, a criterion of 203 µg/l of dichloropropane is suggested (B-26).

References

(1) Worthing, C.R., *Pesticide Manual,* 6th ed., p. 176, British Crop Protection Council (1979).
(2) Spencer, E.Y., *Guide to the Chemicals Used in Crop Protection,* 6th ed., p. 436, London, Ontario, Agriculture Canada (January 1973).
(3) Reese, R.R., U.S. Patent 2,601,322, June 24, 1952, assigned to Jefferson Chemical Co.

DICHLOROPROPANE-DICHLOROPROPENE MIXTURE

Function: Nematocide (1)

Chemical Name: See common name

Formula: $C_3H_6Cl_2$ and $C_3H_4Cl_2$

Trade Name: D-D® (Shell)

Manufacture

This mixture results from the high temperature chlorination of propylene (2).

Process Wastes and Their Control

Product Wastes: The MCA recommends incineration methods of disposal. One source notes that D-D® reacts with dilute inorganic bases, concentrated acids, some metal salts, and active metals, as well as undergoing further halogenation. D-D® was said to be unstable in soil (B-3).

Toxicity

This material is a mixture of four dichloropropanes and dichloropropenes plus related chlorinated C_3 hydrocarbons. The LD_{50} of the mixture is 140 mg/kg which is moderately toxic (B-3).

The actions of dichloropropanes and dichloropropenes on living organisms seems to depend upon the isomer (volatility, solubility, etc.) and the individual organisms. Additionally, judging by the rapid excretion of dichloropropanes and dichloropropenes in rats, it is unlikely that these compounds will remain and accumulate in mammals.

Dichloropropanes and dichloropropenes were both shown to be mutagenic but differed in degree. However, both were shown to have a low tumor causing potential if any at all (B-26).

References

(1) Worthing, C.R., *Pesticide Manual,* 6th ed., p. 177, British Crop Protection Council (1979).
(2) Gross, H.P.A. and Hearne, G., *Ind. Eng. Chem.* 31, 1530 (1939).

1,3-DICHLOROPROPENE

Function: Soil fumigant and nematocide

Chemical Name: 1,3-dichloro-1-propene

Formula: $ClCH=CHCH_2Cl$

Trade Name: Telone® (Dow Chemical)

Manufacture

Made by the high temperature chlorination of propylene to yield allyl chloride as a primary product. The 1,3-dichloropropene is separated from the heavy ends after removal of allyl chloride as an overhead product in distillation.

Process Wastes and Their Control

Air: Air emissions from dichloropropene manufacture have been reported (B-15) to consist of 1.0 kg of hydrocarbons and 0.5 kg of hydrogen chloride per metric ton of pesticide produced.

Toxicity

The acute oral LD_{50} value for rats is 250 to 500 mg/kg (moderately toxic).

Allowable Limits on Exposure and Use

Water: The criterion to protect freshwater aquatic life from 1,3-dichloropropene is 18 μg/l as a 24-hour average. The concentration should not exceed 250 μg/l at any time.

For 1,3-dichloropropene the criterion to protect saltwater aquatic life is 5.5 μg/l as a 24-hour average. The concentration should not exceed 14 μg/l at any time (B-26).

For the protection of human health a criterion of 0.63 μg/l of dichloropropene is suggested (B-26).

References

(1) Worthing, C.R., *Pesticide Manual,* 6th ed., p. 178, British Crop Protection Council (1979).
(2) Spencer, E.Y., *Guide to the Chemicals Used in Crop Protection,* 6th ed., p. 192, London, Ontario, Agriculture Canada (January 1973).
(3) Fairbairn, A.W., Cheyney, H.A. and Cherniavsky, A.J., *Chem. Eng. Progress* 43, 280 (1947).

DICHLORPROP

Function: Herbicide (1)(2)(3)

Chemical Name: 2-(2,4-dichlorophenoxy)propionic acid

Formula:

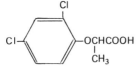

Trade Names: RD-406 (Boots Co., Ltd.)
Cornox RK® (Boots, Co., Ltd.)
Fernoxone®

Manufacture

Dichlorprop may be made by:

    (A) Condensation of 2-chloropropionic acid with 2,4-dichlorophenol or
    (B) Chlorination of 2-phenoxypropionic acid.

Toxicity

The acute oral LD_{50} for rats is 800 mg/kg (slightly toxic).

References

(1) Worthing, C.R., *Pesticide Manual,* 6th ed., p. 179, British Crop Protection Council (1979).

(2) Spencer, E.Y., *Guide to the Chemicals Used in Crop Protection,* 6th ed., p. 193, London, Ontario, Agriculture Canada (January 1973).
(3) Poignant, P. and Jacqumet, H., Canadian Patent 736,882, June 21, 1966.

DICHLORVOS

Function: Contact and stomach insecticide (1)(2)

Chemical Name: 2,2-dichloroethenyl dimethyl phosphate

Formula:

$$(CH_3O)_2\overset{\displaystyle O}{\overset{\|}{P}}OCH{=}CCl_2$$

Trade Names: Nogos® (Ciba-Geigy)
Nuvan® (Ciba-Geigy)
Vapona® (Shell)
Bayer 19149 (Bayer)
Dedevap® (Bayer)
Mafu®
Oko®
Nerkol®

Manufacture (3)(5)

The equation for the production of dichlorvos is as follows:

$$(CH_3O)_3P + Cl_3CCHO \rightarrow (CH_3O)_2POOCH{=}CCl_2 + CH_3Cl$$

The feed materials are chloral and trimethyl phosphite. The reactants are usually employed in about equimolar quantities but lesser amounts of either reactant may be employed. A broadly applicable range of mol ratios of the reactants may be from 1:10 to 10:1. A preferred range is from 2:1 to 1:2. The reaction is exothermic. Temperatures of 10° up to 150°C may be used. The reaction is usually concluded by heating to 50° to 120°C.

The reaction is carried out at essentially atmospheric pressure. The time required for completion of the reaction is short, varying from 10 minutes to an hour or two. The exothermic reaction is usually carried out in the presence of inert solvents to assist in temperature control according to R. Sallman (4). Suitable solvents include benzene, toluene, ether, dioxane or hexane. No catalyst is needed in this process.

The reaction may be carried out in a stirred, jacketed kettle of conventional design. The reaction product can be separated from the reaction mixture, when its separation is desired, by conventional techniques, such as distillation, extraction with selective solvents or the like. For some uses, separation of the product from the crude reaction mixture may be unnecessary.

Process Wastes and Their Control

Air: Air emissions from dichlorvos manufacture have been reported to consist of 1.0 kg hydrocarbon and 0.5 kg of dichlorvos per metric ton of pesticide produced.

Product Wastes: Fifty percent hydrolysis is obtained in pure water in 25 minutes at 70°C, and in 61.5 days at 20°C. A buffered solution yields 50% hydrolysis (37.5°C) in 301 minutes at pH 8, 462 minutes at pH 7, 4,620 minutes at pH 5.4. Hydrolysis yields no toxic residues. Dichlorvos is a hydrolysis product of trichlorfon (B-3).

Allowable Limits on Exposure and Use

Air: The threshold limit value for dichlorvos in air is 1.0 mg/m³ (0.1 ppm) as of 1979. The tentative short term exposure limit is 3.0 mg/m³ (0.3 ppm). These limits bear the notation "skin," indicating that cutaneous absorption should be prevented so that the threshold limit value is not invalidated.

Product Use: The tolerances set by the US EPA for dichlorvos in or on raw agricultural commodities are as follows:

	40 *CFR* Reference	Parts per Million
Cattle, fat	180.235	0.02 N
Cattle, mbyb	180.235	0.02 N
Cattle, meat	180.235	0.02 N
Cucumbers (exp. as naled)	180.235	0.5
Eggs	180.235	0.05 N
Goats, fat	180.235	0.02 N
Goats, mbyp	180.235	0.02 N
Goats, meat	180.235	0.02 N
Horses, fat	180.235	0.02 N
Horses, mbyp	180.235	0.02 N
Horses, meat	180.235	0.02 N
Lettuce (exp. as naled)	180.235	1.0
Milk	180.235	0.02 N
Mushrooms (exp. as naled)	180.235	0.5
Non-per bag rac 6% fat or less (post-h)	180.235	0.5
Non-per bag rac 6% fat or more (post-h)	180.235	2.0
Non-per pkg rac 6% fat or less (post-h)	180.235	0.5
Non-per pkg rac 6% fat or more (post-h)	180.235	2.0
Non-per, bulk stored rac (post-h)	180.235	0.5
Poultry, fat	180.235	0.05 N
Poultry, mbyp	180.235	0.05 N
Poultry, meat	180.235	0.05 N
Radishes	180.235	0.5
Sheep, fat	180.235	0.02 N
Sheep, mbyp	180.235	0.02 N
Sheep, meat	180.235	0.02 N
Swine, edible tissue	180.235	0.1 N
Tomatoes (pre & post-h) (exp. as naled)	180.235	0.5

Issuance of a rebuttable presumption against registration for dichlorvos was being considered by US EPA as of September 1979 on the basis of possible mutagenicity, reproductive and fetotoxic effects, oncogenicity and neurotoxicity.

References

(1) Worthing, C.R., *Pesticide Manual,* 6th ed., p. 180, British Crop Protection Council (1979).
(2) Spencer, E.Y., *Guide to the Chemicals Used in Crop Protection,* 6th ed., p. 194, London, Ontario, Agriculture Canada (January 1973).
(3) Whetstone, R.R. et al, U.S. Patent 2,956,073, October 11, 1960, assigned to Shell Oil Co.
(4) Sallman, R., U.S. Patent 2,861,912, November 25, 1958, assigned to Ciba, Ltd.
(5) Whetstone, R. and Harman, D., U.S. Patent 3,116,201, December 31, 1963, assigned to Shell Oil Co.

DICLORAN

Function: Fungicide (1)(2)(3)

Chemical Name: 2,6-dichloro-4-nitroaniline

Formula:

Trade Names: Botran® (Upjohn)
Allisan® (Boots Pure Drug)
Ditranil®

Manufacture

Dicloran is made by the direct chlorination of p-nitroaniline.

Toxicity

The lowest published lethal dose for rats, LD_{LO} is 1,500 mg/kg (B-5).

Allowable Limits on Exposure and Use

Product Use: The tolerances set by the US EPA for dicloran in or on raw agricultural commodities are as follows:

	40 *CFR* Reference	Parts per Million
Apricots (pre and post-h)	180.200	20.0
Beans, snap	180.200	20.0
Blackberries	180.200	15.0
Boysenberries	180.200	15.0
Carrots (post-h)	180.200	10.0
Celery	180.200	15.0
Cherries, sweet (pre & post-h)	180.200	20.0
Cotton, seed	180.200	0.1
Cucumbers	180.200	5.0
Garlic	180.200	5.0
Grapes	180.200	10.0
Lettuce	180.200	10.0
Nectarines (pre & post-h)	180.200	20.0
Onions	180.200	5.0
Peaches (pre & post-h)	180.200	20.0
Plums (fresh prunes) (pre & post-h)	180.200	15.0
Potatoes (pre-h)	180.200	0.25
Raspberries	180.200	15.0
Rhubarb	180.200	10.0
Sweet potatoes (post-h)	180.200	10.0
Tomatoes	180.200	5.0

References

(1) Worthing, C.R., *Pesticide Manual,* 6th ed., p. 182, British Crop Protection Council (1979).

(2) Spencer, E.Y., *Guide to the Chemicals Used in Crop Protection,* 6th ed., p. 181, London, Ontario, Agriculture Canada (January 1973).
(3) Clark, N.G., Stevenson, H.A., Brookes, R.F. and Hams, A.F., British Patent 845,916, August 24, 1960, assigned to Boots Pure Drug Co., Ltd.

DICOFOL

Function: Acaricide (1)(2)(5)

Chemical Name: 4-chloro-α-(4-chlorophenyl)-α-(trichloromethyl)benzyl alcohol

Formula:

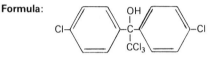

Trade Names: FW-293 (Rohm and Haas)
Kelthane® (Rohm and Haas)

Manufacture (3)(4)

The feed materials for dicofol manufacture are commercial DDT, chlorine and formic acid. In a first step, the DDT is chlorinated according to the following reaction.

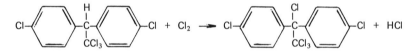

Then, that intermediate is reacted with an aqueous formic acid solution to give the following product.

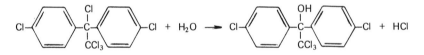

The initial exothermic chlorination step is carried out at 80° to 90°C. The subsequent exothermic hydrolysis step is carried out at temperatures in the range from 125° to 165°C and preferably from 135° to 150°C. While the rate of reaction increases with increased temperatures, side reactions also increase, depending on the particular catalyst employed, until above 160° to 165°C, appreciable amounts of benzil and tar are produced. The reaction temperature is controlled by the rate of addition of reactant water and by the amount of heat supplied, according to H.F. Wilson et al (4).

The chlorination of DDT is conducted at atmospheric pressure, as is the subsequent hydrolysis step. The reaction period in the DDT chlorination step is about 2 hours (3). The hydrolysis reaction requires from 2 to 20 hours, depending on the reaction temperature and the amount of catalyst used. Typical reaction times might be 5 to 6 hours.

The chlorination step is conducted in the liquid phase in ethylene dichloride solution. The hydrolysis reaction is carried out in a mixture consisting of perhaps 33% p-toluene sulfonic acid, 33% sulfuric acid and 33% water. Azodiisobutyronitrile is used in the amount of 1% based on DDT charged as a catalyst in the chlorination step.

An arylsulfonic acid or alkanesulfonic acid of not more than 12 carbon atoms is used as a promoter for the hydrolysis reaction. A stirred, jacketed kettle can be used for the chlorination step. An acid-resistant reactor equipped with a stirrer and reflux condenser is used for the hydrolysis step. After the chlorination step, the ethylene dichloride solvent can be removed under reduced pressure to give a technical grade product which partially solidifies when cooled.

After the completion of the hydrolysis step as indicated by cessation of HCl evolution, the reaction mixture is cooled to 90° to 100°C and a low-boiling solvent such as ethylene dichloride is added along with water. An organic layer and an aqueous acid layer are allowed to form and are separated. The organic layer is washed with a dilute alkaline solution and then with water. It is then freed of solvent by distillation under reduced pressure to give the product. The product may be recrystallized from dilute naphtha solution at 0°C if desired. Yields of 90 to 95% can generally be obtained in the hydrolysis step.

Dicofol may also be made by the chlorination of chlorfenethol.

Another alternate route (6) involves the preparation of a Grignard reagent from p-dichlorobenzene followed by reaction of the p-chlorophenylmagnesium chloride with ethyl trichloroacetate followed by hydrolysis. This is analogous to the preparation of chlorfenethol (which see).

Process Wastes and Their Control

Air: Air emissions from dicofol manufacture have been reported (B-15) to consist of:

Component	Kilograms per Metric Ton Pesticide Produced
Dicofol	0.5
Chlorobenzene	0.5
Chloral	0.5

Water: A process developed by Kennedy (7) effected waste treatment of a stream from dicofol production. Influent waste streams containing 15-35 ppm of chlorinated organics were treated with ion exchange resins. The effluent did not exceed 0.5 ppm and the treatment was found superior to activated carbon treatment.

Product Wastes: This compound is unstable in aqueous alkali and is degraded to dichlorobenzophenone and chloroform:

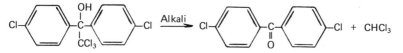

It is unaffected by concentrated H_2SO_4 (B-3).

Allowable Limits on Exposure and Use

Product Use: The tolerances set by the US EPA for dicofol in or on raw agricultural commodities are as follows:

	40 CFR Reference	Parts per Million
Apples	180.163	5.0
Apricots	180.163	10.0

(continued)

	40 *CFR* Reference	Parts per Million
Beans, dry	180.163	5.0
Beans, lima, succulent	180.163	5.0
Beans, snap, succulent	180.163	5.0
Blackberries	180.163	5.0
Boysenberries	180.163	5.0
Bush nuts	180.163	5.0
Butternuts	180.163	5.0
Cherries	180.163	5.0
Chestnuts	180.163	5.0
Cotton, seed	180.163	0.1
Crabapples	180.163	5.0
Cucumbers	180.163	5.0
Dewberries	180.163	5.0
Eggplant	180.163	5.0
Figs	180.163	5.0
Filberts	180.163	5.0
Grapefruit	180.163	10.0
Grapes	180.163	5.0
Hazelnuts	180.163	5.0
Hickory nuts	180.163	5.0
Hops	180.163	30.0
Kumquats	180.163	10.0
Lemons	180.163	10.0
Limes	180.163	10.0
Loganberries	180.163	5.0
Melons	180.163	5.0
Nectarines	180.163	10.0
Oranges	180.163	10.0
Peaches	180.163	10.0
Pears	180.163	5.0
Pecans	180.163	5.0
Peppermint, hay	180.163	25.0
Peppers	180.163	5.0
Pimentos	180.163	5.0
Plums (fresh prunes)	180.163	5.0
Pumpkins	180.163	5.0
Quinces	180.163	5.0
Raspberries	180.163	5.0
Spearmint, hay	180.163	25.0
Squash, summer	180.163	5.0
Squash, winter	180.163	5.0
Strawberries	180.163	5.0
Tangerines	180.163	10.0
Tomatoes	180.163	5.0
Walnuts	180.163	5.0

The tolerances set by the US EPA for dicofol in food are as follows (the *CFR* Reference is to Title 21):

	CFR Reference	Parts per Million
Tea, dried	193.80	45.0

References

(1) Worthing, C.R., *Pesticide Manual,* 6th ed., p. 183, British Crop Protection Council (1979).
(2) Spencer, E.Y., *Guide to the Chemicals Used in Crop Protection,* 6th ed., p. 195, London, Ontario, Agriculture Canada (January 1973).

(3) Wilson, H.F. et al, U.S. Patent 2,812,280, November 5, 1957, assigned to Rohm and Haas Company.

(4) Wilson, H.F. et al, U.S. Patent 2,812,362, November 5, 1957, assigned to Rohm and Haas Company.

(5) Riley, G.C. and Nolan, E.A., U.S. Patent 3,102,070, August 27, 1963, assigned to Rohm and Haas Company.

(6) Metal and Thermit Corp. British Patent 831,421 (March 30, 1960).

(7) Kennedy, D.C., U.S. Patent 4,042,498, August 16, 1977, assigned to Rohm and Haas Company.

DICROTOPHOS

Function: Systemic insecticide and acaricide (1)(2)

Chemical Name: 3-(dimethylamino)-1-methyl-3-oxo-1-propenyl dimethyl phosphate

Formula:

$$(CH_3O)_2\overset{\overset{O}{\|}}{P}OC\!\!=\!\!CH\!-\!\underset{\underset{O}{\|}}{C}\!-\!N(CH_3)_2$$
$$\underset{CH_3}{|}$$

Trade Names: SD-3562 (Shell)
Bidrin® (Shell)
C-709 (Ciba-Geigy)
Carbicron® (Ciba-Geigy)
Ektafos® (Ciba-Geigy)

Manufacture (3)(4)

Seven hundred and seventy grams of 2-chloro-N,N-dimethylacetoacetamide were mixed with 100 ml of glacial acetic acid at 90° to 95°C and 656 grams of trimethyl phosphite was added to the constantly stirred mixture over a period of 1.25 hours. The temperature of the mixture was held at 100°C during the addition and for an additional 2.75 hours. The mixture then was stripped to a kettle temperature of 100°C at 1.8 mm mercury pressure. There was obtained 1,044 grams of product, which contained 82% of the trans isomer of dimethyl 2-(dimethylcarbamoyl)-1-methylvinyl phosphate. Conduct of this same procedure, but omitting the acetic acid, ordinarily has resulted in a product containing about 50% of the trans isomer.

The 2-chloro-N,N-dimethylacetoacetamide is obtained by first reacting diketene with dimethylamine and then reacting the resulting N,N-dimethylacetoacetamide with sulfuryl chloride as follows. Diketene in amount of 126 parts (1.5 mols) was added dropwise during a period of about an hour to a stirred solution of 68 parts of dimethylamine in 800 parts of water while maintaining the temperature of the reaction mixture at 0° to 15°C. The mixture was then stirred for an additional hour, after which the aqueous solution was evaporated to a residue of 180 parts which was distilled. The desired N,N-dimethylacetoacetamide in amount of 142 parts distilled at 80° to 83°C at 0.5 mm pressure.

Sulfuryl chloride in amount of 143.5 parts (1.07 mols) was added dropwise to 138 parts (1.07 mols) of N,N-dimethylacetoacetamide held at about 10°C, after which the mixture was warmed to 50°C to remove the majority of the hydrogen chloride. The residue was then distilled and 156 parts of alpha-chloro-N,N-dimethylacetoacetamide was recovered which distilled at 93° to 98°C at 0.5 mm pressure. The liquid amide was found to contain 22.0% chlorine (theory, 21.7%).

Process Wastes and Their Control

Air: Air emissions from dicrotophos manufacture have been reported (B-15) to consist of 0.5 kg of hydrocarbons and 0.5 kg of dicrotophos per metric ton of pesticide produced.

Product Wastes: Dicrotophos decomposes after 31 days at 75°C or 7 days at 90°C. Hydrolysis is 50% complete in aqueous solutions at 38°C after 50 days at pH 9.1 (100 days are required at pH 1.1). Alkaline hydrolysis (NaOH) yields $(CH_3)_2NH$ (B-3).

Toxicity

The acute oral LD_{50} value for rats is 13-30 mg/kg which is highly toxic.

Allowable Limits on Exposure and Use

Air: The threshold limit value for dicrotophos in air is 0.25 mg/m³ as of 1979. Further, cutaneous absorption should be prevented so that the threshold limit value is not invalidated (B-23).

Product Use: The tolerances set by the US EPA for dicrotophos in or on raw agricultural commodities are as follows:

	40 *CFR* Reference	Parts per Million
Cotton, seed	180.299	0.05

References

(1) Worthing, C.R., *Pesticide Manual,* 6th ed., p. 184, British Crop Protection Council (1979).
(2) Spencer, E.Y., *Guide to the Chemicals Used in Crop Protection,* 6th ed., p. 196, London, Ontario, Agriculture Canada (January 1973).
(3) Tiemah, C.H. and Stiles, A.R., U.S. Patent 3,068,268, December 11, 1962, assigned to Shell Oil Company.
(4) Whetstone, R.R. and Stiles, A.R., U.S. Patent 2,802,855, August 13, 1957, assigned to Shell Development Company.

DICRYL (CHLOROANOCRYL)

Function: Herbicide

Chemical Name: 3',4'-dichloro-2-methylacrylanilide

Formula:

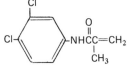

Trade Names: Niagara 4556 (FMC Corp.)
DCMA

Manufacture (1)

Dicryl, N-(3,4-dichlorophenyl)methacrylamide, may be prepared by the reaction of 3,4-dichloroaniline with methacrylyl chloride in the presence of a tertiary amine, for example, as follows. To a stirred solution of 17.8 grams (0.11 mol) of 3,4-dichloroaniline and 10.1 grams (0.1 mol) of triethylamine in 200 ml of benzene was added dropwise 10.5 grams (0.01 mol) of freshyl distilled methacrylyl chloride. The mixture was heated under reflux for one hour with stirring, then allowed to cool to room temperature. To the cooled mixture was added 100 ml of 1% hydrochloric acid and stirring was continued for 0.5 hour.

The benzene layer was separated and the aqueous phase was extracted twice with 50 ml portions of ether. The combined benzene and ether solutions were washed with 100 ml of water, separated and dried over anhydrous potassium carbonate. The solvent was removed by distillation to give 18.5 grams (76%) of solid product which melted at 123° to 125°C. Recrystallization from ligroin-ethanol gave a solid which melted at 124° to 126°C.

Process Wastes and Their Control

Product Wastes: Dicryl can be hydrolyzed by alkali to the more toxic 3,4-dichloroaniline. Therefore, hydrolysis cannot be recommended. Proper incineration would appear to be the method of choice (B-3).

Toxicity

The acute oral LD_{50} value for rats is 1,800 mg/kg (B-5) which is slightly toxic.

Reference

(1) Thompson, J.T., U.S. Patent 3,169,850, February 16, 1965.

DIELDRIN

Function: Insecticide (1)(2)

Chemical Name: 3,4,5,6,9,9-hexachloro-1aα, 2β,2aα,3β,6β,6aα,7β,7aα-octahydro-2,7:3,6-di-methanonaphth[2,3-b] oxirene

Formula:

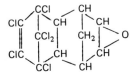

Trade Names: Compound 497 (Julius Hyman and Co.).
Octalox® (Julius Hyman and Co.)
Panoram D-31®

Manufacture

The feed materials for dieldrin manufacture are essentially aldrin and a peracid whereby the aldrin is epoxidized as follows as described by S.B. Soloway (5).

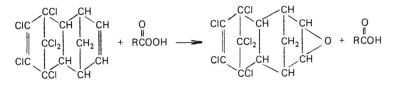

Alternatively the epoxidation can be achieved by the use of hydrogen peroxide in the presence of a tungstic oxide catalyst whereby the reaction is essentially as follows as described by C.W. Smith et al (3) and G.B. Payne and C.W. Smith (7).

$$\text{aldrin} + H_2O_2 \rightarrow \text{dieldrin} + H_2O$$

Operating temperatures below 100°C and generally in the range of 0° to 75°C are suitable for this epoxidation reaction. The reaction is generally conducted at atmospheric pressure. The pressure should be adequate to maintain the liquid phase but at the usual operating temperatures of 20° to 75°C, this is no problem. Typical reaction times for this epoxidation reaction are in the range from 1 to 6 hours. The time necessary is inversely proportional to the reaction temperature employed.

The reaction system using peracids contains two liquid phases, one aqueous phase containing the organic peracid being used as the epoxidizing agent and the other nonaqueous phase containing the aldrin to be epoxidized. The nonaqueous phase may typically utilize benzene as a solvent for the aldrin.

When hydrogen peroxide is used, a solvent capable of dissolving 90% H_2O_2 is used. Tertiary butanol is a particularly desirable solvent. Dioxane, dimethylformamide and sulfolane are other suitable solvents. When using a peracid, the reaction can be carried out noncatalytically or an epoxidation catalyst such as sulfuric or phosphoric acid may be used, as described by C. Yang (4). When using hydrogen peroxide, tungsten trioxide may be used as a catalyst, essentially producing pertungstic acid according to C.W. Smith et al (3).

A reactor system suitable for dieldrin manufacture is described by C. Yang (4). A product and waste schematic flow sheet for dieldrin manufacture is given in Figure 27.

Process Wastes and Their Control

Water: Acids and water-soluble materials from dieldrin manufacture are sent to an evaporation basin (B-10).

Product Wastes: Dieldrin like aldrin (from which it is derived by oxidation) is quite stable to heat (no decomposition at 250°C) and to refluxing aqueous or alcoholic acustic. The instability of dieldrin to acids is similar to that of aldrin. Dieldrin is more resistant to oxidation than is aldrin, but is attacked by ozone. Reaction would be expected to occur at the double bond and at the epoxide to give the tetracarboxylic acid (see aldrin). Dieldrin is 100% degraded by sodium or lithium in liquid ammonia but this is not a practical disposal method. The MCA recommends incineration methods for the disposal of dieldrin (B-3).

Toxicity

The acute oral LD_{50} value for rats is 46 mg/kg (highly toxic).

See "Aldrin" for a more detailed discussion of toxicity.

Allowable Limits on Exposure and Use

Air: The threshold limit value for dieldrin in air is 0.25 mg/m[3] as of 1979. The tentative short term exposure limit is 0.75 mg/m[3]. These limits bear the notation "skin," indicating that cutaneous absorption should be prevented so that the threshold limit value is not invalidated (B-23).

NIOSH has recommended, as of October 1978, that dieldrin be held to the lowest reliably detectable level in air which is 0.15 mg/m[3] on a time-weighted average basis.

Water: In water, EPA has set (B-12) a limit of 0.003 µg/l for the protection of freshwater and

Figure 27: Production and Waste Schematic for Dieldrin

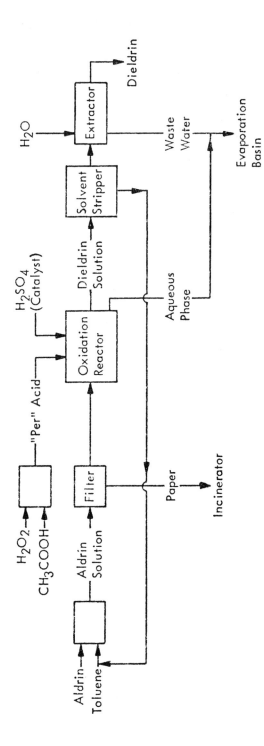

marine aquatic life. They note that persistence, bioaccumulation potential and carcinogenicity of dieldrin caution minimum human exposure.

Subsequently (B-26), EPA has set criteria for dieldrin in water to protect freshwater and saltwater aquatic life (see Aldrin). For the protection of human health from potential carcinogenic effects, dieldrin concentration should be zero. For a cancer risk of 1 in 100,000 the dieldrin concentration should be 4.4×10^{-2} ng/l.

Product Use: The US EPA as of March 18, 1971 cancelled most uses of dieldrin. See "Aldrin" for the list of permitted uses.

The tolerances set by the US EPA for dieldrin in or on raw agricultural commodities are as follows:

	40 *CFR* Reference	Parts per Million
Alfalfa	180.137	0.1
Apples	180.137	0.1
Apricots	180.137	0.1
Asparagus	180.137	0.1
Bananas	180.137	0.1
Barley, grain	180.137	0.02 i
Barley, straw	180.137	0.1 i
Beans	180.137	0.0
Beets, garden	180.137	0.0
Beets, garden, tops	180.137	0.0
Broccoli	180.137	0.1
Brussels sprouts	180.137	0.1
Cabbage	180.137	0.1
Carrots	180.137	0.1
Cauliflower	180.137	0.1
Cherries	180.137	0.1
Clover	180.137	0.0
Collards	180.137	0.0
Corn, forage	180.137	0.0
Corn, grain	180.137	0.0
Corn, pop	180.137	0.0
Cowpeas	180.137	0.0
Cranberries	180.137	0.1
Cucumbers	180.137	0.1
Eggplant	180.137	0.1
Endive (escarole)	180.137	0.0
Grapefruit	180.137	0.05 i
Grapes	180.137	0.1
Horseradish	180.137	0.1
Kale	180.137	0.0
Kohlrabi	180.137	0.0
Lemons	180.137	0.05 i
Lespedeza	180.137	0.0
Lettuce	180.137	0.1
Limes	180.137	0.05 i
Mangoes	180.137	0.1
Mustard, greens	180.137	0.0
Nectarines	180.137	0.1
Oats, grain	180.137	0.02 i
Oats, straw	180.137	0.1 i
Onions	180.137	0.1
Oranges	180.137	0.05 i
Parsnips	180.137	0.1
Peaches	180.137	0.1

(continued)

	40 *CFR* Reference	Parts per Million
Pears	180.137	0.1
Peas	180.137	0.0
Peas, black-eyed	180.137	0.0
Peas, cowpeas, hay	180.137	0.0
Peas, hay	180.137	0.0
Peas, cowpea hay	180.137	0.0
Peppers	180.137	0.1
Pimentos	180.137	0.1
Plums (fresh prunes)	180.137	0.1
Potatoes	180.137	0.1
Quinces	180.137	0.1
Radishes	180.137	0.1
Radishes, tops	180.137	0.1
Rutabagas	180.137	0.0
Rye, grain	180.137	0.02 i
Rye, straw	180.137	0.1 i
Salsify, roots	180.137	0.1
Salsify, tops	180.137	0.0
Sorghum, grain	180.137	0.0
Sorghum, grain, forage	180.137	0.0
Soybeans	180.137	0.0
Soybeans, hay	180.137	0.0
Spinach	180.137	0.0
Squash, summer	180.137	0.1
Strawberries	180.137	0.1
Sweet potatoes	180.137	0.1
Swiss chard	180.137	0.0
Tangerines	180.137	0.05 i
Tomatoes	180.137	0.1
Turnips	180.137	0.0
Turnips, tops	180.137	0.0
Wheat, grain	180.137	0.02 i
Wheat, straw	180.137	0.1 i

References

(1) Worthing, C.R., *Pesticide Manual,* 6th ed., p. 185, British Crop Protection Council (1979).
(2) Spencer, E.Y., *Guide to the Chemicals Used in Crop Protection,* 6th ed., p. 198, London, Ontario, Agriculture Canada (January 1973).
(3) Smith, C.W. et al, U.S. Patent 2,786,854, March 26, 1957, assigned to Shell Development Company.
(4) Yang, C., U.S. Patent 2,873,283, February 10, 1959, assigned to Shell Development Company.
(5) Soloway, S.B., U.S. Patent 2,676,131, April 20, 1954, assigned to Shell Development Company.
(6) Midwest Research Institute, *The Pollution Potential in Pesticide Manufacturing,* Washington, DC, Environmental Protection Agency (June 1972).
(7) Payne, G.B. and Smith, C.W., U.S. Patent 2,776,301, January 1, 1957, assigned to Shell Development Company.

DIENOCHLOR

Function: Acaricide (1)

Chemical Name: 1,1',2,2',3,3',4,4',5,5'-decachlorobi-2,4-cyclopentadien-1-yl

Formula:

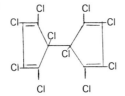

Trade Name: Pentac® (Hooker Chemical Corp.)

Manufacture

Dienochlor may be made by the reductive coupling of hexachlorocyclopentadiene according to the following equation as described by J.T. Rucker (2).

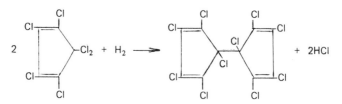

Alternatively, two molecules of hexachlorocyclopentadiene may be coupled by the use of metallic copper (3) according to the following equation.

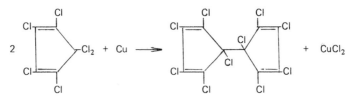

Emphasis will be placed in the description which follows on the reductive coupling route using hydrogen and a catalyst. The most desirable temperature for the hydrogenation reaction is about room temperature. Lower temperatures increase cooling costs and these costs are not compensated by increased yields. Increasing the temperature only increases the yields of by-products.

This reaction is carried out at atmospheric pressure. A reaction time of six hours has been cited by J.T. Rucker (2). The reaction is conducted in the liquid phase. A solvent may or may not be employed. If a solvent is used, it should be inert with respect to the reactants and the reaction products. Among the solvents which may be employed are benzene, toluene, etc.

Noble metal catalysts such as ruthenium, rhodium, palladium, osmium and platinum may be used. Palladium has been found to be most satisfactory. It is most effective when deposited on an inert carrier such as activated alumina, activated charcoal or kieselguhr in a finely-divided state, in a concentration of about 5% by weight of metal on the support. A jacketed stirred reactor equipped with a reflux condenser and means for hydrogen introduction is used for this reaction. The catalyst is first removed from the reaction product by filtration. The filtrate is then cooled to precipitate the product in crystalline form. It may be recrystallized from isopropanol and then from hexane if desired.

The final product has the formula $C_{10}Cl_{10}$, a molecular weight of 475, and a MP of 120° to 122°C.

Process Wastes and Their Control

Product Wastes: This perchlorinated compound is stable to aqueous acids and bases. It loses activity upon heating (50% loss after 6 hours at 130°C) or upon exposure to sunlight or ultraviolet light (B-3).

Toxicity

The acute oral LD_{50} value for rats is 3,150 mg/kg (slightly toxic).

References

(1) Worthing, C.R., *Pesticide Manual*, 6th ed., p. 186, British Crop Protection Council (1979).
(2) Rucker, J.T., U.S. Patent 2,934,470, April 26, 1960, assigned to Hooker Chemical Corp.
(3) Ladd, E.C., U.S. Patent 2,732,409, January 24, 1956, assigned to United States Rubber Co.

DIFENOXURON

Function: Herbicide (1)(2)

Chemical Name: N'-[4-(4-methoxyphenoxy)phenyl]-N,N-dimethylurea

Formula:

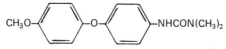 CH$_3$O— —O— —NHCON(CH$_3$)$_2$

Trade Names: C-3470 (Ciba-Geigy Ltd.)
Lironion® (Ciba-Geigy Ltd.)

Manufacture (2)

4-(4-methoxyphenoxy)phenyl isocyanate is dissolved in acetonitrile and stirred into a 40% solution of dimethylamine in water. The product is rapidly precipitated. Stirring is continued for some hours, the product is filtered with suction and washed with large quantities of water and a little dilute acetic acid. The product is dried in vacuuo at 60°C. The product melts at 138° to 139°C.

Toxicity

The acute oral LD_{50} value for rats is more than 7,750 mg/kg which is insignificantly toxic.

References

(1) Worthing, C.R., *Pesticide Manual*, 6th ed., p. 191, British Crop Protection Council (1979).
(2) Ciba, Ltd., British Patent 913,383, December 19, 1962.

DIFENZOQUAT METHYL SULFATE

Function: Herbicide (1)(2)

Chemical Name: 1,2-Dimethyl-3,5-diphenyl-1H-pyrazolium methyl sulfate

Formula:

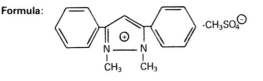

$\cdot CH_3SO_4^{\ominus}$

Trade Names: AC 84,777 (American Cyanamid Co.)
Avenge® (American Cyanamid Co.)
Finaven® (American Cyanamid Co.)

Manufacture (2)

(A) *Preparation of 1-Methyl-3,5-Diphenylpyrazole*—5.0 g (0.022 mols) of di-benzoylmethane in 40 ml of n-propyl alcohol is heated to 80° to 90°C, and 10.5 g of methyl hydrazine in 10 ml of n-propyl alcohol slowly added thereto. The mixture is maintained at 95°C for 30 minutes and then poured into 600 ml of ice water.

An oil separates and turns solid after one-half hour. The solid is filtered, washed with cold water and dried to give 5.2 g of the desired product having a MP of 58° to 59°C.

(B) *Preparation of 1,2-Dimethyl-3,5-Diphenylpyrazolium Methyl Sulfate*—5.0 g of 1-methyl-3,5-diphenylpyrazole is dissolved in 30 ml of dry xylene with heating and constant stirring. The solution is cooled to 60°C, and 2.78 g of dimethyl sulfate is added in 10 ml of xylene. The mixture is then heated to 100°C for 6 hours and allowed to cool. After cooling, the mixture is filtered. The solid which is recovered is stirred with dry acetone and the mixture filtered. This yields 3.91 g of the methyl sulfate, 50.7% yield, having a MP of 146° to 148°C.

Toxicity

The acute oral LD_{50} value for rats is 470 mg/kg which is moderately toxic.

Allowable Limits on Exposure and Use

Product Use: The tolerances set by the US EPA for difenzoquat methyl sulfate in or on raw agricultural commodities are as follows:

	40 *CFR* Reference	Parts per Million
Barley, grain	180.369	0.2
Barley, straw	180.369	20.0
Cattle, fat	180.369	0.05
Cattle, mbyp	180.369	0.05
Cattle, meat	180.369	0.05
Goats, fat	180.369	0.05
Goats, mbyp	180.369	0.05
Goats, meat	180.369	0.05
Hogs, fat	180.369	0.05
Hogs, mbyp	180.369	0.05
Hogs, meat	180.369	0.05
Horses, fat	180.369	0.05
Horses, mbyp	180.369	0.05

(continued)

	40 *CFR* Reference	Parts per Million
Horses, meat	180.369	0.05
Poultry, fat	180.369	0.05
Poultry, mbyp	180.369	0.05
Poultry, meat	180.369	0.05
Sheep, fat	180.369	0.05
Sheep, mbyp	180.369	0.05
Sheep, meat	180.369	0.05
Wheat, straw	180.369	20.0
Wheat, grain	180.369	0.05

References

(1) Worthing, C.R., *Pesticide Manual,* 6th ed., p. 192, British Crop Protection Council (1979).
(2) Walworth, B.L. and Klingsberg, E., U.S. Patent 3,882,142, May 6, 1975, assigned to American Cyanamid Co.

DIFLUBENZURON)
(formerly Difluron)

Function: Insecticide (1)(2)(3)(4)

Chemical Name: N-[[(4-chlorophenyl)amino]carbonyl]-2,6-difluorobenzamide

Formula:

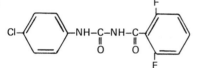

Trade Names: PH60-40 (Philips)
TH6040 (Philips)
Ent 29054 (Philips)
OMS 1804 (Philips)
PDD 6040-1 (Philips)
Dimilin® (Philips)

Manufacture (2)(3)(4)

2,6-difluorobenzamide is reacted with 4-chlorophenyl isocyanate in an organic solvent medium. The product isolated from the reaction mixture has a MP of 210° to 230°C.

Toxicity

The acute oral LD_{50} value for rats is more than 4,640 mg/kg which rates as slightly to insignificantly toxic.

Allowable Limits on Exposure and Use

Product Use: Issuance of a rebuttable presumption against registration was being considered by US EPA as of September 1979 on the basis of possible oncogenicity, hazard to wildlife and other chronic effects.

The tolerances set by the US EPA for diflubenzuron in or on raw agricultural commodities are as follows:

	40 *CFR* Reference	Parts per Million
Cattle, fat	180.377	0.05
Cattle, mbyp	180.377	0.05
Cattle, meat	180.377	0.05
Eggs	180.377	0.05
Goats, fat	180.377	0.05
Goats, mbyp	180.377	0.05
Goats, meat	180.377	0.05
Hogs, fat	180.377	0.05
Hogs, mbyp	180.377	0.05
Hogs, meat	180.377	0.05
Horses, fat	180.377	0.05
Horses, mbyp	180.377	0.05
Horses, meat	180.377	0.05
Milk	180.377	0.05
Poultry, fat	180.377	0.05
Poultry, mbyp	180.377	0.05
Poultry, meat	180.377	0.05
Sheep, fat	180.377	0.05
Sheep, mbyp	180.377	0.05
Sheep, meat	180.377	0.05

References

(1) Worthing, C.R., *Pesticide Manual,* 6th ed., p. 193, British Crop Protection Council (1979).
(2) Wellinga, K. and Mulder, R., U.S. Patent 3,748,356, July 24, 1973, assigned to U.S. Philips Corp.
(3) NV Philips Gloeilampenfabrieken, British Patent 1,324,293, July 25, 1973.
(4) Wellinga, K. and Mulder, R., U.S. Patent 4,013,717, March 22, 1977, assigned to U.S. Philips Corp.

DIMEFOX

Function: Systemic insecticide and acaricide (1)(2)

Chemical Name: Tetramethylphosphoro diamidic fluoride

Formula:
$$[(CH_3)_2N]_2\!-\!\overset{\displaystyle O}{\overset{\|}{P}}\!-\!F$$

Trade Names: Pestox IV® (Pest Control, Ltd.)
Terra Sytam® (Murphy Chemical, Ltd.)
S-14® (Wacker-Chemie)
Hanane®

Manufacture (3)(5)

To a mixture of 50 cc (75 g) chloroform and 157 g methyl dibutylamine and 76.5 g phosphorus oxychloride, was added with agitation and cooling, 45 g of dimethylamine dissolved in 150 cc CHCl$_3$, the temperature being kept at 33°±5°C.

After addition of all the dimethylamine, the mixture was cooled to 25°C, and a solution of 29 g of potassium fluoride in 30 cc of water added.

Potassium chloride was immediately precipitated and the temperature of the reaction mixture rose, over 30 minutes, by 14°C. It was stirred for a further 15 minutes.

The product of reaction was isolated by adding 50 g of sodium hydroxide dissolved in 300 cc of water, separating the organic layer, and largely removing the chloroform by distillation to a liquid temperature of 140°C followed by washing the mixture of methyl dibutylamine and bisdimethylaminofluorophosphine oxide with 2 x 50 cc of water, by which the bisdimethylaminofluorophosphine oxide is sufficiently removed from the tertiary base.

From the aqueous solution of the product the pure bisdimethylaminofluorophosphine oxide was obtained by extraction with chloroform, followed by fractional distillation under reduced pressure.

The use of organophosphorus compounds as activators for bisdimethylaminofluorophosphine oxide (BFPO) has also been described (4).

Product Wastes and Their Control

Product Wastes: This compound is stable in aqueous solutions and is resistant to hydrolysis by alkali. It is hydrolyzed by acids, slowly oxidized by vigorous oxidizing agents, and rapidly oxidized by chlorine. Treatment with acids followed by bleaching powder has been recommended for decomposition. Acid hydrolysis yields amine. Equipment contaminated with dimefox can be decontaminated with hypochlorite (B-3).

Toxicity

The acute oral LD_{50} value for rats is 1-2 mg/kg which is extremely toxic.

References

(1) Worthing, C.R., *Pesticide Manual,* 6th ed., p. 196, British Crop Protection Council (1979).
(2) Spencer, E.Y., *Guide to the Chemicals Used in Crop Protection,* 6th ed., p. 205, London, Ontario, Agriculture Canada (January 1973).
(3) Pound, D.W., British Patent 688,760, assigned to Pest Control Ltd. (March 11, 1953).
(4) Murphy Chemical Co. Ltd., British Patent 741,662 (December 7, 1955).
(5) McCombie, H., Saunders, B.C., Chapman, N.B. and Heap, R., U.S. Patent 2,489,917, November 29, 1949.

DIMETHACHLOR

Function: Herbicide (1)(2)

Chemical Name: 2-chloro-N-(2,6-dimethylphenyl)-N-(2-methoxyethyl) acetamide

Formula:

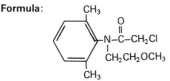

Trade Names: CGA 17020 (Ciba-Geigy Ltd.)
Teridox® (Ciba-Geigy Ltd.)

Manufacture (2)

A solution of 24.2 (0.2 mol) of 2,6-dimethylaniline and 17.6 g (0.24 mol) of methoxy-acetalde-hyde in 150 ml of benzene is treated with 1 ml of 25% trimethylamine solution in methanol and the mixture is refluxed for 5 hours using a steam trap. The reaction mixture is evaporated in vacuo and vacuum distillation of the residue yields the 1-(2'-methoxy-ethylidene-amino)-2,6-dimethylbenzene with a BP of 58° to 61°C at 0.1 torr.

A solution of 16.3 g (0.092 mol) of this intermediate product in 200 ml of absolute ethanol is hydrogenated at 25°C under normal pressure with the addition of 2 g of 5% palladium char-coal. After filtering off the catalyst and evaporating the filtrate in vacuo, vacuum distillation of the residue yields the N-(2'-methoxyethyl)-2,6-xylidine with a BP of 64° to 65°C at 0.2 torr.

A suspension of 4.4 g of N-(2'-methoxyethyl)-2,6-xylidine and 2.6 g of potassium hydrogen carbonate in 30 ml of absolute benzene is treated dropwise with a solution of 2.94 g of chloroacetyl chloride in 10 ml of benzene, whereupon the mixture is subsequently further stirred for 2 hours at 25°C. For processing it is diluted with 100 ml of ether. The organic phase is washed repeatedly with water and dried. The desired N-(2'-methoxy-ethyl)-2,6-di-methyl-chloroacetanilide is obtained pure and in quantitative yield in the form of an oil by evaporating off the solvent. The compound crystallizes on being left to stand at low temper-ature; MP 42° to 45°C.

Toxicity

The acute oral LD_{50} value for rats is 1,600 mg/kg which is slightly toxic.

References

(1) Worthing, C.R., *Pesticide Manual,* 6th ed., p. 197, British Crop Protection Council (1979).
(2) Ciba-Geigy AG, British Patent 1,422,473, January 28, 1976.

DIMETHAMETRYN

Function: Herbicide (1)

Chemical Name: N-(1,2-dimethylpropyl)-N'-ethyl-6-(methylthio)-1,3,5-triazine-2,4-diamine

Formula:

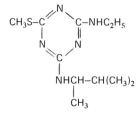

Trade Name: C-18898 (Ciba-Geigy)

Manufacture (2)

The 2,4-dichloro-6-(α,β-dimethylpropylamino)-1,3,5-triazine required as starting material is prepared as follows. Four hundred and sixty parts of cyanuric chloride are dissolved in 2,500 parts by volume of toluene, and while stirring at 0° to 5°C, there are added, drop by drop to this solution, 275 parts of α,β-dimethylpropylamine and then 100 parts of sodium hydroxide dissolved in 500 parts of water. The mixture is stirred at 0°C until it displays a neutral reaction. The organic phase is separated, dried over anhydrous sodium sulfate, filtered and evaporated. The residue is crystallized from hexane. Melting point is 63° to 65°C.

Forty-eight parts of methylmercaptan and 1 g of trimethylamine are introduced into a solution of 199.5 parts of 2,4-dichloro-6-(α,β-dimethylpropylamino)-1,3,5-triazine in 500 parts by volume of toluene and then, while stirring at 0°C, a solution of 40 parts of sodium hydroxide in 200 parts of water is added drop by drop, and the mixture is stirred at 0°C, until it displays a neutral reaction. The organic phase is separated, dried and evaporated. The residue is distilled in a high vacuum. The product boils at 146°C, under a pressure of 0.4 mm Hg.

The resulting 2-chloro-4-(α,β-dimethylpropylamino)-6-methylmercapto-1,3,5-triazine is dissolved in toluene and, while stirring and supplying moderate cooling, aqueous ethylamine of 32% strength is added drop by drop. The mixture is stirred for 14 hours at 40°C, and then diluted with 100 parts by volume of toluene and 100 parts of water. The organic phase is separated, dried and evaporated and the residue distilled to give the dimethametryn product. The product boils at 151° to 153°C under a pressure of 0.5 mm Hg.

Toxicity

The acute oral LD_{50} value for rats is 3,000 mg/kg which is slightly toxic.

References

(1) Worthing, C.R., *Pesticide Manual,* 6th ed., p. 198, British Crop Protection Council (1979).
(2) Nikles, E., U.S. Patent 3,799,925, March 26, 1974, assigned to Ciba-Geigy Corporation.

DIMETHIRIMOL

Function: Systemic fungicide (1)(2)

Chemical Name: 5-butyl-2-(dimethylamino)-6-methyl-4-pyrimidinol

Formula:

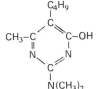

Trade Names: PP-675® (I.C.I.)
Milcurb® (Plant Protection, Ltd.)

Manufacture

Dimethirimol may be made by one of two routes:

    (A) Condensation of 2-butylethyl acetoacetate with N,N-dimethylguanidine sulfate (3); or

    (B) Condensation of dimethylamine acetate with 5-butyl-4-hydroxy-6-methyl-2-methylthiopyrimidine.

Toxicity

The acute oral LD_{50} value for rats is 2,350 mg/kg which is slightly toxic.

References

(1) Worthing, C.R., *Pesticide Manual*, 6th ed., p. 199, British Crop Protection Council (1979).
(2) Spencer, E.Y., *Guide to the Chemicals Used in Crop Protection*, 6th ed., p. 208, London, Ontario, Agriculture Canada (January 1973).
(3) Snell, B.K., Elias, R.S. and Freeman, P.F.H., British Patent 1,182,584, February 25, 1970, assigned to Imperial Chemical Industries, Ltd.

DIMETHOATE (FOSFAMID IN U.S.S.R.)

Function: Insecticide and acaricide (1)(2)

Chemical Name: O,O-dimethyl-S-[2-(methylamino)-2-oxoethyl] phosphorodithioate

Formula:

$$(CH_3O)_2\underset{\underset{S}{\|}}{P}SCH_2\underset{\underset{O}{\|}}{C}N\underset{CH_3}{\overset{H}{<}}$$

Trade Names: EI 12,880 (American Cyanamid Co.)
Cygon® (American Cyanamid Co.)
Dimetate® (American Cyanamid Co.)
L395 (Montecatini)
Fostion MM® (Montecatini)
Rogor® (Montecatini)
De-Fend®
Ferkethion®
Perfekthion® (BASF)
Roxion® (Cela)

Manufacture (3)(4)(5)(6)

The following is a specific example (3) of the conduct of the production of dimethoate. 42.9 g of O,O-dimethyl dithiophosphoric acid were added slowly to a well agitated mixture of 28.7 g of anhydrous sodium carbonate suspended in 100 cc of methyl isobutyl ketone. The mixture was then heated to about 65°C and 25.3 g of N-methyl chloroacetamide were added. The resulting mixture was held at 80°C for one hour with continued stirring, then cooled to room temperature and filtered. The filtrate was washed twice with water, dried over anhydrous sodium sulfate, and filtered. The filtrate was heated under vacuum to remove the methyl isobutyl ketone. The residual product, S-carbamylmethyl O,O-dimethyl dithiophosphate, was a colorless crystalline solid melting at 60° to 62°C.

Process Wastes and Their Control

Air: Air emissions from dimethoate manufacture have been reported (B-15) as follows:

Component	Kilograms per Metric Ton Pesticide Produced
SO_2	0.3
NO_2	0.3
Hydrocarbons	2.0
Dimethoate	0.5

Product Wastes: Dimethoate is thermally unstable and decomposes on heating after first being converted to the more toxic dithio isomer. It is hydrolyzed more rapidly in alkaline medium: 50% hydrolysis at 70°C at pH 9, in 0.8 hour, and 21 hours at pH 2 (B-3).

Toxicity

The acute oral LD_{50} value for rats is about 350 mg/kg which is moderately toxic.

Allowable Limits on Exposure and Use

Product Use: A rebuttable presumption against registration was issued for dimethoate on September 12, 1977 on the basis of oncogenicity, mutagenicity, fetotoxicity, and reproductive effects.

The tolerances set by the US EPA for dimethoate in or on raw agricultural commodities are as follows:

	40 *CFR* Reference	Parts per Million
Alfalfa	180.204	2.0
Apples	180.204	2.0
Beans, dry	180.204	2.0
Beans, lima	180.204	2.0
Beans, snap	180.204	2.0
Broccoli	180.204	2.0
Cabbage	180.204	2.0
Cattle, fat	180.204	0.02 N
Cattle, mbyp	180.204	0.02 N
Cattle, meat	180.204	0.02 N
Cauliflower	180.204	2.0
Celery	180.204	2.0
Collards	180.204	2.0
Corn, fodder	180.204	1.0
Corn, forage	180.204	1.0
Corn, grain	180.204	0.1 N
Cotton, seed	180.204	0.1
Eggs	180.204	0.02 N
Endive (escarole)	180.204	2.0
Goats, fat	180.204	0.02 N
Goats, mbyp	180.204	0.02 N
Goats, meat	180.204	0.02 N
Grapefruit	180.204	2.0
Grapes	180.204	1.0
Hogs, fat	180.204	0.02 N
Hogs, mbyp	180.204	0.02 N
Hogs, meat	180.204	0.02 N
Horses, fat	180.204	0.02 N
Horses, mbyp	180.204	0.02 N
Horses, meat	180.204	0.02 N
Kale	180.204	2.0
Lemons	180.204	2.0
Lettuce	180.204	2.0
Melons	180.204	1.0
Milk	180.204	0.002 N
Mustard, greens	180.204	2.0
Oranges	180.204	2.0
Pears	180.204	2.0
Peas	180.204	2.0
Pecans	180.204	0.1
Peppers	180.204	2.0

(continued)

	40 *CFR* Reference	Parts per Million
Potatoes	180.204	0.2
Poultry, fat	180.204	0.02 N
Poultry, mbyp	180.204	0.02 N
Poultry, meat	180.204	0.02 N
Safflower, seed	180.204	0.1
Sheep, fat	180.204	0.02 N
Sheep, mbyp	180.204	0.02 N
Sheep, meat	180.204	0.02 N
Sorghum, forage	180.204	0.2
Sorghum, grain	180.204	0.1
Soybeans	180.204	0.05 N
Soybeans, forage	180.204	2.0
Soybeans, hay	180.204	2.0
Spinach	180.204	2.0
Swiss chard	180.204	2.0
Tangerines	180.204	2.0
Tomatoes	180.204	2.0
Turnips, roots	180.204	2.0
Turnips, tops	180.204	2.0
Wheat, fodder, green	180.204	2.0
Wheat, grain	180.204	0.04 N
Wheat, straw	180.204	2.0

The tolerances set by the EPA for dimethoate in animal feeds are as follows (the *CFR* Reference is to Title 21):

	CFR Reference	Parts per Million
Citrus pulp, dried (ct.f)	561.170	5.0

References

(1) Worthing, C.R., *Pesticide Manual,* 6th ed., p. 200, British Crop Protection Council (1979).
(2) Spencer, E.Y., *Guide to the Chemicals Used in Crop Protection,* 6th ed., p. 209, London, Ontario, Agriculture Canada (January 1973).
(3) Cassady, J.T., Hoegberg, E.I. and Gleissner, B.D., U.S. Patent 2,494,283, January 10, 1950, assigned to American Cyanamid Co.
(4) Young, R.W. and Clark, E.L., U.S. Patent 3,210,242, October 5, 1965, assigned to American Cyanamid Co.
(5) Montecatini Soc. Generale per L'Industria Mineraria e Chimica, British Patent 791,824, March 12, 1958.
(6) Young, R.W., U.S. Patent 2,996,531, August 15, 1961, assigned to American Cyanamid Co.

1,3-DI-(METHOXYCARBONYL)-1-PROPEN-2-YL DIMETHYL PHOSPHATE

Function: Contact insecticide and acaricide (1)

Chemical Name: Dimethyl 3-[(dimethoxyphosphinyl)oxy]-2-pentendioate

Formula:

$$(CH_3O)_2\overset{\overset{\displaystyle O}{\|}}{P}OC\!=\!CHCOOCH_3$$
$$\underset{\displaystyle CH_2COOCH_3}{|}$$

Trade Names: GC-307 (Allied Chemical Corp.)
Bomyl® (Allied Chemical Corp.)

Manufacture (2)

A dialkyl acetone dicarboxylate is reacted with sodium metal dissolved in alcohol to form an intermediate sodium compound, which is then reacted with dialkyl chlorophosphonate to produce the vinyl ester phosphate. These reactions are represented by the following equations:

$$2RO\!-\!\overset{\overset{\displaystyle O}{\|}}{C}\!-\!CH_2\!-\!\overset{\overset{\displaystyle O}{\|}}{C}\!-\!CH_2\!-\!COOR \ + \ 2Na \ \longrightarrow \ 2RO\!-\!\overset{\overset{\displaystyle O}{\|}}{C}\!-\!CH\!=\!\overset{\overset{\displaystyle ONa}{|}}{C}\!-\!CH_2\!-\!COOR \ + \ H_2$$

$$RO\!-\!\overset{\overset{\displaystyle O}{\|}}{C}\!-\!CH\!=\!\overset{\overset{\displaystyle ONa}{|}}{C}\!-\!CH_2\!-\!COOR \ + \ (R'O)_2POCl \ \longrightarrow \ RO\!-\!\overset{\overset{\displaystyle O}{\|}}{C}\!-\!CH\!=\!\underset{\displaystyle CH_2\!-\!COOR}{\overset{\overset{\displaystyle O}{\|}}{C}\!-\!O\!-\!P\!-\!(OR')_2} \ + \ NaCl$$

According to another method, a dialkyl acetone dicarboxylate is reacted with sulfuryl chloride to produce an intermediate chlorinated compound which is then reacted with a trialkyl phosphite to produce the vinyl ester phosphate. The equations representing these reactions are as follows:

$$RO\!-\!\overset{\overset{\displaystyle O}{\|}}{C}\!-\!CH_2\!-\!\overset{\overset{\displaystyle O}{\|}}{C}\!-\!CH_2\!-\!COOR \ + \ SO_2Cl_2 \ \longrightarrow \ RO\!-\!\overset{\overset{\displaystyle O}{\|}}{C}\!-\!CHCl\!-\!\overset{\overset{\displaystyle O}{\|}}{C}\!-\!CH_2\!-\!COOR \ + \ SO_2 \ + \ HCl$$

$$RO\!-\!\overset{\overset{\displaystyle O}{\|}}{C}\!-\!CHCl\!-\!\overset{\overset{\displaystyle O}{\|}}{C}\!-\!CH_2\!-\!COOR \ + \ (R'O)_3P \ \longrightarrow \ RO\!-\!\overset{\overset{\displaystyle O}{\|}}{C}\!-\!CH\!=\!\underset{\displaystyle CH_2\!-\!COOR}{\overset{\overset{\displaystyle O}{\|}}{C}\!-\!O\!-\!P\!-\!(OR')_2} \ + \ R'Cl$$

The last-described method was employed in the following example illustrating the production of a typical vinyl ester phosphate by the process. In the example, parts are by weight.

Example—Two hundred and six parts of dimethyl acetone dicarboxylate were placed in a reaction vessel. One hundred and thirty-five parts of sulfuryl chloride were added dropwise to the reaction vessel with stirring over a 15 minute period at a temperature of 30° to 40°C. After addition of the sulfuryl chloride, the reaction mixture was heated for about 15 minutes at 80°C to complete the reaction. The chlorinated product was cooled to 20°C, and 124 parts of trimethyl phosphite were added dropwise with stirring and external cooling to maintain the reaction temperature at 35° to 40°C. The reaction mixture was then heated to 55°C and finally heated on a steam bath with a gentle flow of air for 2 hours to remove the methyl chloride formed during the reaction. A yield of 318 parts of a yellow oil was obtained. Upon distillation of the yellow oil, the vinyl ester phosphate was recovered as the fraction boiling between 155° and 164°C at 1.7 mm Hg.

Process Wastes and Their Control

Product Wastes: This compound is hydrolyzed by alkali (50% hydrolyzed at pH 5 after more than 10 days, at pH 6 after more than 4 days, at pH 9 less than 1 day) (B-3).

Toxicity

The acute oral LD_{50} value for rats is 32 mg/kg which is highly toxic.

References

(1) Worthing, C.R., *Pesticide Manual,* 6th ed, p. 201, British Crop Protection Council (1979).
(2) Gilbert, E.E., U.S. Patent 2,891,887, June 23, 1959, assigned to Allied Chemical Corp.

DIMETHYL 4-(METHYLTHIO)PHENYL PHOSPHATE

Function: Insecticide (1)(2)

Chemical Name: See common name

Formula:

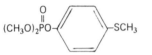

Trade Name: GC-6506 (Allied Chemical Corp.)

Manufacture (2)

O,O-dimethyl-O-(4-methylmercaptophenyl) phosphate may be prepared by reacting 4-methyl-mercaptophenol with dimethyl chlorophosphate in the presence of a basic material, such as potassium carbonate, triethylamine or pyridine. If desired, a catalyst such as copper may be used in order to accelerate the reaction. The reaction is preferably carried out in an inert organic solvent, such as benzene, toluene, xylene or acetone. The preparation of O,O-di-methyl-O-(4-methylmercapotophenyl) phosphate using potassium carbonate as the basic material may be represented by the following equation:

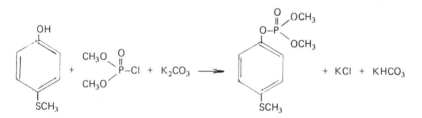

In a typical example of the conduct of this process, 14 parts of 4-methylmercaptophenol and 50 parts of benzene were placed in a reaction vessel. To this solution were added 14 parts of potassium carbonate and about 0.3 part of powdered copper. The resulting mixture was heated to a temperature of 50° to 60°C, and 144.5 parts of dimethyl chlorophosphate were slowly added with stirring.

The mixture was then heated at reflux with continued stirring for about 3½ hours. After this period, the mixture was filtered and the precipitate washed with benzene and then with acetone. Residual solvent was stripped off at reduced pressure, and a liquid residue of 22.5 parts of O,O-dimethyl-O-(4-methylmercaptophenyl) phosphate, corresponding to a yield of about 65% of theory, was obtained.

Process Wastes and Their Control

Product Wastes: The alkaline hydrolysis rate of this compound at pH 9.5 and 37.5°C is 3.6 x 10^{-3} sec^{-1} (B-3).

Toxicity

The acute oral LD$_{50}$ value for rats is about 7 mg/kg which is highly to extremely toxic.

References

(1) Worthing, C.R., *Pesticide Manual,* 6th ed., p. 203, British Crop Protection Council (1979).
(2) Gilbert, E.E. and Otto, J.A., U.S. Patent 3,151,022, September 29, 1964, assigned to Allied Chemical Corp.

DIMETHYL PHTHALATE

Function: Insect repellent (1)(2)

Chemical Name: Dimethyl 1,2-benzenedicarboxylate

Formula:

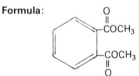

Trade Names: DMP
NTM

Manufacture (3)

The reactants in dimethyl phthalate production are phthalic acid and methanol which react according to the following equation.

$$C_6H_4(COOH)_2 + 2CH_3OH \rightarrow C_6H_4(COOCH_3)_2 + 2H_2O$$

The reaction may be carried out in a plate column-type reactor as described by W.M. Billing (3).

The reaction is carried out at 180° to 350°C and preferably at about 250°C. The reaction pressure is in the range of 50 to 1,000 psi, preferably from 200 to 600 psi. The residence time in the reaction zone is about two hours.

The reaction is carried out in the liquid phase in the absence of any added reaction solvents. The reaction mass actually is in a mixed vapor-liquid phase condition because operating temperature and pressure are interrelated so the alcohol is maintained above its boiling point, with a portion of the vapor dissolved in the liquid phase. This reaction is carried out in the absence of any added catalyst. The products from the reactor may be fed to a separator, such as a centrifugal separator, in which the high boiling ester is separated from any alcohol.

Toxicity

The acute oral LD$_{50}$ value for rats is 8,200 mg/kg which is insignificantly toxic.

Allowable Limits on Exposure and Use

Air: The threshold limit value for dimethyl phthalate in air is 5 mg/m³ as of 1979. The tentative short term exposure limit is 10 mg/m³ (B-23).

Water: In water, a criterion was set by EPA (B-12) at a level of 3 μg/l of phthalate esters for the protection of freshwater aquatic life.

In water, EPA has subsequently suggested (B-26) limits for the protection of human health from the toxic properties of dimethyl phthalate ingested through water and through contaminated aquatic organisms of 160 mg/l.

References

(1) Worthing, C.R., *Pesticide Manual,* 6th ed., p. 204, British Crop Protection Council (1979).
(2) Spencer, E.Y., *Guide to the Chemicals Used in Crop Protection,* 6th ed., p. 219, London, Ontario, Agriculture Canada (January 1973).
(3) Billing, W.M., U.S. Patent 2,813,891, November 19, 1957, assigned to Hercules Powder Company.

DIMETILAN

Function: Stomach insecticide (specific for flies) (1)(2)

Chemical Name: 1-[(dimethylamino)carbonyl]-5-methyl-1H-pyrazol-3-yl dimethylcarbamate

Formula:

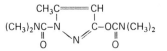

Trade Names: G22870 (Ciba-Geigy)
GS 13332 (Ciba-Geigy)
Snip® (Ciba-Geigy)
Dimetilane®

Manufacture (3)(4)

Dimetilan may be produced by the reaction of methylpyrazolone with two equivalents of dimethyl carbamoyl chloride.

Toxicity

The acute oral LD₅₀ value for rats is 64 mg/kg which is moderately toxic.

References

(1) Worthing, C.R., *Pesticide Manual,* 6th ed., p. 205, British Crop Protection Council (1979).
(2) Spencer, E.Y., *Guide to the Chemicals Used in Crop Protection,* 6th ed., p. 222, London, Ontario, Agriculture Canada (January 1973).
(3) Gysin, H., Margot, A. and Simon, C., U.S. Patent 2,681,879, June 22, 1954, assigned to J.R. Geigy AG.
(4) Grauer, T. and Urwyler, H., U.S. Patent 3,452,043, June 24, 1969, assigned to J.R. Geigy AG.

DINITRAMINE

Function: Herbicide (1)(2)

Chemical Name: N¹,N¹-diethyl-2,6-dinitro-4-(trifluoromethyl)-1,3-benzenediamine

Formula:

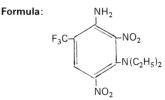

Trade Names: USB 3584 (U.S. Borax and Chemical Corp.)
Cobex® (U.S. Borax and Chemical Corp.)

Manufacture (3)

The 2,4-dihalo-3,5-dinitrobenzotrifluoride starting materials are readily prepared (3) by nitration of the 2,4-dihalobenzotrifluoride with a mixture of fuming nitric and fuming sulfuric acids at a temperature below about 80°C.

A heavy walled glass reaction tube of about 50 ml capacity was charged with 5 g (0.0163 mol) of 2,4-dichloro-3,5-dinitrobenzotrifluoride, 30 ml of ethanol and 10 ml of diethylamine. The reaction tube was sealed and heated in a bath at 94° to 99°C for 46.5 hours. The cooled reaction tube was then opened and the contents evaporated to dryness to give a solid residue. The residue was extracted with 200 ml of refluxing diethyl ether and the insoluble amine hydrochloride was separated therefrom by filtration. The ether extract was evaporated to dryness and the residue dissolved in 40 ml of refluxing 95% ethanol and 10 ml of water. The product was subjected to ammoniation to give dinitramine.

Toxicity

The acute oral LD_{50} value for rats is 3,000 mg/kg which rates as slightly toxic.

Allowable Limits on Exposure and Use

Product Use: The tolerances set by the EPA for dinitramine in or on raw agricultural commodities are as follows:

	40 *CFR* Reference	Parts per Million
Beans, dry	180.327	0.05 N
Beans, garbanzo	180.327	0.05 N
Beans, green	180.327	0.05 N
Beans, lima	180.327	0.05 N
Beans, vines, hay	180.327	0.05 N
Cotton, forage	180.327	0.05 N
Cotton, seed	180.327	0.05 N
Lentils	180.327	0.05 N
Peanuts	180.327	0.05 N
Peanuts, forage	180.327	0.05 N
Peanuts, hulls	180.327	0.05 N
Peanuts, vines, hay	180.327	0.05 N
Peas	180.327	0.05 N
Peas, forage	180.327	0.05 N

(continued)

	40 *CFR* Reference	Parts per Million
Soybeans	180.327	0.05 N
Soybeans, forage	180.327	0.05 N
Sunflower, forage	180.327	0.05 N
Sunflower, seed	180.327	0.05 N

References

(1) Worthing, C.R., *Pesticide Manual,* 6th ed., p. 207, British Crop Protection Council (1979).
(2) Spencer, E.Y., *Guide to the Chemicals Used in Crop Protection,* 6th ed., p. 127, London, Ontario, Agriculture Canada (January 1973).
(3) Hunter, D.L., Woods, W.G., Stone, J.D. and LeFevre, C.W., U.S. Patent 3,617,252, November 2, 1971, assigned to United States Borax and Chemical Corporation.

DINOBUTON

Function: Acaricide and fungicide (1)(2)

Chemical Name: 1-methylethyl-2-(1-methylpropyl)-4,6-dinitrophenyl carbonate

Formula:

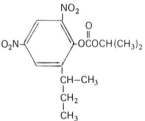

Trade Names: MC 1053 (Murphy Chemical Co., Ltd.)
Acrex® (Murphy Chemical Co., Ltd.)
Sytasol® (Murphy Chemical Co., Ltd.)
UC 19786 (Union Carbide)
Dessin® (Union Carbide)
Dinofen® (J and A Margesin)
Talan® (Monteshell)

Manufacture (3)(4)

The following is a typical example of dinobuton preparation (3). 2,4-Dinitro-6-sec-butyl-phenol (96.6% pure, 49.7 g) was dissolved with stirring in acetone (400 cc). To the solution was added potassium hydroxide (11.2 g) and the mixture was stirred till all reactants were in solution. To this solution was added, all at once, a solution of isopropyl chloroformate (96% pure, 25.3 g) in acetone (100 cc). The mixture was refluxed for 2 hours.

After cooling, the precipitated potassium chloride was filtered off, washed with a little acetone, dried and weighed. Weight of potassium chloride was 14.5 g (98% of theoretical). The acetone was then stripped off. To the residue (71.2 g) was added methanol (100 cc) and the solution stirred. Crystallization started immediately and continued overnight. The crystalline solid was filtered off and washed on the filter with very little methanol to give almost colorless washings.

The white crystalline solid (first crop) weighed 46.6 g (71.5% of theoretical), MP 55° to 57°C. The mother liquor (without the washings) was kept overnight at −9°C. The second crop of a white crystalline solid was filtered off weighing 7.8 g (12% of theoretical), MP 54° to 57°C. On recrystallization from petrol, BP 40° to 60°C, almost white crystals, melting at 56° to 57°C, were obtained. Found: N, 8.46%; $C_{11}H_{18}N_2O_7$ requires N, 8.59%.

Total yield of isopropyl 2,4-dinitro-6-sec-butylphenyl carbonate was therefore 54.4 g (83.5% of theoretical). By repeating this method but using sodium carbonate or potassium carbonate instead of potassium hydroxide, isopropyl 2,4-dinitro-sec-butylphenyl carbonate was obtained in 81% yield and 91% yield respectively. The basic equations underlying the preparation of this compound are as follows:

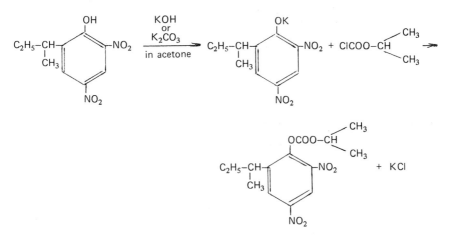

Process Wastes and Their Control

Product Wastes: Dinobuton is easily converted by hydrolysis into the corresponding dinitrophenol (dinoseb) which is even more toxic than the original product (B-3).

Toxicity

The acute oral LD_{50} for rats is 59 mg/kg (B-5) which is moderately to highly toxic. A value of 140 mg/kg (1) has also been published.

References

(1) Worthing, C.R., *Pesticide Manual,* 6th ed., p. 207, British Crop Protection Council (1979).
(2) Spencer, E.Y., *Guide to the Chemicals Used in Crop Protection,* 6th ed., p. 224, London, Ontario, Agriculture Canada (January 1973).
(3) Pianka, M. and Polton, D.J., U.S. Patent 3,234,082, February 8, 1966, assigned to Murphy Chemical Co., Ltd.
(4) Pianka, M. and Polton, D.J., U.S. Patent 3,234,260, February 8, 1966, assigned to Murphy Co., Ltd.

DINOCAP

Function: Acaricide and fungicide (1)(2)

Chemical Name: 4-(methylheptyl)-2,6-dinitrophenyl-1-butenoate

Formula:

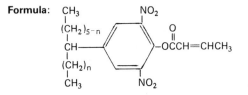

Trade Names: CR-1639 (Rohm and Haas)
Arathane® (Rohm and Haas)
Karathane® (Rohm and Haas)
Iscothan® (Innis, Speiden and Co.)
Crotothane® (May and Baker, Ltd.)
Mildex®
Caprane®

Manufacture

Capryldinitrophenyl crotonate may be prepared (3) from capryl alcohol (octanol-2), and phenol through nitration and esterification. The reaction of capryl alcohol with phenol, as is known, may be effected in the presence of an acidic condensing agent, such as acid clays or an organic sulfonic acid. To a solution of 1.03 kg of acid condensed capryl phenol in 1.875 liters of ethylene dichloride, there was slowly added over the course of 110 minutes 1.125 kg of 70% nitric acid. The temperature was held at 25° to 35°C.

After the acid had been added, the reaction mixture was stirred for one-half hour. It was then washed twice with water in a volume amounting to about half the volume of the reaction mixture. The solvent was then evaporated while the mixture was heated on a water bath under reduced pressure. The residue consisted of capryldinitrophenol together with some caprylmononitrophenol in the form of a dark-red viscous oil.

A solution of 2.08 kg of capryldinitrophenol was prepared with about 1 liter of petroleum ether (boiling range 60° to 100°C) and thereto was added 610 g of refined pyridine. The resulting solution was stirred, and thereto was added dropwise, over the course of 45 minutes, 769 g of crotonyl chloride. The reaction mixture was maintained at 29° to 46°C with the aid of external cooling. Stirring was continued for another 45 minutes and the mixture left standing for 2 days.

About 2 liters of benzene was added and a solution of 1 kg of concentrated hydrochloric acid in 5 liters of water stirred thoroughly therewith. The organic layer was separated from the aqueous layer and washed successively with 5.5 liters of water, twice with 3.5 liters of normal sodium hydroxide solution, and twice with 5 liters of a solution containing 200 g of sodium chloride. The organic layer was then dried over anhydrous sodium sulfate and concentrated on a water bath under reduced pressure. The residue was a viscous dark-brown oil comprising 2.265 kg of technical capryldinitrophenyl crotonate.

An alternative mode of manufacture (4) involves first reacting an alkylphenol with 88 to 98% sulfuric acid in molecular excess at about 25° to 45°C until sulfonation is effected, as indicated by no further heat of reaction or as shown by analysis of the reaction mixture, and adding this sulfonated mixture with agitation to an aqueous 25 to 50% sodium nitrate solution which is at a temperature between 70° and about 100°C at a rate to maintain the resulting mixture with the aid of cooling within this temperature range and preferably between 75° and 85°C.

Layers are allowed to form and are separated. For convenience the organic layer may be taken up in an inert, water-immiscible, volatile organic solvent either before or after its separation to aid in the washing of the product with water. After removal of wash water the solvent is distilled off, if desired, and the product is dried, as by heating under reduced pressure.

There is obtained 95 to 98% pure 2,4-dinitro-6-alkylphenol in a yield of 92 to 96%. Unnitrated phenol is essentially absent as are oxidized and resinous materials. Where further purification is desired, charcoaling or extraction may be practiced or, in the case of solids, recrystallization.

Process Wastes and Their Control

Product Wastes: This fungicide and miticide is chemically similar to the other dinitrophenol derivatives (binapacryl, DNOC, dinoseb, dinobuton, and dinitrocyclohexylphenol). Hydrolysis produces the corresponding dinitrophenol derivative which is more toxic than the original product (B-3).

Toxicity

The acute oral LD_{50} value for rats is 980-1,190 mg/kg which is slightly toxic.

Allowable Limits on Exposure and Use

Product Use: The tolerances set by the EPA for dinocap in or on raw agricultural commodities are as follows:

	40 *CFR* Reference	Parts per Million
Apples	180.341	0.1 N
Apricots	180.341	0.1 N
Blackberries	180.341	0.15 N
Boysenberries	180.341	0.15 N
Caneberries	180.341	0.15 N
Cantaloups	180.341	0.1 N
Cucumbers	180.341	0.1 N
Dewberries	180.341	0.15 N
Gooseberries	180.341	0.15 N
Grapes	180.341	0.1 N
Honeydew	180.341	0.1 N
Loganberries	180.341	0.15 N
Muskmelons	180.341	0.1 N
Nectarines	180.341	0.1 N
Peaches	180.341	0.1 N
Pears	180.341	0.1 N
Pumpkins	180.341	0.1 N
Raspberries	180.341	0.15 N
Squash, summer	180.341	0.1 N
Squash, winter	180.341	0.1 N
Watermelons	180.341	0.1 N

The tolerances set by the EPA for dinocap in animal feeds are as follows (the *CFR* Reference is to Title 21):

	CFR Reference	Parts per Million
Apple pomace, dried	561.200	0.3

References

(1) Worthing, C.R., *Pesticide Manual,* 6th ed., p. 208, British Crop Protection Council (1979).

(2) Spencer, E.Y., *Guide to the Chemicals Used in Crop Protection,* 6th ed., p. 225, London, Ontario, Agriculture Canada (January 1973).

(3) Hester, W.F. and Craig, W.E., U.S. Patent 2,526,660, October 24, 1950, assigned to Rohm and Haas Company.

(4) Clarke, D.g., McKeever, C.H. and Wolffe, E.L., U.S. Patent 2,810,767, October 22, 1957, assigned to Rohm and Haas Company.

DINOSEB

Function: Insecticide and herbicide (1)(2)

Chemical Name: 2-(1-methylpropyl)-4,6-dinitrophenol

Formula:

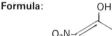
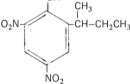

Trade Names: DN 289 (Dow)
Premerge® (Dow)
Gebutox® (Farbwerke Hoechst)
Chemox®
Knox-Weed 55®
Sinox General®
Supersevtox®

Manufacture (3)

The 2,4-dinitro-6-alkylphenols can be prepared by reacting a 6-alkylphenol with concentrated sulfuric acid until the phenol is converted substantially to a sulfonic acid derivative. This reaction product is then dissolved in water, and sufficient nitric acid is gradually added to the solution, with agitation, to convert the sulfonic acid derivative to the desired dinitrophenol compound.

After all of the nitric acid has been added, the reaction mixture is stirred and maintained at a temperature between about 60° and about 80°C for a suitable period of time to insure complete reaction, cooled, and the dinitro derivative separated therefrom, for example, by filtration, extraction with solvent, or decantation.

The crude reaction product so obtained may, if desired, be further purified, as by recrystallization from an organic solvent such as ethyl alcohol, chloroform, chlorobenzene, dilute acetic acid, etc. or by preparing therefrom the corresponding sodium phenolate, which may be recrystallized from water or alcohol and thereafter acidified to recover the desired free phenolic compound.

Process Wastes and Their Control

Air: Air emissions from dinoseb manufacture have been reported (B-15) as follows:

Component	Kilograms per Metric ton Pesticide Produced
Particulates	0.1
SO$_2$	1.2
NO$_x$	0.1
Hydrocarbons	1.0

Toxicity

The acute oral LD$_{50}$ value for rats is 58 mg/kg (moderately toxic).

Allowable Limits on Exposure and Use

Product Use: The tolerances set by the EPA for dinoseb in or on raw agricultural commodities are as follows:

	40 *CFR* Reference	Parts per Million
Alfalfa	180.281	0.1 N
Alfalfa, hay	180.281	0.1 N
Almonds, Hulls	180.281	0.1 N
Apples	180.281	0.1 N
Apricots	180.281	0.1 N
Barley, forage	180.281	0.1 N
Barley, grain	180.281	0.1 N
Barley, straw	180.281	0.1 N
Beans	180.281	0.1 N
Beans, forage	180.281	0.1 N
Beans, hay	180.281	0.1 N
Blackberries	180.281	0.1 N
Blueberries	180.281	0.1 N
Boysenberries	180.281	0.1 N
Cherries	180.281	0.1 N
Citrus fruits	180.281	0.1 N
Clover	180.281	0.1 N
Clover, hay	180.281	0.1 N
Corn, fodder	180.281	0.1 N
Corn, forage	180.281	0.1 N
Corn, fresh (inc. sweet) (k+cwhr)	180.281	0.1 N
Corn, grain	180.281	0.1 N
Cotton, forage	180.281	0.1 N
Cotton, seed	180.281	0.1 N
Cotton, seed, hulls	180.281	0.1 N
Cucurbits	180.281	0.1 N
Currants	180.281	0.1 N
Dates	180.281	0.1 N
Figs	180.281	0.1 N
Filberts	180.281	0.1 N
Garlic	180.281	0.1 N
Gooseberries	180.281	0.1 N
Grapes	180.281	0.1 N
Hops	180.281	0.1 N
Loganberries	180.281	0.1 N
Nectarines	180.281	0.1 N
Nuts, almonds	180.281	0.1 N
Oats	180.281	0.1 N
Oats, grain	180.281	0.1 N
Oats, straw	180.281	0.1 N
Olives	180.281	0.1 N

(continued)

	40 *CFR* Reference	Parts per Million
Onions	180.281	0.1 N
Peaches	180.281	0.1 N
Peanuts	180.281	0.1 N
Peanuts, forage	180.281	0.1 N
Peanuts, hay	180.281	0.1 N
Peanuts, hulls	180.281	0.1 N
Pears	180.281	0.1 N
Peas	180.281	0.1 N
Peas, forage	180.281	0.1 N
Peas, hay	180.281	0.1 N
Pecans	180.281	0.1 N
Plums (prunes)	180.281	0.1 N
Potatoes	180.281	0.1 N
Raspberries	180.281	0.1 N
Rye, forage	180.281	0.1 N
Rye, grain	180.281	0.1 N
Rye, straw	180.281	0.1 N
Soybeans	180.281	0.1 N
Soybeans, forage	180.281	1.0
Soybeans, hay	180.281	1.0
Strawberries	180.281	0.1 N
Trefoil, birdsfoot	180.281	0.1 N
Trefoil, birdsfoot, hay	180.281	0.1 N
Vetch	180.281	0.1 N
Vetch, hay	180.281	0.1 N
Walnuts	180.281	0.1 N
Wheat, forage	180.281	0.1 N
Wheat, grain	180.281	0.1 N
Wheat, straw	180.281	0.1 N

References

(1) Worthing, C.R., *Pesticide Manual,* 6th ed., p. 209, British Crop Protection Council (1979).
(2) Spencer, E.Y., *Guide to the Chemicals Used in Crop Protection,* 6th ed., p. 228, London, Ontario, Agriculture Canada (January 1973).
(3) Mills, L.E. and Fayerweather, B.L., U.S. Patent 2,192,197, March 5, 1940, assigned to Dow Chemical Co.

DINOSEB ACETATE

Function: Herbicide (1)(2)

Chemical Name: 2-(1-methylpropyl)-4,6-dinitrophenyl acetate

Formula:

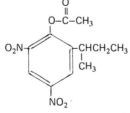

Trade Names: Aretit® (Farbwerke Hoechst)
Ivosit® (Farbwerke Hoechst)
Phenotan® (Farbwerke Hoechst)

Manufacture (3)

The products are prepared in the following manner. Carboxylic acid halides or carboxylic acid anhydrides are caused to act in an anhydrous medium and in a manner which is already known, on dinitro alkylphenols. For example, carboxylic acid halides are caused to react with dinitro alkylphenols in the presence of tertiary amines as catchers for halogen halide, or carboxylic acid anhydrides are boiled with dinitro alkylphenols and catalytic amounts of tertiary amines, or the anhydrous alkali metal salts or alkaline earth metal salts of dinitrophenols, carboxylic acid halides or carboxylic acid anhydrides are caused to react with each other.

In this manner the esters of dinitro-sec-butylphenol and acetic acid, propionic acid, butyric acid, valeric acid and caproic acid are obtained. These esters are syrupy yellow-brown oils which can be distilled in most cases only with decomposition. They are insoluble in water and well soluble in most of the organic solvents. Specifically, dinoseb acetate is made by the reaction of dinoseb with acetic anhydride (or acetyl chloride).

Toxicity

The acute oral LD_{50} value for rats is 60 to 65 mg/kg which is moderately toxic.

References

(1) Worthing, C.R., *Pesticide Manual,* 6th ed., p. 210, British Crop Protection Council (1979).
(2) Spencer, E.Y., *Guide to the Chemicals Used in Crop Protection,* 6th ed., p. 229, London, Ontario, Agriculture Canada (January 1973).
(3) Scherer, O., Reichner, K. and Frensch, H., U.S. Patent 3,130,037, April 21, 1964, assigned to Farbwerke Hoechst AG.

DINOTERB

Function: Herbicide (1)

Chemical Name: 2-(1,1-dimethylethyl)-4,6-dinitrophenol

Formula:

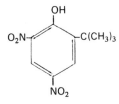

Trade Name: None

Manufacture

Dinoterb is made from tert-butylphenol by nitration to dinitro-tert-butylphenol.

Its use in the form of an alkali metal salt or an ammonium salt or an amine salt has been described (2).

Toxicity

The acute oral LD$_{50}$ value for mice is about 25 mg/kg.

References

(1) Worthing, C.R., *Pesticide Manual,* 6th ed., p. 211, British Crop Protection Council (1979).
(2) Pechiney-Progil, British Patent 1,126,658 (September 11, 1968).

DIOXACARB

Function: Insecticide (particularly effective against cockroaches) (1)(2)(3)

Chemical Name: 2-(1,3-dioxolan-2-yl)phenyl methylcarbamate

Formula:

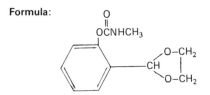

Trade Names: C-8353 (Ciba-Geigy)
Famid® (Ciba-Geigy)
Elocron® (Ciba-Geigy)

Manufacture (3)

Dioxacarb may be made by the condensation of salicylaldehyde with ethylene glycol. That condensation product is in turn reacted with methyl isocyanate to give dioxacarb.

In a typical example, a mixture of 244 parts of salicylaldehyde, 125 parts of ethylene glycol, 1 part of zinc chloride, 1 part by volume of concentrated phosphoric acid and 500 parts by volume of benzene was boiled in a circulating distillation apparatus until water was no longer eliminated. The solution of the product was filtered and evaporated. The residue was distilled under a high vacuum. The dioxolanyl phenol intermediate boiled at 88° to 91°C under 0.04 mm Hg pressure and melted at 68° to 70°C.

A solution of 50 parts of o-(1,3-dioxolan-2-yl)phenol in 300 parts by volume of dry toluene was mixed with about 0.2 part by volume of triethylamine, and 20 parts of methyl isocyanate were dropped into this solution at room temperature. The temperature rose gradually to 31°C. The mixture was kept for one day at room temperature. The product was filtered off and crystallized from toluene; it melted at 111° to 114°C.

Toxicity

The acute oral LD$_{50}$ value for rats is 90 mg/kg (B-5) which is moderately toxic.

References

(1) Worthing, C.R., *Pesticide Manual,* 6th ed., p. 212, British Crop Protection Council (1979).
(2) Spencer, E.Y., *Guide to the Chemicals Used in Crop Protection,* 6th ed., p. 231, London, Ontario, Agriculture Canada (January 1973).
(3) Ciba, Ltd., British Patent 1,122,633, August 7, 1968.

DIOXATHION

Function: Insecticide and acaricide (1)(2)

Chemical Name: S,S'-1,4-dioxane-2,3-diyl O,O,O',O'-tetraethyl di(phosphorodithioate)

Formula:

$$CH_2 \overset{O}{\diagdown} CH-\overset{S}{\overset{\|}{S}P(OC_2H_5)_2}$$
$$CH_2 \underset{O}{\diagup} CH-SP(OC_2H_5)_2$$

Trade Names: AC-528 (Hercules, Inc.)
Delnav® (Hercules, Inc.)
Novadel® (Hercules, Inc.)

Manufacture (3)(4)

Dioxathion may be made by reacting 2,3-dichloro-p-dioxane or 2,3-dibromo-p-dioxane with O,O-diethyl phosphorodithioic acid, the latter being the reaction product of ethanol with phosphorus pentasulfide. O,O-diethyl phosphorodithioic acid may be reacted directly with the 2,3-dihalodioxane or it may be reacted in the form of its salt or in the presence of materials which sequester the hydrogen halide produced in the reaction. The reaction is as follows.

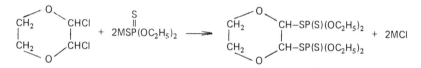

The 2,3-dichlorodioxanes may be made by the chlorination of dioxane. The preparation of dioxathion may be carried out at a temperature in the range from 30° to 110°C; the temperature is not critical but must be below the product decomposition temperature of about 110°C. The reaction is usually carried out at substantially atmospheric pressure. The reaction time may be in the range of 1 to 6 hours.

The reaction is preferably carried out in a solvent which is inert to the reaction, although solvents are not necessary according to R.M. Speck (4). When a solvent is used, volatile aromatic hydrocarbons such as benzene, toluene, xylene or cymene are preferred because they have the desired power for the reaction mixture without high solvent power for the hydrogen chloride produced. Water is usually included because it renders the reaction mixture nonhomogeneous.

The reaction may be carried out noncatalytically and only in the presence of an amine acceptor for the hydrogen chloride formed, as described by W.R. Diveley et al (3).

It is economically advantageous, however, to speed up the reaction by the use of catalytic amounts of zinc, iron or tin chlorides, according to R.M. Speck (4). Zinc and tin chlorides are preferred because they give lighter colored products. From 0.01 to 0.1 mol percent of the catalyst based on the organic chloride reactant is the preferred concentration range.

The catalyst may be added directly as the metal chloride or may be formed in situ by reaction of a metal or metal salt with the hydrogen chloride evolved in the reaction. A conventional stirred, jacketed vessel may be used. It should be equipped with a reflux condenser and may be provided with an inert gas (nitrogen) sparger to assist in removal of the hydrogen chloride evolved.

The reaction product may be used without further purification as insecticides. However, they may also be purified by washing the reaction mixture with water, then with sufficient aqueous sodium hydroxide to neturalize the acids present, and finally with fresh water. The wet product may then be dried over sodium sulfate and any reaction solvent, such as benzene, distilled off.

Process Wastes and Their Control

Product Wastes: This compound is hydrolyzed by alkali and on heating. It is unstable on iron or tin surfaces and with certain carriers, and decomposes at 130° to 140°C yielding $(EtO)_2P(S)SH$ and $RSP(S)OEt)_2$. Dioxathion is oxidized to the corresponding $P=O$ compound (B-3).

Toxicity

The acute oral LD_{50} value for rats is 20 mg/kg (B-5) which is highly toxic.

Allowable Limits on Exposure and Use

Air: The threshold limit value for dioxathion in air is 0.2 mg/m³ as of 1979. It is further noted that cutaneous absorption of dioxathion should be prevented so that the threshold limit value is not invalidated.

Product Use: The tolerances set by the EPA for dioxathion in or on raw agricultural commodities are as follows:

	40 *CFR* Reference	Parts per Million
Apples	180.171	5.0
Cattle, fat of meat	180.171	1.0
Fruits, stone	180.171	0.14 N
Goats, fat of meat	180.171	1.0
Grapefruit	180.171	3.0
Grapes	180.171	2.1
Hogs, fat of meat	180.171	1.0
Horses, fat of meat	180.171	1.0
Lemons	180.171	3.0
Limes	180.171	3.0
Milk	180.171	0.0
Oranges	180.171	3.0
Pears	180.171	5.0
Quinces	180.171	5.0
Sheep, fat of meat	180.171	1.0
Tangerines	180.171	3.0
Walnuts	180.171	0.14 N

The tolerances set by the EPA for dioxathion in animal feeds are as follows (the *CFR* Reference is to Title 21):

	CFR Reference	Parts per Million
Citrus pulp, dehydrated (ct.f)	561.210	18.0

References

(1) Worthing, C.R., *Pesticide Manual,* 6th ed., p. 213, British Crop Protection Council (1979).
(2) Spencer, E.Y., *Guide to the Chemicals Used in Crop Protection,* 6th ed., p. 232, London, Ontario, Agriculture Canada (January 1973).

(3) Diveley, W.R. et al, U.S. Patent 2,725,328, November 29, 1955, assigned to Hercules Powder Company.
(4) Speck, R.M., U.S. Patent 2,815,350, December 3, 1957, assigned to Hercules Powder Company.

DIPHACINONE

Function: Rodenticide (1)(3)

Chemical Name: 2-(diphenylacetyl)-1H-indene-2,3(2H)dione

Formula:

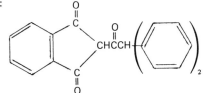

Trade Names: Diphenadione
Diphacin®
PID

Manufacture (2)(4)

6.9 g (0.3 mol) of sodium was dissolved in 70 ml of anhydrous methanol and 250 ml of benzene was added with vigorous stirring. The stirring was continued while the benzene and excess methanol were distilled. The resulting sodium methoxide suspension was diluted with 250 ml of benzene and 46 g (0.24 mol) of dimethyl phthalate. The mixture was stirred and heated with an oil bath to vigorous reflux.

A solution of 21 g (0.1 mol) of diphenyl acetone in 100 ml of benzene was added dropwise over a 1.5 hour period. During this addition 100 ml of distillate was removed by use of a water take-off trap. A solution of 21 g (0.1 mol) of diphenyl acetone and 30 g (0.13 mol) of dimethyl phthalate in 100 ml of benzene was added dropwise over a second 1.5 hour period. Another 100 ml of distillate was removed.

The reaction temperature was raised to 140°C and maintained at this level for one hour. The oil bath was removed and the reaction temperature was allowed to drop to 90°C before 150 ml of ethanol was added. Stirring and refluxing was resumed until all of the dark viscous material was in solution. The solution was poured into a 500 ml beaker and heated on the steam bath until the total volume was about 100 ml. This solution was cooled and stirred in an ice bath at about 25°C while adding 50 ml of concentrated hydrochloric acid in about one-half minute.

The product separated as a light yellow solid and was collected on a filter. The product was agitated and washed for about one minute with 500 ml of water to remove sodium chloride. The resulting suspension was filtered and recrystallized from ethanol to produce 42 g (62% yield based on diphenyl acetone) of 2-diphenylacetyl-1,3-indandione which melted between 146° and 147°C.

Toxicity

The acute oral LD_{50} value for rats is not available but the compound is extremely toxic.

References

(1) Worthing, C.R., *Pesticide Manual*, 6th ed., p. 214, British Crop Protection Council (1979).
(2) Birkenmeyer, R.D. and Speeter, M.E., U.S. Patent 2,827,489, March 18, 1958, assigned to The Upjohn Company.
(3) Correll, J.T., U.S. Patent 2,900,302, August 18, 1959, assigned to The Upjohn Company.
(4) Thomas, D.G., U.S. Patent 2,672,483, March 16, 1954, assigned to The Upjohn Company.

DIPHENAMID

Function: Herbicide (1)(2)

Chemical Name: N,N-dimethyl-α-phenylbenzeneacetamide

Formula:

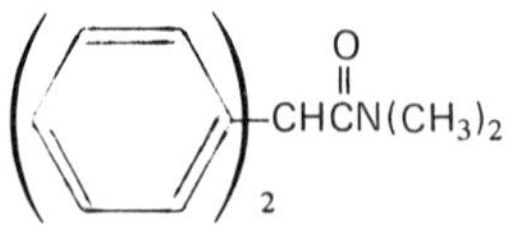

Trade Names: L-34314 (Eli Lilly and Co.)
Dymid® (Eli Lilly and Co.)
Enide® (Upjohn Co.)

Manufacture (3)

The compounds which are the herbicidally active ingredients are prepared by reacting diphenylacetyl chloride or a suitably substituted diphenylacetyl chloride with a secondary amine in the presence of a base. The base is preferably employed in at least equimolar ratio to the diphenylacetyl chloride in order to take up the liberated hydrogen chloride. For this purpose, the base may be merely an additional quantity of the secondary amine. The diphenylacetyl chlorides are prepared by hydrolyzing the corresponding diphenylacetonitrile, and subsequently reacting the hydrolytically produced acid with thionyl chloride.

The following is a specific example of the conduct of the process. 9.66 g of diphenylacetyl chloride were dissolved in 25 ml of benzene. A solution of 5.68 g of dimethylamine in 50 ml of benzene was added dropwise to the acid chloride solution. After the addition had been completed, the reaction mixture was heated at refluxing temperature for an additional 2½ hours. The reaction mixture was cooled, and sufficient chloroform was added to render the mixture homogeneous.

The organic layer was washed with successive 100 ml portions of 10% hydrochloric acid, water, saturated sodium bicarbonate solution and water, and was then dried. The solvents were removed by evaporation in vacuo, and the residue, comprising N,N-dimethyldiphenylacetamide formed in the above reaction, was recrystallized three times from a mixture of benzene and hexane and twice from a mixture of ethyl acetate and hexane. N,N-dimethyldiphenylacetamide thus prepared melted at about 128°C.

Process Wastes and Their Control

Product Wastes: The product is moderately stable to heat and light. It is phytotoxic 10 to 11 months after application (B-3).

Toxicity

The acute oral LD$_{50}$ value for rats is 1,050 mg/kg which is slightly toxic.

Allowable Limits on Exposure and Use

Product Use: The tolerances set by the EPA for diphenamid in or on raw agricultural commodities are as follows:

	40 *CFR* Reference	Parts per Million
Apples	180.230	0.1 N
Cattle, fat	180.230	0.05 N
Cattle, mbyp	180.230	0.05 N
Cattle, meat	180.230	0.05 N
Cotton, forage	180.230	0.2
Cotton, seed	180.230	0.1 N
Goats, fat	180.230	0.05 N
Goats, mbyp	180.230	0.05 N
Goats, meat	180.230	0.05 N
Hogs, fat	180.230	0.05 N
Hogs, mbyp	180.230	0.05 N
Hogs, meat	180.230	0.05 N
Horses, fat	180.230	0.05 N
Horses, mbyp	180.230	0.05 N
Horses, meat	180.230	0.05 N
Milk	180.230	0.01 N
Okra	180.230	0.1 N
Peaches	180.230	0.1 N
Peanuts	180.230	0.1 N
Peanuts, forage	180.230	2.0
Peanuts, hay	180.230	2.0
Peanuts, hulls	180.230	0.5
Potatoes	180.230	1.0
Sheep, fat	180.230	0.05 N
Sheep, mbyp	180.230	0.05 N
Sheep, meat	180.230	0.05 N
Soybeans	180.230	0.1 N
Soybeans, forage	180.230	0.5
Soybeans, hay	180.230	0.5
Strawberries	180.230	1.0
Sweet potatoes	180.230	0.1 N
Vegetables, fruiting	180.230	0.1 N

References

(1) Worthing, C.R., *Pesticide Manual,* 6th ed., p. 215, British Crop Protection Council (1979).
(2) Spencer, E.Y., *Guide to the Chemicals Used in Crop Protection,* 6th ed., p. 233, London, Ontario, Agriculture Canada (January 1973).
(3) Pohland, A., U.S. Patent 3,120,434, February 4, 1964, assigned to Eli Lilly and Company.

DIPHENYL

Function: Fungicide (1)(2)

Chemical Name: 1,1'-biphenyl

Formula:

Trade Name: None

Manufacture (3)

Diphenyl is made by the vapor-phase, noncatalytic pyrolysis of benzene in a tubular steel coil reactor at 600° to 800°C and atmospheric pressure for a contact time of 0.4 to 1.5 seconds. Yields are 85 to 90%.

Toxicity

The acute oral LD_{50} value for rats is 3,280 mg/kg (slightly toxic).

Allowable Limits on Exposure and Use

Product Use: The tolerances set by the EPA for diphenyl in or on raw agricultural commodities are as follows:

	40 *CFR* Reference	Parts per Million
Citrus citron (post-h)	180.141	110.0
Citrus fruits—hybrids thereof (post-h)	180.141	110.0
Grapefruit (post-h)	180.141	110.0
Kumquats (post-h)	180.141	110.0
Lemons (post-h)	180.141	110.0
Limes (post-h)	180.141	110.0
Oranges (post-h)	180.141	110.0
Tangerines (post-h)	180.141	110.0

References

(1) Martin, H., Worthing, C.R., *Pesticide Manual,* 5th ed., p. 217, British Crop Protection Council (January 1977).
(2) Spencer, E.Y., *Guide to the Chemicals Used in Crop Protection,* 6th ed., p. 234, London, Ontario, Agriculture Canada (January 1973).
(3) Degeorges, M.E. et al, U.S. Patent 3,227,525, January 4, 1966, assigned to Societe Progil.

DIPROPETRYN

Function: Preemergence herbicide (1)(2)

Chemical Name: 6-(ethylthio)-N,N'-bis(1-methylethyl)-1,3,5-triazine-2,4-diamine

Formula:

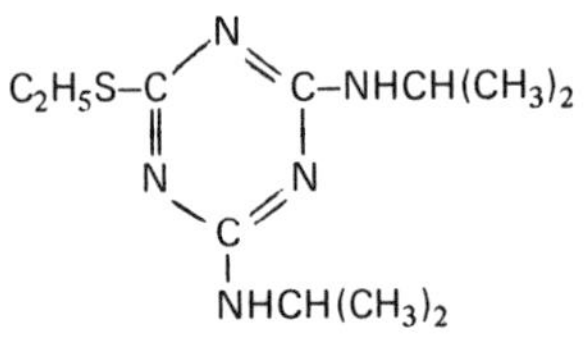

Trade Names: GS 16068 (Ciba-Geigy)
Cotofor® (Ciba-Geigy)
Sancap® (Ciba-Geigy)

Manufacture

Dipropetryn is made by reacting propazine (which see) with thiourea and diethyl sulfate in turn to attach the C_2H_5S- group in place of the $-Cl$.

Toxicity

The acute oral LD_{50} for rats is 3,900 to 4,200 mg/kg which is only slightly toxic.

Allowable Limits on Exposure and Use

Product Use: The tolerances set by the EPA for dipropetryn in or on raw agricultural commodities are as follows:

	40 *CFR* Reference	Parts per Million
Cotton, seed	180.329	0.1

References

(1) Worthing, C.R., *Pesticide Manual,* 6th ed., p. 216, British Crop Protection Council (1979).
(2) Spencer, E.Y., *Guide to the Chemicals Used in Crop Protection,* 6th ed., p. 46, London, Ontario, Agriculture Canada (January 1973).

DIPROPYL ISOCINCHOMERONATE

Function: Fly repellent (1)(2)

Chemical Name: Di-n-propyl 2,5-pyridinedicarboxylate

Formula:

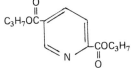

Trade Name: MGK Repellent 326 (McLaughlin Gormley King Co.)

Manufacture (3)(4)(5)

Paraldehyde and ammonia are reacted in an initial step to give 2-methyl-5-ethyl pryidine.

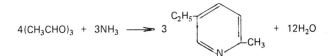

The methylethyl pyridine is then oxidized to pyridine-2,5-dicarboxylic acid. It is, in turn, esterified with propanol to give the product.

Toxicity

The acute oral LD_{50} for rats is above 5,000 mg/kg which is insignificantly toxic.

Product Use: The tolerances set by the EPA for dipropyl isocinchomeronate in or on raw agricultural commodities are as follows:

	40 *CFR* Reference	Parts per Million
Cattle, fat	180.143	0.1 N
Cattle, mbyp	180.143	0.1 N
Cattle, meat	180.143	0.1 N
Goats, fat	180.143	0.1 N
Goats, mbyp	180.143	0.1 N
Goats, meat	180.143	0.1 N
Hogs, fat	180.143	0.1 N
Hogs, mbyp	180.143	0.1 N
Hogs, meat	180.143	0.1 N
Horses, fat	180.143	0.1 N
Horses, mbyp	180.143	0.1 N
Horses, meat	180.143	0.1 N
Milk	180.143	0.004 N
Sheep, fat	180.143	0.1 N
Sheep, mbyp	180.143	0.1 N
Sheep, meat	180.143	0.1 N

References

(1) Worthing, C.R., *Pesticide Manual,* 6th ed., p. 217, British Crop Protection Council (1979).
(2) Spencer, E.Y., *Guide to the Chemicals Used in Crop Protection,* 6th ed., p. 236, London, Ontario, Agriculture Canada (January 1973).
(3) Berger, D.E., U.S. Patent 2,926,074, February 23, 1960, assigned to Phillips Petroleum Company.
(4) Leonard, N.J., U.S. Patent 2,757,120, July 31, 1956, assigned to Phillips Petroleum Company.
(5) Mahan, J.E. and Stansbury, R.E., U.S. Patent 3,067,091, December 4, 1962, assigned to Phillips Petroleum Company.

DIQUAT

Function: Contact herbicide (1)(2)

Chemical Name: 6,7-dihydrodipyrido[1,2-a:2',1'-c]pyrazinediium dibromide

Formula:

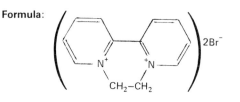

Trade Names: FB/2® (I.C.I. Ltd.)
Reglone® (Plant Protection, Ltd.)
Aquacide® (Chipman Chemical Co., Ltd.)

Manufacture

The oxidative coupling of 2 molecules of pyridine over a heated Raney nickel catalyst gives

2,2'-bipyridyl. It is reacted with ethylene dibromide in water to give diquat. The reaction may be carried out by heating the reagents together, or they may be heated together in the presence of an inert solvent or diluent, preferably a high-boiling polar liquid, for example, nitrobenzene or β-ethoxy-ethanol. The following specific examples (3) illustrate the conduct of this process.

Example 1—One part of 2,2'-dipyridyl and 10 parts of ethylene dibromide are mixed and the mixture is heated under reflux for 20 hours. It is then cooled and filtered and the solid residue is washed with acetone. It is then crystallized from aqueous methanol and there is obtained a quaternary bromide monohydrate which darkens above 300°C and melts at 335° to 340°C.

Example 2—1.6 parts of 2,2'-dipyridyl and 2.1 parts of ethylene dibromide are dissolved in 10 parts of nitrobenzene and the solution is boiled under reflux for 30 minutes. Twenty parts of acetone are then added and the mixture is filtered and the solid residue is washed with acetone and crystallized from aqueous methanol. The product so obtained is identical with the material obtained according to the process of Example 1.

Process Wastes and Their Control

Product Wastes: Diquat is inactivated by inert clay or by anionic surfactants. Therefore, an effective and environmentally safe disposal method would be to mix the product with ordinary household detergent and bury the mixture in clay soil (B-3).

Toxicity

The acute oral LD_{50} for rats is about 230 mg/kg which is moderately toxic.

Allowable Limits on Exposure and Use

Air: The threshold limit value for diquat in air is 0.5 mg/m³ as of 1979. A tentative short term exposure limit is 1.0 mg/m³ (B-23).

Product Use: The tolerances set by the EPA for diquat dibromide in or on raw agricultural commodities are as follows:

	40 *CFR* Reference	Parts per Million
Sugarcane	180.226	0.05

The tolerances set by the EPA for diquat dibromide in food are as follows (the *CFR* Reference is to Title 21):

	CFR Reference	Parts per Million
Water, potable	193.160	0.01 i

References

(1) Worthing, C.R., *Pesticide Manual,* 6th ed., p. 218, British Crop Protection Council (1979).
(2) Spencer, E.Y., *Guide to the Chemicals Used in Crop Protection,* 6th ed., p. 237, London, Ontario, Agriculture Canada (January 1973).
(3) Fielden, R.J., Homer, R.F. and Jones, R.L., U.S. Patent 2,823,987, February 18, 1958, assigned to Imperial Chemical Industries, Ltd.

DISODIUM OCTABORATE

Function: Herbicide (1)

Chemical Name: Disodium borate

Formula: $Na_2B_8O_{13} \cdot 4H_2O$

Trade Name: Polybor® (U.S. Borax and Chemical Corp.)

Manufacture (2)

The feed ingredients typically consist essentially of a mixture of such soluble compounds as boric acid, sodium pentaborate, sodium tetraborate and sodium metaborate, for example, in such proportions as to yield the desired molar ratio Na_2O/B_2O_3. In the special case of ratio 0.20, the solution may contain such a mixture or may be prepared from sodium pentaborate alone. Water is then evaporated very rapidly from that solution, leaving the solutes in solid form.

To insure rapid removal of water from the solution, the latter may be spread at elevated temperature into a thin film, as on a hot plate or on the heated roll of a drum dryer; or is otherwise divided to increase the relative surface area through which evaporation may take place. The preferred method of evaporation is by spraying the solution at moderately elevated temperature into a stream of hot and relatively dry air.

The droplets of the spray become substantially dry, and preferably also are cooled nearly to room temperature, while still carried in the air stream and before striking any solid wall of the enclosure. Each droplet thus produces a generally spherical solid particle containing each ingredient in the initial solution proportions. Although superficially dry, the resulting solid composition may include an appreciable proportion of water, preferably between about 0.5 and 1.0 mol of water per mol of B_2O_3, the exact amount of water and the average size of the particles depending primarily upon the particular spraying and drying conditions used.

Toxicity

The acute oral LD_{50} for guinea pigs is 5,300 mg/kg.

References

(1) Worthing, C.R., *Pesticide Manual,* 6th ed., p. 220, British Crop Protection Council (1979).
(2) O'Brien, P.J. and Connell, G.A., U.S. Patent 2,998,310, August 29, 1961, assigned to United States Borax and Chemical Corporation.

DISUL (SESONE)

Function: Herbicide (1)(2)(3)

Chemical Name: Sodium-2,4-dichlorophenoxyethyl sulfate

Formula:

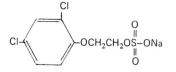

Trade Names: Crag Herbicide 1® (Union Carbide Corp.)
2,4-DES-Sodium
SES

Manufacture

Sodium 2-(2,4-dichlorophenoxy)ethyl sulfate may be prepared (2) by adding 1 gram-mol of 2-(2,4-dichlorophenoxy)ethanol to a mixture of 1.25 mols of ethyl ether and 1.15 mols of chlorosulfonic acid at $-10°$ to $0°C$. The mixture was stirred for two hours at this temperature and then slowly added to 600 cc of a 15% water solution of NaOH (2.25 mols of NaOH) at 25°C.

The ethyl ether was removed by distillation, the residue allowed to cool to room temperature and the solid sodium 2-(2,4-dichlorophenoxy)ethyl sulfate which separated was filtered off and dried. The filtrate was a water solution containing the residues and by-products of the reaction, that is, a little NaOH, and the sodium sulfate and sodium chloride which were formed.

The organic sulfate is substantially insoluble in the aqueous solution of the inorganic materials at a temperature below 30°C. The product obtained by filtration melted at 170°C, was a white crystalline solid and analyzed 98% sodium 2-(2,4-dichlorophenoxy)ethyl sulfate. The process operated at an efficiency better than 95%.

An alternate route is also available (3). 84.5 parts by weight of ethylenechlorohydrin (1.05 mols) were stirred into 128 parts by weight of chlorosulfonic acid (1.1 mols) over a period of 30 minutes at 0°C. The cold sulfonation mixture was thoroughly agitated and mixed and then permitted to pass into 1,100 parts by weight of a 9% aqueous sodium hydroxide solution, while maintaining the temperature at about 0°C. Thereafter, the amount of free alkali in the reaction mixture was determined and found to be 42.3 parts by weight.

After adding 163 parts by weight of 2,4-dichlorophenol (1 mol), the remaining amount of free sodium hydroxide (2.3 parts by weight) was transformed into sodium sulfate by adding a calculated amount of dilute sulfuric acid (29 parts by weight of a 10% aqueous solution of sulfuric acid).

Subsequently, the reaction mixture was boiled until the gradually decreasing alkalinity of the reaction mixture indicated that the reaction was practically completed. This required boiling for about 14 to 16 hours. By evaporating or spray-drying the resulting solution, a solid residue was obtained which contained high percentages of 2,4-dichlorophenoxyethyl sodium sulfate.

If this product is not to be sold as a selective weed killing agent in this form, it may be purified by repeated extraction with hot alcohol, whereby inorganic salts formed during the reaction are removed. In this manner 279 parts by weight of pure crystalline 2,4-dichlorophenoxyethyl sodium sulfate were obtained, corresponding to 90% of the theoretical yield.

Process Wastes and Their Control

Product Wastes: $2,4\text{-}Cl_2C_6H_3OC_2H_4OSO_3Na$, is hydrolyzed by alkali to $NaHSO_4$ and apparently the dichlorophenoxyethanol (B-3).

Allowable Limits on Exposure and Use

Air: The threshold limit value for disul in air is 10 mg/m^3 as of 1979. The tentative short term exposure limit value is 20 mg/m^3 (B-21).

Product Use: The tolerances set by the EPA for disul in or on raw agriculture commodities are as follows:

	40 *CFR* Reference	Parts per Million
Asparagus	180.102	2.0
Peanuts	180.102	6.0
Peanuts, hay	180.102	6.0
Peanuts, hulls	180.102	6.0
Potatoes	180.102	6.0
Strawberries	180.102	2.0

References

(1) Worthing, C.R., *Pesticide Manual,* 6th ed., p. 159, British Crop Protection Council (1979).
(2) Lambreach, J.A., U.S. Patent 2,573,769, November 6, 1951, assigned to Union Carbide and Carbon Corporation.
(3) Gundel, W. and Linden, H., U.S. Patent 2,852,548, September 16, 1958, assigned to Henkel and Cie.

DISULFOTON

Function: Systemic insecticide (1)(2)

Chemical Name: O,O-diethyl S-2-ethylthioethyl phosphorodithioate

Formula:

$$(C_2H_5O)_2\overset{\overset{\displaystyle S}{\|}}{P}SCH_2CH_2SC_2H_5$$

Trade Names: Bayer 19639 (Bayer)
S-276 (Bayer)
Disyston® (Bayer)
Dithio-Systox® (Bayer)
Ekatine®
Frumin®
Solvirex®

Manufacture

The following is the production chemistry.

$$P_2S_5 + 4C_2H_5OH + 2NaOH \xrightarrow{\text{toluene}} 2(C_2H_5O)_2P(S)SNa + H_2S + 2H_2O$$
(DES)

$$PCl_3 + 3HOC_2H_4-S-C_2H_5 \longrightarrow 3ClC_2H_4-S-C_2H_5 + P(OH)_3$$
(CTA)

$$(C_2H_5O)_2P(S)SNa + ClC_2H_4-S-C_2H_5 \longrightarrow (C_2H_5O)_2P(S)-S-C_2H_4-S-C_2H_5 + NaCl$$
Disulfoton

The following is a specific example of disulfoton preparation (3). One hundred and twelve grams (0.5 mol) of potassium O,O-diethyl dithiophosphate are dissolved in 400 cc of ethyl alcohol. Sixty-four grams (0.5 mol) of β-chloroethyl ethyl sulfide are dropped to this solution at 70°C and heated for 2 hours at this temperature.

After cooling the salt is sucked off, the solvent is distilled off, the residue is dissolved in benzene and washed with water. After drying with anhydrous sodium sulfate it is fractionated. One hundred and fifteen grams of O,O-diethyl-S-ethylmercaptoethyl dithiophosphate of boiling point 132° to 133°C at 1.5 mm are obtained. A flow diagram of a commercial disulfoton plant is shown in Figure 28.

Process Wastes and Their Control

Air: Air emissions from disulfoton manufacture have been reported (B-15) as 205 kg SO_2 and 0.5 kg disulfoton per metric ton of pesticide produced.

As regards waste control, H_2S is flared and other materials are vented to the atmosphere, according to one source (B-10).

Water: Triester and organic skimmings are sent to burial, process wastewater to toluene extractor and skimmer with final NaOH and NaOCl treatment before flowing to the wastewater treatment plant. Caustic conditions (lime) remove some of the organophosphates during a retention period of 36 hours. The pH is also adjusted and polyelectrolyte flocculents added. The effluent is discharged into a river with containments in the parts per million range (B-10).

A wastewater treatment system for a plant making both disulfoton and azinphos methyl is described in some detail under azinphos-methyl (which see).

Product Wastes: Disulfoton is resistant to hydrolysis in acid media. Fifty percent hydrolysis at 70°C requires 60 hours at pH 5, and 7.2 hours at pH 9. It is hydrolyzed by refluxing with 1N KOH in i-PrOH in 30 minutes (B-3).

Toxicity

The acute oral LD_{50} value for rats is 2.6 to 8.6 mg/kg which is highly to extremely toxic.

Phorate and disulfoton are widely used agricultural insecticides. They are organophosphorus compounds whose mode of action, inhibition of acetylcholinesterase, is well understood. Both have high acute toxicity in laboratory animals. Because of this and a known mode of action, studies on the possible chronic toxicity of these compounds have been neglected. There is very little toxicologic research on disulfoton reported in the open literature. There is a single subchronic-toxicity study, in which no adverse effects were observed after administration of 0.75 mg/kg/day for 30 days; this indicates that disulfoton would pose less hazard than phorate. Based on subchronic toxicity data, an ADI was calculated at 0.0001 mg/kg/day for both phorate and disulfoton.

Phorate and disulfoton are converted in the environment and in mammalian systems to a series of highly toxic oxidative metabolites, which are known to be more potent cholinesterase inhibitors than the parent compounds. These materials must be considered when evaluating the toxicity of phorate and disulfoton. Therefore, it is proposed that the derived no-adverse-effect dosages of these compounds be considered to include their oxidative metabolites as well.

The most obvious research need for both these compounds is studies on chronic toxicity, including carcinogenicity and teratogenicity. Some of these studies may have been done by the manufacturers; if so, they should be made generally available to assist in the evaluation of toxicology by the scientific community.

There is also a need for corroboration of the no-adverse-effect cholinesterase-inhibition dosage in human subjects in a controlled study with at least two dosages. This would allow the extrapolation of a no-adverse-effect dosage with a higher degree of confidence and a lower uncertainty factor (B-22).

Figure 28: Production and Waste Schematic for Disulfoton

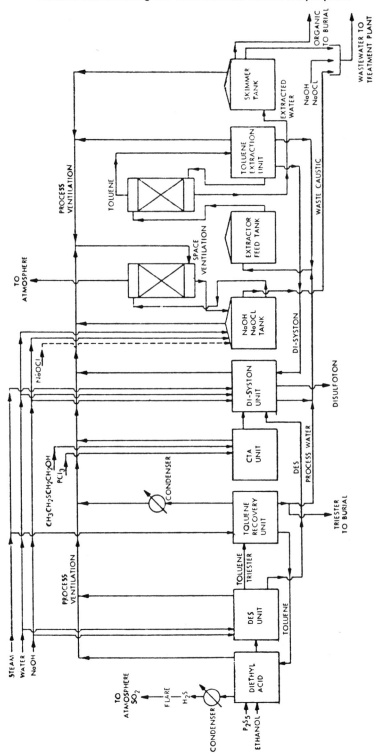

Source: Reference (4)

Allowable Limits on Exposure and Use

Air: The threshold limit value for disulfoton in air is 0.1 mg/m^3 as of 1979. The tentative short term exposure limit is 0.3 mg/m^3. It is noted, further, that cutaneous absorption should be prevented so that the threshold limit value is not invalidated.

Product Use: The tolerances set by the EPA for disulfoton in or on raw agricultural commodities are as follows:

	40 *CFR* Reference	Parts per Million
Alfalfa, fresh	180.183	5.0
Alfalfa, hay	180.183	12.0
Barley, fodder, green	180.183	5.0
Barley, grain	180.183	0.75
Barley, straw	180.183	5.0
Beans, dry	180.183	0.75
Beans, lima	180.183	0.75
Beans, snap	180.183	0.75
Beans, vines	180.183	5.0
Beets, sugar	180.183	0.5
Beets, sugar, tops	180.183	2.0
Broccoli	180.183	0.75
Brussels sprouts	180.183	0.75
Cabbage	180.183	0.75
Cauliflower	180.183	0.75
Clover, fresh	180.183	5.0
Clover, hay	180.183	12.0
Coffee beans	180.183	0.3
Corn, field, fodder	180.183	5.0
Corn, field, forage	180.183	5.0
Corn, grain	180.183	0.3
Corn, pop, fodder	180.183	5.0
Corn, pop, forage	180.183	5.0
Corn, sweet, fodder	180.183	5.0
Corn, sweet, forage	180.183	5.0
Corn, sweet, grain (k+cwhr)	180.183	0.3
Cotton, seed	180.183	0.75
Hops	180.183	0.5
Lettuce	180.183	0.75
Oats, fodder, green	180.183	5.0
Oats, grain	180.183	0.75
Oats, straw	180.183	5.0
Peanuts	180.183	0.75
Peanuts, hay	180.183	5.0
Peanuts, hulls	180.183	0.3
Peas	180.183	0.75
Peas, vines	180.183	5.0
Pecans	180.183	0.75
Peppers	180.183	0.1
Pineapples	180.183	0.75
Pineapples, foliage	180.183	5.0
Potatoes	180.183	0.75
Rice	180.183	0.75
Rice, straw	180.183	5.0
Sorghum, fodder	180.183	5.0
Sorghum, forage	180.183	5.0
Sorghum, grain	180.183	0.75
Soybeans	180.183	0.1
Soybeans, forage	180.183	0.25

(continued)

	40 *CFR* Reference	Parts per Million
Soybean, hay	180.183	0.25
Spinach	180.183	0.75
Sugarcane	180.183	0.3
Tomatoes	180.183	0.75
Wheat, fodder, green	180.183	5.0
Wheat, grain	180.183	0.3
Wheat, straw	180.183	5.0

The tolerances set by the EPA for disulfoton in animal feeds are as follows (the *CFR* Reference is to Title 21):

	CFR Reference	Parts per Million
Pineapple bran	561.160	5.0
Sugar beet pulp, dehydrated	561.160	5.0

References

(1) Worthing, C.R., *Pesticide Manual,* 6th ed., p. 221, British Crop Protection Council (1979).
(2) Spencer, E.Y., *Guide to the Chemicals Used in Crop Protection,* 6th ed., p. 239, London, Ontario, Agriculture Canada (January 1973).
(3) Lorenz, W. and Schrader, G., U.S. Patent 2,759,010, August 14, 1956, assigned to Farbenfabriken Bayer AG.
(4) Midwest Research Institute, *The Pollution Potential in Pesticide Manufacturing,* Washington, DC, Environmental Protection Agency (June 1972).

DITALIMFOS

Function: Fungicide (1)

Chemical Name: O,O-diethyl (1,2-dihydro-1,3-dioxo-2H-isoindol-2-yl)phosphonothioate

Formula:

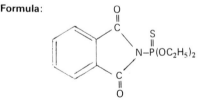

Trade Names: Dowco 199 (Dow Chemical Co.)
Plondrel® (Dow Chemical Co.)

Manufacture

Ditalimfos may be made by the reaction of O,O-diethyl phosphorochloridothionate with potassium phthalimide. The following is a specific example (2) of the process. O,O-diethyl phosphorochloridothioate (18.9 g, 0.1 mol) was added portionwise with stirring to 18.5 g of potassium phthalimide (0.1 mol) dispersed in 175 ml of N-methyl-2-pyrrolidone. The addition was carried out at a temperature of 0° to 2°C.

Twenty-five milliliters of N-methyl-2-pyrrolidone was added to the reaction mixture; thereafter, the mixture was set aside for a period of about 16 hours at 0° to 2°C, with stirring, to ensure completion of the reaction. The reaction mixture was then diluted with water to remove alkali metal chloride of reaction and precipitate the product, and filtered to obtain the O,O-diethyl phthalimidophosphonothioate compound as a crystalline residue. The residue was dried, recrystallized from hexane, and found to melt at 81.5° to 83.5°C.

Toxicity

The acute oral LD_{50} value for rats is about 5,000 mg/kg which is very slightly toxic.

References

(1) Worthing, C.R., *Pesticide Manual,* 6th ed., p. 222, British Crop Protection Council (1979).
(2) Dow Chemical Co., British Patent 1,034,493, June 29, 1966.

DITHIANON

Function: Fungicide (1)(2)(3)(5)

Chemical Name: 5,10-dihydro-5,10-dioxonaphtho[2,3-b]-1,4-dithiin-2,3-dicarbonitrile

Formula:

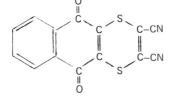

Trade Names: IT931 (E. Merck AG)
MV119A (E. Merck AG)
Delan® (E. Merck AG)
Delan-Col®
Thynon®

Manufacture

CS_2 and NaCN are first reacted to give 1,2-dicyano-1,2-dimercaptoethylene (4). The sodium salt of that product is, in turn, reacted with dichloro-1,4-naphthoquinone to give dithianon.

The following is a specific example (3) of the process. Five grams of sodium cyanide were added to 25 grams of dimethyl sulfoxide. Then 8 grams of carbon disulfide were added to the resulting reaction mixture at 30°C over a period of one-half hour. The mixture was then stirred for 1½ hours at 40°C. Thereafter 50 cc of water were added to the resulting reaction solution and the solution heated for 1 hour at 50°C. The sulfur which had precipitated was filtered off and washed with a little water.

The filtrate was then dropped into a solution of 16 grams of naphthoquinone-(1,4) and 6 grams of glacial acetic acid in 75 cc of methylene chloride over a 30 minute period, the reaction solution being maintained at 20°C by light cooling. The solution was stirred for a further 30 minutes and then 60 grams of $FeCl_3 \cdot 6H_2O$ in 60 cc of water added thereto.

After stirring the reaction mixture for an hour at room temperature the solids were filtered off on a suction filter and then washed thoroughly with water and with methanol. 13.5

grams of 2,3-dicyan-1,4-dithia-anthraquinone remained on the filter corresponding to a yield of 91% of the theoretical. The melting point of the product was 218° to 220°C. About half of the anthraquinone-(1,4) supplied could be recovered from the filtrate.

Toxicity

The acute oral LD_{50} for rats is 638 mg/kg which rates as slightly toxic.

References

(1) Worthing, C.R., *Pesticide Manual,* 6th ed., p. 223, British Crop Protection Council (1979).
(2) van Schoor, A., Jacobi, E., Lust, S. and Flemming, H., U.S. Patent 2,976,296, March 21, 1961, assigned to E. Merck AG.
(3) Jacobi, E., van Schoor, A. and Hahn, H., U.S. Patent 3,030,381, April 17, 1962, assigned to E. Merck AG.
(4) Hahn, H., Mohr, G. and van Schoor, A., U.S. Patent 3,152,169, October 6, 1964, assigned to E. Merck AG.
(5) Merck AG, British Patent 857,383, December 29, 1960.

DIURON

Function: Herbicide (1)(2)(3)

Chemical Name: N'-(3,4-dichlorophenyl)-N,N-dimethylurea

Formula:

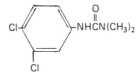

Trade Names: Karmex® (DuPont)
Marmex®

Manufacture (3)(5)

These herbicidally active compounds may be prepared by the reaction of an appropriate mono- or dihalo-substituted phenyl isocyanate with dimethyl or methylethyl amine (3). The following equation showing specific reactants illustrates the reaction.

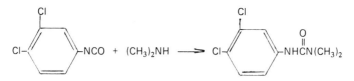

The amine-isocyanate reaction is most readily carried out in the presence of an inert solvent, such as toluene, anisole, benzene, chlorobenzene, or dioxane. No catalyst is needed, and since the reaction is exothermic it is ordinarily unnecessary to supply heat. Thus the reaction is conveniently carried out by first mixing the isocyanate with the inert solvent at room temperature and then gradually adding the secondary amine reactant while permitting the temperature to increase through the range of 25° to 75°C. The tri-substituted urea products are generally quite insoluble in the solvent used and, therefore, precipitate out as formed and are readily separated from the reaction mass.

The halophenyl dialkyl ureas are white crystalline solids. They are insoluble or only slightly soluble in water and cold benzene and, in general, appreciably soluble in dioxane, acetone, ethyl acetate, ethanol and hot benzene. The halophenyl isocyanates used as starting materials can be prepared by heating the haloaryl carbamyl chloride first obtained by treating the haloaryl primary amine with phosgene at ordinary temperature in the presence of an appropriate solvent or reaction media.

The following is a specific example of the conduct of this process (3). 3,4-Dichlorophenyl isocyanate was prepared by passing phosgene into a slurry of 40.8 parts by weight of 3,4-dichloroaniline hydrochloride in about 260 parts by weight of dioxane at 75° to 85°C until a clear solution of the isocyanate was obtained (about 1.3 hours). The excess phosgene and hydrogen chloride were removed from the reaction mass by distillation of approximately 100 parts by weight of the dioxane.

The resulting undistilled mixture was then cooled to 18°C and dimethylamine was passed into it rapidly at 18° to 34°C until present in excess. The mixture was poured into 500 parts by weight of water and the excess amine neutralized by the addition of dilute hydrochloric acid. After cooling, the crystalline 3-(3,4-dichlorophenyl)-1,1-dimethylurea which precipitated was removed by filtration, washed with water and cold ethyl alcohol, then recrystallized from 235 parts by weight of ethyl alcohol.

27.1 parts by weight of recrystallized product was obtained having a MP of 153° to 154°C. Additional 3-(3,4-dichlorophenyl)-1,1-dimethylurea having the same MP was obtained by recrystallization of additional crude product recovered by concentration of the mother liquor. The combined yield amounted to 61% of the theoretical yield.

A production and waste schematic of the commercial diuron process is given in Figure 29. The commercial reaction chemistry is believed (4) to be as follows.

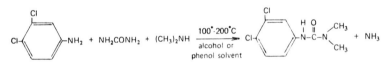

Figure 29: Production and Waste Schematic for Diuron

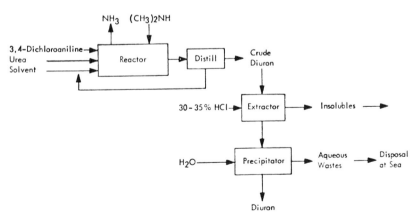

Source: Reference (4)

Process Wastes and Their Control

Air: Air emissions from diuron manufacture have been reported (B-15) to consist of 0.5 kg hydrocarbons and 0.5 kg chlorine per metric ton of pesticide produced.

Water: Aqueous waste contains some quantities of HCl (B-10).

Product Wastes: Diuron, stable under normal conditions, decomposes on heating to 180° to 190°C giving dimethylamine and 3,4-dichlorophenylisocyanate. Treatment at elevated temperatures by acid or base yields dimethylamine and 3,4-dichloroaniline. Hydroysis is not recommended as a disposal procedure because of the generation of the toxic products, 3.4-dichloroaniline and dimethylamine (B-3).

Toxicity

The acute oral LD_{50} value for rats is 3,400 mg/kg which is slightly toxic.

Allowable Limits on Exposure and Use

Air: The threshold limit value for diuron in air is 10 mg/m³ as of 1979 (B-23).

Product Use: The tolerances set by the EPA for diuron in or on raw agricultural commodities are as follows:

	40 *CFR* Reference	Parts per Million
Alfalfa	180.106	2.0
Apples	180.106	1.0
Artichokes	180.106	1.0
Asparagus	180.106	7.0
Bananas	180.106	0.1 N
Barley, forage	180.106	2.0
Barley, grain	180.106	1.0
Barley, hay	180.106	2.0
Barley, straw	180.106	2.0
Blackberries	180.106	1.0
Blueberries	180.106	1.0
Boysenberries	180.106	1.0
Cattle, fat	180.106	1.0
Cattle, mbyp	180.106	1.0
Cattle, meat	180.106	1.0
Citrus, fruits	180.106	1.0
Clover, forage	180.106	2.0
Clover, hay	180.106	2.0
Corn, field, ear	180.106	1.0
Corn, field, fodder	180.106	2.0
Corn, field, forage	180.106	2.0
Corn, grain	180.106	1.0
Corn, pop, ear	180.106	1.0
Corn, pop, fodder	180.106	2.0
Corn, pop, forage	180.106	2.0
Corn, sweet, ear	180.106	1.0
Corn, sweet, fodder	180.106	2.0
Corn, sweet, forage	180.106	2.0
Cotton, seed	180.106	1.0
Currants	180.106	1.0
Dewberries	180.106	1.0
Goats, fat	180.106	1.0
Goats, mbyp	180.106	1.0

(continued)

	40 *CFR* Reference	Parts per Million
Goats, meat	180.106	1.0
Gooseberries	180.106	1.0
Grapes	180.106	1.0
Grasses, Bermuda	180.106	7.0
Grasses, Bermuda, hay	180.106	7.0
Grasses, crops (exc. Bermuda grass)	180.106	2.0
Grasses, hay (exc. Bermuda grass)	180.106	2.0
Hogs, fat	180.106	1.0
Hogs, mbyp	180.106	1.0
Hogs, meat	180.106	1.0
Horses, fat	180.106	1.0
Horses, mbyp	180.106	1.0
Horses, meat	180.106	1.0
Huckleberries	180.106	1.0
Loganberries	180.106	1.0
Nuts	180.106	0.1 N
Oats, forage	180.106	2.0
Oats, grain	180.106	1.0
Oats, hay	180.106	2.0
Oats, straw	180.106	2.0
Olives	180.106	1.0
Papayas	180.106	0.5
Peaches	180.106	0.1 N
Pears	180.106	1.0
Peas	180.106	1.0
Peas, forage	180.106	2.0
Peas, hay	180.106	2.0
Peppermint, hay	180.106	2.0
Pineapples	180.106	1.0
Potatoes	180.106	1.0
Raspberries	180.106	1.0
Rye, forage	180.106	2.0
Rye, grain	180.106	1.0
Rye, hay	180.106	2.0
Rye, straw	180.106	2.0
Sheep, fat	180.106	1.0
Sheep, mbyp	180.106	1.0
Sheep, meat	180.106	1.0
Sorghum, fodder	180.106	2.0
Sorghum, forage	180.106	2.0
Sorghum, grain	180.106	1.0
Sugarcane	180.106	1.0
Trefoil, birdsfoot, forage	180.106	2.0
Trefoil, birdsfoot, hay	180.106	2.0
Vetch, forage	180.106	2.0
Vetch, hay	180.106	2.0
Vetch, seed	180.106	1.0
Wheat, forage	180.106	2.0
Wheat, grain	180.106	1.0
Wheat, hay	180.106	2.0
Wheat, straw	180.106	2.0

The tolerances set by the EPA for diuron in animal feeds are as follows (the *CFR* Reference is to Title 21):

	CFR Reference	Parts per Million
Citrus pulp, dried (ls.f)	561.220	4.0

References

(1) Worthing, C.R., *Pesticide Manual,* 6th ed., p. 224, British Crop Protection Council (1979).
(2) Spencer, E.Y., *Guide to the Chemicals Used in Crop Protection,* 6th ed., p. 241, London, Ontario, Agriculture Canada (January 1973).
(3) Todd, C.W., U.S. Patent 2,655,445, October 13, 1953, assigned to DuPont.
(4) Midwest Research Institute and RvR Consultants, *Production, Distribution, Use and Environmental Impact Potential of Selected Pesticides,* Washington, DC, Council on Environmental Quality (March 15, 1974).
(5) Jones, R.L., U.S. Patent 2,768,971, October 30, 1956, assigned to Imperial Chemical Industries, Ltd.

DNOC

Function: Insecticide (1)

Chemical Name: 2-methyl-4,6-dinitrophenol

Formula:

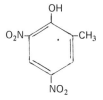

Trade Names: Sinox® (G. Truffaut et Cie)
Elgetol 30® (FMC)
Dinitrol®
Dinitrosol®

Manufacture

DNOC is made by the controlled nitration of o-cresol. Formulation of DNOC in powder or solution form has been described (2).

Toxicity

DNOC usage in the U.S. has declined in recent years because the compound is highly toxic to plants in the growing stage and nonselectively kills both desirable and undesirable vegetation. Additionally, the compound is highly toxic to humans and is considered one of the more dangerous agricultural pesticides (B-26). The acute oral LD_{50} for rats is 25-40 mg/kg.

As regards systemic effects of DNOC, the following has been reported (B-25): DNOC blocks the formation of high energy phosphate compounds, and the energy from oxidative metabolism is liberated as heat. Early symptoms of intoxication by inhalation or skin absorption are elevation of the basal metabolic rate and rise in temperature accompanied by fatigue, excessive sweating, unusual thirst, and loss of weight. The clinical picture resembles in part a thyroid crisis. Weakness, fatigue, increased respiratory rate, tachycardia, and fever may lead to rapid deterioration and death. Bilateral cataracts have been seen following oral ingestion for therapeutic purposes. These have not been seen during industrial or agricultural use.

Allowable Limits on Exposure and Use

Air: The threshold limit value for DNOC in air is 0.2 mg/m³ as of 1979. The tentative short term exposure limit is 0.6 mg/m³. Both limits bear the notation "skin," indicating that cutaneous absorption should be prevented so that the threshold limit value is not invalidated.

Water: In water, EPA has suggested (B-26) limits to protect human health from the adverse effects of dinitro-o-cresol ingested in contaminated water and fish of 12.8 µg/l.

Product Use: The tolerances set by the EPA for DNOC in or on raw agricultural commodities are as follows:

	40 *CFR* Reference	Parts per Million
Apples (blossom stage)	180.344	0.02 N

References

(1) Worthing, C.R., *Pesticide Manual*, 6th ed., p. 225, British Crop Protection Council (1979).
(2) Truffaut, G., British Patent 425,295 (February 28, 1935).

DODEMORPH

Function: Fungicide (1)(2)(3)

Chemical Name: 4-cyclododecyl-2,6-dimethylmorpholine

Formula:

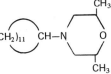

Trade Name: BAS-238F (BASF for acetate salt)

Manufacture (2)(3)

500 parts (by weight) of concentrated sulfuric acid is allowed to flow into 531 parts of N-di-(2-hydroxypropyl)-cyclododecylamine with stirring. Stirring is continued while the mixture is kept at 160°C for an hour under a water jet vacuum. The mixture is then slowly added to 2,500 parts of 25% caustic soda solution while stirring, allowed to settle and the upper layer (371 parts) separated off and dried over 50 parts of 50% caustic soda solution. The oily layer is separated off and distilled under reduced pressure. The boiling point of the product is 161° to 162°C at 1.5 mm Hg. The yield is 331 parts of 2,6-dimethyl-4-cyclododecylmorpholine, that is, 66% of the theory with reference to N-di-(2-hydroxypropyl)-cyclododecylamine.

Toxicity

The acute oral LD_{50} value for rats is 1,800 mg/kg which is slightly toxic.

References

(1) Worthing, C.R., *Pesticide Manual*, 6th ed., p. 226, British Crop Protection Council (1979).

(2) Sanne, W., Koenig, K.H., Pommer, E.H. and Stummeyer, H., U.S. Patent 3,468,885, September 23, 1969, assigned to BASF.

(3) Sanne, W., Koenig, K.H., Pommer, E.H. and Stummeyer, H., U.S. Patent 3,686,399, August 22, 1972, assigned to BASF.

DODINE

Function: Fungicide (1)(2)(3)

Chemical Name: Dodecylguanidine

Formula: $C_{12}H_{25}NHCNH_2$
                      $\overset{\parallel}{NH}$

Trade Names: AC 5223 (American Cyanamid Co.)
                 Cyprex[®] (American Cyanamid Co.)
                 Melprex[®] (American Cyanamid Co.)

Manufacture

Dodine may be made by the reaction of dodecyl chloride with sodium cyanamide followed by treatment of the product of the first reaction with ammonia. Dodine may also be made by the reaction of dodecyl amine with cyanamide in the presence of acetic acid (4). The reaction occurs according to the following equation.

$$C_{12}H_{25}NH_2 + H_2NCN + CH_3COOH \longrightarrow C_{12}H_{25}N\overset{\overset{\textstyle NH}{\parallel}}{H}CNH_2 \cdot CH_3COOH$$

The following process conditions relate to the second process described above. The reaction temperature may be in the range from 80° to 170°C. The upper temperature limit is dependent on the rate at which the product decomposes. Ordinarily, the best results are obtained at 140° to 160°C. At operating temperatures of 120° to 160°C it is usually necessary to conduct this reaction under autogenous pressure in an autoclave. Under most conditions the reaction can be completed in about 6 hours. Naturally, the reaction time required is essentially inversely related to reaction temperature.

The reaction is carried out in aqueous medium having a pH from 9.5 to 11.0. No added catalyst is used in this process. A stirred, jacketed autoclave is used for the conduct of this reaction. The product of this reaction may be isolated as the acid salt. The reaction product in aqueous or alcoholic solution may be reacted with other acids, such as boric, phthalic or maleic acids to give improved fungitoxic and bacteriotoxic compositions as described by G. Lamb (5).

Process Wastes and Their Control

Product Wastes: Dodine is stable under moderately alkaline conditions, but is hydrolyzed by strong alkali, liberating dodecylamine. It is incompatible with lime and anionic surfactants (B-3).

Toxicity

The acute oral LD$_{50}$ value for rats is about 1,000 mg/kg which is slightly toxic.

Allowable Limits on Exposure and Use

Product Use: The tolerances set by the EPA for dodine in or on raw agricultural commodities are as follows:

	40 *CFR* Reference	Parts per Million
Apples	180.172	5.0
Cherries, sour	180.172	5.0
Cherries, sweet	180.172	5.0
Meat	180.172	0.0
Milk	180.172	0.0
Peaches	180.172	5.0
Pears	180.172	5.0
Pecans	180.172	0.3
Strawberries	180.172	5.0
Walnuts, black	180.172	0.3

References

(1) Worthing, C.R., *Pesticide Manual,* 6th ed., p. 227, British Crop Protection Council (1979).
(2) Spencer, E.Y., *Guide to the Chemicals Used in Crop Protection,* 6th ed., p. 245, London, Ontario, Agriculture Canada (January 1973).
(3) Lamb, G., U.S. Patent 2,867,562, January 6, 1959, assigned to American Cyanamid Co.
(4) Paden, J.H. et al, U.S. Patent 2,425,341, August 12, 1947, assigned to American Cyanamid Co.
(5) Lamb, G., U.S. Patent 2,921,881, January 8, 1960, assigned to American Cyanamid Co.

DRAZOXOLON

Function: Fungicide (1)(2)(3)

Chemical Name: 3-methyl-4,5-isoxazoledione 4-[(2-chlorophenyl)hydrazone]

Formula:

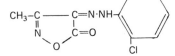

Trade Names: PP-781 (Plant Protection, Ltd.)
Mil-col® (Plant Protection, Ltd.)

Manufacture (3)

Drazoxolon may be prepared by coupling o-chlorobenzenediazonium chloride with ethyl acetoacetate. The following is a specific example (4): A solution of 35 parts of sodium nitrite in 55 parts of water was cooled to about 2°C in a container surrounded by ice. 160 parts of ice and 29.4 parts of sodium bisulfite dissolved in 55 parts of water were then added slowly to the sodium nitrite solution thus maintaining the temperature at from 0° to 3°C. Sulfur dioxide was then blown into the solution until the solution was acid to Congo Red. The resulting mixture was added to a solution of 28.6 parts of ethyl acetoacetate in 75 parts of methanol. This mixture was heated to 50°C and kept at this temperature for 3 hours; it was then cooled to below 10°C.

To this mixture was added a solution of a diazonium compound, the addition taking about 15 minutes and the temperature of the reaction being maintained at from 2° to 6°C. The diazonium compound was prepared by addition of 14 parts of sodium nitrite dissolved in 52 parts of water to 25.5 parts of o-chloroaniline dissolved in 64 parts of concentrated hydrochloric acid, 66 parts of water, and 150 parts of ice, the addition taking place at a temperature below 5°C.

After the addition the reaction mixture was heated to 60°C for 3 hours, it was then filtered while still hot and the resulting solid washed with 2,000 parts of water followed by 5 washings each of 50 parts of methanol. After drying at 60°C 40.4 parts of a product melting between 163° and 166°C and having a content (measured by infrared analysis) of 3-methyl-4-(o-chlorophenylhydrazono)-5-isoxazolone greater than 97% was obtained. This represents a yield of about 85% of theory when calculated on the o-chloroaniline used and about 77% of theory when calculated on the ethyl acetoacetate used.

Toxicity

The acute oral LD_{50} value for rats is 126 mg/kg which rates as moderately toxic.

References

(1) Worthing, C.R., *Pesticide Manual,* 6th ed., p. 228, British Crop Protection Council (1979).
(2) Spencer, E.Y., *Guide to the Chemicals Used in Crop Protection,* 6th ed., p. 246, London, Ontario, Agriculture Canada (January 1973).
(3) Summers, L.A., Freeman, P.F.H., Geohagan, M.J.A. and Turner, J.A.W., British Patent 999,097, July 21, 1965, assigned to Imperial Chemical Industries, Ltd.
(4) Green, M.B. and Roberts, R., British Patent 1,049,103, November 23, 1966, assigned to Imperial Chemical Industries Ltd.

DSMA

Function: Postemergence contact herbicide (1)(2)(4)

Chemical Name: Disodium methylarsonate

Formula: $CH_3AsO(ONa)_2$

Trade Names: Ansar® (Ansul Chemical Co.)
Crab-E-Rad®
Ditac®
Methar®
Sodar®

Manufacture (3)(5)(6)

In carrying out the process (3), a solution of sodium arsenite in water is formed in accordance with the following well-known equation:

$$As_2O_3 + 6NaOH \rightarrow 2Na_3AsO_3 + 3H_2O$$

It should be noted that a sufficient amount of sodium hydroxide is employed, so that its combination with the arsenious oxide so as to form sodium orthoarsenite is assured. As an illustration of the sodium arsenite solution and its constituents, 25 pounds of arsenious oxide and 30.3 pounds of sodium hydroxide in 30 gallons of solution, produces very satisfactory results.

The sodium arsenite solution, after preparation, is placed into a feed tank and from thence is charged at a constant rate of flow to a high tower filled with suitable inert packing. The sodium arsenite solution from the feed tank trickles downwardly through the packed tower and is returned from the bottom of the tower to the feed tank in any suitable manner. As the sodium arsenite solution descends from the top of the tower it is met by an ascending current of methyl chloride in gaseous form which is admitted at the bottom of the tower.

To carry out the methylation of the sodium arsenite solution, the system employed is a closed one and is maintained at substantially 60°C, in any well-known manner, and under a pressure of substantially 60 psi. If desired, the pressure in the system may be obtained by injecting the gaseous methyl chloride under suitable pressure into the tower. Under these conditions, the sodium arsenite solution and the methyl chloride react to form sodium methyl arsonate according to the following equation.

$$CH_3Cl + Na_3AsO_3 \rightarrow CH_3AsO(ONa)_2 + NaCl$$

While a pressure of substantially 60 psi and a temperature of substantially 60°C have been disclosed, it is to be understood that either or both of these may be varied to meet the particular operating conditions encountered. Of course, the temperature as well as the pressure of the system will vary depending upon the size and type of tower employed, the maximum pressure being that at which the methyl chloride gas will condense under operating conditions.

The apparatus is maintained as a closed system at the temperature and pressure mentioned, while the sodium arsenite solution is continuously circulated from the charging tank through the tower and returned to the former, and the methyl chloride gas is admitted at the bottom of the tower. Circulation of the sodium arsenite solution is continued until substantially 90% of it has been converted into sodium methyl arsenate. This point may be determined in any suitable and well-known manner.

It should be noted that with the use of methyl chloride as a methylating agent, a side reaction occurs with the loss of methyl chloride and sodium hydroxide (from the arsenious oxide) as shown by the following equation.

$$CH_3Cl + NaOH \rightarrow CH_3OH + NaCl$$

When the loss of sodium hydroxide has reached a point where its ratio to the arsenious oxide is below that required for the formation of sodium orthoarsenite, methylation is substantially arrested. In order to avoid this difficulty, sodium hydroxide is added to the original charge in excess of that required so as to compensate for its average loss in forming the methyl alcohol on the right hand side of the above equation.

An improved version of the process (5) involves the step of adding sodium hydroxide to the system at such a rate as to maintain in the system a substantially constant ratio of about 6 mols of sodium hydroxide for each mol of arsenious oxide present, thereby most efficiently compensating for the loss of sodium hydroxide through its reaction with methyl chloride.

Process Wastes and Their Control

Air: Air emissions from DSMA manufacture have been reported (B-15) to consist of the following:

Component	Kilograms per Metric Ton Pesticide Produced
Hydrocarbons	1.0
Arsenic trioxide	3×10^{-6}
Methyl chloride	0.05
Acetone	0.05
Methanol	0.05

Water: NaCl appears as a waste product of the methyl-arsenic unit but no data are available as to the dispostion of the waste (B-10).

Toxicity

The acute oral LD_{50} for rats is 1,800 mg/kg (slightly toxic).

Allowable Limits on Exposure and Use

Product Use: Issuance of a rebuttable presumption against registration for DSMA was being considered by EPA as of September 1979 on the basis of possible oncogenicity and mutagenicity.

The tolerances set by the EPA for DSMA in or on raw agricultural commodities are as follows:

	40 CFR Reference	**Parts per Million**
Citrus fruits	180.289	0.35
Cotton, seed	180.289	0.7

The tolerances set by the EPA for disodium methanearsonate in food are as follows (the CFR Reference is to Title 21):

	CFR Reference	**Parts per Million**
Sugarcane molasses	193.284	3.0
Sugarcane sugar	193.284	3.0
Sugar syrup	193.284	3.0

References

(1) Worthing, C.R., *Pesticide Manual,* 6th ed., p. 352, British Crop Protection Council (1979).
(2) Spencer, E.Y., *Guide to the Chemicals Used in Crop Protection,* 6th ed., p. 238, London, Ontario, Agriculture Canada (January 1973).
(3) Miller, G.E. and Seaton, S.G., U.S. Patent 2,442,372, June 1, 1948, assigned to the U.S. Secretary of War.
(4) Schwerdle, U.S. Patent 2,678,265, May 11, 1954, assigned to Vineland Chemical Co.
(5) Miller, G.E. and Reid, E.E., U.S. Patent 2,695,306, November 23, 1954.
(6) Schwerdle, A., U.S. Patent 2,889,347, June 2, 1959.

E

EDIFENPHOS

Function: Fungicide

Chemical Name: O-ethyl S,S-diphenyl phosphorodithioate

Formula:

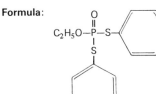

Trade Names: Bayer 78,418
Hinosan® (Bayer)

Manufacture

Edifenphos may be made by the reaction of two molecules of thiophenol with a molecule of ethylphosphorodichloridate.

Toxicity

The acute oral LD_{50} value for rats is 150 mg/kg (B-5) which is moderately toxic.

References

(1) Worthing, C.R., *Pesticide Manual,* 6th ed., p. 229, British Crop Protection Council (1979).
(2) Spencer, E.Y., *Guide to the Chemicals Used in Crop Protection,* 6th ed., p. 260, London, Ontario, Agriculture Canada (January 1973).

ENDOSULFAN

Function: Insecticide (1)(2)

Chemical Name: 6,7,8,9,10,10-hexachloro-1,5,5a,6,9,9a-hexahdyro 6,9-methano-2,4,3-benzodioxathiepin 3-oxide

Formula:

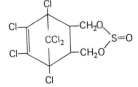

Trade Names: HOE 2671 (Farbwerke Hoechst AG)
Thiodan® (Farbwerke Hoechst AG)
FMC 5462
Thiodan® (FMC)
Cyclodan®
Beosit®
Malix®
Thimul®
Thifor®

Manufacture

The feed materials for endosulfan manufacture are hexachlorocyclopentadiene, cis-2-butene-1,4-diol and thionyl chloride. The initial reaction which takes place is a Diels-Alder condensation,

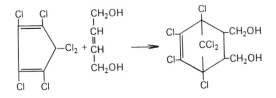

and the second stage is the conversion to a sulfite by means of reaction with thionyl chloride:

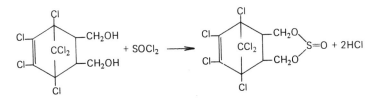

In the first step, the butenediol is added slowly over a period of time to an excess of hexachlorocyclopentadiene. Optimum results are obtained when the addition takes place over one-quarter to one-half of the total reaction time, according to E.J. Geering et al (3).

The presence of calcium carbonate is essential to avoid the acid-catalyzed ring closure:

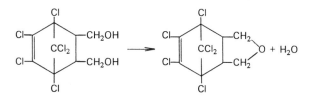

The Diels-Alder condensation step should take place at temperatures in excess of 75°C and preferably between 125° and 250°C in order to give commercially practical reaction rates. The reaction may conveniently be carried out at the reflux temperature of the toluene solvent (111°C).

The temperatures employed in the reaction with thionyl chloride are greater than room temperature and preferably between 50° and about 110°C. The condensation step is carried out at atmospheric pressure, as is the subsequent reaction with thionyl chloride.

The time of contact in the condensation step is not critical and varies with temperature and desired degree of conversion. Generally speaking, high yields are obtained with reaction times in the range of 20 minutes to 20 hours. A total reaction time of 8 hours, 4 hours of which are occupied by reactant addition, is typical. Similarly, the time of reaction with thionyl chloride is not critical and may vary from several minutes to several hours. Two hours might be a typical time.

The reaction of hexachlorocyclopentadiene with butenediol may conveniently be carried out in the presence of a solvent such as toluene. A small amount of an acid acceptor should be present; epichlorohydrin may be used to remove traces of HCl from the hexachlorocyclopentadiene.

Toluene may also be used as the solvent for the thionyl chloride reaction step. Carbon tetrachloride may also be used as noted by H. Frensch et al (4). No catalyst is used in the Diels-Alder condensation step, or in the subsequent conversion of the diol to a sulfite. A stirred, jacketed reaction kettle of conventional design, equipped with a reflux condenser may be used for both steps. The mixture from the condensation reaction is cooled and the crystalline product filtered out and the filter cake washed with toluene. The yield is typically about 85%. After reaction with thionyl chloride, the reaction mixture is cooled, filtered to remove any foreign matter, and stripped to remove any unreacted thionyl chloride. The yield is 99.5% on the Diels-Alder adduct charged.

The product contains two isomers, one melting at 108° to 110°C and the other melting at 208° to 210°C. The pure isomers thus obtained have differing periods of persistence when applied as insecticides on crops and, therefore, are most suitable for differing applications, depending on whether long or short persistence is desired. It is especially important that the low melting isomer should be substantially free of the high melting, more residual isomer in some applications, otherwise intolerable insecticide residues may remain on the treated crops after harvesting. Because the high melting isomer is more stable than the low melting isomer and more stable than endosulfan, there will be certain applications for this isomer where a highly effective, long lasting insecticide is required.

On the other hand, the low melting isomer is more volatile than the high melting isomer when it is desirable to have the insecticide volatilize comparatively quickly. For example, when it appears desirable under certain circumstances to apply a highly effective insecticide on agricultural crops shortly before harvest time, the low melting isomer meets the qualification of having high insecticidal acitivity and at the same time is volatile enough to be gone by harvest time or to be decomposed to nontoxic degradation products a very short time after application.

Figure 30 shows a schematic flow diagram for an isomer separation technique as described by H.L. Schlichting (5). As shown there, 300 parts by weight of endosulfan containing a 68/30 isomer mixture may be mixed with 100 parts by volume of trichloroethylene. The mixture may then be heated and filtered to concentrate the high-melting isomer. The filtrate is then processed to give a nearly pure low-melting isomer.

Figure 30: Schematic Diagram of Endosulfan Isomer Separation Technique

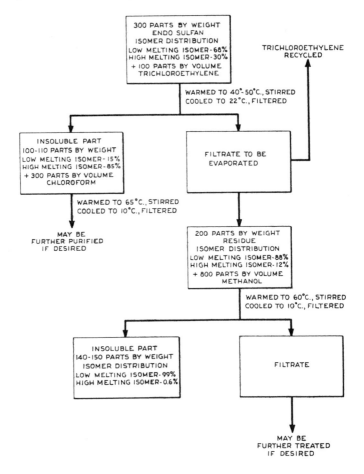

Source: Reference (5).

Process Wastes and Their Control

Air: Air emissions from endosulfon manufacture have been reported (B-15) to consist of 1.0 kg hydrocarbons and 0.5 kg hydrogen chloride per metric ton of pesticide produced.

Product Wastes: Endosulfan is a sulfur-containing compound and unlike most of the hexachlorocyclopentadiene family, is sensitive to moisture, bases and acids. It is slowly hydrolyzed to give SO_2 and the corresponding diol $C_7Cl_6(CH_2OH)_2$. The analytical method is based on reaction with NaOH in methanol: Heat 0.7 g endosulfan with 100 ml methanol and 3 to 4 g NaOH for 2 hours on a hot plate or under reflux. (The SO_2 is largely retained as sodium sulfite.) Endosulfan is said to give calcium sulfate with lime. Endosulfan is stable in sunlight. A recommended method for disposal is burial 18 inches deep in noncropland away from water supplies, but bags can be burned (B-3).

Toxicity

The acute oral LD_{50} value for rats is 80-110 mg/kg which is moderately toxic.

Endosulfan has been demonstrated to be highly toxic to fish and marine invertebrates and is readily adsorbed by sediments. It therefore represents a potential hazard in the aquatic environment (B-26).

Allowable Limits on Exposure and Use

Air: The threshold limit value for endosulfan in air is 0.1 mg/m^3 as of 1979. The tentative short term exposure limit is 0.3 mg/m^3. It is further noted that cutaneous absorption should be prevented so that the threshold limit value is not invalidated.

Water: In water, EPA has set (B-12) a criterion of 0.003 μg/l of endosulfan for the protection of freshwater aquatic life and a limit of 0.001 μg/l for the protection of marine aquatic life.

In water, EPA has subsequently suggested (B-26) a criterion to protect freshwater aquatic life of 0.042 μg/l as a 24-hour average; the concentration should not exceed 0.49 μg/l at any time.

For saltwater aquatic life, no criterion for endosulfan can be derived; there are insufficient data to estimate a criterion.

For the protection of human health from the toxic properties of endosulfan ingested through water and contaminated aquatic organisms, the ambient water criterion is 0.1 mg/l.

Product Use: The tolerances set by the EPA for endosulfan in or on raw agricultural commodities are as follows:

	40 *CFR* Reference	Parts per Million
Alfalfa, fresh	180.182	0.3
Alfalfa, hay	180.182	1.0
Almonds	180.182	0.2 N
Almonds, hulls	180.182	1.0
Apples	180.182	2.0
Apricots	180.182	2.0
Artichokes	180.182	2.0
Barley, grain	180.182	0.1 N
Barley, straw	180.182	0.2 N
Beans	180.182	2.0
Beets, sugar, without tops	180.182	0.1 N
Blueberries	180.182	0.1 N
Broccoli	180.182	2.0
Brussels sprouts	180.182	2.0
Cabbage	180.182	2.0
Carrots	180.182	0.2
Cattle, fat	180.182	0.2
Cattle, mbyp	180.182	0.2
Cattle, meat	180.182	0.2
Cauliflower	180.182	2.0
Celery	180.182	2.0
Cherries	180.182	2.0

(continued)

	40 *CFR* Reference	Parts per Million
Collards	180.182	2.0
Corn, sweet (k+cwhr)	180.182	0.2
Cotton, seed	180.182	1.0
Cucumbers	180.182	2.0
Eggplant	180.182	2.0
Goats, fat	180.182	0.2
Goats, mbyp	180.182	0.2
Goats, meat	180.182	0.2
Grapes	180.182	2.0
Hogs, fat	180.182	0.2
Hogs, mbyp	180.182	0.2
Hogs, meat	180.182	0.2
Horses, fat	180.182	0.2
Horses, mbyp	180.182	0.2
Horses, meat	180.182	0.2
Kale	180.182	2.0
Lettuce	180.182	2.0
Macadamia nuts	180.182	0.2 N
Melons	180.182	2.0
Milk, fat (= N in whole milk)	180.182	0.5
Mustard, greens	180.182	2.0
Mustard, seed	180.182	0.2 N
Nectarines	180.182	2.0
Nuts, filberts	180.182	0.2 N
Oats, grain	180.182	0.1 N
Oats, straw	180.182	0.2 N
Peaches	180.182	2.0
Pears	180.182	2.0
Peas, succulent	180.182	2.0
Pecans	180.182	0.2 N
Peppers	180.182	2.0
Pineapples	180.182	2.0
Plums	180.182	2.0
Potatoes	180.182	0.2 N
Prunes	180.182	2.0
Pumpkins	180.182	2.0
Rape, seed	180.182	0.2 N
Rye, grain	180.182	0.1 N
Rye, straw	180.182	0.2 N
Safflower, seed	180.182	0.2 N
Sheep, fat	180.182	0.2
Sheep, mbyp	180.182	0.2
Sheep, meat	180.182	0.2
Spinach	180.182	2.0
Squash, summer	180.182	2.0
Squash, winter	180.182	2.0
Strawberries	180.182	2.0
Sugarcane	180.182	0.5
Sunflower, seed	180.182	2.0
Sweet potatoes	180.182	0.2
Tomatoes	180.182	2.0
Turnips, greens	180.182	2.0
Walnuts	180.182	0.2 N
Watercress	180.182	2.0
Wheat, grain	180.182	0.1 N
Wheat, straw	180.182	0.2 N

The tolerances set by the EPA for endosulfan in food are as follows (the *CFR* Reference is to Title 21):

	CFR Reference	Parts per Million
Tea, dried	193.170	24.0

References

(1) Worthing, C.R., *Pesticide Manual*, 6th ed., p. 231, British Crop Protection Council (1979).
(2) Spencer, E.Y., *Guide to the Chemicals Used in Crop Protection*, 6th ed., p. 247, London, Ontario, Agriculture Canada (January 1973).
(3) Geering, E.J. et al, U.S. Patent 2,983,732, May 9, 1961, assigned to Hooker Chemical Corp.
(4) Frensch, H. et al, U.S. Patent 2,799,685, July 16, 1957, assigned to Farbwerke Hoechst AG.
(5) Schlichting, H.L., U.S. Patent 3,251,856, May 17, 1966, assigned to Hooker Chemical Corp.

ENDOTHALL (ENDOTHAL in U.K.)

Function: Herbicide (1)(2)

Chemical Name: 7-oxabicyclo[2.2.1]heptane-2,3-dicarboxylic acid

Formula:

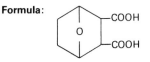

Manufacture

Endothal is made by the condensation of furan and maleic anhydride according to the equation shown below as described by N. Tischler et al (3).

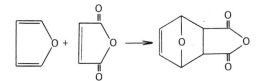

The reaction is conducted at 35°C and atmospheric pressure in the liquid phase. Isopropyl ether may be used as a reaction diluent. No catalyst is used and the reaction time is about 7 hours.

An aluminum reaction vessel equipped with an agitator and a reflux condenser may be used for the conduct of this reaction, as described by N. Tischler et al (5). The sodium salt may be made by reacting the anhydride product of the initial condensation reaction with caustic as described by J.F. Olin (6). The reaction which occurs is as follows:

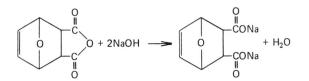

Other salts may be produced by similar means (4).

Process Wastes and Their Control

Product Wastes: Endothal is not readily degraded by common reagents. It is stable in acid and reacts with bases to form salts. It is stable to about 90°C, at which temperature it is slowly converted to the anhydride. Burial of unwanted quantities has also been suggested (B-3).

Toxicity

The acute oral LD_{50} value for rats is 51 mg/kg which is moderately to highly toxic.

Allowable Limits on Exposure and Use

Product Use: Tolerances set by the EPA for endothal in or on raw agricultural commodities are as follows:

	40 *CFR* Reference	**Parts per Million**
Cotton, seed	180.293	0.1
Potatoes	180.293	0.1
Rice, grain	180.293	0.05 N
Rice, straw	180.293	0.05 N

The tolerances set by the EPA for endothal in food are as follows (the *CFR* Reference is to Title 21):

	***CFR* Reference**	**Parts per Million**
Water, potable	193.180	0.2 i

References

(1) Worthing, C.R., *Pesticide Manual,* 6th ed., p. 233, British Crop Protection Council (1979).
(2) Spencer, E.Y., *Guide to the Chemicals Used in Crop Protection,* 6th ed., p. 248, London, Ontario, Agriculture Canada (January 1973).
(3) Tischler, N. et al, U.S. Patent 2,576,080, November 20, 1951, assigned to Sharples Chemicals Inc.
(4) Tischler, N. et al, U.S. Patent 2,576,081, November 20, 1951, assigned to Sharples Chemicals Inc.
(5) Tischler, N. et al, U.S. Patent 2,576,082, November 20, 1951, assigned to Sharples Chemicals Inc.
(6) Olin, J.F., U.S. Patent 2,550,494, April 24, 1951, assigned to Sharples Chemicals Inc.

ENDRIN

Function: Insecticide (1)(2)

Chemical Name: 3,4,5,6,9,9-hexachloro-1aα,2β,2aβ,3α,6α,6aβ,7β,7aα-octahydro-2,7:3,5-dimethanonaphth[2,3-b] oxirene

Formula:

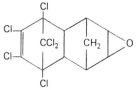

Trade Names: Experimental Insecticide 269 (Julius Hyman and Co.)
   Mendrin®

Manufacture

The various steps in endrin manufacture as as follows:

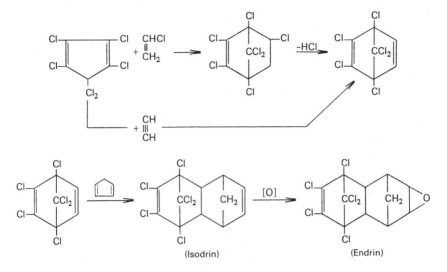

(Isodrin)   (Endrin)

The starting material for endrin manufacture is hexachlorocyclopentadiene. It may be re-acted with vinyl chloride to give a product which is then dehydrochlorinated as described by H. Bluestone (3). That intermediate is condensed with cyclopentadiene to give isodrin and then epoxidized to give endrin.

Alternatively, a slightly more straightforward process may be practiced involving condensa-tion of hexachlorocyclopentadiene with acetylene to give the intermediate which may then be condensed with cyclopentadiene and epoxidized to give endrin.

The acetylene-hexachlorocyclopentadiene condensation is carried out at 150° to 175°C and 2,000 to 4,000 psi pressure in the liquid phase in the absence of a catalyst (4)(5). Reaction time is 0.5 to 3.0 hours.

The second stage condensation with cyclopentadiene is carried out at 50° to 90°C and atmos-pheric pressure. It involves a two-hour reaction period after gradual reactant addition. The reaction mixture from the second stage of endrin preparation is cooled, whereupon the product begins to precipitate out. The mixture may be poured into a boiling acetone-methanol mixture and then crystallized therefrom.

The final epoxidation step is carried out using peracetic acid at 20° to 45°C and atmospheric pressure in the presence of a benzene solvent. The epoxidation step is fairly slow. After slow reactant addition, the mixture is agitated at length and finally heated to 45°C for about 1 hour.

After the final epoxidation step, the reaction mixture may be steam distilled to remove ex-cess peracid, carboxylic acid and benzene solvent. The residue may be ether extracted and the ether solution washed and dried. The ether may then be evaporated to give the final endrin product in crystalline form.

Thus, it has been found that the presence of metals during the final epoxidation step in the preparation of endrin leads to undesired by-products and further to a less satisfactory product (6). This product as compared to technical endrin commercially produced in a metal-free state is relatively unstable. It also contains color bodies, while the aforementioned product not adversely affected by metal in its production is white in color. The instability of this product creates some problems. For example, metal-containing endrin has a melting point much lower than pure endrin (245°C), and since it is a mixed melting point, the value will vary with the actual amount of by-product formed, metal content, and the like. This low melting point creates problems in drying the product and also in grinding, a standard operation in the preparation of commercial dry insecticide formulations.

Further, the presence of these undesired compounds lowers the amount of insecticidally active ingredients in the product, making it necessary to use higher concentration formulations, which may be impractical due to economic considerations or due to physical limitations, such as the inability of the carrier to adsorb the composition.

Unexpectedly, it has been found that this vexing problem can be solved by the addition of dipicolinic acid in the production of endrin (6). Dipicolinic acid, which hereafter will be referred to as DPA, is meta-2,6-pyridine dicarboxylic acid. While the presence of DPA in even low concentrations is highly beneficial, a minimum DPA:metal molar ratio of 1:1 is of even further advantage, and DPA:metal ratio of 2:1 or higher is preferred. Mol ratios of DPA to metal above 4 produces a high grade product containing over 95% endrin. When this mol ratio is between about 1 and 4, the concentration of endrin in the product drops to between 90 and 95%; below a mol ratio of DPA to metal of about 1, the concentration of final product is still lower, dropping to below 80%.

The mode of addition of the DPA to the reaction zone may be performed by any of the methods known to the art for the addition of solid matter such as by the use of conveyor belts, manual addition, feed tanks, etc. It also may be added in its melted state, although this is not a preferred method. Actually the amount of DPA added on a percentage basis is very small in most cases, and hence addition thereof to the reaction zone does not constitute a problem.

If desired, the DPA may be added concurrently with the isodrin or one of the other charged reactants. Also, but not as an alternative to DPA being present during reaction, the DPA may be added to the final product after completion of the reaction and subsequent purification. This would prevent any degradation of the endrin due to metallic contamination during formulation and final use of the endrin. A standard blender for the mixing of solid materials may be used to incorporate this unique stabilizer with the endrin. While the amount of DPA necessary in such a mixture will depend on the amount of metal contamination it will encounter under the normal circumstances, from about 0.5 to 500 ppm DPA (based on 100% endrin) should be adequate although smaller or larger concentrations may be used.

Process Wastes and Their Control

Air: Air emissions from endrin manufacture have been reported to consist of 1.0 kg hydrocarbons and 0.5 kg HCl per metric ton of pesticide produced (B-15).

Water: A typical waste stream from a diolefin based chlorinated hydrocarbon pesticide manufacture will include:

COD	500 mg/l
BOD,	50 mg/l
Total solids	1,000 mg/l
Suspended solids	100 mg/l
pH	2
Chlorides	High
Nitrates	—
Phosphates	—

Product as endrin	100-300 ppb
Toxicity	High
Flow	0.375 gal/lb*

*Gallons treated wastewater/pound product

(This does not include concentrated liquids, tank "bottoms," spent catalysts, etc., which are normally landfilled.)

Waste streams containing peracid, carboxylic acid and benzene are steam washed and discharged (B-10).

Product Wastes: Endrin is a stereoisomer of dieldrin and has similar chemistry. It is stable to alkalis. It is apparently stable with dilute acids, but rearranges with strong acids, acid catalysts or certain metals or when heated above 200°C to give a compound which is insecticidally less active than is endrin.

A disposal procedure recommended by the manufacturer consists of absorption, if necessary, and burial at least 18 inches deep, preferably in sandy soil in a flat or depressed location away from wells, livestock, children, wildlife, etc. (B-3).

Toxicity

The acute oral LD_{50} value for rats is 7.5-17.5 mg/kg which is highly toxic.

In the aquatic environment endrin is acutely toxic to carp at 0.046 μg/l and to the pink shrimp at 0.037 μg/l. It is chronically toxic to the fathead minnow at 0.187 μg/l and at 0.038 μg/l to the grass shrimp. Endrin has been reported to bioconcentrate by factors as high as 15,000 in freshwater fish and 6,400 in marine fish.

Endrin is toxic to mammals, but a no-effect level of 1 mg/kg for the rat and the dog has been established. Quantitative data on endrin toxicity to humans are not available (B-26).

In animals, chronic exposure to endrin may result in damage to the liver, kidneys, heart, brain, lung, adrenal glands, and spleen. Effects secondary to central nervous system disorders have also been observed following chronic exposure of mammals to sublethal doses of endrin. These include behavioral abnormalities, changes in carbohydrate metabolism, and changes in the composition of the blood. Although no malignancies attributable to endrin have been reported, chromosomal abnormalities and teratogenesis have been induced by endrin in several mammalian species (B-26).

Allowable Limits on Exposure and Use

Air: The threshold limit value for endrin in air is 0.1 mg/m³ as of 1979. The tentative short term exposure limit is 0.3 mg/m³. Further, it is noted that cutaneous absorption should be prevented so that the threshold limit value is not invalidated.

Water: In water, EPA has set (B-12) criteria of 0.2 μg/l for domestic water supplies on a health protection basis. They also set a criterion of 0.004 μg/l for the protection of freshwater and marine aquatic life.

In water, EPA has subsequently suggested (B-26) a criterion to protect freshwater aquatic life of 0.0020 μg/l as a 24-hour average; the concentration should not exceed 0.10 μg/l at any time.

For endrin the criterion to protect saltwater aquatic life is 0.0047 μg/l as a 24-hour average and the concentration should not exceed 0.031 μg/l at any time.

For the protection of human health from the toxic properties of endrin ingested through water and contaminated aquatic organisms, the ambient water criterion is determined to be 1 μg/l.

Toxic pollutant effluent standards (40 *CFR* Part 129.102) were promulgated by the EPA. These allowed an effluent concentration of 1.5 μg/l per average working day calculated over a period of one month, not to exceed 7.5 μg/l in any sample representing one working day's effluent. In addition, discharge is not to exceed 0.0006 kg per 1,000 kg of production.

Product Use: A rebuttable presumption notice against registration was issued on July 27, 1976 by EPA on the basis of oncogenicity, teratogenicity, and reductions in endangered species and nontarget species.

In a notice on May 20, 1964, EPA (B-17) had done the following:

(1) Cancelled the use of endrin on:

    (a) Tobacco.

In subsequent notices, issued July 25, 1979, EPA cancelled uses of endrin on:

    (b) Cotton in all areas east of Interstate Highway #35 (includes all states east of the Mississippi River, Arkansas, Louisiana, Missouri, and portions of Texas and Oklahoma).

    (c) Small grains to control all pests other than the army cutworm, the pale western cutworm.

    (d) Apple orchards in Eastern States to control meadow voles.

    (e) Sugarcane to control the sugarcane borer.

    (f) Ornamentals.

Further, the EPA took the following actions:

(2) Denial of applications for new registrations for the above uses [(1) (b) through (e)], as well as for the use of endrin in unenclosed bird perch treatment.

(3) Cancellation of the following registrations of endrin products unless registrants modify the terms and conditions of registration as specified below:

    (a) Use on cotton west of Interstate Highway #35 [must modify label to add statements (5) (a) through (g)].

    (b) Use on small grains to control army cutworms and pale western cutworms [must modify label to add statements (5) (b) through (f) and (h)].

    (c) Use on apple orchard's in Eastern States to control the pine vole and in Western States to control meadow voles [must modify label to add statements (5) (g), (c), (i), (r), (j) through (l)].

    (d) Use on sugarcane to control the sugarcane beatle [must modify label to add statements (5) (g), (c), (m) and (n)].

    (e) Use for conifer seed treatment [must modify label to add statement (5) (o)].

    (f) Use in enclosed bird perch treatments [must modify label to add statements (5) (s), (c) and (p)],

(4) Denial of applications for new registrations for any of the above endrin uses [(3) (a) through (f)], as well as for the following endrin uses unless the applications are modified to meet the terms and conditions specified herein.

    (a) As a tree painting (in Texas) [must modify label to add statements (5) (s) and (c)].

(b) On alfalfa and clover seed crops (in Colorado) [must modify label to add statements (5) (b) through (f)].

(c) On small grains to control grasshopper (in Montana) [must modify label to add statements (5) (b) through (f) and (h)].

(5) Label Statements:

(a) For use in areas west of Interstate Highway #35 only.

(b) *Required Clothing for Female Workers*—Female ground applicators, mixers and loaders and flagpersons must wear long-sleeved shirts and long pants made of a closely woven fabric, and wide-brimmed hats. Mixers and loaders must also wear rubber or synthetic rubber boots and aprons.

(c) *Warning to Female Workers*—The United States Environmental Protection Agency has determined that endrin causes birth defects in laboratory animals. Exposure to endrin during pregnancy should be avoided. Female workers must be sure to wear all protective clothing and use all protective equipment specified on this label. In case of accidental spills or other unusual exposure, cease work immediately and follow directions for contact with endrin.

(d) *Equipment*—

1. Ground Application: For use with boom-nozzle ground equipment. Apply at not less than 5 gallons total mixture, water and chemical, per acre. Do not use nozzle liquid pressure at greater than 40 pounds per square inch (psi). Do not use cone nozzle size smaller than 0.16 gallon per minute (gpm), at 40 psi such as type D2-25 or TX-10, or any other atomizer or nozzle giving smaller drop size.

2. Aerial Application: Do not apply at less than 2 gallons total mixture of water and chemical per acre. Do not operate nozzle liquid pressure over 40 psi or with any fan nozzle smaller than 0.4 gpm or fan angle greater than 65 degrees such as type 6504. Do not use any cone type nozzles smaller than 0.4 gpm nor whirl plate smaller than #46 such as type D4-46 or any other atomizer or nozzle giving smaller drop size. Do not release this material at greater than 10 feet height above the crop.

(e) *Application Restrictions*—Do not apply this product within 1/8 mile of human habitation. Do not apply this product by air within 1/4 mile or by ground within 1/8 mile of lakes, ponds, or streams. Application may be made at distances closer to ponds owned by the user but such application may result in excessive contamination and fish kills. Do not apply when rainfall is imminent. Apply only when wind velocity is between 2 and 10 mph.

(f) *Procedures to Be Followed if Fish Kills Occur or if Ponds Are Contaminated*—In case of fish kills, fish must be collected and disposed of by burial. Ponds in which fish kills have occurred, and user-owned ponds exposed to endrin by application at distances closer than otherwise prohibited, must be posted with signs stating: "Contaminated: No Fishing." Signs must remain for one year after a fish kill has occurred or for six months after lesser contamination unless laboratory analysis shows endrin residues in the edible portion of fish to be less than 0.3 part per million (ppm).

(g) *Prophylactic Use*—Unnecessary use of this product can lead to resistance in pest populations and subsequent lack of efficacy.

(h) *Pests for Which This Product May Be Applied*—This product may be applied to control the following pests only: army cutworm; pale western cutworm; grasshoppers. (Currently grasshoppers may only be included on endrin products for use in Montana.)

(i) *Application Restrictions*—Do not apply this product within 50 feet of lakes, ponds or streams. Do not apply this product within 50 feet of areas occupied by unprotected humans. Do not apply when rainfall is imminent.

(j) *Equipment*—Apply by ground equipment only. Use a very coarse spray with minimum pressure necessary to penetrate ground cover. Do not apply as fine spray. Power air blast equipment must be modified to meet the above application restriction. Consult the state recommendations for acceptable methods of adapting equipment.

(k) *Prophylactic Use*—Unnecessary use of this product can lead to resistance in the vole population and subsequent lack of efficacy.

(l) *Pests for Which This Product May Be Applied*—This product may be applied to control the following pests only:

Eastern United States—Pine vole (*Microtus pinetorum*)
Western United States—Meadow voles (*Microtus* species)

(m) *Application Restrictions*—Apply only with low-pressure ground equipment. Cover furrows with soil promptly after application.

(n) *Pests for Which This Product May Be Applied*—This product may be applied only to control the sugarcane beetle.

(o) *Application Restrictions*—Do not sow treated seed when large numbers of migratory birds are expected.

(p) *Special Warning*—Do not use within one mile of roosting sites or within two miles of nesting sites of peregrine falcons, as identified by the United States Fish and Wildlife Service.

(q) *Required Clothing for Female Workers*—Female applicators, mixers and loaders must wear long-sleeved shirts and long pants made of a closely woven fabric, and wide-brimmed hats. Mixers and loaders must also wear rubber or synthetic rubber boots and aprons.

(r) *Procedures to Be Followed if Fish Kills Occur*—In case of fish kills, fish must be collected promptly and disposed of by burial. Ponds in which fish kills have occurred must be posted "Contaminated: No Fishing." Signs must remain for one year after a fish kill has occurred unless laboratory analysis shows endrin residues in the edible portion of fish to be less than 0.3 ppm.

(s) *Required Clothing for Female Workers*—Female workers handling or applying this product must wear long-sleeved shirts and long pants made of a closely woven fabric, wide-brimmed hat, and rubber or synthetic rubber boots and aprons.

The tolerances set by the EPA for endrin in or on raw agricultural commodities are as follows:

	40 *CFR* Reference	Parts per Million
Beets, sugar	180.131	0.0
Beets, sugar, tops	180.131	0.0
Broccoli	180.131	0.0
Brussels sprouts	180.131	0.0
Cabbage	180.131	0.0
Cauliflower	180.131	0.0
Cotton, seed	180.131	0.0
Cucumbers	180.131	0.0
Eggplant	180.131	0.0
Peppers	180.131	0.0
Potatoes	180.131	0.0
Squash, summer	180.131	0.0
Tomatoes	180.131	0.0

References

(1) Worthing, C.R., *Pesticide Manual,* 6th ed., p. 234, British Crop Protection Council (1979).
(2) Spencer, E.Y., *Guide to the Chemicals Used in Crop Protection,* 6th ed., p. 249, London, Ontario, Agriculture Canada (January 1973).
(3) Bluestone, H., U.S. Patent 2,676,132, April 20, 1954, assigned to Shell Development Co.
(4) Howald, J.M. et al, U.S. Patent 2,813,915, November 19, 1957, assigned to Shell Development Co.
(5) Anderson, J. et al, U.S. Patent 2,900,419, August 18, 1959, assigned to Shell Development Co.
(6) Marks, D.R., U.S. Patent 2,899,446, August 11, 1959, assigned to Velsicol Chemical Corp.

EPN

Function: Insecticide and acaricide (1)(2)

Chemical Name: O-ethyl O-4-nitrophenyl phenyl-phosphonothioate

Formula:

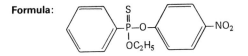

Trade Name: EPN-300® (DuPont)

Manufacture

In an initial step (3), sodium ethylate reacts with phenyl thiophosphonyl dichloride as follows:

$$C_2H_5ONa + C_6H_5\overset{\overset{S}{\|}}{P}Cl_2 \longrightarrow C_6H_5\overset{\overset{S}{\|}}{\underset{OC_2H_5}{P}}Cl + NaCl$$

In a subsequent step, the intermediate product is reacted with sodium-p-nitrophenate as follows:

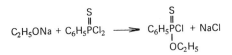

The first step of this process is conveniently carried out at room temperature of about 20°C. The second step is preferably carried out at 100° to 150°C and is conveniently operated at the reflux temperature of chlorobenzene (130°C). Both steps of this process are carried out at atmospheric pressure. A reaction time of about 2 hours at 20°C may be used for the initial condensation step. A reaction time of about 4 hours may be used for the second condensation step in this process.

Both steps of this process are carried out in the liquid phase, the first in benzene solution and the second in chlorobenzene solution. Catalysts are not used in either of the two reaction steps in EPN manufacture.

Stirred, jacketed reaction vessels of conventional design may be used in this process. The second stage requires the incorporation of a reflux condenser in the apparatus. The products cf the first stage reaction are separated and the benzene layer dried and distilled. The second stage reaction product is cooled and filtered to remove suspended sodium chloride. The solvent may then be removed under vacuum to give the product EPN an oily liquid with a light yellow color.

Process Wastes and Their Control

Water: A plant described in a study by Midwest Research Institute under EPA contract (B-14) uses two different aqueous waste disposal systems. Primary production waste was disposed of by off-site deep well injection. Rain runoff, blowdown and washdown waters were discharged into the channel of an industrial area's central waste treatment plant that treats waste from approximately 35 plants of various types.

That plant is shown in Figure 31. It consists of: preaeration, activated sludge treatment with clarification and sludge recycle, chlorination, and final polishing where additional natural biological stabilization occurs. The pretreatment sample was taken of the EPN effluent just prior to discharge into Bio-San channel. There was no information to indicate waste processing prior to this discharge point. The posttreatment sample was taken following chlorination but prior to the final polishing stage.

Figure 31: Wastewater Treatment for EPN Production

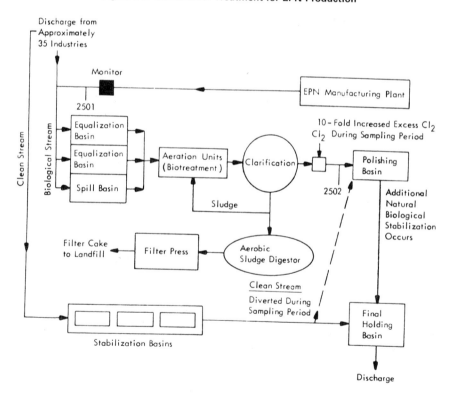

Source: Reference (B-14)

Product Wastes: EPN is incompatible with alkaline pesticides. It is relatively rapidly hydrolyzed in alkaline medium to benzene thiophosphoric acid, alcohol and p-nitrophenol (B-3).

Toxicity

The acute oral LD_{50} value for rats is 8 mg/kg (B-5) which is very highly toxic.

Allowable Limits on Exposure and Use

Air: The threshold limit value for EPN in air is 0.5 mg/m^3 as of 1979. The tentative short term exposure limit is 2.0 mg/m^3. The note is appended that cutaneous absorption should be avoided so that the threshold limit value is not invalidated.

Product Use: A rebuttable presumption against registration was issued on September 19, 1979 by EPA on the basis of neurotoxicity.

The tolerances set by the EPA for EPN in or on raw agricultural commodities are as follows:

	40 *CFR* Reference	Parts per Million
Almonds	180.119	0.5
Apples	180.119	3.0
Apricots	180.119	3.0
Beans	180.119	3.0
Beets, greens (alone)	180.119	3.0
Beets, sugar, without tops	180.119	3.0
Beets, with tops	180.119	3.0
Beets, without tops	180.119	3.0
Blackberries	180.119	3.0
Boysenberries	180.119	3.0
Cherries	180.119	3.0
Citrus fruits	180.119	3.0
Corn	180.119	3.0
Cotton, seed	180.119	0.5
Dewberries	180.119	3.0
Grapes	180.119	3.0
Lettuce	180.119	3.0
Loganberries	180.119	3.0
Nectarines	180.119	3.0
Olives	180.119	3.0
Peaches	180.119	3.0
Pears	180.119	3.0
Pecans	180.119	0.5
Pineapples	180.119	3.0
Plums (fresh prunes)	180.119	3.0
Quinces	180.119	3.0
Raspberries	180.119	3.0
Rutabaga tops	180.119	3.0
Rutabagas, with tops	180.119	3.0
Rutabagas, without tops	180.119	3.0
Soybeans	180.119	0.05 N
Spinach	180.119	3.0
Strawberries	180.119	3.0
Tomatoes	180.119	3.0
Turnips, greens	180.119	3.0
Turnips, with tops	180.119	3.0
Turnips, without tops	180.119	3.0
Walnuts	180.119	0.5
Youngberries	180.119	3.0

References

(1) Worthing, C.R., *Pesticide Manual,* 6th ed., p. 235, British Crop Protection Council (1979).
(2) Spencer, E.Y., *Guide to the Chemicals Used in Crop Protection,* 6th ed., p. 251, London, Ontario, Agriculture Canada (January 1973).
(3) Jelinek, A.G., U.S. Patent 2,503,390, April 11, 1950, assigned to DuPont.

EPTC

Function: Herbicide (1)(2)

Chemical Name: S-ethyl dipropylcarbamothioate

Formula:

$$(CH_3CH_2CH_2)_2N\overset{\overset{\displaystyle O}{\|}}{C}-SCH_2CH_3$$

Trade Names: R-1608 (Stauffer Chemical Co.)
Eptam® (Stauffer Chemical Co.)

Manufacture (3)

The raw materials are di-n-propyl amine, ethyl mercaptan and phosgene. The first step involves the production of di-n-propyl carbamyl chloride according to the reaction:

$$(C_3H_7)_2NH + COCl_2 \rightarrow (C_3H_7)_2NCOCl + HCl$$

The carbamyl chloride is then reacted with sodium ethyl mercaptide to give the product by the reaction:

$$(C_3H_7)_2NCOCl + NaSC_2H_5 \rightarrow (C_3H_7)_2NCOSC_2H_5 + NaCl$$

The following is a specific example of the conduct of the process. Sodium is dispersed in xylene using oleic acid as the stabilizing agent until a particle size of 5 to 200 μ in diameter is obtained. Dispersion equivalent to an amount of 16.9 parts (0.733 mol) of sodium is then transferred to a reactor which has been previously flushed out with argon (or other inert gas such as nitrogen). A solution of 50 parts (0.806 mol) of ethanethiol dissolved in 86 parts of xylene is then gradually added to the sodium dispersion over an interval of 30 minutes. The temperature is maintained at 25° to 36°C by cooling.

The sodium ethylmercaptide forms as finely divided crystals which make an easily stirrable slurry. This suspension is heated to reflux, the heat is turned off, and 120 parts (0.733 mol) of di-n-propylcarbamyl chloride is added over an interval of 17 minutes to the refluxing slurry. The heat of reaction is sufficient to keep the xylene refluxing. After all of the di-n-propylcarbamyl chloride has been added, the mixture is refluxed for an additional 3 hours. It is then cooled, filtered from sodium chloride which has formed during the reaction, and the solvent is removed under reduced pressure. The residual liquid is then distilled under vacuum to give 125.5 parts (90.6% yield) of ethyl N,N-di-n-propylthiocarbamate, BP (31.5 mm) 135.5° to 137.0°C.

Process Wastes and Their Control

Air: Air emissions from EPTC manufacture have been reported to consist of 205 kg SO_2, 1.5 kg hydrocarbons and 0.5 kg EPTC per metric ton of pesticide produced (B-15).

Product Wastes: In addition to the hydrolysis reaction, EPTC has been reported to be oxidatively cleaved to give the amine and ethane sulfonic acid (B-3).

Toxicity

The acute oral LD_{50} value for rats is about 1,650 mg/kg which is slightly toxic.

Allowable Limits on Exposure and Use

Product Use: The tolerances set by the EPA for EPTC in or on raw agricultural commodities are as follows:

	40 *CFR* Reference	Parts per Million
Almonds, hulls	180.117	0.1 N
Asparagus	180.117	0.1 N
Beans, castor	180.117	0.1 N
Citrus fruits	180.117	0.1 N
Cotton, forage	180.117	0.1 N
Cotton, seed	180.117	0.1 N
Flax, seed	180.117	0.1 N
Fruits, small	180.117	0.1 N
Grain crops	180.117	0.1 N
Grasses, forage	180.117	0.1 N
Legumes, forage	180.117	0.1 N
Nuts	180.117	0.1 N
Pineapples	180.117	0.1 N
Safflower, seed	180.117	0.1 N
Strawberries	180.117	0.1 N
Sunflower, seed	180.117	0.1 N
Vegetables, fruiting	180.117	0.1 N
Vegetables, leafy	180.117	0.1 N
Vegetables, pod	180.117	0.1 N
Vegetables, root crop	180.117	0.1 N
Vegetables, seed	180.117	0.1 N

References

(1) Worthing, C.R., *Pesticide Manual,* 6th ed., p. 237, British Crop Protection Council (1979).
(2) Spencer, E.Y., *Guide to the Chemicals Used in Crop Protection,* 6th ed., p. 252, London, Ontario, Agriculture Canada (January 1973).
(3) Tiles, H. and Antognini, J., U.S. Patent 2,913,327, November 17, 1959, assigned to Stauffer Chemical Co.

ERBON

Function: Herbicide (1)(2)

Chemical Name: 2-(2,4,5-trichlorophenoxy)ethyl 2,2-dichloropropionate

Formula:

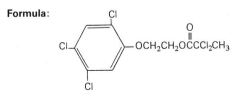

Trade Names: Baron® (Dow Chemical Co.)
Erbon® (Dow Chemical Co.)
Novege®

Manufacture (3)

These compounds may be prepared by reacting [1] α,α-dichloropropionic acid and [2] a haloaryloxy lower alkanol. The reaction may be carried out in the presence of an esterification catalyst such as sulfuric acid and conveniently in a water-immiscible solvent such as toluene, monochlorobenzene or ethylene dichloride. The amount of the reactants to be employed is not critical, some of the desired product being produced with any proportion of ingredients. Good results are obtained when substantially equimolecular proportions of the reactants are employed. The employment of an excess of the ether-alcohol coupled with the removal of the water of reaction as formed, generally results in optimum yields.

In carrying out the reaction, the α,α-dichloropropionic acid, haloaryloxy alkanol and catalyst, if employed, are mixed together and the resulting mixture heated at a temperature of from about 75° to 180°C for a period of time to complete the reaction. Alternatively the α,α-dichloropropionic acid employed may be dispersed in the solvent and the resulting mixture heated at the boiling temperature. During the heating, a mixture of solvent and water of reaction may be continuously distilled out of the reaction vessel, condensed, and the solvent recovered. Additional solvent is introduced into the reaction zone as may be necessary.

Upon completion of the reaction, the desired product may be separated by fractional distillation under reduced pressure. When the desired product precipitates as a crystalline solid in the cooled reaction mixture or in the reaction residue to be obtained following removal of the reaction solvent, this solid product may be separated in an alternative method by filtration and thereafter purified by recrystallization from various organic solvents. In another method of separation and purification, the solvent mixture of the reaction product is neutralized with an alkali such as dilute aqueous sodium carbonate. The resulting mixture divides into aqueous and solvent layers. The solvent layer, which contains the ester reaction product, is separated and washed several times with water to extract the water-soluble salts of catalyst and any unreacted α,α-dichloropropionic acid. The washed mixture is then fractionally distilled under reduced pressure to obtain the desired ester compound. The following is a specific example of the conduct of this process.

Example—383 grams (1.59 mols) of 2-(2,4,5-trichlorophenoxy)ethanol, 239 grams (1.67 mols of α,α-dichloropropionic acid and 0.5 ml of concentrated sulfuric acid were dispersed in 600 ml of ethylene dichloride and the resulting mixture heated for about 21 hours at a temperature of from 91.5° to 95.5°C. The heating was carried out in the usual fashion with continuous distillation of ethylene dichloride together with the water of reaction as formed, separation of the water and recycling of the ethylene dichloride. Following the reaction, the mixture was washed with water and thereafter fractionally distilled under reduced pressure to separate a 2-(2,4,5-trichlorophenoxy)ethyl-α,α-dichloropropionate product as a crystalline solid melting at 49° to 50°C and boiling at 161° to 164°C at 0.5 mm pressure.

Process Wastes and Their Control

Product Wastes: Erbon is said to be nonflammable and stable to ultraviolet light (B-3).

Toxicity

The acute oral LD_{50} value for rats is 1,120 mg/kg which is slightly toxic.

Allowable Limits on Exposure and Use

Product Use: Issuance of a rebuttable presumption against registration for erbon was being considered by EPA as of September 1979 on the basis of possible oncogenicity, teratogenicity and fetotoxicity.

The tolerances set by the EPA for erbon in or on raw agricultural commodities are as follows:

	40 *CFR* Reference	Parts per Million
Citrus fruits (post-h)	180.271	8.0
Cotton, seed	180.271	30.0

References

(1) Martin, H. and Worthing, C.R., *Pesticide Manual,* 5th ed., p. 239, British Crop Protection Council (January 1977).
(2) Spencer, E.Y., *Guide to the Chemicals Used in Crop Protection,* 6th ed., p. 253, London, Ontario, Agriculture Canada (January 1973).
(3) Brust, H.F. and Senkbeil, H.O., U.S. Patent 2,754,324, July 10, 1956, assigned to Dow Chemical Co.

ETACELASIL

Function: Abscission agent (1)(2)

Chemical Name: 6-(2-chloroethyl)-6-(2-methoxyethoxy)-2,5,7,10-tetraoxa-6-silaundecane

Formula: $(CH_3OCH_2CH_2O)_3SiCH_2CH_2Cl$

Trade Names: CGA 13586 (Ciba-Geigy AG)
Alsol® (Ciba-Geigy AG)

Manufacture (2)

2-Chloroethyltrichlorosilane is dissolved in absolute diethyl ether. To this solution is added methoxyethanol which reacts to give etacelasil.

Toxicity

The acute oral LD_{50} for rats is 2,066 mg/kg which is slightly toxic.

References

(1) Worthing, C.R., *Pesticide Manual,* 6th ed., p. 238, British Crop Protection Council (1979).
(2) Agripat, S.A. British Patent 1,371,804, October 30, 1974.

ETHEPHON

Function: Plant growth regulator (1)(2)

Chemical Name: 2-chloroethylphosphonic acid

Formula:

$$\underset{\text{ClCH}_2\text{CH}_2\overset{\displaystyle O}{\overset{\displaystyle \|}{P}}\text{(OH)}_2}{}$$

Trade Names: Ethrel® (Amchem Products, Inc.)
Florel® (Amchem Products, Inc.)
Cepha® (GAF Corp.)

Manufacture

Ethephon is made from tris(2-chloroethyl) phosphite by rearrangement to bis(2-chloroethyl) 2-chloroethyl phosphonate followed by hydrolysis.

Product separation and purification have been covered in some detail (3)(4).

Toxicity

The acute oral LD_{50} value for rats is about 4,000 mg/kg which is very slightly toxic.

Allowable Limits on Exposure and Use

Product Use: The tolerances set by the EPA for ethephon in or on raw agricultural commodities are as follows:

	40 *CFR* Reference	Parts per Million
Alfalfa, hay	180.301	5.0
Apples	180.300	5.0
Blackberries	180.300	30.0
Blueberries	180.300	20.0
Cantaloupes	180.300	2.0
Cherries	180.300	10.0
Coffee beans	180.300	0.1 N
Cranberries	180.300	5.0
Figs	180.300	5.0
Filberts	180.300	0.5
Lemons	180.300	2.0
Peppers	180.300	30.0
Pineapples	180.300	2.0
Pineapples, fodder	180.300	3.0
Pineapples, forage	180.300	3.0
Tangerines	180.300	0.5
Tangerines, hybrids	180.300	0.5
Tomatoes	180.300	2.0

References

(1) Worthing, C.R., *Pesticide Manual,* 6th ed., p. 241, British Crop Protection Council (1979).
(2) Fritz, C.D., Evans, W.F. and Cooke, A.R., U.S. Patent 3,879,188, April 22, 1975, assigned to Amchem Products Inc.
(3) Jacques, A.M.V., U.S. Patent 3,896,163, July 22, 1975, assigned to Amchem Products, Inc.
(4) Jacques, A.M.V., U.S. Patent 3,897,486, July 29, 1975, assigned to Amchem Products, Inc.

ETHIOFENCARB

Function: Insecticide (1)

Chemical Name: 2-ethylthiomethylphenyl methylcarbamate

Formula:

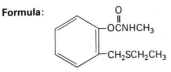

Trade Names: HOX 1901 (Bayer)
Croneton® (Bayer)

Manufacture (2)

Ethiofencarb may be made by the reaction of methyl isocyanate with 2-ethylthiomethyl phenol. 30 g of methylisocyanate are added to 0.5 mol of 2-ethylmercaptomethylphenol in 300 ml of benzene and allowed to stand for 12 hours. The solvent is then distilled. There remain behind 72 g (74% of the theory) of the compound of the above formula; refractive index n_D^{24} = 1.5602.

The N-methyl-O-(2-ethylmercaptomethyl-phenyl)-carbamic acid ester is also obtained in 45 to 55% yield when 2-ethylmercaptomethyl phenol is reacted with an excess of phosgene to give the chloroformic acid ester or with approximately equimolar amounts of phosgene to give the bis-(phenyl)-carbonate, and then treated with methylamine.

The phenol used as starting material of the formula:

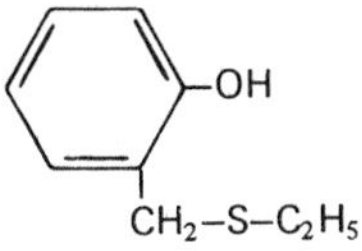

can be prepared as follows:

71 g of freshly prepared 2-chloromethyl-phenol in 300 ml of acetonitrile are reacted with 35 g of ethylmercaptan and 0.5 mol of sodium methylate in 100 ml of acetonitrile at 0° to 10°C. After standing for two hours, pouring into water is effected, followed by taking up with benzene and distillation, BP 105°/5 mm Hg, yield: 50% of the theory.

Toxicity

The acute oral LD_{50} value for rats is 400-500 mg/kg which is moderately to highly toxic.

References

(1) Worthing, C.R., *Pesticide Manual,* 6th ed., p. 243, British Crop Protection Council (1979).
(2) Hoffmann, H., Hammann, I. and Unterstenhofer, G., British Patent 1,298,515, December 6, 1972, assigned to Bayer AG.

ETHIOLATE

Function: Herbicide (1)(2)

Chemical Name: S-ethyl diethylcarbamothioate

Formula:

$$\underset{(C_2H_5)_2NCSC_2H_5}{\overset{O}{\overset{\|}{}}}$$

Trade Names: S-15076
Prefox® (Gulf Oil)

Manufacture

Ethiolate may be made by the reaction of diethylamine with ethyl chlorothioformate in aqueous NaOH solution. After reaction the mixture separates into two layers—a lower NaCl solution layer and an upper light-yellow ethiolate layer.

Toxicity

The acute oral LD_{50} value for rats is 400 mg/kg (moderately toxic).

Allowable Limits on Exposure and Use

Product Use: The tolerances set by the EPA for ethiolate in or on raw agricultural commodities are as follows:

	40 *CFR* Reference	Parts per Million
Corn, fodder	180.343	0.1 N
Corn, forage	180.343	0.1 N
Corn, fresh (inc. sweet) (k+cwhr)	180.343	0.1 N
Corn, grain	180.343	0.1 N

References

(1) Martin, H. and Worthing, C.R., *Pesticide Manual,* 4th ed., p. 246, British Crop Protection Council (November 1974).
(2) Spencer, E.Y., *Guide to the Chemicals Used in Crop Protection,* 6th ed., p. 259, London, Ontario, Agriculture Canada (January 1973).

ETHION

Function: Insecticide and acaricide (1)(2)

Chemical Name: O,O,O',O'-tetraethyl-S,S'-methylene di(phosphorodithioate)

Formula:

$$\underset{(C_2H_5O)_2PSCH_2SP(OC_2H_5)_2}{\overset{S \qquad S}{\overset{\| \qquad \|}{}}}$$

Trade Names: FMC 1240 (FMC Corp.)
Nialate® (FMC Corp.)

Manufacture

The raw materials for ethion preparation are O,O-diethyl phosphorodithioic acid and dibromomethane. These materials react according to the equation:

$$2(C_2H_5O)_2\overset{\overset{S}{\|}}{P}-SM + CH_2Br_2 \longrightarrow (C_2H_5O)_2\overset{\overset{S}{\|}}{P}-S-CH_2-S-\overset{\overset{S}{\|}}{P}(OC_2H_5)_2 + 2MBr$$

The phosphorodithioic acid is usually used in the form of its sodium salt. The reaction is carried out at a temperature of below 50°C during addition of the reactants, followed by heating to 80° to 85°C to complete the reaction. The reaction is carried out at atmospheric pressure.

The reaction period for this process is two or more hours. Reaction times of 4 to 14 hours under reflux are cited by J.R. Willard et al (3). The reaction may be carried out in a liquid two phase system, with an aqueous alkaline phase present and an organic dibromomethane phase present.

Alternatively, the process may be carried out in an organic solvent such as a lower alcohol, dioxane, pyridine or some similar solvent which will not react with the dihalomethane. However, results obtained are not as satisfactory as with the preferred aqueous medium, according to J.R. Willard et al (4). Separation of the product from an organic solvent is substantially more difficult and the reaction times are somewhat longer in organic media.

No catalyst is employed in this reaction. A stirred reaction vessel equipped with heating means and with a reflux condenser may be used to carry out this reaction. The reaction product separates from an aqueous reaction mixture as an oil, is separated from the aqueous salt solution by decantation and is then filtered, washed and dried.

Process Wastes and Their Control

Product Wastes: Analysis of this compound is accomplished by hydrolysis to form diethyl phosphorodithioic acid. Ethion is subject to both acid and alkaline hydrolysis and is slowly oxidized in air (B-3).

Toxicity

The acute oral LD_{50} value for rats is 13 mg/kg (B-5) which is highly toxic.

Allowable Limits on Exposure and Use

Air: The threshold limit value for ethion in air is 0.4 mg/m^3 as of 1979. It is noted that cutaneous absorption of ethion should be avoided so that the threshold limit value is not invalidated.

Product Use: Tolerances set by the EPA for ethion in or on raw agricultural commodities are as follows:

	40 *CFR* Reference	Parts per Million
Almonds	180.173	0.1
Almonds, hulls	180.173	5.0
Apples	180.173	2.0
Apricots	180.173	0.1
Beans	180.173	2.0
Cattle, fat	180.173	2.5
Cattle, mbyp	180.173	0.75
Cattle, meat	180.173	0.75
Cherries	180.173	0.1
Chestnuts	180.173	0.1
Citrus fruits	180.173	2.0
Corn, fodder	180.173	14.0
Corn, forage	180.173	14.0
Corn, grain	180.173	0.1 N
Cotton, seed	180.173	0.5
Cucumbers	180.173	0.5

(continued)

	40 *CFR* Reference	Parts per Million
Eggplant	180.173	1.0
Eggs	180.173	0.2
Filberts	180.173	0.1
Goats, fat	180.173	0.2
Goats, mbyp	180.173	0.2
Goats, meat	180.173	0.2
Grapes	180.173	2.0
Hogs, fat	180.173	0.2
Hogs, mbyp	180.173	0.2
Hogs, meat	180.173	0.2
Horses, fat	180.173	0.2
Horses, mbyp	180.173	0.2
Horses, meat	180.173	0.2
Melons	180.173	2.0
Milk, fat (= N in whole milk)	180.173	0.5 N
Nectarines	180.173	1.0
Onions	180.173	1.0
Peaches	180.173	1.0
Pears	180.173	1.0
Pecans	180.173	0.1
Peppers	180.173	1.0
Pimentos	180.173	1.0
Plums (fresh prunes)	180.173	2.0
Poultry, fat	180.173	0.2
Poultry, mbyp	180.173	0.2
Poultry, meat	180.173	0.2
Sheep, fat	180.173	0.2
Sheep, mbyp	180.173	0.2
Sheep, meat	180.173	0.2
Sorghum, forage	180.173	2.0
Sorghum, grain	180.173	2.0
Squash, summer	180.173	0.5
Strawberries	180.173	2.0
Tomatoes	180.173	2.0
Walnuts	180.173	0.1

The tolerances set by the EPA for ethion in animal feeds are as follows (the *CFR* Reference is to Title 21):

	CFR Reference	Parts per Million
Citrus pulp, dehydrated	561.230	10.0

The tolerances set by the EPA for ethion in food are as follows (the *CFR* Reference is to Title 21):

	CFR Reference	Parts per Million
Raisins	193.190	4.0
Tea, dried	193.190	10.0

References

(1) Worthing, C.R., *Pesticide Manual,* 6th ed., p. 244, British Crop Protection Council (1979).
(2) Spencer, E.Y., *Guide to the Chemicals Used in Crop Protection,* 6th ed., p. 254, London, Ontario, Agriculture Canada (January 1973).
(3) Willard, J.R. et al, U.S. Patent 2,873,228, February 10, 1959, assigned to Food Machinery & Chemical Corp.

(4) Willard, J.R. et al, U.S. Patent 3,014,058, December 9, 1961, assigned to Food Machinery & Chemical Corp.

ETHIRIMOL

Function: Fungicide (1)(2)

Chemical Name: 5-butyl-2-(ethylamino)-6-methyl-4-pyrimidinol

Formula:

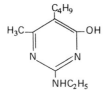

Trade Names: PP149
Milstem® (Plant Protection, Ltd.)
Milgo®
New Milstem®

Manufacture (3)

Ethirimol is made by the reaction of ethylguanidine and 2-n-butylethyl acetoacetate.

Toxicity

The acute oral LD_{50} value for rats is 4,000 mg/kg (B-5) which is very slightly toxic.

References

(1) Worthing, C.R., *Pesticide Manual,* 6th ed., p. 245, British Crop Protection Council (1979).
(2) Spencer, E.Y., *Guide to the Chemicals Used in Crop Protection,* 6th ed., p. 255, London, Ontario, Agriculture Canada (January 1973).
(3) Snell, B.K., Elias, R.S. and Freeman, P.F.H., British Patent 1,182,584, February 25, 1970, assigned to Imperial Chemical Industries Ltd.

ETHOFUMESATE

Function: Herbicide (1)(2)

Chemical Name: 2-ethoxy-2,3-dihydro-3,3-dimethylbenzofuran-5-yl methanesulfonate

Formula:

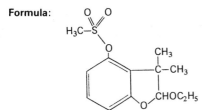

Trade Names: NC8438 (Fisons, Ltd.)
Nortron® (Fisons, Ltd.)

Manufacture

Ethofumesate is made by first reacting 2-methylpropionaldehyde and morpholine. That product is reacted with p-benzoquinone to give 2,3-dihydro-3,3-dimethyl-2-morpholino-benzofuran-5 ol. That compound can be reacted with mesyl chloride (methanesulfonyl chloride, CH_3SO_2Cl) to give ethofumesate (3).

Toxicity

The acute oral LD_{50} value for rats is greater than 6,400 mg/kg which is insignificantly toxic.

Allowable Limits on Exposure and Use

Product Use: The tolerances set by the EPA for ethofumesate in animal feeds are as follows (the *CFR* Reference is to Title 21):

	CFR Reference	Parts per Million
Sugar beet molasses	561.235	0.5

References

(1) Worthing, C.R., *Pesticide Manual,* 6th ed., p. 246, British Crop Protection Council (1979).
(2) Gates, P.S., Gillon, J. and Saggers, D.T., British Patent 1,271,659, April 26, 1972.
(3) Gates, P.S., Gillon, J. and Saggers, D.T., U.S. Patent 3,689,507, September 5, 1972, assigned to Fisons, Ltd.

ETHOHEXADIOL

Function: Insect repellant (1)(2)

Chemical Name: 2-ethyl-1,3-hexanediol

Formula: $CH_3CH_2CHCHOHCH_2CH_2CH_3$
                           $\overset{|}{C}H_2OH$

Trade Names: Rutgers 6-12
Rutgers 6-12®

Manufacture

Ethohexadiol is made by the self-condensation of butyraldehyde to butyraldol in an initial step:

$$2CH_3CH_2CH_2CHO \longrightarrow CH_3CH_2CH_2CHOH\overset{\overset{\displaystyle CHO}{|}}{C}HCH_2CH_3$$

The butyraldol is then hydrogenated to ethohexadiol.

Toxicity

The acute oral LD_{50} value for rats is 2,400 mg/kg (B-5) which is slightly toxic.

References

(1) Worthing, C.R., *Pesticide Manual,* 6th ed., p. 247, British Crop Protection Council (1979).
(2) Wilkes, B.G., U.S. Patent 2,407,205, September 3, 1946, assigned to Carbide & Carbon Chemicals Corp.

ETHOPROP
(ETHOPROPHOS in U.K. and elsewhere)

Function: Nematocide and soil insecticide (1)(2)(4)

Chemical Name: O-ethyl S,S-dipropyl phosphorodithioate

Formula:

$$\underset{(C_3H_7S)_2\overset{\displaystyle O}{\overset{\|}{P}}OCH_2CH_3}{}$$

Trade Names: VC9-104 (Mobil Chemical Co.)
Prophos® (Mobil Chemical Co.)
Mocap® (Mobil Chemical Co.)

Manufacture (3)

(A) S,S-dipropyl phosphorochloridodithioite—607 parts (8 mols) of propyl mercaptan were added to 550 parts (4 mols) of PCl_3 at 25° to 30°C in 1 hour and 55 minutes, while stirring, after which the mixture was allowed to stand overnight. The product was separated by fractionally distilling at 5 mm of Hg up to a BP of 140°C.

(B) S,S-dipropyl O-ethyl-phosphorodithioite—54 parts of the product from (A) were added at 0° to 5°C, while stirring, over a period of 30 minutes to a mixture of ethanol and triethylamine in hexane. The reaction mixture was allowed to stir for an additional hour at 0° to 5°C. The mass was filtered to remove the amine hydrochloride. An atmosphere of N_2 was maintained during reaction, stirring and filtration to prevent oxidation. After filtration to remove the hydrochloride, the hexane was fractionated, first at atmospheric pressure, then to 30 mm at a final pot temperature of 100°C. The residue was distilled at 10 mm of Hg to yield the desired product.

(C) S,S-dipropyl O-ethyl-phosphorodithioate—14 parts of hydrogen peroxide were added to the product from (B) in refluxing benzene. Slight cooling was necessary because of the exothermic nature of the reaction. After the addition was completed, stirring continued for 30 minutes. The mass was cooled and washed with two 50-part portions of 5% sodium hydroxide. Aqueous and organic layers were separated and the solvent was removed at a pressure of 30 mm of Hg and a temperature of 100°C. The resulting residue was distilled at 8 mm of Hg, yielding the product.

Process Wastes and Their Control

Product Wastes: This compound is very stable in acid aqueous media from 25° to 100°C. It is hydrolyzed moderately faster in basic media at 25°C and rapidly at 100°C. Prophos is thermally stable for 8 hours at 150°C (B-3).

Toxicity

The acute oral LD$_{50}$ for rats is 62 mg/kg (moderately toxic).

Allowable Limits on Exposure and Use

Product Use: The tolerances set by the EPA for ethoprop in or on raw agricultural commodities are as follows:

	40 *CFR* Reference	Parts per Million
Bananas	180.262	0.02 N
Beans, lima	180.262	0.02 N
Beans, lima, forage	180.262	0.02 N
Beans, snap	180.262	0.02 N
Beans, snap, forage	180.262	0.02 N
Cabbage	180.262	0.02 N
Corn, fodder	180.262	0.02 N
Corn, forage	180.262	0.02 N
Corn, fresh (inc. sweet) (k+cwhr)	180.262	0.02 N
Corn, grain	180.262	0.02 N
Cucumbers	180.262	0.02 N
Peanuts	180.262	0.02 N
Peanuts, hay	180.262	0.02 N
Pineapples, fodder	180.262	0.02 N
Pineapples, forage	180.262	0.02 N
Potatoes	180.262	0.02 N
Soybeans	180.262	0.02 N
Soybeans, forage	180.262	0.02 N
Soybeans, hay	180.262	0.02 N
Sugarcane	180.262	0.02 N
Sugarcane, fodder	180.262	0.02 N
Sugarcane, forage	180.262	0.02 N
Sweet potatoes	180.262	0.02 N

References

(1) Worthing, C.R., *Pesticide Manual,* 6th ed., p. 248, British Crop Protection Council (1979).
(2) Spencer, E.Y., *Guide to the Chemicals Used in Crop Protection,* 6th ed., p. 261, London, Ontario, Agriculture Canada (January 1973).
(3) Wilson, J.H., Jr., U.S. Patent 3,268,393, August 23, 1966, assigned to Mobil Oil Corp.
(4) Goyette, L.E., U.S. Patent 3,112,244, November 26, 1963, assigned to Virginia-Carolina Chemical Corp.

ETHOXYQUIN

Function: Fungicide and plant growth regulator (1)

Chemical Name: 1,2-dihydro-6-ethoxy-2,2,4-trimethylquinoline

Formula:

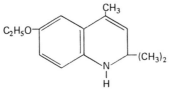

Trade Names: Santoquin®
                Stop Scald®

Manufacture (2)

6-Ethoxy-2,2,4-trimethyl-1,2-dihydroquinoline is made by passing acetone through p-phenetidine (containing 1% of iodine) at 120° to 130°C until about 2½ mols of acetone have been absorbed by each mol of p-phenetidine. The resulting condensation product is fractionally distilled under reduced pressure and 6-ethoxy-2,2,4-trimethyl-1,2-dihydroquinoline is obtained in about 65% yield and is a liquid of BP 140° to 150°C at 3 to 4 mm. About 20% of p-phenetidine is also obtained in the distillation and this recovered material is suitable for using in another condensation.

Toxicity

The acute oral LD_{50} value for rats is 800 mg/kg (B-5) which is slightly toxic.

Allowable Limits on Exposure and Use

Product Use: The tolerances set by the EPA for ethoxyquin in or on raw agricultural commodities are as follows:

	40 *CFR* Reference	Parts per Million
Apples (pre & post-h)	180.178	3.0
Pears (pre & post-h)	180.178	3.0

References

(1) Mowry, D.T. and Schlesinger, A.H., U.S. Patent 2,661,277, December 1, 1953, assigned to Monsanto Chemical Co.
(2) Baird, W., Goldstein, R.F. and Jones, M., British Patent 505,113, May 4, 1939, assigned to Imperial Chemical Industries, Ltd.

ETHYLENE DIBROMIDE

Function: Insecticidal fumigant (1)(2)(3)(4)

Chemical Name: 1,2-dibromoethane

Formula: CH_2BrCH_2Br

Trade Names: Bromofume®
                Dowfume W-85®
                Soilbrome 85®
                Nephis®
                EDB

Manufacture

The direct addition of bromine to ethylene takes place with ease according to the reaction:

$$CH_2=CH_2 + Br_2 \rightarrow BrCH_2CH_2Br$$

This reaction has traditionally been carried out by a batch process involving the bubbling of ethylene gas into liquid bromine. That process was slow and cumbersome and major attention is paid here to newer process variations.

Pure bromine and pure ethylene are desired if the highest quality ethylene dibromide product is wanted. However, bromine containing other substances, such as organic bromides, may be used. Also, ethylene containing methane, ethane, nitrogen and hydrogen may be used. Provision may be made for feeding high-concentration ethylene at one point and dilute ethylene-containing gas at another point as described by A.A. Gunkler et al (5).

The molar ratio of bromine to ethylene may vary from 1.25:1 to 1:2. However, it is commercially desirable to use substantially equimolar proportions or a slight excess, not exceeding 10%, of ethylene according to D.B. Clapp et al (6).

The reaction between ethylene and bromine is highly exothermic and, further, has a negative temperature coefficient so that the reaction rate falls off appreciably as the temperature is increased as pointed out by D.B. Clapp et al (6). Reaction temperatures may range from 10° to 150°C but preferably should not exceed 100°C. The addition of bromine to ethylene is carried out at substantially atmospheric pressure.

Residence times of 16 hours or so were typical for batch operations where ethylene was bubbled into liquid bromine. This has been cut to 2 hours in the continuous system described by D.B. Clapp et al (6). Most commonly, ethylene dibromide itself is used as the reaction medium. No catalyst is used. The reactor is preferably made of corrosion-resistant material such as glass, stoneware or the like. Mild steel lined with polyethylene has also been used.

A vertical column reactor having a lower packed zone and an upper reaction zone containing heat exchange coils has been designed for ethylene dibromide manufacture. A stream of substantially bromine-free product is withdrawn from the bottom of the packed zone which functions primarily as a stripping column for unreacted bromine although some reaction occurs there according to D.B. Clapp et al (6). Ethylene dibromide recovery usually involves stripping of the reactor product with ethylene to remove excess bromine. The product may then be distilled.

Process Wastes and Their Control

Product Wastes: The Manufacturing Chemists Association suggests the following disposal procedures for EDB and other bromine-containing compounds.

- Pour onto vermiculite, sodium bicarbonate or a sand-soda ash mixture (90-10). Mix and shovel into paper boxes. Place in an open incinerator. Cover with scrap wood and paper. Ignite with an excelsior train; stay on upwind side. Or dump into a closed incinerator with afterburner.

- Dissolve in a flammable solvent. Spray into the fire box of an incinerator equipped with afterburner and scrubber (alkali).

Toxicity

The acute oral LD_{50} for rats is 146 mg/kg which is moderately toxic.

The ACGIH (B-23) has cited ethylene dibromide as a "substance with recognized carcinogenic potential awaiting reassignment of TLV pending further data acquisition." In November 1978, NCI called EDB "the most potent cancer-causing substance ever found in the animal test program of NCI." Industry sources rebutted this and a decision by EPA on use restriction is in process.

Prolonged contact of the liquid with the skin may cause erythema, blistering, and skin ulcers. These reactions may be delayed 24 to 48 hours. Dermal sensitization to the liquid may develop. The vapor is irritating to the eyes and to the mucous membranes of the respiratory tract.

Inhalation of the vapor may result in severe acute respiratory injury, central nervous system depression, and severe vomiting. Animal experiments have produced injury to the liver and kidneys. Ethylene dibromide induced squamous cell carcinomas in the stomachs of both rats and mice when administered via chronic oral intubation at maximum tolerated doses (MTD) and at half MTD's (B-25).

Allowable Limits on Exposure and Use

Air: The threshold limit value for ethylene dibromide in air is 155 mg/m^3 (20 ppm) as of 1979. A tentative short term exposure limit is 230 mg/m^3 (30 ppm). However, an intended change designates EDB as a human carcinogen where no exposure by any route (respiratory, skin or oral) should be permitted (B-23).

NIOSH has recommended a lower limit of 1 mg/m^3 (0.13 ppm) based on 15-minute exposure.

Product Use: A rebuttable presumption against registration for EDB was issued on December 14, 1977 by EPA on the basis of oncogenicity, mutagenicity and reproductive effects.

The tolerances set by the EPA for EDB in or on raw agricultural commodities are as follows:

	40 *CFR* Reference	Parts per Million
Asparagus	180.126	10.0
Bananas, cavendish (post-h)	180.146	10.0
Barley, grain (post-h)	180.146	50.0
Barley, grain (post-h)	180.100	Exempt
Beans, lima	180.126	5.0
Beans, string (post-h)	180.146	10.0
Broccoli	180.126	75.0
Cantaloupes (post-h)	180.146	10.0
Carrots	180.126	75.0
Cauliflower	180.126	10.0
Cherries (post-h)	180.146	25.0
Citrus fruits (post-h)	180.146	10.0
Corn, grain (post-h)	180.146	50.0
Corn, grain (post-h)	180.100	Exempt
Corn, sweet	180.126	50.0
Corn, sweet, forage	180.126	50.0
Cotton, seed	180.126	25.0
Cucumbers	180.126	30.0
Cucumbers (post-h)	180.146	10.0
Eggplant	180.126	50.0
Guava (post-h)	180.146	10.0
Lettuce	180.126	30.0
Litchi, fruit (post-h)	180.146	10.0
Litchi, nut (post-h)	180.146	10.0
Longan fruit (post-h)	180.146	10.0
Mangoes (post-h)	180.146	10.0
Melons	180.126	75.0
Melons, bitter (post-h)	180.146	10.0
Oats, grain (post-h)	180.146	50.0
Oats, grain (post-h)	180.100	Exempt
Okra	180.126	50.0
Papayas (post-h)	180.146	10.0

(continued)

	40 *CFR* Reference	Parts per Million
Parsnips	180.126	75.0
Peanuts	180.126	25.0
Peppers	180.126	30.0
Peppers, bell (post-h)	180.146	10.0
Pineapples	180.126	40.0
Pineapples (post-h)	180.146	10.0
Plums (fresh prunes) (post-h)	180.146	25.0
Potatoes	180.126	75.0
Rice, grain (post-h)	180.146	50.0
Rice, grain (post-h)	180.100	Exempt
Rye, grain (post-h)	180.100	Exempt
Rye, grain (post-h)	180.146	50.0
Sorghum, grain (milo) (post-h)	180.146	50.0
Sorghum, grain (milo) (post-h)	180.100	Exempt
Squash, summer	180.126	50.0
Squash, zucchini (post-h)	180.146	10.0
Strawberries	180.126	5.0
Sweet potatoes	180.126	50.0
Tomatoes	180.126	50.0
Wheat, grain (post-h)	180.146	50.0
Wheat, grain (post-h)	180.100	Exempt

The tolerances set by the EPA for inorganic bromides in animal feed resulting from fumigation with ethylene dibromide are as follows (*CFR* Reference is to Title 21):

	CFR Reference	Parts per Million
Barley, milled fractions	561.260	125.0
Citrus pulp, dehydrated	561.260	90.0
Corn, milled fractions	561.260	125.0
Dog food	561.260	400.0
Oats, milled fractions	561.260	125.0
Rice, milled fractions	561.260	125.0
Rye, milled fractions	561.260	125.0
Sorghum, grain, milo, milled fractions	561.260	125.0
Wheat, milled fractions	561.260	125.0

The tolerances set by the EPA for inorganic bromides in food resulting from fumigation with ethylene dibromide are as follows (*CFR* Reference is to Title 21):

	CFR Reference	Parts per Million
Barley, flours	193.250	125.0
Biscuit mixes	193.250	125.0
Bread mixes	193.250	125.0
Breading	193.250	125.0
Cake mixes	193.250	125.0
Cereal grains, milled fractions	193.225	125.0
Cereal flours & related products	193.250	125.0
Cheese, parmesan	193.250	325.0
Cheese, rouquefort	193.250	325.0
Cookie mixes	193.250	125.0
Eggs, dried	193.250	400.0
Figs, dried	193.250	250.0
Foods, processed, except below	193.250	125.0
Herbs, processed	193.250	400.0
Macaroni products	193.250	125.0

(continued)

	40 CFR Reference	Parts per Million
Malt beverages	193.250	25.0
Milo (sorghum), flours	193.250	125.0
Noodle products	193.250	125.0
Oats, flours	193.250	125.0
Pie mixes	193.250	125.0
Rice, cracked	193.250	125.0
Rice, flours	193.250	125.0
Rye, flours	193.250	125.0
Soya, flour	193.250	125.0
Spices, processed	193.250	400.0
Tomato products, conc.	193.250	250.0
Vegetables, dried	193.250	125.0

References

(1) Worthing, C.R., *Pesticide Manual,* 6th ed., p. 249, British Crop Protection Council (1979).
(2) Spencer, E.Y., *Guide to the Chemicals Used in Crop Protection,* 6th ed., p. 253, London, Ontario, Agriculture Canada (January 1973).
(3) Kagy, J.R. and McPherson, R.R., U.S. Patent 2,448,265, August 31, 1948, assigned to Dow Chemical Co.
(4) Bickerton, J.M., U.S. Patent 2,473,984, June 21, 1949, assigned to Innis, Speiden & Co.
(5) Gunkler, A.A. et al, U.S. Patent 2,746,999, May 22, 1956, assigned to Dow Chemical Co.
(6) Clapp, D.B. et al, U.S. Patent 2,914,577, November 24, 1959, assigned to Associated Ethyl Co., Ltd.

ETHYLENE DICHLORIDE

Function: Insecticidal fumigant (1)(2)

Chemical Name: 1,2-dichloroethane

Formula: CH_2ClCH_2Cl

Trade Name: EDC

Manufacture

Ethylene dichloride may be made by the direct addition of chlorine to ethylene according to the reaction:

$$CH_2=CH_2 + Cl_2 \rightarrow ClCH_2CH_2Cl$$

or it may be made by the oxychlorination route (4)(5)(6)(7) following the reaction:

$$CH_2=CH_2 + 2HCl + \tfrac{1}{2}O_2 \rightarrow ClCH_2CH_2Cl + H_2O$$

Process Wastes and Their Control

Product Wastes: The MCA recommends incineration methods for disposal of this compound (B-3).

Toxicity

The acute oral LD_{50} value for rats is 680 mg/kg (B-5) which is slightly toxic.

1,2-Dichloroethane has been shown to be weakly mutagenic in two different mutagenicity screening tests. No data are available on its potential carcinogenicity according to the National Academy of Sciences in 1977 (B-22).

In view of the relative paucity of data on teratogenicity, carcinogenicity, and long-term oral toxicity of 1,2-dichloroethane, estimates of the effects of chronic oral exposure at low levels cannot be made with any confidence. It was recommended that studies to produce such information be conducted before final limits in drinking water were established.

However, subsequently (9), there have been strong indications shown that EDC is a carcinogen.

Allowable Limits on Exposure and Use

Air: The threshold limit value for ethylene dichloride in air is 200 mg/m^3 (50 ppm) as of 1979 but it is intended to change these values to 40 mg/m^3 (10 ppm). The short term exposure limit is 300 mg/m^3 (75 ppm) as of 1979 but it is intended to change this value to 60 mg/m^3 (15 ppm) (B-23).

The current U.S. Federal standard is 400 mg/m^3 (100 ppm) (B-21). NIOSH recommended in 1976 that the permissible exposure limit be reduced to 5 ppm on a 40-hour week average basis with a 15 ppm ceiling over a 15-minute period. Then in 1978, as a result of a special hazard review, NIOSH recommended a 4 mg/m^3 (1 ppm) time weighted average with an 8 mg/m^3 (2 ppm) ceiling for 15-minute exposure. This last limit reflects the limits of current analytical sampling and detection capability and is said to be justified by indications that EDC may be a carcinogen.

In water, EPA has suggested (B-26) a concentration of 1,2-dichloroethane in water, calculated to keep the lifetime cancer risk below 10^{-5}, of 7.0 μg/l.

Product Use: The tolerances set by the EPA for EDC in or on raw agricultural commodities are as follows:

	40 *CFR* Reference	Parts per Million
Barley, grain (post-h)	180.100	Exempt
Corn, grain (post-h)	180.100	Exempt
Oats, grain (post-h)	180.100	Exempt
Rice, grain (post-h)	180.100	Exempt
Rye, grain (post-h)	180.100	Exempt
Sorghum, grain, milo (post-h)	180.100	Exempt
Wheat, grain (post-h)	180.100	Exempt

The tolerances set by the EPA for EDC as a fumigant in food are as follows (*CFR* Reference is to Title 21):

	CFR Reference	Parts per Million
Cereal grains, milled fractions	193.225	125.0
Corn, grits, malted beverages	193.230	125.0
Rice, cracked, malted beverages	193.230	125.0

References

(1) Worthing, C.R., *Pesticide Manual,* 6th ed., p. 250, British Crop Protection Council (1979).

(2) Spencer, E.Y., *Guide to the Chemicals Used in Crop Protection,* 6th ed., p. 264, London, Ontario, Agriculture Canada (January 1973).

(3) Benedict, D.B. et al, U.S. Patent 2,929,852, March 22, 1960, assigned to Union Carbide Corp.

(4) Bohl, L.E. et al, U.S. Patent 3,256,352, June 14, 1966, assigned to Pittsburgh Plate Glass Co.

(5) Schwarzenbek, E.F., U.S. Patent 3,363,010, January 9, 1968, assigned to Pullman, Inc.

(6) Dunn, J.L. et al, U.S. Patent 2,866,830, December 30, 1958, assigned to Dow Chemical Co.

(7) Hirsh, D.H. et al, U.S. Patent 3,042,728, July 3, 1962, assigned to Union Carbide Corp.

(8) National Institute for Occupational Safety and Health, *Criteria for a Recommended Standard: Occupational Exposure to Ethylene Dichloride,* NIOSH Doc. No. 76-139 (1976).

(9) National Institute for Occupational Safety and Health, *Ethylene Dichloride,* NIOSH Current Intelligence Bulletin No. 5, Washington, DC (April 19, 1978).

ETHYLENE OXIDE

Function: Insecticidal fumigant (1)(2)

Chemical Name: 1,2-epoxyethane

Formula:

$$\underset{CH_2\text{------}CH_2}{\overset{O}{\triangle}}$$

Trade Name: Carboxide® (with 90% CO_2) (Union Carbide)

Manufacture (3)

The production of ethylene oxide from ethylene proceeds according to the reaction:

$$2CH_2{=}CH_2 + O_2 \longrightarrow 2CH_2\text{------}CH_2 \;(\overset{O}{\triangle})$$

This competes with the total oxidation reaction:

$$C_2H_4 + 3O_2 \longrightarrow 2H_2O + 2CO_2$$

The alternative and older route to ethylene oxide involves the reaction of ethylene chlorohydrin with calcium hydroxide according to the equation:

$$2HOCH_2CH_2Cl + Ca(OH)_2 \longrightarrow 2CH_2\text{------}CH_2 \;(\overset{O}{\triangle}) + CaCl_2 + 2H_2O$$

but the cost of manufacturing ethylene oxide is higher by the chlorohydrin route and all new plants built in the last 10 years have been based upon the direct oxidation route. Hence, no details of the chlorohydrin route will be given here.

There is a large body of patent and periodical literature dealing with ethylene oxide manufacture. In the interests of brevity, a single reasonably typical process will be cited here (2). The reaction may be carried out in the vapor phase in a multitubular fluidized bed reactor using a silver catalyst. The reactor is operated at 175° to 300°C and 100 to 150 psi. A contact time of about 1 second is employed. Yields are about 75%.

Process Wastes and Their Control

Product Wastes: Ethylene oxide (BP 10.7°C) is flammable and explosive in air at concentrations >3%. It is soluble in water. The aqueous solution is fairly stable, but is slowly hydrolyzed to ethylene glycol (B-3).

Toxicity

The acute oral LD_{50} for rats is 330 mg/kg (B-5) which is moderately toxic.

Breathing high concentrations of ethylene oxide may cause nausea, vomiting, irritation of the nose, throat, and lungs. Pulmonary edema may occur. In addition, drowsiness and unconsciousness may occur. Ethylene oxide has been found to cause cancer in female mice exposed to it for prolonged periods (B-25).

Ethylene oxide is a well-known mutagen in commercial use in plants. No mutagenic effect has been demonstrated in man.

Allowable Limits on Exposure and Use

Air: The threshold limit value for ethylene oxide in air is 90 mg/m³ (50 ppm) as of 1979 with a tentative short term exposure limit of 135 mg/m³ (75 ppm). However, an intended change would lower the TLV to 20 mg/m³ (10 ppm) (B-23).

Product Use: A rebuttable presumption against registration was issured on January 27, 1978 by EPA on the basis of mutagenicity and testicular effects.

The tolerances set by the EPA for ethylene oxide in or on raw agricultural commodities are as follows:

	40 *CFR* Reference	Parts per Million
Copra (post-h)	180.151	50.0
Spices, whole (post-h)	180.151	50.0
Walnuts, black, meat (post-h)	180.151	50.0

The tolerances set by the EPA for ethylene oxide in food are as follows (the *CFR* Reference is to Title 21):

	CFR Reference	Parts per Million
Seasonings, processed natural	193.200	50.0

References

(1) Worthing, C.R., *Pesticide Manual,* 6th ed., p. 236, British Crop Protection Council (1979).
(2) Spencer, E.Y., *Guide to the Chemicals Used in Crop Protection,* 6th ed., p. 265, London, Ontario, Agriculture Canada (January 1973).
(3) Drummond, V.D. et al, U.S. Patent 2,752,363, June 26, 1956, assigned to Vulcan Copper & Supply Co.

N-(2-ETHYLHEXYL)-8,9,10-TRINORBORN-5-ENE-2,3-DICARBOXIMIDE

Function: Insecticidal synergist (1)(2)

Chemical Name: 2-(2-ethylhexyl)-3a,4,7,7a-tetrahydro-4,7-methano-1H-isoindole-1,3(2H)-dione

Formula:

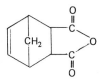

Trade Names: Octacide 264®
　　　　　　　MGK-264® (McLaughlin Gormley King)

Manufacture (3)

This compound can be synthesized in various ways. For example, the respective N-2-ethylhexyl di-keto pyrrolidine may be first formed by condensation of maleic anhydride with 2-ethyl hexylamine, preferably in the presence of an organic solvent of the benzene series, such as benzene, toluene or the xylenes, to form α,α'-di-keto N-2-ethyl-hexyl pyrrolidine.

The latter is condensed with monomeric cyclopentadiene, preferably in the presence of a diluent such as a solvent of the benzene series, or in the presence of water, so as to allow a better control of the substantial reaction heat. However, by-products are formed in these steps, especially in aqueous medium, and the yields obtainable in these steps are not satisfactory. A better method giving excellent yields consists in first condensing maleic anhydride with monomeric cyclopentadiene and then condensing the resulting α,β-(1,4-Δ^2-cyclopentenylene) succinic acid anhydride of the formula:

with 2-ethyl hexylamine.

Process Wastes and Their Control

Product Wastes: Chemically, this is the most stable material of the imide and hydrazide class. However, it should be readily hydrolyzed by base (B-3).

Toxicity

The acute oral LD_{50} value for rats is 3,640 mg/kg which is slightly toxic.

Allowable Limits on Exposure and Use

Product Use: The tolerances set by the EPA for this material in or on raw agricultural commodities are as follows:

	40 *CFR* Reference	Parts per Million
Cattle, fat	180.368	0.02
Cattle, fat	180.367	0.3
Goats, fat	180.367	0.3

(continued)

	40 *CFR* Reference	Parts per Million
Hogs, fat	180.367	0.3
Horses, fat	180.367	0.3
Milk, fat	180.367	0.3
Sheep, fat	180.367	0.3

The tolerances set by the EPA for ethylhexyl trinorbornene dicarboximide in food are as follows (the *CFR* Reference is to Title 21):

	CFR Reference	Parts per Million
Processed foods	193.320	10.0

References

(1) Worthing, C.R., *Pesticide Manual*, 6th ed., p. 251, British Crop Protection Council (1979).
(2) Spencer, E.Y., *Guide to the Chemicals Used in Crop Protection,* 6th ed., p. 378, London, Ontario, Agriculture Canada (January 1973).
(3) Schreiber, A.A., U.S. Patent 2,476,512, July 19, 1949, assigned to Van Dyke and Co.

S-ETHYLSULFINYLMETHYL O,O-DIISOPROPYL PHOSPHORODITHIOATE

Functon: Insecticide (1)(2)

Chemical Name: S-[(ethylsulfinyl)methyl] O,O-bis-(1-methyethyl) phosphorodithioate

Formula:
$$[(CH_3)_2CH]_2OP\overset{\underset{\|}{S}}{}SCH_2\overset{\underset{\|}{O}}{}SC_2H_5$$

Trade Name: Aphidan® (Hokko Chemical Industry Co., Ltd.)

Manufacture

This compound is made starting with O,O-diisopropyl hydrogen phosphorodithioate which is reacted with formaldehyde and ethanethiol to give the O,O-diisopropyl ethylthiomethyl thiophosphate ether. The thioether is oxidized with H_2O_2 to give the final product.

Toxicity

The acute oral LD_{50} for mice is 84.5 mg/kg.

References

(1) Worthing, C.R., *Pesticide Manual,* 6th ed., p. 252, British Crop Protection Council (1979).
(2) Yamamoto, T., Kobayashi, S., Nakatomi, K., Takahashi, S., Veda, H. and Kondo, M., U.S. Patent 3,408,426, October 29, 1968, assigned to Hokko Chemical Industry, Co., Ltd.

ETRIDIAZOLE

Function: Fungicide (1)(2)

Chemical Name: 5-ethoxy-3-trichloromethyl-1,2,4-thiadiazole

Formula:

$$Cl_3C-C \overset{\displaystyle N}{\underset{\displaystyle N \diagdown_S \diagup C-OC_2H_5}{\parallel \qquad \parallel}}$$

Trade Names: OM-2424 (Olin Chemicals)
Terrazole® (Olin Chemicals)
Terracoat®
Terrachlor-Super X®
Truban® (Mallinckrodt, Inc.)

Manufacture (3)(4)

A solution of 65 grams (1.625 mols) of sodium hydroxide in 130 ml of water was added drop-wise with stirring during 2 hours to a mixture of 75 grams (0.378 mol) of trichloroacetamidine hydrochloride, 70 grams (0.376 mol) of trichloromethanesulfenyl chloride and 500 ml of methylene chloride. The temperature was kept between $-4°$ and $1°C$ by cooling in an ice-salt mixture. The methylene chloride layer was separated, washed twice with 50 ml of water and dried with sodium sulfate. After evaporation of the solvent, the residue was vacuum distilled to obtain 50 grams (56% of the theoretical yield) of 3-trichloromethyl-5-chloro-1,2,4-thiadi-azole. It boiled at 73°C at 0.3 mm Hg pressure.

Then, in a second stage of the process (4) a solution of 805 mg (0.035 mol) of sodium in 20 ml of anhydrous ethanol was added within 5 minutes with stirring to a solution of 8.3 grams (0.035 mol) 3-trichloromethyl-5-chloro-1,2,4-thiadiazole in 20 ml of ethanol. After an addi-tional 10 minutes the reaction mixture was neutral and it was then filtered to remove precipi-tated sodium chloride. Excess ethanol was evaporated and the residue was vacuum distilled to obtain 7.2 grams (83% of the theoretical yield) of 3-trichloromethyl-5-ethoxy-1,2,4-thiadi-azole. It boiled at 94.5°C at 1 mm Hg pressure.

Process Wastes and Their Control

Product Wastes: This liquid fungicide is thermally stable to at least 165°C. It is not degraded by UV radiation or oxygen. It is hydrolyzed on contact with alkaline media. The compound contains chlorine and is usually formulated with pentachloronitrobenzene. Therefore, incin-eration will produce an added air pollution hazard because of the production of HCl (B-3).

Toxicity

The acute oral LD_{50} for mice is 2,000 mg/kg.

Allowable Limits on Exposure and Use

Product Use: The tolerances set by the EPA for this product in or on raw agricultural com-modities are as follows:

	40 *CFR* Reference	Parts per Million
Avocados	180.370	0.15

References

(1) Worthing, C.R., *Pesticide Manual,* 6th ed., p. 254, British Crop Protection Council (1979).
(2) Spencer, E.Y., *Guide to the Chemicals Used in Crop Protection,* 6th ed., p. 486, London, Ontario, Agriculture Canada (January 1973).

(3) Schroeder, H.A., U.S. Patent 3,260,588, July 12, 1966, assigned to Olin Mathieson Chemical Corp.
(4) Schroeder, H.A., U.S. Patent 3,260,725, July 12, 1966, assigned to Olin Mathieson Chemical Corp.

ETRIMFOS

Function: Insecticide (1)(2)(3)

Chemical Name: O-(6-ethoxy-2-ethyl-4-pyrimidinyl)-O,O-dimethyl phosphorothioate

Formula:

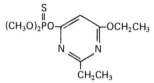

Trade Names: San 197 I (Sandoz, Ltd.)
Ekamet® (Sandoz, Ltd.)

Manufacture

Sodium hydride (0.1 mol) (50% in mineral oil) is added while stirring vigorously and in the absence of moisture to 2-ethyl-4-ethoxy-6-hydroxypyrimidine (0.1 mol) in 300 cc of absolute dimethylformamide. After stirring the mixture at room temperature for about half an hour solids are rapidly removed by suction and O,O-dimethylthionophosphoric acid chloride (0.1 mol) in 100 cc of absolute toluene are added dropwise and while stirring to the filtrate. The mixture is stirred at 45°C for a further 5 hours and is allowed to stand over night at room temperature. The solvent is decanted as far as possible in a vacuum/high vacuum and the residue is taken up in ether and the etheral solution is concentrated by evaporation.

Toxicity

The acute oral LD_{50} for rats is 1,800 mg/kg which is slightly toxic.

References

(1) Worthing, C.R., *Pesticide Manual,* 6th ed., p. 255, British Crop Protection Council (1979).
(2) Milzner, K.H. and Reisser, F., U.S. Patent 3,862,188, January 21, 1975, assigned to Sandoz, Ltd.
(3) Milzner, K.H. and Reisser, F., U.S. Patent 3,928,353, December 23, 1975, assigned to Sandoz, Ltd.

F

FAMPHUR

Function: Insecticide

Chemical Name: O,O-dimethyl O-p-(dimethyl sulfamoyl)phenyl phosphorothioate

Formula:

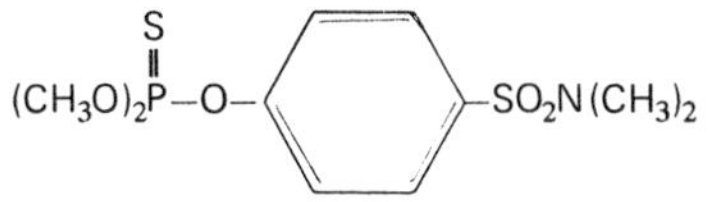

$$(CH_3O)_2\overset{\overset{\textstyle S}{\|}}{P}-O-\langle\bigcirc\rangle-SO_2N(CH_3)_2$$

Trade Names: Cyflee®
Famophos®
Warbex®
AC-38023 (American Cyanamid)

Manufacture

A detailed example of the conduct of the process (1) is as follows. To a mixture of N,N-dimethyl-1-phenol-4-sulfonamide (5.0 grams, 0.025 mol) and sodium hydroxide (1.0 gram, 0.025 mol) in 50 ml of water are added separately and simultaneously O,O-dimethyl phosphorochloridothioate (4.0 grams, 0.025 mol) and a solution of sodium hydroxide (1.0 gram, 0.025 mol) in 25 ml of water.

The mixture is stirred at room temperature for 3 hours, extracted with ether, the ethereal extracts dried over magnesium sulfate, and the ether removed under reduced pressure, yielding 3.8 grams (47%) of solid material. Recrystallization from toluene-hexane yields the pure product MP 52.5° to 53.5°C.

Toxicity

The acute oral LD_{50} value for rats is 35 mg/kg (B-5) which is highly toxic.

Allowable Limits on Exposure and Use

Product Use: The tolerances set by the EPA for famphur in or on raw agricultural commodities are as follows:

	40 *CFR* Reference	Parts per Million
Cattle, fat	180.233	0.1
Cattle, mbyp	180.233	0.1
Cattle, meat	180.233	0.1

Reference

(1) Berkelhammer, G., U.S. Patent 3,005,004, October 17, 1961, assigned to American Cyanamid Company.

FENAC
(CHLORFENAC in U.K. and elsewhere)

Function: Herbicide (1)(2)(3)

Chemical Name: 2,3,6-trichlorobenzeneacetic acid

Formula:

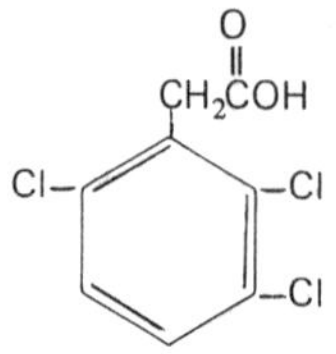

Trade Names: Fenac®
Tri-Fene®
Kanepar®

Manufacture (4)

2,3,6-Trichlorotoluene is first made by the chlorination of toluene.

A quantity of 2,3,6-trichlorotoluene (prepared by the process described by Brimelow et al *JCS*, 1951, 1208) is placed in a standard chlorination vessel and elemental chlorine is introduced into the chlorotoluene under the influence of a 250 watt mercury vapor lamp to produce side-chain chlorination while the reaction mixture is maintained at a temperature between about 90°C and about 130°C. When about 70 to 80 mol percent of the amount of chlorine theoretically necessary to produce monochlorination in the side-chain has been introduced, the passage of chlorine is interrupted and the reaction product is fractionated to recover the 2,3,6-trichlorobenzyl chloride so produced.

To a solution of 37.5 parts of sodium cyanide in 40 parts of water and 150 parts of ethyl alcohol at reflux was added slowly 138 parts of 2,3,6-trichlorobenzyl chloride. After refluxing for 4½ hours, the mixture was filtered, evaporated to remove the alcohol, and the residual solids recrystallized several times from aqueous methanol. The product was a colorless crystalline solid melting at 58.9°C, and identified as 2,3,6-trichlorobenzyl cyanide. The yield under the above described conditions was 70%.

588 parts of the cyanide was heated for one hour with 1,816 parts of 65% aqueous sulfuric acid at reflux temperature. The reaction mixture was then cooled to room temperature, the aqueous acid decanted and the reaction product washed with water. After air-drying, the

product was recrystallized from benzene. The yield was 363 parts of 2,3,6-trichlorophenyl-acetic acid. The MP of the pure colorless crystalline solid is 161°C.

Toxicity

The acute oral LD_{50} for rats is 1,780 mg/kg which is slightly toxic.

Allowable Limits on Exposure and Use

Product Use: The tolerances set by the EPA for fenac in or on raw agricultural commodities are as follows:

	40 *CFR* Reference	Parts per Million
Sugarcane	180.283	0.01

References

(1) Martin, H. and Worthing, C.R., *Pesticide Manual,* 4th ed., p. 98, British Crop Protection Council (November 1974).
(2) Spencer, E.Y., *Guide to the Chemicals Used in Crop Protection,* 6th ed., p. 519, London, Ontario, Agriculture Canada (January 1973).
(3) Tischler, N., U.S. Patent 2,977,212, March 28, 1961, assigned to Heyden Newport Chemical Corporation.
(4) Hooker Chemical Corp., British Patent 860,310 (February 1, 1961).

FENAMINOSULF

Function: Fungicide (1)(2)(3)

Chemical Name: Sodium [4-(dimethylamino)phenyl] diazenesulfonate

Formula:

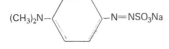

$$(CH_3)_2N-\langle\bigcirc\rangle-N{=}NSO_3Na$$

Trade Names: Bayer 22,555 (Bayer)
Bayer 5072 (Bayer)
Dexon® (Bayer)

Manufacture

4-Dimethylaminoaniline is diazotized and the diazonium chloride reacted with sodium sulfite to give fenaminosulf.

Process Wastes and Their Control

Product Wastes: Fenaminosulf is unstable in alkaline media and is rapidly decomposed in the presence of light. Dilute aqueous solutions are completely decolorized in 30 minutes or less when exposed to ordinary light, concentrated solutions decompose more slowly (B-3).

Toxicity

This is a moderately to highly toxic (LD_{50} = 60), yellow-brown, solid fungicide.

References

(1) Worthing, C.R., *Pesticide Manual,* 6th ed., p. 256, British Crop Protection Council (1979).
(2) Spencer, E.Y., *Guide to the Chemicals Used in Crop Protection,* 6th ed., p. 274, London, Ontario, Agriculture Canada (January 1973).
(3) Urbschat, E. and Frohberger, P.E., U.S. Patent 2,911,336, November 3, 1959, Reissue 24,960, April 4, 1961, assigned to Farbenfabriken Bayer AG.

FENAMIPHOS

Function: Nematocide (1)(2)

Chemical Name: Ethyl-3-methyl-4-(methylthio)phenyl 1-methylethylphosphoroamidate

Formula:

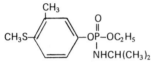

Trade Names: Bay 68,138 (Bayer)
Nemacur® (Bayer)

Manufacture (3)

3-Methyl-4-mercaptomethylphenyl dichlorothionophosphate is made by reacting 3-methyl-4-methylmercaptophenol with phosphorus oxychloride. That intermediate is reacted first with ethanol and then with diisopropylamine to give fenamiphos.

Toxicity

The acute oral LD_{50} for rats is 15-20 mg/kg which is highly toxic.

Allowable Limits on Exposure and Use

Product Use: The tolerances set by the EPA for fenamiphos in or on raw agricultural commodities are as follows:

	40 *CFR* Reference	**Parts per Million**
Bananas	180.349	0.1
Brussels sprouts	180.349	0.1
Cabbage	180.349	0.1
Cotton, seed	180.349	0.05
Peanuts	180.349	0.02
Peanuts, hulls	180.349	0.4
Soybeans	180.349	0.05

References

(1) Worthing, C.R., *Pesticide Manual,* 6th ed., p. 257, British Crop Protection Council (1979).
(2) Spencer, E.Y., *Guide to the Chemicals Used in Crop Protection,* 6th ed., p. 269, London, Ontario, Agriculture Canada (January 1973).
(3) Kayser, H. and Schrader, G., U.S. Patent 2,978,479, April 4, 1961, assigned to Farbenfabriken Bayer AG.

FENARIMOL

Function: Fungicide (1)(2)

Chemical Name: α-(2-chlorophenyl)-α-(4-chlorophenyl)-5-pyrimidinemethanol

Formula:

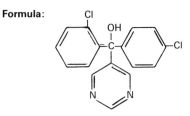

Trade Names: EL-222 (Eli Lilly & Co.)
     Bloc® (Eli Lilly & Co.)
     Rimidin® (Eli Lilly & Co.)
     Rubigan® (Eli Lilly & Co.)

Manufacture (2)

2-Chlorobenzoyl chloride and chlorobenzene are first reacted to give 2,4'-dichlorobenzophenone. Then n-butyllithium and 5-bromopyrimidine were reacted to give 5-lithiopyrimidine. The two intermediates were then reacted to give fenarimol, melting at 117° to 119°C.

Toxicity

The acute oral LD_{50} for rats is 2,500 mg/kg which is slightly toxic.

References

(1) Worthing, C.R., *Pesticide Manual,* 6th ed., p. 258, British Crop Protection Council (1979).
(2) Davenport, J.D., Hackler, R.E. and Taylor, H.M., British Patent 1,218,523, January 6, 1971, assigned to Eli Lilly & Co.

FENBUTATIN OXIDE

Function: Acaricide (1)(2)

Chemical Name: Hexakis(2-methyl-2-phenylpropyl)-distannoxane

Formula:

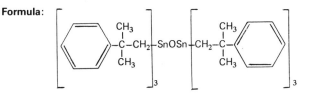

Trade Names: SD 14,114 (Shell)
     Vendex® (Shell)
     Torque® (Shell)

Manufacture (3)

Neophyl chloride (1-chloro-2-methyl-2-phenyl propane) is reacted with magnesium to give the Grignard reagent. It is reacted with tin tetrachloride to give trineophyltin chloride. The chloride is treated with NaOH to give bis(trineophyltin) oxide.

Toxicity

The acute oral LD_{50} for rats is 2,630 mg/kg which is slightly toxic.

Allowable Limits on Exposure and Use

Product Use: The tolerances set by the EPA for fenbutatin oxide in or on raw agricultural commodities are as follows:

	40 *CFR* Reference	Parts per Million
Apples	180.362	4.0
Cattle, kidney	180.362	0.3
Cattle, liver	180.362	0.3
Citrus fruits	180.362	4.0
Goats, kidney	180.362	0.3
Goats, liver	180.362	0.3
Hogs, kidney	180.362	0.3
Hogs, liver	180.362	0.3
Horses, kidney	180.362	0.3
Horses, liver	180.362	0.3
Pears	180.362	4.0
Sheep, kidney	180.362	0.3
Sheep, liver	180.362	0.3

The tolerances set by the EPA for fenbutatin oxide in animal feeds are as follows (the *CFR* Reference is to Title 21):

	CFR Reference	Parts per Million
Apple pomace, dried	561.255	20.0
Citrus pulp, dried	561.255	7.0

References

(1) Worthing, C.R., *Pesticide Manual,* 6th ed., p. 259, British Crop Protection Council (1979).
(2) Horne, C.A., Jr., U.S. Patent 3,657,451, April 18, 1972, assigned to Shell Oil Company.
(3) Zimmer, H., Homberg, O.A. and Jayawant, M., *J. Org. Chem.* 31, 3857-60 (1966).

FENFURAM

Function: Fungicide (1)(2)

Chemical Name: 2-methyl-N-phenyl-3-furancarboxamide

Formula:

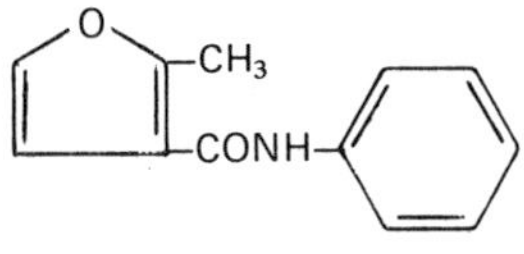

Trade Names: WL 22361 (Shell Research Ltd.)

Manufacture (2)

2-Methylfuran-3-carboxylic acid and thionyl chloride are heated together on a water bath for 2 hours. Excess thionyl chloride is distilled off in vacuum and the residue dissolved in dry benzene and then treated with aniline dissolved in benzene. After standing for 3 days, the mixture was washed with 1:5 hydrochloric acid and the residue remaining after removal of solvent was purified chromatographically over silica using CH_2Cl_2 as eluant. The product re-crystallized from petroleum ether melts at 109° to 110°C.

Toxicity

The acute oral LD_{50} for rats is 12,900 mg/kg which is insignificantly toxic.

References

(1) Worthing, C.R., *Pesticide Manual*, 6th ed., p. 261, British Crop Protection Council (1979).
(2) Ten Haken, P., British Patent 1,215,066, December 9, 1970, assigned to Shell International Research Maatschappij NV.

FENITROTHION

Function: Insecticide (1)(2)

Chemical Name: O,O-dimethyl O-(3-methyl-4-nitrophenyl) phosphorothioate

Formula:

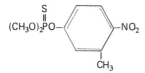

Trade Names: Bayer 48131 (Bayer)
Bayer S 5660 (Bayer)
Bayer S-1102A (Bayer)
Folithion® (Bayer)
AC 47,300 (American Cyanamid Co.)
Accothion® (American Cyanamid Co.)
Cytel® (American Cyanamid Co.)
Cyfen® (American Cyanamid Co.)
Sumithion® (Sumitomo Chemical Co.)
Agrothion®
Dicofen®
Fenstan®
Metathion E-50®
Verthion®

Manufacture

Fenitrothion may be made by the reaction of O,O-dimethylphosphorochloridothioate with sodium 3-methyl-4-nitrophenolate.

Process Wastes and Their Control

Product Wastes: Hydrolysis rate for this compound is lower than that of methyl parathion. Fifty percent hydrolysis at 30°C requires 272 minutes in 0.01N NaOH and 12 minutes in 0.1N NaOH (B-3).

Toxicity

The acute oral LD_{50} for rats is 250 to 500 mg/kg which is moderately toxic.

References

(1) Worthing, C.R., *Pesticide Manual,* 6th ed., p. 262, British Crop Protection Council (1979).
(2) Spencer, E.Y., *Guide to the Chemicals Used in Crop Protection,* 6th ed., p. 276, London, Ontario, Agriculture Canada (January 1973).

FENSON

Function: Acaricide (1)

Chemical Name: p-Chlorophenyl benzene sulfonate

Formula:

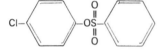

Trade Names: Murvesco® (Murphy Chemicals Ltd.)
 PCI® (Pan Britannica Industries)

Manufacture

Fenson may be made by the interaction of benzene sulfonyl chloride with p-chlorophenol.

Process Wastes and Their Control

Product Wastes: Fenson, ovex and 2,4-dichlorophenylbenzene sulfonate are structural analogues. Alkaline hydrolysis produces compounds of greater toxicity, p-chlorophenol (LD_{50} = 670) or dichlorophenol. The compounds are relatively resistant to oxidation. Incineration of these products would be acceptable if precautions were taken to eliminate air pollution (employ caustic scrubbers to trap HCl and SO_2) (B-3).

Toxicity

The acute oral LD_{50} value for rats is 1,350 mg/kg (B-5) which is slightly toxic.

Reference

(1) Spencer, E.Y., *Guide to the Chemicals Used in Crop Protection,* 6th ed., p. 277, London, Ontario, Agriculture Canada (January 1973).

FENSULFOTHION

Function: Insecticide and nematocide (1)(2)

Chemical Name: O,O-diethyl O-4-(methylsulfinyl)phenyl phosphorothioate

Formula:

Trade Names: Bayer 25141 (Bayer)
Bayer S 767 (Bayer)
Dasanit® (Bayer)
Terracur P® (Bayer)

Manufacture (3)

83 Grams of finely divided potash and 1 gram of pulverized copper are added to a solution of 83.5 grams of p-methylmercaptophenol in 300 cc of benzene. The mixture is heated with stirring up to 65°C and 140 grams of O,O-diethylthionophosphoric acid chloride are slowly added dropwise at this temperature without cooling.

The temperature spontaneously rises to 80°C. The mixture is kept at boiling temperature for another hour, the inorganic salts are filtered off with suction and the benzene solution is washed several times with diluted ammonia and water. After drying over sodium sulfate and distilling off the solvent, 163 grams of the ester are obtained (yield: 98% of the theoretical). The ester boils at 111°C under a pressure of 0.01 mm Hg.

0.5 cc of 50% sulfuric acid is added to a solution in 70 cc of methanol of 36.5 grams (0.125 mol) of the ester obtained according to the preceding paragraph, and 11.25 cc of a 37.8% hydrogen peroxide solution are added dropwise at 40° to 45°C. The reaction proceeds exothermally. After the oxidation is complete, the solution is stirred at 50°C for another hour, neutralized with prepared chalk and filtered. The water and the methanol are azeotropically distilled off with benzene. A light yellow viscous oil remains behind in a quantity of 38 grams which corresponds to a yield of 99% of the theoretical. The sulfoxide boils at 140° to 141°C under a pressure of 0.01 mm Hg.

Process Wastes and Their Control

Air: Air emissions from fensulfothion manufacture have been reported (B-15) to consist of 205 kg SO_2 and 0.5 kg fensulfothion per metric ton of pesticide produced.

Product Wastes: This compound is readily oxidized to the sulfone and apparently isomerized readily to the S-ethyl isomer (B-3).

Toxicity

The acute oral LD_{50} value for rats is 5 to 10 mg/kg which is highly to extremely toxic.

Allowable Limits on Exposure and Use

Air: The threshold limit value for this material in air is 0.1 mg/m³ as of 1979 (B-23).

Product Use: The tolerances set by the EPA for fensulfothion in or on raw agricultural commodities are as follows:

	40 *CFR* Reference	Parts per Million
Bananas	180.234	0.02
Beets, sugar	180.234	0.05

(continued)

	40 *CFR* Reference	Parts per Million
Beets, sugar, tops	180.234	0.05
Cattle, fat	180.234	0.02
Cattle, mbyp	180.234	0.02
Cattle, meat	180.234	0.02
Corn, field, fodder	180.234	1.0
Corn, field, forage	180.234	1.0
Corn, fresh (inc. sweet) (k+cwhr)	180.234	0.1
Corn, grain	180.234	0.1
Corn, pop, fodder	180.234	1.0
Corn, pop, forage	180.234	1.0
Corn, pop, grain	180.234	0.1
Corn, sweet, fodder	180.234	1.0
Corn, sweet, forage	180.234	1.0
Cotton, seed	180.234	0.02
Goats, fat	180.234	0.02
Goats, mbyp	180.234	0.02
Goats, meat	180.234	0.02
Hogs, fat	180.234	0.02
Hogs, mbyp	180.234	0.02
Hogs, meat	180.234	0.02
Horses, fat	180.234	0.02
Horses, mbyp	180.234	0.02
Horses, meat	180.234	0.02
Onions, dry	180.234	0.1
Peanuts	180.234	0.05
Peanuts, hulls	180.234	5.0
Pineapples	180.234	0.05
Pineapples, forage	180.234	0.05
Plantains	180.234	0.02
Potatoes	180.234	0.1
Rutabagas, roots	180.234	0.1
Sheep, fat	180.234	0.02
Sheep, mbyp	180.234	0.02
Sheep, meat	180.234	0.02
Sorghum, forage	180.234	1.0
Sorghum, grain	180.234	0.1
Sorghum, fodder	180.234	1.0
Soybeans	180.234	0.02
Soybeans, forage	180.234	0.1
Sugarcane	180.234	0.02
Sweet potatoes	180.234	0.05
Tomatoes	180.234	0.1

References

(1) Worthing, C.R., *Pesticide Manual,* 6th ed., p. 264, British Crop Protection Council (1979).
(2) Spencer, E.Y., *Guide to the Chemicals Used in Crop Protection,* 6th ed., p. 278, London, Ontario, Agriculture Canada (January 1973).
(3) Schegk, E. and Schrader, G., U.S. Patent 3,042,703, July 3, 1962, assigned to Farbenfabriken Bayer AG.

FENTHION

Function: Insecticide (1)(2)

Chemical Name: O,O-dimethyl O-[3-methyl-4-(methylthio)phenyl] phosphorothioate

Formula:

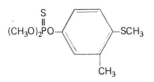

$(CH_3O)_2PO$—⟨benzene ring⟩—SCH_3 with CH_3 substituent

Trade Names: Bayer 29493 (Bayer)
Bayer S-1752 (Bayer)
Baycid® (Bayer)
Baytex® (Bayer)
Entex® (Bayer)
Lebaycid® (Bayer)
Mercaptophos® (Bayer)
Queleton® (Bayer)
Tiguvon® (Bayer)
Figuron®

Manufacture (3)

Fenthion may be made by the condensation of 4-methylmercapto-m-cresol and dimethyl-phosphorochloridothionate.

Process Wastes and Their Control

Product Wastes: Fenthion is more resistant to hydrolysis and heating than methyl para-thion: 50% hydrolysis at 80°C in acid medium requires 36 hours, or 95 minutes in alkaline medium (B-3).

Toxicity

The acute oral LD_{50} value for rats is 215 mg/kg (B-5) which is moderately toxic.

Allowable Limits on Exposure and Use

Product Use: The tolerances set by the EPA for fenthion in or on raw agricultural commodities are as follows:

	40 *CFR* Reference	Parts per Million
Alfalfa, hay	180.214	18.0
Cattle, fat	180.214	0.1
Cattle, mbyp	180.214	0.1
Cattle, meat	180.214	0.1
Grasses, hay	180.214	18.0
Hogs, fat	180.214	0.1
Hogs, mbyp	180.214	0.1
Hogs, meat	180.214	0.1
Milk	180.214	0.01 N
Poultry, fat	180.214	0.1
Poultry, mbyp	180.214	0.1
Poultry, meat	180.214	0.1
Rice	180.214	0.1
Rice, straw	180.214	0.5

References

(1) Worthing, C.R., *Pesticide Manual,* 6th ed., p. 265, British Crop Protection Council (1979).

(2) Spencer, E.Y., *Guide to the Chemicals Used in Crop Protection,* 6th ed., p. 279, London, Ontario, Agriculture Canada (January 1973).

(3) Schegk, E. and Schrader, G., U.S. Patent 3,042,703, July 3, 1962, assigned to Farbenfabriken Bayer AG.

FENURON

Function: Herbicide (1)

Chemical Name: N,N-dimethyl-N'-phenylurea

Formula:

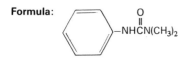

Trade Name: Dybar® (DuPont)

Manufacture

In one process aniline and urea are reacted

$$2C_6H_5NH_2 + H_2NCONH_2 \rightarrow C_6H_5NHCONHC_6H_5 + 2NH_3$$

and the resultant diphenylurea is reacted with dimethylamine in a second step to give fenuron (5).

$$C_6H_5NHCONHC_6H_5 + (CH_3)_2NH \rightarrow C_6H_5NHCON(CH_3)_2 + C_6H_5NH_2$$

The aniline is recycled (2) as shown in Figure 36. The following is one specific example of the conduct of this process.

Urea and aniline in a proportion of 1 mol to 8 mols of aniline are placed in a reaction chamber and heated with stirring to a temperature of 140° to 180°C and maintained at that temperature, about 1½ hours, until there is no further evolution of NH_3 from the mixture. The reaction liquid is cooled and the resulting slurry is then filtered.

The wet symmetrical diphenylurea is washed with methanol and then dried in an oven at 85°C. A small additional amount of symmetrical diphenylurea is isolated from the methanol wash by slow evaporation of the wash. The yield of symmetrical diphenylurea is 94.1% based upon urea. Symmetrical diphenylurea separated from the reaction mixture is fed together with dimethylamine in the proportion of 1 mol diphenylurea and 5.2 mols dimethylamine into the top of a heated vertical tube, passing down through the vertical tube maintained at 400°C and discharging from the bottom thereof.

The cooled liquid reaction products upon analysis show a conversion of 86% of the symmetrical diphenylurea with a yield of 71% of unsymmetrical substituted phenyldimethylurea based on the symmetrical diphenylurea reacted.

In another variant of fenuron manufacture, phenylisocyanate is reacted with dimethylamine (3) according to the equation:

$$C_6H_5NCO + (CH_3)_2NH \rightarrow C_6H_5NHCON(CH_3)_2$$

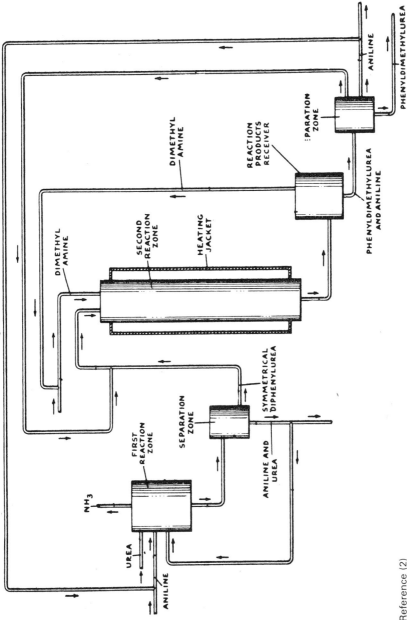

Figure 36: Flow Diagram of Process for Fenuron Manufacture

Source: Reference (2)

In still another variant of fenuron manufacture, aniline, dimethylamine, CO, and sulfur are reacted (4):

$$C_6H_5NH_2 + CO + (CH_3)_2NH + S \rightarrow C_6H_5NHCON(CH_3)_2 + H_2S$$

The following is a specific example of this alternate process. A 1.8 liter stainless steel bomb, having a working pressure of 400 psia and provided with a heating element, was used as a reactor. The bomb was charged with 13.8 grams of sulfur, 20.0 grams of aniline, and 29.0 grams of dimethylamine.

Methanol was added as a diluent, 77 ml being used. Carbon monoxide was charged to the reactor under an initial pressure of 100 psi. The closed bomb was heated at about 100°C for approximately 2 hours with agitation.

The reactor was then emptied and the contents filtered hot. The filtrate was evaporated to a small volume, leaving behind a slurry containing crude substituted urea. The crude slurry was mixed with 400 ml hot water to dissolve unreacted amine and the substituted urea.

This solution was then chilled to precipitate the substituted urea which was then recovered by filtration. Based on the primary amine, a yield of 79.0% of 1,1-dimethyl-3-phenylurea was obtained; MP found: 131° to 133°C; reported MP: 134°C.

Process Wastes and Their Control

Product Wastes: Fenuron decomposes in boiling strong bases or mineral acids. The products, however (aniline and dimethylamine), are both toxic, thus precluding hydrolysis as a disposal method (B-3).

Toxicity

The acute oral LD_{50} value for rats is 6,400 mg/kg (B-5) which is insignificantly toxic.

References

(1) Worthing, C.R., *Pesticide Manual,* 6th ed., p. 268, British Crop Protection Council (1979).
(2) Gilbert, E.E. and Sorma, G.J., U.S. Patent 2,729,677, January 3, 1956, assigned to Allied Chemical and Dye Corporation.
(3) Todd, C.W., U.S. Patent 2,655,447, October 13, 1953, assigned to DuPont.
(4) Applegarth, F., Barnes, M.D. and Franz, R.A., U.S. Patent 2,857,430, October 21, 1958, assigned to Monsanto Chemical Company.
(5) Jones, R.L., U.S. Patent 2,768,971, October 30, 1956, assigned to Imperial Chemical Industries, Ltd.

FENURON-TCA

Function: Herbicide (1)(2)

Chemical Name: Trichloroacetic acid compound with N,N-dimethyl-N'-phenylurea

Formula:

Trade Names: GC-2603 (Allied Chemical)
Urab® (Allied Chemical)

Manufacture (2)

To 5 parts by weight trichloroacetic acid, a colorless solid having a MP of 57.5°C, dissolved in 30 parts by weight xylene was added 5 parts by weight 1-(phenyl)-3,3-dimethylurea, (fenuron) a white solid melting at about 130°C. With stirring, a homogeneous colorless solution resulted. (Solubility of the phenyldimethylurea in xylene alone is about 0.5 gram per 100 grams xylene.) The mixture was drowned in 300 parts by weight petroleum ether. After stirring for a few minutes a white precipitate resulted. This was collected on a filter and washed with petroleum ether to yield 8.7 parts by weight snow white crystals which had a MP of 65° to 68°C. The white crystals were analyzed and found to be the reaction product of phenyldimethylurea and trichloroacetic acid in a 1:1 mol ratio. X-ray spectrographic analysis of the white crystals showed them to have a different diffraction pattern from either of the individual reactants, namely trichloroacetic acid and phenyldimethylurea.

Toxicity

The acute oral LD_{50} for rats is 4,000 to 5,700 mg/kg which is slightly to insignificantly toxic.

References

(1) Worthing, C.R., *Pesticide Manual,* 6th ed., p. 269, British Crop Protection Council (1979).
(2) Gilbert, E.E., Otto, J.A. and Pellerano, S.A., U.S. Patent 2,782,112, February 19, 1957, assigned to Allied Chemical and Dye Corp.

FERBAM

Function: Fungicide (1)(2)

Chemical Name: Tris(dimethylcarbamodithioate-S,S') iron

Formula:

$$\left[(CH_3)_2 N\overset{\overset{\displaystyle S}{\|}}{C} S \right]_3 Fe$$

Trade Names: Fermate® (DuPont)
Ferbeck®
Ferradow®
Karbam Black®

Manufacture

Dimethyl amine and carbon disulfide react to give dimethylthiocarbamate. Reaction of that material with soluble ferric salts gives ferbam.

Process Wastes and Their Control

Product Wastes: Ferbam is hydrolyzed by alkali and is unstable to moisture, lime and heat. Ferbam can be incinerated (B-3).

Toxicity

The acute oral LD_{50} value for rats is over 4,000 mg/kg which is slightly to insignificantly toxic.

Allowable Limits on Exposure and Use

Air: The threshold limit value for ferbam in air is 10 mg/m^3 as of 1979. The tentative short term exposure limit is 20 mg/m^3 (B-23). The existing U.S. federal limit is 15 mg/m^3 (B-21).

Product Use: The tolerances set by the EPA for ferbam in or on raw agricultural commodities are as follows:

	40 *CFR* Reference	Parts per Million
Almonds	180.114	0.1
Apples	180.114	7.0
Apricots	180.114	7.0
Asparagus	180.114	7.0
Beans	180.114	7.0
Beets, greens (alone)	180.114	7.0
Beets, with tops	180.114	7.0
Beets, without tops	180.114	7.0
Blackberries	180.114	7.0
Blueberries (huckleberries)	180.114	7.0
Boysenberries	180.114	7.0
Broccoli	180.114	7.0
Brussels sprouts	180.114	7.0
Cabbage	180.114	7.0
Carrots	180.114	7.0
Cauliflower	180.114	7.0
Celery	180.114	7.0
Cherries	180.114	7.0
Citrus fruits	180.114	7.0
Collards	180.114	7.0
Corn	180.114	7.0
Cranberries	180.114	7.0
Cucumbers	180.114	7.0
Currants	180.114	7.0
Dates	180.114	7.0
Dewberries	180.114	7.0
Eggplant	180.114	7.0
Gooseberries	180.114	7.0
Grapes	180.114	7.0
Guavas	180.114	7.0
Kale	180.114	7.0
Kohlrabi	180.114	7.0
Lettuce	180.114	7.0
Loganberries	180.114	7.0
Mangoes	180.114	7.0
Melons	180.114	7.0
Mustard, greens	180.114	7.0
Nectarines	180.114	7.0
Onions	180.114	7.0
Papayas	180.114	7.0
Peaches	180.114	7.0
Peanuts	180.114	7.0
Pears	180.114	7.0
Peas	180.114	7.0
Peppers	180.114	7.0
Plums (fresh prunes)	180.114	7.0
Pumpkins	180.114	7.0
Quinces	180.114	7.0
Radishes, tops	180.114	7.0
Radishes, with tops	180.114	7.0
Radishes, without tops	180.114	7.0

(continued)

	40 *CFR* Reference	Parts per Million
Raspberries	180.114	7.0
Rutabaga tops	180.114	7.0
Rutabagas, with tops	180.114	7.0
Rutabagas, without tops	180.114	7.0
Spinach	180.114	7.0
Squash	180.114	7.0
Squash, summer	180.114	7.0
Strawberries	180.114	7.0
Tomatoes	180.114	7.0
Turnips, greens	180.114	7.0
Turnips, with tops	180.114	7.0
Turnips, without tops	180.114	7.0
Youngberries	180.114	7.0

References

(1) Worthing, C.R., *Pesticide Manual,* 6th ed., p. 271, British Crop Protection Council (1979).
(2) Tisdale, W.H. and Williams, I., U.S. Patent 1,972,961, September 11, 1934, assigned to DuPont.

FLAMPROP-ISOPROPYL

Function: Herbicide (1)(2)

Chemical Name: DL-N-benzoyl-N-(3-chloro-4-fluorophenyl)alanine 1-methylethyl ester

Formula:

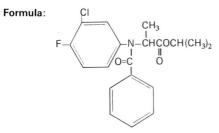

Trade Names: WL 29762 (Shell)
Barnon® (Shell)

Manufacture (3)(4)

(A) Preparation of N-(3-chloro-4-fluorophenyl)-2-aminopropionic acid—3-Chloro-4-fluoro-aniline (600 g) was dissolved in isopropyl alcohol (2.3 liters) and water (122 liters), 2-chloropropionic acid (445 g) and sodium bicarbonate (692 g) were added to the solution which was then heated under reflux with stirring for 24 hours. The mixture was cooled to 60°C and a further charge of 2-chloropropionic acid (445 g) and sodium bicarbonate (692 g) was added. The mixture was then heated under reflux for 72 hours. The cooled mixture was diluted with water (25 liters) and extracted with methylene chloride (7 liters). The aqueous solution was acidified with concentrated hydrochloric acid to pH 4. The crude acid which was precipitated was filtered off, washed with water and dried.

(B) Preparation of isopropyl N-(3-chloro-4-fluorophenyl)-2-aminopropionate—N-(3-chloro-

4-fluorophenyl)-2-aminopropionic acid [725 g, prepared as in (A)] was dissolved in dry iso-propyl alcohol (2.5 liters). The solution was saturated with hydrogen chloride gas and heated under reflux for 4 hours. The volatile components were then distilled off and the residue was washed with ice-cold aqueous sodium bicarbonate solution to yield the required ester.

(C) Benzoylation of isopropyl N-(3-chloro-4-fluorophenyl)-2-aminopropionate—Isopropyl N-(3-chloro-4-fluorophenyl)-2-aminopropionate [840 g, prepared as in (B)] in dry toluene (2 liters) was treated with benzoyl chloride (562 g) and the mixture was heated under reflux for 5 hours. The solvent and excess benzoyl chloride were distilled off and the residue was re-crystallized from benzene to yield isopropyl-N-benzoyl N-(3-chloro-4-fluoro)-2-aminopro-pionate MP 63° to 65°C.

Toxicity

The acute oral LD$_{50}$ for rats is over 3,000 mg/kg which is slightly toxic.

References

(1) Worthing, C.R., *Pesticide Manual,* 6th ed., p. 272, British Crop Protection Council (1979).
(2) Haddock, E., British Patent 1,289,283, September 13, 1972, assigned to Shell International Research NV.
(3) Yates, J. and Payne, D.H., British Patent 1,164,160, September 17, 1969, assigned to Shell International Research NV.
(4) Haddock, E. and Sampson, A.J., British Patent 1,437,711, June 2, 1976, assigned to Shell International Research Maatschappij BV.

FLAMPROP-METHYL

Function: Herbicide (1)(2)

Chemical Name: Methyl N-benzoyl-N-(3-chloro-4-fluorophenyl)-DL-alaninate

Formula:

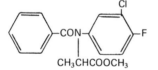

Trade Names: WL-29,761 (Shell Research Ltd.)
Mataven® (Shell Research Ltd.)

Manufacture (2)

(A) See Flamprop-Isopropyl for the preparation of N-(3-chloro-4-fluorophenyl)-2-amino-propionic acid.

(B) Preparation of methyl N-(3-chloro-4-fluorophenyl)-2-aminopropionate—N-(3-chloro-4-fluorophenyl)-2-aminopropionic acid (5.0 g) was dissolved in dry methanol (50 ml). The solution was saturated with hydrogen chloride gas and heated under reflux for 4 hours. The volatile components were then distilled off and the residue was washed with ice-cold aque-ous sodium bicarbonate solution to yield the required ester.

(C) Benzoylation of methyl N-(3-chloro-4-fluorophenyl)-2-aminopropionate—Methyl N-(3-chloro-4-fluorophenyl)-2-aminopropionate [2.32 g, prepared as in (A)] in dry toluene (50 ml) was treated with benzoyl chloride (1.45 g) and the mixture was heated under reflux for 5 hours. The solvent and excess benzoyl chloride were distilled off and the residue was recrystallized from the benzene to yield methyl N-benzoyl-N-(3-chloro-4-fluorophenyl)-2-aminopropionate MP 77° to 79°C.

Toxicity

The acute oral LD_{50} value for rats is 1,210 mg/kg which is slightly toxic.

References

(1) Worthing, C.R., *Pesticide Manual,* 6th ed., p. 274, British Crop Protection Council (1979).
(2) Haddock, E. and Sampson, A.J., British Patent 1,437,711, June 3, 1976, assigned to Shell International Research Maatschappij BV. .

FLUCHLORALIN

Functon: Herbicide

Chemical Name: N-(2-chloroethyl)-2,6-dinitro-N-propyl-4-trifluoromethyl aniline

Formula:

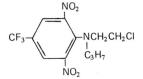

Trade Name: Basalin® (BASF Wyondotte Corp.)

Manufacture

10.1 Parts by weight of N-propyl-N-hydroxyethylamine is added to a solution of 13.5 parts by weight of 2,6-dinitro-4-trifluoromethylchlorobenzene in 50 parts by weight of dry tetrahydrofuran and the whole boiled under reflux for one to two hours. The solvent is then distilled off, the residue treated with ethyl acetate/water, and the organic phase is removed and dried over sodium sulfate. After the ethyl acetate has been distilled off, the intermediate is obtained in 90% yield. The hydroxylethyl intermediate is reacted with thionyl chloride to give fluchloralin.

Toxicity

The acute oral LD_{50} for rats is 1,550 mg/kg (B-5) which is slightly toxic.

Allowable Limits on Exposure and Use

Product Use: The tolerances set by the EPA for fluchloralin in or on raw agricultural commodities are as follows:

	40 *CFR* Reference	Parts per Million
Cotton, seed	180.363	0.05 N
Soybeans	180.363	0.05 N

Reference

(1) Kiehs, K., Koenig, K.H. and Fischer, A., U.S. Patent 3,854,927, December 17, 1974, assigned to BASF.

FLUOMETURON

Function: Herbicide (1)(2)(3)

Chemical Name: N,N-dimethyl-N'-[3-(trifluoromethyl)phenyl] urea

Formula:

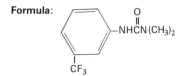

Trade Names: Ciba 2059 (Ciba-Geigy)
Cotoran® (Ciba-Geigy)

Manufacture

Fluometuron can be made by the reaction of dimethylamine on 3-trifluoromethylphenyl isocyanate.

Process Wastes and Their Control

Air: Air emissions from fluometuron manufacture have been reported (B-15) to consist of 0.5 kg hydrocarbons and 0.5 kg fluorine per metric ton of pesticide produced.

Product Wastes: The chemical properties of this compound are similar to other urea derivatives; it is decomposed by strong base or acid at elevated temperature. However, this method is not recommended due to generation of toxic products (B-3).

Toxicity

The acute oral LD_{50} value for rats is 89 mg/kg according to NIOSH (B-5) which is moderately toxic but much higher values of 6,000 to 8,000 mg/kg are given by Worthing (1) which would indicate insignificant toxicity.

Allowable Limits on Exposure and Use

Product Use: The tolerances set by the EPA for fluometuron in or on raw agricultural commodities are as follows:

	40 *CFR* Reference	Parts per Million
Cotton, seed	180.229	0.1 N
Sugarcane	180.229	0.1 N

The tolerances set by the EPA for fluometuron in animal feeds are as follows (the *CFR* Reference is to Title 21):

	CFR Reference	Parts per Million
Sugarcane Bagasse	561.240	0.2

References

(1) Worthing, C.R., *Pesticide Manual,* 6th ed., p. 275, British Crop Protection Council (1979).
(2) Spencer, E.Y., *Guide to the Chemicals Used in Crop Protection,* 6th ed., p. 283, London, Ontario, Agriculture Canada (January 1973).
(3) Martin, H. and Aebi, H., U.S. Patent 3,134,665, May 26, 1964, assigned to Ciba, Ltd.

FLUOROACETAMIDE

Function: Rodenticide

Chemical Name: 2-Fluoracetamide

Formula: FCH_2CONH_2

Trade Names: Compound 1081
Fluorakil 100®
Fussol®

Manufacture

This compound is made by reacting fluoroacetyl chloride and ammonia.

Toxicity

The acute oral LD_{50} for rats is 13 mg/kg which is highly toxic.

Allowable Limits on Exposure and Use

Product Use: A rebuttable presumption against registration was issued by EPA on November 22, 1976 on the basis of toxicity to nontarget species.

As of November 2, 1979, the EPA required that labeling be amended to allow use only inside of sewers against the Norwegian root rat. This use is restricted and may be applied only by a certified applicator or a competent person acting under the instructions and control of a certified applicator. The RPAR was terminated on February 28, 1980.

Reference

(1) Worthing, C.R., *Pesticide Manual,* 6th ed., p. 276, British Crop Protection Council (1979).

FLUORODIFEN

Function: Herbicide (1)(2)(3)

Chemical Name: 2-nitro-1-(4-nitrophenoxy)-4-trifluoromethylbenzene

Formula:

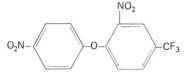

Trade Names: C-6989 (Ciba-Geigy)
Preforan® (Ciba-Geigy)

Manufacture (3)

Fluorodifen may be made by the reaction of p-nitrophenol with 2-nitro-trifluoromethyl chlorobenzene in the presence of caustic. The reaction is carried out for 6½ hours at 175°C.

Process Wastes and Their Control

Product Wastes: Fluorodifen does contain fluorine, and therefore incineration presents the increased hazard of HF in the off-gases (B-3). The Manufacturing Chemists Association suggests that, prior to incineration, fluorine-containing compounds should be mixed with slaked lime plus vermiculite, sodium carbonate or sand-soda ash mixture (90-10).

Toxicity

Fluorodifen, a solid herbicide, is relatively nontoxic. The acute oral LD_{50} for rats is 9,000 mg/kg.

Allowable Limits on Exposure and Use

Product Use: The tolerances set by the EPA for fluorodifen in or on raw agricultural commodities are as follows:

	40 *CFR* Reference	Parts per Million
Peanuts	180.290	0.2 N
Peanuts, hulls	180.290	0.2 N
Peanuts, vines, hay	180.290	0.2 N
Soybeans	180.290	0.1 N
Soybeans, forage	180.290	0.1 N
Vegetables, seed & pod	180.290	0.1 N
Vegetables, seed & pod, forage	180.290	0.1 N

References

(1) Worthing, C.R., *Pesticide Manual*, 6th ed., p. 277, British Crop Protection Council (1979).
(2) Spencer, E.Y., *Guide to the Chemicals Used in Crop Protection,* 6th ed., p. 284, London, Ontario, Agriculture Canada (January 1973).
(3) British Patent 1,033,163, June 15, 1966, assigned to Ciba, Ltd.

FLUOTHIURON

Function: Herbicide (1)(2)

Chemical Name: N-[3-chloro-4-[(chlorodifluoromethyl)-thio]phenyl]-N,N-dimethylurea

Formula:

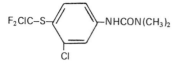

Trade Names: Bay Kue 2079A (Bayer)
Clearcide® (Bayer)

Manufacture (2)

3-Chloro-4-difluorochloromethylmercaptophenyl-isocyanate was obtained by fluorination of 770 g 3-chloro-4-trichloromethylmercaptophenylisocyanate (BP 144° to 147°C/0.8 mm Hg, n_D^{20} 1.6287) with 650 ml of anhydrous hydrofluoric acid at 0° to 20°C. After distillation, there were obtained 502 g of the desired isocyanate of BP 139°C/14 mm Hg, n_D^{20} 1.5650.

10 g 3-chloro-4-difluorochloromethylmercaptophenylisocyanate was added dropwise to 50 ml of a 20% strength aqueous dimethylamine solution. The temperature was kept below 35°C by external cooling. The crystalline product was filtered off with section, and 11 g N-(3-chloro-4-difluoromethylmercaptophenyl)-N',N'-dimethylurea of MP 112°C were obtained.

Toxicity

The acute oral LD_{50} value for rats is 336 to 554 mg/kg which is moderately toxic.

References

(1) Martin, H. and Worthing, C.R., *Pesticide Manual,* 5th ed., p. 107, British Crop Protection Council (January 1977).
(2) Klauke, E., Kuhle, E. and Eue, L., U.S. Patent 3,931,312, January 6, 1976, assigned to Bayer AG.

FLUOTRIMAZOLE

Function: Fungicide (1)(2)

Chemical Name: 1-[diphenyl[3-(trifluoromethyl)-phenyl]methyl]-1H-1,2,4-triazole

Formula:

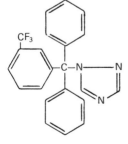

Trade Names: BUE 0620 (Bayer)
Persulon® (Bayer)

Manufacture (2)

1,2,4-Triazole (0.1 mol) is dissolved in 250 ml of absolute acetonitrile, and (0.1 mol) 3-trifluoromethylphenyl-diphenyl-methyl chloride and about 0.11 mol triethyl amine are added thereto, and the mixture is heated to 80°C for 4 hours. After cooling, the amine hydrochloride and, in part, the reaction products, separate. The solvent is distilled off and the residue is washed out with water until Cl ions can no longer be detected. After drying, recrystallization from acetone is effected and a 90% yield of fluotrimazole, melting at 128° to 130°C is obtained.

Toxicity

The acute oral LD_{50} value for rats is over 5,000 mg/kg which is insignificantly toxic.

References

(1) Worthing, C.R., *Pesticide Manual,* 6th ed., p. 278, British Crop Protection Council (1979).
(2) Buchel, K.H., Grewe, F. and Kaspers, H., U.S. Patent 3,682,950, August 8, 1972, assigned to Farbenfabriken Bayer AG.

FLURECOL-BUTYL

Function: Plant growth regulator

Chemical Name: 9-Hydroxy-9H fluorene-9-carboxylic acid butyl ester

Formula:

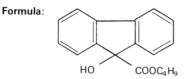

HO   COOC$_4$H$_9$

Trade Name: IT-3233 (E. Merck AG)

Manufacture

Phenanthraquinone is first converted to 9-hydroxyfluorene-9-carboxylic acid which is then esterfied with 1-butanol.

Toxicity

The acute oral LD_{50} for rats is over 5,000 mg/kg which is insignificantly toxic.

References

(1) Worthing, C.R., *Pesticide Manual,* 6th ed., p. 279, British Crop Protection Council (1979).
(2) E. Merck AG, British Patent 1,051,652, December 14, 1966.
(3) E. Merck AG, British Patent 1,051,653, December 14, 1966.
(4) E. Merck AG, British Patent 1,051,654, December 14, 1966.

FLURIDONE

Function: Herbicide (1)(2)

Chemical Name: 1-Methyl-3-phenyl-5-[3-(trifluoromethyl)phenyl]-4(1H)-pyridinone

Formula:

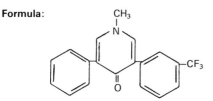

Trade Name: EL-171 (Eli Lilly & Co.)

Manufacture (2)

To a solution of 4 liters of tetrahydrofuran and 284 g of sodium methoxide was added 556 g of 1-(3-trifluoromethylphenyl)-3-phenyl-2-propanone at 10° to 15°C over a 20-minute period. The reaction mixture was stirred for 15 minutes. Then 370 g of ethyl formate was added over a period of 30 minutes and the mixture stirred for 1 hour at 10° to 15°C. To the mixture was added an additional 296 g of ethyl formate over a period of 30 minutes. The reaction mixture was allowed to warm to room temperature and was stirred overnight. Then, a solution of 336 g of methylamine hydrochloride in 1 liter of water was added. The two-phase mixture was stirred at 30°C for 30 minutes. The mixture was then extracted with methylene chloride, and the extracts were combined and concentrated under vacuum, leaving an oily residue which consisted of a mixture containing 1-methylamino-2-phenyl-4-(3-trifluoromethyphenyl)-1-buten-3-one and 1-methylamino-4-phenyl-2-(3-trifluoromethylphenyl)-1-buten-3-one.

The residue was reacted by the same procedure of the previous paragraph. After being dissolved in methylene chloride, the mixture was washed with water and dried. After drying and removal of the solvent, the solid product was found to weigh 430 g, yield 65%. The product was recrystallized from ethyl ether, and the purified product was identified as 1-methyl-3-phenyl-5-(3-trifluoromethylphenyl)-4(1H)-pyridone, MP 153° to 155°C, by infrared, nuclear magnetic resonance, and thin-layer chromatography analyses.

Toxicity

The acute oral LD_{50} value for rats is more than 10,000 mg/kg which is insignificantly toxic.

References

(1) Worthing, C.R., *Pesticide Manual*, 6th ed., p. 280, British Crop Protection Council (1979).
(2) Taylor, H.M., British Patent 1,521,092, August 9, 1978, assigned to Eli Lilly & Co.

FOLPET

Function: Fungicide (1)(2)

Chemical Name: 2-[(trichloromethyl)thio]-1H-isoindole-1,3(2H)-dione

Formula:

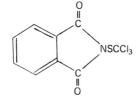

Trade Names: Phaltan® (Chevron Chemical Co.)
Folpan®

Manufacture

Folpet is made by the reaction of perchloromethylmercaptan with sodium phthalimide. See Captan synthesis details for related process steps (3)(4)(5).

Process Wastes and Their Control

Air: Air emissions from folpet manufacture have been reported (B-15) to consist of the following:

Component	Kilograms per Metric Ton Pesticide Produced
Folpet	0.1
Butadiene	1.0
Carbon disulfide	0.5
Water (see Captan)	

Product Waste: Folpet hydrolyzes slowly in water or in preparations not completely dry; stable when dry. It hydrolyzed rapidly at elevated temperatures or in alkaline media (B-3).

Toxicity

The acute oral LD_{50} value for rats is in excess of 10,000 mg/kg which is insignificantly toxic.

The reader is referred to the section of this volume on Captan for a discussion of both folpet and captan (B-22).

Allowable Limits on Exposure and Use

Product Use: Tolerances set by the EPA for folpet in or on raw agricultural commodities are as follows:

	40 CFR Reference	Parts per Million
Apples	180.191	25.0
Avocados	180.191	25.0
Blackberries	180.191	25.0
Blueberries	180.191	25.0
Boysenberries	180.191	25.0
Celery	180.191	50.0
Cherries	180.191	50.0
Citrus fruits	180.191	15.0
Crabapples	180.191	25.0
Cranberries	180.191	25.0
Cucumbers	180.191	15.0
Currants	180.191	25.0
Dewberries	180.191	25.0
Garlic	180.191	15.0
Gooseberries	180.191	25.0
Grapes	180.191	25.0
Huckleberries	180.191	25.0
Leeks	180.191	50.0
Lettuce	180.191	50.0
Loganberries	180.191	25.0
Melons	180.191	15.0

(continued)

	40 *CFR* Reference	Parts per Million
Onions, dry bulb	180.191	15.0
Onions, green	180.191	50.0
Pumpkins	180.191	15.0
Raspberries	180.191	25.0
Shallots	180.191	50.0
Squash, summer	180.191	15.0
Squash, winter	180.191	15.0
Strawberries	180.191	25.0
Tomatoes	180.191	25.0

References

(1) Worthing, C.R., *Pesticide Manual,* 6th ed., p. 281, British Crop Protection Council (1979).
(2) Spencer, E.Y., *Guide to the Chemicals Used in Crop Protection,* 6th ed., p. 285, London, Ontario, Agriculture Canada (January 1973).
(3) Kittleson, A.R., U.S. Patent 2,553,770, May 22, 1951, assigned to Standard Oil Development Corp.
(4) Kittleson, A.R. et al, U.S. Patent 2,553,771, May 22, 1951, assigned to Standard Oil Development Company.
(5) Kittleson, A.R., U.S. Patent 2,553,776, May 22, 1951, assigned to Standard Oil Development Company.

FONOFOS

Function: Insecticide (1)(2)

Chemical Name: O-ethyl S-phenyl ethylphosphonodithioate

Formula:

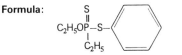

Trade Names: N-2790 (Stauffer Chemical Co.)
Dyfonate® (Stauffer Chemical Co.)

Manufacture (2)

The compounds can be made by the reaction of a compound having the formula

where X is chlorine or bromine, with thiophenol or an alkyl substituted thiophenol, preferably in the presence of an alkali in an organic solvent. The following is a specific example of the conduct of the process.

Thiophenol (11.7 g), O-ethyl-ethylphosphonochloridothioate (16.4 g) and acetone (70 ml) are brought together and Et_3N (10 g) is added in portions while the mixture is efficiently stirred. An exothermic reaction takes place with the prompt precipitation of the amine hydrochloride.

When all the triethylamine is added the mixture is refluxed (58°C) for 20 minutes, then allowed to cool down. Benzene (70 ml) is added and the mixture is successively washed with water, 3% NaOH, and water again. The benzene layer is separated and dried over anhydrous $MgSO_4$. Then the solvent is stripped off to obtain a crude oily product in an 89% yield.

In an alternative route (3) O-ethyl-S-phenyl ethylphosphonodithioate is prepared by isomerization of triethyl phosphite in the presence of an iodine-containing catalyst at 185°C and pressures between about 12 and 30 psig. The resulting diethylethyl phosphonate is reacted with phosgene at a temperature between about 45° and about 55°C and a pressure between about 0 and about 80 psig. The dialkyl phosphono chloridate formed is then reacted with sodium thiophenolate at about 15°C with a pH controlled at about 12.5. Finally the phosphonic acid ester is reacted with phosphorus pentasulfide in the presence of a promoter between about 90° and 130°C. Yield of pure O-ethyl-S-phenyl ethyl phosphonodithioate was 90%, based on diethylethyl phosphonate.

Process Wastes and Their Control

Product Wastes: This phosphono compound is reported to be satisfactorily decomposed by hypochlorite (B-3).

Toxicity

The acute oral LD_{50} value for rats is 8 to 17.5 mg/kg which is very highly toxic.

Allowable Limits on Exposure and Use

Air: The threshold limit value for fonofos in air is 0.1 mg/m³ as of 1979 (B-23).

Product Use: The tolerances set by EPA for fonofos in or on raw agricultural commodities are as follows:

	40 *CFR* Reference	Parts per Million
Asparagus	180.221	0.5
Beans, forage	180.221	0.1 N
Beans, vines, hay	180.221	0.1 N
Beets, sugar, tops	180.221	0.1 N
Corn, field, fodder	180.221	0.1 N
Corn, field, forage	180.221	0.1 N
Corn, fresh (inc. sweet) (k+cwhr)	180.221	0.1 N
Corn, grain	180.221	0.1 N
Corn, pop, fodder	180.221	0.1 N
Corn, pop, forage	180.221	0.1 N
Corn, sweet, fodder	180.221	0.1 N
Corn, sweet, forage	180.221	0.1 N
Peanuts	180.221	0.1 N
Peanuts, forage	180.221	0.1 N
Peanuts, hay	180.221	0.1 N
Peanuts, hulls	180.221	0.1 N
Peas, forage	180.221	0.1 N
Peas, vines, hay	180.221	0.1 N
Peppermint	180.221	0.1 N
Peppermint, hay	180.221	0.1 N
Sorghum, fodder	180.221	0.1 N
Sorghum, forage	180.221	0.1 N
Sorghum, grain	180.221	0.1 N
Soybeans, forage	180.221	0.1 N
Soybeans, hay	180.221	0.1 N

(continued)

	40 *CFR* Reference	Parts per Million
Spearmint	180.221	0.1 N
Spearmint, hay	180.221	0.1 N
Strawberries	180.221	0.1 N
Sugarcane	180.221	0.1 N
Vegetables, fruiting	180.221	0.1 N
Vegetables, leafy	180.221	0.1 N
Vegetables, root crop	180.221	0.1 N
Vegetables, seed & pod	180.221	0.1 N

References

(1) Worthing, C.R., *Pesticide Manual,* 6th ed., p. 282, British Crop Protection Council (1979).
(2) Szabo, K., Brady, J.G. and Williamson, T.B., U.S. Patent 2,988,474, June 13, 1961, assigned to Stauffer Chemical Company.
(3) Pitt, H.M. and Simone, R.A., U.S. Patent 3,642,960, February 15, 1972, assigned to Stauffer Chemical Company.

FORMALDEHYDE

Function: Bactericide and fungicide (1)(2)

Chemical Name: Formaldehyde

Formula:
$$\underset{HCH}{\overset{O}{\underset{\|}{}}}$$

Manufacture

Formaldehyde may be made by various routes: (a) the oxidation of liquified petroleum gas (LPG) (3)

$$C_4H_{10} + \tfrac{3}{2}O_2 \longrightarrow CH_3CHO + CH_2O + CH_3OH$$

(b) the oxidation of methane (4)

$$CH_4 + O_2 \longrightarrow HCHO + H_2O$$

(c) the oxidation of methanol (5)

$$CH_3OH + \tfrac{1}{2}O_2 \longrightarrow HCHO + H_2O$$

Toxicity

The acute oral LD_{50} value for rats is 800 mg/kg which is slightly toxic.

Allowable Limits on Exposure and Use

Air: The threshold limit value for formaldehyde in air is 3 mg/m^3 (2 ppm) as of 1979 (B-23).

Product Use: The tolerances set by the EPA for formaldehyde in or on raw agricultural commodities are as follows:

	40 *CFR* Reference	Parts per Million
Alfalfa, forage	180.103	Exempt
Barley, grain	180.103	Exempt
Bermuda grass, forage	180.103	Exempt
Bluegrass, forage (post-h)	180.103	Exempt
Bromegrass, forage (post-h)	180.103	Exempt
Clover, forage (post-h)	180.103	Exempt
Corn, grain	180.103	Exempt
Cowpea, hay, forage (post-h)	180.103	Exempt
Fescue, forage (post-h)	180.103	Exempt
Lespedeza, forage (post-h)	180.103	Exempt
Lupines, forage (post-h)	180.103	Exempt
Oats, grain (post-h)	180.103	Exempt
Orchard Grass, forage (post-h)	180.103	Exempt
Peanuts, hay, forage (post-h)	180.103	Exempt
Peas, vines, hay, forage (post-h)	180.103	Exempt
Rye grass, forage (post-h)	180.103	Exempt
Sorghum, grain	180.103	Exempt
Soybean, hay, forage (post-h)	180.103	Exempt
Sudan, grass, forage (post-h)	180.103	Exempt
Timothy, forage (post-h)	180.103	Exempt
Vetch, forage (post-h)	180.103	Exempt
Wheat, grain	180.103	Exempt

References

(1) Worthing, C.R., *Pesticide Manual,* 6th ed., p. 283, British Crop Protection Council (1979).

(2) Spencer, E.Y., *Guide to the Chemicals Used in Crop Protection,* 6th ed., p. 287, London, Ontario, Agricutlure Canada (January 1973).

(3) Ramosek et al, U.S. Patent 2,767,203, October 16, 1956, assigned to Stanolind Oil and gas Company.

(4) Magee, E.M., U.S. Patent 3,032,588, May 1, 1962, assigned to Esso Research and Engineering Company.

(5) Allyn, C.L. et al, U.S. Patent 2,849,492, August 26, 1958, assigned to Reichhold Chemicals, Inc.

FORMETANATE

Function: Acaricide (1)(2)(3)(4)

Chemical Name: N,N-dimethyl-N'-dimethyl-N'-(3-[(methylamino)carbonyl]oxy)phenylmethanimidamide

Formula:

Trade Names: Schering 36056 (Schering AG)
Dicarzol® (Schering AG)
EP 332
ENT 27566

Manufacture (3)

3(N,N-dimethylaminomethylenimino)-phenol and methyl isocyanate were dissolved in a mixture of tetrahydrofuran and dimethylformamide, and the solution was left to stand 48 hours at room temperature. It was subsequently heated to a boil for 30 minutes, and the solvent mixture was evaporated in a vacuum.

The oily residue was dissolved in chloroform, and filtered over aluminum oxide (neutral). After evaporation of the chloroform in a vacuum, the residue was crystallized in the presence of a small amount of ether, and was recrystallized from ethyl acetate. The yield was 60%, the MP of the product was 102° to 103°C. The elementary analysis corresponded to that of the compound $C_{11}H_{15}N_3O_2$.

The phenols which were employed as starting materials were prepared as follows. 0.1 mol of a suitably substituted aminophenol was dissolved either in 25 ml dimethylformamide or in another suitable solvent, such as acetonitrile, in the presence of 0.1 mol dimethylformamide; 9.15 ml phosphorus oxychloride were added drop by drop with agitation while the temperature was held at or below 60°C by means of external cooling.

After completion of the phosphorus oxychloride addition, the mixture was agitated for 30 additional minutes at 60°C. When ethanol was added and the mixture was cooled to room temperature, crystallization of the desired hydrochloride usually occurred very rapidly. If necessary, crystallization was made more complete by addition of ethyl ether.

If the hydrochloride obtained was not satisfactorily pure, it was recrystallized from ethanol or from a mixture of ethanol and ether. It was then dissolved or suspended in water, and an equivalent amount of a strong base such as triethylamine, NaOH or Na_2CO_3 was added whereupon the free phenol crystallized.

If necessary the aminophenol was recrystallized from a mixture of tetrahydrofuran and a light petroleum fraction. The pesticidal compounds of the process may also be prepared by reacting a corresponding phenol with a mono-substituted carbamic acid chloride in a manner well known in itself.

Toxicity

The acute oral LD_{50} value for rats is 21 mg/kg which is highly toxic.

Allowable Limits on Exposure and Use

Product Use: The tolerances set by EPA for formetanate hydrochloride in or on raw agricultural commodities are as follows:

	40 *CFR* Reference	Parts per Million
Apples	180.276	3.0
Grapefruit	180.276	4.0
Lemons	180.276	4.0
Limes	180.276	4.0
Nectarines	180.276	4.0
Oranges	180.276	4.0
Peaches	180.276	5.0
Pears	180.276	3.0
Plums (fresh prunes)	180.276	2.0
Tangerines	180.276	4.0

The tolerances set by the EPA for formetanate in animal feeds are as follows (the *CFR* Reference is to Title 21):

	CFR Reference	Parts per Million
Citrus molasses	561.250	10.0

The tolerances set by the EPA for formetanate in food are as follows (the *CFR* Reference is to Title 21):

	CFR Reference	Parts per Million
Prunes, dried	193.220	8.0

References

(1) Worthing, C.R., *Pesticide Manual,* 6th ed., p. 284, British Crop Protection Council (1979).
(2) Spencer, E.Y., *Guide to the Chemicals Used in Crop Protection,* 6th ed., p. 288, London, Ontario, Agriculture Canada (January 1973).
(3) Peissker, H., Jager, A., Steinhausen, W. and Boroschewski, G., U.S. Patent 3,336,186, August 15, 1967, assigned to Schering AG.
(4) Peissker, H., Jager, A., Steinhausen, W. and Boroschewski, G., U.S. Patent 3,542,853, December 8, 1970, assigned to Shering AG.

FORMOTHION

Function: Insecticide and acaricide (1)(2)(3)(4)(5)

Chemical Name: S-[2-(formylmethylamino)-2-oxoethyl]O,O-dimethyl phosphorodithioate

Formula:
$$(CH_3O)_2\overset{\overset{S}{\|}}{P}SCH_2CO\overset{\overset{CH_3}{|}}{N}CH{=}O$$

Trade Names: J-38 (Sandoz, Ltd.)
OMS 968 (Sandoz, Ltd.)
Anthio® (Sandoz, Ltd.)
Aflix® (Sandoz, Ltd.)

Manufacture (3)(4)(5)

N-formyl-N-methylchloroacetamide is first prepared by one of the following techniques (3): (a) 1 mol of an α-chloro acetyl chloride of the formula $CH_2Cl{-}COCl$ is dissolved in trichloroethylene, after which 1 mol of methyl formamide,

$$HCONHCH_3$$

is stirred in dropwise at about 80°C in the course of 1 to 2 hours. Stirring of the reaction mixture is then further continued at the temperature until hydrogen chloride development has practically ceased.

(b) 59 parts (1 mol) of the compound of the formula CH_3NHCHO are dissolved in 100 parts by volume of trichloroethylene and the resultant solution is stirred dropwise in the course of 1 to 2 hours into a solution, heated to 80°C, of 112 parts (1 mol) of chloroacetic acid chloride in 200 parts by volume of trichloroethylene. Stirring is then continued for about one more hour at 80°C. The resultant reaction product of the formula:

$$ClCH_2CON-CH_3$$
$$|$$
$$CHO$$

is recovered from the reaction by distillation. Yield: 85%. Boiling point: 70° to 71°C at 0.4 mm. Then, in a subsequent and final step: 68 parts of the above compound and 90 parts of $(CH_3O)_2PSSNH_4$ are stirred together for 4 hours in 300 parts by volume of chloroform at about 55°C.

The chloroform layer is then washed successively with water, 5% aqueous sodium bicarbonate solution and again with water. After drying the chloroform layer and distilling off the chloroform, there is obtained in about 85% yield the compound of the formula:

$$(CH_3O)_3PSSCH_2CON-CH_3$$
$$|$$
$$CHO$$

in the form of a yellow oil.

Process Wastes and Their Control

Product Wastes: Formothion is similar to dimethoate, but more stable in storage and on heating. It is hydrolyzed by alkali and is incompatible with alkaline pesticides. Analysis is accomplished by a modified hydrolysis (B-3).

Toxicity

The acute oral LD_{50} value for rats is 365 to 500 mg/kg which is moderately toxic.

References

(1) Worthing, C.R., *Pesticide Manual,* 6th ed., p. 285, British Crop Protection Council (1979).
(2) Spencer, E.Y., *Guide to the Chemicals Used in Crop Protection,* 6th ed., p. 289, London, Ontario, Agriculture Canada (January 1973).
(3) Lutz, K. and Schuler, M., U.S. Patent 3,176,035, March 30, 1965, assigned to Sandoz, Ltd.
(4) Lutz, K. and Schuler, M., U.S. Patent 3,178,337, April 13, 1965, assigned to Sandoz, Ltd.
(5) Lutz, K. and Schuler, M., British Patent 900,557, July 11, 1962, assigned to Sandoz, Ltd.

FOSAMINE-AMMONIUM

Function: Herbicide (1)(2)(3)

Chemical Name: Ammonium ethyl(ammocarbonyl) phosphonate

Formula:

$$CH_3CH_2O\overset{\displaystyle O}{\overset{\displaystyle \|}{P}}-CONH_2$$
$$\underset{O^- \ NH_4^+}{|}$$

Trade Names: DXP-1108 (DuPont)
Krenite® (DuPont)

Manufacture (2)(3)

A solution of 48.5 parts of 29% aqueous ammonium hydroxide is stirred and cooled with an external ice bath to 15°C. To the cooled solution 22 parts of diethyl carbomethoxyphosphonate is slowly added over a 10-minute period. The mixture turns cloudy, but clears up after 15 minutes. During this time, the mixture is allowed to warm spontaneously to about 30°C and stirring is continued for 2 hours. The clear solution is stripped under reduced pressure (15 mm of Hg) at a water-bath temperature of 70°C. The residue is a white crystalline solid which is recrystallized from absolute ethyl alcohol, giving 12.3 parts of ammonium ethyl carbamoylphosphonate, MP 173° to 176°C.

Toxicity

The acute oral LD_{50} value for rats is 24,000 mg/kg [for a formulated product whose concentration is not stated in Reference (1)]. It appears to be insignificantly toxic on this basis.

References

(1) Worthing, C.R., *Pesticide Manual,* 6th ed., p. 286, British Crop Protection Council (1979).
(2) Langsdorf, W.P., U.S. Patent 3,627,507, December 14, 1971, assigned to E.I. DuPont de Nemours & Co.
(3) Langsdorf, W.P., U.S. Patent 3,846,512, November 5, 1974, assigned to E.I. DuPont de Nemours & Co.

FOSTHIETAN

Function: Insecticide and nematicide (1)

Chemical Name: Diethyl 1,3-dithietan-2-yl-idenephosphoramidate

Formula:

$$(CH_3CH_2O)_2 \overset{\overset{O}{\|}}{P}N{=}C{\overset{\displaystyle S}{\underset{\displaystyle S}{\diagdown\!\diagup}}}CH_2$$

Trade Name: AC 64,475 (American Cyanamid Co.)

Manufacture (2)

First, diethyl phosphorochloridate and sodium thiocyanate are reacted to give diethoxy phosphinyl isothiocyanate.

To a mixture of 57.6 g (0.512 mol) of potassium tert-butoxide in 650 ml of tert-butanol saturated with hydrogen sulfide is added 100 g (0.512 mol) of diethoxyphosphinyl isothiocyanate in 50 ml of tert-butanol over a one hour period. The reaction is mildly exothermic and maintained at about 25° to 30°C with a water bath. After another hour, the reaction solids are collected and washed twice with tert-butanol/benzene (10:1), then twice with benzene, and finally twice with ether. After vacuum drying overnight, the recovered fluorescent light-yellow powdery salt which weighs 122.7 g (89%) has a MP of 97° to 99°C (dec).

A mixture of 1.50 g (0.0056 mol) of potassium diethoxyphosphinyldithiocarbamate prepared as above, 0.5 ml (0.0058 mol) of methylene bromide, and 0.58 g (0.0069 mol) of sodium bicarbonate in 20 ml of methanol is stirred vigorously overnight. After addition of ether (ca an equal volume), the mixture is filtered and concentrated in vacuo. The residue is remixed with benzene and the mixture is filtered and concentrated to yield 1.30 g of yellow oil. The oil is redissolved in benzene and washed with dilute salt solution three times. The dried organic mixture, on concentration, yields 0.96 g (67%) of crude 2-diethoxyphosphinylimino-1,3-dithiolane identified by comparison of its infrared spectrum with the known compound.

Toxicity

The acute oral LD$_{50}$ value for rats is 5.7 mg/kg which is highly to extremely toxic.

References

(1) Worthing, C.R., *Pesticide Manual,* 6th ed., p. 287, British Crop Protection Council (1979).
(2) Addor, R.W., U.S. Patent 3,476,837, November 4, 1969, assigned to American Cyanamid Company.

FUBERIDAZOLE

Function: Fungicide (1)(2)(4)

Chemical Name: 2-(2-furanyl)-1H-benzimidazole

Formula:

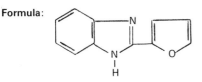

Trade Names: Bayer 33,172 (Bayer)
W VII/117 (Bayer)
Voronit® (Bayer)

Manufacture (3)

Fuberidazole may be made by the condensation of o-phenylenediamine with furfural using cuprous acetate in an oxidizing medium.

Toxicity

The acute oral LD$_{50}$ for rats is 1,100 mg/kg which is slightly toxic.

References

(1) Worthing, C.R., *Pesticide Manual,* 6th ed., p. 288, British Crop Protection Council (1979).
(2) Spencer, E.Y., *Guide to the Chemicals Used in Crop Protection,* 6th ed., p. 290, London, Ontario, Agriculture Canada (1973).
(3) Weidenhagen, R., *Ber. Deut. Chem. Ges.,* 69, 2271 (1936).
(4) Frohberger, P.E. and Weigand, C., U.S. Patent 3,546,813, December 15, 1970, assigned to Farbenfabriken Bayer AG.

FURALAXYL

Function: Fungicide (1)(2)

Chemical Name: Methyl N-(2,6-dimethylphenyl)-N-(2-furanylcarbonyl)-DL-alaninate

Formula:

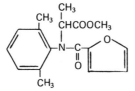

Trade Names: CGA 38140 (Ciba-Geigy AG)
Fongarid® (Ciba-Geigy AG)

Manufacture (2)

(A) 100 g of 2,6-dimethylaniline, 223 g of 2-bromopropionic acid methyl ester and 84 g of NaHCO₃ were stirred for 17 hours at 140°C. The mixture was then cooled, diluted with 300 ml of water and extraction was performed with diethyl ether. The extract was washed with a small amount of water, dried over sodium sulfate, filtered, and the ether evaporated off. After the surplus 2-bromopropionic acid methylester had been distilled off, the crude product was distilled in a high vacuum.

(B) 13 g of furan-2-carboxylic acid chloride were added dropwise with stirring to 17 g of the ester obtained in (A), 2 ml of dimethyl formamide and 150 ml of absolute toluene and the mixture was refluxed for 1 hour. The solvent was evaporated off and the crude product was then crystallized by trituration with petroleum ether; MP 70° to 84°C (ethyl acetate/petroleum ether). The product is the mixture of two pairs of diastereoisomers.

Toxicity

The acute oral LD₅₀ value for rats is 940 mg/kg which is slightly toxic.

References

(1) Worthing, C.R., *Pesticide Manual,* 6th ed., p. 289, British Crop Protection Council (1979).
(2) Ciba-Geigy AG, British Patent 1,498,199, January 18, 1978.

G

GIBBERELLIC ACID

Function: Plant growth regulator (5)

Chemical Name: 2,4a,7-trihydroxy-1-methyl-8-methylenegibb-3-ene-1,10-carboxylic acid
1→4 lactone

Formula:

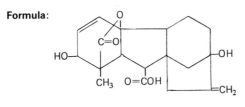

Trade Names: Pro-Gibb®
Gibberellin

Manufacture

Gibberellic acid may be produced (1) by *Gibberella fujikuroi* when grown as a surface culture or as a deep, stirred and aerated culture in a suitable aqueous medium. A suitable medium must contain a carbon source (e.g., sucrose, glucose or glycerol, at 2 to 30% weight per volume), an ammonium salt or a nitrate or a digest of protein (e.g., peptone) as nitrogen source (to give a concentration of N in the medium of 0.01 to 0.5%, a concentration of 0.08% being generally suitable), a magnesium salt (e.g., magnesium sulfate heptahydrate at 0.05% or thereabouts, which also conveniently adds sulfate which is also necessary but which may be added in some other form if another salt of magnesium is used), a phosphate (e.g., potassium dihydrogen phosphate at 0.05 to 0.5%, 0.1% being a suitable concentration), a potassium salt (which may be conveniently the phosphate, see above) and traces (ca. 1 to 10 ppm of metal) of salts of iron, copper, zinc, manganese and a trace (1 to 10 ppm of metal) of a molybdate. Fermentations can be conveniently carried out at a temperature in the range of 25° to 33°C.

The process of obtaining gibberellic acid comprises generally the following steps:

(1) Growth of a mold culture in an aqueous medium.
(2) Filtration and extraction of the medium with an adsorbent.
(3) Elution of the adsorbent.
(4) Purification of the acid obtained by the elution.

It has been found (2) that *Gibberella fujikuroi* when added to sterile-aerated water covered by an oil layer generates in the water phase gibberellin and/or gibberellic acid-like substances capable of altering the rate of plant growth while producing oil-soluble oxygenated compounds which are recoverable from the oil phase. The oil acts as a nutrient for *Gibberella fujikuroi*. Crude oil may be employed as a nutrient and in addition various untreated liquid hydrocarbon fractions such as kerosene, heavy naphthas, aromatic solvent naphthas, petroleum pitches and tars, coke oven pitch and tar and various distillation residues, asphalts and the like are also utilizable. Unsaturation in the hydrocarbon molecule provides an ideal point of attack in the assimilation of the hydrocarbon by the mold.

The use of crude oil as a nutrient is an economic saving over the customarily employed nutrient used previously. In addition, crude oil will form a protective blanket on the surface of the substrate permitting the adaptability of the process to large uncovered earthen or concrete tanks. Alcohols, aldehydes, acids and ketones are produced in the oil phase during the process. These products may be removed from the oil phase by fractionation, solvent extraction, and other well-known methods of separation. To complete the process, the water phase can be treated by various separation methods to form gibberellic acid concentrates or the crystalline acid or acid salts may be recovered. A conventional process involves the extraction of the water phase with an adsorbent such as charcoal, eluting the adsorbent with acetone, concentrating and extracting the eluate with ethyl acetate, extracting the ethyl acetate solution with a phosphate buffer and adjusting the pH with hydrochloric acid, extraction with ethyl acetate and concentrating the solution to obtain gibberellic acid.

Production of gibberellic acid using a glucose medium (3) and separation of the resultant product (4) have also been described.

Process Wastes and Their Control

Product Wastes: The gibberellins are probably nontoxic to humans. Burial or incineration (where permitted) would be effective disposal procedures (B-3).

Toxicity

The acute oral LD_{50} for rats is over 15,000 mg/kg which is insignificantly toxic.

Allowable Limits on Exposure and Use

Product Use: The tolerances set by the EPA for gibberellic acid in or on raw agricultural commodities are as follows:

	40 *CFR* Reference	Parts per Million
Artichokes	180.224	0.15 N
Blueberries	180.224	0.15 N
Citrus fruits	180.224	0.15 N
Fruits, stone	180.224	0.15 N
Grapes	180.224	0.15 N
Hops	180.224	0.15 N
Sugarcane	180.224	0.15 N
Sugarcane, fodder	180.224	0.15 N
Sugarcane, forage	180.224	0.15 N
Vegetables, leafy	180.224	0.15 N

References

(1) Brian, P.W., Radley, M.E., Curtis, P.J. and Elson, G.W., U.S. Patent 2,842,051, July 8, 1958, assigned to Imperial Chemical Industries.

(2) Hitzmans, D.O. and Mills, A.M., U.S. Patent 3,084,106, April 2, 1963, assigned to Phillips Petroleum Co.

(3) Birch, A.J., Nixon, I.S. and Grove, J.F., U.S. Patent 2,977,285, March 28, 1961, assigned to Imperial Chemical Industries Ltd.

(4) Calam, C.T. and Curtis, P.J., U.S. Patent 2,950,288, August 23, 1960, assigned to Imperial Chemical Industries Ltd.

(5) Worthing, C.R., *Pesticide Manual*, 6th ed., p. 291, British Crop Protection Council (1979).

GLYODIN

Function: Fungicide

Chemical Name: 2-heptadecyl-2-imidazoline acetate

Formula:

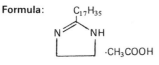

Trade Names: Glyodex®
Glyoxide®
Glyodine®
Crag Fruit Fungicide 341® (Union Carbide)

Manufacture

The feed materials for glyodin manufacture are stearic acid, ethanol, ethylenediamine and acetic acid. The initial reaction is as follows:

$$C_{17}H_{35}COOH + C_2H_5OH \longrightarrow C_{17}H_{35}COOC_2H_5 + H_2O$$

The ester is then reacted with ethylenediamine to give N-(2-aminoethyl)stearamide. The amide is then heated to close the ring and to give an imidazoline.

$$C_{17}H_{35}COOC_2H_5 + H_2NCH_2CH_2NH_2 \longrightarrow C_{17}H_{35}\overset{\overset{\displaystyle O}{\|}}{C}-NHCH_2CH_2NH_2 + C_2H_5OH$$

$$C_{17}H_{35}\overset{\overset{\displaystyle O}{\|}}{C}-NHCH_2CH_2NH_2 \longrightarrow C_{17}H_{35}C\overset{\overset{\displaystyle N-CH_2-CH_2}{\|}}{\underset{\displaystyle \hspace{2.2cm} NH}{\rule{0pt}{0pt}}} + H_2O$$

The final step involves formation of the acetate salt.

The initial esterification step is conducted at 65° to 75°C. The reaction of the ester with ethylenediamine is carried out at about 125°C. The ring closure reaction is carried out at about 200°C. The final step comprises adding molten imidazoline to acetic acid in isopropanol at a rate such that the temperature does not exceed 40°C.

The initial esterification step is conducted under atmospheric pressure, as is the subsequent reaction of the ester with ethylenediamine. The ring closure of the N-aminoethyl stearamide is carried out under reduced pressure of 2 to 10 mm of mercury. The final reaction with acetic acid is carried out at atmospheric pressure. The esterification step is refluxed until water no longer appears as a reaction product. The reaction of the ester with ethylenediamine is carried out over a reaction period of about 8 hours under total reflux conditions according to B.F. Kiff (1).

The esterification step is carried out in the liquid phase, as is the subsequent reaction of the ester with ethylenediamine, and the subsequent ring closure reaction. The reaction of the imidazoline with acetic acid is carried out in essentially anhydrous isopropanol solution. Sulfuric acid may be used as a catalyst in the initial esterification step. No catalyst is used, however, in the subsequent reaction of the ester with ethylenediamine, nor in the ring closure reaction which follows. No catalyst is used in the final reaction step involving the heptadecyl imidazoline and acetic acid.

The esterification step is carried out in a kettle equipped with a reflux column and connected to a decanting device for continuous removal of the water formed during the reaction. A similar type of apparatus is used for the reaction of the ester formed with ethylenediamine.

The ring closure step is carried out in a kettle surmounted by a Claisen-type condenser and then connected to a receiver and a cold trap in series, the entire reaction train being adapted to operation under a vacuum. The final reaction of the heptadecyl imidazoline with acetic acid is simply carried out in a stirred, jacketed kettle. The ester product from the first step is distilled at 150° to 165°C and at an absolute pressure of 1.5 mm of mercury. An ester yield of about 95% is obtained based on acid charged.

At the end of the second stage reaction between ester and ethylenediamine, excess diamine is stripped off by heating from the reaction temperature of 125°C to a temperature of 215°C. In the ring closure step, the imidazoline product is recovered in about 75% yield by distillation from the reaction mass. The final product may simply be filtered to give a clear solution of the imidazoline acetate in isopropanol.

Process Wastes and Their Control

Product Wastes: This light orange solid fungicide is readily decomposed by alkali to produce relatively harmless products (2-aminoethyl stearamide) (B-3).

Toxicity

The acute oral LD_{50} for rats is 6.8 mg/kg (B-5) which is highly toxic.

Allowable Limits on Exposure and Use

Product Use: The tolerances set by the EPA for glyodin in or on raw agricultural commodities are as follows:

	40 *CFR* Reference	Parts per Million
Apples	180.124	5.0
Cherries	180.124	5.0
Peaches	180.124	5.0
Pears	180.124	5.0

Reference

(1) Kiff, B.W., U.S. Patent 2,540,171, February 6, 1951, assigned to Union Carbide and Carbon Corp.

GLYPHOSATE

Function: Herbicide (1)(2)(3)

Chemical Name: N-(phosphonomethyl)glycine

Formula: $(HO)_2POCH_2NHCH_2COOH$

Trade Names: Mon 0573 (Monsanto)
Mon 2139 (Monsanto)
Roundup® (Monsanto)

Manufacture (3)

A mixture of about 50 parts of glycine, 92 parts of chloromethylphosphonic acid, 150 parts of 50% aqueous sodium hydroxide and 100 parts water was introduced into a suitable reaction vessel and maintained at reflux temperature while an additional 50 parts of 50% aqueous sodium hydroxide was added. The pH of the reaction mixture was maintained between 10 and 12 by the rate of addition of the sodium hydroxide. After all of the caustic solution had been added, the reaction mixture was refluxed for an additional 20 hours, cooled to room temperature and filtered. About 160 parts of concentrated hydrochloric acid were then added and the mixture filtered to provide a clear solution which slowly deposited N-phosphonomethylglycine. This material had a MP of 230°C with decomposition.

Toxicity

The acute oral LD_{50} for rats is 4,320 mg/kg which is very slightly toxic.

The tolerances set by the EPA for glyphosate in animal feeds are as follows (the *CFR* Reference is to Title 21):

	CFR Reference	Parts per Million
Citrus pulp, dried	561.253	0.4
Soybean hulls	561.253	20.0
Sugarcane molasses	561.253	2.0

The tolerances set by the EPA for glyphosate in food are as follows (the *CFR* Reference is to Title 21):

	CFR Reference	Parts per Million
Palm, oil	193.235	0.1
Potable water	193.235	1.0
Sugarcane molasses	193.235	2.0

References

(1) Worthing, C.R., *Pesticide Manual,* 6th ed., p. 292, British Crop Protection Council (1979).
(2) Spencer, E.Y., *Guide to the Chemicals Used in Crop Protection,* 6th ed., p. 293, London, Ontario, Agriculture Canada (January 1973).
(3) Franz, J.E., U.S. Patent 3,799,758, March 26, 1974, assigned to Monsanto Co.

GLYPHOSINE

Function: Plant growth regulator (1)(2)

Chemical Name: N,N-bis(phosphonomethyl)glycine

Formula: [(HO)$_2$POCH$_2$]$_2$NCH$_2$COOH

Trade Names: CP-41845 (Monsanto)
Polaris® (Monsanto)

Manufacture (3)

Into a conventional jacketed glass-lined mixing vessel fitted with a water condenser, are blended 750 parts of a 50 weight percent aqueous solution of glycine (H$_2$NCH$_2$COOH), 820 parts of orthophosphorous acid, and 500 parts of concentrated (ca. 38%) hydrochloric acid. The resulting blend, while being continuously stirred, is heated to 100°C. Then over a period of about 30 minutes 1,500 parts of aqueous (37%) formaldehyde solution are added slowly to the blend. The reflux condenser was removed from the mixing vessel and approximately 25% of the volume of the resulting mixture is evaporated over the next 2 hours. A quantitative evaluation of the nuclear magnetic resonance spectrum of the resulting concentrated solution showed that 84% of the orthophosphorous acid has reacted to form the desired N$-$C$-$P linkages, 13% of orthophosphorous acid still remains unreacted, and less than 3% of the orthophosphorous acid has been oxidized to orthophosphoric acid.

The concentrated aqueous solution is then evaporated on a steam bath until a clear syrupy liquid is obtained. This syrupy liquid, when dissolved in hot ethanol and subsequently cooled, precipitates from the aqueous ethanol solution as white crystals. The molecular weight of these crystals, by titration, is 259.2, while, theoretically, it should be 263.1.

Toxicity

The acute oral LD$_{50}$ value for rats is 3,925 mg/kg which is very slightly toxic.

Allowable Limits on Exposure and Use

Product Use: The tolerances set by the EPA for glyphosine in or on raw agricultural commodities are as follows:

	40 *CFR* Reference	Parts per Million
Alfalfa, dry	180.364	0.2
Alfalfa, fresh	180.364	0.2
Almonds, hulls	180.364	1.0
Asparagus	180.364	0.2
Avocados	180.364	0.2
Cattle, fat	180.364	0.2 N
Cattle, kidney	180.364	0.1
Cattle, liver	180.364	0.1
Citrus fruits	180.364	0.2
Coffee beans	180.364	1.0
Cotton, forage	180.364	0.2
Cotton, hay	180.364	0.2
Cotton, seed	180.364	6.0
Cotton, trash	180.364	0.2
Fruits, pome	180.364	0.2
Goats, kidney	180.364	0.1
Goats, liver	180.364	0.1
Grain crops	180.364	0.1 N
Grapes	180.364	0.1
Grasses, forage	180.364	0.2 N
Hogs, kidney	180.364	0.1
Hogs, liver	180.364	0.1
Horses, kidney	180.364	0.1

(continued)

	40 *CFR* Reference	Parts per Million
Horses, liver	180.364	0.1
Nuts	180.364	0.2
Pistachio nuts	180.364	0.2
Poultry, kidney	180.364	0.1
Poultry, liver	180.364	0.1
Sheep, kidney	180.364	0.1
Sheep, liver	180.364	0.1
Soybeans	180.364	6.0
Soybeans, forage	180.364	15.0
Soybeans, hay	180.364	15.0
Sugarcane	180.364	0.1
Sugarcane	180.354	3.0
Vegetables, leafy	180.364	0.2 N
Vegetables, rootcrops	180.364	0.2 N
Vegetables, seed & pod	180.364	0.2 N
Vegetables, seed & pod, forage	180.364	0.2 N
Vegetables, seed, & pod, hay	180.364	0.2 N

References

(1) Worthing, C.R., *Pesticide Manual,* 6th ed., p. 293, British Crop Protection Council (1979).
(2) Hamm, P.C., U.S. Patent 3,556,762, January 19, 1971, assigned to Monsanto Co.
(3) Irani, R.R. and Moedritzer, K., U.S. Patent 3,288,846, November 29, 1966, assigned to Monsanto Co.

GUAZATINE
(GUANOCTINE in U.K.)

Function: Fungicide (1)(2)

Chemical Name: 9-aza-1,17-diguanidino heptadecane

Formula:

$$\left[\begin{array}{c} NH \\ \| \\ H_2NC{-}NH(CH_2)_8 \end{array} \right]_2 NH$$

Trade Names: MC-25 (Murphy Chemical Ltd.)
EM 379 (Murphy Chemical Ltd.)
Panoctine® (Casco Gard)

Manufacture (3)

Guazatine may be made by the reaction of dioctyl triamine with O-methyl isourea.

Toxicity

The acute oral LD_{50} for rats is 230 to 260 mg/kg (as the triacetate) which is moderately toxic.

References

(1) Worthing, C.R., *Pesticide Manual,* 6th ed., p. 294, British Crop Protection Council (1979).
(2) Spencer, E.Y., *Guide to the Chemicals Used in Crop Protection,* 6th ed., p. 295, London, Ontario, Agriculture Canada (January 1973).
(3) Badcock, G.G. and Dyke, W.J.C., British Patent 1,114,155, May 15, 1968, assigned to Evans Medical Ltd.

H

HALACRINATE

Function: Fungicide (1)(2)

Chemical Name: 7-Bromo-5-chloro-8-quinolinyl 2-propenoate

Formula:

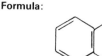

Trade Name: CGA 30599 (Ciba-Geigy)

Manufacture (2)

1 kg of 5-chloro-7-bromo-8-oxyquinoline (3.87 mols) were stirred in 4 liters of benzene and mixed with 0.411 kg of distilled triethylamine (4.06 mols).

While stirring thoroughly 0.353 kg of acrylic chloride (3.90 mols) dissolved in 1 liter of benzene was allowed to run in over the course of half an hour. The mixture was subsequently further stirred for 1 hour at 70°C internal temperature, cooled to room temperature and the crude product was filtered with suction from the precipitated triethylamine hydrochloride.

400 ml of sodium hydroxide solution was added to the filtrate and the mixture stirred for 42 hours, so that both phases were thoroughly mixed with each other. Thereafter the unchanged 5-chloro-7-bromo-8-oxyquinoline was collected by suction filtration from the Na salt (approximately 80 g.).

The phases in the filtrate were separated. The benzene phase was neutralized by stirring with 3 × 300 ml of water, treated with activated charcoal and dried over Na_2SO_4, whereupon the solvent was removed in a rotary evaporator.

1.066 kg (= 88% of theory) of halacrinate was obtained in the form of a crystalline residue. Melting point after recrystallization 100.5° to 101.5°C.

The compound can be recrystallized from cyclohexane (1 g/approximately 4 ml) or from a large quantity of petroleum ether; it occurs in the form of large crystals.

Toxicity

The acute oral LD_{50} value for rats is over 10,000 mg/kg which is insignificantly toxic.

References

(1) Martin, H. and Worthing, C.R., *Pesticide Manual,* 5th ed., p. 292, British Crop Protection Council (January 1977).
(2) Huber-Emden, H., Hubele, A. and Klahre, G., U.S. Patent 3,813,399, May 28, 1974, assigned to Ciba-Geigy AG.

HEPTACHLOR

Function: Insecticide (1)(2)

Chemical Name: 1,4,5,6,7,8,8-heptachloro-3a,4,7,7a-tetrahydro-4,7-methano-1H indene

Formula:

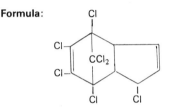

Trade Names: E-3314 (Velsicol Corp.)
Velsicol 104 (Velsicol Corp.)
Drinox®
Heptagran®
Heptalube®

Manufacture

The feed materials for heptachlor manufacture are "chlordene" and chlorine. These are the same feed materials as for chlordane manufacture but in this case the process is substitution chlorination whereas in the case of chlordane it involved addition chlorination of chlordene to chlordane. The reaction for the production of heptachlor is as follows:

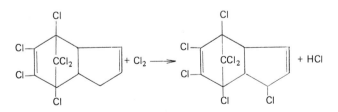

This reaction can be conducted at temperatures within the range of 30° to 100°C. There is evidence which indicates that highest yields are obtained at lower temperatures of perhaps 5°C, however. This reaction is conducted at atmospheric pressure. Reaction periods of 6 to 8 hours may be considered typical.

This reaction is carried out in the liquid phase in benzene solution. The solution may contain 40% of the hexachlorodicyclopentadiene or may be appreciably more dilute. From 0.5 to 5.0% of fuller's earth is used based on the hexachlorodicyclopentadiene present.

This reaction is carried out in a stirred, jacketed reactor of conventional design equipped with a reflux condenser. When operating in the presence of fuller's earth, care should be taken to prevent the access of any lights; the reaction should be conducted in total darkness.

After reaction, the fuller's earth is separated from the reaction mixture. Then the benzene solvent is separated by stripping to give a colorless viscous liquid which does not crystallize. This liquid may be fractionated under vacuum to give a distillate containing the first 25% of the crude. The material is dissolved in methanol and cooled to crystallize out pure heptachlor (3).

Process Wastes and Their Control

Air: Air emissions from heptachlor manufacture have been reported (B-15) to consist of 1.0 kg of hydrocarbons and 0.5 kg hydrogen chloride per metric ton of pesticide produced.

Product Wastes: Heptachlor is said to be stable to at least 160°C and to light, moisture, air and oxidizing agents, acids and apparently alkali, although one account indicates that its susceptibility to alkali is like that of chlordane rather than the aldrin subgroup. Heptachlor is rather easily converted to the epoxide in soil and plants and one would suppose this reaction would occur with peroxides. Heptachlor is decomposed in the presence of iron at 160°C and above. The MCA recommends incineration methods for disposal of heptachlor (B-3).

Toxicity

The acute oral LD_{50} for rats is 100 to 160 mg/kg which is moderately toxic.

Heptachlor has been demonstrated to be highly toxic to aquatic life, to persist for prolonged periods in the environment, to bioconcentrate in organisms at various trophic levels, and to exhibit carcinogenic activity in mice (B-26).

Allowable Limits on Exposure and Use

Air: The threshold limit value for heptachlor in air is 0.5 mg/m^3 as of 1979. The tentative short term exposure limit is 2.0 mg/m^3. It is further noted that cutaneous absorption of heptachlor should be avoided so that the TLV is not invalidated.

Water: In water, EPA has set criteria (B-12) of 0.001 μg/l for the protection of freshwater and marine aquatic life. They went on to point out that persistence, bioaccumulation potential and carcinogenicity of heptachlor indicate minimum human exposure.

Subsequently (B-26), EPA has suggested a criterion of 0.0015 μg/l as a 24-hour average to protect freshwater aquatic life; the concentration should not exceed 0.45 μg/l at any time. To protect saltwater aquatic life, a criterion of 0.0036 μg/l is suggested as a 24-hour average; the concentration should not exceed 0.05 μg/l at any time. For human health protection, the concentration in water should be zero; a cancer risk level of 1 in 100,000 allows a concentration of 0.23 ng/l of heptachlor.

Product Use: In a decision on December 2, 1974 by EPA (B-17), under the provisions of the Administrator's acceptance of the settlement plan to phase out certain uses of the pesticides heptachlor and chlordane, most registered products containing heptachlor will be effectively cancelled, or their application for registration denied by July 1, 1983. A summary of those uses not affected by this settlement, as well as a summary of those uses affected (Phase Out Uses) by the settlement, including the pest to be controlled, the site of application, the use restrictions, and end-use dates, follows:

1. Uses not affected: see chlordane

2. Uses affected (Phase Out Uses):

   (a) Registrations for control of cutworms on field corn will be effectively cancelled or denied by August 1, 1980 in states with EPA approved restricted-use permit programs and immediately in all other states unless and until those states obtain and maintain EPA approved restricted use permit programs. Use shall be by certified applicators only, and shall be applied only by soil broadcast or soil incorporation. The following crops shall not be grown in a field treated with heptachlor in the year of treatment or the following year: legumes (including soybeans, alfalfa, clover, peas, peanuts and other beans); root crops (including potatoes, sugar beets and rutabagas); oil crops (including cotton and safflower); vegetables crops, tobacco or pumpkins. Silage shall not be cut from a field treated with heptachlor in the year of treatment or the following year. Corn which has been treated with heptachlor shall not be followed with any other dairy or meat animal forage crop nor shall livestock be permitted to have access to treated land for a period of two years following treatment.

   (b) Registrations for control of seed corn beetle, seed corn maggot, wireworm, false wireworm, southern corn rootworm and kafir ant on: (1) barley, oats, wheat, rye and corn will be effectively cancelled or denied by September 1, 1982; and (2) sorghum will be effectively cancelled or denied on July 1, 1983. May be used by commercial seed treatment companies only.

   (c) Registrations for control of citrus root weevil larvae and Feller's rose beetle larvae on citrus in the State of Florida will be effectively cancelled or denied by December 31, 1979. Use shall be by certified applicators only, and may be applied by soil incorporation only.

   (d) Registration for control of ants on pineapples in the State of Hawaii will be effectively cancelled or denied by December 31, 1982. To be applied by certified applicators only.

   (e) Registrations for control of narcissus bulb fly on narcissus bulbs will be effectively cancelled or denied by December 31, 1980. Use shall be by certified applicators only. The following protective procedures will be required for persons engaged in treating narcissus with heptachlor:

      1) Wear heavy natural rubber gloves and clean waterproof protective clothes and goggles.

      2) Bathe immediately after work and change all clothing, wash clothing thoroughly with soap and warm water before reuse.

      3) In case of contact, immediately remove contaminated clothes and wash thoroughly with soap and warm water.

      4) Wear a pesticide respirator jointly approved by the Mining Enforcement and Safety Administration and by the National Institute of Occupational Safety and Health under provisions of 30 *CFR* Part II.

      Note: The end-use dates for Phase Out Uses should be those dates listed above, unless the production limitations imposed by FIFRA Docket No. 326 et al is earlier exceeded. Pesticide products in existence 90 days before the effective date of cancellation or denial of a Phase Out Use may: (1) be distributed, sold or otherwise moved in commerce, and used; provided that the pesticide shall not be used inconsistently with its labeling, and (2) may be relabeled by or under the authority of a registrant for another Phase Out Use not already cancelled or denied, and any pesticide product so produced shall not count against the production limitation for the other Phase Out Use.

The tolerances set by the EPA for heptachlor in or on raw agricultural commodities are as follows:

	40 *CFR* Reference	Parts per Million
Alfalfa	180.104	0.0
Apples	180.104	0.0
Barley	180.104	0.0
Beans, lima	180.104	0.0
Beans, snap	180.104	0.1
Beets	180.104	0.0
Beets, sugar	180.104	0.0
Brussels sprouts	180.104	0.0
Cabbage	180.104	0.1
Carrots	180.104	0.0
Cauliflower	180.104	0.0
Cherries	180.104	0.0
Clover	180.104	0.0
Clover, sweet	180.104	0.0
Corn	180.104	0.0
Cotton, seed	180.104	0.0
Cowpeas	180.104	0.0
Grapes	180.104	0.0
Grasses, pasture	180.104	0.0
Grasses, range	180.104	0.0
Kohlrabi	180.104	0.0
Lettuce	180.104	0.1
Meat	180.104	0.0
Milk	180.104	0.0
Oats	180.104	0.0
Onions	180.104	0.0
Peaches	180.104	0.0
Peanuts	180.104	0.0
Peas	180.104	0.0
Peas, black-eyed	180.104	0.0
Pineapples	180.104	0.0
Potatoes	180.104	0.0
Radishes	180.104	0.0
Rutabagas	180.104	0.1
Rye	180.104	0.0
Sorghum, grain (milo)	180.104	0.0
Sugarcane	180.104	0.0
Sweet potatoes	180.104	0.0
Tomatoes	180.104	0.0
Turnips (inc. tops)	180.104	0:0
Wheat	180.104	0.0

References

(1) Worthing, C.R., *Pesticide Manual,* 6th ed., p. 296, British Crop Protection Council (1979).
(2) Spencer, E.Y., *Guide to the Chemicals Used in Crop Protection,* 6th ed., p. 297, London, Ontario, Agriculture Canada (January 1973).
(3) Bluestone, H., U.S. Patent 2,576,666, November 27, 1951, assigned to Julius Hyman and Company.

HEXACHLOROBENZENE

Function: Fungicide (1)(2)

Chemical Name: Hexachlorobenzene

Formula:

Trade Names: Anti-Carie®
                 HCB

Manufacture

Hexachlorobenzene may be made by the liquid phase chlorination of benzene (3) at 180°C and under atmospheric pressure in the presence of an iron oxide catalyst. A stirred, jacketed kettle is used as a reaction vessel and the reaction time is about 4 hours. Yields of 85% are obtained. It may also be made from BHC isomers. Thus, it has been found (4) that hexachlorbenzene is obtained in a simple manner in almost theoretical yields by treating hexachlorcyclohexanes or higher chlorinated cyclohexanes, in particular the stereoisomers of gamma-hexachlorcyclohexane, with chlorides and/or the anhydride of sulfuric acid, advantageously in the presence of catalysts. Suitable catalysts are, for example, the halides of metals, as for example of iron and aluminum, and of nonmetals, as for example phosphorus and sulfur. The halides can also be first formed during the reaction.

The treatment of the initial materials with the chlorides of sulfuric acid or the anhydride of sulfuric acid takes place at elevated temperatures, as for example at 100°C or more, advantageously at about 130° to 200°C. In general normal pressures are used, but higher pressures, as for example about 20 to 250 atmospheres are also suitable.

The process is preferably carried out, for example, by heating the highly chlorinated cyclohexanes with the chlorides of sulfuric acid, such as sulfuryl chloride or chlorsulfonic acid, or the anhydride of sulfuric acid for some time, as for example several hours, advantageously in the presence of a catalyst. The anhydride may be used as such or dissolved in sulfuric acid, for example in the form of oleum.

The amount of the chlorides or the anhydride of sulfuric acid can be different in the case of each individual chlorinated cyclohexane. In general, about 120 to 400 parts, advantageously about 170 parts by weight are used for each 100 parts by weight of the initial material. Inert solvents, such as carbon tetrachloride, chloroform or other halogen hydrocarbons, can be co-employed, but it is then usually necessary to work in closed vessels. The hexachlorbenzene obtained may be readily separated from the reaction mixture after cooling, for example by filtration, and purified from adherent products by washing with solvents such as water.

Process Wastes and Their Control

Air: Air emissions from hexachlorobenzene manufacture have been reported (B-15) to consist of 1.0 kg hydrocarbons and 0.5 kg HCl per metric ton of pesticide produced.

Product Wastes: The MCA recommends incineration methods of disposal (B-3).

Toxicity

The acute oral LD_{50} value for rats is 10,000 mg/kg which is insignificantly toxic.

The acute toxicity of HCB is relatively low, but subchronic or chronic exposure of laboratory animals or humans to HCB results in the development of severe porphyria, especially in females. An ADI was calculated at 0.001 mg/kg/day based on a 10-month feeding study in rats.

A conditional acceptable daily intake of 0.0006 mg/kg/day was derived by the FAO/WHO as the upper limit for residues. The FAO/WHO suggested extreme caution with the compound and indicated that available information is insufficient to establish a firm acceptable intake for HCB.

HCB can be readily determined by electron capture gas chromatography at concentrations as low as 0.0001 ppb.

There are a number of puzzling differences in the highest no-effect and lowest minimal-toxic-effect dosages found for HCB in rats. These differences may be the results of using different rat strains or different HCB formulations in the various studies. They may also result from the use of HCB of uncertain purity. The source of the observed variations should be established. No subchronic- or chronic-toxicity studies have been conducted with HCB in mammalian species other than rats. It is especially important to conduct 2-year feeding experiments and carcinogenicity studies with HCB in two species, because HCB has been found to be extremely toxic on long-term exposure and is on the list of suspected carcinogens (B-22).

Allowable Limits on Exposure and Use

Air: As far as can be determined, the Occupational Safety and Health Administration has not set a standard for occupational exposure of HCB. Russia and Yugoslavia have set the maximum tolerated level of HCB in air at 0.9 mg/m^3 (B-26).

Water: The EPA has set a criterion of 5 ng/l for hexachlorobenzene on the basis of a carcinogenic risk of 1 in 100,000 (B-26).

Product Use: The main agricultural use of HCB is on wheat seed which is intended solely for planting. For this purpose, HCB is mixed with a blue dye, giving the treated wheat a distinct blue color. This coloration is intended as a warning that the seed has been treated with a poison and must not be used for stock or human consumption.

In the wake of widespread HCB contamination of cattle in Louisiana in 1973, and concern over possible contamination of sheep in California, EPA established an interim tolerance of 0.5 ppm (B-25).

HCB has been approved for use as a preemergence fungicide applied to seed grain. The Federal Republic of Germany no longer allows the application of HCB-containing pesticides. The government of Turkey discontinued the use of HCB-treated seed wheat in 1959 after its link to acquired toxic porphyria cutanea tarda was reported. Commercial production of HCB in the United States was discontinued in 1976. The Louisiana State Department of Agriculture has set the tolerated level of HCB in meat fat at 0.3 mg/kg. The NHMRC (Australia) has used this same value for the tolerated level of HCB in cows' milk. WHO has set the tolerated level of HCB in cows' milk at 20 μg/kg in whole milk. The New South Wales Department of Health (Australia) has recommended that the concentration of HCB in eggs must not exceed 0.1 mg/kg. The value of 0.6 μg HCB/kg/day was suggested by FAO/WHO in 1974 as a reasonable upper limit for HCB residues in food for human consumption. The FAO/WHO recommendations for residues in foodstuffs were 0.5 mg/kg in fat for milk and eggs, and 1 mg/kg in fat for meat and poultry.

References

(1) Worthing, C.R., *Pesticide Manual*, 6th ed., p. 298, British Crop Protection Council (1979).
(2) Spencer, E.Y., *Guide to the Chemicals Used in Crop Protection*, 6th ed., p. 298, London, Ontario, Agriculture Canada (January 1973).
(3) Cohen, R.S., U.S. Patent 3,274,269, September 20, 1966, assigned to Dover Chemical Corporation.
(4) Becke, F., and Sperber, H., U.S. Patent 2,792,434, May 14, 1957, assigned to Badische Anilin- and Soda Fabrik AG.

HEXAZINONE

Function: Herbicide (1)(2)

Chemical Name: 3-Cyclohexyl-6-dimethylamino-1-methyl-1,3,5-triazine-2,4-dione

Formula:

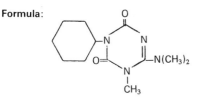

Trade Names: DPX 3674 (DuPont)
Velpar® (DuPont)

Manufacture (2)

(A) Preparation of 1-Methyl-3-cyclohexyl-6-methylthio-s-triazine-2,4(1H,3H)-dione—To a solution of 70 parts of 2-methyl-2-thiopseudourea sulfate in 375 parts of water and 400 parts of toluene at 10°C is added over 1 hour 62.5 parts of cyclohexyl isocyanate. The pH is maintained at 8.5 by addition of 80 parts of 50% aqueous sodium hydroxide over 1 to 1.5 hours. The two-phase system is then reacted with 99 parts of methyl chloroformate and 84 parts of 50% aqueous sodium hydroxide during 1 hour. The temperature is maintained at 25° to 30°C. After the addition is complete the reaction is stirred at 25° to 28°C for an additional 3 hours. The toluene layer is separated and treated to give 76 parts of 1-methyl-3-cyclohexyl-6-methylthio-s-triazine-2,4-(1H,3H)-dione, MP 136° to 138°C.

(B) Preparation of 1-Methyl-3-cyclohexyl-6-dimethylamino-s-triazine-2,4(1H,3H)-dione—A suspension of 300 parts of 1-methyl-3-cyclohexyl-6-methylthio-s-triazine-2,4(1H,3H)-dione in 887 parts of toluene is stirred at 25° to 30°C for 3 hours with 150 parts of dimethylamine. Toluene is distilled from the reaction until a pot temperature of 125°C is attained. The reaction mass is cooled to 50°C at which time 480 parts of hexane is added over 0.66 hours. The slurry at 25°C is filtered to give 282 parts of crystalline 1-methyl-3-cyclohexyl-6-dimethyl-amino-s-triazine-2,4(1H,3H)-dione, MP 110° to 115°C.

Toxicity

The acute oral LD$_{50}$ for rats is 1,690 mg/kg which is slightly toxic.

References

(1) Worthing, C.R., *Pesticide Manual,* 6th ed., p. 299, British Crop Protection Council (1979).
(2) Lin, K., U.S. Patent 3,902,887, September 2, 1975, assigned to E.I. DuPont de Nemours & Company.

HYDROCYANIC ACID

Function: Insecticidal fumigant (1)(2)

Chemical Name: Hydrocyanic acid

Formula: HCN

Trade Names: None

Manufacture

Hydrogen cyanide may be made by various routes.

A. By dehydration of formamide (3)

$$HCONH_2 \rightarrow HCN + H_2O$$

B. From methane, ammonia and air (4)

$$2CH_4 + 2NH_3 + 3O_2 \rightarrow 2HCN + 6H_2O$$

C. From propane and ammonia (5)

$$C_3H_8 + 3NH_3 \rightarrow 3HCN + 7H_2$$

Toxicity

The lowest published acute oral lethal dose, LD_{LO} for rats is 10 mg/kg (B-5).

Allowable Limits on Exposure and Use

Air: The threshold limit value for HCN in air is 11 mg/m^3 (10 ppm) as of 1979. A tentative short term exposure limit is 16 mg/m^3 (15 ppm). It is further noted that cutaneous absorption of HCN should be prevented so that the TLV is not invalidated.

Product Use: The tolerances set by the EPA for HCN in or on raw agricultural commodities are as follows:

	40 *CFR* Reference	Parts per Million
Allspice (post-h)	180.130	250.0
Almonds (post-h)	180.130	25.0
Anise (post-h)	180.130	250.0
Barley (post-h)	180.130	75.0
Basil (post-h)	180.130	250.0
Bay (post-h)	180.130	250.0
Beans, dry (post-h)	180.130	25.0
Buckwheat (post-h)	180.130	75.0
Caraway (post-h)	180.130	250.0
Cashews (post-h)	180.130	25.0
Cassia (post-h)	180.130	250.0
Celery, seed (post-h)	180.130	250.0
Chili (post-h)	180.130	250.0
Cinnamon (post-h)	180.130	250.0
Citrus fruits (post-h)	180.130	50.0
Clove (post-h)	180.130	250.0
Cocoa beans (post-h)	180.130	25.0
Coriander (post-h)	180.130	250.0
Corn (post-h)	180.130	75.0
Cumin (post-h)	180.130	250.0
Dill (post-h)	180.130	250.0
Ginger (post-h)	180.130	250.0
Mace (post-h)	180.130	250.0
Marjoram (post-h)	180.130	250.0
Milo (grain sorghum) (post-h)	180.130	75.0
Nutmeg (post-h)	180.130	250.0
Oats (post-h)	180.130	75.0

(continued)

	40 *CFR* Reference	Parts per Million
Oregano (post-h)	180.130	250.0
Paprika (post-h)	180.130	250.0
Peanuts (post-h)	180.130	25.0
Peas, dried (post-h)	180.130	25.0
Pecans (post-h)	180.130	25.0
Pepper, black (post-h)	180.130	250.0
Pepper, red (post-h)	180.130	250.0
Pepper, white (post-h)	180.130	250.0
Poppy (post-h)	180.130	250.0
Rice (post-h)	180.130	75.0
Rosemary (post-h)	180.130	250.0
Rye (post-h)	180.130	75.0
Sage (post-h)	180.130	250.0
Savory (post-h)	180.130	250.0
Sesame (post-h)	180.130	25.0
Thyme (post-h)	180.130	250.0
Tumeric (post-h)	180.130	250.0
Walnuts (post-h)	180.130	25.0
Wheat (post-h)	180.130	75.0

The tolerances set by the EPA for hydrocyanic acid in food are as follows (the *CFR* Reference is to Title 21):

	CFR Reference	Parts per Million
Bacon, uncooked	193.240	50.0
Cereal flours	193.240	125.0
Cereals, cooked	193.240	90.0
Cocoa	193.240	200.0
Ham, uncooked	193.240	50.0
Sausage, uncooked	193.240	50.0

References

(1) Worthing, C.R., *Pesticide Manual,* 6th ed., p. 300, British Crop Protection Council (1979).
(2) Spencer, E.Y., *Guide to the Chemicals Used in Crop Protection,* 6th ed., p. 300, London, Ontario, Agriculture Canada (January 1973).
(3) Asendorf, E. et al, U.S. Patent 2,904,400, September 15, 1959, assigned to Degussa.
(4) Jenks, W.R. et al, U.S. Patent 3,104,945, September 24, 1963, assigned to DuPont.
(5) Kennedy, D.J. et al, U.S. Patent 3,157,468, November 17, 1964, assigned to Shawinigan Chemicals, Ltd.

2-HYDROXYETHYL OCTYL SULFIDE

Function: Insect repellent (1)(2)(3)

Chemical Name: 2-(octylthio) ethanol

Formula: $CH_3(CH_2)_7SCH_2CH_2OH$

Trade Name: MGK Repellent 874 (McLaughlin Gormley King Co.)

Manufacture (3)

To a three-necked flask (500 ml) equipped with a stirrer and a dry ice-acetone cooled reflux condenser were charged the following ingredients:

n-Octyl mercaptan	269 g
Ethylene oxide	85 g
20% by weight solution of NaOH in MeOH (catalyst)	6 cc

The mercaptan-caustic mixture in the flask was first heated to about 220°F after which the ethylene oxide was added dropwise to the mixture. The reaction which resulted was quite vigorous. After the reaction was complete, the contents of the flask were neutralized with concentrated hydrochloric acid (38%) until a phenolphthalein end point was reached. The purified product which resulted from the reaction amounted to 288.2 grams and had a BP of 98°C at 0.1 mm.

Process Wastes and Their Control

Product Wastes: This material could be disposed of by efficient incineration, if acceptable, by open burning (B-3).

Toxicity

The acute oral LD_{50} for rats is 8,500 mg/kg which is insignificantly toxic.

References

(1) Worthing, C.R., *Pesticide Manual,* 6th ed., p. 301, British Crop Protection Council (1979).
(2) Spencer, E.Y., *Guide to the Chemicals Used in Crop Protection,* 6th ed., p. 301, London, Ontario, Agriculture Canada (January 1973).
(3) Goodhue, L.D. and Cantrel, K.E., U.S. Patent 2,863,799, December 9, 1958, assigned to Phillips Petroleum Co.

I

4-(INDOL-3-YL)BUTYRIC ACID

Function: Plant growth regulator (1)

Chemical Name: 4-(3-indolyl)butyric acid

Formula:

Trade Names: Hormodin®
In Seradix® (May & Baker, Ltd.)
In Rootone® (Amchem Products, Inc.)

Manufacture

The following is a specific example (2) of the conduct of the process. There were charged to a 1 liter, stainless steel rocker autoclave 117 grams (1.0 mol) of indole, 90 grams (1.05 mols) of gamma-butyrolactone and 70 grams (1.06 mols) of potassium hydroxide. The resulting mixture was heated at 280° to 290°C for 19 hours. After cooling to ambient temperature, the light brown-colored solid potassium 4-(3-indolyl)butyrate that had formed was removed from the autoclave and dissolved in 1 liter of water. The aqueous solution was extracted with two 250 ml portions of isopropyl ether. The ether extracts were combined and the ether was evaporated. A residue containing 24 grams of indole was recovered.

The aqueous, ether-washed solution was acidified with concentrated hydrochloric acid to a pH of 2 and contacted with 250 ml of isopropyl ether, whereupon two liquid phases formed, a top, ether-rich phase and a lower aqueous phase, which were separated. The aqueous phase was extracted with 250 ml of isopropyl ether, after which the aqueous phase was discarded. The ether extract was combined with the ether extract obtained previously and washed with three 250 ml portions of water. The ether was evaporated and 188 grams of 4-(3-indolyl)butyric acid, representing a yield of about 92% was recovered. The structure of the 4-(3-indolyl)butyric acid was confirmed by its infrared spectrum. The efficiency of this re-action, based upon indole was 100%.

Toxicity

The acute oral LD$_{50}$ value for mice is 100 mg/kg (B-5).

References

(1) Worthing, C.R., *Pesticide Manual,* 6th ed., p. 303, British Crop Protection Council (1979).
(2) Fritz, H.E., U.S. Patent 3,051,723, August 28, 1962, assigned to Union Carbide Corp.

IODOFENPHOS

Function: Insecticide (1)(2)

Chemical Name: O-2,5-dichloro-4-iodophenyl-O,O-dimethyl phosphorothioate

Formula:

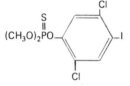

Trade Names: C-9491 (Ciba-Geigy)
Nuvanol N® (Ciba-Geigy)
Alfacron®

Manufacture (3)

24 parts of 2,5-dichloro-4-iodophenol are dissolved in 100 parts of water by means of 3.4 parts of sodium hydroxide. The water is removed in vacuo and the dried sodium salts dis-solved in 100 parts by volume of toluene.

The solution is warmed to 70° to 80°C and 15 parts of dimethyl thiochlorophosphate are added dropwise. The mixture is stirred for a further 8 hours at 70° to 80°C. After cooling the salt is filtered off, and the solution successively washed with water and 1N ice-cooled caustic soda. The solution is dried over sodium sulfate and the solvent is evaporated off in vacuo, at 50°C bath temperature. The residue obtained consists of 24 parts of the condensation prod-uct in the form of a pale yellow oil.

This solidifies, after standing for a while, to crystals melting at 72° to 73°C.

Toxicity

The acute oral LD$_{50}$ for rats is 2,100 mg/kg which is slightly toxic.

References

(1) Worthing, C.R., *Pesticide Manual,* 6th ed., p. 304, British Crop Protection Council (1979).
(2) Spencer, E.Y., *Guide to the Chemicals Used in Crop Protection,* 6th ed., p. 303, London, Ontario, Agriculture Canada (January 1973).
(3) Ciba, Ltd., British Patent 1,057,609, February 1, 1967.

IOXYNIL

Function: Herbicide (1)(2)

Chemical Name: 4-hydroxy-3,5-diiodobenzonitrile

Formula:

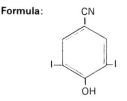

Trade Names: ACP 63-303 (Amchem Products)
Certrol® (Amchem Products)
MB-8873 (May and Baker, Ltd.)
Actril® (May and Baker, Ltd.)
Toxynil®
Acrilawn®

Manufacture (3)

Ioxynil may be made starting with 4-hydroxybenzaldehyde which is iodinated and then converted to the nitrile, or is converted to the nitrile and then iodinated with iodine monochloride.

Process Wastes and Their Control

Product Wastes: Ioxynil, a white crystalline herbicide, is chemically stable, but produces little or no residual activity in soil. It contains substantial quantities of iodine which presents the same incineration hazards as chlorinated pesticides (B-3).

Toxicity

The acute oral LD_{50} for rats is 110 mg/kg which is moderately toxic.

References

(1) Worthing, C.R., *Pesticide Manual,* 6th ed., p. 305, British Crop Protection Council (1979).
(2) Spencer, E.Y., *Guide to the Chemicals Used in Crop Protection,* 6th ed., p. 304, London, Ontario, Agriculture Canada (January 1973).
(3) Auwers, K. and Reis, J., *Ber. Deut. Chem. Ges.* 29, 2355 (1896).

IOXYNIL OCTANOATE

Function: Herbicide (1)(2)

Chemical Name: 4-cyano-2,6-diiodophenyl octanoate

Formula:

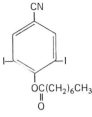

Trade Names: MB 11,461 (May and Baker, Ltd.)
Totril® (May and Baker, Ltd.)

Manufacture (3)

Ioxynil octanoate may be made by the reaction of ioxynil (which see) with n-octanoyl chloride in the presence of dry pyridine. Commencing with 3,5-dichloro-4-hydroxybenzonitrile (19 g), n-octanoyl chloride (18 g) (prepared according to Aschan, *Ber.,* 1898, 31, 2348) and dry pyridine (190 cc), a crude product (29 g) is obtained which is twice recrystallized from ethanol. The resulting solid is washed with saturated aqueous sodium bicarbonate solution (3 × 100 cc) and water to give 3,5-dichloro-4-n-octanoyloxy-benzonitrile (18 g) as a white solid, MP 51° to 53°C.

Toxicity

The acute oral LD_{50} value for rats is 390 mg/kg which is moderately toxic.

References

(1) Worthing, C.R., *Pesticide Manual,* 6th ed., p. 306, British Crop Protection Council (1979).
(2) Spencer, E.Y., *Guide to the Chemicals Used in Crop Protection,* 6th ed., p. 305, London, Ontario, Agriculture Canada (January 1973).
(3) Heywood, B.J. and Leeds, W.G., British Patent 1,067,033, April 26, 1967, assigned to May and Baker, Ltd.

IPRODIONE

Function: Fungicide (1)(2)

Chemical Name: 3-(3,5-Dichloropheny)-N-(1-methylethyl)-2,4-dioxo-1-imidazoline-carboxamide

Formula:

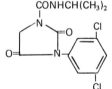

Trade Names: 26 019 RP (Rhone-Poulenc)
ROP 500F (Rhone-Poulenc)
NRC 910 (Rhone-Poulenc)
LEA 2043 (Rhone-Poulenc)
FA 2071 (Rhone-Poulenc)
Rovral® (Rhone-Poulenc)

Manufacture (2)

3-(3,5-Dichlorophenyl)hydantoin, MP 199°C, employed as starting material can be prepared according to the method described by Dhar, *J. Soc. Ind. Research,* 20c, 145 (1961).

Isopropyl isocyanate (4.6 g) and triethylamine (5.5 g) are added to a solution of 3-(3,5-dichlorophenyl)-hydantoin (11 g) in acetone (150 cc). After 30 minutes heating under reflux followed by cooling, the acetone is evaporated under reduced pressure. The residue obtained is washed with petroleum ether (BP 50° to 70°C; 250 cc) and recrystallized from diisopropyl ether to yield 1-isopropylcarbamoyl-3-(3,5-dichlorophenyl)hydantoin (11 g) melting at 136°C.

Toxicity

The acute oral LD_{50} value for rats is 3,500 mg/kg which is slightly toxic.

References

(1) Worthing, C.R., *Pesticide Manual,* 6th ed., p. 307, British Crop Protection Council (1979).
(2) Sauli, M., U.S. Patent 3,755,350, August 28, 1973, assigned to Rhone-Poulenc SA.

ISOCARBAMID

Function: Herbicide (1)(2)

Chemical Name: N-(2-methylpropyl)-2-oxo-1-imidazolinecarboxamide

Formula:

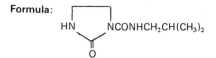

$$HN \quad NCONHCH_2CH(CH_3)_2$$
$$O$$

Trade Names: MNF 0166 (Bayer)
Merpelan AZ® (Bayer) (63% Isocarbamid and 13% Lenacil)

Manufacture (2)

148.5 g imidazolidin-2-one-1-carbonyl chloride are stirred with 100 ml of water. To this mixture there are added dropwise from one dropping funnel 77 g isobutyl amine and, from a second dropping funnel, 100 ml sodium hydroxide solution with a content of 40 g NaOH in such a manner that the pH value does not rise above 10. The temperature is kept to 35° to 40°C by cooling. When the entire amount of the amine has been added dropwise, the pH value is raised to 12 by the remainder of the sodium hydroxide solution. Stirring until cold is effected; the product is filtered off with suction and washed with a little water. The yield of imidazolidin-2-one-1-carboxylic acid isobutyl amide is 159 g (86% of the theory). The MP is 95° to 96°C.

Toxicity

The acute oral LD_{50} for rats is over 2,500 mg/kg which is slightly toxic.

References

(1) Worthing, C.R., *Pesticide Manual,* 6th ed., p. 308, British Crop Protection Council (1979).
(2) Munz, F., Hack, H. and Eue, L., U.S. Patent 3,875,180, April 1, 1975, assigned to Bayer AG.

ISOFENPHOS

Function: Insecticide and acaricide (1)(2)(3)

Chemical Name: 1-Methylethyl 2-[[ethoxy[(1-methylethyl)amino]phosphinothioyl] oxy]benzoate

Formula:

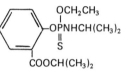

Trade Names: BAY SRA 12869 (Bayer)
Oftanol® (Bayer)

Manufacture (2)

To 180 g ethyl-thiono-phosphoric acid ester dichloride there are added at 40° to 50°C, with stirring, a mixture of 136 g salicylic acid isopropyl ester, 220 g water and 24 g sodium hydroxide; the reaction mixture is then stirred for 4 hours and it is then taken up in 400 ml benzene. The benzene solution is washed twice with water, dried over sodium sulfate and, finally, the product is fractionally distilled. 155 g O-ethyl-O-[(2-carbo-isopropoxy)phenyl]-thiono-phosphoric acid diester monochloride of BP 120°C/0.01 mm Hg are obtained.

To a solution of 162 g (0.5 mol) O-ethyl-O-[(2-carboisopropoxy)phenyl]-thiono-phosphoric acid diester monochloride (prepared as above) in 600 ml benzene there are added, at 20° to 40°C, 75 g isopropylamine dissolved in 75 ml benzene. After subsequent stirring of the reaction mixture for one hour, it is extracted with water, the benzene phase is separated, dried, and evaporated, and the residue is fractionally distilled. The N-isopropylamido-thiono-phosphoric acid O-ethyl-O-[(2-carbo-isopropoxy)phenyl] ester boils at 120°C under a pressure of 0.01 mm Hg. The yield is 140 g (82% of the theory).

Toxicity

The acute oral LD_{50} for rats is 28-38 mg/kg which is highly toxic.

References

(1) Worthing, C.R., *Pesticide Manual,* 6th ed., p. 309, British Crop Protection Council (1979).
(2) Schrader, G., Hammann, I. and Stendel, W., U.S. Patent 3,621,082, November 16, 1971, assigned to Farbenfabriken Bayer AG.
(3) Schrader, G., Hammann, I. and Stendel, W., U.S. Patent 3,755,572, August 28, 1973, assigned to Bayer AG.

ISOPROPALIN

Function: Herbicide (1)(2)

Chemical Name: 4-(1-methylethyl)-2,6-dinitro-N,N-dipropylbenzeneamine

Formula:

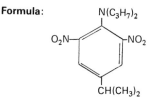

Trade Names: EL 179 (Eli Lilly & Co.)
Paarlan® (Eli Lilly & Co.)

Manufacture

Isopropalin may be made by the reaction of dipropylamine with 1-chloro-4-isopropyl-2,6-dinitrobenzene.

Toxicity

The acute oral LD_{50} value for rats is more than 5,000 mg/kg which is insignificantly toxic.

Allowable Limits on Exposure and Use

Product Use: The tolerances set by the EPA for isopropalin in or on raw agricultural commodities are as follows:

	40 *CFR* Reference	Parts per Million
Peppers	180.313	0.05 N
Tomatoes	180.313	0.05 N

References

(1) Worthing, C.R., *Pesticide Manual,* 6th ed., p. 312, British Crop Protection Council (1979).
(2) Soper, Q.F., U.S. Patent 3,257,190, June 21, 1966, assigned to Eli Lilly & Co.

2-ISOVALERYLINDANE-1,3-DIONE

Function: Insecticide (1)(2)(3)

Chemical Name: 2-(3-methyl-1-oxobutyl)-1H-indene-1,3(2H)dione

Formula:

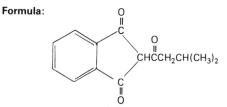

Trade Name: Valone® (Kilgore Chemical Co.)

Manufacture

This material may be prepared by the Claisen condensation reaction between methyl iso-amyl ketone and diethyl phthalate.

Toxicity

The lowest published lethal dose, LD_{LO} for rats is 250 mg/kg (B-5).

References

(1) Martin, H. and Worthing, C.R., *Pesticide Manual,* 5th ed., p. 318, British Crop Protection Council (1979).
(2) Spencer, E.Y., *Guide to the Chemicals Used in Crop Protection,* 6th ed., p. 308, London, Ontario, Agriculture Canada (January 1973).
(3) Kilgore, L.B., U.S. Patent 2,228,170, January 7, 1941.

K

KASUGAMYCIN

Function: Fungicide (1)(2)

Chemical Name: D-3-O-(2-amino-4-[(1-carboxyiminomethyl)amino]-2,3,4,6-tetradeoxy-α-D-arabino-hexapyranosyl)-D-chiro-inositol

Formula:

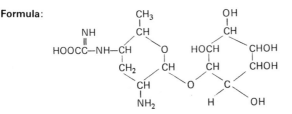

Trade Name: Kasumin® (Hokko Chemical Industry Co.)

Manufacture (3)

Kasugamycin is made by the fermentation of *S. kasugaensis* and is extracted from the fermentation broth with strongly acid ion-exchange resins.

Toxicity

The acute oral LD_{50} value for rats is 22,000 mg/kg which is insignificantly toxic.

References

(1) Worthing, C.R., *Pesticide Manual,* 6th ed., p. 315, British Crop Protection Council (1979).
(2) Spencer, E.Y., *Guide to the Chemicals Used in Crop Protection,* 6th ed., p. 311, London, Ontario, Agriculture Canada (January 1973).
(3) Zaidan Hojin Biseibutsu Kagaku Kenkyukai, British Patent 1,094,566, December 13, 1967.

L

LEAD ARSENATE

Function: Insecticide (1)(2)

Chemical Name: Lead arsenate

Formula: PbHAsO$_4$

Trade Name: Gypsine

Manufacture (3)

Lead arsenate is manufactured by a batch process, similar to that previously discussed for calcium arsenate (which see). Lead oxide is mixed with arsenic acid on a batch basis. The precipitate slurry is drum dried and packaged. Water vapors are collected and vented, and all spills are caught and returned to the process. There are thus no wastewaters associated with lead arsenate production. However, it has been reported that some producers of lead arsenate filter the precipitate slurry prior to drying, to remove undesirable (soluble) reaction side products. For those producers, this filtrate liquid may require treatment prior to discharge, for removal of lead and arsenate.

Process Wastes and Their Control

Air: Air emissions from lead arsenate manufacture have been reported (B-15) to consist of 3×10^{-6} kg arsenic trioxide per metric ton of pesticide produced.

Toxicity

The acute oral LD$_{50}$ value for rats is 100 mg/kg which is moderately toxic.

Allowable Limits on Exposure and Use

Air: The threshold limit value for lead arsenate in air (as Pb) is 0.15 mg/m^3 as of 1979. A tentative short term exposure limit is 0.45 mg/m^3 (B-23).

Product Use: A rebuttable presumption against registration was issued on October 18, 1978 by EPA on the basis of oncogenicity, teratogenicity and mutagenicity.

The tolerances set by the EPA for lead arsenate in or on raw agricultural commodities are as follows:

	40 *CFR* Reference	Parts per Million
Apples	180.194	7.0
Apricots	180.194	7.0
Asparagus	180.194	7.0
Avocados	180.194	7.0
Blackberries	180.194	7.0
Blueberries (huckleberries)	180.194	7.0
Boysenberries	180.194	7.0
Celery	180.194	7.0
Cherries	180.194	7.0
Citrus fruits	180.194	1.0
Cranberries	180.194	7.0
Currants	180.194	7.0
Dewberries	180.194	7.0
Eggplant	180.194	7.0
Gooseberries	180.194	7.0
Grapes	180.194	7.0
Loganberries	180.194	7.0
Mangoes	180.194	7.0
Nectarines	180.194	7.0
Peaches	180.194	7.0
Pears	180.194	7.0
Peppers	180.194	7.0
Plums (fresh prunes)	180.194	7.0
Quinces	180.194	7.0
Raspberries	180.194	7.0
Strawberries	180.194	7.0
Tomatoes	180.194	7.0
Youngberries	180.194	7.0

References

(1) Worthing, C.R., *Pesticide Manual,* 6th ed., p. 316, British Crop Protection Council (1979).
(2) Spencer, E.Y., *Guide to the Chemicals Used in Crop Protection,* 6th ed., p. 312, London, Ontario, Agriculture Canada (January 1973).
(3) Patterson, J.W., *State of the Art for the Inorganic Chemicals Industry: Inorganic Pesticides,* Report EPA-600/2-74-009a, Washington, DC, Environ. Protection Agency (March 1975).

LENACIL

Function: Herbicide (1)(2)

Chemical Name: 3-cyclohexyl-6,7-dihydro-1H-cyclopentapyrimidine-2,4(3H,5H)dione

Formula:

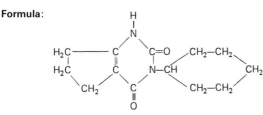

Trade Names: DuPont 634 (DuPont)
Venzar® (DuPont)

Manufacture (2)

A mixture of 343 parts by weight of ethyl-2-cyclopentanone-1-carboxylate, 284 parts by weight of cyclohexylurea, 10 parts by weight of p-toluene sulfonic acid, and, 1,750 parts by weight of xylene is stirred at reflux for 6 hours. During this time the water given off by the reaction is trapped out. The solvent is stripped from the resulting solution at reduced pressure. The residue is then dissolved in 793 parts by weight of absolute ethyl alcohol.

To this solution is added a mixture of 120 parts by weight of sodium methoxide in 400 parts by weight of absolute ethyl alcohol. The mixture is then refluxed for 10 minutes. The solvent is stripped from this mixture and the resulting solid is dissolved in 3,000 parts by weight of water.

This solution is cooled, acidified with excess hydrochloric acid and the solid is filtered off. Recrystallization of this solid from dimethylformamide gives light gray crystals of the desired compound, melting at 310° to 313°C.

Process Wastes and Their Control

Product Wastes: This compound is decomposed by the action of strong alkali (B-3).

Toxicity

The acute oral LD_{50} for rats is more than 11,000 which is insignificantly toxic.

References

(1) Worthing, C.R., *Pesticide Manual,* 6th ed., p. 317, British Crop Protection Council (1979).
(2) Soboszenski, E.J., U.S. Patent 3,235,360, February 15, 1966, assigned to DuPont.

LEPTOPHOS

Function: Insecticide (1)(2)(3)

Chemical Name: O-4-bromo-2,5-dichlorophenyl O-methyl phenylphosphonothioate

Formula:

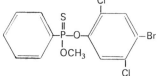

Trade Names: VCS-506 (Velsicol Chemical Corp.)
Abar® (Velsicol Chemical Corp.)
Phosvel® (Velsicol Chemical Corp.)

Manufacture (3)

2,5-Dichloro-4-bromophenol (7.5 g; 0.025 mol) was dissolved in acetone (25 ml) and placed in a three-neck, round-bottom flask equipped with a mechanical stirrer, internal thermometer and reflux condenser. A solution of sodium hydroxide (1 g; 0.025 mol) in water (5 ml) was added to the flask. The contents were stirred, and a solution of O-methyl phenylthiophosphonyl chloride (5.2 g; 0.025 mol) in acetone (25 ml) was slowly added to the flask. The reaction mixture was stirred and heated at reflux for 15 minutes and then cooled. The reaction mixture was filtered and the filtrate distilled in vacuo to remove the acetone.

The residue was extracted with diethyl ether and the extract washed with a 5% aqueous solution of sodium hydroxide (100 ml) and then twice with water. The ether extract was dried over magnesium sulfate, filtered, and the filtrate heated under reduced pressure to remove diethyl ether and recover O-methyl O-2,5-dichloro-4-bromophenyl phenylthiophosphonate as a light yellow liquid having a refractive index at 22°C of 1.6385 which solidifies on standing.

Toxicity

The acute oral LD_{50} value for rats is about 50 mg/kg which is moderately to highly toxic.

Allowable Limits on Exposure and Use

Product Use: The tolerances set by the EPA for leptophos in or on raw agricultural commodities are as follows:

	40 *CFR* Reference	Parts per Million
Beets, sugar, roots	180.345	0.1
Beets, sugar, tops	180.345	1.0
Cattle, fat	180.345	0.05
Cattle, mbyp	180.345	0.05
Cattle, meat	180.345	0.05
Goats, fat	180.345	0.05
Goats, mbyp	180.345	0.05
Goats, meat	180.345	0.05
Hogs, fat	180.345	0.05
Hogs, mbyp	180.345	0.05
Hogs, meat	180.345	0.05
Horses, fat	180.345	0.05
Horses, mbyp	180.345	0.05
Horses, meat	180.345	0.05
Lettuce	180.345	10.0
Sheep, fat	180.345	0.05
Sheep, mbyp	180.345	0.05
Sheep, meat	180.345	0.05
Tomatoes	180.345	2.0

References

(1) Worthing, C.R., *Pesticide Manual,* 6th ed., p. 318, British Crop Protection Council (1979).
(2) Spencer, E.Y., *Guide to the Chemicals Used in Crop Protection,* 6th ed., p. 313, London, Ontario, Agriculture Canada (January 1973).
(3) Richter, S.B., U.S. Patent 3,459,836, August 5, 1969, assigned to Velsicol Chemical Corp.

LINDANE

Function: Insecticide (1)(2)

Chemical Name: $1\alpha,2\alpha,3\beta,4\alpha,5\alpha,6\beta$-hexachlorocyclohexane

Formula:

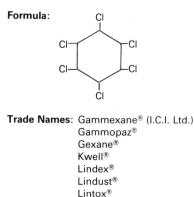

Trade Names: Gammexane® (I.C.I. Ltd.)
Gammopaz®
Gexane®
Kwell®
Lindex®
Lindust®
Lintox®

Manufacture (3)(4)(5)

The production of lindane is not a chemical synthesis operation but a physical separation process. As noted earlier in this volume under BHC Manufacture, it is possible to influence the gamma isomer content of benzene hexachloride to an extent during the synthesis process. Basically, however, one is faced with the problem of separating a 99%-plus purity gamma isomer from a crude product containing perhaps 12 to 15% of the gamma isomer. The separation and concentration process is done by a carefully controlled solvent extraction and crystallization process. A schematic representation of one such process is shown in Figure 32 as described by R.D. Donaldson et al (3). Another description of hexachlorocyclohexane isomer separation is given by R.H. Kimball (4).

Process Wastes and Their Control

Air: Air emissions from lindane manufacture have been reported (B-15) to consist of 1.0 kg hydrocarbons and 0.5 kg HCl per metric ton of pesticide produced.

Product Wastes: The BHC isomers are stable to light, air, and strong acids, but undergo dehydrochlorination upon prolonged heating (e.g., 110°C for 24 hours) or in strongly alkaline solution at room temperature (excepting the β-isomer which constitutes about 6% of BHC) e.g., lindane was 98.5% removed in 6.5 hours at pH 11.5. Studies with BHC in alcoholic alkali (e.g., 1.5N) indicate the reaction is complete in 1 hour and that the product is a mixture of 1,2,4-trichlorobenzene (65 to 85%) with smaller amounts of the 1,2,3-isomer (5 to 18%) and the 1,3,5-isomer (0 to 15%).

Oxidation of lindane was ineffective with Cl_2 or H_2O_2, partially effective with $KMnO_4$, and effective with ozone but the products were not identified.

Aeration of aqueous lindane solutions caused volatilization. Reduction with zinc dust in acid medium converted BHC to benzene.

The MCA recommends incineration methods for the disposal of lindane. A process has been patented for the destructive pyrolysis of benzene hexachloride at 400° to 500°C with a catalyst mixture which contains 5 to 10% of either cupric chloride, ferric chloride, zinc chloride, or aluminum chloride on activated carbon (B-3).

Toxicity

The acute oral LD_{50} value for rats is about 90 mg/kg which is moderately toxic.

Figure 32: Extraction and Separation Process for the Concentration
of the Pure Gamma Isomer from a Crude BHC Feed Material

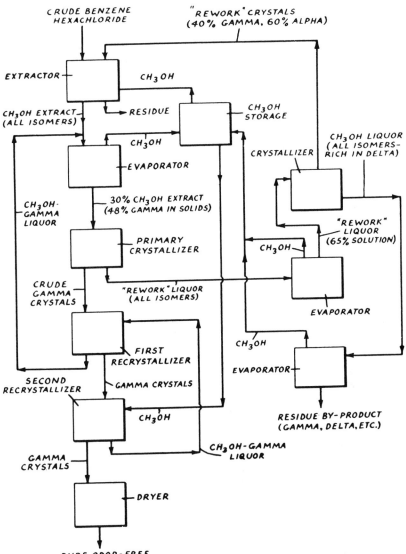

Allowable Limits on Exposure and Use

Air: A threshold limit value for lindane in air is 0.5 mg/m^3 as of 1979. A tentative short term exposure limit is 1.5 mg/m^3. It is further noted that cutaneous absorption of lindane should be prevented so that the TLV is not invalidated.

Water: In water, EPA has set (B-12) the following criteria for lindane:

> 4.0 μg/l for domestic water supply
> 0.01 μg/l for freshwater aquatic life
> 0.004 μg/l for marine aquatic life

Subsequently, EPA has suggested (B-26) for lindane a criterion to protect freshwater aquatic life of 0.21 μg/l as a 24-hour average; the concentration should not exceed 2.9 μg/l at any time.

For saltwater aquatic life, there are insufficient data to estimate a criterion.

For the maximum protection of human health, the concentration of lindane in water should be zero. The concentration for a 1 in 100,000 cancer risk is 54 ng/l (B-26).

Product Use: A rebuttable presumption against registration was issued on February 17, 1977 by EPA on the basis of oncogenicity, fetotoxicity, reproductive effects and acute toxicity.

In an action dated April 28, 1969 (B-17), EPA cancelled the use of lindane in vaporizers.

The tolerances set by the EPA for lindane in or on raw agricultural commodities are as follows:

	40 *CFR* Reference	Parts per Million
Apples	180.133	1.0
Apricots	180.133	1.0
Asparagus	180.133	1.0
Avocados	180.133	1.0
Broccoli	180.133	1.0
Brussels sprouts	180.133	1.0
Cabbage	180.133	1.0
Cattle, fat of meat	180.133	7.0
Cauliflower	180.133	1.0
Celery	180.133	1.0
Cherries	180.133	1.0
Collards	180.133	1.0
Cucumbers	180.133	3.0
Eggplant	180.133	1.0
Goats, fat of meat	180.133	7.0
Grapes	180.133	1.0
Guavas	180.133	1.0
Hogs, fat of meat	180.133	4.0
Horses, fat of meat	180.133	7.0
Kale	180.133	1.0
Kohlrabi	180.133	1.0
Lettuce	180.133	3.0
Mangoes	180.133	1.0
Melons	180.133	3.0
Mushrooms	180.133	3.0
Mustard greens	180.133	1.0
Nectarines	180.133	1.0

(continued)

	40 *CFR* Reference	Parts per Million
Okra	180.133	1.0
Onions, dry bulb only	180.133	1.0
Peaches	180.133	1.0
Pears	180.133	1.0
Pecans	180.133	0.01 N
Peppers	180.133	1.0
Pineapples	180.133	1.0
Plums (fresh prunes)	180.133	1.0
Pumpkins	180.133	3.0
Quinces	180.133	1.0
Sheep, fat of meat	180.133	7.0
Spinach	180.133	1.0
Squash	180.133	3.0
Squash, summer	180.133	3.0
Strawberries	180.133	1.0
Swiss chard	180.133	1.0
Tomatoes	180.133	3.0

References

(1) Worthing, C.R., *Pesticide Manual*, 6th ed., p. 290, British Crop Protection Council (1979).
(2) Spencer, E.Y., *Guide to the Chemicals Used in Crop Protection,* 6th ed., p. 315, London, Ontario, Agriculture Canada (January 1973).
(3) Donaldson, R.D. et al, U.S. Patent 2,767,223, October 16, 1956, assigned to Allied Chemical and Dye Corp.
(4) Kimball, R.H., U.S. Patent 2,767,224, October 16, 1956, assigned to Hooker Electrochemical Co.
(5) Hay, J.K. and Webster, K.C., U.S. Patent 2,502,258, March 28, 1950, assigned to Imperial Chemical Industries, Ltd.

LINURON

Function: Herbicide (1)(2)(3)(4)

Chemical Name: N'-(3,4-dichlorophenyl)-N-methoxy-N-methylurea

Formula:

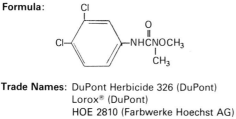

Trade Names: DuPont Herbicide 326 (DuPont)
Lorox® (DuPont)
HOE 2810 (Farbwerke Hoechst AG)
Afalon® (Farbwerke Hoechst AG)

Manufacture (3)(4)

For the preparation of linuron, there are used 10.2 g of O,N-dimethyl-hydroxylamine in 30 ml of benzene and 31.3 g of 3,4-dichlorophenylisocyanate in 100 ml of benzene. The strong reaction heat is reduced by cooling with ice. The yield of the compound N-(3,4-dichlorophenyl)-

N'-(methoxy)-N'-(methyl)urea obtained in the form of crystals amounts to 32 g (77% of the theoretical yield), MP is 92° to 93°C.

Process Wastes and Their Control

Air: Air emissions from linuron manufacture have been reported (B-15) to consist of 0.5 kg hydrocarbons and 0.5 kg chlorine per metric ton of pesticide produced.

Product Wastes: Linuron can be hydrolyzed in alkaline and especially in acidic media. However, this procedure is not recommended due to generation of more toxic products (B-3).

Toxicity

The acute oral LD_{50} value for rats is 4,000 mg/kg which is very slightly toxic.

Allowable Limits on Exposure and Use

Product Use: The tolerances set by the EPA for linuron in or on raw agricultural commodities are as follows:

	40 *CFR* Reference	Parts per Million
Asparagus	180.184	3.0
Barley, forage	180.184	0.5
Barley, grain	180.184	0.25
Barley, hay	180.184	0.5
Barley, straw	180.184	0.5
Carrots	180.184	1.0
Cattle, fat	180.184	1.0
Cattle, mbyp	180.184	1.0
Cattle, meat	180.184	1.0
Celery	180.184	0.5
Corn, field, fodder	180.184	1.0
Corn, field, forage	180.184	1.0
Corn, fresh (inc. sweet) (k+cwhr)	180.184	0.25
Corn, grain	180.184	0.25
Corn, pop, fodder	180.184	1.0
Corn, pop, forage	180.184	1.0
Corn, sweet, fodder	180.184	1.0
Corn, sweet, forage	180.184	1.0
Cotton, seed	180.184	0.25
Goats, fat	180.184	1.0
Goats, mbyp	180.184	1.0
Goats, meat	180.184	1.0
Hogs, fat	180.184	1.0
Hogs, mbyp	180.184	1.0
Hogs, meat	180.184	1.0
Horses, fat	180.184	1.0
Horses, mbyp	180.184	1.0
Horses, meat	180.184	1.0
Oats, forage	180.184	0.5
Oats, grain	180.184	0.25
Oats, hay	180.184	0.5
Oats, straw	180.184	0.5
Parsnips, tops	180.184	0.5
Parsnips, with tops	180.184	0.5
Parsnips, without tops	180.184	0.5
Potatoes	180.184	1.0
Rye, forage	180.184	0.5

(continued)

	40 *CFR* Reference	Parts per Million
Rye, grain	180.184	0.25
Rye, hay	180.184	0.5
Rye, straw	180.184	0.5
Sheep, fat	180.184	1.0
Sheep, mbyp	180.184	1.0
Sheep, meat	180.184	1.0
Sorghum, fodder	180.184	1.0
Sorghum, forage	180.184	1.0
Sorghum, grain (milo)	180.184	0.25
Soybeans, dry	180.184	1.0
Soybeans, forage	180.184	1.0
Soybeans, hay	180.184	1.0
Soybeans, succulent	180.184	1.0
Wheat, forage	180.184	0.5
Wheat, grain	180.184	0.25
Wheat, hay	180.184	0.5
Wheat, straw	180.184	0.5

The tolerances set by the EPA for linuron in animal feeds are as follows (the *CFR* Reference is to Title 21):

	CFR Reference	Parts per Million
Feeds, animal	561.265	0.1

References

(1) Worthing, C.R., *Pesticide Manual,* 6th ed., p. 320, British Crop Protection Council (1979).
(2) Spencer, E.Y., *Guide to the Chemicals Used in Crop Protection,* 6th ed., p. 317, London, Ontario, Agriculture Canada (January 1973).
(3) Scherer, O. and Heller, P., U.S. Patent 2,960,534, November 15, 1960, assigned to Farbwerke Hoechst AG.
(4) Scherer, O. and Heller, P., U.S. Patent 3,079,244, February 26, 1963, assigned to Farbwerke Hoechst AG.

M

MALATHION

Function: Insecticide and acaricide (1)(2)

Chemical Name: Diethyl(dimethoxyphosphinothioyl)thiobutanedioate

Formula:

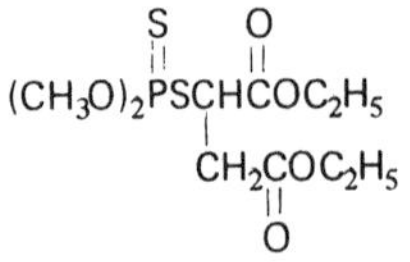

$$(CH_3O)_2PSCHCOC_2H_5 \quad\begin{smallmatrix}S\ O\end{smallmatrix}$$

$$\underset{O}{CH_2COC_2H_5}$$

Trade Names: E14049 (American Cyanamid Co.)
Cythion® (American Cyanamid Co.)
Malathion®
Malatiozol®
Malathiozoo®
Emmaton®
Karbophos®
Chemathion®
Malaspray®

Manufacture (3)(4)(5)

The feed materials for malathion manufacture are O,O-dimethyl phosphorodithioic acid and diethyl maleate or fumarate which react according to the equation:

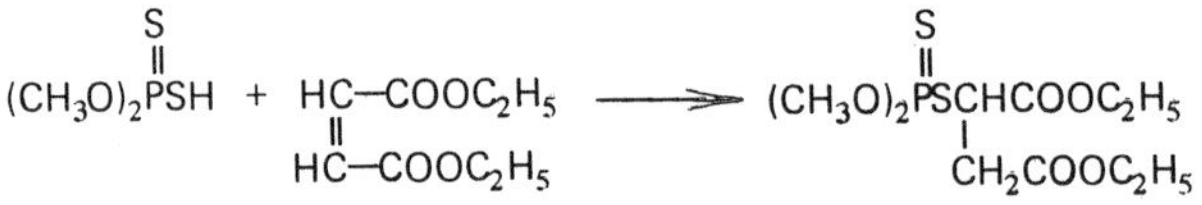

$$(CH_3O)_2PSH \;+\; \begin{matrix}HC-COOC_2H_5\\ \| \\ HC-COOC_2H_5\end{matrix} \;\longrightarrow\; (CH_3O)_2PSCHCOOC_2H_5 \atop CH_2COOC_2H_5$$

An antipolymerization agent such as hydroquinone may be added to the reaction mixture to inhibit the polymerization of the maleate or fumarate compound under the reaction conditions. This reaction is preferably carried out at a temperature within the range of 20° to 150°C. This reaction is preferably carried out at atmospheric pressure. Reaction times of 16 to 24

474

hours have been specified for this reaction by J.T. Cassaday (4). The reaction is preferably carried out in a solvent such as the low molecular weight aliphatic monohydric alcohols, ketones, aliphatic esters, aromatic hydrocarbons or trialkyl phosphates.

The reaction may be accelerated by using an aliphatic tertiary amine catalyst, usually within the range of 0.2 to 2.0% based on the total weight of the reactants. A stirred, jacketed reactor of conventional design may be used. After cooling, the reaction mixture may be taken up in benzene. It is then washed with 10% Na_2CO_3 and with water. The organic layer is dried over anhydrous Na_2SO_4, filtered and concentrated in vacuo to give the final product as residue.

A production and waste schematic for malathion manufacture is shown in Figure 33. Decolorization and deodorization of malathion with a peroxide or hydroperoxide has been described (6).

Figure 33: Product and Waste Schematic for Malathion

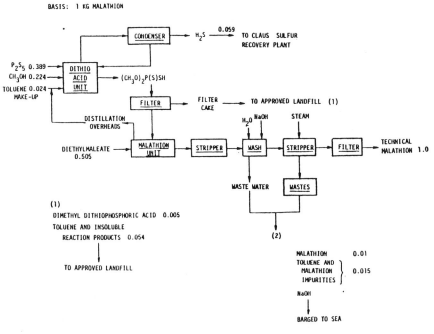

Source: Reference (B-1)

Process Wastes and Their Control

Air: Air emissions from malathion manufacture have been reported (B-15) to consist of the following:

Component	Kilograms per Metric Ton Pesticide Produced
SO_2	0.3
NO_x	0.3
Hydrocarbons	2.0
Malathion	0.5

Water: The liquid stream contains NaCl. The wastewater from the strippers and the waste stream from the NaOH wash is barged to sea at one plant.

The discarded filter cake for a typical malathion plant with an annual production rate of 14,000 tons per year totals 1,826 metric tons per year (0.13 kg/kg malathion). These wastes consist of 1,000 metric tons of filter aid, 756 metric tons of toluene and insoluble reaction products and 70 metric tons of dimethyl dithiophosphoric acid (B-10).

At present, this waste is detoxified with sodium hydroxide and sent to landfill. In the detoxification with sodium hydroxide it is hypothesized that the following reactions take place:

$$4NaOH + 2(CH_3O)_2\overset{\overset{\displaystyle S}{\|}}{P}-SH \longrightarrow 2NaSH + (CH_3O)_2\overset{\overset{\displaystyle S}{\|}}{P}-OH + Na_2S + (CH_3O)_2P(OH)_3$$

This method of disposal can still be considered hazardous, because of the possibility of leakage of materials into groundwater (B-10).

A more desirable alternative treatment process for the spent filter cake from malathion manufacture requires the following steps (see Figure 34): (1) hydrolysis; (2) steam stripping; (3) decantation; (4) composting; and (5) biological treatment.

Figure 34: Improved Process for Filter Cake Treatment—Malathion Manufacture

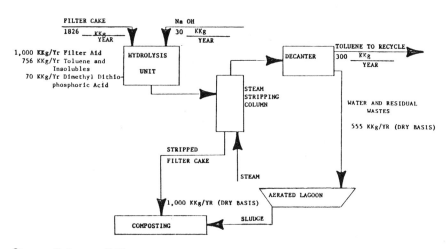

Source: Reference (B-7)

The combined wastes from the washing and stripping steps total approximately 14,350 metric tons per year for typical 14,000 MT malathion plant. The waste stream consists of 350 metric tons per year (0.025 kg/kg malathion) of malathion, toluene, and malathion impurities, and the remaining 14,000 metric tons per year (1 kg/kg malathion) of an estimated 2% sodium hydroxide solution.

At present, the liquid process wastes from malathion manufacture are disposed of by ocean dumping. This method of disposal is unacceptable for several reasons; first U.S. laws and, in the near future, international convention, will make this practice illegal; second, this method does not make any attempt to detoxify the waste, merely to dispose of it; and third, no attempt is made to recover the 140 metric tons per year of malathion and approximately 100 metric tons per year of toluene (B-7).

In order to recover usable products, reduce the volume of the waste stream and, detoxify the wastes, an alternative treatment process has been selected based on sedimentation, resin adsorption with solvent regeneration, vacuum distillation, hydrolysis and finally biotreatment (composting). See Figure 35.

Figure 35: Improved Process for Treatment of Liquid Wastes—Malathion Manufacture

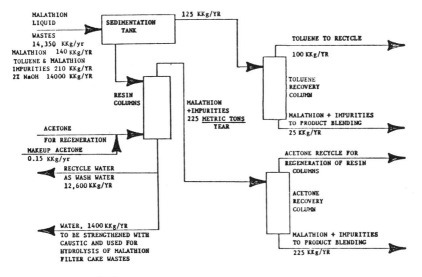

Source: Reference (B-7).

Product Wastes: Malathion is reported to be "hydrolyzed almost instantly" at pH 12; 50% hydrolysis at pH 9 requires 12 hours. Alkaline hydrolysis under controlled conditions (0.5N NaOH in ethanol) gives quantitative yields of $(CH_3O)_2P(S)SNa$, whereas hydrolysis in acidic media yields $(CH_3O)_2P(S)OH$. On prolonged contact with iron or iron-containing material, it is reported to break down and completely lose insecticidal activity (B-3).

Toxicity

The acute oral LD_{50} value for rats is 2,800 mg/kg which is slightly toxic.

Malathion is a widely used organophosphorus insecticide with a wide spectrum of activity and many diverse uses in agricultural, home, and garden applications. It is quite likely that malathion could appear as a contaminant of drinking water, although there are no reports of its having been found as yet. The mode of action of malathion is similar to that of other organophosphorus insecticides—inhibition of acetylcholinesterase. Its acute toxicity, however, is quite low, compared with that of other members of this class of insecticides, primarily because of its facile metabolism in mammalian systems, by carboxylesterase or aliesterase enzymes, to products of decreased or no toxicity. Its toxic potential, however, is illustrated by the possibility of potentiation when degradative enzymes are inhibited by other chemicals.

The chronic-toxicity information available on malathion is surprisingly sparse for a compound that has been extensively used in the past. However, the two rat studies that have been reported, the subchronic administration of malathion to human volunteers, and the establishment of no-adverse-effect dosages in rats and humans on the basis of anticholinesterase activity allows the establishment of a no-adverse-effect concentration for drinking-water with a high degree of assurance of safety. An ADI was calculated at 0.02 mg/kg/day based on these data.

Additional chronic toxicity data are needed for malathion with particular concentration on long-term feeding studies in which teratogenicity, mutagenicity, and carcinogenicity are evaluated. Of particular importance would be a good study of the metabolism and persistence of malathion in water. In view of the extent of past use of malathion, continued monitoring for its presence in food materials and water is necessary (B-22).

Allowable Limits on Exposure and Use

Air: The threshold limit value for malathion in air is 10 mg/m^3 as of 1979. It is noted that cutaneous absorption of malathion should be prevented so that the TLV is not invalidated (B-23).

NIOSH has recommended (B-7) a limit of 15 mg/m^3 on a 10-hour time weighted average basis.

Water: In water, EPA has set (B-12) criteria of 0.1 μg/l for the protection of freshwater and marine aquatic life.

Product Use: The tolerances set by the EPA for malathion in or on raw agricultrual commodities are as follows:

	40 *CFR* Reference	Parts per Million
Alfalfa (pre-h)	180.111	135.0
Almonds (pre & post-h)	180.111	8.0
Almonds, hulls (pre-h)	180.111	50.0
Apples (pre-h)	180.111	8.0
Apricots (pre-h)	180.111	8.0
Asparagus (pre-h)	180.111	8.0
Avocados (pre-h)	180.111	8.0
Barley, grain (pre & post-h)	180.111	8.0
Beans (pre-h)	180.111	8.0
Beets (inc. tops) (pre-h)	180.111	8.0
Beets, sugar, roots (pre-h)	180.111	1.0
Beets, sugar, tops (pre-h)	180.111	8.0
Blackberries (pre-h)	180.111	8.0
Blueberries (pre-h)	180.111	8.0
Boysenberries (pre-h)	180.111	8.0
Broccoli (pre-h)	180.111	8.0
Brussels sprouts (pre-h)	180.111	8.0
Cabbage (pre-h)	180.111	8.0
Carrots (pre-h)	180.111	8.0
Cattle, fat (pre-s)	180.111	4.0
Cattle, mbyp (pre-s)	180.111	4.0
Cattle, meat (pre-s)	180.111	4.0
Cauliflower (pre-h)	180.111	8.0
Celery (pre-h)	180.111	8.0
Cherries (pre-h)	180.111	8.0
Chestnuts (pre-h)	180.111	1.0
Clover (pre-h)	180.111	135.0
Collards (pre-h)	180.111	8.0
Corn (inc. sweet) (k+cwhr) (pre-h)	180.111	2.0
Corn, forage (pre-h)	180.111	8.0
Corn, grain (post-h)	180.111	8.0
Cotton, seed (pre-h)	180.111	2.0
Cowpeas, forage (pre-h)	180.111	135.0

(continued)

	40 *CFR* Reference	Parts per Million
Cowpeas, hay (pre-h)	180.111	135.0
Cranberries (pre-h)	180.111	8.0
Cucumbers (pre-h)	180.111	8.0
Currants (pre-h)	180.111	8.0
Dandelions (pre-h)	180.111	8.0
Dates (pre-h)	180.111	8.0
Dewberries (pre-h)	180.111	8.0
Eggplant (pre-h)	180.111	8.0
Eggs	180.111	0.1
Endive (escarole) (pre-h)	180.111	8.0
Figs (pre-h)	180.111	8.0
Filberts (pre-h)	180.111	1.0
Garlic (pre-h)	180.111	8.0
Goats, fat (pre-s)	180.111	4.0
Goats, mbyp (pre-s)	180.111	4.0
Goats, meat (pre-s)	180.111	4.0
Gooseberries (pre-h)	180.111	8.0
Grapefruit (pre-h)	180.111	8.0
Grapes (pre-h)	180.111	8.0
Grasses (pre-h)	180.111	135.0
Grasses, hay (pre-h)	180.111	135.0
Guavas (pre-h)	180.111	8.0
Hogs, fat (pre-s)	180.111	4.0
Hogs, mbyp (pre-s)	180.111	4.0
Hogs, meat (pre-s)	180.111	4.0
Hops (pre-h)	180.111	1.0
Horseradish (pre-h)	180.111	8.0
Horses, fat (pre-s)	180.111	4.0
Horses, mbyp (pre-s)	180.111	4.0
Horses, meat (pre-s)	180.111	4.0
Kale (pre-h)	180.111	8.0
Kohlrabi (pre-h)	180.111	8.0
Kumquats (pre-h)	180.111	8.0
Leeks (pre-h)	180.111	8.0
Lemons (pre-h)	180.111	8.0
Lentils (pre-h)	180.111	8.0
Lespedeza, hay (pre-h)	180.111	135.0
Lespedeza, seed (pre-h)	180.111	8.0
Lespedeza, straw (pre-h)	180.111	135.0
Lettuce (pre-h)	180.111	8.0
Limes (pre-h)	180.111	8.0
Loganberries (pre-h)	180.111	8.0
Lupine, hay (pre-h)	180.111	135.0
Lupine, seed (pre-h)	180.111	8.0
Lupine, straw (pre-h)	180.111	135.0
Macadamia (pre-h)	180.111	1.0
Mangoes (pre-h)	180.111	8.0
Melons (pre-h)	180.111	8.0
Milk, fat (from appli to dairy cows)	180.111	0.5
Mushrooms (pre-h)	180.111	8.0
Mustard, greens (pre-h)	180.111	8.0
Nectarines (pre-h)	180.111	8.0
Oats, grain (pre & post-h)	180.111	8.0
Okra (pre-h)	180.111	8.0
Onions (pre-h)	180.111	8.0
Onions, green (pre-h)	180.111	8.0
Oranges (pre-h)	180.111	8.0
Papayas (pre-h)	180.111	1.0
Parsley (pre-h)	180.111	8.0
Parsnips (pre-h)	180.111	8.0

(continued)

	40 *CFR* Reference	Parts per Million
Passion fruits (pre-h)	180.111	8.0
Peanuts (pre & post-h)	180.111	8.0
Peanuts, forage (pre-h)	180.111	135.0
Peanuts, hay (pre-h)	180.111	135.0
Pears (pre-h)	180.111	8.0
Peas (pre-h)	180.111	8.0
Peas, vines (pre-h)	180.111	8.0
Peas, vines, hay (pre-h)	180.111	8.0
Pecans (pre-h)	180.111	8.0
Peppermint (pre-h)	180.111	8.0
Peppers (pre-h)	180.111	8.0
Pineapples (pre-h)	180.111	8.0
Plums (pre-h)	180.111	8.0
Potatoes (pre-h)	180.111	8.0
Poultry, fat (pre-s)	180.111	4.0
Poultry, mbyp (pre-s)	180.111	4.0
Poultry, meat (pre-s)	180.111	4.0
Prunes, (pre-h)	180.111	8.0
Pumpkins (pre-h)	180.111	8.0
Quinces (pre-h)	180.111	8.0
Radishes (pre-h)	180.111	8.0
Raspberries (pre-h)	180.111	8.0
Rice, grain (pre & post-h)	180.111	8.0
Rice, wild	180.111	8.0
Rutabagas (pre-h)	180.111	8.0
Rye, grain (pre & post-h)	180.111	8.0
Safflower, seed (pre-h)	180.111	0.2
Salsify (inc. tops) (pre-h)	180.111	8.0
Shallots (pre-h)	180.111	8.0
Sheep, fat (pre-s)	180.111	4.0
Sheep, mbyp (pre-s)	180.111	4.0
Sheep, meat (pre-s)	180.111	4.0
Sorghum, forage (pre-h)	180.111	8.0
Sorghum, grain (pre & post-h)	180.111	8.0
Soybeans, dry (pre-h)	180.111	8.0
Soybeans, forage (pre-h)	180.111	135.0
Soybeans, hay (pre-h)	180.111	135.0
Soybeans, succulent (pre-h)	180.111	8.0
Spearmint (pre-h)	180.111	8.0
Spinach (pre-h)	180.111	8.0
Squash, summer (pre-h)	180.111	8.0
Squash, winter (pre-h)	180.111	8.0
Strawberries (pre-h)	180.111	8.0
Sweet potatoes (pre-h)	180.111	1.0
Swiss chard (pre-h)	180.111	8.0
Tangerines (pre-h)	180.111	8.0
Tomatoes (pre-h)	180.111	8.0
Trefoil, birdsfoot, forage (pre-h)	180.111	135.0
Trefoil, birdsfoot, hay (pre-h)	180.111	135.0
Turnips (inc. tops) (pre-h)	180.111	8.0
Vetch, hay (pre-h)	180.111	135.0
Vetch, seed (pre-h)	180.111	8.0
Vetch, straw (pre-h)	180.111	135.0
Walnuts (pre-h)	180.111	8.0
Watercress (pre-h)	180.111	8.0
Wheat, grain (pre & post-h)	180.111	8.0

The tolerances set by the EPA for malathion in animal feeds are as follows (the *CFR* Reference is to Title 21):

	CFR Reference	Parts per Million
Cattle feed, conc. (nonmed.)	561.270	10.0
Citrus pulp, dehydrated (ct f)	561.270	50.0

The tolerances set by the EPA for malathion in food are as follows (the *CFR* Reference is to Title 21):

	CFR Reference	Parts per Million
Raisins, processed	193.260	12.0
Safflower, oil, refined	193.260	0.6

References

(1) Worthing, C.R., *Pesticide Manual,* 6th ed., p. 321, British Crop Protection Council (1979).
(2) Spencer, E.Y., *Guide to the Chemicals Used in Crop Protection,* 6th ed., p. 318, London, Ontario, Agriculture Canada (January 1973).
(3) Midwest Research Institute, *The Pollution Potential in Pesticide Manufacturing,* Washington, DC, Environmental Protection Agency (June 1972).
(4) Cassady, J.T., U.S. Patent 2,578,652, December 18, 1951, assigned to American Cyanamid Co.
(5) Backlund, G.R., Martino, J.F. and Divine, R.D., U.S. Patent 3,463,841, August 26, 1969, assigned to American Cyanamid Co.
(6) Usui, M., U.S. Patent 2,962,521, November 29, 1960, assigned to Sumitomo Chemical Co.
(7) National Institute for Occupational Safety and Health, *Criteria for a Recommended Standard: Occupational Exposure to Malathion,* NIOSH Doc. No. 76-205, Washington, DC (1976).

MALEIC HYDRAZIDE

Function: Plant growth regulator (1)(2)(4)

Chemical Name: 1,2-dihydro-3,6-pyridazinedione

Formula:

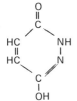

Trade Names: MH-30 (American Cyanamid)
Slo-Gro®
Sucker Stuff®
Super Sucker Stuff®
Retard®
Desprout®
Malazide®
Regulox®

Manufacture (3)

The reaction may be carried out quite simply by stirring solid maleic anhydride into an aqueous solution of a stoichiometric amount of the hydrazine salt. The solution becomes cloudy and a temperature increase is noted. The solution may then be refluxed, or heated at a temperature just below the refluxing temperature, for a short time, cooled, and the product filtered, washed with water and dried. Water serves to moderate the reaction, and to provide for easy heat transfer, ease of mixing, and the like.

Water also results in a better product. While it is much preferred to carry out the reaction in the presence of water, the reaction can be effected by heating the dry components, viz, maleic acid or anhydride and the hydrazine salt of the strong inorganic acid, such a procedure giving the same products; alternatively, these reactants may be heated in a nonsolvent such as benzene to effect the reaction. Water is preferred, however, because of its cheapness.

It is preferred to employ approximately stoichiometric amounts of the reactants, i.e., approximately 1 mol of maleic acid or anhydride per mol of hydrazine in the form of the salt of the strong inorganic acid. Thus, if the monohydrazine salt (obtained from 1 mol of hydrazine per mol of acid) is used, 1 mol of maleic acid or anhydride per mol of such salt is used. In the case of dihydrazine salts (which are made from 2 mols of hydrazine per mol of acid), 2 mols of maleic acid or anhydride per mol of such salt are used.

Generally speaking, the method is carried out by commingling the reactants and thereafter heating to an elevated temperature, typically ranging from 75° to 110°C until the reaction has attained the degree of completion. Upon cooling, the product precipitates from the mixture when an aqueous reaction medium is employed, and is readily recovered therefrom by filtration and washing.

The following is one specific example of the operation of the process. Technical dihydrazine sulfate 810 g (5.0 mols) was dissolved in 3 liters of warm water. This solution was stirred vigorously while maleic anhydride 980 g (10.0 mols) was added rapidly. An immediate reaction set in as evidenced by a cloudy appearance and a rise in temperature from 33°C at the beginning to 56°C at the end of the addition. The reaction mixture was then heated as rapidly as possible by a glass heating mantle, to a temperature just short of reflux. This temperature was maintained for 3½ hours.

The mixture was light yellow at first, darkened somewhat and thickened a great deal at about 66°C. At 72°C it became quite fluid. The color decreased gradually until it almost completely disappeared after about 30 minutes at the highest temperature. Stirring was continued and the flask was cooled to 25°C in a cold water bath. The product was collected by vacuum filtration, washed free of sulfate and dried. The product was snow-white, coarse crystalline solid. Yield 960 g or 85.6% of theory. It melts with decomposition at 305° to 308°C.

Process Wastes and Their Control

Water: H_2SO_4 is a by-product (B-10).

Product Wastes: This product is relatively stable to hydrolysis but decomposes in presence of strong acids and oxidizing agents. Treatment with alkali hydroxides and amines results in the formation of water-soluble salts. Salts of alkaline-earth and heavy metals are practically insoluble in water. Burning with polyethylene has been reported to result in >99% decomposition (B-3).

Toxicity

The acute oral LD_{50} value for rats is 3,800 mg/kg (B-5) which is slightly toxic.

Allowable Limits on Exposure and Use

Product Use: A rebuttable presumption against registration was issued on October 28, 1977 by EPA on the basis of oncogenicity, mutagenicity and reproductive effects.

The tolerances set by the EPA for maleic hydrazide in or on raw agricultural commodities are as follows:

	40 *CFR* Reference	Parts per Million
Onions, dry bulb	180.175	15.0
Potatoes	180.175	50.0

The tolerances set by the EPA for maleic hydrazide in food are as follows (the *CFR* Reference is to Title 21):

	CFR Reference	Parts per Million
Potato chips	193.270	160.0

References

(1) Worthing, C.R., *Pesticide Manual,* 6th ed., p. 322, British Crop Protection Council (1979).
(2) Spencer, E.Y., *Guide to the Chemicals Used in Crop Protection,* 6th ed., p. 355, London, Ontario, Agriculture Canada (January 1973).
(3) Harris, W.D. and Schoene, D.L., U.S. Patent 2,575,954, November 20, 1951, assigned to United States Rubber Co.
(4) Hoffman, O.L. and Schoene, D.L., U.S. Patent 2,614,916, October 21, 1952, assigned to United States Rubber Co.

MANCOZEB

Function: Fungicide (1)(2)(3)

Chemical Name: [[1,2-ethanediylbis(carbamodithioato)] (2-)] manganese zinc salt

Formula:

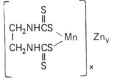

Trade Name: Dithane M-45® (Rohm and Haas Co.)

Manufacture (3)

There were reacted equimolar amounts of disodium ethylenebisdithiocarbamate and manganous chloride in concentrated aqueous solution. A precipitate formed which was washed by decantation three times to give an aqueous slurry containing 38% of hydrated manganese ethylenebisdithiocarbamate. A portion of 800 parts of this slurry was mixed with a separately prepared solution of 24.5 parts of zinc nitrate hexahydrate in 25 parts of water. The resulting mixture was stirred for one-half hour at 30°C and was spray dried, the inlet gas temperature being 280°C and the outlet temperature 125°C. The dry solid which formed was micronized. Product collected amounted to 272 parts.

By analysis this product contained 47.5% of carbon disulfide, 17.3% of manganese, and 1.7% of zinc. It is a complex salt in which maneb is modified by reaction with zinc ions. The disodium ethylenebisdithiocarbamate is nabam (q.v.).

Toxicity

The acute oral LD_{50} value for rats is over 8,000 mg/kg which is insignificantly toxic.

The dithiocarbamate fungicides are low in acute toxicity and do not present alarming properties during long-term administration to experimental animals, except at very high dosages. An acceptable daily intake has been temporarily set by the FAO/WHO (in 1974) at 0.005 mg/kg for both the dimethyldithiocarbamates (including thiram) and the ethylene-bis-dithiocarbamates (FAO/WHO, 1975). That value represents a fivefold lowering from the previous value used by FAO/WHO for all dithiocarbamate fungicides. That decision was based, for the dimethyldithiocarbamates, on the recent evidence of teratogenic and mutagenic effects, as well as the possibility of nitrosation to form carcinogenic nitrosamines. For the EBDC compounds, the ETU problem and its associated teratogenic, mutagenic, and carcinogenic effects prompted the lowered values.

It could be held that ETU in water should be considered independently as a contaminant separate from the parent compounds. In light of the above and taking into account the carcinogenic risk projections it is suggested that very strict criteria be applied when limits for ETU in drinking water are established.

Based on long-term feeding studies results, ADI's were calculated at 0.005 mg/kg/day for maneb, zineb and mancozeb; at 0.0125 mg/kg/day for ziram; and at 0.005 mg/kg/day for thiram (B-22).

Allowable Limits on Exposure and Use

Product Use: A rebuttable presumption against registration was issued on August 10, 1977 by EPA on the basis of oncogenicity, teratogenicity and hazard to wildlife.

The tolerances set by the EPA for mancozeb in or on raw agricultural commodities are as follows:

	40 *CFR* Reference	Parts per Million
Apples	180.176	7.0
Asparagus	180.176	0.1 N
Bananas (pre-h)	180.176	4.0
Bananas, pulp (no peel)	180.176	0.5
Barley, grain	180.176	5.0
Barley, straw	180.176	25.0
Beets, sugar	180.176	2.0
Beets, sugar, tops	180.176	65.0
Carrots	180.176	2.0
Celery	180.176	5.0
Corn, fodder	180.176	5.0
Corn, forage	180.176	5.0
Corn, fresh (inc. sweet) (k+cwhr)	180.176	0.5
Corn, grain (exc. pop)	180.176	0.1
Corn, pop, grain	180.176	0.5

(continued)

	40 *CFR* Reference	Parts per Million
Cotton, seed	180.176	0.5
Crabapples	180.176	10.0
Cranberries	180.176	7.0
Cucumbers	180.176	4.0
Fennel	180.176	10.0
Grapes	180.176	7.0
Kidney	180.176	0.5
Liver	180.176	0.5
Melons	180.176	4.0
Oats, grain	180.176	5.0
Oats, straw	180.176	25.0
Onions, dry bulb	180.176	0.5
Papaya, edible pulp (no peel)	180.176	0.0
Papaya, whole	180.176	10.0
Peanuts	180.176	0.5
Peanuts, vines, hay	180.176	65.0
Pears	180.176	10.0
Quinces	180.176	10.0
Rye, grain	180.176	5.0
Rye, straw	180.176	25.0
Squash, summer	180.176	4.0
Tomatoes	180.176	4.0
Wheat, grain	180.176	5.0
Wheat, straw	180.176	25.0

The tolerances set by the EPA for mancozeb in animal feeds are as follows (the *CFR* Reference is to Title 21):

	CFR Reference	Parts per Million
Barley, milled fractions	561.410	20.0
Oats, milled fractions	561.410	125.0
Rye, milled fractons	561.410	125.0
Wheat, milled fractions	561.410	20.0

References

(1) Worthing, C.R., *Pesticide Manual,* 6th ed., p. 324, British Crop Protection Council (1979).
(2) Spencer, E.Y., *Guide to the Chemicals Used in Crop Protection,* 6th ed., p. 320, London, Ontario, Agriculture Canada (January 1973).
(3) Lyon, C.B., Nemec, J.W. and Unger, V.H., U.S. Patent 3,379,610, April 23, 1968, assigned to Rohm and Haas Co.

MANEB

Function: Fungicide (1)(2)(4)(5)

Chemical Name: [[1,2-ethanediylbis(carbamodithioato)](2-)] manganese

Formula:

Trade Names: Dithane M-22® (Rohm and Haas)
Manzate® (DuPont)
Maneba®
Manebgan®
Manesan®
Sopranebe®
Trimangol®
Vancide®

Manufacture

The reaction chemistry for maneb production is believed to be approximately as shown below, based on discussions with the manufacturers (3).

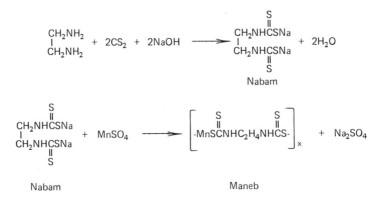

Nabam

Maneb

A production and waste control schematic for maneb is shown in Figure 36. Raw materials include carbon disulfide, ethylenediamine and sodium hydroxide (50%). These are reacted in stainless steel, cooled units. The exothermic reaction is controlled by the feed rate and excess carbon disulfide is distilled out. The sodium hydroxide addition controls pH. The resulting concentrated nabam solution is reacted (within 24 hours) with manganese sulfate, and the desired manganese ethylenebisdithiocarbamate is precipitated. The slurry is washed with water to remove sodium sulfate and dried to less than 1% water content. Process byproducts include sodium sulfate and small amounts of carbon disulfide and sodium hydroxide. Hexamethylene tetramine (urotropin) may be added to the final product to improve its storage stability as described by R.J. Sobatzki (6).

The preparation of maneb has also been described by Flenner (4), as follows: To 600 parts by weight of a 14% aqueous solution of sodium hydroxide at about room temperature, there is added 87 parts by weight of a 69% aqueous solution of ethylene diamine with agitation. A total of 156 parts by weight of carbon disulfide is added slowly with vigorous agitation of the reaction mass, keeping the temperature between 25° to 35°C throughout the reaction period. There is thus formed a solution of disodium ethylenebisdithiocarbamate.

The solution of disodium ethylenebisdithiocarbamate is diluted with 800 parts by weight of water and acetic acid is then added with agitation to adjust the solution to neutral to phenolphthalein indicator. The diluted and neutralized solution is agitated slowly while adding 600 parts by weight of a 20% aqueous solution of manganous chloride. The product, manganous ethylenebisdithiocarbamate, which is a light yellow colored solid, precipitates as the manganous chloride solution is added. The product is removed by filtration and dried at about 50°C in a forced draft oven. There is obtained 205 parts by weight of the product. The dried material is then pulverized to give a finely powdered product.

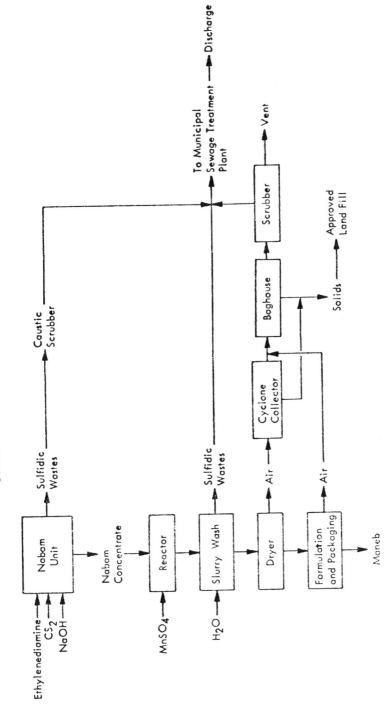

Figure 36: Production and Waste Schematic for Maneb

Source: Reference (3)

Product Wastes and Their Control

Air: Air emissions from maneb manufacture have been reported (B-15) to be as follows:

Component	Kilograms per Metric Ton Pesticide Produced
SO$_2$	205.0
Hydrocarbons	1.5
Maneb	0.5

Air emissions are controlled by cyclone collectors, bag filters, and scrubbers. A small amount of hydrogen sulfide is given off in the reaction; this is collected from process vents and passed through caustic (5%) scrubbers before release to the atmosphere. Emissions from the drying step are also controlled. A cyclone collector, bag filter and scrubber are used to remove particulates from the air. The final packaging operation is dusty; this area must be well ventilated and the airstream is passed through bag filters before release. Collected dusts are landfilled in one case (B-10).

Water: The liquid waste stream contains primarily salts, about 9lb/13 lb of maneb produced. The salts are mostly sodium sulfate, with some manganese sulfate and certain by-products, such as sodium trithiocarbamate, formed by the reaction of carbon disulfide with sodium hydroxide. The liquid stream goes untreated to a large city sewage treatment plant.

Rohm and Haas (and other chemical plants in the vicinity) have an agreement with municipal authorities under which they can discharge their effluent to their plant for treatment. Thus, the plant does not have separate waste treatment facilities and does not require any discharge permits other than agreement from the city authorities to accept their waste streams. The waste stream going to the city treatment plant must contain less than 1% solids. Noncontaminated cooling water is monitored and discharged directly to the river.

Solid Wastes: About 2% of the total weight of product is lost as solids. Solid wastes (broken bags, sweepings, solids collected in the dust collectors, etc.) are discharged at an approved dump site. None of the solid waste, including spillage from broken bags, is recovered.

Waste treatment methods employed in the maneb process do not include incineration (sulfur dioxide and manganese would create air pollution problems), evaporation, deep-well disposal (the salts would tend to clog the deep well) or sea dumping.

Product Wastes: Maneb is unstable to moisture and is hydrolyzed by acids and hot water. It decomposes at about 100°C and may spontaneously decompose vigorously when stored in bulk (B-3).

Toxicity

The acute oral LD$_{50}$ for rats for maneb is 6,750 mg/kg which is insignificantly toxic.

The toxic properties of dithiocarbamate fungicides have been discussed at some length by the National Research Council (B-22) and the reader is referred to the section of this volume on "Mancozeb" for more detail.

Allowable Limits on Exposure and Use

Product Use: A rebuttable presumption against registration was issued on August 10, 1977 on the basis of oncogenicity, teratogenicity and hazard to wildlife.

The tolerances set by the EPA for maneb in or on raw agricultural commodities are as follows:

	40 *CFR* Reference	Parts per Million
Almonds	180.110	0.1
Apples	180.110	2.0
Apricots	180.110	10.0
Bananas, pulp (pre-h)	180.110	0.5
Bananas, whole	180.110	4.0
Beans, dry	180.110	7.0
Beans, succulent	180.110	10.0
Beets, sugar, tops	180.110	45.0
Broccoli	180.110	10.0
Brussels sprouts	180.110	10.0
Cabbage	180.110	10.0
Cabbage, chinese	180.110	10.0
Carrots	180.110	7.0
Cauliflower	180.110	10.0
Celery	180.110	5.0
Collards	180.110	10.0
Corn, sweet (k+cwhr)	180.110	5.0
Cranberries	180.110	7.0
Cucumbers	180.110	4.0
Eggplant	180.110	7.0
Endive (escarole)	180.110	10.0
Figs	180.110	7.0
Grapes	180.110	7.0
Kale	180.110	10.0
Kohlrabi	180.110	10.0
Lettuce	180.110	10.0
Melons	180.110	4.0
Mustard, greens	180.110	10.0
Nectarines	180.110	10.0
Onions	180.110	7.0
Papayas	180.110	10.0
Peaches	180.110	10.0
Peppers	180.110	7.0
Potatoes	180.110	0.1
Pumpkins	180.110	7.0
Rhubarb	180.110	10.0
Spinach	180.110	10.0
Squash, summer	180.110	4.0
Squash, winter	180.110	4.0
Tomatoes	180.110	4.0
Turnips, roots	180.110	7.0
Turnips, tops	180.110	10.0

References

(1) Worthing, C.R., *Pesticide Manual,* 6th ed., p. 325, British Crop Protection Council (1979).
(2) Spencer, E.Y., *Guide to the Chemicals Used in Crop Protection,* 6th ed., p. 321, London, Ontario, Agriculture Canada (January 1973).
(3) Midwest Research Institute and RvR Consultants, *Production, Distribution, Use and Environmental Impact Potential of Selected Pesticides,* Washington, DC, Council on Environmental Quality (March 15, 1974).
(4) Flenner, A.L., U.S. Patent 2,504,404, April 10, 1950, assigned to DuPont.
(5) Golding, D.R.V. and Richards, B.L., Jr., U.S. Patent 2,710,822, June 14, 1955, assigned to DuPont.
(6) Sobatzki, R.J., U.S. Patent 2,974,156, May 7, 1961, assigned to Rohm and Haas Co.

MCPA

Function: Fungicide (1)(2)

Chemical Name: 4-chloro-2-methylphenoxyacetic acid

Formula:

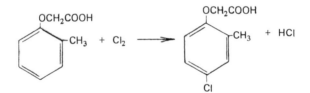

Trade Names: Agroxone® (Plant Protection, Ltd.)
Agritox® (May and Baker, Ltd.)
Cornox M® (The Boots Co., Ltd.)
Mephanac®
Methoxone®

Manufacture

The feed materials for MCPA manufacture may be methylphenoxyacetic acid (cresoxyacetic acid) and chlorine (3). They react according to the equation:

The chlorination reaction may be carried out at temperatures ranging from 60° to 100°C. The reaction is exothermic and only when the chlorine concentration is low or when the reaction nears completion is it necessary to apply external heat. Atmospheric pressure is used and a typical reaction time is one hour. This reaction is carried out in the liquid phase in the presence of 1,2-dichloropropane as the solvent. The advantages obtained from such an operation are substantial. Not only is any problem of equipment corrosion eliminated, but also the recovery of solvent is readily accomplished. Moreover, the 1,2-dichloropropane exhibits a singular combination of desirable properties not shared by other chlorinated aliphatic compounds. More particularly, 1,2-dichloropropane has a relatively low toxicity, which obviously is advantageous. In addition, it is not appreciably lost during the chlorination reaction, yet following the chlorination reaction it is readily removed from the reaction product.

This process may be carried out in the absence of any catalyst or inhibitor, but the use of a small amount of iodine or ferric chloride may be desirable. Generally, iodine is the preferred solvent chlorination inhibitor because it is readily removed following the reaction. Also, it is easily inactivated with sodium thiosulfate in those cases where the filtrate is to be stored or where the solvent is to be distilled.

A chlorinator equipped with cooling means and a reflux condenser may be used for the conduct of this process. When the reaction is completed, the reaction mixture is cooled and the product precipitates out. It is recovered on a filter and washed with cold solvent and dried. Yields of 75 to 80% are obtained.

Alternatively, o-cresol may be chlorinated and the p-chlorocresol reacted with chloracetic acid to give MCPA.

Toxicity

The acute oral LD_{50} for rats is 700 mg/kg which is sightly toxic.

In 90-day feeding studies no-adverse-effect doses were reported at up to 10 mg/kg/day for MCPA in rats, but histopathologic changes in livers and kidneys were reported once at 1.25 mg/kg day. Based on these data an ADI was calculated at 0.00125 mg/kg/day for MCPA.

There is considerable variation in the no-adverse-effect and minimal-toxic-effect dosages found in the various subchronic-toxicity experiments with MCPA. The reasons for these differences are not apparent, and further work is needed to resolve them. There have been no 2-year chronic-toxicity tests with MCPA, and such studies should be undertaken. Moreover, very little is known about the reproductive, mutagenic, and carcinogenic properties of MCPA. Additional research is needed, particularly in view of the reported weak mutagenic acitivity of MCPA.

There appears to be a complete lack of data on human toxicity related to MCPA (B-22).

Allowable Limits on Exposure and Use

Product Use: The tolerances set by the EPA for MCPA in or on raw agricultural commodities are as follows:

	40 *CFR* Reference	Parts per Million
Barley, forage	180.339	20.0
Barley, grain	180.339	0.1 N
Barley, straw	180.339	2.0
Cattle, fat	180.339	0.1 N
Cattle, mbyp	180.339	0.1 N
Cattle, meat	180.339	0.1 N
Flax, seed	180.339	0.1 N
Flax, straw	180.339	2.0
Goats, fat	180.339	0.1 N
Goats, mbyp	180.339	0.1 N
Goats, meat	180.339	0.1 N
Grasses, hay	180.339	20.0
Grasses, pasture	180.339	300.0
Grasses, rangeland	180.339	300.0
Hogs, fat	180.339	0.1 N
Hogs, mbyp	180.339	0.1 N
Hogs, meat	180.339	0.1 N
Horses, fat	180.339	0.1 N
Horses, mbyp	180.339	0.1 N
Horses, meat	180.339	0.1 N
Milk	180.339	0.1 N
Oats, grain	180.339	0.1 N
Oats, forage	180.339	20.0
Oats, straw	180.339	2.0
Peas, vines	180.339	0.1 N
Peas, vines, hay	180.339	0.1 N
Rice, grain	180.339	0.1 N
Rice, straw	180.339	2.0
Rye, forage	180.339	20.0
Rye, grain	180.339	0.1 N
Rye, straw	180.339	2.0
Sheep, fat	180.339	0.1 N
Sheep, mbyp	180.339	0.1 N
Sheep, meat	180.339	0.1 N

(continued)

	40 CFR Reference	Parts per Million
Sorghum, fodder	180.339	20.0
Sorghum, forage	180.339	20.0
Sorghum, grain	180.339	0.1
Vegetables, seed & pod	180.339	0.1
Wheat, forage	180.339	20.0
Wheat, grain	180.339	0.1 N
Wheat, straw	180.339	2.0

References

(1) Worthing, C.R., *Pesticide Manual*, 6th ed., p. 326, British Crop Protection Council (1979).
(2) Spencer, E.Y., *Guide to the Chemicals Used in Crop Protection*, 6th ed., p. 322, London, Ontario, Agriculture Canada (January 1973).
(3) Skeeters, M.J., U.S. Patent 2,740,810, April 3, 1956, assigned to Diamond Alkali Co.

MCPB

Functon: Herbicide (1)(2)(3)

Chemical Name: 4-(4-chloro-2-methylphenoxy)butanoic acid

Formula:

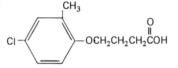

Trade Names: MB 3046 (May and Baker, Ltd.)
Tropotox® (May and Baker, Ltd.)
Cantrol®
Thitrol®

Manufacture

MCPB can be made by the reaction of butyrolactone with sodium 4-chloro-o-cresylate. An alternative route is as follows (4): 50 ml of absolute alcohol were placed in a 250 ml flask fitted with reflux condenser and calcium chloride guard tube and 1.2 grams sodium was added. When the reaction ceased, 7.1 grams of 2-methyl-4-chlorophenol was added followed by 5.2 grams of γ-chlorobutyronitrile. After refluxing for 4 hours, 50 ml of 2N sodium hydroxide solution were added and the mixture refluxed for 6 hours. The alcohol was then removed by distillation and the cold diluted solution was filtered into excess 4N hydrochloric acid.

The buff colored precipitate (7.7 grams) was filtered, washed with water, dissolved in ether and the solution extracted three times with 5% sodium bicarbonate solution. The bulked aqueous layers were warmed to remove traces of ether, cooled and acidified (Congo). The precipitate was filtered and dried. The dry product was crude γ-(2-methyl-4-chlorophenoxy)-butyric acid in the form of a white powder, MP 99° to 101°C. After recrystallization from petroleum ether, 4.5 grams of colorless needles (MP 100° to 101°C) were obtained.

Toxicity

The acute oral LD_{50} value for rats is 680 mg/kg which is slightly toxic.

Allowable Limits on Exposure and Use

Product Use: The tolerances set by the EPA for MCPB in or on raw agricultural commodities are as follows:

	40 *CFR* Reference	Parts per Million
Peas	180.318	0.1 N

References

(1) Worthing, C.R., *Pesticide Manual,* 6th ed., p. 327, British Crop Protection Council (1979).
(2) Spencer, E.Y., *Guide to the Chemicals Used in Crop Protection,* 6th ed., p. 108, London, Ontario, Agriculture Canada (January 1973).
(3) National Research Development Corp., British Patent 758,980, October 10, 1956.
(4) Wain, R.L., U.S. Patent 2,863,753, December 9, 1958, assigned to National Research Development Corp.

MECARBAM

Function: Insecticide and acaricide (1)()(3)

Chemical Name: Ethyl[[(diethoxyphosphinothioyl)thio]acetyl]methylcarbamate

Formula:

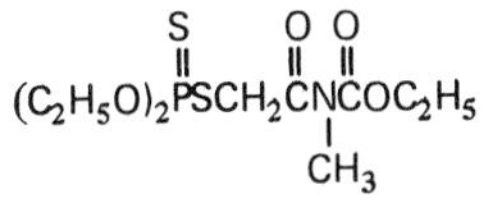

$$(C_2H_5O)_2\overset{\overset{S}{\|}}{P}SCH_2\overset{\overset{O}{\|}}{C}\underset{\underset{CH_3}{|}}{N}\overset{\overset{O}{\|}}{C}OC_2H_5$$

Trade Names: P-474 (Murphy Chemical Ltd.)
MC-47 (Murphy Chemical Ltd.)
Murfotox® (Murphy Chemical Ltd.)
Pestan® (Takeda Chemical Industries)
Afos® (J. and A. Margesin)

Manufacture (3)

Mecarbam may be made by the reaction of O,O-diethyldithiophosphoric acid with an intermediate N-chloroacetyl N-methyl ethyl carbamate which was prepared as follows: Chloroacetyl chloride and N-methyl ethyl carbamate were heated at 100°C during 2 hours. N-chloroacetyl N-methyl ethyl carbamate distilled as a colorless liquid.

Process Wastes and Their Control

Product Wastes: This compound is compatible with all but highly alkaline pesticides. At pH 3, it is hydrolyzed to $(EtO)_2P(S)SCH_2COOH$, CO_2, EtOH, and CH_3NH_2 (B-3).

Toxicity

The acute oral LD_{50} value for rats is 36 to 53 mg/kg which is highly toxic.

References

(1) Worthing, C.R., *Pesticide Manual,* 6th ed., p. 328, British Crop Protection Council (1979).

(2) Spencer, E.Y., *Guide to the Chemicals Used in Crop Protection,* 6th ed., p. 323, London, Ontario, Agriculture Canada (January 1973).
(3) Pianka, M. and Polton, D.J., British Patent 867,780, May 10, 1961, assigned to Murphy Chemical Co., Ltd.

MECOPROP

Function: Herbicide (1)(2)(3)(4)

Chemical Name: 2-(4-chloro-2-methylphenoxy)propanoic acid

Formula:

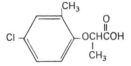

Trade Names: RD4593 (Boots Co., Ltd.)
Iso-Cornox® (Boots Co., Ltd.)
Mecopar®
Mecopex®
Mecoturf®
Vi-Pex®
Compitox®
Clovotox®
Herrifex DS®

Manufacture

Mecoprop may be made by the condensation of 2-chloropropionic acid with 4-chloro-o-cresol. The preparation of various esters has also been described (5).

Process Wastes and Their Control

Product Wastes: 2-CH_3-4-Cl-C_6H_3-OCH(CH_3)-COOH is said to be "stable" to heat and "resistant" to hydrolysis, reduction and atmospheric oxidation (B-3).

Toxicity

The acute oral LD_{50} for rats is 930 mg/kg which is slightly toxic.

References

(1) Worthing, C.R., *Pesticide Manual,* 6th ed., p. 329, British Crop Protection Council (1979).
(2) Spencer, E.Y., *Guide to the Chemicals Used in Crop Protection,* 6th ed., p. 325, London, Ontario, Agriculture Canada (January 1973).
(3) Lush, G.B. and Stevenson, H.A., British Patent 820,180, September 16, 1959, assigned to Boots Pure Drug Co., Ltd.
(4) Lord, J.G., British Patent 822,973, November 4, 1959, assigned to Boots Pure Drug Co., Ltd.
(5) Stevenson, H.A., Brooks, R.F., Lush, G.B. and Fraser, J., British Patent 825,875, December 23, 1959, assigned to Boots Pure Drug Co., Ltd.

MEFLUIDIDE

Function: Plant growth regulator (1)(2)

Chemical Name: N-[2,4-dimethyl-5-[[(trifluoromethyl)-sulfonyl]amino]phenyl]acetamide

Formula:

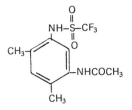

Trade Names: MBR-12325 (3-M Company)
Embark® (3-M Company)

Manufacture (2)

Trimethylamine (174 g, 2.95 mols), trifluoromethanesulfonyl fluoride (129 g, 0.826 mol), ethyl acetate (175 ml) and 5-amino-2,4-dimethylacetanilide (105 g, 0.59 mol) were heated for 1 day in about 1 liter of 10% sodium hydroxide solution. The resulting solution was steam distilled until no basic distillate was obtained. The residual solution was cooled with an ice bath, then filtered to remove insoluble impurities. The filtrate was extracted twice with 500 ml portions of dichloromethane. The aqueous phase was filtered, then the filtrate was cooled with an ice bath and acidified with cold dilute hydrochloric acid to provide a light yellow solid. The product was washed with water and dried to provide 155 g (85%) of 5-acetamido-2,4-dimethyltrifluoromethanesulfonanilide, MP 170° to 176°C. Further purification was effected by recrystallization from acetonitrile to provide product with MP 181° to 184°C.

An alternative work-up of the reaction mixture from the pressure reactor is to pour it into dilute hydrochloric acid, filter the crude product and dry it, or extract rather than filter, followed by drying and isolation of the product.

Toxicity

The acute oral LD_{50} for rats is over 4,000 mg/kg which is very slightly toxic.

References

(1) Worthing, C.R., *Pesticide Manual,* 6th ed., p. 330, British Crop Protection Council (1979).
(2) Fridlinger, T.L., U.S. Patent 3,894,078, July 8, 1975, assigned to Minnesota Mining and Manufacturing Co.

MENAZON

Function: Aphicide (1)(2)

Chemical Name: S-[(4,6-diamino-1,3,5-triazin-2-yl)methyl]O,O-dimethylphosphorodithioate

Formula:

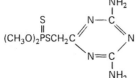

$$(CH_3O)_2\overset{S}{\overset{\|}{P}}SCH_2C$$

Trade Names: PP-175 (I.C.I.)
          Sayfos® (Plant Protection Ltd.)
          Saphizon® (Plant Protection Ltd.)
          Saphicol® (Plant Protection (Ltd.)
          Azidithion®

Manufacture (3)(4)(5)

Menazon may be made by one of two routes:

  (A) The reaction of biguanide with ethyl(dimethoxyphosphinothioylthio)acetate or

  (B) The reaction of sodium O,O-dimethylphosphorodithioate with 4,6-diamino-2-chloromethyl-1,3,5-triazine.

The following is a specific example of the second process above: A solution of 15.8 parts of O,O'-dimethylphosphorodithioic acid in 125 parts of methanol is treated with 5.3 parts of anhydrous sodium carbonate. 15.9 parts of 2-chloromethyl-4,6-diamino-s-triazine are then added and the mixture is heated under reflux for 7 hours. The reaction mixture is then cooled and filtered and the solid residue is washed with water and dried. It is crystallized from aqueous methanol or aqueous β-ethoxyethanol and there is thus obtained 2-dimethoxyphosphinothioylthiomethyl-4,6-diamino-s-triazine, MP 164° to 166°C.

Process Wastes and Their Control

Product Wastes: Menazon is compatible with all but strongly alkaline pesticides. It may be decomposed by reactive surfaces of some "inert" filters (B-3).

Toxicity

The acute oral LD_{50} value for rats is 1,950 mg/kg which is slightly toxic.

References

(1) Worthing, C.R., *Pesticide Manual,* 6th ed., p. 331, British Crop Protection Council (1979).
(2) Spencer, E.Y., *Guide to the Chemicals Used in Crop Protection,* 6th ed., p. 327, London, Ontario, Agriculture Canada (January 1973).
(3) Calderbank, A., Edgar, E.C. and Silk, J.A., U.S. Patent 3,169,904, February 16, 1965, assigned to Imperial Chemical Industries, Ltd.
(4) Calderbank, A., Edgar, E.C. and Silk, J.A., U.S. Patent 3,169,964, February 16, 1965, assigned to Imperial Chemical Industries, Ltd.
(5) Calderbank, A., Edgar, E.C. and Silk, J.A., British Patent 899,701, June 27, 1962, assigned to Imperial Chemical Industries, Ltd.

MEPHOSFOLAN

Function: Insecticide (1)(2)

Chemical Name: Diethyl-4-methyl-1,3-dithiolan-2-ylidenephosphoramidate

Formula:

$$(C_2H_5O)_2\overset{\overset{\displaystyle O}{\|}}{P}N{=}C\overset{\diagup S-CH-CH_3}{\underset{\diagdown S-CH_2}{\Big|}}$$

Trade Names: E147,470 (American Cyanamid)
Cytrolane® (American Cyanamid)
Dithiolane Iminophosphate

Manufacture (3)

Mephosfolan may be made by the reaction of diethylphosphorochloridate and 2-imino-4-methyldithiolane. To make the latter compound, dry hydrogen chloride is passed rapidly into a mixture of 25 parts of chloroform and 0.2 part of ethanol for ten minutes. After addition of 7.8 parts of propane-1,2-dithiol, 5.3 parts of cyanogen chloride are introduced over an 80-minute period. The solids which form are collected and amount to 10.9 parts or 89% of theory of 2-imino-4-methyl-1,3-dithiolane hydrochloride. When the latter is recrystallized from ethanol, it melts at 166° to 170°C with decomposition.

To a mixture of 10.6 parts of sodium bicarbonate and 11.2 parts of O,O-diethylphosphorochloridothioate in 20 parts of water and 10 parts of benzene are added over 15 minutes 10.2 parts of 2-imino-4-methyl-1,3-dithiolane hydrochloride in 10 parts of water. After mixing for nineteen hours at room temperature, the reaction mixture is extracted several times with benzene. The combined benzene layers, after washing, drying and stripping in the usual manner, leave 15.2 parts of yellow oil. Molecular distillation at 114° to 116°C/0.001 mm yields 9.7 parts of pure product, $N_D^{25} = 1.5814$.

Toxicity

The acute oral LD_{50} value for rats is 4 to 9 mg/kg which is highly to extremely toxic.

References

(1) Worthing, C.R., *Pesticide Manual,* 6th ed., p. 332, British Crop Protection Council (1979).
(2) Spencer, E.Y., *Guide to the Chemicals Used in Crop Protection,* 6th ed., p. 201, London, Ontario, Agriculture Canada (January 1973).
(3) American Cyanamid Co., British Patent 974,138 (November 4, 1964).

MERCURIC CHLORIDE

Function: Fungicide (1)(2)

Chemical Name: Mercury chloride

Formula: $HgCl_2$

Trade Name: None

Manufacture (3)

Mercuric chloride is made by the reaction of mercury with nitric acid followed by conversion to the sulfate using H_2SO_4 and then to the chloride by HCl as shown in Figure 37.

Figure 37: Production and Waste Schematic for Mercury Chloride

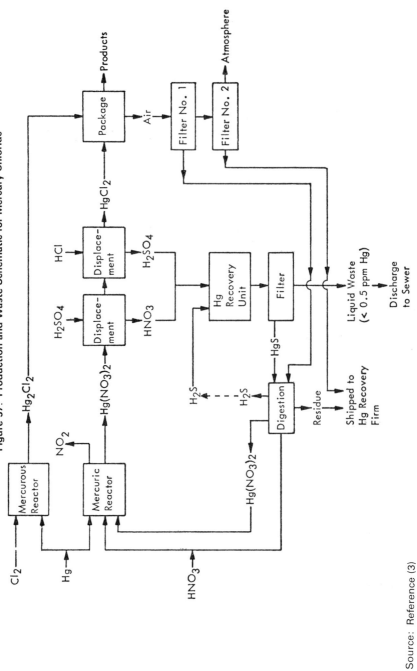

Source: Reference (3)

Toxicity

The acute oral LD_{50} for rats is 1.5 mg/kg which is extremely toxic.

Allowable Limits on Exposure and Use

Air: The threshold limit value for inorganic mercury compounds in air is 0.05 mg/m³ as of 1979. The tentative short term exposure limit is 0.15 mg/m³ (B-23).

The current U.S. Federal ceiling on inorganic mercury is 0.1 mg/m³ (B-21).

Water: In water, EPA has set (B-12) criteria as follows for mercury:

> 2.0 µg/l for domestic water supplies
> 0.05 µg/l for freshwater aquatic life and wildlife
> 0.10 µg/l for marine aquatic life

In water, EPA has subsequently suggested (B-26) a criterion to protect freshwater aquatic life from inorganic mercury of 0.064 µg/l as a 24-hour average; the concentration should not exceed 3.2 µg/l at any time. To protect saltwater aquatic life, 0.19 µg/l is a suggested 24-hour average; the concentration should not exceed 1.0 µg/l at any time. For human health protection from the toxic properties of mercury ingested through water or contaminated aquatic organisms, the criterion is 0.2 µg/l (B-26).

Product Use: In a series of actions (B-17), the first of which was dated March 22, 1972, the EPA cancelled all uses of mercury compounds as pesticides except the following:

1. As a fungicide in the treatment of textiles and fabrics intended for continuous outdoor use.

2. As a fungicide to control brown mold on freshly sawn lumber.

3. As a fungicide treatment to control Dutch elm disease.

5. As an in-can preservative in water-based paints and coatings.

5. As a fungicide in water-based paints and coatings used for exterior application.

6. As a seed disinfectant for treating such farm seeds as wheat, oats, barley, flax, sorghum, and cotton.

7. As a fungicide to treat "summer turf diseases" such as dollar spot, brown patch, leaf spot, copper spot, red thread, fairy ring, fusarium blight, helminthosporium, and snow mold.

8. As a fungicide to control "winter turf diseases" such as *Sclerotinia boreales,* and gray and pink snow mold subject to the following:

   (a) The use of these products shall be prohibited within 25 feet of any water body where fish are taken for human consumption.

   (b) There products can be applied only by or under the direct supervision of golf course superintendents.

   (c) The products will be classified as restricted use pesticides when they are reregistered and classified in accordance with section 4(c) of FEPCA.

Note: For purposes of the settlement agreements, "summer turf diseases" include not only the various pathogens which attack fine turf under temperate climate conditions, but also winter diseases which occur on areas of a golf course exclusive of greens, tees, and aprons; "winter diseases" refer to the forms of snow mold which can attack and damage the fine turf of greens, tees, and aprons.

The settlement allows no more production of mercury after August 31, 1978, for use on "summer turf diseases" or for protection of seeds. However, if the equivalent of 2 years previous production is reached prior to this date, this earlier date becomes the effective date of cancellation.

References

(1) Worthing, C.R., *Pesticide Manual,* 6th ed., p. 333, British Crop Protection Council (1979).
(2) Spencer, E.Y., *Guide to the Chemicals Used in Crop Protection,* 6th ed., p. 329, London, Ontario, Agriculture Canada (January 1973).
(3) Midwest Research Institute, *The Pollution Potential in Pesticide Manufacturing,* Washington, DC, Environmental Protection Agency (June 1972).

MERCUROUS CHLORIDE

Function: Fungicide (1)(2)

Chemical Name: Mercury chloride

Formula: Hg_2Cl_2

Trade Name: Calomel

Manufacture (3)

Mercurous chloride may be made by the direct chlorination of mercury (3). (See Figure 37).

Toxicity

The acute oral LD_{50} value for rats is 210 mg/kg which is moderately toxic.

See Mercuric Chloride for Allowable Limits on Exposure and Use.

References

(1) Worthing, C.R., *Pesticide Manual,* 6th ed., p. 335, British Crop Protection Council (1979).
(2) Spencer, E.Y., *Guide to the Chemicals Used in Crop Protection,* 6th ed., p. 330, London, Ontario, Agriculture Canada (January 1973).
(3) Midwest Research Institute, *The Pollution Potential in Pesticide Manufacturing,* Washington, DC, Environmental Protection Agency (June 1972).

METALAXYL

Function: Fungicide (1)(2)

Chemical Name: Methyl N-(2-methoxyacetyl)-N-(2,6-xylyl)-DL-alaninate

Formula:

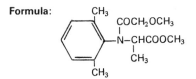

Trade Names: CGA 48 988 (Ciba-Geigy AG)
Ridomil® (Ciba-Geigy AG)

Manufacture (2)

(A) A mixture of 100 g of 2,6-dimethylaniline, 223 g of 2-bromopropionic acid methyl ester and 84 g of NaHCO₃ was stirred for 17 hours at 140°C, then cooled, diluted with 300 ml of water and extracted with diethyl ether. The extract was washed with a small amount of water, dried over sodium sulfate, filtered and the ether evaporated. After the excess 2-bromopropionic acid methyl ester had been distilled off, the crude product was distilled in a high vacuum.

(B) A mixture of 11 g of the ester obtained according to (A), 6.5 g of methoxyacetyl chloride, 2 ml of dimethyl formamide and 250 ml of absolute toluene was stirred at room temperature and refluxed for 1 hour. The solvent was evaporated off and the crude product then distilled in vacuo.

The D-forms of both cis-trans-isomers were obtained by acylating the pure D-form of α-(2,6-dimethylanilino)-propionic acid methyl ester with methoxyacetic acid or with one of the reactive derivatives thereof.

Toxicity

The acute oral LD_{50} value for rats is 30 mg/kg which is highly toxic.

References

(1) Worthing, C.R., *Pesticide Manual,* 6th ed., p. 336, British Crop Protection Council (1979).
(2) Ciba-Geigy AG, British Patent 1,500,581, February 8, 1978.

METALDEHYDE

Function: Molluscicide (1)(2)

Chemical Name: Acetaldehyde (polymer)

Formula:

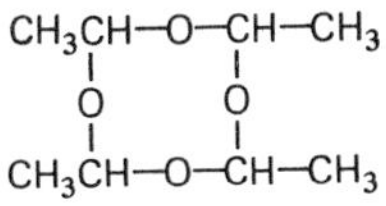

$$CH_3CH-O-CH-CH_3$$
$$\begin{matrix} | & & | \\ O & & O \\ | & & | \end{matrix}$$
$$CH_3CH-O-CH-CH_3$$

Trade Name: Meta®

Manufacture

Metaldehyde is made by the polymerization of acetaldehyde in aqueous ethanol in the presence of an acid catalyst.

Toxicity

The acute oral LD_{50} value for rats is 630 mg/kg (B-5) which is slightly toxic.

Allowable Limits on Exposure and Use

Product Use: In an action dated July 1, 1974 (B-17), the EPA specified that labeling for metaldehyde snail and slug baits must have the following statement on the front panel of the product label: "This pesticide may be fatal to children and dogs or other pets if eaten. Keep children and pets out of treated area."

The tolerances set by the EPA for metaldehyde in food are as follows (the *CFR* Reference is to Title 21):

	CFR Reference	Parts per Million
Strawberries	193.280	0.0

References

(1) Worthing, C.R., *Pesticide Manual,* 6th ed., p. 337, British Crop Protection Council (1979).
(2) Spencer, E.Y., *Guide to the Chemicals Used in Crop Protection,* 6th ed., p. 331, London, Ontario, Agriculture Canada (January 1973).

METHABENZTHIAZURON

Function: Herbicide (1)(2)

Chemical Name: N-2-benzothiazolyl-N,N'-dimethylurea

Formula:

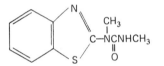

Trade Names: B-74,283 (Bayer)
Tribunil® (Bayer)

Manufacture (3)

Methabenzthiazuron may be made by the reaction of methyl isocyanate with 2-methyl-aminobenzothiazole.

Toxicity

The acute oral LD_{50} value for rats is more than 2,500 mg/kg which is slightly toxic.

References

(1) Worthing, C.R., *Pesticide Manual,* 6th ed., p. 339, British Crop Protection Council (1979).
(2) Hack, H., Eve, L. and Schafer, W., British Patent 1,085,430, October 4, 1967, assigned to Farbenfabriken Bayer AG.
(3) Searle, N.E., U.S. Patent 2,756,135, July 24, 1956, assigned to DuPont.

METHAMIDOPHOS

Function: Insecticide and acaricide (1)(2)

Chemical Name: O,S-dimethyl phosphoramidothioate

Formula:

$$CH_3O \diagdown \overset{\overset{O}{\|}}{P} NH_2$$
$$CH_3S \diagup$$

Trade Names: Bayer 71628 (Bayer)
SRA 5172 (Bayer)
Tamaron® (Bayer)
Ortho 9006 (Chevron Chemical)
Monitor (Chevron Chemical)

Manufacture (3)

Methamidophos may be made by the isomerization of O,O-dimethyl phosphoramidothioate. The following is a specific example (3) of the conduct of the process. A 130-gram portion of O,O-dimethylchlorophosphorothioate dissolved in 600 ml benzene was charged to a flask and cooled in an ice bath. Through this solution were passed 36 grams gaseous ammonia. The temperature was held at 10° to 15°C. The solids were allowed to settle; the solution was filtered and the salt cake was washed with benzene. The solution was then stripped to 60°C at 20 mm Hg.

The stripped product was combined with a 100-ml portion of methyl iodide and refluxed for 6 hours. The mixture was then stripped again to 60°C at 20 mm Hg and the residual oil was dissolved in 570 ml of an 80% dichloromethane-20% hexane solvent with stirring, the solution was filtered and the solids were removed. The solvent was stripped from the filtrate to 60°C at 20 mm Hg leaving 98 grams of O-methyl-S-methyl phosphoroamidothioate. This compound was observed as a pale yellow liquid of moderate viscosity which crystallizes on standing, melting completely at 32°C.

Process Wastes and Their Control

Product Wastes: This compound is stable in the pH range of 3 to 8. Hydrolysis by acids and alkalis increased with higher temperatures (B-3).

Toxicity

The acute oral LD_{50} value for rats is 30 mg/kg which is highly toxic.

Allowable Limits on Exposure and Use

Product Use: The tolerances set by the EPA for methamidophos in or on raw agricultural commodities are as follows:

	40 *CFR* Reference	Parts per Million
Broccoli	180.315	1.0
Brussels sprouts	180.315	1.0
Cabbage	180.315	1.0
Cauliflower	180.315	2.0
Cotton, seed	180.315	0.1 N
Cucumbers	180.315	1.0
Eggplant	180.315	1.0
Lettuce	180.315	2.0
Melons	180.315	0.5
Peppers	180.315	1.0
Potatoes	180.315	0.1 N
Tomatoes	180.315	2.0

References

(1) Worthing, C.R., *Pesticide Manual,* 6th ed., p. 340, British Crop Protection Council (1979).
(2) Spencer, E.Y., *Guide to the Chemicals Used in Crop Protection,* 6th ed., p. 333, London, Ontario, Agriculture Canada (January 1973).
(3) Magee, P.S., U.S. Patent 3,309,266, March 14, 1967, assigned to Chevron Research Co.

METHAM-SODIUM

Function: Soil fungicide, nematocide and herbicide (1)(2)

Chemical Name: Methylcarbamodithioic acid sodium salt

Formula:
$$\underset{CH_3NHCSNa}{\overset{\displaystyle S}{\underset{\|}{}}}$$

Trade Names: N-869 (Stauffer)
Vapam® (Stauffer)
VPM® (DuPont)
Maposol®
Sistan®
Unifume®
Vitafume®
SMDC

Manufacture (3)(4)

Metham-sodium may be made by reacting methylamine and carbon disulfide in the presence of caustic soda.

Process Wastes and Their Control

Product Wastes: This compound reacts slowly with moisture to liberate the toxic gas, methyl isothiocyanate, the active agent in its use as a soil fungicide or sterilant. The metham is stabilized in concentrated aqueous solution, but liberates CH_3NCS upon dilution. Hydrolysis to CS_2 should be an effective disposal method if excess acid (2- to 3-fold) is used (B-3).

Toxicity

The acute oral LD_{50} for rats is 820 mg/kg which is slightly toxic.

References

(1) Worthing, C.R., *Pesticide Manual,* 6th ed., p. 341, British Crop Protection Council (1979).
(2) Spencer, E.Y., *Guide to the Chemicals Used in Crop Protection,* 6th ed., p. 334, London, Ontario, Agriculture Canada (January 1973).
(3) Dorman, S.C. and Lindquist, A.B., U.S. Patent 2,766,554, October 16, 1956, assigned to Stauffer Chemical Co.
(4) Dorman, S.C. and Lindquist, A.B., U.S. Patent 2,791,605, May 7, 1957, assigned to Stauffer Chemical Co.

METHAZOLE

Function: Herbicide (1)

Chemical Name: 2-(3,4-dichlorophenyl)-4-methyl-1,2,4-oxadiazolidine-3,5-dione

Formula:

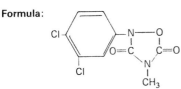

Trade Names: VCS 438 (Velsicol Chemical Corp.)
Probe® (Velsicol Chemical Corp.)
Bioxone®
Paxilon® (Fisons)
Tunic®

Manufacture (2)

Methazole may be made by the reaction of methyl isocyanate with N-(3,4-dichlorophenyl) hydroxylamine to give 1-(3-4-dichlorophenyl)-1-hydroxy-3-methyl urea. The intermediate 1-(3,4-dichlorophenyl)-1-hydroxy-3-methyl urea was dissolved in a cooled (10°C) 2N aqueous sodium hydroxide solution. Ethyl chloroformate was added dropwise at 10° to 15°C with stirring. The stirring was continued for about ½ hour after the addition was completed. The desired compound, which precipitated as formed, was removed by filtration, washed with water and dried.

Toxicity

The acute oral LD_{50} value for rats is 1,350 mg/kg which is slightly toxic.

Allowable Limits on Exposure and Use

Product Use: The tolerances set by the EPA for methazole in or on raw agricultural commodities are as follows:

	40 *CFR* Reference	Parts per Million
Cotton, seed	180.357	0.1

References

(1) Worthing, C.R., *Pesticide Manual,* 6th ed., p. 342, British Crop Protection Council (1979).
(2) Krenzer, J., U.S. Patent 3,437,664, April 8, 1969, assigned to Velsicol Chemical Corp.

METHIDATHION

Function: Insecticide (1)(2)

Chemical Name: S[[5-methoxy-2-oxo-1,3,4-thiadiazol-3(2H)-yl]methyl]O,O-dimethyl phosphorodithioate

Formula:

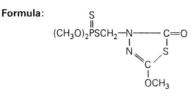

Trade Names: GS 13005 (Ciba-Geigy)
Supracide® (Ciba-Geigy)
Ultracide® (Ciba-Geigy)
Ustracide®

Manufacture (3)(4)

132 parts of 5-methoxy-1,3,4-thiadiazol-2(3H)-one (MP 113° to 114°C, produced by phosgenation of thiocarbazic acid-O-methyl ester) and 130 parts by volume of an about 37% aqueous formaldehyde solution are heated until a clear solution is formed. 130 parts of a 5% aqueous sodium carbonate solution, cooled to 0°C, are then added to this mixture whereupon it crystallizes by scratching with a glass rod. The so-obtained 3-hydroxymethyl compound melts at 84° to 86°C.

A solution of 98 parts of this 3-hydroxymethyl compound in 200 parts by volume of chloroform is quickly added to a well-cooled solution of thionyl chloride in 200 parts by volume of chloroform. The mixture is refluxed for 1 hour. The easily volatile components are then distilled off in a water jet vacuum at a bath temperature of 40° to 50°C and the residue is fractionated under high vacuum. The 3-chloromethyl-5-methylthio-1,3,4-thiadiazol-2(3H)-one obtained is a pale yellow oil and boils at 78°C/0.1 Torr.

32 parts of 3-chloromethyl-5-methoxy-1,3,4-thiadiazol-2(3H)-one are added dropwise to a solution of 40 parts of potassium salt of O,O-dimethyl-dithiophosphoric acid in 200 parts by volume of methanol. The temperature rises temporarily to 35°C. The mixture is then stirred for 4 hours at room temperature and 100 parts of water are then added. The methanol is then distilled off essentially in a water jet vacuum at a bath temperature of about 35°C. The so-obtained O,O-dimethyl-S-[5-methoxy-1,3,4-thiadiazol-2(3H)-onyl-(3)-methyl] dithiophosphate forms a colorless oil. On molecular distillation it boils at 130°C/0.0001 Torr.

Toxicity

The acute oral LD_{50} value for rats is 25 to 43 mg/kg which is highly toxic.

Allowable Limits on Exposure and Use

Product Use: Tolerances set by the EPA for methidathion in or on raw agricultural commodities are as follows:

	40 *CFR* Reference	Parts per Million
Alfalfa	180.298	6.0
Alfalfa, hay	180.298	6.0
Almond hulls	180.298	6.0
Artichokes	180.298	0.05
Clover	180.298	6.0
Clover, hay	180.298	6.0
Cotton, seed	180.298	0.2
Fruits, stone	180.298	0.05
Grapefruit	180.298	2.0

(continued)

	40 *CFR* Reference	Parts per Million
Grasses	180.298	6.0
Grasses, hay	180.298	6.0
Lemons	180.298	2.0
Oranges	180.298	2.0
Peaches	180.298	0.05
Pecans	180.298	0.05
Potatoes	180.298	0.2
Sorghum, fodder	180.298	2.0
Sorghum, forage	180.298	2.0
Sorghum, grain	180.298	0.2
Sunflower, seed	180.298	0.5
Walnuts	180.298	0.05

References

(1) Worthing, C.R., *Pesticide Manual*, 6th ed., p. 343, British Crop Protection Council (1979).
(2) Spencer, E.Y., *Guide to the Chemicals Used in Crop Protection*, 6th ed., p. 335, London, Ontario, Agriculture Canada (January 1973).
(3) Rufenacht, K., U.S. Patent 3,230,230, January 18, 1966, assigned to J.R. Geigy AG.
(4) Rufenacht, K., U.S. Patent 3,240,668, March 15, 1966, assigned to J.R. Geigy AG.

METHIOCARB

Function: Insecticide (1)(2)

Chemical Name: 3,5-dimethyl-4-methylthiophenyl methylcarbamate

Formula:

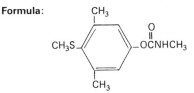

Trade Names: Bayer 37344 (Farbenfabriken Bayer AG)
H-321 (Farbenfabriken Bayer AG)
Mesurol® (Farbenfabriken Bayer AG)
Draza® (Farbenfabriken Bayer AG)
Metmercapturon
Mercaptodimethur

Manufacture (3)

Methiocarb may be made by the reaction of methyl isocyanate with 4-methylthio-3,5-xylenol.

Toxicity

The acute oral LD_{50} value for rats is 100 mg/kg which is moderately toxic.

References

(1) Worthing, C.R., *Pesticide Manual*, 6th ed., p. 344, British Crop Protection Council (1979).

(2) Spencer, E.Y., *Guide to the Chemicals Used in Crop Protection,* 6th ed., p. 328, London, Ontario, Agriculture Canada (January 1973).
(3) Schegk, E., Schrader, G. and Wedemeyer, K.F., U.S. Patent 3,313,684, April 11, 1967, assigned to Farbenfabriken Bayer AG.

METHOMYL

Function: Insecticide (1)(2)

Chemical Name: Methyl-N-[[(methylamino)carbonyl]oxy]ethanimidothioate

Formula:

$$CH_3S-\overset{\overset{\displaystyle CH_3}{|}}{C}=NOCNHCH_3$$
$$\underset{O}{\overset{||}{}}$$

Trade Names: DuPont 1179 (DuPont)
Lannate® (DuPont)

Manufacture (3)(4)

A total of 41 parts of acetonitrile, 250 parts of anhydrous ether and 48 parts of methyl mercaptan are stirred and maintained in an atmosphere of nitrogen as 36.5 parts of dry HCl is gradually added to the mixture. Near the end of the HCl addition the clear solution becomes cloudy and a precipitate begins to form. The white crystals that form are collected by filtration, rinsed with cold ether, and dried in a vacuum oven at room temperature. A total of 83 parts of hygroscopic, white, crystalline methyl thiolacetimidate hydrochloride is obtained, melting at 81° to 89°C. These imidates are reacted with hydroxylamine hydrochloride in the presence of pyridine to give the corresponding reaction intermediate hydroxamates.

To a stirred solution of 105 parts of methyl thiolacetohydroxamate in 400 parts of methylene chloride, at 25°C is added 60 parts of methyl isocyanate. The temperature rises during addition until the solvent begins to reflux. After the evolution of heat diminishes the mixture is heated to reflux and stirred for an additional 45 minutes. After removal of the solvent and drying, methomyl is obtained as a white solid.

Toxicity

The acute oral LD_{50} value for rats is 17 mg/kg (B-5) which is highly toxic.

In view of the relative paucity of data on the mutagenicity, carcinogenicity, and long-term oral toxicity of methomyl, estimates of the effects of chronic oral exposure at low levels cannot be made with any confidence. It is recommended that studies to produce such information be conducted before limits in drinking water can be established. The behavior of either aldicarb (which see) or methomyl in water and the possibility of their appearing in drinking water is not understood and should be the subject of high-priority research. Effects in humans have not been well-documented and efforts should be made in this direction (B-22).

Allowable Limits on Exposure and Use

Air: The threshold limit value for methomyl in air is 2.5 mg/m^3 as of 1979. Skin absorption of methomyl should be prevented so that the TLV is not invalidated (B-23).

Product Use: The tolerances set by the EPA for methomyl in or on raw agricultural commodities are as follows:

	40 *CFR* Reference	Parts per Million
Alfalfa	180.253	10.0
Apples	180.253	1.0
Apples	180.376	0.15
Asparagus	180.253	2.0
Avocados	180.253	2.0
Beans, dry	180.253	0.1 N
Beans, forage	180.253	10.0
Beans, succulent	180.253	2.0
Beets, tops	180.253	6.0
Blueberries	180.253	6.0
Broccoli	180.253	2.0
Brussels sprouts	180.253	2.0
Cabbage, chinese	180.253	5.0
Collards	180.253	6.0
Corn, fodder	180.253	10.0
Corn, forage	180.253	10.0
Corn, fresh (inc. sweet) (k+cwhr)	180.253	0.1 N
Corn, grain (inc. popcorn)	180.253	0.1 N
Cotton, seed	180.253	0.1 N
Cucurbits	180.253	0.2 N
Dandeloins	180.253	
Dandelions	180.253	6.0
Endive (escarole)	180.253	5.0
Grapefruit	180.253	2.0
Grapes	180.253	5.0
Grasses, Bermuda	180.253	10.0
Grasses, Bermuda, hay (dry, dehydrated)	180.253	40.0
Kale	180.253	6.0
Lemons	180.253	2.0
Lettuce	180.253	5.0
Mint, hay	180.253	2.0
Mustard greens	180.253	6.0
Nectarines	180.253	5.0
Oranges	180.253	2.0
Parsley	180.253	6.0
Peaches	180.253	5.0
Peanuts	180.253	0.1 N
Peanuts, hulls	180.253	0.1 N
Peas	180.253	5.0
Peas, vines	180.253	10.0
Peppers	180.253	2.0
Pomegranates	180.253	0.2 N
Sorghum, forage	180.253	1.0
Sorghum, grain	180.253	0.2 N
Soybeans	180.253	0.2 N
Soybeans, forage	180.253	10.0
Spinach	180.253	6.0
Swiss chard	180.253	6.0
Tangerines	180.253	2.0
Tomatoes	180.253	1.0
Turnip greens, tops	180.253	6.0
Vegetables, fruiting	180.253	0.2 N
Vegetables, leafy (exc. broccoli)	180.253	0.2 N
Vegetables, leafy (exc. brs. sprouts)	180.253	0.2 N
Vegetables, leafy (exc. cabbage)	180.253	0.2 N
Vegetables, leafy (exc. cauliflower)	180.253	0.2 N
Vegetables, leafy (exc. celery)	180.253	0.2 N
Vegetables, leafy (exc. endive)	180.253	0.2 N
Vegetables, leafy (exc. lettuce)	180.253	0.2 N
Vegetables, leafy (exc. spinach)	180.253	0.2 N
Vegetables, root crop	180.253	0.2 N

References

(1) Worthing, C.R., *Pesticide Manual,* 6th ed., p. 345, British Crop Protection Council (1979).
(2) Spencer, E.Y., *Guide to the Chemicals Used in Crop Protection,* 6th ed., p. 336, London, Ontario, Agriculture Canada (January 1973).
(3) Buchanan, J.B., U.S. Patent 3,576,834, April 27, 1971, assigned to DuPont.
(4) Buchanan, J.B., U.S. Patent 3,639,633, February 1, 1972, assigned to DuPont.

METHOPRENE

Function: Hormonal insecticide (1)(2)

Chemical Name: Isopropyl (E,E)-11-methoxy-3,7,11-trimethyl-2,4-dodecadienoate

Formula:

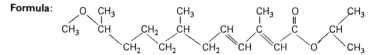

Trade Name: Altosid® (Zoecon Corp.)

Manufacture (1)(2)(3)(4)

To a mixture of 10 g of 7-methoxy-3,7-dimethyloctan-1-al, 17 g of diethyl 3-ethoxycarbonyl-2-methylprop-2-enyl phosphonate (77% trans), and 150 ml of dimethylformamide, under nitrogen, 0°C, with stirring, is added sodium isopropanolate (prepared from 1.5 g of sodium in 150 ml of isopropanol). After addition is complete, the reaction is stirred for 18 hours at room temperature and then worked up by extraction with hexane to yield isopropyl-11-methoxy-3,7,11-trimethyldodeca-2,4-dienoate (mostly trans-2, trans-4), which can be chromatographed and distilled for further purification.

Process Wastes and Their Control

Product Wastes: The manufacturer of this insect growth regulator states that the product is very nonpersistent and of no hazard to wildlife. They suggest disposal by burial, but, if permissible, the product could be burned (B-3).

Allowable Limits on Exposure and Use

Product Use: The tolerances set by the EPA for methoprene in or on raw agricultural commodities are as follows:

	40 *CFR* Reference	Parts per Million
Cattle, fat	180.359	0.3
Cattle, mbyp	180.359	0.1
Cattle, meat	180.359	0.1
Eggs	180.103	Exempt
Fish	180.103	Exempt
Goats, fat	180.103	Exempt
Goats, mbyp	180.103	Exempt
Goats, meat	180.103	Exempt
Grasses, forage	180.103	Exempt

(continued)

	40 *CFR* Reference	Parts per Million
Hogs, fat	180.103	Exempt
Hogs, mbyp	180.103	Exempt
Hogs, meat	180.103	Exempt
Horses, fat	180.103	Exempt
Horses, mbyp	180.103	Exempt
Horses, meat	180.103	Exempt
Legumes, forage	180.103	Exempt
Milk, fat	180.359	0.05
Poultry, fat	180.103	Exempt
Poultry, mbyp	180.103	Exempt
Poultry, meat	180.103	Exempt
Rice	180.103	Exempt
Rice, straw	180.103	Exempt
Sheep, fat	180.103	Exempt
Sheep, mbyp	180.103	Exempt
Sheep, meat	180.103	Exempt
Shellfish	180.103	Exempt

The tolerances set by the EPA for methoprene in food are as follows (the *CFR* Reference is to Title 21):

	CFR Reference	Parts per Million
Water, potable	193.285	Exempt

References

(1) Henrick, C.A. and Siddall, J.B., U.S. Patent 3,904,662, September 9, 1975, assigned to Zoecon Corp.
(2) Henrick, C.A. and Siddall, J.B., U.S. Patent 3,912,815, October 14, 1975, assigned to Zoecon Corp.
(3) Henrick, C.A., U.S. Patent 3,818,047, June 18, 1974.
(4) Henrick, C.A., U.S. Patent 3,865,874, February 11, 1975, assigned to Zoecon Corp.

METHOPROTRYNE

Function: Herbicide (1)(2)

Chemical Name: N-(3-methoxypropyl)-N'-(1-methylethyl)-6-(methylthio)-1,3,5-triazine-2,4-diamine

Formula:

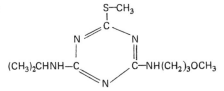

Trade Names: G 36393 (Ciba-Geigy)
Gesaran® (Ciba-Geigy)

Manufacture (3)

46 parts of cyanuric chloride are dissolved in 300 parts of chlorobenzene. Then, at −15° to −5°C, first 14 parts of isopropylamine dissolved in 22 parts of water and then 10 parts of sodium hydroxide dissolved in 40 parts of water are added dropwise. The whole is stirred until the reaction mixture has a neutral reaction, any traces of precipitated 2-chloro-4,6-bis-isopropylamino-s-triazine are removed and then, at room temperature, 23 parts of γ-methoxy-propylamine in 46 parts of water and afterwards 10 parts of sodium hydroxide in 40 parts of water are added. The whole is stirred at 40° to 50°C until the reaction mixture has a neutral reaction. The chlorobenzene is eliminated by steam distillation whereupon the difficultly soluble 2-chloro-4-isopropylamino-6-(γ-methoxy-propylamino)-s-triazine can be filtered off under suction and recrystallized. MP 112° to 114°C.

26 parts of 2-chloro-4-isopropylamino-6-(γ-methoxy-propylamino)-s-triazine are mixed with 100 ml of concentrated aqueous potassium hydrosulfide solution and stirred until a practically clear solution is obtained. The mixture is then neutralized with acetic acid whereby precipitation of the resulting 2-mercapto-4-isopropylamino-6-(γ-methoxy-propylamino)-s-triazine occurs. The crystalline product is separated by filtration and washed with cold water.

2.3 parts of sodium are dissolved in 200 parts of anhydrous methanol and then 25.7 parts of 2-mercapto-4-isopropylamino-6-(γ-methoxy-propylamino)-s-triazine are added. 20 parts of methyl iodide are then added dropwise and the reaction mixture is stirred at 40° to 50°C until it has a neutral reaction. The solvent is then distilled off, the residue is taken up in benzene, the solution is washed with 2N caustic soda lye and with water, the benzene is eliminated and the 2-methylmercapto-4-isopropylamino-6-(γ-methoxy-propylamino)-s-triazine is recrystallized from petroleum ether. MP 68° to 70°C.

Toxicity

The acute oral LD_{50} value for rats is in excess of 5,000 mg/kg which is insignificantly toxic.

References

(1) Worthing, C.R., *Pesticide Manual,* 6th ed., p. 346, British Crop Protection Council (1979).
(2) Spencer, E.Y., *Guide to the Chemicals Used in Crop Protection,* 6th ed., p. 352, London, Ontario, Agriculture Canada (January 1973).
(3) Knusli, E. and Gysin, H., U.S. Patent 3,326,914, June 20, 1967, assigned to J.R. Geigy AG.

2-METHOXY-4H-BENZO-1,3,2-DIOXAPHOSPHORINE-2-SULFIDE

Function: Insecticide (1)(2)

Chemical Name: 2-methoxy-4H-benzo-1,3,2-dioxaphosphorine-2-sulfide

Formula:

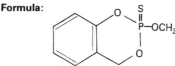

Trade Name: Salithion® (Sumitomo Chemical Co., Ltd.)

Manufacture (3)

To a mixture of 12.3 g (0.1 mol) of o-hydroxybenzyl alcohol and 100 ml of dry toluene at 80°C were added 27.6 g (0.2 mol) of anhydrous potassium carbonate and 0.5 g of copper powder. A solution of 16.5 g (0.1 mol) of methyl phosphorodichloridothionate in 50 ml of dry toluene was slowly added to the resulting mixture while being stirred. After the addition is finished, the mixture was allowed to react at 80° to 90°C for 15 hours. Thereafter the reaction mixture was cooled, and the solid substance was filtered off and washed with toluene. The combined filtrate and toluene washing mixture was washed with a cold 5% sodium hydroxide solution and with water, and then dried. The solvent in the mixture was driven off in vacuo, and the remaining solid was recrystallized from methanol. The yield of the crystals was 13 g, MP 51° to 53°C.

Toxicity

The acute oral LD_{50} for rats is 180 mg/kg which is moderately toxic.

References

(1) Worthing, C.R., *Pesticide Manual,* 6th ed., p. 347, British Crop Protection Council (1979).
(2) Spencer, E.Y., *Guide to the Chemicals Used in Crop Protection,* 6th ed., p. 453, London, Ontario, Agriculture Canada (January 1973).
(3) Sumitomo Chemical Co., Ltd., British Patent 987,378 (March 31, 1965).

METHOXYCHLOR

Function: Insecticide (1)(2)

Chemical Name: 1,1'-(2,2,2-trichloroethylidene)bis(4-methoxybenzene)

Formula:

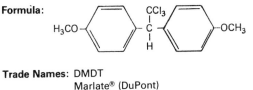

Trade Names: DMDT
Marlate® (DuPont)
Moxie®

Manufacture

Methoxychlor may be made by the condensation of one molecule of chloral with two molecules of anisole in the presence of an acidic condensing agent.

Process Wastes and Their Control

Air: Air emissions from methoxychlor manufacture have been reported (B-15) to consist of 0.5 kg methoxychlor and 0.5 kg anisole per metric ton of pesticide produced.

Product Wastes: $(CH_3OC_6H_4)_2CHCCl_3$ is like DDT in that it is resistant to heat and oxidation. By some accounts it is less readily dehydrochlorinated by alkali than is DDT, but other data indicate little difference. The dehydrochlorination is catalyzed by heavy metals. Methoxychlor is dechlorinated by refluxing with sodium in isopropyl alcohol. It is described as resistant to ultraviolet light, but other studies have shown that it breaks down rapidly under UV in hexane solution. Incineration is recommended by the MCA as the disposal method for methoxychlor (B-3).

Toxicity

The acute oral LD_{50} value for rats is 6,000 mg/kg which is insignificantly toxic.

Methoxychlor, a close relative of DDT, has very low mammalian toxicity. In a 2-year feeding study no adverse effect was observed at 200 ppm in rats. On the basis of these chronic data an ADI was calculated at 0.1 mg/kg/day (B-22).

Allowable Limits on Exposure and Use

Air: The threshold limit value for methoxychlor in air is 10 mg/m^3 as of 1979 (B-23). The current U.S. Federal limit is 15 mg/m^3 (B-21).

Water: In water, EPA has set criteria (B-12) of 100 µg/l for domestic water supplies and 0.03 µg/l of methoxychlor for the protection of freshwater and marine aquatic life.

Product Use: The tolerances set by the EPA for methoxychlor in or on raw agricultural commodities are as follows:

	40 *CFR* Reference	Parts per Million
Alfalfa	180.120	100.0
Apples	180.120	14.0
Apricots	180.120	14.0
Asparagus	180.120	14.0
Barley, grain, stored (post-h)	180.120	2.0
Beans	180.120	14.0
Beets, greens (alone)	180.120	14.0
Beets, with tops	180.120	14.0
Beets, without tops	180.120	14.0
Blackberries	180.120	14.0
Blueberries (huckleberries)	180.120	14.0
Boysenberries	180.120	14.0
Broccoli	180.120	14.0
Brussels sprouts	180.120	14.0
Cabbage	180.120	14.0
Carrots	180.120	14.0
Cattle, fat of meat	180.120	3.0
Cauliflower	180.120	14.0
Cherries	180.120	14.0
Clover	180.120	100.0
Collards	180.120	14.0
Corn	180.120	14.0
Corn, grain, stored (post-h)	180.120	2.0
Cranberries	180.120	14.0
Cucumbers	180.120	14.0
Currants	180.120	14.0
Dewberries	180.120	14.0
Eggplant	180.120	14.0
Goats, fat of meat	180.120	3.0
Gooseberries	180.120	14.0
Grapes	180.120	14.0
Grasses, forage	180.120	100.0
Hogs, fat of meat	180.120	3.0
Horses, fat of meat	180.120	3.0
Kale	180.120	14.0
Kohlrabi	180.120	14.0
Lettuce	180.120	14.0
Loganberries	180.120	14.0

(continued)

	40 *CFR* Reference	Parts per Million
Melons	180.120	14.0
Milk, fat	180.120	1.25
Mushrooms	180.120	14.0
Nectarines	180.120	14.0
Oats, grain, stored (post-h)	180.120	2.0
Peaches	180.120	14.0
Peanuts	180.120	14.0
Peanuts, forage	180.120	100.0
Pears	180.120	14.0
Peas	180.120	14.0
Peas, cowpeas	180.120	100.0
Peppers	180.120	14.0
Pineapples	180.120	14.0
Plums (fresh prunes)	180.120	14.0
Potatoes	180.120	1.0
Pumpkins	180.120	14.0
Quinces	180.120	14.0
Radishes, tops	180.120	14.0
Radishes, with tops	180.120	14.0
Radishes, without tops	180.120	14.0
Raspberries	180.120	14.0
Rice, grain, stored (post-h)	180.120	2.0
Rutabaga tops	180.120	14.0
Rutabagas, with tops	180.120	14.0
Rutabagas, without tops	180.120	14.0
Rye, grain, stored (post-h)	180.120	2.0
Sheep, fat of meat	180.120	3.0
Sorghum, grain, stored (post-h)	180.120	2.0
Soybeans, forage	180.120	100.0
Spinach	180.120	14.0
Squash	180.120	14.0
Squash, summer	180.120	14.0
Strawberries	180.120	14.0
Sweet potatoes (pre & post-h)	180.120	7.0
Tomatoes	180.120	14.0
Turnips, greens	180.120	14.0
Turnips, with tops	180.120	14.0
Turnips, without tops	180.120	14.0
Wheat, grain, stored (post-h)	180.120	2.0
Yams (pre & post-h)	180.120	7.0
Youngberries	180.120	14.0

References

(1) Worthing, C.R., *Pesticide Manual,* 6th ed., p. 348, British Crop Protection Council (1979).
(2) Bousquet, E.W. and Goddin, A.H., U.S. Patent 2,420,928, May 20, 1947, assigned to DuPont.

4-METHOXY-3,3'-DIMETHYLBENZOPHENONE

Function: Herbicide (1)(2)

Chemical Name: (4-Methoxy-3-methylphenyl)-(3-methylphenyl)methanone

Formula:

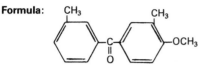

Trade Names: NK-049 (Nippon Kayaku Co.)
Kayametone® (Nippon Kayaku Co.)

Manufacture (2)

13.3 g (0.1 mol) of aluminum chloride is added in small portions to a mixture of 18.3 g (0.15 mol) of o-methylanisole and 20 ml of benzene at a temperature not higher than 30°C. After the addition, 15.5 g (0.1 mol) of m-toluyl chloride is added dropwise to the mixture over 20 minutes while maintaining the temperature at or below 10°C. During this addition, a violent exothermic reaction takes place with hydrogen chloride being evolved. Thereafter, the reaction is continued at a temperature not higher than 20°C for 2 hours to give the reaction mixture which is a red viscous liquid. The reaction is further continued by keeping the mixture at 25° to 30°C for 1 hour. When the resulting reaction mixture is poured into a mixture of water (100 ml) and ice (100 g) containing a small amount of concentrated hydrochloric acid, the mixture is decomposed with evolution of heat whereby an organic substance is separated out. This organic substance is then extracted with 50 ml of benzene and the extract is washed twice with 5% sodium hydroxide and then with water, dried over anhydrous sodium sulfate and concentrated in vacuo to give 23.3 g (crude yield 97.1%) of a crude crystalline product. The crude product is then recrystallized from 30 ml of methanol to give 21.5 g (yield 89.5%) of a white crystalline product. MP 62° to 62.5°C.

Toxicity

The acute oral LD_{50} for rats is more than 4,000 mg/l which is very slightly toxic.

References

(1) Worthing, C.R., *Pesticide Manual,* 6th ed., p. 349, British Crop Protection Council (1979).
(2) Yamada, O., Kurozumi, A., Ishida, S., Futatsuya, F., Ito, K. and Yamamoto, H., U.S. Patent 3,873,304, March 25, 1975, assigned to Nippon Kayaku K.K.

2-METHOXYETHYL MERCURIC CHLORIDE

Function: Seed treatment (1)

Chemical Name: Chloro(2-methoxyethyl) mercury

Formula: $CH_3OCH_2CH_2HgCl$

Trade Names: Agallol®
Aretan®
Ceresan Universal Nassbeize®

Manufacture (2)

This compound may be made by passing ethylene into a solution of mercuric acetate in methanol, followed by acidification with HCl to precipitate the chloride product.

Toxicity

The acute oral LD_{50} value for rats is 30 mg/kg which is highly toxic.

References

(1) Worthing, C.R., *Pesticide Manual,* 6th ed., p. 350, British Crop Protection Council (1979).
(2) Schoeller, W. et al, *Ber. Deut. Chem. Ges.* 46, 2864 (1913).

2-METHOXYETHYLMERCURY SILICATE

Function: Seed treatment (1)

Chemical Name: (2-methoxyethyl)(trihydrogen orthosilicato)mercury

Formula:

$$CH_3OCH_2CH_2-Hg-O-\underset{\underset{OH}{|}}{\overset{\overset{OH}{|}}{Si}}-OH$$

Trade Name: Ceresan Universal Trockenbeize® (Bayer)

Manufacture

This material may be prepared by passing ethylene into mercuric acetate in methanol (2) followed by precipitation of the silicate.

Toxicity

The acute oral LD_{50} for rats is 1,140 mg/kg (B-5) which is slightly toxic.

References

(1) Worthing, C.R., *Pesticide Manual,* 6th ed., p. 351, British Crop Protection Council (1979).
(2) Schoeller, W. et al, *Ber. Deut. Chem. Ges.* 46, 2864 (1913).

METHYL BROMIDE

Function: Insecticidal and acaricidal fumigant (1)

Chemical Name: Bromomethane

Formula: CH_3Br

Trade Names: Embafume®
Metafume®
Metabrom®

Manufacture (2)

Methyl bromide is made by the reaction of bromine on methanol in the presence of sulfur as shown in Figure 38 and as portrayed by the following equation:

$$6CH_3OH + 3Br_2 + S \rightarrow 6CH_3Br + H_2SO_4 + 2H_2O$$

Figure 38: Production and Waste Schematic for Methyl Bromide (Dow Chemical)

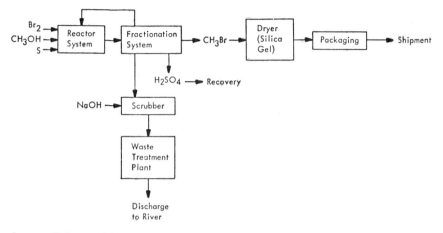

Source: Reference (2)

Process Wastes and Their Control

Air: Air emissions from methyl bromide manufacture have been reported (B-15) to consist of 1.0 kg hydrogen bromide, 0.5 kg bromine and 0.5 kg methyl bromide per metric ton of pesticide produced.

The fumes are scrubbed with NaOH, treated and discharged. The wastewater from the scrubbing process may possibly contain Na_2SO_4.

Product Wastes: The MCA recommended disposal procedures are the same as for EDB, but only the latter is at all practical, i.e., spray the gas into the fire box of an incinerator equipped with an afterburner and scrubber (alkali). However, incineration of such a toxic gas by the layman appears too hazardous to be recommended. Methyl bromide is apparently degraded rapidly in the sunlight in air and a preferable disposal procedure for the layman would be to release small amounts slowly to the atmosphere in a well-ventilated outdoor location. The cylinders can also be returned to the manufacturers (B-3).

Toxicity

Methyl bromide is a very toxic gas (BP = 3.6°C) at room temperature, and the concentration should not be allowed to exceed 20 ppm in air used for breathing. It is ordinarily marketed as a condensed liquid (which may have a warning agent added) in a pressurized container.

Allowable Limits on Exposure and Use

Air: The threshold limit value for methyl bromide in air is 60 mg/m^3 (15 ppm) as of 1979. Intended changes will result in a TLV of 20 mg/m^3 (5 ppm) and a short term exposure limit of 60 mg/m^3 (15 ppm). It is further noted that cutaneous absorption of methyl bromide should be prevented so the TLV is not invalidated (B-23).

The current U.S. Federal limit is 80 mg/m^3 (20 ppm) (B-21).

Water: For methyl bromide the EPA (B-26) has set up a criterion to protect freshwater aquatic life of 140 μg/l on a 24-hour average; the concentration should never exceed 320 μg/l at any time. To protect saltwater aquatic life, a criterion of 170 μg/l is suggested as a 24-hour average; the concentration should never exceed 380 μg/l.

For the protection of human health from the toxic properties of methyl bromide ingested through water and through contaminated aquatic organisms, a criterion level of 2 μg/l is suggested by EPA (B-26).

Product Use: The tolerances set by the EPA for methyl bromide in or on raw agricultural commodities are as follows:

	40 *CFR* Reference	Parts per Million
Alfalfa, hay (post-h)	180.123	50.0
Almonds (post-h)	180.123	200.0
Apples (post-h)	180.123	5.0
Apricots (post-h)	180.123	20.0
Artichokes, Jerusalem (post-h)	180.123	30.0
Asparagus (post-h)	180.123	100.0
Avocados (post-h)	180.123	75.0
Barley (post-h)	180.123	50.0
Beans (post-h)	180.123	50.0
Beans, green (post-h)	180.123	50.0
Beans, lima (post-h)	180.123	50.0
Beans, snap (post-h)	180.123	50.0
Beets, garden, roots (post-h)	180.123	30.0
Beets, sugar, roots (post-h)	180.123	30.0
Brazil nuts (post-h)	180.123	200.0
Bush nuts (post-h)	180.123	200.0
Butternuts (post-h)	180.123	200.0
Cabbage (post-h)	180.123	50.0
Cantaloupes (post-h)	180.123	20.0
Carrots (post-h)	180.123	30.0
Cherries (post-h)	180.123	20.0
Chestnuts (post-h)	180.123	200.0
Cipollini, bulb (post-h)	180.123	50.0
Citrus citron (post-h)	180.123	30.0
Cocoa beans (post-h)	180.123	50.0
Coffee beans (post-h)	180.123	75.0
Copra (post-h)	180.123	100.0
Corn (post-h)	180.123	50.0
Corn, pop (post-h)	180.123	240.0
Corn, sweet (k+cwhr) (post-h)	180.123	50.0
Cotton, seed (post-h)	180.123	200.0
Cucumbers (post-h)	180.123	30.0
Cumin, seed (post-h)	180.123	100.0
Eggplant (post-h)	180.123	20.0
Filberts (hazelnuts) (post-h)	180.123	200.0
Garlic (post-h)	180.123	50.0
Ginger, roots (post-h)	180.123	100.0
Grapefruit (post-h)	180.123	30.0
Grapes (post-h)	180.123	20.0
Hickory nuts (post-h)	180.123	200.0
Honeydew (post-h)	180.123	20.0
Horseradish (post-h)	180.123	30.0
Kumquats (post-h)	180.123	30.0
Lemons (post-h)	180.123	30.0
Limes (post-h)	180.123	30.0
Mangoes (post-h)	180.123	20.0
Muskmelons (post-h)	180.123	20.0

(continued)

	40 *CFR* Reference	Parts per Million
Nectarines (post-h)	180.123	20.0
Nuts, cashews (post-h)	180.123	200.0
Nuts, walnuts (post-h)	180.123	200.0
Oats (post-h)	180.123	50.0
Okra (post-h)	180.123	30.0
Onions (post-h)	180.123	20.0
Oranges (post-h)	180.123	30.0
Papayas (post-h)	180.123	20.0
Parsnips, roots (post-h)	180.123	30.0
Peaches (post-h)	180.123	20.0
Peanuts (post-h)	180.123	200.0
Pears (post-h)	180.123	5.0
Peas (post-h)	180.123	50.0
Peas, black-eyed (post-h)	180.123	50.0
Pecans (post-h)	180.123	200.0
Peppers (post-h)	180.123	30.0
Pimentos (post-h)	180.123	30.0
Pineapples (post-h)	180.123	20.0
Pistachio nuts (post-h)	180.123	200.0
Plums (fresh prunes) (post-h)	180.123	20.0
Pomegranates (post-h)	180.123	100.0
Potatoes (post-h)	180.123	75.0
Pumpkins (post-h)	180.123	20.0
Quinces (post-h)	180.123	5.0
Radishes (post-h)	180.123	30.0
Rice (post-h)	180.123	50.0
Rutabagas (post-h)	180.123	30.0
Rye (post-h)	180.123	50.0
Salsify, roots (post-h)	180.123	30.0
Sorghum, grain (milo) (post-h)	180.123	50.0
Soybeans (post-h)	180.123	200.0
Squash, summer (post-h)	180.123	30.0
Squash, winter (post-h)	180.123	20.0
Squash, zucchini (post-h)	180.123	20.0
Strawberries (pre-h)	180.123	30.0
Sweet potatoes (post-h)	180.123	75.0
Tangerines (post-h)	180.123	30.0
Timothy, hay (post-h)	180.123	50.0
Tomatoes (post-h)	180.123	20.0
Turnips, roots (post-h)	180.123	30.0
Watermelons (post-h)	180.123	20.0
Wheat (post-h)	180.123	50.0
Yams (post-h)	180.123	30.0

The tolerances set by the EPA for inorganic bromides in animal feed resulting from fumigation with methyl bromide are as follows (*CFR* Reference is to Title 21):

	CFR Reference	Parts per Million
Barley, milled fractions	561.260	125.0
Citrus pulp, dehydrated	561.260	90.0
Corn, milled fractions	561.260	125.0
Dog food	561.260	400.0
Oats, milled fractions	561.260	125.0
Rice, milled fractions	561.260	125.0
Rye, milled fractions	561.260	125.0
Sorghum, grain, milo, milled fractions	561.260	125.0
Wheat, milled fractions	561.260	125.0

The tolerances set by the EPA for inorganic bromides in food resulting from fumigation with methyl bromide are as follows (*CFR* Reference is to Title 21):

	CFR Reference	Parts per Million
Barley, flours	193.250	125.0
Biscuit mixes	193.250	125.0
Bread mixes	193.250	125.0
Breading	193.250	125.0
Cake mixes	193.250	125.0
Cereal flours & related prods.	193.250	125.0
Cereal grains, milled fractions	193.225	125.0
Cheese, parmesan	193.250	325.0
Cheese, rouquefort	193.250	325.0
Cookie mixes	193.250	125.0
Corn grits, malted beverages	193.230	125.0
Eggs, dried	193.250	400.0
Figs, dried	193.250	250.0
Herbs, processed	193.250	400.0
Macaroni products	193.250	125.0
Malt beverages	193.250	25.0
Milo (sorghum), flours	193.250	125.0
Noodle products	193.250	125.0
Oats, flours	193.250	125.0
Pie mixes	193.250	125.0
Rice, cracked	193.250	125.0
Rice, cracked, malted beverages	193.230	125.0
Rice, flours	193.250	125.0
Rye, flours	193.250	125.0
Soya, flour	193.250	125.0
Spices, processed	193.250	400.0
Tomato products, conc.	193.250	250.0
Vegetables, dried	193.250	125.0

References

(1) Worthing, C.R., *Pesticide Manual,* 6th ed., p. 354, British Crop Protection Council (1979).
(2) Midwest Research Institute, *The Pollution Potential in Pesticide Manufacturing,* Washington, DC, Environmental Protection Agency (June 1972).

METHYL CARBOPHENOTHION

Function: Acaricide and insecticide (1)(2)

Chemical Name: S-[[(4-chlorophenyl)thio] methyl]O,O-dimethyl phosphorodithioate

Formula:

Trade Names: R-1492 (Stauffer)
Tri-Me® (Stauffer)
Methyl Trithion® (Stauffer)

Manufacture (3)

The feed materials for methyl carbophenothion manufacture are p-chloromethylchlorophenyl sulfide and sodium dimethyl phosphorodithioate which react according to the equation:

$$(CH_3O)_2\overset{\overset{S}{\|}}{P}SNa + ClCH_2SC_6H_4Cl \longrightarrow (CH_3O)_2\overset{\overset{S}{\|}}{P}SCH_2SC_6H_4Cl + NaCl$$

In practice, a slight (20%) molar excess of the alkali dialkyl phosphorodithioate is used. The reaction is conducted at the reflux temperature of isopropanol (82.5°C). This reaction is conducted at atmospheric pressure. A reaction period of 2 hours under reflux is used. The reaction is conducted in the liquid phase in the presence of isopropanol as a solvent. No added catalyst is used in this condensation reaction. A jacketed kettle equipped with a reflux condenser may be used for the conduct of this reaction.

After completion of the reaction, the bulk of the isopropanol solvent may be distilled off and the residue treated with water and petroleum ether. After separation, a lower aqueous layer is discarded and the petroleum ether layer is water washed, dried and filtered. Finally, the petroleum ether is distilled off to give a light yellow colored liquid product in 60 to 65% yield.

Process Wastes and Their Control

Product Wastes: This compound is resistant to hydrolysis because of low water solubility. Fifty percent hydrolysis at an unspecified temperature requires 32 days at pH 8.3 (initial methyl trithion concentration of 0.6 ppm); 83 days at pH 4.6 (initial concentration of 0.6 ppm); and 64 days in distilled H_2O (initial concentration of 1.0 ppm) (B-3).

Toxicity

The acute oral LD_{50} for rats is 157 mg/kg which is moderately toxic.

References

(1) Worthing, C.R., *Pesticide Manual,* 6th ed., p. 111, British Crop Protection Council (1979).
(2) Spencer, E.Y., *Guide to the Chemicals Used in Crop Protection,* 6th ed., p. 343, London, Ontario, Agriculture Canada (January 1973).
(3) Fancer, L.W., U.S. Patent 2,793,224, May 21, 1957, assigned to Stauffer Chemical Co.

1,1-METHYLENEDI(THIOSEMICARBAZIDE)

Function: Rodenticide (1)(2)

Chemical Name: 2,2'-methylenebis(hydrazinecarbothioamide)

Formula:

$$NH_2-\overset{\overset{S}{\|}}{C}-NH-NH-CH_2-NH-NH-\overset{\overset{S}{\|}}{C}-NH_2$$

Trade Names: NK-15561 (Nippon Kayaku Co. Ltd.)
Kayanex® (Nippon Kayaku Co. Ltd.)

Manufacture (2)

137 parts of thiosemicarbazide and 1,500 parts of acetonitrile were mixed with agitation and

then 91 parts (about 1.5 times the theoretical amount) of 37% formalin were added into the mixture. The resultant mixture was reacted for 2 hours under reflux. Thereafter, the reacted mixture was cooled to 50°C and the resultant precipitated crystals were filtered. The thus obtained crystals were washed eight times with 50 parts of acetonitrile and were dried to yield 88.6 parts (91.5% of the theoretical amount) of white crystals. The white crystals have a decomposition point of 173° to 175°C and the results of elemental analysis in % are as follows: Calculated—C, 18.54; H, 5.19; N, 43.26; S, 33.01. Found—C, 18.50; H, 5.14; N, 43.26; S, 32.78.

The acid salt of the abovementioned compound can be obtained by adding an acid such as hydrochloric acid, sulfuric acid, etc., by the use of a usual method. Moreover, a small amount of formaldehyde thiosemicarbazone was collected from the filtrate.

Toxicity

The acute oral LD_{50} value for mice is about 30 mg/kg. No value for rats was available. It appears to be quite toxic.

References

(1) Worthing, C.R., *Pesticide Manual,* 6th ed., p. 355, British Crop Protection Council (1979).
(2) Yamamoto, H., Koike, K., Ohgushi, K. and Tokumitsu, I., U.S. Patent 3,826,641, July 30, 1974.

METHYL ISOTHIOCYANATE

Function: Nematocide (1)(2)

Chemical Name: Isothiocyanatomethane

Formula: $CH_3-N=C=S$

Trade Name: Trapex® (Schering AG)

Manufacture

Methyl isothiocyanate may be made by reacting sodium methyldithiocarbamate with ethyl chlorocarbonate.

Toxicity

The acute oral LD_{50} value for rats is 175 mg/kg which is moderately toxic.

References

(1) Worthing, C.R., *Pesticide Manual,* 6th ed., p. 356, British Crop Protection Council (1979).
(2) Pieroh, E.A. and Werres, H., U.S. Patent 3,113,908, December 10, 1963, assigned to Schering AG.

METIRAM

Function: Fungicide (1)(2)(3)

Chemical Name: Zineb (mixture with ethylene thiuram disulfide)

Formula:

$$[(-CH_2-N(H)-C(S)-S-S-C(S)-N(H)-CH_2-)_n(CH_2-N(H)-C(S)-S-Zn-S-C(S)-N(H)-CH_2-)_m]_x$$

n:m = 1:3   x is unknown

Trade Names: FMC 9102
Polyram® (BASF)

Manufacture

Metiram is made from ethylenebisdithiocarbamaic acid by oxidation using H_2O_2 and H_2SO_4 in the presence of zinc chloride.

Toxicity

The acute oral LD_{50} value for rats is more than 10,000 mg/kg which is insignificantly toxic.

Allowable Limits on Exposure and Use

Product Use: A rebuttable presumption against registration was issued on August 10, 1977 by EPA on the basis of oncogenicity, teratogenicity and hazard to wildlife.

References

(1) Worthing, C.R., *Pesticide Manual,* 6th ed., p. 550, British Crop Protection Council (1979).
(2) Spencer, E.Y., *Guide to the Chemicals Used in Crop Protection,* 6th ed., p. 350, London, Ontario, Agriculture Canada (January 1973).
(3) Flieg, O. and Windel, H., U.S. Patent 3,248,400, April 26, 1966, assigned to Badische Anilin- Soda-Fabrik AG.

METOBROMURON

Function: Herbicide (1)(2)(3)

Chemical Name: N'-(4-bromophenyl)-N-methoxy-N-methylurea

Formula:

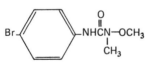

Trade Names: C-3126 (Ciba-Geigy)
Patoran® (Ciba-Geigy)

Manufacture

Metobromuron may be made by bromination of 3-phenyl-1-methoxy-1-methylurea (3): 27 g of N-phenyl-N'-methoxy-N'-methylurea are dissolved in 100 cc of glacial acetic acid, 12.3 g of anhydrous sodium acetate are added, and the bromination is carried out at 70°C with 26.3 g Br_2 in glacial acetic acid.

The yield of crude N-4-bromophenyl-N'-methyl-N'-methoxyurea is 34 g. The crude product melts at 91° to 94°C, and, when recrystallized from cyclohexane, the product melts at 95° to 96°C.

Alternatively, metobromuron may be made from p-bromophenyl isocyanate and O,N-dimethylhydroxylamine.

Process Wastes and Their Control

Product Wastes: See "Linuron" for disposal procedures (B-3).

Toxicity

The acute oral LD_{50} value for rats is 2,000 to 3,000 mg/kg which is slightly toxic.

Allowable Limits on Exposure and Use

Product Use: The tolerances set by the EPA for metobromuron in or on raw agricultural commodities are as follows:

	40 *CFR* Reference	Parts per Million
Potatoes	180.250	0.2

References

(1) Worthing, C.R., *Pesticide Manual,* 6th ed., p. 359, British Crop Protection Council (1979).
(2) Spencer, E.Y., *Guide to the Chemicals Used in Crop Protection,* 6th ed., p. 351, London, Ontario, Agriculture Canada (January 1973).
(3) Martin, H., Aebi, H. and Ebner, L., U.S. Patent 3,223,721, December 14, 1965, assigned to Ciba, Ltd.

METOLACHLOR

Function: Plant growth regulator and herbicide (1)(2)

Chemical Name: 2-Chloro-N-(2-ethyl-6-methylphenyl)-N-(2-methoxy-1-methylethyl) acetamide

Formula:

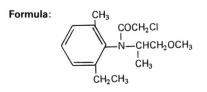

Trade Names: CGA 24705 (Ciba-Geigy Ltd.)
Dual® (Ciba-Geigy Ltd.)

Manufacture (2)

(A) A mixture of 540 g (4.0 mols) of 2-ethyl-6-methylaniline and 306 g (2.0 mols) of 2-bromo-1-methoxypropane is heated for 40 hours at reduced pressure (10 mm Hg) to 120°C with stirring. After it has cooled, the light red, viscous solution is diluted with 200 ml of water and made alkaline with 210 ml of concentrated sodium hydroxide solution. The precipitated product is taken up in ether, and the ethereal solution is washed neutral with water, dried, and evaporated.

Distillation of the residue yields pure-2-ethyl-6-methyl-N-[1'-methoxy-prop-2'-yl]-aniline which boils at 64° to 66°C/0.07 Torr.

(B) A solution of 9.7 g (0.047 mol) of the intermediate described under (A) and 5.05 g (0.05 mol) of triethylamine in 30 ml of benzene is treated dropwise with a solution of 5.65 g (0.05 mol) of chloroacetyl chloride in 10 ml of absolute benzene, and the mixture is further stirred for 2 hours at room temperature. The reaction mixture is diluted with ether, the solution repeatedly washed with water, and dried. The solvent mixture is evaporated off in vacuo to give pure-2-ethyl-6-methyl-N-[1'-methoxy-prop-2'-yl]-chloroacetanilide.

Toxicity

The acute oral LD_{50} value for rats is 2,780 mg/kg which is slightly toxic.

Allowable Limits on Exposure and Use

Product Use: The tolerances set by the EPA for metolachlor in or on raw agricultural commodities are as follows:

	40 *CFR* Reference	Parts per Million
Cattle, mbyp	180.368	0.02
Cattle, meat	180.368	0.02
Corn, grain (exc. pop)	180.368	0.1
Eggs	180.368	0.02
Goats, fat	180.368	0.02
Goats, mbyp	180.368	0.02
Goats, meat	180.368	0.02
Hogs, fat	180.368	0.02
Hogs, mbyp	180.368	0.02
Hogs, meat	180.368	0.02
Horses, fat	180.368	0.02
Horses, mbyp	180.368	0.02
Horses, meat	180.368	0.02
Milk	180.368	0.02
Poultry, fat	180.368	0.02
Poultry, mbyp	180.368	0.02
Poultry, meat	180.368	0.02
Sheep, fat	180.368	0.02
Sheep, mbyp	180.368	0.02
Sheep, meat	180.368	0.02
Soybeans	180.368	0.1

Effective February 27, 1980, the EPA added the following limits on tolerances for metolachlor in or on the following raw agricultural commodities (Reference Title 40, *CFR* 180.368):

Commodity	Parts per Million
Corn, forage and fodder	1.0
Peanuts	0.1
Peanut, forage and hay	3.0
Peanut, hulls	1.0
Sorghum, forage and fodder	2.0
Sorghum grain	0.3
Soybean, forage and fodder	2.0

References

(1) Worthing, C.R., *Pesticide Manual,* 6th ed., p. 360, British Crop Protection Council (1979).
(2) Vogel, C. and Aebi, R., U.S. Patent 3,937,730, February 10, 1976, assigned to Ciba-Geigy Corp.

METOXURON

Function: Herbicide (1)

Chemical Name: N'-(3-chloro-4-methoxyphenyl)-N,N-dimethylurea

Formula:

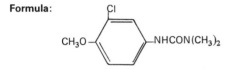

Trade Names: Herbicide 6602 (Sandoz, Ltd.)
Dosanex® (Sandoz, Ltd.)

Manufacture (2)

Metoxuron may be made by the reaction of dimethylamine with 3-chloro-4-methoxyphenylisocyanate. 73.4 g (0.4 mol) of 3-chloro-4-methoxyphenylisocyanate (BP 79° to 80°C at 0.2 mm of Hg) are added dropwise at 5° to 10°C during ¾ to 1 hour to 45 g of a 40% by weight aqueous dimethylamine solution (corresponding to 0.4 mol of dimethylamine); an exothermic reaction takes place and, as the reaction proceeds, the product is continuously precipitated in crystalline form. The crystalline precipitate is filtered with suction, washed with a little water and dried at 50° to 70°C in the vacuum of a water pump. The compound is obtained in the form of colorless crystals having a MP of 123° to 125°C with a yield of 82.5 to 87 g (90 to 95% of theory). After crystallization from ethanol, colorless crystals of MP 126° to 127°C are obtained.

The compound is likewise obtained when gaseous dimethylamine is introduced to saturation point into a solution of 0.1 mol of 3-chloro-4-methoxyphenylisocyanate in 200 ml of anhydrous ether at 20°C while stirring well and cooling, the resulting reaction product being filtered off with suction and dried.

Toxicity

The acute oral LD_{50} value for rats is 3,200 mg/kg which is slightly toxic.

References

(1) Worthing, C.R., *Pesticide Manual,* 6th ed., p. 361, British Crop Protection Council (1979).
(2) Schuler, M., British Patent 1,165,160, September 24, 1969, assigned to Sandoz, Ltd.

METRIBUZIN

Function: Herbicide (1)(2)(3)

Chemical Name: 4-Amino-6-(1,1-dimethylethyl)-3-methylthio-1,2,4-triazin-5(4H)one

Formula:

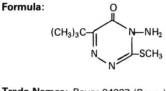

Trade Names: Bayer 94337 (Bayer)
Bayer 6159H (Bayer)
Bayer 6443H (Bayer)
DIC 1468 (Bayer)
Sencor® (Bayer)
Sencorex® (U.K.) (Bayer)
Sencoral® (France) (Bayer)
Lexone® (DuPont)

Manufacture (2)

(A) Preparation of 3-mercapto-4-amino-5-imino-6-tert-butyl-1,2,4-triazine—142.5 g (1 mol) thiocarbohydrazide hydrochloride were dissolved in 500 ml dimethyl sulfoxide and reacted in one hour with 166 g (1 mol) 2-tert-butylimino-3,3-dimethylbutyronitrile, with cooling to 20°C. Stirring was afterwards effected for one hour before the reaction product was precipitated with one liter of ice water. After recrystallization from ethanol, 3-mercapto-4-amino-5-imino-6-tert-butyl-1,2,4-triazine of the MP 181°C was obtained. Yield: 150 g (75% of the theory).

(B) Hydrolysis—10 parts by weight 3-mercapto-4-amino-5-imino-6-tert-butyl-1,2,4-triazine were heated to 100°C for 2 hours with 80 parts by weight ethanol and 100 parts by weight 1-normal hydrochloric acid. After cooling to 20°C, the crystalline reaction product separated, was filtered off with suction and was dried. There were obtained 9.3 parts by weight 3-mercapto-4-amino-6-tert-butyl-1,2,4-triazine-5-one of the MP 215° to 217°C. Yield: 93% of the theory.

This hydrolysis can also be carried out in a one-pot process immediately following the reaction of 2-tert-butyl-imino-3,3-dimethylbutyronitrile with thiocarbohydrazide hydrochloride without intermediate isolation of the 3-mercapto-4-amino-5-imino-6-tert-butyl-1,2,4-triazine.

(C) Methylation—4 parts by weight 3-mercapto-4-amino-6-tert-butyl-1,2,4-triazine-5-one were dissolved in a mixture of 11 parts by weight 2-normal sodium hydroxide solution and 4 parts by weight methanol, and 3.2 parts by weight methyl iodide were added at 0°C. The reaction mixture was subsequently stirred at 20°C for a further 4 hours. The reaction product crystallized out, was filtered off with suction, dried and recrystallized from benzene. There were obtained 3.52 parts by weight 3-methylmercapto-4-amino-6-tert-butyl-1,2,4-triazine-5-one of the MP 126° to 127°C. Yield: 82% of the theory.

Toxicity

The acute oral LD_{50} value for rats is 2,200 to 2,345 mg/kg which is slightly toxic.

Allowable Limits on Exposure and Use

Product Use: The tolerances set by the EPA for metribuzin in or on raw agricultural commodities are as follows:

	40 *CFR* Reference	Parts per Million
Alfalfa, green	180.332	2.0
Alfalfa, hay	180.332	7.0
Asparagus	180.332	0.05
Barley, grain	180.332	0.75
Barley, straw	180.332	1.0
Cattle, fat	180.332	0.7
Cattle, mbyp	180.332	0.7
Cattle, meat	180.332	0.7
Corn, fodder	180.332	0.1
Corn, forage	180.332	0.1
Corn, fresh (inc. sweet k+cwhr)	180.332	0.05
Corn, grain	180.332	0.05
Eggs	180.332	0.01
Goats, fat	180.332	0.7
Goats, mbyp	180.332	0.7
Goats, meat	180.332	0.7
Grasses	180.332	2.0
Grasses, hay	180.332	7.0
Hogs, fat	180.332	0.7
Hogs, mbyp	180.332	0.7
Hogs, meat	180.332	0.7
Horses, fat	180.332	0.7
Horses, mbyp	180.332	0.7
Horses, meat	180.332	0.7
Lentils, dry	180.332	0.05
Lentils, forage	180.332	0.5
Lentils, vine hay	180.332	0.05
Milk	180.332	0.05
Peas	180.332	0.1
Peas, dry	180.332	0.05
Peas, forage	180.332	0.5
Peas, vinehay	180.332	0.05
Potatoes	180.332	0.6
Poultry, fat	180.332	0.7
Poultry, mbyp	180.332	0.7
Poultry, meat	180.332	0.7
Sainfoin	180.332	2.0
Sainfoin, hay	180.332	7.0
Sheep, fat	180.332	0.7
Sheep, mbyp	180.332	0.7
Sheep, meat	180.332	0.7
Soybeans	180.332	0.1
Sugarcane	180.332	0.1
Tomatoes	180.332	0.1
Wheat, forage	180.332	2.0
Wheat, grain	180.332	0.75
Wheat, straw	180.332	1.0

The tolerances set by the EPA for metribuzin in animal feeds are as follows (the *CFR* Reference is to Title 21):

	CFR Reference	Parts per Million
Potatoes, processed (inc. chips)	561.41	3.0
Sugarcane bagasse	561.41	0.5
Sugarcane molasses	561.41	0.3
Tomato, pomace, dry	561.41	2.0

The tolerances set by the EPA for metribuzin in food are as follows (the *CFR* Reference is to Title 21):

	CFR Reference	Parts per Million
Potatoes, processed (inc. pot. chips)	193.25	3.0
Sugarcane molasses	193.25	2.0

References

(1) Worthing, C.R., *Pesticide Manual.* 6th ed., p. 362, British Crop Protection Council (1979).
(2) Jautelat, M., Kabbe, H.J. and Ley, K., U.S. Patent 3,752,808, August 14, 1973, assigned to Bayer AG.
(3) Fawzi, M.M., U.S. Patent 3,905,801, September 16, 1975, assigned to E.I. DuPont de Nemours & Co.

MEVINPHOS

Function: Insecticide and acaricide (1)(2)

Chemical Name: Methyl 3-[(dimethoxyphosphinyl)oxy]-2-butenoate

Formula:

$$(CH_3O)_2\overset{\overset{O}{\|}}{P}OC=\overset{\overset{O}{\|}}{C}HCOCH_3$$
$$\underset{CH_3}{|}$$

Trade Names: OS-2046 (Shell Chemical)
Phosdrin® (Shell Chemical)

Manufacture (3)

One convenient way of preparing the compound is to react trimethyl phosphite with methyl α-chloroacetoacetate, the reaction proceeding in accordance with the following equation:

$$CH_3CO \cdot CHClCOOCH_3 + (CH_3O)_3P \longrightarrow (CH_3O)_2 \overset{\overset{O}{\|}}{P} - O - \overset{\overset{CH_3}{|}}{C} = CH - COOCH_3 + CH_3Cl$$

In a typical example of this operation, 100 g of trimethyl phosphite was added dropwise to 121.5 g of methyl α-chloroacetoacetate over the course of a 1-hour reaction period during which the temperature rose gradually from room temperature to about 85°C. The resulting reaction mixture was then distilled in vacuo, there being recovered, as the fraction boiling between 106° and 107.5°C at 1 mm Hg, 123 g of a liquid product.

Process Wastes and Their Control

Air: Air emissions from mevinphos manufacture have been reported (B-15) to be 0.5 kg hydrocarbons and 0.5 kg mevinphos per metric ton of pesticide produced.

Product Wastes: Mevinphos is 50% hydrolyzed in aqueous solutions at an unspecified temperature in 1.4 hour at pH 11, 35 days at pH 7, and 120 days at pH 6. Decomposition is rapidly accomplished by lime sulfur (B-3).

Toxicity

The acute oral LD_{50} value for rats is 3 to 12 mg/kg which is highly to extremely toxic.

Allowable Limits on Exposure and Use

Air: The threshold limit value for mevinphos in air is 0.1 mg/m^3 (0.01 ppm) as of 1979. The tentative short term exposure limit is 0.3 mg/m^3 (0.03 ppm). It is noted that cutaneous absorption should be prevented so that the TLV is not invalidated.

Product Use: The tolerances set by the EPA for mevinphos in or on raw agricultural commodities are as follows:

	40 *CFR* Reference	Parts per Million
Alfalfa	180.157	1.0
Apples	180.157	0.5
Artichokes	180.157	1.0
Beans	180.157	0.25
Beets, garden (inc. tops)	180.157	1.0
Birdsfoot, trefoil, forage	180.157	1.0
Birdsfoot, trefoil, hay	180.157	1.0
Broccoli	180.157	1.0
Brussels sprouts	180.157	1.0
Cabbage	180.157	1.0
Carrots	180.157	0.25
Cauliflower	180.157	1.0
Celery	180.157	1.0
Cherries	180.157	1.0
Clover	180.157	1.0
Collards	180.157	1.0
Corn, forage	180.157	1.0
Corn, grain	180.157	0.25
Corn, pop, forage	180.157	1.0
Corn, pop, grain	180.157	0.25
Corn, sweet, forage	180.157	1.0
Corn, sweet, grain	180.157	0.25
Cucumbers	180.157	0.25
Eggplant	180.157	0.25
Grapefruit	180.157	0.25
Grapes	180.157	0.5
Kale	180.157	1.0
Lemons	180.157	0.25
Lettuce	180.157	0.5
Melons	180.157	0.5
Mustard, greens	180.157	1.0
Okra	180.157	0.25
Onions, green	180.157	0.25
Oranges	180.157	0.25
Parsley	180.157	1.0
Peaches	180.157	1.0
Pears	180.157	0.5
Peas	180.157	0.25
Peas, vines	180.157	1.0
Peppers	180.157	0.25
Plums	180.157	1.0
Potatoes	180.157	0.25
Raspberries	180.157	1.0
Sorghum, forage	180.157	1.0
Sorghum, grain	180.157	1.0

(continued)

	40 *CFR* Reference	Parts per Million
Spinach	180.157	1.0
Squash, summer	180.157	0.25
Strawberries	180.157	1.0
Tomatoes	180.157	0.25
Turnips	180.157	0.25
Turnips, tops	180.157	1.0

References

(1) Worthing, C.R., *Pesticide Manual,* 6th ed., p. 363, British Crop Protection Council (1979).
(2) Spencer, E.Y., *Guide to the Chemicals Used in Crop Protection,* 6th ed., p. 353, London, Ontario, Agriculture Canada (January 1973).
(3) Stiles, A.R., U.S. Patent 2,685,552, August 3, 1954, assigned to Shell Development Co.

MEXACARBATE

Function: Insecticide and acaricide (1)(2)

Chemical Name: 4-dimethylamino-3,5-dimethylphenyl methylcarbamate

Formula:

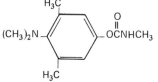

Trade Names: Dowco 139 (Dow Chemical Co.)
Zectran® (Dow Chemical Co.)

Manufacture (3)(4)

The following is a specific example (3) of the process: 80 g of sodium bicarbonate and 80 ml (0.86 mol) of dimethyl sulfate were added to a solution of 32 g (0.23 mol) of 4-amino-3,5-xylenol in 117 ml of water. An exothermic reaction took place during which time the temperature of the reaction mixture was maintained between 25° and 30°C with external cooling. The mixture was then allowed to stand overnight at room temperature whereupon the desired 4-dimethylamino-3,5-xylenol precipitated as a white solid. The latter was recovered by filtration, washed with pentane and air dried to obtain a purified 4-dimethylamino-3,5-xylenol melting from 85° to 92°C.

Then 25.5 ml (24.6 g; 0.45 mol) of methyl isocyanate and 5 drops of triethylamine catalyst were added to a solution of 71.5 g (0.45 mol) of 4-dimethylamino-3,5-dimethylphenol (MP 85° to 92°C) in 500 ml of hexane. The mixture was allowed to react at a temperature of about 35°C for a period of 8 hours and thereafter concentrated by heating on the steam bath to remove most of the hexane. The mixture was then cooled in an ice bath and seeded with crystals obtained from the reaction mixture to precipitate the desired 4-dimethylamino-3,5-dimethylphenyl methylcarbamate product. The latter was recovered by filtration and found to be a crystalline solid melting from 80° to 83°C.

Toxicity

The acute oral LD_{50} value for rats is between 15 and 63 mg/kg which is highly toxic.

Allowable Limits on Exposure and Use

Product Use: The tolerances set by the EPA for mexacarbate in or on raw agricultural commodities are as follows:

	40 *CFR* Reference	Parts per Million
Cherries	180.320	25.0
Corn, fodder	180.320	0.03
Corn, forage	180.320	0.03
Corn, fresh (inc. sweet) (k+cwhr)	180.320	0.03
Corn, grain, field	180.320	0.03
Corn, grain, pop	180.320	0.03
Peaches	180.320	15.0

References

(1) Martin, H. and Worthing, C.R., *Pesticide Manual,* 4th ed., p. 359, British Crop Protection Council (1974).
(2) Spencer, E.Y., *Guide to the Chemicals Used in Crop Protection,* 6th ed., p. 354, London, Ontario, Agriculture Canada (January 1973).
(3) Shulgin, A.T., U.S. Patent 3,084,098, April 2, 1963, assigned to Dow Chemical Co.
(4) Dow Chemical Co., British Patent 925,424, May 8, 1963.

MIREX

Function: Insecticide (particularly effective against fire ants) (1)(2)(3)(5)

Chemical Name: 1,1a,2,2,3,3a,4,5,5,5a,5b,6-dodecachlorooctahydro-1,3,4-metheno-1H-cyclobuta(cd)pentalene

Formula:

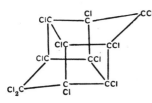

Trade Names: GC 1283
Dechlorane®

Manufacture (4)(6)

Mirex may be made by the dimerization of hexachlorocyclopentadiene in the presence of an aluminum chloride catalyst. The following is a specific example (4) of the conduct of this process.

One mol of hexachlorocyclopentadiene, dissolved in 7.0 mols of hexachlorobutadiene, and two-tenths mol of substantially anhydrous aluminum chloride is charged into a reactor provided with an agitator. The charge in the reactor is agitated, heated to a temperature of about 110°C and is maintained at that temperature for a period of five hours.

The product recovered is water-washed at room temperature to effect a removal of aluminum chloride, separated, and the bottom organic layer dried with anhydrous sodium sulfate. The dried product is fractionated under reduced pressure in an efficient column to remove hexachlorobutadiene and unreacted hexachlorocyclopentadiene. The residue is then cooled and a 45% yield of $C_{10}Cl_{12}$ recovered, based on the hexachlorocyclopentadiene charged to the process. After purification by recrystallization from benzene, the snow-white crystals of $C_{10}Cl_{12}$ product sublime above 240°C and analyzes 77.8% chlorine (theory 77.98% chlorine).

Process Wastes and Their Control

Product Wastes: This highly chlorinated compound is unaffected by mineral acids (HCl, HNO_3 and H_2SO_4). It would be expected to be extremely resistant to oxidation except at the high temperature of an efficient incinerator (B-3).

Toxicity

The acute oral LD_{50} value for rats is about 300 mg/kg which is moderately toxic.

Allowable Limits on Exposure and Use

Water: In water, EPA has set criteria (B-12) of 0.001 $\mu g/l$ of mirex for the protection of freshwater and marine aquatic life.

Product Use: The tolerances set by the EPA for mirex in or on raw agricultural commodities are as follows:

	40 *CFR* Reference	Parts per Million
Cattle, fat of meat	180.251	0.1 N
Eggs	180.251	0.1 N
Goats, fat of meat	180.251	0.1 N
Hogs, fat of meat	180.251	0.1 N
Horses, fat of meat	180.251	0.1 N
Milk, fat (= N in whole milk)	180.251	0.1 N
Poultry, fat of meat	180.251	0.1 N
Raw commodities not listed in 180.251	180.251	0.01 N
Sheep, fat of meat	180.251	0.1 N

As a consequence of EPA actions on October 20 and December 29, 1976 (B-17), all registered products containing mirex were effectively cancelled on December 1, 1977. (A technical mirex product made by Hooker Chemical Company is unaffected by this Settlement Agreement. However, since mirex produced under this registration may be used only in the formulation of other pesticide products, the registration was useless after December 1, 1977.) All existing stocks of mirex within the continental U.S. will not be sold, distributed, or used after June 30, 1978. A summary of mirex products, their uses, registration numbers, effective cancellation dates, and termination of uses of existing stocks as specified in the Mirex Settlement Agreement follows:

(1) Fire Ant Bait, Reg. No. 38962-1, used for controlling fire ants through broadcast and mound treatment; Fire Ant Bait "150," Reg. No. 38962-2, used for controlling fire ants through ground broadcast and mound treatment; Mirex Special Concentrate (25%), Reg. No. 38962-6, used for formulating products; Mirex Pelleted Bail 450, Reg. No. 38962-8, used for controlling Texas leaf cutting ants, harvester ants, and fire ants; A-C Mirex Pelleted Bait 450, Reg. No. 38962-9, used for controlling Texas leaf cutting ants, harvester ants, and fire ants; and Yellowjacket Stopper, Reg. No. 38962-10, used for controlling yellowjackets are cancelled; existing stocks could be sold, distributed, and used until December 31, 1977.

(2) Granular Bait 2X (now 10:5), Reg. No. 38962-3, used for controlling fire ants in the Federal-State cooperative program for imported fire ant suppression was effectively cancelled December 1, 1977; existing stocks were to be applied aerially only through December 31, 1977, and applied through ground broadcast and mound treatment only until June 30, 1978.

(3) Granular Bait 4X, Reg. No. 38962-4, used for controlling fire ants in the Federal-State cooperative program for imported fire ant suppression is cancelled and existing stocks may no longer be used except that 1,000 pounds of this bait can be used as a control in experiments until June 30, 1978.

(4) Harvester Bait 300, Reg. No. 38962-5, may only be used for the control of the pheidole ant, Argentine ant, and fire ant on pineapples in Hawaii. The effective date of cancellation for these uses was December 1, 1977; existing stocks as of December 1, 1977 may not be applied aerially after December 31, 1977, but may be sold and used (other than aerially) indefinitely.

(5) Mirex Technical Concentrate (80%), Reg. No. 38962-7, used for formulating products, was effectively cancelled on December 1, 1977; no stocks of Technical Mirex existing on December 1, 1977, shall be distributed, sold, or used.

The application of mirex is subject to the following restrictions:

(1) Aerial application: No longer permitted.

(2) Ground application.

    a. Permissible in all areas of infestation provided that there is no ground application to aquatic and heavily forested areas or areas where run-off or flooding will contaminate such areas.

    b. Treatment shall be confined to areas where the imported fire ants are causing significant problems.

Note: The following definition is included in order to clarify the ground application restrictions. Aquatic areas: encompasses without limitation estuaries, rivers, streams, wetlands (those land and water areas subject to inundation by tidal, riverine, or lacustrine flowage), lakes, ponds, and other bodies of water.

References

(1) Martin, H. and Worthing, C.R., *Pesticide Manual,* 5th ed., p. 368, British Crop Protection Council (January 1977).

(2) Spencer, E.Y., *Guide to the Chemicals Used in Crop Protection,* 6th ed., p. 359, London, Ontario, Agriculture Canada (January 1973).

(3) Gilbert, E.E., U.S. Patent 2,671,043, March 2, 1954, Reissue 24,750, assigned to Allied Chemical and Dye Corp.

(4) Johnson, A.N., U.S. Patent 2,996,553, August 15, 1961, assigned to Hooker Chemical Corp.

(5) Greenbaum, S.B., U.S. Patent 3,220,921, November 30, 1965, assigned to Hooker Chemical Corp.

(6) Gilbert, E.E., U.S. Patent 2,702,305, February 15, 1955, Reissue 24,397, November 26, 1957, assigned to Allied Chemical and Dye Corp.

MOLINATE

Function: Herbicide (1)(2)

Chemical Name: S-ethyl hexahydro-1H-azepine-1-carbothioate

Formula:

$$CH_2CH_2CH_2 \diagdown \atop CH_2CH_2CH_2 \diagup N - \overset{\overset{O}{\parallel}}{C}SC_2H_5$$

Trade Names: R-4572 (Stauffer Chemical Co.)
Ordam® (Stauffer Chemical Co.)
Hydram®

Manufacture (3)

A 500-cc flask was provided with stirrer, thermometer and addition funnel. A solution of 14.0 g (0.35 mol) of sodium hydroxide in 200 cc of water was charged to the flask and this was followed by the addition of 31.7 g (0.32 mol) of hexamethyleneimine in 100 cc of n-pentane. To the vigorously stirred mixture with cooling was then added 37.4 g (0.30 mol) of ethyl chlorothiolformate. The temperature of the reaction mixture was maintained at 15° to 20°C.

After the addition was completed, the mixture was stirred for an additional 5 minutes and then phase separated. The upper organic phase was washed with two 50-cc portions of dilute hydrochloric acid (5 cc of concentrated hydrochloric acid made up to a volume of 55 cc with water) and with three 50-cc portions of water. It was then dried over anhydrous magnesium sulfate, filtered and concentrated on the stream bath. The residual liquid was then fractionally distilled under vacuum. There was obtained 51.9 g (92.3% yield) of molinate, BP (10 mm) 136.5° to 137°C.

Process Wastes and Their Control

Product Wastes: This cyclic thiocarbamate, $\overline{(CH_2)_6} - N - CO - SC_2H_5$, is stable to hydrolysis by water, but is hydrolyzed by H_2SO_4 (B-3).

Toxicity

The acute oral LD_{50} value for rats is 720 mg/kg which is slightly toxic.

Allowable Limits on Exposure and Use

Product Use: The tolerances set by the EPA for molinate in or on raw agricultural commodities are as follows:

	40 *CFR* Reference	Parts per Million
Rice	180.228	0.1 N
Rice, straw	180.228	0.1 N

References

(1) Worthing, C.R., *Pesticide Manual,* 6th ed., p. 364, British Crop Protection Council (1979).
(2) Spencer, E.Y., *Guide to the Chemicals Used in Crop Protection,* 6th ed., p. 360, London, Ontario, Agriculture Canada (January 1973).
(3) Tilles, H. and Curtis, R., U.S. Patent 3,198,786, August 3, 1965, assigned to Stauffer Chemical Co.

MONALIDE

Function: Herbicide (1)(2)

Chemical Name: N-(4-Chlorophenyl)-2,2-dimethyl-pentanamide

Formula:

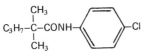

Trade Names: Schering 35,830 (Schering AG)
Potablan® (Schering AG)

Manufacture (2)

To a solution of 63.5 g (0.5 mol) of p-chloroaniline and 70.0 ml (0.05 mol) of triethylamine in 500 ml of toluene there are added dropwise with stirring at room temperature within about 15 minutes 74.0 g (0.5 mol) of α-α-dimethyl-valeryl chloride. In this operation the temperature rises to 60° to 70°C. After a further 2 hours stirring (without additional heating or cooling) 100 ml of water are added to dissolve the deposited triethylamine hydrochloride. Then, with stirring, the toluene is distilled off with steam under vacuum. The resulting deposit of α,α-dimethyl-valeric acid-(4-chloroanilide) is filtered with suction, washed with water and dried at about 50°C. Yield: 115 to 120 g (96 to 100% of the theoretical), MP 85° to 86°C. After recrystallization from petroleum ether (60° to 80°C), the product obtained had the MP 86° to 88°C.

Toxicity

The acute oral LD$_{50}$ value for rats is over 4,000 mg/kg which is very slightly toxic.

References

(1) Worthing, C.R., *Pesticide Manual,* 6th ed., p. 365, British Crop Protection Council (1979).
(2) Schering, A.G., British Patent 971,819, October 7, 1964.

MONOCROTOPHOS

Function: Insecticide (1)(2)

Chemical Name: Dimethyl-1-methyl-3-(methylamino)-3-oxo-1-propenyl phosphate

Formula:

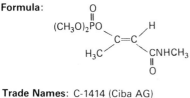

Trade Names: C-1414 (Ciba AG)
Nuvacron (Ciba AG)
SD9129 (Shell)
Azodrin (Shell)

Manufacture

Monocrotophos may be made by the reaction of trimethyl phosphite with 2-chloro-N-methyl-3-oxobutyramide.

Process Wastes and Their Control

Air: Air emissions from monocrotophos manufacture have been reported (B-15) to be 0.5 kg hydrocarbons and 0.5 kg monocrotophos per metric ton of pesticide produced.

Product Wastes: Treatment of this product with water produces hydrolysis at a rate that is almost independent of pH in the pH range of 1 to 7 (50% hydrolyzed at pH 1 to 7 in 22 to 23 days at 38°C). At higher pH's the rate of hydrolysis rapidly increases (B-3).

Toxicity

The acute oral LD_{50} value for rats is 14 to 23 mg/kg which is highly toxic.

Allowable Limits on Exposure and Use

Air: The threshold limit value for monocrotophos in air is 0.25 mg/m^3 as of 1979 (B-23).

Product Use: The tolerances set by the EPA for monocrotophos in or on raw agricultural commodities are as follows:

	40 *CFR* Reference	Parts per Million
Cotton, seed	180.296	0.1
Peanuts	180.296	0.05
Peanuts, hulls	180.296	0.5
Potatoes	180.296	0.1
Sugarcane	180.296	0.1
Tomatoes	180.296	0.5

The tolerances set by the EPA for monocrotophos in food are as follows (the *CFR* Reference is to Title 21):

	CFR Reference	Parts per Million
Tomato products, conc.	193.151	2.0

References

(1) Worthing, C.R., *Pesticide Manual*, 6th ed., p. 366, British Crop Protection Council (1979).
(2) Spencer, E.Y., *Guide to the Chemicals Used in Crop Protection*, 6th ed., p. 361, London, Ontario, Agriculture Canada (Janaury 1973).

MONOLINURON

Function: Herbicide (1)(2)

Chemical Name: N'-(4-chlorophenyl)-N-methoxy-N-methylurea

Formula:

Trade Names: HOE 2747 (Farbwerke Hoechst)
Aresin® (Farbwerke Hoechst)
Arresin®
Afesin®
Gramonol®

Manufacture (3)

50 g of freshly distilled para-chlorophenyl isocyanate are dissolved in 100 ml of benzene and, while shaking, mixed with a solution of 20 g of O,N-dimethyl hydroxylamine in 100 ml of benzene. The heat produced is compensated by cooling with ice water. After standing for 24 hours at room temperature, the reaction mixture is advantageously cooled with a freezing mixture of ice and sodium chloride in order to complete the crystallization.

Monolinuron is obtained in the form of crystals and is then isolated by filtering with suction. After drying under reduced pressure, the colorless substance melts at 76° to 78°C. The yield amounts to 69 g (98.5% of the theoretical yield).

Toxicity

The acute oral LD_{50} value for rats is 2,250 mg/kg which is slightly toxic.

References

(1) Worthing, C.R., *Pesticide Manual,* 6th ed., p. 367, British Crop Protection Council (1979).
(2) Spencer, E.Y., *Guide to the Chemicals Used in Crop Protection,* 6th ed., p. 362, London, Ontario, Agriculture Canada (January 1973).
(3) Scherer, O. and Heller, P., U.S. Patent 3,079,244, February 26, 1963, assigned to Farbwerke Hoechst AG.

MONURON

Function: Herbicide (1)(2)

Chemical Name: N'-(4-chlorophenyl)-N,N-dimethylurea

Formula:

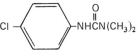

Trade Names: Telvar® (DuPont)
Karmex W®
CMU
Chlorfenidim

Manufacture (3)(4)

The feed materials for monuron manufacture are p-chloroaniline, phosgene and dimethylamine. These may be combined according to differing routes as follows. One route involves reaction of the p-chloroaniline with phosgene to give p-chlorophenyl carbamyl chloride which is turn may be dehydrochlorinated to give p-chlorophenyl isocyanate. The isocyanate is then reacted with dimethylamine to give the final product according to the following reaction sequence:

$$p\text{-}ClC_6H_4NH_2 + COCl_2 \rightarrow p\text{-}ClC_6H_4NHCOCl + HCl$$

$$p\text{-}ClC_6H_4NHCOCl \rightarrow p\text{-}ClC_6H_4NCO + HCl$$

$$p\text{-}ClC_6H_4NCO + HN(CH_3)_2 \rightarrow p\text{-}ClC_6H_4NHCON(CH_3)_2$$

Another route involves reaction of dimethylamine with phosgene to give an intermediate which is reacted with p-chloroaniline according to the reaction sequence:

$$(CH_3)_2NH + COCl_2 \rightarrow (CH_3)_2NCOCl + HCl$$

$$p\text{-}ClC_6H_4NH_2 + (CH_3)_2NCOCl \rightarrow p\text{-}ClC_6H_4NHCON(CH_3)_2 + HCl$$

A third possibility involves the chlorination of phenyl isocyanate to p-chlorophenyl carbamyl chloride (5), which is then dehydrochlorinated and then reacted with dimethylamine as in the first reaction sequence above.

Emphasis will be put here on the use of p-chlorophenyl isocyanate as a raw material for reaction with dimethylamine to give monuron. The use of phenyl isocyanate would give fenuron under similar conditions and the use of 3,4-dichlorophenyl isocyanate would give diuron under analogous conditions.

The following is a specific example (3) of the conduct of this process: To a stirred solution of dimethylamine in dry dioxane was added dropwise p-chlorophenyl isocyanate with the evolution of heat. The temperature was maintained at 20° to 30°C. After cooling the reaction mixture to 15°C, the solid, white crystalline product which separated was filtered and dried in a vacuum (30 mm) oven at 50°C.

Process Wastes and Their Control

Air: Air emissions from monuron manufacture have been reported (B-15) to be 0.5 kg hydrocarbons and 0.5 kg chlorine per metric ton of pesticide produced.

Product Wastes: See Diuron for disposal procedures (B-3).

Toxicity

The acute oral LD_{50} value for rats is 3,600 mg/kg which is slightly toxic.

Allowable Limits on Exposure and Use

Product Use: Issuance of a rebuttable presumption against registration for monuron was being considered by EPA as of September 1979 on the basis of possible oncogenicity.

Some monuron products have been voluntarily cancelled according to a EPA notice dated August 16, 1977.

References

(1) Worthing, C.R., *Pesticide Manual,* 6th ed., p. 368, British Crop Protection Council (1979).
(2) Spencer, E.Y., *Guide to the Chemicals Used in Crop Protection,* 6th ed., p. 365, London, Ontario, Agriculture Canada (January 1973).
(3) Todd, C.W., U.S. Patent 2,655,444, October 13, 1953, assigned to DuPont.
(4) Todd, C.W., U.S. Patent 2,655,445, October 13, 1953, assigned to DuPont.
(5) Julian, F.M. et al, U.S. Patent 2,974,163, March 7, 1961, assigned to DuPont.

MONURON-TCA

Function: Herbicide (1)(2)

Chemical Name: 3-(4-chlorophenyl)-1,1-dimethyluronium trichloroacetate

Formula:

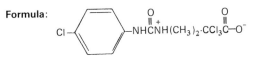

Trade Names: GC-2996 (Allied Chemical)
Urox® (Allied Chemical)

Manufacture (3)

The reaction may be readily accomplished by mixing the diamide and halogenated acid preferably in the presence of a liquid aromatic hydrocarbon solvent such as benzene, toluene or xylene in which the acid is soluble, at about room temperature. The reaction product may be separated from the reaction mixture by crystallization and more easily by adding a large excess of paraffin hydrocarbon such as petroleum ether which causes the reaction product to separate out of solution from the xylene. The following is a specific example of the conduct of this process.

Sixty parts by weight of xylene and 6.5 parts by weight trichloroacetic acid, and 6.3 parts by weight of 1-(4-chlorophenyl)-3,3-dimethylurea (monuron) having a MP of about 170°C, when mixed formed a homogeneous clear solution. (Solubility of chlorophenyldimethylurea is about 0.2 g/100 g xylene). The mixture was drowned in 400 parts by weight petroleum ether. A white solid precipitated immediately. Collected on a filter, washed with petroleum ether and air dried, the resultant solid constituted 10.3 parts by weight. The glistening snow-white solid melted at 78° to 81°C. Percent trichloroacetic acid: Found = 45.0. Calculated for trichloroacetic acid-chlorophenyldimethylurea compound (1:1 mol ratio) = 45.1%.

Process Wastes and Their Control

Product Wastes: For disposal procedures, see Diuron (B-3).

Toxicity

The acute oral LD_{50} for rats is 2,300 mg/kg (B-5) which is slightly toxic.

References

(1) Worthing, C.R., *Pesticide Manual,* 6th ed., p. 369, British Crop Protection Council (1979).
(2) Spencer, E.Y., *Guide to the Chemicals Used in Crop Protection,* 6th ed., p. 366, London, Ontario, Agriculture Canada (January 1973).
(3) Gilbert, E.E., Otto, J.A. and Pellerano, S.A., U.S. Patent 2,782,112, February 19, 1957, assigned to Allied Chemical and Dye Corp.

MSMA

Function: Herbicide (1)(2)

Chemical Name: Monosodium methanearsonate

Formula:

$$CH_3-\overset{\overset{\displaystyle O}{\|}}{\underset{\underset{\displaystyle OH}{|}}{As}}-ONa$$

Trade Names: Ansar® (Ansul Chemical Co.)
Bueno® (Diamond Shamrock Corp.)
Daconate® (Diamond Shamrock Corp.)
Phyban®
Silvisar®
Weed-E-Rad®

Manufacture (3)(4)(5)(6)

$$As_2O_3 + 6NaOH \longrightarrow 2Na_3AsO_3 + 3H_2O$$

$$Na_3AsO_3 + CH_3Cl \longrightarrow CH_3AsO(ONa)_2 + NaCl$$

$$2CH_3AsO(ONa)_2 + H_2SO_4 \longrightarrow 2CH_3\overset{\overset{\displaystyle OH}{|}}{\underset{\underset{\displaystyle O}{\|}}{As}}-ONa + Na_2SO_4$$

The first step of the process is performed in a separate, dedicated building. The drums of arsenic trioxide are opened in an air-evacuated chamber and automatically dumped into the 50% caustic. A dust collection system is employed. The drums are washed carefully with water, the washwater is added to the reaction mixture, and the drums are crushed and are sold as scrap steel. The intermediate sodium arsenite is obtained as a 25% solution and is stored in large tanks prior to further reaction. In the next step, the 25% solution of sodium arsenite is treated with methyl chloride to give the disodium salt, DSMA. DSMA must be used at higher application rates and is not as soluble as MSMA, and has therefore never become as popular as MSMA.

In order to obtain MSMA, the solution is partially acidified with sulfuric acid and the resulting solution is concentrated by evaporation. The active ingredient is sold at a number of concentrations, but ~58% is the maximum concentration that can be prepared without encountering an undesirable increase in viscosity.

As the aqueous solution is being concentrated, a mixture of sodium sulfate and sodium chloride precipitates out, about 0.5 lb/100 lb of active ingredient. These salts, a troublesome disposal problem because they are contaminated with arsenic, are removed by centrifugation, washed in a 5-stage countercurrent washing cycle, and then disposed of in a dump. A production and waste scheme for MSMA manufacture is shown in Figure 39 (3).

Process Wastes and Their Control

Air: Air emissions from MSMA manufacture have been reported (B-15) to be as follows:

Component	Kilograms per Metric Ton Pesticide Produced
Hydrocarbons	1.0
Arsenic trioxide	3×10^{-8}
Methyl chloride	0.05
Acetone	0.05
Methanol	0.05

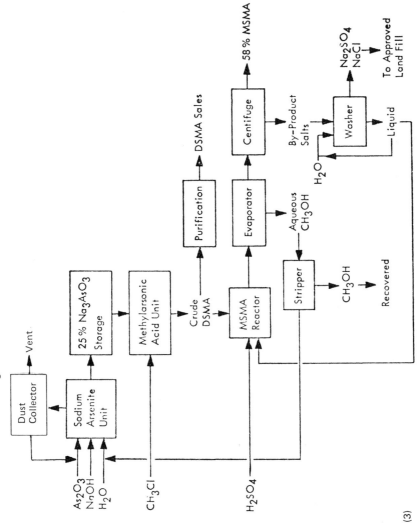

Figure 39: Production and Waste Schematic for MSMA

Source: Reference (3)

Air pollution control in MSMA production involves the following (B-15):

Control Device	Emissions Controlled
Baghouse	Arsenic trioxide
Water scrubber	Arsenic trioxide
Acidifier vent scrubbers	

Water: Discharge at one plant to two equalization ponds contains 0.7 to 0.8 ppm arsenic as well as NaCl and Na_2SO_4. The by-product salts, Na_2SO_4 and NaCl, are a disposal problem because they contain arsenic. One manufacturer washes the salts in a five-stage countercurrent washing cycle and then disposes of them in an approved dump (B-10).

Toxicity

The acute oral LD_{50} value for rats is 900 mg/kg which is slightly toxic.

Allowable Limits on Exposure and Use

Product Use: Issuance of a rebuttable presumption against registration for MSMA was being considered by EPA as of September 1979 on the basis of possible oncogenicity and mutagenicity.

References

(1) Worthing, C.R., *Pesticide Manual,* 6th ed., p. 352, British Crop Protection Council (1979).
(2) Spencer, E.Y., *Guide to the Chemicals Used in Crop Protection,* 6th ed., p. 364, London, Ontario, Agriculture Canada (January 1973).
(3) Midwest Research Institute and RvR Consultants, *Production, Distribution, Use and Environmental Impact Potential of Selected Pesticides,* Washington, DC, Council on Environmental Quality (March 15, 1974).
(4) Miller, G.E. and Seaton, S.G., U.S. Patent 2,442,372, June 1, 1948, assigned to the U.S. Secretary of War.
(5) Miller, G.E. and Reid, E.E., U.S. Patent 2,695,306, November 23, 1954.
(6) Schwerdle, A., U.S. Patent 2,889,347, June 2, 1959.

N

NABAM

Function: Fungicide (1)(2)

Chemical Name: Disodium 1,2-ethanediylbis(carbamodithioate)

Formula:

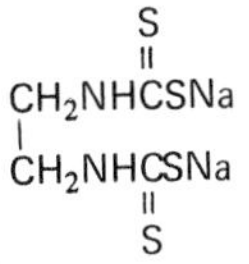

Trade Names: Dithane D-14® (Rohm & Haas)
Parzate® (DuPont)

Manufacture (3)

A mixture of 250 parts of 72% ethylene diamine and 480 parts of a 50% aqueous sodium hydroxide solution was stirred and externally cooled while 478 parts of carbon disulfide was added dropwise below the surface of the mixture. The addition of carbon disulfide required two hours, during which time the temperature was held below 25°C. The reaction mixture was further stirred for two hours at the end of which time it set to a solid. This was recrystallized from 1,800 parts of anhydrous ethanol to give 868 parts of an air-dried solid which melted within the range of 85° to 110°C and which consisted essentially of the disodium salt of ethylene bisdithiocarbamic acid. A stable anhydrous form of nabam may be prepared as described by Nemec et al (4).

The reaction is preferably carried out in a packed column as described by F. Schaefer et al (5). The carbon disulfide vaporizes and distills in the lower part and the ethylene diamine solution and caustic are introduced into the ascending vapors; the resulting salt solution is removed from the base of the column. It is a feature of this process that the product is completely free of CS_2.

It is claimed that this type of operation permits a 40-fold reduction in reactor volume for equal throughput when compared with prior art processes due to the increase in reaction velocity. Further, the absence of moving parts is an added safety feature when working with carbon disulfide, which has a flash point of −30°C and an agitation temperature of only 102°C. The product is withdrawn directly from the base of the reactor as noted above in the form of a pure sodium salt solution. This may be used as is, or may be reacted with other metallic salts to produce other fungicidal materials such as zineb, as described in a subsequent section of this volume.

The ammonium salt may be prepared using ammonium hydroxide in place of sodium hydroxide (6).

Process Wastes and Their Control

Air: Air emissions from nabam manufacture have been reported (B-15) to consist of 205 kg SO_2, 1.5 kg hydrocarbons and 0.5 kg nabam per metric ton of pesticide produced.

Water: A process developed by Stevenson and Harkness (7) is a purification process for industrial effluent containing dissolved dithiocarbamate. Before discharging the effluent, it is mixed at pH 6 to 8 with a soluble heavy metal salt and heavy metal dithiocarbamate thereby precipitated and the precipitate is separated from the effluent. By the process poisoning of fish in water courses into which the effluents are discharged can be avoided as can malfunctioning of sewage treatment processes handling the effluents.

Toxicity

The acute oral LD_{50} value for rats is 395 mg/kg which is moderately toxic.

Allowable Limits on Exposure and Use

Product Use: A rebuttable presumption against registration was issued on August 10, 1977 on the basis of oncogenicity, teratogenicity and hazard to wildlife.

References

(1) Worthing, C.R., *Pesticide Manual,* 6th ed., p. 370, British Crop Protection Council (1979).
(2) Spencer, E.Y., *Guide to the Chemicals Used in Crop Protection,* 6th ed., p. 368, London, Ontario, Agriculture Canada (January 1973).
(3) Hester, W.F., U.S. Patent 2,317,765, April 27, 1943, assigned to Rohm & Haas Co.
(4) Nemec, J.W. and Schechter, S.J., U.S. Patent 3,050,439, August 21, 1962, assigned to Rohm & Haas Co.
(5) Schaefer, F. et al, U.S. Patent 3,210,409, October 5, 1965, assigned to Farbenfabriken Bayer AG.
(6) Fike, E.A., U.S. Patent 2,844,623, July 22, 1958, assigned to Roberts Chemicals, Inc.
(7) Stevenson, A. and Harkness, N., U.S. Patent 3,966,601, June 29, 1976, assigned to Robinson Brothers Ltd., England.

NALED

Function: Insecticide (1)(2)

Chemical Name: 1,2-dibromo-2,2-dichloroethyl dimethyl phosphate

Formula:

$$(CH_3O)_2POCHBrCBrCl_2$$
$$\overset{O}{\overset{\|}{}}$$

Trade Names: RE-4355 (Chevron Chemical Co.)
Dibrom® (Chevron Chemical Co.)
Bromex

Manufacture

The insecticide naled is made by addition of bromine to O,O-dimethyl O-(2,2-dichlorovinyl phosphate) which is the insecticide dichlorvos (q.v.). The reaction is essentially:

$$(CH_3O)_2POCH=CCl_2 + Br_2 \longrightarrow (CH_3O)_2POCHBrCCl_2Br$$

The rate of addition of bromine during the reaction materially affects the yield and purity of the product. Whenever the amount of free bromine in the reaction mixture exceeds 4% by weight of the total bromine added, a reduction in yield and purity result, according to J.N. Ospenson et al (3). The reaction is preferably conducted in the range from 0° to 30°C. Below 0°C, one obtains low yields and purities. Temperatures above 30°C are conducive to undesirable side reactions. This reaction is carried out at atmospheric pressure. The bromine may be added slowly over a period of 10 to 11 hours. The use of an inert aliphatic solvent is preferred to obtain maximum yields of the desired product and to minimize side reactions.

While the preferred solvent is carbon tetrachloride, a number of inert polar solvents such as diethyl ether, acetic acid, chloroform, etc., have been used with satisfactory results. The use of photochemical catalysis is necessary to obtain any appreciable yields of the desired product. When the reaction is carried out in the absence of light, 75% absorption of bromine was obtained but only a 10 mol percent yield of the desired product was obtained. In the presence of ultraviolet light, on the other hand, 100% bromine adsorption was obtained, giving a 93 mol percent yield of desired product.

A Pfaudler glass-lined kettle equipped with turbine agitator and baffles may be used. It should be equipped with facilities for the vacuum stripping of low boiling materials. Further, it should be jacketed for heating and cooling. Finally, quartz mercury vapor light sources should be installed within a water-cooled immersion well in the reactor. Upon completion of the reaction, the product may be stripped under reduced pressure (50 millimeters of mercury) at a maximum temperature of 80°C. Under these conditions, all the solvent that can be vaporized is removed, as well as any excess bromine. The yield of product is substantially quantitative and the low-viscosity liquid product crystallizes to a white crystalline solid having a purity of 90 to 93%.

Process Wastes and Their Control

Air: Air emissions from naled manufacture have been reported (B-15) to consist of 0.5 kg hydrocarbons, 0.5 kg bromine and 0.5 kg naled per metric ton of pesticide produced.

Product Wastes: This pesticide is more stable to hydrolysis than dichlorvos (50% hydrolysis at pH 9 at 37.5°C in 301 minutes). It is unstable in alkaline conditions, in presence of iron, and is degraded by sunlight. About 10% hydrolysis per day is obtained in ambient water (B-3).

Toxicity

The acute oral LD_{50} value for rats is 430 mg/kg which is moderately toxic.

Allowable Limits on Exposure and Use

Air: The threshold limit value for naled in air is 3 mg/m³ as of 1979. The tentative short term exposure limit is 6 mg/m³ (B-23).

Product Use: The tolerances set by the EPA for naled in or on raw agricultural commodities are as follows:

	40 *CFR* Reference	Parts per Million
Almonds	180.215	0.05 N
Almonds, hulls	180.215	0.05 N
Beans, dry	180.215	0.5
Beans, succulent	180.215	0.5
Beets, sugar, roots	180.215	0.5
Beets, sugar, tops	180.215	0.5
Broccoli	180.215	1.0
Brussels sprouts	180.215	1.0
Cabbage	180.215	1.0
Cattle, fat	180.215	0.05 N
Cattle, mbyp	180.215	0.05 N
Cattle, meat	180.215	0.05 N
Cauliflower	180.215	1.0
Celery	180.215	3.0
Collards	180.215	3.0
Cotton, seed	180.215	0.5
Cucumbers	180.215	0.5
Eggplant	180.215	0.5
Eggs	180.215	0.05 N
Goats, fat	180.215	0.05 N
Goats, mbyp	180.215	0.05 N
Goats, meat	180.215	0.05 N
Grapefruit	180.215	3.0
Grapes	180.215	0.5
Grasses, forage	180.215	10.0
Hogs, fat	180.215	0.05 N
Hogs, mbyp	180.215	0.05 N
Hogs, meat	180.215	0.05 N
Hops	180.215	0.5
Horses, fat	180.215	0.05 N
Horses, mbyp	180.215	0.05 N
Horses, meat	180.215	0.05 N
Kale	180.215	3.0
Legumes, forage	180.215	10.0
Lemons	180.215	3.0
Lettuce	180.215	1.0
Melons	180.215	0.5
Milk	180.215	0.05 N
Mushrooms	180.215	0.5
Oranges	180.215	3.0
Peaches	180.215	0.5
Peas (succulent)	180.215	0.5
Peppers	180.215	0.5
Poultry, fat	180.215	0.05 N
Poultry, mbyp	180.215	0.05 N
Poultry, meat	180.215	0.05 N
Pumpkins	180.215	0.5
Rice	180.215	0.5
Safflower, seed	180.215	0.5
Sheep, fat	180.215	0.05 N
Sheep, mbyp	180.215	0.05 N
Sheep, meat	180.215	0.05 N
Spinach	180.215	3.0
Squash, summer	180.215	0.5
Squash, winter	180.215	0.5
Strawberries	180.215	1.0

(continued)

	40 *CFR* Reference	Parts Per Million
Swiss chard	180.215	3.0
Tangerines	180.215	3.0
Tomatoes	180.215	0.5
Turnips, tops	180.215	3.0
Walnuts	180.215	0.5

References

(1) Worthing, C.R., *Pesticide Manual,* 6th ed., p. 371, British Crop Protection Council (1979).
(2) Spencer, E.Y., *Guide to the Chemicals Used in Crop Protection,* 6th ed., p. 369, London, Ontario, Agriculture Canada (January 1973).
(3) Ospenson, J.N. et al, U.S. Patent 2,971,882, February 14, 1961, assigned to California Research Corp.

NAPHTHALENE

Function: Insecticidal fumigant (1)(2)

Chemical Name: Naphthalene

Formula:

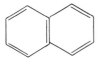

Trade Name: None

Manufacture

Naphthalene may be made by one of two routes: (a) crystallization from coal tar; and (b) hydrocracking of alkylnaphthalenes in petroleum (3).

Toxicity

The acute oral LD_{50} value for rats is 2,200 mg/kg which is slightly toxic.

Naphthalene has been shown to be toxic to microorganisms and has been reported to reduce photosynthetic rates in algae. It has also been reported to be acutely toxic to various invertebrate and vertebrate species of aquatic organisms. In laboratory animals and humans, naphthalene has been linked to blood disorders and is suspected of traversing the placental membrane in humans following naphthalene ingestion by the mother (B-26).

The harmful effects of naphthalene have been summarized as follows (B-25):

Local—Naphthalene is a primary irritant and causes erythema and dermatitis upon repeated contact. It is also an allergen and may produce dermatitis in hypersensitive individuals. Direct eye contact with the dust has produced irritation and cataracts.

Systemic—Inhaling high concentrations of naphthalene vapor or ingesting may cause intravascular hemolysis and its consequences. Initial symptoms include eye irritation, headache, confusion, excitement, malaise, profuse sweating, nausea, vomiting, abdominal pain, and irritation of the bladder.

There may be progressive jaundice, hematuria, hemoglobinuria, renal tubular blockage, and acute renal shutdown. Hematologic features include red cell fragmentation, icterus, severe anemia with nucleated red cells, leukocytosis, and dramatic decreases in hemoglobin, hematocrit, and red cell count. Individuals with a deficiency of glucose-6-phosphate dehydrogenase in erythrocytes are more susceptible to hemolysis by naphthalene.

Allowable Limits on Exposure and Use

Air: The threshold limit value for naphthalene in air is 50 mg/m^3 (10 ppm) as of 1979. The tentative short term exposure limit is 75 mg/m^3 (15 ppm).

Water: In water, EPA has suggested (B-26) a criterion for the protection of human health from the toxic properties of naphthalene ingested through water and through contaminated aquatic organisms of 143 μg/l in water.

References

(1) Worthing, C.R., *Pesticide Manual,* 6th ed., p. 372, British Crop Protection Council (1979).
(2) Spencer, E.Y., *Guide to the Chemicals Used in Crop Protection,* 6th ed., p. 370, London, Ontario, Agriculture Canada (January 1973).
(3) Emmerson, H.R., U.S. Patent 3,193,594, July 6, 1965, assigned to Union Oil Co. of Calif.

1,8-NAPHTHALIC ANHYDRIDE

Function: Seed protectant against carbamate herbicides (1)(2)

Chemical Name: 1H,3H-naphthol[1,8-cd]pyran-1,3-dione

Formula:

Trade Name: Protect® (Gulf Oil Co.)

Manufacture

1,8-naphthalic anhydride may be made by oxidizing acenaphthene.

Toxicity

The acute oral LD$_{50}$ value for rats is 12,300 mg/kg which is insignificantly toxic.

References

(1) Worthing, C.R., *Pesticide Manual,* 6th ed., p. 373, British Crop Protection Council (1979).
(2) Hoffman, O.L., U.S. Patent 3,564,768, February 23, 1971, assigned to Gulf Research and Development Co.

NAPHTHYLACETIC ACID

Function: Plant growth regulator (1)

Chemical Name: α-naphthaleneacetic acid

Formula: CH$_2$COOH

Trade Names: NAA
NAA-800® (Amchem Products, Inc.)
Fruitone-N® (Amchem Products, Inc.)
Rootone® (Amchem Product Inc.)
Phyomone® (I.C.I. Ltd.)

Manufacture

One manufacturing process (2) involves preparing 1-naphthaleneacetic acid by the reaction of naphthalene with chloroacetic acid in the presence of potassium bromide as a catalyst. The amount of potassium bromide appears to possess some function in the reaction since it has been found that the yield of 1-naphthalene acetic acid increases to some extent as the amount of the potassium bromide is increased up to about one-tenth mol of potassium bromide per mol of total reactants. Additions of potassium bromide in excess of this amount do not appear to greatly increase the yield of the desired end product which rises to a maximum of from 25 to 35% yield.

Although it is desirable to employ materials which are relatively free of moisture, small amounts of water in the reaction mixture will not be found to be deleterious. In general, the product is obtained by heating technical grades of naphthalene and chloroacetic acid in open reflux equipment with the indicated amount of potassium bromide and at temperatures from 210° to 235°C and subsequently maintaining the reaction mixture at the boiling temperature of the mixture for 24 hours. The chloroacetic acid and potassium bromide were heated together until the chloroacetic acid melted and the temperature of the mixture was maintained above the melting point of chloroacetic acid for about a half hour or so before the naphthalene was added. Generally, yields are slightly better when proceeding in this fashion rather than in heating the three components together. The naphthalene and chloroacetic acid were reacted with a slight excess of the latter.

The desired end product was separated from the reaction mixture by heating the same with approximately 10% aqueous caustic soda, filtering the mixture and acidifying the filtrate to precipitate the crude 1-naphthaleneacetic acid which usually has the form of a dark grey solid. This product was then dried, ground to a fine powder, and extracted with a suitable solvent, in this instance, petroleum ether. The final product was obtained by crystallizing it from the solvent from which it was then removed and dried. The product so obtained possessed a melting point in the range of 127° to 129°C, whereas the crude product melted in the range from 98° to 129°C.

An alternative procedure (3) involves the saponification of the corresponding nitrile. 20 g of naphthyl-1-acetonitrile are stirred with 100 cc of concentrated hydrochloric acid for 5 hours at 60° to 70°C, 50 cc of water are then added. The mixture is refluxed with stirring for 10 hours. After cooling, the mixture is extracted with sodium carbonate. The alkaline extract is acidified and extracted with benzene. The benzene layer is distilled to dryness leaving naphthyl-1-acetic acid. The acid is recrystallized from benzene-petroleum ether or water.

Toxicity

The acute oral LD_{50} value for rats is 1,000 mg/kg which is slightly toxic.

Allowable Limits on Exposure and Use

Product Use: The tolerances set by the EPA for NAA in or on raw agricultural commodities are as follows:

	40 *CFR* Reference	Parts per Million
Apples	180.155	1.0
Olives	180.155	0.1 N
Pears	180.155	1.0
Pineapples	180.155	0.05 N
Quinces	180.155	1.0

References

(1) Worthing, C.R., *Pesticide Manual,* 6th ed., p. 376, British Crop Protection Council (1979).
(2) Southwick, P.L., U.S. Patent 2,655,531, October 13, 1953, assigned to Food Machinery and Chemical Corp.
(3) Wenner, W., U.S. Patent 2,489,348, November 29, 1949, assigned to Hoffmann-La Roche, Inc.

NAPROPAMIDE

Function: Herbicide (1)(2)(3)

Chemical Name: N,N-diethyl-2-(1-naphthalenyloxy) propanamide

Formula:

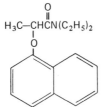

Trade Names: R-7465 (Stauffer Chemical Co.)
Devrinol® (Stauffer Chemical Co.)

Manufacture (3)

α-Naphthol (144 g, 1.00 mol), N-N-diethyl α-bromopropionamide (208 g, 1.00 mol), and a 25% methanol solution of sodium methoxide (216 g, 1.00 mol) were mixed together and heated to reflux with stirring for 2½ hours. The reaction mixture was cooled to room temperature and diluted with water (1,000 ml). The product was extracted with chloroform (2 × 200 ml) and the chloroform extract was dried over magnesium sulfate and evaporated in vacuo to yield 212 g of a dark oil. On standing, crystals started to form and the remainder was crystallized from n-pentane and washed with n-pentane to yield 190 g of light brown solid, MP 63° to 64°C. Alternatively napropamide may be made from the reaction of naphthyloxypropionic acid with diethylamine (1)(2).

Toxicity

The acute oral LD_{50} for rats is over 5,000 mg/kg which is insignificantly toxic.

Allowable Limits on Exposure and Use

Product Use: The tolerances set by the EPA for napropamide in or on raw agricultural commodities are as follows:

	40 *CFR* Reference	Parts per Million
Citrus fruits	180.328	0.1 N
Figs	180.328	0.1 N
Fruits, pome	180.328	0.1 N
Fruits, small	180.328	0.1 N
Fruits, stone	180.328	0.1 N
Nuts	180.328	0.1 N
Nuts, almonds, hulls	180.328	0.1 N
Vegetables, fruiting	180.328	0.1 N

References

(1) Worthing, C.R., *Pesticide Manual,* 6th ed., p. 378, British Crop Protection Council (1979).
(2) Spencer, E.Y., *Guide to the Chemicals Used in Crop Protection,* 6th ed., p. 371, London, Ontario, Agriculture Canada (January 1973).
(3) Tilles, H., Baker, D.R. and Dewald, C.L., U.S. Patent 3,480,671, November 25, 1969, assigned to Stauffer Chemical Co.

NAPTALAM

Function: Herbicide (1)(2)(3)(4)(5)(6)

Chemical Name: 2-[(1-naphthalenylamino)carbonyl] benzoic acid

Formula:

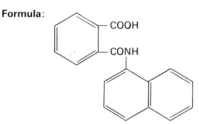

Trade Names: 6Q8 (Uniroyal, Inc.)
Alanap® (Uniroyal, Inc.)
Dyanap®
Naptalal®
Panala®
NPA

Manufacture (3)(4)

Naptalam may be prepared by the reaction of phthalic anhydride with alpha-naphthylamine according to the equation:

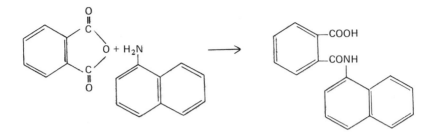

Equimolar amounts of phthalic anhydride and the primary monoaryl amine are stirred together in a solvent such as benzene, xylene, or kerosene. The reaction takes place rapidly at room temperature, giving the desired product.

Process Wastes and Their Control

Product Wastes: Naptalam reacts with strong acids and strong alkalies which hydrolyze the product to α-naphthylamine, which is much more toxic than the original product, and phthalic acid (or its salt) (B-3).

Toxicity

Naptalam is relatively nontoxic (acute oral LD_{50} = 8,200 mg/kg) to warm blooded animals such as rats.

Allowable Limits on Exposure and Use

Product Use: The tolerances set by the EPA for naptalam in or on raw agricultural commodities are as follows:

	40 *CFR* Reference	Parts per Million
Cantaloupes	180.297	0.1 N
Cranberries	180.297	0.1 N
Cucumbers	180.297	0.1 N
Muskmelons	180.297	0.1 N
Peanuts	180.297	0.1 N
Peanuts, hay	180.297	0.1 N
Soybeans	180.297	0.1 N
Soybeans, hay	180.297	0.1 N
Watermelons	180.297	0.1 N

References

(1) Worthing, C.R., *Pesticide Manual*, 6th ed., p. 379, British Crop Protection Council (1979).
(2) Spencer, E.Y., *Guide to the Chemicals Used in Crop Protection*, 6th ed., p. 372, London, Ontario, Agriculture Canada (January 1973).
(3) Smith, A.E. and Hoffman, O.L., U.S. Patent 2,556,664, June 12, 1951, assigned to United States Rubber Co.
(4) Smith, A.E. et al, U.S. Patent 2,556,665, June 12, 1951, assigned to United States Rubber Co.
(5) Smith, A.E., Feldman, A.W. and Stone, G.M., U.S. Patent 2,736,646, February 28, 1956, assigned to United States Rubber Co.
(6) Smith, A.E., Feldman, A.W. and Stone, G.M., U.S. Patent 2,736,647, February 28, 1956, assigned to United States Rubber Co.

NEBURON

Function: Herbicide (1)

Chemical Name: N-butyl-N'-(3,4-dichlorophenyl)-N-methylurea

Formula:

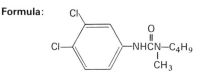

Trade Name: Kloben® (DuPont)

Manufacture (2)

Methyl n-butylamine was added gradually with stirring to 3,4-dichlorophenyl isocyanate in toluene at room temperature. The mixture was heated to reflux temperature and maintained at that temperature for thirty minutes. On cooling, white crystalline neburon precipitated and was separated and dried.

Toxicity

The acute oral LD_{50} for rats is more than 11,000 mg/kg which is insignificantly toxic.

References

(1) Worthing, C.R., *Pesticide Manual,* 6th ed., p. 380, British Crop Protection Council (1979).
(2) Todd, C.W., U.S. Patent 2,655,444, October 13, 1953, assigned to DuPont.

NICLOSAMIDE

Function: Molluscicide (1)(2)(3)

Chemical Name: 5-chloro-N-(2'-chloro-4'-nitrophenyl)-2-hydroxybenzamide

Formula:

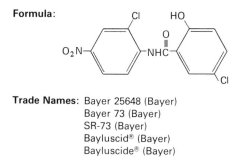

Trade Names: Bayer 25648 (Bayer)
Bayer 73 (Bayer)
SR-73 (Bayer)
Bayluscid® (Bayer)
Bayluscide® (Bayer)

Manufacture

Equimolecular amounts of 5-chloro-salicylic acid and 2-chloro-4-nitro-aniline are dissolved in 250 ml of xylene. While boiling, there are introduced slowly 5 g of PCl_3. Heating is continued for 3 further hours. The mixture then is allowed to cool down and the crystals which separate are filtered off with suction. The crude niclosamide may be recrystallized from ethanol, melting at 233°C.

Process Wastes and Their Control

Product Wastes: Niclosamide is reported to be highly stable to heat (B-3).

Toxicity

The acute oral LD_{50} for rats is 250 mg/kg (B-5) which is moderately toxic.

References

(1) Worthing, C.R., *Pesticide Manual,* 6th ed., p. 381, British Crop Protection Council (1979).
(2) Schraufstatter, E. and Gonnert, R., U.S. Patent 3,079,297, February 26, 1963, assigned to Farbenfabriken Bayer AG.
(3) Strufe, R., Schraufstatter, E. and Gonnert, R., U.S. Patent 3,113,067, December 3, 1963, assigned to Farbenfabriken Bayer AG.

NICOTINE

Function: Insecticide

Chemical Name: (S)-3-(1-methyl-2-pyrrolidinyl)pyridine

Formula:

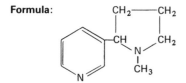

Trade Name: Black Leaf 40®

Manufacture

Nicotine may be prepared from waste tobacco by: (a) steam distillation in the presence of alkali; and (b) solvent extraction with trichloroethylene in the presence of alkali followed by extraction from the solvent with dilute H_2SO_4 (3).

Process Wastes and Their Control

Product Wastes: The Manufacturing Chemists Association has suggested incineration as a method of disposal for nicotine (B-3).

Toxicity

Nicotine is highly toxic (LD_{50} = 55 mg/kg) and is reported to be hazardous to wildlife (B-3).

At high doses, nicotine is quite toxic and lethal. Nicotine is metabolized readily, principally to cotinine. Nicotine is teratogenic in mice only at high doses. Evidence on carcinogenicity is equivocal, but it is a cocarcinogen.

In view of the relative paucity of data on the carcinogenicity and long-term oral toxicity of nicotine, estimates of the effects of chronic oral exposure at low levels cannot be made with any confidence. It is recommended that studies to produce such information be conducted before limits in drinking water are established (B-22).

Allowable Limits on Exposure and Use

Air: The threshold limit value for nicotine in air is 0.5 mg/m^3 as of 1979. The tentative short term exposure limit is 1.5 mg/m^3. It is further noted that cutaneous absorption should be prevented so that the TLV is not invalidated (B-23).

Product Use: The tolerances set by the EPA for nicotine in or on raw agricultural commodities are as follows:

	40 *CFR* Reference	Parts per Million
Apples	180.167	2.0
Apricots	180.167	2.0
Artichokes	180.167	2.0
Asparagus	180.167	2.0
Avocados	180.167	2.0
Beans	180.167	2.0
Beets, greens (alone)	180.167	2.0
Beets, with tops	180.167	2.0
Beets, without tops	180.167	2.0
Blackberries	180.167	2.0
Boysenberries	180.167	2.0
Broccoli	180.167	2.0
Brussels sprouts	180.167	2.0
Cabbage	180.167	2.0
Cauliflower	180.167	2.0
Celery	180.167	2.0
Cherries	180.167	2.0
Citrus fruits	180.167	2.0
Collards	180.167	2.0
Corn	180.167	2.0
Cranberries	180.167	2.0
Cucumbers	180.167	2.0
Currants	180.167	2.0
Dewberries	180.167	2.0
Eggplant	180.167	2.0
Eggs	180.167	1.0
Gooseberries	180.167	2.0
Grapes	180.167	2.0
Kale	180.167	2.0
Kohlrabi	180.167	2.0
Lettuce	180.167	2.0
Loganberries	180.167	2.0
Melons	180.167	2.0
Mushrooms	180.167	2.0
Mustard, greens	180.167	2.0
Nectarines	180.167	2.0
Okra	180.167	2.0
Onions	180.167	2.0
Parsley	180.167	2.0
Parsnips, greens (alone)	180.167	2.0
Parsnips, with tops	180.167	2.0
Parsnips, without tops	180.167	2.0
Peaches	180.167	2.0
Pears	180.167	2.0

(continued)

	40 *CFR* Reference	Parts per Million
Peas	180.167	2.0
Peppers	180.167	2.0
Plums (fresh prunes)	180.167	2.0
Poultry, fat	180.167	1.0
Poultry, mbyp	180.167	1.0
Poultry, meat	180.167	1.0
Pumpkins	180.167	2.0
Quinces	180.167	2.0
Radishes, tops	180.167	2.0
Radishes, with tops	180.167	2.0
Radishes, without tops	180.167	2.0
Raspberries	180.167	2.0
Rutabagas, tops	180.167	2.0
Rutabagas, with tops	180.167	2.0
Rutabagas, without tops	180.167	2.0
Spinach	180.167	2.0
Squash	180.167	2.0
Squash, summer	180.167	2.0
Strawberries	180.167	2.0
Swiss chard	180.167	2.0
Tomatoes	180.167	2.0
Turnips, greens	180.167	2.0
Turnips, with tops	180.167	2.0
Turnips, without tops	180.167	2.0
Youngberries	180.167	2.0

References

(1) Worthing, C.R., *Pesticide Manual,* 6th ed., p. 382, British Crop Protection Council (1979).
(2) Spencer, E.Y., *Guide to the Chemicals Used in Crop Protection,* 6th ed., p. 373, London, Ontario, Agriculture Canada (January 1973).
(3) Von Bethmann, M.F., Lipp, G. and Bayer, H., U.S. Patent 3,396,735, August 13, 1968, assigned to Eresta Warenhandel GmbH.

NITRALIN

Function: Herbicide (1)(2)(3)

Chemical Name: 4-(methylsulfonyl)-2,6-dinitro-N,N-dipropylaniline

Formula:

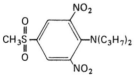

Trade Names: SD-11,831 (Shell Research, Ltd.)
Planavin® (Shell Research Ltd.)

Manufacture (2)

The starting material is 4-chloro-3-nitrophenyl methylsulfone. In an initial step, 6 parts of 4-chloro-3-nitrophenyl methylsulfone was added portionwise to a mixture of 49 parts of fuming H_2SO_4 (30%) and 16 parts of red fuming HNO_3 (d. 1.60) at 30°C. The reaction was heated with stirring to 120° to 140°C for 3 hours, then poured over ice. The separated solid was filtered, water-washed, and dried. The product was recrystallized from dimethyl formamide solution by adding alcohol to give 4.5 parts of pale yellow crystals, MP 201° to 203°C.

Then, in a second step, 15 parts of 4-chloro-3,5-dinitrophenyl methylsulfone was suspended in 120 parts of methanol and 15 parts of dipropylamine were added. The reaction mixture was heated for two hours, chilled, and the separated solid filtered. The product was washed and recrystallized to give 18 parts of golden crystals, MP 150° to 151°C (91% yield).

Process Wastes and Their Control

Product Wastes: Nitralin has a half-life in the soil of 30 to 50 days. It decomposes vigorously and explosively at 225°C and is flammable. It is evidently unstable toward strong bases (B-3).

Toxicity

The acute oral LD_{50} value for rats is over 2,000 mg/kg which is slightly toxic.

Allowable Limits on Exposure and Use

Product Use: The tolerances set by the EPA for nitralin in or on raw agricultural commodities are as follows:

	40 *CFR* Reference	Parts per Million
Almonds	180.237	0.1 N
Almonds, hulls	180.237	0.1 N
Broccoli	180.237	0.1 N
Brussels sprouts	180.237	0.1 N
Cabbage	180.237	0.1 N
Cauliflower	180.237	0.1 N
Cotton, seed	180.237	0.1 N
Cucurbits	180.237	0.1 N
Fruits, pome	180.237	0.1 N
Fruits, stone	180.237	0.1 N
Grapes	180.237	0.1 N
Legumes, forage	180.237	0.1 N
Peanuts	180.237	0.1 N
Safflower, seed	180.237	0.1 N
Soybeans, dry	180.237	0.1 N
Vegetables, fruiting	180.237	0.1 N
Vegetables, seed & pod	180.237	0.1 N

References

(1) Worthing, C.R., *Pesticide Manual*, 6th ed., p. 383, British Crop Protection Council (1979).
(2) Soloway, S.B. and Zwahlen, K.D., U.S. Patent 3,227,734, January 4, 1966, assigned to Shell Oil Co.
(3) Shell International Research, N.V., Belgian Patent 672,199, November 10, 1965.

NITRAPYRIN

Function: Soil bactericide (stabilizes nitrogen fertilizers) (1)(2)(3)

Chemical Name: 2-chloro-6-trichloromethylpyridine

Formula:

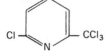

Trade Names: Dowco 163 (Dow Chemical Co.)
N-Serve® (Dow Chemical Co.)

Manufacture (4)(5)

Acetaldehyde and ammonia are first condensed to give methyl pyridine (α-picoline). Then, liquid α-picoline hydrochloride is produced as an essential intermediate from α-picoline and gaseous hydrogen chloride under mild conditions and that intermediate is then contacted and reacted with gaseous chlorine at elevated temperatures in the absence of added water to give nitrapyrin (4).

Process Wastes and Their Control

Product Wastes: The manufacturer of this nitrification inhibitor suggests that unwanted quantities can be disposed of by burial in a sanitary landfill (B-3).

Toxicity

The acute oral LD_{50} for rats is in the range of 1,100 to 1,200 mg/kg which is slightly toxic.

Allowable Limits on Exposure and Use

Air: The threshold limit value for nitrapyrin in air is 10 mg/m^3 as of 1979. The tentative short term exposure limit is 20 mg/m^3 (B-23).

Product Use: The tolerances set by the EPA for nitrapyrin in or raw agricultural commodities are as follows:

	40 *CFR* Reference	Parts per Million
Cattle, fat	180.350	0.05 N
Cattle, mbyp	180.350	0.05 N
Cattle, meat	180.350	0.05 N
Corn, fodder	180.350	0.5
Corn, forage	180.350	0.5
Corn, fresh (inc. sweet) (k+cwhr)	180.350	0.1 N
Corn, grain	180.350	0.1 N
Cotton, seed	180.350	1.0
Goats, fat	180.350	0.05 N
Goats, mbyp	180.350	0.05 N
Goats, meat	180.350	0.05 N
Hogs, fat	180.350	0.05 N
Hogs, mbyp	180.350	0.05 N
Hogs, meat	180.350	0.05 N
Horses, fat	180.350	0.05 N
Horses, mbyp	180.350	0.05 N
Horses, meat	180.350	0.05 N
Poultry, fat	180.350	0.05 N
Poultry, mbyp	180.350	0.05 N
Poultry, meat	180.350	0.05 N

(continued)

	40 *CFR* Reference	Parts per Million
Sheep, fat	180.350	0.05 N
Sheep, mbyp	180.350	0.05 N
Sheep, meat	180.350	0.05 N
Sorghum, fodder	180.350	0.5
Sorghum, forage	180.350	0.1 N
Sorghum, grain	180.350	0.1 N
Wheat, forage	180.350	0.5
Wheat, grain	180.350	0.1 N
Wheat, straw	180.350	0.5

References

(1) Worthing, C.R., *Pesticide Manual*, 6th ed., p. 384, British Crop Protection Council (1979).
(2) Spencer, E.Y., *Guide to the Chemicals Used in Crop Protection*, 6th ed., p. 122, London, Ontario, Agriculture Canada (January 1973).
(3) Goring, C.A.I., U.S. Patent 3,135,594, June 2, 1964, assigned to Dow Chemical Co.
(4) Taplin, W.H., III, U.S. Patent 3,424,754, January 28, 1969, assigned to Dow Chemical Co.
(5) Dow Chemical Co., British Patent 957,276, May 6, 1964.

NITRILACARB

Function: Insecticide and acaricide (1)(2)(3)

Chemical Name: (4,4-Dimethyl-5-methylaminocarbonyloxyimino)pentanenitrile zinc chloride salt

Formula: $N{\equiv}CCH_2CH_2C(CH_3)_2CH{=}NOCONHCH_3 \cdot ZnCl_2$

Trade Names: AC-85,258 (American Cyanamid Co.)
Accotril® (American Cyanamid Co.)

Manufacture (2)(3)

The free base is produced as follows: A solution of 32 g of hydroxylamine hydrochloride in 45 ml of water was added to a solution containing 44.0 g of 2,2-dimethyl-4-cyanobutyralde-hyde, BP 82° to 85°C (2.0 mm), and 36 g of pyridine in 450 ml of alcohol gave a mildly exo-thermic reaction. The solution was warmed to 45°C to complete the reaction, concentrated to remove alcohol, diluted with water and the oxime isolated by extraction with benzene, washing and evaporation. The oxime (37.0 g) was obtained as a straw colored oil which solidified at room temperature.

To a stirred mixture of 37.0 g of 2,2-dimethyl-4-cyanobutyraldoxime in 175 ml of benzene was added to 17.5 g of methyl isocyanate followed by 0.2 ml of triethylamine. A very mildly exothermic reaction occurred. Infrared absorption spectra indicated reaction was complete in 3.5 hours at room temperature. After partial concentration of the mixture under vacuum to remove any unreacted methyl isocyanate, it was diluted with benzene and ether, washed with water and with saturated salt solution and dried with magnesium sulfate. The carbamate was recovered by evaporation of solvent and crystallized from ether to obtain 42 g of white solid, MP 42.5° to 43°C. This free base may then be combined with zinc chloride.

Toxicity

The acute oral LD_{50} value for rats is 9 mg/kg which is highly toxic.

References

(1) Worthing, C.R., *Pesticide Manual,* 6th ed., p. 385, British Crop Protection Council (1979).
(2) Addor, R.W. and Allman, D.E., U.S. Patent 3,621,049, November 16, 1971, assigned to American Cyanamid Co.
(3) Addor, R.W. and Allman, D.E., U.S. Patent 3,681,505, August 1, 1972, assigned to American Cyanamid Co.

NITROFEN

Function: Herbicide (1)(2)(3)

Chemical Name: 2,4-dichloro-1-(4'-nitrophenoxy) benzene

Formula:

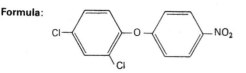

Trade Names: FW-925 (Rohm and Haas Co.)
TOK E-25® (Rohm and Haas Co.)
Niclofen®
Nip®

Manufacture (3)

A mixture of 86.5 parts by weight (0.53 mol) of 2,4-dichlorophenol, 27 parts (0.41 mol) of aqueous 85% potassium hydroxide solution, 50 parts (0.32 mol) of p-chloronitrobenzene and 0.5 part of copper powder is heated for eight hours at 200°C. The reaction mixture is poured into ice water and the resulting mixture is extracted with ethylene dichloride. Layers form and are separated. The solvent layer is treated with charcoal, filtered, and washed with potassium hydroxide solution to remove free 2,4-dichlorophenol. Solvent is distilled off and the product is fractionally distilled. After a forerun of chiefly chloronitrobenzene, a main fraction, 2,4-dichlorophenyl-4'-nitrophenyl ether, is taken at 176° to 180°C at 0.9 mm in a yield of 60%. The distilled product can be recrystallized from a benzene-hexane mixture and then melts at 62° to 65°C.

Process Wastes and Their Control

Product Wastes: The dichloro aromatic constituent of the nitrofen molecule precludes rapid degradation in the environment (B-3).

Toxicity

The acute oral LD_{50} for rats is 740 mg/kg (B-5) which is slightly toxic although the literature contains the statement that nitrofen is relatively nontoxic (an LD_{50} = 3,050 is quoted) and is nonirritating to the skin.

Allowable Limits on Exposure and Use

Product Use: The tolerances set by the EPA for nitrofen in or on raw agricultural commodities are as follows:

	40 *CFR* Reference	Parts per Million
Beets, sugar, roots	180.223	0.05 N
Beets, sugar, tops	180.223	0.05 N
Broccoli	180.223	0.75
Brussels sprouts	180.223	0.75
Cabbage	180.223	0.75
Carrots	180.223	0.75
Cattle, fat	180.223	0.05 N
Cattle, mbyp	180.223	0.05 N
Cattle, meat	180.223	0.05 N
Cauliflower	180.223	0.75
Celery	180.223	0.75
Eggs	180.223	0.05 N
Goats, fat	180.223	0.05 N
Goats, mbyp	180.223	0.05 N
Goats, meat	180.223	0.05 N
Hogs, fat	180.223	0.05 N
Hogs, mbyp	180.223	0.05 N
Hogs, meat	180.223	0.05 N
Horseradish	180.223	0.05 N
Horses, fat	180.223	0.05 N
Horses, mbyp	180.223	0.05 N
Horses, meat	180.223	0.05 N
Kohlrabi	180.223	0.75
Milk, fat	180.223	0.5
Milk, whole	180.223	0.02 N
Onions, dry bulb	180.223	0.75
Onions, green	180.223	0.75
Parsley	180.223	0.75
Poultry, fat	180.223	0.2
Poultry, mbyp (exc. fat)	180.223	0.05 N
Poultry, meat	180.223	0.05 N
Rice	180.223	0.1
Rice, straw	180.223	0.1
Sheep, fat	180.223	0.05 N
Sheep, mbyp	180.223	0.05 N
Sheep, meat	180.223	0.05 N
Taro (corm)	180.223	0.02

References

(1) Worthing, C.R., *Pesticide Manual,* 6th ed., p. 386, British Crop Protection Council (1979).
(2) Spencer, E.Y., *Guide to the Chemicals Used in Crop Protection,* 6th ed., p. 376, London, Ontario, Agriculture Canada (January 1973).
(3) Wilson, H.F. and McRae, D.H., U.S. Patent 3,080,225, March 5, 1963, assigned to Rohm and Haas Co.

NORBORMIDE

Function: Rodenticide (1)

Chemical Name: 3a,4,7,7a-tetrahydro-5-(hydroxyphenyl-2-pyridylmethyl)-7-(phenyl-2-pyridyl-methylene)-4,7-methano-1H-isoindole-1,3(2H)-dione

Formula:

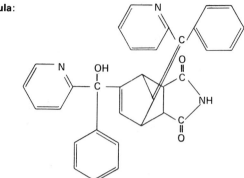

Trade Names: McN 1025 (McNeil Laboratories, Inc.)
Shoxin® (McNeil Laboratories, Inc.)
Raticate® (McNeil Laboratories, Inc.)

Manufacture (2)

Cyclopentadiene and 2-benzoylpyridine are first condensed. That product is reacted with maleimide to give norbormide.

As described by McNeil Laboratories in more detail (3), the following is the first step: To a cold (ice-bath) solution of sodium ethoxide which is prepared by dissolving 0.8 part by weight of sodium metal in 380 parts by volume of absolute ethanol is added 64 parts by weight of phenyl-2-pyridylketone and then 40 parts by weight of freshly distilled cyclopentadiene over twenty minutes. The reaction solution is kept under nitrogen and darkens rapidly during the addition. It is stirred at ice-bath temperature for two hours after which crystals of product begin to appear. After the mixture is stirred for fifteen hours under nitrogen at ice-bath temperature, the crystalline product is separated by filtration. The filtrate contains 6-phenyl-6-(2-pyridyl)-fulvene. The solid product amount to 49 parts by weight of orange crystals melting at 144° to 168°C. One recrystallization from absolute ethanol gives pure α-phenyl-α-[6-phenyl-6-(2-pyridyl)-2-fulvenyl]-2-pyridine methanol as orange prisms, MP 164° to 170°C.

Then, the following details the second step: A 5.4 parts by weight sample of a mixture of geometric isomers of α-phenyl-α-[6-phenyl-6-(2-pyridyl)-2-fulvenyl]-2-pyridine methanol and 1.26 parts by weight of maleimide are combined in 25 parts by volume of benzene and the solution is refluxed for three hours. After thirteen hours standing at room temperature the solution is refluxed for 90 minutes, then cooled in an ice bath and filtered to give 5.4 parts by weight of white solid 5-[α-hydroxy-α-phenyl-α-(2-pyridyl)methyl]-7-(phenyl-2-pyridylmethyl-ene)-5-norbornene-2,3-dicarboximide. The filtrate is concentrated under reduced pressure to give a second crop, MP 190° to 198°C. The first crop is recrystallized twice from ethyl acetate to yield a white, crystalline low melting isomer of 5-[α-hydroxy-α-phenyl-α-(2-pyridyl)methyl]-7-(phenyl-2-pyridyl-methylene)-5-norbornene-2,3-dicarboximide   hemihydrate MP 193.5° to 194.5°C.

The second crop of crystals from the reaction mixture is recrystallized three times from ethyl acetate to give white, crystalline high melting isomer of 5-[α-hydroxy-α-phenyl-α-(2-pyridyl) methyl]-7-(phenyl-2-pyridylmethylene)-5-norbornene-2,3-dicarboximide MP 213° to 214°C.

Process Wastes and Their Control

Product Wastes: Norbormide is stable at room temperature when dry but is hydrolyzed by alkali (B-3).

Toxicity

The acute oral LD_{50} for rats is 52 mg/kg which is moderately to highly toxic.

However, norbormide is a very selective rodenticide; it is lethal to all members of the genus *Rattus* but is relatively nonlethal to other animals (B-3).

References

(1) Worthing, C.R., *Pesticide Manual,* 6th ed., p. 388, British Crop Protection Council (1979).
(2) Mohrbacher, R.J., et al, *J. Org. Chem.* 31, 2141 (1966).
(3) McNeil Laboratories, British Patent 1,059,405 (February 22, 1967).

NORFLURAZON

Function: Herbicide (1)

Chemical Name: 4-chloro-5-methylamino-2-(3-trifluoromethylphenyl)-3-(2H)-pyridazinone

Formula:

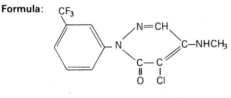

Trade Names: H-52,143 (Sandoz, Ltd.)
H-9789 (Sandoz, Ltd.)
Zorial® (Sandoz, Ltd.)
Evitol® (Sandoz, Inc.)

Manufacture (2)

The 1-(3-trifluoromethylphenyl)-4,5-dichloropyridazone-(6) used as starting material may be produced as follows: 57.5 g of mucochloric acid are dissolved in 100 cc of absolute alcohol and 60 g of m-trifluoromethylphenylhydrazine are added thereto. The solution is then evaporated to dryness at reduced pressure and heated for 40 minutes under reflux with the addition of 120 cc of glacial acetic acid and 120 cc of acetic anhydride. Subsequently evaporation to dryness at reduced pressure is again effected, after recrystallizing from 90% aqueous alcohol colorless crystals having a MP of 92° to 94°C are obtained. The yield of 1-(3-trifluoro-methylphenyl)4,5-dichloropyridazone-(6) amounts to 94 g. 20 g of 1-(3-trifluoromethylphenyl)-4,5-dichloropyridazone-(6) are added to a solution of 15 g of methylamine in 175 cc of alcohol and are heated to 85°C to 24 hours. Precipitation with water is then effected, the precipitate is filtered off and dried. 17 g of 1-(3-trifluoromethylphenyl)-4-methylamino-5-chloropyridazone-(6), having a MP of 183° to 185°C, are obtained by recrystallization from alcohol.

Toxicity

The acute oral LD_{50} value for rats is more than 8,000 mg/kg which is insignificantly toxic.

Allowable Limits on Exposure and Use

Product Use: The tolerances set by the EPA for norflurazon in or on raw agricultural commodities are as follows:

	40 *CFR* Reference	Parts per Million
Cotton, seed	180.356	0.1
Cranberries	180.356	0.1

References

(1) Worthing, C.R., *Pesticide Manual,* 6th ed., p. 389, British Crop Protection Council (1979).
(2) Ebner, C. and Schuler, M., U.S. Patent 3,644,355, February 22, 1972, assigned to Sandoz, Ltd.

NORURON (NOREA)

Function: Herbicide

Chemical Name: N'-(hexahydro-4,7-methanoindan-5-yl)-N,N-dimethyl urea

Formula:

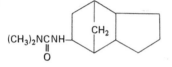

Trade Names: Hercules 7531 (Hercules, Inc.)
Herban® (Hercules, Inc.)

Manufacture (1)

Cyclopentadiene and cyclopentene are first subjected to a Diels-Alder reaction to give nor-bornene. The norbornene is then reacted with thiocyanic acid. The reaction of HSCN is effected by contacting the norbornene with HSCN generated in situ from a salt of thiocyanic acid and a mineral acid in aqueous medium, at ordinary temperature below 150°C. The reaction of the resultant isothiocyanate compound with dimethyl amine, also takes place on contacting the two reagents at ordinary temperatures below 150°C. The chlorinolysis reaction of the thiourea with chlorine and water takes place in aqueous medium at temperatures up to about 100°C and requires as a minimum one mol of chlorine and one of water per mol of thiourea whereby sulfur and hydrochloric acid are produced as by-products. An excess of chlorine is preferred, and a total of four mols of chlorine will convert the thiourea to the urea with hydrochloric acid and sulfuric acid as the by-products. The reaction is carried out in acid medium and may be started in aqueous acid if desired, in which case the urea compound is taken into solution more readily than when only by-product acid is depended on for acidity.

The following is a specific example of the isothiocyanate preparation (2): A solution of 23.5 parts of concentrated sulfuric acid and 6.5 parts water was added dropwise over about two hours to a well-stirred mixture of 28.2 parts of norbornylene, 29.1 parts crushed potassium thiocyanate and 130 parts toluene at 35° to 40°C. The reaction was completed by continuing the stirring at 35° to 40°C for an additional 3 hours. After further standing for 16 hours, 250 parts water was added and the organic layer was separated from the aqueous layer, washed with water and dried over sodium sulfate. The solvent was distilled off, the product distilled to recover 22.15 parts 2-norbornyl isothiocyanate boiling at 74° to 80°C at 0.3 mm.

Process Wastes and Their Control

Product Wastes: Norea can be hydrolyzed at elevated temperature by acids or bases. This procedure is not recommended due to generation of the toxic material, dimethylamine (B-3).

Toxicity

The acute oral LD_{50} value for rats is 2,000 mg/kg which is slightly toxic.

Allowable Limits on Exposure and Use

Product Use: The tolerances set by the EPA for noruron in or on raw agricultural commodities are as follows:

	40 *CFR* Reference	Parts per Million
Cotton, seed	180.260	0.2 N
Potatoes	180.260	0.2 N
Sorghum, cane	180.260	0.2 N
Sorghum, forage	180.260	0.2 N
Sorghum, grain	180.260	0.2 N
Soybeans	180.260	0.2 N
Spinach	180.260	0.2 N
Sugarcane	180.260	0.2 N

References

(1) Diveley, W.R. and Pombo, M.M., U.S. Patent 3,150,179, September 22, 1964, assigned to Hercules Powder Co.
(2) Buntin, G.A. and Diveley, W.R., U.S. Patent 3,304,167, February 14, 1967, assigned to Hercules, Inc.

NUARIMOL

Function: Fungicide (1)(2)

Chemical Name: α-(2-chlorophenyl)-α-(4'-fluorophenyl)-5-pyrimidinemethanol

Formula:

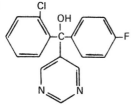

Trade Names: EL-228 (Eli Lilly & Co.)
Trimidal® (Eli Lilly & Co.)
Triminol® (Eli Lilly & Co.)

Manufacture (2)

To 300 ml of anhydrous ether maintained in an atmosphere of dry nitrogen gas in a suitably equipped 3-neck round-bottom reaction flask cooled to −118°C by an alcohol-liquid nitrogen cooling bath, were added 170 ml (0.3 mol) of a 15% solution of butyl lithium in hexane. Cooling and stirring and the dry nitrogen atmosphere were continued while a solution of 0.3 mol of 5-bromopyrimidine in 150 ml of dry tetrahydrofuran was added and the whole stirred for about 2 hours. The temperature of the reaction mixture was lowered to −125°C and a solution of 0.3 mol of 4-fluoro-2'-chlorobenzophenone in 150 ml of dry tetrahydrofuran was added slowly while maintaining the temperature of the mixture at about −120°C.

The reaction product mixture was stirred overnight and warmed to ambient room temperature. The reaction product mixture was neutralized by the addition of a saturated aqueous solution of ammonium chloride. The neutralized mixture was extracted with ether and the combined ether extracts dried over anhydrous potassium carbonate, filtered, and concentrated to dryness in vacuo and the residue dissolved in benzene. The benzene solution was chromatographed over 1,500 g of silica gel, elution being accomplished with an ethyl acetate-benzene mixture, using a gradient elution technique. The fraction obtained using a solvent containing 30:50 ethyl acetate-benzene was concentrated to dryness at reduced pressure, yielding a product having a MP of about 126° to 127°C after recrystallization from ether. The product was identified by elemental analyses and NMR spectrum as α-(4-fluorophenyl)-α-(2'-chlorophenyl)-5-pyrimidinemethanol.

Toxicity

The acute oral LD_{50} value for rats is 1,250 to 2,000 mg/kg which is slightly toxic.

References

(1) Worthing, C.R., *Pesticide Manual,* 6th ed., p. 390, British Crop Protection Council (1979).
(2) Davenport, J.D., Hackler, R.E. and Taylor, H.M., British Patent 1,218,623, January 6, 1971, assigned to Eli Lilly & Co.

O

OCTAHYDRODIBENZOFURANCARBALDEHYDE

Function: Insect repellent (1)(2)

Chemical Name: 1,5a,6,9,9a,9b-hexahydro-4a(4H)-dibenzofurancarboxaldehyde

Formula:

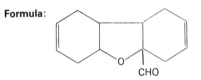

Trade Name: MGK Repellent 11® (McLaughlin Gormley King Co.)

Manufacture (3)(4)

This product is produced (3) from furfural and 1,3-butadiene using amounts of butadiene varying from 5 to 50 weight percent of the monomer charge. It is preferred to utilize anhydrous furfural, in order to minimize furfural homopolymerization and the possible formation of lactones (4). The process may be operated at 90° to 150°C (3) or at 190° to 220°C (4). Pressures of 75 to 135 psi are preferred; higher pressures increase reaction rate but favor butadiene dimer formation over the formation of the desired cotrimer.

Toxicity

The acute oral LD_{50} value for rats is 2,500 mg/kg (slightly toxic).

References

(1) Worthing, C.R., *Pesticide Manual,* 6th ed., p. 391, British Crop Protection Council (1979).
(2) Spencer, E.Y., *Guide to the Chemicals Used in Crop Protection,* 6th ed., p. 43, London, Ontario, Agriculture Canada (January 1973).
(3) Hillyer, J.C. and Nicewander, D.A., U.S. Patent 2,683,151, July 6, 1954, assigned to Phillips Petroleum Company.
(4) Hillyer, J.C. et al, U.S. Patent 2,898,347, August 4, 1959, assigned to Phillips Petroleum Company.

OMETHOATE

Function: Insecticide and acaricide

Chemical Name: O,O-dimethyl S-[2-methylamino)-2-oxo-ethyl] phosphorothioate

Formula:

$$(CH_3O)_2\overset{\overset{O}{\|}}{P}SCH_2\overset{\overset{O}{\|}}{C}NHCH_3$$

Trade Names: Bayer 45,432 (Bayer)
Bayer S-6876 (Bayer)
Folimat® (Bayer)

Manufacture

Omethoate may be prepared by the reaction of O,O-dimethylphosphorothioic acid with 2-chloro-N-methyl acetamide.

Toxicity

The acute oral LD_{50} value for rats is 50 mg/kg which is moderately/highly toxic.

References

(1) Worthing, C.R., *Pesticide Manual,* 6th ed., p. 392, British Crop Protection Council (1979).
(2) Spencer, E.Y., *Guide to the Chemicals Used in Crop Protection,* 6th ed., p. 379, London, Ontario, Agriculture Canada (January 1973).

ORYZALIN

Function: Herbicide (1)(2)

Chemical Name: 4-(dipropylamino)-3,5-dinitrobenzene sulfonamide

Formula:

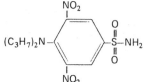

Trade Names: EL-119
Dirimal® (Eli Lilly &Co.)
Surflan® (Eli Lilly & Co.)
Ryzelan®

Manufacture

For example (2), 3,5-dinitro-4-[di(n-propyl)amino] benzenesulfonyl chloride and an excess of concentrated NH_4OH, the excess acting as solvent is heated to reflux temperature for several hours and the mixture cooled, poured onto crushed ice, and filtered. The filtrate is concentrated to dryness in vacuo and recrystallized from an acetone-petroleum ether mixture to yield 3,5-dinitro-N¹,N¹-di(n-propyl)sulfanilamide, MP 137° to 138°C.

Process Wastes and Their Control

Product Wastes: This material is reported (B-3) to disappear rapidly when mixed with soil maintained under flooded conditions.

Toxicity

This yellow-orange solid herbicide is relatively nontoxic (LD_{50} = 10,000). It presents no "undue hazard" to fish.

Allowable Limits on Exposure and Use

Product Use: The tolerances set by the EPA for oryzalin in or on raw agricultural commodities are as follows:

	40 *CFR* Reference	Parts per Million
Cotton, seed	180.304	0.05
Soybeans	180.304	0.1

References

(1) Worthing, C.R., *Pesticide Manual,* 6th ed., p. 393, British Crop Protection Council (1979).
(2) Soper, Q.F., U.S. Patent 3,367,949, February 6, 1968, assigned to Eli Lilly & Co.

OVEX (CHLORFENSON IN U.K.)

Function: Acaricide (1)(2)(3)

Chemical Name: 4-chlorophenyl 4-chlorobenzene sulfonate

Formula:

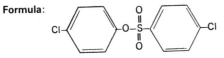

Trade Names: K-6451® (Dow Chemical Co.)
Ovotran® (Dow Chemical Co.)
Ovochlor®
Estonmite®
Mitran®
Orthotran®
Ovotox®
Ovomite®

Manufacture (4)(5)

Ovex is made by the condensation of p-chlorophenyl sulfonyl chloride with p-chlorophenol in alkaline solution.

Process Wastes and Their Control

Product Wastes: See Fenson for disposal methods (B-3).

Toxicity

The acute oral LD_{50} value for rats is 2,000 mg/kg which is slightly toxic.

Allowable Limits on Exposure and Use

Product Use:

The tolerances set by the EPA for ovex in or on raw agricultural commodities are as follows:

	40 *CFR* Reference	Parts per Million
Apples	180.134	3.0
Grapefruit	180.134	5.0
Lemons	180.134	5.0
Oranges	180.134	5.0
Peaches	180.134	3.0
Pears	180.134	3.0
Plums (fresh prunes)	180.134	3.0
Tangerines	180.134	5.0

References

(1) Worthing, C.R., *Pesticide Manual*, 6th ed., p. 99, British Crop Protection Council (1979).
(2) Spencer, E.Y., *Guide to the Chemicals Used in Crop Protection*, 6th ed., p. 381, London, Ontario, Agriculture Canada (January 1973).
(3) Hummer, R.W., and Kenaga, E.E., U.S. Patent 2,528,310, assigned to Dow Chemical Co.
(4) Slagh, H.R. and Britton, E.C., *Jour. Am. Chem. Soc.,* 72, 2808 (1950).
(5) Diamond Alkali Co., British Patent 747,368, April 4, 1956.

OXADIAZON

Function: Herbicide (1)(2)

Chemical Name: 3-[2,4-dichloro-5-(1-methylethoxy)phenyl]-5-(1,1-dimethylethyl)-1,3,4-oxadiazol-2(3H)-one

Formula:

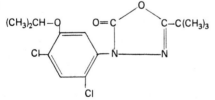

Trade Names: RP 17,623 (Rhone Poulenc)
Ronstar® (Rhone Poulenc)

Manufacture (3)

The initial 1-trimethylacetyl-2-(2,4-dichloro-5-isopropyloxyphenyl)hydrazine (MP 100°C) is obtained by the action of trimethylacetyl chloride on 2,4-dichloro-5-isopropyloxyphenyl-hydrazine (MP 71°C) in benzene in the presence of triethylamine. A solution obtained from 1-trimethylacetyl-2-(2,4-dichloro-5-isopropyloxyphenyl)hydrazine (28.5 g) and a 20% solution

(132 ml) of phosgene in toluene is heated gradually to about 100° to 110°C until the evolution of gas ceases. After concentration of the toluene solution under reduced pressure, the residual solid is recrystallized from ethanol. 5-tert-butyl-3-(2,4-dichloro-5-isopropyloxyphenyl)-1,3,4-oxadiazol-2-one (23 g) is thus obtained, MP 90°C.

Toxicity

The acute oral LD_{50} value for rats is over 8,000 mg/kg which is insignificantly toxic.

Allowable Limits on Exposure and Use

Product Use: The tolerances set by the EPA for oxadiazon in or on raw agricultural commodities are as follows:

	40 *CFR* Reference	Parts per Million
Brazil nuts	180.346	0.05 N
Bush nuts	180.346	0.05 N
Butternuts	180.346	0.05 N
Cashews	180.346	0.05 N
Cattle, fat	180.346	0.01 N
Cattle, mbyp	180.346	0.01 N
Cattle, meat	180.346	0.01 N
Chestnuts	180.346	0.05 N
Crabapples	180.346	0.05 N
Filberts	180.346	0.05 N
Fruits, stone	180.346	0.05 N
Goats, fat	180.346	0.01 N
Goats, mbyp	180.346	0.01 N
Goats, meat	180.346	0.01 N
Hazelnuts	180.346	0.05 N
Hickory nuts	180.346	0.05 N
Hogs, fat	180.346	0.01 N
Hogs, mbyp	180.346	0.01 N
Hogs, meat	180.346	0.01 N
Horses, fat	180.346	0.01 N
Horses, mbyp	180.346	0.01 N
Horses, meat	180.346	0.01 N
Macadamia nuts	180.346	0.05 N
Milk, fat (= N in whole milk)	180.346	0.1 N
Pears	180.346	0.05 N
Pecans	180.346	0.05 N
Pistachio nuts	180.346	0.05 N
Quinces	180.346	0.05 N
Rice, grain	180.346	0.05 N
Rice, straw	180.346	0.2 N
Sheep, fat	180.346	0.01 N
Sheep, mbyp	180.346	0.01 N
Sheep, meat	180.346	0.01 N
Walnuts	180.346	0.05 N

References

(1) Worthing, C.R., *Pesticide Manual,* 6th ed., p. 394, British Crop Protection Council (1979).
(2) Spencer, E.Y., *Guide to the Chemicals Used in Crop Protection,* 6th ed., p. 382, London, Ontario, Agriculture Canada (January 1973).
(3) Metivier, J. and Boesch, R., U.S. Patent 3,385,862, May 29, 1968, assigned to Rhone-Poulenc SA.

OXAMYL

Function: Insecticide and nematocide (1)(2)

Chemical Name: Methyl-2-(dimethylamino)-N-[[(methylamino)carbonyl]oxy]-2-oxoethan-iminothioate

Formula:

$$(CH_3)_2N\overset{\overset{\displaystyle O}{\|}}{C}C=NO\overset{\overset{\displaystyle O}{\|}}{C}NHCH_3$$
$$|$$
$$SCH_3$$

Trade Names: DPX-410 (DuPont)
Vydate® (DuPont)

Manufacture (3)(4)

A solution of 206 parts of methyl glyoxalate oxime in 1,500 parts of water is treated at 0° to −5°C with 145 parts of chlorine for about 37 minutes. An oil of 1-(methoxycarbonyl)-formhydroxamyl chloride is formed, which can be isolated by extraction with methylene chloride, evaporation of the methylene chloride and recrystallization from benzene, MP 63° to 65°C. To the suspension of 1-(methoxycarbonyl)formhydroxamyl chloride in water as obtained above are added at 0° to −5°C 105 parts of methyl mercaptan followed by 130 parts of 50% aqueous sodium hydroxide mixed with 330 parts of water. The resulting two-phase reaction mass is then extracted with methylene chloride. Removal of the solvent under reduced pressure leaves the solid methyl 1-methoxycarbonyl-N-hydroxy-thioformimidate, which can be recrystallized from benzene, MP 63° to 64°C.

The crude methyl 1-methoxycarbonyl-N-hydroxythioformimidate obtained above is dissolved in 250 parts of methanol and 180 parts of anhydrous dimethylamine dissolved in the solution below 30°C. This solution is allowed to stand at ambient temperature overnight. Removal of excess dimethylamine and the solvent under reduced pressure leaves 240 parts (74%) of methyl 1-(dimethylcarbamoyl)-N-hydroxythioformimidate.

To a suspension of 70 parts of methyl 1-(dimethylcarbamoyl)-N-hydroxythioformimidate and one-half part of Dabco in 350 parts of acetone at 40°C is added slowly 27 parts of methyl isocyanate. The temperature of the reaction rises to 58°C. After the temperature has subsided to 25°C, the solvent is evaporated under reduced pressure and the resulting residue recrystallized. Recrystallization from benzene gives one form of methyl 1-(dimethylcarbamoyl)-N-methylcarbamoyloxy)-thioformimidate of MP 109° to 110°C. Recrystallization from water gives the other form of MP 101° to 103°C.

Toxicity

The acute oral LD_{50} value for rats is 5.4 mg/kg which is highly to extremely toxic.

Allowable Limits on Exposure and Use

Product Use: The tolerances set by the EPA for oxamyl in or on raw agricultural commodities are as follows:

	40 *CFR* Reference	Parts per Million
Apples	180.303	2.0
Celery	180.303	3.0
Citrus fruits	180.303	3.0

(continued)

	40 *CFR* Reference	Parts per Million
Pineapples	180.303	1.0
Pineapples, forage	180.303	10.0
Potatoes	180.303	0.1
Tomatoes	180.303	2.0

The tolerances set by the EPA for oxamyl in animal feeds are as follows (the *CFR* Reference is to Title 21):

	CFR Reference	Parts per Million
Pineapple, bran	561.285	6.0

References

(1) Worthing, C.R., *Pesticide Manual,* 6th ed., p. 395, British Crop Protection Council (1979).
(2) Spencer, E.Y., *Guide to the Chemicals Used in Crop Protection,* 6th ed., p. 345, London, Ontario, Agriculture Canada (January 1973).
(3) Buchanan, J.B., U.S. Patent 3,530,220, September 22, 1970, assigned to DuPont.
(4) Buchanan, J.B., U.S. Patent 3,658,870, April 25, 1972, assigned to DuPont.

OXINE-COPPER

Function: Fungicide (1)(2)

Chemical Name: Bis(8-quinolinolate-N^1,O^x) copper

Formula:

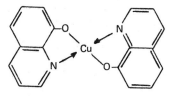

Trade Name: None

Manufacture

Oxine-copper is made by the reaction of 8-hydroxyquinoline with copper sulfate.

Toxicity

The acute oral LD_{50} value for rats is more than 10,000 mg/kg which is insignificantly toxic.

References

(1) Worthing, C.R., *Pesticide Manual,* 6th ed., p. 396, British Crop Protection Council (1979).
(2) Spencer, E.Y., *Guide to the Chemicals Used in Crop Protection,* 6th ed., p. 383, London, Ontario, Agriculture Canada (January 1973).

OXYCARBOXIN

Function: Fungicide (1)(2)(3)(4)(5)

Chemical Name: 5,6-dihydro-2-methyl-N-phenyl-1,4-oxathin-3-carboxamide 4,4-dioxide

Formula:

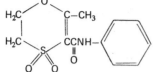

Trade Names: F-461 (Uniroyal Inc.)
Plantvax® (Uniroyal Inc.)

Manufacture (3)(4)(5)

Oxycarboxin may be made by the oxidation of carboxin (which see) with hydrogen peroxide.

Process Wastes and Their Control

Product Wastes: Oxycarboxin is hydrolyzed by very acidic or basic conditions to yield the more toxic aniline (B-3).

Toxicity

The acute oral LD_{50} value for rats is 2,000 mg/kg which is slightly toxic.

References

(1) Worthing, C.R., *Pesticide Manual,* 6th ed., p. 397, British Crop Protection Council (1979).
(2) Spencer, E.Y., *Guide to the Chemicals Used in Crop Protection,* 6th ed., p. 384, London, Ontario, Agriculture Canada (January 1973).
(3) Kulka, M., Thiara, D.S. and Harrison, W.A., U.S. Patent 3,399,214, August 27, 1968, assigned to Uniroyal Inc.
(4) von Schmeling, B., Kulka, M., Thiara, D.S. and Harrison, W.A., U.S. Patent 3,402,241, September 17, 1968, assigned to Uniroyal Inc.
(5) von Schmeling, B., Kulka, M., Thiara, D.S. and Harrison, W.A., U.S. Patent 3,249,499, May, 3, 1966, assigned to United States Rubber Company.

OXYDEMETON-METHYL

Function: Insecticide (1)(2)

Chemical Name: S-[2-(ethylsulfinyl)ethyl] O,O-dimethyl phosphorothioate

Formula:

$$(CH_3O)_2\overset{\overset{O}{\|}}{P}SCH_2CH_2\overset{\overset{O}{\|}}{S}C_2H_5$$

Trade Names: B-21,097 (Bayer)
R2170 (Bayer)
Metasystox-R® (Bayer)

Manufacture (3)(4)

The feed materials for oxydemeton-methyl manufacture are essentially the dimethyl analog of the insecticide demeton and hydrogen peroxide. The reaction which occurs is as follows:

$$(CH_3O)_2 \overset{\overset{\displaystyle O}{\parallel}}{P} SCH_2CH_2SC_2H_5 + H_2O_2 \longrightarrow (CH_3O)_2 \overset{\overset{\displaystyle O}{\parallel}}{P} \overset{\overset{\displaystyle O}{\parallel}}{S} CH_2CH_2SC_2H_5 + H_2O$$

The concentration of hydrogen peroxide in the reaction mixture must not be too low or the reaction rate will be economically inadequate. A considerable excess of hydrogen peroxide can be tolerated, provided the temperature is controlled during reaction.

20 grams of β-ethylmercapto-ethyl-thionophosphoric acid-dimethyl ester in 20 ml of methanol are oxidized with 8.5 ml of 36% hydrogen peroxide which are added dropwise to the solution at 35° to 40°C. After the reaction has subsided during which the reaction mixture must be slightly cooled, stirring is continued at 40°C for 2 hours. In the clear neutral solution a little free hydrogen peroxide is still detectable but disappears on subsequent fractionating.

There are obtained 14 g of sulfoxide having a BP of 95° to 96°C at a pressure of 0.01 mm Hg in a yield of 66% of the theoretical. The compound is only water soluble to some extent. O,O-dimethylphosphorochloridothioate and ethyl 2-hydroxyethyl sulfide (from ethyl mercaptan and ethylene oxide) are the raw materials for the demeton-methyl.

Toxicity

The acute oral LD_{50} value for rats is 65 to 80 mg/kg which is moderately toxic.

Allowable Limits on Exposure and Use

Product Use: The tolerances set by the EPA for oxydemeton-methyl in or on raw agricultural commodities are as follows:

	40 *CFR* Reference	Parts per Million
Alfalfa, chaff, for seed	180.330	11.0
Alfalfa, green	180.330	5.0
Alfalfa, hay for seed	180.330	11.0
Apples	180.330	2.0
Beans, lima	180.330	0.5
Beans, lima, forage	180.330	2.0
Beans, snap	180.330	0.5
Beans, snap, forage	180.330	2.0
Beets, sugar	180.330	0.3
Beets, sugar, tops	180.330	0.5
Blackberries	180.330	2.0
Broccoli	180.330	1.0
Brussels sprouts	180.330	1.0
Cabbage	180.330	1.0
Cattle, fat	180.330	0.01
Cattle, mbyp	180.330	0.01
Cattle, meat	180.330	0.01
Cauliflower	180.330	1.0
Clover, chaff, for seed	180.330	11.0

(continued)

	40 *CFR* Reference	Parts per Million
Clover, green	180.330	5.0
Clover, hay, for seed	180.330	11.0
Corn, fodder	180.330	3.0
Corn, forage	180.330	3.0
Corn, fresh (inc. sweet) (k+cwhr)	180.330	0.5
Corn, grain	180.330	0.5
Cotton, seed	180.330	0.1
Cucumbers	180.330	1.0
Eggplant	180.330	1.0
Filberts	180.330	0.05
Goats, fat	180.330	0.01
Goats, mbyp	180.330	0.01
Goats, meat	180.330	0.01
Grapefruit	180.330	1.0
Grapes	180.330	0.1
Hogs, fat	180.330	0.01
Hogs, mbyp	180.330	0.01
Hogs, meat	180.330	0.01
Horses, fat	180.330	0.01
Horses, mbyp	180.330	0.01
Horses, meat	180.330	0.01
Lemons	180.330	1.0
Lettuce, head	180.330	2.0
Melons	180.330	0.3
Milk	180.330	0.01
Mint, hay	180.330	12.5
Onions, dry bulb	180.330	0.05
Oranges	180.330	1.0
Pears	180.330	0.3
Peas	180.330	0.3
Peas, forage	180.330	2.0
Peas, hay	180.330	8.0
Peppers	180.330	0.75
Plums (fresh prunes)	180.330	1.0
Potatoes	180.330	0.1
Pumpkins	180.330	0.3
Raspberries	180.330	2.0
Safflower	180.330	1.0
Sheep, fat	180.330	0.01
Sheep, mbyp	180.330	0.01
Sheep, meat	180.330	0.01
Sorghum, forage	180.330	2.0
Sorghum, grain	180.330	0.75
Squash, summer	180.330	1.0
Squash, winter	180.330	0.3
Strawberries	180.330	2.0
Turnips	180.330	2.0
Turnips, tops	180.330	2.0
Walnuts	180.330	0.3

The tolerances set by the EPA for oxydemeton-methyl in animal feeds are as follows (the *CFR* Reference is to Title 21):

	CFR Reference	Parts per Million
Sorghum, milled fractions (exc flour)	561.234	2.0

References

(1) Worthing, C.R., *Pesticide Manual,* 6th ed., p. 398, British Crop Protection Council (1979).

(2) Spencer, E.Y., *Guide to the Chemicals Used in Crop Protection,* 6th ed., p. 385, London, Ontario, Agriculture Canada (January 1973).

(3) Muhlmann, R., Lorenz, W. and Schrader, G., U.S. Patent 2,963,505, December 6, 1960, assigned to Farbenfabriken Bayer AG.

(4) Lane, D.W.J., U.S. Patent 2,791,599, May 7, 1957, assigned to Pest Control, Ltd.

OXYTHIOQUINOX
(QUINOMETHIONATE IN U.K.)
(CHINOMETHIONAT, INTERNATIONAL)

Function: Acaricide (1)(2)

Chemical Name: 6-methyl-1,3-dithiolo[4,5-b] quinoxalin-2-one

Formula:

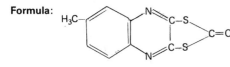

Trade Names: B-36,205 (Bayer)
Bayer Ss 2074 (Bayer)
Morestan® (Bayer)
Quinomethionate
Chinomethionat

Manufacture

According to D.C. Morrison and A. Furst (3), 2,3-dimercaptoquinoxaline may be obtained by reacting 2,3-dichloroquinoxaline with thiourea and subsequent alkaline splitting of the bis-thiouronium salt thus formed, or also by the action of phosphorus pentasulfide on 2,3-dihydroxyquinoxaline.

The product of 2,3-dimercaptoquinoxaline and its nuclear substitution products is expediently carried out in an especially simple manner by treating 2,3-dichloroquinoxaline or its derivatives with aqueous solutions of potassium or sodium hydrosulfide at an elevated temperature in general below 100°C (4). This practical process gives quantitative yields.

The following is a specific example of the conduct of the subsequent acylation step. Into a solution of 19.4 parts of 2,3-dimercaptoquinoxaline and 12 parts of sodium hydroxide in 150 parts of water, gaseous phosgene is introduced with ice cooling at 5° to 10°C until the solution shows an acid reaction. The excess phosgene is removed by blowing in air, the separated product is filtered off with suction, washed and dried. After boiling the filter cake with dioxane, the yield is 8 parts, i.e., 33% of the theoretical, of oxythioquinox.

Process Wastes and Their Control

Product Wastes: This compound is similar in chemical behavior to thioquinox, although it is more stable to oxidation. It is subject to alkaline hydrolysis (B-3).

Toxicity

The acute oral LD_{50} for rats is 2,500 to 3,000 mg/kg which is slightly toxic. However (B-3), this compound should not be allowed to contaminate streams, lakes, or ponds. Some phytotoxicity has been observed with certain varieties of roses.

Allowable Limits on Exposure and Use

Product Use: The tolerances set by the EPA for oxythioquinox in or on raw agricultural commodities are as follows:

	40 *CFR* Reference	Parts per Million
Apples	180.338	0.05 N
Cattle, fat	180.338	0.05 N
Cattle, mbyp	180.338	0.05 N
Cattle, meat	180.338	0.05 N
Citrus fruits	180.338	0.5
Goats, fat	180.338	0.05 N
Goats, mbyp	180.338	0.05 N
Goats, meat	180.338	0.05 N
Hogs, fat	180.338	0.05 N
Hogs, mbyp	180.338	0.05 N
Hogs, meat	180.338	0.05 N
Horses, fat	180.338	0.05 N
Horses, mbyp	180.338	0.05 N
Horses, meat	180.338	0.05 N
Macadamia nuts	180.338	0.1 N
Milk	180.338	0.01 N
Pears	180.338	0.05 N
Sheep, fat	180.338	0.05 N
Sheep, mbyp	180.338	0.05 N
Sheep, meat	180.338	0.05 N
Walnuts	180.338	0.1 N

References

(1) Worthing, C.R., *Pesticide Manual,* 6th ed., p. 463, British Crop Protection Council (1979).
(2) Spencer, E.Y., *Guide to the Chemicals Used in Crop Protection,* 6th ed., p. 387, London, Ontario, Agriculture Canada (January 1973).
(3) Morrison, D.C. and Furst, A., *J. Org. Chem.,* 21, 470 (1956).
(4) Sasse, K., Wegler, R. and Unterstenhofer, G., U.S. Patent 3,091,613, May 28, 1963, assigned to Farbenfabriken Bayer AG.

P

PARAQUAT

Function: Herbicide (1)(2)(3)

Chemical Name: 1,1'-dimethyl-4,4'-bipyridinium dichloride

Formula:

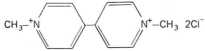

Trade Names: PP-148 (Plant Protection, Ltd.)
PP-910 (Plant Protection, Ltd.)
Dextrone®
Gramoxone®
Weedol®
Esgram®

Manufacture

Paraquat is made by the reaction of 4,4'-bipyridyl with methyl chloride in water. The 4,4'-bipyridyl is made by the reaction of pyridine with sodium in liquid ammonia to give 4,4'-tetrahydrobipyridyl which is oxidized by air to give 4,4'-bipyridyl.

Process Wastes and Their Control

Product Wastes: Paraquat is rapidly inactivated in soil. It is also inactivated by anionic surfactants. Therefore an effective and environmentally safe disposal method would be to mix the product with ordinary household detergent and bury the mixture in clay soil (B-3).

Toxicity

The acute oral LD_{50} value for rats is 150 mg/kg which is moderately toxic.

Paraquat is a highly effective, general herbicide that is acutely toxic to man and animals in its concentrated form (20% liquid concentrate). Oral exposure to high doses of paraquat frequently results in death, which is usually due to progressive fibrosis and epithelial proliferation in the lungs. However, in 2-year feeding studies in rats, paraquat did not produce any significant abnormalities.

Paraquat is rapidly inactivated by contact with clay particles in soil and is firmly bound physically. In this form, it is biologically inactive and apparently does not have any immediate or prolonged harmful effects. Thus, it is unlikely that paraquat would be found in large amounts in drinking water. Based on a 2-year feeding study in rats, an ADI was calculated at 0.0085 mg/kg/day (B-22).

The clinical picture following accidental or suicidal ingestion of paraquat is as follows (B-9). Paraquat ingestions are frequently fatal. Their management is unsatisfactory and largely symptomatic. Three clinical stages follow ingestion of as little as one ounce of paraquat.

The first is a gastrointestinal phase with burning in the mouth and throat, nausea, vomiting, abdominal pain, and diarrhea.

Several days after exposure, signs of hepatic and renal toxicity appear. These are due to central zone necrosis of the liver and acute tubular necrosis of the kidney.

Ten to twenty days after ingestion, progressive proliferative changes develop in the lungs. Hyperplastic changes in the terminal bronchioles occur with alveolar fibroblastic proliferation. Loss of lung surfactant has been demonstrated. Within a few days, death from respiratory failure occurs.

Allowable Limits on Exposure and Use

Air: The threshold limit value for respirable sizes of paraquat in air is 0.1 mg/m^3 as of 1979 (B-23). NIOSH cites a permissible exposure limit of 0.5 mg/m^3 (B-21).

Product Use: Issuance of a rebuttable presumption against registration for paraquat was being considered by EPA as of September 1979 on the basis of possible chronic effects; reduced fertility; environmental effects; and data gaps (oncogenicity and mutagenicity).

The tolerances set by the EPA for paraquat in or on raw agricultural commodities are as follows:

	40 *CFR* Reference	Parts per Million
Alfalfa	180.205	5.0
Almonds, hulls	180.205	0.5
Apples	180.205	0.05 N
Apricots	180.205	0.05 N
Avocados	180.205	0.05 N
Bananas	180.205	0.05 N
Barley, grain	180.205	0.05 N
Beans, guar	180.205	0.5
Beets, sugar	180.205	0.5
Beets, sugar, tops	180.205	0.5
Cattle, fat	180.205	0.01 N
Cattle, mbyp	180.205	0.01 N
Cattle, meat	180.205	0.01 N
Cherries	180.205	0.05 N
Citrus fruits	190.205	0.05 N
Clover	180.205	5.0
Coffee beans	180.205	0.05 N
Corn, fodder	180.205	0.05 N

(continued)

	40 *CFR* Reference	Parts per Million
Corn, forage	180.205	0.05 N
Corn, fresh (inc. sweet) (k+cwhr)	180.205	0.05 N
Corn, grain	180.205	0.05 N
Cotton, seed	180.205	0.5
Eggs	180.205	0.01 N
Figs	180.205	0.05 N
Fruits, passion	180.205	0.2
Fruits, small	180.205	0.05 N
Goats, fat	180.205	0.01 N
Goats, mbyp	180.205	0.01 N
Goats, meat	180.205	0.01 N
Grasses, pasture	180.205	5.0
Grasses, range	180.205	5.0
Guavas	180.205	0.05 N
Hogs, fat	180.205	0.01 N
Hogs, mbyp	180.205	0.01 N
Hogs, meat	180.205	0.01 N
Hops, fresh	180.205	0.1
Hops, vines	180.205	0.5
Horses, fat	180.205	0.01 N
Horses, mbyp	180.205	0.01 N
Horses, meat	180.205	0.01 N
Lettuce	180.205	0.05 N
Melons	180.205	0.05 N
Milk	180.205	0.01 N
Nectarines	180.205	0.05 N
Nuts	180.205	0.05 N
Oats, grain	180.205	0.05 N
Olives	180.205	0.05 N
Papayas	180.205	0.05 N
Peaches	180.205	0.05 N
Pears	180.205	0.05 N
Peppers	180.205	0.05 N
Pineapples	180.205	0.05 N
Plums (fresh prunes)	180.205	0.05 N
Potatoes	180.205	0.5
Poultry, fat	180.205	0.01 N
Poultry, mbyp	180.205	0.01 N
Poultry, meat	180.205	0.01 N
Rye, grain	180.205	0.05 N
Safflower, seed	180.205	0.05 N
Sheep, fat	180.205	0.01 N
Sheep, mbyp	180.205	0.01 N
Sheep, meat	180.205	0.01 N
Sorghum, forage	180.205	0.05 N
Sorghum, grain	180.205	0.05 N
Soybeans	180.205	0.05 N
Soybeans, forage	180.205	0.05 N
Sugarcane	180.205	0.5
Sunflower, seed	180.205	2.0
Tomatoes	180.205	0.05 N
Trefoil, birdsfoot	180.205	5.0
Wheat, grain	180.205	0.05 N

The tolerances set by the EPA for paraquat dichloride in animal feeds are as follows (the *CFR* Reference is to Title 21):

	CFR Reference	Parts per Million
Sunflower seed, hulls	561.289	6.0

The tolerances set by the EPA for paraquat bis(methylsulfate) in food are as follows (the *CFR* Reference is to Title 21):

	CFR Reference	Parts per Million
Hops, dried	193.331	0.2

References

(1) Worthing, C.R., *Pesticide Manual,* 6th ed., p. 399, British Crop Protection Council (1979).
(2) Spencer, E.Y., *Guide to the Chemicals Used in Crop Protection,* 6th ed., p. 389, London, Ontario, Agriculture Canada (January 1973).
(3) Brian, R.C., Driver, G.W., Homer, R.F. and Jones, R.L., U.S. Patent 2,972,528, February 21, 1961, assigned to Imperial Chemical Industries, Ltd.

PARATHION

Function: Insecticide (1)(2)

Chemical Name: O,O-diethyl O-4-nitrophenyl phosphorothioate

Formula:

$$(C_2H_5O)_2\overset{\overset{\text{S}}{\|}}{P}O-\!\!\langle\bigcirc\rangle\!\!-NO_2$$

Trade Names: E605 (Bayer)
Folidol® (Bayer)
Bladan® (Bayer)
ACC-3422 (American Cyanamid)
Thiophos® (American Cyanamid)
Niran® (Monsanto)
Fosferno (Plant-Protection, Ltd.)
Alkron®
Alleron®
Etilon®
Danthion®
Parawet®
Phoskil®
Nitrostigmine®

Manufacture

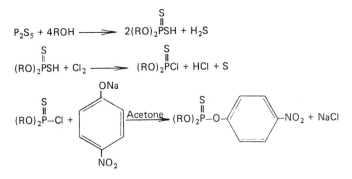

$$P_2S_5 + 4ROH \longrightarrow 2(RO)_2\overset{\overset{\text{S}}{\|}}{P}SH + H_2S$$

$$(RO)_2\overset{\overset{\text{S}}{\|}}{P}SH + Cl_2 \longrightarrow (RO)_2\overset{\overset{\text{S}}{\|}}{P}Cl + HCl + S$$

In the first step, ethyl dithiophosphone acid may be prepared (4) by reacting 208 pounds of anhydrous ethyl alcohol with 248 pounds of phosphorus pentasulfide (91% P_2S_5) crushed to pass a four-mesh screen. The reaction is preferably carried out in a glass lined kettle provided with an aluminum reflux condenser. The alcohol is charged into the kettle, heated to 35°C and the phosphorus pentasulfide added slowly through a suitable feed pipe. Since the reaction is exothermic, care is used to add the phosphorus pentasulfide at such a rate as to prevent undue rise in temperature, cooling means being used if necessary. The temperature is kept below 85°C and preferably around 45°C during the mixing of the ingredients which takes a period of about 1½ hours. The temperature of the mix is then raised to about 85°C and maintained at about that point for 2 to 3 hours after which the products are allowed to cool and the diethyl dithiophosphoric acid is removed. The crude acid obtained weighs over 400 pounds and contains about 88% pure acid.

The chlorination step follows (5) and may use elemental chlorine, sulfur monochloride or sulfur dichloride. An alternative route has been described (6). A production and waste schematic for the overall process of parathion manufacture is shown in Figure 40.

To 83.1 pounds of thiophosphoryl chloride was added, over a period of 2 to 3 hours, 350.5 pounds of an approximately 20% solution of sodium ethylate in ethyl alcohol, maintaining the temperature below 10°C. To this mixture was then added 68.2 pounds of p-nitrophenol, and to this resulting solution was added, over a period of 1 to 2 hours, 166.9 pounds of an approximately 20% solution of sodium ethylate in ethyl alcohol, maintaining the temperature below 10°C. This reaction mass was agitated by mechanical stirring for approximately 18 hours at the end of which time the reaction was complete. A larger portion of the ethyl alcohol in the batch was distilled from the stirred batch at atmospheric pressure. To the reaction mixture was then added 47 gallons of water in order to dissolve sodium chloride and to separate the O,O-diethyl O-p-nitrophenyl thiophosphate from the residual alcohol. The resulting mixture was pumped through a precoated filter in order to remove gums. The filtrate was allowed to separate into an upper aqueous layer which was decanted from the lower layer of O,O-diethyl O-p-nitrophenyl thiophosphate.

The lower layer was washed twice by stirring with 18 gallons of a 3% aqueous solution of sodium carbonate which was followed by one 15 gallon water wash. The wet lower layer was then subjected to a steam distillation to remove triethyl thiophosphate. The distillation was continued until no more oil was noticed in the distillate.

The water layer was decanted from the cooled material in the still pot. The organic layer was dried by heating to 110°C under 100 mm vacuum for ½ hour. The finished product, O,O-diethyl O-p-nitrophenyl thiophosphate, was obtained in an overall yield of 70% based on thiophosphoryl chloride.

Process Wastes and Their Control

Air: Air emissions from parathion manufacture have been reported (B-15) to consist of the following:

Component	Kilograms per Metric Ton Pesticide Produced
SO_2	410.0
Parathion	1.0
Ethanol	1.0
Ammonia	0.7

Gaseous emissions from parathion production contain significant amounts of mercaptans including hydrogen sulfide (B-10).

Figure 40: Parathion Manufacture

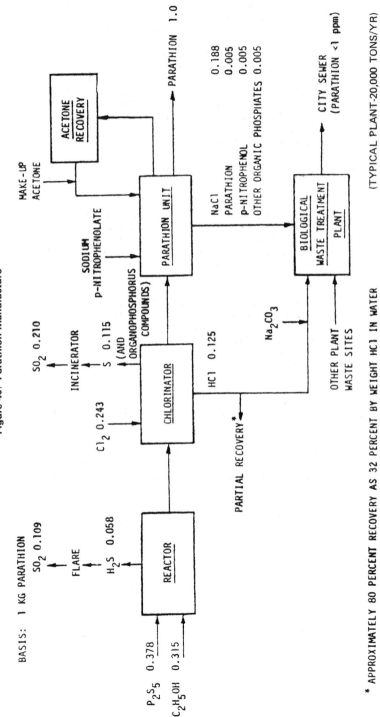

BASIS: 1 KG PARATHION

Air pollution control in parathion manufacture involves the following (B-15):

Control Device	Emissions Controlled	Reported Efficiency, %
Incinerator	Hydrogen sulfide, sulfur, mercaptan	—
Water scrubber	Phosphorus pentoxide, hydrogen chloride	95
Brink® mist eliminator	Phosphorus pentoxide (for visibility)	99.9

Water: The waste streams from parathion manufacture may contain sulfur, HCl, sodium chloride, sodium carbamate, trimethyl thiophosphate, and other organics including para-nitrophenol and small amounts of product (B-10). A typical waste stream can be characterized by:

COD	3,000 mg/l
BOD	700 mg/l
Total solids	27,000 mg/l
pH	2.0
Acidity	3,000 mg/l
Sodium	6,000 mg/l
Chlorides	7,000 mg/l
Phosphates	250 mg/l
Nitrates	20 mg/l
Sulfates	3,000 mg/l
Parathion	20 mg/l
Calcium	High

A wastewater treatment system for a parathion plant is shown in Figure 41. The chlorinated acid wastewater from parathion production was sampled at location 2201 (pretreatment). Neutral wastewater from parathion production was sampled at location 2202 (midtreatment). The combined wastewater passes through aerobic biodegradation lagoons, clarifiers and finally at point 2203 (posttreatment) enters the city sewer system.

Twenty-four hour composite samples were taken at the pretreatment and midtreatment sampling locations. An 8-hour composite was made from grab samples taken at the post-treatment location. The pH's at the pretreatment, midtreatment and posttreatment samples were 0.7, 6.4, and 7.1, respectively. The flow at the pretreatment location is not measured. The flows at midtreatment and posttreatment are considered classified information by the production plant personnel.

It is difficult to determine the effect of the waste treatment process between the pretreatment and midtreatment samples because the volume of the pretreatment effluent was not measured and the water is diluted with nonacid wastes before the midtreatment sample location. The wastewater is not diluted beyond this point; the volumes of water leaving the midtreatment sampling location and the posttreatment location are monitored and are of nearly equal volume. Concentration differences between midtreatment and posttreatment can be attributed to the effects of the biotreatment process. This process can remove organophosphorus compounds by several mechanisms which include: aerobic biodegradation, air oxidation, adsorption onto the biomass, neutral hydrolysis, and vaporization.

Solid Waste: Concentrated residues and tank bottoms contain large amounts of intermediates and some product. This portion of the waste may be a slurry with a paste-like consistency which poses severe handling problems but seldom enters the wastewater stream (B-10).

The elemental sulfur which is formed in the chlorinator is assumed to appear in the form of "microspheres." These microspheres will tend to encapsulate both the starting material (diethyl dithiophosphoric acid), the chlorination product (diethyl chlorothiophosphate), and other side reaction materials. For a plant producing 20,000 metric tons per year of parathion the sulfur sludge from the chlorinator is 2,300 metric tons per year (B-17).

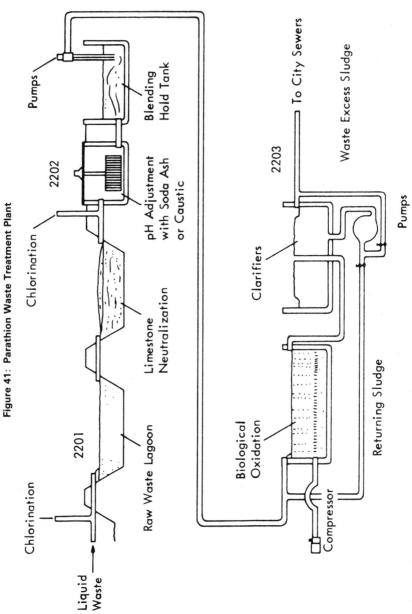

Figure 41: Parathion Waste Treatment Plant

Source: Reference (B-14)

A sulfur balance indicates that this sulfur sludge is composed of 93% (2,140 metric tons per year) elemental sulfur with the remaining 7% (160 metric tons per year) being organophosphorus compounds. Because the organophosphorus compounds are toxic, the sludge is classed a potentially hazardous discharge.

The present method of disposal of the waste chlorinator sludge from parathion manufacture is incineration without controls for abatement of SO_2 and phosphorus oxide emissions. This results in emissions estimated at approximately 4,000 cfm (based on 300 days per year and 8 hours per day) of which about 20% is SO_2. The magnitude of the emissions of SO_2 and phosphorus oxides coupled with the extreme toxicity of the sludge if not incinerated result in a totally unacceptable means of disposal.

Goals established for an alternative treatment were: (1) to separate the sulfur, for recovery purposes, from the organophosphates and (2) to detoxify the organophosphorus compounds. The following sequence of treatment processes (see Figure 42) can accomplish these purposes: (1) heated sedimentation; (2) ultrafiltration; (3) filtration; and (4) composting with lime and/or limestone.

Figure 42: Improved Process for Treatment of Sulfur Sludge from Chlorination Unit from Parathion Manufacture

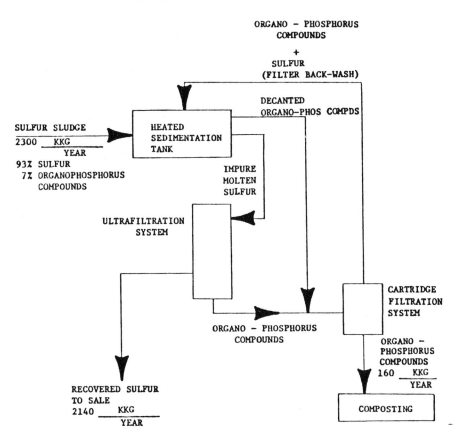

Source: Reference (B-7)

Product Wastes: Parathion is relatively resistant to hydrolysis: 50% hydrolysis at 70°C and pH 9 requires 2.7 hours; 17 to 20 hours at 70°C and pH 1 to 5; and 690 days at 20°C and pH 1 to 5. This compound is subject to reduction by metals in acid media (as Zn in 9 parts HOAc and 1 part HCl) to yield, O,O-diethyl O-(4-aminophenyl)thiophosphate (aminoparathion) which is nontoxic to animals and does have an insecticidal effect. One manufacturer recommends the use of a detergent in a 5% trisodium phosphate solution for parathion disposal and cleanup problems (B-3).

Toxicity

The acute oral LD_{50} value for rats is 2 mg/kg (B-5) which is extremely toxic.

The toxic properties of parathion and parathion-methyl are very similar and the reader is referred to the section on Parathion-Methyl for a comprehensive statement by the National Research Council (B-22) dealing with both compounds.

Allowable Limits on Exposure and Use

Air: The threshold limit value for parathion in air is 0.1 mg/m^3 as of 1979. The tentative short term exposure limit is 0.3 mg/m^3. It is further noted that cutaneous absorption should be prevented so that the TLV is not invalidated (B-23).

NIOSH has recommended (7) a value of 0.05 mg/m^3 on a 10-hour time weighted average basis.

Water: In water, EPA has set (B-12) criteria of 0.05 μg/l for the protection of freshwater and marine aquatic life.

Product Use: The tolerances set by the EPA for parathion in or on raw agricultural commodities are as follows:

	40 *CFR* Reference	Parts per Million
Alfalfa, fresh	180.121	1.25
Alfalfa, hay	180.121	5.0
Almonds	180.121	0.1 N
Almonds, hulls	180.121	3.0
Apples	180.121	1.0
Apricots	180.121	1.0
Artichokes	180.121	1.0
Avocados	180.121	1.0
Barley	180.121	1.0
Beans	180.121	1.0
Beets, greens (alone)	180.121	1.0
Beets, sugar	180.121	0.1 N
Beets, sugar, tops	180.121	0.1 N
Beets, with tops	180.121	1.0
Beets, without tops	180.121	1.0
Blackberries	180.121	1.0
Blueberries (huckleberries)	180.121	1.0
Boysenberries	180.121	1.0
Broccoli	180.121	1.0
Brussels sprouts	180.121	1.0
Cabbage	180.121	1.0
Carrots	180.121	1.0
Cauliflower	180.121	1.0
Celery	180.121	1.0
Cherries	180.121	1.0

(continued)

	40 *CFR* Reference	Parts per Million
Citrus fruits	180.121	1.0
Clover	180.121	1.0
Collards	180.121	1.0
Corn	180.121	1.0
Corn, forage	180.121	1.0
Cotton, seed	180.121	0.75
Cranberries	180.121	1.0
Cucumbers	180.121	1.0
Currants	180.121	1.0
Dates	180.121	1.0
Dewberries	180.121	1.0
Eggplant	180.121	1.0
Endive (escarole)	180.121	1.0
Figs	180.121	1.0
Filberts	180.121	0.1 N
Garlic	180.121	1.0
Gooseberries	180.121	1.0
Grapes	180.121	1.0
Grasses, forage	180.121	1.0
Guavas	180.121	1.0
Hops	180.121	1.0
Kale	180.121	1.0
Kohlrabi	180.121	1.0
Lettuce	180.121	1.0
Loganberries	180.121	1.0
Mangoes	180.121	1.0
Melons	180.121	1.0
Mustard, greens	180.121	1.0
Mustard, seed	180.121	0.2
Nectarines	180.121	1.0
Oats	180.121	1.0
Okra	180.121	1.0
Olives	180.121	1.0
Onions	180.121	1.0
Parsnips, greens alone	180.121	1.0
Parsnips, with tops	180.121	1.0
Parsnips, without tops	180.121	1.0
Peaches	180.121	1.0
Peanuts	180.121	1.0
Pears	180.121	1.0
Peas	180.121	1.0
Peas, forage	180.121	1.0
Pecans	180.121	0.1 N
Peppers	180.121	1.0
Pineapples	180.121	1.0
Plums (fresh prunes)	180.121	1.0
Potatoes	180.121	0.1 N
Pumpkins	180.121	1.0
Quinces	180.121	1.0
Radishes, tops	180.121	1.0
Radishes, with tops	180.121	1.0
Radishes, without tops	180.121	1.0
Rape, seed	180.121	0.2
Raspberries	180.121	1.0
Rice	180.121	1.0
Rutabaga tops	180.121	1.0
Rutabagas, with tops	180.121	1.0
Rutabagas, without tops	180.121	1.0
Safflower, seed	180.121	0.1 N

(continued)

	40 *CFR* Reference	Parts per Million
Sorghum	180.121	0.1 N
Sorghum, fodder	180.121	3.0
Sorghum, forage	180.121	3.0
Soybeans	180.121	0.1
Soybeans, hay	180.121	1.0
Spinach	180.121	1.0
Squash	180.121	1.0
Squash, summer	180.121	1.0
Strawberries	180.121	1.0
Sugarcane	180.121	0.1 N
Sugarcane, fodder	180.121	0.1 N
Sugarcane, forage	180.121	0.1 N
Sunflower, seed	180.121	0.2
Sweet Potatoes	180.121	0.1 N
Swiss Chard	180.121	1.0
Tomatoes	180.121	1.0
Turnips, greens	180.121	1.0
Turnips, with tops	180.121	1.0
Turnips, without tops	180.121	1.0
Vetch	180.121	1.0
Walnuts	180.121	0.1 N
Wheat	180.121	1.0
Youngberries	180.121	1.0

As a consequence of EPA action on May 28, 1976, it was specified that:

1. Registration of ethyl parathion should be limited to those packed in one gallon containers or larger.

2. Manufacturers and formulators of registered ethyl parathion should be in compliance with the standardized safety label that was enclosed with EPA Notice PR 71-2 dated May 28, 1976.

References

(1) Worthing, C.R., *Pesticide Manual,* 6th ed., p. 401, British Crop Protection Council (1979).
(2) Spencer, E.Y., *Guide to the Chemicals Used in Crop Protection,* 6th ed., p. 390, London, Ontario, Agriculture Canada (January 1973).
(3) Midwest Research Institute, *The Pollution Potential in Pesticide Manufacturing,* Washington, DC, Environmental Protection Agency (June 1972).
(4) Christmann, L.J., U.S. Patent 1,893,018, January 3, 1933, assigned to American Cyanamid Company.
(5) Hechenbleikner, I., U.S. Patent 2,482,063, September 13, 1949, assigned to American Cyanamid Company.
(6) Dvornikoff, M.N. and Young, E.J., U.S. Patent 2,663,721, December 22, 1953, assigned to Monsanto Chemical Company.
(7) National Institute for Occupational Safety and Health, *Criteria for a Recommended Standard: Occupational Exposure to Parathion,* NIOSH Doc. No. 76-190 Washington, DC (1976).

PARATHION-METHYL
(METAPHOS IN U.S.S.R.)

Function: Insecticide (1)(2)

Chemical Name: O,O-dimethyl O-4-nitrophenyl phosphorothioate

Formula:

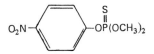

Trade Names: Dalf® (Bayer)
Folidoc M® (Bayer)
Metacide® (Bayer)
Bladan M® (Bayer)
Nitrox 80® (Bayer)
Metron®
Patron M®
Tekwaisa®

Manufacture (3)(4)

The feed materials for methyl parathion manufacture are as follows: O,O-dimethyl phosphorothionochloridate and sodium p-nitrophenoxide. These materials react according to the equation:

$$(CH_3O)_2\overset{\overset{S}{\|}}{P}Cl \ + \ NaOC_6H_4NO_2 \ \longrightarrow \ (CH_3O)_2\overset{\overset{S}{\|}}{P}OC_6H_4NO_2 \ + \ NaCl$$

The exothermic reaction may be conducted at 80° to 100°C according to G. Schrader (5). However, it is preferred to operate at a much lower temperature of from −10° to +10°C according to M.N. Dvornikoff et al (6). The condensation reaction in parathion methyl manufacture is carried out at atmospheric pressure. A reaction time of about 5 hours is required at 80° to 95°C and a time of about 18 hours is required when operating at lower temperatures near 0°C. The reaction is conducted in the liquid phase. One of various organic solvents may be present. An alcoholic medium is specified by M.N. Dvornikoff et al (6). However, an inert solvent such as benzene or chlorobenzene is preferred by G. Schrader (5).

Copper powder may be used as a catalyst or the reaction may simply be conducted in a copper reaction vessel, thereby shortening reaction time appreciably. A small amount of potassium bromide can be used as an effective cocatalyst. Aliphatic amine catalysts have been cited for use in this reaction by A.D.F. Toy et al (7). At least 0.25% such materials as triethyl amine, tributyl amine, N-ethyl morpholine and hexamethylene tetramine may be used.

The reaction is carried out in a stirred, jacketed vessel. Although the reaction is conducted at essentially atmospheric pressure, closed vessels must be used because by-product gases are both toxic and nuisances. Yields of methyl parathion are 90% or higher. The reaction product mixture may be pumped through a precoated filter to remove gummy impurities. The filtrate may then be separated into an aqueous and an oily layer. The lower oily layer may then be washed with a dilute sodium carbonate solution, then with water; it may then be steam distilled to remove trimethyl thiophosphate. After cooling and settling, the organic layer can be dried by heating under vacuum to give the product.

Process Wastes and Their Control

Air: Air emissions from methyl parathion manufacture have been reported (B-15) to consist of the following:

Component	Kilograms per Metric Ton Pesticide Produced
SO$_2$	410.0
Parathion methyl	1.0
Methyl alcohol	1.0
Ammonia	0.05

By-product H$_2$S from this process is flared (B-10).

Water: Liquid wastes from one plant, including by-product HCl-NaCl, go to biological treatment facilities and then into a sewer system. Sulfur and waste solvents are burned. The only solid waste is sludge from the biological oxidation system which is recycled and discharged at a slow rate into the sewer. Spills are washed into the waste treatment system (B-10).

Product Wastes: This compound is subject to hydrolysis: 50% in 210 minutes in 0.01N NaOH at 30°C, and 32 minutes in 1N NaOH at 15°C. Heating to 140° to 160°C sometimes takes place explosively. (See also Parathion) (B-3.)

Toxicity

The acute oral LD$_{50}$ for rats is 14 mg/kg which is highly toxic.

Parathion and parathion methyl are highly toxic organophosphorus insecticides that are widely used in commercial agriculture. Acute toxicity of both compounds is very high, but chronic toxicity does not appear to be a major consideration. The mode of action of these compounds is well known to be inhibition of acetylcholinesterase. Subchronic and chronic studies with the compounds have been primarily concerned, therefore, with measuring the decrease in cholinesterase enzymes as a result of oral treatment or incorporation in the diet of a variety of animals.

An ADI was calculated at 0.0043 mg/kg/day for both parathion and parathion methyl based on human data on parathion.

The obvious scarcity of data on the toxicity of parathion methyl indicates a pressing need for research. It appears that the assumption has been made that parathion methyl is toxicologically the same as parathion and that extrapolations have been made from parathion toxicology to parathion methyl. The data on teratogenic effects of parathion methyl, however, indicate that this is not an acceptable procedure in this case. Furthermore, in the last several years, parathion methyl has greatly surpassed parathion in total volume of use, making the need for specific data on parathion methyl even more pressing. The first priority in developing new information must be on the possibility of teratological effects of parathion methyl (B-22).

Allowable Limits on Exposure and Use

Air: The threshold limit value for parathion methyl is 0.2 mg/m^3 as of 1979. The tentative short term exposure limit is 0.6 mg/m^3. It is noted that cutaneous absorption should be prevented so that the TLV is not invalidated (B-23). NIOSH has recommended the same TLV value (B-23), although no OSHA standard exists.

References

(1) Worthing, C.R., *Pesticide Manual,* 6th ed., p. 402, British Crop Protection Council (1979).
(2) Spencer, E.Y., *Guide to the Chemicals Used in Crop Protection,* 6th ed., p. 349, London, Ontario, Agriculture Canada (January 1973).
(3) Midwest Research Institute and RvR Consultants, *Production, Distribution, Use and Environmental Impact Potential of Selected Pesticides,* Washington, DC, Council on Environmental Quality (March 15, 1974).

(4) Schneller, G. and Smith B., *Ind. Eng. Chem.* 41, 1027 (1949).
(5) Schrader, G., U.S. Patent 2,624,745, January 6, 1953, assigned to Farbenfabriken Bayer.
(6) Dvornikoff, M.N. et al, U.S. Patent 2,663,721, assigned to Monsanto Chemical Co.
(7) Toy, A.D.F. et al, U.S. Patent 2,471,464, May 31, 1949, assigned to Victor Chemical Works.
(8) National Institute for Occupational Safety and Health, *Criteria for a Recommended Standard: Occupational Exposure to Methyl Parathion,* NIOSH Doc. No. 77-106, Washington, DC (1977).

PARIS GREEN

Function: Insecticide (1)(2)

Chemical Name: Bis(acetato) (hexametaarsenitotetracopper)

Formula:

$$(CH_3CO)_2Cu \cdot 3Cu(AsO_2)_2$$

Trade Name: None

Manufacture (3)(4)

Paris green may be made by the interaction of sodium arsenite ($NaAsO_2$), copper sulfate and acetic acid.

A process using acetic acid, ammonia, copper metal and arsenic trioxide as raw materials has been described (3).

Toxicity

The acute oral LD_{50} for rats is 22 mg/kg which is highly toxic.

References

(1) Worthing, C.R., *Pesticide Manual,* 6th ed., p. 403, British Crop Protection Council (1979).
(2) Spencer, E.Y., *Guide to the Chemicals Used in Crop Protection,* 6th ed., p. 392, London, Ontario, Agriculture Canada (January 1973).
(3) Serciron, M., U.S. Patent 2,159,864, May 23, 1939.
(4) Krefft, O., U.S. Patent 2,268,123, December 30, 1941, assigned to Chemical Marketing Co., Inc.

PEBULATE

Function: Herbicide (1)(2)

Chemical Name: S-propyl butylethylcarbamothioate

Formula:

$$C_4H_9 \diagdown \atop C_2H_5 \diagup N\overset{\overset{O}{\|}}{C}SC_3H_7$$

Trade Names: R-2061 (Stauffer Chemical Co.)
           Tillam® (Stauffer Chemical Co.)

Manufacture

It has been found (3) that carbamyl chlorides can be reacted directly with mercaptans to produce the desired thiocarbamic esters if zinc chloride is employed as a catalyst. The reaction can be conducted without the use of a solvent, and since no sodium salt is involved, there is no filtration or solids handling problem. A batch or continuous process can be used.

When it was attempted to conduct the reaction without a catalyst, it was found that the reaction would not go at any appreciable rate below 170°C, and that if the reaction were conducted at a higher temperature, there was a substantial decomposition reducing the yield. Moderately elevated temperatures are useful in carrying out the reaction. Generally speaking, the temperature should be maintained in the range of 20° to 160°C.

The amount of zinc chloride which is used is not critical as any amount will tend to increase the reaction rate, but in order to secure practical results, it is best to use from 0.5 to 20 grams per gram mol of carbamyl chloride and preferably between 2 and 10 grams per gram mol. The amount of zinc chloride used depends on the reaction temperature and the desired rate of reaction. Normally, the mercaptan is employed in an excess over the stoichiometric amount with the carbamyl chloride. Any excess mercaptan used is recovered and recycled. The following is a specific example of the conduct of this process.

To a flask fitted with a reflux condenser, thermometer, and heating mantle are added 163.5 g (1 mol) of N-ethyl-n-butylcarbamyl chloride, 9.14 g (1.2 mols) of n-propylmercaptan and about 5 g of anhydrous zinc chloride. The temperature is held between 70° and 160°C for 2 hours. The excess mercaptan is distilled out during the final period of reaction. The zinc chloride is removed by washing with dilute hydrochloric acid. The weight of recovered product was 183 g (90% yield).

An alternative method has also been described (4)(5). In this procedure, 10 g (0.72 mol) of n-propyl chlorothiolformate was dissolved in 125 cc of ethyl ether and the solution was cooled in an ice bath to 5°C. To this mixture was slowly added 14.9 g (0.148 mol) of N-ethyl-n-butylamine.

The precipitated amine hydrochloride was then removed by filtration and the filter cake was washed with a little ether. The washings were combined with the original filtrate and the ether solution of the product was concentrated on a steam bath. The residual liquid was transferred to a 50 cc of round-bottom flask and the solvent was completely removed at 25°C and 25 mm Hg pressure by means of a Rinco rotating film evaporator. There was obtained 14.0 g (96% yield) of pebulate.

Toxicity

The acute oral LD_{50} value for rats is 1,120 mg/kg which is slightly toxic.

Allowable Limits on Exposure and Use

Product Use: The tolerances set by the EPA for pebulate in or on raw agricultural commodities are as follows:

	40 *CFR* Reference	Parts per Million
Beets, sugar, roots	180.238	0.1 N
Beets, sugar, tops	180.238	0.1 N
Tomatoes	180.238	0.1 N

References

(1) Worthing, C.R., *Pesticide Manual,* 6th ed., p. 404, British Crop Protection Council (1979).
(2) Spencer, E.Y., *Guide to the Chemicals Used in Crop Protection,* 6th ed., p. 393, London, Ontario, Agriculture Canada (January 1973).
(3) Campbell, R.G. and Klingman, G.E., U.S. Patent 2,983,747, May 9, 1961, assigned to Stauffer Chemical Co.
(4) Tilles, H. and Antognini, J., U.S. Patent 3,175,897, March 30, 1965, assigned to Stauffer Chemical Co.
(5) Tilles, H. and Antognini, J., U.S. Patent 3,185,720, May 25, 1965, assigned to Stauffer Chemical Co.

PENDIMETHALIN
(formerly Penoxalin)

Function: Herbicide (1)(2)

Chemical Name: N-(1-Ethylpropyl)-3,4-dimethyl-2,6-dinitrobenzeneamine

Formula:

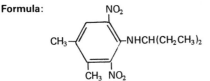

Trade Names: AC 92,553 (American Cyanamid)
Prowl® (American Cyanamid)
Stomp® (American Cyanamid)
Herbadox® (American Cyanamid)

Manufacture (2)

A mixture of 4-chloro-3,5-dinitro-o-xylene (0.61 mol), 3-pentaneamine (1.82 mols), and xylene (1,400 ml) is brought to reflux. After refluxing overnight, the reaction mixture is cooled and filtered. The precipitate is washed with petroleum ether. The filtrate and washings are combined, washed with 500 ml of 10% hydrochloric acid, and finally with 2 liters of water. The organic layer is separated and dried. Removal of the drying agent and the solvent leaves an orange oil which crystallizes with the addition of methanol. A yellow orange solid with MP 56° to 57°C is collected in about 85% yield.

Toxicity

The acute oral LD_{50} value for rats is 1,050 to 1,250 mg/kg which is slightly toxic.

Allowable Limits on Exposure and Use

Product Use: The tolerances set by the EPA for pendimethalin in or on raw agricultural commodities are as follows:

	40 *CFR* Reference	Parts per Million
Corn, fodder	180.361	0.1
Corn, forage	180.361	0.1
Corn, grain	180.361	0.1
Cotton, seed	180.361	0.1
Soybeans	180.361	0.1
Soybeans, forage	180.361	0.1
Soybeans, hay	180.361	0.1

References

(1) Worthing, C.R., *Pesticide Manual,* 6th ed., p. 405, British Crop Protection Council (1979).
(2) Lutz, A.W. and Diehl, R.E., U.S. Patent 3,920,742, November 18, 1975, assigned to American Cyanamid Co.

PENTACHLOROPHENOL

Function: Insecticide, fungicide and herbicide (1)(2)(3)

Chemical Name: Pentachlorophenol

Formula:

Trade Names: Dowicide 7® (Dow Chemical Co.)
Santophen 20® (Monsanto)
Santobrite (Monsanto)
Chlorophen (Reichhold Chemicals)
Glazd Penta®
Sinituho®

Manufacture

Pentachlorophenol may be made by one of two routes: the chlorination of phenol (3)(4)(5) or the hydrolysis of hexachlorobenzene (6)(7)(8).

The chlorination of phenol proceeds as follows:

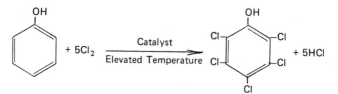

The chlorination is performed at substantially atmospheric pressure. The temperature of the phenol in the primary reactor at the start is in the range of 65° to 130°C (preferably 105°C) and is held in this range until the melting point of the product reaches 95°C. About 3 to 4 atoms of chlorine are combined at this point, and the temperature is progressively increased to maintain a temperature of about 10°C over the product melting point, until the reaction is completed in 5 to 15 hours. The mixture is a liquid, and a solvent is not required, but the catalyst concentration is critical; about 0.0075 mol of anhydrous aluminum chloride is usually used per mol of phenol.

The off-gas from the chlorination reactor (largely HCl during the initial reaction and chlorine near the conclusion) is sent to a scrubber-reactor system containing excess phenol. It is held at a temperature such that the chlorine is almost completely reacted to give the lower chlorinated phenols, which may be either separated, purified and sold, or returned to be used in the primary pentachlorophenol reactor. The residual gas is substantially pure HCl. Figure 43 is a production and waste schematic for the process.

Figure 43: Production and Waste Diagram for Pentachlorophenol Manufacture

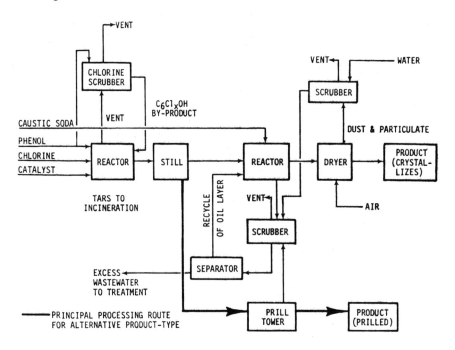

Source: Reference (B-4)

The second alternative route involves the hydrolysis of hexachlorobenzene according to the equation:

$$C_6Cl_6 + NaOH \rightarrow C_6Cl_5OH + NaCl$$

Figure 44 is a flow diagram of the process with emphasis on product recovery. An improved hydrolysis process using dimethyl sulfoxide (DMSO) as a solvent has been developed (8).

Figure 44: Process Diagram for Pentachlorophenol Manufacture from Hexachlorobenzene

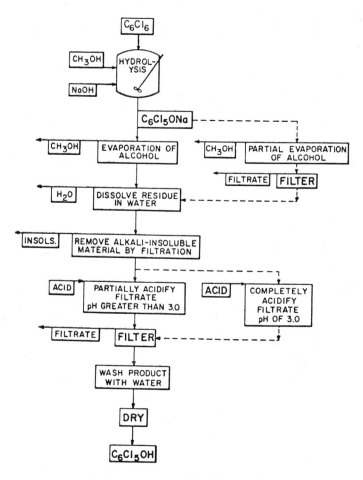

Source: Reference (7)

Process Wastes and Their Control

Air: Air emissions from the manufacture of pentachlorophenol and its sodium salts have been reported (B-15) to consist of the following:

Component	Kilograms per Metric Ton Pesticide Produced
Pentachlorophenol	0.6
Sodium pentachlorophenate	2.2
Phenol	1.0
Hydrogen chloride	1.0
Chlorine	0.5

Air pollution control in pentachlorophenol manufacture involves the following (B-15):

Control Devices	Emissions Controlled	Reported Efficiency, %
Packed and venturi scrubber	Chlorine, phenol, acids	90 to 100
Baghouse	Pentachlorophenol	95 to 99
Mechanical seals	(For pentachlorophenol reactor)	—

Water: The wastes from this process are cycled to the secondary reactor where more phenol is added. The products of the second reactor are partially chlorinated phenols and HCl. The phenols are recycled to the primary reactor, and the HCl is recovered and cycled through a recovery process (B-10).

Wastewater containing lower chlorinated phenols may be sent to a treatment facility.

Pentachlorophenol may be removed from wastewater by liquid-liquid extraction with methylene chloride (9).

Product Wastes: Pentachlorophenol is nearly insoluble in water, but soluble in most organic solvents, while the sodium salt is soluble in water, but is insoluble in oils. Photolysis of aqueous PCP by sunlight produces a series of degradation products in which chlorine atoms are replaced by hydroxyl groups followed by air-oxidation to quinones and other reaction products: Principal products are $C_6Cl_4(OH)_2$, $C_6Cl_2O_2(PH)_2$ and the coupled compounds $C_6Cl_5OC_6Cl_2O_2(OH)$, $C_6Cl_4(OH)OC_6Cl_2O_2(OH)$, and $C_6Cl_4(OH)OC_6Cl_3O_2$.

Under ultraviolet irradiation in methanol solution, PCP underwent reductive dechlorination to give tetrachlorophenol (the 2,3,5,6 isomer only), but irradiation of an aqueous suspension of PCP gave humic acid (an unchlorinated polymeric material).

PCP is nonflammable alone, but the MCA recommends incineration methods of disposal (B-3).

Toxicity

The acute oral LD_{50} value for rats is 210 mg/kg which is moderately toxic.

There are substantial disagreements in the results of several of the subacute and chronic toxicity experiments with PCP perhaps because of the use of inadequately characterized PCP preparations in these studies. In addition, 2-year chronic toxicity experiments in one or more species have not yet been conducted with this extensively used chemical. High doses (>5 mg/kg/day) of PCP have been shown to be teratogenic in rats and hamsters administered during susceptible days of gestation. There is also a need for an adequate determination of the carcinogenic potential of this chemical.

On the basis of the available chronic toxicity data an ADI for pentachlorophenol has been calculated to be 0.003 mg/kg/day (B-22).

PCP and its sodium salt, sodium pentachlorophenate (Na-PCP), have been demonstrated to be highly toxic to man, mammals, and aquatic life. As a result of their toxic nature, accumulation in tissues, and widespread industrial and agricultural applications, both PCP and Na-PCP pose a potential threat to terrestrial and aquatic life (B-26).

Allowable Limits on Exposure and Use

Air: The threshold limit value for pentachlorophenol in air is 0.5 mg/m^3 as of 1979. The tentative short term exposure limit is 1.5 mg/m^3. It is further noted that cutaneous absorption of pentachlorophenol should be prevented so that the TLV is not invalidated.

Water: In water, EPA has suggested (B-26) a criterion to protect freshwater aquatic life of 6.2 μg/l as a 24-hour average and the concentration should not exceed 14 μg/l at any time. For pentachlorophenol the criterion to protect saltwater aquatic life is 3.7 μg/l as a 24-hour average and the concentrations should not exceed 8.5 μg/l at any time.

For the protection of human health from the toxic properties of pentachlorophenol ingested through water and through contaminated aquatic organisms, the ambient water quality criterion is determined to be 140 μg/l (B-26).

Product Use: A rebuttable presumption against registration was issued on October 18, 1978 by EPA on the basis of fetotoxicity and teratogenicity.

References

(1) Worthing, C.R., *Pesticide Manual,* 6th ed., p. 406, British Crop Protection Council (1979).
(2) Spencer, E.Y., *Guide to the Chemicals Used in Crop Protection,* 6th ed., p. 395, London, Ontario, Agriculture Canada (January 1973).
(3) Midwest Research Institute and RvR Consultants, *Production, Distribution, Use and Environmental Impact Potential of Selected Pesticides,* Washington, DC, Council on Environmental Quality (March 15, 1974).
(4) Stoesser, W.C., U.S. Patent 2,131,259, September 17, 1938, assigned to Dow Chemical Company.
(5) Shelton, F.J. et al, U.S. Patent 2,947,790, August 2, 1960, assigned to Reichhold Chemicals, Inc.
(6) Smith, F.B. et al, U.S. Patent 2,107,650, February 8, 1938, assigned to Dow Chemical Company.
(7) Rosen, I., U.S. Patent 2,812,366, November 5, 1957, assigned to Diamond Alkali Company.
(8) Cohen, R.S., U.S. Patent 3,481,991, December 2, 1969, assigned to Dover Chemical Corporation.
(9) Winn, W.D., U.S. Patent 3,931,001, January 6, 1976, assigned to Dow Chemical Company.

PERFLUIDONE

Function: Herbicide (1)(2)

Chemical Name: 1,1,1-Trifluoro-N-[2-methyl-4-(phenylsulfonyl)phenyl]methane-sulfonamide

Formula:

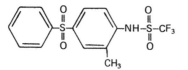

Trade Name: MBR-8251 (3M Co.)

Manufacture

The starting material, 4-phenylthio-o-toluidine is reacted with trifluoromethane sulfonyl chloride. The intermediate so produced is oxidized by reaction with hydrogen peroxide to give perfluidone, a white crystalline product melting at 142° to 144°C.

Toxicity

The acute oral LD_{50} value for rats is 633 mg/kg which is slightly toxic.

Allowable Limits on Exposure and Use

Product Use: The tolerances set by the EPA for perfluidone in or on raw agricultural commodities are as follows:

	40 *CFR* Reference	Parts per Million
Cotton, seed	180.165	0.01

References

(1) Worthing, C.R., *Pesticide Manual,* 6th ed., p. 408, British Crop Protection Council (1979).
(2) Minnesota Mining and Manufacturing Co., British Patent 1,306,564, February 14, 1973.

PERMETHRIN

Function: Insecticide (1)(2)

Chemical Name: (3-Phenoxyphenyl)methyl-1-(2,2-dichloroethenyl)-2,2-dimethyl-cyclopropanecarboxylate

Formula:

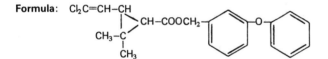

Trade Names: NRDC-143 (National Research Dev. Corp.)
FMC-33,297 (FMC Corp.)
PP-557 (I.C.I.)
Ambush® (I.C.I.)
Ambushfog® (I.C.I.)
Perthrine® (I.C.I.)
Kafil® (I.C.I.)
WL 43479 (Shell)
Talcord® (Shell)
Outflank® (Shell)
Stockade® (Shell)
Coopex® (Wellcome Fdn.)
Qamlin® (Wellcome Fdn.)
Stomoxin® (Wellcome Fdn.)
Perigen® (Wellcome Fdn.)

Manufacture (2)

The starting acid was prepared by a variant on the conventional chrysanthemic acid synthesis using ethyldiazoacetate in which, in this case, 1,1-dichloro-4-methyl-1,3-pentadiene was reacted with ethyldiazoacetate in the presence of the copper catalyst and the resulting ethyl (±)-cis-trans-2,2-dimethyl-3-(2,2-dichlorovinyl)-cyclopropane carboxylate hydrolyzed to the free acid. The cis- and trans-isomers can be separated from one another by selective crystallization from n-hexane in which the cis-isomer is more soluble. The isomeric mixture was dissolved in hexane at room temperature and cooled to 0° or −20°C, when the trans-isomer precipitates. This precipitate was ground up, washed with a small volume of hexane at room temperature and the residue recrystallized again from hexane at 0° to −20°C, to give the trans-isomer as a residue. The cis-isomer is recovered from the hexane solution.

The starting acid was then reacted with 3-phenoxybenzyl alcohol to give permethrin.

Toxicity

The acute oral LD_{50} value for rats ranges from 430 to 4,000 mg/kg (1) which is slightly toxic.

Allowable Limits on Exposure and Use

Product Use: The tolerances set by the EPA for permethrin in or on raw agriculture commodities are as follows:

	40 *CFR* Reference	Parts per Million
Cattle, fat	180.378	0.05
Cattle, mbyp	180.378	0.05
Cattle, meat	180.378	0.05
Cotton, seed	180.378	0.05
Eggs	180.378	0.05
Goats, fat	180.378	0.05
Goats, mbyp	180.378	0.05
Goats, meat	180.378	0.05
Hogs, fat	180.378	0.05
Hogs, mbyp	180.378	0.05
Hogs, meat	180.378	0.05
Horses, fat	180.378	0.05
Horses, mbyp	180.378	0.05
Horses, meat	180.378	0.05
Milk	180.378	0.05
Poultry, fat	180.378	0.05
Poultry, mbyp	180.378	0.05
Poultry, meat	180.378	0.05
Sheep, fat	180.378	0.05
Sheep, mbyp	180.378	0.05
Sheep, meat	180.378	0.05

References

(1) Worthing, C.R., *Pesticide Manual,* 6th ed., p. 409, British Crop Protection Council (1979).
(2) Elliott, M., Janes, N.F. and Pulman, D.A., British Patent 1,413,491, November 12, 1975, assigned to National Research Development Corp.

PHENMEDIPHAM

Function: Herbicide (1)(2)(3)(4)

Chemical Name: 3-[(methoxycarbonyl)amino] phenyl(3-methylphenyl)carbamate

Formula:

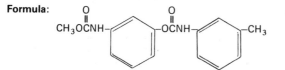

Trade Names: Schering 38,584
Schering 4075
Betanal® (Schering AG)

Manufacture

Phenmedipham may be made by the reaction of m-tolyl isocyanate with 3-methoxycarbonyl-aminophenol. The intermediate 3-methoxycarbonylaminophenol (methyl m-hydroxycar-banilate) was prepared as follows (3). To a solution of 43.6 grams of m-aminophenol in 250 ml of dry acetonitrile was added 18.9 grams of methyl chloroformate at a rate which maintained the temperature of the solution below 40°C.

The resultant mixture was stirred for 1 hour at room temperature and for 2 hours at reflux temperature, then cooled. The precipitate of m-aminophenol hydrochloride was removed by filtration and the filtrate was evaporated to dryness. The crude product was recrystallized from water to yield 11.0 grams of methyl m-hydroxycarbanilate melting at 94° to 96°C.

Toxicity

The acute oral LD_{50} value for rats is more than 8,000 mg/kg which is insignificantly toxic.

Allowable Limits on Exposure and Use

Product Use: The tolerances set by the EPA for phenmedipham in or on raw agricultural commodities are as follows:

	40 *CFR* Reference	Parts per Million
Beets	180.278	0.2 N
Beets, sugar, roots	180.278	0.1 N
Beets, sugar, tops	180.278	0.1 N

References

(1) Worthing, C.R., *Pesticide Manual*, 6th ed., p. 413, British Crop Protection Council (1979).
(2) Spencer, E.Y., *Guide to the Chemicals Used in Crop Protection*, 6th ed., p. 401, London, Ontario, Agriculture Canada (January 1973).
(3) Wilson, K.R. and Hill, K.L., U.S. Patent 3,404,975, October 8, 1969, assigned to FMC Corp.
(4) Boroschewski, G., Arndt, F. and Rusch, R., U.S. Patent 3,692,820, September 19, 1972, assigned to Schering AG.

PHENOTHRIN

Function: Insecticide (1)(2)

Chemical Name: (3-Phenoxyphenyl)methyl 2,2-dimethyl-3-(2-methyl-1-propenyl) cyclopropanecarboxylate

Formula:

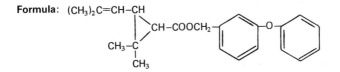

Trade Names: S-2539 (Sumitomo Chemical Co.)
S-2539 Forte (Sumitomo Chemical Co.)
Sumithrin® (Sumitomo Chemical Co.)

Manufacture (2)

To a solution of 0.05 mol of 3-phenoxybenzyl alcohol in 3 times by volume of dry benzene, 0.075 mol of pyridine is added. To the solution, there is added a solution containing 0.053 mol of chrysanthemum carboxylic acid chloride in 3 times by volume of dry benzene to react with generation of heat. After allowing to stand overnight with tight sealing, the reaction mixture is treated with a slight amount of water to dissolve the pyridine hydrochloride precipitate, and the aqueous layer formed is removed. The organic layer is successively washed with an aqueous solution containing 5% by weight of hydrochloric acid, a saturated aqueous solution of sodium hydrogen carbonate, and a saturated aqueous solution of sodium chloride, dried over anhydrous sodium sulfate, and distilled to remove benzene. The residual liquid is subjected to silica gel chromatography to recover the purified ester in the form of pale yellow oil.

Toxicity

The acute oral LD_{50} value for rats is more than 500 mg/kg which is slightly toxic.

References

(1) Worthing, C.R., *Pesticide Manual,* 6th ed., p. 414, British Crop Protection Council (1979).
(2) Itaya, N., Kamoshita, K., Mizutani, T., Kitamura, S., Nakai, S., Kameda, N., Fujimoto, K. and Okuno, Y., U.S. Patent 3,666,789, May 30, 1972, assigned to Sumitomo Chemical Co., Ltd.

PHENTHOATE

Function: Insecticide (1)(2)(3)

Chemical Name: Ethyl α-[(dimethoxyphosphinothioyl)thio] benzeneacetate

Formula:

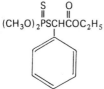

$$(CH_3O)_2PSCHCOC_2H_5$$

Trade Names: L-561 (Montecatini)
Cidial® (Montecatini)
Elsan® (Montecatini)
B-18,510 (Bayer)
Papthion® (Sumitomo Chemical Co.)

Manufacture (3)

38 grams crude dimethyldithiophosphoric acid dissolved in 150 cc acetone are neutralized with 13 grams anhydrous sodium carbonate. After stirring for half an hour, 48.6 grams ethyl alpha-phenyl-alpha-bromoacetate are added. The mixture is refluxed for one hour and most

of the acetate is distilled off at normal pressure on a boiling water bath. After cooling, the mass is treated with ice water and extracted with carbon tetrachloride. The extract is shaken with a 5% Na_2CO_3 solution until the pH is adjusted to 7, dried over $MgSO_4$ and the solvent removed under vacuum on a boiling water bath. The residue consists of 49 grams of a yellow oily phenthoate product.

Process Wastes and Their Control

Product Wastes: Phenthoate is stable in acid and neutral media. In a buffered solution of pH 9.7, approximately 25% is hydrolyzed after 20 days (B-3).

Toxicity

The acute oral LD_{50} value for rats is 300 to 400 mg/kg which is moderately toxic.

References

(1) Worthing, C.R., *Pesticide Manual,* 6th ed., p. 415, British Crop Protection Council (1979).
(2) Spencer, E.Y., *Guide to the Chemicals Used in Crop Protection,* 6th ed., p. 207 and p. 617, London, Ontario, Agriculture Canada (January 1973).
(3) Fusco, R., Losco, G. and Perini, M., U.S. Patent 2,947,662, assigned to Montecatini SA.

PHENYLMERCURY ACETATE

Function: Fungicide (1)(2)

Chemical Name: (acetate-O)phenylmercury

Formula:

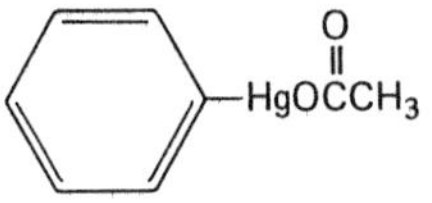

Trade Names: Tag HL 331® (Chevron Chemical Co.)
Gallotox®
Liquiphene®
Nylmerate®
Phix®
Scutl®
Agrosan D® (I.C.I.)
Ceresol® (I.C.I.)
Harvesan Plus (I.C.I.)
Mist-O-Matic (Murphy Chemical Ltd.)
Leytosan (Steetley Chemicals Ltd.)

Manufacture (3)

PMA may be made by the reaction of benzene with mercuric acetate. The following is a typical example (4) of the conduct of the process. Mercuric oxide or mercuric acetate may be placed in an autoclave.with an excess of thiophene-free benzene and glacial acetic acid. This mixture may be autoclaved at a pressure of about 6 psi and at a temperature of about 100°C for a sufficient time to complete the reaction, about 3 or 4 hours. This results in a reaction mixture containing in solution not only the desired phenyl mercury acetate but also mercurated bodies which can be regarded as impurities, among them, polymercury benzene compounds.

When the reaction is complete the pressure may be relieved and the benzene distilled off. This distillation should be carried on long enough to remove all traces of benzene. This may be readily accomplished by maintaining the temperature at about 100°C, since benzene boils at about 80°C and acetic acid boils at 118°C. It is important that all the benzene should be removed since it is a solvent for the polymercury benzene compounds which are formed as well as phenylmercury acetate in the reaction. When the benzene has been entirely removed it will be found that the phenylmercury acetate will remain in solution in the acetic acid while the impurities (polymercury benzene compounds) will be precipitated.

The resulting mixture may then be cooled and filtered whereby a solution of phenylmercury acetate in acetic acid will remain. Prior practice indicates that the acetic acid should be removed by evaporation of the liquid to dryness but it has been found that such procedure results in a poor yield and, what is more important, a deterioration of the product. Instead of following this procedure, the acetic acid solution is mixed with a large quantity of cold water, at least twice the volume of the solution and preferably several times the volume thereof.

Inasmuch as phenylmercury acetate is insoluble in cold water and in dilute acetic acid it is precipitated almost quantitatively. While a certain portion of the product may be precipitated by the addition of a lesser quantity of cold water, a reduction below two volumes of cold water to each volume of solution will result in a decrease in yield of material of equal purity. The dilute acetic acid may now be decanted and the precipitate washed with cold water until the mass becomes neutral and the last trace of acetic acid is removed. The phenylmercury acetate may now be dissolved in boiling water, filtered and allowed to crystallize. In this way a finished product is secured which has a constant and sharp MP of 148°C.

Figure 45 is a flow diagram of the manufacture of the most widely used organic mercurial— phenylmercuric acetate. Here yields are high and mercury losses are small, and most of the mercury that is lost is recycled from the waste effluent. (It should be noted that the very high value of the waste mercury is in itself an incentive to recover it; in this case the recovered mercury represents a savings of over 2% on raw material costs alone. In contrast, in the chlor-alkali industry the amounts of mercury losses are much larger, but the additional cost per kilogram of product produced is insignificant.) The finished phenylmercuric acetate is sold either as a dry powder (100%) or as a dilute suspension.

Figure 45: Mercury Balance for Synthesis of Phenylmercuric Acetate
(Using 1973 Technology)

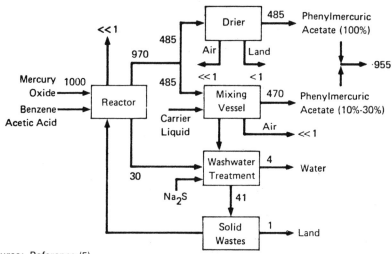

Toxicity

The acute oral LD_{50} value for rats is 30 mg/kg (B-5) which is highly toxic.

References

(1) Worthing, C.R., *Pesticide Manual,* 6th ed., p. 416, British Crop Protection Council (1979).
(2) Spencer, E.Y., *Guide to the Chemicals Used in Crop Protection,* 6th ed., p. 404, London, Ontario, Agriculture Canada (January 1973).
(3) Kobe, K.A. and Leuth, P.F., *Ind. Eng. Chem.,* 34, 309 (1942).
(4) Rentschler, M.J., U.S. Patent 2,050,018, August 4, 1936, assigned to The Hamilton Laboratories, Inc.
(5) Office of Toxic Substances, *Materials Balance and Technology Assessment of Mercury and Its Compounds on National and Regional Bases,* Report EPA-560/3-75-007, Wash., DC, Environmental Protection Agency (October 1975).

2-PHENYLPHENOL

Function: Fungicide (1)(2)

Chemical Name: 1,1'-biphenyl-2-ol

Formula:

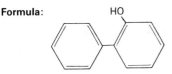

Trade Names: Dowicide I®
Nectryl®
Topane®

Manufacture

2-Phenylphenol may be recovered from the reaction products in the Dow phenol process (3) involving the reaction of monochlorobenzene with calcium hydroxide. The following is a specific example (4) of the conduct of this process.

Example 1—16,300 pounds of crude o-phenylphenol, prepared by the caustic-chlorobenzene process, was mixed with 120 pounds of flake caustic and 1,500 pounds of water. The resulting mixture was heated to between 95° and 105°C and 45.6 cubic feet of air per minute (calculated at standard pressure and temperature) was passed through the solution for 8 hours. The pressure in the reactor was 40 to 50 psig. Upon completion of this operation the reaction mixture was fractionally distilled under reduced pressure to obtain 83.4% of the starting crude phenol product as a purified o-phenylphenol having a freezing point of 57°C and a boiling point of 165°C at 25 mm pressure. The product obtained was found to be superior in color and light stability as compared against a control of o-phenylphenol which was not oxidized but purified by distillation.

An alternative route (5) starts with diphenylene oxide (dibenzofuran) which may be made, in turn, by heating phenol with lead oxide. The following is a specific example of the conduct of this process.

Example 2—168 grams of diphenylene oxide, 230 cc of ligroin and 46 grams of sodium metal are introduced into a V_2A stirrer-type autoclave and, after forcing in hydrogen to a pressure of 40 atmospheres, heated to 190°C while stirring rapidly. The hydrogen pressure initially rises owing to the raising of the temperature but then gradually falls. After 6 hours, the calculated amount of 0.5 mol of hydrogen has been taken up. The autoclave is then cooled and opened, whereupon the ligroin is distilled off from the yellow reaction mixture which has been obtained and the remaining reaction product is introduced in portions into methanol in order to decompose the free sodium metal which is still present.

Upon being diluted with water, a small amount (7 grams) of unmodified diphenylene oxide (MP 82°C) is precipitated from the alcoholic solution, the oxide being filtered off. By acidifying the alkaline filtrate with hydrochloric acid, 2-hydroxydiphenyl is precipitated as a yellow oil, which solidifies slowly and can be filtered with suction.

155 grams of alkali-soluble product are obtained, from which it is possible to obtain by fractionation 125 grams of pure 2-hydroxydiphenyl [BP (15 mm Hg) 152° to 154°C, MP 56°C]. 30 grams of a phenol mixture boiling at higher temperature remain in the distillation residue.

Toxicity

The acute oral LD_{50} value for rats is 2,480 mg/kg which is slightly toxic.

Allowable Limits on Exposure and Use

Product Use: The tolerances set by the EPA for 2-phenylphenol in or on raw agricultural commodities are as follows:

	40 *CFR* Reference	Parts per Million
Apples (post-h)	180.129	25.0
Cantaloupes (post-h)	180.129	125.0
Cantaloupes (post-h) (edible portion)	180.129	10.0
Carrots (post-h)	180.129	20.0
Cherries (post-h)	180.129	5.0
Citrus citron (post-h)	180.129	10.0
Cucumbers (post-h)	180.129	10.0
Grapefruit (post-h)	180.129	10.0
Kumquats (post-h)	180.129	10.0
Lemons (post-h)	180.129	10.0
Limes (post-h)	180.129	10:0
Nectarines (post-h)	180.129	5.0
Oranges (post-h)	180.129	10.0
Peaches (post-h)	180.129	20.0
Pears (post-h)	180.129	25.0
Peppers, bell (post-h)	180.129	10.0
Pineapples (post-h)	180.129	10.0
Plums (fresh prunes) (post-h)	180.129	20.0
Sweet potatoes (post-h)	180.129	15.0
Tangerines (post-h)	180.129	10.0
Tomatoes (post-h)	180.129	10.0

References

(1) Worthing, C.R., *Pesticide Manual*, 6th ed., p. 419, British Crop Protection Council (1979).
(2) Spencer, E.Y., *Guide to the Chemicals Used in Crop Protection*, 6th ed., p. 405, London, Ontario, Agriculture Canada (January 1973).
(3) Grebe, J.J. et al, U.S. Patent 2,275,044, March 3, 1942, assigned to Dow Chemical Co.
(4) Widiger, A.H., U.S. Patent 3,087,969, April 30, 1963, assigned to Dow Chemical Co.
(5) Muller, K.W. and Delfs, D., U.S. Patent 2,862,035, November 25, 1958, assigned to Farbenfabriken Bayer AG.

PHORATE

Function: Insecticide (1)(2)(6)

Chemical Name: O,O-diethyl S-ethylthiomethyl phosphorodithioate

Formula:

$$C_2H_5O \diagdown \overset{\overset{S}{\|}}{\underset{\diagup}{P}}-S-CH_2-S-C_2H_5$$
$$C_2H_5O$$

Trade Names: El 3911 (American Cyanamid)
Thimet® (American Cyanamid)
Granutox®

Manufacture (3)(4)(5)(7)

$$P_2S_5 + 4C_2H_5OH \longrightarrow 2(C_2H_5O)_2\overset{\overset{S}{\|}}{P}SH + H_2S$$

$$(C_2H_5O)_2\overset{\overset{S}{\|}}{P}SH + H_2C{=}O \longrightarrow (C_2H_5O)_2\overset{\overset{S}{\|}}{P}S-CH_2OH$$

$$(C_2H_5O)_2\overset{\overset{S}{\|}}{P}-S-CH_2OH + C_2H_5SH \longrightarrow (C_2H_5O)_2\overset{\overset{S}{\|}}{P}-SCH_2SC_2H_5$$

Alternatively, O,O-diethyl dithiophosphoric acid may be reacted with chloromethyl ethyl sulfide (5) as in the following example.

Example—98 grams of O,O-diethyl dithiophosphoric acid (95%) are dissolved with 45 grams of sodium bicarbonate in 250 cc of water. 55 grams of chloromethyl ethyl sulfide are dropped thereto at 20° to 25°C. The solution which has turned acid to Congo red paper, is rendered neutral again by adding some sodium bicarbonate and stirred for 3 hours more at 20° to 25°C. After this time the solution has become clear. The oil is dissolved in benzene and the benzene solution is washed with water and dried with anhydrous sodium sulfate. Figure 46 is a production and waste scheme for phorate manufacture.

Process Wastes and Their Control

Air: Air emissions from phorate manufacture have been reported (B-15) to consist of 205 kg SO_2 and 0.5 kg phorate per metric ton of pesticide produced.

Product Wastes: Phorate is relatively unstable to hydrolysis, and is incompatible with alkaline pesticides. Fifty percent hydrolysis required 2 hours at pH 8 and 70°C, and 9.6 hours at pH 1 to 5 and 40°C. Phorate is easily oxidized to the corresponding sulfoxide, which is more resistant to hydrolysis (B-3).

Toxicity

The acute oral LD_{50} value for rats is 1.6 to 3.7 mg/kg which rates as extremely toxic.

A comprehensive statement on the toxic properties of phorate from a National Research Council study (B-22) may be found in the section on the related compound disulfoton.

Allowable Limits on Exposure and Use

Air: The threshold limit value for phorate in air is 0.05 mg/m³ as of 1979. A tentative short term exposure limit is 0.2 mg/m³. Further, cutaneous absorption of phorate should be prevented so that the TLV is not invalidated.

Figure 46: Production and Waste Schematic for Phorate

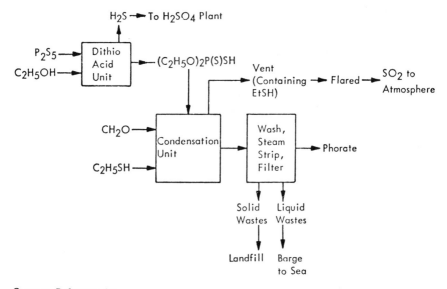

Source: Reference (7)

Product Use: The tolerances set by the EPA for phorate in or on raw agriculture commodities are as follows:

	40 *CFR* Reference	Parts per Million
Alfalfa, fresh	180.206	0.5
Alfalfa, hay	180.206	1.0
Barley, grain	180.206	0.1
Barley, straw	180.206	0.1
Beans	180.206	0.1
Beans, vines	180.206	0.5
Beets, sugar, roots	180.206	0.3
Beets, sugar, tops	180.206	3.0
Cattle, fat	180.206	0.05 N
Cattle, mbyp	180.206	0.05 N
Cattle, meat	180.206	0.05 N
Corn, forage	180.206	0.5
Corn, grain	180.206	0.1
Corn, sweet (k+cwhr)	180.206	0.1
Cotton, seed	180.206	0.05
Eggs	180.206	0.05 N
Goats, fat	180.206	0.05 N
Goats, mbyp	180.206	0.05 N
Goats, meat	180.206	0.05 N
Grasses, Bermuda, straw	180.206	0.5
Hogs, fat	180.206	0.05 N
Hogs, mbyp	180.206	0.05 N

(continued)

	40 *CFR* Reference	Parts per Million
Hogs, meat	180.206	0.05 N
Hops	180.206	0.5
Horses, fat	180.206	0.05 N
Horses, mbyp	180.206	0.05 N
Horses, meat	180.206	0.05 N
Lettuce	180.206	0.1
Milk	180.206	0.02 N
Peanuts	180.206	0.1
Peanuts, hay	180.206	0.3
Peanuts, vines	180.206	0.3
Potatoes	180.206	0.5
Poultry, fat	180.206	0.05 N
Poultry, mbyp	180.206	0.05 N
Poultry, meat	180.206	0.05 N
Rice	180.206	0.1
Sheep, fat	180.206	0.05 N
Sheep, mbyp	180.206	0.05 N
Sheep, meat	180.206	0.05 N
Sorghum, fodder	180.206	0.1
Sorghum, grain	180.206	0.1
Soybeans	180.206	0.1
Sugarcane	180.206	0.1
Tomatoes	180.206	0.1
Wheat, fodder, green	180.206	1.5
Wheat, grain	180.206	0.05
Wheat, straw	180.206	0.05

The tolerances set by the EPA for phorate in animal feeds are as follows (the *CFR* Reference is to Title 21):

	CFR Reference	Parts per Million
Sugar beet pulp, dried	561.290	1.0

References

(1) Worthing, C.R., *Pesticide Manual,* 6th ed., p. 420, British Crop Protection Council (1979).

(2) Spencer, E.Y., *Guide to the Chemicals Used in Crop Protection,* 6th ed., p. 407, London, Ontario, Agriculture Canada (January 1973).

(3) Hook, E.O. and Moss, P.H., U.S. Patent 2,586,655, February 19, 1952, assigned to American Cyanamid Company.

(4) Hook, E.O. and Moss, P.H., U.S. Patent 2,596,076, May 6, 1952, assigned to American Cyanamid Company.

(5) Lorenz, W. and Schrader, G., U.S. Patent 2,759,010, August 14, 1956, assigned to Farbenfabriken Bayer AG.

(6) Oros, N.R. and Vartanian, R.D., U.S. Patent 2,970,080, January 31, 1961, assigned to American Cyanamid Company.

(7) Midwest Research Institute, *The Pollution Potential in Pesticide Manufacturing,* Washington, DC, Environmental Protection Agency (June 1972).

PHOSALONE

Function: Insecticide and acaricide (1)(2)(3)

Chemical Name: S-[(6-chloro-2-oxo-3(2H)-benzoxazolyl)methyl] O,O-diethyl phosphorodithioate

Formula:

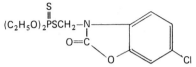

Trade Names: RP-11,974 (Rhone-Poulenc)
Zolone® (Rhone-Poulenc)
Rubitox® (Rhone-Poulenc)

Manufacture (3)

Urea and o-aminophenol are condensed in a first step to give benzoxazolone. In a second step, benzoxazolone is chlorinated to give 6-benzoxazolone. Then chloromethylation with formaldehyde and HCl gives 6-chloro-3-chloromethylbenzoxazolone. Finally, O,O-diethyl-phosphorodithiolate is condensed with the 6-chloromethylbenzoxazolone to give phosalone.

Process Wastes and Their Control

Product Wastes: Phosalone is readily hydrolyzed in alkaline medium. The principal hydrolysis products are 6-chlorobenzoxazolone, $(EtO)_2P(O)OH$, and formaldehyde (B-3).

Toxicity

The acute oral LD_{50} value for rats is 120 to 170 mg/kg which is moderately toxic.

Allowable Limits on Exposure and Use

Product Use: The tolerances set by the EPA for phosalone in or on raw agricultural commodities are as follows:

	40 *CFR* Reference	Parts per Million
Almonds	180.263	0.1 N
Almonds, hulls	180.263	50.0
Apples	180.263	10.0
Apricots	180.263	15.0
Artichokes	180.263	25.0
Brazil nuts	180.263	0.05 N
Bush nuts	180.263	0.05 N
Butternuts	180.263	0.05 N
Cashews	180.263	0.05 N
Cattle, fat	180.263	0.25
Cattle, mbyp	180.263	0.25
Cattle, meat	180.263	0.25
Cherries	180.263	15.0
Chestnuts	180.263	0.05 N
Citrus fruits	180.263	3.0
Filberts	180.263	0.05 N
Goats, fat	180.263	0.25
Goats, mbyp	180.263	0.25
Goats, meat	180.263	0.25
Grapes	180.263	10.0
Hazelnuts	180.263	0.05 N

(continued)

	40 *CFR* Reference	Parts per Million
Hickory nuts	180.263	0.05 N
Hogs, fat	180.263	0.25
Hogs, mbyp	180.263	0.25
Hogs, meat	180.263	0.25
Horses, fat	180.263	0.25
Horses, mbyp	180.263	0.25
Horses, meat	180.263	0.25
Macadamia nuts	180.263	0.05 N
Nectarines	180.263	15.0
Peaches	180.263	15.0
Pears	180.263	10.0
Pecans	180.263	0.05 N
Plums (fresh prunes)	180.263	15.0
Potatoes	180.263	0.1 N
Sheep, fat	180.263	0.25
Sheep, mbyp	180.263	0.25
Sheep, meat	180.263	0.25
Walnuts	180.263	0.05 N

The tolerances set by the EPA for phosalone in animal feeds are as follows (the *CFR* Reference is to Title 21):

	CFR Reference	Parts per Million
Apple pomace, dried	561.300	85.0
Citrus pulp, dried	561.300	12.0
Grape pomace, dried	561.300	45.0

The tolerances set by the EPA for phosalone in food are as follows (the *CFR* Reference is to Title 21):

	CFR Reference	Parts per Million
Prunes, dried	193.340	40.0
Raisins	193.340	20.0
Tea, dried	193.340	8.0

References

(1) Worthing, C.R., *Pesticide Manual,* 6th ed., p. 421, British Crop Protection Council (1979).
(2) Spencer, E.Y., *Guide to the Chemicals Used in Crop Protection,* 6th ed., p. 408, London, Ontario, Agriculture Canada (January 1973).
(3) British Patent 1,005,372, September 22, 1965, assigned to Rhone-Poulenc, SA.

PHOSFOLAN

Function: Insecticide (1)(2)

Chemical Name: Diethyl 1,3-dithiolan-2-ylidene-phosphoramidate

Formula:

Trade Names: El 47,031 (American Cyanamid)
Cyolane® (American Cyanamid)
Cyolan® (American Cyanamid)
Cylan® (American Cyanamid)

Manufacture

Phosfolan may be made by the reaction of diethyl phosphorochloridate with 2-imino-1,3-dithiolane. In the first step of the process (3), 2-imino-1,3-dithiolane hydrochloride may be produced as described in the following example.

Into a mixture of 210 parts of toluene and 2.0 parts of ethanol in a suitable three-neck flask, equipped with a stirrer, ice-water cooled condenser, thermometer and gas inlet tube, is introduced hydrogen chloride as a gas for 6 minutes at a moderate rate. Titration of a 5.0 part aliquot of the toluene mixture in a homogeneous methanol-water mixture shows the hydrogen chloride concentration to be 0.12 molar. Ethane-1,2-dithiol (108.6 parts) is then added. The gas inlet tube is raised above the liquid surface and gaseous cyanogen chloride addition is begun from a tared cylinder. The rate of addition is 0.37 part per minute and solid product begins to appear after about 25 minutes. The initial reaction temperature is 21°C and rises spontaneously to 50°C after 80 minutes.

A room temperature water bath is placed in contact with the reaction flask so that the heat of reaction is controlled and is further maintained at between about 40° and 50°C. Over a period of 3½ hours, 76 parts of cyanogen chloride are added. After removal of the cooling bath and an additional 42 minute reaction period, infrared examination of the toluene mixture shows that less than about 5% of ethane-1,2-dithiol can be detected.

To insure that the reaction has gone to completion, the reaction slurry is left undisturbed for an additional 45 minutes. Then the thick slurry is diluted with toluene and poured into a Buchner funnel. The solids are successively washed with additional portions of toluene and petroleum ether, respectively. After vacuum drying, 179.3 parts of finely divided 2-imino-1,3-dithiolane hydrochloride, having a melting point of 212° to 216°C with decomposition, are obtained. The yield corresponds to 91.2% of theory based on ethane-1,2-dithiol. In the foregoing example, similar results are obtained where cyanogen bromide is substituted for cyanogen chloride.

Then, in a second step, the product from the first step is reacted with $(RO)_2PSCl$ as follows. To a stirred mixture of 44.0 parts of 2-imino-1,3-dithiolane hydrochloride and 53.4 parts of O,O-diethylphosphorochloridothioate in 100 parts of water and 200 parts of benzene are added 51.0 parts of sodium acetate in 100 parts of water over a 30 minute period. After stirring for 60 hours, the benzene layer is separated and the water layer extracted once with ether. The combined ether-benzene mixture is washed with dilute sodium carbonate solution, then with water and dried. The solvent and some unreacted chloridothioate are stripped off under reduced pressure in a film type evaporator. On standing the product crystallizes. Washing off the solids with petroleum ether (30° to 60°C) yields 40.6 parts, or 56.3% of theory, of the iminophosphate having a melting point of 36.5° to 38.8°C.

Process Wastes and Their Control

Product Wastes: Although phospholan is stable under neutral or acid conditions, it is hydrolyzed by alkali, and is nonpersistent in soil (B-3).

Toxicity

The acute oral LD_{50} value for rats is 8 to 9 mg/kg which is very highly toxic.

References

(1) Worthing, C.R., *Pesticide Manual,* 6th ed., p. 422, British Crop Protection Council (1979).
(2) Spencer, E.Y., *Guide to the Chemicals Used in Crop Protection,* 6th ed., p. 200, London, Ontario, Agriculture Canada (January 1973).
(3) Addor, R.W., U.S. Patent 3,197,481, July 27, 1965, assigned to American Cyanamid Co.

PHOSMET

Function: Acaricide and insecticide (1)(2)

Chemical Name: S-[(1,3-dihydro-1,3-dioxo-2H-isoindol-2-yl)methyl] O,O-dimethyl phosphorodithioate

Formula:

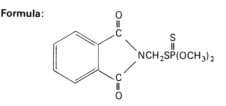

Trade Names: R-1504 (Stauffer Chemical Co.)
Imidan® (Stauffer Chemical Co.)
Prolate®

Manufacture (3)

Phthalimide is the starting material for phosmet manufacture. It is reacted with formaldehyde and HCl to give N-chloromethylphthalimide as the intermediate. Then, in a typical example, 12.3 grams (0.06 mol) of N-chloromethylphthalimide, 12.6 grams (0.07 mol) of sodium O,O-dimethyl dithiophosphate, and 50 cc of methanol were mixed and refluxed on a steam bath for 3½ hours. The reaction mixture was cooled and shaken with 100 cc of benzene, then filtered. The filtrate was transferred to a separatory funnel and washed three times with water. After drying over anhydrous carbonate, the dry benzene solution was filtered, then evaporated on the steam bath. The product, a yellow liquid, weighed 10.2 grams.

Process Wastes and Their Control

Product Wastes: Phosmet is reported to be rapidly decomposed with hypochlorite. Fifty percent hydrolysis at room temperature in a buffered solution of initial phosmet concentration of 20 ppm requires 13 days at pH 4.5, <12 hours at pH 7, and <4 hours at pH 9.3 (B-3).

Toxicity

The acute oral LD_{50} value for rats is 230 to 300 mg/kg which is moderately toxic.

Allowable Limits on Exposure and Use

Product Use: The tolerances set by the EPA for phosmet in or on raw agricultural commodities are as follows:

	40 *CFR* Reference	Parts per Million
Alfalfa	180.261	40.0
Almonds, hulls	180.261	10.0
Apples	180.261	10.0
Apricots	180.261	5.0
Blueberries	180.261	10.0
Cattle, fat	180.261	0.2
Cattle, mbyp	180.261	0.2
Cattle, meat	180.261	0.2
Cherries	180.261	10.0
Citrus fruits	180.261	5.0
Corn, field, fodder	180.261	10.0
Corn, field, forage	180.261	10.0
Corn, fresh	180.261	0.5
Corn, fresh (inc. sweet) (k+cwhr)	180.261	0.5
Corn, grain	180.261	0.5
Corn, pop, fodder	180.261	10.0
Corn, pop, forage	180.261	10.0
Corn, sweet, fodder	180.261	10.0
Corn, sweet, forage	180.261	10.0
Cranberries	180.261	10.0
Goats, fat	180.261	0.2
Goats, mbyp	180.261	0.2
Goats, meat	180.261	0.2
Grapes	180.261	10.0
Hogs, fat	180.261	0.2
Hogs, mbyp	180.261	0.2
Hogs, meat	180.261	0.2
Horses, fat	180.261	0.2
Horses, mbyp	180.261	0.2
Horses, meat	180.261	0.2
Kiwi fruit	180.261	25.0
Nectarines	180.261	5.0
Nuts	180.261	0.1 N
Peaches	180.261	10.0
Pears	180.261	10.0
Peas	180.261	0.5
Peas, forage	180.261	10.0
Peas, hay	180.261	10.0
Plums (fresh prunes)	180.261	5.0
Potatoes	180.261	0.1
Sheep, fat	180.261	0.2
Sheep, mbyp	180.261	0.2
Sheep, meat	180.261	0.2
Sweet potatoes (post-h)	180.261	10.0

References

(1) Worthing, C.R., *Pesticide Manual,* 6th ed., p. 423, British Crop Protection Council (1979).
(2) Spencer, E.Y., *Guide to the Chemicals Used in Crop Protection,* 6th ed., p. 409, London, Ontario, Agriculture Canada (January 1973).
(3) Fancher, L.W., U.S. Patent 2,767,194, October 16, 1956, assigned to Stauffer Chemical Co.

PHOSPHAMIDON

Function: Insecticide (1)(2)

Chemical Name: 2-chloro-3-(diethylamino)-1-methyl-3-oxo-1-propenyl dimethyl phosphate

Formula:

$$(CH_3O)_2\overset{\overset{O}{\|}}{P}\overset{\overset{}{|}}{C}\!\!=\!\!=\!\!\overset{\overset{O}{\|}}{C}\overset{\overset{}{|}}{C}N(C_2H_5)_2$$
$$\quad\quad\quad CH_3\quad Cl$$

Trade Names: Ciba 570 (Ciba-Geigy)
Dimecron® (Ciba-Geigy)

Manufacture (3)

Acetoacetic acid diethylamide is the starting material for phosphamidon manufacture. It is treated with sulfuryl chloride (SO_2Cl_2) to give α,α-dichloroacetoacetic acid diethylamide. Then, in a typical example, dichloroacetoacetic acid diethylamide (boiling at 92.5° to 93°C under 0.18 mm pressure) and trimethyl phosphite are mixed together at room temperature. In order to initiate the reaction, the mixture is heated to 90°C. At that temperature a violent evolution of gas sets in, and the reaction solution heats up to 160°C without external heating. When the reaction is finished the product is subjected for a short time longer at 95°C to the reduced pressure of a water jet pump. There remains behind a pale yellow oily product which can be purified by distillation in a high vacuum. It boils at 162°C under 1.5 mm pressure.

Process Wastes and Their Control

Air: Air emissions from phosphamidon manufacture have been reported to consist of 0.5 kg hydrocarbons and 0.5 kg phosphamidon per metric ton of pesticide produced (B-15).

Product Wastes: Fifty percent hydrolysis at 23°C requires 13.8 days at pH 7, and 2.2 days at pH 10 (B-3).

Toxicity

The acute oral LD_{50} value for rats is 17 to 30 mg/kg which is highly toxic.

Allowable Limits on Exposure and Use

Product Use: The tolerances set by the EPA for phosphamidon in or on raw agricultural commodities are as follows:

	40 *CFR* Reference	Parts per Million
Apples	180.239	1.0
Broccoli	180.239	0.5
Cantaloupes	180.239	0.25
Cauliflower	180.239	0.5
Cotton, seed	180.239	0.1
Cucumbers	180.239	0.5
Grapefruit	180.239	0.75
Lemons	180.239	0.75
Oranges	180.239	0.75
Peppers	180.239	0.5
Potatoes	180.239	0.1
Sugarcane	180.239	0.1
Tangerines	180.239	0.75
Tomatoes	180.239	0.1
Walnuts	180.239	0.1
Watermelons	180.239	0.25

References

(1) Worthing, C.R., *Pesticide Manual,* 6th ed., p. 424, British Crop Protection Council (1979).

(2) Spencer, E.Y., *Guide to the Chemicals Used in Crop Protection,* 6th ed., p. 410, London, Ontario, Agriculture Canada (January 1973).

(3) Beriger, E. and Sallmann, R., U.S. Patent 2,908,605, October 13, 1959, assigned to Ciba, Ltd.

PHOXIM

Function: Insecticide (1)(2)

Chemical Name: α-[[(diethoxyphosphinothioyl)oxy]imino]benzeneacetonitrile

Formula:

$$\underset{(C_2H_5O)_2PON=C}{\overset{\underset{\parallel}{S} \quad \overset{CN}{|}}{}}$$

Trade Names: Bay 77,488 (Bayer)
Bay 5621 (Bayer)
Baythion® (Bayer)
Volaton® (Bayer)
Valexon®

Manufacture (3)(4)

61 grams (0.36 mol) of the sodium salt of α-oximinophenylacetic acid nitrile (α-cyanoben-zaldoxime) [prepared according to M.R. Zimmermann, *Journal fur praktische Chemie* (2), 66, 359/1902] are suspended in 280 cc acetone and 57 grams (0.3 mol) O,O-diethylthionophos-phoric acid ester chloride are added dropwise to this suspension at 30° to 35°C, while cool-ing. After stirring the mixture for 1 hour, it is poured into water, the precipitated oil is taken up with benzene, the benzene solution washed with water and a 2N sodium hydroxide solu-tion, the organic phase dried over anhydrous sodium sulfate and the solvent distilled off. As the residue there are obtained 85 grams (95.5% of theory) O,O-diethylthionophosphoryl-α-oximinophenylacetic acid nitrile in the form of a pale yellow oil of refractive index n^{22} 1.5395 and density γ^{20} 1.176 g/cc. The compound can only be distilled in small amounts, even un-der strongly reduced pressure, and then boils at 102°C/0.01 mm Hg.

Toxicity

The acute oral LD_{50} for rats is about 2,000 mg/kg which is slightly toxic.

References

(1) Worthing, C.R., *Pesticide Manual,* 6th ed., p. 425, British Crop Protection Council (1979).

(2) Spencer, E.Y., *Guide to the Chemicals Used in Crop Protection,* 6th ed., p. 413, London, Ontario, Agriculture Canada (January 1973).

(3) Lorenz, W., Fest, C., Hammann, I., Federmann, M., Flucke, W. and Stendel, W., U.S. Pa-tent 3,591,662, July 6, 1971, assigned to Farbenfabriken Bayer AG.

(4) Lorenz, W., Fest, C. Hammann, I., Federmann, M., Flucke, W. and Stendel, W., U.S. Pa-tent 3,689,648, September 5, 1972, assigned to Farbenfabriken Bayer AG.

PICLORAM

Function: Herbicide (1)(2)

Chemical Name: 4-amino-3,5,6-trichloropyridine-2-carboxylic acid

Formula:

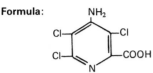

Trade Names: Tordon® (Dow Chemical Company)

Manufacture (5)

α-Picoline (α-methylpyridine) is the starting material which is in turn made from acetaldehyde and ammonia. It is chlorinated to give 2,3,4,5-tetrachloro-6-(trichloromethyl)pyridine. Then the tetrachloro intermediate and excess anhydrous ammonia at −80°C were sealed in a pressure vessel and thereafter heated for 1 hour at 100°C. At the end of this period, the reaction mixture was cooled to −80°C, the pressure released and then allowed to warm to room temperature to evaporate unreacted ammonia. As a result of these operations, there remained as residue the desired 4-amino-3,5,6-trichloro-2-(trichloromethyl)pyridine compound (3). The latter was recrystallized from 60% aqueous acetic acid to obtain a purified compound.

Then in a final conversion step the trichloromethyl intermediate is hydrolyzed to the carboxylic acid product, picloram (4). The following is a typical example of that operation. A solution of 80% (weight percent) sulfuric acid, prepared by dissolving 278 ml of 95% sulfuric acid in 122 ml of water, was added to 278 grams (0.88 mol) of 4-amino-2,3,5-trichloro-6-(trichloromethyl)pyridine and the resulting mixture heated to 120°C. As a result of these operations, a reaction commenced as evidenced by foaming brought on by the evolution of hydrogen chloride by-product from the reaction mixture.

The mixture was removed from the heat until the foaming subsided, thereafter heated with increase in temperature until a temperature of 160°C was reached and maintained thereat for two hours. The resulting mixture was then cooled to room temperature and then added to 1,500 ml of ice water to precipitate the desired 4-amino-3,5,6-trichloropicolinic acid product. The latter was recovered by filtration, and purified by dissolving in dilute aqueous sodium hydroxide, treating the alkaline solution with activated carbon, filtering the treated solution through diatomaceous earth and reacidifying to obtain a purified product melting at 218° to 219°C. The yield of the purified product was 192.0 grams or 96% of theoretical.

Process Wastes and Their Control

Product Wastes: This chlorinated brush killer is usually formulated with 2,4-D and the disposal problems are similar (B-3). Incineration at 1000°C for 2 seconds is required for thermal decomposition. Alternatively, the free acid can be precipitated from its solutions by addition of a mineral acid. The concentrated acid can then be incinerated and the dilute residual solution disposed in an area where several years persistence in the soil can be tolerated.

Toxicity

The acute oral LD_{50} value for rats is 8,200 mg/kg which is insignificantly toxic.

Picloram is reported to have low toxicity to wildlife and "no problem exists when used according to label directions." The product is highly phytotoxic, accumulates in the soil, and is persistent (B-3).

Allowable Limits on Exposure and Use

Air: The threshold limit value for pichloram in air is 10 mg/m^3 as of 1979. The tentative short term exposure limit is 20 mg/m^3 (B-23).

Product Use: The tolerances set by the EPA for picloram in or on raw agriculture commodities are as follows:

	40 *CFR* Reference	Parts per Million
Barley, forage, green	180.292	1.0
Barley, grain	180.292	0.5
Barley, straw	180.292	1.0
Cattle, fat	180.292	0.2
Cattle, kidney	180.292	5.0
Cattle, liver	180.292	0.5
Cattle, mbyp (exc. kidney, liver)	180.292	0.2
Cattle, meat	180.292	0.2
Eggs	180.292	0.05
Goats, fat	180.292	0.2
Goats, kidney	180.292	5.0
Goats, liver	180.292	0.5
Goats, mbyp (exc. kidney, liver)	180.292	0.2
Goats, meat	180.292	0.2
Grasses, forage	180.292	80.0
Hogs, fat	180.292	0.2
Hogs, kidney	180.292	5.0
Hogs, liver	180.292	0.5
Hogs, mbyp (exc. kidney, liver)	180.292	0.2
Hogs, meat	180.292	0.2
Horses, fat	180.292	0.2
Horses, kidney	180.292	5.0
Horses, liver	180.292	0.5
Horses, mbyp (exc. kidney, liver)	180.292	0.2
Horses, meat	180.292	0.2
Milk	180.292	0.05
Oats, forage, green	180.292	1.0
Oats, grain	180.292	0.5
Oats, straw	180.292	1.0
Poultry, fat	180.292	0.05
Poultry, mbyp	180.292	0.05
Poultry, meat	180.292	0.05
Sheep, fat	180.292	0.2
Sheep, kidney	180.292	5.0
Sheep, liver	180.292	0.5
Sheep, mbyp (exc. kidney, liver)	180.292	0.2
Sheep, meat	180.292	0.2
Wheat, forage, green	180.292	1.0
Wheat, grain	180.292	0.5
Wheat, straw	180.292	1.0

The tolerances set by the EPA for picloram in animal feeds are as follows (the *CFR* Reference is to Title 21):

	CFR Reference	Parts per Million
Barley, milled fractions (exc. flour)	561.305	3.0
Oats, milled fractions (exc. flour)	561.305	3.0
Wheat, milled fractions (exc. flour)	561.305	3.0

The tolerances set by the EPA for picloram in food are as follows (the *CFR* Reference is to Title 21):

	CFR Reference	Parts per Million
Barley, flour	193.350	1.0 i
Barley, milled fractions (exc. flour)	193.350	3.0
Wheat, flour	193.350	1.0 i
Wheat, milled fractions (exc. flour)	193.350	3.0

References

(1) Worthing, C.R., *Pesticide Manual,* 6th ed., p. 426, British Crop Protection Council (1979).
(2) Spencer, E.Y., *Guide to the Chemicals Used in Crop Protection,* 6th ed., p. 414, London, Ontario, Agriculture Canada (January 1973).
(3) Redemann, C.T., U.S. Patent 3,234,229, February 8, 1966, assigned to Dow Chemical Co.
(4) Johnston, H. and Tomita, M.S., U.S. Patent 3,285,925, November 15, 1966, assigned to Dow Chemical Co.
(5) Dow Chemical Co., British Patent 957,831, May 13, 1964.

PINDONE

Function: Insecticide and rodenticide (1)(2)(3)

Chemical Name: 2-(2,2-dimethyl-1-oxopropyl)-1H-indene-1,3(2H)dione

Formula:

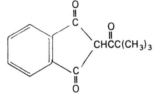

Trade Names: Pivalyl Valone® (Kilgore Chemical Co.)
Pival® (Kilgore Chemical Co.)
Pivalyn® (Kilgore Chemical Co.)
Tri-Ban®

Manufacture (4)

Pindone may be made by the condensation of pinacolone (methyl tert-butyl ketone) with diethyl phthalate, using sodium in benzene as the condensing agent.

Toxicity

The acute oral LD_{50} value for rats is 280 mg/kg (B-5) which is moderately toxic but an LD_{50} value for rats by injection of 50 mg/kg has also been reported (1).

Allowable Limits on Exposure and Use

Air: The threshold limit value for pindone in air is 0.1 mg/m³ as of 1979. The tentative short term exposure limit is 0.3 mg/m³ (B-23).

References

(1) Worthing, C.R., *Pesticide Manual,* 6th ed., p. 428, British Crop Protection Council (1979).
(2) Spencer, E.Y., *Guide to the Chemicals Used in Crop Protection,* 6th ed., p. 415, London, Ontario, Agriculture Canada (January 1973).
(3) Kilgore, L.B., U.S. Patent 2,228,170, January 7, 1941.
(4) Kilgore, L.B. et al, *Ind. Eng. Chem.* 34, 494 (1942).

PIPERONYL BUTOXIDE

Function: Insecticidal synergist (1)(2)(3)(4)

Chemical Name: 5-[2-(2-butoxyethoxy)ethoxymethyl]-6-propyl-1,3-benzodioxole

Formula:

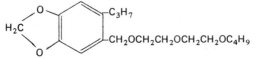

Trade Names: Butoxide
NIA-5273 (FMC Corp.)
Butacide® (FMC Corp.)
Butocide®

Manufacture (3)(4)

Piperonyl butoxide is obtained when the chloromethyl derivative of dihydrosafrole is reacted with the sodium salt of diethylene glycol monobutyl ether (butyl Carbitol) to form an ether derivative. This is illustrated by the following equation:

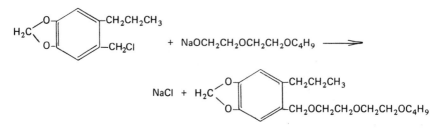

The chloromethyl dihydrosafrole referred to can be readily prepared by the reaction of formaldehyde and hydrochloric acid on dihydrosafrole. The preparation of this product is illustrated by the following example, the parts being by weight.

Example 1—162 parts of dihydrosafrole, 150 parts of 40% formaldehyde solution, and 500 parts of concentrated hydrochloric acid are mixed and agitated for a period of about 36 hours at a temperature below 20°C. The oily bottom layer is separated and the water layer is extracted with benzol. The benzol solution and the separated oil are combined and neutralized with sodium bicarbonate solution. The solution is dried, the benzol is distilled off, and the remaining oil is distilled in vacuo. The chloromethyl derivative distilled at about 128°C at 4 mm pressure. It is a colorless oil heavier than water. The chlorine content was determined by saponification and gave a saponification value of 265 (calculated 262). The preparation of the butyl Carbitol derivative is illustrated by the following example.

Example 2—22 parts of sodium hydroxide pellets were added to 162 parts of butyl Carbitol in 90 parts of benzene, and the mixture was refluxed using a water trap until no more water was collected (approximately 13 hours). The solution was cooled and under continued cooling, 106 parts of chloromethyl compound in 45 parts of benzene were added. After standing overnight, the mixture was refluxed for 4 hours. The salt which separates is removed by washing with water, the separated benzol solution is dried, and the benzol is distilled off. The remaining oil may be distilled in vacuo. Some of the excess butyl Carbitol distills over, then the final product distills at about 195°C at 2 mm pressure. It is a colorless oil, soluble in benzene, isopropanol and most organic solvents.

Toxicity

The acute oral LD_{50} value for rats is about 7,500 mg/kg which is insignificantly toxic.

Allowable Limits on Exposure and Use

Product Use: Issuance of a rebuttable presumption against registration was being considered by EPA as of September 1979 on the basis of possible cocarcinogenicity.

The tolerances set by the EPA for piperonyl butoxide in or on raw agricultural commodities are as follows:

	40 *CFR* Reference	Parts Per Million
Almonds (post-h)	180.127	8.0
Apples (post-h)	180.127	8.0
Barley (post-h)	180.127	20.0
Beans (post-h)	180.127	8.0
Birdseed mixtures (post-h)	180.127	20.0
Blackberries (post-h)	180.127	8.0
Blueberries (huckleberries) (post-h)	180.127	8.0
Boysenberries (post-h)	180.127	8.0
Buckwheat (post-h)	180.127	20.0
Cattle, fat	180.127	0.1 N
Cattle, mbyp	180.127	0.1 N
Cattle, meat	180.127	0.1 N
Cherries (post-h)	180.127	8.0
Cocoa beans (post-h)	180.127	8.0
Copra (post-h)	180.127	8.0
Corn (post-h)	180.127	20.0
Cotton, seed (post-h)	180.127	8.0
Crabapples (post-h)	180.127	8.0
Currants (post-h)	180.127	8.0
Dewberries (post-h)	180.127	8.0
Eggs	180.127	1.0
Figs (post-h)	180.127	8.0
Flax, seed (post-h)	180.127	8.0
Goats, fat	180.127	0.1 N
Goats, mbyp	180.127	0.1 N
Goats, meat	180.127	0.1 N
Gooseberries (post-h)	180.127	8.0
Grapes (post-h)	180.127	8.0
Guavas (post-h)	180.127	8.0
Hogs, fat	180.127	0.1 N
Hogs, mbyp	180.127	0.1 N
Hogs, meat	180.127	0.1 N
Horses, fat	180.127	0.1 N
Horses, mbyp	180.127	0.1 N
Horses, meat	180.127	0.1 N

(continued)

	40 *CFR* Reference	Parts per Million
Loganberries (post-h)	180.127	8.0
Mangoes (post-h)	180.127	8.0
Milk, fat	180.127	0.25
Muskmelons (post-h)	180.127	8.0
Oats (post-h)	180.127	8.0
Oranges (post-h)	180.127	8.0
Peaches (post-h)	180.127	8.0
Peanuts, shell removed (post-h)	180.127	8.0
Pears (post-h)	180.127	8.0
Peas (post-h)	180.127	8.0
Pineapples (post-h)	180.127	8.0
Plums (fresh prunes (post-h)	180.127	8.0
Potatoes (post-h)	180.127	0.25
Poultry, fat	180.127	3.0
Poultry, mbyp	180.127	3.0
Poultry, meat	180.127	3.0
Raspberries (post-h)	180.127	8.0
Rice (post-h)	180.127	20.0
Rye (post-h)	180.127	20.0
Sheep, fat	180.127	0.1 N
Sheep, mbyp	180.127	0.1 N
Sheep, meat	180.127	0.1 N
Sorghum, grain (post-h)	180.127	8.0
Tomatoes (post-h)	180.127	8.0
Walnuts (post-h)	180.127	8.0
Wheat (post-h)	180.127	20.0

The tolerances set by the EPA for piperonyl butoxide in animal feeds are as follows (the *CFR* Reference is to Title 21):

	CFR Reference	Parts per Million
Feeds, dried (from cotton bags)	561.310	10.0
Feeds, dried (from paper bags)	561.310	10.0

The tolerances set by the EPA for piperonyl butoxide in food are as follows (the *CFR* Reference is to Title 21):

	CFR Reference	Parts per Million
Cereal grains, milled fractions (exc. flour)	193.360	10.0
Dried foods (2-ply bag-cloth & waxpaper)	193.360	10.0
Dried foods in paper bags (multiwalled)	193.360	10.0
Processed food	193.360	10.0

References

(1) Worthing, C.R., *Pesticide Manual,* 6th ed., p. 429, British Crop Protection Council (1979).
(2) Spencer, E.Y., *Guide to the Chemicals Used in Crop Protection,* 6th ed., p. 417, London, Ontario, Agriculture Canada (January 1973).
(3) Wachs, H., U.S. Patent 2,485,681, October 25, 1949, assigned to U.S. Industrial Chemicals, Inc.
(4) Wachs, H., U.S. Patent 2,550,737, May 1, 1951, assigned to U.S. Industrial Chemicals, Inc.

PIPEROPHOS

Function: Herbicide (1)(2)

Chemical Name: S-[2-(2-methyl-1-piperidinyl)-2-oxoethyl] O,O-dipropyl phosphorodithioate

Formula:

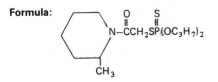

Trade Names: C-19,490 (Ciba-Geigy, Ltd.)
Rilof® (Ciba-Geigy, Ltd.)

Manufacture (2)

A solution of 32.0 grams of N-chloracetyl-2-methyl piperidine (0.184 mol) in 200 ml of acetone is added dropwise while stirring to a solution of 45.0 grams of the sodium salt of O,O-di-n-propyl-dithiophosphoric acid (0.191 mol) in 250 ml of acetone. During the addition, the temperature rises to 32°C and NaCl separates. The batch is then stirred for one hour at 50°C. The NaCl which has separated is then filtered off, and the solvent expelled under a pressure of 10 mm of Hg at 50°C. the desired compound remains behind as a yellowish, viscous oil which is practically pure according to analysis.

Toxicity

The acute oral LD_{50} value for rats is 324 mg/kg which is moderately toxic.

References

(1) Worthing, C.R., *Pesticide Manual,* 6th ed., p. 430, British Crop Protection Council (1979).
(2) British Patent 1,255,946, December 1, 1971, assigned to Ciba-Geigy, AG.

PIRIMICARB

Function: Aphicide (1)(2)

Chemical Name: 2-(dimethylamino)-5,6-dimethyl-4-pyrimidinyl dimethylcarbamate

Formula:

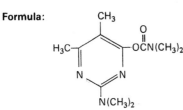

Trade Names: PP 062 (I.C.I.)
Pirimor® (Plant Protection, Ltd.)
Aphox®

Manufacture (3)(4)

A mixture of 2-dimethylamino-4-hydroxypyrimidine (20.0 grams), anhydrous potassium carbonate (20.0 grams) and dimethylcarbamoyl chloride (13.0 cc) in dry acetone (150 cc) was refluxed for 4 hours, cooled to 20°C and the insoluble portion filtered off and washed with acetone (50 cc). The washings and filtrate were combined and evaporated under reduced pressure, and the residual oil was dissolved in chloroform (3 volumes) and washed first with 1% sodium hydroxide solution, and then with water until the washings were neutral. The chloroform extract was dried over anhydrous sodium sulfate, and the solvent removed under reduced pressure. The residue was distilled, and 2-dimethylamino-4-dimethylcarbamoyl-oxypyrimidine was obtained. Synthesis of pirimicarb is accomplished in an analogous manner.

Toxicity

The acute oral LD$_{50}$ value for rats is 147 mg/kg which is moderately toxic.

Allowable Limits on Exposure and Use

Product Use: The tolerances set by the EPA for pirimicarb in or on raw agricultural commodities are as follows:

	40 *CFR* Reference	Parts per Million
Potatoes	180.365	0.1

References

(1) Worthing, C.R., *Pesticide Manual,* 6th ed., p. 432, British Crop Protection Council (1979).
(2) Spencer, E.Y., *Guide to the Chemicals Used in Crop Protection,* 6th ed., p. 420, London, Ontario, Agriculture Canada (January 1973).
(3) Baranyovits, F.L.C., Ghosh, R., Bishop, N.D., Freeman, P.F.H. and Jones, W.G.M., U.S. Patent 3,493,574, February 3, 1970, assigned to Imperial Chemical Industries, Ltd.
(5) Baranyovits, F.L.C., Ghosh, R., Bishop, N.D., Freeman, P.F.H., and Jones W.G.M., U.S. Patent 3,577,543, May 4, 1971, assigned to Imperial Chemical Industries, Ltd.

PIRIMIPHOS-ETHYL

Function: Insecticide (1)(2)

Chemical Name: O-[2-(diethylamino)-6-methyl-4-pyrimidinyl] O,O-diethyl phosphorothioate

Formula:

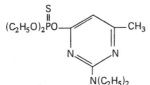

Trade Names: PP-211 (I.C.I.)
Primicid® (Plant Protection, Ltd.)

Manufacture (3)(4)(5)

Pirimiphos-ethyl may be made from N,N-diethylguanidine which is condensed with ethyl acetoacetate and then reacted with O,O-diethyl phosphorochloridothioate.

Toxicity

The acute oral LD$_{50}$ value for rats is 140 to 200 mg/kg which is moderately toxic.

References

(1) Worthing, C.R., *Pesticide Manual,* 6th ed., p. 433, British Crop Protection Council (1979).
(2) Spencer, E.Y., *Guide to the Chemicals Used in Crop Protection,* 6th ed., p. 421, London, Ontario, Agriculture Canada (January 1973).
(3) McHattie, G.V., British Patent 1,019,227, February 2, 1966, assigned to Imperial Chemical Industries, Ltd.
(4) Snell, B.K. and Sharpe, S.P., British Patent 1,205,000, September 9, 1970, assigned to Imperial Chemical Industries, Ltd.
(5) McHattie, G.V., U.S. Patent 3,287,453, November 22, 1966, assigned to Imperial Chemical Industries, Ltd.

PIRIMIPHOS-METHYL

Function: Insecticide and acaricide (1)(2)

Chemical Name: O-[2-(diethylamino)-6-methyl-4-pyrimidinyl] O,O-dimethyl phosphoro-thioate

Formula:

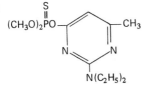

Trade Names: PP-511 (I.C.I.)
Actellic® (Plant Protection, Ltd.)
Actellifog®

Manufacture (3)(4)(5)

Primiphos-methyl may be made by the condensation of N,N-diethylguanidine with ethyl acetoacetate. That intermediate is then reacted with O,O-dimethylphosphorochloridothioate. In a specific example of the latter step (5), 2-diethylamino-4-methyl-6-hydroxy primidine (0.06 mol) was mixed with 0.7 gram (0.07 mol) anhydrous potassium carbonate in ethyl acetate (130 ml) and dimethyl chlorothiophosphate (0.06 mol) added slowly. The solution was refluxed overnight, cooled and evaporated to dryness under reduced pressure. The residue was then taken up into toluene, washed free of unreacted hydroxypyrimidine with cold 3% aqueous sodium hydroxide followed by water until the washings were neutral.

After drying over anhydrous magnesium sulfate and removing the solvent the crude product was obtained. This was heated to 75°C under a pressure of 0.2 mm mercury for 2 hours to remove unreacted chlorothiophosphate and a final yield of O-(2-diethylamino-4-methyl-6-pyrimidinyl) O,O-dimethyl phosphorothioate obtained that was 80% of the theoretical yield.

Toxicity

The acute oral LD$_{50}$ value for rats is 2,050 mg/kg (B-5) which is slightly toxic.

References

(1) Worthing, C.R., *Pesticide Manual,* 6th ed., p. 434, British Crop Protection Council (1979).
(2) Spencer, E.Y., *Guide to the Chemicals Used in Crop Protection,* 6th ed., p. 422, London, Ontario, Agriculture Canada (January 1973).
(3) McHattie, G.V., British Patent 1,019,227, February 2, 1966, assigned to Imperial Chemical Industries, Ltd.
(4) Sharpe, S.P. and Snell, B.K., U.S. Patent 1,204,552, September 9, 1970, assigned to Imperial Chemical Industries, Ltd.
(5) Sharpe, S.P. and Snell, B.K., U.S. Patent 3,651,224, March 21, 1972, assigned to Imperial Chemical Industries, Ltd.

POLYOXIN-B

Function: Fungicide (1)

Chemical Name: 5-[[2-amino-5-O-(aminocarbonyl)-2-deoxy-L-xylonoyl]amino]-1,5-dideoxy-1-[3,4-dihydro-5-(hydroxymethyl)-2,4-dioxo-1(2H)-pyrimidinyl]-β-D-allofuranuronic acid

Formula:

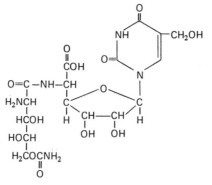

Trade Name: None

Manufacture

Polyoxin-B may be isolated from culture solutions of *Streptomyces cacaoi* var. asoenis which give a series of related antibiotics.

Toxicity

The acute oral LD_{50} value for mice is 800 mg/kg (B-5).

Reference

(1) Worthing, C.R., *Pesticide Manual,* 6th ed., p. 435, British Crop Protection Council (1979).

PROFENOFOS

Function: Insecticide and acaricide (1)(2)

Chemical Name: O-(4-Bromo-2-chlorophenyl) O-ethyl-S-propyl phosphorothioate

Formula:

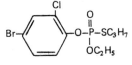

Trade Names: CGA-15324 (Ciba-Geigy Ltd.)
Curacron® (Ciba-Geigy Ltd.)

Manufacture (2)

31.1 grams (0.15 mol) 2-chloro-4-bromophenol were dissolved in 150 ml benzene and 15.1 grams triethylamine were then added. At 10° to 15°C, 32.0 grams of O-ethyl-S-n-propyl-chlorothiophosphate were added dropwise with continuous stirring. Stirring was then continued for 12 hours at room temperature. The mixture was washed with water, 3% Na_2CO_3 solution and water and dried over anhydrous sodium sulfate. The benzene was distilled off and the residue purified by distillation (at 130°C/0.01 torr). Yield: 49 grams; refractive index n_D^{20} 1.5466.

Toxicity

The acute oral LD_{50} value for rats is 358 mg/kg which is moderately toxic.

Allowable Limits on Exposure and Use

Product Use: The tolerances set by the EPA for profenofos in animal feeds are as follows (the *CFR* Reference is to Title 21):

	CFR Reference	Parts per Million
Cottonseed, hulls	561.53	6.0
Soapstock	561.53	15.0

References

(1) Worthing, C.R., *Pesticide Manual,* 6th ed., p. 438, British Crop Protection Council (1979).
(2) Beriger, E. and Drabek, J., U.S. Patent 3,992,533, November 16, 1976, assigned to Ciba-Geigy Corp.

PROFLURALIN

Function: Herbicide (1)(2)(3)

Chemical Name: N-(cyclopropylmethyl)-2,6-dinitro-N-propyl-4-(trifluoromethyl)-benzeneamine

Formula:

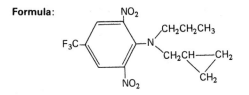

Trade Names: CGA-10,832 (Ciba-Geigy, Ltd.)
Pregard® (Ciba-Geigy, Ltd.)
Tolban® (Ciba-Geigy, Ltd.)

Manufacture (3)(4)

In 600 ml of benzene was dissolved n-propylamine (130 g, 2.2 mols, 181 ml). This well-stirred solution was cooled to about 10°C, and a solution of cyclopropanecarboxylic acid chloride (104.5 g, 1.0 mol) in 100 ml of benzene was added dropwise over a period of 1½ hours. During this time, white solids formed in the reaction medium. The mixture was heated at reflux for about 3 hours, cooled to room temperature, and filtered to remove the insoluble white n-propylamine hydrochloride salt. The filtrate was washed successively with 150 ml portions of water, 5% HCl, water, 10% NaHCO$_3$ and water. The solution, after being dried with anhydrous MgSO$_4$, was stripped of solvent to yield a crude N-n-propylcyclopropane carboxamide oil (108.6 g) which crystallized to a solid upon standing. A sample was recrystallized from petroleum ether at subzero temperatures to give material melting at 34° to 35°C.

In a reaction flask was placed a suspension of lithium aluminum hydride (30 g, 0.786 mol) in anhydrous ether (1 liter). A nitrogen blanket was held over the mixture throughout the reaction. To the stirred mixture was added dropwise a solution of N-n-propylcyclopropanecarboxamide (50 g, 0.393 mol) in 300 ml of dry ether over a period of 2 hours. Gentle refluxing of the solvent was evident during the addition.

After the addition was complete, the mixture was heated at reflux for approximately 10 hours. The solution was cooled with an ice bath while 6% aqueous sodium hydroxide was cautiously added dropwise to destroy the unreacted lithium aluminum hydride. The resulting white aluminate salts were removed by filtration and the clear etheral filtrate was dried with potassium hydroxide pellets. After the ether was removed by distillation, the N-cyclopropylmethyl-N-n-propylamine was distilled at atmospheric pressure and found to have a boiling range of 136° to 138°C. The yield was 27 g.

In a reaction flask was placed 3,5-dinitro-4-chlorobenzotrifluoride (85.4 g, 0.316 mol) and triethylamine (40.5 g, 0.40 mol), dissolved in 800 ml of benzene. The solution was cooled to approximately 10°C with stirring and N-cyclopropylmethyl-N-n-propylamine (0.347 mol), dissolved in 100 ml of benzene, was added dropwise over 1½ hours. The mixture was refluxed for 2½ hours and allowed to cool. Filtration of the solids (amine hydrochloride) gave a red-colored filtrate which was washed with 5% HCl and water. After drying the solution with K$_2$CO$_3$, the solvent was removed at reduced pressure to yield a yellow solid which was recrystallized from cold hexane and which melted at 27° to 28°C.

Toxicity

The acute oral LD$_{50}$ value for rats is about 10,000 mg/kg which is insignificantly toxic.

Allowable Limits on Exposure and Use

Product Use: The tolerances set by the EPA for profluralin in or on raw agricultural commodities are as follows:

	40 *CFR* Reference	Parts per Million
Alfalfa, forage	180.348	0.1 N
Alfalfa, hay	180.348	0.1 N
Cattle, fat	180.348	0.02 N
Cattle, mbyp	180.348	0.02 N
Cattle, meat	180.348	0.02 N

(continued)

	40 *CFR* Reference	Parts per Million
Cotton, seed	180.348	0.1 N
Eggs	180.348	0.02 N
Goats, fat	180.348	0.02 N
Goats, mbyp	180.348	0.02 N
Goats, meat	180.348	0.02 N
Hogs, fat	180.348	0.02 N
Hogs, mbyp	180.348	0.02 N
Hogs, meat	180.348	0.02 N
Horses, fat	180.348	0.02 N
Horses, mbyp	180.348	0.02 N
Horses, meat	180.348	0.02 N
Milk	180.348	0.02 N
Poultry, fat	180.348	0.02 N
Poultry, mbyp	180.348	0.02 N
Poultry, meat	180.348	0.02 N
Safflower, seed	180.348	0.1 N
Sheep, fat	180.348	0.02 N
Sheep, mbyp	180.348	0.02 N
Sheep, meat	180.348	0.02 N
Soybeans, hay	180.348	0.3 N
Sunflower, seed	180.348	0.1 N
Vegetables, seed & pod (dry/succulent)	180.348	0.1 N
Vegetables, seed & pod, fodder	180.348	0.1 N
Vegetables, seed & pod, forage	180.348	0.1 N

References

(1) Worthing, C.R., *Pesticide Manual,* 6th ed., p. 439, British Crop Protection Council (1979).
(2) Spencer, E.Y., *Guide to the Chemicals Used in Crop Protection,* 6th ed., p. 223, London, Ontario, Agriculture Canada (January 1973).
(3) Maravetz, L.L., U.S. Patent 3,546,295, December 8, 1970, assigned to Esso Research and Engineering Co.
(4) Maravetz, L.L., U.S. Patent 3,672,864, June 27, 1972, assigned to Esso Research and Engineering Co.

PROMECARB

Function: Insecticide (1)(2)

Chemical Name: 3-methyl-5-(1-methylethyl)phenyl methylcarbamate

Formula:

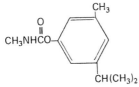

Trade Names: Schering 34,615 (Schering AG)
Carbamult® (Schering AG)
Minacide®

Manufacture

Promecarb may be made by the reaction of 5-isopropyl-3-methylphenol and methyl isocyanate. Alternatively the 5-isopropyl-3-methylphenol may be converted to the chloroformic acid derivative and reacted with methylamine (2). The following is a specific example of the latter process.

Into a stirred solution of 243 g (2.45 mols = 175 ml) phosgene in 500 ml toluene was added dropwise and at a reaction temperature below 15°C a solution of 300 g 3-methyl-5-isopropyl-phenol and 242 g (2 mols) dimethylaniline in 500 ml toluene. Thereafter the mixture was stirred further for 30 minutes at a temperature of about 10°C, then for 1 hour at a tempera-ture of about 50°C. Subsequently the mixture was cooled, whereupon the precipitated di-methylaniline hydrochloride was separated by suction. The toluene-filtrate, after drying over sodium sulfate, was distilled under vacuum. The 3-methyl-5-isopropyl-phenyl ester of chloroformic acid (383 g) boils at $BP_{0.002}$, 53°C.

Into a stirred solution of 10.65 g (0.05 mol) of this chloroformic acid ester in 25 ml toluene there was added dropwise at a temperature of 5° to 8°C a solution of 4.65 g (0.15 mol = 3.5 ml) methylamine in 25 ml toluene and thereafter stirred further for about 15 minutes. Then the reaction mixture was diluted with ether. The ether solution was washed with water. After drying over sodium sulfate and filtering the ether solution was concentrated. The residue (10.2 g, MP 77° to 79°C) was recrystallized and yielded 6.33 g of 3-methyl-5-isopropyl-phenyl-ester of N-methylcarbamic acid: MP 87° to 87.5°C

Toxicity

The acute oral LD_{50} value for rats is 60 to 90 mg/kg which is moderately toxic.

References

(1) Worthing, C.R., *Pesticide Manual,* 6th ed., p. 441, British Crop Protection Council (1979).
(2) Czyzewski, A. and Jager, A., U.S. Patent 3,167,472, January 26, 1965, assigned to Schering AG.

PROMETON

Function: Herbicide (1)(2)(3)

Chemical Name: 6-methoxy-N,N'-bis-(1-methylethyl)-1,3,5-triazine-2,4-diamine

Formula:

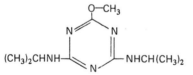

Trade Names: G-31,435 (Ciba-Geigy)
In Primatol® (Ciba-Geigy)
Primatol O® (Ciba-Geigy)
Gesafram®

Manufacture

Prometon is made by the reaction of propazine (which see) with methanol in the presence of one equivalent of sodium hydroxide.

Toxicity

The acute oral LD$_{50}$ value for rats is 3,000 mg/kg which is slightly toxic.

References

(1) Worthing, C.R., *Pesticide Manual*, 6th ed., p. 442, British Crop Protection Council (1979).
(2) Spencer, E.Y., *Guide to the Chemicals Used in Crop Protection*, 6th ed., p. 425, London, Ontario, Agriculture Canada (January 1973).
(3) Gysin, H. and Knusli, E., U.S. Patent 2,909,420, October 20, 1959, assigned to J.R. Geigy AG.

PROMETRYNE

Function: Herbicide (1)(2)(3)

Chemical Name: N,N'-bis(1-methylethyl)-6-(methylthio)-1,3,5-triazine-2,4-diamine

Formula:

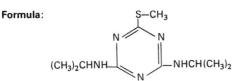

Trade Names: G-34,161 (Ciba-Geigy)
Gesagard® (Ciba-Geigy)
Caparol® (Ciba-Geigy)
Primatol Q®

Manufacture

Prometryne may be made by reacting propazine (which see) with methyl mercaptan.

Toxicity

The acute oral LD$_{50}$ value for rats is 3,150 to 5,230 mg/kg which is slightly toxic.

Allowable Limits on Exposure and Use

Product Use: The tolerances set by the EPA for prometryne in or on raw agricultural commodities are as follows:

	40 *CFR* Reference	Parts per Million
Celery	180.222	0.5
Corn, field, fodder	180.222	0.25
Corn, field, forage	180.222	0.25
Corn, fresh (inc. sweet) (k+cwhr)	180.222	0.25
Corn, grain	180.222	0.25
Corn, pop, fodder	180.222	0.25
Corn, pop, forage	180.222	0.25
Corn, sweet, fodder	180.222	0.25
Corn, sweet, forage	180.222	0.25
Cotton, forage	180.222	1.0
Cotton, seed	180.222	0.25

References

(1) Worthing, C.R., *Pesticide Manual,* 6th ed., p. 443, British Crop Protection Council (1979).
(2) Spencer, E.Y., *Guide to the Chemicals Used in Crop Protection,* 6th ed., p. 426, London, Ontario, Agriculture Canada (January 1973).
(3) Gysin, H. and Knusli, E., U.S. Patent 2,909,420, October 20, 1959, assigned to J.R. Geigy AG.

PRONAMIDE
(PROPYZAMIDE IN U.K., JAPAN)

Function: Herbicide (1)(2)(3)(4)

Chemical Name: 3,5-dichloro-N-(1,1-dimethylpropynyl) benzamide

Formula:

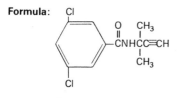

Trade Names: RH 315 (Rohm & Haas)
                 Kerb® (Rohm & Haas)
                 Pronamide

Manufacture (3)(4)

A suspension of 40 parts of 3,5-dichlorobenzoic acid in 160 parts of thionyl chloride was heated under reflux for 4 hours with solution of the solid occurring. Excess thionyl chloride was removed under reduced pressure and the residue distilled to give 39.6 parts (90% yield) of 3,5-dichlorobenzoyl chloride, BP 71° to 75°C (1 mm).

Pronamide was then prepared by adding 23.2 parts (0.28 mol) of 3-amino-3-methylbutyne, 40.4 parts of 24.8% aqueous sodium hydroxide (0.25 mol) and 125 parts of Esso octane (a commercial product containing 18% paraffins, 60% naphthenes and 22% aromatic hydrocarbons and having a boiling range of 102° to 113°C) to a 500-ml flask equipped with a stirrer, thermometer and addition funnel.

While stirring and cooling there was added 52.3 parts (0.25 mol) of 3,5-dichlorobenzoyl chloride over a period of 34 minutes at below 20°C. A thick white slurry resulted. The mixture was stirred 3 hours as the temperature rose to room temperature. The solid was filtered off and washed with 100 parts of 50°C water. The isolated solid was dried at 0°C at 20 mm (Hg) pressure for 14 hours to give 62 parts (95% yield) of N-(1,1-dimethylpropynl)-3,5-dichlorobenzamide as a white solid which melted at 155° to 158°C. The solid was found to be 95% pure by gas-liquid chromatography.

Toxicity

The acute oral LD_{50} value for rats is 5,620 mg/kg (B-5) which is insignificantly toxic.

Allowable Limits on Exposure and Use

Product Use: A rebuttable presumption against registration was issued on May 20, 1977 by EPA on the basis of oncogenicity.

The tolerances set by the EPA for pronamide in or on raw agricultural commodities are as follows:

	40 *CFR* Reference	Parts per Million
Alfalfa	180.317	5.0
Alfalfa, forage, fresh	180.317	10.0
Alfalfa, hay, fresh	180.317	10.0
Blackberries	180.317	0.05 N
Blueberries	180.317	0.05 N
Boysenberries	180.317	0.05 N
Cattle, fat	180.317	0.02 N
Cattle, kidney	180.317	0.2
Cattle, liver	180.317	0.2
Cattle, mbyp (exc. kidney, liver)	180.317	0.02 N
Cattle, meat	180.317	0.02 N
Clover	180.317	5.0
Eggs	180.317	0.02 N
Endive (escarole)	180.317	2.0
Goat, kidney	180.317	0.2
Goat, liver	180.317	0.2
Goats, fat	180.317	0.02 N
Goats, mbyp (exc. kidney, liver)	180.317	0.02 N
Goats, meat	180.317	0.02 N
Hog, kidney	180.317	0.2
Hog, liver	180.317	0.2
Hogs, fat	180.317	0.02 N
Hogs, mbyp (exc. kidney, liver)	180.317	0.02 N
Hogs, meat	180.317	0.02 N
Horse, kidney	180.317	0.2
Horse, liver	180.317	0.2
Horses, fat	180.317	0.02 N
Horses, mbyp (exc. kidney, liver)	180.317	0.02 N
Horses, meat	180.317	0.02 N
Lettuce	180.317	2.0
Milk	180.317	0.02 N
Poultry, fat	180.317	0.02 N
Poultry, kidney	180.317	0.2
Poultry, liver	180.317	0.2
Poultry, mbyp (exc. kidney, liver)	180.317	0.02 N
Poultry, meat	180.317	0.02 N
Raspberries	180.317	0.05 N
Sainfoin	180.317	5.0
Sheep, fat	180.317	0.02 N
Sheep, kidney	180.317	0.2
Sheep, liver	180.317	0.2
Sheep, mbyp (exc. kidney, liver)	180.317	0.02 N
Sheep, meat	180.317	0.02 N
Trefoil	180.317	5.0
Vetch, crown	180.317	5.0

References

(1) Worthing, C.R., *Pesticide Manual,* 6th ed., p. 453, British Crop Protection Council (1979).
(2) Spencer, E.Y., *Guide to the Chemicals Used in Crop Protection,* 6th ed., p. 427, London, Ontario, Agriculture Canada (January 1973).

(3) Horrom, B.W., Crovetti, A.J. and Viste, K.L., U.S. Patent 3,534,098, October 13, 1970, assigned to Rohm & Haas Co.
(4) Horrom, B.W., Crovetti, A.J. and Viste, K.L., U.S. Patent 3,640,699, February 8, 1972, assigned to Rohm & Haas Co.

PROPACHLOR

Function: Herbicide (1)(2)(3)

Chemical Name: 2-chloro-N-(1-methylethyl)-N-phenylacetamide

Formula:

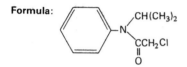

Trade Names: CP 31393 (Monsanto)
Ramrod® (Monsanto)

Manufacture (3)

Propachlor may be made by reacting α-chloroacetyl chloride with an excess of N-isopropyl aniline. In a typical example, a reaction flask was charged with N-isopropyl aniline, 20% sodium hydroxide solution and ethylene dichloride. The flask and its contents were cooled to −10°C and chloroacetyl chloride gradually added over a period of 15 minutes. After stirring until the mixture was warmed to 5°C, the resulting organic phase was separated and washed with dilute hydrochloric acid and water. Upon the evaporation of the ethylene dichloride a solid product was obtained, which was recrystallized from aqueous ethanol. The white crystalline product had a melting point of 67° to 76°C.

Process Wastes and Their Control

Air: Air emissions from propachlor manufacture have been reported (B-15) to consist of 0.5 kg hydrocarbons per metric ton of pesticide produced.

Product Wastes: Alkaline hydrolysis would yield N-isopropylaniline (B-3).

Toxicity

The acute oral LD_{50} value for rats is 780 mg/kg which is slightly toxic.

Propachlor is the most toxic of the group consisting of alachlor, butachlor and propachlor, although all appear to be fairly well tolerated by mammals (B-22).

The maximal tolerated dosage of propachlor without adverse effect is reported as 133.3 mg/kg/day in both rats and dogs. Other workers reported slight organ pathology in rats, mice, and rabbits at 100 mg/kg/day or higher; this agrees approximately with the former data.

The existing toxicity data for these compounds are largely those produced by the manufacturer for registration purposes. Based on the available data, ADI's were calculated at 0.1, 0.1 and 0.01 mg/kg/day for alachlor, propachlor, and butachlor, respectively.

Apparently, no long-term toxicity studies have been completed that would contribute information on reproductive effects or carcinogenic potential of these acetanilides or their degradation products, which include aniline derivatives. These studies are needed (B-22).

Allowable Limits on Exposure and Use

Product Wastes: The tolerances set by the EPA for propachlor in or on raw agricultural commodities are as follows:

	40 *CFR* Reference	Parts per Million
Beets, sugar, roots	180.211	0.2
Beets, sugar, tops	180.211	1.0
Cattle, fat	180.211	0.02 N
Cattle, mbyp	180.211	0.02 N
Cattle, meat	180.211	0.02 N
Corn, forage	180.211	1.5
Corn, grain	180.211	0.1 N
Corn, sweet (k+cwhr)	180.211	0.1 N
Cotton, seed	180.211	0.1 N
Eggs	180.211	0.02 N
Goats, fat	180.211	0.02 N
Goats, mbyp	180.211	0.02 N
Goats, meat	180.211	0.02 N
Hogs, fat	180.211	0.02 N
Hogs, mbyp	180.211	0.02 N
Hogs, meat	180.211	0.02 N
Horses, fat	180.211	0.02 N
Horses, mbyp	180.211	0.02 N
Horses, meat	180.211	0.02 N
Milk	180.211	0.02 N
Peas (pod removed)	180.211	0.02
Peas, forage	180.211	1.5
Poultry, fat	180.211	0.02 N
Poultry, mbyp	180.211	0.02 N
Poultry, meat	180.211	0.02 N
Sheep, fat	180.211	0.02 N
Sheep, mbyp	180.211	0.02 N
Sheep, meat	180.211	0.02 N
Sorghum, forage	180.211	3.0
Sorghum, grain	180.211	0.25

References

(1) Worthing, C.R., *Pesticide Manual,* 6th ed., p. 444, British Crop Protection Council (1979).
(2) Spencer, E.Y., *Guide to the Chemicals Used in Crop Protection,* 6th ed., p. 428, London, Ontario, Agriculture Canada (January 1973).
(3) Hamm, P.C. and Speziale, A.J., U.S. Patent 2,863,752, December 9, 1958, assigned to Monsanto Chemical Co.

PROPANIL

Function: Herbicide (1)(2)(4)

Chemical Name: N-(3,4-dichlorophenyl) propanamide

Formula:

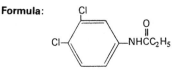

Trade Names: FW-734 (Rohm & Haas)
Stam F-34® (Rohm & Haas)
B-30,130 (Bayer)
Surcopur® (Bayer)
Rogue® (Monsanto)
DPA

Manufacture (3)(4)

Propanil may be made by the reaction of 3,4-dichloroaniline with propionic acid in the presence of thionyl chloride; SO_2 and HCl are split off in the course of the reaction. Alternatively (3), propionyl chloride may be reacted with the dichloroaniline. Generally the acid halide is added with stirring at room temperature to an inert solvent such as toluene, benzene or dioxane containing an acid acceptor such as triethylamine, pyridine or sodium carbonate, and the properly substituted aniline. To insure complete interaction the entire reaction is heated briefly at 75° to 100°C. Essentially pure acyl anilide product can be obtained by distilling off the solvent and washing the residue with water until free of the acid acceptor salt.

Process Wastes and Their Control

Air: Air emissions from propanil manufacture have been reported (B-15) to consist of 0.5 kg hydrocarbons per metric ton of pesticide produced.

Product Wastes: Hydrolysis in acidic or basic media yields the more toxic substance, 3,4-dichloroaniline, and is not recommeded (B-3).

Toxicity

The acute oral LD_{50} value for rats is 1,285 to 1,485 mg/kg which is slightly toxic.

Propanil is well tolerated by experimental animals on a chronic basis, and there is little or no indication of mutagenic or oncogenic properties of the compound. The highest no-adverse-effect concentration of propanil based on reproduction in the rat and acute, subchronic, and chronic studies in rats and dogs is 400 ppm in the diet. Based on these data an ADI was calculated at 0.02 mg/kg/day (B-22).

Allowable Limits on Exposure and Use

Product Use: The tolerances set by the EPA for propanil in or on raw agricultural commodities are as follows:

	40 *CFR* Reference	Parts per Million
Cattle, fat	180.274	0.1 N
Cattle, mbyp	180.274	0.1 N
Cattle, meat	180.274	0.1 N
Eggs	180.274	0.05 N
Goats, fat	180.274	0.1 N
Goats, mbyp	180.274	0.1 N
Goats, meat	180.274	0.1 N

(continued)

	40 *CFR* Reference	Parts per Million
Hogs, fat	180.274	0.1 N
Hogs, mbyp	180.274	0.1 N
Hogs, meat	180.274	0.1 N
Horses, fat	180.274	0.1 N
Horses, mbyp	180.274	0.1 N
Horses, meat	180.274	0.1 N
Milk	180.274	0.05 N
Poultry, fat	180.274	0.1 N
Poultry, mbyp	180.274	0.1 N
Poultry, meat	180.274	0.1 N
Rice	180.274	2.0
Rice, straw	180.274	75.0
Sheep, fat	180.274	0.1 N
Sheep, mbyp	180.274	0.1 N
Sheep, meat	180.274	0.1 N

The tolerances set by the EPA for propanil in animal feeds are as follows (the *CFR* Reference is to Title 21):

	CFR Reference	Parts per Million
Rice bran	561.150	10.0
Rice hulls	561.150	10.0
Rice polishings	561.150	10.0
Rice, milled fractions	561.150	10.0

References

(1) Worthing, C.R., *Pesticide Manual,* 6th ed., p. 446, British Crop Protection Council (1979).
(2) Spencer, E.Y., *Guide to the Chemicals Used in Crop Protection,* 6th ed., p. 429, London, Ontario, Agriculture Canada (January 1973).
(3) Fielding, M.J. and Stoddard, D.L., U.S. Patent 3,108,038, October 22, 1963, assigned to DuPont.
(4) British Patent 903,766, August 22, 1962, assigned to Rohm & Haas Company.

PROPARGITE

Function: Acaricide (1)(2)(3)(4)(5)

Chemical Name: 2-[4-(1,1-dimethylethyl)phenoxy] cyclohexyl 2-propynyl sulfite

Formula:

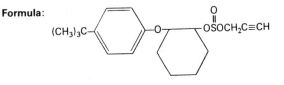

Trade Names: DO-14 (Uniroyal, Inc.)
Omite® (Uniroyal, Inc.)
Comite®

Manufacture (3)(4)(5)

2-(p-tert-butylphenoxy)cyclohexanol was made as follows: p-tert-butylphenol (711 grams, 4.74 mols) and 10.6 grams sodium hydroxide were combined and heated to 150°C. Cyclohexene oxide, i.e., 1,2-epoxycyclohexane (465 grams, 4.74 mols) was added dropwise during 1 hour, maintaining the reaction temperature at 150° to 160°C. After the addition was completed, the mixture was stirred at this temperature for 30 minutes. Xylene (500 ml) was added and the mixture was cooled to 100°C and neutralized with 12.8 grams concentrated sulfuric acid. Some of the xylene was distilled off in order to azeotrope out the water formed in the neutralization. The resulting xylene solution of the product is suitable for use in making the chlorosulfinate.

An aliquot of this solution was heated under reduced pressure to remove the xylene and other volatiles. Analysis of the residue showed that the product was obtained in 98.3% yield. The product can be recrystallized from hexane and melts at 93° to 95°C.

2-(p-tert-butylphenoxy)cyclohexyl chlorosulfinate was prepared as follows. The xylene solution of 2-(p-tert-butylphenoxy)cyclohexanol (1,653 grams of solution containing 1,025 grams, 4.13 mols) was warmed to 60°C whereupon a clear solution was obtained. Thionyl chloride (540 grams, 4.54 mols) was added during 30 minutes with enough cooling to cause the temperature to drop continuously. The mixture was allowed to stir at 5° to 10°C for 2½ hours and then stored at room temperature for 15 hours. The volatile materials were removed by warming the mixture to 40°C (5 mm) to obtain 2-(p-tert-butylphenoxy) cyclohexyl chlorosulfinate.

Propargyl alcohol (254 grams, 4.54 mols), 326 grams (4.13 mols) pyridine and 250 ml xylene were combined and the solution was cooled to 0° to 5°C. The chlorosulfinate was added during 1 hour, keeping the reaction temperature between 5° and 15°C. The mixture was stirred for 30 minutes and was washed with 2 liters of water. The aqueous layer was extracted with ether and the ether layer combined with the product. The crude propargite product was washed with 1 liter saturated salt solution. The solvents were removed and the residue heated to 90°C (0.3 mm). Yield was 1,348 grams (93.3%).

Process Wastes and Their Control

Product Wastes: Incineration of this compound would produce sulfur dioxide which, in small quantities, would not constitute a serious air pollution hazard (B-3).

Toxicity

The acute oral LD_{50} value for rats is 2,200 mg/kg which is slightly toxic.

Allowable Limits on Exposure and Use

Product Use: The tolerances set by the EPA for propargite in or on raw agricultural commodities are as follows:

	40 *CFR* Reference	Parts per Million
Almonds	180.259	0.1
Almonds, hulls	180.259	55.0
Apples	180.259	3.0
Apricots	180.259	7.0
Beans, dry	180.259	0.2
Beans, succulent	180.259	20.0
Cattle, fat	180.259	0.1

(continued)

	40 *CFR* Reference	Parts per Million
Cattle, mbyp	180.259	0.1
Cattle, meat	180.259	0.1
Corn, fodder	180.259	10.0
Cotton, forage	180.259	10.0
Corn, grain	180.259	0.1
Cotton, seed	180.259	0.1
Cranberries	180.259	10.0
Eggs	180.259	0.1
Figs	180.259	3.0
Goats, fat	180.259	0.1
Goats, mbyp	180.259	0.1
Goats, meat	180.259	0.1
Grapefruit	180.259	5.0
Grapes	180.259	10.0
Hogs, fat	180.259	0.1
Hogs, mbyp	180.259	0.1
Hogs, meat	180.259	0.1
Hops	180.259	15.0
Horses, fat	180.259	0.1
Horses, mbyp	180.259	0.1
Horses, meat	180.259	0.1
Lemons	180.259	5.0
Milk, fat (0.08 in whole milk)	180.259	2.0
Mint	180.259	50.0
Nectarines	180.259	4.0
Oranges	180.259	5.0
Peaches	180.259	7.0
Peanuts	180.259	0.1
Peanuts, forage	180.259	10.0
Peanuts, hay	180.259	10.0
Peanuts, hulls	180.259	10.0
Pears	180.259	3.0
Plums (fresh prunes)	180.259	7.0
Potatoes	180.259	0.1
Poultry, fat	180.259	0.1
Poultry, mbyp	180.259	0.1
Poultry, meat	180.259	0.1
Sheep, fat	180.259	0.1
Sheep, mbyp	180.259	0.1
Sheep, meat	180.259	0.1
Sorghum, fodder	180.259	10.0
Sorghum, forage	180.259	10.0
Sorghum, grain	180.259	10.0
Strawberries	180.259	7.0
Walnuts	180.259	0.1

The tolerances set by the EPA for propargite in animal feeds are as follows (the *CFR* Reference is Title 21):

	CFR Reference	Parts per Million
Apple pomace, dried	561.330	80.0
Citrus pulp, dried	561.330	40.0
Grape pomace, dried	561.330	40.0

The tolerances set by the EPA for propargite in food are as follows (the *CFR* Reference is to Title 21):

	CFR Reference	Parts per Million
Figs, dried	193.370	9.0
Hops, dried	193.370	30.0
Raisins	193.370	25.0

References

(1) Worthing, C.R., *Pesticide Manual,* 6th ed., p. 447, British Crop Protection Council (1979).
(2) Spencer, E.Y., *Guide to the Chemicals Used in Crop Protection,* 6th ed., p. 68, London, Ontario, Agriculture Canada (January 1973).
(3) Covey, R.A., Smith, A.E. and Hubbard, W.L., U.S. Patent 3,272,854, September 13, 1966, assigned to United States Rubber Co.
(4) Covey, R.A., Smith, A.E. and Hubbard, W.L., U.S. Patent 3,311,534, March 28, 1967, assigned to United States Rubber Co.
(5) Covey, R.A., Smith, A.E. and Hubbard, W.L., U.S. Patent 3,463,859, August 26, 1969, assigned to Uniroyal, Inc.

PROPAZINE

Function: Herbicide (1)(2)(3)

Chemical Name: 6-chloro-N,N'-bis(1-methylethyl)-1,3,5-triazine-2,4 diamine

Formula:

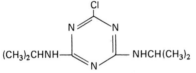

Trade Names: G-30,028 (Ciba-Geigy)
Gesamil® (Ciba-Geigy)
Milogard® (Ciba-Geigy)
Primatol P®

Manufacture

Cyanuric chloride is first made from chlorine and hydrocyanic acid. It is then reacted with two equivalents of isopropylamine in the presence of an acid acceptor to give propazine.

Process Wastes and Their Control

Air: Air emissions from propazine manufacture have been reported (B-15) to consist of 1.5 kg of hydrocarbons per metric ton of pesticide produced.

Toxicity

The acute oral LD_{50} value for rats is 7,700 mg/kg which is insignificantly toxic.

Atrazine, propazine, and simazine all appear to have low chronic toxicity. The only good carcinogenicity feeding study done on these compounds did not reveal a significant increase in cancer incidence over controls. On the basis of these chronic studies, an ADI was calculated for each of these compounds. The ADI for atrazine is 0.0215 mg/kg/day, for propazine 0.0464 mg/kg/day, and for simazine 0.215 mg/kg/day (B-22).

Allowable Limits on Exposure and Use

Product Use: The tolerances set by the EPA for propazine in or on raw agricultural commodities are as follows:

	40 *CFR* Reference	Parts per Million
Sorghum, fodder	180.243	0.25 N
Sorghum, forage	180.243	0.25 N
Sorghum, grain	180.243	0.25 N
Sorghum, sweet	180.243	0.25 N

References

(1) Worthing, C.R., *Pesticide Manual,* 6th ed., p. 448, British Crop Protection Council (1979).
(2) Spencer, E.Y., *Guide to the Chemicals Used in Crop Protection,* 6th ed., p. 430, London, Ontario, Agriculture Canada (January 1973).
(3) Gysin, H. and Knusli, E., U.S. Patent 2,891,855, June 23, 1959, assigned to J.R. Geigy AG.

PROPETAMPHOS

Function: Insecticide (1)(2)

Chemical Name: (E)-1-methylethyl 3-[[(ethylamino)methoxyphosphinothioyl]oxy]-2-butenoate

Formula:

$$CH_3CH_2NH-\overset{\overset{\displaystyle S}{\|}}{P}-O-\overset{\overset{\displaystyle H-C-COOCH(CH_3)_2}{\|}}{C}-CH_3$$
$$\underset{OCH_3}{|}$$

Trade Names: SAN 52 1391 (Sandoz AG)
Safrotin® (Sandoz AG)

Manufacture (2)

185.5 grams (1 mol) of tri-n-butylamine are added to a mixture of 144.2 grams (1 mol) of acetoacetic acid isopropyl ester and 178 grams (1.05 mols) of thiophosphoryl chloride while stirring and cooling to about −3°C during the course of 1 hour. The solution which turns viscous is stirred at 0° to 3°C for a further half hour, whereby the solution is mixed as thoroughly as possible. As soon as a viscous oil starts to separate, 20 cc of toluene are added 1.2 liters of petroleum ether are added, the mixture is stirred for a short period and the tributyl ammonium chloride is filtered off. The filtrate is concentrated in a vacuum at a bath temperature of 30°C. 128 grams (4.0 mols) of methanol are added to the residue as rapidly as possible at 0°C and the mixture is stirred at 0°C for 4 hours. 226 grams of a 70% aqueous solution of ethylamine (3.5 mols) are then added at −5°C within one-half hour, and the mixture is stirred at 0°C for a further one-half hour.

The reaction mixture is extracted twice in a separatory funnel with 120 cc amounts of approximately 10% hydrochloric acid (25 cc of concentrated hydrochloric acid with 100 grams of crushed ice). The separation of the phases during the first washing with aqueous hydrochloric acid is effected without the addition of petroleum ether; petroleum ether is preferably added during the second washing with aqueous hydrochloric acid in order to complete the separation. The petroleum ether phase is again washed with water and with hydrogen carbonate solution, is dried with sodium sulfate, and the solvent is removed. The resulting O-(1-

carboisopropoxy-1-propen-2-yl)-O-methyl-N-ethylphosphorothioamidate has a BP of 87° to 89°C/5 × 10^{-3} mm of Hg, n_D^{20} = 1.495. Analysis: $C_{10}H_{20}NO_4PS$, molecular weight: 281.3.

Toxicity

The acute oral LD$_{50}$ value for rats is 82 mg/kg which is moderately toxic.

References

(1) Worthing, C.R., *Pesticide Manual*, 6th ed., p. 449, British Crop Protection Council (1979).
(2) Leber, J.P. and Lutz, K., U.S. Patent 3,758,645, September 11, 1973, assigned to Sandoz, Ltd.

PROPHAM

Function: Plant growth regulator (1)(2)(4)

Chemical Name: 1-methylethyl phenylcarbamate

Formula:

Trade Names: IPC
Chem Hoe®

Manufacture

Propham may be made by (a) condensing phenyl isocyanate and isopropanol or (b) condensing aniline and isopropyl chloroformate (3). According to the prior art, isopropyl chloroformate is reacted with aniline, usually in the presence of a base such as an alkali metal hydroxide, in order to produce the carbamate in solid state. Such product is contaminated with various impurities including water, alkali metal chloride, alkali metal hydroxide or other base, aniline, etc. The product of this reaction may be purified by filtering the solid from the reaction mixture, washing with cold water, and washing with dilute HCl to reduce the pH of the product to approximately 7 or 8 approximate neutrality. This product may be purified by recrystallization from ethyl alcohol.

In this process, however, the product which is to be filtered from the reaction mixture is in the form of a lumpy, waxy curd, part of which floats on the top of the reaction mixture and part of which settles to the bottom. In such a state, or condition, filtration is a difficult, inefficient and incomplete method of separating the product from the reaction mixture. In addition, it appears that there is entrapped in this lumpy waxy curd a slight amount of aniline which detracts from the purity of the product.

In accordance with an improved process (3), a method of preparing isopropyl N-phenyl carbamate has been discovered. This method comprises reacting an isopropyl haloformate, such as isopropyl chloroformate, iodoformate, or bromoformate, with aniline at a temperature above the freezing point of the reaction mixture, but below 10° to 15°C, usually 0° to 10°C, and in the presence of a base such as an alkali metal hydroxide, carbonate and/or bicarbonate. This reaction mixture containing the product in an aqueous solution normally is acidified to neutralize the base, and the mixture, or at least solid isopropyl N-phenyl carbamate obtained therefrom, is heated to a temperature above the melting temperature of the carbamate, but below the boiling temperature of the water phase of the mixture, usually 85° to 100°C. Upon such heating, the carbamate melts and two immiscible liquid layers are formed, one being an aqueous layer and the other an organic layer containing the product.

After separating the two layers, it is then advantageous to extract the aniline by decantation, centrifugation or other convenient method, from the product-containing organic layer. This is done by one of several methods. One method comprises adding water to the product layer and steam distilling. Another method comprises adding to the product layer a quantity of water, then adding a sufficient amount of a mineral acid, usually as a solution containing, say 5 to 30% by weight of a mineral acid such HCl, to adjust the pH of the solution to about 3 or 4 or other acid concentration at which the aniline is transformed into a water-soluble salt. Following separation of aniline by a liquid phase separation, the product may be converted to a pulverulent granular solid by atomizing and chilling.

A preferred method of procedure is to add isopropyl chloroformate to a mixture of aniline and the base, employing substantially one mol of each reactant. Equally good results are obtained, however, when an excess, for example, up to 100% or more by molar ratio of aniline is used. When such an excess of aniline is used, a correspondingly smaller amount of base may be used.

It is also intended that the process may be practiced by adding aniline to a mixture of the base and isopropyl chloroformate, or by adding the base to a mixture of aniline and isopropyl chloroformate. Various basic compounds may be used in the practice of the process. Inorganic bases such as the oxides, hydroxides, carbonates, the bicarbonates of sodium, potassium, calcium, barium, strontium and magnesium, or other alkaline earth metal or alkali metal, organic bases such as pyridine, dimethyl aniline, and quaternary ammonium bases such as trimethyl phenyl ammonium hydroxide are included among these.

Frequently it is desirable to have present an excess, for example, up to 5 to 20% by molar ratio of the basic compound. This excess may be provided by use of an additional base or simply by using an additional amount of the same base. A suitable procedure to employ when this excess is desired is to add isopropyl chloroformate and the base simultaneously to a mixture of aniline and an amount of the same base or a different base.

Toxicity

The acute oral LD_{50} value for rats is about 5,000 mg/kg which is slightly to insignificantly toxic.

References

(1) Worthing, C.R., *Pesticide Manual*, 6th ed., p. 450, British Crop Protection Council (1979).
(2) Spencer, E.Y., *Guide to the Chemicals Used in Crop Protection*, 6th ed., p. 431, London, Ontario, Agriculture Canada (January 1973).
(3) Allen, E.M., U.S. Patent 2,615,916, October 28, 1952, assigned to Columbia-Southern Chemical Corp.
(4) Sexton, W.A. and Templeman, W.G., British Patent 574,995, January 30, 1946, assigned to Imperial Chemical Industries, Ltd.

PROPINEB

Function: Fungicide (1)(2)

Chemical Name: N,N'-[1-methyl-1,2-ethanediyl)bis[carbamodithioato(2-)]]zinc

Formula:

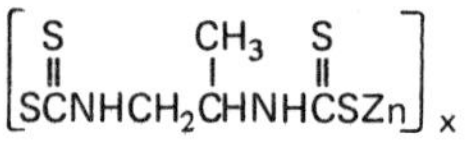

$$\left[\begin{array}{c} S \quad\quad CH_3 \quad S \\ \| \quad\quad\quad | \quad\quad \| \\ SCNHCH_2CHNHCSZn \end{array} \right]_x \qquad x > 1$$

Trade Names: B-46,131 (Bayer)
LH 30/Z (Bayer)
Antracol® (Bayer)

Manufacture

Propylenebisdithiocarbamates are made by reacting 1,2-propylenediamine and carbon bisulfide in the presence of NaOH. They are reacted in turn with zinc salts (zinc nitrate, e.g.) to give propineb.

Toxicity

The acute oral LD_{50} value for rats is 8,500 mg/kg which is insignificantly toxic.

References

(1) Worthing, C.R., *Pesticide Manual,* 6th ed., p. 451, British Crop Protection Council (1979).
(2) Spencer, E.Y., *Guide to the Chemicals Used in Crop Protection,* 6th ed., p. 432, London, Ontario, Agriculture Canada (January 1973).

PROPOXUR

Function: Insecticide (1)(2)(3)

Chemical Name: 2-(1-methylethoxy)phenyl methylcarbamate

Formula:

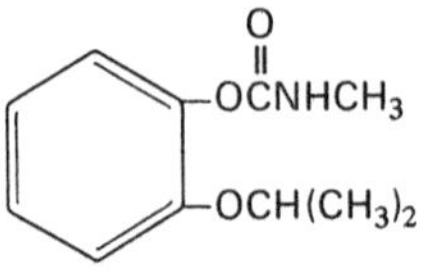

Trade Names: B-39,007 (Bayer)
58 12.315 (Bayer)
Baygon® (Bayer)
Blattanex® (Bayer)
Unden® (Bayer)
Sendran®
Suncide®
Tendex®
Undene®

Manufacture (3)

The following is a specific example of the preparation of propoxur. 23.25 grams of o-iso-propoxyphenol (0.15 mol) are treated in 15 ml of anhydrous dioxane with 8.63 grams of methyl isocyanate (0.152 mol) and 2 drops of triethylamine as catalyst. The solutions become warm soon, and upon cooling the product precipitates as a colorless crystalline substance. In order to complete the precipitation, the reaction mixture is stirred with 50 ml of petroleum ether. The crystalline product is filtered off, washed with ligroin and with water in order to remove any urea by-products, and dried at 50°C in a vacuum. After recrystallization from benzene, the product melts at 91°C. Yield: 26.3 grams, corresponding to 84% of the theoretical.

Toxicity

The acute oral LD_{50} for rats is 90 to 128 mg/kg which is moderately toxic.

Allowable Limits on Exposure and Use

Air: The threshold limit value for propoxur in air is 0.5 mg/m^3 as of 1979. The tentative short term exposure limit is 2.0 mg/m^3 (B-23).

References

(1) Worthing, C.R., *Pesticide Manual*, 6th ed., p. 452, British Crop Protection Council (1979).
(2) Spencer, E.Y., *Guide to the Chemicals Used in Crop Protection,* 6th ed., p. 433, London, Ontario, Agriculture Canada (January 1973).
(3) Bocker, E., Delfs, D., Unterstenhofer, G. and Behrenz, W., U.S. Patent 3,111, 539, November 19, 1963, assigned to Farbenfabriken Bayer AG.

PROTHIOCARB

Function: Fungicide (1)(2)

Chemical Name: S-Ethyl [3-(dimethylamino)propyl]carbamothioate

Formula: $(CH_3)_2NCH_2CH_2CH_2NHCOSCH_2CH_3$

Trade Names: SN-41,703 (Schering AG)
Previcur® (Schering AG)-U.K.
Dynone® (Schering AG)-South Africa

Manufacture (1)(2)

Prothiocarb is made by the reaction of N,N-dimethylpropane diamine with S-ethyl chlorothioformate. As solvents, there may be used, for example, inert organic liquids, such as ether or hydrocarbons. Suitable acid acceptors are tertiary organic amines, namely triethylamine or pyridine, etc., inorganic bases, namely alkali hydroxides or carbonates, etc., or also the amine required for the reaction, which is then used in correspondingly higher quantity. The reaction proceeds smoothly in a temperature range from about 0° to 100°C, but can be carried out also at higher or lower temperatures. The conversion may be conducted as a one-phase or two-phase reaction, the latter with the use of nonmiscible liquids, such as water and an organic solvent.

Toxicity

The acute oral LD_{50} value for rats is 1,300 mg/kg which is slightly toxic.

References

(1) Worthing, C.R., *Pesticide Manual,* 6th ed., p. 454, British Crop Protection Council (1979).
(2) Hoyer, G.A. and Pieroh, E.A., U.S. Patent 3,513,241, May 19, 1970, assigned to Schering AG.

PROTHOATE

Function: Acaricide and insecticide (1)(2)(3)

Chemical Name: O,O-diethyl S-[2-(1-methylethyl)amino-2-oxoethyl] phosphorodithioate

Formula:

$$(C_2H_5O)_2\overset{\underset{\parallel}{S}}{P}SCH_2\overset{\underset{\parallel}{O}}{C}NHCH(CH_3)_2$$

Trade Names: L-343 (Montecatini)
FAC® (Montecatini)
EI 18,682 (American Cyanamid)
Fostion®

Manufacture (4)

Prothoate may be made by condensing sodium diethylphosphorodithioate with N-isopropyl chloroacetamide. 1.58 kg of crude O,O-diethyl-hydrogen phosphorodithioate are dissolved in 2.5 liters of acetone (commercial grade), then 450 g of anhydrous Na_2CO_3 are added at a temperature not higher then 30°C, while stirring. Stirring is continued for half an hour, then the solution is cooled to 10°C. 920 g N-isopropylchloroacetamide dissolved in 1,000 cc acetone are added rapidly while cooling moderately to maintain the reaction mass at a temperature of about 10°C. After some time a precipitate consisting of NaCl is formed; stirring is continued and the above temperature is maintained for 12 hours, then the acetone is distilled off at normal pressure (about 65°C).

After distillation of the solvent is completed, the reaction mass is cooled and dipped into 6 liters of cold water. The oily phase is separated and shaken with a sodium bicarbonate solution if the pH is acidic, or with a carbon dioxide solution if the pH is alkaline, till the pH is 7. Finally, the mixture is allowed to separate completely and the oil is clarified by filtration through a pleated filter. 1,950 g of crude O,O-diethyl-S-(N-monoisopropyl)-carbamylmethyl-phosphorodithioate are thus obtained. The product comes as a straw-colored oil, d_4^{20} 1.1561 at 20°C, BP 152° to 155°C at 0.1 to 0.2 mm Hg, refractive index n_D^{20} 1.5182, insoluble in water and soluble in numerous organic solvents like monohydric aliphatic alcohols such as ethanol and methanol.

Process Wastes and Their Control

Product Wastes: Prothoate is analogous to dimethoate, but more stable in storage. It is stable in neutral, moderately acid or slightly alkaline media. Prothoate decomposed in approximately 48 hours at pH 9.2 and 50°C (B-3).

Toxicity

The acute oral LD_{50} value for rats is 8 to 9 mg/kg which is very highly toxic.

References

(1) Worthing, C.R., *Pesticide Manual*, 6th ed., p. 456, British Crop Protection Council (1979).
(2) Spencer, E.Y., *Guide to the Chemicals Used in Crop Protection,* 6th ed., p. 439, London, Ontario, Agriculture Canada (January 1973).
(3) Cassady, J.T., Hoegberg, E.I. and Gleissner, B.D., U.S. Patent 2,494,283, January 10, 1950, assigned to American Cyanamid Co.
(4) Montecatini Soc. Generale per L'Industria Mineraria e Chimica, British Patent 791,824 (March 12, 1958).

PYRACARBOLID

Function: Fungicide (1)(2)(3)

Chemical Name: 3,4-dihydro-6-methyl-N-phenyl-2H-pyran-5-carboxamide

Formula:

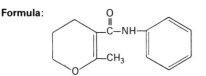

Trade Names: HOE 2989 (Farbwerke Hoechst)
HOE 6052 (Farbwerke Hoechst)
HOE 6053 (Farbwerke Hoechst)
HOE 137640F (Farbwerke Hoechst)
Sicarol® (Farbwerke Hoechst)

Manufacture (3)

14.2 grams of 2-methyl-5,6-dihydropyran-3-carboxylic acid were dissolved in 70 ml of absolute benzene and 8.5 ml of thionyl chloride were added at 22°C. After standing for 2½ hours
the reaction mixture was concentrated to one-fourth of the original volume under reduced
pressure. The crude acid chloride thus obtained was dropped over a period of 40 minutes
into a solution of 9.3 grams of aniline and 9.5 grams of pyridine in 100 ml of benzene, which
solution was stirred at −5°C. The reaction mixture was then stirred for another hour without
cooling and finally poured onto ice. The benzenic solution was separated and washed with
water. 16.3 grams of 2-methyl-5,6-dihydropyran-3-carboxylic acid anilide were obtained
from the dried benzenic solution, corresponding to a yield of 75% of the theoretical. After recrystallization from methanol/water with the addition of active charcoal the compound
melted at 108° to 109°C.

Toxicity

The acute oral LD$_{50}$ vaue for rats is 15,000 mg/kg which is insignificantly toxic.

References

(1) Worthing, C.R., *Pesticide Manual,* 6th ed., p. 457, British Crop Protection Council (1979).
(2) Spencer, E.Y., *Guide to the Chemicals Used in Crop Protection,* 6th ed., p. 440a, London,
Ontario, Agriculture Canada (January 1973).
(3) Scherer, O. and Heubach, G., U.S. Patent 3,632,821, January 4, 1972, assigned to Farbwerke Hoechst AG.

PYRAZOPHOS

Function: Fungicide (1)(2)

Chemical Name: Ethyl 2-[(diethoxyphosphinothioyl)oxy]-5-methylpyrazolo[1,5-a]pyrimi-
dine-6-carboxylate

Formula:

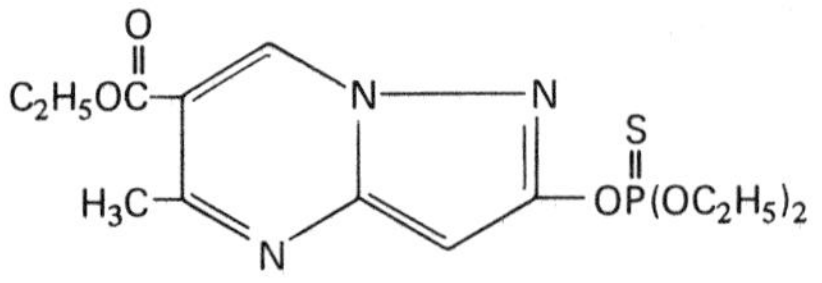

$$C_2H_5O\overset{\overset{O}{\|}}{C}\text{—} \quad \text{(pyrazolopyrimidine ring)} \quad H_3C\text{—} \quad \text{—}OP(OC_2H_5)_2 \;\; \overset{S}{\|}$$

Trade Names: HOE 2783 (Farbwerke Hoechst)
Afugan® (Farbwerke Hoechst)
Curamil® (Farbwerke Hoechst)

Manufacture (2)(3)

A mixture of 24.3 grams (0.1 mol) of the sodium salt of 2-hydroxy-5-methyl-6-carbethoxy-pyrazolopyrimidine, 18.9 grams (0.1 mol) O,O-diethyl-thiophosphoryl chloride and 200 cc acetone was stirred at about 50°C for 8 to 10 hours. From the filtrate which had been freed from NaCl, 37 grams 2-(O,O-diethyl-thiono-phosphoryl)-5-methyl-6-carbethoxypyrazolo-pyrimidine were obtained in the form of a yellow oil (MP 38° to 40°C).

Toxicity

The acute oral LD_{50} value for rats is 140 mg/kg (B-5) which is moderately toxic.

References

(1) Worthing, C.R., *Pesticide Manual,* 6th ed., p. 458, British Crop Protection Council (1979).
(2) Scherer, O. and Mildenberger, H., U.S. Patent 3,496,178, February 17, 1970, assigned to Farbwerke Hoechst AG.
(3) Scherer, O. and Mildenberger, H., U.S. Patent 3,632,757, January 4, 1972, assigned to Farbwerke Hoechst AG.

PYRETHRINS OR PYRETHRUM

Function: Insecticide (1)(2)

Chemical Name: 4-hydroxy-3-methyl-2-(2,4-pentadienyl)-2-cyclopentene-1-one and 2-(2-bu-tenyl)-4-hydroxy-3-methyl-2-cyclopentene-1-one esters with 2,2-dimethyl-3-(2-methyl-1-propenyl)cyclopropanecarboxylic acid and 3-(3-methoxy-2-methyl-3-oxo-1-propenyl) 2,2-dimethylcyclopropanecarboxylic acid

Formula:

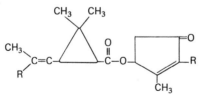

Trade Name: None

Manufacture

Chrysanthemum flowers $\xrightarrow{\text{Solvent}}$ Crude Extract $\xrightarrow[\text{with } CH_3OH]{\text{Extract}}$ $\xrightarrow[\text{with } C_6H_{14}]{\text{Extract}}$ Pyrethrum concentrate

A pyrethrum refinery and waste schematic is shown in Figure 47 (3).

Figure 47: Pyrethrum Refinery and Waste Schematic

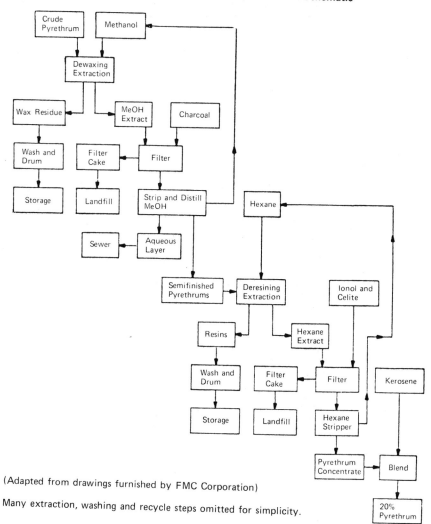

(Adapted from drawings furnished by FMC Corporation)

Many extraction, washing and recycle steps omitted for simplicity.

Source: Reference (3).

Process Wastes and Their Control

Air: Air emissions from pyrethrum production have been estimated (B-15) to be 0.5 kg of particulates and 0.5 kg of hydrocarbons per metric ton of pesticide produced.

Product Wastes: Pyrethrin is highly unstable in the presence of light, moisture, and air. It is rapidly oxidized and inactivated by air. Most of the insecticidal activity of the product is destroyed by minor changes in the molecule.

Pyrethrin products are not apt to be disposal candidates. It could be dumped into a landfill, or buried in noncrop land away from water. In each of these cases it would be better to mix the product with lime. Incineration would be an effective disposal procedure where permitted. If an efficient incinerator is not available, the product should be mixed with large amounts of combustible material and contact with the smoke should be avoided (B-3).

Toxicity

The acute oral LD_{50} value for rats is 600 to 900 mg/kg which is slightly toxic.

Allowable Limits on Exposure and Use

Air: The threshold limit value for pyrethrum in air is 5 mg/m^3 as of 1979. The tentative short term exposure limit is 10 mg/m^3 (B-23).

Product Use: The tolerances set by the EPA for pyrethrum in or on raw agricultural commodities are as follows:

	40 *CFR* Reference	Parts per Million
Almonds (post-h)	180.128	1.0
Apples (post-h)	180.128	1.0
Barley (post-h)	180.128	3.0
Beans (post-h)	180.128	1.0
Birdseed mixtures (post-h)	180.128	3.0
Blackberries (post-h)	180.128	1.0
Blueberries (huckleberries) (post-h)	180.128	1.0
Boysenberries (post-h)	180.128	1.0
Buckwheat (post-h)	180.128	3.0
Cattle, fat	180.128	0.1 N
Cattle, mbyp	180.128	0.1 N
Cattle, meat	180.128	0.1 N
Cherries (post-h)	180.128	1.0
Cocoa beans (post-h)	180.128	1.0
Copra (post-h)	180.128	1.0
Corn (post-h)	180.128	3.0
Cotton, seed (post-h)	180.128	1.0
Crabapples (post-h)	180.128	1.0
Currants (post-h)	180.128	1.0
Dewberries (post-h)	180.128	1.0
Eggs	180.128	0.1 N
Figs (post-h)	180.128	1.0
Flax, seed (post-h)	180.128	1.0
Goats, fat	180.128	0.1 N
Goats, mbyp	180.128	0.1 N
Goats, meat	180.128	0.1 N
Gooseberries (post-h)	180.128	1.0
Grapes (post-h)	180.128	1.0
Guavas (post-h)	180.128	1.0
Hogs, fat	180.128	0.1 N
Hogs, mbyp	180.128	0.1 N
Hogs, meat	180.128	0.1 N
Horses, fat	180.128	0.1 N
Horses, mbyp	180.128	0.1 N
Horses, meat	180.128	0.1 N
Loganberries (post-h)	180.128	1.0
Mangoes (post-h)	180.128	1.0
Milk, fat	180.128	0.5 N
Muskmelons (post-h)	180.128	1.0

(continued)

	40 *CFR* Reference	Parts per Million
Oats (post-h)	180.128	1.0
Oranges (post-h)	180.128	1.0
Peaches (post-h)	180.128	1.0
Peanuts, shell removed (post-h)	180.128	1.0
Pears (post-h)	180.128	1.0
Peas (post-h)	180.128	1.0
Pineapples (post-h)	180.128	1.0
Plums (fresh prunes) (post-h)	180.128	1.0
Potatoes (post-h)	180.128	0.05
Poultry, fat	180.128	0.2
Poultry, mbyp	180.128	0.2
Poultry, meat	180.128	0.2
Raspberries (post-h)	180.128	1.0
Rice (post-h)	180.128	3.0
Rye (post-h)	180.128	3.0
Sheep, fat	180.128	0.1 N
Sheep, mbyp	180.128	0.1 N
Sheep, meat	180.128	0.1 N
Sorghum, grain (post-h)	180.128	1.0
Tomatoes (post-h)	180.128	1.0
Walnuts (post-h)	180.128	1.0
Wheat (post-h)	180.128	

The tolerances set by the EPA for pyrethrum in animal feeds are as follows (the *CFR* Reference is to Title 21):

	CFR Reference	Parts per Million
Feeds, dried (from cotton bags)	561.340	1.0
Feeds, dried (from paper bags)	561.340	1.0

The tolerances set by the EPA for pyrethrum in food are as follows (the *CFR* Reference is to Title 21):

	CFR Reference	Parts per Million
Cereal grains (milled fractions)	193.390	1.0
Dried foods (2-ply bag-cloth & waxpaper)	193.390	1.0
Dried foods in paper bags (multiwalled)	193.390	1.0
Processed food	193.390	1.0

References

(1) Worthing, C.R., *Pesticide Manual,* 6th ed., p. 459, British Crop Protection Council (1979).
(2) Spencer, E.Y., *Guide to the Chemicals Used in Crop Protection,* 6th ed., p. 442, London, Ontario, Agriculture Canada (January 1973).
(3) Midwest Research Institute, *The Pollution Potential in Pesticide Manufacturing,* Wash., DC, Environmental Protection Agency (June 1972).

PYRIDINITRIL

Function: Fungicide (1)

Chemical Name: 2,6-dichloro-4-phenylpyridine-3,5-dicarbonitrile

Formula:

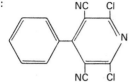

Trade Names: IT3296 (E. Merck AG)
Cilvan® (E. Merck AG)

Manufacture (2)

Benzaldehyde is first condensed with two molecules of ethylcyanoacetate and then with ammonia. The intermediate thus produced is chlorinated to give pyridinitril. The chlorination step has been described as follows (2): 23.7 g dry 3,5-dicyano-6-hydroxy-4-phenyl-2-pyridone [produced according to *Journal of the Chemical Society* (London) vol. 117 (1920), p. 1473] or 25.5 g of its ammonium salt in 150 ml chlorobenzene are reacted with 41.7 g phosphorus pentachloride and boiled two hours. The dark red reaction mixture is poured over ice, the organic phase separated, washed to neutrality with sodium bicarbonate solution, dried over sodium sulfate and concentrated. The yield is 12.0 g (44% of the theoretical) 2,6-dichloro-3,5-dicyano-4-phenyl pyridine melting at 192° to 196°C. After recrystallization from benzene or ethanol, colorless crystals melting at 203° to 204°C are obtained.

Toxicity

The acute oral LD_{50} value for rats is over 5,000 mg/kg which is insignificantly toxic.

References

(1) Martin, H. and Worthing, C.R., *Pesticide Manual,* 5th ed., p. 454, British Crop Protection Council (January 1977).
(2) Mohr, G., Niethammer, K., Lust, S. and Schneider, G., U.S. Patent 3,468,895, September 23, 1969, assigned to E. Merck AG.

Q

QUINACETOL SULFATE

Function: Fungicide (1)(2)

Chemical Name: 1-(8-Hydroxy-5-quinolinyl)ethanone (2:1) sulfate

Formula:

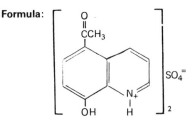

Trade Names: G-20,072 (Ciba-Geigy)
Fongoren® (Ciba-Geigy)

Manufacture (3)

8-Hydroxyquinoline is reacted with acetyl chloride to give 5-acetyl-8-hydroxyquinoline which combines with H_2SO_4 to give quinacetol sulfate.

Toxicity

The acute oral LD_{50} value for rats is 1,600 to 2,200 mg/kg which is slightly toxic.

References

(1) Martin, H. and Worthing, C.R., *Pesticide Manual,* 5th ed., p. 455, British Crop Protection Council (January 1977).
(2) Hodel, E. and Gatzl, K., U.S. Patent 3,759,719, September 18, 1973, assigned to Ciba-Geigy Corp.
(3) Matsumura, K., *J. Am. Chem. Soc.,* 52, 4433-36 (1930).

QUINALPHOS

Function: Insecticide and acaricide

Chemical Name: O,O-diethyl O-2-quinoxalinyl phosphorothioate

Formula:

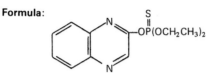

Trade Names: Sandoz 6538 e.c. (Sandoz, Ltd.)
Sandoz 6626 g (Sandoz, Ltd.)
Ekalux® (Sandoz, Ltd.)
Bayer 77,049 (Bayer)
Bayrusil® (Bayer)

Manufacture (1)

Quinalphos may be made by the condensation of o-phenylenediamine with chloroacetic acid. The intermediate so produced is then reacted with O,O-diethylphosphorochloridothioate to give quinalphos.

Toxicity

The acute oral LD_{50} value for rats is 60 to 140 mg/kg which is moderately toxic.

Reference

(1) Worthing, C.R., *Pesticide Manual,* 6th ed., p. 462, British Crop Protection Council (1979).

QUINTOZENE

Function: Fungicide (1)(2)

Chemical Name: Pentachloronitrobenzene

Formula:

Trade Names: Brassicol® (Farbwerke Hoechst)
Tritisan® (Farbwerke Hoechst)
Folosan® (Olin Mathieson Chemical Corp.)
Terraclor® (Olin Mathieson Chemical Corp.)
Botrilex®
Tilcarex® (Bayer)

Manufacture

By the chlorination of nitrobenzene at 60° to 70°C using iodine as a catalyst.

Process Wastes and Their Control

Product Wastes: It has been observed that the product decomposes readily when burned with polyethylene. The compound is highly stable in soil, as would be expected on the basis of the polychlorinated aromatic structure (B-3).

Toxicity

Pentachloronitrobenzene is a solid soil fungicide and is relatively nontoxic (acute oral LD_{50} value for rats = 12,000 mg/kg) but may cause skin irritation on repeated contact (B-3).

Although without effect in the rat and dog, PCNB appears to be carcinogenic in two strains of mice. In light of the above and taking into account the carcinogenic risk projections, it is suggested that very strict criteria be applied when limits for PCNB in drinking water are established.

A lower limit of detection of PCNB by gas chromatography was 0.01 ppm. A temporary acceptable daily intake of PCNB has been established for humans by the WHO at 0.001 mg/kg.

Most of the subchronic- and chronic-toxicity studies on PCNB have used technical-grade material, which normally contains about 1.8% HCB, but in some cases as much as 11% HCB. It is therefore not clear whether HCB and other impurities significantly contribute to the observed toxicity of PCNB. Moreover, some of the studies have involved PCNB formulations containing relatively low concentrations of the fungicide. The subchronic and chronic studies, particularly the latter, should be repeated in two species with pure PCNB. Such studies are particularly warranted, because of the suspected carcinogenicity of PCNB. Additional long-term oncogenic studies should also be conducted in susceptible strains of mice and other experimental animals. In addition, the FAO/WHO has recommended further short-term studies to elucidate the difference in teratogenic activity between rats and mice; studies to explain the effects on the liver and bone marrow of dogs; and further studies on the toxicity of PCNB metabolites (B-22).

Allowable Limits on Exposure and Use

Product Use: A rebuttable presumption against registration was issued on October 13, 1977 by EPA on the basis of oncogenicity.

The tolerances set by the EPA for PCNB in or on raw agricultural commodities are as follows:

	40 *CFR* Reference	Parts per Million
Cotton, seed	180.291	0.1 N

References

(1) Worthing, C.R., *Pesticide Manual,* 6th ed., p. 465, British Crop Protection Council (1979).
(2) Spencer, E.Y., *Guide to the Chemicals Used in Crop Protection,* 6th ed., p. 445, London, Ontario, Agriculture Canada (January 1973).

R

RESMETHRIN

Function: Insecticide (1)(2)(3)(4)(5)(6)

Chemical Name: [5-phenylmethyl-3-furan] methyl-2,2-dimethyl-3-(2-methyl-1-propenyl) cyclopropane carboxylate

Formula:

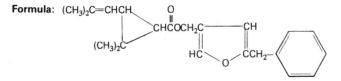

Trade Names: NRDC 104 (Nat. Research Dev. Corp.)
SBP 1382 (Penick)
FMC 17,370 (FMC)
Benzyfuroline
Chryson® (Sumitomo Chemical Co. Ltd.)
Synthrin®

Manufacture

Resmethrin is made by the esterification of 5-benzyl-3-furylmethyl alcohol with chrysanthemum monocarboxylic acid chloride. In a specific example (6), 5-benzyl-3-furylmethyl alcohol (1.1 mol) as a 10% solution in ether is added with cooling and stirring to (±)-cis,trans-chrysanthemoyl chloride (1.0 mol) as a 10% solution in benzene.

Pyridine (1.0 mol) is added and the mixture set aside overnight. Water is then added to the organic layer, which is washed with dilute sulfuric acid, with saturated aqueous potassium hydrogen carbonate and twice with saturated sodium chloride and is finally dried, evaporated and distilled to give 5-benzyl-3-furylmethyl-(±)-cis,trans-chrysanthemate; BP is 169° to 172°C at 2.5×10^{-2} mm.

The preparation of the 5-benzyl-3-furylmethyl alcohol involves a multistep synthesis (6) as follows. Tetracarboxyfuran is first prepared as follows. Bromine (19 ml) in chloroform is added over 1 hour to ethyl sodiooxalacetate (175 g) in chloroform (40 ml) at 0° to 10°C. After washing with water (4 × 300 ml) the chloroform solution is dried (Na_2SO_4) and evaporated. The residues from four such experiments after recrystallization from ethanol (800 ml) at −20°C give the required product (317 g). Concentration of the mother liquor and cooling to −20°C gives a further 80 g of product. Total yield 397 g (63%), MP 79°C.

The above tetracarboxylate (100 g) is added with stirring during 5 minutes to sulfuric acid (300 ml). After being heated to 50°C for 5 minutes, the product is cooled to room temperature and added to ice (1,000 g). The combined product from four such experiments is taken into ether. The ether is washed twice with saturated sodium chloride solution, dried (Na_2SO_4) and evaporated. The residue is refluxed with glacial acetic acid (600 ml), constant boiling hydrobromic acid (400 ml) and water (200 ml) for 5 hours and then evaporated.

The residue is recrystallized from acetic acid (350 ml) and chloroform (600 ml) then washed with chloroform and dried in vacuo. Yield, 216 g, MP 233° to 238°C (decomposed).

3-Furoic acid is prepared in the next step. Tetracarboxyfuran (20 g) is heated with copper powder (1 g) in a fused salt bath. Evolution of carbon dioxide is controlled by removal of the heating bath when necessary (temperature 250° to 290°C). At 290°C this product (5.8 g) distills. Repetition gives a total of 40 g of product which is recrystallized from water to give pure acid (30 g), MP 114° to 118°C.

Methyl-3-furoate is next prepared. 3-Furoic acid (41.5 g) is refluxed with methanol (190 ml) and sulfuric acid (3.75 ml) for 4 hours. Most of the methanol is evaporated, the residue poured into water and the ester taken into ether. After washing (saturated potassium hydrogen carbonate, and saturated sodium chloride) and drying (Na_2SO_4) the solvent is evaporated and the product distilled, BP 80°C at 30 mm, n_D^{20} 1.4640 (33 g).

Methyl-5-chloromethyl 3-furoate is the product of the next step. Dry hydrogen chloride is passed into a mixture of methyl 3-furoate (15 g), paraformaldehyde (4.2 g) and zinc chloride (4.2 g) in chloroform (90 ml) maintained at 20° to 25°C for 1½ hours. The product is shaken with water, more chloroform added and the combined organic layers washed (3× H_2O), dried (Na_2SO_4) and evaporated. Distillation of the residue gives methyl-5-chloromethyl 3-furoate (9.4 g), BP 80° to 108°C, n_D^{20} 1.5003 to 1.5072, MP 42° to 51°C.

Methyl 5-benzyl-3-furoate is next prepared. Aluminum trichloride (9.84 g freshly sublimed) is added with stirring to methyl 5-chloromethyl 3-furoate (10.9 g) in benzene while the temperature is maintained below 20°C for 50 minutes. Water (100 ml) is added dropwise with cooling (to below 30°C) and the product taken up in ether, washed twice with 10% sodium hydroxide and twice with sodium chloride and finally dried (Na_2SO_4), evaporated and distilled. The methyl 5-benzyl-3-furoate (8.7 g) has BP 127° to 135°C at 0.4 mm, MP 52° to 53°C.

5-Benzyl-3-furylmethyl alcohol finally results when methyl 5-benzyl-3-furoate (6.64 g) in dry ether (90 ml) is added dropwise to lithium aluminum hydride (1.35 g) in ether (135 ml). The mixture is then warmed on steam, cooled and decomposed by addition of water. The ethereal solution is dried (Na_2SO_4), evaporated and distilled to give 5-benzyl-3-furylmethyl alcohol (4.84 g), BP 151° to 156°C at 1.5 mm, MP 36° to 39°C.

Process Wastes and Their Control

Product Wastes: This product, a pyrethrum analog, decomposes fairly rapidly on exposure to air and light. See pyrethrum for disposal recommendations (B-3).

Toxicity

The acute oral LD_{50} value for rats is 1,500 mg/kg (B-5) which is slightly toxic.

References

(1) Worthing, C.R., *Pesticide Manual,* 6th ed., p. 467, British Crop Protection Council (1979).
(2) Spencer, E.Y., *Guide to the Chemicals Used in Crop Protection,* 6th ed., p. 446, London, Ontario, Agriculture Canada (January 1973).
(3) Elliott, M. and Janes, N.F., British Patent 1,168,799, October 29, 1969, assigned to National Research Development Corp.
(4) Elliott, M. and Janes, N.F., British Patent 1,168,798, October 29, 1969, assigned to National Research Development Corp.
(5) Elliott, M., British Patent 1,168,797, October 29, 1969, assigned to National Research Development Corp.
(6) Elliott, M., U.S. Patent 3,465,007, September 2, 1969, assigned to National Research Development Corp.

RONNEL (U.S. and Canada)
FENCHLORFOS (Elsewhere)

Function: Insecticide (1)(2)

Chemical Name: O,O-dimethyl O-2,4,5-trichlorophenyl phosphorothioate

Formula:

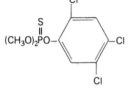

Trade Names: Dow ET-57 (Dow Chemical Co.)
Dow ET-14 (Dow Chemical Co.)
Trolene® (Dow Chemical Co.)
Korlan® (Dow Chemical Co.)
Nankor®
Viozene®

Manufacture (3)(4)

The feed materials for ronnel manufacture are 2,4,5-trichlorophenyl dichlorothiophosphate and sodium methoxide. The reaction proceeds as follows:

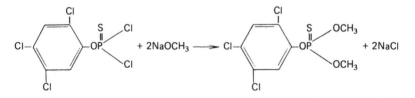

Good results are obtained when using essentially stoichiometric proportions of the reactants. The reaction may be accomplished by adding either reactant gradually to the other reactant dispersed in the reaction solvent.

The O-(2,4,5-trichlorophenyl)dichlorothiophosphate as a starting material may be prepared by reacting a molecular excess of thiophosphoryl chloride ($PSCl_3$) with an alkali metal 2,4,5-trichlorophenate. Good results are obtained when employing from 2 to 4 mols of thiophosphoryl chloride per mol of 2,4,5-trichlorophenate. The phenate, preferably as the sodium salt, is added portionwise with stirring to the thiophosphoryl chloride and the mixture subsequently warmed for a short time to complete the reaction. The crude reaction mixture is then filtered and the filtrate fractionally distilled under reduced pressure to separate the product.

This reaction is preferably carried out at a temperature from 10° to 85°C. The reaction is somewhat exothermic, and temperature control is maintained by regulation of the rate of addition of the reactants, as well as by the addition and subtraction of heat, if required. This reaction is carried out at atmospheric pressure. This reaction is quite rapid and a reaction period of 10 minutes is quite adequate following a 5-minute period of reactant addition.

This reaction is carried out in an organic solvent and conveniently in the alcohol employed in the preparation of the alcoholate, according to C.L. Moyle (3)(4). No catalyst is employed in this reaction. A stirred, jacketed reaction vessel of conventional design may be used for the conduct of this reaction.

Upon completion of the reaction, the solvent is removed from the crude mixture by evaporation and the residue dispersed in a nonreactive organic solvent such as methylene dichloride, carbon tetrachloride, or diethyl ether. The resultant mixture may be successively washed with dilute aqueous sodium hydroxide and water, and dried with anhydrous sodium sulfate. The separation of the product is then accomplished by evaporation of the solvent. If desired, the product may be further purified by fractional distillation under reduced pressure.

Process Wastes and Their Control

Air: Air emissions from ronnel manufacture have been estimated (B-15) to be as follows:

Component	Kilograms per Metric Ton Pesticide Produced
SO_2	0.3
NO_x	0.3
Hydrocarbons	2.0
Ronnel	0.5

Product Wastes: This compound is stable in both neutral and acid media at 60°C. It is incompatible with alkaline pesticides and is hydrolyzed in weakly alkaline medium to the mono O-methyl compound. Hydrolysis in strongly alkaline media yield mainly $(CH_3O)_2P(S)OH$, and 2,4,5-trichlorophenol (LD_{50} 150 to 250 mg/kg). Ronnel is saponified with concentrated HCl to produce 2,4,5-trichlorophenol (B-3).

Toxicity

The acute oral LD_{50} value for rats is 1,740 mg/kg which is slightly toxic.

Allowable Limits on Exposure and Use

Air: The threshold limit value for ronnel in air is 10 mg/m³ as of 1979 (B-23).

Product Use: The tolerances set by the EPA for ronnel in or on raw agricultural commodities are as follows:

	40 *CFR* Reference	Parts per Million
Bananas, pulp	180.177	0.0
Bananas, whole	180.177	0.5
Cattle, fat	180.177	10.0
Cattle, mbyp	180.177	4.0
Cattle, meat	180.177	4.0
Eggs	180.177	0.03
Goats, fat	180.177	10.0
Goats, mbyp	180.177	4.0
Goats, meat	180.177	4.0
Hogs, fat	180.177	3.0
Hogs, mbyp	180.177	2.0
Hogs, meat	180.177	2.0
Milk, fat (= N in whole milk)	180.177	1.25
Poultry, fat	180.177	0.01
Poultry, mbyp	180.177	0.01
Poultry, meat	180.177	0.01
Sheep, fat	180.177	10.0
Sheep, mbyp	180.177	4.0
Sheep, meat	180.177	4.0

Issuance of a rebuttable presumption against registration for ronnel was being considered by EPA as of September 1979 on the bases of possible oncogenicity, teratogenicity and feto-toxic effects.

References

(1) Worthing, C.R., *Pesticide Manual,* 6th ed., p. 260, British Crop Protection Council (1979).
(2) Spencer, E.Y., *Guide to the Chemicals Used in Crop Protection,* 6th ed., p. 447, London, Ontario, Agriculture Canada (January 1973).
(3) Moyle, C.L., U.S. Patent 2,599,515, June 3, 1952, assigned to Dow Chemical Co.
(4) Moyle, C.L., U.S. Patent 2,599,516, June 3, 1952, assigned to Dow Chemical Co.

ROTENONE

Function: Insecticide and acaricide (1)(2)

Chemical Name: [2R-(2a,6aα,12aα)]-1,2,12,12a-tetrahydro-8,9-dimethoxy-2-(1-methyl-ethenyl)[l]benzopyrano[3,4-b]furo[2,3-h]benzopyran-6(6aH)-one

Formula:

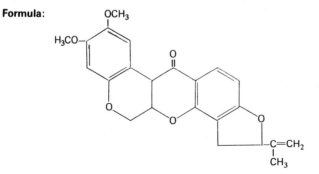

Trade Names: Derrin
Tubatoxin
Nicouline

Manufacture

Rotenone is made by extraction of *Derris* or *Lonchocarpus* roots and crystallization from the extracts.

Process Wastes and Their Control

Product Wastes: Rotenone is decomposed by light and alkali to less insecticidal products. It is readily detoxified by the action of light and air. It is also detoxified by heating; 2 hours at 100°C results in 76% decomposition. Oxidation products are probably nontoxic. Incineration has been recommended as a disposal procedure. Burial with lime would also present minimal danger to the environment (B-3).

Toxicity

The acute oral LD_{50} value for rats is 132 mg/kg (B-5) which is moderately toxic.

Allowable Limits on Exposure and Use

Air: The threshold limit value for commercial rotenone in air is 5 mg/m^3 as of 1979. The tentative short term exposure limit is 10 mg/m^3 (B-23).

Product Use: Issuance of a rebuttable presumption against registration for rotenone was being considered by EPA as of September 1979 on the basis of possible oncogenicity.

References

(1) Worthing, C.R., *Pesticide Manual,* 6th ed., p. 468, British Crop Protection Council (1979).
(2) Spencer, E.Y., *Guide to the Chemicals Used in Crop Protection,* 6th ed., p. 167, 448, London, Ontario, Agriculture Canada (January 1973).

RYANODINE

Function: Insecticide (1)(2)(3)(4)

Chemical Name: Ryania

Formula: Ryanodine, the principal insecticidal component of ryania has the formula $C_{25}H_{35}NO_9$; it is an ester of pyrrole-2-carboxylic acid.

Trade Names: Ryanex®
Ryanicide®
Ryanexcel®

Manufacture (3)(4)

The ground stem of *Ryania speciosa,* a shrub native to Trinidad and the Amazon Basin, is extracted with water, chloroform or methanol to give ryania. Material for use in insecticidal dusts and sprays may also be obtained (4) by merely comminuting, grinding, pulverizing or otherwise finely dividing various parts such as leaves, stems and roots of plants of the genus *Ryania*. When the comminuted plant materials are mixed with solid diluents such as powdered talc or with a liquid diluent or carrier such as water, dusts and sprays are obtained which are both safe to use and highly effective for insect control.

Process Wastes and Their Control

Product Wastes: Ryania is considerably more stable to heat and atmospheric oxidation than pyrethrin. It is stable to light and does not decompose on storage. Incineration (see pyrethrum) or burial with lime are effective disposal methods (B-3).

Toxicity

The acute oral LD_{50} value for rats is 750 to 1,000 mg/kg which is slightly toxic.

References

(1) Worthing, C.R., *Pesticide Manual,* 6th ed., p. 469, British Crop Protection Council (1979).
(2) Spencer, E.Y., *Guide to the Chemicals Used in Crop Protection,* 6th ed., p. 450, London, Ontario, Agriculture Canada (January 1973).
(3) Folkers, K., Rogers, E. and Heal, R.E., U.S. Patent 2,400,295, May 14, 1946, assigned to Merck and Company, Inc.
(4) Heal, R.E., U.S. Patent 2,590,536, March 25, 1952, assigned to Merck and Company, Inc.

S

SABADILLA

Function: Insecticide (1)(2)(3)(4)

Chemical Name: Veratrine

Formula: A mixture of alkaloids

Trade Name: Cevadilla

Manufacture

The sabadilla plant, and particularly the seed thereof, has toxic properties which are not in the natural state of the plant sufficiently potent to insects to make a suitable commercial insecticide, but which are activated or rendered highly potent by the application of heat.

The seed is finely powdered or ground or comminuted, and then heated sufficiently above normal temperature and for a sufficient period of time to develop or enhance the toxic potency to its maximum, and then allowed to cool down to normal temperature. A temperature of about 150°C for a period of about one hour has been found to be most effective.

The application of heat to the powdered seed may be accomplished generally in two ways and with any suitable apparatus. If the material is to be used as an insecticide in dry or dusting powder form, then the powder itself can be heated in a suitable chamber and allowed to cool, after which it is ready for use. On the other hand, if the material is to be used as the toxic ingredient of a liquid or spray type of insecticide, the powdered seed is mixed with a solvent or vehicle such as kerosene, and the mixture heated to the desired degree and for the proper length of time.

The extract thus formed is then filtered either hot or cooled. Any of the solvents, such as petroleum hydrocarbon of the kerosene type, commonly used for commercial spray insecticides, may be used as a solvent or vehicle for the powdered sabadilla seed.

Process Wastes and Their Control

Product Wastes: The alkaloids present in this product are rapidly destroyed by the action of light. Careful incineration is an effective disposal procedure. Burial with lime is not recommended because the product is relatively stable toward alkali and is frequently formulated with lime (B-3).

Toxicity

The acute oral LD_{50} value for rats is 4,000 mg/kg which is very slightly toxic.

References

(1) Martin, H. and Worthing, C.R., *Pesticide Manual,* 5th ed., p. 464, British Crop Protection Council (January 1977).
(2) Spencer, E.Y., *Guide to the Chemicals Used in Crop Protection,* 6th ed., p. 451, London, Ontario, Agriculture Canada (January 1973).
(3) Allen, T.C. and Dicke, R.J., U.S. Patent 2,348,949, May 16, 1944, assigned to Wisconsin Alumni Research Foundation.
(4) Allen, T.C. and Harris, H.H., U.S. Patent 2,390,911, December 11, 1945, assigned to Wisconsin Alumni Research Foundation.

SCHRADAN

Function: Insecticide (1)(2)

Chemical Name: Octamethyldiphosphoramide

Formula:
$$[(CH_3)_2N]_2 \overset{\overset{O}{\|}}{P}-O-\overset{\overset{O}{\|}}{P}[N(CH_3)_2]_2$$

Trade Names: Pestox 3® (Fisons, Ltd.)
Sytam® (Murphy Chemical Co.)
OMPA

Manufacture (3)

Tetramethylphosphorodiamic chloride is first produced by the reaction of dimethylamine and phosphoryl chloride, $POCl_3$. Two molecules of that intermediate are then condensed with an alkali metal hydroxide or an amine as an alkaline agent. The following is an example of the second or condensation step.

To 94.5 g (0.554 mol) bis(dimethylamido)phosphoryl chloride there were added 10 g (0.55 mol, 100% excess) water. The temperature rose to 39°C. Then 56.5 g (0.559 mol, 1% excess) of triethylamine were added in two minutes. The temperature was maintained between 40° and 45°C for two hours. Then 100 ml carbon tetrachloride were added and the slurry stirred for one hour. It was filtered to remove the solid amine hydrochloride salt. The carbon tetrachloride was removed by distillation. The liquid residue, octamethylpyrophosphoramide, weighed 73.2 g representing a 92.5% yield, with a refractive index of 1.4620 at 25°C. Upon distillation the crude liquid yielded 84% of substantially pure product.

Process Wastes and Their Control

Product Wastes: This compound is stable in aqueous solutions. Fifty percent hydrolysis at 25°C requires 100 years in neutral solution, 70 days in 1N NaOH, and 200 minutes in 1N HCl. In water, an excess of chlorine leads to complete breakdown to nontoxic compounds (B-3).

Toxicity

The acute oral LD_{50} value for rats is 9 mg/kg which is highly toxic.

Allowable Limits on Exposure and Use

Product Use: As noted by EPA on May 28, 1976 (B-17), schradan (OMPA) was the subject of voluntary cancellation of all products.

The tolerances set by the EPA for OMPA in or on raw agricultural commodities is as follows:

	40 *CFR* Reference	Parts per Million
Walnuts, English	180.166	0.75

References

(1) Worthing, C.R., *Pesticide Manual,* 6th ed., p. 470, British Crop Protection Council (1979).
(2) Spencer, E.Y., *Guide to the Chemicals Used in Crop Protection,* 6th ed., p. 454, London, Ontario, Agriculture Canada (January 1973).
(3) Toy, A.D.F. and Costello, J.F., Jr., U.S. Patent 2,717,249, September 6, 1955, assigned to Victor Chemical Works.

SECBUMETON

Function: Herbicide (1)(2)

Chemical Name: N-ethyl-6-methoxy-N'-(1-methyl-propyl)-1,3,5-triazine-2,4-diamine

Formula:

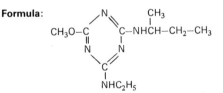

Trade Names: GS-14,254 (Ciba-Geigy)
Etazine® (Ciba-Geigy)
Sumitol® (Ciba-Geigy)

Manufacture

Secbumeton is made by reacting cyanuric chloride with sec-butylamine and ethylamine in sequence and then with methanol and caustic.

Toxicity

The acute oral LD_{50} value for rats is 2,680 mg/kg which is slightly toxic.

Allowable Limits on Exposure and Use

Product Use: The tolerances set by the EPA for secbumeton in or on raw agricultural commodities are as follows:

	40 *CFR* Reference	Parts per Million
Sugarcane	180.323	0.25 N

References

(1) Worthing, C.R., *Pesticide Manual,* 6th ed., p. 471, British Crop Protection Council (1979).
(2) Spencer, E.Y., *Guide to the Chemicals Used in Crop Protection,* 6th ed., p. 61, London, Ontario, Agriculture Canada (January 1973).

SESAMEX

Function: Insecticidal synergist (1)(2)

Chemical Name: 5-[1-[2-(2-ethoxyethoxy)ethoxy]ethoxy]-1,3-benzodioxole

Formula:

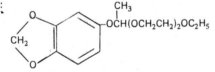

Trade Names: Sesoxane® (Shulton, Inc.)

Manufacture (3)

Sesamex may be made by the acetylation of piperonal as a first step followed by hydrolysis and condensation of the hydrolysis product with vinyl ethyl Carbitol. Alternatively, 3,4-methylenedioxyphenol may be reacted with 2-(2-ethoxyethoxy) ethylene to give sesamex.

Toxicity

The acute oral LD_{50} value for rats is 2,000 to 2,270 mg/kg which is slightly toxic.

References

(1) Worthing, C.R., *Pesticide Manual,* 6th ed., p. 472, British Crop Protection Council (1979).
(2) Spencer, E.Y., *Guide to the Chemicals Used in Crop Protection,* 6th ed., p. 455, London, Ontario, Agriculture Canada (January 1973).
(3) Beroza, M., U.S. Patent 2,832,792, April 29, 1958, assigned to the Secretary of Agriculture.

SIDURON

Function: Herbicide (1)(2)

Chemical Name: N-(2-methylcyclohexyl)-N'-phenylurea

Formula:

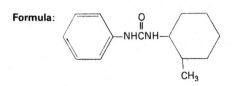

$$CH_3$$

Trade Names: DuPont 1318 (DuPont)
Tupersan® (DuPont)

Manufacture (3)(4)

A solution of 2-methylcyclohexylamine (cis,trans-mixture) (22.6 parts by weight) in 200 parts of toluene is treated with 23.8 parts by weight of phenylisocyanate over 44 minutes. The temperature increases from 23° to 41°C. After stirring for two hours the solution is evaporated in vacuum on a steam bath. The residue crystallizes on cooling. After washing with n-pentane, the products melt at 120° to 122°C. It is a mixture of cis- and trans-1-(2-methylcyclohexyl)-3-phenylurea.

Process Wastes and Their Control

Product Wastes: Siduron is slowly decomposed by acids or bases liberating aniline. This procedure is not recommended for disposal (B-3).

Toxicity

The acute oral LD_{50} value for rats is 7,500 mg/kg which is insignificantly toxic.

References

(1) Worthing, C.R., *Pesticide Manual,* 6th ed., p. 473, British Crop Protection Council (1979).
(2) Spencer, E.Y., *Guide to the Chemicals Used in Crop Protection,* 6th ed., p. 457, London, Ontario, Agriculture Canada (January 1973).
(3) Luckenbaugh, R.W., U.S. Patent 3,309,192, March 14, 1967, assigned to DuPont.
(4) E.I. DuPont de Nemours and Co., British Patent 1,028,818, May 11, 1966.

SILVEX (U.S.A.)
FENOPROP (U.K.)
2,4,5-TP (France and U.S.S.R.)

Function: Herbicide (1)(2)(3)

Chemical Name: 2-(2,4,5-trichlorophenoxy) propionic acid

Formula:

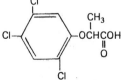

Trade Names: Kuron® (ester) (Dow)
Kurosal® (K salt) (Dow)
Aqua-Vex® (K salt)
O-X-D®

Manufacture

Silvex may be made by the reaction of 2,4,5-trichlorophenol with sodium α-chloropropionate.

Process Wastes and Their Control

Air: Air emissions from silvex manufacture have been reported (B-15) to consist of 1.0 kg phenol, 0.5 kg trichlorophenol and 1.0 kg hydrogen chloride per metric ton of pesticide produced.

Toxicity

The acute oral LD_{50} value for rats is 650 mg/kg which is slightly toxic.

Allowable Limits on Exposure and Use

Water: In water, EPA has set (B-12) a criterion of 10 μg/l for water for domestic water supply on a health basis.

Product Use: The tolerances set by the EPA for silvex in or on raw agricultural commodities are as follows:

	40 *CFR* Reference	Parts per Million
Pears (post-h)	180.340	0.05

A rebuttable presumption against registration for silvex was issued on April 21, 1978 by EPA on the basis of oncogenicity, teratogenicity and fetotoxicity.

In a notice dated September 28, 1970 (B-17), EPA noted that chlorodioxin contaminants were not allowed in silvex.

In a notice dated March 15, 1979, the EPA suspended all pesticide products containing silvex for forestry uses, rights-of-way uses, pasture uses, home and garden uses, commercial/or-namental turf uses, and aquatic weed control/ditch bank uses.

In a further notice dated July 17, 1979, the EPA stated that the only allowable uses for silvex are on rice, rangeland, sugarcane (field and stubble), preharvest fruit drop of apples, prunes, and pears, and noncrop uses. Noncrop uses of silvex include use on or around non-crop sites, including fencerows, hedgerows, fences (not otherwise included in suspended uses, e.g., rights-of-way, pasture); industrial sites or buildings (not otherwise included in suspended uses, e.g., rights-of-way, commercial/ornamental turf); storage areas, waste areas, vacant lots, and parking areas. The following definitions are included in order to help clarify the suspension orders for 2,4,5-T and silvex.

Range is nonpasture grazing land on which forage is produced through native species, or on which introduced species are managed as native species. This precludes land on which regular cultivation practices of the nature contained in the pasture definition are followed.

Pasture is land producing forage for animal consumption, harvested by grazing, which has annual or more frequent cultivation, seeding, fertilization, irrigation, pesticide application and other similar practices applied to it. Fencerows enclosing pastures are included as part of the pasture.

References

(1) Worthing, C.R., *Pesticide Manual,* 6th ed., p. 263, British Crop Protection Council (1979).
(2) Spencer, E.Y., *Guide to the Chemicals Used in Crop Protection,* 6th ed., p. 458, London, Ontario, Agriculture Canada (January 1973).
(3) Williams, B.M., U.S. Patent 2,749,360, June 5, 1956, assigned to Dow Chemical Co.

SIMAZINE

Function: Herbicide (1)(2)(3)

Chemical Name: 6-chloro-N,N'-diethyl-1,3,5-triazine-2,4-diamine

Formula:

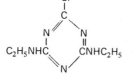

Trade Names: G-27,692 (Ciba-Geigy)
Gesatop® (Ciba-Geigy)
Princep® (Ciba-Geigy)
In Primatol® (Ciba-Geigy)
Gesapun®

Manufacture (4)

Simazine is made by the reaction of cyanuric chloride with two equivalents of ethylamine in the presence of an acid acceptor.

Process Wastes and Their Control

Air: Air emissions from simazine manufacture have been reported (B-15) to be 1.5 kg hydrocarbons per metric ton of pesticide produced.

Toxicity

The acute oral LD_{50} value for rats is over 5,000 mg/kg which is insignificantly toxic.

Atrazine, propazine, and simazine all appear to have low chronic toxicity. The only good carcinogenicity feeding study done on these compounds did not reveal a significant increase in cancer incidence over control. On the basis of these chronic studies, an ADI was calculated for each of these compounds. The ADI for atrazine is 0.0215 mg/kg/day, for propazine 0.0464 mg/kg/day, and for simazine 0.215 mg/kg/day (B-22).

Allowable Limits on Exposure and Use

Product Use: The tolerances set by the EPA for simazine in or on raw agricultural commodities are as follows:

	40 *CFR* Reference	Parts per Million
Alfalfa	180.213	15.0
Alfalfa, forage	180.213	15.0
Alfalfa, hay	180.213	15.0
Almond hulls	180.213	0.25
Almonds, hulls & nuts	180.213	0.25
Apples	180.213	0.25
Artichokes	180.213	0.5
Asparagus	180.213	10.0
Avocados	180.213	0.25
Bananas	180.213	0.2
Blackberries	180.213	0.25
Blueberries	180.213	0.25
Boysenberries	180.213	0.25
Cattle, fat	180.213	0.02 N
Cattle, mbyp	180.213	0.02 N
Cattle, meat	180.213	0.02 N
Cherries	180.213	0.25
Corn, fodder	180.213	0.25
Corn, forage	180.213	0.25
Corn, fresh (inc. sweet) (k+cwhr)	180.213	0.25
Corn, grain	180.213	0.25
Cranberries	180.213	0.25
Currants	180.213	0.25
Dewberries	180.213	0.25
Eggs	180.213	0.02 N
Fish	180.213	12.0
Goats, fat	180.213	0.02 N
Goats, mbyp	180.213	0.02 N
Goats, meat	180.213	0.02 N
Grapefruit	180.213	0.25
Grapes	180.213	∪.25
Grasses	180.213	15.0
Grasses, Bermuda	180.213	15.0
Grasses, Bermuda, forage	180.213	15.0
Grasses, Bermuda, hay	180.213	15.0
Grasses, forage	180.213	15.0
Grasses, hay	180.213	15.0
Hogs, fat	180.213	0.02 N
Hogs, mbyp	180.213	0.02 N
Hogs, meat	180.213	0.02 N
Horses, fat	180.213	0.02 N
Horses, mbyp	180.213	0.02 N
Horses, meat	180.213	0.02 N
Lemons	180.213	0.25
Loganberries	180.213	0.25
Milk	180.213	0.02 N
Nuts, filberts	180.213	0.25
Nuts, macadamia	180.213	0.25
Nuts, pecans	180.213	0.1 N
Nuts, walnuts	180.213	0.25
Olives	180.213	0.25
Oranges	180.213	0.25
Peaches	180.213	0.25
Pears	180.213	0.25
Plums	180.213	0.25
Poultry, fat	180.213	0.02 N
Poultry, mbyp	180.213	0.02 N
Poultry, meat	180.213	0.02 N
Raspberries	180.213	0.25

(continued)

	40 *CFR* Reference	Parts per Million
Sheep, fat	180.213	0.02 N
Sheep, mbyp	180.213	0.02 N
Sheep, meat	180.213	0.02 N
Sugarcane	180.213	0.25
Walnuts	180.213	0.25

The tolerances set by the EPA for simazine in animal feeds are as follows (the *CFR* Reference is to Title 21):

	CFR Reference	Parts per Million
Sugarcane molasses	561.350	1.0

The tolerances set by the EPA for simazine in food are as follows (the *CFR* Reference is to Title 21):

	CFR Reference	Parts per Million
Sugarcane molasses	193.400	1.0
Sugarcane syrup	193.400	1.0
Water, potable	193.400	0.01

References

(1) Worthing, C.R., *Pesticide Manual,* 6th ed., p. 474, British Crop Protection Council (1979).
(2) Spencer, E.Y., *Guide to the Chemicals Used in Crop Protection,* 6th ed., p. 459, London, Ontario, Agriculture Canada (January 1973).
(3) Gysin, H. and Knusli, E., U.S. Patent 2,891,855, June 23, 1959, assigned to J.R. Geigy AG.
(4) Pearlman, W.M. and Banks, C.K., *Jour Am Chem Soc* 70, 3726 (1948).

SODIUM ARSENITE

Function: Insecticide and herbicide (1)(2)

Chemical Name: Sodium metaarsenite

Formula: $NaAsO_2$

Trade Names: Altas A®
Chem Sen®
Penite®
Kill-All®

Manufacture (3)

Sodium arsenite is manufactured as a liquid product by batch mixing caustic, water and arsenic trioxide. The product is sold in the liquid form, and there is thus no liquid waste associated with the process. All cleanup water from the batch mix vats is saved to mix with the next batch.

Toxicity

The acute oral LD_{50} value for rats is 41 mg/kg (B-5) which is highly toxic.

Allowable Limits on Exposure and Use

Product Use: A rebuttable presumption against registration was issued on October 18, 1978 by EPA on the basis of oncogenicity, teratogenicity and mutagenicity.

In a notice dated August 1, 1967, EPA (B-17) stated that sodium arsenite was unacceptable for home use if compound is in excess of 2.0% unless the following warning statements appear on the label: "Do not use or store in or around the home" and "Do not allow domestic animals to graze treated area."

Two products containing sodium arsenite were voluntarily cancelled according to a EPA notice dated March 11, 1977.

The tolerances set by the EPA for sodium arsenite in or on raw agricultural commodities are as follows:

	40 *CFR* Reference	Parts per Million
Cattle, fat (pre-s min 14 days)	180.335	0.7
Cattle, kidney (pre-s min 14 days)	180.335	2.7
Cattle, liver (pre-s min 14 days)	180.335	2.7
Cattle, mbyp (exc. kidney, liver)	180.335	0.7
Cattle, meat (pre-s min 14 days)	180.335	0.7
Horses, fat (pre-s min 14 days)	180.335	0.7
Horses, kidney (pre-s min 14 days)	180.335	2.7
Horses, liver (pre-s min 14 days)	180.335	2.7
Horses, mbyp (exc. kidney, liver)	180.335	0.7
Horses, meat (pre-s min 14 days)	180.335	0.7

References

(1) Worthing, C.R., *Pesticide Manual,* 6th ed., p. 476, British Crop Protection Council (1979).
(2) Spencer, E.Y., *Guide to the Chemicals Used in Crop Protection,* 6th ed., p. 460, London, Ontario, Agriculture Canada (January 1973).
(3) Patterson, J.W., *State of the Art for the Inorganic Chemicals Industry: Inorganic Pesticides,* Report EPA-600/2-74-0092, Wash., DC, Environmental Protection Agency (March 1975).

SODIUM CHLORATE

Function: Herbicide

Chemical Name: Sodium chlorate

Formula: $NaClO_3$

Trade Names: Atlacide®
Chlorax®
Dropleaf®
Fall®
Shed-A-Leaf®
Tumbleleaf®

Manufacture (3)(4)

Sodium chlorate production basically involves electrolysis of concentrated sodium chloride solution to produce a mixture of sodium chlorate, sodium hypochlorite and residual sodium chloride. As shown in Figure 48, rock salt plus makeup water is added to a mother liquor representing recycled electrolyte solution from which sodium chlorate has been crystallized.

Figure 48: Sodium Chlorate Production Schematic

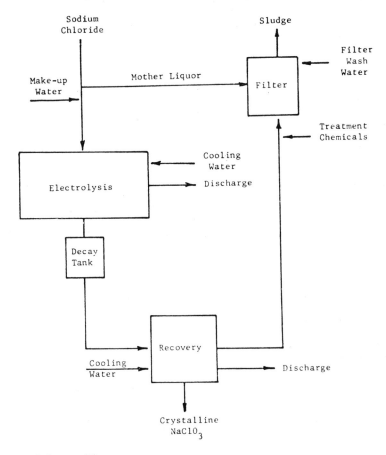

Source: Reference (3)

The brine solution is decanted to the electrolytic cells. At one plant, muriatic acid is added for pH control, and sodium dichromate to improve cell efficiency. The cells are cooled by a once-through noncontact water flow. After completion of electrolysis, the sodium chlorate mother liquor is decanted to a hypochlorite decay tank, where (at one plant) barium chloride is added to precipitate (as barium sulfate) sulfate impurities in the mother liquor. The solution is then pumped to a precipitation tank, where soda ash is added. This precipitates calcium as $CaCO_3$, and excess barium as $BaCO_3$.

At the second plant, sulfate impurity in the sodium chloride is removed by lime addition to the mother liquor to precipitate calcium sulfate, with residual calcium precipitated as calcium carbonate by addition of soda ash. In both plants, the solution is then pressure filtered to remove the mixture of precipitates previously formed, plus graphite fragments from the cell electrodes. The pressure filter is backwashed several times daily, with cooling water, and the backwash discharged.

At one plant, the filtrate then goes to a steam heated evaporator, to drive off water vapor and precipitate NaCl. The sodium chlorate-rich supernatant is decanted to a chlorate crystallization process, while the NaCl slurry is vacuum filtered. Crystalline NaCl is returned to the dissolver, and filtrate pumped to the chlorate crystallizer. The chlorate solution is cooled by refrigeration, or by reducing the pressure, with water vapor released. This vapor, plus that driven off in the steam heated NaCl crystallization chamber is condensed using cooling water at 2,500 gpm, and returned to the dissolver, along with a portion of the cooling water for makeup.

The sodium chlorate slurry is then centrifuged to separate the crystals from the mother liquor. The mother liquor is returned to the electrolytic cells, and the crystalline sodium chlorate dried in rotary driers. This completes the manufacturing process, and the product is loaded into drums or tank cars for shipping.

Process Wastes and Their Control

Product Wastes: Water-soluble chlorate compounds are of moderate toxicity, but are very strong oxidizing agents and should not be permitted to contact organic matter especially in the absence of water. They can be chemically reduced to less dangerous products. The following procedure has been suggested (B-3): Add to a large volume of concentrated solution of reducer (hypo, a bisulfite, or a ferrous salt) and acidify with 3M H_2SO_4. When reduction is complete add soda ash or dilute hydrochloric acid to the solution. Wash into drain with large excess of water.

Toxicity

The acute oral LD_{50} value for rats is 1,200 mg/kg which is slightly toxic.

Allowable Limits on Exposure and Use

Product Use: The tolerances set by the EPA for sodium chlorate in or on raw agricultural commodities are as follows:

	40 *CFR* Reference	Parts per Million
Chili peppers	180.102	Exempt
Corn, fodder	180.102	Exempt
Corn, forage	180.102	Exempt
Corn grain	180.102	Exempt
Cotton, seed	180.102	Exempt
Rice	180.102	Exempt
Rice, straw	180.102	Exempt
Safflower, seed	180.102	Exempt
Sorghum, fodder	180.102	Exempt
Sorghum, forage	180.102	Exempt
Sorghum, grain	180.102	Exempt
Soybeans	180.102	Exempt
Sunflower, seed	180.102	Exempt

References

(1) Worthing, C.R., *Pesticide Manual,* 6th ed., p. 477, British Crop Protection Council (1979).
(2) Spencer, E.Y., *Guide to the Chemicals Used in Crop Protection,* 6th ed., p. 461, London, Ontario, Agriculture Canada (January 1973).

(3) Patterson, J.W., *State of the Art for the Inorganic Chemicals Industry: Inorganic Pesticides,* Report EPA-600/2-74-0092, Wash., DC, Environmental Protection Agency (March 1975).
(4) Holmes, A.J., U.S. Patent 3,043,757, July 10, 1962, assigned to Olin Mathieson Chemical Corp.

SODIUM FLUORIDE

Function: Insecticide

Chemical Name: Sodium fluoride (1)(2)

Formula: NaF

Trade Names: Florocid®
Karidium®
Villiaumite®
Zymafluor®

Manufacture

Sodium fluoride may be made by:

(A) Neutralization of HF with NaOH or
(B) The reaction of fluorspar (CaF_2) and NaOH (3)

Process Wastes and Their Control

Product Wastes: Sodium fluoride, cryolite (sodium aluminofluoride, Na_3AlF_6) and sodium fluosilicate (Na_2SiF_6) are sources of toxic fluoride ions which cannot be detoxified. Thus, precautions must be taken to insure that these materials do not enter a water supply in large amounts, but small amounts could be added to wastewaters, since fluoride is already present at trace levels in natural waters. A suggested disposal method converts the soluble fluoride ions to insoluble calcium fluoride (LD_{50} = 5,000 compared to 200 for the soluble fluorides), a naturally occurring mineral (fluorspar) which can safely be added to a landfill. The method is as follows (B-3): Add slowly to a large container of water. Stir in slight excess of soda ash. If fluoride is present add slaked lime also. Let stand 24 hours. Decant or siphon into another container and neutralize with 6M HCl before washing down with large excess of water. The sludge may be added to landfill.

Toxicity

The acute oral LD_{50} value for rats is 180 mg/kg (B-5) which is moderately toxic.

Allowable Limits on Exposure and Use

Product Use: In a notice dated June 1, 1970, the EPA (B-17) cancelled home use of sodium fluoride if the product contains more than 40% of sodium fluoride.

References

(1) Worthing, C.R., *Pesticide Manual,* 6th ed., p. 479, British Crop Protection Council (1979).
(2) Spencer, E.Y., *Guide to the Chemicals Used in Crop Protection,* 6th ed., p. 463, London, Ontario, Agriculture Canada (January 1973).
(3) Fredrickson, R.E. et al, U.S. Patent 2,985,508, May 23, 1961, assigned to Dow Chemical Co.

SODIUM FLUOROACETATE

Function: Rodenticide (1)(2)

Chemical Name: Sodium fluoroacetate

Formula:

$$CH_2FCONa$$
with O double-bonded above the C.

Trade Name: Compound 1080®

Manufacture

Sodium fluoroacetate may be made by one of two routes:

(A) Condensing methyl chloroacetate with potassium fluoride and then converting the fluoroester to the sodium salt with NaOH.

(B) Reacting CO, HF and formaldehyde at high pressures to give fluoroacetic acid which is converted by NaOH to sodium fluoroacetate.

Process Wastes and Their Control

Product Wastes: The compound is unstable at temperatures above 110°C and decomposes at 200°C. Thus, careful incineration has been suggested as a disposal procedure by the Manufacturing Chemists Association. According to their procedure, the product should be mixed with large amounts of vermiculite, sodium bicarbonate and sand-soda ash. Slaked lime should also be added to the mixture. Two incineration procedures for this mixture are suggested. The better of these procedures is to burn the mixture in a closed incinerator equipped with an afterburner and an alkali scrubber. The other procedure suggests that the mixture be covered with scrap wood and paper in an open incinerator. (The incinerator should be lighted by means of an excelsior train.) (B-3)

Toxicity

This extremely dangerous rodenticide has a very high acute oral toxicity (LD_{50} for rats is 0.2 mg/kg) and is also absorbed through unbroken skin. It is extremely hazardous to all wildlife. The pure salt is very soluble in water (insoluble in organic solvents) but is normally used as a bait at low concentrations (B-3).

Allowable Limits on Exposure and Use

Air: The threshold limit value for sodium fluoroacetate in air is 0.05 mg/m^3 as of 1979. The tentative short term exposure limit is 0.15 mg/m^3. Further, it is noted that cutaneous absorption should be prevented so that the TLV is not invalidated.

Product Use: A rebuttable presumption against registration was issued on December 1, 1976 on the basis of reductions in nontarget and endangered species and because there is no human antidote.

In a notice dated March 9, 1972 (B-17), the EPA cancelled the use of sodium fluoroacetate in mammalian predator control. Label should have instructions for predator use blocked out.

References

(1) Worthing, C.R., *Pesticide Manual,* 6th ed., p. 480, British Crop Protection Council (1979).
(2) Spencer, E.Y., *Guide to the Chemicals Used in Crop Protection,* 6th ed., p. 464, London, Ontario, Agriculture Canada (January 1973).

SODIUM METABORATE

Function: Herbicide (1)

Chemical Name: Sodium metaborate

Formula: $Na_2B_2O_4 \cdot 4H_2O$

Trade Name: Monobor®

Manufacture

Sodium metaborate may be made by the reaction of sodium tetraborate (borax) and sodium hydroxide (3).

Sodium metaborate is commonly used in admixture with sodium chlorate (2). Monobor chlorate is manufactured by mixing caustic solution, borax and crystalline sodium chlorate. The reaction is exothermic, and the resultant slurry is cooled on water-chilled rollers, to form flakes. The flakes are sized by screening. Dust (collected by bag collectors) and fines are returned to the continuous mixer. Oversized flakes are granulated, and returned to the screening operation.

Figure 49 is a schematic representation of the production of Monobor chlorate.

Figure 49: Monobor Chlorate Production Schematic

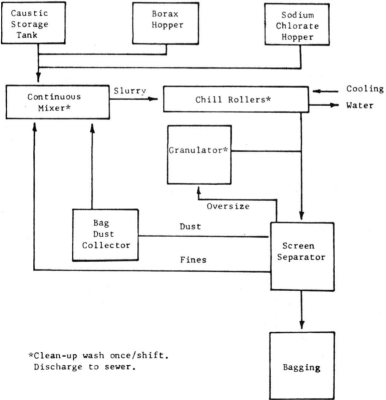

Source: Reference (3)

Toxicity

The acute oral LD_{50} value for rats is 2,330 mg/kg which is slightly toxic.

References

(1) Worthing, C.R., *Pesticide Manual,* 6th ed., p. 481, British Crop Protection Council (1979).
(2) Mitchell, E.M. and Yannacakis, J., U.S. Patent 3,032,405, May 1, 1962, assigned to United States Borax and Chemical Corp.
(3) Patterson, J.W., *State of the Art for the Inorganic Chemicals Industry: Inorganic Pesticides,* Report EPA-600/2-74-0092, Wash., DC, Environmental Protection Agency (March 1975).

SODIUM 2,4,5-TRICHLOROPHENATE

Function: Fungicide

Chemical Name: Sodium 2,4,5-trichlorophenate

Formula:

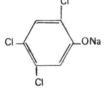

Trade Names: Dowicide B®
Dowicide 25®

Manufacture (1)

The process comprises adding 2,4,5-trichlorophenol to an aqueous solution or suspension of an alkali metal hydroxide, maintaining the mixture at the proper reaction temperature for a suitable period of time, and cooling the reaction product whereby hydrated crystals of the phenolates are obtained.

In carrying out the foregoing method for the preparation of the various alkali metal trichlorophenolates, the compounds are obtained directly by crystallization in a hydrated form, generally the pentahydrate. However, upon exposure to the air at ordinary room temperatures the pentahydrate compounds lose about 4 molecules of water of crystallization. The resultant monohydrate may then be readily dehydrated to the anhydrous compounds, such as by heating them to a temperature between 40° and 60°C at atmospheric pressure, or by drying in vacuo over sulfuric acid or calcium chloride at room temperature.

The compounds are white, crystalline substances which do not melt, but decompose at varying temperatures upon heating; e.g., the sodium salt decomposes at 280°C. The compounds are partially decomposed in aqueous solution by carbon dioxide with the formation of the metal carbonates and trichlorophenol. The alkali metal phenolates are soluble in water, alcohol and acetone.

Toxicity

The acute oral LD_{50} value for rats is 1,620 mg/kg (B-5) which is slightly toxic.

Allowable Limits on Exposure and Use

Product Use: A rebuttable presumption against registration was issued on August 2, 1978 by EPA on the basis of oncogenicity and fetotoxicity.

Reference

(1) Mills, L.E., U.S. Patent 1,991,329, February 12, 1935, assigned to Dow Chemical Company.

SOLAN (Canada and U.S.A.)
PENTANOCHLOR (Elsewhere)

Function: Herbicide (1)(2)(3)(4)

Chemical Name: N-(3-chloro-4-methylphenyl)-2-methylpentanamide

Formula:

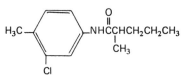

Trade Names: Mephamide
NIA-4512 (FMC Corp.)
Herbon Solan®

Manufacture (3)(4)

To a stirred suspension of 1,909 parts of 3-chloro-4-methylaniline and 742 parts of sodium carbonate in 3,000 parts of benzene were added 1,810 parts of 2-methylpentanoyl chloride, 4 hours being required to complete the addition. The mixture was held at the reflux temperature for 1 hour. Allowed to stand overnight, the mixture was then raised to the reflux temperature and maintained at that temperature for 3 hours. The mixture was cooled, an additional 2,000 parts of benzene added and the mixture washed with 2,000 parts of 2% hydrochloric acid solution. The acid wash was decanted and the benzene layer concentrated under reduced pressure to give 2,726 parts of pale, yellow solid which melted at 74° to 77°C. Recrystallization increased the melting point to 79° to 80°C.

Process Wastes and Their Control

Product Wastes: This material, stable at room temperature, is hydrolyzed under basic conditions to give the substituted aniline (B-3).

Toxicity

The acute oral LD_{50} value for rats is more than 10,000 mg/kg which is insignificantly toxic.

References

(1) Worthing, C.R., *Pesticide Manual,* 6th ed., p. 407, British Crop Protection Council (1979).
(2) Spencer, E.Y., *Guide to the Chemicals Used in Crop Protection,* 6th ed., p. 465, London, Ontario, Agriculture Canada (January 1973).
(3) Willard, J.R. and Dorschner, K.P., U.S. Patent 3,020,142, February 6, 1962, assigned to FMC Corp.

(4) Dorschner, K.P., Gates, R.L. and Willard, J.R., British Patent 869,169, May 31, 1961, assigned to Food Machinery and Chemical Corp.

STIROFOS (U.S.A.)
TETRACHLORVINPHOS (U.K. and International)

Function: Insecticide (1)(2)

Chemical Name: 2-chloro-1-(2,4,5-trichlorophenyl)-ethenyl dimethyl phosphate

Formula:

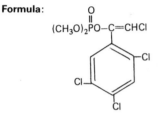

Trade Names: SD-8447 (Shell)
Gardona® (Shell)
Rabon®

Manufacture (3)

Stirofos is made by the reaction of trimethyl phosphite with pentachloroacetophenone. The acetophenone is prepared in a first step as follows. To 88 parts of aluminum chloride was added 109 parts of 1,2,4-trichlorobenzene. To this slurry, with stirring, was added 88 parts of dichloroacetyl chloride over a period of 10 minutes. The mixture then was heated slowly to 90°C where it was held for 4 hours. Decomposition of the complex was effected by pouring the reaction mixture onto a mixture of ice and hydrochloric acid.

The resulting mixture was extracted with ether and the organic phase thus obtained was washed successively with dilute hydrochloric acid, water, dilute sodium bicarbonate solution and finally with saturated salt (NaCl) solution. The solvent was evaporated and the residue was distilled to give 134 parts (77% yield) of 2,2,2',4',5'-pentachloroacetophenone, a colorless liquid, boiling point 103° to 105°C (0.05 torr).

Then the stirofos is prepared in a succeeding step as follows. To 50 parts of 2,2,2',4',5'-pentachloroacetophenone was added, over one-half hour, 25.5 parts of trimethyl phosphite. The temperature was kept between 30° and 50°C during the addition. After the addition was complete, the mixture was heated to 110°C for one-half hour. The reaction mixture was then cooled to room temperature and treated with ether to induce crystallization. The crystals that formed were cooled in ice and filtered to give 46 parts of a first crop, melting point: 97° to 98°C. The mother liquors afforded 3 parts of a second crop on a similar treatment with ether and pentane to raise the total yield to 79%.

2,2,2',4',5'-pentachloroacetophenone is ordinarily prepared by Friedel-Crafts ketone synthesis, viz by reaction of 1,2,4-trichlorobenzene with dichloroacetyl chloride in the presence of aluminum chloride followed by decomposition of the resulting complex with ice and hydrochloric acid. Invariably some of the isomeric 2,2,2',3',6'-pentachloroacetophenone is formed. The presence of this isomer is highly undesirable, since it too reacts with trimethyl phosphite to form 2-chloro-1-(2,3,6-trichlorophenyl)vinyl dimethyl phosphate. Both this phosphate and its ketone precursor produce a very marked hormonal effect on certain crops such as cotton, grapes, tobacco, and melons. This effect is similar to that exhibited by hormonal-type weed killers and may cause retardation in growth or may kill sensitive crops. This undesirable hormonal effect is greatly reduced or not shown if the phytotoxic isomer is present in concentrations less than one percent by weight. It has been found that it is very difficult to separate the 2',4',5'-ketone isomer from the 2',3',6'-ketone isomer by distillation.

An improved synthesis process (4) involves the reaction of trimethyl phosphite selectively with 2,2,2',4',5'-pentachloroacetophenone in a mixture of isomeric pentachloroacetophenones by carrying out the reaction in a liquid alkane as the reaction medium.

Process Wastes and Their Control

Product Wastes: Fifty percent hydrolysis (at 50°C) is obtained in 1,300 hours at pH 3 and in 80 hours at pH 10.5 (B-3).

Toxicity

The acute oral LD_{50} value for rats is 4,000 to 5,000 mg/kg which is very slightly toxic.

Allowable Limits on Exposure and Use

Product Use: The tolerances set by the EPA for stirofos in or on raw agricultural commodities are as follows:

	40 *CFR* Reference	Parts per Million
Alfalfa	180.252	110.0
Apples	180.252	10.0
Cattle, fat	180.252	1.5
Cherries	180.252	10.0
Corn, field, fodder	180.252	110.0
Corn, field, forage	180.252	110.0
Corn, fresh (k+cwhr)	180.252	10.0
Corn, grain	180.252	10.0
Corn, pop, fodder	180.252	110.0
Corn, pop, forage	180.252	110.0
Corn, sweet (k+cwhr)	180.252	10.0
Corn, sweet, fodder	180.252	110.0
Corn, sweet, forage	180.252	110.0
Cranberries	180.252	10.0
Eggs	180.252	0.1
Hogs, fat	180.252	1.5
Horses, fat	180.252	0.5
Milk, fat	180.252	0.5 N
Peaches	180.252	0.1
Pears	180.252	10.0
Poultry, fat	180.252	0.75
Sheep, fat	180.252	0.5
Tomatoes	180.252	5.0

References

(1) Worthing, C.R., *Pesticide Manual,* 6th ed., p. 504, British Crop Protection Council (1979).
(2) Spencer, E.Y., *Guide to the Chemicals Used in Crop Protection,* 6th ed., p. 490, London, Ontario, Agriculture Canada (January 1973).
(3) Phillips, D.D. and Ward, L.F., Jr., U.S. Patent 3,102,842, September 3, 1963, assigned to Shell Oil Co.
(4) Ramey, D.E., U.S. Patent 3,553,297, January 5, 1971, assigned to Shell Oil Co.

STREPTOMYCIN

Function: Bactericide (1)(2)

Chemical Name: O-2-deoxy-2(methylamino)-α-L-glucopyranosyl-(1→2)-O-5-deoxy-3C-
formyl-α-L-lyxofuranosyl-(1→4)-N,N'-bis(aminoiminomethyl)-D-streptamine

Formula:

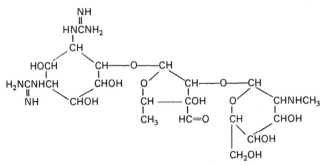

Trade Names: Agrimycin®
Agristrep®

Manufacture

Recovered from the fermentation broth of *Streptomyces griseus*

Process Wastes and Their Control

Product Wastes: Streptomycin is unstable to heat and does not accumulate in the soil. Therefore, disposal by incineration or burial should not result in harm to the environment (B-3).

Toxicity

The acute oral LD_{50} value for rats is 9,000 mg/kg which is insignificantly toxic.

Allowable Limits on Exposure and Use

Product Use: The tolerances set by the EPA for streptomycin in or on raw agricultural commodities are as follows:

	40 *CFR* Reference	Parts per Million
Celery	180.245	0.25 N
Fruits, pome	180.245	0.25 N
Peppers	180.245	0.25 N
Potatoes	180.245	0.25 N
Sweet potatoes, seed (post-h)	180.245	0.02 N
Tomatoes	180.245	0.25 N

References

(1) Martin, H. and Worthing, C.R., *Pesticide Manual,* 5th ed., p. 478, British Crop Protection Council (January 1977).
(2) Spencer, E.Y., *Guide to the Chemicals Used in Crop Protection,* 6th ed., p. 467, London, Ontario, Agriculture Canada (January 1973).

STRYCHNINE

Function: Rodenticide (1)(2)

Chemical Name: Strychnidin-10-one

Formula:

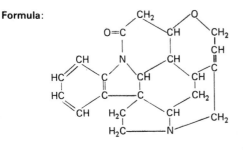

Trade Names: Kwik-Kil®
Mouse-Tox®
Ro-Dex®

Manufacture

Strychnine is extracted from Strychnos seeds with benzene but may be synthesized (3) by a complex synthesis.

Process Wastes and Their Control

Product Wastes: Careful incineration has been recommended for disposal; two procedures are suggested (B-3): (1) Pour or sift onto a thick layer of sand and soda ash mixture (90-10). Mix and shovel into a heavy paper box with much paper packing. Burn in incinerator. Fire may be augmented by adding excelsior and scrap wood. Stay on the upwind side. (2) Waste may be dissolved in flammable solvent (alcohols, benzene, etc.) and sprayed into fire box of an incinerator with afterburner and scrubber.

Toxicity

Strychnine is the most toxic product (acute oral LD_{50} = 5 mg/kg for rats) in the botanical group (B-3).

Allowable Limits on Exposure and Use

Air: The threshold limit value for strychnine in air is 0.15 mg/m³ as of 1979. The tentative short term exposure limit is 0.45 mg/m³ (B-23).

Product Use: A rebuttable presumption against registration was issued on December 1, 1976 on the basis of reductions in nontarget and endangered species.

In a decision dated March 9, 1972, EPA (B-17) cancelled the use of strychnine in mammalian predator control. Label should have instructions for predator use blocked out.

References

(1) Worthing, C.R., *Pesticide Manual,* 6th ed., p. 482, British Crop Protection Council (1979).
(2) Spencer, E.Y., *Guide to the Chemicals Used in Crop Protection,* 6th ed., p. 468, London, Ontario, Agriculture Canada (January 1973).
(3) Woodward, R.B. et al, *Jour. Am. Chem. Soc.* 76, 4749 (1954).

SULFALLATE (CDEC in U.S.A.)

Function: Herbicide (1)(2)

Chemical Name: 2-chloro-2-propenyl diethylcarbamodithioate

Formula:

$$(C_2H_5)_2NC\overset{\overset{\displaystyle S}{\|}}{S}CH_2\overset{\overset{\displaystyle Cl}{|}}{C}{=}CH_2$$

Trade Names: CP 4742 (Monsanto)
Vegadex® (Monsanto)

Manufacture (3)(4)(5)

Sulfallate may be made by adding 2,3-dichloropropene with stirring to an equimolar amount of sodium diethyldithiocarbamate containing a few drops of dodecylbenzene sulfonate wetting agent. Within 20 minutes, a temperature rise from 30° to 45°C is noted. The mixture is then heated to 50° to 60°C for 4 hours. After cooling to room temperature, the reaction mixture separates into 2 layers. The organic layer is washed with warm water and dried over sodium sulfate. Any unreacted 2,3-dichloropropene is removed in a vacuum at room temperature.

Process Wastes and Their Control

Air: Air emissions from sulfallate manufacture have been reported as follows (B-15):

Component	Kilograms per Metric Ton Pesticide Produced
SO_2	205.0
Hydrocarbons	1.5
Sulfallate	0.5

Product Wastes: The only dithiocarbamate used as a herbicide, sulfallate is hydrolyzed very slowly in either very weak acid, pH 5, or very weak base, pH 8. It is hydrolyzed rapidly in boiling caustic and is completely decomposed by strong oxidizing agents (B-3).

Toxicity

The acute oral LD_{50} value for rats is 850 mg/kg which is slightly toxic.

Allowable Limits on Exposure and Use

Product Use: The tolerances set by the EPA for CDEC in or on raw agricultural commodities are as follows:

	40 *CFR* Reference	Parts per Million
Beans, lima	180.247	0.2 N
Beans, snap	180.247	0.2 N
Beans, vines	180.247	0.2 N
Broccoli	180.247	0.2 N
Brussels sprouts	180.247	0.2 N
Cabbage	180.247	0.2 N
Cantaloupes	180.247	0.2 N
Cauliflower	180.247	0.2 N
Celery	180.247	0.2 N

(continued)

	40 *CFR* Reference	Parts per Million
Chicory	180.247	0.2 N
Collards	180.247	0.2 N
Corn (k+cwhr)	180.247	0.2 N
Corn, fodder	180.247	0.2 N
Corn, forage	180.247	0.2 N
Corn, grain	180.247	0.2 N
Cucumbers	180.247	0.2 N
Endive (escarole)	180.247	0.2 N
Kale	180.247	0.2 N
Lettuce	180.247	0.2 N
Mustard, greens	180.247	0.2 N
Okra	180.247	0.2 N
Potatoes	180.247	0.2 N
Soybeans	180.247	0.2 N
Soybeans, forage	180.247	0.2 N
Soybeans, hay	180.247	0.2 N
Spinach	180.247	0.2 N
Tomatoes	180.247	0.2 N
Turnips	180.247	0.2 N
Turnips, greens	180.247	0.2 N
Watermelons	180.247	0.2 N

References

(1) Worthing, C.R., *Pesticide Manual,* 6th ed., p. 483, British Crop Protection Council (1979).
(2) Spencer, E.Y., *Guide to the Chemicals Used in Crop Protection,* 6th ed., p. 470, London, Ontario, Agriculture Canada (January 1973).
(3) Harman, M.W. and D'Amico, J.J., U.S. Patent 2,854,467, September 30, 1958, assigned to Monsanto Chemical Co.
(4) Harman, M.W. and D'Amico, J.J., U.S. Patent 2,919,182, December 29, 1959, assigned to Monsanto Chemical Co.
(5) Harman, M.W. and D'Amico, J.J., U.S. Patent 2,744,898, May 8, 1956, assigned to Monsanto Chemical Co.
(6) D'Amico, J.J. and Harman, M.W., British Patent 769,222, assigned to Monsanto Chemical Co. (March 6, 1957).

SULFOTEPP

Function: Insecticide

Chemical Name: Tetraethylthiodiphosphate

Formula:

$$(C_2H_5O)_2\overset{\overset{S}{\|}}{P}-O-\overset{\overset{S}{\|}}{P}(OC_2H_5)_2$$

Trade Names: E-393 (Bayer)
Bladafum® (Bayer)
ASP-47 (Victor Chemical Works)
Dithio®
Dithione®
Dithiotepp®
Lethalaire G-57®
Sulphatepp®
Thiotepp®
Tinotepp®

Manufacture

Sulfotepp may be made by:

(A) The action of sulfur on TEPP (which see) or
(B) The interaction of O,O-diethyl phosphorochloridothioate with aqueous sodium carbonate in the presence of pyridine.

Process Wastes and Their Control

Product Wastes: This material is resistant to hydrolysis (B-3).

Toxicity

The acute oral LD_{50} value for rats is about 5 mg/kg which is very highly toxic.

Allowable Limits on Exposure and Use

Air: A permissible exposure limit for sulfotepp (TEDP) is 0.2 mg/m³ (B-21).

References

(1) Worthing, C.R., *Pesticide Manual,* 6th ed., p. 484, British Crop Protection Council (1979).
(2) Spencer, E.Y., *Guide to the Chemicals Used in Crop Protection,* 6th ed., p. 471, London, Ontario, Agriculture Canada (January 1973).
(3) Toy, A.D.F., *Jour. Am. Chem. Soc.* 73, 4670 (1951).

SULFOXIDE

Function: Insecticidal synergist (1)(2)(3)

Chemical Name: 5-[2-(octylsulfinyl)propyl]-1,3-benzodioxole

Formula:

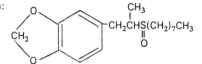

Trade Names: Sulfox-cide® (S.B. Penick and Co.)
Sulfoxyl®

Manufacture

In a first step, octyl mercaptan is reacted with iso-safrole to give a thioether. The thioether is oxidized with hydrogen peroxide to sulfoxide. In preparing such compounds, it is found convenient to dissolve the intermediate sulfide in acetic acid and to add an excess of 30% hydrogen peroxide to the acetic acid or acetone solution. After several hours the mixture is poured into water whereupon the sulfoxide separates as an oil and is isolated by extraction with suitable solvent (benzene or ether) which is later removed.

Process Wastes and Their Control

Product Wastes: Sulfoxide is resistant to alkaline hydrolysis. Oxidation to the sulfone produces an oxidation-resistant compound of similar moderate toxicity. Incineration or open burning with proper precautions is suggested (B-3).

Toxicity

The acute oral LD_{50} value for rats is 2,000 to 2,500 mg/kg which is slightly toxic.

References

(1) Worthing, C.R., *Pesticide Manual,* 6th ed., p. 485, British Crop Protection Council (1979).
(2) Spencer, E.Y., *Guide to the Chemicals Used in Crop Protection,* 6th ed., p. 472, London, Ontario, Agriculture Canada (January 1973).
(3) Synerholm, M.E., U.S. Patent 2,486,445, November 1, 1949, assigned to Boyce Thompson Institute for Plant Research.

SULFUR

Function: Fungicide and acaricide (1)(2)

Chemical Name: Sulfur

Formula: S

Trade Names: None

Manufacture (3)

Bulk sulfur, brought in by hopper cars, is crushed by teethed rollers to 10 mesh size, and transported by conveyor belts to holding bins. Additional processing is shown in Figure 50. The crushed sulfur is placed, in charges of 750 to 800 pounds, into a batch mixer. Clay, if required, is added at 7 to 50% of the finished product. Wetting agents may also be added to allow field spraying of the pesticide as a liquid suspension.

After batch mixing, the preparation is conveyed to an enclosed grinding mill to further reduce the size of the sulfur. In order to prevent explosions, an oxygen-lean atmosphere must be maintained in the grinding mill. This is accomplished by using CO_2 rich flue gas from the plant steam boiler. After grinding, the mixture goes to a final post-blender, and then is bagged for shipment. When shifting from one product to another, it is necessary to clean out the mix and grinding system. As a first step, the caked material in the units is chipped out by hand. Then, charges of clay are run through the batch mixer, grinding mill and post-blender. This clay is stockpiled, and later used when the same product is again being formulated.

Toxicity

The acute oral LD_{50} for rats is not available but sulfur is relatively nontoxic to mammals.

References

(1) Worthing, C.R., *Pesticide Manual,* 6th ed., p. 486, British Crop Protection Council (1979).
(2) Spencer, E.Y., *Guide to the Chemicals Used in Crop Protection,* 6th ed., p. 474, London, Ontario, Agriculture Canada (January 1973).
(3) Patterson, J.W., *State of the Art for the Inorganic Chemical Industry: Inorganic Pesticides,* Report EPA-600/2-74-0092, Wash., DC, Environmental Protection Agency (March 1975).

Figure 50: Sulfur Pesticide Production

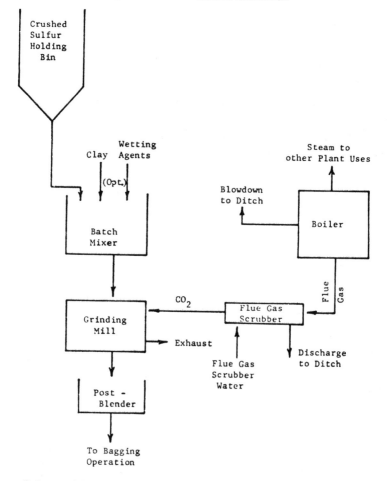

Source: Reference (3)

SULFURIC ACID

Function: Herbicide (1)(2)

Chemical Name: Sulfuric Acid

Formula: H_2SO_4

Trade Names: None

Manufacture

Sulfuric acid is made (3) by the oxidation of sulfur dioxide to sulfur trioxide followed by absorption of the sulfur trioxide in water.

Toxicity

The acute oral LD_{50} value for rats is 2,140 mg/kg (B-5) which is slightly toxic.

Allowable Limits on Exposure and Use

Air: The threshold limit value for H_2SO_4 in air is 1.0 mg/m^3 as of 1979 (B-23).

Product Use: The tolerances set by the EPA for sulfuric acid in or on raw agricultural commodities are as follows:

	40 *CFR* Reference	Parts per Million
Garlic	180.101	Exempt
Onions	180.101	Exempt

References

(1) Worthing, C.R., *Pesticide Manual,* 6th ed., p. 487, British Crop Protection Council (1979).
(2) Spencer, E.Y., *Guide to the Chemicals Used in Crop Protection,* 6th ed., p. 475, London, Ontario, Agriculture Canada (January 1973).
(3) Sittig, M., *Sulfuric Acid Manufacture and Effluent Control,* Park Ridge, NJ, Noyes Data Corp. (1971).

SULFURYL FLUORIDE

Function: Insecticidal fumigant (1)(2)(3)

Chemical Name: Sulfuryl fluoride

Formula:

$$\begin{array}{c} O \\ \| \\ F-S-F \\ \| \\ O \end{array}$$

Trade Name: Vikane® (Dow Chemical Co.)

Manufacture

Sulfuryl fluoride may be made by:

    A. Burning fluorine in sulfur dioxide or
    B. By heating barium fluorosulfonate (4):

$$Ba(SO_3F)_2 \rightarrow BaSO_4 + SO_2F_2$$

Process Wastes and Their Control

Product Wastes: The following disposal method has been suggested for this gaseous inorganic fumigant: allow gas to flow into a mixed solution of caustic soda and slaked lime. After neutralization, the solution, which is relatively harmless, may be washed down the drain. The precipitated calcium fluoride may be buried or added to a landfill (see sodium fluoride). Small amounts could also be released directly to the atmosphere without serious harm (B-3).

Toxicity

The acute oral LD_{50} value for rats is 100 mg/kg (B-5) which is moderately toxic.

Allowable Limits on Exposure and Use

Air: The threshold limit value for sulfuryl fluoride in air is 20 mg/m³ (5 ppm) and the tentative short term exposure limit is 40 mg/m³ (10 ppm) (B-23).

References

(1) Worthing, C.R., *Pesticide Manual*, 6th ed., p. 488, British Crop Protection Council (1979).
(2) Spencer, E.Y., *Guide to the Chemicals Used in Crop Protection*, 6th ed., p. 476, London, Ontario, Agriculture Canada (January 1973).
(3) Kenaga, E.E., U.S. Patent 2,875,127, February 24, 1959, assigned to Dow Chemical Co.
(4) Traube, W., Hoerenz, J. and Wunderlich, F., *Ber.* 52, 1272 (1919).

SULPROFOS

Function: Insecticide

Chemical Name: O-Ethyl O-(4-methylthiophenyl)-S-propyl phosphorodithioate

Formula:

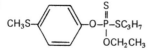

Trade Names: Bay NTN 9306 (Bayer AG)
Bolstar® (Bayer AG)
Helothion® (Bayer AG)

Manufacture

Sulprofos is made by the condensation of 4-methylthiophenol with O,O-dimethyl phosphoro-chloridothioate.

Toxicity

The acute oral LD_{50} value for rats is 304 mg/kg which is moderately toxic.

Allowable Limits on Exposure and Use

Product Use: The tolerances set by the EPA for sulprofos in or on raw agricultural commodities are as follows:

	40 *CFR* Reference	Parts per Million
Cattle, fat	180.374	0.01
Cattle, mbyp	180.374	0.01
Cattle, meat	180.374	0.01
Cotton, seed	180.374	0.5
Eggs	180.374	0.001

(continued)

	40 *CFR* Reference	Parts per Million
Goats, fat	180.374	0.01
Goats, mbyp	180.374	0.01
Goats, meat	180.374	0.01
Hogs, fat	180.374	0.01
Hogs, mbyp	180.374	0.01
Hogs, meat	180.374	0.01
Horses, fat	180.374	0.01
Horses, mbyp	180.374	0.01
Horses, meat	180.374	0.01
Milk	180.374	0.001
Poultry, fat	180.374	0.01
Poultry, mbyp	180.374	0.01
Poultry, meat	180.374	0.01
Sheep, fat	180.374	0.01
Sheep, meat	180.374	0.01
Sheep, mbyp	180.374	0.01

The tolerances set by the EPA for sulprofos in food are as follows (the *CFR* Reference is to Title 21):

	CFR Reference	Parts per Million
Cottonseed, oil	193.212	1.0

Reference

(1) Worthing, C.R., *Pesticide Manual,* 6th ed., p. 489, British Crop Protection Council (1979).

SWEP

Function: Herbicide (1)(2)(3)

Chemical Name: Methyl N-(3,4-dichlorophenyl)carbamate

Formula:

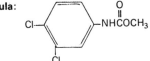

Trade Name: FMC 2995 (FMC Corp.)

Manufacture (3)

Swep may be made by (a) reacting 3,4-dichloroaniline with methyl chloroformate, or by (b) reacting methanol with 3,4-dichlorophenyl isocyanate.

Method a: In a flask were placed 40.5 g of 3,4-dichloroaniline, 23.8 g of methyl chloroformate and 250 ml of benzene. While stirring the cloudy solution at room temperature, 30.3 g of dimethylaniline was added dropwise. The solution was refluxed for one hour, cooled to room temperature, washed with 5% hydrochloric acid and with water. A white solid crystallized at once, and was separated by filtration and dried, to yield 21 g of methyl N-(3,4-dichlorophenyl)carbamate, MP 112° to 114°C.

Method b: In a flask fitted with stirrer, thermometer, condenser and powder funnel was placed 750 ml of methanol. While stirring at room temperature, a total of 470 g of 3,4-dichlorophenyl isocyanate was added, slowly enough to maintain the temperature of the exothermic reaction below 60°C. The clear solution was then refluxed for 30 minutes, and allowed to cool. On cooling a white solid crystallized to a dense cake. This cake was triturated in hexane, collected, washed and dried, to yield 474 g of methyl N-(3,4-dichlorophenyl)-carbamate, MP 108° to 111°C.

Toxicity

The acute oral LD_{50} for rats is 552 mg/kg which is slightly to moderately toxic.

References

(1) Martin, H., Worthing, C.R., *Pesticide Manual,* 5th ed., p. 486, British Crop Protection Council (January 1977).
(2) Spencer, E.Y., *Guide to the Chemicals Used in Crop Protection,* 6th ed., p. 477, London, Ontario, Agriculture Canada (January 1973).
(3) Willard, J.R. and Dorschner, K.P., U.S. Patent 3,116,995, January 7, 1964, assigned to FMC Corp.

T

2,4,5-T

Function: Herbicide (1)(2)

Chemical Name: 2,4,5-trichlorophenoxyacetic acid

Formula:

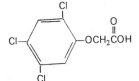

Trade Names: Weedone® 2,4,5-T (Amchem Products)
Esteron®
Reddox®
Trinoxol®

Manufacture (3)(4)

The product 2,4,5-T is made by the reaction of 2,4,5-trichlorophenol with monochloroacetic acid according to the reaction:

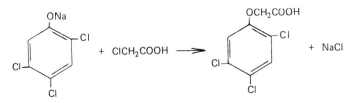

The manufacture of 2,4,5-T is carried out at a temperature from 100° to 180°C depending on the reaction medium used and the pressure employed. It is preferred to use as high a temperature as possible in order to effect rapid reaction.

This reaction is carried out at essentially atmospheric pressure. Reaction periods of from 1.5 to 3.0 hours may be used in 2,4,5-T manufacture.

The reaction is preferably carried out in the presence of an inert organic liquid boiling above 100°C such as amyl alcohol. Lower boiling liquid media require pressure operation in order to attain economically desirable reaction temperatures and reaction rates.

A production and waste schematic for 2,4,5-T manufacture is shown in Figure 51.

Figure 51: Production and Waste Schematic for 2,4,5-T (Dow Process)

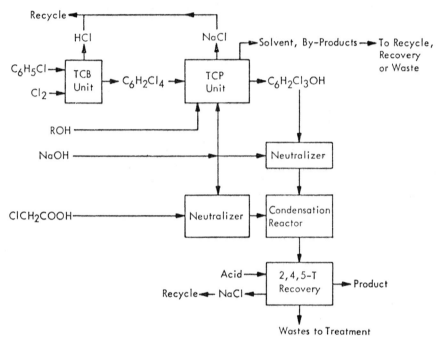

Source: Reference (3)

Process Wastes and Their Control

Water: The acid waste from a 2,4,5-T plant had the following analysis (B-10):

pH	7.9
COD	25,700 mg/l
BOD$_5$	16,680 mg/l*
Chlorides	69,000 mg/l
Total solids	172,467 mg/l
Total volatile solids	18,150 mg/l
Suspended solids	700 mg/l
Volatile suspended solids	242 mg/l
Total Kjeldahl nitrogen	40 mg/l
Odors	Not offensive

*Using 50-50 dilution of 2,4,5-T and acclimated seed (below toxic levels)

Product Wastes: The reactions of 2,4,5-Cl$_3$C$_6$H$_2$-O-CH$_2$COOH are essentially the same as those of 2,4-D. The MCA notes two disposal procedures for 2,4,5-T: (1) Mix the excess sodium carbonate, add water and let stand for 24 hours before flushing down the drain with excess water; and (2) pour onto vermiculite and incinerate with wood, paper, and waste alcohol (B-3).

Toxicity

The acute oral LD$_{50}$ value for rats is 300 mg/kg which is moderately toxic.

Although pure 2,4,5-T is moderately toxic, contamination of the herbicide with TCDD, which is very toxic, greatly increases the toxicity. No-adverse-effect doses were: for 2,4,5-T, 10 mg/kg/day in dogs and mice and up to 30 mg/kg/day in rats; and for TCDD, 0.01 μg/kg/day in rats. Based on these data ADI's were calculated at 0.1 mg/kg/day for 2,4,5-T and 10^{-4} μg/kg/day for TCDD.

There are substantial differences in the reported toxicity of 2,4,5-T, probably because of varying degrees of contamination with TCDD number of the subchronic, carcinogenicity, etc., studies should be repeated with 2,4,5 very high purity. Apparently, no adequate 2-year chronic-toxicity studies have been conducted with 2,4,5-T, and 2-year feeding studies are needed. The data available are largely from relatively short-term exposure experiments; these data, however, are fairly consistent. An exception is the Russian study in rats that reported toxic effects in mothers and their pups at extremely low maternal doses of 2,4,5-T butyl ester and a no-adverse-effect dosage only one-thousandth as high as that found by other investigators. The 2,4,5-T butyl ester used by Konstantinova may have been heavily contaminated with TCDD, but the reason for this large discrepancy is still unexplained and should be resolved (B-22).

Allowable Limits on Exposure and Use

Air: Threshold limit value for 2,4,5-T in air is 10 mg/m^3 as of 1979. The tentative short term exposure limit is 20 mg/m^3 (B-23).

Product Use: A rebuttable presumption against registration was issued on April 21, 1978 by EPA on the basis of oncogenicity, teratogenicity and fetotoxicity.

In a notice dated September 28, 1970, EPA stated that chlorodioxin contaminants were not allowed in 2,4,5-T (B-17).

In a notice dated April 20, 1970 regarding 2,4,5-T, the EPA suspended the following list of uses: (1) all uses in lakes, ponds, or on ditch banks and (2) liquid formulation for use around the home, recreation areas and similar sites (B-17).

In a notice dated May 1, 1970, the EPA cancelled the following list of uses: (1) all granular formulations for use around the home, recreation areas and similar sites and (2) all uses on food crops intended for human consumption (B-17). (Note: use on rice not finally cancelled.)

In a notice dated March 15, 1979 the EPA suspended all uses of 2,4,5-T in forestry, rights-of-way and pastures.

In a further notice on July 17, 1979 the EPA stated that the only allowable uses for 2,4,5-T are for rice, rangeland and noncrop uses. Noncrop uses include uses at airports; fences, hedgerows (not otherwise included in suspended uses e.g., rights-of-way, pasture); lumber yards; refineries; nonfood crop areas; storage areas; wastelands (not otherwise included in suspended uses, e.g., forestry); vacant lots; tank farms; industrial sites and areas (not otherwise included in suspended uses, e.g., rights-of-way). For definitions of range and pasture see "Silvex."

References

(1) Worthing, C.R., *Pesticide Manual,* 6th ed., p. 490, British Crop Protection Council (1979).
(2) Spencer, E.Y., *Guide to the Chemicals Used in Crop Protection,* 6th ed., p. 478, London, Ontario, Agriculture Canada (January 1973).
(3) Midwest Research Institute, *The Pollution Potential in Pesticide Manufacturing,* Washington, DC, Environmental Protection Agency (June 1972).
(4) Gilbert, E.E. et al, U.S. Patent 2,830,083, April 8, 1958, assigned to Allied Chemical Corp.

2,3,6-TBA

Function: Herbicide (1)(2)(3)(4)(7)

Chemical Name: 2,3,6-trichlorobenzoic acid

Formula:

Trade Names: HC-1281 (Heyden Chemical Corporation)
Trysben® (DuPont)

Manufacture

The feed materials for 2,3,6-trichlorobenzoic acid manufacture are benzoyl chloride, chlorine and water. In a first stage reaction, the benzoyl chloride is chlorinated to give trichlorobenzoyl chloride according to the reaction:

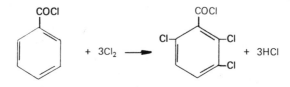

Then, in a second stage, the trichlorobenzoyl chloride is hydrolyzed to trichlorobenzoic acid according to the equation:

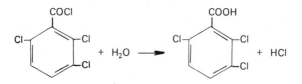

as described by R.L. Brown et al (5) and by T.A. Girard (6).

The chlorination step is carried out at 100° to 150°C. The hydrolysis step may be carried out at 75° to 170°C. The chlorination reaction is carried out at atmospheric pressure, as is the hydrolysis.

A reaction period of 8 to 14 hours may be employed to convert the benzoyl chloride to 2,3,6-

trichlorobenzoyl chloride. The hydrolysis step is relatively much more rapid. The chlorination reaction is carried out by passing chlorine into molten benzoyl chloride in the absence of any added solvent. Similarly, the hydrolysis step proceeds essentially in the melt phase, the only water added being that required stoichiometrically plus perhaps a slight excess to insure completeness of the reaction.

From 1 to 5% of a metal halide such as anhydrous ferric chloride, anhydrous antimony trichloride, or mixtures thereof, may be used to catalyze the chlorination step. Catalytic amounts of iodine, in addition to the above catalysts, also assist the chlorination. From 0.4% to 3.0% of sulfuric acid based on the weight of the reaction mass is used as a catalyst for the hydrolysis reaction. A stirred, jacketed kettle may be used for the chlorination step, and for the hydrolysis step as well.

The product mixture from the chlorination step need not be separated prior to the hydrolysis step; since no solvent was used, no solvent removal is necessary.

The product of the hydrolysis step is a molten mass consisting of various polychlorobenzoic acid isomers. It contains as impurities the metal halide catalyst and sulfuric acid. The product may be used as such in herbicidal compositions; alternatively, it can be purified by crystallization from water or from an organic solvent such as chloroform.

Alternatively, 2,3,6-TBA may be prepared (3) by nuclear-chlorinating toluene, o-chlorotoluene, m-chlorotoluene or various dichlorotoluenes or mixtures thereof to form trichlorotoluene. The mixed isomers of the trichlorotoluene fraction preferably, but not necessarily, after being separated, can be converted to trichlorobenzoic acid by side-chain chlorination (8) of the trichlorotoluene to form trichlorobenzyl chloride, which can be esterified with an alkali metal salt of an aliphatic acid to form the corresponding trichlorobenzyl ester of the acid. This ester is oxidized to trichlorobenzoic acid using, for example, nitric acid as an oxidizing agent. Under optimum conditions, using o-chlorotoluene as a starting material, the trichlorobenzoic acid will contain about 50 to 75% of the 2,3,6-isomer.

In still another alternative scheme, 2,3,6-TBA may be made from tetrachlorobenzenes as described by D.X. Klein et al (9). This process can be represented by the following equations:

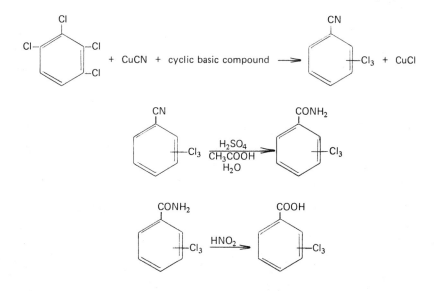

The following is a specific example of the conduct of this alternative process. A mixture of 21.6 grams (0.10 mol) of 1,2,3,4-tetrachlorobenzene, 14.3 grams (0.05 mol) of cuprous bromide, 17.9 grams (0.10 mol) of cuprous cyanide, and 39.0 grams (0.30 mol) of quinoline was heated with stirring at approximately 180°C for 4.5 hours. After cooling to 90°C and dilution with 150 ml of benzene, the reaction mixture was heated at reflux temperature for one hour and then filtered to separate the copper salts from the organic material. The salts were washed with hot benzene. The benzene solution was combined with the benzene washings and then washed with water, acidified with 10% hydrochloric acid, and washed again with water until neutral. After removal of the benzene by distillation, 18.0 grams of crude trichlorobenzonitrile was obtained.

To the crude trichlorobenzonitrile was added a mixture of 250 grams of concentrated sulfuric acid, 63 grams of glacial acetic acid, and 69 grams of water. The hydrolysis was carried out by heating the reaction mixture with stirring at 110°C for 30 minutes and then at 158°C for 4 hours. The acidic reaction mixture which contained trichlorobenzamides dissolved in acetic acid was cooled to 10°C in an ice bath and treated with a solution consisting of 4 grams of sodium nitrite in 10 ml of water.

This solution was added slowly through a tube extending below the surface of the reaction mixture. During the addition the reaction mixture was stirred rapidly. The reaction mixture was warmed at 70°C on a steam bath to complete the reaction. It was then cooled, diluted with 200 ml of water, and extracted with two 75 ml portions of benzene. After filtration the benzene solution was washed with water and then with 5% sodium hydroxide solution. The alkaline solution was filtered and acidified with hydrochloric acid.

The crude trichlorobenzoic acid that precipitated was purified by means of a procedure which involved dissolving it in dilute sodium hydroxide solution, adjusting the pH to 7 with hydrochloric acid, treating the neutral solution at its boiling point with activated carbon and then filtering, cooling and acidifying the solution with hydrochloric acid. The precipitated acid was separated and the above described purification procedure repeated. After drying, 6.7 grams (0.029 mol) of trichlorobenzoic acid was obtained which contained 63% of the 2,3,6-isomer, 27% of the 2,3,4-isomer, 6.5% of the 2,3,5-isomer, and 2.4% of the 2,4,5-isomer. From the benzene solution from which the trichlorobenzoic acid had been extracted was recovered 6 grams (0.028 mol) of 1,2,3,4-tetrachlorobenzene.

Process Wastes and Their Control

Product Wastes: 2,3,6-Trichlorobenzoic acid is usually used in the form of its dimethyl amine salt which is said to be chemically stable, but to undergo some decomposition when its aqueous solutions are evaporated to dryness and to lose amine at high pH. The acid is slightly soluble in water, and is stable in air up to at least 60°C, and to light (B-3).

Toxicity

The acute oral LD_{50} value for rats is 1,500 mg/kg which is slightly toxic.

Allowable Limits on Exposure and Use

Product Use: Products containing 2,3,6-TBA were voluntarily cancelled on the basis of oncogenicity according to a notice by EPA dated February 9, 1978.

References

(1) Worthing, C.R., *Pesticide Manual,* 6th ed., p. 493, British Crop Protection Council (1979).
(2) Spencer, E.Y., *Guide to the Chemicals Used in Crop Protection,* 6th ed., p. 515, London, Ontario, Agriculture Canada (January 1973).
(3) Girard, T.A., DiBella, E.P. and Sidi, H., U.S. Patent 2,848,470, August 19, 1958, assigned to Heyden Newport Chemical Corp.

(4) Tischler, N., U.S. Patent 3,081,162, March 12, 1963, assigned to Heyden Newport Chemical Corp.

(5) Brown, R.L. et al, U.S. Patent 2,890,243, June 9, 1959, assigned to DuPont.

(6) Girard, T.A., U.S. Patent 2,975,211, March 14, 1961, assigned to Heyden Newport Chemical Corp.

(7) Beatty, R.H., U.S. Patent 2,978,838, April 11, 1961, assigned to Amchem Products, Inc.

(8) Girard, T.A., U.S. Patent 2,980,732, April 18, 1961, assigned to Heyden Newport Chemical Corp.

(9) Klein, D.X. and Girard, D.A., U.S. Patent 3,009,942, November 21, 1961, assigned to Heyden Newport Chemical Corp.

TCA

Function: Herbicide (1)(2)(3)

Chemical Name: Sodium trichloroacetate

Formula: CCl_3COONa

Trade Names: Nata® (Farbwerke Hoechst)
Natal®
Tecane®

Manufacture

Trichloroacetic acid may be made by the reaction of acetic acid and chlorine as described by J.A. Sonia et al (4) as follows:

$$CH_3COOH + 3Cl_2 \rightarrow CCl_3COOH + 3HCl$$

The introduction of chlorine may be commenced at room temperature. It has been found more economical, however, to heat the reaction mixture up to a temperature of at least 30°C before introducing the chlorine. The temperature is then raised to between 190° and 220°C by external heating and maintained there during at least the last half of the reaction period.

Pressures of from 10 to 70 psi gauge have been found satisfactory but a preferred operating range is from 25 to 45 psi gauge pressure. Reaction periods ranging from 8 to 100 hours and typically about 20 hours may be used in the chlorination of acetic acid. The reaction is carried out in the liquid phase in the absence of added solvents. Chemical catalysts such as phosphorus halides or metal halides may be used to assist the progress of the reaction. Further, actinic light containing substantial proportions of wavelengths from 2,800 to 5,400 angstrom units may be used to promote the chlorination.

A pressure chlorinator provided with heating means, a brine-cooled reflux condenser and a light well is used for the reaction. A small amount (about 1%) of water is added to the reaction mixture to convert trichloroacetic anhydride and trichloroacetyl chloride present to trichloroacetic acid. The crude mixture of reaction products can also be converted directly to substantially pure sodium trichloroacetate by treating the mixture with an excess of caustic soda in the form of a 50% solution of sodium hydroxide at a temperature of not more than 50°C.

A variant of TCA manufacture developed by Eaker (5) is based on the discovery that a mixture containing from about 15 to 75% glacial acetic acid and about 85 to 25% acetic anhydride can be chlorinated until the reaction product reaches approximately the trichloroacetic acid stage, that is approximately 100% trichloroacetic acid and compounds hydrolyzable to trichloroacetic acid, by carrying out the chlorination process in the presence of an acid of phosphorus and preferably in the presence of a catalytic amount of an acid of phosphorus. By a catalytic amount is meant an amount up to about 2% by weight of the sum of the weight of glacial acetic acid and acetic anhydride.

A production and waste schematic for TCA manufacture is shown in Figure 52.

Process Wastes and Their Control

Air: Air emissions from TCA manufacture have been reported (B-15) to consist of 1.0 kg hydrocarbons and 0.5 kg hydrogen chloride per metric ton of pesticide produced.

Figure 52: General Process Flow Diagram for Halogenated Aliphatic Acid Production Facilities

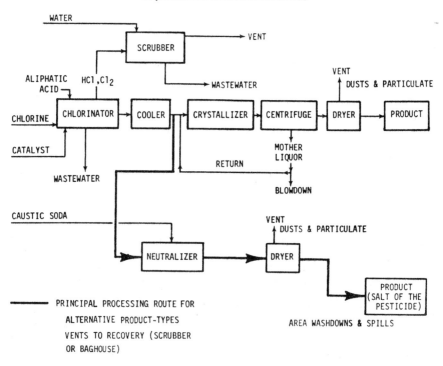

Source: Reference (B-4)

Product Wastes: The TCA sodium salt tends to decarboxylate to give chloroform under strongly alkaline conditions, e.g., treatment with 30% NaOH, and is known to undergo slow decomposition in dilute solutions (B-3).

Toxicity

The acute oral LD_{50} value for rats is 3,320 mg/kg which is slightly toxic.

Allowable Limits on Exposure and Use

Product Use: The tolerances set by the EPA for TCA in or on raw agricultural commodities are as follows:

	40 CFR Reference	Parts per Million
Beets, sugar, roots	180.310	0.5
Sugarcane	180.310	0.5

References

(1) Worthing, C.R., *Pesticide Manual,* 6th ed., p. 494, British Crop Protection Council (1979).
(2) Spencer, E.Y., *Guide to the Chemicals Used in Crop Protection,* 6th ed., p. 513, London, Ontario, Agriculture Canada (January 1973).
(3) Bousquet, E.W., U.S. Patent 2,393,086, January 15, 1946, assigned to DuPont.
(4) Sonia, J.A. et al, U.S. Patent 2,674,620, April 6, 1954, assigned to Hooker Electrochemical Co.
(5) Eaker, C.M., U.S. Patent 2,832,803, April 29, 1958, assigned to Monsanto Chemical Co.

TCMTB

Function: Fungicide (1)

Chemical Name: 2-(Thiocyanomethyl)-benzothiazole

Formula:

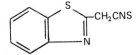

Trade Name: Busan 72® (Buckman Labs.)

Manufacture (2)

162.4 grams of chloromethyl thiocyanate was added to a solution of 258.5 grams of 2-mercaptobenzothiazole and 102.8 grams of NaOEt in 400 ml absolute EtOH which was cooled <40°C while the reaction took place. The mixture was kept at 35° to 40°C overnight and 15 days at room temperature and filtered. The filtrate plus CH_2Cl_2 extractions from the filter cake gave 118 g of 2-(thiocyanomethyl)-benzothiazole. This was combined with 242.1 grams of product recovered from the alcohol phase. The product could not be distilled without decomposition but was identified by a strong infrared band at 4.62 μ.

Toxicity

The acute oral LD_{50} value for rats is 1,590 mg/kg (B-5) which is slightly toxic.

Allowable Limits on Exposure and Use

Product Use: The tolerances set by the EPA for TCMTB use in or on raw agricultural products are as follows:

	40 CFR Reference	Parts per Million
Barley, fodder	180.288	0.1 N
Barley, forage	180.288	0.1 N
Barley, grain	180.288	0.1 N

(continued)

	40 *CFR* Reference	Parts per Million
Barley, straw	180.288	0.1 N
Beets, sugar, roots	180.288	0.1 N
Beets, sugar, tops	180.288	0.1 N
Corn, fodder	180.288	0.1 N
Corn, forage	180.288	0.1 N
Corn, grain	180.288	0.1 N
Cotton, forage	180.288	0.1 N
Cotton, seed	180.288	0.1 N
Oats, fodder	180.288	0.1 N
Oats, forage	180.288	0.1 N
Oats, grain	180.288	0.1 N
Oats, straw	180.288	0.1 N
Rice, grain	180.288	0.1 N
Rice, straw	180.288	0.1 N
Safflower, fodder	180.288	0.1 N
Safflower, forage	180.288	0.1 N
Safflower, seed	180.288	0.1 N
Sorghum, fodder	180.288	0.1 N
Sorghum, forage	180.288	0.1 N
Sorghum, grain	180.288	0.1 N
Wheat, fodder	180.288	0.1 N
Wheat, forage	180.288	0.1 N
Wheat, grain	180.288	0.1 N
Wheat, straw	180.288	0.1 N

References

(1) Packer, K., *Nanogen Index*, Freedom, CA, Nanogens International (1975).
(3) Buckman Laboratories, British Patent 1,129,575, October 9, 1968.
(3) Buckman, S.J., Pera, J.D. and Raths, F.W., U.S. Patent 3,463,785, August 26, 1969, assigned to Buckman Laboratories, Inc.

TDE

Function: Insecticide (1)(2)

Chemical Name: 2,2-bis(p-chlorophenyl)-1,1-dichloroethane

Formula:

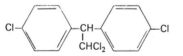

Trade Names: Rhothane® (Rohm and Haas)
DDD

Manufacture

Ethanol may be chlorinated to give 2,2-dichlorovinylethyl ether, $CCl_2=CHOC_2H_5$, (2). This is then condensed with two mols of chlorobenzene to give TDE.

Process Wastes and Their Control

Product Wastes: The chemical properties of $(ClC_6H_4)_2CHCHCl_2$ are similar to DDT, but it is apparently dehydrochlorinated slightly more slowly by alkali. It is reduced by sodium in iso-propyl alcohol (B-3).

Toxicity

The acute oral LD_{50} value for rats is 113 mg/kg which is moderately toxic.

Allowable Limits on Exposure and Use

Product Use: In an action on March 18, 1971, EPA (B-17) cancelled all uses of this product which is a metabolite of DDT.

The tolerances set by the EPA for TDE in or on raw agriculture commodities are as follows:

	40 *CFR* Reference	Parts per Million
Apples	180.187	7.0
Apricots	180.187	7.0
Beans	180.187	7.0
Blackberries	180.187	3.5
Blueberries (huckleberries)	180.187	7.0
Boysenberries	180.187	3.5
Broccoli	180.187	1.0
Brussels sprouts	180.187	1.0
Cabbage	180.187	1.0
Carrots	180.187	1.0
Cauliflower	180.187	1.0
Cherries	180.187	3.5
Citrus fruits	180.187	3.5
Corn, sweet (k+cwhr)	180.187	3.5
Cucumbers	180.187	7.0
Dewberries	180.187	3.5
Eggplant	180.187	7.0
Grapes	180.187	7.0
Kohlrabi	180.187	1.0
Lettuce	180.187	1.0
Loganberries	180.187	3.5
Melons	180.187	7.0
Nectarines	180.187	7.0
Peaches	180.187	7.0
Pears	180.187	7.0
Peas	180.187	1.0
Peppers	180.187	7.0
Plums (fresh prunes)	180.187	3.5
Pumpkins	180.187	7.0
Quinces	180.187	7.0
Raspberries	180.187	3.5
Rutabagas, roots	180.187	1.0
Rutabagas, tops	180.187	7.0
Spinach	180.187	1.0
Squash	180.187	7.0
Squash, summer	180.187	7.0
Strawberries	180.187	3.5
Tomatoes	180.187	7.0
Turnips, roots	180.187	1.0

References

(1) Spencer, E.Y., *Guide to the Chemicals Used in Crop Protection,* 6th ed., p. 154, London, Ontario, Agriculture Canada (January 1973).
(2) Meitzner, E.F. et al, U.S. Patent 2,464,600, March 15, 1949, assigned to Rohm and Haas Co.

TECNAZENE

Function: Fungicide (1)(2)

Chemical Name: 1,2,4,5-tetrachloro-3-nitrobenzene

Formula:

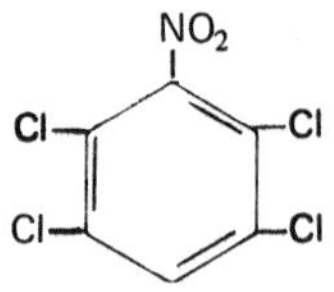

Trade Names: TCNB
Fusarex® (Bayer)
Folosan DB-905® (Bayer)
Fumite®

Manufacture

Tecnazene is made by the nitration of 1,2,4,5-tetrachlorobenzene. The 1,2,4,5-tetrachlorobenzene is made (3) by the chlorination of trichlorobenzenes which are produced by the dehydrochlorination (cracking) of unwanted BHC isomers.

Process Wastes and Their Control

Product Wastes: The disposal of this crystalline solid fungicide presents difficulties similar to PCNB (B-3). (See Quintozene.)

Toxicity

The acute oral LD_{50} value for rats is about 250 mg/kg (B-5) which is moderately toxic.

Allowable Limits on Exposure and Use

Product Use: The tolerances set by the EPA for tecnazene in or on raw agricultural commodities are as follows:

	40 *CFR* Reference	Parts per Million
Potatoes (post-h)	180.203	25.0

References

(1) Worthing, C.R., *Pesticide Manual*, 6th ed., p. 496, British Crop Protection Council (1979).
(2) Spencer, E.Y., *Guide to the Chemicals Used in Crop Protection*, 6th ed., p. 480, London, Ontario, Agriculture Canada (January 1973).
(3) Weimer, P.E. et al, U.S. Patent 2,767,226, October 16, 1956, assigned to Ethyl Corp.

TEMEPHOS

Function: Larvicide (1)(2)(3)

Chemical Name: O,O'-(thiodi-4,1-phenylene) O,O,O',O'-tetramethyl di(phosphorothioate)

Formula:

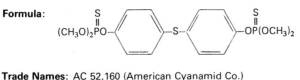

Trade Names: AC 52,160 (American Cyanamid Co.)
Abate® (American Cyanamid Co.)
Abathion® (American Cyanamid Co.)
Abat® (American Cyanamid Co.)
Swebate® (American Cyanamid Co.)
Nimitex® (American Cyanamid Co.)
Biothion® (American Cyanamid Co.)

Manufacture (3)

The compound O,O,O',O'-tetramethyl-O,O'-thiodi-p-phenylene phosphorothioate may be prepared by reacting the diphenol

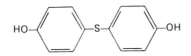

with at least 2 mols of O,O-dimethyl phosphorohalidothioate represented by the formula

where X is halogen and preferably chlorine.

The following is a specific example of the preparation of temephos in a nonaqueous solvent system. This compound is prepared by reacting 11 grams (0.05 mol) of 4,4'-thiodiphenol and 5.5 grams (0.1 mol) of sodium methoxide slurried in 400 ml of methyl ethyl ketone and refluxing for 15 minutes, distilling to remove methanol, diluting with 100 ml of methyl ethyl ketone and adding 16.2 grams (0.11 mol) of O,O-dimethyl phosphorochloridothioate dissolved in 50 ml of methyl ethyl ketone over 5 minutes under reflux. The mixture is then refluxed for 1.5 hours.

The solids are filtered and the solvent removed in vacuo. The residue is dissolved in chloroform and the solution washed with 5% sodium hydroxide, 5% hydrochloric acid, water and saturated sodium chloride solution. The washed solution is then dried and concentrated in vacuo to give 18.3 grams of oil. Purification by washing with hexane and chromatography on acid washed alumina gives pure temephos.

In contrast, the following is a specific example of the preparation of temephos in an aqueous reaction system. 4,4'-dihydroxy diphenyl sulfide (12.4 grams) is dissolved in 57 grams of 10% aqueous sodium hydroxide and the pH adjusted to 10 to 11. To this is added 36 grams of O,O-dimethyl phosphorochloridothioate with stirring and the temperature held at about 40°C, with external cooling as required, for about 4 hours. Additional 25% aqueous sodium hydroxide is added as required to maintain the initial pH. The reaction mixture is extracted with toluene, and toluene solution dried. After distillation of the solvent, the product is obtained as 24 grams (91%) yield of a viscous oil.

Process Wastes and Their Control

Product Wastes: No hydrolysis was observed after several hours at 40°C and pH 11, or at pH 8 and room temperature for several weeks. Essentially complete hydrolysis occurred upon heating in concentrated KOH for 20 minutes (B-3).

Toxicity

The acute oral LD_{50} value for rats is 1,000 mg/kg (B-5) which is slightly toxic.

Allowable Limits on Exposure and Use

Air: The threshold limit value for this material in air is 10.0 mg/m³ as of 1979. The short term exposure limit has been set tentatively at 20 mg/m³ (B-23).

Product Use: The tolerances set by the EPA for temephos in or on raw agricultural commodities are as follows:

	40 *CFR* Reference	Parts per Million
Citrus, fruits	180.170	0.1 N

References

(1) Worthing, C.R., *Pesticide Manual,* 6th ed., p. 497, British Crop Protection Council (1979).
(2) Spencer, E.Y., *Guide to the Chemicals Used in Crop Protection,* 6th ed., p. 493, London, Ontario, Agriculture Canada (January 1973).
(3) Lovell, J.B. and Baer, R.W., U.S. Patent 3,317,636, May 2, 1967, assigned to American Cyanamid Co.

TEPP

Function: Aphicide and acaricide (1)(2)

Chemical Name: Tetraethyl diphosphate

Formula:

$$(C_2H_5O)_2\overset{\displaystyle O}{\overset{\|}{P}}-O-\overset{\displaystyle O}{\overset{\|}{P}}(OC_2H_5)_2$$

Trade Names: Nifos T® (Monsanto)
Vapotone (Chevron Chemical Co.)
Tetron®

Manufacture

TEPP may be made by the reaction of triethyl phosphate and phosphorus oxychloride, $POCl_3$, as described by N.E. Willis (3):

$$5(C_2H_5O)_3P{=}O + POCl_3 \longrightarrow 3(C_2H_5O)_2\overset{O}{\overset{\|}{P}}-O-\overset{O}{\overset{\|}{P}}(OC_2H_5)_2 + 3C_2H_5Cl$$

Another method involves the interaction of triethyl phosphate and phosphorus pentoxide as described by W.H. Woodstock (4):

$$4(C_2H_5O)_3P{=}O + P_2O_5 \longrightarrow 3(C_2H_5O)_2\overset{O}{\overset{\|}{P}}-O-\overset{O}{\overset{\|}{P}}(OC_2H_5)_2$$

Still another method involves the reaction of ethanol with $POCl_3$ according to R.L. Metcalf (5).

$$15C_2H_5OH + 6POCl_3 \longrightarrow 3(C_2H_5O)_2\overset{\overset{O}{\|}}{P}-O-\overset{\overset{O}{\|}}{P}(OC_2H_5)_2 + 3C_2H_5Cl + 15HCl$$

Still another route to TEPP involves the reaction of triethyl phosphate with diethyl phosphorochloridate as described by S. Hall and M. Jacobson (6).

$$(C_2H_5O)_3P{=}O + (C_2H_5O)_2PCl \longrightarrow (C_2H_5O)_2\overset{\overset{O}{\|}}{P}-O-\overset{\overset{O}{\|}}{P}(OC_2H_5)_2 + C_2H_5Cl$$

A further possibility for TEPP manufacture involves the controlled hydrolysis of diethyl phosphorochloridate using pyridine to tie up the HCl produced as pyridine hydrochloride. As described by A.D.F. Toy (7) this process follows the equation:

$$2(C_2H_5O)_2\overset{\overset{O}{\|}}{P}Cl + H_2O + 2C_5H_5N \longrightarrow (C_2H_5O)_2\overset{\overset{O}{\|}}{P}-O-\overset{\overset{O}{\|}}{P}(OC_2H_5)_2 + 2C_5H_5N{\cdot}HCl$$

The separation of TEPP from its reaction mixtures has been described by N.E. Willis (3). The mixture is agitated with 9% aqueous NaCl to selectively hydrolyze higher polyphosphates and salt out the TEPP. This aqueous solution is then extracted with hexane to remove triethyl phosphate. The TEPP is then extracted from the aqueous layer with chlorobenzene. This solution is washed with sodium carbonate and fractionated to remove the solvent.

Process Wastes and Their Control

Air: Air emissions from TEPP manufacture have been reported (B-15) to consist of 0.5 kg hydrocarbons and 0.05 kg TEPP per metric ton of pesticide produced.

Product Wastes: TEPP is 50% hydrolyzed in water in 6.8 hours at 25°C, and 3.3 hours at 38°C; 99% hydrolysis requires 45.2 hours at 25°C, or 21.9 hours at 38°C. Hydrolysis of TEPP yields nontoxic products (B-3).

Toxicity

The acute oral LD_{50} value for rats is 0.5 mg/kg (B-5) which is extremely toxic.

Allowable Limits on Exposure and Use

Air: The threshold limit value for TEPP in air is 0.05 mg/m^3 (0.004 ppm) as of 1979. The tentative short term exposure limit is 0.2 mg/m^3 (0.01 ppm). Further, it is noted that cutaneous absorption should be prevented so that the TLV is not invalidated.

Product Use: The tolerances set by the EPA for TEPP in or on raw agricultural commodities are as follows:

	40 *CFR* Reference	Parts per Million
Alfalfa, fresh	180.347	0.01 N
Alfalfa, hay	180.347	0.01 N
Apples	180.347	0.01 N
Cabbage	180.347	0.01 N
Cauliflower	180.347	0.01 N
Oranges	180.347	0.01 N
Peaches	180.347	0.01 N
Potatoes	180.347	0.01 N

References

(1) Worthing, C.R., *Pesticide Manual,* 6th ed., p. 498, British Crop Protection Council (1979).
(2) Spencer, E.Y., *Guide to the Chemicals Used in Crop Protection,* 6th ed., p. 481, London, Ontario, Agriculture Canada (January 1973).
(3) Willis, N.E., U.S. Patent 2,523,243, September 19, 1950, assigned to Monsanto Chemical Co.
(4) Woodstock, W.H., U.S. Patent 2,402,703, June 25, 1946, assigned to Victor Chemical Works.
(5) Metcalf, R.L., *Organic Insecticides,* New York, Interscience Publishers (1955).
(6) Hall, S. and Jacobson, M., *Ind. Eng. Chem., Ind. Ed.,* 40, 694 (1948).
(7) Toy, A.D.F., *Jour. Am. Chem. Soc.* 70, 3882 (1948) and 72, 2065 (1950).

TERBACIL

Function: Herbicide (1)(2)(3)

Chemical Name: 5-chloro-3-(1,1-dimethylethyl)-6-methyl-2,4(1H,3H)-pyrimidinedione

Formula:

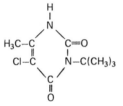

Trade Names: DuPont Herbicide 732 (DuPont)
Sinbar® (DuPont)

Manufacture

Terbacil may be made by the chlorination of 3-tert-butyl-6-methyluracil.

Process Wastes and Their Control

Air: Air emissions from terbacil production have been reported (B-15) to consist of 0.5 kg hydrocarbons, 1.0 kg HCl and 0.5 kg Cl_2 per metric ton of pesticide produced.

Toxicity

The oral LD_{50} value for rats is more than 5,000 mg/kg which is insignificantly toxic.

Allowable Limits on Exposure and Use

Product Use: The tolerances set by the EPA for terbacil in or on raw agricultural commodities are as follows:

	40 *CFR* Reference	Parts per Million
Alfalfa, forage	180.209	5.0
Alfalfa, hay	180.209	5.0

(continued)

	40 *CFR* Reference	Parts per Million
Apples	180.209	0.1
Blueberries	180.209	0.1
Caneberries	180.209	0.1
Cattle, fat	180.209	0.1
Cattle, mbyp	180.209	0.1
Cattle, meat	180.209	0.1
Citrus fruits	180.209	0.1
Goats, fat	180.209	0.1
Goats, mbyp	180.209	0.1
Goats, meat	180.209	0.1
Hogs, fat	180.209	0.1
Hogs, mbyp	180.209	0.1
Hogs, meat	180.209	0.1
Horses, fat	180.209	0.1
Horses, mbyp	180.209	0.1
Horses, meat	180.209	0.1
Milk, fat (= 0.1 in whole milk)	180.209	0.5
Peaches	180.209	0.1
Pears	180.209	0.1
Pecans	180.209	0.1
Sainfoin, forage	180.209	5.0
Sainfoin, hay	180.209	5.0
Sheep, fat	180.209	0.1
Sheep, mbyp	180.209	0.1
Sheep, meat	180.209	0.1
Sugarcane	180.209	0.1

References

(1) Worthing, C.R., *Pesticide Manual,* 6th ed., p. 499, British Crop Protection Council (1979).
(2) Spencer, E.Y., *Guide to the Chemicals Used in Crop Protection,* 6th ed., p. 482, London, Ontario, Agriculture Canada (January 1973).
(3) Loux, H.M., U.S. Patent 3,235,357, February 15, 1966, assigned to DuPont.

TERBUFOS

Function: Insecticide (1)(2)(3)(4)

Chemical Name: S-[[(1,1-dimethylethyl)thio]methyl] O,O-diethyl phosphorodithioate

Formula:

$$(CH_3CH_2O)_2 \overset{\displaystyle S}{\overset{\|}{P}}-SCH_2SC(CH_3)_3$$

Trade Names: AC-92,100 (American Cyanamid Co.)
Counter® (American Cyanamid Co.)

Manufacture (2)

Tertiary butyl mercaptan, formaldehyde, and diethyl dithiophosphoric acid are reacted together to produce S-tert-butylmercaptomethyl O,O-diethyl dithiophosphate.

Toxicity

The acute oral LD_{50} value for rats is about 1.6 mg/kg which is extremely toxic.

Allowable Limits on Exposure and Use

Product Use: The tolerances set by the EPA for terbufos in or on raw agricultural commodities are as follows:

	40 *CFR* Reference	Parts per Million
Beets, sugar, roots	180.352	0.05 N
Beets, sugar, tops	180.352	0.1
Corn, field, fodder	180.352	0.05
Corn, field, forage	180.352	0.05
Corn, grain	180.352	0.05 N
Corn, pop, fodder	180.352	0.5
Corn, pop, forage	180.352	0.5
Corn, sweet (k+cwhr)	180.352	0.05 N
Corn, sweet, fodder	180.352	0.5
Corn, sweet, forage	180.352	0.5

References

(1) Worthing, C.R., *Pesticide Manual,* 6th ed., p. 500, British Crop Protection Council (1979).
(2) Hook, E.O. and Moss, P.H., U.S. Patent 2,596,076, May 6, 1952, assigned to American Cyanamid Co.
(3) Lindsay, A.D., U.S. Patent 4,059,700, November 22, 1977, assigned to American Cyanamid Co.
(4) Gordon, F.M., U.S. Patent 4,065,558, December 27, 1977, assigned to American Cyanamid Co.

TERBUMETON

Function: Herbicide (1)(2)

Chemical Name: N-(1,1-dimethylethyl)-N'-ethyl-6-methoxy-1,3,5-triazine-2,4-diamine

Formula:

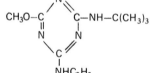

Trade Names: GS-14,259 (Ciba-Geigy)
Caragard® (Ciba-Geigy)

Manufacture

Terbumeton may be made by reacting terbuthylazine (which see) with sodium methoxide.

Toxicity

The acute oral LD_{50} value for rats is 480 to 650 mg/kg which is slightly to moderately toxic.

References

(1) Worthing, C.R., *Pesticide Manual,* 6th ed., p. 501, British Crop Protection Council (1979).

(2) Spencer, E.Y., *Guide to the Chemicals Used in Crop Protection,* 6th ed., p. 62, London, Ontario, Agriculture Canada (January 1973).

TERBUTHYLAZINE

Function: Herbicide (1)(2)

Chemical Name: 6-chloro-N-(1,1-dimethylethyl)-N'-ethyl-1,3,5-triazine-2,4-diamine

Formula:

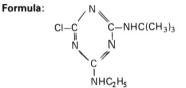

Trade Names: GS-13,529 (Ciba-Geigy)
Gardoprim® (Ciba-Geigy)

Manufacture

Cyanuric chloride as a starting material is reacted first with an equivalent of tert-butylamine and in a second step with an equivalent of ethylamine, both in the presence of NaOH to produce terbuthylazine.

Toxicity

The acute oral LD_{50} value for rats is about 2,000 mg/kg which is slightly toxic.

Allowable Limits on Exposure and Use

Product Use: The tolerances set by the EPA for terbuthylazine in or on raw agricultural commodities are as follows:

	40 *CFR* Reference	Parts per Million
Corn, fodder	180.333	0.1 N
Corn, forage	180.333	0.1 N
Corn, grain	180.333	0.1 N
Sorghum, forage	180.333	0.1 N
Sorghum, grain	180.333	0.1 N

References

(1) Worthing, C.R., *Pesticide Manual,* 6th ed., p. 502, British Crop Protection Council (1979).
(2) Spencer, E.Y., *Guide to the Chemicals Used in Crop Protection,* 6th ed., p. 60, London, Ontario, Agriculture Canada (January 1973).

TERBUTRYNE

Function: Herbicide (1)(2)(3)

Chemical Name: N-(1,1-dimethylethyl)-N'-ethyl-6-(methylthio)-1,3,5-triazine-2,4-diamine

Formula:

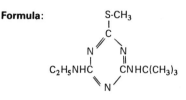

Trade Names: GS-14,250 (Ciba-Geigy)
Igrane® (Ciba-Geigy)
Preban® (Ciba-Geigy)

Manufacture (3)

Terbutryne may be made by reacting terbuthylazine (which see) with methyl mercaptan in the presence of an equivalent amount of caustic.

Toxicity

The acute oral LD_{50} value for rats is 2,000 to 3,000 mg/kg which is slightly toxic.

Allowable Limits on Exposure and Use

Product Use: The tolerances set by the EPA for terbutryne in or on raw agricultural commodities are as follows:

	40 CFR Reference	Parts per Million
Barley, fodder	180.265	0.1 N
Barley, grain	180.265	0.1 N
Barley, green	180.265	0.1 N
Barley, straw	180.265	0.1 N
Sorghum, grain	180.265	0.1 N
Wheat, fodder	180.265	0.1 N
Wheat, grain	180.265	0.1 N
Wheat, green	180.265	0.1 N
Wheat, straw	180.265	0.1 N

References

(1) Worthing, C.R., *Pesticide Manual,* 6th ed., p. 503, British Crop Protection Council (1979).
(2) Spencer, E.Y., *Guide to the Chemicals Used in Crop Protection,* 6th ed., p. 484, London, Ontario, Agriculture Canada (January 1973).
(3) Knusli, E., Schappi, W. and Berrer, D., U.S. Patent 3,145,208, August 18, 1964, assigned to J.R. Geigy AG.

TERPENE POLYCHLORINATES

Function: Insecticide and acaricide (1)

Chemical Name: Not precise; see formula under Manufacture below.

Formula: $C_{10}H_{10}Cl_8$

Trade Names: 3960-X14 (B.F. Goodrich Co.)
Strobane® (Heyden-Newport Chemical Co.)

Manufacture

The raw materials are a terpene mixture (camphene plus pinenes) and chlorine. The reactions involved are approximately as follows:

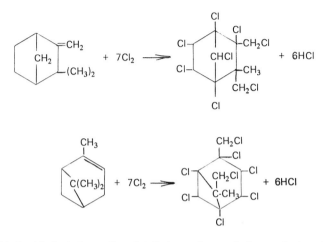

It is possible to chlorinate turpentine directly to produce a similar product as described by J.T. Smith (2).

The halogenation is exothermic and requires no heat in the initial stages of the reaction but it may be desirable to warm the mixture to 80° to 150°C or slightly higher during the later stages of the reaction when substitution is fairly slow. The chlorination temperature is not critical as long as it is held below the decomposition point of the chlorinated mixture which is about 200°C. This reaction is carried out at essentially atmospheric pressure. This reaction is fairly slow and takes several hours to complete.

No solvent or diluent is necessary for the liquid-phase chlorination reaction. However, if desired, solvents inert toward chlorine may be used. Carbon tetrachloride or other highly chlorinated solvents are examples of suitable materials.

Actinic energy in the form of ultraviolet rays or sunlight is the preferred catalyst for the reaction. Other catalysts such as red phosphorus, PCl_5 and organic peroxides can also be employed, according to A.L. Schultz et al (3). A glass-lined reaction vessel equipped with stirrer, heating and cooling jacket and a light well is suitable for this reaction.

A dark stage reactor may be used to initiate the chlorination reaction when using turpentine as the feed material. This may be followed by an actinic reactor as shown in Figure 53 and as described by J.T. Smith (2).

After the chlorine content of the chlorinated product reaches the desired level (66% chlorine), the HCl and any unreacted chlorine are swept from the product by a stream of nitrogen or air.

The crude product from the surge tank in Figure 53 may be sprayed first into a tower containing sodium carbonate solution and then into a settling tank. The product is then fed to a dehydrating tower where the last traces of moisture are removed by heating prior to sending the product to storage. About 2% by weight of alpha-pinene is then added to the chlorinated product to stabilize it against decomposition in storage.

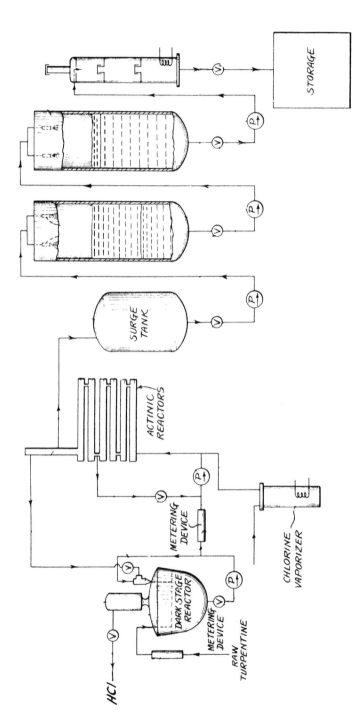

Figure 53: Flow Diagram of Process for the Chlorination of Turpentine

Source: Reference (2)

Process Wastes and Their Control

Product Wastes: Like toxaphene, this material is unstable in the presence of alkalis and organic bases and is slowly dehydrochlorinated at 100°C. It is said to resemble closely the mixture of compounds present in toxaphene (B-3).

Toxicity

The acute oral LD_{50} value for rats is 200 mg/kg (B-5) which is moderately toxic.

Allowable Limits on Exposure and Use

Product Use: As noted by EPA on June 28, 1976 (B-17) all products containing strobane were subject to voluntary cancellation.

The tolerances set by EPA for terpene polychlorinates in or on raw agricultural commodities are as follows:

	40 *CFR* Reference	Parts per Million
Cotton, seed	180.164	5.0

References

(1) Spencer, E.Y., *Guide to the Chemicals Used in Crop Protection,* 6th ed., p. 485, London, Ontario, Agriculture Canada (January 1973).
(2) Smith, J.T., U.S. Patent 3,287,241, November 22, 1966, assigned to Mission Chemical Co.
(3) Schultz, A.L. et al, U.S. Patent 2,767,115, October 16, 1956, assigned to B.F. Goodrich Co.

TETRADIFON

Function: Acaricide (1)(2)

Chemical Name: 1,2,4-trichloro-5[(4-chlorophenyl)sulfonyl] benzene

Formula:

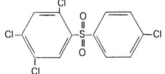

Trade Names: Tedion V-18® (N.V. Philips-Duphar)
NIA-5488 (FMC Corp.)

Manufacture

Tetradifon is made by the reaction of chlorosulfonic acid plus 1,2,4-trichlorobenzene and benzene. The first (sulfonation) step follows the equation:

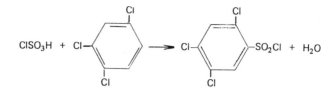

and the second (condensation) step follows the equation:

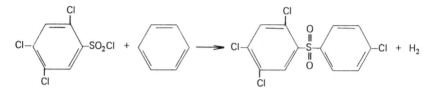

The sulfonation step may be conducted at 90°C. The condensation step may be carried out at 80° to 90°C. Both the sulfonation and condensation steps of this process are carried out at atmospheric pressure. A reaction period of about 3 hours may be used for the sulfonation step. A reaction period of from 1.5 to 3.0 hours may be used for the condensation step. Both the sulfonation and condensation steps of this process are carried out in the liquid phase in the absence of any solvent materials other than the reactants themselves. No catalyst is employed in the sulfonation step.

Condensation agents such as aluminum chloride, zinc chloride, boron trifluoride, or stannic chloride are used in the second stage of the process, however, as described by J. Meltzer (3). Stirred, jacketed reaction vessels of conventional design may be employed for both stages of this process. The second, or condensation stage must be conducted in a closed vessel from which moisture can be effectively excluded.

The sulfonation product from the first step of this process may be poured onto ice in good old time-honored tradition for the handling of such product. The product may then be recrystallized from petroleum ether to give crystals melting at 65° to 67°C.

The product of the condensation step may again be poured onto ice to quench the condensing agent. The precipitated product may be dissolved in ethyl acetate and then reprecipitated by the addition of petroleum ether. The final product is a solid melting at 127° to 129°C.

Process Wastes and Their Control

Product Wastes: Tetradifon is quite stable in dilute alkali and strong acid solutions and is even resistant to the action of mineral acids and alkalis upon prolonged heating. It is not oxidized by chromium trioxide in boiling acetic acid, but is dechlorinated by sodium in isopropyl alcohol or (rapidly) by sodium biphenyl solution (B-3).

Toxicity

The sulfur atom in this miticide, p-chlorophenyl-2,4,5-trichlorophenyl sulfone, is quite inactive and the compound is very nontoxic; the acute oral LD_{50} value for rats is 14,700 mg/kg.

Allowable Limits on Exposure and Use

Product Use: The tolerances set by the EPA for tetradifon in or on raw agricultural commodities are as follows:

	40 *CFR* Reference	Parts per Million
Apples	180.174	5.0
Apricots	180.174	5.0
Cherries	180.174	5.0
Citrus citron	180.174	2.0
Crabapples	180.174	5.0
Cucumbers	180.174	1.0
Figs	180.174	6.0
Grapefruit	180.174	2.0
Grapes	180.174	5.0
Hops, fresh	180.174	30.0
Lemons	180.174	2.0
Limes	180.174	2.0
Meat	180.174	0.0
Melons	180.174	1.0
Milk	180.174	0.0
Nectarines	180.174	5.0
Oranges	180.174	2.0
Peaches	180.174	5.0
Pears	180.174	5.0
Peppermint	180.174	100.0
Plums (fresh prunes)	180.174	5.0
Pumpkins	180.174	1.0
Quinces	180.174	5.0
Spearmint	180.174	100.0
Squash, winter	180.174	1.0
Strawberries	180.174	5.0
Tangerines	180.174	2.0
Tomatoes	180.174	1.0

The tolerances set by the EPA for tetradifon in food are as follows (the *CFR* Reference is to Title 21):

	CFR Reference	Parts per Million
Figs, dried	193.420	10.0
Hops, dried	193.420	120.0
Tea, dried	193.420	8.0

References

(1) Worthing, C.R., *Pesticide Manual,* 6th ed., p. 505, British Crop Protection Council (1979).
(2) Spencer, E.Y., *Guide to the Chemicals Used in Crop Protection,* 6th ed., p. 491, London, Ontario, Agriculture Canada (January 1973).
(3) Meltzer, J. et al, U.S. Patent 2,812,281, November 5, 1957, assigned to North American Philips Co.

TETRAMETHRIN

Function: Insecticide (1)(2)(3)

Function: (1)(2)(3)

Chemical Name: (1,3,4,5,6,7-hexahydro-1,3-dioxo-2H-isoindol-2-yl)methyl 2,2-dimethyl-3-(2-methyl-1-propenyl)cyclopropane carboxylate

Formula:

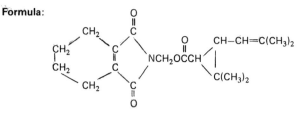

Trade Names: FMC-9260 (FMC Corp.)

Neopynamin® (Sumitomo Chemical Co.)

Manufacture (3)

A mixture of 18.1 grams of N-hydroxymethyl-1-cyclohexene-1,2-dicarboximide and 48 ml of dry pyridine was added with 50 ml of dry toluene and the mixture was cooled with ice. A solution of 19 grams of chrysanthemoyl chloride in 50 ml of dry toluene was dropped to the mixture while being stirred. The reaction proceeded exothermically, whereby pyridine-hydrochloric acid salt was isolated in the mixture. The reaction vessel was tightly closed and allowed to stand overnight.

The excess pyridine was neutralized with 5% hydrochloric acid, and the resulting two layers were separated from each other. The organic layer was washed with a saturated sodium bicarbonate solution and then with a saturated sodium chloride solution, and dried over sodium sulfate. Evaporation of the solvent in vacuo and recrystallization of the residue from ligroin yielded 28.5 grams of N-(chrysanthemoxymethyl)-1-cyclohexene-1,2-dicarboximide (MP 100° to 107°C, colorless leaflets).

Process Wastes and Their Control

Product Wastes: This compound is stable under normal conditions. Alkaline hydrolysis should readily decompose the material (B-3).

Toxicity

The acute oral LD_{50} for rats is over 4,640 mg/kg which is very slightly toxic.

References

(1) Worthing, C.R., *Pesticide Manual,* 6th ed., p. 506, British Crop Protection Council (1979).

(2) Spencer, E.Y., *Guide to the Chemicals Used in Crop Protection,* 6th ed., p. 492, London, Ontario, Agriculture Canada (January 1973).

(3) Kato, T., Ueda, K., Horie, S., Mizutani, T., Fujimoto, K. and Okuno, Y., U.S. Patent 3,268,398, August 23, 1966, assigned to Sumitomo Chemical Co.

O,O,O′,O′-TETRAPROPYL DITHIOPYROPHOSPHATE

Function: Insecticide (1)(2)(3)

Chemical Name: Tetrapropyl thiodiphosphate

Formula:

$$(C_3H_7O)_2\overset{\overset{S}{\|}}{P}-O-\overset{\overset{S}{\|}}{P}(OC_3H_7)_2$$

Trade Names: ASP-51 (Stauffer Chemical Co.)
Aspon® (Stauffer Chemical Co.)
NPD® (DuPont)

Manufacture (3)

Pyridine or another amine and water react in theoretical portions with part of the dipropyl chlorothionophosphate to give tetrapropyl thiodiphosphate and pyridine hydrochloride. The pyridine hydrochloride immediately reacts with the inorganic base to regenerate pyridine and water which react with a further portion of the dipropyl chlorothionophosphate. These reactions are illustrated in the following equations in which R is C_3H_7:

$$(1) \quad 2(RO)_2\overset{\overset{S}{\|}}{P}-Cl + H_2O + 2C_5H_5N \longrightarrow (RO)_2\overset{\overset{S}{\|}}{P}-O-\overset{\overset{S}{\|}}{P}(OR)_2 + 2C_5H_5N \cdot HCl$$

$$(2) \quad 2C_5H_5N \cdot HCl + Na_2CO_3 \longrightarrow 2C_5H_5N + H_2O + CO_2 + 2NaCl$$

Very small amounts of pyridine or other amine will suffice to promote the reaction, but for practical purposes, it is generally preferred to employ not less than 5% by weight of the theoretical amount for commercial production. While pyridine is the preferred catalyst, it is possible to employ tertiary amines such as alpha-picoline and triethylamine with a limited degree of success where low yields and quality of the product are not objectionable. The reaction may be carried out in the presence of an inert solvent, if desired, or may be carried out without the aid of a solvent.

The following is a specific example of the conduct of this process. To 216.5 grams (1 mol) of di-n-propyl chlorothionophosphate was added a mixture of 81.4 grams (1.03 mols) pyridine and 9.3 grams (0.515 mol) water. The temperature was maintained at 36° to 38°C by periodic cooling. The mixture was stirred overnight and 150 cc water added. The oily layer was separated and washed four times with 50 cc of water each. It was heated to 110°C under less than 1 mm Hg to remove unreacted $(n-C_3H_7O)_2PSCl$. The residue weighed 161 grams (81% yield).

Process Wastes and Their Control

Product Wastes: This material is stable in water at 100°C for at least 24 hours. However, it decomposes without explosive hazard at 149°C (B-3).

Toxicity

The acute oral LD_{50} for rats is 450 mg/kg (B-5) which is moderately toxic.

References

(1) Worthing, C.R., *Pesticide Manual,* 6th ed., p. 507, British Crop Protection Council (1979).
(2) Spencer, E.Y., *Guide to the Chemicals Used in Crop Protection,* 6th ed., p. 494, London, Ontario, Agriculture Canada (January 1973).
(3) Toy, A.D.F., U.S. Patent 2,663,722, December 22, 1953, assigned to Victor Chemical Works.

TETRASUL

Function: Acaricide (1)(2)(3)

Chemical Name: 1,2,4-trichloro-5-[4-(chlorophenyl)thio]benzene

Formula:

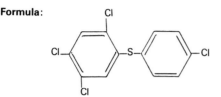

Trade Names: V-101 (N.V. Philips-Duphar)
Animert V-101® (N.V. Philips-Duphar)

Manufacture

Sodium 2,4,5-trichlorothiophenolate and p-chloronitrobenzene are first condensed to give 2,4,5-trichloro-4'-nitrodiphenylsulfide. Then, in a typical example, 15 grams of 2,4,5-trichloro-4'-nitrodiphenylsulfide (0.045 mol) and 45 grams of iron powder (0.8 gram-atom) were suspended in 300 ml of water. After the addition of 0.3 ml of acetic acid, the mixture was refluxed, while stirring, for 5 hours. Then, after cooling, 1.5 grams of sodium hydroxide was added and the precipitate was filtered off. The filtrate was extracted with benzene. After drying, filtering and thickening, 12 grams (88%) of 2,4,5-trichloro-4'-aminodiphenylsulfide with a melting point of 125° to 126°C was obtained.

Five grams of 2,4,5-trichloro-4'-aminodiphenylsulfide (0.0164 mol) obtained was dissolved in 60 ml of acetic acid, which solution was added in drops at a temperature of 5° to 10°C while stirring, to a solution of 1.25 grams of sodium nitrite (0.0181 mol) in 12.5 ml of concentrated sulfuric acid. The reaction mixture had a dark brown color and after the whole was added, the cooling bath was removed. When the temperature had increased to 15°C, the reaction mixture was poured out into a solution of 3.2 grams of cuprochloride in 32 ml of concentrated hydrochloric acid. Nitrogen was gradually liberated.

After a few hours the reaction was heated for a short time and diluted with 1 liter of water. After filtering, the yield of crude, nitrogen-free product was 4.82 grams or 91%; MP 72° to 86°C. After recrystallization from ethanol, 2.7 grams (51%) of 2,4,5,4'-tetrachlorodiphenylsulfide having a MP of 85° to 86°C was obtained.

Process Wastes and Their Control

Product Wastes: This miticide can be oxidized to its sulfone, tetradifon (which see) and poses similar disposal problems. It is apparently degraded in sunlight (B-3).

Toxicity

The acute oral LD_{50} value for rats is 3,960 mg/kg (B-5) which is slightly toxic.

Allowable Limits on Exposure and Use

Product Use: The tolerances set by the EPA for tetrasul in or on raw agriculture commodities are as follows:

	40 *CFR* Reference	Parts per Million
Apples	180.256	0.1 N

References

(1) Worthing, C.R., *Pesticide Manual,* 6th ed., p. 508, British Crop Protection Council (1979).

(2) Spencer, E.Y., *Guide to the Chemicals Used in Crop Protection,* 6th ed., p. 495, London, Ontario, Agriculture Canada (January 1973).

(3) Uhlenbroek, J.H. and Meltzer, J., U.S. Patent 3,054,719, September 18, 1962, assigned to North American Philips Co.

THIABENDAZOLE

Function: Fungicide (1)(2)

Chemical Name: 2-(4-thiazolyl)-1H-benzimidazole

Formula:

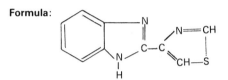

Trade Names: Bovizole®
Eprofil®
Equizole®
Mertect® (Merck and Co.)
Mintezol®
Tecto®
Thiaben®
Thibenzole®

Manufacture (3)

The following are examples of alternative modes of preparation of thiabendazole:

(a) Three grams of thiazole-4-carboxylic acid hydrobromide (14.3 mmol) and 2 grams of o-phenylenediamine (18.5 mmol) are mixed and added to 60 grams of polyphosphoric acid. The mixture is heated slowly with stirring to 240°C and maintained at this temperature for 3 hours. The hot solution is then poured onto about 200 grams of ice. A taffy-like mass separates which dissolves on stirring. The mixture is filtered and the filtrate neutralized with 30% sodium hydroxide. At a pH of about 6, 2-(4'-thiazolyl)-benzimidazole precipitates. It is filtered, washed with water, and dried in air, MP 296° to 298°C. This product is extracted with boiling ethanol. Some benzene is added to the extract and the solution boiled to remove traces of water. On concentration of the solution to a small volume and cooling, 2-(4'-thiazolyl)-benzimidazole crystallizes, MP 301° to 302°C.

(b) 22.6 grams of thiazole-4-aldehyde in 25 ml of methanol is added to a suspension of 22 grams of o-phenylenediamine in 75 ml of nitrobenzene. The resulting mixture is stirred at room temperature for a few minutes and then heated slowly to 210°C for one minute. During the heating period, the methanol is removed by distillation. The reaction mixture is then cooled with stirring to about 10°C whereupon the 2-(4'-thiazolyl)-benzimidazole crystallizes. It is filtered off and washed with ether. Any nitrobenzene remaining with the product is removed by recrystallization of the benzimidazole from alcohol.

(c) 15 grams of 4-carbethoxy isothiazole is added at room temperature to 11 grams of o-phenylenediamine in 150 grams of polyphosphoric acid. The mixture is stirred and the temperature raised to 125°C for 2 hours. It is then heated at 175°C for an additional 2 hours. The mixture is then poured into 1 liter of ice water and neutralized to a pH of about 6 with sodium hydroxide whereupon 2-(4'-isothiazolyl)-benzimidazole precipitates. The product is filtered off and extracted with hot acetone. The acetone extracts are treated with decolorizing charcoal and the filtrate obtained after removal of the charcoal is concentrated to dryness in vacuo to give the desired product.

(d) 13 grams of 4-thiazolyl acid chloride and 13 grams of o-nitroaniline are stirred together in 35 ml of pyridine at room temperature for about 12 hours. At the end of this time, the mixture is quenched in ice water and the solid nitroanilide recovered by filtration and washed with dilute sodium carbonate solution. The solid is suspended in 150 ml of glacial acetic acid, and 80 ml of 6N hydrochloric acid added to the suspension. 60 grams of zinc dust is added in small portions to the acetic mixture. After the zinc addition is complete, the reaction is essentially filtered and the filtrate neutralized with concentrated ammonium hydroxide to precipitate 2-(4'-thiazolyl)-benzimidazole. The product is purified by recrystallization from ethyl acetate. The acid chloride employed as starting material is obtained by treating 4-carboxy thiazole with thionyl chloride by known methods.

Process Wastes and Their Control

Product Wastes: Thiabendazole is a fungicide which is reported to be stable in both acid and alkaline solutions (B-3).

Toxicity

The acute oral LD_{50} value for rats is 3,300 mg/kg which is slightly toxic.

Allowable Limits on Exposure and Use

Product Use: The tolerances set by the EPA for thiabendazole in or on raw agriculture commodities are as follows:

	40 *CFR* Reference	Parts per Million
Apples (post-h)	180.242	10.0
Banana, pulp (pre & post-h)	180.242	0.4
Bananas (pre & post-h)	180.242	3.0
Beets, sugar, tops	180.242	10.0
Beets, sugar, without tops (pre & post-h)	180.242	6.0
Cattle, fat	180.242	0.1
Cattle, mbyp	180.242	0.1
Cattle, meat	180.242	0.1
Citrus fruits (post-h)	180.242	10.0
Eggs	180.242	0.1
Goats, fat	180.242	0.1
Goats, mbyp	180.242	0.1
Goats, meat	180.242	0.1
Hogs, fat	180.242	0.1
Hogs, mbyp	180.242	0.1
Hogs, meat	180.242	0.1
Horses, fat	180.242	0.1
Horses, mbyp	180.242	0.1
Horses, meat	180.242	0.1
Milk	180.242	0.1
Pears (post-h)	180.242	10.0

(continued)

	40 *CFR* Reference	Parts per Million
Potatoes (pre & post-h)	180.242	3.0
Poultry, mbyp	180.242	0.1
Polutry, meat	180.242	0.1
Sheep, fat	180.242	0.1
Sheep, mbyp	180.242	0.1
Sheep, meat	180.242	0.1
Soybeans	180.242	0.1
Squash, hubbard	180.242	1.0
Sweet potatoes, seed (post-h)	180.242	0.02 N
Wheat, grain	180.242	0.1
Wheat, straw	180.242	0.2

The tolerances set by the EPA for thiabendazole in animal feeds are as follows (the *CFR* Reference is to Title 21):

	CFR Reference	Parts per Million
Apple pomace, dried (post-h)	561.380	33.0
Beets, sugar, dried and/or dehydrated	561.380	3.5
Citrus molasses (post-h)	561.380	20.0
Citrus pulp, dried (post-h)	561.380	35.0
Potato processing waste (pre & post-h)	561.380	30.0

References

(1) Worthing, C.R., *Pesticide Manual,* 6th ed., p. 509, British Crop Protection Council (1979).
(2) Spencer, E.Y., *Guide to the Chemicals Used in Crop Protection,* 6th ed., p. 496, London, Ontario, Agriculture Canada (January 1973).
(3) Sarett, L.H. and Brown, H.D., U.S. Patent 3,017,415, January 16, 1962, assigned to Merck and Company.

THIAZFLURON

Function: Herbicide (1)(2)(3)

Chemical Name: N,N'-dimethyl-N-[5-trifluoromethyl)-1,3,4-thiadiazol-2-yl] urea

Formula:

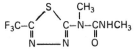

Trade Names: GS-29,696 (Ciba-Geigy)
Erbotan® (Ciba-Geigy)

Manufacture (3)

(a) 46.8 grams of trifluoroacetic anhydride are added dropwise at 0°C to a suspension of 27.3 grams of 4-methyl-thiosemicarbazide in 300 ml of ether. The mixture is then stirred for 8 hours at room temperature. After cooling, the reaction product is separated and washed with a little cold ether. The 1-trifluoroacetyl-4-methyl-thiosemicarbazide thus obtained melts at 163° to 164°C.

(b) 25 grams of 1-trifluoroacetyl-4-methyl-thiosemicarbazide are introduced in portions within 15 minutes into 125 grams of polyphosphoric acid at 80°C. The reaction mixture is then heated to 120°C and stirred for about 30 minutes at this temperature. The reaction mixture, previously cooled to about 70°C, is then poured into 500 ml of ice-water, and the product is precipitated by adding aqueous concentrated ammonia solution, separated and washed with water. After recrystallizing from ethanol/water, the 2-methylamino-5-trifluoromethyl-1,3,4-thiadiazole has a MP of 115° to 116°C.

(c) 12 grams of 2-methylamino-5-trifluoromethyl-1,3,5-thiadiazole are dissolved at about 50° to 60°C in 30 ml of dimethyl formamide, and 4 grams of methyl isocyanate are then added. Following completion of the reaction, the N-[5-trifluoromethyl-1,3,4-thiadiazolyl(2)]-N,N'-dimethyl urea separates in crystalline form. Melting point 134°C.

Toxicity

The acute oral LD_{50} value for rats is 278 mg/kg which is moderately toxic.

References

(1) Worthing, C.R., *Pesticide Manual,* 6th ed., p. 510, British Crop Protection Council (1979).
(2) Spencer, E.Y., *Guide to the Chemicals Used in Crop Protection,* 6th ed., p. 221, London, Ontario, Agriculture Canada (January 1973).
(3) Agripat, S.A., British Patent 1,254,468, November 24, 1971.

THIOFANOX

Function: Insecticide, acaricide and nematocide (1)

Chemical Name: 3,3-dimethyl-(methylthio)-2-butanone-O-[(methylamino)carbonyl] oxime

Formula:

$$(CH_3)_3CC{=}NOCNHCH_3$$
with O double-bonded at the carbonyl and CH_2SCH_3 substituent

Trade Names: Decomox® (Diamond Shamrock Corp.)

Manufacture (2)

The starting material is pinacolone (3,3-dimethyl-2-butanone); bromopinacolone and methanethiol give 3,3-dimethyl-1-methylthio-2-butanone, which is first converted to the oxime by reaction with hydroxylamine. The oxime is then treated in one of two ways: (a) with phosgene to give the oxime chloroformate followed with methyl amine or (b) with methyl isocyanate.

Toxicity

The acute oral LD_{50} value for rats is 8.5 mg/kg which is very highly toxic.

References

(1) Worthing, C.R., *Pesticide Manual,* 6th ed., p. 514, British Crop Protection Council (1979).
(2) Magee, T.A., U.S. Patent 3,875,232, April 1, 1975, assigned to Diamond Shamrock Corp.

THIOMETON

Function: Insecticide and acaricide (1)

Chemical Name: S-2-ethylthioethyl O,O-dimethyl phosphorodithioate

Formula:

$$\begin{array}{c} S \\ \parallel \\ (CH_3O)_2PSCH_2CH_2SC_2H_5 \end{array}$$

Trade Names: Bayer 23,129 (Farbenfabriken Bayer)
Ekatin® (Sandoz AG)
Thiomethon

Manufacture (2)

Thiometon may be made by the reaction of: (a) 2-ethylthioethyl chloride with sodium O,O-dimethylphosphorodithioate or (b) 2-ethylthioethanol with O,O-dimethyl hydrogen phosphorodithioate in the presence of p-toluene sulfonyl chloride.

Toxicity

The acute oral LD_{50} value for rats is 120 to 130 mg/kg which is moderately toxic.

References

(1) Worthing, C.R., *Pesticide Manual,* 6th ed., p. 515, British Crop Protection Council (1979).
(2) Lutz, K., Schuler, M. and Jucker, O., Swiss Patent 319,579, February 28, 1957, assigned to Sandoz AG.

THIONAZIN

Function: Insecticide and nematocide (1)(2)

Chemical Name: O,O-diethyl O-pyrazinyl phosphorothioate

Formula:

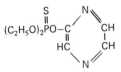

Trade Names: Experimental Nematocide 18,133
Nemaphos® (American Cyanamid Co.)
Zinophos® (American Cyanamid Co.)
Cynem® (American Cyanamid Co.)

Manufacture (3)(4)(5)(6)

To a slurry of 11.8 grams (0.1 mol) of the sodium salt of 2-hydroxypyrazine in 150 cc of N-methyl-2-pyrrolidone, 18.9 grams (0.1 mol) of O,O-diethyl phosphorochloridothioate was added with stirring. The temperature of the mixture rose immediately to 50°C, and stirring was continued at about 40°C for three hours. The reaction mixture was filtered and the precipitate was washed with a small portion of N-methyl-2-pyrrolidone. The combined filtrates were concentrated to remove the solvent and the resulting residue was dissolved in 100 cc of toluene. The toluene solution was washed with 10% aqueous sodium carbonate and then with saturated sodium chloride solution to neutrality. After drying over anhydrous magnesium sulfate, the toluene solution was concentrated in vacuo and the residue was filtered through a Hyflo magnesium sulfate mat to give 19.9 grams (73% of theory) of product, a clear amber-colored liquid.

Process Wastes and Their Control

Product Wastes: Hydrolysis yields the sodium salt of 2-pyrazinol. The reaction is used in analysis of thionazin; the hydrolysis product fluoresces (B-3).

Toxicity

The acute oral LD$_{50}$ for rats is 12 mg/kg which is highly toxic.

Allowable Limits on Exposure and Use

Product Use: The tolerances set by the EPA for thionazin in or on raw agriculture commodities are as follows:

	40 *CFR* Reference	Parts per Million
Beans, snap	180.264	0.1 N
Beans, snap, vines	180.264	0.1 N
Beets, sugar, roots	180.264	0.1 N
Beets, sugar, tops	180.264	0.1 N
Broccoli	180.264	0.1 N
Brussels sprouts	180.264	0.1 N
Cabbage	180.264	0.1 N
Cauliflower	180.264	0.1 N
Corn, field, fodder	180.264	0.1 N
Corn, field, forage	180.264	0.1 N
Corn, fresh (inc sweet) (k+cwhr)	180.264	0.1 N
Corn, grain	180.264	0.1 N
Corn, pop, fodder	180.264	0.1 N
Corn, pop, forage	180.264	0.1 N
Corn, sweet, fodder	180.264	0.1 N
Corn, sweet, forage	180.264	0.1 N
Cotton, seed	180.264	0.1 N
Mint	180.264	0.1 N
Strawberries	180.264	0.1 N

References

(1) Worthing, C.R., *Pesticide Manual,* 6th ed., p. 516, British Crop Protection Council (1979).
(2) Spencer, E.Y., *Guide to the Chemicals Used in Crop Protection,* 6th ed., p. 498, London, Ontario, Agriculture Canada (January 1973).
(3) Dixon, J.K., DuBreuil, S. and Boardway, N.L., U.S. Patent 2,918,468, December 22, 1959, assigned to American Cyanamid Co.
(4) Gordon, F.M., U.S. Patent 2,938,831, May 31, 1960, assigned to American Cyanamid Co.
(5) Miller, B. and Forbes, M.R., U.S. Patent 3,091,614, May 28, 1963, assigned to American Cyanamid Co.
(6) Gagliardi, G.N., U.S. Patent 3,340,262, September 5, 1967, assigned to American Cyanamid Co.

THIOPHANATE

Function: Fungicide (1)(2)

Chemical Name: Diethyl[1,2-phenylenebis(iminocarbonothioyl)] biscarbamate

Formula:

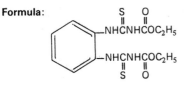
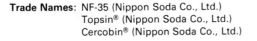

Trade Names: NF-35 (Nippon Soda Co., Ltd.)
Topsin® (Nippon Soda Co., Ltd.)
Cercobin® (Nippon Soda Co., Ltd.)

Manufacture

Preparation of compounds of this general formula, although not this specific compound, have been detailed by Nippon Soda Co., Ltd (3).

Preparation of thiophanate is as follows (4)(5): 59.0 g (0.54 mol) of ethyl chloroformate were added to 54.4 g (0.56 mol) of potassium thiocyanate in 300 ml of acetone at room temperature under agitation, and the mixture was heated and kept at a room temperature of 35° to 45°C on a water bath for one hour. Then the mixture containing the resulting ethoxycarbonylisothiocyanate was cooled and kept at a temperature at 10° to 20°C on an ice water bath under agitation. 15.5 g (0.143 mol) of o-phenylenediamine were dropped into the mixture, while it was maintained at a temperature of 10° to 20°C on an ice water bath under agitation. Then the reaction mixture was kept at room temperature for one hour, and allowed to stand to precipitate a large quantity of crystals. The reaction mixture was filtered, and the recovered crystals were washed with water and dried.

47 g of crystals were obtained. The crystals were light yellow, and had a decomposition point of 190° to 191°C. Colorless plates having the decomposition point of 194°C were obtained by recrystallization from acetone.

Toxicity

The acute oral LD_{50} value for rats is over 10,000 mg/kg which is insignificantly toxic.

References

(1) Worthing, C.R., *Pesticide Manual,* 6th ed., p. 517, British Crop Protection Council (1979).
(2) Spencer, E.Y., *Guide to the Chemicals Used in Crop Protection,* 6th ed., p. 499, London, Ontario, Agriculture Canada (January 1973).
(3) Nippon Soda Co., Ltd., British Patent 1,214,415, December 2, 1970.
(4) Noguchi, T., Kohmoto, K., Yasuda, Y., Hashimoto, S., Kato, K., Miyazaki, K. and Takiguchi, D., U.S. Patent 3,745,187, July 10, 1973, assigned to Nippon Soda Co., Ltd.
(5) Kohmoto, K. and Miyazaki, K., U.S. Patent 3,769,308, October 30, 1973.

THIOPHANATE-METHYL

Function: Fungicide (1)(2)

Chemical Name: Dimethyl[1,2-phenylenebis(iminocarbonothioyl)] biscarbamate

Formula:

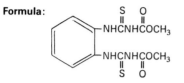

Trade Names: NF-44 (Nippon Soda Co., Ltd.)
Topsin-Methyl® (Nippon Soda Co., Ltd.)
Cercobin-Methyl® (Nippon Soda Co., Ltd.)
Mildophane® (May and Baker Co., Ltd.)

Manufacture

Methyl chloroformate is reacted with potassium thiocyanate to give methyl isothiocyano-formate which is then reacted with o-phenylenediamine to give the thiophanate.

Toxicity

The acute oral LD_{50} value for rats is 7,500 mg/kg (B-5) which is insignificantly toxic.

Allowable Limits on Exposure and Use

Product Use: A rebuttable presumption against registration was issued on December 1, 1977 by EPA on the basis of mutagenicity and reduction in nontarget species.

The tolerances set by the EPA for thiophanate-methyl in or on raw agricultural commodities are as follows:

	40 *CFR* Reference	Parts per Million
Apricots (pre and post-h)	180.371	15.0
Bananas (pre-h)	180.371	2.0
Bananas, pulp	180.371	0.2
Cherries (pre and post-h)	180.371	15.0
Nectarines (pre and post-h)	180.371	15.0
Peaches (pre and post-h)	180.371	15.0
Plums (pre and post-h)	180.371	15.0
Prunes (pre and post-h)	180.371	15.0
Strawberries (pre-h)	180.371	5.0

References

(1) Worthing, C.R., *Pesticide Manual,* 6th ed., p. 518, British Crop Protection Council (1979).
(2) Spencer, E.Y., *Guide to the Chemicals Used in Crop Protection,* 6th ed., p. 500, London, Ontario, Agriculture Canada (January 1973).

THIRAM

Function: Fungicide (1)(2)(3)

Chemical Name: Tetramethyl thioperoxydicarbonic diamide

Formula:

$$\underset{(CH_3)_2 NCSSCN(CH_3)_2}{\overset{\overset{S}{\|}\quad\overset{S}{\|}}{}}$$

Trade Names: Arasan® (DuPont)
Tersan® (DuPont)
Pomarsol® (Bayer)
Nomersan® (Plant Protection, Ltd.)
Tuads®
Thylate®
Delsan®
Spotrete®
Thiurad®
Thiuramyl®
Tiuramyl®

Manufacture

Thiram may be made first by the reaction of carbon disulfide and dimethylamine in the presence of NaOH to give sodium dimethyldithiocarbamate. Two molecules of the intermediate are then oxidatively coupled using H_2O_2.

Process Wastes and Their Control

Product Wastes: Thiram can be dissolved in alcohol or other flammable solvent and burned in an incinerator with an afterburner and scrubber (B-3).

Toxicity

The acute oral LD_{50} value for rats is 560 mg/kg (B-5) which is slightly toxic.

The toxic properties of dithiocarbamate fungicides have been discussed at some length by the National Research Council (B-22) and the reader is referred to the section of this volume on "Mancozeb" for more detail.

Allowable Limits on Exposure and Use

Air: The threshold limit value for thiram in air is 5 mg/m³ as of 1979. The tentative short term exposure limit is 10 mg/m³ (B-23).

Product Use: The tolerances set by the EPA for thiram in or on raw agricultural commodities are as follows:

	40 *CFR* Reference	Parts per Million
Apples	180.132	7.0
Bananas, pulp	180.132	1.0
Bananas, with peel (pre & post-h)	180.132	7.0
Celery	180.132	7.0
Onions, dry bulb	180.132	0.5
Peaches	180.132	7.0
Strawberries	180.132	7.0
Tomatoes	180.132	7.0

References

(1) Worthing, C.R., *Pesticide Manual,* 6th ed., p. 519, British Crop Protection Council (1979).
(2) Spencer, E.Y., *Guide to the Chemicals Used in Crop Protection,* 6th ed., p. 502, London, Ontario, Agriculture Canada (January 1973).

(3) Tisdale, W.H. and Williams, I., U.S. Patent 1,972,961, September 11, 1934, assigned to DuPont.

TOXAPHENE (CAMPHECHLOR IN U.K.)

Function: Insecticide (1)(2)(3)(4)

Chemical Name: Toxaphene

Formula:

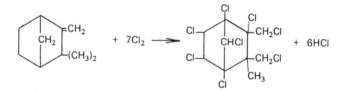

Trade Names: Hercules 3956 (Hercules, Inc.)
Alltox®
Estonox®
Chem-Phene®
Geniphene®
Gy-phene®
Phenacide®
Phenatox®
Toxadust®
Toxaspra®

Manufacture

The raw materials for toxaphene manufacture are camphene and chlorine and the reaction is approximately as follows:

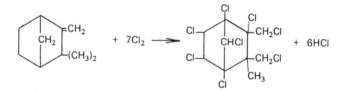

The initial reaction involving addition to the double bond is rapid while the second involving substitution proceeds with more difficulty, as pointed out by M.A. Phillips (5). The reaction temperature initially rises, due to the heat of reaction, to 85° to 90°C, and some cooling may be required. It then drops and may be 50° to 75°C at the end of the reaction. This chlorination reaction is carried out at atmospheric pressure. This reaction takes from 15 to 30 hours to reach completion. This reaction is carried out in the liquid phase using about 5 parts of carbon tetrachloride solvent per part of camphene feed, as described by G.A. Buntin (3)(4). Ultraviolet light is the catalyst for this reaction.

Lead-lined, glass-lined or nickel-clad vessels may be used for this reaction. The vessel should be equipped with a heat-exchange jacket, a reflux condenser and a well for the ultraviolet lamp. The carbon tetrachloride solvent is removed from the reaction product by distillation under reduced pressure after HCl and excess chlorine have been blown out. The residue from the distillation is allowed to solidify. A production and waste schematic for toxaphene manufacture is shown in Figure 54 (6).

Figure 54: Production and Waste Schematic for Toxaphene

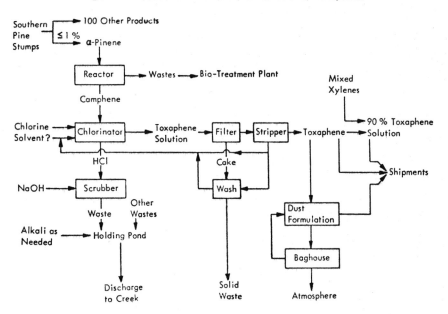

Source: Reference (6)

Process Wastes and Their Control

Air: Air emissions from toxaphene production have been reported (B-15) to consist of the following:

Component	Kilograms per Metric Ton Pesticide Produced
HCl	2.65
Cl$_2$	0.25
Toluene	1.0
Toxaphene	5×10^{-6}

Air pollution control in toxaphene manufacture involves the following (B-15):

Control Device	Emissions Controlled	Reported Efficiency
Alkali and water scrubber	Solvent vapor, hydrogen chloride, chlorine	—
Stripping	Solvent vapor, hydrogen chloride, chlorine	—
Limestone adsorption	Solvent vapor, hydrogen chloride, chlorine	100
Baghouse	Toxaphene	—

The chlorinator waste HCl gas passes through a water absorber and the resulting muriatic acid is recovered or neutralized and sent to wastewater treatment. The toxaphene product then goes either to a solution or to a dust formulation step. Emissions from the dust formulation are vented to a baghouse, with the captured dust then recycled to the formulation step (B-10).

Product Wastes: Toxaphene is said to dehydrochlorinate in the presence of alkali, upon prolonged exposure to sunlight, and at temperatures of about 155°C. Reduction with sodium in isopropyl alcohol is the analytical method for total chloride (B-3).

Toxicity

The acute oral LD_{50} value for rats is 80 to 90 mg/kg which is moderately toxic.

Toxaphene is a widely used organochlorine insecticide that apparently has not caused a great deal of environmental harm, although it has been used in agriculture for many years. Because it is a complex mixture of uncharacterized camphene derivatives, very little is known about its metabolism in plants or other higher organisms. Considerable information is available, however, on its toxicity in laboratory animals and various aquatic organisms. An ADI of 0.00125 mg/kg/day was calculated on the basis of the chronic toxicity data (B-22).

A summary of the results of examination of over 100,000 samples of raw agricultural commodities by the FDA between 1963 and 1969 shows that toxaphene residues are seldom present. Thus, the possibility that large quantities of toxaphene residues could be found in drinking water is not great.

Toxaphene has demonstrated carcinogenic effects in laboratory animals. In addition, toxaphene is highly toxic to many aquatic invertebrate and vertebrate species and has been shown to cause the "broken back syndrome" in fish fry. These observations, together with reported bioconcentration factors as high as 91,000 indicate that toxaphene poses a threat to living organisms, particularly in the aquatic environment (B-26).

Allowable Limits on Exposure and Use

Air: The threshold limit value for chlorinated camphenes in air has been set at 0.5 mg/m^3 as of 1979. The tentative short term exposure limit is 1.0 mg/m^3 (B-23).

Water: In water, EPA set criteria (B-12) for toxaphene of 5 μg/l for domestic water supply and 0.005 μg/l for the protection of freshwater and marine aquatic life.

Subsequently, EPA has suggested (B-26) limits to protect freshwater aquatic life of 0.007 μg/l as a 24-hour average and the concentration should not exceed 0.47 μg/l at any time.

For toxaphene the criterion to protect saltwater aquatic life is 0.019 μg/l as a 24-hour average and the concentration should not exceed 0.12 μg/l at any time.

For the maximum protection of human health from the potential carcinogenic effects of exposure to toxaphene through ingestion of water and contaminated aquatic organisms, the ambient water concentration is zero. Concentrations of toxaphene estimated to result in additional lifetime cancer risks ranging from no additional risk to an additional risk of 1 in 100,000 have been determined by the EPA. The agency is considering setting criteria at an interim target risk level in the range of 10^{-5}, 10^{-6}, or 10^{-7} with corresponding criteria of 4.7×10^{-4} μg/l, 4.7×10^{-5} μg/l, and 4.7×10^{-6} μg/l, respectively (B-26).

Product Use: A rebuttable presumption against registration was issued on May 25, 1977 by EPA on the basis of oncogenicity and reductions in nontarget species.

In a notice dated February 14, 1969, the EPA (B-17) cancelled all uses of toxaphene products bearing directions for use on lettuce and cabbage except the following:

(1) Cabbage at application rates of 4.0 pounds actual/acre must have the warning statement "Do not apply after heads start to form."

(2) Lettuce at application rates of 5.0 pounds actual/acre must have the warning statement "Do not apply after seedling stage on leaf lettuce. Do not apply after heads begin to form on head of lettuce."

The tolerances set by the EPA for toxaphene in or on raw agricultural commodities are as follows:

	40 *CFR* Reference	Parts per Million
Apples	180.138	7.0
Apricots	180.138	7.0
Bananas	180.138	3.0
Bananas, pulp	180.138	0.3
Barley	180.138	5.0
Beans	180.138	7.0
Blackberries	180.138	7.0
Boysenberries	180.138	7.0
Broccoli	180.138	7.0
Brussels sprouts	180.138	7.0
Cabbage	180.138	7.0
Carrots	180.138	7.0
Cattle, fat of meat	180.138	7.0
Cauliflower	180.138	7.0
Celery	180.138	7.0
Citrus fruits	180.138	7.0
Collards	180.138	7.0
Corn	180.138	7.0
Cotton, seed	180.138	5.0
Cranberries	180.138	7.0
Cucumbers	180.138	7.0
Dewberries	180.138	7.0
Eggplant	180.138	7.0
Goats, fat of meat	180.138	7.0
Hazelnuts	180.138	7.0
Hickory nuts	180.138	7.0
Hogs, fat of meat	180.138	7.0
Horseradish	180.138	7.0
Horses, fat of meat	180.138	7.0
Kale	180.138	7.0
Kohlrabi	180.138	7.0
Lettuce	180.138	7.0
Loganberries	180.138	7.0
Nectarines	180.138	7.0
Oats	180.138	5.0
Okra	180.138	7.0
Onions	180.138	7.0
Parsnips	180.138	7.0
Peaches	180.138	7.0
Peanuts	180.138	7.0
Pears	180.138	7.0
Peas	180.138	7.0
Pecans	180.138	7.0
Peppers	180.138	7.0
Pimentos	180.138	7.0
Pineapples	180.138	3.0
Quinces	180.138	7.0
Radishes, tops	180.138	7.0
Radishes, with tops	180.138	7.0
Radishes, without tops	180.138	7.0
Raspberries	180.138	7.0
Rice	180.138	5.0
Rutabagas	180.138	7.0
Rye	180.138	5.0
Sheep, fat of meat	180.138	7.0
Sorghum, grain	180.138	5.0
Soybeans, dry form	180.138	2.0
Spinach	180.138	7.0

(continued)

	40 *CFR* Reference	Parts per Million
Strawberries	180.138	7.0
Sunflower seeds	180.138	0.1
Tomatoes	180.138	7.0
Walnuts	180.138	7.0
Wheat	180.138	5.0
Youngberries	180.138	7.0

The tolerances set by the EPA for toxaphene in food are as follows (the *CFR* Reference is to Title 21):

	CFR Reference	Parts per Million
Soybean, oil, crude	193.450	12.0

References

(1) Worthing, C.R., *Pesticide Manual,* 6th ed., p. 76, British Crop Protection Council (1979).
(2) Spencer, E.Y., *Guide to the Chemicals Used in Crop Protection,* 6th ed., p. 506, London, Ontario, Agriculture Canada (January 1973).
(3) Buntin, G.A., U.S. Patent 2,565,471, August 28, 1951, assigned to Hercules Powder Co.
(4) Buntin, G.A., U.S. Patent 2,657,164, October 27, 1953, assigned to Hercules Powder Co.
(5) Phillips, M.A., *Brit. Chem. Eng.,* 10, No. 8, 550-51 (August 1965).
(6) Midwest Research Institute, *The Pollution Potential in Pesticide Manufacturing,* Washington, DC, Environmental Protection Agency (June 1972).

TRIADIMEFON

Function: Fungicide (1)(2)

Chemical Name: 1-(4-chlorophenoxy)-3,3-dimethyl-1-(1H-1,2,4-triazol-1-yl)-2-butanone

Formula:

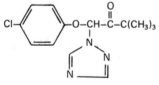

Trade Names: Bay Meb 6447 (Bayer AG)
Bayleton® (Bayer AG)

Manufacture (2)

35.8 grams (0.2 mol) of α-bromo-pinacolone in 50 ml of ethyl acetate were added dropwise to sodium 4-chlorophenolate which was prepared from 0.2 mol of 4-chlorophenol and 4.6 grams (0.2 mol) of sodium in 130 ml of absolute alcohol, and the mixture was heated to the boil overnight. Thereafter the sodium bromide produced was filtered off hot, the filtrate was distilled in vacuo and the solid residue was recrystallized from a little ligroin.

1-(4'-chlorophenoxy)-3,3-dimethyl-butan-2-one (73% of theory) was obtained.

6 ml (0.11 mol) of bromine were added to 0.1 mol of 1-(4'-chlorophenoxy)-3,3-dimethyl-butan-2-one and the mixture was heated under reflux to 140°C for 1 hour. The resulting oily resi-

due was taken up with petroleum ether, whereupon it crystallized; the solid residue was filtered off and well rinsed.

1-bromo-1-(4'-chlorophenoxy)-3,3-dimethyl-butan-2-one (89% of theory) was obtained.

0.033 mol of 1-bromo-1-(4'-chlorophenoxy)-3,3-dimethyl-butan-2-one and 9.9 grams (0.15 mol) of 1,2,4-triazole were dissolved in 80 ml of acetonitrile and heated under reflux for 48 hours. Thereafter the solvent was distilled off in vacuo, the residue was taken up with 150 ml of water and the aqueous solution was extracted by shaking three times with 40 ml of methylene chloride at a time. The organic phase was thereafter twice washed with 150 ml of water at a time, dried over sodium sulfate and distilled.

The oil obtained as residue was fractionally recrystallized from a little ether, whereby triadimefon, melting at about 82°C was obtained.

Toxicity

The acute oral LD_{50} value for rats is 560 to 570 mg/kg which is moderately toxic.

References

(1) Worthing, C.R., *Pesticide Manual,* 6th ed., p. 523, British Crop Protection Council (1979).
(2) Meiser, W., Buchel, K.H. and Kramer, W., U.S. Patent 3,912,752, October 14, 1975, assigned to Bayer AG.

TRIALLATE

Function: Herbicide (1)(2)(3)(4)

Chemical Name: S-(2,3,3-trichloro-2-propenyl)bis(1-methylethyl)carbamothioate

Formula:

$$[(CH_3)_2CH]_2NCSCH_2CCl=CCl_2$$

with $\overset{O}{\underset{\|}{}}$ above the C of NCSCH

Trade Names: CP 23426 (Monsanto)
Avadex BW® (Monsanto)
Fargo® (Monsanto)

Manufacture (3)(4)

To a stirred solution of 202.4 grams (2.0 mols) of diisopropylamine in 1,000 ml of dry ethyl ether at −10° to 0°C there was bubbled in carbon oxysulfide until the gain in weight was 120 grams. This addition required 30 minutes and the mixture was then stirred at −10° to 0°C for an additional 90 minutes. Thereupon 145.4 grams (1.0 mol) of 1,1,2,3-tetrachloropropene was added in one portion and the reaction mixture stirred at 25° to 30°C for 24 hours. The by-product salt was removed by filtration and the excess ether removed in vacuo. The residue was distilled in vacuo and the fraction boiling at 148° to 149°C at 9 mm collected.

Toxicity

The acute oral LD_{50} value for rats is 1,471 mg/kg (B-5) which is slightly toxic.

Allowable Limits on Exposure and Use

Product Use: Issuance of a rebuttable presumption against registration for triallate was being considered by EPA as of September 1979 on the basis of possible mutagenicity.

The tolerances set by the EPA for triallate in or on raw agricultural commodities are as follows:

	40 *CFR* Reference	Parts per Million
Barley, grain	180.314	0.05 N
Barley, straw	180.314	0.05 N
Lentils	180.314	0.05 N
Lentils, forage	180.314	0.05 N
Lentils, hay	180.314	0.05 N
Peas	180.314	0.05 N
Peas, forage	180.314	0.05 N
Peas, hay	180.314	0.05 N
Wheat, grain	180.314	0.05 N
Wheat, straw	180.314	0.05 N

References

(1) Worthing, C.R., *Pesticide Manual*, 6th ed., p. 524, British Crop Protection Council (1979).
(2) Spencer, E.Y., *Guide to the Chemicals Used in Crop Protection*, 6th ed., p. 507, London, Ontario, Agriculture Canada (January 1973).
(3) Harman, M.W. and D'Amico, J.J., U.S. Patent 3,330,643, July 11, 1967, assigned to Monsanto Co.
(4) Harman, M.W. and D'Amico, J.J., U.S. Patent 3,330,821, July 11, 1967, assigned to Monsanto Co.

TRIAMIPHOS

Function: Fungicide (1)(2)(3)(4)

Chemical Name: p-(5-Amino-3-phenyl-1H-1,2,4-triazol-1-yl)-N,N,N',N'-tetramethyl phosphonic diamide

Formula:

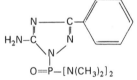

Trade Names: WP-155 (N.V. Philips-Duphar)
Wepsyn® (N.V. Philips-Duphar)

Manufacture

In a solution of 2.3 grams of sodium (0.1 mol) in 100 ml of methanol is dissolved 16.0 grams of 5-phenyl-3-aminotriazol-1,2,4 (0.1 mol). The methanol is evaporated in vacuo and the remaining salt is suspended in 80 ml of acetonitrile. To the suspension is added 20 grams of bis(N,N-dimethylamido)-phosphorylchloride. The mixture is then heated while stirring, for 2.5 hours at 80° to 85°C. The solution is filtered hot and the filtrate is concentrated by evaporation. The residue, a crystalline product, is recrystallized from a mixture of ethanol and water (1:3). Yield: 23.4 grams (80%), MP 167° to 168°C.

Toxicity

The acute oral LD_{50} value for rats is 20 mg/kg which is highly toxic.

References

(1) Worthing, C.R., *Pesticide Manual,* 6th ed., p. 525, British Crop Protection Council (1979).
(2) Spencer, E.Y., *Guide to the Chemicals Used in Crop Protection,* 6th ed., p. 508, London, Ontario, Agriculture Canada (January 1973).
(3) Koopmans, M.P., Meltzer, J., Huisman, H.O., van den Bos, G. and Wellinga, K., U.S. Patent 3,121,090, February 11, 1964, assigned to North American Philips Co., Inc.
(4) Koopmans, M.P., Meltzer, J., Huisman, H.O., van den Bos, G. and Wellinga, K., U.S. Patent 3,220,922, November 30, 1965, assigned to North American Philips Co., Inc.

TRIAZOPHOS

Function: Insecticide and acaricide (1)(2)(3)

Chemical Name: O,O-diethyl O-(1-phenyl-1H-1,2,4-triazol-3-yl) phosphorothioate

Formula:

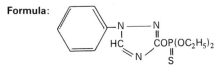

Trade Names: HOE 2960 (Farbwerke Hoechst)
Hostathion® (Farbwerke Hoechst)

Manufacture (3)

32.2 grams (0.2 mol) of 1-phenyl-3-hydroxy-1,2,4-triazole was suspended in 250 ml of acetone and 38 grams (0.2 mol) of O,O-diethyl-thiophosphoryl chloride were added thereto. Subsequently, 22 grams (0.22 mol) of triethylamine were added dropwise and the mixture was stirred for 6 hours at about 50°C. After cooling, the triethylamine hydrochloride was filtered off and the solvent was evaporated from the filtrate. There were obtained 60 grams of 1-phenyl-3-(O,O-diethylthionophosphoryl)-1,2,4-triazole as a light brown oil.

Toxicity

The acute oral LD_{50} value for rats is 66 mg/kg which is moderately toxic.

References

(1) Worthing, C.R., *Pesticide Manual,* 6th ed., p. 526, British Crop Protection Council (1979).
(2) Spencer, E.Y., *Guide to the Chemicals Used in Crop Protection,* 6th ed., p. 508a, London, Ontario, Agriculture Canada (January 1973).
(3) Scherer, O. and Mildenberger, H., U.S. Patent 3,686,200, August 22, 1972, assigned to Farbwerke Hoechst AG.

S,S,S-TRIBUTYL PHOSPHOROTRITHIOATE

Function: Defoliant (1)(2)

Chemical Name: S,S,S-tributyl phosphorotrithioate

Formula: $(C_4H_9S)_3P=O$

Trade Names: B-1776 (Chemagro Corp.)
DEF Defoliant® (Chemagro Corp.)
Fosfall A®
Degreen®
For Fall A®

Manufacture

The raw materials for manufacture are n-butyl mercaptan and phosphorus trichloride. These react to produce a trithiophosphite which is then air oxidized (5) to give a phosphorotrithioate. The reactions may be expressed as follows:

$$PCl_3 + 3C_4H_9SH \longrightarrow (C_4H_9S)_3P + 3HCl$$

$$(C_4H_9S)_3P + \tfrac{1}{2}O_2 \longrightarrow (C_4H_9S)_3PO$$

Actually, however, the reaction is not that simple as pointed out by K.H. Rattenbury et al (3). The reaction of phosphorus trichloride with butyl mercaptan to form S,S,S-tributyl trithiophosphite is one which is irreversible with respect to butyl mercaptan and hydrogen chloride but reversible with respect to phosphorus trichloride. The following reactions are proceeding:

(1) $\qquad PCl_3 + C_4H_9SH \longrightarrow C_4H_9SPCl_2 + HCl \uparrow$

(2) $\qquad C_4H_9SPCl_2 + C_4H_9SH \longrightarrow (C_4H_9S)_2PCl + HCl \uparrow$

(3) $\qquad (C_4H_9S)_2PCl + C_4H_9SH \longrightarrow (C_4H_9S)_3P + HCl \uparrow$

while at the same time the following competing reactions are preventing the reaction from going to completion:

(4) $\qquad 2(C_4H_9S)_3P + PCl_3 \longrightarrow 3(C_4H_9S)_2PCl$

(5) $\qquad (C_4H_9S)_2PCl + PCl_3 \longrightarrow 2C_4H_9SPCl_2$

(6) $\qquad (C_4H_9S)_3P + C_4H_9SPCl_2 \longrightarrow 2(C_4H_9S)_2PCl$

An alternative scheme is to react the mercaptan first with metallic sodium to give the RSNa compound which is then reacted with POCl₃ to give the trithiophosphite final product as described by F.X. Markley (4).

Process Wastes and Their Control

Product Wastes: This compound is slowly hydrolyzed under alkaline conditions (B-3).

Toxicity

The acute oral LD$_{50}$ value for rats is 325 mg/kg which is moderately toxic.

Allowable Limits on Exposure and Use

Product Use: Issuance of a rebuttable presumption against registration for this compound was being considered by EPA as of September 1979 on the basis of possible neurotoxicity.

The tolerances set by the EPA for this material in or on raw agricultural commodities are as follows:

	40 *CFR* Reference	Parts per Million
Cattle, fat	180.272	0.02 N
Cattle, mbyp	180.272	0.02 N
Cattle, meat	180.272	0.02 N
Cotton, seed	180.272	4.0
Goats, fat	180.272	0.02 N
Goats, mbyp	180.272	0.02 N
Goats, meat	180.272	0.02 N
Milk	180.272	0.002 N
Sheep, fat	180.272	0.02 V
Sheep, mbyp	180.272	0.02 N
Sheep, meat	180.272	0.02 N

The tolerances set by the EPA for tributyl phosphorotrithioate in animal feeds are as follows (the *CFR* Reference is to Title 21):

	CFR Reference	Parts per Million
Cotton, seed, hulls	561.390	6.0

References

(1) Worthing, C.R., *Pesticide Manual,* 6th ed., p. 527, British Crop Protection Council (1979).
(2) Spencer, E.Y., *Guide to the Chemicals Used in Crop Protection,* 6th ed., p. 510, London, Ontario, Agriculture Canada (January 1973).
(3) Rattenbury, K.H. et al, U.S. Patent 2,943,107, June 28, 1960, assigned to Chemagro Corp.
(4) Markley, F.X., U.S. Patent 2,965,467, December 20, 1960, assigned to Pittsburgh Coke and Chemical Co.
(5) Warner, P.F. et al, U.S. Patent 3,174,989, March 23, 1965, assigned to Phillips Petroleum Co.
(6) Clark, J.W., U.S. Patent 3,178,468, April 13, 1965, assigned to Phillips Petroleum Co.

TRIBUTYL PHOSPHOROTRITHIOITE

Function: Defoliant (1)(5)

Chemical Name: Tributylphosphorotrithioite

Formula: $(C_4H_9S)_3P$

Trade Name: Folex® (Mobil Chemical Co.)

Manufacture

The raw materials for manufacture are butyl mercaptan and phosphorus trichloride which react as follows:

$$3C_4H_9SH + PCl_3 \rightarrow (C_4H_9S)_3P + 3HCl$$

as described, for example, by W.W. Crouch et al (2). A ratio of 5 or 6 mols of mercaptan per mol of PCl_3 gives superior yields, it is stated. An alternative route is by the reaction of elemental phosphorus with a dialkyl sulfide according to the reaction:

$$3C_4H_9SSC_4H_9 + 2P \rightarrow 2(C_4H_9S)_3P$$

as described, for example, by D.R. Stevens et al (3).

Temperatures from 10° to 135°C may be used in the conduct of the reaction of PCl_3 and butyl mercaptan. The reaction of dibutyl disulfide and yellow phosphorus is carried out at 150° to 250°C and preferably at 170° to 210°C. The reaction temperature must be high enough to cause solution of the phosphorus in the disulfide-solvent liquid phase. The rate of reaction is, of course, favorably influenced by the use of higher temperatures; however, it is necessary to operate below the decomposition temperatures of the product trithiophosphites (about 300°C).

The reaction of PCl_3 with butyl mercaptan is carried out at atmospheric pressure or under slightly reduced pressure to assist in the removal of the hydrogen chloride evolved. The reaction of a dialkyl disulfide and phosphorus is carried under such pressure that both the disulfide and the solvent are in the liquid phase. When using decalin as the solvent, the pressure may be essentially atmospheric.

Reaction times ranging from 8 to 20 hours may be used in the reaction of butyl mercaptan with PCl_3. The time required for the reaction of dialkyl disulfides and yellow phosphorus is dependent on the temperature and the degree of agitation used. About 50 hours may be required at 150°C, about 2 hours at 200°C.

The reaction of PCl_3 with butyl mercaptan may advantageously be carried out in the absence of added solvents as described by W.W. Crouch et al (2). The reaction between dialkyl disulfides and yellow phosphorus is preferably carried out in the presence of a cycloparaffinic or a methylcycloparaffinic solvent. In contrast, the reaction does not proceed to any practical extent in the presence of a paraffinic solvent such as kerosene, nor in the presence of an alkyl aromatic where the alkyl group is ethyl or higher. In the latter cases, the predominant reaction seems to be the decomposition of the disulfide to the corresponding mercaptans. From 120 to 250 volume percent of cycloparaffin solvent based on the sulfide is the preferred amount. No catalyst is employed in the reaction of mercaptan and PCl_3 or in the reaction of dialkyl disulfide and phosphorus.

A closed agitated kettle equipped with a reflux condenser may be used for the reaction of butyl mercaptan with PCl_3. A glass-lined kettle equipped with a Dowtherm jacket and a reflux condenser is used for the reaction of dialkyl disulfide with phosphorus.

The tributyl trithiophosphate product is a low-melting solid and is recovered as a still residue after distillation of any unreacted mercaptan remaining after the reaction with PCl_3. Alternatively, the product may be distilled as an overhead product from any heavier byproducts which may be present.

The mixture of products from the reaction of dialkyl disulfide and phosphorus is passed through a heat exchanger and then enters a fractionating column operated at about 1 mm of mercury to avoid decomposition of the product. A decalin-disulfide fraction is taken overhead and recycled. A heavy bottoms fraction is removed and the desired product stream taken off at the midpoint of the column, as shown in Figure 55 (4).

Process Wastes and Their Control

Air: Air emissions have been reported (B-15) to consist of 205 kg SO_2 per ton of pesticide produced.

Product Wastes: This product is slowly hydrolyzed in water to form butyl mercaptan. The rate of hydrolysis is increased by alkali. It is slowly oxidized by oxygen in air to the phosphate, a reaction which is accelerated by heating (B-3).

Toxicity

The acute oral LD_{50} value for rats is about 1,270 mg/kg which is slightly toxic.

Figure 55: Process Plant for the Manufacture of Tributyl Phosphorotrithioite

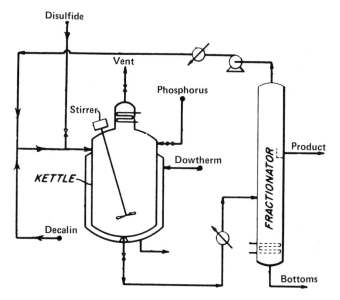

Source: Reference (4)

Allowable Limits on Exposure and Use

Product Use: Issuance of a rebuttable presumption against registration was being considered by EPA as of September 1979 on the basis of possible neurotoxicity.

The tolerances set by the EPA in or on raw agricultural commodities are as follows:

	40 *CFR* Reference	Parts per Million
Cotton, seed	180.186	0.25

References

(1) Worthing, C.R., *Pesticide Manual,* 6th ed., p. 528, British Crop Protection Council (1979).
(2) Crouch, W.W. et al, U.S. Patent 2,682,554, June 29, 1954, assigned to Phillips Petroleum Co.
(3) Stevens, D.R., et al, U.S. Patent 2,542,370, February 20, 1951, assigned to Gulf Research and Development Co.
(4) McLeod, G.D. et al, U.S. Patent 2,819,290, January 7, 1958, assigned to Standard Oil Co., Indiana.
(5) Goyette, L.E., U.S. Patent 2,955,803, October 11, 1960, assigned to Virginia-Carolina Chemical Corp.

TRICHLORFON (TRICHLORPHON IN U.K.)

Function: Insecticide (1)(2)(3)

Chemical Name: Dimethyl 2,2,2-trichloro-1-hydroxyethyl phosphonate

Formula:

$$\begin{matrix} & O \\ & \| \\ (CH_3O)_2 & PCHCCl_3 \\ & | \\ & OH \end{matrix}$$

Trade Names: B-15,922 (Bayer)
L 13/59 (Bayer)
Dipterex® (Bayer)
Neguvon® (Bayer)
Tugon® (Bayer)
Dylox® (Chemagro)
Dyrex®
Anthon®
Proxol®
Masoten®

Manufacture (3)

The following is a specific example of the conduct of the process. 75 grams of chloral are dropped into 60 grams of dimethyl phosphite at an initial temperature of 25°C. The temperature slowly rises to 50°C and is kept at 50° to 60°C by external cooling. After cooling the oil is dissolved in benzene, and the benzene solution is washed with a sodium bicarbonate solution and dried with anhydrous sodium sulfate.

After the solvent has been distilled off, an oil is left which almost completely solidifies. After aspirating and short washing with an icy mixture of ether and petroleum ether, the trichloro-α-hydroxyethylphosphonic dimethyl ester is obtained in the form of colorless needles with a MP of 81°C. Yield 90 grams.

Process Wastes and Their Control

Product Wastes: Trichlorfon initially hydrolyzes to the more toxic compound dichlorvos (LD_{50} of 56 mg/kg) at pH 8 and 37.5°C. Trichlorfon is essentially 100% hydrolyzed in approximately 24 hours to nontoxic products (B-3).

Toxicity

The acute oral LD_{50} value for rats is 560 to 630 mg/kg which is slightly to moderately toxic.

Allowable Limits on Exposure and Use

Product Use: Issuance of a rebuttable presumption against registration for trichlorfon was being considered by EPA as of September 1979 on the basis of possible oncogenicity, teratogenicity and mutagenicity.

The tolerances set by the EPA for trichlorfon in or on raw agricultural commodities are as follows:

	40 *CFR* Reference	Parts per Million
Alfalfa	180.198	60.0
Alfalfa, hay	180.198	90.0
Artichokes	180.198	0.1 N
Bananas, pulp	180.198	0.2
Bananas, whole	180.198	2.0
Barley, forage	180.198	50.0
Barley, grain	180.198	0.1 N

(continued)

	40 *CFR* Reference	Parts per Million
Barley, straw	180.198	1.0
Beans, dry	180.198	0.1 N
Beans, lima	180.198	12.0
Beans, lima, shelled	180.198	0.1 N
Beans, lima, vines	180.198	12.0
Beans, lima, vines, hay	180.198	12.0
Beans, snap	180.198	0.1 N
Beans, vines	180.198	1.0
Beets	180.198	0.1 N
Beets, sugar	180.198	0.1 N
Beets, sugar, tops	180.198	12.0
Brussels sprouts	180.198	0.1 N
Cabbage	180.198	0.1 N
Carrots	180.198	0.1 N
Cattle, fat	180.198	0.1 N
Cattle, mbyp	180.198	0.1 N
Cattle, meat	180.198	0.1 N
Cauliflower	180.198	0.1 N
Citrus fruits	180.198	0.1 N
Clover	180.198	60.0
Clover, hay	180.198	90.0
Collards	180.198	0.1 N
Corn, fodder	180.198	30.0
Corn, forage	180.198	30.0
Corn, fresh (inc sweet) (k+cwhr)	180.198	0.1 N
Corn, grain (inc pop)	180.198	0.1 N
Cotton, seed	180.198	0.1 N
Cowpeas	180.198	0.1 N
Cowpeas, vines	180.198	1.0
Flax, seed	180.198	0.1 N
Flax, straw	180.198	1.0
Goats, fat	180.198	0.1 N
Goats, mbyp	180.198	0.1 N
Goats, meat	180.198	0.1 N
Grasses, pasture	180.198	60.0
Grasses, pasture, hay	180.198	90.0
Grasses, range	180.198	240.0
Grasses, range, hay	180.198	240.0
Horses, fat	180.198	0.1 N
Horses, mbyp	180.198	0.1 N
Horses, meat	180.198	0.1 N
Lettuce	180.198	0.1 N
Milk	180.198	0.01 N
Oats, forage	180.198	50.0
Oats, grain	180.198	0.1 N
Oats, straw	180.198	1.0
Peanuts	180.198	0.05 N
Peanuts, hulls	180.198	4.0
Peanuts, vines, hay	180.198	4.0
Peas, cowpeas	180.198	0.1 N
Peppers	180.198	0.1 N
Pumpkins	180.198	0.1 N
Safflower, seed	180.198	0.1 N
Sheep, fat	180.198	0.1 N
Sheep, mbyp	180.198	0.1 N
Sheep, meat	180.198	0.1 N
Tomatoes	180.198	0.1 N
Wheat, forage	180.198	50.0
Wheat, grain	180.198	0.1 N
Wheat, straw	180.198	1.0

The tolerances set by the EPA for trichlorfon in animal feeds are as follows (the *CFR* Reference is to Title 21):

	CFR Reference	Parts per Million
Citrus pulp, dried	561.190	2.5

References

(1) Worthing, C.R., *Pesticide Manual,* 6th ed., p. 531, British Crop Protection Council (1979).
(2) Spencer, E.Y., *Guide to the Chemicals Used in Crop Protection,* 6th ed., p. 511, London, Ontario, Agriculture Canada (January 1973).
(3) Lorenz, W., U.S. Patent 2,701,225, February 1, 1955, assigned to Farbenfabriken Bayer AG.

TRICHLORONATE

Function: Insecticide (1)

Chemical Name: O-ethyl O-2,4,5-trichlorophenyl ethylphosphonothioate

Formula:

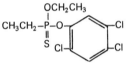

Trade Names: Bayer 37 289 (Bayer AG)
S-4400 (Bayer AG)
Agrisil® (Bayer AG)
Agritox® (Bayer AG)
Phytosol® (Bayer AG)

Manufacture

The starting material is 2,4,5-trichlorophenol which is reacted with O-ethyl ethylphosphono-chloridothioate to give the trichloronate product.

Allowable Limits on Exposure and Use

Product Use: The tolerances set by the EPA in January 1980 (45 *FR* 6103) for trichloronate in or on raw agricultural commodities are as follows:

	40 *CFR* Reference	Parts per Million
Alfalfa	180.198	60
Alfalfa, hay	180.198	90
Artichokes	180.198	0.1 N
Bananas (nmt 0.2 ppm will be present after the peel is removed)	180.198	2
Barley, forage	180.198	50
Barley, grain	180.198	0.1 N
Barley, straw	180.198	1
Beans dried	180.198	0.1 N

(continued)

	40 *CFR* Reference	Parts per Million
Beans, lima	180.198	12
Beans, lima, shelled	180.198	0.1 N
Beans, lima, vine, hay	180.198	12
Beans, lima, vines	180.198	12
Beans, snap	180.198	0.1 N
Beans, vines	180.198	1
Beets	180.198	0.1 N
Beets, sugar	180.198	0.1 N
Beets, sugar, tops	180.198	12
Birdsfoot trefoil, hay	180.198	90
Brussels sprouts	180.198	0.1 N
Cabbage	180.198	0.1 N
Carrots	180.198	0.1 N
Cattle, fat	180.198	0.1 N
Cattle, mbyp	180.198	0.1 N
Cattle, meat	180.198	0.1 N
Cauliflower	180.198	0.1 N
Citrus fruit	180.198	0.1 N
Clover	180.198	60
Clover, hay	180.198	90
Collards	180.198	0.1 N
Corn, fodder	180.198	30
Corn, forage	180.198	30
Corn, fresh (inc sweet) (k+cwhr)	180.198	0.1 N
Corn, grain	180.198	0.1 N
Cottonseed	180.198	0.1 N
Cowpeas	180.198	0.1 N
Cowpeas, vines	180.198	1
Flax, straw	180.198	1
Flaxseed	180.198	0.1 N
Goats, fat	180.198	0.1 N
Goats, mbyp	180.198	0.1 N
Goats, meat	180.198	0.1 N
Grass, pasture	180.198	60
Grass, pasture, hay	180.198	90
Grass, range	180.198	240
Grass, range, hay	180.198	240
Horses, fat	180.198	0.1 N
Horses, mbyp	180.198	0.1 N
Horses, meat	180.198	0.1 N
Lettuce	180.198	0.1 N
Milk	180.198	0.01 N
Oats, forage	180.198	50
Oats, grain	180.198	0.1 N
Oats, straw	180.198	1
Peanuts	180.198	0.05 N
Peanuts, vine, hay	180.198	4
Peanuts, vine, hulls	180.198	4
Peppers	180.198	0.1 N
Pumpkins	180.198	0.1 N
Safflower seed	180.198	0.1 N
Sheep, fat	180.198	0.1 N
Sheep, mbyp	180.198	0.1 N
Sheep, meat	180.198	0.1 N
Tomatoes	180.198	0.1 N
Wheat, forage	180.198	50
Wheat, grain	180.198	0.1 N
Wheat, straw	180.198	1

Reference

(1) Worthing, C.R., *Pesticide Manual,* 6th ed., p. 530, British Crop Protection Council (1979).

TRICYCLAZOLE

Function: Fungicide (1)(2)

Chemical Name: 5-methyl-1,2,4-triazolo-[3,4-b]-benzothiazole

Formula:

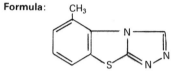

Trade Name: EL-291 (Eli Lilly & Co.)

Manufacture (2)

A solution of 97 to 100% formic acid was added to 2-hydrazino-4-methylbenzothiazole and the mixture was refluxed for 24 hours with stirring. The mixture was then cooled to room temperature and poured into water. The product precipitated and was separated by filtration to give crystals melting at 187° to 188°C.

Toxicity

The acute oral LD_{50} value for rats is about 290 mg/kg which is moderately toxic.

References

(1) Worthing, C.R., *Pesticide Manual,* 6th ed., p. 533, British Crop Protection Council (1979).
(2) Eli Lilly & Co., British Patent 1,419,121, December 24, 1975.

TRIETAZINE

Function: Herbicide (1)(2)

Chemical Name: 6-chloro-N,N',N'-triethyl-1,3,5-triazine-2,4-diamine

Formula:

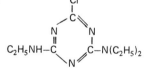

Trade Names: G-27,901 (J.R. Geigy SA)
NC-1667 (Fisons Ltd.)
Gesafloc®

Manufacture

Trietazine may be made by reacting cyanuric chloride, first with an equivalent of ethylamine and then with an equivalent of diethylamine.

Toxicity

The acute oral LD_{50} value for rats is 2,800 to 4,000 mg/kg which is slightly toxic.

References

(1) Worthing, C.R., *Pesticide Manual,* 6th ed., p. 535, British Crop Protection Council (1979).
(2) Gysin, H. and Knusli, E., U.S. Patent 2,891,855, June 23, 1959, assigned to J.R. Geigy AG.

TRIFENMORPH

Function: Molluscicide (1)

Chemical Name: 4-(triphenylmethyl) morpholine

Formula:

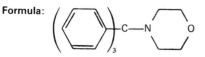

Trade Names: WL 8008 (Shell Research Ltd.)
Frescon® (Shell Research Ltd.)

Manufacture (2)

(A) Preparation of Trityl Chloride-Aluminum Chloride Complex—179 grams (1.16 mols) of dry carbon tetrachloride was added with stirring over two hours at a temperature of from 60° to 65°C to a mixture of 133.5 grams (97%, 0.97 mol) of anhydrous $AlCl_3$ and 560 grams (7.78 mols) of thiophene-free, dry benzene. After all the carbon tetrachloride had been added, stirring was continued for an hour at a temperature of 60°C. Subsequently, dry nitrogen was passed through until no more HCl escaped.

(B) Preparation of N-Tritylmorpholine—A solution of 130.7 grams (1.5 mols) of morpholine in 500 ml of CCl_4 was mixed dropwise, over a period of 30 minutes with stirring at a temperature of 30°C with external cooling, with the trityl chloride-$AlCl_3$ complex solution (prepared from 0.485 mol of $AlCl_3$). The reaction mixture was subsequently stirred for 45 minutes at 30°C.

Transferred to a separatory funnel, the reaction mixture separated into two phases, the lower phase containing the trityl derivative dissolved in CCl_4, the upper phase containing morpholine hydrochloride and a morpholine-$AlCl_3$ complex. After separation of the lower phase the upper phase was extracted twice, each time with 250 ml of CCl_4, and the CCl_4-extracts were combined with the bulk of the trityl derivative. The CCl_4-solution was contacted with 150 cc of dilute hydrochloric acid and subsequently washed with water to remove unconverted morpholine. After the solution had been dried over $CaCl_2$ the solvent was evaporated off and the resultant crystalline product was stirred for an hour at room temperature with three times the weight of methanol. After the methanol had been removed by suction and the residue had been dried, 146 grams of N-triphenylmethyl morpholine (98%) was obtained (92.3% of theory, based on a yield of complex of 97%). Melting point 175°C.

Toxicity

The acute oral LD_{50} value for rats is 121 mg/kg (B-5) which is moderately toxic. Higher values of 450 to 2,200 mg/kg have also been cited (1).

References

(1) Worthing, C.R., *Pesticide Manual*, 6th ed., p. 536, British Crop Protection Council (1979).
(2) Adams, C.R., Arpe, H.J., Schulze-Steinen, N.J., Falbe, J.F., Edwards, A.C. and Tetteroo, H., U.S. Patent 3,577,413, May 4, 1971, assigned to Shell Oil Co.

TRIFLURALIN

Function: Herbicide (1)(2)(4)

Chemical Name: 2,6-dinitro-N,N-dipropyl-4-(trifluoromethyl)benzenamine

Formula:

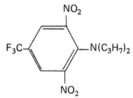

Trade Names: L-36,352 (Eli Lilly & Co.)
Treflan® (Eli Lilly & Co.)

Manufacture (3)(4)

A specific example of the preparation of trifluralin is as follows (4). Fifty grams of 4-chloro-3,5-dinitrobenzoic acid were reacted with 50 grams of sulfur tetrafluoride in an autoclave at 120°C for 7 hours to form 4-trifluoromethyl-2,6-dinitrochlorobenzene as an intermediate. Evaporation of the reaction mixture to dryness left a solid residue comprising 4-trifluoro-methyl-2,6-dinitrochlorobenzene, which was purified by recrystallization from a hexane-benzene solvent mixture. Crystalline 4-trifluoromethyl-2,6-dinitrochlorobenzene thus prepared melted at about 53° to 57°C.

Eight and one-tenth grams of the 4-trifluoromethyl-2,6-dinitrochlorobenzene intermediate were then mixed with 10 ml of di-n-propylamine, and the reaction mixture was heated at about 100°C for about 2 hours. The reaction mixture was diluted with ether and was filtered to remove di-n-propylamine hydrochloride formed as a by-product in the reaction. The filtrate was washed with dilute hydrochloric acid and then was evaporated to dryness in vacuo. The residue containing N,N-di-n-propyl-4-trifluoromethyl-2,6-dinitroaniline (trifluralin) was recrystallized from hexane and melted at about 41° to 43°C.

Figure 56 is a process and waste diagram for trifluralin manufacture by another method.

Process Wastes and Their Control (3)

Air: Air emissions from trifluralin production have been reported (B-15) to be as follows:

Component	Kilograms per Metric Ton Pesticide Produced
Particulates	1.0 (nitrates, sulfates and chlorides)
NO_x	1.0
HCl	3.0
HF	0.3
SO_2	1.2

Air pollution control in trifluralin production involves the following (B-15):

Control Device	Reported Efficiency
1- and 2-stage venturi scrubber and Tri-mer wet scrubber	90%

Water: The wastewaters contain NaCl. The wastewater stream from the decanter is sent to activated carbon adsorption and then to biological treatment (B-10).

Solid Waste: The spent activated carbon, from a typical plant producing 10,000 metric tons of trifluralin per year, amounts to a total of 1,150 metric tons per year. Of the total spent carbon waste stream 600 metric tons per year is spent carbon, 457 metric tons is unreacted intermediates and solvent, and 93 metric tons is trifluralin and related compounds (B-7).

At present the existing treatment method for the spent activated carbon from trifluralin manufacture is storage in plastic-lined steel drums. The plastic lining is necessary due to the fluoride content of the components adsorbed within the carbon and their corrosiveness on bare steel. This method of disposal is only a short term solution due to the accumulation of drums over the years and the danger of leaks and/or spills from the drums.

Figure 56: Trifluralin Manufacture

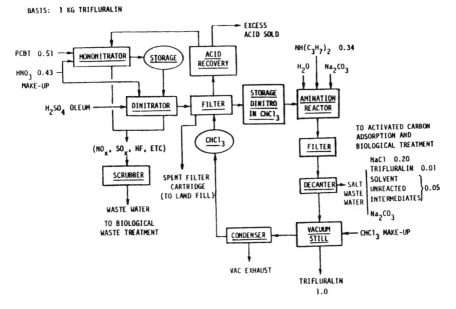

Source: Reference (B-1)

To effectively treat the activated carbon wastes from trifluralin manufacture it is necessary to first separate the activated carbon from the wastes and then to treat the separated wastes. Based on these criteria, the selected treatment scheme (see Figure 57) for the trifluralin wastes requires the following unit operations: 1) grinding; 2) solvent extraction; 3) centrifugation; 4) vacuum stripping and distillation; 5) composting; and 6) chemical landfill.

Figure 57: Treatment of Spent Activated Carbon from Adsorption in Trifluralin Manufacture

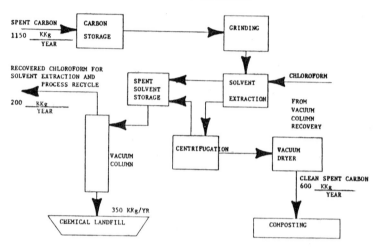

Source: Reference (B-7)

Production Wastes: Trifluralin is susceptible to photochemical decomposition. If incinerated, the same precautions should be taken as recommended for fluorodifen (which see) (B-3).

Toxicity

Trifluralin is relatively nontoxic (acute oral LD_{50} for rats $\geq$10,000 mg/kg).

Allowable Limits on Exposure and Use

Product Use: Issuance of a rebuttable presumption against registration for trifluralin was being considered by EPA as of September 1979 on the basis of possible oncogenicity and mutagenicity.

The tolerances set by the EPA for trifluralin in or on raw agricultural commodities are as follows:

	40 *CFR* Reference	Parts per Million
Alfalfa, hay	180.207	0.2 N
Beans, mung, sprouts	180.207	2.0
Carrots	180.207	1.0
Citrus fruits	180.207	0.05 N
Corn, field, fodder	180.207	0.05 N
Corn, field, forage	180.207	0.05 N

(continued)

	40 *CFR* Reference	Parts per Million
Corn, field, grain	180.207	0.05 N
Cotton, seed	180.207	0.05 N
Cucurbits	180.207	0.05 N
Fruits, stone	180.207	0.05 N
Grapes	180.207	0.05 N
Hops	180.207	0.05 N
Legumes, forage	180.207	0.05 N
Nuts	180.207	0.05 N
Peanuts	180.207	0.05 N
Peppermint, hay	180.207	0.05 N
Safflower, seed	180.207	0.05 N
Spearmint, hay	180.207	0.05 N
Sugarcane	180.207	0.05 N
Sunflower, seed	180.207	0.05 N
Vegetables, fruiting	180.207	0.05 N
Vegetables, leafy	180.207	0.05 N
Vegetables, root crop (exc carrots)	180.207	0.05 N
Vegetables, seed and pod	180.207	0.05 N
Wheat, grain	180.207	0.05 N
Wheat, straw	180.207	0.05 N

The tolerances set by the EPA for trifluralin in food are as follows (the *CFR* Reference is to Title 21):

	CFR Reference	Parts per Million
Peppermint oil	193.440	2.0
Spearmint oil	193.440	2.0

References

(1) Worthing, C.R., *Pesticide Manual,* 6th ed., p. 537, British Crop Protection Council (1979).
(2) Spencer, E.Y., *Guide to the Chemicals Used in Crop Protection,* 6th ed., p. 524, London, Ontario, Agriculture Canada (January 1973).
(3) Midwest Research Institute, *The Pollution Potential in Pesticide Manufacturing,* Washington, DC, Environmental Protection Agency (June 1972).
(4) Soper, Q.F., U.S. Patent 3,257,190, June 21, 1966, assigned to Eli Lilly & Co.

TRIMEDLURE

Function: Hormonal insect attractant (1)

Chemical Name: tert-butyl 4(or 5)-chloro-2-methylcyclohexane carboxylate

Formula:

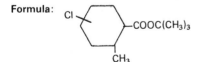

Trade Name: Pherocon MFF® (Zoecon Corp.)

Manufacture (1)

A mixture of 56 grams of trans-6-methyl-3-cyclohexene-carboxylic acid (2), 60 ml dioxane and 148 ml of concentrated hydrochloric acid in a sealed bottle was heated at 95°C for six hours with vigorous agitation. About 300 ml of water was added and after cooling, the crystallized product was filtered off and the filtrate extracted with benzene. Removal of the solvent gave 68.7 grams (97%) of the 4(or 5)-chlorohexahydro-o-toluic acid, MP 73° to 83°C.

The 4(or 5)-chlorohexahydro-o-toluic acid, prepared as described above, was converted to the ester via the acid halide route as follows. A total of 68 grams of the acid was dissolved in 100 ml of benzene and 42.6 ml of thionyl chloride (phosphorus trichloride or tribromide may be used in place of thionyl chloride) was added. The mixture was allowed to stand overnight at room temperature. The benzene and thionyl chloride were removed under water pump vacuum and the residue, dissolved in 50 ml benzene, was added slowly to a cold, stirred solution of 40 ml tert-butyl alcohol, 100 ml benzene and 40 ml pyridine. After standing one hour the mixture was worked up by adding 100 ml ether, washing with water, 10% Na_2CO_3, dilute HCl, and again with water. The solvent was evaporated and the residue distilled at 90° to 92°C, at 0.6 mm. The yield was 65.4 grams of the ester having a pleasant fruity odor.

Toxicity

The acute oral LD_{50} value for rats is 4,556 mg/kg (B-5) which is very slightly toxic.

References

(1) Beroza, M., Green, N. and Gertler, S.I., U.S. Patent 3,016,329, January 9, 1962.
(2) Diels, O. and Alder, K., *Ann.* 470, 88-92 (1929).

TRIMETHYLPHENYL METHYLCARBAMATE

Function: Insecticide

Chemical Name: 3,4,5-trimethylphenyl methylcarbamate

Formula:

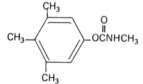

Trade Names: SD-8530 (Shell)
Landrin® (Shell)

Manufacture (1)

A solution of 27.2 parts of 3,4,5-trimethylphenol, 14.3 parts of methyl isocyanate and 0.3 part by volume of triethylamine in 125 parts by volume of anhydrous ether was allowed to stand at room temperature (25°C) over a period of 3 days.

The final reaction mixture was mixed with 500 parts by volume of ether, the resulting mixture was mixed with Norite activated carbon and refluxed, then filtered and the resulting solution was concentrated to about 300 parts by volume on the steam bath. Upon chilling to −20°C, 29 parts (75% yield) of 3,4,5-trimethylphenyl methylcarbamate, as white crystals melting at 122° to 123°C, were obtained.

Toxicity

The acute oral LD$_{50}$ value for rats is 208 mg/kg which is moderately toxic.

Allowable Limits on Exposure and Use

Product Use: The tolerances set by the EPA for this compound in or on raw agricultural commodities are as follows:

	40 *CFR* Reference	Parts per Million
Corn, field, grain	180.305	0.1 N
Corn, fodder	180.305	0.1 N
Corn, forage	180.305	0.1 N
Corn, pop, grain	180.305	0.1 N

Reference

(1) Kuderna, J.G. and Phillips, D.D., U.S. Patent 3,130,122, April 21, 1964, assigned to Shell Oil Co.

TRIPHENYLTIN ACETATE (U.S.)
FENTIN ACETATE (U.K.)

Function: Fungicide (1)(2)(5), acaricide (4)

Chemical Name: (Acetyloxy) triphenylstannane

Formula:

Trade Names: VP-1940 (Farbwerke Hoechst)
HOE 2824 (Farbwerke Hoechst)
Brestan® (Farbwerke Hoechst)

Manufacture

The reaction of stannic chloride, SnCl$_4$, on tetraphenyltin yields fentin chloride. The fentin chloride is reacted with sodium acetate to give fentin acetate. The fentin chloride intermediate may also be made by the Grignard reaction of SnCl$_4$ and magnesium on chlorobenzene (3).

Toxicity

The acute oral LD$_{50}$ value for rats is 125 mg/kg which is moderately toxic.

References

(1) Worthing, C.R., *Pesticide Manual,* 6th ed., p. 266, British Crop Protection Council (1979).
(2) Spencer, E.Y., *Guide to the Chemicals Used in Crop Protection,* 6th ed., p. 280, London, Ontario, Agriculture Canada (January 1973).

(3) Midwest Research Institute and RvR Consultants, *Production, Distribution, Use and Environmental Impact Potential of Selected Pesticides,* Washington, DC, Council on Environmental Quality (March 15, 1974).
(4) Taylor, J.L., U.S. Patent 3,268,395, August 23, 1966, assigned to Thompson-Hayward Chemical Co.
(5) Brueckner, H. and Haertel, K., U.S. Patent 3,499,086, March 3, 1970, assigned to Farbwerke Hoechst AG.

TRIPHENYLTIN HYDROXIDE (U.S.)
FENTIN HYDROXIDE (U.K.)

Function: Fungicide (1)(2)(4), acaricide (5)

Chemical Name: Triphenyltin hydroxide

Formula:

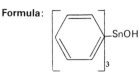

Trade Names: Du-Ter® (N.V. Philips-Duphar)
Erithane®
Farmatin®
Profarma Tinspray®
Tubatin®
Vitospot®

Manufacture

Stannic chloride and phenyl magnesium chloride (the latter prepared from magnesium and chlorobenzene) react to give triphenyl tin chloride which is in turn reacted with sodium hydroxide to give triphenyltin hydroxide (3). Figure 58 is a production and waste schematic for this process.

**Figure 58: Production and Waste Schematic
for Organotin Hydroxide Manufacture**

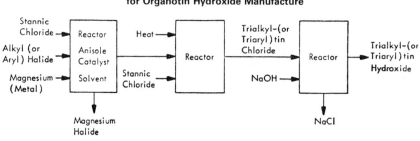

Source: Reference (3)

Process Wastes and Their Control

Product Wastes: It has been suggested that tin organic compound wastes be dumped into landfills (B-3).

Toxicity

The acute oral LD_{50} value for rats is 108 mg/kg which is moderately toxic.

Allowable Limits on Exposure and Use

Product Use: The tolerances set by the EPA for triphenyltin hydroxide in or on raw agriculture commodities are as follows:

	40 *CFR* Reference	Parts per Million
Beets, sugar, roots	180.236	0.1 N
Carrots	180.236	0.1 N
Cattle, kidney	180.236	0.05 N
Cattle, liver	180.236	0.05 N
Goats, kidney	180.236	0.05 N
Goats, liver	180.236	0.05 N
Hogs, kidney	180.236	0.05 N
Hogs, liver	180.236	0.05 N
Horses, kidney	180.236	0.05 N
Horses, liver	180.236	0.05 N
Peanuts	180.236	0.05 N
Peanuts, hulls	180.236	0.4
Pecans	180.236	0.05 N
Potatoes	180.236	0.05 N
Sheep, kidney	180.236	0.05 N
Sheep, liver	180.236	0.05 N

References

(1) Worthing, C.R., *Pesticide Manual,* 6th ed., p. 267, British Crop Protection Council (1979).
(2) Spencer, E.Y., *Guide to the Chemicals Used in Crop Protection,* 6th ed., p. 282, London, Ontario, Agriculture Canada (January 1973).
(3) Midwest Research Institute and RvR Consultants, *Production, Distribution, Use and Environmental Impact Potential of Selected Pesticides,* Washington, DC, Council on Environmental Quality (March 15, 1974).
(4) Duyfjes, W. and De Lange, W., U.S. Patent 3,140,977, July 14, 1964, assigned to North American Philips Co., Inc.
(5) Taylor, J.L., U.S. Patent 3,268,395, August 23, 1966, assigned to Thompson-Hayward Chemical Co.

U

UNDECYLENIC ACID

Function: Herbicide (1)(2)(3).

Chemical Name: 10-Undecenoic acid

Formula:

$$CH_2=CH(CH_2)_8\overset{\overset{\displaystyle O}{\|}}{C}OH$$

Trade Names: Declid®
Desenex®
Renselin®
Sevinon®

Manufacture

Undecylenic acid may be produced in low yields by the pyrolysis of ricinoleic acid from castor oil. The process is carried out under reduced pressure; yields may be improved by conducting the destructive distillation in the presence of rosin.

Toxicity

The acute oral LD_{50} for rats is 2,500 mg/kg (B-5) which is slightly toxic.

References

(1) Martin, H. and Worthing, C.R., *Pesticide Manual,* 5th ed., p. 531, British Crop Protection Council (January 1977).
(2) Spencer, E.Y., *Guide to the Chemicals Used in Crop Protection,* 6th ed., p. 528, London, Ontario, Agriculture Canada (January 1973).
(3) Zimmermann, P.W. and Hitchcock, A.E., U.S. Patent 2,626,862, January 27, 1953, assigned to Boyce Thompson Institute for Plant Research, Inc.

V

VALIDAMYCIN A

Function: Fungicide (1)(2)

Chemical Name: D-1,5,6-trideoxy-3-O-β-D-glucopyranosyl-5-(hydroxymethyl)-1-[[4,5,6-trihydroxy-3-(hydroxymethyl)-2-cyclohexen-1-yl]amino]-D-chiroinositol

Formula:

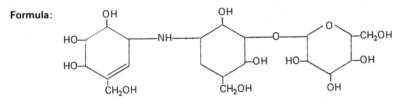

Trade Names: Validacin® (Takeda Chemical Industries, Ltd.)

Manufacture

Validamycin A is made by the fermentation of *Streptomyces hygroscopicus* var. limoneus.

Toxicity

The acute oral LD_{50} value for rats is in excess of 20,000 mg/kg which is insignificantly toxic.

References

(1) Worthing, C.R., *Pesticide Manual,* 6th ed., p. 541, British Crop Protection Council (1979).
(2) Spencer, E.Y., *Guide to the Chemicals Used in Crop Protection,* 6th ed., p. 529, London, Ontario, Agriculture Canada (January 1973).

VAMIDOTHION

Function: Insecticide and acaricide (1)(2)

Chemical Name: O,O-dimethyl-5-(2-[[1-methyl-2-(methylamino)-2-oxoethyl]thio]ethyl) phosphorothioate

Formula:

$$(CH_3O)_2PSCH_2CH_2SCHCNHCH_3$$

Trade Names: RP-10465 (May and Baker, Ltd. and Rhone-Poulenc)
NPH-83 (May and Baker, Ltd. and Rhone-Poulenc)
Kilval® (May and Baker, Ltd. and Rhone-Poulenc)
Trucidor® (Rhone-Poulenc)

Manufacture (3)

In an initial step, 2-chloroethanol (ethylene chlorohydrin) is reacted with ethyl 2-mercapto propionate giving ethyl-2-(2-hydroxyethylthio)propionate. That intermediate is, in turn, treated successively with methylamine and thionyl chloride to give 2-(2-chloroethylthio)-N-methylpropionamide. Condensation with O,O-dimethylphosphorothioate gives vamidothion.

Toxicity

The acute oral LD_{50} value for rats is 65 to 105 mg/kg which is moderately toxic.

References

(1) Worthing, C.R., *Pesticide Manual,* 6th ed., p. 542, British Crop Protection Council (1979).
(2) Spencer, E.Y., *Guide to the Chemicals Used in Crop Protection,* 6th ed., p. 530, London, Ontario, Agriculture Canada (January 1973).
(3) Metivier, J., U.S. Patent 2,943,974, July 5, 1960, assigned to Societe des Usines Chimiques Rhone-Poulenc.

VERNOLATE

Function: Herbicide (1)(2)

Chemical Name: S-propyl dipropylcarbamothioate

Formula:

$$(C_3H_7)_2NCSC_3H_7$$

Trade Names: R-1607 (Stauffer Chemical Co.)
Vernam® (Stauffer Chemical Co.)
PPTC

Manufacture (3)

Sodium is dispersed in xylene using oleic acid as the stabilizing agent until a particle size of 5 to 200 microns in diameter is obtained. Dispersion equivalent to an amount of 9.2 parts (0.40 mol) of sodium, is then transferred to a reactor which has been previously flushed out with argon (or other inert gas such as nitrogen). A solution of 38 parts (0.50 mol) of 1-pro-panethiol (propyl mercaptan) dissolved in 86 parts of xylene is then gradually added to the sodium dispersion over an interval of 30 minutes. The temperature is maintained at 25° to 36°C by cooling.

The sodium propyl mercaptide forms as finely divided crystals which make an easily stirrable slurry. This suspension is heated to reflux, the heat is turned off, and 65 parts (0.40 mol) of di-n-propylcarbamyl chloride is added over an interval of 17 minutes to the refluxing slurry. The heat of reaction is sufficient to keep the xylene refluxing. After all of the di-n-propyl-carbamyl chloride has been added, the mixture is refluxed for an additional 3 hours. It is then cooled, filtered from sodium chloride which has formed during the reaction, and the solvent is removed under reduced pressure. The residual liquid is then distilled under vacuum to give 70.3 parts (86.5% yield) of vernolate.

Process Wastes and Their Control

Air: Air emissions from vernolate production have been reported (B-15) to be 205 kg SO_2, 1.5 kg hydrocarbons and 0.5 kg vernolate per metric ton of pesticide produced.

Toxicity

The acute oral LD_{50} value for rats is 1,780 mg/kg which is slightly toxic.

Allowable Limits on Exposure and Use

Product Use: The tolerances set by the EPA for vernolate in or on raw agricultural commodities are as follows:

	40 *CFR* Reference	Parts per Million
Corn, fodder	180.240	0.1 N
Corn, forage	180.240	0.1 N
Corn, fresh (inc sweet) (k+cwhr)	180.240	0.1 N
Corn, grain	180.240	0.1 N
Peanuts	180.240	0.1 N
Peanuts, forage	180.240	0.1 N
Peanuts, hay	180.240	0.1 N
Potatoes	180.240	0.1 N
Soybeans	180.240	0.1 N
Soybeans, forage	180.240	0.1 N
Soybeans, hay	180.240	0.1 N
Sweet potatoes	180.240	0.1 N

References

(1) Worthing, C.R., *Pesticide Manual,* 6th ed., p. 543, British Crop Protection Council (1979).
(2) Spencer, E.Y., *Guide to the Chemicals Used in Crop Protection,* 6th ed., p. 531, London, Ontario, Agriculture Canada (January 1973).
(3) Tilles, H. and Antognini, J., U.S. Patent 2,913,327, November 17, 1959, assigned to Stauffer Chemical Co.

W

WARFARIN

Function: Rodenticide (1)(2)

Chemical Name: 4-hydroxy-3-(3-oxo-1-phenylbutyl)-2H-1-benzopyran-2-one

Formula:

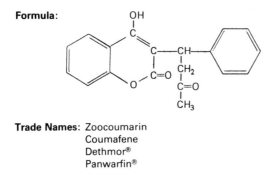

Trade Names: Zoocoumarin
Coumafene
Dethmor®
Panwarfin®

Manufacture (3)

About 0.1 mol each of 4-hydroxycoumarin and benzalacetone are dissolved, in any desired order, in about three times their combined weight of pyridine. The solution is refluxed for about twenty-four hours, and then allowed to cool; after which it is poured into about fifteen volumes of water, and acidified to about pH 2 by the addition of hydrochloric acid. An oil separates, and on cooling and standing overnight solidifies. The solid product is recovered, as by filtration, and recrystallized from ethanol. The yield is about 44%. The recrystallized product melts at about 161°C. Crystalline warfarin sodium may be prepared on a large scale in dry stable form by a process described by Schroeder et al (4) and by Link (5).

Toxicity

The acute oral LD_{50} value for rats is 3 mg/kg (B-5) which is extremely toxic.

Allowable Limits on Exposure and Use

Air: The threshold limit value for warfarin in air is 0.1 mg/m³ as of 1979. The tentative short term exposure limit is 0.3 mg/m³ (B-23).

References

(1) Worthing, C.R., *Pesticide Manual,* 6th ed., p. 545, British Crop Protection Council (1979).
(2) Spencer, E.Y., *Guide to the Chemicals Used in Crop Protection,* 6th ed., p. 534, London, Ontario, Agriculture Canada (January 1973).
(3) Stahmann, M.A., Ikawa, M. and Link, K.P., U.S. Patent 2,427,578, September 16, 1947, assigned to Wisconsin Alumni Research Foundation.
(4) Schroeder, C.H. and Link, K.P., U.S. Patent 2,765,321, October 2, 1956, assigned to Wisconsin Alumni Research Foundation.
(5) Link, K.P., U.S. Patent 2,777,859, January 15, 1957, assigned to Wisconsin Alumni Research Foundation.

X

3,4-XYLYL METHYLCARBAMATE

Function: Insecticide (1)(2)

Chemical Name: 3,4-Dimethylphenyl methylcarbamate

Formula:

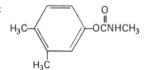

Trade Names: S-1046 (Sumitomo Chemical Co., Ltd.)
Meobal® (Sumitomo Chemical Co., Ltd.)

Manufacture

The material may be made by one of two routes: (a) condensation of 3,4-xylenol with methyl isocyanate or (b) condensation of 3,4-dimethylphenyl chloroformate with methylamine.

Toxicity

The acute oral LD_{50} value for rats is 380 mg/kg which is moderately toxic.

References

(1) Worthing, C.R., *Pesticide Manual,* 6th ed., p. 546, British Crop Protection Council (1979).
(2) Spencer, E.Y., *Guide to the Chemicals Used in Crop Protection,* 6th ed., p. 218, London, Ontario, Agriculture Canada (January 1973).

3,5-XYLYL METHYLCARBAMATE

Function: Insecticide (1)

Chemical Name: 3,5-Dimethylphenyl methylcarbamate

Formula:

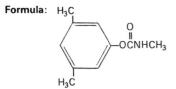

Trade Names: H-69 (Hokko Chemical Industry Co. and Hodogaya Chemical Co., Ltd.)
Macbal® (Hokko Chemical Industry Co. and Hodogaya Chemical Co., Ltd.)

Manufacture

This compound may be made by one of two routes: (a) reaction of 3,5-xylenol with methyl isocyanate or (b) reaction of 3,5-dimethylphenyl chloroformate with methylamine.

Toxicity

The acute oral LD_{50} value for rats is 542 mg/kg which is slightly toxic.

Reference

(1) Worthing, C.R., *Pesticide Manual,* 6th ed., p. 547, British Crop Protection Council (1979).

Z

ZINC PHOSPHIDE

Function: Rodenticide

Chemical Name: Zinc phosphide

Formula: Zn_3P_2

Trade Names: Mous-Con®
Kilrat®
Rumetan®

Manufacture (1)(2)

Zinc phosphide may be made by the direct combination of zinc and phosphorus.

Toxicity

The acute oral LD_{50} value for rats is about 46 mg/kg which is highly toxic.

Allowable Limits on Exposure and Use

Product Use: The tolerances set by the EPA for zinc phosphide in or on raw agricultural commodities are as follows:

	40 *CFR* Reference	Parts per Million
Grapes	180.284	0.01
Grasses, rangeland	180.284	0.1
Sugarcane	180.284	0.01

References

(1) Worthing, C.R., *Pesticide Manual,* 6th ed., p. 548, British Crop Protection Council (1979).
(2) Spencer, E.Y., *Guide to the Chemicals Used in Crop Protection,* 6th ed., p. 535, London, Ontario, Agriculture Canada (January 1973).

ZINEB

Function: Fungicide (1)(2)(3)

Chemical Name: [[1,2-ethanediylbis(carbamodithioato)] (2-)] zinc

Formula:

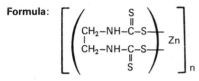

Trade Names: Dithane Z-78® (Rohm and Haas)
Parzate Zineb® (DuPont)

Manufacture

Made by the precipitation of nabam (which see) with zinc sulfate.

Process Wastes and Their Control

Air: Air emissions from zineb production have been estimated (B-15) to be 205 kg SO_2, 1.5 kg hydrocarbons and 0.5 kg zineb per metric ton of pesticide produced.

Product Wastes: Zineb is unstable toward mositure, and acid hydrolysis yields CS_2. Heating above 120°C causes decomposition to a carbonaceous product (B-3).

Toxicity

The acute oral LD_{50} value for rats is over 5,200 mg/kg which is insignificantly toxic.

The toxic properties of dithiocarbamate fungicides have been discussed at some length by the National Research Council (B-22) and the reader is referred to the section of this volume on Mancozeb for more detail.

Allowable Limits on Exposure and Use

Product Use: A rebuttable presumption against registration was issued on August 10, 1977 by EPA on the basis of oncogenicity, teratogenicity and hazard to wildlife.

The tolerances set by the EPA for zineb in or on raw agricultural commodities are as follows:

	40 *CFR* Reference	Parts per Million
Apples	180.115	2.0
Apricots	180.115	7.0
Beans	180.115	7.0
Beets, garden, roots only	180.115	7.0
Beets, tops	180.115	25.0
Blackberries	180.115	7.0
Boysenberries	180.115	7.0
Broccoli	180.115	7.0
Brussels sprouts	180.115	7.0
Cabbage	180.115	7.0
Cabbage, chinese	180.115	25.0
Carrots	180.115	7.0
Cauliflower	180.115	7.0
Celery	180.115	5.0

(continued)

	40 *CFR* Reference	Parts per Million
Cherries	180.115	7.0
Citrus fruits	180.115	7.0
Collards	180.115	25.0
Corn, grain	180.115	0.1
Corn, sweet (k+cwhr)	180.115	5.0
Cranberries	180.115	7.0
Cucumbers	180.115	4.0
Currants	180.115	7.0
Dewberries	180.115	7.0
Eggplant	180.115	7.0
Endive (escarole)	180.115	10.0
Gooseberries	180.115	7.0
Grapes	180.115	7.0
Guavas	180.115	7.0
Hops	180.115	60.0
Kale	180.115	10.0
Kohlrabi	180.115	7.0
Lettuce	180.115	10.0
Loganberries	180.115	7.0
Melons	180.115	4.0
Mushrooms	180.115	7.0
Mustard, greens	180.115	10.0
Nectarines	180.115	7.0
Onions	180.115	7.0
Parsley	180.115	7.0
Peaches	180.115	7.0
Peanuts	180.115	7.0
Pears	180.115	7.0
Peas	180.115	7.0
Peppers	180.115	7.0
Plums (fresh prunes)	180.115	7.0
Pumpkins	180.115	7.0
Quinces	180.115	7.0
Radishes, tops	180.115	7.0
Radishes, with tops	180.115	7.0
Radishes, without tops	180.115	7.0
Raspberries	180.115	7.0
Romaine	180.115	25.0
Rutabaga tops	180.115	7.0
Rutabagas, with tops	180.115	7.0
Rutabagas, without tops	180.115	7.0
Salsify	180.115	7.0
Spinach	180.115	10.0
Squash	180.115	4.0
Squash, summer	180.115	7.0
Strawberries	180.115	7.0
Swiss chard	180.115	25.0
Tomatoes	180.115	4.0
Turnips, greens	180.115	7.0
Turnips, with tops	180.115	7.0
Turnips, without tops	180.115	7.0
Wheat	180.115	1.0
Youngberries	180.115	7.0

References

(1) Worthing, C.R., *Pesticide Manual,* 6th ed., p. 549, British Crop Protection Council (1979).
(2) Spencer, E.Y., *Guide to the Chemicals Used in Crop Protection,* 6th ed., p. 536, London, Ontario, Agriculture Canada (January 1973).
(3) Heuberger, J.W., U.S. Patent 2,457,674, December 28, 1948, assigned to Rohm and Haas Co.

ZIRAM

Function: Fungicide (1)(2)

Chemical Name: Bis(dimethylcarbamodithioato-S,S')zinc

Formula:

$$\left[(CH_3)_2N\overset{\overset{\displaystyle S}{\|}}{C}-S \right]_2 Zn$$

Trade Names: Milbam® (DuPont)
Zerlate® (DuPont)
Nibam®
Fuklasin®
Cuman®

Manufacture

Ziram may be made by the precipitation of sodium dimethyldithiocarbamate with zinc sulfate. The sodium dimethyldithiocarbamate is, in turn, made by the interaction of carbon disulfide and dimethylamine in caustic solution.

A more direct route (3) involves the reaction of an amine, carbon disulfide and a water-insoluble metal oxide or hydroxide, such as zinc oxide or hydroxide. The reaction is preferably conducted in the absence of a mutual solvent for the reactants. The direct reaction method for the preparation of the insoluble metal salts of the substituted dithiocarbamic acids has the following distinct advantages: a shorter reaction time; economy in the expenditure of raw materials; reduced number of essential steps; improved utilization of the heat of reaction; and much larger yield of finished product per unit of reactor volume.

The method of practicing the direct process is to place a water-insoluble metal oxide or hydroxide in a jacketed mixer, and then to add an approximately equivalent quantity of an amine as quickly as possible. After several minutes of very intensive mixing, an approximately equivalent quantity of carbon disulfide is poured into the mixer and intensively mixed with the amine and metal oxide or hydroxide. The temperature will begin to rise and may be controlled by circulating hot or cold water through the jacket for the purpose of maintaining a temperature optimum for promotion of the reaction to the right. The reaction temperature will range from 0° to 150°C, and the particular temperature used will depend somewhat upon the amine and metal oxide or hydroxide used in the reaction. The pressure to be maintained during the reaction will ordinarily vary between subatmospheric and 100 psi, and the pressure has a tendency to drop during the course of the reaction.

As noted above, the reaction is best conducted with approximately stoichiometric equivalents of the reactants, but in view of the fact that carbon disulfide and some of the amines are volatile, it is sometimes desirable to use a slight excess of these compounds in order to compensate for vapor losses sustained. The reaction time varies with the temperature and particular dithiocarbamate being formed, but usually ranges from one-half hour to 8 hours. After completion of the reaction, the water of reaction and any excess of carbon disulfide or amine are removed by vacuum distillation.

Process Wastes and Their Control

Product Wastes: Ziram has a chelate structure and is more stable than most dithiocarbamates; it has good stability to moisture and is not decomposed readily by dilute acids. It is decomposed by strong acids and when heated with caustic alkalis, and can decompose violently upon prolonged heating at 170° to 180°C (B-3).

Toxicity

The acute oral LD_{50} value for rats is 1,400 mg/kg which is slightly toxic.

The toxic properties of dithiocarbamate fungicides have been discussed at some length by the National Research Council (B-22) and the reader is referred to the section of this volume on Mancozeb for more detail.

Allowable Limits on Exposure and Use

Product Use: The tolerances set by the EPA for ziram in or on raw agricultural commodities are as follows:

	40 *CFR* Reference	Parts per Million
Almonds	180.116	0.1
Apples	180.116	7.0
Apricots	180.116	7.0
Beans	180.116	7.0
Beets, greens (alone)	180.116	7.0
Beets, with tops	180.116	7.0
Beets, without tops	180.116	7.0
Blackberries	180.116	7.0
Blueberries (huckleberries)	180.116	7.0
Boysenberries	180.116	7.0
Broccoli	180.116	7.0
Brussels sprouts	180.116	7.0
Cabbage	180.116	7.0
Carrots	180.116	7.0
Cauliflower	180.116	7.0
Celery	180.116	7.0
Cherries	180.116	7.0
Collards	180.116	7.0
Cranberries	180.116	7.0
Cucumbers	180.116	7.0
Dewberries	180.116	7.0
Eggplant	180.116	7.0
Gooseberries	180.116	7.0
Grapes	180.116	7.0
Kale	180.116	7.0
Kohlrabi	180.116	7.0
Lettuce	180.116	7.0
Loganberries	180.116	7.0
Melons	180.116	7.0
Nectarines	180.116	7.0
Onions	180.116	7.0
Peaches	180.116	7.0
Peanuts	180.116	7.0
Pears	180.116	7.0
Peas	180.116	7.0
Pecans	180.116	0.1
Peppers	180.116	7.0
Pumpkins	180.116	7.0
Quinces	180.116	7.0
Radishes, tops	180.116	7.0
Radishes, with tops	180.116	7.0
Radishes, without tops	180.116	7.0
Raspberries	180.116	7.0
Rutabaga tops	180.116	7.0
Rutabagas, with tops	180.116	7.0

(continued)

	40 *CFR* Reference	Parts per Million
Rutabagas, without tops	180.116	7.0
Spinach	180.116	7.0
Squash	180.116	7.0
Squash, summer	180.116	7.0
Strawberries	180.116	7.0
Tomatoes	180.116	7.0
Turnips, greens	180.116	7.0
Turnips, with tops	180.116	7.0
Turnips, without tops	180.116	7.0
Youngberries	180.116	7.0

References

(1) Worthing, C.R., *Pesticide Manual,* 6th ed., p. 551, British Crop Protection Council (1979).
(2) Spencer, E.Y., *Guide to the Chemicals Used in Crop Protection,* 6th ed., p. 537, London, Ontario, Agriculture Canada (January 1973).
(3) Olin, J.F. and Deger, T.E., U.S. Patent 2,492,314, December 27, 1949, assigned to Sharples Chemicals, Inc.

RAW MATERIALS INDEX

This entire volume has been cross indexed by raw material. The only exceptions are those few materials which are extracted from natural roots and the like. Also those pesticides which are produced by fermentation are listed separately.

The question arises of course of how far back to go in the raw material chain. It has been the attempt, where information was available, to go back to reasonably simple raw materials such as methyl isocyanate.

FERMENTATION

Bacillus thuringiensis
Cycloheximide
Gibberellic acid

Kasugamycin
Polyoxin-B

Streptomycin
Validamycin A

CHEMICALS

Acenaphthene
 1,8-Naphthalic anhydride

Acetaldehyde
 Acrolein
 Metaldehyde
 Nitrapyrin
 Picloram

Acetic acid
 Chloroacetic acid
 Glyodin
 Paris green
 PMA
 TCA

Acetic anhydride
 Dinoseb acetate
 Sesamex

Acetoacetanilide
 Carboxin

Acetoacetic acid diethylamide
 Phosphamidon

Acetoacetic acid isopropyl ester
 Propetamphos

Acetone
 Bendiocarb
 Ethoxyquin

Acetonitrile
 Methomyl
 Phenmedipham

Acetyl chloride
 Acephate
 Quinacetol sulfate

Acetylene
 Acrylonitrile
 Aldrin
 Endrin

Acrolein
 Acrylonitrile

Acrolein cyanohydrin
 Acrylonitrile

Atrazine
  Ametryne

Azelaic acid
  Codlelure

Barium fluorosulfonate
  Sulfuryl fluoride

Benzalacetone
  Warfarin

Benzaldehyde
  Pyridinitril

Benzazimide
  Azinphos-ethyl
  Azinphos-methyl

Benzene
  Anthraquinone
  Benzene hexachloride
  Dichlorobenzene
  Diphenyl
  Hexachlorobenzene
  Phenylmercury acetate
  Tetradifon
  Trifenmorph

Benzene hexachloride
  Hexachlorobenzene
  Lindane
  Tecnazene

Benzene sulfonic acid
  4-Chlorophenyl phenyl sulfone

Benzene sulfonyl chloride
  Fenson

p-Benzoquinone
  Ethofumesate

Benzoyl chloride
  Benzoximate
  Benzoylprop-ethyl
  Flamprop-isopropyl
  Flamprop-methyl
  2,3,6-TBA

2-Benzoyl pyridine
  Norbormide

5-Benzyl-3-furylmethyl alcohol
  Resmethrin

Benzyl isopropylamine
  Butam

Benzyl mercaptan
  Benzyl dibutylthiocarbamate

Biguanide
  Menazon

Bis(N,N-dimethylamido)phosphoryl chloride
  Triamiphos

Borax
  Disodium octaborate
  Sodium metaborate

Bromine
  Bromacil
  Bromoxynil
  Bronopol
  Chlorbromuron
  Chlorophacinone
  Codlelure
  Dibromochloropropane
  Ethylene dibromide
  Methyl bromide
  Metobromuron
  Naled
  Thiofanox
  Triadimefon

Bromochloromethane
  Chlormephos

2-Bromo-1-methoxypropane
  Metolachlor

p-Bromophenyl isocyanate
  Metobromuron

α-Bromo-pinacolone
  Triadimefon

2-Bromopropionic acid methyl ester
  Furalaxyl
  Metalaxyl

5-Bromopyridine
  Ancymidol
  Nuarimol

5-Bromopyrimidine
  Fenarimol

Bromoxynil
  Bromoxynil octanoate

Butadiene
  Anthraquinone
  Captafol
  Captan
  Octahydrodibenzofurancarbaldehyde

n-Butanol
  Butachlor
  Dibutyl phthalate
  Flurecol-butyl

2-Butanol
  sec-Butylamine

t-Butanol
  Trimedlure

Maneb
  Metham-sodium
  Nabam
  Propineb
  Thiram
  Ziram

Carbon monoxide
  Carbon disulfide
  Fenuron
  Sodium fluoroacetate

Carbon tetrachloride
  Trifenmorph

Carbonyl sulfide
  Diallate
  Triallate

Carboxin
  Oxycarboxin

Catechol
  Carbofuran

Chloral
  Chloralose
  DDT
  Dichlorvos
  Methoxychlor
  Trichlorfon

Chlorfenethol
  Dicofol

Chlorine
  Allyl alcohol
  Benzene hexachloride
  Captan
  Carbon tetrachloride
  Chlordane
  Chloroacetic acid
  Chloroneb
  Chloronitropropane
  Chloropicrin
  Chlorothalonil
  Chlorthal-dimethyl
  Crimidine
  2,4-D
  Dalapon
  Dialifos
  Dichlobenil
  Dichlone
  Dichlorobenzene
  Di-(2-chloroethyl ether)
  1,1-Dichloro-1-nitroethane
  3,6-Dichloropicolinic acid
  1,2-Dichloropropane
  Dichloropropane-dichloropropene
  1,3-Dichloropropene
  Dichlorprop
  Dicloran
  Dicofol

Dioxathion
Ethylene dichloride
Fenac
Heptachlor
Hexachlorobenzene
MCPA
Mercurous chloride
Methyl parathion
Nitrapyrin
Oxamyl
Parathion
Pentachlorophenol
Phosalone
Picloram
Pyridinitril
Quinotozene
2,3,6-TBA
TCA
TDE
Terbacil
Terpene polychlorinates
Toxaphene

Chloroacetamide
  Dimethoate

Chloroacetic acid
  2,4-D
  MCPA
  Naphthylacetic acid
  Quinalphos
  2,4,5-T

Chloroacetyl chloride
  Alachlor
  Allidochlor
  Butachlor
  Dimethachlor
  Formothion
  Mecarbam
  Metolachlor
  Propachlor

N-chloroacetyl-2-methyl piperidine
  Piperophos

m-Chloroaniline
  Barban
  Chlorpropham

o-Chloroaniline
  Anilazine

p-Chloroaniline
  Chlorfenprop-methyl
  Monalide
  Monuron

p-Chlorobenzaldehyde
  1,1-Di(4-chlorophenyl)-2-nitropropane/butane

Chlorobenzene
  Bufencarb

Chlorophacinone
4-Chlorophenyl phenyl sulfone
DDT
1,1-Di-(4-chlorophenyl)-2-nitropropane/bu-
tane
2-Phenylphenol
TDE
Triphenyltin acetate
Triphenyltin hydroxide

p-Chlorobenzene diazonium chloride
Chlorfenprop-methyl
Drazoxolon

p-Chlorobenzotrifluoride
Trifluralin

o-Chlorobenzoyl chloride
Fenarimol

p-Chlorobenzylacetone
Coumachlor

p-Chlorobenzyl chloride
Chlorbenside

5-Chloro-7-bromo-8-oxyquinoline
Halacrinate

2-Chloro-4-bromophenol
Profenofos

2-Chlorobutanone
Butocarboxim

4-Chloro-2-butynol
Barban

4-Chloro-2-butynyl chloroformate
Barban

γ-Chlorobutyronitrile
MCPB

4-Chloro-o-cresol
Mecoprop

1-Chloro-2,4-dinitrobenzene
Bromofenoxim

4-Chloro-3,5-dinitrobenzoic acid
Benfluralin
Trifluralin

4-Chloro-3,5-dinitro-o-xylene
Pendimethalin

N-β-chloroethyl benzene sulfonamide
Bensulfide

Chloroethyltrichlorosilane
Etacelasil

3-Chloro-4-fluoroaniline
Flamprop-isopropyl
Flamprop-methyl

1-Chloro-4-isopropyl-2,6-dinitrobenzene
Isopropalin

3-Chloro-4-methoxyphenyl isocyanate
Metoxuron

2-Chloro-N-methyl acetamide
Omethoate

3-Chloro-4-methyl aniline
Solan

p-Chloromethylchlorophenyl sulfide
Methyl carbophenothion

3-Chloro-4-methyl-7-hydroxycoumarin
Coumaphos

2-Chloro-N-methyl-3-oxobutyramide
Monocrotophos

4-Chloro-2-methylphenyl isothiocyanate
Chloromethiuron

Chloromethyl phosphonic acid
Glyphosate

Chloromethyl thiocyanate
TCMTB

2-Chloro-4-nitroaniline
Niclosamide

p-Chloronitrobenzene
Nitrofen
Tetrasul

2-Chloro-6-nitrobenzonitrile
Dichlobenil

4-Chloro-3-nitrophenyl methyl sulfone
Nitralin

p-Chlorophenol
Dichlorophen
Fenson
Ovex
Triadimefon

4-p-Chlorophenoxyphenyl isocyanate
Chloroxuron

m-Chlorophenyl isocyanate
Barban
Chlorbufam

p-Chlorophenyl isocyanate
Buturon

Diflubenzuron
Monolinuron

p-Chlorophenyl mercaptan
Chlorbenside

N-(3-Chlorophenyl)-N'-methoxy-N'-methylurea
Chlorbromuron

p-Chlorophenyl sulfonyl chloride
Ovex

N-(2-chlorophenyl) thiourea
Benazolin

2-Chloropropionic acid
Benzoylprop-ethyl
Dichlorprop
Flamprop-isopropyl
Flamprop-methyl
Mecoprop

5-Chlorosalicylic acid
Niclosamide

Chlorosulfonic acid
Disul
Tetradifon

3-Chloro-4-trichloromethylmercapto-
phenyl isocyanate
Fluothiuron

Chromic acid
Copper zinc chromate

Chrysanthemum monocarboxylic acid
Allethrin
Phenothrin
Resimethrin
Tetramethrin

Copper
Copper carbonate
Copper oxychloride
Copper sulfate
Cuprous oxide
Paris green

Copper sulfate
Oxine-copper
Paris green

o-Cresol
DNOC
MCPA

Crotonyl chloride
Dinocap

Cupric chloride
Copper oxychloride

Cyanamide
Dodine

Cyanoacetyl-3-ethylurea
1-(2-cyano-2-methoxyiminoacetyl)-3-
ethylurea

α-Cyanobenzaldoxime
Phoxim

Cyanogen chloride
Mephosfolan
Phosfolan

4-Cyanophenol
Cyanofenphos
Cyanophos

Cyanuric chloride
Anilazine
Atrazine
Aziprotryn
Cyanazine
Cyprazine
Desmetryne
Dimethametryn
Methoprotryne
Propazine
Secbumeton
Simazine
Terbuthylazine
Trietazine

Cyclohexane
Chloranil

Cyclohexene oxide
Propargite

Cyclohexyl chloride
Cyhexatin

Cyclohexyl isocyanate
Hexazinone

Cyclohexyl urea
Lenacil

Cyclooctylamine
Cycluron

Cyclopentadiene
Aldrin
Chlordane
Endrin
N-(2-Ethylhexyl)-8,9,10-trinorborn-5-ene-2,3-
dicarboximide
Heptachlor
Norbormide
Noruron

Cyclopentene
Noruron

3,4-Dichlorophenyl isocyanate
  Linuron
  Neburon
  Swep

3,5-Dichlorophenyl isocyanate
  3-(3,5-Dichlorophenyl)-1-isopropyl-
  carbamoylhydantoin

2,3-Dichloro-1-propene
  Sulfallate

α,α-Dichloropropionic acid
  Erbon

Dichloropropyl dimethylamine
  Cartap

2,3-Dichloroquinoxaline
  Oxythioquinox

2,2-Dichlorovinyl ethyl ether
  1,1-Dichloro-2,2-bis(4-ethylphenyl)ethane

Dichlorvos
  Naled

Diethylamine
  Dinitramine
  Ethiolate
  Napropamide
  Trietazine

Diethylaniline
  Alachlor
  Butachlor

N,N-diethyl-α-bromopropionamide
  Napropamide

Diethyl carbomethoxyphosphonate
  Fosamine-ammonium

Diethyl 3-ethoxycarbonyl-2-methylprop-2-
  enyl phosphonate
  Methoprene

N,N-diethyl guanidine
  Pirimiphos-ethyl
  Pirimiphos-methyl

Diethyl maleate
  Malathion

Diethyl phosphite
  Demeton

Diethyl phosphorochloridate
  Fosthietan
  Mephosfolan
  Paraoxon
  Phosfolan
  Tepp

O,O-diethyl phosphorochloridothioate
  (O,O-diethyl thiophosphoric acid chloride)
  Bromophos-ethyl
  Chlorphoxim
  Chlorpyrifos
  Chlorthiophos
  Coumaphos
  Demeton
  Diazinon
  Dichlofenthion
  Ditalimfos
  Fensulfothion
  Phoxim
  Pirimiphos-ethyl
  Pyrazophos
  Quinalphos
  Sulfotepp
  Thionazin
  Triazophos

Diethyl phosphorodithioic acid
  (O,O-diethyl-dithiophosphoric acid)
  Azinphos-ethyl
  Carbophenothion
  Chlormephos
  Dialifos
  Dioxathion
  Ethion
  Mecarbam
  Phosalone
  Prothoate
  Terbuphos

Diethyl phthalate
  Pindone
  2-Valerylindane-1,3-dione

Diethyl sulfate
  Chlorobenzilate
  Dipropetryn

2,6-Difluorobenzamide
  Diflubenzuron

Dihydrazine sulfate
  Maleic hydrazide

Dihydrosafrol
  Piperonyl butoxide

N-di-(2-hydroxypropyl)cyclododecylamine
  Dodemorph

Diisobutylamine
  Butylate

Diisopropylamine
  Diallate
  Fenamiphos
  Triallate

O,O-diisopropyl hydrogen phosphorodithioate
  Bensulfide

Chlorpyrifos-methyl
Cyanophos
Demeton-methyl
Dimethyl-4-(methylthio)phenyl phosphate
Etrimfos
Famphur
Fenitrothion
Fenthion
Iodofenphos
Methamidophos
Methyl parathion
Oxydemetron-methyl
Pirimiphos-methyl
Sulprofos
Temephos

O,O-dimethyl phosphorothioic acid
Omethoate
Vamidothion

Dimethyl phthalate
Diphacinone

Dimethyl propylamine
Dimethametryn

Dimethyl sulfamoyl chloride
Bupirimate

Dimethyl sulfate
Carbendazim
Dicamba
Mexacarbate

α,α-Dimethylvaleryl chloride
Monalide

4,6-Dinitro-2-sec-butylphenol
Binapacryl
Dinobuton

3,5-Dinitro-4-chlorobenzotrifluoride
Profluralin

3,5-Dinitro-4-[di-(n-propyl)amino]benzene
sulfonyl chloride
Oryzalin

2,6-Dinitro-4-trifluoromethylchlorobenzene
Fluchloralin

Dinoseb
Dinoseb acetate

Dioctyl triamine
Guazatine

Dioxane
Dioxathion

Diphenyl acetone
Diphacinone

Diphenylacetyl chloride
Diphenamid

Diphenylene oxide
2-Phenylphenol

Di-n-propylamine
EPTC
Isopropalin
Nitralin
Trifluralin

Di-n-propyl carbamyl chloride
Vernolate

Di-n-propyl chlorothionophosphate
O,O,O',O'-tetrapropyl dithiopyrophosphate

Di-n-propyl dithiophosphoric acid
Piperophos

Dodecylamine
Dodine

Dodecyl chloride
Dodine

Ethanedithiol
Phosfolan

Ethanol
Benzoylprop-ethyl
Disulfoton
Ethoprop
Etridiazole
Fenamiphos
Glyodin
Parathion
Phorate
TDE
TEPP

Ethirimol
Bupirimate

Ethyl acetate
Chlorfenethol

Ethyl acetoacetate (Acetoacetic ester)
Bromacil
Carboxin
Crimidine
Dehydroacetic acid
Diazinon
Drazoxolon
Pirimiphos-ethyl
Pirimiphos-methyl

Ethylamine
Atrazine
Carbetamide
Cyanazine

Phorate

β-Ethylmercaptoethyl thiocyanate
    Demeton

2-Ethyl-6-methylaniline
    Metolachlor

O-ethyl-methylpropylthionophosphoric acid
chloride
    Butamiphos

Ethyl-α-phenyl-α-bromoacetate
    Phenthoate

O-ethyl phenyl phosphonochloridothionate
    Cyanofenphos
    O-2,4-dichlorophenyl-O-ethyl phenylphos-
    phonothioate

Ethyl phosphorodichloridate
    Edifenphos
    Isofenphos

O-ethyl-S-n-propyl chlorothiophosphate
    Profenofos

Ethylpropyl phenol
    Bufencarb

Ethylthioethanol
    Disulfoton
    Thiometon

2-Ethylthiomethyl phenol
    Ethiofencarb

Fenuron
    Fenuron-TCA

Fluorine
    Sulfuryl fluoride

Fluoroacetyl chloride
    Fluoroacetamide

4-Fluoro-2'-chlorobenzophenone
    Nuarimol

Fluorodichloromethyl sulfenyl chloride
    Dichlofluanid

Fluorspar
    Sodium fluoride

Formaldehyde
    Alachlor
    Azinphos-ethyl
    Azinphos-methyl
    Butachlor
    Carbophenothion
    Dazomet
    Dichlorophen

S-ethylsulfinylmethyl O,O-diisopropyl
    phosphorodithioate
    Glyphosine
    Methidathion
    1,1-Methylenedi(thiosemicarbazide)
    Phorate
    Phosalone
    Phosmet
    Piperonyl butoxide
    Sodium fluoroacetate
    Terbufos

Formamide
    Hydrocyanic acid

Formic acid
    Amitrole
    Dicofol
    Tricyclazole

Furan
    Endothall

Furan-2-carboxylic acid chloride
    Furalaxyl

Furfural
    Fuberidazole
    Octahydrodibenzofurancarbaldehyde

Glucose
    Chloralose

Glycine
    3-(3,5-Dichlorophenyl)-1-isopropyl-
    carbamoylhydantoin
    Glyphosate
    Glyphosine

Hexachlorobenzene
    Pentachlorophenol

Hexachlorocyclopentadiene
    Aldrin
    Chlordane
    Chlordecone
    Dienochlor
    Endosulfan
    Endrin
    Heptachlor
    Mirex

Hexachloro-p-xylene
    Chlorthal-dimethyl

Hexadecyl alcohol
    Cycloprate

Hexamethyleneimine
    Molinate

2-Hydrazino-4-methylbenzothiazole
    Tricyclazole

Atrazine
Aziprotryn
Cyprazine
Desmetryne
Isofenphos
Methoprotryne
Propazine

n-Isopropyl aniline
Phopachlor

N-isopropyl chloroacetamide
Prothoate

Isopropyl chloroformate
Chloropropham
Dinobuton
Propham

Isopropyl isocyanate
3-(3,5-Dichlorophenyl)-1-isopropyl-
carbamoylhydantoin
Iprodione

Isopropyl sulfamoyl chloride
Bentazon

Isosafrole
Sulfoxide

Lactonitrile
Acrylonitrile

Lead oxide
Lead arsenate

Lithium
Fenarimol

Lithium aluminum hydride
Profluralin

Magnesium
Chlorfenethol
Fenbutatin oxide
Triphenyltin acetate
Triphenyltin hydroxide

Maleic anhydride
Captafol
Captan
Endothall
N-(2-ethylhexyl)-8,9,10-trinorborn-5-ene-
2,3-dicarboximide
Maleic hydrazide

Maleimide
Norbormide

Manganous chloride
Mancozeb

Manganous sulfate
Maneb

2-Mercaptobenzothiazole
TCMTB

2-Mercaptoethanol
Carboxin

Mercuric acetate
2-Methoxyethyl mercuric chloride
2-Methoxyethyl mercury silicate
Phenylmercury acetate

Mercuric oxide
PMA

Mercury
Mercuric chloride
Mercurous chloride

Mesityl oxide
Butopyronoxyl

Mesyl chloride
Ethylfumesate

Methacrylyl chloride
Dicryl

Methallyl chloride
Carbofuran

Methane
Carbon disulfide
Carbon tetrachloride
Formaldehyde
Hydrocyanic acid

Methanol
Alachlor
Chlorflurecol-methyl
Chlorthal-dimethyl
Chlorthiophos
Codlelure
Crufomate
Demephion
Diazinon
Dimethyl phthalate
Flamprop-methyl
Formaldehyde
Malathion
Methyl bromide
Prometon
Propetamphos
Secbumeton
Swep

Methoxyacetaldehyde
Dimethachlor

Methoxyacetyl chloride
Metalaxyl

4-Methoxybenzoyl cyclopropane
Ancymidol

Methyl isoamyl ketone
  2-Isovalerylindane-1,3-dione

Methyl isocyanate
  Aldicarb
  Aminocarb
  Bendiocarb
  Benzthiazuron
  Bufencarb
  Butocarboxim
  Carbofuran
  Dioxacarb
  Ethiofencarb
  Formetanate
  Methabenzthiazuron
  Methazole
  Methiocarb
  Methomyl
  Mexacarbate
  Nitrilacarb
  Oxamyl
  Promecarb
  Propoxur
  Thiazfluron
  Thiofanox
  Trimethylphenyl methyl carbamate
  3,4-Xylyl methyl carbamate
  3,5-Xylyl methyl carbamate

3-Methyl-5-isopropyl phenol
  Promecarb

Methyl isothiocyanate
  Metham-sodium

O-methyl isourea
  Guazatine

Methyl lactate
  Carbetamide

Methyl magnesium bromide
  Chlorfenethol

Methyl mercaptan
  Ametryne
  Butocarboxim
  Demephion
  Desmetryne
  Dimethametryn
  Methomyl
  Oxamyl
  Prometryne
  Terbutryne
  Thiofanox

4-Methylmercapto-n-cresol
  Fenthion

3-Methyl-4-mercaptomethyl phenol
  Fenamiphos

4-Methylmercapto phenol

Dimethyl 4-(methylthio)phenyl phosphate
  Fensulfothion

4-Methylmercapto-3,5-xylenol
  Methiocarb

3-Methyl-4-methylmercaptophenol
  Fenamiphos

O-methyl methyl phosphonodithioic acid
  Mecarphon

2-Methyl-2-methylthiopropionaldehyde N-
  methylcarbamoyloxime
  Aldoxycarb

3-Methyl-6-nitrophenol
  Butamiphos

2-Methyl pentanoyl chloride
  Solan

Methyl phenoxy acetic acid
  MCPA

O-methyl phenylthiophosphonyl chloride
  Leptophos

O-methyl phosphorothionodichloridate
  2-Methoxy-4H-benzo-1,3,2-dioxaphosphorin-
  2-sulfide

Methyl phthalate
  Chlorophacinone

2-Methyl propionaldehyde
  Ethofumesate

Methyl propynol
  Chlorbufam

3-Methyl-5-pyrazolone
  Dimetilan

4-Methylthiophenol
  Sulprofos

2-Methyl-2-thiopseudourea sulfate
  Hexazinone

4-Methyl thiosemicarbazide
  Thiazfluron

4-Methylthio-3,5-xylenol
  Methiocarb

Monuron
  Monuron-TCA

Morpholine
  Ethofumesate
  Trifenmorph

Phenanthraquinone
  Flurecol-butyl

p-Phenetidine
  Ethoxyquin

Phenol
  2,4-D
  Dinocap
  Pentachlorophenol

3-Phenoxybenzyl alcohol
  Permethrin
  Phenothrin

2-Phenoxypropionic acid
  Dichlorprop

Phenyl acetone
  Chlorophacinone

5-Phenyl-3-amino-1,2,4-triazole
  Triamiphos

o-Phenylene diamine
  Carbendazim
  Fuberidazole
  Quinalphos
  Thiabendazole
  Thiophanate
  Thiophanate-methyl

Phenyl hydrazine
  Chloridazon

1-Phenyl-3-hydroxy-1,2,4-triazole
  Triazophos

Phenyl isocyanate
  Carbetamide
  Desmedipham
  Fenuron
  Propham
  Siduron

3-Phenyl-1-methoxy-1-methylurea
  Metobromuron

N-phenyl-N'-methoxy-N'-methylurea
  Metobromuron

O-phenyl phosphorochloridate
  Diamidaphos

Phenyl thiophosphonyl dichloride
  EPN

4-Phenylthio-o-toluidine
  Perfluidone

Phosgene
  Benodanil
  Bentazon

Benzyl dibutyl thiocarbamate
Bromacil
Bufencarb
Butacarb
Buthidazole
Butylate
2-sec-Butyl phenyl methylcarbamate
Carbaryl
EPTC
Methidathion
Monuron
Oxadiazon
Oxythioquinox
Pirimicarb
Promecarb

Phosphorus
  Aluminum phosphide
  Merphos
  Zinc phosphide

Phosphorus oxychloride
  Chlordimeform
  Crufomate
  Dimefox
  Fenamiphos
  Schradan
  TEPP
  S,S,S-tributyl phosphorotrithioate

Phosphorus pentasulfide
  Disulfoton
  Malathion
  Parathion
  Phorate

Phosphorus pentoxide
  TEPP

Phosphorus trichloride
  Disulfoton
  Ethoprop
  Merphos
  S,S,S-tributyl phosphorotrithioate

Phthalic anhydride
  Anthraquinone
  Dibutyl phthalate
  Dimethyl phthalate
  Naptalam

Phthalimide
  Phosmet

Picoline
  3,6-Dichloropicolinic acid

Pinacolone
  Thiofanox

Pinene
  Terpene polychlorinates

Piperonal
  Sesamex

Piperonylic acid
  Butyl Carbitol piperonylate

Pivalyl chloride
  Butam

Potassium cyanide
  Dichlobenil

Potassium fluoride
  Dimefox
  Sodium fluoroacetate

Potassium hydrosulfide
  Methoprotryne

Potassium hydroxide
  Methyl tetrachloro-N-methoxy-N-methyl
    terephthalamate

Potassium permanganate
  Demeton S-methyl sulfone

Potassium phthalimide
  Ditalimfos

Potassium thiocyanate
  Isobornyl thiocyanoacetate
  Noruron
  Thiophanate
  Thiophanate-methyl

Propane-1,2-dithiol
  Mephosfolan

Propanol
  Dipropyl isocinchomeronate

Propargyl alcohol
  Propargite

Propazine
  Dipropetryn
  Prometon
  Prometryne

Propionic acid
  Dalapon
  Propanil

n-Propylamine
  Profluralin

Propyl chlorothiolformate
  Pebulate

Propylene
  Acrolein
  Acrylonitrile
  Allyl alcohol

1,2-Dichloropropane
Dichloropropane-dichloropropene
1,3-Dichloropropene

1,3-Propylene diamine
  Propineb

Propylene oxide
  Allyl alcohol
  Butylphenoxy isopropoxy isopropyl chloro-
    ethyl sulfite
  Butylphenoxy methylethyl chloroethyl sulfite

N-propyl-N-hydroxyethylamine
  Fluchloralin

Propyl mercaptan
  Ethoprop
  Pebulate
  Vernolate

Pyridine
  Diquat
  Paraquat

Pyrogallol
  Bendiocarb

Quinone
  Chloroneb

Ricinoleic acid
  Undecylenic acid

Salicylaldehyde
  Dioxacarb

Salicylic acid isopropyl ester
  Isofenphos

Saligenol
  2-Methoxy-4H-benzo-1,3,2-dioxaphos-
    phorine-2-sulfide

Sodium
  Benzyl dibutylthiocarbamate
  Buthiobate
  Butylate
  Cyhexatin
  EPTC
  Paraquat
  2-Phenyl phenol
  Triadimefon
  Triamiphos
  Vernolate

Sodium acetate
  Triphenyltin acetate

Sodium arsenite
  Paris green

Sodium azide
  Aziprotryn

Sodium carbonate
  Copper carbonate
  Phenthoate
  Sulfotepp

Sodium chloride
  Calcium cyanide
  Cryolite
  Sodium chlorate

Sodium 4-chloro-o-cresylate
  MCPB

Sodium-α-chloropropionate
  Silvex

Sodium cyanamide
  Dodine

Sodium cyanide
  Dithianon
  Fenac

Sodium 2,5-dichloro-4-iodophenolate
  Iodofenphos

Sodium 2,4-dichlorophenolate
  O-2,4-dichlorophenyl-O-ethyl
    phenylphosphonothioate

Sodium diethyldithiocarbamate
  Sulfallate

Sodium ethoxide
  EPN

Sodium hydrosulfide
  Oxythioquinox

Sodium hydroxide
  Allyl alcohol
  Cacodylic acid
  Cyhexatin
  Dalapon
  Dicamba
  Disulfoton
  DSMA
  Fenbutatin oxide
  MCPB
  Metham-sodium
  MSMA
  Nabam
  Pentachlorophenol
  Phenylphenol
  Propineb
  Propyzamide
  Sodium arsenite
  Sodium fluoride
  Sodium fluoroacetate
  Sodium metaborate
  Sodium 2,4,5-trichlorophenate
  TCA
  Thiram

Triphenyltin hydroxide
  Ziram

Sodium metaborate
  Disodium octaborate

Sodium methoxide
  Diphacinone
  Ronnel
  Temephos
  Terbumeton

Sodium methyl dithiocarbamate
  Methyl isothiocyanate

Sodium methyl mercaptide
  Aldicarb
  Aziprotryn

Sodium O-methyl methyl phosphonodithioate
  Mecarphon

Sodium 3-methyl-4-nitrophenolate
  Fenitrothion

Sodium nitrite
  Aldicarb
  Chlorfenprop-methyl
  1-(2-Cyano-2-methoxyiminoacetyl)-3-
    ethylurea
  Drazoxolon
  Tetrasul

Sodium p-nitrophenoxide
  EPN
  Methyl parathion
  Parathion

Sodium phthalimide
  Folpet

Sodium sulfite
  Fenaminosulf

Sodium thiocyanate
  Cartap
  Fosthietan

Stannic chloride
  Triphenyltin acetate
  Triphenyltin hydroxide

Stannic oxide
  Bis(tributyltin) oxide

Stearic acid
  Glyodin

Succinic anhydride
  Daminozide

Sulfur
  Carbon disulfide

p-Toluene sulfonic acid
   Amitraz

m-Toluyl chloride
   4-Methoxy-3,3'-dimethyl benzophenone

m-Tolyl isocyanate
   Phenmedipham

1,2,4-Triazole
   Fluotrimazole
   Triadimefon

Tri-n-butyl phosphine
   Chlorphonium

Trichloroacetamidine hydrochloride
   Etridiazole

Trichloroacetic acid
   Fenuron-TCA
   Monuron-TCA

1,2,4-Trichlorobenzene
   Dicamba
   Stirofos
   Tetradifon

Trichloroethylene
   Chloroacetic acid

Trichloromethanesulfenyl chloride
   Etridiazole

2,4,5-Trichlorophenol
   Ronnel
   Silvex
   Sodium 2,4,5-trichlorophenate
   2,4,5-T
   Trichloronate

2-(2,4,5-Trichlorophenoxy) ethanol
   Erbon

2,4,5-Trichlorophenyl methyl sulfide
   Chlorthiophos

1,2,3-Trichloropropene
   Diallate

3,5,6-Trichloro-2-pyridinol
   Chlorpyrifos
   Chlorpyrifos-methyl

2,4,5-Trichlorothiophenol
   Tetrasul

Triethyl phosphate
   TEPP

Triethyl phosphite
   Chlorfenvinphos

Trifluoroacetic anhydride
   Thiazfluron

Trifluoromethanesulfonyl fluoride
   Mefluidide
   Perfluidone

3-Trifluoromethylphenyl-diphenyl-methyl
      chloride
   Fluotrimazole

m-Trifluoromethylphenyl hydrazine
   Norflurazon

3-Trifluoromethylphenyl isocyanate
   Fluometuron

1-(3-Trifluoromethylphenyl)-3-phenyl-2-
      propanone
   Fluridone

Trimethylacetyl chloride
   Oxadiazon

Trimethylamine
   Chlormequat

3,4,5-Trimethyl phenol
   Trimethylphenyl methylcarbamate

Trimethyl phosphite
   Crotoxyphos
   Dichlorvos
   Dicrotophos
   1,3-Di-(methoxycarbonyl)-propen-2-yl
      dimethyl phosphate
   Mevinphos
   Monocrotophos
   Phosphamidon
   Stirofos

Tris(2-chloroethyl) phosphite
   Ethephon

Urea
   AMS
   Crimidine
   Diuron
   Fenuron
   Phosalone
   Potassium cyanate

Vinyl ethyl Carbitol
   Sesamex

N-vinyl phthalimide
   Dialifos

m-Xylene
   Chlorothalonil

TRADE NAMES INDEX

DIC 1468 - 528
Dicarzol - 430
Dicestal - 285
Dicofen - 407
Didroxane - 285
Difolatan - 132
Dilan - 288
Dimecron - 619
Dimetate - 314
Dimetilane - 320
Dimilin - 309
Dimite - 176
Dinitrol - 352
Dinitrosol - 352
Dinocide - 243
Dinofen - 322
Dipel - 75
Diphacin - 333
Diphenadione - 333
Di-phenthane 70 - 285
Dipofene - 185
Dipterex - 746
Dirimal - 570
Ditac - 356
Dithane D-14 - 545
Dithane M-22 - 486
Dithane M-45 - 483
Dithane Z-78 - 769
Dithio - 689
Dithiolane Iminophos-
    phate - 497
Dithione - 689
Dithio-Systox - 342
Dithiotepp - 689
Ditranil - 295
DMDT - 513
DMOC - 155
DMP - 319
DN 289 - 326
DO-14 - 641
Dosanex - 527
Dowco 132 - 216
Dowco 139 - 532
Dowco 163 - 560
Dowco-169 - 260
Dowco 179 - 199
Dowco 199 - 346
Dowco 213 - 226
Dowco 214 - 202
Dowco 290 - 289
Dow ET-14 - 662
Dow ET-57 - 662
Dowfume - 290
Dowfume W-85 - 389
Dowicide 7 - 598
Dowicide 25 - 682
Dowicide B - 682

Dowicide I - 609
Dowpon - 235
DPA - 640
DPX-410 - 574
DPX-3217 - 221
DPX-3674 - 451
Drat - 188
Drawin 755 - 116
Draza - 507
Drepamon - 96
Drinox - 445
Dropleaf - 676
Dual - 525
DuPont 634 - 466
DuPont 1179 - 508
DuPont 1318 - 671
DuPont 1991 - 83
DuPont Herbicide 326 -
    471
DuPont Herbicide 732 - 712
DuPont Herbicide 976 -
    100
Dursban - 199
Du-Ter - 758
DXP-1108 - 433
Dyanap - 553
Dybar - 412
Dyfonate - 427
Dylox - 746
Dymid - 334
Dynone - 649
Dyrene - 57
Dyrex - 746

E-393 - 689
E605 - 584
E-3314 - 445
E14049 - 474
E147,470 - 497
EDB - 389
EDC - 393
Ekalux - 658
Ekamet - 400
Ekatin - 729
Ekatine - 342
Ektafos - 299
EI 12,880 - 314
EI 18,682 - 650
EI 47,031 - 616
EL-110 - 81
EL-119 - 570
EL-171 - 425
EL-179 - 461
EL-222 - 405
EL-228 - 567
EL-291 - 750
EL-531 - 56

EL-3911 - 611
Elgetol 30 - 352
Elimite - 163
Elocron - 330
Elsan - 606
Elvaron - 277
EM 379 - 443
Embafume - 517
Embark - 495
Embutox - 240
Emmaton - 474
Enavene - 94
Endosan - 98
Enide - 334
ENT 27566 - 430
Ent 39054 - 309
Entex - 411
EP 332 - 430
Ephidan - 398
EPN-300 - 373
Eprofil - 725
Eptam - 376
Eptapur - 119
Equizole - 725
Erbon - 378
Erbotan - 727
Erithane - 758
Esgram - 581
Esteron - 697
Estonmite - 571
Estonox - 734
Etazine - 669
Ethide - 285
Ethrel - 380
Ethylan - 282
Ethyl Guthion - 68
Etilon - 584
Euparen - 277
Evik - 51
Evitol - 565
Exotherm Termil - 193
Experimental Insecticide
    269 - 367
Experimental Nematocide
    18,133 - 729

F-461 - 576
FA 2071 - 458
FAC - 650
Fall - 676
Famid - 330
Famophos - 401
Faneron - 102
Fargo - 739
Farmatin - 758
FB/2 - 338
Fenac - 402

Saphizon - 496
Sarolex - 260
Saviac - 188
Sayfos - 496
SBP 1382 - 660
Schering 4075 - 605
Schering 34,615 - 633
Schering 35,830 - 537
Schering 36056 - 430
Schering 36,268 - 174
Schering 38,584 - 605
Scorch - 59
Scutl - 607
SD-3562 - 299
SD-4294 - 215
SD-7859 - 158
SD-8447 - 684
SD-8530 - 756
SD9129 - 537
SD-11,831 - 558
SD 14,114 - 405
Semeron - 255
Sencor - 528
Sencoral - 528
Sencorex - 528
Sendran - 648
Seradix - 455
SES - 341
Sesoxane - 670
Sevin - 138
Sevinon - 760
Shed-A-Leaf - 676
Shoxin - 564
Sicarol - 651
Silvisar - 125, 542
Sinbar - 712
Sinituho - 598
Sinox - 352
Sinox General - 326
Sistan - 504
Slo-Gro - 481
SMDC - 504
SN-41,703 - 649
Snip - 320
Sodar - 356
Soilbrome 85 - 389
Soil Fungicide 1823 - 186
Soilgrin - 39
Solvirex - 342
Somio - 158
Sopranebe - 486
Spectracide - 260
Spergon - 160
Sporeine - 75
Spotrete - 733
SR-73 - 555
SR406 - 134

SRA 5172 - 503
S-Seven - 286
Stam F-34 - 640
Standak - 38
Stockade - 603
Stomoxin - 603
Stomp - 597
Stop Scald - 389
Strobane - 716
Sucker Stuff - 481
Suffix - 94
Sulfenimide - 132
Sulfox-cide - 690
Sulfoxyl - 690
Sulphatepp - 689
Sulphenone - 189
Sumithion - 407
Sumithrin - 606
Sumitol - 669
Suncide - 648
Supersevtox - 326
Super Sucker Stuff - 481
Supracide - 506
Surcopur - 640
Surflan - 570
Sutan - 121
Swebate - 709
Synklor - 166
Synthrin - 660
Systox - 251
Sytam - 668
Sytasol - 322

T 1258 - 157
Tag HL 331 - 607
Taktic - 53
Talan - 322
Talcord - 603
Tamaron - 503
TBTO - 99
TCNB - 708
Tecane - 703
Tecto - 725
Tedion V-18 - 719
Tekwaisa - 593
Telone - 291
Telvar - 539
Temik - 36
Tendex - 648
Teniathane - 285
Teniatol - 285
Tenoran - 194
Teridox - 311
Termil - 193
Terrachlor-Super X - 399
Terraclor - 658
Terracoat - 399

Terracur P - 409
Terra Sytam - 310
Terrazole - 399
Tersan - 733
Tersan SP - 186
Tetron - 710
TH6040 - 309
Thiaben - 725
Thibenzole - 725
Thifor - 360
Thimet - 611
Thimul - 360
Thiodan - 360
Thiomethon - 729
Thiophos - 584
Thiotepp - 689
Thitrol - 492
Thiurad - 733
Thiuramyl - 733
Thuricide - 75
Thylate - 733
Thynon - 347
Tiguvon - 411
Tilcarex - 658
Tillam - 596
Tinotepp - 689
Tinox - 250
Tiuramyl - 733
TOK E-25 - 562
Tolban - 632
Tomorin - 211
Topane - 609
Topsin - 731
Topsin-Methyl - 732
Torak - 256
Torque - 405
Totril - 458
Toxadust - 734
Toxaspra - 734
Toxynil - 457
Trapex - 523
Treflan - 752
Triatox - 53
Tribactur - 75
Tri-Ban - 623
Tribunil - 502
Tricornox - 80
Tri-Fene - 402
Trimangol - 486
Tri-Me - 521
Trimidal - 567
Triminol - 567
Trinoxol - 697
Trithion - 153
Tritisan - 658
Trivex-T - 89
Trolene - 662

BIBLIOGRAPHY

(B-1) Gruber, G.I. (TRW Systems Group), *Assessment of Industrial Hazardous Waste Practices: Organic Chemicals, Pesticides and Explosives Industries,* Report EPA/530/SW-118c, Washington, DC, U.S. Environmental Protection Agency (April 1975).

(B-2) Ghassemi, M. and Quinlivan, S. (TRW Systems Group), *A Study of Selected Landfills Designed as Pesticide Disposal Sites,* Report EPA/530/SW-114c, Washington, DC, U.S. Environmental Protection Agency (November 1975).

(B-3) Lawless, E.W., Ferguson, T.L. and Meiners, A.F. (Midwest Research Institute), *Guidelines for the Disposal of Small Quantities of Used Pesticides,* Report EPA-670/2-75-057, Washington, DC, U.S. Environmental Protection Agency (June 1975).

(B-4) U.S. Environmental Protection Agency, *Development Document for Interim Final Effluent Limitations Guidelines for the Pesticide Chemicals Manufacturing Point Source Category,* Report No. EPA-440/1-75-060d, Washington, DC (November 1976).

(B-5) Fairchild, E.J., Ed., *Agricultural Chemicals and Pesticides: A Subfile of the Registry of Toxic Effects of Chemical Substances,* DHEW (NIOSH) Publication No. 77-180, Washington, DC, National Institute for Occupational Safety and Health (July 1977).

(B-6) National Institute for Occupational Safety and Health, *Health and Safety Guide for Pesticide Formulators,* DHEW (NIOSH) Publication No. 77-100, Washington, DC (May 1977).

(B-7) Genser, J.M., Zipperstein, A.H., Klosky, S.P. and Farber, P.S. (Processes Research, Inc.), *Alternatives for Hazardous Waste Management in the Organic Chemical, Pesticides and Explosives Industries,* Report EPA 530/SW-151e, Washington, DC, U.S. Environmental Protection Agency (September 1977).

(B-8) Lawless, E.W., Von Rumker, R., Kelso, G.L., Lawrence, K.A., Maloney, J.D. and Thompson, E.R. (Midwest Research Institute), *A Technology Assessment of Biological Substitutes for Chemical Pesticides,* Report NSF-RA-770508, Washington, DC, National Science Foundation (December 1977).

(B-9) National Institute for Occupational Safety and Health, *Occupational Diseases, A Guide to Their Recognition,* DHEW (NIOSH) Publication No. 77-181, Washington, DC (June 1977).

(B-10) Parsons, T.B., Ed. (Radian Corp.), *Industrial Process Profiles for Environmental Use;* Chapter 8: "Pesticides Industry," Report EPA-600/2-77-023h, Washington, DC, U.S. Environmental Protection Agency (January 1977).

(B-11) Wilkinson, R.R., Kelso, G.L. and Hopkins, F.C. (Midwest Research Institute), *State of the Art Report: Pesticide Disposal Research,* Report EPA-600/2-78-183, Washington, DC, U.S. Environmental Protection Agency (September 1978).

(B-12) U.S. Environmental Protection Agency, *Quality Criteria for Water,* Washington, DC (July 1976).

(B-13) U.S. Environmental Protection Agency, *Development Document for Effluent Limitations Guidelines for the Pesticide Chemicals Manufacturing Point Source Category,* Report EPA-440/1-78-060e, Washington, DC (April 1978).

(B-14) Marcus, M., Spigarelli, J. and Miller, H. (Midwest Research Institute), *Organic Compounds in Organophosphorus Pesticide Manufacturing Wastewaters,* Report EPA-600/4-78-056, Washington, DC, U.S. Environmental Protection Agency (September 1978).

(B-15) Archer, S.R., McCurley, W.R. and Rawlings, G.D. (Monsanto Research Corp.), *Source Assessments: Pesticide Manufacturing Air Emissions: Overview and Prioritization,* Report EPA-600/2-78-004d, Washington, DC, U.S. Environmental Protection Agency (March 1978).

(B-16) Kelso, G.L., Wilkinson, R.R., Malone, J.R., Jr. and Ferguson, T.L. (Midwest Research Institute), *Development of Information on Pesticides Manufacturing for Source Assessments,* Report EPA-600/2-78-100, Washington, DC, U.S. Environmental Protection Agency (May 1978).

(B-17) U.S. Environmental Protection Agency, *Suspended and Cancelled Pesticides,* Washington, DC, Publication No. OPA 159/9, Office of Public Awareness (October 1979).

(B18) National Institute for Occupational Safety and Health, *Criteria for a Recommended Standard: Occupational Exposure During the Manufacture and Formulation of Pesticides,* DHEW (NIOSH) Publication No. 78-174, Washington, DC (July 1978).

(B-19) Office of Technology Assessment, *Pest Management Strategies in Crop Protection,* Publication No. OTA-F-98, Washington, DC (October 1979).

(B-20) SCS Engineers, *Disposal of Dilute Pesticide Solutions,* Report No. SW-174c, Washington, DC, U.S. Environmental Protection Agency (May 1979).

(B-21) National Institute for Occupational Safety and Health, *Pocket Guide to Chemical Hazards,* DHEW (NIOSH) Publication No. 78-210, Washington, DC (September 1978).

(B-22) National Research Council, *Drinking Water and Health,* Washington, DC, National Academy of Sciences (1977).

(B-23) American Conference of Governmental Industrial Hygienists, *Threshold Limit Values for Chemical Substances in Workroom Air,* Cincinnati, Ohio, ACGIH (1979).

(B-24) Cardarelli, N.F. and Walker, K.E., *Development of Registration Criteria for Controlled Release Pesticide Formulations,* Report EPA-540/9-77-016, Washington, DC, U.S. Environmental Protection Agency (January 1978).

(B-25) Sittig, M., *Hazardous and Toxic Effects of Industrial Chemicals,* Park Ridge, NJ, Noyes Data Corp. (1979).

(B-26) Sittig, M., Ed., *Priority Toxic Pollutants—Health Impacts and Allowable Limits,* Park Ridge, NJ, Noyes Data Corp. (1980).

(B-27) U.S. Environmental Protection Agency, *Computer Printout of Pesticide Tolerances,* Washington, DC, Office of Toxic Substances (September 14, 1979).

(B-28) Lee, R.E. and Aspelin, A.L., *Economic Trends and Outlook of the Pesticide Industry: Need for Exclusive Amendments to FIFRA,* Report No. EPA-540/9-78-006, Washington, DC, U.S. Environmental Protection Agency (February 15, 1978).

(B-29) Scopes, N., Ed., *Pest and Disease Control Handbook,* British Crop Protection Council (1979).

(B-30) Sittig, M., "The Toxic Hazards of Industrial Chemicals—In the Workplace, for Factory Neighbors and for Consumers," paper presented at 49th Congress of the A.N.Z.A.A.S., Auckland, New Zealand (January 23, 1979).

(B-31) Wells, W.A., "The Rebuttable Presumption Against Registration (RPAR) Process," paper presented at Executive Enterprises Seminar, DuPont Plaza Hotel, Washington, DC (April 25, 1979).

PRIORITY TOXIC POLLUTANTS
Health Effects and Allowable Limits 1980

Edited by Marshall Sittig

Enivronmental Health Review No. 1

This book is a practical manual with its contents arranged alphabetically in encyclopedic form. It is intended to provide specific information on the 65 priority toxic pollutants (actually reflecting 129 individual compounds), their derivatives or degradation products, or intermedia transfers.

Compelled by a consent decree obtained in a federal court by public interest groups, the Environmental Protection Agency has promulgated criteria on allowable limits and profuse guidelines on how to interpret them and also how to determine further such criteria.

Here, in ready reference form, are the essentials of the criteria and guidelines which so far have been published for the control of priority toxic pollutants. Each typical pollutant is characterized with descriptions of its:

Occurrence
Physical Properties
Chemical Properties
Uses
Toxic Effects
Current Levels of Exposure

Special Groups at Risk
Existing Guidelines & Standards
Summary of Proposed Criteria
Bases for the Human Health Criteria
Pertinent References

A list of the 65 original priority pollutants follows here, however all of the 129 compounds are covered in this book.

1. Acenaphthene
2. Acrolein
3. Acrylonitrile
4. Aldrin/Dieldrin
5. Antimony & Compounds
6. Arsenic & Compounds
7. Asbestos
8. Benzene
9. Benzidine
10. Beryllium & Compounds
11. Cadmium & Compounds
12. Carbon Tetrachloride
13. Chlordane
14. Chlorobenzenes
15. Chlorinated Ethanes
16. Chloroalkyl Ethers
17. Chlorinated Naphthalenes
18. Chlorinated Phenols
19. Chloroform
20. 2-Chlorophenol
21. Chromium & Compounds
22. Copper & Compounds
23. Cyanides
24. DDT & Metabolites
25. Dichlorobenzenes
26. Dichlorobenzidine
27. Dichloroethylene
28. 2,4-Dichlorophenol
29. Dichloropropane/propene
30. 2,4-Dimethylphenol
31. Dinitrotoluenes
32. Diphenylhydrazines
33. Endosulfan
34. Endrin
35. Ethylbenzene
36. Fluoranthene
37. Haloethers
38. Halomethanes
39. Heptachlor
40. Hexachlorobutadiene
41. Hexachlorocyclohexane
42. Hexachlorocyclopentadiene
43. Isophorone
44. Lead & Compounds
45. Mercury & Compounds
46. Naphthalene
47. Nickel & Compounds
48. Nitrobenzene
49. Nitrophenols
50. Nitrosamines
51. Pentachlorophenol
52. Phenol
53. Phthalate Esters
54. Polychlorinated Biphenyls
55. Polynuclear Aromatics
56. Selenium & Compounds
57. Silver & Compounds
58. Tetrachlorodibenzo-p-dioxin
59. Tetrachloroethylene
60. Thallium & Compounds
61. Toluene
62. Toxaphene
63. Trichloroethylene
64. Vinyl Chloride
65. Zinc & Compounds

ISBN 0-8155-0797-6   370 pages

ALLAN LEVINE

King

WILLIAM LYON
MACKENZIE KING
A LIFE GUIDED BY
THE HAND OF DESTINY

Douglas & McIntyre
D&M PUBLISHERS INC.
Vancouver/Toronto

Douglas & McIntyre
An imprint of D&M Publishers Inc.
2323 Quebec Street, Suite 201
Vancouver BC Canada V5T 4S7
www.douglas-mcintyre.com

Cataloguing data available from Library and Archives Canada
ISBN 978-1-55365-560-2 (cloth)
ISBN 978-1-55365-908-2 (ebook)

Editing by Trena White
Copyediting by Lara Kordic
Jacket and text design by Jessica Sullivan
Front jacket photograph © Corbis
Back jacket photograph by Yousuf Karsh, Laurier House Collection,
Library and Archives Canada, 1986-204 NPC, C-090385
Printed and bound in Canada by Friesens
Text printed on acid-free paper

We gratefully acknowledge the financial support of the Canada Council
for the Arts, the British Columbia Arts Council, the Province of British
Columbia through the Book Publishing Tax Credit and the Government
of Canada through the Canada Book Fund for our publishing activities.

With the generous support of the Manitoba Arts Council
With the generous support of the City of Winnipeg through
the Winnipeg Arts Council

*For my "dear" mother and my loyal hound, Maggie
(Mackenzie King would have understood)*

My belief in myself leads me to hope that this will some day be realized. The public will admire me for the courage & spirit I show in sacrificing a certainty for a great uncertainty... Lastly there is the purpose of God in all, the realization of the dream of my life, the page unfolds as by the hand of Destiny. From a child I have looked forward to this hour as that which should lead me into my life's work. I have believed my life's work lies there, and now I am led to the threshold by the Invisible Hand.

THE DIARY OF WILLIAM LYON MACKENZIE KING, *July 25, 1907*

The secret of his unique accomplishment is difficult to identify, for Mr. King was a character of baffling complexity.

WINNIPEG FREE PRESS, *July 24, 1950*

Table of Contents

Introduction:
By the Hand of Destiny

*I am taking up this diary again as a means of keeping true
to my true purpose. It has kept me in the path from drift-
ing more than I otherwise might have, it has helped to
clear me in my thought and convictions, and it has been
a real companion and friend.*—THE DIARY OF WILLIAM
LYON MACKENZIE KING, *January 1, 1902*

LIKE MOST DAYS, January 15, 1929, started for William
Lyon Mackenzie King, the fifty-four-year-old prime minister of Canada,
with the recording of yet another "vision." Early that morning, Willie
King, or "Rex," the nickname used by his small circle of friends, called
from his second-floor bedroom in Laurier House in Ottawa for his personal secretary, Howard Measures. He had had another vivid dream,
and he needed to immediately dictate it before it passed into the realm
of the unknown. Measures obediently took down every loony word, and
later that day the story of the dream was dispatched to King's fortune
teller, Mrs. Rachel Bleaney, in Kingston, for her wise interpretation.

In his vision, King found himself in a large building with a never-
ending hallway. He began to climb the flight of stairs before him. Suddenly, he came across His Majesty, King George V, lying on a couch,
motionless. "There was no sign of breathing and he might have been
thought to be dead," as King explained. He moved on to another room
and found a large blue envelope with a red seal on it containing the
Royal morning mail. Mackenzie King did not see his parents, both long
dead, but he was conscious of their presence.

Mrs. Bleaney soon sent the prime minister her dream analysis.
Something unexpected was going to happen to him, she predicted.

According to the Kingston soothsayer, "Dreaming of royalty generally means some unexpected honour which I firmly believe is soon about to be bestowed on you; yet I feel you are likely to be very grave and anxious over a very important question that may arise in Parliament." She foresaw that a trip to Britain was imminent. "Your dear ones were certainly with you in this dream, but there seems a little cloud that they could not show themselves as clear as you could have liked, but you will have some more strange yet wonderful visions. You may have three ones after the other, but your dear mother is your leading guide and always will be, and although you were lonely and felt this in your heart, she is coming again in a most remarkable and beautiful way to prepare you for something of a most wonderful and remarkable nature as you will be taking the most important part in a large public event." Mrs. Bleaney signed the letter, "Your most sincere and true spiritual friend and adviser."

There was, in fact, no grand trip to Britain that year nor any honours bestowed upon him. But even if Mrs. Bleaney was off the mark, King held her predictions in high regard.

What are we to make of this seemingly cockeyed vision and the hundreds of other such records of Mackenzie King's dreams, séances and communications with his "dear mother" and all those others in the "Great Beyond," as he reverentially referred to the afterlife?

Regardless of how it sometimes seems, Mackenzie King was not the only fervent believer and practitioner of spiritualism in the first half of the twentieth century, just the only Canadian prime minister to embrace its mystical tenets and ethereal customs. Did that make him unfit to be the leader of the country? Not in the least. There is no denying that his behaviour was bizarre and unconventional, as is his current popular reputation. Yet somehow he was able to tap his intellectual dexterity and make the prime minister's role his own for a generation. "He wasn't a crackpot," asserted Walter Turnbull, who served as King's principal secretary during the early forties, in a 1975 interview. "He had this idea, as all leaders do, that he was the smartest man around. His ego was very big... but he was one of the sanest people I ever knew in my life."

The mediums King sought out constantly reassured him about both personal and political issues. Sometimes, though not often, he was guided by what they told him; most of these prognostications and supernatural messages were exactly what his psychics knew he needed to hear.

Consider another example. In February 1939, seven months before the outbreak of the Second World War, King was understandably in a tense frame of mind. Early that month, he invited the English clairvoyant and palm reader Mrs. Quest Brown to Laurier House for an "interview." She helped him make sense of things. "You are obviously torn spiritually between duty and preserving your own inner integrity— this aggressive and destructive battle is weakening your resistance—it is an insidious poison that is gradually sapping your strength," she told him. "You have done so much for others and there is in every one's life a point where one analyses success at such a price is of doubtful merit. You can still fulfill your destinies from the maddening crowd. You have many years in front of you in which to achieve all you desire for Canada and possibly this must be achieved in the peace and silence of your own inner life." King, not surprisingly, concurred with this insightful assessment of his character. "Mrs. Brown . . . has diagnosed absolutely correctly the real conflict that I experienced over many years," he wrote in a personal note about the palm reading. "My exhaustion and fatigue has arisen out of it, the eternal effort to preserve an inner peace with the terrible pressure of obligations arising out of my environment and duties." Such was the daily self-sacrifice that he believed he made for the country.

The Mackenzie King list of weirdness is a long one. King was fascinated by his convoluted dreams of waiting endlessly for trains, floating in the air and meeting the dead and the forgotten. He had an obsessive-compulsive fixation with the numbers and position of the hands on the clock. He saw magical images in tea leaves and, in what must rank at the top of the "bizarrest of the bizarre" category, the mysterious shapes formed in his shaving cream lather. My favourite of dozens of examples is from his last year in office. On January 20, 1948, he visualized in his morning lather the symbols of the Cold War, a polar bear and an eagle. The bear "seemed to be moving into space and crushing down somewhat the one outspread wing of the eagle." A dog appeared, perhaps symbolizing Britain or Canada, which helped push the bear off the eagle.

King devoted innumerable hours to scribbling or dictating pages and pages of commentaries about his nocturnal experiences. At his magical Kingsmere estate in the Gatineau near Ottawa, he constructed romantic ruins adorned with an eclectic "collection of stoneware deer, cocks, gnomes, donkeys and rabbits," as Malcolm MacDonald, the British high commissioner to Ottawa from 1941 to 1946, described it. "In these surroundings," added MacDonald, "most of [King's] great decisions on policy are taken, often with his household servants, a favourite Irish terrier and that menagerie of petrified creatures out-of-doors as his only companions."

Most famously of all, King heeded table rappings as messages from the other world and participated in countless séances, all so that he could communicate with those departed souls he missed most: his parents, John and Isabel; sister Bella and brother Max; his mentor, Prime Minister Wilfrid Laurier; and those loyal Irish terriers, Pat I and II. (Pat III outlived him.) He often shared these other-world encounters with his closest friend and soulmate, the woman he surely wished he had married, Mrs. Joan Patteson, who figures largely in this biography. Still, if any part of the story of King's life demands historical perspective, his embrace of spiritualism is it.

The interesting question is why he was a believer. And that really is not difficult to answer. "Life is contingent and filled with uncertainties, the most frightening of which is the manner, time, and place of our own demise," says Michael Shermer, the American science writer and publisher and editor of *Skeptic* magazine, who has made a career debunking psychics, UFOlogists, alien abductees, creationists, Holocaust deniers and cosmologists. Anyone, he adds, who has suffered a tragic loss—and in King's case it was the death of four family members between 1915 and 1922—is particularly susceptible to reassurances from psychics. Instant gratification was what King sought, and instant gratification was what he got. "Under the pressure of reality we become credulous," continues Shermer. "We seek reassuring certainties from fortune-tellers and palm-readers, astrologers and psychics. Our critical faculties break down under the onslaught of promises and hopes offered to assuage life's great anxieties." Why do smart people believe in weird things? Because, Shermer explains, smart people "are skilled at defending beliefs they arrived at for non-smart reasons." But as with so much of King's life, there are multifaceted complexities to unravel before this is entirely understandable.

FROM 1919 TO his retirement in 1948, Mackenzie King humbly led and occasionally lorded over the Liberal Party. For nearly two and a half decades during that time period, he also served as prime minister—1921 to 1926, 1926 to 1930 and 1935 to 1948. It is hard to imagine any politician ever surpassing King's twenty-two years in Canada's highest office. He was motivated by a powerful drive to honour the memory of his embattled rebel grandfather, William Lyon Mackenzie, and combat the nefarious Family Compact of the 1830s, which in his imagination had morphed into his detested Tory opponents. Urging him on was his overwhelming belief that his life was guided by God's will or the "Hand of Destiny," as he deemed it. "I may be prepared to play that strong part in the history of Canada," he wrote on a trip to Britain in March 1908, "which may help to bring the people of the country nearer to the principles of Righteousness which exalteth a nation and which it is the purpose of God in my life that I shall do."

Thirty-four years later, this perception remained intact. "Personally, I do not believe in a man's life and career being a matter of some chance or accident," he recorded in his diary at the end of October 1942. "It is the result of many forces, some known, some unknown—many beyond his [control];... these forces are linked with others that have gone before, which run into the far distant past. How far, none of us know. They may well be into the centuries that human life is fulfilling its purpose in terms of eternal laws. I am sure those which accord with right are fulfilling God's purpose in the world. Those which ally themselves with evil are helping to make clear that evil cannot and will not endure. In this sense there is some truth to the old doctrine of predestination."

So determined, and with the Almighty cheering him on, he reshaped Canada's status within the British Empire, advancing the country's independence, redefined the role of the governor general and federalism, initiated social welfare legislation and brilliantly guided the country through the Second World War.

Mackenzie King was the master juggler. He did what he had to do to survive and rule another day, whether it was keeping Quebec happy and assuring national unity—coincidentally guaranteeing the success of the Liberals and himself at the polls—cozying up to the western Progressives during the twenties, proclaiming victory in his celebrated constitutional dispute with Governor General Lord Byng, rising again like a phoenix in 1935 after five years in the wilderness of opposition or managing the World War II conscription crises with a calculated precision. "He succeeded with hardly a mistake... in giving expression by way of that curious cloudy rhetoric of his, to what lay in the Canadian subconscious mind," concluded the historian Frank Underhill in his 1950 post-mortem of King's career. "Mr. King, for twenty-five years, was the leader who divided us least."

In 1997, when twenty-five scholars, all specialists in Canadian political history, were asked to rank the country's prime ministers in order of greatness, Mackenzie King came out on top. The academics pointed to King's success in maintaining national unity, "well-articulated goals in domestic and foreign policy" and "a solid record of achievement, not least in getting elected and staying in power." Judgments about greatness are admittedly subjective. In King's case, this verdict has more to do with his tremendous impact on the history of the country and far less on his personality. (In the most recent academic ranking of Canada's prime ministers published in *Maclean's* June 20, 2011, issue, King has dropped to number three in the list after Wilfrid Laurier and John A. Macdonald.)

Short, squat, stocky and pudgy and a life-long bachelor, devout Christian and ardent spiritualist, King was also about the quirkiest political leader of his era or any other. During his life, he was regarded as a dreary and cautious plodder always standing back to see which way the political winds were blowing before committing himself. "We will frankly confess that Mr. King is a puzzle to us," wrote the Conservative partisan journalist Arthur Ford in 1943. "He has none of the qualifications that one would expect in a prime minister, who has been able to hold office in a country like Canada for over a decade and a half... He lacks colour; he has never been a popular public figure even with his own followers; except on rare occasions his speeches are dull and prolix... And yet it is foolish to dismiss his continued success as due to just plain rabbit's luck. It takes more astuteness to maintain power as long as Mr. King has." Conservative Party guru Dalton Camp was more succinct: King, he wrote, was "a pale, colourless little man, clad in the aura of indestructibility."

Since King's death in 1950 and following the controversial decision by his literary executors to grant access to his renowned diary, his multitude of strange idiosyncrasies have been exposed to ridicule and treated with disbelief (as will be seen, this was not entirely unwarranted). In short, he has become the Rodney Dangerfield of Canadian political leaders. Despite his extraordinary record of accomplishments, Mackenzie King can't get any respect.

King had not been in the ground long before the assault of humiliation started, and it has not really stopped. "How shall we speak of Canada, Mackenzie King dead?" the first line of the 1954 satirical poem "W.L.M.K" by Montreal constitutional lawyer Frank Scott begins. "The Mother's boy in the lonely room with his dog, his medium and his ruins?... He skilfully avoided what was wrong without saying what was right, and never let his on the one hand know what his on the other hand was doing." Twenty years later, poet Dennis Lee included this ditty in his children's poem *Alligator Pie*: "William Lyon Mackenzie King sat in the middle & played with string and he loved his mother like anything—William Lyon Mackenzie King."

Novelist Robertson Davies took a shot at King in his 1972 novel, *The Manticore*. "The Right Honourable William Lyon Mackenzie King was undoubtedly an odd man, but subsequent study has led me to the conclusion that he was a political genius of an extraordinary order," argues lawyer David Staunton, Davies's flawed protagonist. A friend of the family, Dunstan Ramsay, begs to differ: "You'd better face it, Boy," he declares. "Mackenzie King rules Canada because he himself is the embodiment of Canada—cold and cautious on the outside, dowdy and pussy in every overt action, but inside a mass of intuition and dark intentions. King is Destiny's child. He will probably always do the right thing for the wrong reasons." Similarly, in his witty and satirical novel *Joshua Then and Now,* Mordecai Richler has lots of fun at King's expense, mocking his life and career and describing him as "mean-spirited, cunning, somewhat demented, and a hypocrite on a grand scale."

More disparaging still was the portrait of King in *Willie: A Romance,* the first volume of Heather Robertson's fictional trilogy about his life, published in 1983. She depicted him rather unfairly as immoral and depraved. Likewise, in Donald Brittain's 1988 television miniseries, *The King Chronicle,* he was "an unctuous fraud, fat, devious and dull." And, in Allan Stratton's award-winning 1981 play *Rexy!,* King and Joan Patteson's séances become light farcical fare.

The true-life assessments were more layered but just as harsh. There is no way around it: King was disdained by almost everyone who laboured for and served him. After speaking with many of his former secretaries and staff in the 1980s, political scientist Paul Roazen reluctantly conceded, "I am not sure I met anyone who genuinely seemed to like him; on the whole I found it painful to have to interview some of these people, since they found him so easy to hate." King was a miserable employer, who felt that anyone who was lucky enough to work for him should be honoured to do so. Whether it was the under-secretary of state for external affairs or his cook, he expected this small army of bureaucrats and domestics to wait on him hand and foot seven days a week and almost twenty-four hours a day.

His staff, however, in the words of civil servant and diplomat Hugh Keenleyside, who pretty much summed up the general feeling, thought King "selfish and inconsiderate." And that was putting it politely. In late April 1941, Grant Dexter, the *Winnipeg Free Press*'s well-connected Ottawa correspondent, reported back to his esteemed editor, John Dafoe, about an enlightening conversation he had with Norman Lambert, one of the Liberal Party's key organizers, who detested King despite the fact that he had appointed Lambert to the Senate. "It is amusing to hear [Lambert] explain that he simply can't stand the worm at close quarters," Dexter noted with gossipy glee, "bad breath, a fetid, unhealthy sinister atmosphere, like being close to some filthy object.

But get off a piece and he looks better and better. He thinks he will make a speech eulogizing Willie in the Senate." A month earlier, Lester Pearson, who was then working for the Department of External Affairs in London and was familiar with King's eccentricities, was horrified when he heard that he might have to return to Ottawa to serve as the prime minister's private secretary. "God protect me from that!" he noted in his diary.

When Mackenzie King wanted someone to like him, or if he believed an individual in his immediate circle had made a decision that might negatively affect him, he could be obnoxiously charming and transparent. "He really is excessively friendly," remarked Sir Francis Floud, the British High Commissioner to Canada in the mid-thirties. "My wife says after a conversation with him she feels as if the cat had licked her all over and she ought to go and have a bath."

Leonard Brockington, the witty and outgoing first head of the Canadian Broadcasting Corporation, had been conscripted into King's prime minister's office (PMO) as a special adviser during the early years of the Second World War, and he wasn't especially pleased about it. (The term "PMO" was technically not used for the prime minister's office until the 1960s and is used here and throughout the book merely as an abbreviation.) One day, he decided he was going to submit his resignation, except that King did not want him to leave (even though he was quite critical of Brockington in his diary). Brockington and his wife were brought out to Kingsmere for tea, where King proceeded to tell Mrs. Brockington what a gem her husband was. Later, when Harry Ferns, who also worked briefly in the PMO in the forties, asked Brockington about the visit, he aptly commented that King had "covered me with whipped cream and bullshit." Brockington stayed in his position for a while longer.

His tactics to win over Eleanor "Nell" Martin, the wife of Paul Martin, Sr., were more debonair. Soon after they were married, in the late thirties, the young novice MP and his wife were invited to Laurier House for dinner. Nell Martin, as her son former prime minister Paul Martin recalls, was "attractive and did not have much interest in politics." Before the dinner, Martin, Sr., beseeched Nell to be on her best behaviour. They arrived and Martin made the introductions. "How do you do, Mr. Prime Minister," said Nell. "My husband thinks you're a very great man." King eyed her. "And what do you think, Mrs. Martin?" he inquired. "Well," she replied, "I'm going to need some convincing." Martin later remembered that he "just about dropped from embarrassment and fidgeted for the rest of the meal." Determined to make the best impression on Nell, the prime minister turned up unannounced at the Martins' flat the following day to escort Mrs. Martin for a stroll. All went well, and King admired Nell Martin and vice versa from that day onwards.

Mackenzie King was intelligent, politically astute, strategic, compassionate, insightful, deferential and respectful. He had a "tender human side," asserted Fred McGregor, one of his secretaries, who had more positive memories about him. Blair Fraser, an influential Ottawa correspondent for *Maclean's* magazine's from 1943 to 1960, only ever received one phone call from King. It was after Fraser had put an ad in the newspaper that his Irish terrier—the same breed as King's dogs, the three Pats—had run off and was lost. As a devoted dog owner, the prime minister was understandably concerned. Late one night, Fraser recalled, the telephone rang. "'Fraser: This is Mackenzie King speaking. Have you found your dog?'" Fraser was taken aback by the call and for a time believed it had been a set-up by someone in the press gallery. "When I finally plucked up the courage to tell one of his secretaries about it," he added, "he said, 'Oh that was the PM all right. He does that kind of thing all the time.'"

That kind and thoughtful aspect of his character was frequently overshadowed by his other side, the one that was self-righteous, egotistical, petty, vain, moralistic, paranoid, selfish, self-centred and vindictive. To compensate for any perceived deficiencies, he went to painstaking lengths to ensure every part of his life or any public statement, from the kitchen utensils his cook used to a major speech in the House of Commons, was perfect. It could drive those around him to distraction, a fact to which he was oblivious. His self-consciousness was legendary. The renowned Ottawa-based photographer Yousuf Karsh took King's picture at Kingsmere in 1940 and found him a difficult subject to capture. "King wished me to depict the man he visualized himself to be," Karsh later recalled. Karsh was not thrilled with the outcome of the sitting, but King was delighted with the iconic images of him and Pat. "You will never know what this means to me," King wrote eight years later, "particularly in the light of the problem that pictures of myself seem to have presented to photographers in the past."

Mackenzie King was famously fastidious. When he travelled, he would always pack two pairs of extra shoelaces, just in case there was a footwear emergency. Through the acquisition of Laurier House in Ottawa and the Kingsmere property, as well as a series of gifts from such wealthy supporters as the tea merchant Peter Larkin, King was a fairly rich man by the time he regained the prime ministership in

1935. Yet he was thrifty to the point of the ridiculous. Pencils in his office had to be cut in thirds, and each little part was then used until it was whittled away to nothing. He rarely carried around money, as Gordon Robertson, who worked in the PMO during the last years of King's career, discovered when he functioned as the prime minister's "banker" on official trips to the United States and Europe. "I soon learned," Robertson remembered, "that if I did not have his signature for any money I gave him or any account I paid for him, at the moment of the transaction, there was little prospect of having the charge accepted later as something that should go on his expense account. After a few costly mistakes, I had slips prepared, small enough to carry in my pocketbook, with space for a date and reading, 'Received from R.G. Robertson the sum of...,' and a line below for the prime minister's signature."

You do not have to be a psychiatrist to get inside King's head or diagnose the reasons for the long list of personal flaws. Each one was related to the debilitating and oppressive insecurity that stifled him and shaped his touchy personal relationships with his colleagues and friends—a major theme that runs through this biography. Indeed, if we do take the learned opinion of the experts in this matter, it is immediately apparent that King was a walking textbook example of an insecure passive-aggressive male.

Of the eleven behaviours of such individuals identified by Dr. Scott Wetzler, a professor of psychiatry at the Albert Einstein College of Yeshiva University in New York, King daily manifested at least nine of them. They include speaking ambiguously or cryptically as a means of engendering a feeling of insecurity in others, fearing competition, making chaotic situations, sulking and feeling victimized instead of recognizing one's own weaknesses. "Problems arise with the passive-aggressive man because of his fatal flaw: an indirect and inappropriate way of expressing hostility, hidden under the guise of innocence, generosity or passivity," writes Wetzler. "If what he says or does confuses you, or, more likely, angers you, this is why. You're not the only one to react this way. It's what passive-aggression is all about." Mackenzie King in a nutshell: a devout humanitarian version of Richard Nixon.

Factor in King's obsessive-compulsive tendencies—the constant checking of the clock for no apparent reason and the saving of every scrap of paper he received, right down to his dental X-rays and the Pats' dog tags—and you have one difficult prime minister. King was "a literary magpie of unprecedented proportions," as Peter C. Newman described him in a 1976 essay. "After King died, his notes on Edward VIII's abdication were found in a piano bench at Laurier House." (The collection of King's papers kept in Ottawa at Library and Archives Canada includes more than two million documents and 25,000 photographs. It measures 315.89 metres or 1,036.38 feet, almost three Canadian football fields long.)

King made a point of noting in his diary every compliment he received and each time his caucus or an audience cheered him on, in addition to every slight and sarcastic remark he regarded as offensive. And he always blew these usually insignificant matters out of all proportion. No wonder he was "fatigued" much of the time: he expended a huge amount of emotional energy on minor issues and sheer nonsense. It started early. In 1904, when Prime Minister Laurier did not properly compliment him on a speech he delivered at the Canadian Club, he was troubled about it for a week.

At Christmas, he tallied up all those people who sent him cards and, more significantly, those who did not. In December 1926, some months after his nasty public row with Lord Byng over a dissolution of Parliament that he demanded and the governor general did not wish to grant, a Christmas card from Lord and Lady Byng (who now detested King more than her husband did) was not forthcoming. "The Byngs have not sent me a Xmas card," he noted, "pretty nice sort of treatment after the four years of close relationship we had." When Lord Byng died in Britain in June 1935, King was thrown into a crisis. He was uncertain whether he should send a telegram of condolence, which he finally did, and was incapable of deciding what to say publicly about the former governor general's passing. He was in an anxious tizzy about it for days, reviewing the entire 1926 episode again with the necessary rationalizations for his actions.

No one had it harder or felt sorrier for himself than Mackenzie King did. A friendly phone call from Kathleen King, the wife of his nephew Arthur (the son of his brother, Max) was not just a friendly phone call. "I think this is one of the very few times that anyone related by marriage or any other way has made an enquiry concerning my health," he noted in October 1943 with typical self-absorption. "It shows how completely isolated one has become in giving one's years as well as days in public affairs and allowing all elements of home and its association to slip by unshared." Violet Markham, the British social reformer and socialite who was friends with King for more than four decades, remembered that she "begged him to take life more easily and not allow affairs of state to submerge him so completely. No human frame could stand with impunity the strain he put upon it, but he had little power either to relax or amuse himself."

In his memoirs, historian Charles Stacey recalled that during a guided tour he gave King through the battlefields of Normandy in August 1946, there was a terribly uncomfortable moment for the prime minister when the group stood at the spot where Canadian soldiers had been murdered by the Nazis and temporarily buried. King always feared not doing the right thing or following the correct protocol, so he did nothing. "What does one do at an empty grave?" Stacey remembered wondering. "We military people waited for the Prime Minister to set us some example. We then realized he was waiting for an example from us; in fact he was craning forward to see what those of us on each side of him were doing. We saluted; and King then removed his hat."

King was so out of tune with his own persona that it borders on the absurd. During the second conscription crisis in 1944, he had recruited General Andrew McNaughton to replace James Ralston, his problematic minister of defence. But McNaughton failed to win a seat in the House of Commons. In his diary, King, who mentally catalogued every caustic remark ever directed at him, castigated the general for being "much too strong in his suspicions and dislikes and hatreds... McNaughton was not a good man in politics for that reason. Sir Wilfrid was right when he said that it does not do to cherish resentments in public life." Similarly, after King had died, Alex Hume, the veteran *Ottawa Citizen* reporter who had known the prime minister for years, recounted that King had once told him—with a straight face, no less—that "it is a great mistake to take anything in public life in a personal way to embitter one." If only he had paid attention to that advice he would have saved himself from so much aggravation, but it was not to be.

THE WINDOW INTO King's turbulent personality and his tortured soul is the diary he kept almost religiously from the time he was eighteen in 1893 to his death in 1950. It is the treasure trove of his triumphs, anxieties, sexual proclivities and chronic guilt, which this book is framed around. "This diary is to contain a very brief sketch of the events, actions, feelings and thoughts of my daily life," the first entry began on September 6, 1893. "It must above all be a true and faithful account. The chief object of my keeping this diary is that I may be ashamed to let even one day have nothing worthy of its showing, and it is hoped that through its pages the reader may be able to trace how the author has sought to improve his time." Running about 30,000 pages (7.5 million words), it is one of the greatest historical documents in Canadian history. King's diary, journalist and critic Robert Fulford once opined,

"might turn out to be the only Canadian work of our century that someone will look at in five hundred years."

Until about 1935, King scrawled the diary in his own difficult-to-
read handwriting, usually every evening before he "turned the lights
out," as innumerable entries end. In later years, he dictated much of
it to his loyal secretaries, mainly the unflappable and faithful Edouard
Handy. After King's death, the handwritten diary was transcribed by
Fred McGregor, one of the literary executors of King's estate, so that
his first official biographer, political scientist R.M. Dawson, could use
it. It was a major operation. Over the course of many months, McGregor
read the diary into a Dictaphone, and a trio of stenographers then typed
it out. He showed incredible fortitude in deciphering King's handwriting, and historians and other researchers everywhere owe him a debt
of gratitude. If he had not taken on this painstaking assignment, the
vast majority of the early portion of the diary would have been surely
inaccessible.

Before King became prime minister in 1921, the entries tended to
be briefer, and he often missed days at a time; uncharacteristically,
there are only four records in 1913. Once he assumed office, the entries
become an extended day-to-day chronicle of every person he saw and
spoke to, every meeting he attended, every social event he was invited
to, every backhanded comment he received, every séance he participated in—and his love for "dear mother" (the phrase appears more than
a thousand times in fifty-seven years). He occasionally recognized the
unhealthy consequences of keeping a daily record of his life. "The process of writing a little of every action is I think bad," he suggested in a
lucid moment in April 1900, "it helps to make one too introspective, too
conscious of self in everything, it is too a restraint on freedom which is
sometimes irksome." But he continued with the diary the next day and
the next day after that. There are notations on the hands of the clock,
his over-the-top love for his dogs, his intense affection for Joan Patteson
and other women, his elusive quest to find a wife and the prime ministerial bowel movements (like his clock checking, another tell-tale sign
of obsessive-compulsive disorder).

King obsessed about checking the clock and found hidden but never
explained meaning in the position of the hands. Typical was this entry
from November 5, 1941. He was having dinner with Joan and Godfroy
Patteson at their home. "It was quite curious how the hands of the clock
asserted themselves in connection with the evening," he wrote. "They
were exactly together near 20 to 8 when I looked at the clock at the

head of the stairs and drew [Joan's] attention to it. Next time I looked at the clock, it was exactly quarter to nine... When we started out for the walk, [the hands of the clock] were in a perfectly straight line, at quarter past nine, each one of these glances were unpremeditated and unforeseen events brought them exactly in sequence."

Ever since Charles Stacey structured his fascinating 1976 study of the diary around an entry the twenty-three-year-old King wrote in mid-February 1898 about leading a "double life," with alleged sordid tales of King's encounters with Toronto prostitutes in the 1890s and his sexual hang-ups, historians, journalists and memoir writers have embraced that characterization and tried to separate the successful politician from the zany, eccentric spiritualist. "Much of what Stacey wrote about King, and much of what 'the psychologists and pornographers' have found in the private King, is at best 'half true,'" argues historian Michael Bliss. Civil servant par excellence Gordon Robertson, who was later one of King's literary executors, dismisses King's "diary confessions and his idiosyncrasies" as not being "of any serious consequence". So, too, did Grattan O'Leary of the *Ottawa Journal*. Stacey recalled that after the publication of his book, "many respectable Canadians felt that the conspiracy of silence [of King's diaries and private life] should have continued." He was accused of "fabricating the whole thing," and the University of Toronto president "received angry letters from alumni (often doubtless good Liberals) demanding that [he] be fired or deprived of [his] tenure."

This "double life" approach might make King's life story more palatable, but it is entirely incorrect. No understanding of his public career and private interests can be achieved without considering the whole crazy package. Michael Bliss was accurate about one thing: King's life was not a "Jekyll-and-Hyde epic" (as he himself described it in 1895) and is not treated in this biography as one. You cannot write about King the prime minister without reference to his zealous faith in God, the séances, his off-the-wall adoration of his mother, his talking to his dogs (at least they did not talk back) and all the other inanities found in the diary. It is equally futile to attempt to prove—as is the wont of many historians—that King tended to disregard the advice he received from fortune tellers and mediums, thereby diminishing their significance in his life. Sometimes he did; sometimes he did not. However, that is not the point. All of his spiritualist experiences, his other superstitions and his multi-paranoid reactions imprinted on his consciousness, shaping his thoughts and feelings in a thousand different ways no biographer can completely unravel.

Over the centuries, diaries have been kept for any number of reasons. There is the life-and-times variety popularized by the British politician Samuel Pepys, the most famous of seventeenth-century diarists; the official yet personal statesman diary like that of U.S. presidents Harry Truman and Ronald Reagan; the poignant account of young Anne Frank penned while she was hiding from the Nazis in an attic in Amsterdam during the Second World War; and the confessional diaries left by the Puritans, who felt the need to cleanse their spirits and account for their "acts of salvation" in case the devil showed up on their doorsteps.

Keeping a diary is also supposed to make you a happier person, though that likely did not apply to Mackenzie King. His diary tends to fall into the Puritan category, with elements of the "official states-man" type, particularly for the lengthy moment-by-moment entries recorded during the war years. As might be expected, he portrayed himself always in the best possible light, blaming others for the conflicts and problems he routinely encountered. In its self-reflection, its rules to live by, the devotion to social reform, and the painful guilt over sexual urges (and actions), King's diary resembled that of the politician he worshipped most, the nineteenth-century British Liberal leader William Ewart Gladstone. "Oh God I want to be like that man," King wrote in August 1900, "and my whole nature thirsts for the opportunity to do the work even as he did his, to be strong in my ideals, true to the most real purposes, and firm in the struggle for the right and true."

Reading King's diary is an experience in itself—and I make no claim that I have read each volume in its entirety; it is doubtful if anyone ever has or could. Dip into it at any point and the most common reaction is: "You can't make this stuff up." It is akin to a journey down Alice's rabbit hole or, as York University political scientist Reginald Whitaker once fittingly described it, "a descent into delirium." It offers a unique entrance into the complex mind of the most complex politician Canada has ever produced. Even if similar diaries existed for other Canadian prime ministers, rest assured none would be as spellbinding or as warped as that of Mackenzie King.

Take as only one of thousands of examples the entry in King's diary for September 15, 1939, two weeks after the start of the Second World War. The day began for the sixty-four-year-old prime minister of Canada with a "vision" of listening to Lady Jessie Foster, Sir George Foster's wife, reciting a James Russell Lowell poem. The dream aroused him. King woke up hot, tearing at his bed clothes, and called for his personal physician. The diary continued: "When I spoke of the effect of too much bed clothing and the feeling of being consumed of the kind of [sexual] fire which it brought with it, he told me [those were] perfectly natural phenomena [and] to do what I could in keeping my bedding light not too warm. He spoke of the relation of the brain to the sexual organs, and mentioned not being surprised if the one seemed to stimulate the other. He thought at my age, it was marvelous that I had the evidences of young manhood which I appear to have."

A Bible reading from the "little books" was next, followed by a stern lecture to his hapless servants, whom he warned "to follow literally the terms of their engagement" or they risked having the police register their comings and goings from Laurier House. One of the maids, Carmen, had taken off without telling anyone. Another, Jean, wanted to spend that evening at the movies and play bridge. She threatened to quit if she could not do this. King was beside himself. "I asked her if she thought that was a nice way to behave when everyone was trying their best to serve their country. She said she knew that we were all having a hard time but that she liked movies and bridge, and was more concerned about them. I pointed out that the police had been keeping record of time; and she had been coming in last night—a quarter past twelve—some time, one o'clock. Said I did not think this was right, particularly in the P.M.'s residence. She agreed to be in by 12. I told her there might be special occasions when she could stay out later, but I have to leave the matter to talk over with Mrs. [Joan] Patteson and would decide later."

This negligible episode he connected to the war: "Is it any wonder that nations like Germany with great discipline and the rest of it soon play the devil with some of the free countries with their little appreciation of what freedom really means?" he asked. "How ludicrous that one with the burdens I have at the moment should be worried with problems of this kind!" How "ludicrous" indeed that he allowed himself to be so burdened is the real question. He was incapable of seeing the world through any other lens.

A Victorian Presbyterian in mind, spirit and heart—with a seventeenth-century strict Calvinism thrown in for good measure—King diligently struggled to make himself a moral person and purge vice whenever it crossed his path. He detested card games, hated smoking, especially by women, and did not allow it at his cabinet meetings (though he tolerated Winston Churchill's cigars), and castigated himself each and every time he wasted time or took a drink of alcohol.

King's nineteenth-century attitude towards sex could occupy an entire book and is considered in due course in more detail in the early chapters of this biography. A life-long bachelor, was he, in fact, a virgin (a possibility suggested to me by historian and King expert Jack Granatstein)? Not likely, but if he did have intimate sexual relations with women, they were very infrequent. However, some years ago, Paul Roazen interviewed King's valet—presumably John Nicol, though he does not identify him—who told Roazen that on one occasion he had to "get women's 'pomade' off King's trousers ('pomade' was used then to keep a woman's hair in place) and the butler was said to have been shocked to find King in a compromising situation in his study." When King's physician suggested to him in November 1916 that he find the right woman to settle down with, his reply as he recorded it in his diary was: "I wanted to devote my life to my work and my country and sought some one with like principles and aims."

At the age of nineteen, King composed his twelve commandments of life, which directed him towards salvation:

1. Fear God.
2. Honour thy Father and thy Mother.
3. Love one another.
4. Guard your lips as they speak the truth and pure word.
5. Keep your mind pure.
6. Do all you can to help others especially the poor and the sick.
7. Let your Life bear witness for Christ.
8. Be honest and upright in all things.
9. Seek to live a better life every day.
10. Do not waste Time. "Be up and doing."
11. Learn ever the great lessons of Life.
12. Be nearer Home each day than ever before.

HE QUICKLY FOUND that no matter how much he prayed—and he began each day on his knees—or how much he was prepared to sacrifice himself to the teachings of Jesus Christ, these rules were impossible to obey at all times and in all cases.

Like a fierce game of tug-of-war, Mackenzie King's various personality traits wrenched and tormented his inner soul for much of his life. "Hand of Destiny" and guardian angels or not, he was still able to summon the strength to steer his life in a positive direction to attain political glory by the age of forty-six. Then, in the mid-twenties, when he confronted opposition within his own party and critics in the press and was ultimately defeated in the federal election of 1930, he did not surrender. Instead, he regrouped and in 1935 recaptured the country, firmly ensconcing himself in the prime minister's office for the next thirteen years. Surrounding himself with capable and sometimes brilliant cabinet ministers, paying close attention to the needs of Quebec, keeping national unity as his main focus, usually proceeding with great circumspection and showing an amazing willingness for flexibility and compromise, he cemented his reputation as one of the most accomplished politicians in Canadian history. In the process, he also established the Liberal Party as the country's "natural governing party" (at least until the late eighties).

Without question, he was the country's most important prime minister and certainly its most peculiar. Making sense of the deep and often moving inner conflicts that haunted King, his quirks and his incredible ability to conquer and control his own fate is the aim of the story told in the following pages.

MACKENZIE KING did not want his entire diary made public, let alone put on the World Wide Web for everyone with an Internet connection to examine, as they can today. After he retired he planned to use the diary in the writing of his memoirs. That's why he spent so much energy keeping a record of his daily life, especially during the Second World War. But the instructions he included in his will signed at the end of February 1950 were ambiguous. He wanted the diary destroyed, "except those parts which I have indicated are and shall be available for publication or use." What exactly did that mean? Did he truly want the diligent work of a lifetime burned?

There was a strong indication of his thinking in an entry of March 4, 1950. On that day, King had received a letter from his close friend (and one-time employer) John D. Rockefeller, Jr., urging him to add a codicil in his will, "allowing at least [Fred] McGregor to go through my diaries and not to have them destroyed." Rockefeller viewed his request as "an obligation" King had in lieu of $100,000 the Rockefeller Foundation had advanced King for his memoirs project. Rockefeller's letter set King off, enough for him to consider briefly returning the foundation's gift. "I think he is quite mistaken about any obligation to the Foundation. Indeed, the agreement with the Foundation was that there was no obligation of any kind in connection with the writing of the Memoirs, and secondly, I think there was a distinct statement on my part that I intended to destroy the diaries though the suggestion was made by the Chairman [of the foundation] that I might consider making some extracts, etc. The letter threw me into a profuse perspiration. I could not but feel that there has been correspondence behind my back on this matter of the diaries and that this was evidence of pressure being brought to bear." Later that day, he discussed the matter over with Joan

Patteson, who agreed with his interpretation of Rockefeller's letter and confirmed something King already knew: returning the foundation's funds to Rockefeller would hurt him deeply.

Joan's advice about the diary was more telling and no doubt related to the fact that she assumed she played a starring role in it. She felt that King "owed" it to himself "not to make the diary available to anyone" but himself. He agreed:

> In the first place the diary has been kept for motives quite other than those connected with the war. It has been intended as a guide to myself and as an aide-memoire, etc. I cannot, however, deny that it contains material about proceedings at the Cabinet Table and as I have taken an oath not to disclose anything to anyone save one of the Cabinet, as to what has transpired there. I think that of itself would make it impossible for me to leave the diary open to anyone save someone of the Cabinet. Besides I think as regards the Foundation, a distinct understanding was reached that the diaries should be destroyed.

MOREOVER, ROOSEVELT AND Churchill, he concluded, had confided in him private matters that neither would have trusted him with if they had known this information was to be made public. As fate would have it, according to a notation and signature at the end of this entry, it was only read by Jack Pickersgill, one of the literary executors of King's papers, on November 13, 1960.

In the months and years immediately after King's death, Pickersgill and the three other literary executors—Fred McGregor, Norman Robertson, then the secretary to the cabinet, and W. Kaye Lamb, the dominion archivist—were faced with a troubling conundrum. They realized that the diary was unique but also dangerous. Even a brief perusal revealed that it was much more than the record of King's political career; it was also an intimate chronicle of his private thoughts and his journey into spiritualism. Thus, they had a big problem: how to honour King's wishes without destroying this most significant of Canadian historical documents.

From a legal perspective, whether or not the diary was preserved or destroyed was up to the executors. That was, at any rate, the learned opinion of Frederick P. Varcoe, the deputy minister of justice. King's family supported preservation. As Colonel Charles B. Lindsey (the son of King's first cousin George Lindsey) reported to McGregor in early 1952, the family members "are unanimous in the opinion that the interests of Mr. King's memory and the country would be best preserved by preserving the diaries in their entirety." The colonel did recommend, however, that restrictions be placed on the personal entries. Edouard Handy, King's secretary, was of the same mind, as was the press, once journalists and editors got wind of its existence. "Frankness is the spice of literature and memoirs and if that is taken away Mr. King in print will be duller, less understandable than he ever was in life," argued the *Ottawa Journal* in March 1952, not realizing just how stunningly correct that assessment was to prove.

Joan Patteson, however, had not changed her mind and was upset by the executors' procrastination. "Mr. McGregor and I have not been at one over the diaries," she wrote to Violet Markham in 1951. "Rex made me promise to see that Mr. [McGregor] burned them without reading— and while I saw the need perhaps of reading them, the dying request was sacred to me—and I think is what I have suffered most. I have no knowledge of their contents—but have wished that his wishes could be carried out."

Despite his enormous frustration with King's personal foibles, Norman Robertson, among the executors, was the most insistent about protecting the Liberal leader's reputation. As such, he argued fervently that the diary be kept under wraps. He did not even want to permit political scientist Robert MacGregor Dawson—whom the executors had appointed in 1951 to write the official biography of King—to read the diary. At first, Fred McGregor and Edouard Handy agreed to go through the diary and pick out what they believed was relevant, but that proved an impossible task. Dawson finally asked Violet Markham to intercede with Robertson on his behalf, which she did. This, plus the more open-minded attitude of the other executors, gave Dawson, as well as historian Blair Neatby, who completed the second and third volumes of the biography after Dawson's death in 1958, nearly full use of the diary.

Dawson, whose volume covered King's life to 1923, made judicious use of the diary. There was no mention of King's work with saving prostitutes in Toronto in the 1890s, and scant attention was paid to his love affairs of the heart, his obsession with his mother or the beginning of his friendship with Joan Patteson. In 1960, Pickersgill began publishing the first of four mammoth volumes of *The Mackenzie King Record*, selective and sanitized excerpts from the diaries from the beginning of the Second World War to King's retirement in 1948. At the time, political scientist James Eayrs was working as a research assistant for Dawson and had read parts of the whole diary for the war years. He wasn't impressed with Pickersgill's editing. "I regard the decision to publish the diaries in the bowdlerized and truncated format in the *Mackenzie*

King Record," he later recalled in a 1976 interview, "as the Pacific scandal of Canadian letters."

Nevertheless, these publications solved the problem of access. Now every historian in Canada wanted permission to use the diary, and the executors and the national archives found it increasingly difficult to refuse their requests. In late 1971, the diary up to 1931 was opened to researchers, and within nine years the entire fifty-seven years of King's personal writings in all their glorious detail were available.

Pickersgill, McGregor and Lamb initially agreed with Robertson that King's separate spiritualism records (mainly written in the 1930s), as well as his séance transcripts, should be destroyed. That action was initially delayed because biographer Blair Neatby wanted to research this material. And in the end, only a few spiritualist notebooks were burned by Pickersgill and Gordon Robertson in 1977.

Gordon Robertson tells that story like this. Norman Robertson (no relation) died in 1968, and Gordon was asked to replace him as one of the literary executors. A number of King's "scribblers" were still to be examined, most of which contained detailed descriptions of the séances he participated in and the strange writings of his various mediums. The notebooks were, Robertson recalls, "incoherent scribbles" and "meaningless gibberish" with no historic value. The literary executors had made the decision to get rid of this material some time ago, but they still understandably hesitated for many years about destroying any of King's papers. "King would never have directed that any part be preserved," writes Robertson. "Jack and I burned the scribblers, page by page, in the Pickersgills' fireplace." The bulk of King's spiritualist writings and correspondence, which were preserved, was only made available to researchers in 2001 as per the edict of the literary executors.

University of Toronto historian Charles Stacey, no fan of King, started reading the diary while he was writing a book on Canadian foreign affairs. "I found little on external policy," Stacey recalled, "and a great deal on King's relation with women. I was drawn, as they say, to read on." He figured that if he did not delve into this fantastic story, someone else would. This new line of research ultimately led in 1976 to the publication of his book *A Very Double Life: The Private World of Mackenzie King.* With its unbelievable tales mined from the diary depicting King's world, it was a minor sensation. Nothing was ever the same for King's legacy after that. Once the salacious details of the diary were publicized, the family members were scandalized, according to Toronto architect Harry Morison Lay, King's sister Jennie's grandson.

Since the publication of Stacey's book, King's diary and voluminous papers have been probed, dissected and analyzed in hundreds of scholarly studies and doctoral dissertations. The specialized work of each—most notably that of Robert MacGregor Dawson, Blair Neatby, Jack Granatstein, Norman Hillmer and Charles Stacey—were of enormous assistance to me in completing this biography. And yet almost all of these comprehensive books, articles and theses remain largely inaccessible to the general public, at whom this book is directed. For them, King remains a curious enigma. He is the chubby face on the fifty-dollar bill and the subject of novels, plays and children's books, which have painted him as crazy Willie. He was "the loner," as one recent book for young people by Toronto writer Nate Hendley put it, "Wily Willie a dull leader for exciting times... who preferred the company of his dogs to humans."

Of William Lyon Mackenzie King, his lifelong friend Violet Markham had this to say in her memoirs: "No figure in contemporary history has excited more diverse views, or given rise to such contradictory estimates of character. I sometimes thought his personality might be likened to a set of Chinese boxes which fit so surprisingly into each other, each box different in size and colour and yet making a perfect whole."

And that was only the half of it.

I

His Mantle
Has Fallen on Me

*I feel that I have a great work to do in this life. I believe
that in some sphere I shall rise to be influential and helpful ... I believe it may be a professor of Political Economy,
an earnest student of social questions. Or it may be in
public life, parliament perhaps.*—THE DIARY OF
WILLIAM LYON MACKENZIE KING, *August 27, 1895*

IN SEPTEMBER 1947, near the end of his life, William Lyon
Mackenzie King paid a final visit to Kitchener, Ontario, where he was
born on December 17, 1874. As he toured the grounds of his family's
home there, called "Woodside," he was bursting with emotion and sen-
timental memories. "I was delighted to see in front of the house, there
was still the appearance of the old lawn," he recorded that evening in
his diary, an entry that runs six pages. "I missed many of the old pine
trees ... I walked through some of [the long grass] to the spot where, as
children, we all mourned the loss and burial of the little dog Fanny. I
recalled many of the incidents related to old Bill, our horse; thought
of the days we played cricket on the lawn; of sliding down a hill in the
winter and of my brother, Max sliding down on one or two occasions in
a barrel."

The visit also triggered King's spiritual demons. In an unsettling
dream he had the previous night, he saw his deceased parents standing
on a platform by the train, "looking very much concerned and frightened," sharing in their son's "distress and mental condition" at the prospect of returning to Woodside. But as with so many other interactions
in Mackenzie King's life that caused him needless anxiety, the day-long
visit to Kitchener turned out to be quite enjoyable.

Some years earlier, out of respect for King, the North Waterloo Liberal Association purchased the Woodside property, which had badly deteriorated. The group had started raising funds to restore the house where the King family had lived from 1886 to 1893 to its former quaint Victorian glory. The work had not begun, yet as he walked through the house and gardens he was delighted about the plans to transform Woodside into a national historic site.

Mackenzie King no doubt would be pleased if he could see Woodside today. The winding stone walkway leads visitors into a tidy and orderly home, complete with a piano in the parlour, where each Sunday the King family gathered to sing Presbyterian hymns. His father's library is stacked with books by Dickens, Carlyle and Browning (and resembles King's own third-floor library at Laurier House); a cast iron cook stove dominates the kitchen, where feasts for the holidays were prepared; and there is a classic oak fold-top desk in one of the bedrooms, where young Willie did his homework and contemplated the many career paths open to him. In the back yard is a rare tulip tree and lily pond, among other foliage.

You instantly feel as if you are stepping back in time and into the setting for a melodramatic Dickens or Thackeray novel. As Charlotte Gray has noted in her biography of King's mother, Isabel, Woodside "made the perfect backdrop for that middle-class Victorian ideal, an idyllic family life." That's just the way King remembered it: "It is perfectly true to say that all of the time of my boyhood, the early years of my life—the years that left the most abiding of all impressions and most in the way of family associations were those lived at Woodside," he wrote in 1944. (In order to hold on to those boyhood memories, King paid for trees from Woodside to be transplanted on the grounds of his Kingsmere estate in the Gatineau Hills. He also had fond memories of Berlin's street oil lamps, and he had several shipped over and installed at Kingsmere.)

As pleasant as that sounds, the reality of his family life was somewhat more complicated, frequently characterized by emotional turbulence, unending financial difficulties and volatile family dynamics.

WHEN THE AREA around Kitchener, about ninety kilometres west of Toronto, was first settled in the early nineteenth century, the town sprang up literally in the middle of nowhere, according to the city's historians, where "the Great Road from Dundas cut across the farm of Joseph Schneider, a Mennonite pioneer who arrived in Upper Canada in 1807." Other Mennonite farmers followed from Pennsylvania, and a large number of German immigrants travelled from Hamburg to New

York and north for available land. Many, too, were craftsmen and artisans, and a tiny village began to take shape in the 1830s. The name Berlin, used first in 1833 (and the town's name until anti-German feelings during the First World War necessitated a name change to Kitchener in 1916), was thus an appropriate choice for the gradually growing town, which was incorporated in 1853 (in Waterloo County) with a population hovering around 750 people. In 1856, the Grand Trunk Railway connected Berlin with Toronto—you were more likely to hear German spoken on a train trip than English—which led to further expansion. By the time Mackenzie King was born, Berlin's population was 2,800, of which 73 per cent were German-speaking (and of those, 29 per cent had been born in Germany). For good reason, Berlin came to be known far and wide as Canada's "German capital." Or, as the *Toronto Mail* put it in an 1886 retrospective, Berlin was "a patch of old Germany set down in the garden of Ontario."

The town was also a pleasant place to raise a family. True, it had its share of taverns and drunken brawls, and there was at least one brothel. Yet making your way around the town, you were more likely to encounter a stray cow than a dangerous criminal or bothersome streetwalker. Berlin's leading citizens prided themselves on upholding the highest standards of nineteenth-century morality, and that meant (in theory, at any rate) that public drunkenness, profane or blasphemous language, gambling, prostitution and horse racing in the streets were verboten. Tramps and vagabonds were "encouraged" to leave Berlin as quickly as possible.

Long before multiculturalism was defined as the quintessential Canadian characteristic, Berlin's German and Anglo citizens learned to co-exist in an atmosphere of tolerance and brotherhood, even when it came to toasting German militarism. On May 2, 1871, thousands from Berlin and nearby villages and farms turned out for a public celebration of Germany's quick victory over France in the short-lived Franco-Prussian War. There were German songs, concerts and revelry. One of only a few non-German speakers to address the crowd that day was lawyer John King, Mackenzie King's father. He conceded that he did not completely understand German heritage and nationalistic pride, but he was as exhilarated as anyone there about Germany's rise to power.

John King had found his way back to Berlin, where he had grown up, soon after he completed his law degree at the University of Toronto. His father John, Sr., born in North East Scotland in 1814, had come to Upper Canada as a young soldier in the service of the Crown. In 1837, while Mackenzie King's more famous grandfather, William Lyon

Mackenzie, was leading a failed rebellion in Toronto, John King, Sr., was on duty at Fort Henry in Kingston. He married Christina Macdougall, a Scottish immigrant, but then contracted tuberculosis and died in 1843 at the age of twenty-nine. Four months later, Christina King gave birth to their son John in Toronto. Without a husband, she and her infant son ended up in Berlin, where her brother, Dougall Macdougall, was a successful newspaper publisher and a prominent member of Berlin's English and Scottish elite. In an era when most people, and especially newspaper proprietors, took great pride in publicly promoting their partisan political views, Macdougall was a die-hard Reformer and educated his impressionable and intelligent nephew accordingly.

While he was studying to be a lawyer in Toronto during the mid-1860s, John King was introduced to Isabel Grace Mackenzie, the thirteenth and youngest child of William Lyon Mackenzie and his wife, Isabel Baxter. Isabel Grace was born in 1843 in New York City while her renowned (or notorious, depending on your point of view) father was in exile in the United States. Mackenzie had spent nearly a year in a New York state jail in 1839 and 1840 for violating U.S. neutrality laws—he had used the pages of one of his numerous short-lived newspapers to vehemently condemn the pro-British policies of President Martin Van Buren—and thereafter had a terrible time making a living as a journalist and customs clerk. He was finally granted amnesty and returned with his family to Toronto in 1850. (In a beautiful example of historical irony, Mackenzie wanted his two sons, Bill and George, to attend Upper Canada College, the boy's school founded in 1829 for the sons of the Family Compact, Mackenzie's Tory enemies.)

DURING THE LAST decade of his life, Mackenzie was elected as Reformer to the Province of Canada assembly and served for several somewhat unhappy terms. He remained so disillusioned with pre-Confederation politics that in 1858 he proposed Canadian independence from Britain and possible annexation to the United States. A team player he was not, and he could never accept that party-style politics always involves compromises and trade-offs. He also embarked on a few more journalistic ventures. None of his newspapers were profitable, and he fended off creditors to the end of his days, yet another hardship his wife and children had to endure. Still, the "Little Rebel" died with his tarnished reputation slightly restored.

The Mackenzies were firm parents. Presbyterian dictates were followed to the letter; life was about penitence for past, present and future sins. Young Isabel was as fiery and moody as her father and expected a lot from the people around her, even if she rarely reciprocated. She never let go of the bitter memories of growing up poor or the stigma that her father was an outlaw. A highly sensitive woman, family discussions about her father literally could make her ill. At other times, she argued long and loud, trying to convince herself as much as anyone else that he was a folk hero like Robin Hood, who had stood up to tyrants and defended the democratic rights of the people. In later years, the members of the extended family (including Willie) preached that without Mackenzie's 1837 rebellion, responsible government would never have been achieved in Upper Canada in 1848 and 1849.

Such revisionist thinking ignored the historical truth that this constitutional transfer of power from the appointed governor and council to the elected assembly resulted more from a shift in British economic policies than from any Canadian reform or radical push. But myths died hard, and within the King family William Lyon Mackenzie's rightful place in history was a point of honour that could never be questioned.

When Isabel, or Bell, as she was also called, and John King were courting, she was a lively "handsome" woman, with, as Charlotte Gray puts it, "enough dignity and self-assurance to stare down anybody who made slighting remarks about her father." Historians generally have not been kind to Isabel, regarding her as self-centred, possessive, manipulative and even spiteful. Her biographer portrays a more complex character. Isabel could indeed manipulate people and situations to her own advantage, yet she was also a loving mother who wanted only the best for her children. In truth, she expected far too much out of life and her future husband, holding unrealistic expectations that would never be realized. For Isabel, the glass was usually half-empty, and consequently she was regularly disappointed when things did not work out the way she wanted. One of the few positive exceptions was her elder son, Willie.

JOHN AND ISABEL were married on December 12, 1872. John was called to the bar in 1869 and briefly served as the assistant law clerk in the Ontario Legislative Assembly. When his Toronto law work did not produce the results he desired, he relocated to Berlin, hoping that his uncle's business and political connections would serve him well. He even contemplated running for public office, a suggestion that strained his relationship with his wife, whose memories of her father's political hardships were still fresh. Isabel hated to leave her family in Toronto and the nearly two decades she spent in Berlin were mixed.

Almost immediately, Isabel became pregnant and gave birth to a daughter, Isabel Christina Grace (Bella), in mid-November 1873. Then, just over a year later on a miserable snowy and bitterly cold December night, she had a son. John, who was attending a Reform Association political meeting in the nearby town of Linwood, arrived home in time to witness the birth of his second child, who was a month early. As much as Isabel had conflicting emotions about her father, she insisted that the newborn boy be anointed with his name. That moniker was to engender a sense of both pride and angst in William Lyon Mackenzie King as he strived to live up to the mighty legacy he believed it carried. Two more children followed: Janet (Jennie), in 1876, and Dougall Macdougall, who would be called Max, in 1878. Only Jennie was to outlive her older brother, Willie.

At the time of Jennie's birth, the Kings still rented a small cottage at 43 Benton Street, not too far from Berlin's main business area on King Street. The growth of the family necessitated several moves into larger quarters until the Kings settled into Woodside and its ten rooms on Spring Street in 1886.

John King rented Woodside from Frederick Colquhoun, a lawyer and friend whom he had known for many years, as John's uncle and mother lived next door. Woodside's large grounds permitted a garden and a few farm animals—including old Bill, the horse Mackenzie King remembered so fondly nearly five decades later. There was a cricket pitch where John King instructed his sons on the finer points of the game and an area for exciting croquet matches.

Young Willie, or Billie, as he was also called, had a mop of brown hair and a stocky build. He was a fairly happy child, not yet weighed down by the ambition, religious earnestness, insecurities and incessant worries that characterized his adult life. At the same time, it was not long before his siblings started referring to him as "little old grandpa," because of the self-righteous manner he adopted and his haranguing of them about what he perceived as their unacceptable character flaws. Bella, ever serious, usually adhered to her brother's reprimands, whereas Jennie, who strived to be independent, and Max, fun loving but plagued with illness, sloughed them off with a shrug or laugh. Somewhat isolated at Woodside, they truly loved and were dependent on one another, and the memories of those years remained with each of them for the rest of their lives.

Willie was educated at local Berlin schools and was a good student (better at classics than mathematics). He enjoyed debating and sports and was also a bit of a mischief maker; he loved to make his sisters laugh in church and wasn't above stealing apples from any available tree. He was not a tall boy, and years later he recalled with a laugh how the principal of his Berlin school, James Connor, scolded him: "King if you were not so small, I would like to pound you." He was fairly popular among his classmates. In short, Willie was a typical nineteenth-century Canadian youngster, with possibly one exception. Growing up in Berlin, he and his sisters and brother learned about diversity at an early age. Surrounded in their classes with German-speaking children and exposed to German newspapers, music and food, they appreciated German culture and language. (In fact, King later noted that he disapproved of Berlin's name change to Kitchener.)

As the elder boy, Willie King had a special place in the family hierarchy, and he learned as a teenager that much was expected of him. Jennie resented the attention that Willie received from their parents. "It seems to me you have more love and understanding than the rest of us here," she wrote to him in a letter of April 1898 while he was away at university. Jennie's challenge to her brother's pre-eminence in the family and her inclination "to see her brother as he was, rather than as he wanted to be" (to use political scientist Joy Esberey's apt phrase) led to some heated arguments between them.

With Max, Willie adopted a more paternalistic attitude. "As for Max, I have such hopes for him," he suggested to Jennie early in 1898. "I would have him a physician and writer as well, and in both the top of his profession. This can be, this must be and this he will be." (Max did indeed become a doctor and writer.) Willie was not above lecturing Max about the importance of religion, of being guided by Christ's teachings and, as he regularly emphasized it, "the need for being a good man." These were the same stringent principles he tried to adhere to and demanded of himself.

The focal point of the household was Isabel King. She was a loving but strict and occasionally needy mother. Prayers, chores and school work were regimented, and obedience was demanded from both her and John. Isabel doted on all of her children, though her relationship with Willie mushroomed into something unique, even for the late Victorian age. She insisted that they learn how to play the piano, and singing hymns and other songs together was a favourite King family pastime. Each child's birthday was treated with reverence, a custom Mackenzie King continued his entire life.

During the twenty years Isabel spent in Berlin, she relished her upper-middle-class small-town status as the wife of one of Berlin's leading professionals. Yet much to her chagrin, a lack of money was a constant concern. John King's inability to make a decent living and provide for Isabel in the custom she desired was arguably the most significant issue in their marriage, pushing them apart and distressing their children.

John King likely would have been content to be a legal scholar, whiling away his time researching arcane cases in a law library and teaching inquisitive students. He was a decent man with a good heart, who adored his wife and family and was highly respected in the Berlin and Toronto professional communities. He was well read in literature and science and served on the University of Toronto Senate for nearly four decades. Through the years, particularly after the family left Berlin for Toronto in 1893 (where John hoped to find better business opportunities), he was a frequent contributor to the editorial pages of the *Globe,* edited by his friend John Willison. More notably, he became a recognized expert on the law of defamation. Among his numerous publications were one textbook on defamation law and another one on criminal libel, both of which were used in Canadian law schools. When Ontario newspaper publishers found themselves in libel lawsuits, they inevitably turned to John King for legal representation.

Since Isabel absolutely refused to permit John to run for political office once he was in Toronto, he set his sights on a judgeship, full-time employment as a professor of law at the University of Toronto or the coveted principal's position at Toronto's Osgoode Law School, where he lectured. Much to his and the family's great disappointment, none of these materialized. Try as he might, he did not have the cut-throat mentality required to sustain a successful and lucrative law practice. To be fair, this was not for lack of trying. In Berlin, for example, John King had represented Consolidated Bank and its successor, the Canadian Bank of Commerce. He was a county solicitor and was occasionally hired as a Crown counsel. For a brief period, he had a law office in Galt. He also represented the Canadian Press Association at an annual retainer of $640. Like other businessmen, he suffered from an "erratic" cash flow that made it difficult to keep up with his monthly bill payments.

John accepted his lot in life much better than his wife did. Isabel was a difficult, if not impossible, woman to satisfy. She accepted without reservation the narrow world view of Ontario's Anglo elite, who arrogantly professed that wealth and moral superiority went hand in hand. Since women like Isabel would never have dreamed of working outside the home, John King was his family's sole provider.

In Toronto, John and Isabel, mainly because of John's political connections, were associated with such luminaries of high society as lawyer

William Mulock, soon to be a member of Parliament and cabinet minister in Wilfrid Laurier's government who would play a pivotal role in Mackenzie King's career path; Professor Goldwin Smith, the historian, critic and prolific writer who held court at his manor, the Grange (now part of the Art Gallery of Ontario), and advised young Willie on religion and life; and even Sir Oliver Mowat, the Father of Confederation and premier of Ontario. One of the highlights of the social season was the St. Andrew's Day Ball. Isabel and her daughters wanted nothing more than to attend in style and be seen. But come each December, John disappointed them, since he could never afford to purchase the tickets.

In a very good year, John King might have brought in from his law work, teaching and freelance writing about $2,000, and that was rare. The family's expenses for their house—rent at 147 Beverley Street in Toronto, where the Kings lived for nine years during the 1890s, was approximately $90 a month, and coal during the winter was $100 or more—his law office, the children's education, clothing, streetcars and entertainment, among a slew of other day-to-day expenditures, almost always exceeded John's income. His solution to make up any shortfalls was to borrow money against a $5,000 insurance policy he had taken out with London and Lancashire Life Assurance Company. From 1883 to 1905, this amounted to a total of $1,829, of which he had paid back only $102.

John felt sorry for himself. "I came into this world with a highly sensitive nature, and I cannot get rid of it," he wrote to Willie in December 1899. "Extreme sensitiveness will mar anyone's success in life, and I know it has marred mine. If I had coarser feelings and a harder nature, careless of riding rough shod over everybody and everything, I would have been better off today." He deeply resented, as well, Isabel's contention that the Kings were paupers. "I am constantly reminded at home that we are poor, that we can't get this, that or the other thing," he told Willie several years later. "All this is most galling to me, and makes me most unhappy."

When he was still in his early twenties, Mackenzie King had an air of self-importance and decided that his father required his psychological and financial support. "I have awakened to the fact that father greatly needs my assistance and needs to be looked after," he jotted in his diary in mid-June 1895. "I have resolved to try and get him out of his money difficulties which are at present wearing out his life." Willie was true to his word. In later years, once he was working full-time, King sent his parents and siblings money in amounts ranging from $5 to $100. From January to December 1901, he advanced his father $1,069.65, listing each payment in his financial record book. In November 1900, after his grandmother Christina died, it was left to Willie to cover the tab for the funeral.

Thereafter there was a regular flow of money from King to his family. But as was his nature, he tightly controlled the payments instead of merely sending them a monthly allowance. Hence, in a humiliating exercise, Isabel and John had to go on bended knee to Willie with pathetic requests for funds to cover everything from butter and eggs to a new stove. "Now Billie, I have a request to make," one of Isabel's letters from December 1907 began. "I have run myself into debt." After detailing the various household expenses incurred, she added with the right amount of guilt sure to tug at her son's heart, "I wearied over the thought of these things until I made myself sick. Now I have come to you to ask to lend me $75 and I will be sure to save the money and pay you in the beginning of the year." Willie immediately dispatched his distraught mother the $75 and added an extra $5 for her as a gift. "You must always feel and know that I understand the situation," he replied to her, "and am only too ready to save you anxiety."

Like many sons, Willie held his father in high esteem, particularly when he was younger. He never forgot how John had supported him while he was a university student, or the many financial sacrifices he had made for him. Over time, however, he became far less tolerant of his father's inability to make money and of his pseudo-English-gentry lifestyle. "The last word I should say against anyone is father, but I lose all patience with him," he recorded in his diary at the end of September 1898. "It has been a great disappointment... that he has not made more of himself... I cannot understand why he dawdles away his time and his life. He spends too much time on newspapers and small talk, lives as tho' he were a gentlemen [sic] of leisure, when he is in debt. Irregularity is present in all household management. Meanwhile, mother suffers and her life is not made full and beautiful as it should be... My hope was that father would be a great man, inspiring his sons to noble effort... but he made his mistake in the days of prosperity, lived beyond his means [,] has since been in debt, become indifferent, yet retained thro' all an intense love for his children and for none more than for me."

In subsequent letters, he urged his long-suffering father to finally complete his textbook on defamation (which was not published until 1907). King was always enormously respectful of his father's literary skills and writing talent. Nevertheless, John was hurt by his son's insinuation that he was not working hard enough or contributing as best he could to the family's finances. He still believed, against all odds, that a full-time law school teaching appointment was about to happen. "You children could never have the comfort and happiness which you have had," he wrote to Willie, "if I had been a mean or miserly father. My family is my greatest pride and joy—a blessing for which I am daily thankful." There was not much that Willie could say to that declaration of love. Yet financial crises were the norm in the King household for the remainder of John and Isabel's lives, and an unfortunate reality that caused their eldest son annoyance and grave concern.

DURING ONE OF Mackenzie King's many visits to London as prime minister, he took a cup of tea on the terrace of the House of Commons. He was joined by Mrs. Clementine Churchill and Brendan Bracken, Winston Churchill's former parliamentary private secretary and minister of information during the Second World War. Mrs. Churchill teased King about remaining a bachelor. King took the jesting in good humour for a few minutes and then pulled out his pocketbook, which contained a photograph. "This is the only woman I ever loved," he proclaimed. Both Bracken and Mrs. Churchill expected King to show them a portrait of his mother. To their surprise, the picture was of Clementine's good friend, Miss Horatia Seymour, whose father, Horace, had been Prime Minister William Gladstone's private secretary and who then lived at Wellstreet Cottage on the Churchills' Chartwell estate. Mrs. Churchill later asked Horatia about King, and Horatia told her that in the spring of 1908, when King was on a mission to England as deputy minister of labour, he had courted her and had almost proposed marriage. "I liked him very much," said Horatia, "but he would have made a rather dull husband."

There were, as will be told, many other women in King's life and many whom he also wanted to marry, though never did. But as Bracken, Mrs. Churchill and almost anyone else who was acquainted with King knew, none came close to matching the intensity of affection he had for his mother during her life and, even more curiously, after she was gone. At Laurier House, her portrait, a painting that King had commissioned in 1901, was kept like a shrine on a stand by the fireplace and illuminated by a light that was never shut off. (Violet Markham later referred to it as "that mausoleum of horrors.") King cherished this painting and others of her, even if, in his opinion, they did not quite capture his mother's real essence. "No portrait of her will ever reveal [her] spiritual beauty and true loveliness," he wrote to a friend in January 1918, a month after Isabel died. "I have seen nothing on earth so wholly divine as I have seen her look on one or two occasions."

Did Mackenzie King suffer from one of the worst cases of Oedipus complex ever chronicled, as is generally believed? "Men are more likely to confess to a predilection for pornography than admit to a close relationship with their mother," British novelist William Sutcliffe recently suggested. "There isn't much left that the modern man is made to feel ashamed of, yet confessing to your friends that you sometimes call your mum for a chat is something few do... Women—with good reason—run a mile from a man who loves his mother too much."

In Victorian Canada, as in England, however, this was much less of an embarrassing issue. Women were revered and idolized as "naturally pious, moral and nurturing," as the "angel in the house" expected to play a key role in their sons' lives. (They were also forced to be subservient to their husbands.) Novels of the period such as *Margaret Ogilvy*, J.M. Barrie's 1896 loving biographical fiction of his own mother, William Makepeace Thackeray's *Vanity Fair* (1848) and Anthony Trollope's *Barchester Towers* (1857) were testaments to women's delicacy, beauty, inimitability and emotional passion.

By the end of the nineteenth century, fears were raised about the dangers of close mother-son bonding creating excessively "feminine men." Boys were encouraged to be tougher and manlier—think Teddy Roosevelt rather than Oscar Wilde. The mother-son relationship was further transformed with the eventual acceptance of Sigmund Freud's psychoanalytical theories, which postulated that doting mothers could inflict "psychosexual damage" on their unsuspecting sons. The favourable reception accorded to Freud's work, including his notions about the Oedipus complex and its ramifications for interpreting mother-son love meant, however, that Isabel and Willie's relationship forever would be judged as excessive and downright creepy.

Neither Isabel nor Willie regarded the bonds that linked them as abnormal. They were connected by a powerful affection that at times overwhelmed them both, a spiritual, almost religious, union of their minds and shared desires with lofty ambitions to restore the family honour—in particular to resuscitate William Lyon Mackenzie's tainted historical reputation. Still, reading the gushing depictions of Isabel in King's diary, or some of her letters to him or other members of the family about him, is not only uncomfortably revealing of their mother-son intimacy but also makes it difficult to dispute that this relationship was beyond the norm even given Victorian sensibilities.

"She is like a little girl, I do love her so much," he recorded in his diary early in January 1899. "She is so bright, cheerful, good, happy and lovely. She has so much grit and courage in her. I have met no woman so true and lovely a woman in every way as my mother." Some months later, he observed, "If I can only win such a wife as I have such a mother, how infinitely happy!" While he was travelling in London in January 1900, he wrote to her, "I have no less than five photographs of you in my room, three on the mantle over the grate, one hanging in a frame, and one in the cedar of Lebanon frame." And then there is this poetic affirmation from September 4, 1900: "She is, I think, the purest and sweetest soul that God ever made. She is all tenderness and love, all devotion, knows nothing of selfishness and thinks only of others. Her heart has explained to me the mystery of God's creation and she lives in the light of his love. It is like coming near to an angel to be with her, and where she has been a holy calm and purity seems to remain. The more I think and see of her the more I love her and the greater do I believe her to be. She is as young too in heart and feeling as a girl of 15, in beauty she is wonderfully fair. Everyone looks with admiration on her."

Isabel reciprocated this love. Once King acquired a small property in the Gatineau Hills in December 1901, which was in time developed into his Kingsmere estate, Isabel regularly visited him each summer, a cherished time for them both. "Billie likes all my clothes," Isabel reported back to her daughters in Toronto in 1901, "but he is especially struck on my hat-cloud [a thick white veil to protect her from dust] and lace underclothes. He thinks they are a perfect success." In any age, observes Charlotte Gray, "the idea of a son admiring his mother's underclothes is stunning... and the undertones of their relationship are clearly erotic. The passion that was suppressed by the banality of Isabel's marriage was given full rein with Willie. Yet both Isabel and Willie would have been profoundly shocked to see their chaste ardour in anything but the most spiritual light."

As true as that might have been, there can be no doubt that Isabel's strong presence in King's life—later chapters will tell of other "Oedipus moments"—and his intense memories of her in his dreams and during spiritualist encounters with the "other side" shaped his distinctive character and personality and likely prevented him from getting married. That was the assessment of his close confidante Violet Markham. "It would be most unfair to hold Mrs. King responsible for the cult into which her son's love developed, indeed, I hesitate to cast any shadow of criticism on a unique relationship," she wrote in her 1956 memoir volume about her famous friends. "But in the jargon of the psychiatrists, it is undeniable that the mother-complex was a misfortune for Mackenzie King... The mother-cult stood between him and the normal ties of wife and child, which can humanize and soften the often inhuman job of politics. A character naturally introspective was flung back more and more upon itself, and as time went on a man sufficiently gay and sociable in youth became increasingly a recluse."

IN 1891, A large sign reminded gentlemen entering Wardell Hall at the University of Toronto that they should "Please Not Spit on the Floor." Freshmen were routinely hazed and tormented. And, more significantly, the university administration was wholly answerable to the Ontario government for its policies and faculty hirings. That year, Mackenzie King left the warm confines of Woodside in Berlin to pursue his education at the faculty of arts at University College at the University of Toronto.

He arrived with high expectations for himself and a willingness to work hard to achieve his career goals, whatever they might be, as he began to dutifully record in his diary. One day he wanted to become a Presbyterian minister, the next a lawyer, the day after that a professor of political economy, and three days after that a politician. He decided to let the Lord show him the way. He was ambitious, though probably no more so than any other up-and-comer, and easily impressed by the rich, famous and powerful. He dined at the home of George W. Ross, the provincial minister of education, regularly sought advice from Goldwin Smith and ensured that he was introduced to the American social reformer Jane Addams when she spoke at the university in 1895 (and whom he quickly worshipped and later saw again in Chicago).

During the summer of 1895, when he was twenty, he fell madly in love (or lust, more accurately) with an attractive seventeen-year-old, Kitty Riordan, whom he met on a holiday in Muskoka, approximately 160 kilometres north of Toronto. She was one of many young women to capture King's libidinous attention during his university years. These infatuations followed a similar pattern: King, who had a genial nature, enjoyed discussing the great issues of the day and Western civilization with his female objects of desire, only to find a brief time later that his intellectual and spiritual interests did not match theirs.

It was a similar story with Miss Riordan. Smitten for a few precious weeks, he read to her, as was his usual dating custom, excerpts from Edward Gibbon's *The History of the Decline and Fall of the Roman Empire*. Somehow she endured one three-hour session, which covered sixty pages, before they went walking in the woods and rowing on the lake. Then, just as quickly as it had begun, King ended it, insisting to himself and her that she was too young for him. Instead, he desired a "pure friendship" with her. Kitty's feelings were naturally hurt, and she scorned him, which confused and upset him.

King's superficial tendencies were tempered by a sincere Christian impulse for social reform and a desire to improve the world and the lot of his fellow man; all this was typically wrapped up in his sombre self-reflection. On one occasion, in early December 1895, he wrote a letter to a university friend about his hopes and dreams for himself and the world, and he could not control his emotions. "I cried very hard when writing it," he noted, "due a little to physical fatigue, and also to a consciousness of our weakness and inability to do much in removing the evil of this world."

Willie's family cheered his election as class president during his freshman year. According to a report in the *Varsity,* the university student newspaper, King had cleverly introduced himself as the son of "Senator Rex" from Berlin. (John King was a member of the University of Toronto Senate.) The nickname stuck and was used by his closest friends for the rest of his life. "Rex" was an avid, if average, athlete (he had a few moments of glory in track) and an excellent debater; he participated in a few memorable bouts against fellow student Arthur Meighen, who was to become his great Tory enemy in the twenties and during the early years of the Second World War.

King also possessed a particular style that made other students take notice. "There were several more brilliant personages than King in the student body [of the mid-1890s]," recalled B.K. Sandwell, the acerbic writer and *Saturday Night* magazine editor in 1953. "About King however there hung already a sort of aura of destiny, which to me was inexplicable by anything in his personality or performance, and which probably emanated from his own absolute conviction that he had what in religious circles is designated a 'call.'" Lawyer and writer Arthur Chisholm, however, who also attended the University of Toronto in the early 1890s, was always astounded by King's later achievements. "Who in the hell would have thought that Billy King would have done what he has done," he commented to a friend, another former University of Toronto graduate from 1895. "It really makes a man doubt a lot of things."

King took courses in history, political science, law and the new field of political economy and did well. He became assistant editor of the *Varsity* and immediately earned a reputation for his verbose writing and speaking style, an affliction for which he would soon become famous.

King's most memorable moment as an undergraduate and his first real taste of political action occurred in his final year, during the great student strike of 1895. To protest the university's firing of William Dale, a popular professor of Latin, and the administration's censorship of the *Varsity* and the Political Science Association, the students voted to boycott classes until the Ontario government convened a public inquiry into the university administration's questionable actions. The day Dale's dismissal was announced, the students voted to walk out of classes. At a boisterous rally, King delivered a passionate speech denouncing "the age-old cult of tyranny," as journalist Hector Charlesworth, who attended the gathering, later remembered. In his diary, King enthusiastically noted that when he first heard the news about Dale, "I was that excited that I could not keep still, my blood fairly boiled. I scarcely ate any lunch."

King's commitment to the boycott, however, was not total. He decided to attend a few lectures and tried to convince other students to do so as well. His seemingly hypocritical actions were lampooned in the *Varsity* in a cartoon depicting him as the "King of Clubs." On the top head of the card, the figure declared, "Let us boycott lectures," while on the bottom the figure said, "Let us return to lectures." King, who enjoyed the attention, was not troubled by the satire, a commentary foretelling his celebrated status as a politician who straddled the middle. Recounting this episode decades later, King's fellow student B.K. Sandwell was less complimentary. "The sad thing is that the truth about King is so bad it can't be printed even yet," he wrote to the daughter of one of the strike leaders, "for all the evidence goes to show that after making a most stinging speech at the first Strike meeting he ratted on the whole business and attended lectures as usual." That was an exaggeration, to be sure. Sandwell wrote those words in February 1949, and his judgment was no doubt clouded by King's eminent stature at the time. In 1895, King was merely an ambitious university student trying to figure out his life and be true to his high sense of morality while not showing too much contempt for the academic establishment he wished to be a part of.

The strike lasted only a short time before the government agreed to convene a commission led by Manitoba chief justice Thomas Taylor to investigate the cause of the problems, and the students returned to classes. The commission's final report sided with the Ontario government and the university administration's handling of Dale's dismissal, though the commissioners suggested that the students might have been treated with more "tact." During the hearings, which were held for several weeks in April, King was called as a witness. On the stand, he was asked whether he had any gripe with the president of the university, James Loudon. He hedged for a few moments, before being compelled to answer the question. In words he long regretted, he replied truthfully to the commission that Loudon "had always treated him with great respect, as a gentleman." The audience at the hearing erupted in mocking laughter at King's obsequious response.

There might have been one other factor to explain King's deference to the university administration. His father was a close friend of William Mulock, a Liberal MP and then the vice-chancellor of the university, whom King enjoyed visiting and had done so frequently in the winter of 1895. Mulock's influence on King was strong, and he surely would have been reluctant to jeopardize such a relationship—a relationship that was to alter King's life when Mulock summoned him to Ottawa in 1900—by attacking the university president.

SEVERAL KEY INFLUENCES shaped young Mackenzie King as a university student: a personal quest for achievement; his desire to do right by his family and restore any lost honour; an "ideal of service" and "moral obligation" to the larger community that framed the guiding philosophy of the University of Toronto's department of political economy; and his fervently held Christian principles to lead a decent life. Each of these, in retrospect, played a pivotal role in developing King's righteous sense of duty and, in combination, propelled him forward to his ultimate destiny in Ottawa.

His often unrestrained youthful ambition was a direct consequence of his family's high expectations for him to make something of himself. This was in turn tied to the perceived tarnished legacy of his grandfather. The Mackenzie factor, itself laced with religious overtones and grand proclamations of "Divine Providence," is especially relevant in any broader understanding of King's life. That was also the conclusion of those who knew King well. In a 1953 review of Bruce Hutchison's biography of King, *The Incredible Canadian,* Senator Thomas Crerar, who was associated with King for three decades (including more than ten years as a Liberal cabinet minister), pointed to one major reason for King's tremendous success: his "determination to restore and bring to a pinnacle the prestige of the family name."

It was while he attended university that he truly became conscious of his ancestral namesake; by 1896, he started signing his correspondence "W.L. Mackenzie King" rather than Willie King. Like other members of his family, he concluded that the Upper Canadian rebel was single-handedly responsible for the country's democracy. In this highly subjective interpretation of history, King found one of his life's missions. (He also memorized the words to the 1837 Royal Proclamation that put a price of £1,000 on Mackenzie's head, and he later hung up a copy of the wanted poster at Laurier House.)

In mid-June 1895, he began reading his uncle Charles Lindsey's 1862 biography of William Lyon Mackenzie, a book he would return to many times. As soon as he glanced through the introduction, he imagined he could feel his grandfather's "blood coursing through my veins." A few days after that he added, "Reading the life of my dear grandfather, I have become a greater admirer of his than ever, prouder of my own mother and the race from which I am sprung. Many of his principles I pray I have inherited... I can feel his inner life in myself. I have a greater desire to carry on the work he endeavoured to perform, to better the condition of the poor, denounce corruption, the tyranny of abused power and uphold right and honourable principles." Again, three years later, when he was moving closer to a career in public life, he asserted that his grandfather's "mantle has fallen upon me, and it shall be taken up and worn. I never felt it could be done before. I see it now... His voice, his words shall be heard in Canada again and the cause he so nobly fought for shall be carried on." Woe to anyone who dared challenge King's version of Mackenzie's life and legacy.

KING HAD TO deal with much more complex problems than his grandfather. The last decade of the nineteenth and the beginning of the twentieth century was an exhilarating as well as frightful time to be young. Dramatic social and economic forces were transforming Canada, the United States and the Western world, often at an alarming and uncontrollable pace. Inventive technology brought skyscrapers, faster communication, industrial factories, automobiles and movies. But with urbanization and modernism in North America also came non-English-speaking immigrants, slums, crime and prostitution, along with suffragettes, Darwinists and socialists, who, it seemed, were determined to turn the world upside down. From the perspective of the traditionalists, those who feared this rapid change, immorality and wickedness were everywhere. "Underneath the seemingly moral surface of our national life," declared a Canadian Salvation Army journal in 1887, "there is a terrible undercurrent of unclean vice with all its concomitant evils of ruined lives, desolated hearth-stones, prostituted bodies, decimated conditions, and early dishonoured graves."

As a young man with a bright future ahead of him, King had one foot in the traditionalists' camp and the other in that of the modernists, who embraced the revolutionary changes as progress. Like a good Victorian traditionalist, he prayed daily, read the scripture regularly, confessed puritan-style in his diary to repent for his real and imagined sins and sought Christ's guidance to ward off the temptation presented by the modern world. "I hope that my life may be a pure and holy one devoted to Christ alone," he professed on October 15, 1893, two months shy of his nineteenth birthday. Such diary entries suggest that King existed in a fire-and-brimstone environment, where everything was black and white or good and evil. "I have learned with the weariness of battle," he admitted to his close friend Violet Markham some years later, "how fraught with meaning and understanding of human life are the words: 'The Spirit is willing, but the flesh is weak!'" His predilection to accept the New Testament dictum that "He who is not with me is against me" (Matthew 12:30) likely explains his stubborn inclination to view any challenges to his opinions or authority as personal attacks as opposed to the reasoned criticisms they often were.

King, as he might well have, did not cower before or feel overly threatened by science or secularism. At university, he read Darwin's *On the Origin of Species* "carefully and rather skeptically," as he noted on July 1, 1895, and was not shocked by its apparent challenge to traditional religious thought. On the contrary, he insisted that Christianity would be strengthened rather than weakened by modernist theories and trends. He was certain, he wrote, that "the reformation in religion which the next century will witness will be its establishment on a scientific basis—its greatest advance. The inevitability of law in the spiritual as well as in the material world will be shown." He was, in the words of historian Ramsay Cook, a "modernist pilgrim."

This world view was given further credibility by his introduction in his political economy lectures (primarily by Professor William J. Ashley, who founded the department) to the British economic historian Arnold Toynbee. Toynbee had been deeply troubled by the heartlessness of industrialized capitalism in London and elsewhere. He beseeched his devoted cadre of male upper- and middle-class followers at Balliol College at Oxford University to make a real difference, to lead a "useful life" and to work to improve the lot of labourers and the destitute. After Toynbee died in 1883 of meningitis, his writings were collected in the volume *Lectures on the Industrial Revolution of the 18th Century in England,* which influenced a generation of students such as King, who were searching for a deeper purpose in their lives. Toynbee's call to action led to the university settlement house movement and the opening of Toynbee Hall in London's impoverished East End, where his social reform ideas were put into practice.

As King read Toynbee's *Industrial Revolution* in the summer of 1894, he was in a state of euphoria. "I was simply enraptured by his writings," he wrote in his diary, "and believe I have at last found a model for my future work in life." A year and a half later, he listened intently to a visiting scholar describe Toynbee and life at Oxford. "The reference he made to Toynbee, brought all the blood to my face in a rush," he recorded. "I felt it almost as a personal reference." King became somewhat of an expert on Toynbee, speaking about his life and beliefs in Toronto and wherever he travelled. Toynbee's philosophy, a secular social gospel, laid the foundations for King's "Industry and Humanity" approach to solving labour disputes less than a decade later—an ardent conviction that trade unions were "inevitable" (as Professor Ashley had told him) and that they had to be managed "according to the principles of charity, equality and justice." More than four decades after he first read Toynbee, this moment of awakening to his life's calling still resonated with him. "Today I had a curious experience," he reflected in early December 1939. "Out walking, I thought of Arnold Toynbee and of how, when I had read his *Industrial Revolution,* I was so overcome with emotion at finding my ideal of a man and the purpose of life in the kind of work which had set for himself, that I recall kneeling down and praying very earnestly that I might be like him."

KING HEEDED TOYNBEE'S call to do something useful. That meant venturing beyond the genteel university lecture halls and into the raw, mean streets of late-nineteenth-century Toronto, where disease and depravity were the norm. "Toronto the Good," as the upstanding city fathers liked to think of it, was by the 1890s a fantasy. "Houses of ill fame in Toronto?" journalist C.S. Clark sarcastically asked in his 1898 study of the city's dark side. "Certainly not! The whole city is an immense house of ill-fame, the roof of which is the blue canopy of heaven during the summer months."

At first, King began visiting the Hospital for Sick Children every Sunday afternoon for several hours, where he told stories to the young patients. He participated in the university's city mission committee and vowed to do "good work." Later, during the summer of 1895 and then as a part-time job, he worked as a journalist for several different Toronto newspapers covering funerals, cultural events and police court. He felt the police court beat offered him a window into "the shadowy side of life."

It was his gallant efforts to rescue prostitutes in downtown Toronto—similar to his idol, British Prime Minister William Gladstone, who undertook the same work in London's notorious Whitechapel district—that has attracted the most attention and snickering over the years. Despite biographer Charles Stacey's assertions to the contrary, there was probably nothing smutty about it—or about most of what transpired, at any rate. Early in February 1894, a young and innocent Mackenzie King decided to volunteer with the Haven, a local Toronto charity that offered a wide range of services and assistance to homeless and troubled women, unwed mothers and prostitutes. His first attempt to save a girl from a "wicked life," as he described it, was unsuccessful. He prayed with her and had long talks with her and her family, but his efforts were to no avail.

Another encounter with a young prostitute named Edna, probably about eighteen years old, was more promising. He and a university friend, David Duncan, met Edna and a girl named Jennie on King Street and accompanied them back to their room. They talked until two o'clock in the morning, and later over a steak dinner both King and Duncan expressed how bad they felt about the girls' plight. Three days later, King returned and heard further "sad stories" about Edna and Jennie's troubled lives. He spoke for upwards of four hours, as only he could, about the "love of the Saviour." He was devastated again by the girls' experiences.

"Poor little girls, my heart nearly broke as I talked with them," he wrote. "They do feel an awful consciousness of their sin and yet feel it holds them very fast. We all three kneeled down and had a little prayer together... They said they loved to hear any one speak thus to them and they hated to see me go away. I know they did poor girls, they kissed me as [though] I had done them a great kindness... May God hear my prayer when I ask him that I may live to see the day when both these girls are serving Him."

There were more prayers and scripture and poetry readings, and within a week Edna consented to relocate to the Haven and leave the street. (Jennie would not.) Edna told King that God had sent him to her in answer to her mother's prayers. Naturally, he was elated at his achievement of "bringing this tossed little ship [Edna] into a quiet harbour." During the next few months, King visited Edna at the shelter many times. Eventually, his relationship with her ended, and he gave no indication whether or not she returned to the street—or was "overcome with temptation," in King's words. He later turned his attention to other girls, such as Maud "Maggie" Taylor, who assured him of "God changing her heart." He was able to reunite her with her family.

Did Mackenzie King at nineteen and twenty years old pay for the services of prostitutes himself, as Stacey alleged? Were the countless references in his diaries during this period to "strolls," as in "I must go out and stroll round," proof of encounters with ladies of the evening on Toronto's King Street? "Maybe" might be the best answer. "Tried to avoid temptation, but seemed set on it, and was not willing to resist," he wrote on April 25, 1895. "Managed to waste $2 willfully and thought-lessly," he noted a month later, "but I seemed to have my mind set on wasting it so it went." In May 1897, he was delighted with himself: "Tonight for once I completely beat the devil. I felt very tempted to go out and waste time, instead stayed in."

We will never know exactly what King meant by these confessional passages and whether or not he was a young adult "john." As Charlotte Gray suggests, he might have been merely drinking, gambling or watch-ing burlesque shows. Was young King sexually excited by the "wicked" women he encountered? Did he ever masturbate—as Stacey also not-so-subtly insinuates—after seeing them and then feel terribly guilty about it? These are other issues entirely.

Willie King and his male university friends grew up in an era when sex was considered both "natural" as well as "sinful." As for masturba-tion, the common view was that it produced insanity, immorality and deviant behaviour. Arthur Beall, who worked as a "purity agent" for the Ontario branch of the Women's Christian Temperance Union and later as a morality lecturer for the provincial department of education for more than two decades starting in 1911, took special care warn-ing young boys on the evils of masturbation. A eugenics proponent, he admonished his charges to protect their "life fluid" since the conse-quences were severe—"mental bankruptcy" and an insane asylum. At the end of each lecture, he would have the boys stand and declare, "The more you use the penis muscle, the weaker it becomes; but the less you use the penis muscle, the stronger it becomes!"

Who, then, could blame Mackenzie King or any other young male of his generation for regarding masturbation with disdain and fear? King detested wasting time; the early years of his diary are filled with such phrases as, "worse than wasted" or "tonight was practically wasted." In the twelve "do's" and "don'ts" he listed for the new year on the January 1, 1894, number ten was "Do not waste any Time. 'Be up and doing.'" Number five was "Keep your mind pure." So, for an individual who desired to lead a pious Christian life, time wasted on sin and impurity was absolutely deplorable.

"I committed a sin today which reminded me of my weakness and so aroused me that I went over and had a long talk with [J.L.] Murray [a classmate]. We had a short prayer which I hope will be answered," he confessed. "I have to live a Christian life." Some months later, he admit-ted, "I cried after coming home tonight. I feel sorry for something I did last night. What sort of man am I to become?" His sense that he was weak and impious continued for many years, leading to that well-known, but misinterpreted, reflection in his diary of February 13, 1898. "There is no doubt I lead a very double life," he wrote. "I strive to do right and continually do wrong. Yet I do not do the right I do to make a cloak for evil. The evil that I do is done unwillingly[;] it comes of the frailty of my nature. I am sorry for it."

A month earlier, King had been travelling on an overnight train from Toronto to Boston. Because of the flimsy structure of the berths, he spied a young woman, who, as he described it, was "making her toilet." This presented him with a big dilemma. "What man could have resisted the temptation to look?" he later reflected. "I should have, and might have but did not. However I fought hard against possible temptations last night and won completely. I am determined that this year shall not witness the stains of old. This resolve I have made above every other, to aim ever at becoming purer in thought and word and act." The lessons one learns in youth are never forgotten. Remarkably, King was still consulting his physicians about "evil passions" and "natural phenomena" as late as 1939.

KING RECEIVED HIS bachelor of arts (honours) from the University of Toronto in the spring of 1895. He contemplated accepting a $320 fellowship to continue his studies at the University of Chicago, but pressure from his parents, who did not want him to leave Toronto, convinced him to decline it. King opted instead to take a one-year law program available in those years, and received his bachelor of laws in June 1896. He worked at his father's law firm for several months before determining that a career as a lawyer was definitely not for him.

His heart was set on pursuing graduate studies in economics at the University of Toronto. Those ambitions were initially thwarted by the new head of the department of political economy, James Mavor, who had not been overly impressed with King (and perhaps still blamed him for the student strike) and refused him the needed scholarship. King, who had genuinely liked and respected Mavor and believed the feelings were mutual, was crushed. He never forgave him, or the University of Toronto for that matter, for this slight. Across the top of his diary on June 18, 1896, he scrawled a description of Mavor given to him by a friend who had known the professor while he was teaching in Glasgow: "His presumption is limitless, his ability is practically nil and his power for irritating humanity egregious."

Faced with few other options, he went back to the University of Chicago, which again offered him a $320 graduate fellowship for a year of study he could use towards a master of arts degree in Toronto. He accepted, and although he had to endure his mother crying as he left for the train station, he arrived in Chicago in the fall of 1896 ready to pursue an academic career. At first, King believed he would combine his studies with more hands-on community service, which would uplift his spirit. He immediately went to Hull House, a settlement house one of his idols, Jane Addams, had established in a run-down immigrant neighbourhood in Chicago's rough-and-tumble west side. "To me she is Christlike," he later wrote of Addams in his diary. King had expected Hull House to have a "bare floor and walls," and he was not disappointed by the residence he was offered or the local clientele he had to counsel. As he wrote to his close friend Bert Harper in December about the area around Hull House, "misery and wretchedness, vice and degradation, abomination and filthiness are the characteristic features on all sides."

His stint at Hull House lasted only two months. The settlement house was a fair distance from the University of Chicago, and quite quickly the back and forth travel took its toll on King, who was by then taking several classes and researching his MA thesis on the International Typographical Union. That, plus the fact that he became disillusioned with the difficult chore of "uplifting and bettering society," led to his decision to leave. Even a personal appeal from Jane Addams to remain a resident could not change his mind.

From a young age, King was a clever rationalizer, and he had no problem convincing himself that he had made the right choice. He explained it like this in his diary on January 7, 1897: "My decision I think is wise . . . I have to consider my life's usefulness and at present the University seems my field of work. I ought above all to be faithful to it. I had hoped to combine both but I find I cannot and the work I accomplish for the neighbourhood here is not worth the sacrifice, at least it does not appear so . . . I felt it my duty to go; duty brought me here and takes me away again."

True to his word, he embraced his academic studies, especially the sessions he spent with the ingenious and radical economist and sociologist Thorstein Veblen, whom King held in great esteem. Veblen opened King's thinking to the intellectual and human side of socialism, so much so that he began to accept the truth of many of its tenets. Still, socialism never really dented King's liberal principles. Later, he concluded that "true progress" was and always would be the result of intelligence, hard work and the leadership of a wealthy and educated elite.

Veblen also accepted two articles based on King's thesis about the International Typographical Union for the *Journal of Political Economy,* which Veblen edited—an impressive accomplishment for a graduate student. This was King's first published work on labour unions and demonstrated his progressive thinking on the issue.

AS IMPORTANT AS King's academic work was, he had not forsaken his quest for true love. Few of his contemporaries likely could have matched his over-the-top romantic flamboyance. To wit, from March of 1898: she had, King wrote, "all the goodness, purity and greatness that God ever gave to woman, something higher and nobler than is the heritage of the children of this world alone, something deeper and more profound than even the chosen few possess... I love you... because of your self but I love you also because of your devotion to others... You are the only woman with whom it has ever seemed to me that the ocean of life might be crossed in perfect happiness and peace."

It is hard to imagine that such a mushy declaration of adoration would have won over anyone. Yet a few weeks later, twenty-five-year-old Mathilde Grossert, a lovely German-born nurse King had met in Chicago a year earlier while he was convalescing at St. Luke's Hospital with a bout of typhoid, sent him a telegram on April 7 that her answer to his declaration of love and near–marriage proposal was "yes." King was ecstatic. "I felt my heart bound as I read the words," he wrote.

Mathilde had come to Chicago with relatives she had been living with in England and trained as a nurse. She had caught King's eye working on the ward, and they began dating when his health improved. He was instantly enamoured by what he described as her "beautiful Christian character." Seeing her in church, he remarked that "she looked pure and lovely." In the summer of 1897, they exchanged love letters for several months, though his family did not know yet of their relationship. King's correspondence to her is rife with religious imagery and tender, if not quixotic, avowals of love. One example, a letter from August 30, 1897, will have to suffice: "May you reflect into many hearts the image of that self-sacrificing love to which we all owe our happiness of today and of Eternity, and may your life appear to many as it has appeared to me a precious star of hope and love commissioned by Heaven to point men's hearts ever towards the Master's throne a light to dispel the gloom to much of earth's suffering and distress, a comforter of much affliction and a bearer of such happiness and peace." (Over the years, members of Mathilde Grossert's family, valuing their privacy and her memory, have consistently refused to permit any of her letters to King to be published.)

As much as King seemed to have been in love with Mathilde, he was also carrying on a romance with a young Chicago woman named Rosa Humpal. According to his diary entries, he ended his relationship with Rosa on June 23, 1897. "I told her that I did not know what life or love meant," he recorded. "I will never forget our Goodbye. She cried and really seemed very much in love." Curiously, or perhaps to relieve his guilt, he wrote Rosa a month later, while he was also corresponding with Mathilde. "I will never forget our little romance," his letter to Rosa began. "It was the one bit of true poetry which I carried away with me from the great commercial city. I have you, as I last saw you, standing alone in the window, pictured for all time in my gallery of beautiful women."

By January 1898, King had decided that he loved Mathilde "with all my heart" and that she was as close to "my ideal woman" as he was ever to find. By the spring, he was ready to move the relationship to the next level, even if Mathilde was not as certain. Then King made a fateful error that changed everything. Before he had received Mathilde's positive telegram of April 7, in which she finally confirmed that she felt the same way about him as he did about her, he excitedly wrote to his parents and family with what he anticipated they would accept as good news.

He was terribly mistaken. Upon reading that their Willie, only twenty-three, wanted to marry a German-born nurse—nursing was not as respected a profession then as it is today—sent the King household into a state of panic. One after the other, his mother and father and then Jennie berated him for his selfishness in harsh words that made him "sick at heart." Isabel's letter arrived first, oozing with guilt. She did not want him to live a lonely life, but she also admonished him since his marriage would mean that he could not become the family financial provider she had expected of him. "I have built castles without number for you," she wrote. "Are all these dreams but to end in dreams? I am getting old now Willie and disappointment wearies and the heart grows sick. Sometimes when I hear you talk so much what you would do for those that suffer I think charity begins at home."

His sister Jennie was next with a six-page letter expressing concern about his future career path if he was to tie himself down with a wife and reminding him of the many sacrifices his parents had made for his education. "Why did Mother do without a servant all last year?" she pointedly asked him. "You ought to think of all of these things and remember that by helping your parents, you are doing what will be a blessing to you all your life." This was followed a few days later by a letter from John King, who reminded his son more formally that "his first duty is those at home" and harped as Jennie had that marriage would be a detriment to his career.

King predictably could not sleep very well. He started doubting himself and how he felt about Mathilde. He noted in his diary that all he felt like doing was crying, and he did. Still, for a few more days, he stood fast, writing home that his love—"deep, deep love, deep and true," as he put it—for Mathilde was indeed real. But the pressure on him to recant was unbearable, and few mothers could guilt their sons as Isabel King could. Her treatment of Willie, as well as of the entire family, was desperate, most unfair and clearly life altering. Had King married Mathilde Grossert, he might not have gone into Canadian politics and become prime minister; he also might have been a lot happier.

Near the end of April, he journeyed to Chicago to see Mathilde. Although she was still passionate about him—"I have had [Mathilde] in my arms. I have kissed her lips, I have seen she loves me and I am still the same"—he nevertheless pulled back, deciding that he was not "worthy" to marry her. During a taxi ride, he told her about his parents' objections and read her portions of his diary. They agreed to take a step back. King returned to Boston, and the relationship burned itself out during the next four months. By the time it was really over, in mid-August, King felt like a new man with a clear mind, baldly lying to himself that he had not been happy since he had told her he loved her. He only worried that she "may be caused the pain and pangs of disappointment."

Both of them survived this tumultuous episode of young love, though Mathilde did so better than King. Two years later, she married her cousin George Barchet and moved to rural Maryland, where she lived for much of her life. In 1901, King returned Mathilde's letters to her, at the request of her husband, but burned most of the ones he wrote her, which she had returned to him. Thereafter, King remained in contact with her for the next five decades, almost until the day he died, exchanging Christmas cards and birthday wishes. King followed the lives of her children and corresponded with them as well; her son, Stephen Barchet, was a star halfback for the navy football team in 1922 and rose to the rank of Rear Admiral. King personally visited with Mathilde and her family on at least three occasions and sent her $250 when a fire destroyed part of their farm.

After he visited her in early May 1940, he wrote to her fondly about their past, enclosing a photograph of him with his beloved Irish terrier, Pat. "It gave me a great happiness to see you again in your own home and to have the hour or more which we were privileged to share together... The last glimpse I had of you was as you walked around the turn of the cottage, clothed in white, falling as it did as a mantle across the white dress you were wearing. It was a radiant picture of one arrayed in white. It brought back many thoughts of the St. Luke uniform which you wore many years ago and which was so very becoming to you."

LOVESTRUCK OR NOT, King kept up with his academic studies. He had an uncanny ability to compartmentalize his often topsy-turvy emotional state from his workaholic intellect, a talent that served him well in his political career. In March 1897, he submitted his thesis on the International Typographical Union to the political economy department at the University of Toronto. Professor James Mavor accepted it with the following comment: "Pass: A very competent paper. The only drawback is that there are no references to authorities. The Reports of the various organizations ought at least to have been referred to." King received his master of arts at the spring convocation. His family wanted him to remain in Toronto, and he applied for a scholarship to pursue a doctoral degree. Again, Mavor thwarted King's dreams of studying at U of T; despite accepting King's MA thesis, the university refused to award him any funding. This snub once more cut King to the bone, causing him much hardship and bitterness. "Divine Providence," as he described it, came in mid-July with notification that Harvard University was offering him a $250 scholarship—not quite enough to survive, but better than nothing. More to the point, he felt vindicated. "I will work to my worth," he claimed, "and Mavor and Toronto will regret its action."

Before he left for Massachusetts, he accepted a freelance job from the *Mail and Empire,* a newspaper with strong Conservative Party connections, to write a series of articles about the city's social and economic problems. The unsigned lengthy pieces, which appeared in the paper in September and October, were eye opening and methodical exposés detailing the seamier side of urbanization, including slum housing, sweatshops and immigration. They embodied King's social reform impulses, which had been such a significant aspect of his university education, and revealed his higher-level thinking about labour issues confronting the new modern world.

He spent hours researching and interviewing the police and religious leaders, as well as workers and street people. The fact was little had changed in Canada in the past decade, since the 1889 Royal Commission on the Relations of Labour and Capital had uncovered the extent of appalling labour abuse and exploitation. "What a day I have had today and how have I witnessed the oppression of man over his fellows," he noted in his diary early in mid-September 1897. He was escorted on tours through sweatshops in Toronto's immigrant quarter, St. John's Ward (colloquially, "the Ward"), by Benjamin Gurofsky, whom King thought "a very decent jew [*sic*]." They visited with a Polish-Jewish family, spoke with Italian workers, examined slum housing and witnessed first hand the grinding poverty and hardship that took such a brutal toll on the lives of so many. "What a story of Hell," he concluded. "My mind all ablaze."

King discovered that uniforms manufactured for the federal post office department were produced in a Toronto sweatshop, where immigrant workers were paid pennies an hour in horrendous conditions for toiling close to twelve hours a day. As a courtesy, and probably because he did not want to embarrass the Liberals in a partisan Tory newspaper, King decided to tell his mentor, William Mulock, who had been appointed postmaster general in Wilfrid Laurier's cabinet, what his investigations had uncovered rather than write about them. Mulock, who had a well-deserved reputation for his brusque manner, was genuinely shocked at King's report. Within days, he enlisted King for a fee of $100 and another hundred for expenses, to study the federal connection with sweatshops and make recommendations to reform it. Although King might not have guessed at the time, this was a pivotal door-opening moment in his life. Mulock also immediately ensured that mail bags and postal uniforms were no longer made at any sweatshops where workers were abused and announced a new set of mandatory federal guidelines that all contractors had to obey. These rules were soon implemented as government policy for all federal manufacturing contracts and were extended in the Fair Wages Resolution passed in July 1900.

At twenty-three, King rightly felt satisfied at these developments. The report he prepared for Mulock while he was at Harvard—which showed without question the government's culpability in supporting sweatshops—led to other assignments from Mulock as well as similar work in Boston from the Consumers' League of Massachusetts. He told his parents that he was like "the power behind the throne," adding, "This has turned out to be the first influential part I have played in the history of Canadian politics." Despite his proclivity for a more sedate life as a university professor, he started contemplating once again a career in politics, in which he believed he could make a real difference in the lives of ordinary people. Considering that politicians and business owners of this period had such primitive understanding of labour issues, if they thought about them at all, King's expertise in this new area was propitious. Still, he hedged about what was to be his true calling.

At Harvard, King reconnected with his former University of Toronto professor W.J. Ashley. He became acquainted with such intellectual luminaries as the economist Frank Taussig, who tutored King on the complexities of free trade; Dr. William Cunningham, another Toynbee reformer; and Charles Eliot Norton, the septuagenarian fine arts scholar. Much to King's delight, Norton offered to let King live in his home, known as Shady Hill, during the summer of 1898 while he and his family were away in the country. King considered this a "great honour" and rejoiced that he was "in the home of the greatest man in America." He later used the name Shady Hill for one of his cottages on his Kingsmere estate.

About the only troubling issue to arise that year occurred when his older sister, Bella, wanted to train as a nurse at Boston's Children's Hospital. King was not thrilled with the idea of having his sister so close by. He regarded her as an intruder in the private and cozy academic world he had constructed. After much hand wringing, King obtained the application information Bella wanted, though he still tried to convince her in a rambling ten-page letter that she would be better off remaining in Toronto and finding a husband. John King agreed, but Isabel supported her daughter's wishes, despite the family's lowly view on nursing. Bella arrived in Boston in July 1898 and enrolled in the hospital program. She lasted less than six months. A high-strung person, she may have found the intensive training and night shifts too difficult, and King did not provide his sister with the emotional support she required; in fact, he undermined her hopes and dreams as much as he could. Predictably, when his mother accused him of acting selfishly towards Bella, King was adamant that he had done nothing wrong.

By then, King had received a master of arts degree from Harvard, his second MA and fourth university degree in three years, a notable achievement by any standard. With another scholarship worth $450, he began work on his doctorate and planned to write his dissertation on labour and sweatshop issues related to the clothing industry. Then a year later, as he was completing his PhD oral exam, his scholarship was renewed, with funds included to travel abroad to conduct research.

As he embarked on a journey to England in September 1899, he was as affected with his accomplishments as ever. "I go away not without many feelings of sadness," he wrote to his parents, "yet with a heart filled with courageous hope and gladness, a mind newly fired with earnest determination and a body reinvigorated by rest and recreation. I have no fears, but many hopes, and go out as did the knights in search of the Holy Grail."

Once in London, he did not exactly find the Holy Grail, but he did get to meet some activists who had personally known his hero, Arnold Toynbee. While in London, King lived at Passmore Edwards Settlement (now the Mary Ward Centre) located in Bloomsbury, a middle-class neighbourhood not too far from the London School of Economics, where he studied. It suited his refined tastes much more than Chicago's Hull House had. It took him five months to venture to Toynbee Hall in the downtrodden East End. After receiving a tour of the grounds, he noted in his diary that "there was not much to see."

He was consistently in awe of the intellectual elite, whose exclusive club he so eagerly wanted to join. Having tea and conversation in the country with the novelist Mary Ward (Mrs. Humphrey Ward to her readers) and on another occasion with Toynbee's sister, Gertrude, stirred his soul. Yet as pleasant and stimulating as these encounters were, they also brought out his mounting insecurities, never too far below the surface of his character. He wrote in his diary that he felt "unholy, unclean, so unworthy."

His enlightening discussions with socialists and key founders of the Fabian Society, Sidney and Beatrice Webb, made less of an impression on him. King could not support state socialism then or decades later when he confronted the Fabian-influenced leaders of the Canadian socialist party, the Co-operative Commonwealth Federation (CCF). His attendance at a dinner party that included thirty members of the Fabian Society troubled his puritan sensibilities. "Two of the women smoked, some of the men stayed near to the wine and whiskey, others talked. I did not care for the crowd at all," he noted. In a stunning display of immature arrogance, he believed them all to be uncouth, lacking education and "sore headed" about some misfortune or other to have befallen them.

In the spring of 1900, King left for an extended trip to France, Belgium, Germany, Italy and other European countries. When he was in Rome on June 26, a telegram arrived for him. The cable was from William Mulock, offering him full-time employment in Ottawa as the editor of the new federal government–run *Labour Gazette* at an annual salary of $1,500 with a promise of a raise if things worked out. This unexpected development naturally excited King, but initially he was overcome with indecision. By this time, he had seemingly given up on a career outside of academia and was awaiting news of a lecturing position at Harvard. That job offer eventually came in a letter on June 29. King could teach part-time at Harvard and continue researching his thesis for a stipend of $400. Other contract teaching work Harvard had arranged for him outside the university would bring him another $200.

During the next week and a half, as King continued his tour of Italy, he agonized about which career path he should take. In a nineteenth-century version of "helicopter parenting," John King interceded on his son's behalf with Professor Frank Taussig, with whom Mackenzie King would be working in the economics department. The Harvard teaching job was a term position only at this stage, Taussig explained, and suggested to the elder King that his son accept the government's offer. Back in London, Mackenzie King was still uncertain what to do and spoke about his dilemma to anyone who would hear him out. At long last, on July 9 he sent a telegram to Mulock with his acceptance, a life-altering decision if there ever was one, and he knew it.

"This has been an important day for me," he concluded, "it marks a crisis in my life, a choice made which is to make the future course of my life follow a different path of practical activity than it otherwise would have done." After breakfast that day, he ventured down to the British Museum to read back copies of the British *Labour Gazette* in order to prepare for his new calling. In the days ahead, he made plans for his return to Canada, confident that his momentous decision was "a victory of reason over inclination."

A position in Ottawa also appealed to his patriotism, or so he had convinced himself. "I am a Canadian by blood and very much so I fear and my influence should count for more in Canada than elsewhere... I was fast becoming an American not voluntarily but I was being won over... If I can find in Canada the sphere I desire and can be much service in, I would rather my life should be lived there than anywhere else."

2

The Peacemaker

It is necessary to be fearless, and enmities must come sooner or later to the man who seeks reform, so that one has to find one's reward in the sense of justice and fair play in one's own conscience.—THE DIARY OF WILLIAM LYON MACKENZIE KING, *September 13, 1907*

BURIED IN the thousands of newspaper clippings squirrelled away by Mackenzie King over his life is one from the Toronto-based *Globe* of August 18, 1900, announcing his appointment as the twenty-five-year-old editor-in-chief of the new *Labour Gazette*. There is an accompanying black-and-white photograph of him dressed in a stylish black frock coat, crisp white shirt and a black silk tie. He has the distinguished presence of a man ready to take on the world. It is easy to imagine the young ladies of Ottawa swooning over this portrait of the city's newest eligible bachelor.

Thereafter, the *Globe* was enamoured with King. A full-page photo from March 1905 entitled "Four Friends" showed King, then the president of the Canadian Club of Ottawa, with three other "distinguished young Canadians": novelist Norman Duncan, poet and civil servant Wilfred Campbell and lawyer Henry Burbidge, one of King's close companions. All four men, the newspaper pointed out, were graduates of the University of Toronto, and three (King, Duncan and Campbell) "have won reputations far beyond the bounds of the Dominion." A year later, King Edward VII rewarded Mackenzie King's government service with a Companion of the Most Distinguished Order of St. Michael and St. George, or CMG. King's fellow civil servants, some much older and more experienced, deeply resented this singular achievement for such an upstart. "His rise to prominence has been as rapid as it has been sure," the *Globe* extolled. "Perhaps the word 'thorough' explains his advancement better than anything else." A year after that, on July 6, 1907, the *Globe's* Saturday magazine ran another full-page photograph of the dapper King. By then, he was the deputy minister of labour, and an accompanying biographical story about him proclaimed him to be "Canada's Industrial Peacemaker." His rise to prominence truly was rapid and unprecedented.

Ottawa high society, such as it was, might have been ready to welcome young Mr. King when he arrived in the capital on July 1900, but he was not as certain. "Gloomy" was his first impression of Ottawa. The city, in his view, was "small and like... a provincial town, not interesting, but tiresome... Ottawa is not a pretty place save about the parliament buildings." Liberal Prime Minister Wilfrid Laurier, who was then four years into his first term, some years earlier had proclaimed his intentions of developing Ottawa into the "Washington of the North," a capital with style, grace and class. Instead, Laurier confided to his close friend (and likely lover) Mrs. Emilie Lavergne that the city was "dull."

A more fitting nickname for Ottawa was the "Pittsburgh of the North." At the beginning of the twentieth century, the city was still ruled by lumber barons. One large enterprise, Eddy's, churned out more than twenty thousand matches each day. The lumber companies, not the federal government, were Ottawa's main employers. The entire civil service in 1900 consisted of only 1,129 people, overwhelmingly male, and even by 1911 that number had only increased to 3,219 out of a total population of 87,000.

King started his new career in Ottawa less than three months after the Great Fire of April 26, 1900, which destroyed much of Hull and the area of Ottawa near the waterfront and beyond. More than fifteen thousand people lost their homes. Rebuilding took a long time, and this accounted for the "gloomy" appearance of the city that troubled him. Ottawa roads were muddy in the spring or after a heavy rain and cluttered with horses, wagons and electric streetcars powered by overhead wires. The countess of Aberdeen, the energetic wife of the governor general, the earl of Aberdeen, complained in 1893 that the streetcars "frighten the horses and they come along so swiftly and silently that they are a real danger."

King's first impressions notwithstanding, the city was not entirely dead. Sparks Street already had a bit of style and fashioned itself as

Ottawa's "Broadway." The downtown avenue boasted the city's most popular pub with the eye-catching name, the Bucket of Blood. Farther east in Lower Town could be found an assortment of rogues, thieves and prostitutes, with its hub in today's trendy ByWard Market. Closer to the parliament buildings was the "private capital." This was the world of Windsor uniforms and elegant gowns that catered to the upper classes: the prime minister and his cabinet, politicians from the opposition benches and lumber merchants invited to the lavish balls hosted by the governor general and his wife at Rideau Hall, and in the winter to skating parties on the Rideau Canal. It was only a matter of time before Mackenzie King—ambitious, serious and socially opportunistic like his mother—was heralded as a key member of this high society.

SIX WEEKS INTO his new job as editor of the nascent *Labour Gazette,* King started having serious doubts about his career choice. He missed the academic life at Harvard and was keen to complete his PhD dissertation—something that he was not able to do for nine more years. As the months passed, he gradually grew into the position and began to see himself as the only person in the federal government who truly understood the needs of the working class. "I have a nobler conception of my own usefulness to the world and to the cause of labour itself," he told his father in March 1901. Initially, King only wanted to ensure that Canadian workers were treated fairly. "What matter the credit, so long as the end is achieved?" he asked himself in early August 1900. He was being disingenuous, since to Mackenzie King, credit always mattered. Less than four years later, his assessment of his role had changed. "I think I can take credit or blame for [the department's creation]," he wrote, "and know that had I never lived the Department of Labour would not exist today."

He was overstating the situation, yet he was not completely incorrect either. Historian Paul Craven's astute observation that "King was the making of the Labour Department as much as the Labour Department was the making of King" rings true. Until 1909, when the department officially became an independent entity, it remained a secondary responsibility of the postmaster general; in 1900, this cabinet position was held by King's boss William Mulock. From the start, King was determined to give labour issues a higher profile and ultimately envisioned a separate federal labour department, possibly with him heading it as a cabinet minister. "If I can save enough to draw an income from saving I may be able to take the step into active public life for which my whole nature and ambition longs," he wrote in mid-February 1902.

"God has his own plans. He is defining His way very plainly now. He will do so in the Future."

Mulock had hired King to oversee the publication of the *Gazette* as a forum for labour issues, to collect statistics on wages and strikes, as well as to appease working-class voters by keeping them firmly in the Liberal Party camp. It was to be a small operation within the department and as favourable to the Liberals as possible. In the early twentieth century, the affairs of government revolved around a tightly controlled, almost religiously zealous network of businesses, newspapers and, above all, patronage. Those in power rewarded their loyal supporters and intentionally ignored their opponents. Animosity between the two sides was bitter and petty. While King was to eventually grasp and master this fact of life in political Ottawa, he was stunned when he showed Mulock the first issue of the *Labour Gazette*. Any story or statistic that even hinted that the Conservatives had done something positive was deleted. In one case, the fact that a judge known to be a Tory supporter had delivered a pro-labour decision was unacceptable to Mulock. King was forced to remove the reference.

Soon after this encounter, King travelled with Mulock on the train to Toronto and was unable to hide his disgust at the political reality he found himself in. "I was scarcely able to talk all the way," he wrote that night in his diary. It was not long before King discovered that in politics, perception was everything. In September 1900, after only a few months in his position, he was appointed the deputy minister of labour, though he did not receive the commensurate salary of $3,200 until February 1902. King's appointment as deputy minister was highlighted in future newspaper profiles about him. "He is not yet thirty and is a Deputy Minister," reported the *Ottawa Journal* in December 1904. "You must live in Ottawa to appreciate the meaning of that title—it is a word to be breathed with bated breath. But he earned his rise."

He started to personally intervene in labour disputes outside of Ottawa. In October 1900, about a month before a federal election that Laurier and the Liberals would win, he visited Valleyfield, Quebec, where a dispute over wages turned into a violent confrontation between workers and management at a mill company. The mayor of the town, with the support of the mill owners, had called in the militia to restore order, which only made the situation worse. Like a white knight, King rode into the town on his horse. Using the personal charm and common-sense psychological appeal that were to be his trademarks, he had the troops sent away and negotiated an end to the strike. On his way to the train station, however, he was seen chatting happily with the Liberal candidate in the area, George Loy. In a letter to his parents, he suggested that his good work in Valleyfield likely would benefit the Liberals in the coming election and that he hoped Loy would defeat the Conservative candidate, Joseph Bergeron, which was exactly what happened on November 7, 1900.

King probably should have been more discreet. Some months later, the Quebec Conservative Frederick Monk rose in the House of Commons and accused King of being a Liberal "political agent," who advocated on behalf of Loy during the election campaign. Monk was vilified for attacking a civil servant, and King was defended in the House by Liberals as well as independent MP Arthur Puttee, but he learned a valuable lesson about even innocent public displays of partisan support.

ANY *GAZETTE* CORRESPONDENTS or assistants King wanted to hire had to be vetted according to Liberal patronage rules. The one exception was his old university friend Henry Albert "Bert" Harper, who was working in Ottawa as a parliamentary reporter for the *Montreal Herald*. By mid-August 1900, King had received permission to appoint Harper as his chief assistant. After King became deputy minister, Harper was promoted to associate deputy. The two men soon moved in together to a house on Maria Street (now Laurier Avenue West) and then, in September 1901, into a larger apartment on Somerset Street.

King never had a truer or more loyal companion than Harper, who fancied pince-nez eyeglasses of the style made famous by Theodore Roosevelt. The duo of Rex and Bert was connected intellectually and spiritually. Harper, who also kept a diary, shared King's noble Christian vision of saving the world and was as devoted to self-improvement. "No man who desires to make progress in the world can hope to do so if he squanders his evenings," he lectured King, who wholeheartedly concurred with that dictum. They enjoyed hiking in the Gatineau, strolling the capital and reading literary classics aloud to each other curled up by the fire. A favourite book was *Idylls of the King,* a lyrical narrative about the legends of King Arthur by the English poet Lord Alfred Tennyson. For both King and Harper, reading was for stimulating learning rather than entertainment.

No two heterosexual males likely adored each other more. In 1901, Harper started dating King's sister Jennie. Harper began his letters to King with "My Dear Rex" and often signed them, "With much love, Ever affectionately, Bert." When they were apart, Harper genuinely pined for King and vice versa. "I miss you, Rex, very much," wrote Harper in November 1901 while King was out of town. "The meaning of an individual is sometimes emphasized when the individual is absent from the associations which are eloquent of his individuality."

They differed on only one point: Harper did not covet the newspaper attention that King did. On March 30, 1901, the *Ottawa Journal* ran a complimentary story about the labour department, complete with photographs of Deputy Minister King and Associate Deputy Minister Harper. (King did not like the photo of himself and thought he looked like a "negro convict.") King later admitted in his diary that he regretted that he had co-operated with the *Journal,* since he realized it would give the Conservative opposition more ammunition to attack his and Harper's non-partisan credibility. "The deeds of a man are less great when told and where they are poor deeds at best are much better left alone," King wrote. At the same time, he ran out and purchased fifty copies of the newspaper, "why I do not know," he added. Harper knew why. "The information [in the *Journal*] is I think pretty much as Rex dictated it," he noted in his diary the same day. "For myself I don't think it is proper, nor even wise for young men to keep blowing their trumpets too loudly. Vanity is Rex's great weakness however and will I fear remain an alloy in his character to the close."

These personal differences hardly mattered. King and Harper were happiest when they were off on some adventure together. Some months earlier, in October 1900, the two companions, with their bicycles in tow and wearing knickerbockers and leather boots, boarded an early morning train for the short trip to Chelsea, Quebec. From the Chelsea station, they rode into the autumn trees in Gatineau—"The trees were one mass of colour... they seemed full of music, it was like listening to a great world chorus to see the beautiful hues of colour," King poetically put it later—and eventually found themselves on the shore of Lake Kingsmere. "The view," he wrote to his father, "was one of the most beautiful I have ever seen."

On the summit of King Mountain beside the lake, the two friends could see the parliament buildings in the distance, "like small tents on the horizon." Over a lunch of cold roast chicken ("which we pulled limb from limb") and grapes, they read aloud from Hamilton Wright Mabie's *Essays on Nature and Culture.* The day, scenery and moment were almost too much for King, who regarded Mabie's volume as "the most beautiful soul inspiring book that has ever come into my hands." The two friends decided to bike the twenty-four kilometres back to Ottawa, but they swore they would return soon.

The following summer, King and Harper stayed near Lake Kingsmere at Mrs. McMinn's boarding house. Also vacationing there were the Very Reverend Dr. William T. Herridge, the minister of St. Andrew's Presbyterian Church in Ottawa, his captivating wife, Marjorie, and their four children. King and Mrs. Herridge (or "Mrs. H.") instantly bonded on an emotional and possibly sexual level. She introduced him to the literary writings of the English poet Matthew Arnold; the "beauty" of Arnold's poetry inspired King. He soon purchased a lot by Lake Kingsmere for $200, a momentous decision in his life if ever there was one. This was the last summer he and Bert Harper were to spend together.

ON DECEMBER 6, 1901, Mackenzie King was on a train heading back east from B.C. on a work-related trip. Late that same afternoon, Bert Harper decided to join a skating party on the Ottawa River. Among the group of four was Elizabeth "Bessie" Blair, "a pretty social butterfly," as *Saturday Night* magazine's social columnist Amaryllis called her, and the nineteen-year-old daughter of Andrew Blair, Laurier's minister of railways and canals. Out on the river, the foursome was trying to catch up with another, larger party led by Lord and Lady Minto; the earl of Minto had replaced the earl of Aberdeen as governor general in 1898. Bessie's sister May was one of the skaters with the Mintos. Old timers warned the skaters to be cautious. Who knew what "treachery was lurking beneath that broad expanse of ice that looked so fair and firm?" Amaryllis later recounted. "No one heeded those wise people."

Harper's group was still on the river when the sun went down around 5:30 PM. They started for home, but near Kettle Island the river ice gave way and Bessie Blair fell into the water. She laughed and cried out to her companions, "Don't mind me, I can swim." Harper, ever gallant, tried to pull Bessie out of the water. Unable to do so, he tore off his gloves and coat and jumped into the icy water. According to one version, he cried out, "What else can I do?" The strong river current pulled them both under, and they drowned.

A telegram was dispatched to King, but the train conductor forgot to give it to him. The next day, as soon as he arrived in Toronto at Union Station, he saw the newspaper story with tragic details about the accident. He visited with his family briefly (Jennie, too, was heartbroken) and then departed for Ottawa, where he helped arrange his friend's funeral at St. Andrew's Presbyterian Church.

King was consumed with grief over the loss of his friend. His diary at the beginning of January 1902 is full of "Dear Bert" thoughts. "He has gone quickly the soul of the man I loved," he noted in one entry, "as I have loved no other man, my father and brother alone excepted." He consoled himself with the thought that God must have some other purpose for him, which was the reason he now had been abandoned.

Soon after the funeral, William Mulock and King started a fundraising campaign to build a memorial to Harper. To kick-start the fund, Mulock gave $100 and King donated $50. The statue honouring Harper was to be of Sir Galahad (based on a famous painting by George F. Watts), inspired by the poem "The Holy Grail" in Tennyson's *Idylls of the King*, which both Harper and King had loved so much. The statue was completed by Ernest Wise Keyser, a young sculptor from Baltimore. King was not only impressed by Keyser's talent, but he also believed it was significant that Keyser shared Harper's birthday.

On November 18, 1905, a cold Ottawa day, recently appointed Governor General Earl Grey, with a sombre Mackenzie King by his side, unveiled the statue where it still stands in front of the parliament buildings' entrance on Wellington and Metcalfe Streets. Beneath the figure of Sir Galahad, Tennyson's words are cut into stone: "If I lose myself, I save myself!'." About six months later, King published a slim volume about Harper's short, yet gallant, life, entitled *The Secret of Heroism*.

At 161 pages, the book is a heartfelt biography of Harper with excerpts from his diary and letters. "The quality of a man's love will determine the nature of his deeds; occasion may present opportunity, but character alone will record the experience," wrote King. "To a life given over to the pursuit of the beautiful and true, the immortal hour only comes when conduct at last rises to the level of aim, and the ideal found its fulfillment in the realm of the actual."

Harper was portrayed as he was depicted in the statue, a self-sacrificing Sir Galahad. King intended the book to inspire future generations of young men and included Harper's reflections on nature, life and religion. (As Stacey caustically points out, King also did not refrain from incorporating those of Harper's letters which praised his own noble character and strength.) King's prose was heavy and overwrought. Ethel Chadwick, a young woman in the same social circles, aptly described the writing as "very high-flown... As one would expect anything of King's to be." But the critics in London, Toronto and across Canada admired and even raved about the book and its sensitive author.

Harper's death at the age of twenty-six haunted King for decades. On the eighth anniversary of the accident, in 1909, the day he also deliberately chose to make his first speech in the House of Commons, he wrote that the absence of Harper in his life "was irreparable." He left ten white roses at the base of the statue before he entered Parliament. "It was beautiful to leave it there to look at when I came out," he wrote. Thereafter, he acknowledged the anniversary of Harper's death for many years; the last notation in King's diary about Harper was on December 6, 1949, nearly five decades after he had died. Mackenzie King never had another close male friend like Bert Harper. His loss left a real and painful void in his life.

IN THE IMMEDIATE aftermath of Harper's death, it was Marjorie Herridge, sixteen years his senior, who provided King with much-needed emotional support. With an eye-pleasing figure and wavy brown hair, she was definitely not the stereotypical preacher's wife. She had emotional and physical needs that her husband ignored. As King's friendship with her progressed, he stopped referring to her as "Mrs. H." in his diary and called her "the Child." In her letters to him, he was "Boy, dear," "Boy dear heart," and "Boy darling," who could be "bad and naughty."

King began spending a lot of time with Marjorie, which did not go unnoticed. He found the quiet snickering and gossip "most impertinent" and dismissed the "meanness" of the comments, pointing out that "Mrs. H. has twice the head and heart of these people." Depressed about Harper's passing, King visited Marjorie almost every day for months at the Herridges' home, the manse adjacent to St. Andrew's Church. Yet the more he got to know Dr. Herridge, the less he liked him. King regarded Herridge as "conceited and vain" and believed that he and Marjorie had a loveless marriage; at least, she had intimated that to him.

On the night of February 25, 1902, he went for a late evening walk with Marjorie. Around midnight, when King escorted her back to the manse, they were unaware that Dr. Herridge had been out searching for his wife and had been unable to find her. He was angry by the time he returned home. When they finally arrived, King and Marjorie found the door to the manse locked. She rang the bell, and Dr. Herridge opened it. Stepping back, he glared at them. "You may go your own way, I scorn you both," he announced. King was shocked and left, humiliated and cross. At this point in his relationship with Marjorie, there is no evidence that it was anything but a platonic friendship.

The next morning, he and Marjorie met and had what he called a "deeply earnest talk." Marjorie told him that her husband's actions were "insane." King blamed himself for keeping her out so late and for spending too many evenings with her, though he did not think that this excused Dr. Herridge's insult to them both. King wanted to apologize to Herridge, but since Herridge had slammed the door of the manse on him, King felt he could not do so or attend church. He and Marjorie had a long talk about their friendship and "its propriety." This caused them, he recorded, "to speak of things known to the other but unspoken before." She spoke of Dr. Herridge's "indifference" to her, "the want of any true love between them" and his continual work-related trips to England and Europe for long periods of time. She felt alone and abandoned. "All of which to me," he wrote in an attempt to justify his actions, "all gave a tacit understanding that so long as honour could suffer no reproach there could no objection to a closer friendship—where it was as ours has been, on intellectual and spiritual lines—only of the purest and of the best—than might be enjoyed by others less fortunate in the purpose of their lives."

King saw Dr. Herridge the following evening. He told King that he had been annoyed with his wife. King insisted that the fault was his. Dr. Herridge said he was concerned about what the neighbours would say if they saw her out so late with another man. Much to King's dissatisfaction, no apology was forthcoming. On February 28, King was still in emotional turmoil from this incident. "I long for rest and peace, to be alone, to talk with Mrs. H." he wrote. "What a curse the world is! That our hearts have to distort themselves to conform to its conventionalities, or rather to escape its lying slander and vicious tongue."

On March 2, "a strange day, a beautiful day," Dr. Herridge came to see King. He apologized for his behaviour and told King that he was glad his wife had him as a friend. Marjorie had refused to get out of bed, and Dr. Herridge requested that King visit her. He did so, and Marjorie's mental health immediately improved. Nothing, however, had been resolved.

That summer, the two of them were together again at Mrs. McMinn's boarding house at Kingsmere. Dr. Herridge was away on another trip, and Marjorie was alone. King briefly stopped making entries in his diary, and it is possible that their relationship became more intimate; by September 21, 1902, she had become "the Child." "The story of my life for the present is the story of its relation to the Child," he explained. "Our summer has been lived together, lived to ourselves, and not the fall and winter has [sic] come and we are to live apart, and the duties of the life rather than its pleasures are to receive their emphasis. What is to be the outcome of this love, the love which binds her to me and me to, that is the problem now. She loves me more deeply than ever before if that is possible and can do less well without me. I have reason to love her as I never had reason to before."

Charles Stacey and Charlotte Gray, among others, suggest that there is no hard proof that their relationship was ever sexual. "There would have been many barriers," argues Gray. "The mores of the time, and the uneasy and guilt-ridden attitude to sex that Willie had displayed since student days." More to the point, would King have risked his entire future as a politician by having an affair with a minister's wife, knowing that could become public? It seems implausible. And yet, the pet nicknames they had for each other, "the Child" and "Boy darling," imply a relationship beyond one that was platonic. King, too, was capable of exhibiting unrestrained passion.

He continued to spend a great deal of time with Marjorie as well as Dr. Herridge during the next year, and his diary is peppered with references to her. During the late summer of 1904, he and Marjorie, who was left alone at Kingsmere with her children, discovered John Morley's biography of William Gladstone, King's political hero, and they read it to each other.

That year, King built a small cottage on the property he had acquired in 1901 on a spot he called Kingswood, the first piece of what would become his vast Kingsmere estate. (The Herridges had bought land close to Kingswood for a house they named "Edgmoor.") Kingswood was (and is) a quaint summer retreat with a few rooms, a verandah and a fireplace that he modelled on the kind in Shakespeare's home at Stratford-upon-Avon. The fireplace was dedicated in Bert Harper's memory. At a special ceremony, Marjorie Herridge read favourite passages from Arnold. (As fate would have it, the fireplace did not work properly, and the cottage filled with smoke.) King adored the place and spent as much time as he could there. He wrote lovingly, "The bare floor, the timbered walls and ceilings, the brick hearth and fireplace are more beautiful to my eyes than the best interiors of the city and there is nothing to equal the out-of-doors." King's parents and family regularly visited and stayed with him for weeks at a time.

His relationship with Marjorie, however, was far from smooth. One evening, King lost his temper when their cozy solitude was interrupted by a friend of Marjorie's, who had dropped by uninvited. King uncharacteristically took a tantrum, throwing a stone at the door and yelling about the friend's intrusion. "It was a display of anger," he confessed, "to which I am not often given." Overcome with remorse, he apologized to Marjorie the following morning for his unwarranted behaviour and expressed "the anguish in my soul for the longing I have to be a true and noble man, to live the life God would have me live and to fulfill the purpose he has in me." At that, they both knelt and prayed side by side.

Life in Ottawa went on, and if there was an affair it was probably long over by 1906. That spring, as King immersed himself in Canada's labour problems Marjorie travelled by herself to visit her brother in Scotland. Her letters to King during her seven-month absence are indicative of her deep love for him. Whether it was reciprocated is unknown. A few typical notes from Marjorie read as follows: "Friday May 18th The Hillside: Darling boy, I have come out here just to be alone and speak to you... of the glory of the field." On June 18, she wrote, "Boy darling... I had been walking on air ever since I got it... but you are a bad and naughty boy to say nobody loves you. Do you mean to call me a nobody? If I had you here you'd soon know whether I was nobody or not..." On December 7, from the Cleveland Hotel in London, she relates, "My dear Boy, I could not tell you how often I have written a letter to you and not sent them. I said too much or too little... I thank you with all my heart for your cable your kind letter and your beautiful gift. They mean so much to me." Is this not the correspondence between two people who have been physically intimate?

King and Marjorie's close relationship continued until 1914. As his life situation changed, he came to regard her as overly selfish. She made the fateful and unforgivable error of forgetting his birthday in December 1914. Five years later, Marjorie suffered a mental breakdown and was admitted into a psychiatric hospital in Guelph, Ontario. The Herridges' son, Gordon, was killed in a shooting accident in 1922, which further impaired her emotional state. Thereafter, she refused to return to the family's cottage, Edgmoor, because the memories of her lost son were too painful. King eventually acquired this property as part of his country estate and renamed it Moorside.

The day Marjorie died in March 1924, Prime Minister Mackenzie King was unmoved. "The past ten years our friendship has meant little," he recorded in his diary. "My first ten years in Ottawa it was in some respects, apart from home ties and affiliations the most real expression of my life—full of happiness and pain. Selfishness marred it in the end, selfishness and self-indulgence. I could not believe I could have felt so little the word of her death. Of course it is now a release a great release, but a sort of bitterness has since entered my nature, a feeling that she had perhaps injured my life a little through selfishness, perhaps the greater injury was mine to her in ever having been such friends. But it is over now." In the following years, as he delved into spiritualism, he had several visions of Marjorie Herridge and was reminded of much happier times he had enjoyed with "the Child."

King's relationship with Marjorie during his early years in Ottawa, whether it was pleasant or disconcerting, was a distraction from the task at hand. His real objective was to advance the stature of the labour department and in the process himself. From his first day on the job, King was imbued with Arnold Toynbee's "gospel of duty." Whatever small-scale operation Postmaster General William Mulock had in mind, King saw his task in much grander terms: he would transform the federal government into the "impartial umpire" to resolve disputes peacefully between management and labour. As a professional conciliator, he brought a Christian-inspired humanity to the negotiating table.

Still, introducing conciliation that was fair and equitable was a difficult assignment, even for someone as diligent as Mackenzie King. This was an age when workers were frequently treated no better than eighteenth-century Russian serfs. Long hours and exploitation of men, women and children at factories and elsewhere was the norm. Management regarded trade unions as evil and their leaders as socialists, anarchists and foreign agitators. Collective bargaining was deemed destructive.

In 1907, when King investigated Bell Telephone in Toronto, he discovered that little had changed since his sweatshop investigations a decade earlier. The exploitation he found sickened him. "I feel at times as though I would sail without mercy into the company so hideously inhuman and selfish its whole policy seems to have been," he recorded. "The image is constantly before me of some hideous octopus feeding upon the life blood of young women and girls, the more I go into the evidence, the more astounded I am at the revelations it unfolds."

By the time King entered the fray, workers in Canada and the United States had become increasingly impatient and were often more radical in their demand for union recognition, fair wages and better working conditions. No longer were they willing to be treated like a commodity at the mercy and whim of their taskmaster employers. From the 1890s onwards, that meant frequent ugly and violent strikes across the country in a wide range of industries. Most capitalists would have none of it. Leonard Barrett, the vice-president and general manager of the Vulcan Iron Works in Winnipeg, echoing the sentiments of many of his fellow business owners, bluntly put the labour issue into perspective as follows: "God gave me this plant, and by God I'll run it the way I want to!" Often, the federal and provincial governments felt the same way, or at any rate recognized that a truly equal relationship between management and labour was impossible without the business community's full co-operation and consent.

King came to his job full of youthful idealism. He maintained that exploitation of working men had to stop and that they were entitled to the same rights of citizenship (and leisure time) as were the owners of industry. He was never a fan of strikes, but he also did not want striking workers punished if their demands were just. An advocate for nineteenth-century liberalism, King did not oppose unions, but he did fight labour organizers who tried to compel every worker in a shop or factory to join a particular union, whether they wanted to or not. Individual liberty for capitalists and workers alike was sacred. The more he waded into the labour-management quagmire, the more both sides frustrated him. On the one hand, he quickly decided that some union leaders were "tyrannical"; yet, on the other hand, he conceded that he understood why working men found socialism so appealing. "They are driven to it," he wrote to his father at the end of May 1903. Eventually, he arrived at the conclusion that the larger interests of the public (or the "community," as he called it) outweighed the narrower and selfish interests of business and labour.

The first issues of the *Labour Gazette* King produced received generally positive reviews from Laurier and his cabinet, the Trades and Labour Congress of Canada (TLC)—the chief labour organization in the country at the turn of the century, which in 1902 came under the control of the American Federation of Labor, led by Samuel Gompers—and the Canadian Manufacturers' Association (CMA). Only the CMA felt that King was focusing too much on the needs of the workers and ignoring the significant role played by business in labour matters.

He soon made a name for himself in Ottawa as essentially the country's first labour conciliator. Behind the scenes, he was responsible for several key pieces of labour legislation, most notably the Industrial Disputes Investigation Act (IDIA), passed in March 1907. Among other things, IDIA banned strikes and lockouts in public utilities, railways and the mining industry during government-appointed investigations into labour disputes. The investigating committee's findings were not binding, but King was certain, as historian Paul Craven explains, "that the combined forces of public opinion and reason would persuade [management and labour] to settle." Real life was more complicated than King's idealistic theories.

Until IDIA was passed, King operated under the older Conciliation Act of 1900. He successfully intervened in forty-two labour disputes in manufacturing, transportation, mining and construction businesses, everywhere from a tool factory in Dundas, Ontario, to a coal mine in Rossland, British Columbia. Still, between 1900 and 1907 there were more than 1,300 strikes and lockouts registered in Canada, so his track record was admittedly modest at best.

King's main goal in solving any dispute was to get the workers back on the job. He strongly believed that, notwithstanding the greed of business owners and the radicalism of union leaders, confrontation was contrary to human nature. Most people, rich or poor, capitalist or worker, could be appeased by an approach based on a winning personality, a sensitivity to the needs of both sides in the conflict and a healthy dose of reason. He was indeed on the right track: modern mediation techniques emphasize strong interpersonal skills, empathy, intelligence and flexibility—and, to a large degree, King possessed these attributes. His later reputation as a prime minister who could smooth over cabinet disputes was mirrored on his earlier success as a labour conciliator. On the eve of the First World War, he maintained that his technique could solve any problem. "There is a relation, a psychological relation between efforts made in one direction and similar efforts made in another," he explained in a speech he delivered in Detroit in 1914. "Accustom men's minds in the industrial world to remedy industrial wrongs by appealing to reason rather than force, and you have helped to create a sentiment which will also play its part in a much larger way in international affairs."

While this sounds reasonable, the reality of King's labour magic was not as transparent. He was successful if the two parties in the dispute were willing to listen to him. With a coal mining magnate and union hater like James Dunsmuir, who also served as the premier of B.C. from 1900 to 1902, negotiation was futile. King regarded Dunsmuir as a "selfish millionaire" and a "tyrannical autocrat," who "has undertaken to make serfs of a lot [of] free men." But he also innately understood, even if he did not always admit it, that industrialists such as Dunsmuir held all of the cards. In any dispute, most of the arm twisting King did to resolve it was done to the workers, who were usually given a "take-it-or-leave-it" option to accept whatever the owners were offering them. King almost always refused to press management to recognize unions, except this was nearly always one of the key issues in any strike. He was not blind to the damage to goodwill and labour standards caused by owners who used strikebreakers, yet King chose to ignore this questionable business tactic as well.

In 1904, in order to settle one of numerous nasty coal mining strikes in B.C., King had to convince the United Mine Workers of America (UMWA) local to give up its demands for an eight-hour day, which was contrary to a provincial labour law for miners; pay $1 a day for their own transportation to the mine located on Protection Island close to Nanaimo; and worst of all, surrender on the issue of collective bargaining by having each miner negotiate his own separate contract. With their backs against the wall—the owners of the mine threatened to shut it down for good—the miners and their union representatives reluctantly accepted King's terrible negotiated deal.

A similar scenario unfolded in Alberta in 1906 with a coal mining strike in Lethbridge. The miners wanted better wages, improved working conditions and recognition for UMWA. Their employer, the Alberta Railway and Irrigation Company, played hardball with a lockout and used inexperienced strikebreakers who had to be protected by the Royal North-West Mounted Police. The company supplied much of the fuel to rural Alberta and Saskatchewan. By November, with a cold winter approaching, Saskatchewan premier Walter Scott, fearing the worst, demanded that the federal government do something. Again, Mackenzie King spoke to the head of UMWA, John Mitchell, as well as Winnipeg-based business tycoon Augustus Nanton, the company's hard-nosed managing director. "I was desirous of being quite fair and impartial and was not unfriendly to either; also to let them see that I appreciated clearly their side of the case," King later reported. "It is one of the first essentials in conciliation negotiations to let each party see clearly that you understand all the things that have aggravated, angered and incensed them; that you appreciate all the obstacles with which he has to contend, leaving to a secondary stage in the proceedings the weakest parts in his position." In the end, the miners had to give in on a number of key issues. They did receive a modest wage hike and a step forward towards collective bargaining, though not full recognition of UMWA.

King always claimed that his experience in Lethbridge and the horrid thought that Canadians might freeze as a result of a labour strike led the Liberal government to pass the Industrial Disputes Investigation Act the following year. Quebec MP Rodolphe Lemieux, who had replaced Mulock (and Allen Aylesworth, who had filled in temporarily for Mulock) as postmaster general and taken over the labour portfolio, received credit for IDIA, but it was King's baby. King's detractors have long argued that he borrowed the key ideas about arbitration boards and "cooling-off periods" from Australia and New Zealand as well as from Louis-Nazaire Bégin, the archbishop of Quebec, who in 1900 had created a board of arbitration to solve problems in the boot and shoe industry in Quebec City. King, too, was greatly influenced by his former professors, William Ashley and James Mavor, who had taught him that the state had an important role to play in the economy and that the "common good" sometimes took precedence over "individual rights."

King deserves credit for bringing it all together, a point that he made privately in his diary as the bill was being sent to the printer in January 1907. "The measure is my own," he wrote. "The ideas embodied in it were mine before they had found expression in legislation elsewhere. So far as I know no country has legislation just on the lines of the main features of the bill... the prohibition of strikes or lockouts prior to and during an investigation... So far as Canada is concerned it does not owe its origins to other countries, their methods or ideas... God has given me the opportunity, has used me as his instrument to frame the measure, and I have done it along lines and in a manner which I believe will further His Will among men."

Laurier, who generally did not give the highest priority to labour matters, did not pay much attention to the bill, which upset King. After the prime minister finally read it, he told King in front of Lemieux and Aylesworth that it was "a splendid piece of legislation." Conservative leader Robert Borden congratulated him as well. King was as giddy as a schoolboy. Thereafter, he spoke publicly about it and was not shy about promoting himself. The *Globe* dubbed him "Canada's Industrial Peacemaker," and his mother started calling him "the Official Harmonizer." Another Ontario newspaper published a light poem entitled "Official Umpire." The ditty went as follows:

> Mary and John could not agree
> Quarrelled with pertinacity.
> Neighbours said, "What a cruel thing!"
> And sent for Lyon Mackenzie King.

Interest in King's work from the United States, Britain, Australia and New Zealand elevated his reputation further and primed his leap to politics.

Beyond the publicity and promotion King received from IDIA, what was its true impact on labour problems? The act was not perfect, nor did it suddenly transform management's inherent distaste for unions. Yet between 1907 and 1911, IDIA was utilized in 101 disputes and brought about a successful resolution in 90 per cent of the cases. In certain instances, according to one study, direct negotiations between union leaders and management might have produced more positive results, yet where workers were not as united and owners resistant, IDIA proved its worth. IDIA shaped future Canadian labour relations, though it enforced compulsory investigation rather than compulsory arbitration, which neither business nor labour would have accepted at the time in any event.

Despite King's best efforts, serious problems persisted. From 1895 to 1914, the Canadian militia had to be called out more than twenty times to quell labour confrontations. There were "bloody battles" between miners and their employers on Cape Breton Island, in Springhill, Nova Scotia, and on Vancouver Island. Unions and workers were also subject to intimidation, spies and blacklists, tactics that continued to 1919, the most volatile and violent year in Canadian labour history, and after. "Whatever King's impenetrable doctrines of conciliation amounted to," states historian Ian McKay, "they barely concealed the crucial new fact that, in defence of capitalism, the state was prepared to kill."

Outside of his office, King moved within the highest social circles of the capital. Any gossip about his relationship with Marjorie Herridge aside, Governor Generals Minto and Grey and their wives found the young deputy minister bachelor intelligent and suitable company. Earl Grey, in particular, was smitten with King and constantly promoted him to Laurier as a political leader on the rise. Grey had used his influence among the royals to secure King his coveted CMG badge in 1906. (The governor general believed fervently in the nobility of British imperialism, but Laurier wished, as he told King in December 1909, "Earl Grey would mind his own business.") King soon wound up on invitation lists for more dances and dinner parties than he cared to attend. "This last week was like the ragged edge of Hell with its fringe of wasted nights," he scribbled in his diary on February 18, 1901. "I will have no more of it." He hated the useless small talk and the late evenings. Ever ambitious, however, he could turn down few opportunities to hobnob with the rich and powerful.

At one of Earl Grey's parties in December 1905, King met Violet Markham, who would become his lifelong friend and correspondent. She was a handsome, charming and compassionate woman, two years older than King. By the time of her visit to Ottawa, she already had a reputation as a dedicated English social reformer, even though, as a member of the Liberal Party, she had initially opposed women's suffrage. Her father, Charles Markham, was a successful coal mining industrialist, and she was independently wealthy throughout her life. She later married James Carruthers, a First World War veteran and avid racehorse owner. Until her death in 1959, there were few welfare initiatives in which she did not participate. King always made a point of visiting Violet on his various trips to Britain, and they spent a lifetime writing each other and debating the significant events of the day.

That memorable day in 1905, Grey introduced King to Violet with the words, "You must meet Mackenzie King, he will be the Prime Minister of Canada some day." Markham was intrigued. "I found myself talking to a pleasant-looking man of medium build, with a round face and abundant fair hair, who smiled and shook his head at the Governor-General's introduction," she later recalled. She thought he was a bit young to hold the title deputy minister of labour, never mind the idea that he would be the prime minister one day. And yet, he did impress her. She regarded Rex, as she soon called him, "a most charming and able young man, full of the right ideals."

Someone was always trying to find Mackenzie King a wife. For a brief moment, King had ideas about squiring Lady Ruby, one of Lord Minto's lovely daughters, but he had no chance. Lady Minto had already written her sister in England to "scan the peerage" for potential husbands for her daughter. In 1907, Lady Ruby married Rowland Thomas Baring, second earl of Cromer.

Laurier and his wife, Zoë, tried to be matchmakers. In the back of King's mind was always the possibility of jumping into politics to continue his grandfather's work. It was another instance of the "hand of destiny" dictating his future, a future also nurtured by his ambition and encouraged by a variety of federal politicians.

As King contemplated a political career, Laurier tried to convince him that a wife was an asset, not a liability, as King insisted. During the summer of 1907, the Lauriers introduced King to Martha Sheriff, twenty-four years old, recently widowed and very wealthy. She was the daughter of George Fulford, one of Laurier's prominent financial supporters and a senator who had made his fortune hawking Pink Pills for Pale People and other dubious patented medicines. (Fulford was killed in a car accident while on a business trip in Massachusetts in 1905. At the time of his death, he was one of the largest shareholders of General Electric stock.) Martha was a great catch; besides being attractive, starting on her thirtieth birthday she was to receive an annual income of $30,000 from her father's estate.

King and Martha met several more times during August, and she taught him how to play poker. Still, he found excuses not to pursue a relationship. She had told him that money was not everything, yet he considered her wealth a detriment rather than a plus. "Unless she can surpass in my judgment any woman I have yet seen as the one with whose nature my nature could best blend," he noted, "I will not allow wealth, position or aught else to tempt me." Several months later, when Martha was staying with the Lauriers again, they urged King to see her, but he had lost interest. In 1909, Martha married a second time, and Wilfrid Laurier walked her down the aisle. A year after that, Martha died in childbirth.

King's reluctance to pursue Martha was shaped by his own multi-faceted self-image. No one could ever accuse him of not thinking highly of himself or his talents. He might have been rocked by a range of insecurities and neuroses, but in public he often reeked of swagger and ambition, combined with a genuine desire to serve his country. Less than a year after he started working in Ottawa, he had a prime seat in the gallery of the House of Commons to hear Laurier pay tribute to the late Queen Victoria. He gushed in his diary about the prime minister's "beautiful oration" and his "pure and deep" thought. Presumptuous as it may have been, he imagined himself at Laurier's side. "As I see the 'calibre' of other men there," he wrote later that evening. "I have no fears as to my abilities if opportunity presented to serve the country well in parliament. For Grandfather's sake I would like to lead, for his sake and the sake of the principles he stood for."

The idea was firmly planted in his head. Some months later, he was travelling with his mother back to Ottawa from Kingsmere by horse and buggy. One of Isabel's relatives had died, and King was taking her to the train station for the trip back to Toronto. He held her tightly for much of the journey. He described the moment like this: "Partly because of the association, partly to comfort Mother, partly to express my own ambition and partly because my soul was large, my spirit strong and resolve great, I whispered to Mother that I believed, that if opportunity came in the future I might become the Premier of this country. She pressed my hand and said nothing, then said that perhaps I might."

King never lacked personal drive; that he possessed always. Nor did his fussy and formal public demeanour hinder his political aspirations. That style played well in Edwardian Canada; his press clipping files from those years are stocked with stories about King's "inspiring" and "eloquent" speeches, mainly on labour matters. Audiences genuinely found him an appealing speaker. In his corner, too, at least by his count, were the twin spiritual powers of God and destiny. Still, turning his political dream into reality proved more challenging than he anticipated.

In 1907, King's public stature got a boost from his investigations into immigration problems and concerns, specifically those dealing with immigrants from China and Japan. In the early twentieth century, few issues frightened the white Anglo-Saxon Canadian majority more than unwanted immigrants from the non-English, non-Christian world. Laurier's minister of interior from 1897 to 1905, Clifford Sifton, had sought out immigrants, even ones from Eastern Europe, provided that they were prepared to work hard, settle the land and adopt prescribed "Canadian" customs and values. To Sifton's dissatisfaction, however, many of the newcomers were not all "stalwart peasants in sheepskin coats," as he desired, but also Jews, Slavs and Asians who remained in cities, contributing, it was argued, to the rising rates of poverty and crime. Popular writers, journalists, religious leaders and politicians warned that the country was being occupied by "foreign trash," "heathens," "vermin" and "foreign scum."

At the top of the list of undesirables were the Chinese, followed closely by the Japanese. Both groups represented the "unassimilable Yellow Peril." Chinese labourers had sacrificed themselves to build the Canadian Pacific Railway—it is said that one Chinese worker died for every mile of track laid—and immigrants from the Far East were routinely used in British Columbia and elsewhere as cheap strikebreakers. Yet that did not mean they were welcome to stay on. "I have very little hope of any good coming to this country from Asiatic immigration of any kind," Laurier put it succinctly in a letter of April 1899.

Mackenzie King and a majority of politicians, no matter the party or affiliation, would have wholeheartedly endorsed such a sentiment. King worried about "race degeneration" as much as the next white Canadian eugenicist in the years before the First World War. Slightly more enlightened, he did temper that fear with the notion "that the future of white races will depend largely on the extent to which the black and yellow races become educated and prove their capacity for education." Yet it was "natural," he argued in a 1908 report on Oriental and Indian immigration, that "Canada should desire... to remain a white man's country." Was King a racist? By today's standards, the answer would be a loud yes. But the same assessment could be levelled at such diverse leaders as Laurier, U.S. president Theodore Roosevelt and even the saintly social gospel advocate and labour leader James S. Woodsworth. Each was guided by standards and values that dictated the white race supreme and assimilation the norm. Tolerant Canadian-style multiculturalism was decidedly not an option.

Whatever the complaints about the Chinese and Japanese destroying the cultural fabric of the country, it was not easy for them to reach Canada under any circumstances. Chinese immigrants faced a notorious head tax that was increased to $500 in 1903. Due to Britain's ongoing efforts to enhance its relationship with Japan, immigrants from Japan did not have to pay a head tax, but they confronted the same prejudice and discrimination, especially in B.C., where the Asian populations tended to be slightly larger than in the rest of the country.

Vancouver-based immigration companies, some with Japanese connections, figured out ways to bring in Japanese non-unionized labour. More Japanese immigrants reached B.C. via Hawaii in early 1907 following an outbreak of serious illness on the Hawaiian Islands. The

East Indian population on the west coast surged as well, leading to heightened tension. That July, disgruntled members of the Vancouver Trades and Labour Congress formed the independent Asiatic Exclusion League. Vancouver Liberal MP Robert Macpherson hysterically warned Laurier that the province was "slipping into the hands of Asiatics."

The fear and hatred boiled over on the evening of September 7, 1907. An anti-immigration parade and rally promoted by the Asiatic Exclusion League got out of control. A riot broke out as the white crowd marched into Vancouver's Chinatown and Little Tokyo, smashing windows and destroying property. The Japanese fought back, and it took hours for the Vancouver police to quell the violence. Amazingly, no one was killed in the melee, but the property damage was in the thousands of dollars.

Since this was perceived to be strictly a labour matter, Mackenzie King was dispatched to Vancouver to investigate the causes of the riot. At first, he thought the violence was the result of "race agitation," but later he decided that it had been caused primarily by an increase in the number of Asian immigrants. He was probably right on both counts. He awarded the Japanese victims of the riot a total of $9,100 to cover the costs of property damages. Chinese victims were initially excluded from the government compensation, though eventually they received $26,000.

Several Chinese opium dealers in Vancouver submitted property claims to King, which led him to investigate the extent of the opium trade in the city. He visited opium factories and dens and was shocked by what he found. In 1908, acting on King's recommendations, Parliament passed the Opium Act, which prohibited "the importation, manufacture and sale of opium for other than medicinal purposes." Canada was the first country in the Western world to enact such legislation. In 1909, King was Canada's representative in Shanghai at the international convention on opium. Three years later, as minister of labour, he introduced an even more restrictive measure, the Opium and Narcotic Drug Act, to prohibit the improper use of opium and other drugs, but notably not tobacco or alcohol. Possession of opium was now a criminal offence. Police were given wide powers of search and seizure and immediately targeted Chinese immigrants, who were believed to be the main users of the drugs. For many years, as historian Stephanie Bangarth points out, convictions were listed in official reports in two columns, one for "Chinese," the second one for "Others."

Further inquiries King made into the methods by which Japanese immigrants were enticed to come to Canada convinced him that the Japanese government was guilty of violating international immigration agreements. He had drawn the wrong conclusions from the evidence he discovered at Vancouver immigration agency offices. But his work in B.C. still received extensive press coverage. Almost overnight, he became the country's Asian immigration expert and was responsible (along with his superior, Rodolphe Lemieux) for new and tougher regulations governing Japanese and East Indian immigration to Canada.

In 1909, Harvard University, in a creative interpretation of its own academic regulations, accepted King's expanded 1908 report on Oriental and Indian immigration as his doctoral dissertation. He remains the only Canadian prime minister to hold such a high degree.

King's stringent stand on Asian immigration involved him briefly in an intrigue with U.S. president Theodore Roosevelt in 1908. Roosevelt, who was also no supporter of Japanese newcomers, perceived Japanese imperialistic designs in the Pacific region, possibly even the conquest of California and B.C. This diplomatic adventure took King back and forth from Ottawa to the White House and then to England as Roosevelt slyly manipulated King into lobbying British officials to enact an Anglo-American agreement to block Japanese immigration. To ramp up the urgency of his back-channel task, King was deliberately led to believe that the possibility of a war between the United States and Japan was imminent. Edward Grey, the British foreign secretary, eventually clarified the situation for King: Japan had no plans for either a war with the U.S. or a large-scale invasion of North America.

King's own sense of self-importance would never have convinced him that he had been used as an unwilling pawn in Roosevelt's political machinations. And even if he had accepted what had transpired, he would not have been troubled by it. King derived tremendous personal satisfaction whenever he was permitted to enter the puffed-up world of the British and American elites. It delighted him to no end to be treated as an equal by the U.S. president and his secretary of state, Elihu Root, and to discuss the great issues of the day with such British luminaries as Winston Churchill, Andrew Bonar Law, David Lloyd George and Ramsay MacDonald.

More crucially, King's journey to England allowed him to work out his own feelings on British imperialism and Canada's role in the empire—a weighty issue that was later to occupy much of his attention as prime minister. Like Laurier, King strongly believed that Canada owed Britain, the mother country, economic and military loyalty, but always Canadian interests had to be first. In King's mind, Canada had to mature from the junior partner it was in the nineteenth century to a supportive yet autonomous player in the affairs of the British Empire.

If anything, King was realistic. "There is in England a real 'Governing' class, in the sense that it seeks to control and actually does control and guide the national interests both in England and in the Dominions beyond," he recorded in his diary. "The English mind has been so long trained to this way of looking at the world that I can see wherein it will be many years before it will ever come to fully appreciate what self-government means." He also grasped the significance and danger posed to Canadian unity by the anti-imperialist campaign mounted by French-Canadian nationalist Henri Bourassa in his newspaper *Le Devoir*. This, too, was to cause him much concern in the future.

KING RETURNED FROM England confident that he had proven to Laurier that he was ready for the House of Commons and a cabinet position as the country's first true minister of labour. Still, the prime minister refused to be pinned down on where King should run. Instead, Laurier strung him along, dangling the proverbial political carrot just out of King's reach. He knew which of King's buttons to push. In the spring of 1906, King had been offered a full-time professorial job at Harvard's new Graduate School of Business Administration. He was momentarily tempted to leave Ottawa and accept the academic position. "Sir Wilfrid told me," he had written to his parents, "not to think of it, that a man with the blood I had in my veins, my talents, etc. had a great future in this country and that I would be lost as a professor."

A year later, an article published in the Toronto *News* (whose editor was King family friend John Willison) suggested that in the next election King was to run as the Liberal candidate in the Toronto riding of North Oxford. That might have been fine, but the writer also touted him as the future of the party and criticized Laurier's aging cabinet members. Local Liberal Party brokers in North Oxford were not pleased with the rumour that King intended to run in their constituency without their consent. And Liberal cabinet ministers were insulted at King's impertinence. He was forced to make the rounds in Ottawa apologizing, and Laurier had no choice but to delay any announcement about King's political future. That sent King into a slight panic, though he again took comfort that God and destiny were on his side. He also learned an important lesson about talking too much to reporters and thereafter tended to be overly cautious in his dealings with the press.

In early September 1908, Laurier called an election. By then King had smoothed things over, and it was agreed that he was to run in his hometown riding of North Waterloo, then held by the Conservatives. King secured the Liberal nomination at a party gathering in Berlin and immediately resigned as deputy minister of labour. The *Globe,* among other supporters, hailed this development not as one of opportunism, but rather as one of high principles and strong convictions.

At a crowded outdoor event in Berlin on September 24, with King proudly on the platform and his family in the audience, Laurier announced to great cheers that he would create a separate labour department and more or less implied that if victorious, King would become its first minister. Again, the *Globe* and other Liberal papers applauded King's speech that day. "The brilliant young administrator... [had made a start on] a political career which it is safe to say will be crowned with much distinction." For the duration of the campaign, the adjectives "brilliant" and "young" were regularly used in any story about King.

A week before the vote, King was indeed brilliant and at ease in a debate in Berlin. With six thousand people watching, he defended his work in the labour department and bested Alexander Wright, the editor and publisher of the journal *Labour Reformer* and a supporter of King's Conservative opponent, Richard Reid (a high school principal who had once taught King). "It was a most conclusive victory," declared the Liberal-leaning *London Advertiser,* "even the most rabid Conservative having to admit it."

It was the same story the day of the general election: Laurier and the Liberals won 133 seats to the 85 won by Robert Borden and the Conservatives. Mackenzie King won in North Waterloo by a margin of 271 votes. The *Montreal Herald* claimed that the "splendid result" in North Waterloo was due to King's contagious enthusiasm, "splendid executive ability" and "extraordinary" capacity for work. Owing to the parliamentary schedule, and a delay in creating the new labour department, King was not sworn in as a cabinet minister until June 2, 1909. He had to wait another five months, until November 11, to take his rightful place in the House of Commons.

In the days leading up to his inaugural appearance in Parliament, King became embroiled in a protocol argument with the governor general's staff revolving around his mother. It was tradition that the governor general and his wife hosted the State Drawing Room ceremony in the Senate chamber on the Saturday night following the opening of Parliament—in this case, on November 13. Custom had it that cabinet members and other high dignitaries were invited to bring their spouses or unmarried daughters to be presented first to Their Excellencies. King, ever the loyal son, insisted that he escort his mother. The governor general's military secretary, Sir Jonathan Hanbury-Williams, attempted without success to explain the protocol to King. He would not be placated, arguing that his mother was too frail to stand with the larger crowd of journalists and other invited guests and that if his request was not granted he might not attend. King told him in no uncertain terms that "my mother was more to me than all else in this world and that having asked her to come with the sensitive nature she had, I could not let her be the last to be presented, it would look as though I was ashamed of her." King also was dismayed to learn that Lord Grey, whom King regarded as a friend, had instructed his secretary to "stick to his rules."

He decided to show the governor general and everyone else concerned a lesson: "that there is something greater and 'truer' and deeper than 'rules' where 'rules' make a man defy the commandment to honour his father and mother," he scribbled angrily in his diary. "My mother will be more beautiful than any woman there, and will see whether 'rules' or 'worth' will have any precedence after all." In an act of spite and to draw attention, King decided to wear his Windsor uniform complete with his CMG and wait with his family so that they could enter all together. His lively conspiratorial imagination blamed "Toryism" and the "Family Compact" for striking at the descendants of William Lyon Mackenzie once more.

That Saturday evening, King did as he promised, purposefully delaying his entrance and marching in at the front of a large crowd entering the Senate chamber. "I received a pretty cold look from both Their Excellencies," he noted with glee, "who must have realized that my action was intended as a protest."

In contrast, the ceremony in the House of Commons two days earlier had been much smoother and a grand moment for King and his parents. With John and Isabel peering down from the gallery, King was led into the House by Laurier on one arm and on the other by George Graham, the minister of railways. As the House swelled with applause, King shook hands with the speaker and took his seat on the government benches. He was overcome with emotion, all duly recorded in his diary. "The first faces I saw were those of my mother and father and my heart uttered a prayer of gratitude to God for his goodness in having spared them to witness this day," he wrote. "They both look very happy and were so fine looking I was more proud of them than of all else. Next to my pride in them, is my pride in the thought that I should be one of Sir Wilfrid's chosen friends."

After the session ended, he joined his parents in the speaker's apartment for a private reception. He was delighted at how enchanted his mother appeared and was caught up in the historical significance of the event. "They all remarked on her beauty and said 'how distinguished.' Surely there is reward in this for her as well as me, reward for the sacrifices her father made and for what she has had to make in consequence. If her father could only have been present, too, I would have asked for nothing more. He would have felt a recompense for all his struggle. His life, his work are all kept alive in this way, and that I feel to be my chiefest [*sic*] part. Laurier in introducing me gave the name in full 'William Lyon Mackenzie King' and in my ears it seemed to fill the entire room."

The next day, he uttered his first words in Parliament: "Mr. Speaker, I beg to leave to move to lay on the table of the House the report of the Department of Labour for the Fiscal Year ended March 31, 1908." King thought it was a "beautiful" moment that the first words he spoke were about "work which has been largely of my own making." He conceded that others had also contributed to the development of a separate labour department, but without his efforts, he felt it never would have happened. "To present this as the Minister of the Department," he added, "after all that it has meant to gain such recognition is itself a triumph which is worthy the effort of a life. It has all worked out as foreseen years ago."

3

Industrial Prophet

*Of one thing I am certain, and that is that the accepted
practices incidental to the accumulation of wealth under
existing conditions are wholly wrong, and that my part
in life must be along the direction of exposing what is
in error fundamentally, and in helping bring about such
changes in the relations of Labor and Capital as will
secure greater justice to the worker.*—THE DIARY OF
WILLIAM LYON MACKENZIE KING, *January 25, 1919*

MACKENZIE KING'S stint as labour minister and his first
term as an MP lasted slightly more than two years. He dedicated himself
to improving the lot of working men across the country but confronted a
variety of obstacles. His efforts to establish a uniform eight-hour day as
well as enhance workmen's compensation were limited mainly because
these matters fell within provincial jurisdiction.

The most significant event of King's tenure as labour minister, which
also had mixed results, occurred during the Grand Trunk Railway (GTR)
strike of 1910. King's work to settle the strike was later celebrated by
William Murray, a successful Hamilton merchant known for his colourful poetry and called "The Bard of Athol Bank" (the name of his estate).
"Congratulations, with a ring; Once more to great Mackenzie King," the
poem began. "We all today his health have drunk, and I create him
Lord Grant Trunk." King was touched by the sentiment; the reality was
not so black and white.

The GTR strike pitted the railway men's American-controlled union,
which wanted to alter their pay structure to reflect the miles they travelled, against the company's tough American executive, Charles M.
Hays, and an international board of directors. Hays detested unions
and collective bargaining and refused to budge or co-operate with a

government-appointed conciliation board (headed by King's friend Joseph Atkinson, the editor of the *Toronto Star*) investigating the dispute. King was compelled to intervene personally in the negotiations, and Hays strung him along like a puppet, pushing (or rather, manipulating) him to get the union to compromise. King's thirty-eight-page report to Laurier at the conclusion of the strike suggested that Hays had bargained in good faith, and that King had arranged a settlement that both sides agreed to. This is a highly selective interpretation, as several labour historians have convincingly shown. King did in the end effect a deal, though strictly on Hay's terms; the union received a wage increase only because Hays wanted to give one. On such contentious matters as strikebreakers, pension rights and reinstatement of strikers, the union got next to nothing. The striking men who returned to work and lost their pension rights did not get them back until the GTR was nationalized in the early twenties.

Privately, King was furious at Hay's actions with respect to the pension issue. His assessment of the executive's mindset was astute and revealed the insight he had gained during the past decade in Ottawa. "Railroading in the United States is a business which with a certain school of men is run on certain principles," he reported to Laurier. "One is that human life, to say nothing of human feelings, is not to be considered, either as respects its loss through accident or its massacre as a means to an end. The end is power of money as against all other powers in the world. To admit the solidarity of labour in any industrial struggle is to admit something more powerful than money, and that must not be done, no matter how great or tremendous the cost. Mr. Hays has seen himself in this struggle as the chief representative of that school."

King's ego prevented him from seeing the resolution of the GTR strike as anything but a success. If there were flaws in the negotiated settlement, it was the fault of "unscrupulous corporations," as he confided to one friend. In truth, IDIA did not give King the effective and firm tools he required to solve a heated confrontation with a businessman like Hays, who was so inflexible.

Labour voters in North Waterloo knew better. Laurier called another election for September 21, 1911, over his reciprocity or free trade deal with the United States. King was attacked from all sides. He had to defend his role in the GTR strike with his riding's working-class constituents and faced anger from the local business community, which opposed the controversial trade agreement. The Conservatives, including his North Waterloo Tory opponent, William Weichel, a former mayor of Waterloo, kept both issues at the forefront of the campaign.

The 1911 federal election campaign was one of the most notable in Canadian history for the near hysteria over the American threat to Canada it engendered (which was repeated during the 1988 federal election over the Canada-U.S. Free Trade Agreement). The charge was loudly made, especially in Ontario, that the reciprocity treaty was merely the first step towards possible annexation of Canada by the United States— though there was never a real possibility of that happening. Nevertheless, Laurier and the Liberals, who had already been criticized for their decision to create a small ("tin-pot") navy, rather than making the emergency donation to Britain that the Conservatives wanted, were accused of abandoning the empire. On top of that, eighteen influential pro-Liberal Toronto businessmen, including former Liberal cabinet minister Clifford Sifton, publicly rebuked Laurier for his economic policies and declared their support for the Tories.

On election day, Robert Borden and his Conservatives were victorious. Mackenzie King lost his own seat by 315 votes and was now without a job. He did learn from that election, as Jack Pickersgill has observed, "how deep-seated was the Canadian fear of the United States and how easily it could be aroused." Yet in the immediate aftermath, he did not believe that reciprocity was the sole reason for the Liberal loss.

In a long letter King wrote to Violet Markham in mid-December, he pointed to the general ignorance about reciprocity and the more serious impact in Quebec of the French-Canadian nationalist movement, which had opposed Laurier's stand on the Canadian navy. In Ontario, however, "it was largely the Anti-Catholic and Anti-French prejudice" that caused people to perceive Laurier as too much of a Quebecker. "This and an immense campaign fund [of the Conservatives] contributed to by all the interests," he wrote. "These were the real factors." Historians might quibble with his post-election evaluation, given the major focus on free trade, in addition to the desertion of key Ontario Liberals to explain the change in power in Ottawa. Many, like King, also emphasize the nationalist backlash in Quebec against the navy as a primary issue.

For his part, King had some time to contemplate the fickleness of the electorate and the vagaries of Canadian politics. Three days after his defeat, he wrote in his diary that his "sense of relief from the strain of office is almost indescribable... I cannot but believe that defeat is all for the best, and that out of this will come opportunities of a fuller and better life." If anyone had the ability to bounce back, it was Mackenzie King.

WHAT NOW? KING might have been truly delighted by his new-found freedom, but into the wee hours of the night he still scribbled feverishly in his diary about the ramifications of his defeat. He had responsibilities and obligations. For at least the previous five years, he had rightly regarded himself as the head of his family and certainly their main source of income. When his brother, Max, who had become a physician, fell ill while working in a mining camp in Arizona, it was older brother Willie who made the arrangements to bring him back to Toronto. When Jennie renewed a friendship with Harry Lay, a recent widower (Jennie and Lay had first dated in 1899), and accepted his proposal in February 1906, King was angry that his advice on the marriage had not been sought. This issue, as well as others, strained his relationship with Jennie for years.

Now out of work, how was he to continue to support his parents and family? His father was suffering from glaucoma and his mother, demanding as ever, was aging rapidly. And the health of two of his siblings, Bella and Max, was poor and deteriorating fast. In the fall of 1911, King found himself for the first time in more than a decade without a solid paying job or the status of possessing an office on Parliament Hill. At first, he did what came naturally: he fretted and worried incessantly about his future. Then, he took comfort that God's plan for him would proceed accordingly. It was out of his hands, or so he believed.

Joe Atkinson, editor of the *Toronto Star* and an avid Liberal Party supporter and confidant, offered King a job writing editorials at an annual salary of $3,000. King turned him down despite his concern over his personal finances, opting instead to write occasionally for the New York–based magazine *Outlook*. What he truly desired was another chance to win a seat in the House of Commons, but that depended on Laurier. The stately Liberal leader, who after fifteen years in power returned reluctantly to the Opposition benches, refused to put King at the head of the line. Laurier remarked to Rodolphe Lemieux's wife, Berthe, that King should be more patient and that at thirty-six years old he was "in too much of a hurry." As the wonder boy of Ottawa, King, who as one Toronto journalist recalled had an "irrepressible vitality" in those days, also came across as being slightly arrogant, a trait that did not endear him to other members of the Liberal caucus.

No matter the reason, King, whose natural hypersensitivity was now especially heightened, was upset at what he perceived to be Laurier's lack of concern over his career aspirations, and it would not be the last time he felt this way. "One feels one never gets very close to him and that binding sympathy is hardly part of his nature and yet he is loyal to his friends," he noted in his diary about the Liberal chieftain in mid-December 1911. "He can be indifferent to all personal considerations as few men can. I feel a little hurt and disappointed in this, as it makes me realize he is somewhat indifferent as to whether I get into the House or not."

Two years later, he was anointed the Liberal candidate for the Toronto riding of North York, the same area his grandfather had represented. But the outbreak of the First World War in August 1914 would prolong Parliament, and King would not get an opportunity to run again until 1917. Another possibility arose when Atkinson and one of his influential friends, tea magnate Peter Larkin, dangled the leadership of the Ontario Liberal Party in front of him. The position came with an annual salary of $5,000, a $1,500 indemnity and a railway pass. As tempting as this offer was, King felt that his grasp of provincial issues was not sufficient. And more to the point, he was certain that his destiny was in the federal arena. "Somehow I believe God has a great work for me in this Dominion," he jotted, "maybe at some time, to be its prime minister, but to attain this end I must feel myself worthy of it." For the moment, he helped set up a national Liberal information office, started writing Liberal propaganda and in September 1913 edited the first issue of the *Canadian Liberal Monthly*.

His personal finances were better off than he was willing to admit. His savings exceeded $10,000, though he remained distressed about his predicament. In England, his friend Violet Markham sensed the despair in his letters. She sent him £200 after he lost the election and then at the end of December 1911 offered him £300 a year for the next three years (the equivalent today of about $25,000 annually). "So dear Rex," she told him, "for three years feel you are free to work and to rest, to lie fallow and study and read and think without your life being obsessed by the harassing cares of daily bread."

It wasn't books and bread that King worried about so much as finding a wife. His loneliness during this period (and, to a certain extent, for the rest of his life) was often palpable; the anguished references in his letters to his closest friends and in his diary were constant. At first, he was philosophical about it: "I feel I should seek to become married. If I can only find the one that will be the helpmate needed through life I will certainly marry," he recorded on September 24, 1911. Nearly nine years later, with nothing settled and no woman on the horizon, he was more despondent. After visiting his friend Sam Jacobs in Montreal in January 1919 (a prominent Canadian Jewish leader, Jacobs became an MP in 1917 and then served under King until 1938) and watching him interact with his young son, King was envious. "To go into politics without marrying would be folly," he wrote. "I cannot live that cruel life without a home and someone to love and be loved by... Marry then I must."

No one could accuse King of being passive in his search for a spouse. He dated, and then dated some more. On each occasion, however, either he or the young lady walked away. King knew what he wanted. His ideal woman had to be beautiful, of sound moral character and possessing that rare quality of decency he had seen in only a few people. His mother, of course, was his archetype. Isabel, noble and pure in King's eyes, remained for all time the one woman against whom all others were judged; the pedestal he had placed her on reached the heavens. King's quest for a wife was thus doomed; nevertheless, he persisted. He especially favoured the daughters of wealthy Canadian or American businessmen. The list included Miss Fowler, whose father, Thomas, was the president of a New York–based railway; Miss Howard of Montreal, the granddaughter of Lord Strathcona; Miss McCook, also of New York, the daughter of the noted New York lawyer Colonel John J. McCook; Miss Mather of Cleveland, whose father Samuel was a coal and iron magnate; and Miss Stirling, the daughter of a prominent Chicago lawyer. He even had momentary thoughts of pursuing a daughter of tycoon and philanthropist Andrew Carnegie.

The pattern was similar with each. After one or two dates, King fell head over heels in love, certain that he had found his perfect companion. Typical was Miss Fowler, who had worked as a volunteer in Labrador with the medical missionary Dr. Wilfred Grenfell. After spending an evening with her at a dinner party hosted by Governor General Lord Grey, King noted that "she possesses all one can desire." But she soon disappointed him when after a few weeks she did not read the copy of his book, *The Secret of Heroism,* that he had given her as a gift.

Miss Howard was too "nervous" and made him feel ill at ease. "There is something lacking," as he put it. With Miss Mather, he waited too long to make his feelings known, and by the time he did she was engaged to someone else. His reaction was classic King ambivalence. "It was a shock for a moment or two," he wrote when he learned of Miss Mather's plans. "I felt that I had allowed to pass one of the great opportunities of my life, but also that, perhaps, it was as well, as I had never been too sure that it was the path to happiness."

KING MAY NOT have found the perfect woman, but his job prospects suddenly improved with a unique offer from the unlikeliest of employers. In early-twentieth-century America there was no family bigger, richer, more envied and more reviled than the Rockefellers. The patriarch, John D. Rockefeller, Sr., had been one of the key founders of the Standard Oil Company in 1870 and gradually gained control of it as demand for oil, gasoline and kerosene increased to previously unimagined levels. So, too, did Rockefeller's wealth; in time, he became America's first billionaire and certainly one of the richest people in the world. Muckraking journalists like Upton Sinclair and Ida Tarbell—whose celebrated 1904 history of Standard Oil exposed many of Rockefeller's questionable business practices—attacked him for being a bloated, overbearing plutocrat, and U.S. anti-trust laws broke up the corporation in 1911 into more than three dozen different parts. But that hardly bothered Rockefeller; he maintained a controlling interest in a majority of the new companies. "God gave me my money," he once said, adding, "I believe the power to make money is a gift from God... to be developed and used to the best of our ability for the good of mankind." Such a philosophy was instilled in his children, most notably his only son, John D. Rockefeller, Jr., who inherited his father's mantle as head of the family firm. He established the Rockefeller Foundation in 1913, one of the great and most generous philanthropic organizations of the past century.

Rockefeller, Sr., and his wife, Laura, subscribed religiously to Andrew Carnegie's level-headed "Gospel of Wealth," which deemed that one should live a modest lifestyle, provide reasonably yet not excessively for children and channel surplus revenue into philanthropy for the good of the community at large. A pious and God-fearing woman, Laura Rockefeller drilled into John, Jr., and his four older sisters that the only thing that mattered was duty and hard work. "I am so glad my son has told me what he wants for Christmas," she said famously one December when John was still a young boy, "so now it can be denied him."

Following a liberal arts education at Brown University, young John was given an office at Standard Oil's headquarters on Broadway in New York, where he was left to fend for himself. After his father retired in 1897, he was, at the age of twenty-three, technically in charge, though few decisions were made without his father's approval. Junior, who might have been taken for a country pastor, was "doggedly earnest," as one of his biographers put it. He did not drink, smoke or play cards. He worshipped his mother, was tight with his money, keeping track of every penny he spent (something demanded by his father), and was obsessed with fulfilling his Christian duty. In short, he was a near clone of Mackenzie King, except in his case, his mother had insisted he marry his sweetheart, Abby Aldrich, the daughter of a senator, in 1901—and he naturally obeyed. Once their lives intersected in 1914, Rockefeller, Jr.

(hereafter referred to as Rockefeller), and King immediately connected on a spiritual level that cemented their close friendship for more than three decades.

The path that brought King into Rockefeller's tight-knit circle began more than two thousand kilometres away, in the coal mines of southern Colorado. On September 23, 1913, close to nine thousand miners walked off their jobs at the behest of the United Mine Workers of America, a hard-line union demanding recognition as the men's chief representative, higher wages and better working conditions. The miners left the company camps and relocated with their families to nearby tent colonies set up by UMWA. Coal mining in Colorado was in the hands of an elite group of companies, the largest of which was the Colorado Fuel and Iron Company (CFI). The Rockefellers owned 40 per cent of CFI's shares, but they purposely left the day-to-day management of their Colorado operations to the Denver-based head office. In turn, CFI's two key executives, CEO Lamont Bowers, a septuagenarian old-school businessman right out of a Dickens novel (and a Rockefeller relative), and company president Jesse Welborn, who shared Bowers's virulent anti-union views, left the real running of the mines to a collection of unscrupulous overseers, supervisors, pit bosses and store managers.

The miners, many of whom were non-English-speaking immigrants, were at the mercy of favouritism, bullying, corruption and arbitrary rule, as Mackenzie King eventually learned from his personal investigations. The men and their families were pretty well hostages in desolate and unkempt company-owned mining camps, where every aspect of their miserable lives—whom they could speak to, where they could purchase their food and when they could come and go—was rigidly controlled. Any talk of union organization was ruthlessly suppressed. Back in New York, in a classic example of wilful blindness, Rockefeller knew none of this, or chose to accept the biased reports sent to him by Bowers and Lamont that the men were, in fact, satisfied, and the strike was the work of subversive union leaders.

As the strike dragged on and feelings on both sides grew more hostile, a violent tragedy was inevitable. The miners armed themselves and faced off against the combined force of private detectives hired by CFI and the other companies as well as the Colorado National Guard, few of whom had any sympathy or respect for the strikers or their labour and legal rights. The bloodiest moment occurred near the town of Ludlow, three hundred kilometres south of Denver. In the wee hours of April 20, 1914, someone, either a striker or a state guardsman—it is not known for certain—fired a shot. Before the shooting stopped that day, ten men and a child were dead. During the melee, the members of the state militia exacted their revenge on the miners by setting the tent colony on fire. A day later, there was a gruesome discovery: two women and eleven children had desperately tried to escape the carnage by hiding in a bunker dug under one of the tents. They died of asphyxiation and quickly became the victims of the "Ludlow Massacre," as UMWA dubbed it. The miners exploded in rage, attacking and taking over the town of Trinidad. In Washington, President Woodrow Wilson had no choice but to send in federal troops to re-establish order, a military action that took ten more days and led to at least fifty more deaths.

The strike lasted another seven and a half months until UMWA had no more financial resources to carry on the battle. Initially, little was gained. UMWA was not recognized or accepted as the miners' collective bargaining agent. Moreover, many of the striking miners lost their jobs. As far as the union leaders and many federal politicians were concerned, however, the person most responsible for the turmoil in Colorado and the one with blood on his hands was John D. Rockefeller, Jr. He was denounced in the press, angry mobs picketed in front of the company office on Broadway and the saintly Helen Keller described Rockefeller as a "monster of capitalism."

Rockefeller was at first bewildered by these verbal assaults. In public statements and in later testimony before a congressional committee and the United States Commission on Industrial Relations—chaired by Senator Frank Walsh of Missouri, a lawyer who fancied himself a champion of the underdog and was famous for once defending the notorious outlaw Jesse James—he stubbornly maintained that he had nothing to do with the management of the company. He said he felt badly about what had transpired at Ludlow and elsewhere, yet he insisted the blame lay elsewhere. That shirking of responsibility haunted him for a long time. "Those who listened to him," wrote the journalist Walter Lippmann after witnessing one of Rockefeller's public performances, "would have forgiven him much if they had felt they were watching a great figure, a real master of men... But in John D. Rockefeller, Jr., there seemed to be nothing but a young man having a lot of trouble, very much harassed and very well meaning. No sign of the statesman, no quality of leadership in large affairs, just a careful, plodding, essentially uninteresting person... He has been thrust by accident of birth into a position where he reigns but does not rule."

If anyone needed a new public relations strategy in 1914, it was John D. Rockefeller. He and his advisers decided they required professional help. They recruited Ivy Lee, a confident thirty-seven-year-old press agent with Southern roots and a flair for the dramatic. While Lee is credited with being one of the pioneers of the modern public relations industry, his cute tactics—mass mailings that cleverly distorted the story of the Colorado strike and promoted the positive image of the Rockefellers—did not always have the desired effect. Still, Rockefeller was impressed enough to hire his services on a full-time basis.

Next, they went looking for a labour expert, someone who truly understood the new dynamics of management-worker relations and possessed the talent to translate intellectual theory into practical action. Was it possible, Rockefeller and his entourage wondered, to implement a labour policy that would placate their many workers without necessarily accepting a radical union leadership? Jerome Greene, a key Rockefeller adviser, wrote a letter to Charles W. Eliot, the president of Harvard University, that in June 1914 led the Rockefeller Foundation to Mackenzie King's doorstep.

"HOW TERRIBLY BROKEN down on every side is the house of life around me!" King exclaimed in the pages of his diary in mid-May 1914. Only weeks later, waiting for destiny to embrace him, he received an invitation to attend an all-expenses-paid meeting in New York City at Rockefeller's West 54th Street mansion. He was elated and curious about the opportunity presented to him, but equally cautious. Right from the start, King set out his personal philosophy, which was to guide his work for the foundation during the next four years. He explained, as he later recalled, "my experience in personally intervening in industrial disputes had led me to believe that purely economic questions were easily adjusted, that it was the personal antagonisms and matters arising out of prejudice and bitterness and individual antipathies which were the ones which caused the most concern."

At the same time, he worried incessantly about how this public connection with the Rockefellers and U.S. big money interests would impact on his political career, which he expected would eventually turn around for the better. He sought advice from all of the usual suspects: William Mulock, Wilfrid Laurier, Violet Markham and his family. His brother, Max, now stricken with tuberculosis, wrote him from a sanatorium in Denver with a dire warning: "With your clean-cut conscience, it cannot possibly be lasting and it will brand you with the labouring classes for the rest of your life, whether you deserve it or not, as being a tool of the Rockefellers." Violet Markham was more circumspect. "It all depends on the terms, Rex," she wrote. "Nobody must be able to say Rockefeller has bought you and you will speak with his voice in future."

That last bit of wisdom made the most sense to King. Given that another election was at least two years away, he could not possibly turn down a consulting job that paid an annual salary of $12,000. A compromise was worked out. King was hired by the Rockefeller Foundation, rather than Standard Oil, to undertake a major study of industrial relations. He did not have to move to New York, but could travel back and forth from Ottawa. And most important of all, should a federal election be called, he was free to return to his political life. Therefore, he could, and did, claim to anyone who inquired that his new employment was akin to private academic research. He was, he insisted, definitely not the flunky of the most celebrated capitalist in the world. Yet it was also impossible for King, who aimed to please and was acutely aware of Rockefeller's wishes, not to try to win acceptance. Understandably, then, he was extremely careful not to reveal information—and in his investigations he found plenty of it—that confirmed that Rockefeller was indeed a cold-hearted capitalist. King was so paranoid about compromising Rockefeller's position that he used coded telegrams and communicated with Rockefeller and his chief advisers through letters from his young secretary, Fred McGregor, to Rockefeller's executive assistant. On one occasion, before Rockefeller had to testify before the U.S. Commission on Industrial Relations in late May 1915, King gave instructions that Rockefeller was not to ask him what he had learned during his recent trip to Colorado until after his testimony was given. That way, Rockefeller could plead ignorance if the truth ever got out.

Although Rockefeller always addressed King as "Mr. King," and King used the same formalities, the two men instantly bonded. They were the same age, shared the same nineteenth-century moral mindset, were determined to fulfill their Christian duty to mankind and saw the world through a narrow religious lens. The "unseen hand" of destiny governed their actions and, in the words of American historian Howard Gitelman, they "imagined their lives directed by Providence." Reflecting on their relationship many years later, Rockefeller said about King, "Seldom have I been so impressed by a man at first appearance." Rockefeller's official biographer suggested that King was "the closest friend he ever had." Likewise, King raved in his diary about Rockefeller, saying that he knew "of no man living whose character I more admire."

Still, as the labour expert, King could see what Rockefeller did not; neither was he shy about speaking his mind to him in a way few others did. At a meeting in early January 1915, King laid out the problem in black and white. "I repeated... to him," he remembered, "that there appeared no alternative as far as he was concerned to his being either the storm centre of a great revolution in this country or the man who by his fearless stand and position would transfuse a new spirit into industry." Quite soon, the two men were exchanging views on a wide range of personal issues beyond questions of industrial disputes, and King, of all people, lectured Rockefeller about his "excessive seriousness," among other character flaws. "The truth is," King wrote in a moment of honest reflection, "I see in Mr. R. precisely the same mistakes which I have heard others complain of in myself."

Working with McGregor, King enthusiastically dived into his new assignment. By this point, the First World War had been waging in Europe for about seven months. Like most Canadians, King had deplored the war's outbreak. It was, he wrote in his diary in August 1914, "tragic beyond words." In the days leading up to Britain's declaration of war against Germany—which automatically pulled Canada, a member of the British Empire, into the conflict—King did what he could. He urged for unity and non-partisanship and counselled Laurier, astutely fearing that the Liberal leader's all-consuming concerns about Quebec would cause him no end of problems. Yet apart from raising money for the Canadian Patriotic Fund, the bloody conflict remained an intellectual issue removed from King's day-to-day consideration.

In the spring of 1915, his assignment for the Rockefeller Foundation was naturally his main focus. He avoided unnecessary social engagements as much as possible. One of the few leisure activities he enjoyed was horseback riding, during which he often communed with the horse ("I talked with the horse most of the way, and enjoyed the ride exceedingly," was the diary entry for November 25, 1914). A bibliography with ten thousand entries was compiled for him by Harvard economics professor Robert Foerster. In the first few months of his contract, King spent far too much of his time creating his convoluted diagrams, drafted by a local Ottawa architect, and attempting to divine the multidimensional interrelationships between the various segments of society.

Rockefeller valued King's astute advice in ultimately solving the Colorado coal strike; the business tycoon had to be more empathetic to the real problems experienced by the miners. The sage, if straightforward, counsel offered by King dramatically reshaped Rockefeller's Neanderthal views about organized labour and convinced him that he did indeed have a legitimate responsibility to know how CFI was being run. King reminded Rockefeller that in the public eye nothing seemingly was beyond his control. Of course," he added, "we know that this is not true, but the fact that it is not does not lessen the conviction in the public mind, and that conviction is a factor of which constant account has to be taken." During his testimony before the U.S. Commission on Industrial Relations in January 1915, Rockefeller initially suggested that there was nothing wrong with a seven-day work week until King passed him a note. He then said that he favoured six days of work and one day of rest.

King has been accused of being anti-union, though it was more a case of objecting to compulsory union membership. "True unionism is not an end itself," he declared in 1915, "it is a means to an end. It is a means of obtaining and improving standards for the working class." He had felt for many years that strikes with the main purpose of forcing union recognition were counterproductive. In this, he may well have been correct, but he still failed to grasp, as he had during his work with the federal labour department, just how much power management wielded over labour in early-twentieth-century North America. Nor did he truly understand that only through third-party collective bargaining would workers' rights be protected. Unable to make any real contribution to Canada's war effort, King imagined his solution to labour disputes as integral to eventually solving disputes between nations.

For King, it all came down to personal relationships and instilling trust and respect on both sides of the negotiating table. A few months after he started working for the Rockefeller Foundation, King gave an interview to the popular investigative (and muckraking) journal *Everybody's Magazine,* which was no fan of the Rockefellers. Perhaps he was trying to disarm his critics. In any event, he used the opportunity to lecture his new employer and other men of business that their continued abuse of labour was detrimental to their own moral and financial welfare. "You make a mistake in thinking these men are ruled by self-interest alone," he said prophetically. "They also have self-respect. One of their leaders writes you a letter, and you do not answer it, and you expect to meet them afterward on a plane of sweet reasonableness."

In Colorado, even once the strike had ended in mid-December 1914, instilling that trust was easier said than done. To start with, King had Rockefeller fire CEO Lamont Bowers, who never would have altered his rigidly held views that labour must be subservient to the needs of management. On successive trips to Colorado, King charmed and cajoled CFI president Jesse Welborn until he adopted a slightly more enlightened attitude. As Fred McGregor later recalled about King's adventure in Denver and mining towns of the southern part of the state, "he had a knack of establishing friendly relations with and winning the confidence of the men and women he talked with."

King spoke and socialized with a large number of miners and their families, walked with McGregor past the saloons and brothels in the town of Trinidad and even ventured down into a coal mine for a first-hand look at the perilous working conditions the men endured. In the privacy of his diary, his social conscience nagged at him. In truth, he identified much more with the miners' plight than with the position of management, the "Tories" of the piece, who had been oppressive and arbitrary. "One could not help feeling as one looked at the huge seams of coal," he later reflected about his tour of the mine, "that this wealth of nature was never intended to be privately owned, but was intended in reality for society as a whole." But he also knew which master he served and refused to pass judgement about Rockefeller's woeful ignorance of the situation. He decided that the miners would rather be accorded just consideration by the company and guaranteed a feasible grievance mechanism they could use than be members of a union detested by the owners. His goal, he maintained in testimony before the U.S. Commission on Industrial Relations, was not to thwart the cause of unionism in the United States, but only to resolve a bitter labour confrontation in the most expedient and reasonable manner he could devise.

The scheme King came up with was an employee representation or a company union, admittedly a half-measure that shut out UMWA. This innovation permitted the miners to voice their opinions without fear of retribution, or that was the idea. He had borrowed ideas from British labourites and adapted them. Known as the "Rockefeller" or "Colorado Plan," the plan permitted miners to elect representatives to a joint management-worker board, which in theory ensured that the men's labour rights were protected. The company, in turn, agreed to abide by all federal and state regulations and allow the miners a degree of freedom on everything from a complaints department to the purchasing of food. Most importantly, no miner would be fired for opting to join a union. At the same time, CFI executives and King were receiving regular reports about union activities and reaction to the proposal from a spy planted inside the UMWA local. If King was troubled by this sneaky business, he did not write about it and likely rationalized that he was fighting for a far greater good.

Organized labour deeply resented the plan. Many of its leaders, including the outspoken American labour giant Samuel Gompers, regarded the scheme as overly paternalistic. Labour leaders had no official role to play, apart from attempting to have their own men elected as representatives (something the spy confirmed was in the works). At the same time, the UMWA executive adopted a better-than-nothing attitude and welcomed the industrial peace it restored. In the trenches, CFI's miners, beaten down by the strike and persuaded that Rockefeller was a businessman they could trust, endorsed it. King's Colorado Plan became fashionable among American big business.

Mackenzie King succeeded because he knew what had to be done to win over the miners: he had to show them that John D. Rockefeller was "almost" one of them. Shrewdly, he set up a ninety-minute, well-publicized meeting between Rockefeller and the legendary Mary Harris Jones, otherwise known as "Mother Jones," an eighty-three-year-old radical labour agitator and crusader for children's rights. UMWA and some journalists criticized her for fraternizing with the enemy. Following their chat, she told the press that she had "entirely changed her opinion" of Rockefeller and that she and her fellow unionists had been "misrepresenting him terribly." That she later retracted these positive sentiments after tremendous pressure was beside the point; the public relations benefit was dynamite.

Even more brilliant was King's management of Rockefeller's three-week tour of Colorado in late September and early October 1915, designed to convince CFI's miners of his sincerity and the equity of the employee representation plan. A gaggle of reporters followed Rockefeller's every move, and the trip received coverage in the *New York Times* and other major newspapers. King ensured that he met the miners and their families in a friendly atmosphere. During one memorable social gathering at a schoolhouse, Rockefeller played his part by dancing with the miners' wives, to great fanfare. The highlight of the visit was when King and Rockefeller donned dusty overalls and helmets with headlamps for an excursion down into a mine. King had Rockefeller pose with a pickaxe, which left the miners in awe. "You are not as bad as you are painted," one of the miners remarked to Rockefeller as he was departing.

King later admitted in a letter to Rockefeller's wife, Abby, that he directed the tour like a "campaign manager" during an election and had come up with "all kinds of stunts." Indeed, he left nothing to chance, instructing Rockefeller what to say and when and where to say it. In characteristic fashion, he claimed that such blatant manipulation was not his true intention. His real purpose was to show the miners that John D. Rockefeller, Jr., was not their enemy. Their subsequent vote in support of the employee representation plan proved to King that his strategy was sound. As for Rockefeller, he recognized that the entire Colorado Plan had emanated from King's genius, to his ultimate benefit. "I was merely King's mouthpiece," he admitted many years later. "I needed education. No other man did so much for me ... I needed guidance. He had an intuitive sense of the right thing to do—whether it was a man who ought to be talked with or a situation that ought to be met."

KING WAS ENORMOUSLY proud of his accomplishments in Colorado, but not for a minute did he regard his work there as his true purpose with the Rockefeller Foundation. He was intent on producing a well-researched academic study and book that could illuminate, and perhaps revolutionize, thinking about labour relations specifically and human relations in general. It was a tall order for anyone to achieve, even someone as intelligent and driven as Mackenzie King. On top of this, he faced an array of trying personal and professional issues.

His emotional roller coaster began on April 4, 1915. The terrible news arrived by a telegram while he was in Colorado visiting his brother, Max, who was convalescing in a sanatorium. On Easter Sunday, King learned that his dear sister Bella, only forty-two, had died, likely from complications caused by a stomach rupture. Next to depart was his father, John King, who had gone blind from glaucoma; he died on August 30, 1916. Willie took some solace that the who's who of the Toronto legal and political society attended his funeral.

His mother, whose health was failing, was always a concern. After being shuffled around between Max and Jennie, Isabel finally moved in with Willie in December 1916. He would care for her with all the tenderness any son could offer during the last difficult year of her life. It was during this period, shortly before his father's death, that King became increasingly depressed.

Unwilling to accept that both his sister and father were gone, King's passionate convictions of a spiritualistic afterlife, steeped in his ardent religious beliefs, were awakened. Spiritualism's promise of maintaining a supernatural link with his cherished family members and others became one of the driving forces of King's life, providing him with a comfort he craved, but he could never be completely satisfied. "I have never felt more strongly that Death is not an end but a beginning," he wrote to Max soon after he had buried their father, "that the grave is not a fettering but a releasing power to those who go by that way to the Great Unknown and to those they leave behind."

He was as reflective some months later, after another visit to the family plot at the Mount Pleasant Cemetery in Toronto. Something mystical happened, he recalled in his diary on November 25, 1916.

It seemed impossible that his mortal remains could be lying there. To me the spiritual presence of both Bell and himself was more real than their graves which my eyes were witnessing. I stood with one hand on the side of the cross by father's grave, the sun from the West came out from behind a silver cloud in great brightness and lighted all the side of the cross symbolic of immortality. As I turned around the shadow of the cross stretched far behind, and my shadow stretched across father's grave. It was apparent that there could have been no shadow but for the light and the objects between. Here was the whole parable of life, the individual and the cross, the material things left behind in shadow, the immortal relieved in light.

The book he planned to complete for the Rockefellers—soon titled *Industry and Humanity*—weighed heavily on him as well. Whether he was in his apartment at the Roxborough in Ottawa or at his cottage at Kingsmere—in the summer of 1916 he took the phone out of the cottage and, as Fred McGregor remembered, "he lived like a hermit"—he struggled each and every day with the structure and writing for more than two years. Each page had a hundred revisions and he was still not satisfied. Many evenings he wrote in his diary about being discouraged. "The truth is I am not suited to theoretical work, but to practical," he finally conceded in early January 1917, "and need the active touch with men and affairs to give vitality to what I write."

The worst of the depression had hit him a few months earlier, in June 1916. The work on *Industry and Humanity* almost stopped. His diary keeping was sporadic, "an indication itself of a troubled mind," rightly observed McGregor. Out at Kingsmere, after a satisfactory trip to New York with his mother and Jennie, he described his dark mood: "At times I have been depressed and disheartened and unequal to the task. My thoughts have turned away from the ideal I have cherished, and I have found myself in an encounter with my own nature such as I have never known before. It has been at times as though a fire would devour me, and I have been unable to get rest by night or day."

Like other men of his generation, King never would have accepted that his nervous anxiety was anything but a physical ailment, perhaps caused by the excessive pressure on his spine that he claimed troubled him. Following an extensive examination, his Ottawa physician assured King that his spine was uninjured. The doctor told him he had a "severe form of neurasthenia," a nineteenth-century catch-all diagnosis for nervousness and insomnia, possibly related to an inability to comprehend human sexuality. In October, King was back in New York not feeling much better. He shared his emotional distress with John D. Rockefeller, Jr., who quickly saw to it that King had an appointment with Dr. Lewellys Franklin Barker, a leading neurologist at Johns Hopkins Hospital in Baltimore. Barker, who had been born in Canada and worked with the Canadian-born physician and medical educator Sir William Osler, would be King's personal physician for the next twenty-seven years.

Barker reassured King that he was generally physically sound, as a battery of tests with other specialists in Baltimore was to show. In an era when people believed having a cavity might cause insanity, King saw a dentist and had a tooth pulled. Several polyps were also removed from his adenoids in a minor operation. King later described the experience of being put to sleep with gas as akin to entering the world of the supernatural. "My mind was a total blank, there seemed only darkness, oblivion and as I began to come to I thought I was being hurried along in an elevator, and unstrapped," began the long diary entry.

After listening to King ramble on about his anxious "feelings," and observing his hypochondriac tendencies, Dr. Barker decided to send him to confer with his colleague Dr. Adolf Meyer, one of the leading specialists in "mental hygiene." King greatly appreciated Dr. Barker's compassion. "I came back from Barker's," he later wrote in his hotel room, "and knelt down and thanked God for the answer to my prayer, to find a means of freeing me from this nervous fear, and for Mr. Rockefeller as a friend. I also felt that father and Bell were watching over and helping to protect me and to bring this about."

A Swiss-born therapist who was influenced by Freud and rose to become one of the pioneers of psychoanalysis in the United States, the affable Meyer immediately bonded with King. "I felt this man had a soul which could understand mine," King noted after a long session with Meyer. He poured out his deepest secrets. "I outlined to him the strain I have had, the worries and morbid symptoms," he wrote. "I outlined the conflict in my thoughts between spiritual aspiration and material struggles and conflicts, the fight with myself." Presumably, the last point dealt with one of King's earlier sexual hang-ups, masturbation. Dr. Meyer assured him that "all the phenomena" he had described, the "evil passion" to use King's solemn term, was, in fact, "natural." Meyer also strongly recommended that King keep in touch with his friends and read each night before going to bed in order to relax.

With that settled, King left the hospital feeling better than he had in a long time. "My mind is greatly relieved tonight," he recorded, "and I feel this visit to have been one of the providential events of my life. I had come to the point where I thought my work for the future would be undermined by this nervous dread. Now I believe it will be greater than ever before." Meyer's professional assessment of King rings true. As he reported to Dr. Lewellys Barker, "The problem of our patient seems to be distributed about as follows: first, a perfectly obvious elimination of natural sex life from the intensely religious and spiritual affection, which only once became focused away from his mother... Second, the strain from illness and deaths in the family... Third, a most intense feeling of responsibility and also a feeling of jeopardizing his own position politically."

King heeded Meyer's advice about staying connected to his friends and the larger world. He tried to watch what he ate, as Barker recommended, yet his weight tended to fluctuate over the rest of his adult years. The depression was never truly gone, however. "Some days I have complete command of myself," he confessed in a letter to his ailing brother, Max, in February 1921, "but there are times when I seem to lose this grip entirely, and to be overcome with depression which is next to impossible to throw off. I am beginning to discover that highly emotional nature not unlike mother's in some particulars, and that I shall have to learn to guard against all its dangers." Max, who during the last year of his life completed a pop psychology book, entitled *Nerves and Personal Power*, believed Willie's nervousness arose "from a psycho-neurotic condition, some complex due to suppressed emotions." Max, who at the time was emaciated from tuberculosis, assured him that "some things I had believed to be real are not such but wholly due to effect of thought and will."

In early 1917, his mother's health problems consumed King and made any real headway on his labour study for the Rockefellers difficult. He ensured that Isabel had proper home medical care and doted over her every moment he could. For weeks, she was in pain and delirious each night. The beginning of the end for Isabel came in mid-January, when serious heart problems left her in and out of a coma. In a moment of clarity, she told King, as he dutifully recorded on January 24, "it might be all over" and she would no longer be a burden to him. He was beside himself at the thought of her demise. "I have faith and believe," he wrote, "she is going to be spared to me yet a while... Thank God she is with me here... She has spoken often of her love for me, has said while she loved all the others deeply, there was something between each of us 'deeper' than all the rest." The next day, in fact, she nearly died as her doctor had expected. Willie "kissed her dear face a hundred times" and "burst into tears" when she recognized him.

King was relieved that his mother lived to celebrate her seventy-fourth birthday on February 6, for which he thanked God profusely. When Isabel's doctor remained pessimistic about her chances for survival, King became so angry that he fired him. The doctor had not acknowledged Isabel's birthday, and worse, King insisted that he had over-prescribed drugs for her. This event stimulated King's faith in a purely spiritualist solution to his mother's illness and the power of his own love for her. "More and more," he wrote, "I am coming to the Christian Science point of view, the doing away with drugs and the like and putting one's faith in living according to Christ's teachings in all things. All man can do is clear away the underbrush, roll away the stones, the door may open, and the breath of life come in."

As his mother lingered between life and death, he grew more anxious—so much so that he even had some sharp words in his diary about John D. Rockefeller, Jr., who he felt did not truly appreciate the significance of the study he was trying to prepare. At a lengthy meeting in New York in early March 1917, Rockefeller questioned whether completing the book was worthwhile, compared with more practical work that could be accomplished. There was a heated discussion about King's future with the Rockefellers as well. It was King's impression that Rockefeller expected that he would stay on and become an integral part of his executive team, maybe even joining the Standard Oil board. King, who saw himself above the down-and-dirty fray of big business, was aghast at that thought. His dreams of political glory had not been diminished in the least. "He is prepared to crowd and bring pressure to get his own way," King wrote of their conversation, "and to be a little petulant when his wishes are not quickly met—in other words, that he is selfish in his own wishes, like others, not all philanthropist, and that too much cannot be taken for granted on the basis of friendship... He is a dominating type, though he seeks not to be."

When Isabel seemed to improve slightly in October, King took his first step back into a permanent life in politics, though that is not how it first appeared to him. As Canadian casualties on the battlefields in Europe mounted and voluntary enlistment numbers failed to keep up, Prime Minister Robert Borden felt he had no choice but to introduce conscription for military service. He asked Sir Wilfrid Laurier to support conscription and join him in a coalition or union government, but the Liberal Party leader refused. He also refused Borden's request for another extension of Parliament. Laurier knew he could never sell compulsory service in Quebec. This stalemate eventually forced the wartime election of December 17, 1917.

Many English Canadian Liberals were torn between supporting the country in a time of crisis and supporting Laurier, their aging leader. The Borden-led union government cabinet announced on October 12 had a dozen Conservatives, nine key Liberals and one labour minister.

Newton Rowell, the leader of the Ontario Liberal Party, became the president of the Privy Council and such stalwart Liberal newspapers as the *Manitoba Free Press,* the *Globe* and the *Toronto Star* ultimately backed Borden and the unionists. In Winnipeg, the editor of the *Free Press,* John Dafoe, a long-time friend of Laurier, was especially outraged by the Liberal leader's position, going so far as to permit a headline in his newspaper declaring that "a vote for Laurier is a vote for the Kaiser."

Throughout this high drama in Ottawa, King tried on several occasions to convince Laurier to consent to another parliamentary extension and avoid what he rightly anticipated was to be the most divisive election in Canada's history. King was not entirely opposed to conscription, but he maintained that the risk to national unity did not justify the controversial action. When Laurier refused to change his mind, King eventually made a decision that was to serve him well two years later: he remained loyal to Laurier and ran for the Liberals in North York. In his imagination, he was like "Sir Galahad," fighting for the "cause of Right and principle." He swore to himself that if "fate" delivered for him a victory, he would end once and for all his association with the American corporate world. It is "Faith and Fear," he reasoned, "and I am putting my all on Faith, with just enough worldly wisdom to see where my livelihood for the next year or two remains secured. I have never had a mental rest since the association with the Rockefellers was formed, [though] it has been providential as a means to an end."

For a few brief weeks, he thought he might win, despite the fact that he had spent little time in the riding during the past four years. He was further encouraged after he addressed supportive crowds from the same platforms as his grandfather had half a century earlier. But it was not to be for him or most Liberals outside of Quebec. His opponents insinuated that he was in the pockets of the Rockefellers and unpatriotic for not supporting the soldiers on the front, hardly a fair criticism. He lost the election on December 17 by more than a thousand votes; it was a lousy forty-second birthday. Nationally, the union government captured 153 seats to the Liberals' 82—62 of which were from Quebec. King's worst fears were realized. "My whole argument [during the campaign] was one of conciliation and of tolerance," he explained in a letter to Violet Markham a few weeks later, "conciliation between races and religions in this country, and Canadian unity founded upon this basis." Such sentiments were to guide his political career in the months and years ahead, as he strived to reunite the Liberal Party and the country.

Isabel King managed to hang on until the election. "Poor soul," she mumbled when Jennie informed her that Willie had lost. She died a few hours later. King never forgave himself for not being there for her. "This will be a sadness to me," he confided to Violet Markham. He was only comforted by his memory of when they were last together. "I can see now so plainly," he added, "my dear mother's eyes after she had given me her last kiss, and waved to me with her hand the final farewell from the pillow on which she lay."

Thereafter, Isabel King was to haunt her son's thoughts and dreams until the day he died, nurturing him emotionally and instilling him with confidence when he needed it most. "Her spirit is everywhere about," he wrote to Max, "and nowhere am I more conscious of its living presence than here in the room where she fought so bravely to the very last breath, and where God came to take her to Himself." By the middle of February 1918, he was writing in his diary the first of innumerable entries about dreaming and talking to his mother. In King's vivid imagination, she was not "worn by disease" but "radiantly beautiful," as he remembered her. On this initial occasion, he spoke to her about the "other world" and asked if she was still alive and near him. "When I said this," King recalled, "her lips opened and she said, 'I am alive,' but it seemed as though it was forbidden her to say more. I felt she had told me all she could. All day the dream has been with me." These spiritual encounters with his mother reinforced King's "pathological obsession" with her memory, as Charlotte Gray suggests, making the idea of marriage to a lesser woman than her impossible. Isabel's hold on King was to be stronger in death than it had been when she was alive.

That fixation manifested itself almost from the day she died. King absolutely refused to discuss the final calculations of her estate, the cost of the funeral or the price of the monument at the Mount Pleasant Cemetery with either Max or Jennie. Much to his siblings' surprise, Willie was the sole beneficiary, though the estate did not amount to much. Nonetheless, Jennie was riled, believing that Willie had manipulated his mother into changing her will to his advantage. King was obstinate, embarrassingly so when he suggested to Max that since Jennie had not shared in the financial obligation while their mother was alive as he had, she had no right to try "to profit" when the obligation no longer existed. Max, who also depended on his brother's financial handouts and kindness, was in a difficult position—as executor, King forgave a debt of $4,700 that Max owed to their parents' estate—yet he took Jennie's side in the heated argument. For Willie to suggest that Jennie had some ulterior greedy motives, Max scolded him, was "unfair, unbrotherly and unmanly in the extreme." For almost a year, it caused a rift between the three of them, before all was forgiven.

WITH HIS POLITICAL career temporarily on hold and his mother in the world of the spirits, King spent the rest of 1918 toiling away on his book, completing it at long last by mid-September while he was at Kingsmere. On that day, when everything was done, other than a few revisions, he glanced up at the clock on the cottage shelf. To his surprise and delight, as he later recorded, "both hands [were] over the hour twelve." What precise meaning King gave to this is unclear. He did regard such coincidences, as MacGregor Dawson pointed out, "as confirmation of the rightness of the decision reached or the action taken at the particular moment." This ongoing fascination with the hands of the clock and other numerological quirks, which he was to comment on frequently, soothed his many insecurities. And if King was edgy about how his book was about to be received, who could blame him?

Had anyone else other than someone of Mackenzie King's stature authored *Industry and Humanity,* it would have been long forgotten, cast aside as an ill-conceived literary disappointment from another era. It is difficult to argue with one critic who suggested after it first was released that King was "simply encumbered by the weight and copiousness of his own thoughts." Harvard's Charles Eliot, who had recommended King to the Rockefeller Foundation, was much kinder, describing it as "rather diffuse." Rockefeller and his advisers had been politely skeptical when they received drafts of the first few chapters, believing that the writing was too abstract; they were even less impressed when they read the finished product. "I think I get the idea," Starr Murphy of the Rockefeller Foundation told King after wading through an early draft, "but confess to a certain confusion of mind similar to that produced by attempting to read metaphysics, which always was too deep for me." "Who do you expect will read it?" Rockefeller had pointedly asked King over dinner in New York about eight months before the book was published. King did his best to provide an answer, yet was later understandably annoyed by his friend's cynicism.

King had envisioned his work as a kind of roadmap through the murky waters of industrial capitalism, based on his experiences as a labour conciliator and what he regarded as his profound understanding of the human condition. His thesis, as he explained in a report to the Rockefeller Foundation in May 1916, was that "the ends of industry must be made subservient to the ends of humanity, not humanity made subservient to industry." The trick, he argued, was to find the mutual harmonious interconnections between the four parties to industry: labour, capital, management and community. As he maintained that he had proved through his labour work during the last decade, class conflict could be effectively reduced "through rational investigation, the accumulation of knowledge, and improved personal relations."

Within the pages of *Industry and Humanity,* too, if one looked hard enough, was King's inherent liberal social reform philosophy that the state had an obligation to take care of its citizens. His model was the British Labour Party's program, which advocated old-age pensions, unemployment and health insurance, and the public ownership of such vital industries as natural resources and transportation. Within a year, King was to ensure that most of these ideas were included in the Canadian Liberal Party's platform as well.

Industry and Humanity was by no means a master "blueprint" (as Bruce Hutchison postulated), though soon after he completed the manuscript King did infer that it entailed "the work of Liberalism for the future." Two months after that, he suggested, in a moment of spiritual reflection, that the key themes advanced in the book would be his life's work—"that seems inevitable, that seems the leading of God." Clearly, the study inspired him to pursue one of his personal goals: the establishment of the welfare state. The trouble in 1918 was that whatever message King was attempting to present in the chapters of *Industry and Humanity* got lost in his preachy and moralizing prose, which reduced complex societal problems to a Biblical battle between good and evil. Or, in King's ominous terminology, borrowed from a speech given by Louis Pasteur in 1888, a mighty contest between the "Law of Blood and Death" and the "Law of Peace, Work, and Health."

Beyond the angels-and-demons rhetoric, King was a man ahead of his time. Co-operation and democracy rather than conflict and dictatorship are now the norm in the workplace as well as in the international arena. Survival of the fittest and the glorification of war, which social Darwinists so righteously held up as the height of humanity during the late nineteenth and early twentieth century, but which King boldly rejected, has given way, in principle at least, to negotiation and compromise. King's assessment that industry can only function properly when workers feel they have control over their own lives, remains a truism. So, too, does his contention, widely accepted today by modern institutional theorists, that for an organization or institution to be effective, it must consider carefully and communicate openly with all of its various stakeholders. This was the reason why King was wary of Ivy Lee's more contrived public relations campaign, which purposely ignored person-to-person contact and respect.

From a sales perspective, King proved his critics wrong. *Industry and Humanity* was published by Houghton Mifflin Company of Boston. The 560-page book quickly went through three more printings over a twelve-month period and eventually sold about 10,000 copies. King himself bought more than a hundred books (the retail price was $3) and sent them far and wide to family, friends and associates. By 1924, the book had reached its tenth edition and had also been translated into French, with increased sales no doubt stimulated by King's ascension to the prime minister's office. This was an impressive accomplishment for any writer.

King's thoughts on labour left an indelible impression on Rockefeller. He emerged as one of the most vocal advocates for a reformed and more enlightened relationship between American business and its employees. Sounding very much like King, he declared that "the soundest industrial policy is that which has constantly in mind the welfare of the employ-ees as well as the making of profits, and which, when human consid-erations demand it, subordinates profits to welfare. Industrial relations are essentially human relations." King's political and literary success also enhanced Rockefeller's opinion of the book. "It is highly gratifying," Rockefeller wrote to King at the end of January 1924, "to think of the influence which you are thus exerting all over the world in developing a saner and more human attitude on the part of industrial leaders toward their employees and confidence and cooperation with their employers on the part of employees generally."

King undoubtedly concurred. Three decades after the publication of *Industry and Humanity,* he was still touting his perceptive insight. "Industry's basic function is to be the servant of Humanity. So should nationality. Never vice versa," he explained to a writer from *Forbes* mag-azine in a June 1947 interview. "Do you know that now, thirty years later," King added, "I honestly think there is a lot of wisdom in this book, if I do say so myself." He then proceeded to read a few pages to the bemused journalist.

THE PUBLICATION OF *Industry and Humanity* had not solved the far more serious dilemma confronting King in 1919: where did his future lie? He had accepted a retainer of $6,000 a year from the Rockefell-ers to continue as a labour relations consultant and was hired on simi-lar contracts from an assortment of high-powered U.S. corporations. His income for 1918 was approximately $22,000, more money than he ever had earned, and he felt guilty about it; in particular, guilty that his father and Max had had to struggle so much to make a decent living, whereas he had now squirrelled away close to $80,000 in savings.

For King, life was not merely about acquiring wealth. He could have continued developing his lucrative business as a consultant or taken a full-time executive position with the Rockefellers. In early January 1919, another enticing offer was dangled in front of him. Andrew Carnegie, impressed by King's work with the Rockefellers, wanted him to become the director of the Carnegie Corporation, his philanthropic institution. The starting salary was $25,000 a year, and as director he would be responsible for allocating funds from a $150-million endowment. There was also the possibility of writing Carnegie's biography for an additional $100,000. King was tempted, no doubt. Rockefeller, who regarded King as his friend and loyal subject, was more than a little put off by Carnegie's advancements and matched the offer.

At the time, King was courting Jean Greer, the forty-two-year-old daughter of the New York episcopal bishop David Greer. They had met in early 1918. He wooed her with letters and flowers. "I see in her all that I most wish to find in a woman—more than I believed I should ever find in anyone," he noted. He believed she had been sent to him by his departed mother, and in the pages of his diary during February 1918 he had elevated her to near sainthood. On one significant outing, King and Jean had sat together in the front pew of the grand Cathedral Church of St. John the Divine on Amsterdam Avenue in Harlem. "It made me very happy to be sitting at her side, though God knows I felt unworthy enough as I realized what a noble and Christian woman she is," he reflected afterwards. "There was nothing she said or did that was not pleasing to me." Alas, Miss Greer did not reciprocate King's feelings. She felt that he was "rushing things a little hard," as he learned. King decided that she was not ready for marriage. But the hard truth was she did not want to marry him. On June 18, 1919, she wed Franklin W. Robinson, a music instructor.

Had things turned out differently, King might well have abandoned his political aspirations for her and moved to New York. (Thirty years later, he was glancing through the pages of his diary when he came across a reference to Miss Greer. "There is a case of a door closing, all to one's advantage—refulfilling the purpose of my life," he wrote on March 13, 1949. "Had I married her as I evidently had been prepared to do I would have become in all probability a citizen of the U.S. in association with large interests, not P.M. of Canada fulfilling my grandfather's life.") When Miss Greer turned him down, his attention fittingly enough shifted back to the scene in Ottawa.

Since the debacle of the 1917 election, he had met many times with Laurier, who hinted that he thought King might replace him as party leader. Nothing, however, was ever put in writing or said publicly. His desire for Laurier's approval knew no bounds, and when the aging leader's focus was on someone else in the caucus, King pouted like a child. "I confess at times," he wrote in mid-September 1918 following a Liberal Party planning session, "I feel Sir Wilfrid lacks a sense of honour in many things, that he is just playing the game to suit himself and is now more or less indifferent to real reform and the country's needs. He seems too tired to do anything that requires initiatives and progress." Within two months, Laurier announced his intention to resign as leader, boosting King's political hopes and plans. The new leader was to be chosen at an unprecedented three-day convention in Ottawa, the first of its kind in Canadian political history, when Liberals from across the country were to gather in a show of strength and unity. King was told by several supporters that he had Quebec Liberals on his side because of his stand on conscription, and the union government and was convinced "that leadership of the party is mine if I care to go in for it."

On February 17, 1919, King was in Youngstown, Ohio, on business when he learned that Laurier had suffered a fatal stroke. By the time his train reached Buffalo, his leader and mentor was dead. With King's team of key advisers in place—including Peter Larkin, the Salada Tea merchant who was to become one of his closest friends and benefactors, and Joe Atkinson of the influential *Toronto Star*—his confidence was as high as ever, but he remained publicly aloof.

King's two main opponents for party leadership were to be two longtime Liberal Party stalwarts, George Graham and William S. Fielding, both of whom had endorsed Borden's union government. King was certain, as he reasoned with himself, that the Liberals of Quebec would never accept as a leader any man who "betrayed" Laurier. "All the Liberal members of the union government, and Graham and Fielding, have only themselves to thank for the position in which they find themselves today," he wrote. "They failed in a moment of crisis, at a time of great need. They left their leader when the popular tide was rising against him."

Still, King was unsettled by gossip that Laurier had told his wife days before he died that the only way to rebuild the fractured party was for Fielding to succeed him. The night before the final vote at the convention, King spoke to Lady Laurier, who confirmed that this story about Fielding was accurate. Consequently, she could not in good conscience offer King the public declaration of support he desired. No matter, King thought, his own private discussions with Laurier had confirmed in his own mind that he was the Liberal chief's first choice to replace him.

Leading up to the August convention, King decided to take the high road and do little campaigning. He delivered a few speeches, but carried on with his work as a labour consultant. His destiny, he decided, was out of his control and in the hands of the Lord. Besides, in his view, overt displays of partisanship were beneath him and undignified. In moments of doubt, his far greater concern was whether he was truly capable of assuming the leadership without a wife and companion by his side. "Unmarried," he confided to Max in mid-April, "I hesitate to take on the awful burden of leadership of a party and the conflicts of public life alone."

In May, King stuck to his plans and travelled to England to conduct industrial work and to visit with friends. He told Fred McGregor to book their passage home so that he would land in Canada only ten days before the convention. As a precaution, McGregor decided to reserve them alternative passage on a ship leaving England two days earlier, which they finally decided to take. It was a propitious turn of events. Had they travelled on their original reservation, a labour strike in New York would have prevented them from reaching Ottawa until well after the conference had started. King would not have had the opportunity to shine during committee work or, more importantly, make his stirring keynote speech on labour and social welfare issues, the implacable Tory enemy and Laurier's legacy. These factors ultimately propelled him to victory.

The leadership came down to a contest between the elder statesman Fielding and the youthful but experienced King, with George Graham and the sober Scot from Nova Scotia, Daniel McKenzie, who had been serving as the interim leader of the party since Laurier's death, trailing behind. A majority of the country's newspaper editors expected Fielding to win, and several had given him editorial support. The same criticisms were levelled at King again, especially in Ontario: that he was an errand boy of the Rockefellers and that he had shirked his duty to his country during the war.

On August 7, the day of the vote, King meditated and prayed as the balloting proceeded. It took much of the day and three ballots, yet before the night was over William Lyon Mackenzie King was the new leader of the Liberal Party. Quebec delegates, who respected King's loyalty to Laurier two years earlier (and disdained Fielding for his quasi-support of the union government), had likely been the difference, a fact King never forgot or took for granted. One Quebec MP, lawyer Ernest Lapointe, especially impressed him with a "smashing speech." King and Lapointe would soon be the closest of political confidants, and Lapointe would serve as King's loyal French-Canadian lieutenant for the next twenty years.

Standing before the cheering and sweaty throng of more than a thousand delirious Liberals in Howick Hall in the Ottawa Exhibition Grounds, King reflected on his family and the winding path that had brought him to this momentous point in his life. "The majority was better than I had anticipated," he later recorded. "I was too heavy of heart and soul to appreciate the tumult of applause, my thoughts were of dear mother and father and little Bell all of whom I felt to be very close to me, of grandfather and Sir Wilfrid also. I thought: it is right, it is the call of duty. I have sought nothing, it has come. It has come from God. The dear loved ones know and are about, they are alive and with me in this great everlasting Now and Here. It is His work I am called, and to it I dedicate my life."

The next morning, still in a state of awe, he accepted congratulations from Prime Minister Borden and Fielding, whom King felt was "touchy and excited," but not "unfriendly." There were many things to do, including finding a safe constituency he could run in, reuniting the party and solving the problem that troubled him most—finding the "right woman" to share his life and work with. Yet the trepidation of the evening earlier had given way to more a positive outlook. "My happiness continues to be in the happiness of my friends and the great opportunity ahead," he concluded in an eleven-page diary entry about the convention. "The people want clean and honest government, ideals in politics, a larger measure of Social Reform. I am unknown to the people as yet, but they will soon know and will recognize. The Liberal Party will yet rejoice in its entirety at the confidence they have placed in me. They have chosen better than they knew though I say it myself. I say it in all sincerity, believing in my firm determination to win by the right. May God keep me ever near his side and guide me aright."

4

A Not Unacceptable Party Leader

I am happy that I have been true to the Liberal tradition, true to the platform of [the] 1919 Convention, true to the pledges I gave the electors in 1921 and true to the people—the producers and consumers.—THE DIARY OF WILLIAM LYON MACKENZIE KING, *April 10, 1924*

OTHER THAN perhaps Stéphane Dion and Michael Ignatieff in recent years, no leader of the Liberal Party had his work cut out for him quite like Mackenzie King did in 1919. Immediately, doubts were raised about his potential to succeed. "King is by no means a howling success," the Saskatchewan MP James Calder told Robert Borden. "He has left a distinctively weak impression on the House and the country." Conservative veteran Sir George Foster was more succinct and sharp in the pages of his diary: "He is not cut out for a leader, that is certain."

Trying to ignore the rumblings, King pushed on. His first task was to win a seat in the House of Commons, and that was done in quick order in a by-election in October. He opted for a safe seat in Prince Edward Island, which he won by acclamation, but he was determined to return to North York in Toronto in the next general election. His popularity, nevertheless, remained low.

Right from the start, his speeches, almost always too long, abounded "in platitudes, abstract terms, and high sounding and often empty phrases," MacGregor Dawson asserted. In public, at least, he lacked a sense of humour, feeling that it was more dignified to present himself like a Sunday school teacher. This strategy did not always work against

him. Despite his nineteenth-century stiffness and frequent awkward-
ness, his loyal Liberal supporters usually stuck by him. It should not
be forgotten that between 1908 and 1945, King ran in a total of ten
general elections and six by-elections and was elected twelve times, losing only four contests. Nonetheless, his opponents, especially in the
early twenties, regarded him as "an opportunist without principles" and
never missed a chance to point out any inconsistencies in his attitudes
and actions. Such Conservative-leaning newspapers as the *Ottawa
Journal,* the Toronto-based *Mail and Empire* (the *Mail* and the *Empire*
merged in 1895) and the *Winnipeg Tribune* hammered away at King's
inadequacies daily.

Few political journalists disliked him as much as John Stevenson,
for a time the Ottawa correspondent of the *Toronto Star.* King "is as
full of noble sentiments as a new calved cow is full milk," he wrote in
January 1920 to John Dafoe of the *Manitoba Free Press,* "but he is short
on concrete plans for our regeneration." For three years, King continually complained to Joe Atkinson about Stevenson's unfair attacks on
him, but the *Star* editor was reluctant to lose the talented journalist, a
good indication that the new Liberal leader could not even count on the
unconditional support of the most partisan Liberal newspaper in the
country. Stevenson was finally fired in 1923, after writing several anti-
King articles for another publication against Atkinson's direct orders.
That only embittered him further. He quickly found other work writing for American and British publications, which permitted him to continue his "blood feud" with King.

King was also concerned with his waning support from other supposedly Liberal papers. There was Dafoe himself to worry about. The great
Winnipeg editor was, as Pierre Berton fittingly described him, a big
man with "a shock of red hair, a shaggy moustache, and a long, flabby
face with a nose to match." His opinions from Manitoba carried a lot
of weight in Ottawa. In 1919, he had still not forgiven the Liberals for
their opposition to conscription; the *Free Press* had backed the Progressives, the new farmers' party, and pulled the strings behind the scenes.
At the posh Manitoba Club, Dafoe was the unofficial Pooh-Bah of the
"Sanhedrin." This was an elite group of Winnipeg politicians, professionals and business leaders, which included, besides Dafoe, Thomas Crerar, the reluctant leader of the Progressive Party; Bert Hudson, another
prospective Progressive MP; Frank Fowler, a wealthy grain executive;
Herbert J. Symington, a lawyer and railway executive; and James Coyne,
a lawyer. Privately, Dafoe did not have much faith in King's abilities
and believed that he was the Liberals' "poorest asset." The Winnipeg

editor eventually came around when he got to know King better, but he probably never completely accepted King as representing the future of the country.

The *Globe* was even more of a headache. King regarded the *Globe* as the most influential newspaper in Canada. It had been one of his earliest supporters but was now also less than impressed with him, though for odd reasons. In the 1920s, the *Globe*'s publisher was William Jaffray, a man more peculiar than King. A moral crusader and a puritan, Jaffray refused thousands of dollars of advertising for tobacco, liquor and girdles. In 1920, he banned the publication of horse-racing statistics in the *Globe* and insisted that King push for an amendment to the Criminal Code making racing against the law. Racing was too popular for any Liberal or Conservative government to pass such heavy-handed legislation. When King could not meet Jaffray's demands, he became the publisher's mortal enemy. Journalist Melville Rossie, who worked for the *Globe* in the twenties, once told King that Jaffray "was always trying to poison my mind against you." Likewise, Arthur Irwin, who worked with Rossie and had a long career with *Maclean's*, the National Film Board and then as a Canadian diplomat, recalled that anytime he said anything critical of King, Jaffray "just beamed."

PART OF KING'S early problems had to do with his preachy style of conciliation. He wanted to apply the same negotiating techniques he had perfected around the table in Colorado and elsewhere to his control of the Liberal caucus and the country. He did have the right idea: that there was some higher purpose and "greater good" that could unite all Canadians. The pursuit of that worthy objective became the distinguishing feature of his long political stewardship. But in a country such as Canada, finding common ground has never been easy. And, in his first years as leader, King found that strong regional differences, polar opposite economic priorities and huge egos made fulfilling such a lofty goal challenging.

Back in the House of Commons, after nearly a decade-long absence, King was by his own admission rusty. Not wanting to burn any bridges or offend any voters, he came across as overly cautious (another trademark), and as a political leader he was afraid to take a risk or commit to a course of action. This criticism was not without some justification, but he was a quick study and gradually learned how to say a lot without backing himself or the party into a corner. Still, alone at night, he wondered if he was ready for the enormous responsibility and regularly berated himself for not having an absolute grasp of every issue to arise.

The country in 1919 and for the next few years was in a restless mood. Inflation was high; labour, as seen in the upheaval during the Winnipeg General Strike, which had nearly shut down the city for six weeks in May and June 1919, was bitter and unwilling to settle for the status quo. French-Canadian nationalists, in light of the 1917 conscription debate and bitter election that followed, wondered if there was a place for them in the country. Western farmers found a political voice and demanded reduced tariffs, as loudly as the business community in Ontario and Quebec demanded keeping the duties high.

The farmers were unhappy with the two traditional political parties. They contemplated a new kind of movement, in which everything did not revolve around partisanship but instead met the needs of common people. The focus of their resentment was on the least sexy political policies one could imagine: freight rates, the nationalization of railways, grain trade regulations and above all the detested tariff on imported goods, the pinnacle of Canadian economic policy since John A. Macdonald and the Conservatives made it their own during the 1878 election. Thereafter, eastern Canadian businesses elevated the protective tariff to sacred status, whereas western farmers grumbled that they had to sell their grain on an open market but purchase their tools and manufactured goods on a closed one. By the end of the First World War, Prairie farmers' patience had run out.

"Their stand on the tariff was not just an expression of the farmers' interest in cheaper imported implements," explains University of Manitoba historian Gerald Friesen, but as Progressive platforms declared, it "was a defence of 'the people of Canada' against the 'protected interests'; the 'privileged class' must not be permitted to become richer at the expense of the 'poor' or 'the masses'; and these privileged citizens should not continue to subvert the old-line parties by contributing 'lavishly' to campaign funds, thus lowering 'the standard of public morality.'"

This social gospel–style appeal resonated across the Prairies and into Ontario and for a brief time created the Progressive Party and movement, billed as Canada's first successful third option. Remarkably, the Progressive and United Farmers Parties took provincial elections first in Ontario in 1919, then in Alberta in 1921, followed by Manitoba in 1922. On the federal scene, King could not ignore them. He quickly calculated that the farmers, who were liberal in political philosophy, were a much greater political threat to him than the Tories, since in any three-way race the liberal-leaning vote could be split between the United Farmers and the Liberals, allowing the Conservatives to win.

King had one thing going for him: the farmers were squabbling amongst themselves. On one side were the moderates in Manitoba, led for a time by Thomas Crerar, the tall, affable president of the United Grain Growers. He had been a cabinet minister in the union government until he resigned in June 1919 and was aided by John Dafoe's potent pen. The more radical "anti-capitalist" element of the movement was based in Alberta and guided by Henry Wise Wood, president of the United Farmers of Alberta. He advocated for a completely new (and impractical) "group government," based on class, occupation and co-operation. This unwillingness to play politics as it has to be played in Ottawa, with the ruthless go-for-the-jugular abandon necessary to succeed, eventually killed the United Farmers Party. Yet its persistence made King's life miserable for a few years, since he had the most to lose. He rightly regarded the Progressives as "Liberals in a hurry" and sympathized with their anti-tariff policies and social reform and humanitarian outlook. Indeed, King was in many ways as much of a Progressive as Crerar, which explains why he paid so much attention to the party.

King recognized the allure of farming and its connection with virtue: "The farmers' movement is a people's movement and as such the truest kind of Liberalism," he wrote. "Liberalism to hold its own must make clear that it stands for the essential reforms the Farmers and Labor are advocating. I have always done that and indeed as I see the situation, I am the one man who by natural sympathies and past record can bring these three groups into alliance and to a single front." He was at one with their economic zealousness, dismissing protection as an "appeal to the immediate selfish interest of individuals and communities in a manner which causes them to disregard other people and other parts of the country and to discount future consequences."

He tried to portray himself as a man farmers could trust. His speeches "glorified" agriculture. "What we want is a bold peasantry; what we want is men who love the soil, who love contact with nature. We need them . . . if we are to maintain our human society in a proper degree of strength and vigour." And in his diary he swore that he was "determined the Liberal Party shall stand true to the farmer's interests for it is only by developing our agricultural wealth that Canada will ever become a great and prosperous country."

MANY TIMES DURING the period from 1920 to 1925, King reached out to the Progressive Party's leaders, first Crerar and then Robert Forke, offering all sorts of incentives for them to return to the Liberal fold.

Crerar and Liberal MP Ernest Lapointe, King's new ally from Quebec, were friends. "If the three of us ever get together on the one platform," King predicted in December 1919, "we will be able to sweep the country." That was wishful thinking. Far too many Progressives did not trust King or want to be members of a political party that was dominated by protectionist easterners. That also went for Saskatchewan Liberals, led by Premier William Martin, who in 1920 separated themselves from King and the federal party. On a western tour that year, no provincial Liberal would appear on the same platform as the federal leader.

King was a pragmatic and partisan politician, and he could do the arithmetic. In 1920, the four western provinces had fifty-six seats compared with Quebec's sixty-five. Moreover, in the most recent federal election in 1917, sixty-two of Quebec's seats had been captured by Laurier and his loyal Liberal cadre. As much as King wanted to win back Crerar and the farmers, he could not ignore Quebec and its pro-tariff voters. This meant placating Sir Lomer Gouin, the sixty-year-old Napoleonic Quebec premier, who had firmly held the job since 1905 before jumping to federal politics and being elected to the House of Commons in 1921. An ardent protectionist, Gouin—whom Ernest Lapointe regarded as "an unregenerate Tory in the guise of a Liberal"—could barely hide his contempt for Mackenzie King. During the Liberal convention in Ottawa in August 1919, Gouin had held court at the Château Laurier hotel, where, amid the cigar smoke, he strode about with great purpose and pontificated on the state of the Liberal Party and the country.

Heading into the 1921 federal election, King, despite contemporary assertions to the contrary, had one other key advantage: the Conservatives had chosen Arthur Meighen to be their leader after Robert Borden stepped down. "It was too good to be true," King scribbled in his diary when he first heard the news. That overconfidence vanished the first time the two leaders faced off against each other in the House. King had known Meighen when they were both students at the University of Toronto and didn't much like him. The feelings were very mutual. "There was never anything to match Meighen's hatred for King," remembered *Ottawa Citizen* pundit Alex Hume. Meighen "laced it to King to no end, and King was palsied with fear in the face of Meighen's withering attacks." Adds Meighen's sympathetic biographer Roger Graham, "He pursued King so relentlessly, with such savage, ironic scorn, that the pursuit sometimes embarrassed even his own supporters." Meighen was urged by his advisers and friends to pay less attention to King and rebuild his own party. He ignored that political sagacity and learned the hard way that negative campaigning does not always resonate with voters, Canadian voters especially.

For someone as sensitive as King, who had an instinctive longing to be liked by everyone he came into contact with, Meighen's attacks on his character rankled. His contempt for Meighen was "too great for words," as he put it. King soon regarded him as the Tory devil incarnate, an agent of the dastardly Family Compact and a man not to be taken lightly. He was, King noted, the "very antithesis" of himself "in thought and feeling on political matters." Ernest Lapointe, never one to mince words, told his Liberal colleague Chubby Power that Canada was "too damn good a country to let Arthur Meighen run it."

A lawyer, Meighen was smart, dedicated and highly principled. He was, said businessman and Tory loyalist Sir Joseph Flavelle, "destructive in debate . . . [and] crushing in criticism" and carried himself in the House of Commons much more skilfully than King did. Yet he failed to grasp an important lesson about Canadian politics that King innately understood: extreme and inflexible policies will never sustain long-term victory at the polls. "In a country like ours," King reasoned, "it is particularly true that the art of government is largely one of seeking to reconcile rather than to exaggerate differences—to come as near as may be possible to the happy mean."

Meighen refused to understand that French Canadians would never forgive him for drafting the conscription bill and for his staunch and unyielding faith in the pre-eminence of the British Empire; labourites detested him for jailing the leaders of the Winnipeg General Strike; and westerners disliked him for his absolute refusal to lower the tariff. "Meighen was incapable of neutrality on any important issue and it was impossible to be neutral about him," Roger Graham observed. "One had to be with him or against him, because on every question he took his stand and defined it with unhesitating conviction." It was not by accident that Meighen titled his collection of speeches and essays, published in 1949, *Unrevised and Unrepentant*.

DURING THE 1921 campaign, Meighen's intransigence worked in King's favour. Even if the Conservative leader—for this election the party went by the name the "National Liberal and Conservative Party," but that dubious moniker fooled no one—would not have admitted it, he had lost Quebec and the west months before a single vote was cast. The contest ultimately came down to a battle for Ontario's eighty-two seats. King at his wishy-washy best had been trying to sell himself as pro-tariff man in the east and a closet free trader in the west. "The issue, so far as the Liberal party's attitude on the tariff is concerned," he declared, "is not and never has been in this country, between free trade and protection; it has been between a tariff imposed primarily for purposes of protection, and a tariff imposed primarily for purposes of revenue." Each time King spoke about a "tariff for revenue," Meighen's head spun, and rightly so. "Those words are just the circular pomposity of a man who won't say what he means," he stated. "He might as well say he favours a perambulating tariff, or an atmospheric tariff, or a dynamic tariff."

On election day, December 6, 1921, King voted a few minutes after eight in the morning. Back at his apartment, "all was so peaceful and quiet," he later recalled. He knelt in prayer before "dear Mother's picture" and "thanked God for his protecting providence through all." The rest of the day, he calculated and re-calculated how he believed the results were to go. He felt certain that before nightfall, he would be the prime minister following in the footsteps of Gladstone and Laurier. "But," he caught himself, "it is all as God wills."

That evening, he had an early dinner with his new friends, Joan and Godfroy Patteson, who also lived in the Roxborough Apartments, and then headed back to his office to await news of the election returns. Hourly telephone calls from Toronto were positive; King had won his seat in North York by more than a thousand votes. By midnight, the political universe had almost unfolded as King predicted. The Conservatives had been "routed," he gleefully recorded. Rather amazingly, the Tories won only fifty seats, of which more than half were in Ontario. Meighen had lost his own seat in Manitoba. King had expected the Progressives to be the spoilers. But the farmers surprised him and everyone else, including themselves, with an impressive 64-seat victory, the second-highest total for any party. Discontent in the west had led for the first time in Canadian history to the presence of a true third party in the House of Commons. King and the Liberals took 117 seats, which included every one of Quebec's 65 seats, evidence that intense hatred for Meighen and the Tories trumped King's vague assertions about reducing the tariff. With five seats going to labour and independent politicians, King was just shy of a majority. However, he was confident that the Progressives would never side with the Conservatives against the Liberals, and so his government was relatively secure for the time being.

At one-thirty in the morning, he and the Pattesons, who had spent the evening with him, returned to the Roxborough in a taxi and had a celebratory feast until three o'clock. Before the new prime minister tried to fall asleep, he prayed once again before the portrait of his mother and asked the Lord to make him "worthy." In this heightened spiritual moment, King recalled that he had revealed to the Pattesons that in the cherished painting of his mother the book that was opened on her lap was the second volume of Morley's biography of Gladstone. "This was her secret," he wrote, "and mine and it has been known to no one till I told it to Joan and Godfroy tonight."

THE CLOSE BONDS between Mackenzie King and Joan and Godfroy Patteson had been forged three years before that fateful election night. In early October 1918, King hosted a small dinner party at the Hull Golf Club, a favourite out-of-the-way dining spot for Ottawa's elite. The occasion was another celebration of the completion of his book *Industry and Humanity*. There were eight invited guests in all, including King's new neighbours at the Roxborough Apartments, Godfroy and Joan Patteson. After dinner, there was a "dance and chat" in front of the fire. King accompanied the Pattesons back to the Roxborough after ten, and they visited with him for another hour. "It was quite a happy evening," King noted.

This was the first appearance of Joan Patteson in King's diary, but it was certainly not the last. Indeed, for the rest of his life, Joan, and Godfroy to a lesser extent, were to be his closest companions. For the first year or so, he referred to them more formally as "Mr. and Mrs. Patteson"; by the summer of 1920, he had begun calling them both by their first names. In 1918, the Pattesons and their twenty-two-year-old son, John (also known as Jack), had recently moved to Ottawa from London, Ontario. Godfroy was a bald and genial fifty-one-year-old manager at the Molson Bank (soon to merge with the Bank of Montreal). He was the son of the lawyer and civil servant Thomas Patteson, the first editor of the Conservative party organ, the Toronto *Mail*, from 1872 to 1877, and a confidant of John A. Macdonald. Joan, forty-nine and five years older than King, was a slender and pretty woman. She had a good sense of how things were in the world.

Joan MacWhirter and Godfroy Patteson were married on November 21, 1895. Joan Patteson's granddaughter Joan McCallum, who lives in Oakville, Ontario, today recalls her grandmother's inner beauty and says she was "a strict Presbyterian who had her rules." Her younger sister, the well-known British interior designer Mary Fox Linton, sums up the scene in Ottawa and at Kinsgmere she remembers as a child: "My grandfather was there, my grandmother was there and so was Uncle Rex."

WE ONLY HAVE Mackenzie King's version of his and Joan's intimate thirty-year relationship. And the reality he chronicled in his diary, especially when it came to his interactions with women, was often distorted by imaginary demons. Joan wrote plenty of letters to King, and they are as dignified as one might expect from a woman of her era. Had the two of them met earlier, she might well have become his ideal wife, as he long believed. As it was, Joan became King's best friend. Having lost a baby girl, she shared his fascination with spiritualism; Godfroy was not as keen on communicating with the "Great Beyond." Joan planned and hosted King's official dinner parties, acted as a sounding board for his speeches and major decisions and listened sympathetically to him gripe about his multitude of prime ministerial problems. She was discreet and protected his privacy and deepest and darkest secrets.

Most evenings, even after he relocated into Laurier House and the Pattesons moved to a house on Elgin Street, King and Joan spoke or saw each other. Sometimes it was distracting. "It is so delightful to have the time free and quiet, and I am very fond of Mrs. Patteson," he wrote in early May 1920, "but I realize it is not in the interest of my work, and perhaps it is better for each of us to see less of each other."

From 1920 onward, Joan Patteson shared Mackenzie King's life as much as it was possible for a married woman to do so. When Walter Turnbull came to work for King in the late 1930s, he was cautioned "to be nice to Mrs. Patteson because she had great influence with the Prime Minister." He later remarked, "I was nice to her because I liked her. I really genuinely liked her. She was a lady in every sense of the word."

There is no evidence that the relationship between King and Joan was anything other than platonic, with the exception of a possible passionate episode during a few weeks in September 1920. Again, all we have to base this on is King's vivid insinuation of these events—his *Bridges of Madison County* moment, if you will. As King was readying himself for the decisive federal election of 1921, he had become infatuated with Joan. He was more than a little jealous of Godfroy. "She has a nimble mind, a fine appreciation of literature, keen intelligence and wit, a most delightful intellectual companion, she could rise to the greatest heights," he pointed out in the entry for September 6, "but in Godfrey [*sic*] she has not her equal intellectually. His tastes are on a different level. Her nature needs the spiritual, and that side is far from developed in him." Back in Ottawa the next day, he described his emotional turmoil. "It seems so hard to overcome oneself, and this morning it was nearly desperation... These storms of passion—for that is what they are, are madness and wrong. They 'rock the mind' and must cease.

We both have the strength enough to see that and we will help each other to what is best for each, hard as the struggle may be—and it is hard in this lonely solitary life."

We will never know if he and Joan succumbed to temptation, since King sliced the next five days of entries from his diary. When the diary continued on September 14, King was out at Kingsmere in prayer. "Now I must bend all my energies to the great work God has given me to do. He added the following day, "Joan and I pledged our lives to united effort and service today. God grant we may be given strength to endure." As Charles Stacey suggests, whatever took place could not have been that devastating, for on the evening of September 16 King recorded that he spent the evening with Joan and Godfroy reading aloud passages from William Wordsworth's poems "Ode to Duty" and "Character of the Happy Warrior." In a quiet moment, King added that he and Joan "spoke of the fight we would make together for what is best for both of us. The supreme effort to make our lives count in human service... It was one of the happiest evenings we have ever shared together."

Less than a week before the election, King was in Toronto shopping. He stopped by Ryrie-Birks on Yonge and Adelaide Streets, then the city's most fashionable jewellery store, to purchase Joan a $125 bracelet, a fair bit of money for someone as frugal as King was. Romantic as ever, he had it inscribed with words from Tennyson's lengthy poem "The Holy Grail": "M.J.P. from W.L.M.K, A strength was in us from the vision, The Campaign of 1921."

Throughout the twenties, King admired Joan from a respectful distance, confiding only to his diary what his heart truly felt. On August 4, 1923, the Pattesons, among other guests, were dining at his Kingsmere cottage. "I never saw Joan look more beautiful," King wrote. "She was a very distinguished lady in appearance, manner, and in all respects sweet and lovely. I felt very proud of her. All the party talked till after midnight then I walked home with Godfroy and Joan and stayed on there another hour and a half, having one of the most beautiful talks I have ever had with Joan." Occasionally, however, King, who became more self-centred the older he got, reacted negatively to what he perceived to be Joan's selfishness and demanding behaviour when she wanted to see him but he did not have the time to spend with her. It was reminiscent of his attitude to Marjorie Herridge when she also made demands on his time.

In 1924, the Pattesons were responsible for bringing into King's life another significant being: a cute and cuddly six-month-old Irish terrier he named Pat. Joan and Godfroy also took a dog from the same litter, Derry, and the two "brudders," as King liked to refer to them, were the best of canine pals. Any "dog person" will readily identify with King's instant attachment to the animal. He truly loved "little Pat" as a parent would a child. He doted over the dog, talked and read to it—Pat usually wagged his tail when he liked the news King told him—shared late night snacks of oatmeal cookies and Ovaltine and recorded the pooch's various antics in his diary almost daily. Through his regular flow of letters to Violet Markham in England, Pat frequently exchanged pleasantries with Jane, Markham's Labrador—"the whimsy of canine correspondence," Markham calls it. "I hope Jane still remembers me, I think of her often," he wrote to Markham in May 1939. "Only last night I was telling my old dog Pat about the chats we had together, as he was lying at my feet."

If, God forbid, Pat suffered any mishap or ailment, King took it very seriously. On September 29, 1924, a member of King's household staff named Dewet let Pat out and did not watch him closely enough, and the dog ran off. King was heartbroken when he found out. "Poor little thing, so timid and alone. It has saddened me beyond words." The next day, he was compelled to leave Ottawa for business at Port Arthur and Fort William in Northwestern Ontario. "The grief I feel for little Pat is only equaled by the indignation I fell at Dewet's stupidity and disobedience of orders," he wrote. By the time King arrived at his destination on October 1, a telegram had reached him with the good news that Pat had been found. He was an emotional wreck. "All night I thought of him," that day's diary entry began, "and was very sad and lonely because of his loss, indeed he has filled my mind and thoughts to the exclusion of all else, even now that I know he has been found I cannot throw off the feeling of sadness which settled over me when I thought he was gone. I prayed earnestly for his recovery or being recovered last night, fearing my faith was not what it was [as] a child. Dear Mother was very near me in my sleep. I saw and kissed her face. I believe my prayers were heard."

A WEEK AFTER the 1921 election, Canada's tenth prime minister celebrated his forty-seventh birthday. His athletic stature of his university years had vanished and given way to a stouter, portly physique. His hair had thinned into a few wispy strands he combed across the top of his head, and his face was rounder and paler. He was also a new homeowner. A month earlier, Lady Zöe Laurier had died. In her will, she had bequeathed to King her three-storey yellow brick house in Ottawa's fashionable Sandy Hill neighbourhood, about a twenty-minute brisk walk from the parliament buildings. Canada did not yet have an official prime ministerial residence—and would not have one until the federal government acquired 24 Sussex Drive in the late forties—and the Lauriers had lived in this house since the party, at Sir Clifford Sifton's behest, had bought it in 1897 for $9,500. King was the honoured recipient not so much because Lady Laurier was fond of him, but because he was the current leader of the Liberals, and she felt the house should be returned to the party that had financed it. If it was later suggested that the house was owned by the Liberal Party, King immediately pointed out to any journalist or politician who got it wrong that Lady Laurier's will specifically noted him by name. Initially, however, he wasn't so sure the house was for him or if he could handle its expensive upkeep.

In its day, before Laurier's defeat in 1911, the house was stocked with fine French furniture, chandeliers and paintings. Laurier had a library and office on the second floor where he worked and deliberated. But by 1921, the property was in disrepair. The furnace, plumbing and roof all needed upgrading, and the cost was estimated to exceed $30,000. King's anxiety level heightened when he contemplated such a large bill.

In early January 1922, King was invited to dinner by Laurier's nephew Robert and his wife, who were still living in the house. "Tonight I had dinner at the Laurier Mansion—my own home," he wrote later that night in his diary. "I confess it is a bare bleak place in size and furnishings, not at all to my liking in any particular. Will require to be completely gone over. I have little or no sentiment about it. There are flimsy cheap things for the most part in the way of decoration . . . I shall certainly not try to carry it myself."

King's white knight or "fairy godfather," as he later referred to him, was Peter Larkin, the wealthy sixty-seven-year-old founder of the Salada Tea Company, then the third-largest tea business in the world. Larkin, who was born in Montreal, had come from modest means. He started working for grocery merchants as a lad of thirteen before eventually establishing Salada and expanding it until it became one of the chief importers of Indian tea. According to former Canadian diplomat Roy McLaren, Larkin's "most profitable innovation was to have his tea packed in small watertight packages lined with foil rather than in large, bulk chests. His tea, as a result, was fresher and sold better." A lifelong Liberal supporter, Larkin had contributed generously to both Laurier and King's election campaign expenses over the years, had been one of the original investors in the *Toronto Star* in 1899 and had assisted William Mulock in setting up a $100,000 fund for the Lauriers. He now took responsibility for Mackenzie King's financial welfare.

As a first step, Larkin set up a fund of $40,000 deposited in the Old Colony Trust in Boston that was to be used to renovate Laurier House.

King suggested that he deed the house to the trust for the Liberal Party, but Larkin vetoed that idea. He insisted that King maintain ownership and use the money as he saw fit. Work on the house commenced almost immediately, and an elevator was later installed. King's attitude towards his new home changed. "Its imposing and dignified interior makes it a worthy memorial to Sir Wilfrid and Lady Laurier," he wrote to Larkin at the end of November 1922. "No one can enter the house without being impressed with its character as a great residence." He moved into the house in the second week of January 1923. "Am writing for the first time in my new library at Laurier House and spending the first night beneath my own roof. It is quiet and peaceful and there is a feeling of space and repose in this wonderful beautiful room with its shelves of books and raftered ceiling."

Today, visitors can wander around Laurier House, a national historic site (on Laurier Avenue East) and get a real sense of what King's life must have been like there. It has a fairly large main floor drawing room and formal dining room and is cluttered with nineteenth-century-style furniture. The house is still, as Violet Markham depicted it in her 1956 memoirs, "a remarkable Edwardian period piece... [with] florid furnishings... [and] daunting portraits of former Governor-Generals, Prime Ministers and their ladies which hung on the walls."

The house was far too big for a bachelor. King had a small staff, including a butler, cook, maid, chauffer and gardener, who tended to his various needs and kept him company. If he was not entertaining guests, he stayed mostly on the third floor, where his library and office were located. It is a room with a fireplace, comfortable chairs and a couch. Hundreds of books line the shelves on three sides from floor to ceiling, just as he described it on his first night. The focus of the library, however, remains the painting of his mother still kept as a shrine on an easel to the right of the fireplace. The entire setting strongly resembles John King's library at Woodside. Nearby there is a breakfast room where King and his dog, Pat, preferred to eat. He worked prodigiously at his desk in the library, and his secretaries used a crammed space across the hall. There was also a small room (where the air-conditioning unit is now kept) that King used for his séances and table-rapping sessions.

As a sign of gratitude for his political and financial support, King appointed Peter Larkin as Canada's high commissioner in London in February 1922. It was somewhat of a ceremonial position, which Larkin served admirably until his death in 1930. From London, Larkin sent King gifts of furniture, lampshades he picked up on a visit to Paris, brass candle holders, a curio table and a dining room carpet, which King told him was "too lovely for words." He also dispatched butlers and tried to help King iron out his recurring servant problems. "It is really rather ludicrous for a Prime Minister to be writing to a High Commissioner on such a topic as this," King began one letter about his Laurier House staff issues, "but somehow you have assumed the role of a fairy godfather."

Larkin had always told King not to worry about money, and he wasn't kidding. As he had done for Laurier, he also did for King. In late 1925, a year of much turmoil in King's life, Larkin, with the assistance of Senator Arthur Hardy, Sir Herbert Holt, who was the president of the Royal Bank, and several other wealthy donors, established an impressive personal fund for King. During the next decade, it grew to $225,000 and was also deposited in the Old Colony Trust for King's personal use. (A fund of $100,000 was raised for Ernest Lapointe as well.) With his modest prime ministerial salary of $10,000, a $4,000 expense allowance and his own annually growing savings and bonds—by August 1930, King estimated that his savings and bonds were approximately $237,000—in addition to his property in Ottawa and the Gatineau, he was worth at the end of the twenties an estimated $500,000 ($6.4 million in 2010 dollars). Such party funds were routine in those years and date back to John A. Macdonald's day. King was enormously grateful, but was anxious that its existence be kept strictly private. "Larkin has been goodness itself," he wrote at the end of 1927. "He has secured me financially for the rest of my life. I have no longer reason to be anxious on that score, that means everything."

DESPITE HIS GREAT personal victory of 1921 and sound financial footing, the early twenties also had a moment of heartbreaking sorrow for Mackenzie King: the death of his beloved brother, Max, on March 18, 1922, at the age of forty-four. Max and Willie had remained close, even though Max's various maladies had kept him in Colorado for so many years. Max was one of the few people who could speak to King with forthright honesty and with a critical voice. During the election campaign, King, ever the hypochondriac, had in a long letter to Max blamed excessive fatigue when reporting to him about a particularly poor speech he had delivered in Toronto. Max did not believe a word of it. "Let me impress upon you again, my dear Will," he gently scolded him, "the necessity for disassociating irrational auto suggestion with regard to your own health from work toward the greater purpose which you have in view... You could not find one well-informed and honest physician on this continent who would tell you that you are run down physically and to suggest such ideas to yourself is simply to labour under the handicap of self-deception."

A month after King's election victory, Max, who had suffered from tuberculosis and muscular dystrophy, took a turn for the worse. King dropped everything and hurried to Denver. For several days, he visited with Max's wife, May, and her two boys and tried to comfort his ailing brother, who was now paralyzed on his right side. Each night, King lay in his bed listening to Max's deathly coughing. He questioned God's purpose, but ultimately concluded that "we can only trust and hold firm our faith." On the last day of his visit, King promised Max that he would ensure that his family's material needs would be looked after, a promise he kept. The two brothers also spoke of funeral arrangements and of bringing Max back to Toronto. Max, in turn, swore to King that even in death he would be with him always. "I was glad of this," King noted. "I would count on his help from the Great Beyond." Their final parting was as painful as anything King had endured. He knew that he would never see his brother alive again. "I forgot all my pain at parting his poor little frail paralyzed body into my arms," he continued, "called him my dear boy, my dear, dear brother, [and] told him I loved him with all my heart."

When he finally sat down in his train compartment for the return trip to Ottawa, he "cried like a child." He could not help listing the various gifts and assistance for nursing care ($560 in total) he had doled out, but then he caught himself. "The mere mention of money in this connection is hateful," that day's diary entry concluded, "it is only for memo purposes. Where money and love and life enter, there is no relation possible between the first and the last two. I pray God this sad experience may make me a better, stronger, and a purer and greater man. My God bless my dear brother, his wife and little boys, his home and all it means."

Less than two months later, May sent a telegram to Ottawa with the terrible news. "He loved you with all the strength of his great quiet heart," she told King a few days later. "You have indeed been our 'our best man'... [T]he last message he dictated was to you, his last promise was to me, to be with us always and that is to me no word can tell." With thoughts that Max must be now free of pain in the "here after," King boarded the train again for the long journey back to Denver for the funeral. In a classic display of self-centredness so characteristic of him, he wrote that he had no regrets. "I have been at Max's side without fail for 9 years, since 1913, and have failed him in no great matter." In

Denver, King comforted May and the children and covered expenses totalling $1,000 (as he meticulously noted). Following the funeral, at which King delivered a moving eulogy—at least, in his opinion—about Max's courage, the casket was loaded on to the train so that Max could be brought back to Toronto.

There are many "what ifs" in King's life, and the death of Max is certainly one of them. Had his brother lived, he clearly would have been a positive influence. "King was at the beginning of a long period in office when his high position tended to make him more unapproachable and less subject to criticism on personal matters," writes MacGregor Dawson, "and once Max had gone there was no one who could take his place. Many of the habits and idiosyncrasies which Mackenzie King as a lonely bachelor would in time develop might well have perished at an early stage had Max's judgment and common sense been available to act as a corrective."

THE ELECTION VICTORY in December 1921 had not quieted King's critics, who felt that with such strong personalities as Sir Lomer Gouin and William Fielding in his cabinet, the new prime minister was out of his league. He was castigated and ridiculed as the "boy premier." One cutting political cartoon depicted King riding on the back of a horse-drawn wagon with his feet dangling. Sitting up on top were Fielding and Gouin holding the reins and presumably calling the shots.

The reality was far less one-sided. During his first term in office, King handled himself better than he has received credit for. True, he was overly anxious when he had to stand in the House of Commons as prime minister in the spring of 1922. "To my great surprise," he recalled later that day, "the impression on the whole seems to have been good... With a little less stage fright and one day's preparation I might have done very much better. Still I might have done very much worse." Gradually, King whittled away at Progressive support, took a decisive position on Canada's autonomy within the British Empire, pushed forward a reasonable, if not entirely exciting, legislative plan, manoeuvred around a dozen different political sinkholes and ever so gingerly established his dominance over the Liberal Party. As historians Jack Granatstein and Norman Hillmer argue, right from his first day on the job, King "specialized in timing—in knowing when to act and more importantly when not to act."

King himself used a medical analogy to describe his unique style of management. "You keep the disease from developing," he later explained to a *Time* magazine writer. "The important thing is not what action you take to make desirable events happen, but the action you take to keep bad ones from happening… It's the result that counts, not the figure you cut while getting there." Around the cabinet table, he never dominated discussion. He let everyone have his say and usually arrived at a consensus, or at least a position he could live with.

As King had predicted, Thomas Crerar had no desire to be leader of the Opposition—he left that role to Meighen—or to be the prime minister, for that matter; all he truly wanted to do was to influence policy in a way that would take into account the western Canadian perspective. Initially, King thought he had convinced Crerar; Bert Hudson, a Liberal MP with strong connections to the Progressive Party; and Ontario premier Ernest Drury of the United Farmers to join his cabinet. In the end, however, their vocal followers would not stand for them to be part of a government that included the likes of Sir Lomer Gouin and other Montreal protectionists.

Crerar lasted only a year in Ottawa. Falling grain prices, business obligations and the tragic death of his eight-year-old daughter, Audrey, from diphtheria in October 1922 compelled him to return to Winnipeg. There, he resumed his position as the manager of the United Grain Growers. For the next few years, the Progressives' fate was in the hands of Robert Forke, who was a pleasant man but even less effective a leader than Crerar and more disposed to work with King. Upon hearing the news that Forke was taking over, a delighted King jotted in his diary: "This means complete co-operation." He was not far wrong. Except for a handful of disgruntled Progressive radicals and a few Labourites, Forke offered the Liberals near universal support.

Sir Lomer Gouin was King's biggest headache. As the senior Quebec politician (in his own mind, at any rate), the former premier expected an important cabinet portfolio to be dropped into his lap by divine right. In a series of talks with King, he first suggested the justice portfolio, which King had promised to Ernest Lapointe. Then, Gouin decided that he wanted to be the president of the Privy Council. This position had some clout and would have given Gouin too much power, from King's point of view. King smartly kept that job for himself, as well as external affairs, which he regarded as especially crucial since it allowed him to put his own stamp on the country's relations with Britain. He handled the situation with the deftness of a seasoned political veteran. He had already decided that Lapointe was to be his Quebec lieutenant and the one person who could guarantee the party's and his own continual electoral success in the province. "But for you, I would never have been Prime Minister, nor would I have been able to hold the office, as I have held it through the years," King would confess to Lapointe on November 19, 1941, a week before the great Quebec politician died.

Born in rural Quebec in 1876 to a farm family of modest means, Lapointe had worked hard to become a respected lawyer in Rivière-du-Loup by the time he was twenty-two. An admirer of Laurier, he sought out Liberal clients and campaigned tirelessly for the party. At the urging of local party stalwarts, he first ran successfully for Parliament in the 1904 federal election. He arrived in Ottawa unable to speak a word of English. His fellow Liberal Quebec MP Jacques Bureau took Lapointe under his wing and pushed him to learn the language. It took a long time, but eventually he became bilingual. In January 1916, he stunned Laurier with his first English-language speech about Quebec's role in the First World War. Laurier was "simply amazed" at Lapointe's eloquence.

It was not surprising that King and Lapointe found each other and forged a powerful partnership and twenty-year friendship. They both were suspicious of big business and high tariffs and championed the underdog. Lapointe, however, was not a social reformer (or, at least, he did not think of himself as one as King did) and was more fervent than the prime minister in his view that Canada had to assert its autonomy within the British Empire. Lapointe delivered Quebec to King, election after election, though that might have happened regardless of his influence. French-Canadian voters had few options in those years; the Liberals were "it" and routinely won close to 90 per cent of Quebec's sixty-five seats. King quickly came to trust Lapointe's political judgment on all things to do with the province.

At a meeting on December 10, 1921, King and Lapointe had their first of many heart-to-heart conversations. "I told him I regarded him as nearest to me and would give him my confidence in full now and always," King recalled of the discussion. "We would work out matters together. I regarded him as the real leader in Quebec." Lapointe had indicated he wanted the justice portfolio, and King concurred that "he is worthy of justice, is just and honourable at heart [and is] a beautiful Christian character."

Three days later, King met with Gouin for dinner at the Rideau Club, and over oysters on the half shell the jousting for a cabinet position was waged. A week later, King convinced Lapointe to bide his time and accept the more junior cabinet job overseeing marine and fisheries. Lapointe was angry, yet after more prodding consented to the change. Gouin was told that he could have justice—take it or leave it. Almost immediately, gossip circulated among the press gallery and Ottawa political watchers that a conspiracy against King was afoot. The prime minister was warned by several of his loyal supporters, according to what John Dafoe reported to Clifford Sifton, "that if he did not 'get' Sir Lomer, it was only a question of time until Sir Lomer would 'get' him. They told King that it was quite obvious that Sir Lomer Gouin regarded himself as the real head of the administration." Dafoe doubted, however, that King had the courage to take on Gouin, or "bell the cat," as he put it.

Well before a suggestion of a Gouin-led coup was in the works, King recognized his predicament, though he never believed that his position as prime minister was threatened. "I seem to have run counter to or offended the Montreal group at every opportunity," he conceded in early February 1922. "First one then another, the very men I must do my utmost to placate. I have little hope of being able to hold them, they do not want to belong to the Liberal Party, but I do not want to give them cause for complaint or offense." In April, King held his nose and ordered Liberals to oppose a motion by a Progressive MP that would have forbidden cabinet ministers like Gouin to hold directorships in corporations. Westerners shook their heads in dismay at King's kowtowing to Montreal "big business," yet he really had no choice.

For what seemed like a long while, Gouin and Fielding, the Liberal finance minister who also refused to budge on the tariff or make concessions to western Canada over control of its natural resources, made King's life slightly miserable. "We took up the natural resources matter," King complained after one heated cabinet discussion in late April 1922, "but could get nowhere with Fielding who is like a dog in a manger when it comes to making any allowance on an equitable basis." Then, in November, when Gouin and Daniel McKenzie, the solicitor general from Nova Scotia, fought King over another concession to the western provinces that had been agreed to by the cabinet, King uncharacteristically lost his temper. "I pointed out their attitude was putting me in a false light and I was unwilling to be 'humiliated' by any going back on what had already agreed." The following day, King decided he had to lay down the law: "I opened fire pretty strongly... taking exception to colleagues... entering on discussions with which they were not familiar, embarrassing a situation and going back on what was already agreed upon. I said I would not stand for procedure of that kind, would follow out pledges given the electorate in good faith or leave it to others to carry on the Government, if individuals could not agree with policy decided upon they could withdraw from Government. It was the farthest I have gone at any time." There were further disagreements over King's decision to nationalize the Grand Trunk Railway (which became part of the Canadian National Railway), his desire to lower the tariff and the (partial) restoration of the Crowsnest Pass railway rates (which had for a time significantly reduced railway rates paid by western farmers), also favoured by the Progressives but not by Fielding and the Gouin group.

Advancing age finally took care of King's dilemma. At the end of 1923, the seventy-five-year-old Fielding, crippled from a stroke, was compelled to retire. Soon after, health problems forced Gouin to resign his position, too. "I feel a great load off to be rid of Gouin," he admitted in his diary. "He just represented interests. He never moved to Ottawa. The real Liberals of Quebec are all pleased." That opened the door for King to elevate Lapointe to justice and achieve wider consensus on a tariff reduction, which would hopefully win over more Progressives and guarantee him a majority in the next election—or so he believed. He was also assured that Meighen and the Conservatives would continue to promote protectionism at every opportunity.

ON THE AFTERNOON of Saturday, September 16, 1922, King was about to enter the Quaker-inspired Temple of Peace at Sharon, Ontario, north of Toronto, to deliver a speech, when he was approached by a reporter from the *Toronto Star*. He handed King a dispatch from the British government. It was an invitation for Canada to assist British troops who were being threatened by the Turkish army at Chanak (Çanakkale) near the Dardanelles, which had been deemed neutral territory under British control. Until this moment, King had not heard a word about this and refused to comment. He proceeded to speak to the audience inside the temple, but was justifiably angry at the British government's "impertinence."

The following day, back in Ottawa, he received the official request from Winston Churchill, then the secretary of state for the colonies, for Canada to send troops immediately. It was already front-page news across the country: Canada might go to war again to support the mother country. This was all too much for King. "I confess it annoyed me," he wrote. His policy from this day forward, as he made clear in his diary and later to his cabinet ministers and the House of Commons, was that only Parliament would and could decide whether Canada entered another war. He questioned the colonial secretary's common sense and judgment. "It is a serious business having matters in [the] hand of a man like Churchill—the fate of the empire!" he wrote. (On a visit to Canada in 1932, Churchill admitted to King that he should not have issued the public declaration as he had.)

As the Chanak crisis played itself out and the tension in Asia eased, King felt even more secure that his decision had been the correct one. Much of the Liberal press agreed, as did Queen's University political economist Oscar Skelton, who told King that by his stand he had made history. "Never again will a Canadian government be stampeded against its better judgment into giving blank cheques to British diplomacy," Skelton wrote in mid-October, somewhat exaggerating the situation, "now that your government has set this example of firm and self-respecting deliberation." King was already impressed and influenced by the professor's work—Skelton had favourably reviewed *Industry and Humanity*—and his progressive attitude toward Canadian foreign policy.

The Conservative Party and the press, however, played the loyalty card for all it was worth, accusing King and the Liberals of deserting Britain in its alleged hour of need. In words that were to haunt him a few years later, Arthur Meighen declared to an audience of Tory businessmen in Toronto in September 1922: "Let there be no dispute as to where I stand. When Britain's message came... Canada should have said: 'Ready, aye, Ready; we stand by you.'"

Meighen's predecessor, Robert Borden, would have argued that the Canadian blood spilled on the battle fields of Vimy Ridge, Passchendaele and the Somme, as well as subsequent understandings reached during the imperial war conference and peace negotiations, implied that Canada was now "nearly" an autonomous country, yet still morally and psychologically linked to the Britain. (Canadian officials signed the Treaty of Versailles in 1919, but under the signatures of the British delegation.) What precisely this subtle change was to mean for Canada and the other dominions and how the process was to work were the questions that faced King during the twenties.

Trickier still was the haughty British imperial mindset that had not yet grasped or accepted the new realities of dominion partnership and autonomy as quickly as King and several other leaders were about to demand. On many issues, King had taken a dithering approach lest he offend any of his fickle supporters, but not when it came to Canada's role in the world. His deepest personal and political instincts had long confirmed—and in this he was following the lead of Laurier, who had laid the foundation for such change at colonial and imperial conferences before the First World War—that only the Canadian people, through Parliament, would determine any future co-operation with Britain and the empire. From a purely politically practical point of view—and an opportunistic one, as his enemies liked to point out—he also understood that those sixty-five precious seats in Quebec would be more easily won with a "Canada First" foreign policy. This did not mean that King wanted to sever the British connection or that he was not enamoured with British life and society, which he surely was. Nor did it mean that starting in the early twenties he sold out Canada's British heritage to the Americans and continentalism, another historical missile that has been lobbed at his legacy. As he clarified in 1922, he merely wanted the British to stop treating the dominions "as adolescent nations" (in Michael Bliss's words), and accord them the respect they deserved and had earned. The "kindergarten school diplomacy," as the Liberals caustically branded it, had to end.

King's resolve to advance Canadian interests came soon enough. In March 1923, he insisted, despite somewhat hysterical British opposition, that Canada had the right to negotiate and sign a treaty with the United States by itself on matters that were of an entirely Canadian nature. In his case, the issue was trade and regulation over the Pacific halibut fishery. King persisted, and after he threatened to appoint a Canadian diplomatic representative to Washington, D.C., and further undermine the empire hierarchy, the British relented. Granatstein and Hillmer explain the significance of this moment in their survey of Canadian-American relations: "For the first time a treaty had been entirely negotiated and signed by and for Canadians. Generations of school children ever since have memorized the name and date of the 1923 Halibut Treaty as part of the litany of Canadian nation building. Only in Canada, readers might think, and they would be right."

Next, in the early fall, King was on his way to make his case at the imperial conference in London. Instinctively he understood that this was to be one of the defining moments of his first term in office, and he wanted to be at his best. Smartly, he sought out professional help from two people who thought about Canada's relations with Britain as he did and who opposed a common imperial foreign policy, Oscar Skelton from Queen's University and John Dafoe of the *Manitoba Free Press*. The soft-spoken but highly focused Skelton was invited to accompany him as an "expert adviser"; the invitation altered Skelton's life. Within two years, he was King's under-secretary of state for external affairs. He served in the department for the Liberal and Conservative governments until his death in 1941, shaping the country's foreign policy more than any other public servant in Canadian history.

In September 1923, after reading one of Skelton's long, articulate memos on preserving Canada's international control, King wrote in his diary that the professor had "an unusually clear mind and brain, his work is excellent." His friends and foes alike believed that Skelton was virulently anti-British. This was not quite the case, argues his biographer Norman Hillmer. Skelton, he says, "was an anti-imperialist but he was not anti-empire. He described the British Empire as the most wonderful and extraordinary experiment in political organization in world history, 'absolutely unique, unparalleled, unprecedented.'" What Skelton objected to was the same nagging business that troubled King, though Skelton articulated it more clearly and intelligently: imperialistic arrogance and pomposity often concealed the Brits' more calculating motives to expand and centralize their control. Skelton was, to borrow historian W.L. Morton's distinction, "British" and proud of it; yet he was never "English." Skelton never advocated Canada's separation from the British, but he did spend his life promoting Canadian pride in the country's history and traditions.

King rightly believed that if he made any attempt to assert Canada's autonomy, it was bound to receive bad press from British and Canadian Conservative newspapers. He urged several Liberal newspapers to send journalists to cover the conference and presumably to slant the story in a friendlier and pro-Liberal fashion. Such a request was also made of Sir Clifford Sifton, the owner of the *Manitoba Free Press,* which under Dafoe had become close to serving as a Progressive Party organ and had been fairly critical of King and the Liberals. Nonetheless, King was aware that both Sifton and Dafoe were spirited Canadian nationalists on the subject of the empire, and he was well aware that Dafoe was an editor with real authority. As Frank Underhill once put it, "for the past generation now it has been generally true that what the *Free Press* thinks today, western Canada will think tomorrow and the intelligent part of eastern Canada will think a few years hence."

Dafoe had accompanied Robert Borden to Europe in 1919 and dispatched favourable reports back to Canada. Sifton was immediately keen on the idea of Dafoe travelling with King to the imperial conference in London. "You could undoubtedly have great influence with King," he wrote to Dafoe, "and might conceivably exert a determining influence on vital matters." Dafoe was not as certain. "I have very little confidence in King," he told Sifton. "I am afraid his conceit in his ability to take care of himself is equaled only by his ignorance and I should not be surprised if he should find himself trapped. I have no intention of becoming an unofficial member of a board of strategy to assist him while the Conference is on." Despite these critical reservations, Dafoe agreed to accompany the prime minister, Skelton and the members of the small Canadian delegation. Dafoe might not have wanted to become King's unofficial adviser, but once they were on the ship in the middle of the Atlantic, that's exactly what happened. Next to Skelton, there was likely no one else whose advice King valued more.

Once in London, King was "filled with terror," in Skelton's estimation, and full of self-doubt as to whether he was up for the challenge to assert Canada's contentious position. He desperately needed distractions. Based at the posh Ritz Hotel on Piccadilly, he shopped for suits, shoes and a new silk top hat and sifted through the more than fifty social invitations he found waiting for him when he arrived. There was lunch and a pleasant conversation with King George v at Buckingham Palace, dinner with Lord and Lady Astor (among the other invited guests were J.M. Barrie, the author of *Peter Pan,* and Leo Strachey, the editor of the *Spectator*), and a glorious weekend in the country with the former governor general of Canada, the duke of Devonshire, and his wife at their magnificent estate Chatsworth House in Derbyshire. On October 10, he was also sworn in as a member of the British Privy Council, an honour given to past Canadian prime ministers. He was moved beyond words.

At the conference proceedings, where on one occasion King had, in his words, "an unpleasant hour or two," he stubbornly refused to accept any hint of imperial centralization under which Canadian foreign interests might be solely dictated by Britain. "Our attitude is not one of unconditional isolation, nor is it one of unconditional intervention," he declared in a statement that exemplifies the ambiguity he is famous for. (Or, put another way by Leo Amery, then first lord of the admiralty in charge of the British navy, "I formed the impression that King's chief aim was to avoid committing himself to anything.") No one present actually disagreed with King; still, pressure was mounted by such leaders as Stanley Bruce, the intractable thirty-nine-year-old Australian prime minister, who was keen on a unified imperial foreign policy, as well as General Jan Smuts of South Africa, whose agenda was about his own personal aggrandizement as a leading light of an anticipated imperial cabinet. "Mackenzie King, you are a very terrible person," Smuts chided him. "You are giving an awful lot of trouble." Stanley Bruce's adviser Richard Casey (and a future governor general of Australia) was more aggravated with King privately and rather hysterically accused him of wrecking the British Empire. "His efforts to make capital out of his domestic nationalism," said Casey, "are analogous to a vandal who pulls down a castle in order to build a cottage."

Lord Curzon, the pretentious British foreign secretary and former viceroy of India, would have definitely agreed. His patience with colonial insolence had run out before the conference had even started.

Curzon's idea of dominion co-operation was for Britain to dictate the rules of engagement and the colonies to salute smartly in agreement. He angrily dismissed King as "obstinate, tiresome and stupid, and... nervously afraid of being turned out of his own Parliament when he gets back." Yet as several historians have pointed out, both King and Curzon wanted the same thing, but for opposite reasons: on principle, King did not want to be tied to a British-dictated foreign policy he felt he could not control; Curzon had absolutely no interest in a consultative British colonial board that would govern the empire's foreign affairs. The truth was that in his words and actions in London in 1923, King had hardly disowned the empire or Canada's commitment to the empire's welfare. When there was real danger, Canada would be there as it had in 1914 and as it would be in 1939 with the outbreak of the Second World War.

For the time being, however, King returned to Canada confident that his personal reputation and political worth had much improved. Even a curmudgeon like Dafoe had grudgingly conceded that, "as for King, my regard for him has perceptibly increased by what I saw of him in London. He is an abler man than I thought; he has more courage that I gave him credit for." He added, however, what may be the one of the great backhanded compliments in Canadian political history: "In the right setting and with the right men behind him, King would be a not unacceptable party leader."

A CANADIAN PRIME MINISTER's performance at an international event, even a respectable and self-righteous one as King's was in 1923, only produces so much positive karma with the electorate back home. This was especially the case in an era when Canadians could only read about their leader's nationalistic stand, rather than hear about it on the radio or watch him in action on television. King and his disjointed Liberals had provided adequate government, yet by 1924 had still not sufficiently altered the regional divisions that had confronted them in 1921. King's courting of the Progressives and the west, including his commitment to reduce the tariff—which he was finally able to do once Fielding and Gouin were out of the picture—he thought had had the desired effect of improving his chances in the Prairies. Luckily for King, Arthur Meighen remained a marked man in Quebec—Hugh Graham, Lord Atholstan, the cranky and wealthy owner of the *Montreal Star,* had carried out a harsh vendetta against the Conservative leader, demeaning him at every opportunity—and King was confident that Lapointe had control of the province. A Liberal sweep seemed certain again, though there were rumblings that Lapointe's mentor, Jacques Bureau, the minister of customs and excise, had serious problems within his department that he was not properly addressing. And if political polling had existed in the early twenties, it would have likely shown that in Ontario King's so-so performance, his sober personal style and his attempt to appease both protectionists and free traders had not exactly bowled over the province.

King especially obsessed over William Jaffray and the weak coverage he was receiving in the *Globe*. King tried everything to make things right with the puritanical publisher. To appease him, the Liberals had actually pushed through an amendment to ban the publication of racing results, only to have it rejected by the Senate. King tried again two years later, yet again the Senate refused to give its consent. Jaffray blamed King and ensured that during 1924 and 1925, *Globe* editorials and even news stories were highly critical of the prime minister and the Liberals. "If we could only get back the Globe," King told Peter Larkin in September 1924, "we would be able to keep the party in power for many years to come."

In this desperate quest, he enlisted the assistance of John Lewis and Hector Mackinnon, two *Globe* editors and Liberal loyalists. When Lewis was promoted to managing editor in late 1924 and wrote to King promising that he could count on his help, the prime minister was overjoyed. "Your letter has come like breaking of a fresh dawn upon the political horizon," he replied to Lewis. But King's faith was shattered several months later. In June 1925, as an election loomed and the Senate failed to pass the second attempt to ban the publication of horse-racing stats, Jaffray began meddling with Lewis's pro-Liberal editorials. Lewis soon resigned, along with Ross Munro and Melville Rossie, two other devoted Liberal editors. King offered Lewis a Senate appointment and hired him to write election propaganda, which included penning a laudatory short biography of Mackenzie King. The party also covered the salaries of Rossie and Munro for the rest of 1925.

There were high-level discussions with Larkin, Lapointe and Liberal campaign organizer Andrew Haydon. The plan was to make Jaffray an offer to buy the *Globe,* but nothing came of it. Instead, a meeting was arranged between King and Jaffray in September 1925, about a month before the election, to see if a peace accord could be negotiated. King, ever the conciliator, believed that if he was given the opportunity he could win over anyone. Not this time. Jaffray actually questioned King's commitment to Christianity, and his resentment could not be quashed.

Hector Mackinnon, the *Globe*'s city editor, was secretly keeping King up-to-date on Jaffray's pre-election editorials. When Jaffray killed a series of articles written by Mackinnon and Arthur Irwin defending King's "tariff for revenue," both journalists also resigned.

A *Globe* editorial of September 1, 1925, said it all about the support King could expect from the newspaper. "The *Globe* is an exponent of Liberal principles," the long piece began. "It is not and has not been since Confederation the organ of the party. We have criticized and shall continue to criticize the Government of Mr. King when that Government fails, as we believe it has failed." Much to Mackenzie King's dismay, that critical judgment was also shared by thousands of other Canadian voters in the fall of 1925.

5

Vindication and Victory

I confess, I never saw such a combination of adverse circumstances, [and] such a formidable array of difficulties.—THE DIARY OF WILLIAM LYON MACKENZIE KING, *August 15, 1925*

ON THE FIRST DAY of March 1925, a petite forty-eight-year-old woman from Kingston, Mrs. Rachel Bleaney, and her twenty-seven-year-old son, George, arrived at Laurier House to see the prime minister. George had been only sixteen when he enlisted at the beginning of the First World War and was now out of work and ailing. Mrs. Bleaney wondered if King might be able to assist him. What made this visit so memorable, an encounter to which King devoted five pages in his diary, was that Rachel Bleaney was a fortune teller and a woman with apparently astounding powers of clairvoyance. She was born in Milford, England, in 1877. Other than that, little is known of her background or true mystical talents. King had met her four years earlier, but he mentions her by name for the first time on this day.

Once the conversation about her son's problems had ended (King promised to look into his situation), Mrs. Bleaney read the prime minister's fortune as well as that of Joan Patteson, who with Godfroy was invited to join the intimate gathering. It was, King later recorded, "a truly remarkable experience." He jotted down every utterance made by Mrs. Bleaney and was overwhelmed by what he heard. There were words of comfort about his mother and Max and about the challenges that would confront him in the next election. Her declarations, he added, were "amazingly true of the past and if true as to the future will be astonishing beyond words, because so bold and daring in what it prophesizes."

King was assured that there was before him "a straight path" and that his vital work was to continue for many years. Mrs. Bleaney warned him that were men who were conspiring to prevent him from accomplishing his task, yet they were not to succeed. "You will win a higher place than you have at the present," she continued, "there is struggle, not so much." Most notably, she predicted that he would win the next election and be a stronger and wiser leader. Finally, she assured him that marriage was in his future.

"People do not always understand you," stated Mrs. Bleaney, sounding more like a therapist, "because you do not fight back, they would like you to fight, would like to provoke you to fight, but you do not let them rouse you to a temper, you pass by, you do not let them anger you. Yet you don't forget. You would do no one any justice, therefore you feel injustice, though you may not show it. That is why many don't understand you. You don't react as they expect, that is your strength."

Despite Mrs. Bleaney's words of comfort, King remained troubled. He was upset that Saskatchewan premier Charles Dunning, a politician he wanted in Ottawa, had decided to delay his move into federal politics, and was profoundly troubled by his inability to win over William Jaffray at the *Globe*. The puritanical publisher was driving him to distraction. Then, more bad news. During the summer, the Liberals were defeated in two provincial elections: first in Nova Scotia on July 16, where they had held power since 1896, and a month later in New Brunswick, where Liberals had governed for nearly a decade. The losses shook King's confidence to the bone.

"I feel . . . in a personal way that I might be unequal to the strain of an uncertain fall, a trying preparation for the last session, and the turnout of the [House of Commons] and a general election in the spring of next year," he wrote on August 12. "It would be a sort of hell on earth for six months or more and more than one man could endure." In this moment of doubt, King's subconscious defence mechanism kicked in. Perhaps his political career was to be brief. Would that be so bad, he wondered? "If we are defeated," he added, "if I should be defeated, it would give me a chance to get a much needed rest, to get my house—the house of my mind and heart—in order, to see where I stand in the sum of things."

A few days later, he faced his cabinet with the gloomy report: the Liberal election coffers were low; the *Globe* and the *Manitoba Free Press* had deserted the party and little campaign literature had been produced. After the meeting had concluded, Ernest Lapointe remarked that it was "like attending our funeral." Nevertheless, within weeks, King decided that he had no choice, and with Lapointe's encouragement, called an election for October 29. In the meantime, he shored up his cabinet. Vincent Massey, the philanthropist and head of his family's thriving agricultural implement business, Massey-Harris, was seeking a career change. He announced he would run in the constituency of Durham, northeast of Toronto. Massey was a die-hard protectionist, and King was certain that his presence would help in Ontario and Quebec. Massey himself had no political or ideological problems running for a party with a low-tariff platform, but he was also an easy target for his opponents, who labelled him a "renegade Conservative." This marked the beginning of King's shaky twenty-five-year relationship with Massey, one that was to strain both men.

Far worse for the Liberal campaign were the stories circulating around Ottawa and complaints from Canadian businessmen about smuggling and rum running along the U.S. border. The activity was mainly a by-product of the American prohibition of alcohol, which was now in its fifth year and an utter failure. The same U.S. congressmen and senators who had voted in favour of the Eighteenth Amendment instituting Prohibition had foolishly refused to fund its enforcement. Most states along the coast spent more on monitoring fishing and hunting regulations than they did on helping watch their borders. The Great Lakes were filled with speedboats transporting Canadian booze into the U.S. and cheap American clothing, tobacco, radio and other goods back into Canada. "Rum running," the *Financial Post* had stated tongue-in-cheek in 1921, "has provided a tidy bit towards Canada's favourable balance of trade." Corruption was rife among Canada's customs agents, who could be easily bought off, as could their American counterparts. Many of the Canadian agents, especially those in Quebec, were in cahoots with the smugglers. This crooked scheme was probably costing Canada about $50 million in missed duties. One of the key culprits was Joseph Bisaillon, the chief preventive officer for the department of customs in Montreal. For a mid-level bureaucrat, Bisaillon had a bank account that was bursting, and he owned property in Canada and the U.S.

Instead of taking real action, King opted for a band-aid approach, even though he was well aware the customs department was, to use his term, "a sink of iniquity." In his view, the problem was the aging and incompetent minister in charge of the department, Jacques Bureau. Asked in the House of Commons in February 1925 about the smuggling,

Bureau had sheepishly declared, "I don't believe any human force can stop it." (Bisaillon was actually sending his boss, Bureau, regular stocks of seized whiskey, and the minister's chauffeur was driving a smuggled car he had been able to buy cheaply.) King wanted to fire Bureau outright, but Lapointe, ever the loyal friend, objected strenuously. King acquiesced and appointed Bureau to the Senate. During a meeting at Laurier House on the evening of September 1, 1925, when this matter was discussed with Lapointe, Senator Andrew Haydon—a Liberal Party stalwart—and Quebec Liberal cabinet minister Arthur Cardin, Bureau turned up drunk. "I swear I would like to drop him from the Senate as well," King wrote later that night. Bureau was replaced in the cabinet by George Boivin, a lawyer from the Eastern Townships who had been an MP since 1911. One of his first acts was to get rid of Bisaillon. Yet that did not stop the smuggling or the corruption.

King's political instincts were fuzzy during the six weeks of campaigning. He had blind faith that his "tariff for revenue" would resonate with voters across the country, somewhat of a miscalculation, as events were to show. He also promised to reform the Senate, an issue that had most Canadians scratching their heads. As he made his way across the country, he slightly tailored his message depending on the audience. In rural Ontario and the west, he was at one with the farmers, whereas in Montreal he declared that he could protect Canadian business interests. To all, however, he stressed that only the Liberals could keep the country united.

The crowds were lukewarm, but his strategy seemed effective. In Quebec, the Conservatives were in such disarray that Meighen was forced to stay out of the province. The Tories' campaign there was in the independent hands of E.L. Patenaude, a former member of Borden's cabinet, who had resigned over conscription in 1917. He distanced himself from Meighen at every opportunity. Even with the partisan editorial support from the *Montreal Star* and *Gazette,* which rarely mentioned Meighen's name, Patenaude was no match for Lapointe's Liberal machine.

The 1925 campaign was the first on which King and other political leaders could speak on the radio. Approximately 100,000 Canadians owned a radio (within five years that number would rise to 500,000), and many others were avid listeners on the Canadian National Railway's (CNR) popular train-radio-equipped cars. The new invention proved intoxicating. Churches were soon forced to cancel Sunday night services because the only pastime anyone was interested in was "listening in" to such American shows as *The Happiness Boys, Moran and Mack* (the *Two Black Crows*) and especially *Amos 'n' Andy,* about two black country bumpkins trying to find their way in the big city. In Toronto, as early as the spring of 1923, young Foster Hewitt, a cub reporter with the *Toronto Star,* reluctantly agreed to call the play at a local senior league hockey game. Hewitt was soon hooked, and "He shoots! He scores!" entered the Canadian lexicon.

Today it is almost painful to listen to King's high-pitched, whiny voice on old taped broadcasts. He was certainly never a star like U.S. president Franklin Roosevelt, whose relaxed, intimate and personal charm made him the first "radio president." But neither King nor his supporters, nor the press, for that matter, seemed to recognize his lack of talent. Like any technological invention that has changed the world, from the telegraph to the Internet, initial assessments of the radio and its earliest practitioners were understandably enthusiastic and naive. In June 1930, the *Globe* inexplicably suggested that there was no public speaker anywhere "equal on the air" to Mackenzie King, and the Conservative *Ottawa Journal* later praised King for his "soothing, pleasant voice."

King was in awe that people from California to New York could also hear him on the radio as he broadcast from Regina and Calgary. "I confess," he wrote to Violet Markham two weeks before the election, "the experience of speaking to these vast invisible throngs has been one which has caused me to feel, more than I ever did before, the sacredness of the trust which a public man bears when he attempts to carry the lamp of leadership across an entire continent."

King was feeling particularly upbeat when he filled in Markham about the campaign and his chances on election day. "I believe that the Liberal Party stands today more united than it has been at any time since the Reciprocity campaign of 1911," he added. "I think I have been able to heal many wounds and scars by asserting at the outset the right of Liberals to an independent judgment on great national issues, and by making my position clear in disclosing no ill-will by a fair and impartial attitude towards all who at any time differed with me in opinion... I have no fear that we shall not be returned as an Administration."

The spirits, he believed, were also on his side. After visiting a railway shop in Kingston for some glad-handing, he went to see Mrs. Fenwick, a friend and a fellow "believer," who had invited the clairvoyant Rachel Bleaney to join them. (In a telling example of journalistic practices of the day, not a single reporter noted that nine days before the vote the prime minister had stopped in to see the local fortune teller.) The encounter was "one of the most remarkable—if not the most remarkable interview I have ever had," he wrote in another lengthy diary entry. (He was to use the same exaggerated adjective year after year.) Once more, Mrs. Bleaney told him exactly what he wanted and needed to hear: that he would win the election. According to the "little woman," there were two "sinister figures" hovering above him, which King decided were Meighen and his chief adviser, Senator Gideon Robertson. They were trying to "destroy" him, but he was assured that they would not be successful. He had protectors from the "Great Beyond"; his mother and Max were watching over him. Have no fear, they both told him, all would turn out as he wished. King was overwhelmed. "I can never not believe in spiritualism-so-called after today's experience," he concluded. Before the session was finished, King asked Mrs. Bleaney about Joan Patteson. "She said, yes, you are like a brother and sister. She is very fond of you, interested in all you do." The response pleased him.

That night, King spoke to a boisterous crowd at a local theatre and was heckled about his loyalties to the British Empire. He ignored the unwelcome interruption and proudly defended the imperial connection and the flag, delivering, in his opinion, "the best speech I have made in the campaign." A day before the vote, he received "a very remarkable letter" from Mrs. Bleaney. She wrote that she had been in the audience at the theatre and had seen beside him on the stage "the figure of a man," a ghost presumably, with a bandage over his eye, who had held out his hand and smiled warmly. Initially, King thought the figure might be Laurier, but then he decided that the bandage over the eye symbolized blindness. It was his father, whose "faith and courage" confirmed again that King was about to win a great victory. "I had been waiting to receive evidence of his presence," King wrote with enormous satisfaction that night. "No one can make me believe this is not all real, that this little woman is not a clairvoyant revealing the presence of those I most deeply love and who are near to me at this time ... I believe it is all true. That the dear loved ones are round about it, that in this fight I am mostly an instrument to work out the will of God."

With his faith so confirmed, the results of the election the next day staggered him. The Liberals were reduced to 99 seats, 59 of which were won in Quebec. King lost his own seat in North York, as had eight other cabinet ministers. The Progressives had dropped to only twenty-four seats, all except two of which were from the Prairies. The Conservatives had won a total of 116 seats, only 4 in Quebec (with 34 per cent of the popular vote), but 68 of 82 in Ontario, which ultimately decided the contest.

When the results had finally sunk in, King had mixed emotions. He wrote of "feeling regret" at losing in North York, though he was confident he would find and win in another riding in the near future. He rationalized his own loss to the Tory candidate, Thomas Lennox, as a result of "what money and whiskey can do." Indeed, his immediate reaction, which he set down in his diary and in a nine-page letter to Peter Larkin a month later, was to blame Meighen's dishonesty, the "foolish Progressives" and the "millions of dollars" of Tory money he insisted had been used to discredit him and the Liberals across the country—"money from the big interests, seeking further protection, and lack of organization on our part." According to King, the Tories had stooped to the lowest of tactics, spreading stories about the immorality of Liberal candidates. "Do you wonder why I have to have contempt for Meighen which is too great for words?" he asked Larkin.

The *Globe,* on the other hand, which had deserted the Liberals, declared that the election was a "severe want of confidence in the administration." King cursed the paper. "There is hardly a doubt," he told Larkin, "that the . . . loss of Ontario . . . was due to the attitude of the 'Globe,' which bewildered many of our own party who felt that there must be something radically wrong or we would not have been stabbed front and back by that journal as we were." There was, he added with dismay, a report from one of the teachers at the Toronto Institute of the Blind, "who being herself a good Liberal, read aloud the 'Globe' to the inmates of the institution all through the campaign with the result that when the day of polling came these poor physically blind young women each and every one voted against the Liberal party because the 'Globe' had told them it was the only thing to do."

The following day, as King contemplated his next move, he found in the morning mail a "very remarkable letter" from Mrs. Bleaney. A week earlier, he had had a "vivid dream" and had sent Mrs. Bleaney a lengthy commentary describing in excruciating detail every aspect of this vision. He was on the deck of a ship, surrounded by his family members, who were singing. The louder the music became, the darker the image became. And so it went. In Mrs. Bleaney's similarly long interpretation of the dream, penned a day before the election, the ship King imagined represented Parliament. She had predicted a victory for the Liberals all along, but now saw in his dreams the signs of a temporary setback: "of seeming defeat, of the party being in a stronger position a little later on, of a clear way ahead after these troubled times, of the time it would take for people to understand, of great spiritual power to come later on."

KING DID THE math. The twenty-four Progressives, two Labour MPS from Winnipeg—J.S. Woodsworth and Abraham Heaps—and a handful of independents were not about to support a Meighen government. The Liberals could easily hold on to power and maintain the required confidence of the House with their votes. King canvassed his ministers and advisers, and the majority recommended he carry on, as was his traditional parliamentary right. More or less confident in this course of action, King went to see the governor general, Lord Byng.

Field Marshal Julian Hedworth George Byng, First Viscount Byng of Vimy GCB, GCMG, MVO, or "Bungo" to his closest friends (his two older brothers were known as "Byngo" and "Bango"), had taken up his post in Ottawa accompanied by his wife, Lady Evelyn Byng, in the summer of 1921. From the tip of his shiny black boots to his thick moustache, Lord Julian Byng was all soldier. A First World War commander and the esteemed leader of the great victory at Vimy Ridge in April 1917, Byng was well respected by the many Canadians who served under him. He had a military bearing but was friendly, and he was a natural choice as governor general. Early in his term, he and his wife took a train across the country, drawing large, loving crowds. They also became rabid hockey enthusiasts, so much so that in 1925 Lady Byng donated a trophy to the National Hockey League, which still bears her name, given each year to the most gentlemanly player. In retrospect, this honour was somewhat ironic, considering that Lady Byng was known to have had a particularly nasty temper and could swear like a drunken sailor. She also didn't care much for Mackenzie King.

King was genuinely fond of Lord Byng. "I cannot tell you how pleased I am with him as a man and how certain I am that his administration will prove worthy," he confided to Violet Markham at the end of September 1921. "I took an immense liking to him from the moment we met and in each conversation I had since. I have felt a sense of oneness in aim and point of view, which is really remarkable. If by chance I should come to be his adviser in the near future, I shall expect that the association will be one of the happiest and best of my life." At the same time, he found Lady Byng, whom he regarded as "an intellectual face, not a beauty at all, but a sensible woman," difficult to carry on a conversation with at a dinner party. The one sensitive subject they might have broached, though not with other guests within earshot, was spiritualism. Lady Byng was a believer and corresponded with King about it in the late spring of 1923. In one letter, she advised him to be cautious in attempting communication "from the Beyond."

One of King's first informal meetings with Lord Byng took place on September 2, 1921. In words that were to hold much more significant meaning five years later, the governor general told King that he was not a constitutional expert and that he expected "little difficulty." He added, "Mr. King, I want to put myself in your hands. I have only one object that is to be of what service I can, I have no axe to grind only to do what I can where opportunity offers." At the celebration hosted by the Byngs at Government House following King's victory that December, Lady Byng pulled the new prime minister aside. "She congratulated me," King recalled, "said I would find Lord Byng very good at keeping everything secret, that having been in military life, they were not in politics."

By the fall of 1925, Lord Byng's political expertise had apparently improved. On October 30, Byng and King met together in the governor general's library to confer on the government's plans. "Well, I can't tell you, my dear friend, how sorry I am for you," Byng said, lighting his pipe. The two men sat facing each other beside the fireplace. They chatted about the election results for a few more minutes, before they arrived at the crux of the matter. King informed Byng that he did not believe Arthur Meighen, despite having won more seats than the Liberals, would be able to form a government that could survive. Byng interrupted King's constitutional lecture, pointing out that the voting results surely indicated Canadians' unhappiness with King's administration, especially since he and so many other cabinet ministers had been defeated. King countered with the argument that in effect he had only lost the vote in Ontario and that he did not regard the results as a defeat of his government.

Lord Byng cut to the chase. He instructed King that there were three options. The first was dissolution and another election, which he hoped King would not request since he could not in good conscience grant him one at this time. And rightly so. The second was that King resign, and Byng would then ask Meighen to form a government. The third possibility was that King would continue. "I shall of course agree to whatever you say to the last two," said Byng, "but Mr. King as a friend of yours may I say I hope you will consider very carefully the wisdom of the second course. See the position, you will be at the mercy of the Progressives, of Woodsworth, and another Labour man [Heaps], you will have to go to them, or be at their bidding, the country will say you are caring only for office and the fruits of office, they will tear at you ... Remember, my dear friend, it hurts me to say these things. I hope you know I am your friend." King assured Byng that he had no thought of carrying on "by kow-towing to the Progressives." His motives, he said, were politically altruistic: His Majesty's government must carry on, and he could not see how Meighen could do that.

More discussion ensued before King departed completely at ease with the situation, which he coloured with a spiritual hue, linking everything that had transpired to Mrs. Bleaney and his dream. "I felt a great mental relief as his Excellency talked, and was indeed grateful for his genuine friendship," he wrote later that day, "he is a true friend, a man of very great spiritual strength and power. I can see the whole vista clear—exactly as in the interpretation of my dream—how strange. My nature and reason revolt against 'spiritualism' and all the ilk, but not against the things of the spirit, the belief in spiritual guidance, through institutions. It is the material manifestations I feel [cheery] about, on the other hand when in faith and prayer I have asked for them, and they come in such an unmistakable manner, are they not to be accepted in all faith and humility—just at this time when guidance from on High is needed. It is all very beautiful as well as very lonely. I only feel that in so many ways it is not deserved. Were my life more spiritual I would feel more justified."

For two sleepless nights and busy days King went back and forth in his own mind over what decision to make. Lapointe and Charles Murphy, who had been recently appointed to the Senate, counselled him to continue, whereas Vincent Massey, who had lost his bid for a seat, thought he should permit Meighen to try to govern. King listened to this various reasonable advice but was guided by a higher power. "I have had great comfort and help from the spiritual influences about me," he wrote on October 31. "I cannot do other than regard all Mrs. Bleaney has told me as revelation." That evening, he and the Pattesons sang hymns. Leafing through a book of prayers, he found an old telegram he had sent to his sister Jennie back in 1916, after their father had passed away, with the suggestion that their mother come to live with him. In King's mind, especially when he was under stress, such occurrences were never coincidences; everything had a much deeper meaning. "It was like their desire to have me have a word with them. I read the hymn at the page, to my amazement it was of sowing the seed from coast to coast from Labrador to Pacific Coast and all about the interests of the miners, the lumbermen the fishermen, farmers, commerce, etc.—just the things I have been talking about and the people I have been talking to." He dreamed that night that his brother, Max, "was helping to make the way plain."

Even this spiritual guidance could not clarify the path to take. During the morning of November 2, he drafted a letter to the governor general tendering his resignation and asking Byng to call on Meighen to form a government. He never finished it. He met his cabinet in mid-afternoon and everyone, other than Vincent Massey, was in favour of continuing. Three hours later, having convinced himself that his duty to the country was to remain in office, he returned to Government House. Lord Byng was not pleased. He told King that he believed Meighen had earned the right to try to govern. The governor general asked King to give the decision more thought.

The next day, the two men met again. Byng was more direct. He said he was prepared to accept King's advice on remaining as prime minister, but that he did not agree with it. "I immediately expressed surprise," King recollected the conversation, "and asked how his Excellency could accept my advice and still express a different view." They agreed to disagree on this point of constitutional practice, though King argued that as governor general Lord Byng "was not allowed to have views, but to accept or reject the advice of his Ministers." Byng "repeated his view that the elections signified a Conservative wave." Again, King left Government House without arriving at a satisfactory resolution to this political conundrum.

King conferred with lawyer and Canadian nationalist and constitutional expert John S. Ewart, whose opinions he valued. Ewart was firm in his view that Byng "had no right to express his opinion" or ask King to do anything. It was for King "to acquaint him with the general political situation" and to tender him advice accordingly. This was rather a harsh interpretation of constitutional tradition—even today, the governor general is free to express an opinion to a prime minister in private, at any rate—but King and the Liberals certainly had the right to face Parliament and a vote of confidence. Before the night was over, King went back to see Byng and told him of his decision to remain prime minister. Byng consented, but suggested that in any press release King indicate that the governor general had presented other alternatives. King correctly refused to publicize their private discussions, and repeated that as governor general, Byng "had no opinions."

Another round of cabinet discussions took place the next day, followed by more haggling with Byng over the wording of the press release and more advice from Ewart. At long last, King issued a statement of his decision to stay in office. In what would develop into the "King-Byng Affair," Lord Byng and his secretary Arthur Sladen swore that the governor general had informed King that in the event of a non-confidence vote, he would not be granted a dissolution and another election until Meighen had a chance of forming a government. Byng also told Charles Murphy and his former aide-de-camp (and future governor general) Georges Vanier about this, as well as his wife. "Bungo," Lady Byng inquired, "did you get Mackenzie King's promise in writing?" The governor general assured his suspicious spouse that he and King had made a "gentleman's agreement."

With the matter seemingly settled, King was enormously satisfied with how this delicate situation had been handled. He was not bitter about Byng's position but did interpret the governor general's actions from his unique suspicious perspective. "Lord Byng has certainly tried to be fair and just and has been fair and just," he observed. "The natural Tory could not [help assert] itself in the feeling that the Government should resign and let the Tories come in, but this I truly believe, was meant as much, if not more in my own interest than from any love of or desire to help Meighen." Besides, "through all this," he added, "the soundness of all I have been told as coming from the loved ones is amazing." Four days later, still feeling apprehensive even if he did not quite admit it to himself, he reiterated the same point. "I feel I am being guided from above, that dear Mother and Max and Father and Bell the whole family in Heaven are guiding and directing me." Little did King know that this round with Lord Byng was only the beginning of a much longer and more heated bout.

ARTHUR MEIGHEN WAS incredulous that Mackenzie King refused to surrender power. "The popular majority against the government is overwhelming," he declared within days of King's decision. "To cling to office under such circumstances is usurpation of power and contempt of the popular will." Maybe so, but there was not much the Conservative leader could do about it until the House reconvened in early January. Meighen did not help himself or his popularity when on November 16, 1925, in a speech delivered at a banquet in Hamilton, he stated that as a general rule a decision by Parliament to go to war should be put before the people in a general election. He had not quite suggested, as his army of critics asserted, that a referendum be called; he accepted that national emergencies arise and that in such occasions governments had to act quickly. But it was too late; his "Heresy at Hamilton," as it was labelled, shocked his friends and supporters. Grattan O'Leary later recalled that in 1922 Meighen "said our answer should have been 'Ready, aye, Ready.' But now he was saying, 'Ready, aye, Ready... but after a general election.'" Meighen attempted in vain to clarify his position and argued that he had been misinterpreted. Yet as his biographer Roger Graham observes, Meighen had "caused disruption in the party at the very moment when it seemed to stand on the threshold of power, a disruption which someone more sensitive to the intricately varied currents of public opinion might have anticipated and avoided."

Mackenzie King had his own problems and did not pay much attention to Meighen's statement—other than to point out in his diary that it "would do him harm in Great Britain, but they will swallow anything a Tory has to say." He had decided, contrary to the judgment of several key supporters, that he would not seek another seat until Parliament met. That, he told himself, would truly be a "usurpation of power" and unacceptable. There were loud whispers from Vancouver to Ottawa that King should be replaced by Charles Dunning, the premier of Saskatchewan about to join the federal Liberals. Dunning, however, quickly quashed the rumours; he was keen to assume the leadership, but decided against taking such a radical act. King discounted the conspiracy gossip, if it can be even called that, under any circumstances.

During this period, King remained cool and confident. He was sure of himself and handled each problem that confronted him with the deftness of a seasoned political veteran. He certainly could have disappeared from the Canadian political scene in 1925 and been remembered today as an also-ran. Instead, he dug deep and managed an impossible situation as cleverly and as cunningly as he had ever managed any political issue. It would not be an overstatement to say that King cemented his prolonged political future in 1925 and 1926.

His overly sensitive nature was more acute during this time as well. His leadership difficulties aside, he became irritated when he attended a hockey game on the evening of November 19, 1925, at the invitation of Lord and Lady Byng, only to have the governor general and his wife snub him. In between the first and second periods, King spoke to Lord Byng but was not asked to join him and his wife in their private box. King took such slights, whether real or imagined, very personally and was perturbed. "After all I am the prime minister of the country and His Excellency is a visiting governor," he complained in his diary. "The fact that at the moment there is a difficult situation only makes me more indignant that they had not the courtesy to recognize the situation as meriting a little graciousness. The box was full of a lot of people from England who adopt a sort of superior air towards those of us in Canada. I confess to a certain feeling of genuine indignation at the action of their Excellencies tonight—Toryism."

This was followed a few weeks later by an uncomfortable dinner at Government House at which King felt he was once again ignored. He was seated beside Eva Sandford, Lady Byng's lady-in-waiting. What transpired during dinner is anyone's guess. Miss Sanford later claimed that King pinched her thigh twice in the most ungentlemanly fashion. To ward off his unwanted advances, she was forced to kick him hard in the leg with her high heel. There is no reference to this incident in King's diary, and it is difficult to imagine he would have lost control like that in front of the Byngs. Lady Byng, not surprisingly, believed Eva, and once King had departed she told her husband in a harsh tone that she never wanted to invite King for dinner again. Thereafter, she tried to avoid him as much as possible. Soon, according to Sandford, all of Ottawa high society was laughing behind King's back about the alleged pinch.

IN EARLY JANUARY 1926, Lord Byng choked his way through the Speech from the Throne. Without a seat in the Commons, King left the election of the speaker to Ernest Lapointe and remained hidden away in his office. He had canvassed his own MPs and such influential westerners as Charles Dunning and John Dafoe, to ensure that enough of the Progressives and independents would support the Liberals. Holding the balance of power in a parliamentary structure can be intoxicating for any group of politicians who have no hope in hell of ever achieving power. Moreover, how in good conscience could Progressives like Robert Forke or social gospellers such as J.S. Woodsworth vote against King and the Liberals and hand over power to Meighen and the Tories? That scenario would have been much, much worse.

Mackenzie King instantly understood the impossible dilemma the westerners faced and used their predicament to his and the Liberals' advantage. He invited Forke (as well as another Progressive MP) into the cabinet, but as in the past the Alberta members, who still did not trust King and the partisanship of the government, vetoed the move. He did, however, finally bring Charles Dunning from Regina to Ottawa and into the cabinet as minister of railways and canals. Suddenly the Liberals were prepared to establish a farm-loan plan, find the money to complete the Hudson Bay Railway project (linking The Pas to Churchill in northern Manitoba), give the Alberta government control of the province's natural resources and create a neutral tariff advisory board.

The price for the two key votes of Woodsworth and his fellow labourite Abraham Heaps was a national pension plan, social legislation King had long ago endorsed, though he had not included such a scheme in his Speech from the Throne. Once Woodsworth's and Heaps's intentions were known, outlined in a letter they sent to both the Liberals and the Conservatives, King met with the two MPs in his office and then again for dinner at Laurier House, before finally giving them the assurances they were demanding. He also invited Woodsworth to join the cabinet as labour minister, but the social-gospel preacher quickly rejected that idea. Woodsworth knew he could never be part of the Liberal caucus and did not have enough confidence in King to work that closely with him.

By the end of January, with a lot of input from Heaps, a bill was drafted and revised several times. The proposed pension of a maximum of $20 per month for British subjects over the age of seventy, based on an income-means test, was dependent on provincial partnership that was not yet forthcoming. Negotiations still had to take place. The bill passed the House in March without opposition from the Tories, despite Quebec Conservative Robert White's assertion that this modest pension (the average Canadian earned about $1,000 annually in 1926) "would necessarily burden the taxpayers and destroy thrift." The pension scheme was then defeated by Conservative senators. Still, for the time being, King had the votes of Woodsworth and Heaps in his pocket.

BY MID-FEBRUARY 1926, King was back in his rightful place in the House after winning a seat by a margin of five thousand votes in Prince Albert, Saskatchewan. Premier Jimmy Gardiner ensured that the supposedly independent farmer who rudely challenged him— King maintained, as he informed Violet Markham, that his opponent was "supported by Tory money"—did not really have a chance. Since a "Prince Albert" was also a frock coat, made stylish by Queen Victoria's husband, the joke of the day was that King had "borrowed a Quebec fur coat to go west and now he has returned in the spacious folds of his Prince Albert." King, more seriously, took the win to mean that he was becoming a "spiritual westerner" and set the plans in motion to award his new constituency a national park. He felt reinvigorated by his victory and at peace with himself and his actions. "While there has been much that has been trying and fatiguing in all I have had a great silent joy in seeing matters shape themselves as my vision had foreseen they would," he added in his letter to Markham of February 26. "Through it all too I have been conscious of a spiritual power and guidance of which some day I will tell you—a truly remarkable experience and related very closely to my dear mother."

For six months, with the able assistance of the spirits, King and the Liberals survived one vote after the other as Meighen and the

Conservatives sought to defeat them. King had convinced himself, too, that the Byngs were in league with the Tories and conspiring against him.

King was finally undermined by the exposé of the full extent of the scandal in the customs department, which he had not properly dealt with a year earlier. He had hoped in vain that George Boivin, who had replaced Jacques Bureau as minister of the department, would clean up the corruption. (According to Vancouver Conservative Harry Stevens, Bureau had removed from his office nine full filing cabinets worth of damning evidence.) Instead, Boivin had made the situation amazingly worse. In the words of journalist Ralph Allen, Boivin approached his new job "like a man who has inherited a concession." Ongoing smuggling had cost the federal government more than $200 million in lost duty charges. And, as was revealed in mid-June by a special parliamentary committee (of which Harry Stevens was a key member), Boivin had engaged the services of bootleggers and, rather unbelievably, had sprung one of them—Moses Aziz, from the town of Caraquet, New Brunswick—from jail. Aziz was also a talented Liberal fundraiser and the local candidate, Jean George Robichaud, had desperately required his services heading into the October 1925 election. Without conferring with either the minister of justice, Ernest Lapointe or King, Boivin, who was a lawyer, wired the customs office in New Brunswick directing them to release Aziz from custody. Robichaud won his seat, and that was that, until this shocking story of judicial interference exploded in the House of Commons and in Mackenzie King's face.

The committee's report strongly recommended an overhaul of the customs department and a slew of firings. Boivin was to be officially censured. Sensing the worst, King had wanted Boivin to resign and possibly run again in a by-election once things had cooled off, but Lapointe and the other Quebec members would not let their friend be "sacrificed," no matter how poor his judgment had been. King was dumbfounded by this attitude, a peculiarity, he reasoned, of the odd workings of the French mind and soul. "It did not seem to matter that [Boivin] had permitted criminal acts," he wrote. "The French Canadian has a different view of these matters than the Anglo-Saxon. There is something fine in its chivalrous side, but from the point of view of morality it is open to question." Boivin stayed in the cabinet, though the strain of this episode was too much for him. Six weeks later, he died of appendicitis at the age of forty-three.

On June 22, Stevens introduced an amendment to the special committee's report to censure Boivin's actions as "unjustifiable" and the Liberal government's management of the customs department as "wholly indefensible." The Progressives, who were stunned by the stench of corruption and ineptitude emanating from the customs department, could not in good conscience oppose the amendment. Boivin defended himself pathetically. If passed, Stevens's amendment, akin to a vote of confidence, would have forced King to resign and seek a dissolution from the governor general.

J.S. Woodsworth was faced with the toughest decision. He was disgusted by the Liberals' actions. "What is the penalty for debauching a government department?" he asked, with reference to Jacques Bureau. "A senatorship." But how in the same instant could he vote to defeat King in favour of Meighen? Under the Tories, he rightly feared his hard-fought-for old-age-pension legislation would have been lost for good. Woodsworth took the next best course of action: he proposed a sub-amendment that removed Boivin's name from Stevens's indictment and sought the appointment of a royal commission to get to the bottom of corruption in the customs department. Another three days of rancorous debate followed in the House, which included a spur-of-the-moment two-hour speech by King laying out the situation as he saw it. "Our men were greatly pleased and apparently satisfied, which is the main thing just at the present," he later wrote, patting himself on the back for his efforts.

On June 25, Woodsworth's sub-amendment, which would have bought King some much-needed breathing space, failed to pass by two votes. He scrambled for another forty-eight hours, desperately trying to come up with another amendment more Progressives would support. The best all of that backroom negotiating produced, however, was an adjournment of the House at five o'clock in the early morning on Saturday June 26. King was back in his office in the East Block by noon, with his mind made up about what action to take next. He would see Lord Byng and seek a dissolution of Parliament and another election. At an afternoon cabinet meeting, everyone around the table concurred. From King's perspective, neither the Liberals nor the Conservatives nor any other party or grouping could maintain the confidence of the House. An election was the only answer, and, of course, it would save King from a vote of censure.

King's thinking on this as he set it down in his diary was as follows: "That it was the constitutional duty of the Governor General to give a dissolution on the advice of the Prime Minister; that this had not been refused in 100 years in British history nor since Confederation in Canadian history; that it was not advisable to ask him to call on Mr. Meighen as I did not think Mr. Meighen could carry on." In his understanding of constitutional precedents, King was relying on a memorandum prepared for him by Oscar Skelton, which included the expert, and certainly biased, opinion of John Ewart. King was about to discover that Lord Byng interpreted the situation and his constitutional power quite differently.

Late in the afternoon, King discreetly paid Byng the first of several unpleasant visits that were to take place that weekend. (Transcribed in King's diary, the three-day period from June 26 to June 28, 1926, takes up twenty-nine single-spaced typed pages, or about 22,000 words.) After King and the governor general had snacked on tea and bread with butter, the prime minister outlined what had transpired in the House of Commons and then requested that he be granted a dissolution so that another election could be called. Like a desperate schoolboy pleading for a higher grade, King argued that if Byng did not grant the dissolution, he would be setting himself up for unwarranted criticism.

Byng then responded, setting in motion the legendary King-Byng Affair. The governor general pointed out that since October 1925, King (contrary to Byng's desire) had had the opportunity to govern, and that it would be most unfair to call another election without giving Meighen and the Conservative Party, which had the most seats in the House, a chance to form a government. He hoped King would not think he was being "pig-headed," but he had given this matter a lot of thought. In this, Byng was being consistent with what he had told King back in October. No amount of complaining based on scholarly exposition or British constitutional precedence that King threw at Byng during the next day and a half was going to alter this truth.

King believed to his dying day that he had been right and Byng had been wrong; that Byng's great error had been in believing he was the "umpire" between the Liberals and the Conservatives when his duty was to heed the prime minister's advice. The majority of historians and political scientists, including Eugene Forsey, who in the early forties wrote his doctoral thesis on the conflict, profoundly disagreed with King's self-serving rationale. "I had not, even then," Forsey wrote in his memoirs six and a half decades later, "the slightest doubt that Lord Byng's refusal of Mr. King's request for a dissolution of Parliament was completely constitutional, and indeed essential to the preservation of parliamentary government." As is now widely accepted, King was within his right to seek a dissolution, though it would have carried much more weight if he had done so after, not before, the vote on Harry Stevens's motion to censure his government. King had "played fast and loose with the traditions and practices of parliamentary government in his efforts to escape [the vote of censure]," suggests Jack Granatstein. Lord Byng, as Forsey and others have argued, properly refused King's request. Throughout their spirited discussions, Byng, as King noted in his diary, was "very perturbed at my stand and drew himself up in a tense position many times." Indeed, Byng and his wife never forgave King for his behaviour. The governor general's position throughout this ordeal was consistent. "If dissolution can be obtained each time a [prime minister] fears an adverse vote in the House," Byng explained to Georges Vanier, "it is the negation of Parliament's authority."

In a last-ditch effort to save his political skin, King begged the governor general to seek advice from the British government. Byng was rightly taken aback by this odd request, coming from a prime minister who defended Canada's right to be an autonomous nation. As King grasped frantically at straws, he now was inexplicably recommending that the British secretary of state for the dominion affairs, Leo Amery, interfere in Canada's political affairs. Amery called King's suggestion "preposterous." (King had assumed that Amery would agree with him under any circumstances.) Another tense meeting between King and Byng took place on the afternoon of Sunday, June 27, with King once more arguing his case like a lawyer with a client on death row, and the governor general, who "seemed to get a little impatient," as King recalled, not budging from his original position.

That night in the library of Laurier House, King realized that in the morning he would have to do the unthinkable: submit his resignation as prime minister. In need of companionship as well as spiritual guidance, he telephoned Joan Patteson to bid her goodnight and seek comfort. They spoke for a few minutes of the hymns they adored and of God's presence in their lives. When he hung up the phone, King glanced at the painting of his mother and thought again of the biography of Gladstone opened on her lap in the portrait. He sat by the piano and played "O God of Bethel" and "Lead Kindly Light," his mother's favourite hymns.

> I felt notwithstanding I had received what seemed to me messages from dear Mother through the bible, that I must again take it down again to receive a message from her. I took out the large volume, put my hand on the page and finger at certain words—all without my knowledge of what book I was opening to say nothing of the chapter and verse—when I looked down here is what I read. 'The God of our

fathers hath chosen thee, that thou shouldst know His Will and see the Just One and shouldst hear the voice of his mouth For thou shalt be witness unto all men of what thou hast seen and heard.' I could scarcely believe my eyes. My fingers were at the words, "the God of our fathers."

On Monday, with nothing changed, King did what he felt he was commanded to do, and at a final meeting with Byng he resigned as prime minister. Byng was not at all happy that he had been forced into such an uncomfortable situation. In King's version of their conversation, Byng began, "Mr. King, I am an old man and you are a young man. You have many years before you; I have not many years left. I would like to enjoy the quiet life, and this free from ignominy. [King notes that ignominy was not the word, and a less strong one was used.] You are full of ambition and I would like to see you get on; you love power." King replied, "His Excellency was mistaken in that regard, that power was quite secondary to the principles which I felt it my duty to defend and stand for." They shook hands, still friends, or so was King's view, and he departed. King then entered the Commons and stunned everyone with the news of what he had done. Strangely, he wrote that he had "never felt freer or happier" when he announced his resignation and the House adjourned. Bursting with self-righteous elation, he made his way back to Laurier House to rest, give a brief interview to Charles Bowman of the *Ottawa Citizen* and contemplate all that had transpired.

One last point needs to be made about this showdown. In his exasperation with Byng, King did not account for one key factor that might have altered his career path for good. Thomas Crerar wrote of it in a 1950 review of Bruce Hutchison's biography of King, *The Incredible Canadian.* "If Lord Byng had given Mr. King dissolution when he asked for it," Crerar commented, "the election would have been fought upon the record of the Government in the Customs administration and, to me at any rate, there is no doubt the Government would have gone down." In this turn of events, King was most fortunate, though he hardly appreciated it at the time.

Meanwhile, Lord Byng reported the situation to King George v and Leo Amery. "Mr. King, whose bitterness was very marked Monday," Byng wrote to Amery, "will probably take a very vitriolic line against myself—that seems only natural. But I have to wait the verdict of history to prove my having adopted a wrong course and this I do with an easy conscience that, right or wrong, I have acted in the interests of Canada, and have implicated no one else in my decision."

Lady Byng was much angrier and far less diplomatic. Her hatred for King reached a new level. In a letter of July 16 to her friend John Buchan, Lord Tweedsmuir, who was to become the governor general a decade later, she had promised upon their return to England to tell him more about

> what a scurvy cad Mr. K. is and always has been… He has come out in his true colours as totally regardless of Empire, Crown and everything but his own 'place in the sun'. As for his treatment of [Julian] all through that recent period it was disgusting beyond words. For three solid days he came up and insulted, bullied, threatened him, with everything he could think of; in the hopes of bringing an utterly upright man of honour to his own despicable depths of moral degradation. Power is his watchword—the power of M.K.—and there is not one other thing in the whole world that counts in his sight. A true Judas Iscariot, he tired to betray [Julius] in the House with lying protestations of a sham affection that had NEVER really existed, and which was only an incidental handle to him for attempting his own advancement.

A scurvy cad or not, Mackenzie King was right about one thing: Arthur Meighen did not have the votes to survive. Partly because he believed he was entitled to the office and partly because he could not embarrass the governor general and refuse his request, Meighen accepted Byng's invitation to form a government. "No decision he ever made was so often and unreservedly criticized as this one—later on when things turned out badly," wrote Meighen's biographer. The Liberal press, including the *Manitoba Free Press,* railed against him for being power hungry; privately, however, John Dafoe argued that King's resignation meant that his political career was likely over.

Meighen was caught in a trap. In those days, the prime minister and the members of his cabinet were required upon appointment to resign their seats and seek re-election from their constituents as a sign that they could administer the government's affairs. Usually this was a formality, and opposition members did not run against the ministers. (This practice was done away with in 1931.) In any event, Meighen had a big problem. If he and the handful of men he appointed to his new cabinet resigned, his government could not survive a vote of confidence since it would be in a minority position. To get around this, Meighen himself resigned, but appointed a cabinet of ministers without portfolios. This ploy permitted the members to take on the acting responsibilities of a cabinet minister, yet for the moment to remain in their seats. King and the Liberals pounced all over this manoeuvre, declaring that as "acting ministers" these men had no right to administer the country's business. Within a few days, the Progressives joined with the Liberals, and the Conservative government was defeated.

Temporarily out of the House, Arthur Meighen could only watch in dismay as he lost power. He immediately saw Lord Byng, and this time the governor general granted a dissolution. A federal election was called for September 14. Neither Byng nor Meighen were pleased with this result, but their memory of what had transpired was selectively different. Some months later, when he was back in London, Lord Byng told D.B. MacRae, an editor then working for the *Manitoba Free Press,* that Meighen "had insisted that he could carry on and avoid a general election." However, when Byng saw Meighen's cabinet, he knew all was lost. They were, he told MacRae, "the worst looking lot he had ever seen assembled around Parliament." All Byng could think then was "God help Canada."

Meighen, however, in a later discussion with Winston Churchill, said that when his government had failed, he had recommended to Byng that he should ask King to form a new government. He added, "in giving this advice to Lord B., he had said that, in his opinion, he was acting against the interest of his own Party, who strongly desired a dissolution; if the Party learnt that he had given such advice, he would forfeit the leadership of it." In short, Meighen had been a marked man, no matter what action had been taken.

King was naturally as proud as a peacock and feeling completely vindicated. Recounting the events to Violet Markham, he declared, with as much hyperbole as he could muster, that "the whole proceeding is as unconstitutional as anything that has ever happened in British history." Whether the Canadian public truly understood the heady constitutional issues is highly debatable. It didn't much matter to King; he knew that he had found his election issue.

In the west, King also benefited from the renewed editorial support of Dafoe's *Free Press.* Notwithstanding his lack of faith in King's leadership skills, the Winnipeg editor decided that too much was at stake and hammered away throughout the ensuing campaign about Lord Byng's alleged unacceptable behaviour and Arthur Meighen's imperialistic tendencies. He also did not believe that Meighen, despite what he had declared in his Hamilton speech about calling an election before Canada went to war, could be trusted to protect the autonomy of the dominion.

Only King, Dafoe reluctantly concluded, could do that. As he put it to Clifford Sifton, "we have indirectly done King a great service because our fight in the west was more against Meighen and his policies than for King."

Meighen was furious. For all time, he regarded the constitutional issue as a bogus scam by the Liberals to avoid responsibility for the customs scandal. That may well have been the case, yet Meighen still campaigned on a high-tariff policy, alienating western voters, and French Canadians still regarded him with disdain. For such a bright man, he just did not know how to handle people.

One episode says it all. In Prince Albert, Saskatchewan, King's only opponent, a Conservative, was a thirty-one-year-old lawyer named John Diefenbaker. The future Tory prime minister, whom the journalist Bruce Hutchison described as a "frail, wraithlike person," already had "a voice of vehement power and rude health [that] blared like a trombone." Nothing was too low for King, recalled Diefenbaker. The Saskatchewan Liberals who were supporting King continually raised the spectre of the Wartime Elections Act and scared the Ukrainians in the riding by suggesting that a Conservative government might deny them their civil rights again, as had happened in 1917. Yet far worse, in Diefenbaker's opinion, was Meighen's declared opposition to old-age pensions. Any suggestions in favour of pensions, as Diefenbaker proposed, were immediately dismissed as "socialist caricatures of fact." At a campaign stop in Saskatoon at a local Methodist church, one elderly gentleman asked Meighen why he was opposed to a $240 old-age pension. "For ten minutes," Diefenbaker remembered, "Meighen took apart this old man as only he could. Support in the audience went down the drain. The only Conservative elected in the Prairies that year was the Honourable R.B. Bennett. And they tell me that the Meighen name is magic in Canadian politics."

On election day, King defeated Diefenbaker by more than 4,000 votes, and across the country the Liberals amazingly won a majority of 124 seats—this total included 8 Liberal-Progressives led by Robert Forke, who became a member of the new cabinet—to Meighen and the Conservatives' 91. Meighen lost his own seat in Portage la Prairie, Manitoba. "Mr. King's luck is extraordinary," proclaimed the *Canadian Forum*. In retrospect it wasn't luck so much as putting before voters one choice worse than the other. In this Catch-22, King edged out Meighen. In words with more meaning than he could have ever imagined, Dafoe told John Willison a few days later that "the angels are certainly on the side of Willie King. He has a finer opportunity now than he had in 1921, and I hope he will be equal to it. I am beginning to think that probably he will measure up to his opportunities this time."

King knew about the "angels" watching over him the minute Meighen had set the date of the election. Relaxing at Kingsmere on July 4, he walked in the garden and sang "O God of Bethel" to himself. He returned to his cottage in happy spirits, reflecting on when "dear mother made her great flight" into Heaven. Just then, a "beautiful bird" perched on a bird bath. "I watched it," King wrote, "and noticed particularly the scarlet head. Instantly I thought as I often think of the time mother was partially unconscious and spoke of herself as a bird... It was at this moment I must be assured of her presence, which I am. I keep thinking of what Mrs. Bleaney said about winning a great victory. I believe it is at hand. Already a great victory has been won in justification of my stand but it is the victory with the people... [that is] the battle now ahead. I believe the people are with me."

His mother was with him for the next six weeks. By ten in the evening on September 14, as the returns came in, he knew he was about to become prime minister again. He was not able to return to Laurier House until 4:30 AM. Once there, he describes the moving scene: "I kissed the lips of the marble bust of dear Mother and came to my library and prayed very earnestly for strength and guidance." He read his bible, a moving passage from Acts 1, which he had also read at three in the morning on June 25. This was fate, King believed. "I went to bed greatly pleased with the result," he added, "and with little Pat a companion at my side. He seemed to understand as well."

MEIGHEN SPENT THE remainder of his life trying to figure out where he had gone wrong in 1926. Within days, he resigned the Conservative leadership. It was left for his friend Grattan O'Leary to remind an Ottawa audience on September 26 that in its handling of the customs scandal, "not in the memory of any living Canadian has a Government been convicted of so much vulgar evil." Other than the small group at the Ottawa Canadian Club, no one seemed to notice.

Lord Byng, whose term as governor general ended in early August, stayed around Ottawa for another month and a half. Soon after the election, he and his wife were toasted at a farewell dinner at the Ottawa Country Club. It was as awkward an evening as one could imagine. On arrival, the guests were presented to Lord and Lady Byng. Lord Byng was "quite friendly," King remembered. Lady Byng, however, "looked as though it was the trial of her life to shake me by the hand; the expression on her face was a terrifying one and I noticed a like embarrassment on the faces of Miss Sanford and Mrs. Sladen." Later, he tried making small talk with Lady Byng, but "evidently it was a great trial to talk with me," added King. Accompanied by Lapointe, James Robb and other cabinet ministers, the prime minister sat on one side of the room, the deposed Conservatives led by Meighen and Bennett on the other. During the dinner, Meighen did politely acknowledge King, but P.D. Ross, the publisher of the Conservative *Ottawa Journal,* did not. According to King, Ross was "the meanest cuss in the city"; he refused to shake King's hand.

The Byngs left Ottawa a few days later. At the railway station, there was another uncomfortable moment. "This seemed pretty much the climax of an extraordinary situation," noted King. He watched as Byng reviewed the Guard of Honour and then said his farewells. He shook King's hands "warmly," with no malice. Lady Byng was another matter; she played on King's deepest insecurities. "She looked at me like someone from the Chamber of Horrors," continued King, "all that was said was 'Mr. King' and [we] shook hands. I knew the words were used instead of 'Prime Minister' which she should have said . . . His Excellency got a good send off, but Her Excellency received not a cheer. She looked terrible, a wreck of a woman. It was a strange farewell. I was mighty glad when it was over."

There was more to this prickly relationship. The Byngs returned to Ottawa for a visit in mid-May 1932, when R.B. Bennett was prime minister and King was the leader of the Opposition. King was invited to a luncheon at Government House in honour of the Byngs and hosted by the governor general, the earl of Bessborough. Lord Byng was delighted to see King again; Lady Byng had forgiven nothing. She "just put out her hand and shook hands without a word on the part of either of us," according to King, "a cold icy sort of stare on her face." During the meal, she avoided King.

After lunch, King and Lord Byng were left alone. King told him how glad he was to see him again, and he nodded in approval. After those pleasantries, they "drifted into a conversation on the incident," as King described it. Byng suggested to King that it was King and not him who "would go down in history" as being constitutionally correct. Byng insisted that he had done what he had because of the arrangement he believed they had made about permitting Meighen to try to form a government if King failed. And, he said he would do the same thing again. They reviewed in more detail those days in June 1926 and agreed in a show of friendship to disagree about what precisely had been stated and acted upon. Overcome with emotion, King started crying. Byng, ever the gentleman, was indeed polite, but inside he was, according to his wife, seething with anger towards King. Gathering himself together, King said his farewells to Lady Byng, and this time she shook his hands as if she meant it. He even detected "a sort of kindly smile." But his heart was embittered towards her. "She is a viper," he wrote, "and responsible for most of the wrong that has been done."

Lord Byng died at the age of seventy-two on June 6, 1935. King hesitated to send Lady Byng a telegram of condolence. But after deliberating on this, he concluded "that while her attitude in the matter might be a malicious or vindictive one, it was not for me to do other than in my own eyes appeared to be the right thing in the circumstances to do." Much to King's surprise, she sent him a simple acknowledgement the following day. As this was an election year, he was most concerned that any public statement he made be framed with the right sensitivity. Again, he reviewed in his own mind all that had gone on between them. And King had a long memory. "I felt no difficulty over the question of the constitutional issue ... I did feel that Lord Byng's actions at the time, and subsequently, were far from what they should have been, and that he had allowed impressions to be spread concerning my attitude in the matter which were not right." For this King blamed, as he always had, Lady Byng's poisoned feelings towards him and unnamed Tory schemers from Montreal. "There is no doubt," he added with an air of authority, "that [Lord] Byng was being influenced by the Conservative group from the beginning of 1926 on."

Three days later, there was a memorial service for Lord Byng at the Capitol Theatre. King sat by himself in the front box, deep in thought and obsessed with reviewing yet once more the events of a decade earlier.

Throughout the service I felt as if it were the ending of a chapter which had been a very important one ... My conscience was perfectly clear. I felt no bitterness, resentment or disappointment of any kind, but rather that my stand had been completely vindicated. I felt too a sense of triumph in that I had been able to carry the whole controversy through in a manner which prevented the Tory party from using it to their own ends ... I felt that what I had really done by my course of action ... was to save Lord Byng in the public eyes from many of the consequences of his own mistakes. I can honestly say that my attitude towards him has been more than chivalrous throughout.

By the end of the service, King felt a "complete serenity" about the past. "The verdict of history," he suggested, "had been made clear."

Late that evening, the Pattesons and King summoned the spirit of Lord Byng at one of their mystical séances around his "little table," at which the dear and departed communicated through a series of knockings. From the "Great Beyond," Byng asked for King's forgiveness: "I am sorry for any wrong I have done you." King granted Byng's wishes with the utmost gratitude: "So long as you can see into my heart and know the truth," he whispered, "that is all I want." Meanwhile, some weeks later, Lady Byng wrote to R.B. Bennett expressing her true feelings. She was as angry as she had been in 1926. "I don't suppose that fat horror King will give you any chance of a letup in worry—little beast. How I hate him for the way he treated Julian. He is the one person in the whole world to whom I would do whatever harm I could! . . . I loathe liars and traitors. He is both."

There was one last scene in this melodrama. In early December 1940, King attended another dinner at Government House given by the governor general, the earl of Athlone. Upon arriving, he was most surprised to find Lady Byng present. She looked, King wrote, "very hard and worn." They chatted pleasantly during and after dinner. King, ever conscious of his public persona, purposely went out of his way to act in a polite manner. Lady Byng knew how to hold a grudge. In her memoirs, *Up the Stream of Time,* published in 1946, she purposely did not mention Mackenzie King or the constitutional confrontation with her husband.

6

A Place in History

*It is just 5 years ago today since I became Prime Minister.
There is the hiatus of 80 days but it will all be but forgotten. Meanwhile I have won my own place in my own fight
and stand on my own.*—THE DIARY OF WILLIAM LYON
MACKENZIE KING, *December 29, 1926*

ONE SUMMER NIGHT in 1927, Mackenzie King went to
the "moving pictures." A newsreel of the recent Dominion Day festivities preceded the main feature, and he watched himself projected on
to the big screen. It was a painful experience. "I do not like my appearance anywhere," he later admitted in his diary, "a little fat round man,
no expression of lofty character, a few glimpses here and there of the
happier self." He was being a bit tough on himself, though it was true
that into his second term as prime minister his hair had considerably
thinned and he had gained some weight, tipping the scale at over 200
pounds. His doctors regularly urged him to watch his diet, exercise and
stay under 180 pounds. He did walk from Laurier House to Parliament
Hill whenever he could and regularly exercised at Kingsmere, traips-
ing through the maze of paths in the woods accompanied by his loyal
pooch, Pat.

At middle age, the pace of his career and life sometimes took its
toll—or so he frequently complained. Hypochondria does not quite
describe his condition. "I tire a little more easily, perhaps, that may well
be because I'm getting older," he admitted to Peter Larkin in early 1928.
"I am seeking not to overdo, and to meet each day's obligations with as
much equanimity as I can muster." When he couldn't quite get a speech

right, he fussed about "brain fatigue." He was constantly "too fatigued to do [his] work properly," and in late October 1929, he declared, "my brain was fagged out and I did not get the sleep necessary to rest me."

During the last week of February 1927, accompanied by his close friends Senators Andrew Haydon and Wilfrid Laurier McDougald—a wealthy forty-six-year-old physician who craved status and acceptance in the Montreal business community and was a big financial supporter of the Liberal Party—he embarked on what was supposed to be a pleasant holiday to Atlantic City, New Jersey, to enjoy the sea air. Instead, he immediately came down with a sinus infection and nearly fainted several times. He wrote his physician and friend in Baltimore, Dr. Lewellys Barker, with the grim news. Normally, he insisted, he "had the constitution of a lion," but the "stormy session of Parliament" he had just endured had been "a very great strain" and he "was pretty well exhausted at its close."

The Ritz-Carlton Hotel, where King was staying, arranged for a doctor and nurse to tend to him while he recovered. Blood was taken, X-rays were snapped and nostrils were examined and suctioned of foreign matter. He dutifully recorded the entire nausea-ridden ordeal in his diary, right down to his red blood cell count (4,930,000) and his urine analysis ("no albumen, no sugar, no acetone bodies... Clear, amber colour"). Within a few days, he was up and about again, dining with the senators and discussing politics. One night, the trio had a serious talk about psychic phenomena and exchanged tales of bizarre stories they had heard. "This led me," King wrote, "to tell both Haydon and McDougald the story of the last two campaigns which I shall never but believe were all guided and directed from beyond."

Back in Ottawa, life was more mundane. Unless there was something pressing or a special event he had to attend, King's routine ran like clockwork. He liked to sleep until nine o'clock and then begin the day with a prayer or a reading from Mary Tileston's 1884 collection of Biblical and literary wisdom, *Daily Strength for Daily Needs*. It was "intended for a daily companion and counsellor," according to Tileston, who hoped her readers found the strength "to perform the duties and to bear the burdens of each day with cheerfulness and courage." King tried his best.

He spent most mornings in his Laurier House library, dealing with his large volume of correspondence and perusing the newspapers and piles of clippings provided for him. During his second term, he was assisted by his personal secretary, Howard Measures. Meetings and more correspondence followed in the afternoon. His evenings were often filled with social occasions, dinners and other engagements—dancing late into the night at the Ottawa Country Club remained a favourite pastime—which he generally complained about, but these outings did occupy him in a positive way. Still, he was happy to spend some alone time with Pat or visit with Joan and Godfroy Patteson, who remained at the centre of his life.

Like any prime minister, King experienced a whirlwind of speeches, travel and attention to 1,001 details—but not always what one might think. Often, he obsessed over insignificant problems about the comings and goings of his servants at Laurier House and the minutiae involved in staging a public event—who was standing where, who was seated together at dinner—which could have been handled by any number of people on his staff. In January 1928, for example, King hired a new cook, who had worked at the Hunt Club. He thought she would be a fine employee, except for one issue that he noted in his diary: the woman "was about the ugliest looking person" he'd ever seen. On another occasion, he took the time to write to the managing director of a local Ottawa tour company to point out that there was an unacceptable error in its brochure, which stated that Laurier House belonged to the Liberal Party, rather than to King himself. The beleaguered manager replied to the prime minister the next day with apologies, promising to correct the pamphlet as soon as the remaining batch of five thousand had been used up.

Kingsmere, too, was almost a daily fixation. Between 1924 and 1926, King spent $9,000 on three large parcels of land, expanding his property by 175 acres This brought his total holdings to more than 200 acres and made him, in his view, a "landed proprietor." Then, early in 1927, he decided to reinvent himself as a gentleman farmer. His enterprising idea was to raise sheep, chickens, cows and horses and cultivate vegetable and fruit gardens to supply him with fresh food and perhaps turn a small profit. Near his Gatineau property was a nineteenth-century farm that had once belonged to a settler named Henry Fleury. It had fallen into disrepair, and the current owner, Mrs. T. McGillvary, resided in Los Angeles. She had indicated a willingness to sell her estate. He hemmed and hawed for about a month until he finally made Mrs. McGillvary an offer of $3,000. As with most major decisions in King's life, a spiritual aura comforted him, reinforcing his faith that he was taking the correct course of action. "I believe the property is good value, having an eye to the future," he noted. "The view is fine from the moor and I got much pleasure from the sight of the fence, the long stretch. I really believe those hills mean health and strength to me and ultimate salvation of soul as well as body. I seem to experience this month a sensitiveness to invisible psychic forces. It would seem as if the spiritual strength and the material forces alike were each nearer. I pray for the strength to rise only to the highest and the best."

He had to up the price by $1,000, but by mid-April "The Farm," as he dubbed it, was his. It was also necessary to spend a bit more money to purchase some adjacent cottages to expand his pasture. One owner wanted $1,500 for a property called Mainguy House that was valued at a third of that. King reasoned that "it is not worth $500 to me, save to prevent Jews or other undesirable people from getting in." He remained troubled by this expenditure, but justified it several times on the grounds that "the alternative lay between controlling and developing the situation, and having the joys of the property imperiled by the presence of undesirable neighbours, Jews of a low order and others." (Two years earlier, King had received news from the Kingsmere property association that three Jews had rented cottages in the area and one had "intimated" he might purchase a lot.) Once renovated, this new cottage became the Pattesons' summer residence, Shady Hill.

To manage the new acquisition, he found a local named McBurnie, a "fine-looking Scotsman" who, with his wife, agreed for a salary of $60 per month plus free rent to move into the house and become King's tenant farmer. Before he knew it, there were sheep, livestock and chickens running around. The one frustration he had was with the *Ottawa Citizen, Montreal Gazette* and several other newspapers, which had learned of his new enterprise and reported it. The *Ottawa Journal* was the most cutting in an editorial of September 1927: "Now, human nature being what it is, and there being a common notion that premiers, having a supposedly hard job, are presumed to be occasionally in their offices... [w]e're envious of a man who can run a whole party, and a country, and appease people who want senatorships, and judgeships, and breakwaters, and a host of other things—all from a cottage."

King guarded his private life with a protective zealousness of a mother lioness, and journalists of this era were usually deferential. Even Grant Dexter of the *Manitoba Free Press,* John Dafoe's loyal man in Ottawa from 1923 onward and a knowledgeable insider and gossip, drew King's ire for writing in a feature story that the prime minister had taken a bath after his government delivered its budget in 1930. Certainly no reporter speculated in print about King's relationship with Joan Patteson and knew less, if anything, about his spiritualist adventures. And even if they had, they would have ignored them. In those days, claimed Ottawa journalist Alex Hume, "you never covered a prime minister's private life." Hence stories about King's latest property purchase touched a raw nerve. "This is calculated to do great harm in circles that have been helping me with Laurier House and in creating suspicion in minds of public of graft," he wrote angrily. "It takes the pleasure a little out of what I am attempting."

In late January 1928, King agreed to co-operate with *Toronto Star* journalist Frederick Griffin, who visited him at Kingsmere. He had Pat do a number of tricks for Griffin and showed him the farm and the estate. The story that appeared in the *Toronto Star Weekly*, with a large photo spread, was quite complimentary. A few weeks later, King wrote the magazine's editor praising the article, but was very concerned that a "horde of intruders who have no kindly purpose at heart, and are actuated only of curiosity and mischief" were about to descend on his tranquility. They never did.

While the farm gave King much peace and quiet, the sheep and other animals didn't give him pleasure at all. At first, it was a "delight," he wrote, "to see the sheep browsing about the grounds... [and] the little lambs skipping and at play—jumping in the air all four feet at once." Then as the sheep roamed into his neighbours' property, demolished gardens and attracted stray dogs, he became less enamoured with them. Within a year, he had had enough of the entire affair. "The whole livestock end has been a disappointment, failure and annoyance," he grumbled in early June 1928. He sold all of the animals to local farmers, took a bit of a financial loss (he believed that his tenant and friend McBurnie was "swindling" him by switching sheep, taking the healthier ones for his own flock and leaving King with the poorer animals) and opted to concentrate on pastures, fruits and vegetables.

More satisfying was his plan to turn the old Herridge cottage, Moorside, which he had acquired in 1924, into his permanent summer residence. Those renovations proceeded throughout the summer of 1928. As he noted, when the project was still in its early stages: "The more I see of the views from all sides of the house the more wonderful I feel the bargain I have there. It is really an amazing possession to own a mountain and distant valley view to say nothing of the forests in the immediate foreground. There is still much to be done to the place but it easily could be one of the most attractive country residences in Canada." Finally, on August 2, he was able to sleep at Moorside in the willow room with the doors open onto the balcony and was moved beyond words. "The moonlight through the willow branches was as beautiful as anything in Italy or Greece," he wrote the next morning. "I thanked God devoutly for possessing so wonderful a home in so beautiful a spot." In the years ahead, Moorside was to stimulate and tug at King's romantic nature like nothing else in his life. His decision to move into Moorside necessitated relocating the Pattesons into Shady Hill. They had been using Moorside as their summer home. "I can see wherein I have perhaps seemed selfish in 'evicting' Joan and Godfroy after they had worked so hard on the place," he conceded in a diary entry of July 6, 1929, "but I am sure it was right and in my present position necessary."

APART FROM HIS property, Mackenzie King also had "King and country" to concern himself with. His Majesty's government was calling for another Imperial Conference, and the dicey and ever-shifting relationship between Britain and her former colonies was about to be transformed further. Accompanying the prime minister on board the fashionable ss *Megnatic*—operated by the White Star Line of *Titanic* fame—which departed Quebec City on October 9, 1926, were Peter Larkin, the Canadian high commissioner returning from a visit home to his position at the new Canada House on Trafalgar Square; Minister of Justice Ernest Lapointe, King's French-Canadian lieutenant; Vincent Massey, Canada's newly appointed minister in Washington, D.C., who was to function as an unofficial adviser; Dr. Oscar Skelton, the brains of the external affairs department, whose job, he told his wife, was to "stiffen King's backbone"; as well as an assortment of secretaries and family members. The ocean crossing was rough, and King spent much of the time ill or sleeping in his cabin. When he felt better, he dealt with his voluminous correspondence and, much to Skelton's annoyance, occupied the time of his assistant, Marjorie McKenzie, who was coerced into taking the prime minister's dictation.

Once they docked in Liverpool and made their way to the London Ritz, where they resided for the next month, King embarked on a nonstop whirlwind of social activities. "King is in the thick of the social battle," D.B. MacRae, who was covering the conference for the *Manitoba Free Press,* wryly reported to John Dafoe, "luncheon with a duke at noon, dinner at night with some Lord, perhaps a dance at night, and weekends in the country." The dinners and parties were an excellent tonic. King stayed out late each night and slept in the next morning, generally frustrating Skelton and Lapointe, who waited impatiently for him to make an appearance, which he usually did by ten o'clock.

He was not quite so on edge as he had been at his inaugural imperial conference in 1923. A photograph taken in front of Canada House before the 1926 conference began shows King looking happy and content, standing alongside Lapointe, Larkin and Massey. The Canadians were all sporting long dark overcoats, gloves, leather and oxford shoes (everyone but King also wore spats). Other than Massey, who wore a bowler, the men doffed silk top hats, and each wielded a pointed walking stick. King appeared as if he had just made his grand exit from a session of the House of Lords.

The *London Daily News* in a pre-conference profile described Canada's fifty-two-year-old prime minister as "a short stocky man with a genial, youthful presence, looking not more than 35, immaculately tailored, and exuding goodwill." Kevin O'Higgins, the young vice-president of the executive council of the Irish Free State, was less complimentary. As he wrote to his wife, "Mackenzie King has 'disimproved' since 1923—gone fat and American and self-complacent."

King went about his business in his own self-centred way. One day, forty journalists were waiting for him at Canada House for a pre-arranged press conference; he was an hour late and then telephoned to say that he could not make it. His major preoccupation was sitting for a portrait for the well-known Irish artist Sir William Orpen—a gift from the wealthy expat Canadian press baron, Max Aitken, Lord Beaverbrook, who was making his mark on Fleet Street. King had initially "hesitated," as he put it, when Beaverbrook first suggested the idea, but quickly decided this was too good an opportunity to pass up. Always thinking of posterity, he reasoned, "It might be well to have the picture presented to Laurier House in my lifetime to be given to the nation afterwards. The difficulty is going to be securing the necessary time for sitting." He somehow found the time and made certain everyone knew, D.B. MacRae snidely noted in another report, "that [the portrait] is of the utmost importance."

King also found time to meet with Sir Oliver Lodge, the renowned physicist, who had a keen interest in spiritualism. A believer in life after death, Lodge declared after attending a series of séances in the early 1890s that "things hitherto held impossible do actually occur." As he chronicled in his 1916 book, *Raymond, or Life and Death,* his son who had been killed during the First World War had allegedly returned. King's meeting with Lodge was held at the home of old friends, Lord Edward and Lady Dorothy Grey of Fallodon. Lord Grey had served as British foreign secretary for eleven years, from 1905 to 1916. King gave Lodge and the Greys a full account of his recent readings by Mrs. Bleaney and reiterated his feelings that his family members were guiding his actions. "When I had concluded," he wrote in his diary, "all three expressed the opinion that the experience had been a very remarkable one and appreciated my having told them of it." Lodge confirmed King's spiritual moment by confiding that his son Raymond had told him from the "Great Beyond" that he would meet a "prominent man who would tell him some interesting things." King was that man. He was reassured more than ever that, as Lodge explained it to him, "those beyond are trying to have us know of their presence and their interest in our lives, but we have to be receptive and faith is a means to that end." This was a more significant encounter for King than anything that occurred at the conference.

THE TROUBLEMAKER AT the Imperial Conference of 1926 was General J.B.M. ("Barry") Hertzog, a lawyer, judge, proud defender of the Boers at the turn of the century, and for the past two years the nationalist South African prime minister. He remained a "militant prophet of the people," in the words of one biographer. Even before the first session had begun, Hertzog arrived demanding a public declaration of the dominions' independence. Australia and New Zealand, ever reliant on the imperial navy, and the British balked. William Cosgrave, the president of the executive council of the Irish Free State, sided with the South Africans. Canada occupied the proverbial middle. For Mackenzie King, "independence" was a loaded term, linked as it was in his mind with the history of the United States and the American Revolution.

In the three years since the last conference, King's attitude towards the Brits had not really changed. Like most Anglo-Canadians, he still regarded the British connection as the defining ingredient of the country's national identity. "'Autonomy' was the popular description for the contradictory Canadian condition, part free and part not," says Norman Hillmer. "It was used to suggest various degrees—depending on the circumstances, the speaker, and the political tilt—of national breathing room, while the country remained all the while tucked warmly inside the British womb." Still, King was as paranoid as ever about the perceived British intent to create a centralized imperial structure in which the dominions would be like puppets on a string held by a conniving master of the empire. Ominous motives and conspiracies were lurking everywhere in King's vivid imagination.

Back in June 1925, Peter Larkin had proposed to Leo Amery, the secretary of state for the dominions, that he should arrange regular meetings with all the high commissioners so that they could discuss mutual issues and share ideas. It was a reasonable suggestion, except that Larkin forgot to clear it with King. When he finally did, he promised that the prime minister could rely on his discretion and assured him he would not consent to anything without consulting him about it. That still did not sit well with King, who refused to permit Larkin to meet as part of a high commissioners' group, which just might discuss and decide policy, no matter how innocuous. King insisted that only he could speak for Canada, which was precisely the reason he was his own external affairs minister for twenty-two years.

Amery found the episode amusing. King "at once suspected," he recalled, "a sinister design on my part of gradually working towards some sort of Imperial policy situation in London... so poor Larkin had to come and tell me, very apologetically, that he was not allowed to see me except by himself. The rest of us continued our friendly and useful talks." In reality, British government leaders, at least since 1923, had reluctantly accepted the evolution of the dominions' autonomy, even though some of them, like Amery and foreign secretary Sir Austen Chamberlain, still pined for a centralized imperial foreign policy with the British calling the shots. The real difference of opinion now was about nuance, process and saving face.

The British could do little when King, at Skelton's behest, informed them that Canada would not commit itself to the Locarno Treaties, already signed by the British, which would for a time guarantee peace in Europe. King reassured the British that "if the situation arose Canada would do her part" to defend the mother country and the empire. Nor did the British care when King announced Canada's intention to open a legation in Washington, D.C., with Vincent Massey as its first representative. (The issue had been on the table for a number of years in different forms.) All the British requested was that Massey keep the British ambassador in Washington informed so that they would not be ignorant of mutual concerns, and the British ambassador would do likewise.

King had one main goal at the conference, and that was to redefine the role of the governor general. In light of his dispute with Lord Byng, he no longer felt it appropriate for the governor general to function as the British government's agent. The time had come to designate the position solely as the representative of the Crown. With very little discussion, the delegates agreed to this revision to the office of the governor general in Canada as well as in the other dominions. King also had an answer as to how communication with the British government could be more easily facilitated. He suggested that a British high commissioner who could confer frequently with the Canadian prime minister be appointed in Ottawa. The British complied with this request in April 1928.

Most of the time at the conference, however, was taken up trying to appease Hertzog. Smartly, Amery gave the challenging task of arriving at an amenable solution to seventy-eight-year-old statesman Lord Arthur Balfour. Balfour, who had served as prime minister at the turn of the century, was more famous for his declaration of 1917 as foreign secretary promising a Jewish homeland in Palestine. He was at once pleasant and affable, but also arrogant, aloof and ambitious enough to take on Hertzog. "Little as the general public may suspect it, the charming, gracious, and cultured Mr. Balfour is the most egotistical of men," observed the astute journalist Harold Begbie a few years earlier. Or, put more bluntly by politician George Wyndham, who served in Balfour's cabinet, "The truth about Arthur Balfour is this: he knows there's been one ice-age, and he thinks there's going to be another."

Relishing the chance to be the mediator, King did his part, assisted in no small way by Skelton, who did much of the work and received none of the credit. For weeks, Massey recalled, "impromptu informal meetings between various representatives of both sides" took place "at all hours" at the Ritz. At long last, a second "Balfour Declaration" was issued, proclaiming to the satisfaction of a majority of the prime ministers that Great Britain and the dominions "are autonomous Communities within the British Empire, equal in status, in no way subordinate one to another in any aspect of their domestic or external affairs, though united by a common allegiance to the Crown, and freely associated as members of the British Commonwealth of Nations." In essence, the declaration reaffirmed that "every self-governing member of the Empire is now master of its destiny." Everyone but King George v was pleased. "Poor old Balfour has given away my Empire," His Majesty retorted when he read the document. Lapointe told MacRae of the *Manitoba Free Press* (who was being briefed daily by Skelton) that this was "more than [we] ever expected to get."

Later, Amery and officials from the foreign office hailed King as the "umpire of the conference" for his efforts in bringing the two sides together. Sir Maurice Hankey, the secretary of the British cabinet, wrote to his friend William L. Grant, the principal of Upper Canada College, in Toronto, that "no other man ... had been so influential and helpful in keeping Hertzog in line as had Mackenzie King." Oscar Skelton was slightly more modest, telling his wife that the declaration, which he regarded as "epoch-making," was "mainly thanks to the pertinacity of the Irish and ourselves." Most succinct was MacRae, who informed Dafoe that "King has spent the bulk of his time here eating with duchesses. Skelton carried the bulk of the load and seems to have the brains."

King thought otherwise. He was naturally delighted with the results of his work in London, though whether King or his advisers accepted it or not, Canada was still very much governed by British authority. It could not amend its own constitution, the British North America Act (BNA), and would not be able to for another five and a half decades. Nor were King and Lapointe prepared to abolish legal appeals to the Judicial Committee of the Privy Council, which the British had made possible at the 1926 conference. This meant that in the foreseeable future (until 1949) a British court could and did overrule Canada's Supreme Court.

As he was apt to do, King accepted the plaudits he received in the Liberal press—the *Manitoba Free Press* dubbed the declaration the "Charter of Dominion Independence"—and ignored the growing criticism in the Conservative papers for its allegedly amputating Canada's link with Britain. After reading MacRae's final report on the conference, Dafoe was not as bowled over by King's performance. He told Clifford Sifton:

> It is evident from the known circumstances and also from the tenor of Mr. MacRae's letter, that Mr. King, if he could have had his way, would have permitted the matter to drift. We shall owe the declaration... chiefly to South Africa and probably Ireland... Of course in my comments I shall not try to pluck any laurels from W.L.M.K's brow. I have just seen the cable from British United Press in [which] it is explained that [King] by his masterly statesmanship did the whole thing. Perhaps this will go down in history, [which] as Henry Ford once observed is mostly "bunk."

The declaration was not "bunk" to King. He privately oozed about his performance in London, disregarding the timely assistance of the Irish delegates and more significantly, according to historian John MacFarlane, the counsel he received from Skelton and Lapointe, who "pushed [him] in the direction of independence from Britain faster and further than King would have gone on his own." As he confided to his diary some weeks after his return to Ottawa, "The Imperial Conference has helped to give me a place in History."

Two months later, King was required to do a bit of damage control. The negative coverage in the Toronto Tory press as well as in the *Globe*, which was still regarded as a Liberal newspaper despite its critical stand on King, had had an impact. British Torontonians were spooked into believing that Canada had forsaken the empire. Nothing could have been further from the truth, King declared at a banquet at the King Edward Hotel before an audience of five hundred supporters, with his full cabinet in attendance. His two-hour speech extolling the Britain-Canada connection was also broadcast on the radio. "I felt immensely relieved when it was over," he noted later. "God's help was present. He carried me through." The new reality of the Commonwealth, as even King admitted, was more effectively explained by Lapointe in his briefer remarks. Canada was not now and never would be a republic, he stated, and the British monarchy "more than ever would be the keystone of the structure of the empire since each Dominion owed allegiance to him." He gently castigated those "timid souls" for having an "inferiority complex, the subordinate state of mind" about Canada's right to equality and autonomy within the Commonwealth.

That right to equality was on display two weeks later as Vincent Massey settled into his role as Canada's representative in Washington, D.C. He met President Calvin Coolidge and arranged for the purchase of a magnificent and pricey Beaux Arts–style mansion on Massachusetts Avenue to serve as Canada's legation. R.B. Bennett, six months away from being elected the new leader of the Conservative party, decried Massey's appointment and the opening of the legation. "It is but the doctrine of separation," he claimed, "it is but the evidence in many minds of the end of our connection with the empire." He felt similarly about King's decision to follow up Washington with Canadian legations in Paris and Tokyo.

For King and Skelton, the legations were a symbolic statement that Canada had arrived as a nation, albeit one still linked politically and psychologically to Britain. But that was all. Power and decision making remained firmly in the prime minister's hands. At this stage, King trusted Vincent Massey almost as much as he did Peter Larkin in London, and when anything truly critical arose he handled it himself with a personal cable to the British prime minister or a face-to-face meeting with the American secretary of state or president.

The British complained from time to time that it was really Skelton, the under-secretary of state for external affairs, setting Canadian foreign policy and that Skelton had this mysterious and almost malevolent "power" over King. "We would have less difficulty," grumbled Sir Austen Chamberlain, the British foreign secretary in 1929, "If King was not so weak"—and, by implication, Skelton was not so influential. The Brits were wrong. King did indeed trust and heed Skelton's judgment probably more than he did Lapointe's. He regarded the professor as "a really great man and a true soul" and sought his advice on a wide range of domestic and foreign issues. You could count on the fingers of one hand the number of times they had serious disagreements in over fifteen years. But it should not be forgotten, as Norman Hillmer explains, that King "kept the external affairs portfolio firmly in his grasp. Always cautious, he guarded the prerogatives of office jealously, and he was careful to read what went out over his signature." Still, if the British or anyone else for that matter wanted to blame Skelton for this or that policy decision, King—as Skelton pointed out on more than one occasion—did not seem to mind.

King's reluctance to surrender control was also at the root of his lack of support for the League of Nations. With his ardent faith in conciliation, King should have been one of the League's great proponents. The League, however, made him nervous, especially Article x of its covenant, which committed its members to collective security and possible military action against any aggressor. (Because of the League's many weaknesses, the commitment outlined in Article x never amounted to much.) Such a critical decision, King had long argued, was best left to individual governments. Parliament, not a European "think tank," would decide when Canada went to war.

Throughout the twenties, Canadian representatives, following King's instructions, had periodically tried to amend Article x, but they never got very far. King applauded the League's noble aim of disarmament, yet he rejected any policy or protocol that compelled Canada to do anything contrary to its wishes or interests. King also fretted about what he perceived were the high costs associated with the League. To King's chagrin, the auditor-general's report for 1924 had highlighted about $8,000 in expenses for Canada's two delegates to the fifth assembly.

Ernest Lapointe saw the League and Canada's participation in a different light. To him, the League provided Canada with another opportunity to parade its so-called independence. In mid-August 1927, Lapointe, who was in Geneva, wanted Canada to seek election for a non-permanent seat on the League's council for a three-year term. He cabled King with his proposal, stressing that such a move would be "a natural step and crowning point of our policy. We owe it to Canadian prestige and development. Am not afraid of any dangerous aspect, quite the contrary. Let Canada again lead the Dominions." King was not as enthusiastic. He feared that as a council member Canada would become entangled in European matters that were beyond its control, and worse, it would be pitted against the interests of Britain or France.

The matter was left unresolved until Lapointe returned from his travels. Then, he and King discussed it again and once more disagreed, more sharply this time than they had about anything else. In the heat of the moment, Lapointe threatened to resign if King did not comply with his wishes—at least according to the version of the confrontation Lapointe gave to Walter Riddell, Canada's advisory officer to the League of Nations. King immediately backed down and consented to the request: Canada could stand for election. Yet he was not happy about it. "I think it is a mistake, as being unnecessary, unduly pressing our individual status as a nation, and in differences," he wrote with obvious frustration. "But a cleavage with Lapointe on a matter which he feels deeply would be more unfortunate in the long run." The British, too, were uneasy about King's decision for the same reason as he was: there was a potential for a conflict between the mother country and her grown-up colony. Within a few days, and much to Lapointe's satisfaction, the deed was done and Canada was chosen to hold a non-permanent seat on the League of Nations Council.

The various developments at the legations and even in Geneva excited King's imagination and stirred his ego. Reviewing the modest budget estimates for external affairs in the spring of 1928, he gushed about "a new Canada emerging" with a new status. "We are certainly making history," he added. "I am convinced the period of my administration will live in this particular as an epoch in the history of Canada that was formative and memorable." Several months later, however, he brooded about attending a meeting of the League Council, since it meant "giving up Kingsmere for the summer, which is a great sacrifice." Skelton convinced him otherwise, and they both set sail for Plymouth, England, on August 17 on the brand new French ocean liner SS *Île de France*.

King loved the conveniences the ship offered, but he was initially "shocked" by its modern design. For the first few days, the journey was pleasant and King spent his time working as well as gossiping and generally enjoying himself. By the third day, the ocean had become rougher, and he spent most of his time in his stateroom feeling seasick. At midnight on August 22, he was deep in thought about landing at Plymouth and missing his family and Joan Patteson. "[I] began to sing about 'Light of the lonely Pilgrim's heart'... I thought of Plymouth... and myself as the lonely pilgrim," he wrote. He then began to sing different hymns for each of his departed "loved ones," his parents and Bell and Max. The prayers brought to him "assurances of the loved ones round guiding and protecting my life at this time I believe they are very near to me. I think much of Kingsmere the lighted house on the hill, Joan at the door. It all stands out so clearly in my mind."

Once in England, they made their way to Paris so that King could participate in the official signing ceremony of the Briand-Kellogg Pact, a multinational agreement negotiated by the French foreign minister Aristide Briand and the U.S. secretary of state Frank Kellogg, denouncing war as an "instrument of national policy." King was not overly excited about the agreement, other than that it had American involvement, a positive sign for future international events, and its objectives were commendable. Moreover, it asked nothing of him or Canada, which was all for the best. With so many world leaders attending the signing, King did revel in the moment, so laden as it was with spiritual significance. As he wrote his name and came to "Mackenzie," "dear mother's spirit seemed to come between me and the paper to almost illuminate it," he recalled later. "I thought of only her as I wrote, then in signing 'King' I felt much the same about father but not so strongly. As Briand read—'The flags of all the Nations' over the foreign office today, I thought of what Mrs. Bleaney told she saw two years ago 'The flags of all nations, that I would be taking part in, etc—It came to me strongly on the moment."

Then it was back to earth in Geneva. King was honoured by being selected as one of the League's six vice-presidents, until then a largely ceremonial position held by leaders of the Great Powers. The election, he conceded, "discloses the position Canada has gained in League matters." At the same time, he had no desire to participate. He enjoyed the glamour, glitz and gourmet food offered at the League's social events. But, he said, "[I was] reluctant to do any speaking at this Assembly. It all looks like lecturing European countries and I have really no message." With prodding and a lot of input from Skelton, King did take his turn at the podium and offered a few choice words of wisdom about the League, Canada and opportunities for peace—even managing to throw in a few nuggets from *Industry and Humanity*. He departed from Geneva "immensely relieved it was over," as was Oscar Skelton, who had had his hands full keeping Mackenzie King happy.

King spent another few weeks touring France and Italy and personally met with the Fascist dictator Benito Mussolini in Rome. It was an unplanned meeting, but he admitted that he had "become enthused" at the manner in which Italy had been given order and discipline. On the way to Mussolini's office, King's car was stopped. "I saw what government by dictatorship meant," he wrote in his diary, "but when one hears how he came with his blackshirts to [King Victor Emmanuel III], offered his services to clean up the government and House of Representatives filled with communists... cleaned the streets of beggars and the houses of harlots, one becomes filled with admiration." He and

Mussolini had a pleasant chat about government, Canada and, ironically enough, the League of Nations, which Mussolini would in due course thumb his nose at after he invaded Abyssinia (Ethiopia) in 1935. After the conversation ended, King noted, "I wished him well and the necessary strength to carry on his work," adding that "there were evidences of sadness and the tenderness as well as great decision in his countenance. His hand was much softer than I expected to find it."

DURING THIS PERIOD, matters at home also required King's attention. None, in his view, was as important as the celebrations surrounding Canada's Diamond Jubilee, the sixtieth anniversary of Confederation, on July 1, 1927. He doted like a mother hen over every aspect of the festivities that were to take place on Parliament Hill. He spent hours upon hours working on his speech for the dedication of the Peace Tower carillon, the fifty-three musical bells still regularly heard in Ottawa today. All the while, he was cognizant of the jubilee's significance and its boost for national unity, his great purpose in life, as well as his own historical legacy.

King wasn't the only one to be caught up in the hoopla or the moment. There was, as Ralph Allen, *Maclean's* editor during the fifties and sixties, has described it, "an air of unaffected, unsophisticated joy. The country was alive with strawberry festivals, ice-cream festivals, fowl suppers, three-legged races, egg-and-spoon races, baseball games between the fat men and the thin men, historical pageants featuring weirdly made-up and costumed Iroquois and fur trades, hymns, sermons, oratorical contests, millions of Union Jacks, tens of thousands of lithographed portraits of King George and Queen Mary." Most of us today might cringe at the excessive patriotic displays, but back then such celebrations were welcomed.

For the big day, the Royals' two sons, Edward, the Prince of Wales, and his brother George, the Duke of York, were in Ottawa, along with British prime minister Stanley Baldwin. Thousands of people were in the audience, at the time the largest crowd ever to gather on Parliament Hill. King was certain the day would "go down as a memorable occasion." At the commemoration of the Peace Tower, he spoke about the horrific Parliament fire of 1916 and the tragic loss of sixty thousand Canadian soldiers during the war. King was moved by his own words, as he had been when he had drafted his speech a week and a half earlier. The writing had made him cry. "I was so overcome with the beauty of the thought expressed and the truth underlying it," he had noted. Once he had concluded, the carillon rang out with "God Save the King" and

"O Canada," and King was annoyed that a plane circled overhead, "making a frightful noise and spoiling [the] first impression of the carillon."

Another splendid event followed later in the afternoon, with a choir singing and another speech by King, this time about the true meaning of Confederation. (In 1927, King published his two speeches from the Diamond Jubilee, along with other non-political speeches, in a 274-page book entitled *The Message of the Carillon*.) Wearing his fancy embroidered Windsor uniform, he resembled a squat and pudgy bellhop at a five-star hotel. The program and speeches were broadcast over a makeshift radio network. King had memorized his speech, but at the last second opted to use his notes, which he later deeply regretted. The pressure and excitement of speaking across the dominion "and beyond"—as far as England and Brazil—had got to him. Still, he recognized later that the broadcast was "the great feature of the day," adding with just a bit of exaggeration that "never before was the human voice heard at one and the same time over such an extent of the world's surface by so many people. It was the beginning of Canada's place in the world, as a world power."

Wreaths were laid on the monuments on Parliament Hill as well at the foot of the recently completed statue of Laurier (facing the Château Laurier). "It was a proud moment," King reflected, "almost a great spiritual triumph. I look up to his face in pride and joy, recalling I had once told him that that was the place in which a monument would stand. It has all come about more wonderfully than words can express." The marvellous day ended with a state dinner, and King did not make it back to Laurier House until 1:30 in the morning. "I have reason to be profoundly grateful," he scribbled before he fell asleep, "that the day passed off so well—that I was as well prepared as I was in the message to the people, and for the vastness of the opportunity given me—in being prime minister of Canada at this time."

There were still more thrills and sadness the following day. In 1927, the most famous person in the world was arguably a shy and skinny American pilot. Colonel Charles Lindbergh's solo thirty-four-hour flight in May from New York to Paris aboard his airplane, the *Spirit of St. Louis,* captured imaginations everywhere. More than 250,000 newspaper stories were written about him, and he was acclaimed in hundreds of radio broadcasts. He was hailed as a courageous hero and a symbol of man's inventiveness and goodness. "What is the greatest story of all time?" the *New York Times* editors asked in one of their numerous articles on Lindbergh's exploits. "Adam eating the apple? The landing of the Ark? The discovery of Moses in the bulrushes?... But Lindbergh's flight, the suspense of it, the daring of it, the triumph and the glory of it—these are that stuff that makes immortal news." Thanks to the efforts of Vincent Massey, Lindbergh flew into Ottawa (landing at the Hunt Club) to toast Canada's sixtieth birthday. The only dark cloud over the visit was the accidental death of another U.S. pilot accompanying Lindbergh, a Lieutenant Johnson, whose plane crashed when he attempted a risky manoeuvre.

Mackenzie King was naturally caught up in the swirl of activity around Lindbergh. "A more beautiful character I have never seen," he said of the celebrity aviator. "He was like a young god who had appeared from the skies in human form—all that could be desired in youthful appearance, in manner, in charm, in character, as noble a type of the highest manhood as I have ever seen." The guest of honour was feted at a luncheon at Government House (which King felt was organized poorly), taken around Ottawa for a tour of sports facilities and taken to a reception at Laurier House late in the afternoon. As it turned out, Lindbergh was certain he was related to King's uncle Charles Lindsey. He delighted King with his knowledge about William Lyon Mackenzie and Lindsey's biography of the rebel. Another state dinner followed in the evening, which King, despite the presence of Lindbergh, found exhausting. He allowed himself a few glasses of champagne, which he conceded "helped me tonight." He said a few words about Canada-U.S. relations and heaped more praise on Lindbergh's exploits.

Lindberg stayed that night with King at Laurier House and remained in Ottawa for the funeral of the young pilot Johnson. King decided to go all out, and the service was held on Parliament Hill. The flag on the main tower was lowered, the carillon rang out with the funeral march and the Mounties escorted the casket to Union Station past crowds of onlookers. Lindbergh swooped down in his plane and dropped flowers on the train. King found the ceremony stirring and valuable. "I was never so proud of my country," he wrote. "I can think of nothing calculated more to further international amity and to knit together in lasting friendship [the] U.S. and British Empire than national impressions of feeling such as we witnessed today. The death of Johnson was a tragic event but it proved to be the most memorable... It was a sacrifice. In the mystery of divine providence, it would seem to be part of the great plan."

Summing up the past few days, King could not help but dwell on himself and his family:

When I think of dear grandfather and dear mother—it was as if God were bringing to a crowning fruition grandfather's work of nation-building in Canada... How strange all should centre about Laurier

House and myself, even to Col. Lindbergh being my guest there! I shall ever believe it was all the guidance of those I love the most, the fruit of their sacrifice, the crowning glory of all their suffering... Pray God out [of] all the humiliation, the repentance of my failures and weaknesses and shortcomings, there may arise, as with the memorial tower, a larger freedom and spiritual power, to give to this nation the soul that God would have me give, through the spirit of love that comes from him, through the loved ones gone before.

AS IMPRESSIVE AND daring as Lindbergh's heroics and the airplane were, the real invention that changed the world during the twenties was the automobile. Thanks to the creative genius of Henry Ford and others, owning a car was no longer a luxury but a necessity. Between 1908, when Ford's Model-T hit the market, and 1927, sales in Canada exceeded $1 million. Cars linked isolated rural communities with cities, impacted women's fashion and social activities—teenagers quickly discovered all sorts of deeds were possible in the back seat of a Ford or Buick—and dramatically improved public health by getting thousands of horses (and the daily deluge of manure) off the streets. But cars also needed government regulation—in 1918 in Toronto, twenty-eight people died in traffic accidents; ten years later, that number exceeded two hundred—and adequate roads and highways.

According to the BNA, which divvied up jurisdictions and powers between the federal and provincial governments, the latter were responsible for road construction, education and social welfare, including unemployment, whose costs were rising. The Fathers of Confederation, most of whom had put their faith in a federal system with a strong central government, had not anticipated provincial needs or expenses. Starting in the 1890s, a series of pro-provincial constitutional court decisions had shifted some power but little money away from Ottawa— a trend that was temporarily reversed during the First World War. By the late twenties, the provinces were desperately short of cash, since their main and limited source of revenue was derived from motor vehicle licences and gasoline and liquor taxes. Ontario and Quebec, moreover, both targeted hydroelectric power development to nourish and sustain their industrial growth. There was a jurisdictional problem, however: Section 91 of the BNA gave the federal government control over navigation of Canadian waterways, which meant that a clash between Ottawa and the two largest provinces was inevitable.

Meanwhile, Mackenzie King's government was still struggling to pay off the debt for the First World War and dealing with the high expense of railway nationalization. It was in no position to provide the provinces with much financial help. "We got into a good discussion [about federal-provincial relations]," King noted of one cabinet meeting at the end of October 1927, "and I was glad to see the [cabinet was] against federal grants, other than continuation [of the] subsidy to [the] Maritimes." The prime minister himself, who tracked every penny he ever spent, was not about to hand over money to the premiers. Besides, he believed, they were either not as badly off as they claimed, or they were being overly extravagant. Good government demanded prudence and balanced budgets. Increasing subsidies to provinces, King and his colleagues long maintained, would be like giving liquor to an alcoholic: it would never be enough. In either case, he had his own problems to worry about.

Earlier in 1927, King had taken great pride in delivering on his promise to enact the old-age pension legislation, which J.S. Woodsworth and Abraham Heaps had proposed during the chaos of the 1926 legislative session. "The old age pension is now law," King wrote. "For better or worse I can claim much credit for this. It will at all events always be part of liberal administration." What he failed to add was that the $20 pensions paid out to needy Canadians seventy years or older were to be shared with the stressed-out provinces, which the Liberals forgot to consult. Provincial leaders could only shrug; none among them desired to be political curmudgeons who refused to give a few dollars to the elderly.

The mood, therefore, was somewhat tense when King invited the premiers to Ottawa during the first week of November for the third Dominion-Provincial Conference since Confederation. (The first had been in 1906 and the second in 1918.) King, who counted on his labour conciliation skills to soothe warring parties, was more than a little troubled by the unlikely alliance that had been forged between Ontario and Quebec. The cry for provincial rights, which echoed throughout the Railway Committee Room in the parliament buildings, where the closed sessions were convened, emanated from a truly odd couple.

Howard Ferguson, the premier of Ontario, was a Tory, a Protestant Orangeman and an imperialist with small-town charm. The kids called him "Uncle Howard" in Kemptville, where he had been raised and where it was said that Ferguson "always knew what was going on in the barbershop." Ferguson, as a backbencher and then as minister of education, had also been a champion of Regulation 17, the controversial legislation passed in 1912 that severely limited the use of French as a language of instruction in Ontario schools. Franco-Ontarians and Quebeckers had shouted bloody murder at this violation of their educational and cultural rights.

Once Ferguson became premier in 1923 and sought out an alliance with Quebec in the fight against the big, bad feds, the French-language ban was more or less reversed. That act of contrition was enough for Ferguson's counterpart in Quebec, Louis-Alexandre Taschereau, a gaunt sixty-year-old Liberal. Taschereau was an admirer of Wilfrid Laurier and had served under Sir Lomer Gouin before being elected premier in 1920. Like almost every Quebec provincial leader since then, Taschereau stood up for Quebec's economic and social rights. And that meant battling the King government on railway policy, old-age pensions, hydroelectric power and discussion of constructing a seaway in the St. Lawrence River—in short, any and all issues that he felt jeopardized Quebec's interests.

King was prepared to take on Ferguson and his Tory mentality; he never liked him personally, and later in May 1930 when Ferguson denounced King during the federal election campaign, he wrote in his diary that the Ontario premier "is by nature a skunk." But alienating Taschereau was far more dangerous, considering the Liberal government's dependence on Quebec. Perhaps for this reason, he allowed Ernest Lapointe to lead the federal charge during the conference. Whether it was about reforming the Senate, amending the constitution—a contentious conundrum that was to confound prime ministers and premiers for the next six decades—or enhancing provincial taxing powers, the conference of November 1927 proved that Ottawa and the provinces were going to have to agree to disagree on many substantive issues. One of the few victories for Ferguson and Taschereau was King's decision, contrary to Lapointe's wishes, to seek the Supreme Court's ruling on the federal government's claim that it had a right to control the waterways. The court's decision was not delivered until February 1929 and, though vague, tended to favour Ontario and Quebec's interpretation that, as King put it, "power belongs primarily to the provinces."

Considering King's aversion to confrontation, he was happy to take an ostrich-like approach and was happier still when the conference ended without anything really being accomplished. As he spoke about the British Empire and the importance of Canadian unity, his thoughts turned to what Mrs. Bleaney had told him about the great work that lay before him. "The conference broke up with words of appreciation from every province represented," he wrote in his diary on the last day. "All seemed to feel it had been a great success and to have appreciated the spirit, what was arranged of a social character... I believe it has been the greatest possible success and I am truly thankful for the Divine Guidance which helped this to end."

Nevertheless, King also had to concede that the feud between Ernest Lapointe and Louis-Alexandre Taschereau had been particularly nasty, and he was compelled to address the situation a few months later. "May I hasten to say to you very confidentially," he wrote to the Quebec premier at the end of February 1928, "that personally I greatly deplore what on occasions has appeared to me to be a lack of whole-hearted confidence, and to some extent, a degree of estrangement between some of my Quebec colleagues and yourself. Just what the reasons may be, I am at a loss to understand, and I hope that anything of the kind, if it exists, will not be permitted to continue." This was wishful thinking at best, as events subsequently have shown. Differences over the interpretation of Quebec's political and economic interests and how they were to be safeguarded were the bane of Canadian politics in King's day and have been ever since.

LIKE A LOT of other North Americans, Mackenzie King was naturally inclined in 1929 to be optimistic. Prosperity was in the air, and business opportunities seemed limitless. Bankers and politicians alike predicted that 1929 would be the best year yet. No one worried that most stocks in the Toronto, New York and other exchanges were overinflated and being propped up by speculators who were borrowing money at an alarming rate to purchase even more stocks. Or that Canada's export market was at the whim of primarily American protectionist trade policies; that Prairie farmers were growing too much wheat while European markets for grain were shrinking; and that too many people from Vancouver to Halifax were looking for work.

Instead, on January 10, 1929, the prime minister of Canada was entertaining his fortune teller, Mrs. Bleaney, who had taken the bus in from Kingston to Ottawa for an "interview." That afternoon, she spent an hour and a half with King and Joan Patteson at Laurier House, reassuring King yet again of the "help" he had been receiving from his loved ones "from beyond." After the session had finished, he dictated a lengthy memo to Howard Measures describing the experience in painstaking detail. In her trance, Mrs. Bleaney had spoken to him about Woodside, his parents and his siblings. What impressed him, he expounded, "is not what she says as to external events, but her delineation of character, of purpose, and of spiritual forces and their influence. What she seems to do to greatest effect and in a truly remarkable way, is to bring before one those who have passed away, and this in a manner which leaves no room for speculation as to the person of whom she is speaking." As for the future, Mrs. Bleaney predicted he would be married in a year or two to a woman not presently in Ottawa, but that they would not have children. "You did not marry," she told him in words he must have agreed with, "because of the sacrifices you were making for others." In his diary entry about this encounter, he added, "I am wholly convinced in the reality of the spiritual world as contrasted with the material universe we know and which is all subject to change."

Even in his positive frame of mind, there were a few issues in the "material universe" that did, in fact, make him apprehensive. In October 1927, at a convention in Winnipeg, Richard Bedford Bennett had replaced Arthur Meighen as the leader of the Conservative party and immediately become King's number one nemesis. A lawyer and wealthy bachelor, he lived like an emperor in a suite of rooms at the Château Laurier, though he was generous to many charities. Only his small circle of friends and colleagues knew of his good nature and charm. He lacked a sense of humour, had little patience, did not appreciate journalists (even partisan Conservative ones), could be rude and spiteful and trusted hardly anyone. Bennett, said journalist Grattan O'Leary, "was not above asking the opinions of others. He was only above accepting them." Or, as his most recent biographer, John Boyko, adds, Bennett "acted as though he was the smartest person in the room because he usually was."

John Dafoe, who became embroiled in a bitter feud with Bennett, believed King got lucky in 1927. "[Bennett] will have difficulties in a country like Canada, [which is] essentially Liberal," he suggested to Clifford Sifton a few days after the Conservative convention, "in contrast to Mr. King who appears to be that phenomenon—a Liberal statesman who grows more Liberal as his tenure of office lengthens." A year later, the Winnipeg editor was still thinking positively. In words more prophetic than he could possibly have known, he told journalist John Stevenson, "I don't think Bennett can turn King out unless he has some luck on his side through the emergence of difficult questions beyond King's capacity to handle."

Mackenzie King, who had studied Bennett on the offensive in the House of Commons, knew better. Bennett might not have been as intelligent or as sharp a debater as Meighen was, but he had political "smarts." King regarded him as a Tory demagogue and an opponent who could not be lightly dismissed. "I should have preferred any of the others on personal grounds," King wrote after he heard of Bennett's victory in Winnipeg. "Bennett's manner is against him, his money is an asset, and he has ability. He will be a difficult opponent, apt to be very unpleasant, and give a nasty tone to political affairs. I shall just have to try to be all the pleasanter and more diligent."

King eyed political events in the United States with a similar rising anxiety. The election in 1928 of Herbert Hoover, the former secretary of commerce, as president made King toss and turn at night. Hoover was a die-hard Republican with a progressive slant: he put his faith in the ability of government to regulate the economy, and raising tariffs on agricultural products was high on his list. As the U.S. initiated what was to be enacted in June 1930 as the Smoot-Hawley Tariff Act, King resisted getting into a trade war with the Americans. "It is not a red-blooded attitude that is needed so much at this time," King declared in the spring of 1929, a month after Hoover officially took office, "as a cool-headed attitude." He maintained confidence in his own ability to dissuade the new president from his protectionist plans, and if not, he and his finance minister had a few other ideas to combat the tariffs and the growing anti-American sentiments in the country.

His personal finances caused him concern as well. In January of 1929, King had learned from Peter Larkin that Senator Wilfrid Laurier McDougald, whom he regarded with a mixture of awe and wariness, was among the elite group of loyal Liberals who had contributed to his retirement fund at the Old Colony Trust in Boston. In McDougald's case, the total amount was $25,000 donated in two installments during 1927 and 1928. "I felt the embarrassment to which this may give rise," King had written in his diary at the time.

That night, he prepared a memo to Larkin requesting another letter assuring him "that each contributor understood that [the fund] was to give me political independence." King was clearly troubled. "I should have preferred not knowing by whom the contributions have been made and that the names should be solely in Mr. Larkin's keeping," he added. Then, in one of his more wily rationalizations, he continued: "However, one cannot look a gift horse too closely in the mouth and I rather hesitate to send what I had written but feel I must. It may after all be the way of rich men in matters of the kind: be all for the best, but political independence alone justifies accepting anything—without some means independence is next to impossible. Not a cent has been asked for by me." King was always touched by the interest wealthy men had for his welfare, rarely stopping to ask himself whether or not it was only because he was the prime minister of Canada. He maintained an almost childlike view of certain individuals, Senator McDougald and Senator Andrew Haydon being good examples, and never liked to ask too many questions.

Had King probed just a bit more, he would have learned that his two friends McDougald and Haydon were entangled in the various Chinese puzzle layers of the Beauharnois Light, Heat and Power Company, the very same conglomerate whose plans for a massive hydroelectric power development on the St. Lawrence River had been approved by the Liberal cabinet in an order-in-council of March 8, 1929. This decision had been taken after much discussion and lobbying, not surprisingly, by McDougald and possibly Haydon.

By this point, McDougald, through his investment in a holding company called the Sterling Industrial Corporation—which was more of a phantom entity than real—had become a major shareholder in Beauharnois, after Beauharnois had bought out Sterling. This deal was contingent, however, on federal approval of the Beauharnois development. No matter, the value of the shares skyrocketed even before a shovel had hit the ground. As soon as the project had been announced, stock promoters anticipated huge profits, and a frenzy of buying and selling of the company's shares kept driving up the price.

Another key player in Sterling was John Ebbs, Andrew Haydon's law partner. Thereafter, the Ebbs-Haydon Ottawa firm handled nearly all of Beauharnois's legal work as well as that of McDougald. And, finally, for good measure, in June 1928 the law firm (without Haydon's knowledge, according to his version of these events) had been promised a lobby fee of $50,000 by Beauharnois's engineer and chief promoter Robert Sweezy—again, if the federal approval was given. "It is human nature to work harder at a price," Sweezy later said about this contract.

Hence, McDougald and Haydon, as well as Quebec senator Donat Raymond, who had also purchased Beauharnois shares (though his were held in trust) and who was a close associate of Quebec premier Taschereau, all had much to gain if the Beauharnois deal was finalized. Well before any of this information was publicly revealed, King had wrestled with the project. He had not wanted to upset Quebec or Ontario and had ensured that the matter was handled with due diligence; or, as he had put it following yet another caucus meeting at which this matter was raised, "with great care and no haste, but sure and certain steps all the way." There is no reason to doubt King's cautious attitude or his desire to take the correct action. But it is equally hard to believe that he was not the least bit swayed by his various confidants in the Senate, all of whom pushed Beauharnois at every opportunity. Regardless of this behind-the-scenes wheeling and dealing, in early 1929 in King's mind at any rate there was little more to say about the Beauharnois project. Or, so he believed. The project was soon to become one of King's worst political nightmares.

James Robb, who had been King's trusted minister of finance since 1926, died of pneumonia on November 11, 1929, and was replaced by Charles Dunning. In a classic case of the pot calling the kettle black, King believed that Dunning, who had been touted to replace King as Liberal leader, was an arrogant, vain hypochondriac who loved feeling sorry for himself. A few days before Dunning was sworn in as the minister of finance, King wrote in his diary that he had spoken to the forty-four-year-old about his "neurasthenic condition and dropping references to his health ... and coming out strong as to be in best of shape ... both to help in own mental attitude and to have a right attitude towards the public." Once assured that Dunning was in the right frame of mind, King welcomed him to his new position. "It is a relief to have him with his ability and his Western (freer trade) point of view." King was doubly pleased that after eight years of cajoling, he had also convinced Thomas Crerar to join the cabinet as minister of railways and canals. With Dunning, Crerar and Robert Forke, King rightly felt that he had at long last vanquished the Progressive movement. Lapointe had a tight grip on Quebec, and James L. Ralston, the former lieutenant-colonel who had been decorated for bravery during the war and who had joined the cabinet in late 1926 as minister of national defence, had matters well looked after in the Maritimes.

ON BLACK FRIDAY, October 29, 1929, the day the bottom fell out of the stock market, Mackenzie King was preparing a speech he was to deliver in Winnipeg. He attended a routine cabinet meeting and had dinner with Joan Patteson. The day's financial turmoil in New York did not make a blip on his radar. Asked by the press to comment, King stated with confidence that "while no doubt a number of people have suffered owing to the sharp decline in stocks, the soundness of Canadian securities generally is not affected. Business [had] never [been] better, nor faith in Canada's future more justified." He stuck to that positive outlook in his year-end diary assessment:

> I thanked God who has watched over and protected me with the guardian angels he has placed about me—dear Mother, Father, Bell and Max. I am sure they are all near by that they are directing and guiding my path ... I am glad that through Providence of God I have been spared to live until 1930, to have been in office from the first to the last year of the second [sic] decade of the twentieth century. It is something which I cannot be robbed come what may. Within that

time has come growth in public confidence in all parts of the country and all parts of the Empire... In Canada, the Government has grown in strength and confidence... The session of [Parliament] was on the whole a good one, but much has been accomplished... Physically I am in better shape I believe than at any time since I came into office... [And] I believe, which is most important of all, that I have developed a little more character, have better control of myself than in any previous year. That after all has been the supreme endeavour... I believe God has a great work for me still to do. I have no fears as to the future.

In his entire career up to this point, he had never been so wrong about any other political or economic crisis.

King kept all his money in bonds or cash and was not personally impacted by the stock market crash. The same could not be said for, among thousands of others, R.B. Bennett, whose fortune nosedived though he easily absorbed the loss, and Ernest Lapointe, who saw some of the $100,000 he had received from prominent Liberals significantly decline. (Lapointe had trouble paying his taxes in 1933 and 1934.)

It is easy to disparage King for being so naive, for not comprehending the magnitude of what had happened on Wall Street, for not seeing the warning signs of the Great Depression and for not pursuing policies that would have prevented the economic crisis from becoming worse. Such criticism, however, is merely based on convenient hindsight. The fact was that in early 1930 few business and political leaders grasped what was happening. In Britain, the new Labour Party prime minister, Ramsay MacDonald, who took office in early June 1929, followed a conservative financial policy that had little impact on the economic downturn and ignored the sage advice he received about deficit spending from the noted economist John Maynard Keynes. Likewise, in the United States, President Herbert Hoover advocated higher tariffs, which triggered an anti-American protectionist backlash in Europe that compounded the financial disaster. Then he instituted pro-labour policies that kept wages high. According to recent research by the California-based economist Lee Ohanian, this actually increased unemployment because businesses could not afford to pay their workers what they were required to, and it prolonged the Depression. Many of Canada's most astute business moguls, such as the president of the Canadian Pacific Railway, Edward Beatty, regarded the downturn in the economy as a temporary "adjustment."

With a PhD in political economy, could Mackenzie King have been more in tune with the world around him? This was unlikely for several reasons. First, King's view of capitalism and his steadfast faith in government's role as a referee was "orthodox and even old-fashioned," as Blair Neatby argued. King, he wrote, "believed, as did most of his compatriots, that private enterprise was the root of all economic activity." Second, King failed to accept that the federal government had the capacity to guide all facets of the Canadian economy, including responsibility for unemployment relief, which under the BNA was relegated to the municipalities and provinces. And, finally, King was a victim of his own self-centred partisan nature. It was not the first time his partisanship had interfered in his judgement, but in the spring of 1930, it had dire consequences.

By the end of February, the coffers of Canadian cities were already under pressure as relief costs were mounting. And provincial governments, which were responsible for the municipalities and whose own revenues were declining, ignored any pleas for aid. Stories in the Canadian press started to suggest that the looming economic crisis was anything but normal. Mackenzie King was oblivious to these developments. When a Prairie delegation led by Winnipeg mayor Ralph Webb met with the prime minister on February 26 to discuss the rising unemployment and requests for public assistance in their cities, King inexplicably decided that it was "clearly a Tory device to stir up propaganda against the Government ... and to place the Government in an embarrassing position." Rather than try to arrive at a solution, King lectured the mayor and his colleagues that his administration "would assist with our own funds in our own way, the provinces could do likewise and the municipalities likewise." The responsibility for unemployment insurance, he maintained, lay with the provinces.

His view had not changed a month later when Abraham Heaps, the popular and compassionate labourite from the North End immigrant quarter of Winnipeg, together with J.S. Woodsworth, urged the Liberals to introduce unemployment insurance. Several opposition MPs taunted King by quoting from the pages of *Industry and Humanity,* in which he had passionately advocated for such social legislation. "Insurance against unemployment," King had written a decade earlier, "recognizes that an isolated human being, not less than a machine, must be cared for when idle. It recognizes also that nothing is so dangerous to the standard of life, or so destructive of minimum conditions of healthy existence, as widespread or continued unemployment."

In private, King convinced himself that he was not indifferent to the plight of unemployed Canadians and that his government was indeed addressing the problem, though in his opinion the situation was not as calamitous as Heaps and Woodsworth had stated. Alone in his study at Laurier House, he had picked up *Industry and Humanity* and considered quoting from it for the speech he was preparing on unemployment. No one could accuse him of being a hypocrite, he reasoned. Yet when he rose to speak on April 3, he got caught up in the moment. He gave a "fighting speech" that received a "fine ovation," but, as he put it, "I made a slip." That was an understatement.

He began by reiterating his view that unemployment was the responsibility of the provinces and municipalities. As to a federally funded relief program, he declared, "I submit that there is no evidence in Canada today of an emergency situation which demands anything of that kind." Furthermore, the premiers had not officially requested federal money and, he added for good measure, the premier of Ontario, Howard Ferguson, had stated publicly that his province did not have an unemployment problem. "So far as giving money from this federal treasury to provincial governments is concerned," King asserted, "in relation to this question of unemployment as it exists today, I might be prepared to go to a certain length possibly in meeting one or two western provinces that have Progressive premiers at the head of their governments." He was interrupted by shouting. "But," he added over the din, "I would not give a single cent to any Tory government."

R.B. Bennett and his Conservative colleague Harry Stevens were on their feet, yelling, "Shame!" King wasn't done yet. "My honourable friend is getting very indignant," he declared, glaring at the Conservative leader. "Something evidently has got under his skin. May I repeat what I have said? With respect to giving moneys out of the federal treasury to any Tory government in this country for these alleged unemployment purposes, with these governments situated as they are today, with policies diametrically opposed to those of this government, I would not give them a five cent piece." The House exploded in wild applause and uproar.

Isaac MacDougall, a Conservative from Cape Breton Island, asked him if any clauses in the BNA stopped the government from assisting the provinces on unemployment. King answered that it did not. "Then why don't you do it?" MacDougall wondered. "Because," replied the prime minister in a telling response, "we have other uses for our money, other obligations."

King realized later that night that his "five cent piece" remark would be taken out of context. He worried in his diary that "to read the speech I would seem I was indifferent to the condition of the unemployed when really it was feeling against the hypocrisy of the Opposition and [the] self-righteousness of Woodsworth and the Labour Farmer group which made me speak as I did." Except no one but he was listening to his own narcissistic justification for speaking as he did.

Later, he blamed the whole thing on booze. According to John Diefenbaker, King told him at a luncheon that on the day he had made his celebrated remark, he had had lunch at the Royal Ottawa Golf Club and had drunk a few glasses of alcohol. One of the guests at the luncheon had been none other than the premier of Ontario, Howard Ferguson. As King related it to Diefenbaker—and also to Jack Pickersgill, who became King's assistant in the late thirties—he had asked Ferguson how he would handle the demands of the Conservative provincial governments then asking for federal money. Ferguson was said to have replied, "Prime Minister, I wouldn't give a five cent piece to any of them." The phrase stuck in King's mind, and later that afternoon he used it in the House of Commons, much to his regret. "If it had not been for the drinks at lunch," King said to Diefenbaker, "I would never have got into that mess."

Within days, the Conservatives and the press were all over him for his caustic comment. "I can see wherein the Tories intend to misrepresent it as meaning not a cent for unemployment and not a cent to a Tory province for anything. It may afford a chance on the public platform for me to show how I am seeking to guard expenditure, and I believe will appeal to the people when limited to unemployment, as most persons get nothing therefrom."

The real reason for King's partisan jibe was probably that he was gearing up mentally and physically for an election campaign—a campaign that he knew was to be nasty. The Liberal Party, sorely without a proper organization or adequate funds, was not prepared for such a contest, but that was not about to stop him. King did not anticipate the economic situation becoming worse. Yet he decided that it would be strategically wiser to hold an election in 1930 rather than risk one in 1931, when his five-year Parliamentary mandate would end and the economy might indeed have declined further. During the twenties, King hadn't been much interested in building up a Liberal election operation and had let the internal workings of the party slide. The only loyal party member who seemed to care was Andrew Haydon. He had resigned as head of the national organization in 1922 (and was not replaced), yet he still had been plugging away, doing whatever he could to raise money. King trusted Haydon's political instincts, and despite his poor health Haydon would officially run the 1930 campaign. As fate was to have it, money would not be a problem for the Liberals in 1930. The party's campaign was primarily financed by the hundreds of thousands of dollars donated to it by Robert Sweezy of the Beauharnois Light, Heat and Power Company, who was forever grateful for the Liberals' support. His financial support, however, was soon to cause King much grief.

Meanwhile, King was alarmed by the Conservatives' success in creating a well-oiled propaganda machine overseen with military efficiency by Major-General Alexander McRae. Bennett and Hugh Graham of the *Montreal Star* financed the Tory campaign with more than $160,000 split between them, though it was McRae's organizational genius that put the fund to such good use. One of McRae's creations was the weekly Tory publication, the *Canadian*. One issue of the newsletter included a political cartoon that King found tasteless and had a reference to Joan Patteson. It proved to him, he wrote in his diary, "what is to be expected, a vile campaign of slander, insinuation, lying and what not … I can see they will try to make something yet of my friendship for Joan. I shall have to be circumspect in that particular."

To counter the expected increase in the American tariff aimed especially at Canadian agricultural products, King and Dunning had determined the best course of action was to implement an imperial preference with lower duties on imported goods from Britain and the other dominions. From King's perspective, it was a brilliant move: "Switch trade from U.S. to Britain, that will be the cry and it will sweep this country I believe. We will take the flag once more out of the Tory hands."

John Dafoe urged King not to call an election until 1931, so as to ensure that a Liberal prime minister rather than a Tory one would be able to attend the next imperial conference. Once Dunning's budget was presented, Dafoe was even more convinced that King should delay so that the country could see the full benefits of the plan. But King wasn't listening. In fact, he had in all likelihood made up his mind by February, before Dafoe had written to him and before Dunning's budget was unveiled.

A key factor that sealed King and the Liberals' fate in 1930 was Rachel Bleaney, the soothsayer from Kingston. On February 8, Mrs. Bleaney arrived in Ottawa to grant the prime minister another "interview." She told him that June and September were the "important months of the probability of an election this year." She repeated her earlier prediction that King was to be married by the time he was fifty-seven and advised that Laurier's instructions to him of many years earlier were clearer than ever. He was "to keep on with the work I am doing, the all important development … would suffer if other hands took it over." His family was there, as would be expected, watching over him, as were two new spirits, the recently departed Peter Larkin, who had died of a heart attack on February 3, and James Robb. The crux of her message was this, as he later recorded: "Spoke of Bennett as a hen picking up all scraps everywhere so fast, thought he would last as leader. Felt I would come back certainly to power. Spoke of holding up till 1935 to get matters settled, living till 82 always busy." A week later, Mrs. Bleaney wrote to King with more mystical encouragement that an election was imminent. He informed his cabinet of his election decision during the third week of March 1930, and the House was dissolved on the second-last day of May. The election was set for July 28.

IN MID-APRIL, to prepare himself for the coming fight, King, fatigued as ever, had opted for a much-needed trip to sunny Bermuda. Joining him were his usual travelling companions, Andrew Haydon, who was slowing down and starting to show his sixty-three years of age, and W.L. McDougald, who had arrived in Bermuda a day and a half earlier. King enjoyed the two-day journey to the island, consulting with Haydon about the forthcoming campaign and reading the intriguing and gossipy autobiography of Frances "Daisy" Greville, the countess of Warwick. During the late 1880s and 1890s, the married countess was the mistress of Albert, the Prince of Wales and future King Edward VII. King found her book, *Life's Ebb and Flow,* published in 1929, impossible to put down. "I can see in it all a real desire to change the social order on the part of one who was brought up in the midst of privilege," he noted.

This was supposed to be an unofficial private holiday, but once the ship landed at Hamilton, Bermuda, King and his party were greeted by the mayor and received cheers from the assembled crowd. The trio made its way to the lavish Bermudiana Hotel, famous for its gardens and view of Hamilton's harbour. Walking into the hotel, King felt as if he was arriving on a "magic carpet." After they settled in, the three men went for a walk through the gardens. "It was like mid-summer, I have never seen anything lovelier, real enchantment," King recalled. "How I wished Joan had been along—or above all else dear Mother to share such a lovely spot with." McDougald had ensured that King was given a suite befitting a prime minister.

For the next six days, King was entertained by Bermuda's governor, Sir Louis Jean Bols; visited with such friends as Sir Henry Thornton, the chairman of the board and president of the Canadian National Railway, who was also vacationing; and toured an experimental farm and the island's celebrated limestone caves with stalactites. By the ocean, he witnessed the "most lovely beach for bathing." His delicacies were slightly troubled by "men, women and girls [who] were bathing in abbreviated suits." Still, he decided that they were "rather pretty to look at."

McDougald had to leave early. Before he departed, he settled King's and Haydon's hotel bills for a total cost of approximately $400, which King considered "mighty gracious." He found McDougald a curious man, adding most ironically in view of his own insecurities, "all he seems to seek in return is recognition and friendship which he craves." From Bermuda, King travelled to New York, where he met up with McDougald and his wife, Mary, and stayed for another two days at the Ritz-Carlton. McDougald had again demonstrated his devotion to King by arranging for a suite of "palatial" rooms for him and Haydon. King revelled in the surroundings, complete with the "Adams drawing room," though too much of a good thing also gnawed at his Christian conscience.

Back in Ottawa, he flagellated himself for "eating too much, drinking not to excess, very little in fact, but on the holiday better to have drunk nothing at all, not exercising enough and letting my thoughts wander too much. I have to let myself get out of hand in the matter of play and self indulgence in one form or another. My power is a spiritual one. I must get back to that at all costs and with God's wish I shall." Subsequent events were to prove to King that his self-indulgences indeed carried a heavier price than merely feeling fat and guilty.

THE VACATION IN Bermuda did not have the desired effect. The two-month campaign was a slog for King; the phrases "heavy sledding," "exhausted," "tired," and on one occasion, "very tired, so very tired" show up in his diary during June and July 1930. Try as he might, he could not compete with R.B. Bennett's brash electioneering. In early June, Major-General McRae put the Tory leader on a private train car for a 14,000-mile journey that took him to cities and towns across the country. He delivered close to 70 speeches in which he promised to save Canada from Mackenzie King and the worsening economic crisis. There were already nearly 400,000 people out of work, and the price of wheat was in free fall, so Bennett found a ready audience that was prepared to give him the benefit of the doubt.

In Winnipeg, where he opened the campaign, R.B. did not speculate about King's relationship with Mrs. Patteson as King had feared; Bennett had his own secrets with the ladies to keep private. But he did make his famous utterance—the speech was also broadcast on the radio—that he would use tariffs "to blast a way into the markets" of the world. The crowd went crazy. Out at Kingsmere for the evening, King listened to the speech on the radio. "I really blushed," he wrote, "such demagoguery, declamation and ranting there was nothing constructive, nothing really destructive in any concrete fashion of the Government's record or policy... The whole speech was clap trap from beginning to close." King didn't get it then, and he never got it for the duration of the campaign.

On the defensive and unwilling to accept the reality of the unemployment problem, the Liberals pretended that the main issues were trade, taxation and, believe it or not, tariff preferences the Liberals had granted New Zealand butter, which were hurting the Quebec dairy industry. Asked in Edmonton about unemployment, King incredibly replied, "I believe there are people who are unemployed because they do not want to work." On another occasion after he gave a similar response on the number of jobless Canadians, the crowd yelled its disapproval. King pointed at one of the hecklers and accused him of being a "slacker."

In Ontario, Premier Howard Ferguson conveniently forgot that he had (at least, according to King's version) suggested to the prime minister the "five cent piece" comment. Instead, Ferguson seethed about King's partisanship and campaigned for Bennett with the declaration in mid-May that, "Mr. King is the issue." Sensitive as ever, King was "hurt" by Ferguson's "personal attack and bitterness" towards him. "It is contemptible Tory tactics," he wrote. Soon, he was referring in his diary to "Howard Ferguson and his machine methods and [New York–style ruthless] Tammany politics."

A week before the vote, King spoke to a more supportive crowd at the Ottawa arena. It was one of the few campaign events where he was not drowned out by loud jeering. He proudly noted that he had received a rousing ovation, "the finest welcome I have received in my public career." Suffering from a bad cold on July 27, he relaxed out at Kingsmere, and Joan Patteson made him a hot drink to soothe his sore throat. At night, he felt better, and he and Joan sat on the verandah of her cottage, Shady Hill, where they gazed at the stars. Anticipating the results, he claimed he felt "quite indifferent" to the outcome of the election. Back in Ottawa late that evening, he went up to his library, knelt before the painting of his mother and "thanked God for his protection and guidance and the dear loved ones who had helped me so much." He read from the Book of Revelations and immediately "felt the presence" of those in the "Great Beyond." This convinced him that despite everything "forces will help us to victory: Good record of Government, good issue the budget... Bennett not so good a leader as Meighen, a better press than ever before, War veterans pretty much on side, Progressives and Liberals not

divided but working together, in Ontario a common enemy Ferguson...
My faith is stronger and I believe we will win with a good majority."

King's dreams and hopes died the following day. The spirits and Mrs.
Bleaney had been wrong. Bennett and the Conservatives won a majority
with 134 seats and 49 per cent of the popular vote. King and the Liberals won nearly 44 per cent of the popular vote, 1 per cent higher than in
1926, so it was not exactly a rout. But as with many elections in Canada,
the popular vote did not translate into a corresponding number of seats.
The Liberals won a total of 90. The party lost 4 seats in Ontario, 10
in the four western provinces and, most shocking of all, 20 in Quebec,
where the Conservatives won 24 seats, the most they had held since
1911. Memories of Louis Riel, conscription and Arthur Meighen had
gradually faded or at least were temporarily forgotten. Ernest Lapointe's
mythical domination of the province was shown to be vulnerable. "The
authority and prestige of Lapointe in this campaign," later recalled Quebec Liberal MP Charles "Chubby" Power, "was lower than it was before
or afterwards, and there was a consequent deterioration in the strength
of his leadership in the Quebec district." Lapointe himself was "humiliated" by what had transpired, and his health got worse in the months
after the election.

King, too, was "surprised" and "astonished" by the vote. But as with
any major change in his life, he was equally relieved. "The load is heavy
and I would gladly do literary work," he wrote on July 29. "I shall be glad
to throw on to Bennett's shoulders the formation of the government and
finding a solution for unemployment. My guess is he will go to pieces
under the strain." Here was a prediction that did come true. King, of
course, had no clue of just how difficult the next five years were to be
for Bennett and millions of Canadians. Because he was given a lucky
reprieve from confronting the Depression, a legend was soon born in the
annals of the Liberal Party that he had masterminded his own defeat to
seek a much-needed respite. While he did everything to encourage such
notions, in the months ahead, the loss to Bennett gnawed at him.

Late one Sunday night in early November, he was cleaning up his
correspondence and came across Mrs. Bleaney's February predictions
of an election victory. "I feel I must have another talk with Mrs. B.," he
jotted down in his diary. "May get something helpful, but what is most
needed is to reconstruct my own thought and life. I read the introduction and chapter of 'The Life of the Spirit and the Life of Today' by
Evelyn Underhill... It was what I most need and am deeply grateful for.
I need to confirm and deepen my own belief."

7

Scandal and Resurrection

Personally I feel and believe that the time is not far distant when I will be called upon again to take hold. If I can only be given the health and strength, which means if I can only control my own self, walk humbly, keep to the right in everything, power will come from above.—THE DIARY OF WILLIAM LYON MACKENZIE KING, *November 25, 1931*

PERCEPTION IS EVERYTHING in politics, and so it was to prove in 1931 for Mackenzie King. The Beauharnois hydroelectric project was transformed into the Beauharnois Scandal that summer, once the company's letters, reports and financial information fell into the hands of MP Robert Gardiner, the leader of the small United Farmers of Alberta group in the House of Commons and no fan of the Liberals and King. That ultimately led Prime Minister Bennett to appoint a five-man committee, which included Gardiner, to thoroughly investigate the Beauharnois deal. It did not take long for the committee's members to uncover all of the sordid details: exorbitant profits on a project not yet begun (a $190,000 investment in Beauharnois was worth $790,000 eighteen months later), shady financial transactions, the links between the company and Senators McDougald, Haydon and Raymond and the munificent donation to the Liberal election campaign coffers.

Like a deer caught in headlights, King was stunned by this exposé. "I confess I am amazed at some of the things that are being disclosed of which I have known nothing," he wrote on July 10, 1931. All three Liberal-appointed senators were summoned to testify before the committee. Raymond had little to tell, and Haydon had had a heart attack and was unable to testify. McDougald, after some prodding, did answer the committee's numerous queries, but not until July 20, which caused King great anxiety.

From the start of the committee's investigation, King felt that there was an implication fostered by McDougald that he had a special association with King. That may well have been so, but King, growing tenser by the day in the heat of July, was prepared to disavow such a relationship. "It would seem that McDougald had tried to leave the impression that his friendship with me gave him an influence in Ottawa he did not possess," King wrote angrily in his diary. "I have nothing to hide and have no reason to shield McDougald in any way, beyond his having been a friend of the party in its time of need during 1921. The extent to which he had misled others in relation to myself, he himself will have to answer for. It would look as though he had deliberately 'used' me to further his ends... The truth is Beauharnois meant nothing to me in any way till [Frank] Jones and his group applied for the approval... I did not know of any connection with McDougald in the matter."

There was yet more to come. Hidden in the mass of paper the Commons committee scrutinized was a bill from McDougald's office for the $852.32 incurred on the trips to Bermuda and New York in the spring of 1930. In other words, it appeared as if the Beauharnois company had paid for King's vacation. When Ian Mackenzie of Vancouver, the Liberal member on the committee, informed King about this he was "amazed" and furious. He explained his recollection of the trip to Mackenzie and others, which he was compelled to do several more times in the course of the investigation. In King's version, he and Haydon had gone together to Bermuda and McDougald had "joined" them there, and without his knowledge, McDougald had paid the hotel bill. King went to see Haydon, who he said recalled these events the same way. "This voucher is being passed around among the Tories," King added. "It upset me somewhat to have anything of the kind started. I felt incensed at McDougald. It looked as though he were trying to use me, have the [Beauharnois] Company feel he had influenced me."

At a private meeting with Bennett, King explained McDougald's actions and pleaded innocence. He asked that the bills not be used as evidence, imploring that "it would be most unfair" to him and "to the high office of Prime Minister." Bennett told King he was deeply hurt that King had recently called him a "dictator." King defended himself by implying Bennett had been rude to him. Such were the juvenile personal sensitivities that shaped Canadian politics. At the end of the conversation, the two leaders shook hands, and King believed that all was forgotten and forgiven. No such luck. Robert Gardiner insisted that a reference to the voucher be included in the committee's final report.

When the story of the scandal was made public, King decided to adopt a proactive strategy and issued a statement in the House of Commons in which he denied any knowledge of what had transpired and said he knew nothing of McDougald's submission of the bill to Beauharnois, had not talked about Beauharnois with either McDougald or Haydon and had been surprised that McDougald had settled the Bermuda hotel account. He failed to address the obvious question that if he was so troubled by McDougald's payment, why had he not given him the money back? The hotel bill was clearly an error, and most journalists interpreted it that way. The *Kingston Whig-Standard,* a fairly independent newspaper for the era, did not believe, for example, "that there is a right-thinking man in the country who holds for an instant to the thought that the former prime minister of Canada, Rt. Hon. Mackenzie King, is in any sense involved in the nasty mess in connection with the Beauharnois scandal, or that he has in any way been a party to the shocking practices which been exposed in relation to Beauharnois and its financing."

As calm as King appeared when he made his remarks in the House, he was in a real panic over McDougald's coming appearance before the committee. Nightly, his dreams were filled with images of his mother and impending doom. In one, mother and son were in a desperate search for luggage at a crowded train station. King's interpretation of it was that "dear mother was making it plain she was near me at this time of anxiety, guiding and helping me." This anxiety was heightened by his tremendous fear that McDougald would reveal the closely guarded secret of the Larkin trust fund. Predictably, his hostility towards McDougald intensified. He felt that his one-time friend "has so little sense and is so devoid of the ethical side of things that one dreads the outcome of any testimony he may give." King added with some embellishment, "This is a real Gethsemane through which one is being called upon to pass at this time."

The "sensation" and "suspicion" revolving around the inquiry was driving him to distraction. "I can see the Tory party intend to do their worst to destroy me, if they can—they will try to link up all this with McDougald and McDougald with myself," he wrote on July 17. "All of this has made me very depressed and sick at heart. But I have the knowledge that my course was straight." The next day was more of the same, with King's legacy and reputation as a social reformer as well as the future of the Liberal Party now in limbo.

This has been one of the hardest days of my life. I have felt sick at heart and mortified at the revelations of yesterday reflecting as they do on the administration which bears my name... For years to come, there will be repercussions and one does not yet know to what further lengths the matter may go. I have worried over every possession I have, wondering if poverty would not be better than the wealth that has become if anything too great. It has seemed to me that all I have may be misunderstood, misinterpreted, or what is worse might lead to a sort of separation from the poor and simple and humble and honest folk of the world.

That evening he was compelled to arrange a dinner (with Joan Patteson's help), when he believed he should have been "in prayer in sackcloth and ashes."

Another sleepless night followed. On Sunday, July 19, King "felt the blessing of a quiet Sabbath," yet he still was in a tizzy about whether McDougald, who was scheduled to testify the next morning, would mention the Larkin fund. He determined that if required he would rise to the occasion and explain himself and the history of the financial assistance he had received. That thought relaxed him, and he again felt his mother's presence. "I enter the coming week," he wrote, "with a feeling that while the ordeal is one of great trial, it will not be without its blessing and evidences of God's mercy."

The committee room was overflowing on July 20. Much to King's relief, McDougald was "calm and collected" and answered all of the questions posed to him about his Beauharnois stock purchase in a professional manner. Best of all, he declared that he had nothing to do with the election campaign funds given by Robert Sweezy to Haydon, nor did he reveal anything about the Liberal fund set up for King's use. Furthermore, he explained that the hotel bill was a secretarial error and that Beauharnois never paid for King's travel expenses. It was close to one o'clock in the morning when King finally climbed into bed. He did so "grateful to Almighty God for his protecting Providence" and prayed that "it may continue tomorrow and through the years to come."

Unlike the recent sponsorship scandal and corrupt misuse of public money in Quebec, no crimes were committed with Beauharnois, and no one was sent to jail. But guilty parties were named and punishment meted out accordingly. The official report was full of condemnation— for the promoters of the project, their financial manipulations and overt attempt to buy Liberal Party support, as well as for Senators McDougald and Haydon for their conduct and blatant conflict of interest. The report was slightly less harsh on Donat Raymond. Nothing was written about King's actions or that of any other politicians involved in the project. As the *Financial Times* so aptly observed, "The report obviously condemns the conduct of those who operated the toll-gates on the progress of this great project rather than those who paid the tolls."

In an effort to cleanse his and the party's soul, King responded to the damaging issues raised in the report in a heartrending three-and-a-half-hour speech he delivered in the House of Commons. He was contrite, but to a point. No untoward influence had been brought to bear on him, he told himself, and, in a classic case of wilful blindness, he claimed that he had no knowledge of the hundreds of thousands of dollars of campaign funds Sweezy and Beauharnois sent to the Liberals. "Individual members of the Liberal party may have done what they should not have done. The party is not thereby disgraced," he declared in conclusion. "The party is not disgraced but it is in the valley of humiliation. I tell people of this country that as its leader I feel humiliated, and I know my following feel humiliated... But we are going to come of that valley, not in any boasting way but with a determination to see that so far the cause of Liberalism in this country is concerned, it will advance to higher and stronger and better ground that it ever has occupied in the past."

At home later, he conceded, "This has been one of the most trying if not the most trying days of my life, apart from those which have related personal sorrow through illness or death of those I love." He took comfort that he had sensed the "nearness of dear mother." Indeed, that morning when "Little Pat," who had been sick, licked King's outstretched hand, he interpreted the dog's actions as if his mother's spirit was inhabiting Pat's body. "He has so reminded me of mother in all his illness," King added about the Irish terrier, "his great patience etc. that it has been like her spirit sent to comfort."

The Beauharnois ordeal was not quite over. Under substantial pressure, Robert Sweezy was forced to resign as president of the company. McDougald also lost his position as head of the Beauharnois board of directors. He was compelled to relinquish the lucrative shares he had received for the Sterling Industrial Corporation yet still remained a major shareholder in the project. The real trouble for McDougald was in the Senate, where the former Conservative prime minister Arthur Meighen had been appointed government leader in February 1932. Confrontational and partisan as ever, Meighen wanted to expel McDougald as well as Haydon and Raymond, which was easier said than done. Legal opinions varied on how difficult that might be under the circumstances. A Senate inquiry was scheduled for the end of October 1931, and again King's anxiety level started to rise in anticipation of the battle.

At a tense meeting in the library of Laurier House, a bitter McDougald maintained that he had done nothing wrong. King disagreed and thought McDougald was being disingenuous. He pushed him to resign his Senate seat, which "would be best in the end." Righteous as ever, King did not want his retirement fund tainted with what he regarded as dirty money. He informed McDougald that he intended to return his contribution to the Larkin fund. Understandably hurt by this suggestion, McDougald felt as if King was divorcing him from his life, which was precisely what happened, though King denied it at the time. A few weeks later when King visited McDougald at his home in Montreal, he did indeed return $15,000 in the form of a personal cheque (which still left $10,000 unaccounted for). Unknown to McDougald, King had already given orders to Charles Stewart, the former premier of Alberta and federal Liberal cabinet minister, who was briefly employed to organize the Liberal Party, that no further donations be accepted from McDougald. King had even drafted a resignation letter for McDougald. A polite visitor, he stayed for dinner with McDougald and his wife, but the senator was, not surprisingly, in a bitter mood. Having turned fifty years old in the summer, he spoke angrily to King about "the worst year of his life."

King departed from Montreal by train for Halifax assured that McDougald would no longer be a problem for him. That night, he did not sleep well but had a "wonderful vision" of the "angelic loveliness" of "dear mother." His other family members also appeared, and his brother, Max handed him a telegram. When he awoke, he decided that the act of returning the money to McDougald represented his courageous effort to separate himself "from material wealth" and that his family supported what he had done. "What is most remarkable," he recalled, "is that mother appeared as dead or dying in a dream I had when on [the] way to Bermuda and Bermuda was related to [McDougald] ... It is true that spirits are guiding me. This is as real as anything in my life—It has been worth everything."

Having reflected on his position further, McDougald became more defiant and was determined not to resign. He faced the Senate inquiry committee and once more answered questions without saying anything revealing. The same thing went for Raymond. Andrew Haydon's health, however, was worse. He had not recovered from his heart attack and was unable to appear. Instead, he wrote out a statement, with input from King and his attorneys, and the committee questioned him twice at his home with a physician present. When it was over, the Senate committee's report censured both McDougald and Haydon for acting in a manner "unfitting and inconsistent" with their positions as senators. Raymond was merely criticized for his conduct as a Liberal Party fundraiser and for accepting money from a corporation with which he had financial ties. "Two Liberal Senators entangled in the Beauharnois mesh are going to the guillotine as surely as night follows day," began the *Globe*'s headline story about the attack on McDougald and Haydon carried out in the Senate. "Of that there cannot be the slightest doubt ... It was an indictment of political wrongdoing, a condemnation of prostitution of public office for private gain, the like of which probably never has been heard in the legislative halls of Canada."

King saw the report and its disparaging conclusions through an oversized partisan lens, but he was genuinely sympathetic to Haydon's predicament. McDougald, however, got what he deserved, in King's opinion; King embraced Haydon's characterization of McDougald as "selfish and ruthless." He was more upset at the insinuation in the Senate report that the March 1929 order-in-council supporting the Beauharnois project was passed as a result of insider lobbying by the three senators. Under any circumstances, McDougald had no more room to manoeuvre. Following a negotiated deal, which guaranteed an end to the investigations, he finally resigned his Senate seat. That barely satisfied Meighen and Bennett, though they agreed—once Meighen read a signed medical report about Haydon's condition—to leave Haydon and Raymond alone. Had Haydon miraculously recovered, Meighen, as nasty as ever, was quite prepared to go after him again. He never got a chance. The strain of the investigation had been too much for him; Haydon died on November 10, 1932, a broken man.

Grant Dexter of the *Winnipeg Free Press* wrote a tribute extolling Haydon's virtues and suggesting, with a fairly liberal interpretation of the facts, that he had been made a scapegoat in the scandal. King, somewhat selfishly, was more circumspect. At first he was reluctant to issue a public statement about Haydon's passing, but after some prodding from Joan Patteson he did so, uncertain why he had hesitated to begin with. Haydon, he decided, was a martyr, and he might as well say so. He berated himself for not speaking up in the House more clearly in Haydon's defence. At the funeral, King was one of the pallbearers. McDougald also attended, though King avoided him. "It seemed like [the demon] Mephisto coming out of the church," he wrote about seeing his former friend. All that had happened, King believed, was the result of McDougald's "meanness and selfishness." On a solitary walk later that day, King had a "wonderful experience," his first contact with Haydon's spirit from the "beyond." An apparition of his mother had made it clear to him that she and Haydon were now together watching over him. He even felt Haydon pressing on his right shoulder, as he so carefully described: "I said yes Andrew, I know it is you continue to press and I could feel the pressure as of a bag of air—a spiritual body, against my shoulder."

As for the "Mephisto" McDougald, King cut off all ties with him. The former senator had nothing more to do with the Liberal Party or it with him. Following the 1935 federal election, when King became prime minister once again, McDougald sent him a note of congratulations, pointing out that he had deliberately detached himself from the party as King had requested. During the next few years, McDougald, whose health was not good, spoke with King from time to time seeking either reinstatement to the Senate, which he maintained King had promised him, or an appointment to a public board. At one meeting in May 1936 at the Château Laurier, King "begged" McDougald in front of his wife to "forget about public life for some time." His cabinet ministers vehemently opposed the idea, and so did he. McDougald tried again in 1938, arguing that he should be reappointed to the Senate—"I have tried on several occasions when in Ottawa to get an appointment with Your Majesty without success," he wrote with a hint of sarcasm—but King ignored the request. When McDougald finally succumbed to his illness in June 1942 at the age of sixty, King politely refused to be a pallbearer. A decade had passed since the inquiries into Beauharnois, but he still had not forgiven his former friend. "I cannot but feel it is a great release for a life that had been a tragedy," he concluded in his diary, "and might so easily have been so different except for desire for recognition . . . He lacked principle and understanding. It is well that he is at rest."

AS EXPLOSIVE AND juicy as the details of the Beauharnois Scandal were, its exposure and investigations did not end partisanship, Senate appointments for party hacks, questionable lobbying practices or the incestuous relationship between Canadian big business and the major political parties. Moreover, the Beauharnois hydroelectric project continued, and by the time its third phase was completed in 1961 the scandal was forgotten.

From start to finish, Mackenzie King had maintained his innocence (reminiscent of the ignorant German guard Sergeant Schultz on the 1960s television comedy *Hogan's Heroes*: "I see nothing!"). King was quick to dispatch McDougald and Haydon to the lions, yet he told himself and anyone else who would listen that his hands were clean. Was this just another example of his hypocrisy? In assessing King's culpability, University of Saskatchewan historian Theodore Regehr, who in 1990 undertook the most thorough examination of the scandal, rightly argues that "a detailed review of the chronology of events, and the influences brought to bear at critical times, leads to one inescapable conclusion. Without the use of political influence the Beauharnois promoters would not have had a full and sympathetic hearing, and their plans would not have been approved on time. Mackenzie King and his cabinet colleagues decided the case on its own merits, not as a result of a direct bribe. But they seriously considered the merits of the case as quickly as they did only because of the political influence and pressure that was exerted."

The fallout from Beauharnois had convinced King that the Liberals desperately needed an official and properly structured party apparatus. This was not a task he especially relished undertaking, but it was essential nonetheless. "I have grown nearly desperate over the matter of party organization," he wrote to Tom Crerar in early November 1931. "I really do not know what I shall do." In the immediate aftermath of the 1930 election, the Liberals were in constant need of money and with the ominous shadow of the Beauharnois Scandal lingering, there was hardly a lineup of party loyalists eager to replace Andrew Haydon as the next official "bagman."

King detested this unsavoury side of political fundraising—a difficult chore at the best of times, but particularly during one of the worst economic upheavals in modern history—as beneath the dignity of a prime minister or a leader of the Opposition. "I hope you realize how unpleasant it is to me, as the leader of the Party, to have to communicate with any of our friends with respect to its financial affairs," he complained to Senator Frédéric Béique, an old friend from Quebec, "but apparently that is expected of me." King half-expected wealthy senators to contribute as payback for their patronage appointments and encouraged others acting in his name to canvass bankers and Montreal and Toronto corporations. King himself, not wishing to establish a precedent or stir up any further nasty rumours about financial matters, contributed a modest $200 to the party in 1931, despite the fact that he could have afforded much more.

After a few false starts, King focused his energies on wooing Vincent Massey to head a national Liberal organization. There was never any love lost between the two men; King did not like Massey's snooty manner and wealth, though he generally appreciated his intellect and talents. Besides, in 1931, Massey was out of work. In the weeks before the 1930 election, he had agreed to leave his post in Washington, D.C., and replace the late Peter Larkin as high commissioner in London, a position that appealed much more to his anglophile sensibilities. When the Conservatives defeated the Liberals, however, Bennett quickly made it clear to Massey that he had no intention of keeping a King appointee in Britain.

King made his first overture in the summer of 1931, soon after the House of Commons committee had delivered its report on Beauharnois. On August 14, Massey and his wife, Alice, were in Ottawa and dined with King at Laurier House. James Thomas, the British secretary of state for dominion affairs, had offered Massey the position of governor of Western Australia, and Massey wanted King's opinion on whether he should accept it. Given King's desire to see Massey as head of the new Liberal organization, he quickly dissuaded him. Over dinner, Alice mentioned that Massey was planning on starting a study group on liberalism and Liberal Party policies. Massey added that he wanted to "help" in any way he could. Considering all that had transpired during the summer with Beauharnois, this offer of assistance touched a raw nerve in King, who was feeling more sorry for himself than usual. He spoke tersely to the Masseys about his tremendous self-sacrifice. "I opened up straight from the shoulder," he recalled in his diary, "said I hoped they had not misunderstood but that I was getting tired of others speaking of helping *me*. It was the party that needed help, Liberalism that needed help. I had been doing all in my power to that end but had been left very much alone, that I had made up my mind unless there were more in the way of divisions of labour, the party would have to find another leader. That I would not and could not do the work of organization or publicity."

The Masseys agreed with King; however, that was not the true message of this tirade. King knew Massey well enough—he was, King wrote, "very selfish, very ambitious"—and was quite certain that if he dangled the high commission in London as a reward, assuming that King returned to power to grant it, Massey would eventually agree to create and lead the national organization until at least the next election. Before he and Alice departed that evening, Massey asked King whether if he was to take the job in Australia for three years he would still be appointed high commissioner once the Liberals were back in office. King replied sharply that he likely would not, that he felt "that those who left the party in its time of need would get no recognition later on... [that] if he were away it would be impossible to appoint him to London later on." It was blatant blackmail, but in his desperate state of mind he didn't care.

Ten days later, King spent the night at Batterwood, the Masseys' estate near Port Hope, Ontario. Massey informed King that he was turning down the offer in Australia and that once he and Alice had returned from a long vacation to China and Japan, he intended to tackle the organization of the party. Forgetting his arm twisting of August 14, King was delighted with this news. He "felt a mental relief and the dawn of a new day." By then, King had also recruited Senator Arthur Hardy, who agreed to raise a minimum of $4,000 a year for the party, and had convinced his former secretary Norman Rogers, who was now teaching political science at Queen's University in Kingston, to undertake some research and policy development.

Because of other commitments, Massey, much to King's frustration, delayed beginning his new task. Just before he left on his Asia trip, Massey received a twelve-page handwritten letter from King, with another strong dose of guilt. "I have loved the cause," King wrote, "I am proud of the party, and honoured by its Leadership, but my love has been wounded, my pride injured, and honour tarnished by the degree to which I have been left alone in the service of the party, and the exposition of the cause." Once the Masseys had returned from their travels, King dispatched yet another missive, this time more upbeat with his vision for the new organization. "I can see it all so clearly," he wrote in mid-December 1931. "A National Association from coast to coast. Dignified headquarters at the Capital. Study groups, speakers' committees, Liberal Clubs, scattered at regular intervals of space across the continent. Above all, a great body of public opinion slowly mobilizing itself—enlightened and increasingly powerful, restive until it has overthrown the powers that be."

As a Christmas present that year, King gave Massey a book that had inspired this latest quest, a well-received biography of Sir Robert Hudson by journalist John A. Spender. During a career that spanned nearly thirty years, Hudson had nurtured the British Liberal Party's National Federation and masterminded the Liberals' electoral success from 1906 to 1922. In words that King had heeded, Spender wrote that "the best leadership and the greatest cause may miss or fall short of their mark, if not backed by good organization." That bit of wisdom primarily meant having sufficient funds in the party's bank account. To that end, King threatened at a meeting in Ottawa of the national Liberal organization committee in late November 1931 that unless $50,000 was raised immediately—he conceived of $200 plus from each of the country's 245 constituencies—he would tender his resignation. This was a bluff, if ever there was one, for later that day he wrote with spiritual zeal about his destiny to regain his position as prime minister.

Fundraising was a slog, and by March 1932, only about $20,000 of the proposed $50,000 had come in. King found the entire situation "gloomy and depressing," and he was far from happy with Massey, who wanted, as he phrased it, "position and gravy without drudgery and risk." Another eight months passed. Finally in November, assured that he would not have the detested burden of raising money and would in due course also have enough cash at his disposal, Massey became the president of the National Liberal Federation (NLF). The smartest thing he did was to hire Norman Lambert to act as the NLF's general secretary (at a salary of $10,000 a year on a three-year contract). Lambert, rather than Massey, would be the true future and brains behind the NLF.

At forty-seven, Lambert, who was born in the small town of Mount Forest, Ontario, 160 kilometres northwest of Toronto, had worked briefly as a reporter for the *Globe* before moving to Winnipeg to serve as the secretary of the Canadian Council of Agriculture and help edit the popular farm journal the *Grain Growers' Guide*. During the twenties, he moved into business working as the general manager of Manitoba Maple Leaf Milling Company, but he also began writing for the *Globe* again as a Prairie correspondent. He had a well-deserved reputation for being a skilled problem solver, which was to prove a vital asset in his new role.

King had the utmost faith in Lambert's organizational talents and political insight. "He has a fine nature, kind unselfish and big, has too much knowledge of western conditions and economic matters generally," wrote King in the summer of 1933 when Lambert accompanied him on a trip west. And Lambert instinctively understood that he had to treat King with kid gloves and deference, even if King in time grated on his nerves. Lambert had the unfortunate luck of getting caught in the middle of the power and ego struggle that erupted between King and Massey.

In November 1932 Mackenzie King wrote in his diary about the NLF with unrestrained joy. He was sure that the organization would "do more to change the trend of politics in Canada for the better than anything thus far accomplished. The Tories may imitate it in part but it will stand for Liberalism worked out as its policies should be on a democratic basis... I regard today as one of the milestones in the history of Canadian Liberalism. I was glad to find myself paralleling Gladstone's action at the time the National Liberal Federation was formed in Great Britain." Quite soon, the NLF was publishing its own newsletter, the

Liberal Monthly, though lack of funds prevented it from being issued on a regular basis for many years. More significantly, Lambert's grassroots organizational work across the country, along with a growing disenchantment with the Bennett government's inability to solve the Depression, contributed to the Liberals' winning ten out of fifteen by-elections held between 1930 and 1935, and most by substantial margins. Provincial Conservative governments also toppled like a row of dominoes.

Still, King and Massey's vision for the NLF were miles apart. King's own insecurities and his image of himself as the uncontested ayatollah of the Liberal Party precluded him from being instructed on policy, or far worse, being upstaged by the likes of Vincent Massey. King was always willing to take advice from his caucus, but definitely not from the NLF president—despite the fact that he acknowledged Massey's "quite exceptional organizing abilities" in making the NLF work. To aggravate the situation further, Massey became a believer in new U.S. president Franklin Roosevelt's state interventionist New Deal policies as a way to stimulate the economy, whereas King initially did not. King's faith in liberalism was supreme, and he maintained that the Depression would end as soon as business confidence recovered. In September 1933, after listening to a talk about the New Deal by the Canadian economist and educator T.W. MacDermot, he observed, "I confess I thought less of the 'new deal' as I heard it explained. The mad desire to bring about State control and interference beyond all bounds makes one shudder."

Massey's first salvo was in a speech he delivered at the end of March 1933 in Windsor, Ontario. This was soon after the official launch of the Co-operative Commonwealth Federation with J.S. Woodsworth as the party's first leader, with its mandate to advocate for a nationalized economy, welfare state and the "eradication of capitalism." (One Catholic journal in Saskatchewan, where the CCF was later to have its greatest electoral success, described the party's platform as "inspired by the Old Jew Karl Marx, the father and author of the Communist Manifesto.") King regarded the CCF as far too radical to work with. He admired Woodsworth for his "real conviction and integrity," but too much central planning, in his view, smacked of socialism, which was entirely unacceptable. King distinguished between "socialist aims and socialist policies"; as Neatby pointed out, "Socialism was not the answer but King was ready to admit that it could be attractive to well-intentioned men."

Massey, surprisingly for a business tycoon, was more willing to listen to what the radical left-wingers had to say. He entitled his talk in Windsor "New Liberalism," and he argued, "It is the aim of the new Liberalism to attack the questions of the present hour relentlessly honestly

and thoroughly, and to offer a comprehensive plan for reconstruction...
I am one of those who would like to see the closest teamwork between
members of my party and those men and women in the third party who
are prepared to co-operate."

This was all much too much for King. When he read the speech
a few days later, he felt "indignant" and that it was a deliberate personal attack on his views. Massey's "talk about 'the new Liberalism' and
a new platform was gratuitous," King wrote angrily. "His references to
the C.C.F. were mistaken politically, his references to myself were little
short of contemptuous... He is a place seeker and time server despite
all his high flown motives... Massey has made no end of trouble. He
is incapable of working with any one else." Massey, however, felt at the
time that he had made a "pretty good speech," and three decades later
still wasn't sure what all the "shouting was about." For the moment,
King yelled a lot in his diary about Massey, but did not personally confront him.

By the time he spoke in Windsor, Massey had initiated an ambitious plan to convene a Liberal summer conference at Trinity College
School in Port Hope. His idea, not without merit, was to bring together
a diverse group of Canadian, British and American speakers, along with
the Liberal caucus, for an intellectual few days in which the partici-
pants would learn, debate and ultimately fashion a progressive Liberal
Party platform for the future. On the speakers' list were, among others, *Winnipeg Free Press* editor John Dafoe; Sir Herbert Samuel, the
leader of the British Liberal Party; and Americans Averell Harriman, a
Democrat businessman, and Raymond Moley, a Columbia University
law professor and part of FDR's "brain trust," which designed the New
Deal. Massey pitched the conference to former Liberal cabinet members, stressing that the gathering would be "unofficial" but, he believed,
"helpful" in the long run to the party. When the NLF was unable to
finance the conference, Massey covered the costs himself. Most of the
heavyweights of the party, even those who at first had been skeptical,
participated.

"Mackenzie King himself was far from being enthusiastic," was how
Massey politely described the Liberal leader's reaction. In truth, King
was initially seething about the event, dismissing the prominent speakers as a "court of notables to circle around Vincent and Alice in [the]
name of Liberalism and Liberal reform. These things get down to the
purely social and 'vanity of vanities.'" Still, after a discussion with the
British Liberal Sir Walter Layton, who was visiting Ottawa and who had

organized a similar conference for the British Liberal Party, King was prepared to let Massey's gathering go on. But he wasn't thrilled about it or Massey.

Out at Kingsmere on July 7, it was raining when Massey arrived at six o'clock for dinner. Thankfully, Joan Patteson was there to keep the peace. Massey provided more details about the conference. King, feeling ornery, accused his guest of becoming too much of an economic nationalist. "There was nothing very congenial about our conversation," King later conceded, "nothing uncongenial but a sort of lack of sympathetic co-operation. The rain and coolness of the night may have played its part. What I feel most is being so much alone and lacking some of the 'solid' environment needed to interest a man like Massey."

A month later, there was more trouble. Speaking at a public affairs conference at a camp near Orillia, Massey praised FDR, who had "shown the only way forward" and advocated for a somewhat undefined "new nationalism" that would make "each nation an efficient member of the society of nations." Massey made a case for centralized planning, almost along the lines that Bennett was about to push for. King was furious. He told Lambert (who hated the speech as much as King did) that he no longer wanted Massey to speak publicly. And if he did so, Massey should know that his future appointment in London was at risk.

At dawn on September 3, the day he was leaving Ottawa for the Port Hope conference, King had an unsettling dream. He saw, as he depicted it, "first one snake in grass come out of a marshy ground—a little long one—and a littler later a larger one. It is the first time I have dreamed of anything of the kind [and] concluded it was a warning to be careful. I wondered if it could mean that by any chance my hosts at Port Hope were not to be trusted."

King was driven to the train station later that morning. "Poor little Pat" was heartbroken at his master's departure. He arrived at Batterwood, where he found the house full of guests and the Masseys "pretty fatigued and without charm." From his perspective, Massey remembered King at the conference as "a headmaster wanting to find out what the young people were up to."

Given King's negative feelings about central planning, one of the conference's key themes, he did not enjoy himself at many of the sessions. He considered Averell Harriman an excellent speaker and was intrigued by University of London professor T.E. Gregory's lectures on U.S. currency policies. The social activities were much more to his liking, especially when he received a "great ovation" at the concluding dinner and the delegates serenaded him with "For He's a Jolly Good Fellow." At the same time, King griped, "The whole proceeding of this conference is very 'amateurish'[;] everything is a new discovery which fools proclaim from the housetops and concerning which wise men have long known and been silent."

Judging from his diary entries, King's major concern during the few days at Batterwood was his inability to have a bowel movement. Even the beer he drank before he went to bed did not have the desired effect. "Imagine it is due to amount of food, etc, toxin poison in the system," he noted. On September 6, his constipation problems prevented him from attending a morning session, so he went for a walk in the woods with the eminent Chinese philosopher Hu Shih, who had been lecturing that summer at the University of Chicago (and was later China's ambassador to the United States from 1938 to 1942). They had a "most interesting talk about conditions in China that had led to the revolution" under Sun Yat-sen in 1912.

Throughout the conference, King eyed Massey suspiciously. One day, he was standing with Vincent and Alice on their driveway when he spotted a garter snake slithering under a stone. This immediately triggered thoughts about his recent dream featuring the two snakes. It was "hard to know what to believe," he felt, "but I trust the visions in their warning. I noticed Alice seemed anxious to draw me on the subject of planning. Vincent is a believer in it. I am not, but in conscious direction of affairs . . . I certainly think the anti-planners have the best of it thus far. I dread the thought of what may come out of the U.S. experiment. I am beginning to think Roosevelt is a little like Bennett in his outlook, methods, etc."

Massey understood that he had "incurred" King's "displeasure" by his use of the word "planning," even though he insisted that he never embraced it in the socialist sense. The event, as he later told a friend, "exceeded my wildest expectations." Yet a year later when Massey broached the possibility of holding a second conference, King absolutely refused to give his consent. That didn't mean he could not take credit for the gathering. In December 1933, he bragged about the conference to Violet Markham, telling her that "the addresses and discussion were all of a very high order and have since been published in a volume entitled 'The Liberal Way.'"

Massey soon indicated his desire to go on a speaking tour out west. He and King had a row about that as well. King told him that he had no desire to deal with his "mistaken utterances" about central planning or anything else. He strongly suggested that Massey confine himself to organizational work and leave speaking to the experts. The exchange left King most unhappy. "He is so inconsiderate and selfish," King wrote of Massey after yet another meeting at Laurier House ended badly, "just a slave driver used to his wealth and 'efficiency', riding rough shod over others."

After that, King continually reminded Massey that he should expend his energies raising money for the coming election—a task that King had promised him in 1931 he would not have to do—or forget about a posting in London. Massey got the message. By August 1934, he was prepared to concede to King that the NLF should concentrate more on organizational work and less on policy development, as King wanted. Not taking any chances, King constantly reinforced this definition of roles and the future of the high commissioner's job to Norman Lambert, who was then compelled to remind Massey about it. With his wife also advising him to acquiesce to King's demands, Massey halfheartedly played the part of the subservient NLF president. But one gets the impression that he always thought he was a better man than King, and the tension between them flared up from time to time.

DESPITE THE PERSONAL animosity between them, Vincent Massey was not King's true political enemy. His hostility was aimed more directly at R.B. Bennett. Mackenzie King had never accepted Bennett as the prime minister of Canada. Personally, he regarded the Tory leader as a domineering and rude bully, cad, egotist and dictator—to list only the most colourful terms he used in his diary to describe him. On Remembrance Day 1932, King could barely watch Bennett lay a wreath at the official ceremony. He looked, King later wrote, "like a butcher, a tyrant and bully which he is." Quebec Liberal MP Chubby Power concurred. During a heated exchange with Bennett in February 1932, in the midst of the Beauharnois debate, Power declared that in the House of Commons the Conservative leader "often exhibits the manners of a Chicago policeman and the temperament of a Hollywood actor."

The role of leader of the Opposition in Canada's parliamentary system is a second-rate job no matter how you look at it. You have to maintain your dignity and show voters you are ready to be the prime minister, but you spend most of your day complaining in public. Or, as Edward Stanley, the earl of Derby, who briefly served as prime minister of Britain in the 1850s, once quipped, "the duty of an Opposition is very simple... to oppose everything, and propose nothing." King rarely felt that debating was his "forte" and had "misgivings and fears" about taking on Bennett in the House. He found his new duties "more exacting" than he had as prime minister since, as he told Violet Markham, he had to depend much more on himself.

It greatly troubled King that one of Bennett's first actions was to raise tariffs, reversing everything the Liberals had accomplished in the preceding decade. Then, to make matters worse, Bennett attended the imperial conference of 1930 in London, where he tried with limited success to intimidate the British Labour government into accepting a *quid pro quo* imperial preferential tariff. He also saw to it that the autonomy of the dominions—which King, in his view, had so courageously fought for—was soon formalized in the Statute of Westminster of 1931 and reaped the accolades from that achievement, which King felt should have gone to him. The following year, Bennett hosted the imperial economic conference in Ottawa and again assertively pressed for a system of preferential tariffs within the empire. The conference was less than a stellar success, despite some supportive editorial comment about the Conservative prime minister in the press. British officials regarded Bennett's "very aggressive tone" as a colonial affrontery, and only a modest reciprocal tariff deal between Canada and Britain on wheat, lumber and several other commodities at the last second saved the gathering from being declared a failure. Bennett, who had managed to alienate just about everyone in Ottawa, even many of those in his own cabinet, blamed the inflexible Brits and the Liberal Canadian press, led by the *Toronto Star* and *Winnipeg Free Press,* for nearly wrecking the conference.

King was certain that sooner or later, as he wrote to Toronto investment banker Alfred Ames in early 1931, Bennett would "destroy himself by the extremes to which he is apt to go in legislation as well as promises than by anything an Opposition can do in 'tearing him to pieces.' The extreme man is always more or less dangerous, but nowhere more so than in politics." To this King could have added, the cautious man in politics may survive, but will be criticized for avoiding new ideas and for being afraid to take a risk. Many high-ranking Liberals and Liberal newspaper owners and editors whispered among themselves that King's day was over.

Charles Dunning, who had been defeated in the 1930 election, remarked to Norman Lambert that when he thought about returning to political life (which he did in 1936) he quickly reminded himself that he would have to confront King, "that charming, polite, hospitable and inert mass." Robert Cromie, the handsome and high-strung publisher of the *Vancouver Sun,* a traditionally Liberal newspaper, didn't like King or his trade and tariff policies and did not try to hide it. King found him impossible to placate. After Beauharnois, Cromie used his paper to make his point. "He is set. He is stodgy. He has got himself into a rut," a *Sun* editorial declared in early October 1932. "Withdrawn within himself and living the intellectual life of a recluse, he is out of touch with the trends and with the people. He is living in a groove as deep and as narrow as a political grave." King was furious and thereafter treated Cromie as part of the Tory conspiracy.

Up until then, King had been trying his best to shrug off the negative criticism fired at him. "Opposition," he concluded in the summer of 1932, "breeds unrest and discontent, it was ever thus." Instead, he focused his attention on the Conservative enemy. The imperial economic conference "exasperated" King so much that he could "hardly trust [himself] to speak or write about it," he told Violet Markham in early 1933. "The conference instead of [making] a path to peace and progress, became one more stumbling block in that path." But he had astutely kept quiet when the conference was in session, lest he be accused of stifling Canadian trade.

Bennett, who kept tight control of every aspect of his government, was running himself ragged trying to solve the multitude of economic problems and despair caused by the Depression. It was a tall order for most world leaders, and it literally wore him out, a fact not lost on King. "I have never seen Bennett look more completely exhausted and worried than he did today," King wrote in his diary at the end of November 1932. "Bennett looked like a man who had been on a drunk the night before, was very tired, he was very nervous throughout the day." Within a year, Bennett was seriously thinking about retirement. By that point, his popularity had dropped just as King predicted it would. Bennett contemplated censoring the press and instituted harsh military regulations in relief camps, where the unemployed were treated like prisoners. There were no opinion polls in the 1930s, but Michael Bliss suggests that "Bennett's standing may have been worse than even Brian Mulroney's low of twelve per cent sixty years later."

Because Mackenzie King regarded Bennett as such an authoritarian, he failed to appreciate, as many Canadians did, the Conservative government's various proactive economic and social policies. Many did not have the desired effect and, like the relief camps, were handled poorly. Yet Bennett should at least receive high marks for effort. As listed by historian Larry Glassford, between 1930 and 1934, the Conservatives:

provided unprecedented sums for emergency relief, propped up the prairie wheat pools, increased coal subsidies, provided a temporary bonus on wheat exports, increased the federal share of old age pensions, nationalized radio broadcasting, negotiated a treaty with the United States to construct a St. Lawrence deep waterway, hosted an imperial economic conference, negotiated new trade treaties, set up relief camps, forced the two great railways into cooperation, helped to create international silver and wheat cartels, drafted legislation for unemployment insurance, established a central bank, set up natural product marketing boards, eased credit problems for farmers, and launched a major public works program.

The trouble was that unemployment and misery persisted, usually made worse by Bennett's rigid stubbornness.

ON DECEMBER 17, 1934, Mackenzie King turned sixty years old. It was a red-letter day for him. He was up at eight o'clock in the morning and prayed for "Divine Guidance" during the coming year, which he rightly anticipated was to be a significant one in his life. Naturally, too, he was certain his mother's spirit was close to him. He greeted Pat, "kissing the little fellow in his little basket as he lay there and wagged his tail." Next he phoned "dear Joan, who was the first to extend loving good wishes." That birthday greeting was followed by similar wishes from the members of his staff, which greatly pleased King. His sister Jennie had sent him a "beautiful black bowl with silver fish." This was no mere coincidence, he thought, since the "sea and the fish seem to me like the universe of God and the souls that inhabit it, all moving with ease and grace through their ethereal abode." Joan (no mention of Godfroy) got him exquisite ormolu (gilt bronze) candlesticks with goats' hoofs and heads, which delighted him "beyond words."

For the rest of the day, a slew of telegrams and cards were delivered to Laurier House, each one dutifully acknowledged and commented on in his long diary entry. Even his nemesis, Prime Minister Bennett, phoned to congratulate him and commiserated about getting old (Bennett was 64). King's birthday dinner was an intimate affair with both Pattesons. Later that evening, he and Joan spent an hour at the "little table" where the spirits communed with them. The parade of loved ones from the "Great Beyond" "came trouping in," a who's who from King's life: all of the departed members of his family, Marjorie Herridge, Wilfrid and Lady Laurier, Andrew Haydon, Bert Harper, Peter Larkin, William Gladstone, former Canadian prime minister Alexander Mackenzie, former Liberal Party leader Edward Blake and many more. King was a man who plainly wanted to be loved, whether by those who were living or dead.

At sixty, he was mentally ready for another election campaign and physically he was much improved from a year earlier. Out at Kingsmere the previous fall, he had strained his knee and found it difficult to walk. He travelled to Baltimore to undergo a complete examination from Dr. Lewellys Barker and learned that he had arthritis. Barker lectured him that his current weight of 207 pounds was not healthy and that he must watch his diet. His ever-increasing weight had caused the back pain that King was also then experiencing.

He took the doctor's advice to heart, which led to a most banal correspondence, which only someone as obsessive as King could have written. "I have acquired a habit of eating an apple, grapes, or a pear the last thing at night," King explained, "and I usually eat a little fruit at dinner, though between meals I eat very little. It may be that fruit should be eliminated from the list, especially at night, though there may be certain fruits which are prejudicial while others may be advantageous." To this Barker replied, "Your habit of eating fruit is one which I approve, especially an apple or a pear the last thing at night and fruit with lunch or with dinner or with both, either raw or cooked." King was exceptionally disciplined. By February 1934, he had lost nearly thirty pounds and by June another ten pounds. "I have reached the point," King proudly informed Barker, "where, when lying in bed, I am conscious at times of the bones in my body." Newspapers soon commented on the weight loss, and his opponents went so far as to speculate that he was suffering from a disease. "That is the kind of propaganda one expects," King added. Within a few years, he had regained nearly all the weight he had worked so hard to shed.

By mid-1934, King was ready for the next campaign. The Liberal political team was ready and so was the National Liberal Federation. Massey and Lambert had raised a war chest of $1.2 million in Montreal and Toronto alone. (Of this, $626,000 came from Montreal, compared with the Conservative Party, which collected a respectable $458,500 from its supporters in the city). James Ralston had the situation under control in the Maritimes, where he had worked with the NLF on young Liberal clubs, and Ernest Lapointe and Chubby Power had Quebec primed for a massive Liberal vote. As early as December 1933, King had been certain of a Liberal victory. The arrival on the political scene of the CCF had not concerned him, since he rightly considered the new party too radical for most western Progressives.

For almost the next two years, King waited. Bennett, who saw the writing on the wall, delayed calling an election as long as was constitutionally possible. In the interim, he had to deal with his volatile minister of trade commerce, Harry Stevens, who crusaded to expose price gouging and excessive profits in Canadian retail stores, factories and other businesses. Unable to control Stevens or appease him with a royal commission on price spreads, Bennett eventually had to fire him. Stevens soon quit the Conservatives altogether and ultimately established his own political party, the Reconstructionist Party, which in some ridings split the Tory vote in the 1935 election.

In what can only be regarded as a last-ditch effort to save himself and the party, Bennett suddenly and without any public warning transformed himself into an FDR reformer. (Or, if we are to accept historian John Boyko's assessment, Bennett acted in a consistent manner "with the Tory principles he had espoused throughout his political career.") For four years, the Conservative prime minister had rarely listened to anyone. Now, he followed the advice of his brother-in-law, Bill Herridge, Canada's representative in Washington, D.C., who had watched in awe as Roosevelt worked his interventionist New Deal magic—relief, reform, recovery—to confront the Depression. That FDR's policies did not immediately cure the ailing economy was beside the point; it was the decisive political action that attracted Herridge's attention. On the evening of January 2, 1935, without even a word to his cabinet, Bennett delivered the first of five remarkable radio broadcasts (written by Herridge and Bennett's executive assistant, Rod Finlayson) introducing Canadians to his own version of the New Deal—"government intervention... government control and regulation... [and] the end of laissez-faire."

Charles Cahan, the old-school Conservative secretary of state from Montreal (with the fitting nickname "Dino," for dinosaur), was stunned. He could not believe, as he later put it, that Bennett could "espouse the economic fallacies of Karl Marx." Most Tory newspapers were shocked, too, at this embrace of what they regarded as socialism, and Liberal organs treated this radical shift as a cynical political stunt.

Mackenzie King could barely suppress his disgust at Bennett's performance and his blatant about-face. "In its egotist style it is nauseating," he wrote after reading a press release about the first broadcast. A week later, following broadcast number four, the level of disdain he felt had only increased. "I felt humiliated to think of the country being in the hands of such a man," he wrote on January 9. "I uttered spontaneously the words 'what a buffoon.' It was really pathetic, the absolute rot and gush as he talked—platitudes—unction and what not, a mountebank and hypocrite, full of bombast and egotism… sickening and disgusting." Once Bennett brought his New Deal into the House of Commons, not much happened, except the passage of the Employment and Social Insurance Act, but you had to be employed to actually benefit from it. Bennett's program, declared Ralston, was nothing but "a series of measures which prove to only a hollow echo of the flow of fulsome rhetoric with which they were announced."

The situation only got worse for the Conservative chief. He had a mild heart attack in early March and was absent from the House for two months. Then he took a trip to England to mark the Silver Jubilee of King George V. By the time he returned, he faced another serious problem: thousands of desperate and hungry unemployed men, led by the Communist Workers' Unity League, who were extremely frustrated with the authoritarian rule and terrible conditions in the relief camps, began trekking their way to Ottawa. After a futile attempt at negotiating a peaceful settlement, Bennett sent the Royal Canadian Mounted Police to stop the march at Regina. A violent confrontation ensued on July 1, in which a plainclothes policeman was killed, dozens of people wound up in the hospital and downtown Regina looked like it had been hit by a tornado.

Fearful of being stained with a communist brush, King stayed quiet about the riot, despite being urged by Saskatchewan premier Jimmy Gardiner and Winnipeg editor John Dafoe to speak out. He regretted his cautious attitude a few days later, feeling he missed a terrific opportunity to distinguish between the communists and the "helplessness" of the young unemployed men. "I should have made one of the great speeches of my life on the Nemesis of Bennett's action in Regina… I was not prepared as I should have been for that, had I been faithful unto this end, I would have had a veritable crown of life."

The stormy 1935 session of Parliament ended on July 5. Bennett still delayed for another month before he finally set the date of the election for October 14. As it did around the world, the Depression wreaked havoc on politics in Canada. Desperate times propelled the oddest and, in many ways, the most dynamic collection of charismatic authoritarian orators who ever held political office in this country. The list included Duff Pattullo, the sixty-year-old leader of the provincial Liberal Party in B.C., who promised "socialized capitalism" as he came to power in November 1933; William "Bible Bill" Aberhart, the true master of the radio, who indoctrinated Albertans with his Social Credit voodoo and became premier of the province in 1935; the brash and erratic Ontario

Liberal premier Mitch Hepburn, who was never too specific about how he planned to solve the Depression, just that he could do it; and in Quebec, *Le Chef,* Maurice Duplessis, the leader of the provincial Conservatives, who joined with a group of left-leaning Liberals led by Paul Gouin to form the Union Nationale, became premier in 1936 and was the most dominant political personality in the province until his death in 1959. Each was to make Mackenzie King's life more difficult.

Voter behaviour at the national level, however, was exactly the opposite. Weary of Bennett's dictatorial style and sensing the danger of ordaining a political messiah as prime minister, Canadians opted for Mackenzie King, the least charismatic authoritarian orator on the ticket. In the 1935 federal election, an unprecedented 894 candidates contested the 245 seats in the House of Commons. In some ridings, there were upwards of 5 and 6 people running. It was a veritable ideological smorgasbord. Besides the Conservatives and the Liberals, there were candidates from the CCF, Reconstructionist Party, Labour, Liberal-Labour, Liberal-Progressive, Social Credit, United Farmers of Ontario-Labour, Socialist and Communist.

During the spring and summer preceding the election, King fussed over a number of writing projects. Several radio broadcasts were planned for early August. A new edition of *Industry and Humanity,* mainly revised by King's hard-worked secretary Edward Pickering, was published by Macmillan in July to draw attention to King as a social reformer. And John Lewis's 1925 biography of King, with additions by King himself and Norman Rogers of Queen's University, was released again. King spent hours deleting sections, writing new ones, correcting facts, selecting photographs and approving the jacket and lettering on the cover. This volume was designed to be effective campaign propaganda, notwithstanding King's contention, as he told the publisher George Morang, that it was "far from being a political pamphlet. It comes pretty nearly being a first-class biography." Morang was amazed at King's effort. "Permit me to say that I do, indeed, think the book an achievement," the publisher wrote to King. "Your thoroughness has amazed me; after over forty years of experience I am not easily startled by the eccentricities of genius."

Despite enormous pressure to be more daring, King was determined to stick to a safe and traditional platform, stressing moderation, faith in laissez-faire capitalism, a balanced budget and social legislation such as unemployment insurance. King had been advocating unemployment insurance for the past two decades and ignored the fact that the Conservatives had already implemented at least a version of it. The same went for the creation of a central bank.

He had announced his strategy to his caucus back at the end of June. "I did not intend during the campaign," he recalled in his diary, "to propose any plan or any 'ism' as a cure all; that these various plans are only possible of execution by means of dictatorship more in the way of depriving the individual of all of his liberties; that I thought we should make our fight for liberty for which the Liberal party stands and show how strongly the trend had been towards dictatorship in Canada... I pointed out we must tell the electorate they should look for no speedy cure for ills for which the present government had been responsible." As Escott Reid, then the national secretary for the Canadian Institute of International Affairs, explained in his analysis of the campaign, King innately understood that "the Liberal party's policy of having no policy" was a much smarter avenue to pursue.

It needs to be said, as well, that during the summer and fall of 1935, King was heavily influenced by a plethora of messages he was regularly receiving from the spirits, mainly through the magic of table rapping. From his mother, grandfather and Laurier to William Gladstone, St. Luke and St. John, each consistently sustained in his mind the prophecy of his forthcoming victory and thereby reinforced his strategy. "Sir Wilfrid says that you will win handsomely," his father's spirit related to him through knocks on the table at one sitting. "He thinks that that you will carry the country from East to West."

His mind was thus made up early about the strategy to employ as he clarified in three radio talks broadcast before the campaign got truly underway. King was never a radio performer like William Aberhart or FDR—he could never relax or be informal enough to connect with his listeners—but he stuck to the script. Bennett was the problem, he declared. Bennett was the dictator who had used Section 98 of the Criminal Code, which prohibited "unlawful association," to throw innocent Canadians in jail. Bennett had wasted Canadians' money on ill-conceived relief projects. Bennett had ruined the country and only the Liberals could save it. "What this country needs is not the fist of the pugilist," King declared repeatedly during the campaign, "but the hand of the physician."

As might be expected, the Conservative press was at its sarcastic best in pointing out that King's speeches lacked substance. After the first broadcast, the editors of the *Ottawa Journal* caustically observed, "We have before us the full text of Mr. King's speech. In no single line of it does he propose anything, offer anything, suggest anything, calculated to remedy anything. True, he speaks of the 'adoption of Liberal policies.' But what are these polices? Is there a Liberal in Canada today, or a Conservative, or anybody else, who can honestly say he is in a position to define what Liberal policies consist of?" Similarly, in an editorial the Toronto *Mail and Empire* highlighted the fact that King had managed to avoid mentioning the word "Depression" in any of his talks.

The Liberal press, however, had heard much more upbeat broadcasts from a man with a vision. (Paul Martin, Sr., too, running in his first federal campaign, credited King's radio appeal with helping him win his seat in Windsor, Ontario.) For these editors, as the Liberals' catchy slogan for the campaign soon declared: "It's King or Chaos." Another party ad was even more to the point: "Bennett let us down. Drive him out." That was King's point, too. "You have now had five years of the Bennett government," he told his eager audiences. "I wonder if any of you are as well off now as when it started?"

The Conservatives' unimaginative "Vote for Bennett" did not measure up. The Tories had more success with a series of clever radio dramatic programs featuring "Mr. Sage," played in the first two shows by professional radio actor Rupert Lucas, a "typical" Canadian, who was highly critical of King and the Liberals. "Mr. King's so fearful," declared Mr. Sage, "that he does anything at all that he thinks will please the crowd." The programs, which aired on the Canadian Radio Broadcasting Commission (now the CBC), were not identified as a Conservative Party production, a fact an outraged Norman Lambert immediately pointed out to Hector Charlesworth, the head of the CRBC. King was understandably angry and denounced the broadcasts as "scurrilous," "libelous" and "insidious."

During the last week of the 1935 campaign, King travelled in his private train car through southern Ontario, arriving in Toronto on October 5. Vincent Massey showed up in the morning to speak with him, and King had to be roused from his sleep by his valet. Massey had been out speaking to Liberal candidates, recommending that they should be more forthright in challenging Bennett, advice that angered King. He was in a particularly foul mood when he saw Massey that day, and as Massey later recalled, their conversation on the train "was far from pleasant." King's diary entry of the confrontation captures the full vituperation of the moment: "[I] told him he had caused more pain and concern than anyone or all else in the party... that I had never had my privacy invaded as he had, for years... It was always Rex must do this, etc, also his talk about helping me and 'the cause' was all nonsense, it was himself and London that alone kept him to the party, that I had to tell him, it was only in this way he could hope to be appointed... It was a scathing review of his selfish actions, including telling him frankly he had been quite wrong in his view on most things. He was quite crushed, perhaps I went too far but it was the 'last straw'." Massey wisely opted not to see King until after the election.

By the time of a mammoth rally at Maple Leaf Gardens three days later, King had calmed down. Speaking before a cheering audience exceeding seventeen thousand supporters and surrounded by his Liberal team, King, despite being tired and uncomfortable from a tight-fitting shirt band, gave a rousing partisan speech that was broadcast across the country. Liberal premiers from British Columbia to Prince Edward Island had also addressed the crowd by the live radio feed, "an amazing feat," King thought. He did not feel he was up to the task, but, as he noted, "others were kind enough to assure me, it was a great performance."

On the day of the election, King, accompanied by "little Pat joyously at my heels," voted early in the morning. It was crisp but beautiful in Ottawa, "like a heavenly vision." He had prayed and prayed some more that he "might be an instrument to do [God's] holy will." At 6:30 PM, the returns from the Maritimes were broadcast, and they were positive. King sensed something remarkable was about to happen. The Pattesons arrived with their dog, Derry, who presumably romped around Laurier House with Little Pat. King insisted that Joan sit in the chair in which his mother had once been photographed, so that she could listen and critique an interview he planned to give the press. The evening was indeed glorious. King and the Liberals swept the country with 173 seats. In Quebec and Ontario, the party won 111 of the 147 seats up for grabs, or 75.5 per cent of the total. Canadians had clearly forgiven Mackenzie King for any perceived misconduct resulting from the Beauharnois Scandal. The Conservatives were humiliated and reduced to a rump of 40. Other than Bennett, who won his Calgary riding, most of his cabinet was defeated. The various independent parties won another 32 seats.

The Liberals had won the greatest electoral victory in Canadian history to that point, and it was a crowning moment in King's life. True, with 44.8 per cent of the popular vote, the Liberals had actually received a higher percentage in 1930 when they had reached 45.2 per cent of the total, but the number of parties in 1935 accounted for that discrepancy. Historians and journalists harped then and later that the results were more of a vote against Bennett than a vote of confidence in

King—and to a certain extent that is an accurate assessment. Nonetheless, in 1935 King's astute skills as a political strategist, his reading of the frightened and frustrated electorate, his ownership of the middle right of the political spectrum, and his decision to be as vague as possible about his policies all played a part as well in his regaining power. This strategy might not have made King the most inspiring leader in the election (or in subsequent ones), but it is hard to dispute his effectiveness. After all, the most basic goal of politics is to win and hold office, and that King knew how to do very well.

With the victory secure, he gloated, ensuring that in his official statement he gave Bennett "the broadside he deserved." "I waited five years to do that," he wrote, "the people needed to be instructed and to see the significance—the dangers that threatened responsible government—the dangers of dictatorship—the dangers of economic nationalism and imperialism ... and above all the need for a Christian spirit not materialism and selfishness as a guide in national affairs." All that was left to do before he got a couple hours of sleep was to kneel in prayer before his mother's painting, to thank God again for his "mercy and guidance" and "to kiss the photos of all the loved ones."

8

A Romantic Among the Spirits

What [the spirit] told me in addition was more wonderful still—a consecration to God's services in the service of Humanity—God['s] grace having saved me from my sins and His love chosen me to help to work out his will 'on earth as it is in Heaven'... There have been forecastings of this all long the way, but nothing so direct and immediate and now that it has come, I can hardly believe it.—THE DIARY OF WILLIAM LYON MACKENZIE KING, *January 27, 1934*

HE HAD DEMONSTRATED his wily and masterful political skills during the 1935 election. Still, at his raw core, Mackenzie King, whether he was prime minister or opposition leader, was also a great and flawed romantic. He idolized women, yet harped about their imperfections. He adored poetry, art, literature and landscaping—all of which were reflected in the famous "Abbey Ruins" he assembled at Kingsmere. On his various travels through Western Europe, he embraced every ounce of history he could absorb at the Louvre, Arc de Triomphe and the Coliseum. He imagined himself to be Leonardo da Vinci, Julius Caesar and the fourth-century pope Saint Miltiades. He was constantly reading and rereading books and fancied himself one of the "great men" possessing a "sincerity of vision" portrayed by the nineteenth-century writer Thomas Carlyle in *On Heroes, Hero-Worship and the Heroic in History*. The quixotic poetry of William Wordsworth, Matthew Arnold and Alfred Tennyson absorbed his mind, touched his soul and permeated his nightly visions.

King had never given up on his dream of seeking the perfect wife and having a family. But the sainted image of his mother, hovering overhead on an angelic cloud, tainted him, and he wondered whether he would ever find a woman as virtuous as her. That was merely a convenient excuse to camouflage his basic selfishness. Quite simply, into his late fifties, King was incapable of completely sharing his busy life with anyone.

Moreover, he had Joan Patteson, his closest companion, who served—minus the sex—as a surrogate wife. Rarely a day went by when she was not mentioned in his diary. He frequently had dinner with her and the ever-accommodating Godfroy. There were long walks in the evening, tea and conversation until the wee hours of the night, strolling at Kingsmere, reading Carlyle or Tennyson aloud and, most significantly of all, sharing a bond in the world of the spirits. Even when he was travelling, King wrote letters to Joan and found the time to shop for her. He searched the stores in London for the right dress and in Paris he bought her a black evening gown. She hosted his dinner parties and vetted his speeches. The Pattesons' connection to King elevated Joan and Godfroy into the highest echelons of Ottawa upper-crust society.

The Pattesons were truly his anchor, a fact not lost on him. When Joan became ill and had to be hospitalized during the summer of 1930, King was genuinely distressed for her and himself. "I wondered what I would do with [Kingsmere] if anything were to happen to Joan," he wrote. "The internal pain she continues to have, gives me anxiety. I pray it is nothing pernicious... She is the life of the place to me and without her Nature would lose its glory." Fifteen years later, while he was in San Francisco at the planning conference for the United Nations, he confessed to her the same sentiments: "I have never realized before how completely alone I am in all that pertains to my personal affairs. Except for you and Godfroy, I should not know where to turn in regard to any matter relating to Laurier House or Kingsmere when absent from Ottawa."

There were other occasions, when King railed against Godfroy for being "a very stupid man," and occasionally he resented his dependence on the Pattesons. One of the main reasons he had insisted that Joan and Godfroy use Shady Hill as their summer home was because it was a slight distance from Moorside and therefore separate in his mind. He always concerned himself about comments that incorrectly judged his relationship with Joan. "There will not be the close identification of my life with theirs," he wrote in 1929. "I will be able to get back to the old days where there can be companionship without absorption, and where I can be apart and alone with God with nature and the loved ones who still lead me on."

Joan cared deeply for Mackenzie King's welfare, and his opinion truly mattered to her. One night, she showed King a new "ultra-modern" dress she had purchased, and he did not like it. "I spoke out in a way which made her cry," he remembered. "I simply cannot understand women in the matter of dress." Joan often broached the subject of King's bachelor status, "this living alone business," as he termed it. "She has come to see that it is better for me to marry if I can find the right one and she as bravely said she will do her part," he wrote in a diary entry of January 1926. "She thinks I should marry someone young enough to have children. I would like a son and daughter if they were to be strong and noble characters." He was more circumspect two and a half years later. "I wish I were married, that would make my life infinitely happier," he wrote. I need a wife, and I pray God that I may have one to love who is wholly my own."

It is easy to question how effective King would have been as a parent, but he loved young children and embraced the Pattesons' three grand-daughters. Two of them, the eldest, Joan, and the youngest, Mary, have only fond memories of him. In September 1931, Joan and the middle sister, Ann, both under ten years old, were visiting with their grandparents from their home in Britain. Grandmother Joan and the girls stopped by Laurier House. King was enamoured with them, but their presence also triggered his feelings about his own life and what he had missed. He described what he called "a very pretty scene" as follows:

> A fire was burning in the fireplace, little Pat was moving about wagging his tail. Joan played the piano. I had afternoon tea at a little table... They are dear little children full of love and tenderness and I am very fond of each of them and their love is something sweet to behold. The great mistake I have made in my life is not having married, and having a family. I would have loved to have children growing up around me, bringing new joy and new thoughts into my home and life. It would have developed me more and brought me more into touch with the people around me. As it is I am becoming a very solitary and at times lonely man, living in the past, and hard to move into any environment other than one I am in.

This possibly is the most honest assessment King ever provided in his diary about his own life.

INTO HIS FIFTIES and early sixties, women other than Joan did continue to fascinate him. While he was in Paris at the end of August 1928 to attend the ceremonies around the signing of the Kellogg-Briand Pact, he encountered a lovely New York socialite who for a time piqued his interest. When Beatrix Henderson Robb died at the age of eighty-two in 1957, the *New York Times* aptly described her as a "club woman." She also served as a trustee of the Museum of the City of New York. At the time King met her, she was a vivacious fifty-three-year-old divorcee, having split from her husband, the banker and insurance executive Nathaniel Thayer Robb.

For about a month, in between his various official duties, she and King flitted around Paris and then Geneva. They dined alone at the best restaurants as well as with large groups of friends and dignitaries at the luxurious Hôtel de Crillon. One evening in Paris, King escorted Beatrix to see a performance of the grand opera *Samson et Dalila*. Other nights, they drove through the streets of Paris, past the Arc de Triomphe and the Place de la Concorde. "It was a wonderful evening," King wrote about another outing. Back home, he corresponded with Beatrix and telephoned her frequently and kept in contact with her for the rest of his life. A visit to New York usually included a dinner date with her. But that was far as his relationship went. There was no talk of marriage, and the absence of any remorseful or romantic diary entries suggests that King had no regrets. He simply enjoyed Beatrix's company, and the feelings must have been mutual.

SLIGHTLY MORE COMPLICATED was his friendship, also renewed in Paris in 1928, with the charming and accomplished Princess Cantacuzène, otherwise known as Julia Grant. They had first become acquainted nearly three decades earlier in the summer of 1899. During his years at Harvard, King had hobnobbed with the American elite in Newport, Rhode Island. He had been hired to tutor (in French and German) two university students, Bob and Peter Gerry, the sons of a wealthy lawyer, Elbridge T. Gerry, the head of a celebrated and quintessentially American family. When the lessons were finished for the day, the Gerrys introduced King, then twenty-five, into Newport society, which was how he met Julia Grant.

She, too, was a member of U.S. political aristocracy. Julia had been born in the White House in 1876 while her grandfather, Ulysses Grant, was serving as president. (When she was a young child, Ulysses called her "my pet.") On a trip to Europe in 1893, Julia met the dashing Prince Mikhail Cantacuzène, of a noble imperial family, who was then working at the Russian embassy in Rome. They fell in love and were married in Newport at the end of September 1899. The couple, who soon had three children, lived in St. Petersburg. Prince Cantacuzène ("Mike" to Julia) became Czar Nicholas II's chief of staff and was promoted to the rank of general during the First World War. When the Bolshevik revolution broke out, Mikhail, Julia and their family escaped—Julia hid a stash of jewels in her coat—and returned to the United States. Julia wrote several well-received books about her Russian experiences, including *Revolutionary Days: Recollections of Romanoffs and Bolsheviki, 1914–1917,* published in 1920, and contributed many articles to the *New York Times* and *Saturday Evening Post.* For a brief time, Prince Cantacuzène was involved in counter-revolutionary activities from his base in Washington, D.C., and then Florida. He and Julia divorced in October 1934, though as her relationship with King unfolded it was plain that her marriage was in trouble long before that.

King's encounter with her in Paris in 1928 was fortuitous and appealing. He found she had "a fine mind" and was "quite fascinating," "sweet" and "altogether a very interesting and charming person." They corresponded fairly regularly after that, until the end of his life. Their relationship heated up in early 1932. He wrote to thank her for a Christmas present she had sent him and wondered if "some of the thoughts" he had "were shared by others." She replied a week later that she was interested in becoming his "friend." A deeper meaning seemed obvious. Like a prepubescent teenager, King at fifty-seven years of age quickly got cold feet. He finally wrote to her after a three-month delay, indicating his desire to "share" with her life's experiences, "which may reveal to each of us worlds of strength and beauty that lie about us." He scribbled in his diary that Julia was an "exceptional person," who had an "understanding of life and that between us we may disclose powers that are latent but unknown." He was certain that she would "view everything as I do and that together we may help each other to greater heights and knowledge."

He now could not stop thinking about her. "I fear that I have aroused part of my nature which should be subdued by writing the letter I did to Princess Cantacuzène—that may be the fire that shuts us out of paradise. It may, on the other hand, be part of a divine fire which controlled means power and understanding." He decided to leave it to the Almighty to figure it out. Yet two nights later he tossed and turned about how Julia might interpret his letter. These thoughts about her, he wrote, "became absorbing and overwhelming later in the night." To his delight and relief, she wired him with positive words the next morning. "There is something to be fathomed in our relationship," he decided, sounding very much as he did thirty years earlier when he had first met her. Amazingly, his understanding of his own sexuality had progressed very little since he was a young man. "I am wondering if what I experienced last night was self-intoxication as it were, my own thoughts," he added, "or influence of spirits round about, brought near by my own thoughts, or nature's urge swept free as it were after being dammed. This I have to learn."

The intensity of King's feelings for Julia frightened him, but he pushed on. She wrote to him of the "occult influences" that impacted on their world, a sure way to heighten his interest. Letters and telegrams flowed back and forth between Ottawa and New York. Then, suddenly, having hyped their possible future together in his own mind, he now doubted the entire situation. On June 3, he spent half a day writing Julia a letter, stamped and sealed it and gave it to his secretary to mail. A few minutes later, he took the letter back and tore it up. The following day, he drafted another version and tried to "refrain from going farther than I should." He added a note of caution in this letter, pointing out to her that "this celestial fire is fire and could be devastating as well as revealing." By the end of the month, with his anxiety level high, he had decided that anything beyond a close friendship with Julia was not feasible. After writing her a more formal letter, he felt that he was "getting back to a saner frame of mind." He conceded to himself that he had "gone too far" in his earlier letters to her and that he "should have had more control of his feelings."

They continued to write to each other, and the so-called "celestial fire" smouldered. In late September, he was troubled again and wanted "to bring [the relationship] to a close." Two weeks later, he changed gears yet again, writing to her of the "the tenderest, the strongest, the purest, the most sacred feelings that a man can have towards the one he holds in reverence." Three weeks after that, he changed his mind, deciding again, as he wrote in his diary, to "bring it to a close, as something which is certain to lead to a misunderstanding. The experiment has gone far enough to prove much which before I did not understand." Poor Julia's head must have been spinning.

King had encouraged her interest in spiritualism, and in early November she visited Detroit to see the medium Mrs. Etta Wriedt. Julia later told King that at her "sitting" with Mrs. Wriedt, Isabel King had made an unusual appearance and gave her blessing to his "friendship" with her. King could not ignore his mother's wishes, but he also had no desire to become more serious with Julia. The answer to this puzzle had, in fact, come to him the night before in another strange dream, "a remarkable vision," though he had not understood its true meaning until he read Julia's letter. Now it all made sense.

He had fallen asleep on November 6, sensing that Julia was trying to communicate with him. At five o'clock in the morning, he awoke suddenly from a strange dream. He and his mother were in Berlin, Ontario, on their way to the train station. He had a train to catch, but he was arguing with his mother and, as he wrote, "resenting her effort to hold me by pleading for me to love her, or to be absorbed in her love, it was a selfish love I felt that was seeking its own—the kind I have experienced with others but never with her." When he awoke, he knew for certain that his mother was there guiding him "on the journey through life which the train and railway always seems to signify." What he experienced of resentment, he realized, was not directed at his mother, but "towards a selfish love that seeks to claim and to hold and to bind me." He added in convoluted logic, "it was clear to me she was making clear to me that carnal love was wrong, that it separated one from the divine and spiritual, and that what I had been experiencing was that; that she was making clear that if I felt resentment towards her for any kind of love that love must be wrong." His dream, therefore, discounted what Julia had heard his mother say about their relationship. "Without the dream," he concluded, "and all it signified I might have been misled."

Four days later, there was another letter from Julia with romantic subtleties, followed by another bizarre dream starring his mother on a train. "To me it is quite clear," he explained to himself. "The reading of the letter from Julia Grant aroused feelings in my nature before I went to sleep, which I should not have permitted to be aroused. I should have sought in every way not to let them come near me." As he read his daily passage in the Bible from the Book of Judges, he decided that he was Samson and Julia was Delilah. And Samson had to beware the sinful power of the beautiful Delilah, lest she crush him.

The romance between King and Julia, if there ever was any, fizzled after this. They remained in touch, but his letters grew mundane, mainly about politics and his various health concerns. He saw her whenever he was in New York, and in the summer of 1939, a month and a half before the Second World War broke out, she visited him in Ottawa, with her French maid in tow. He brought her out to Kingsmere for the weekend and took great delight in showing her the grounds of the estate, despite the fact that her eyesight had become poor. They dined with Joan and Godfroy, the Skeltons, Lester "Mike" Pearson, who had been by then at the department of external affairs for a decade, and other friends. In private, they spoke of his life as prime minister, the situation in Europe, those in the "Great Beyond" and their "respective views of life and beliefs and of the future and its demands." Almost as a form of closure, King told her that at age sixty-five, he doubted that he would ever marry.

Once Julia departed, King felt an enormous sense of relief "to be back again in the quiet and alone." Julia's visit was enjoyable, but it was, he noted, "a bit of a strain throughout." Joan Patteson, who had acted as hostess during Julia's stay, was as always his refuge. "She is so beautiful in her nature and so humble, so tender, so shrinking from any kind of aggressiveness—a beautiful soul," King wrote, suggesting perhaps that Joan was jealous of his friendship with Julia. "The last few days have been a little trying for her but she has risen above all feelings that were natural enough. I would not let her be hurt for worlds—and she has not been—in thought or word or deed."

JULIA GRANT MAY not have been the right match for him, yet that did not stop King from being drawn like a magnet to wealthy and beautiful women, especially ones who, like Joan, were married and therefore unavailable. His romantic nature thrived on introductions, letters with hidden meanings and nearly always innocent rendezvous. His inflated sense of himself as a lover was thereby reinforced, but so, ultimately, was his bachelorhood.

In the fall of 1934, a year before the decisive federal election of October 1935, King was off to Europe for a much-needed extended vacation. He spent several weeks in London, visiting with Lord Beaverbrook, Violet Markham and many other acquaintances, and then he journeyed to Paris and Rome. His imagination was working overtime. His diary during this period is filled with strange visions and weird and wonderful dreams, which he carefully recorded. Little of it made any sense and he interpreted the dreams in any way that suited his state of mind at any particular moment. In Mackenzie King's world, nothing ever happened by accident.

One night in Paris, he had attended the opera by himself. (One of the dancers had white hair curls, which made him think of the back of his mother's head.) Following the performance, he hailed a taxi and stopped off at the Arc de Triomphe. In the dark, he stood alone before the flame of the Tomb of the Unknown Soldier. His life flashed before his eyes, and he was overwhelmed with emotion. "I prayed earnestly to God to help me overcome my faults and limitations till I came to the light eternal," he wrote later that night. "I prayed to God to bless dear father and mother and Bell and Max, dear old grandfather... Sir Wilfrid and Lady Laurier, Mr. and Mrs. Larkin... I especially named dear Joan, and that we may come more and more together into the fuller light of eternal day—and also named Beatrix Robb and Julia Grant, asking God to help and comfort and bless them in their lives. I thanked God for having helped me to come to the point of Love, of having conquered my own desires to the extent I have."

Once he recovered from this mystical moment, he departed Paris for Rome. His visit there had a specific purpose: to commission a marble bust of his father and a portrait or bust of himself. Again, this was connected to another recent dream in which Lord Strathcona (Donald Smith) appeared to him. To King—and his logic and dream interpretations are often difficult to follow—this vision signified that it was important to immortalize men while they were still alive, "to get something enduring," as he put it. In this way he had convinced himself that it was not vain to have a painting or bust done of him, but a matter of immortality.

He soon met with his old friend, the sculptor Giuseppe Guastalla, whom he had hired back in 1917 to create a bust of his mother (at a total cost of $1,200). That meticulous work had taken nearly four years, but the finished product thrilled King. At the studio, the details of John King's and his own bust were worked out. In negotiating a price, King wanted to be fair, yet he pointed out (with a straight face) that he "was not a man of means." After some haggling, he and Guastalla settled on a price of about $2,000 for both pieces.

King began sitting for his bust a few days later. What made this dreary task much more interesting was the presence of Guastalla's next-door neighbour. Signora Giorgia Borra De Cousandier, who could speak English, was a twenty-four-year-old blonde Madonna who caught the fifty-nine-year-old Mackenzie King's attention. Her husband was in the iron business, and they had a young daughter. She impressed him immediately with her education, wit and style. "She was quite a beautiful type," King wrote, "a lovely nature, so we soon formed a delightful friendship"—which continued for the next sixteen years.

Years later, and despite the age difference, Giorgia claimed that had she not been married when they first met, King would have taken her back to Canada "as a wife." Given King's penchant for cold feet in all of his relationships with women, that would have been an unlikely scenario. Two days after their first encounter, he was smitten. "She is one of the most charming persons I have ever met and one of the most beautiful... She had read half of 'The Message of the Carillon' [King's book of speeches from 1927] last night and spoke particularly of the address on Sir Wilfrid Laurier and liking the description of Canada." A Sunday drive around the city with her and her husband convinced King that Giorgia was "one of the finest women" he had ever known. "Meeting her," he cooed, "has been one of the great joys of the visit to Rome."

Once he returned to power, Italy and Giorgia were very much on King's mind, especially during the crisis over Mussolini's unwarranted attack on Ethiopia in October 1935. King and Oscar Skelton, who was still guiding the external affairs department, believed that economic sanctions imposed by the League of Nations on Italy would be counterproductive. Military sanctions were out of the question for both, as they were for Ernest Lapointe and other French-Canadian members of the cabinet. (Lapointe threatened to resign, again, if military sanctions were supported.) Other English-speaking ministers, including Norman Rogers, now the minister of labour, and James Ilsley, minister of national revenue, supported backing the League's actions. With the memory of conscription still vivid, King was diplomatic as ever and conscious of the English-French split on the question. His personal view, however, was the same as Lapointe's. He stubbornly maintained the idealistic—if naive for the 1930s—position he had advanced in *Industry and Humanity* so many years earlier: that "reason rather than force should prevail." Pursuing "perilous policies for peace," as he called it, was "living dangerously." He settled on supporting the League's call for economic sanctions against Italy, yet doing it as unobtrusively as possible.

In Geneva, Walter Riddell, Canada's officer at the League, did not quite get the ambiguous message about the Liberal government's policy. In a moment of League enthusiasm and possibly to follow the lead of the British, who were steadfast in favour of economic sanctions, he stood up and proposed adding oil to the list of products not to be exported to Italy. Newspapers across Europe and North America were soon writing about the tough "Canadian proposal." Concerned about the American reaction, since the U.S. was the chief supplier of oil to Italy, King privately scolded Riddell and hoped the proposal would quietly disappear. When it did not, the government issued a public statement disavowing Riddell's actions. (In the end, the League of Nations, fearful of further conflict, caved in to Mussolini. Italy occupied Ethiopia, which was merged into Italian East Africa.) Canadian historians have long attributed King's actions during this crisis to his pacifist inclinations, his (and Skelton's) suspicions of the British efforts to impose a single foreign policy among the dominions, as well as his desire to keep Lapointe and Quebec happy. Yet his intense feelings for Giorgia Borra De Cousandier might have also been in play.

On November 28, the day that King approved the public denunciation of Riddell—whose diplomatic career was forever scarred by this episode—he also wrote a comforting letter to Giorgia. "It has broken my heart, as I know it has yours, to see your country involved in war and in the differences which have arisen out of relations with the League of Nations," he told her. "I feel an interest, and I might say, an affection for your country second to none of any other country in the world. I need scarcely say that interest and feelings were in no way minimized by my visit of a year ago. I shall always remember, as among the happiest days of my life, the weeks I spent in Rome & the associations of Mr. Guastalla's studio."

In the ensuing years, as King's worst fears about another world war were realized, he kept in touch with Giorgia and tried to see her whenever he was in Europe. In April 1937, she honoured King by naming him the godfather of her third child (the infant died in childbirth). She had written to him of the child being an "angel" and that the role of godfather carried with it "great spiritual value," something King regarded as highly significant. He wrote to her from Ottawa. "When your letter came, telling me of your serious illness, and the loss of your little son, I was almost broken hearted. I cannot tell you how deeply touched I was at the thought of Mr. [Borra de Cousandier] and yourself inviting me to be the Godfather to the little one. No expression of good will and affection could have been greater than that."

His last letter to her before the Second World War prevented him from corresponding with her further for several years was written in mid-January 1940. He decided that after nearly six years it was time he called her by her first name, and he suggested that she be "equally direct." She had sent him a photograph of a portrait of herself, which he cherished. "I love your picture," he wrote. "I rather hesitated to tell you all I feel about it for... it might make your husband jealous. Please know that, come what may my feelings for you will always be the same. Nothing in the world could change them."

Giorgia divorced her husband in 1942 and spent part of her time translating into Italian a brief biographical portrait of King written by the well-respected German journalist Emil Ludwig. King regarded this as a wonderful tribute. When he was in London in October 1945 on post-war business, he reached her in Italy by phone with tender words of love. "I wanted to hear your voice and tell my love to you," he wrote. He was now seventy and she was thirty-five. Another year would pass before they finally connected again in Paris, where King was attending the peace conference. He had not seen her in more than a decade, but he still found her endearing. "It is deeply touching to see the expression of friendship that is embodied there," he recorded in his diary after spending the afternoon with her at the Louvre and drinking tea at the Hôtel de Crillon. "Indeed, it is a surprise to myself to realize that she had seen the character that she saw in me at that time."

Giorgia, too, was touched by King's affection for her.

> He lavished such tenderness and attentions on me that on more than one occasion it brought tears to my eyes . . . we dined in his private room, but were not left completely alone; the waiter, with some excuse or other, was always making brief appearances in the room. It was at this juncture that he told me how he had been watched all his life and why, at times, in his letters he had to be careful and not say as much as his heart would like. He told me also how he had always become embittered that the political situation had prevented him from coming to Italy . . . We passed the days happily together and only seldom was he forced to leave me for his official duties.

One night, Giorgia enjoyed the racy entertainment in Montmartre, an outing that King avoided, as historian Peter Waite observes. This led to an amusing conversation between King and Giorgia, which she later described: "[He] asked me in all secrecy and a little hesitatingly if it were true that in Paris there were places where some women danced in the nude. I was much affected by that question, thinking how many times he in his life had been to Paris and how in his existence, also in some little diversion, what he had known of frivolity, had been precluded to him."

When they parted, neither of them knew that it would be the last time they would see each other. In 1947, King arranged for Giorgia to work at the new Canadian legation in Rome, but due to an illness that job did not last long. They wrote back and forth during the last years of King's life. "Our friendship is one which I treasure greatly, and ever will," he wrote to her on July 19, 1950, three days before he died. "I am only sorry there has been through the years, so little chance for me to see anything of you or of the beautiful Italy I knew."

NOT THAT KING'S romantic spirit and sense of history needed any boosting. But on his trip to Italy in 1934 and other travels through England, he was overwhelmed by the magnificent architecture, gorgeous landscaping, Renaissance statues and fountains, lush Victorian gardens and ancient ruins he saw. These breathtaking sights triggered in him, as they did in many Europeans and North Americans of this era, an intense desire to recreate that sense of history.

ON NOVEMBER 6, 1934, King dined in Rome at Le Jardin de Russie, a café overlooking the luxurious Hotel de Russie's famous terraced gardens, and was moved. "Bits of statuary, walks—ivy—old statuary—stone seats and balustrade, all most effective," he recorded in his diary. "A small circular pool with a fountain figure in the centre gave me a good idea for Kingsmere... I will try to make that into an Italian garden with... geraniums, foliage leaves... fountains, etc. I love the Italian garden and villas."

Even before his Italian adventure, he started his collection of Greek urns, marble bird baths, statues and cast iron animals. There was a sundial from Boston, dating from the American Revolutionary period, which he came to regard as "a mighty emblem of faith," and a one-hundred-pound bronze ship's bell rescued from Sachs Bros.—"Jews who deal in wastepaper [and] rags." After much thought and conferring with Joan Patteson, each item eventually found its rightful place on the property. Such decisions were not made lightly; sometimes King trusted his own instincts or Joan's, and other times he waited for a symbol from the Heavens. He believed the grounds at Kingsmere were enchanted and sacred, reverberating "with spiritual emanations," according to landscaping historian Edwina von Baeyer. "King's gardens," she adds, "were animated by the presence of God, the spirits of the dearly departed, and a multitude of auspicious signs and messengers from Heaven... Trees were more than intimate scenery, birds were 'messengers from heaven,'" and the ruins he constructed during the thirties "were the foundation for another Westminster Abby [sic]."

While he was in Paris in the fall of 1934, he ordered a statue of the Greek mythical figure Pan, as well as a 150-year-old stone faun. The front of the faun in King's ever-vivid imagination was "full of pain, sadness and equally a sort of slyness and downright wickedness, but mostly a noble nature rising out of the beast." After he and Joan had found the perfect spot for the faun, they chatted on the phone about their momentous decision. King had been looking at a book of Italian ruins when Joan had telephoned him. When he hung up and returned to the table, the book he had been reading was open to a page with a picture of the statue "The Dancing Faun of Pompeii." To King this was not a coincidence, but "evidence" of something magical. The next day, having read an article in the *New York Times* about "sex problems," which he found comforting in explaining "the conflict within one's self," he was restless. "It seems to me," he reasoned, "'the Faun' has come to help me to understand the mixture in man of idealism, brute instincts and sadness and joy."

Some months later, he had Joan's son Jack, who lived in London, ship him two discarded gargoyles that had been removed from Big Ben. Initially, he placed the gargoyles, appropriately called "the Pagan and the Christian," on the base of the steps of Moorside—like his house and country residence, no item or spot on the Gatineau property was not blessed with a suitable and symbolic name—but then decided to relocate them to the base of one of the ruins he was constructing. The gargoyles, he believed, had marked his cottage with "ill omens... symbolic of the evil spirit that was here while they were at the door."

In the same crate that Jack Patteson had shipped the gargoyles, he had included stone carvings from the Westminster parliament buildings as well as two stone angels he had acquired. The angels were not only perfect for the quixotic mood King was trying to create, but they also possessed spiritual significance. "As I stood beside [one of the angels] and removed the wool... from the face and saw the beauty of the face and the curls, I was startled for a moment, the resemblance was so like that of dear mother. It was as if she were lying there quietly and resting. I said nothing. Then Joan said, 'How like your mother' and remarked on how strange it made her feel, as if it were Mother lying there. I felt her presence in a very real way, the more I looked at the figure the more it seemed like her." This angel and its partner were clearly more than ornaments for the garden. King's immediate thought was to build a chapel for them since they were "too precious" to be left outside. He was certain that God would give him a sign for where to place them. And sure enough, the Lord did. As he walked in the front door of Moorside, he looked at the large bench he had in his hallway, and it struck him. This was the Biblical "mercy seat," the seat of grace. "The more I thought of it, " he later reflected, " the more it seemed to me this was the place for the two angels on either side of the Mercy seat, that seat is carved with figures symbolic of celestial harmonies... It will be a daily inspiration to see these figures there—a note to the house—the guardian angels."

The "Abbey Ruins," or "The Temple," began to take shape in King's mind in the early 1930s. King was stirred by the ruins and gardens at Hawarden, William Gladstone's estate in Wales, and even more so by Abbotsford, Sir Walter Scott's mammoth home in Melrose, Scotland, southeast of Edinburgh. In this, he was hardly unusual; picturesque gardens, temples and ruins were long considered to hold creative and social symbolism. In June 1937, he and Joan toured Abbotsford (the Pattesons had accompanied King to London, where he was attending the coronation of King George VI), and King was "immediately struck at the great similarity of the scenery to that of Kingsmere."

One day in May 1935, King had been driving home from Kingsmere when on Daly Avenue he came across a stone house in the process of being demolished and had an epiphany. Still standing was a bay window, which, he wrote, "stood out like the front of some old great Temple." He had his driver stop the car, got out and examined it. There and then, he decided to buy the window, presuming it could be had for a "fair price," and transported to Kingsmere "as a ruin to place on one of the hills." All through the rest of May, the bay window, rather than the coming election, was the focus of King's undivided attention. He was delighted that the price for purchasing, trucking it out to Kingsmere and then reinstalling it was only $250. There were a thousand other details to attend to—or at least so he conjured up, driving the contractor he had hired batty—before the window could be set in its proper place on the top of the hill not far from Moorside. That difficult decision was soon sanctified, he felt, by a "vision" in which he kissed "dear mother... with feelings in my heart known only when I kissed her lips in life," followed by a Bible reading from Joshua (15:2) with the words "the bay that looks southward." The bay window became the first piece in King's ruins project.

In quick order, the owners of the Daly Avenue house, the Shenkmans ("the Jew and Jewess," akin to Moses and Joshua in one of King's dreams), offered King free of charge whatever stone remained. He was so thrilled that he and Joan consulted books on architecture and mythology and started devising plans for a chapel or library. Wisely, he enlisted the services of a talented Ottawa architect, J. Albert Ewart— he had designed the Booth Building and Ottawa Electric Building on Sparks Street and the Metropolitan Life Building on Wellington Street, among many others—to guide his dream and construction. "[Ewart] caught the vision at once," King recorded after he had showed him the site for the first time. That vision, as King explained it to the architect, was nothing less than "a combination of the Parthenon at Athens, [and] a cathedral or Abby [sic] [like] Westminster." Thinking about this discussion later, King extended this tie with the past further, fancying himself a warrior and righteous hero. "There is an association in my life with Prince Charlie," he decided, "and that period in which I must yet discover and which may explain much—some Jacobean blood—as well as Cromwellian spirit—these forces are all exerting their influence I am sure."

Money halted the dream of a Westminster Abbey in the Gatineau almost as soon as it had begun. King's practical and political side won out, and he determined that in the midst of the Depression spending thousands of dollars on an extravagant indulgence like the ruins might well be interpreted incorrectly by Canadian voters. "I am anxious," he wrote, "there should be no room for suspicion let alone criticism of expenditures which seem to be considerable." Instead, Ewart designed a stone wall that, when finished two weeks later, nicely incorporated the bay window and a doorway. Accentuated by wandering vines, it reminded him of a "Greek Temple on a hill." The effect was truly magical, in his opinion. "Nothing could be more exquisite" he wrote, "than the pictures that are framed by the doorway and the window of the landscape beyond, and particularly of individuals approaching either side of the open spaces." The site was officially christened "The Abbey." Its religious connotation thus denoted the land holy, linking it forever in King's mind with the Middle Ages and with British and European culture and history.

On July 5, 1935, the day the sixth session of the seventeenth Parliament ended, King was able to get out to Kingsmere before dark. He and Joan walked in the moonlight like pilgrims into the Abbey Ruins. He told her that they must "keep it always a very sacred place, let no word or thought enter there which was not the holiest and best." She did not argue. "We sat for a while and looked to the distant valley L'Infini [what King called the view through the bay window], and I thought to consecrate myself to the years ahead on which I have now just entered," he wrote. "When we walked around to the front of the window, the valley beneath was filled with fire flies whose little lights were sparking everywhere. It was like a world of spirits. Joan quoted the lines, 'How bright those angel spirits shine.' It was truly like being translated into the spirit world."

The Abbey Ruins remained as they were for the next year as King resumed the office of prime minister. In March 1936, it was brought to his attention that the British North American Bank Note Company building not far from his Parliament Hill office was being torn down. Ultimately, its four Roman-style pillars were acquired and transformed by Ewart into King's splendid "Arc de Triomphe," a "thing of great beauty," as he declared. In early July 1936, he held a dinner party at Moorside with the Pattesons, the Conservative MP and statesman Sir George Perley and his wife, and a few other friends and government associates. After dinner, the men were talking indoors and Joan led the women towards the ruins. "When I saw them in front of the arches I was in raptures," King wrote, barely able to contain himself, "the beauty of the figures in their light garments about the stone was like a Roman or a Grecian tableau. Sir George, then seventy-eight, was visibly dumb-founded with the beauty of the Arc de Triomphe."

There was still more. Other stone pieces and pillars obtained from the demolished bank were formed into a vista he called the "Window on the Forest." In an Ottawa salvage yard, he discovered stones, pillars and part of a fireplace from the old Parliament Buildings dumped there after the 1916 fire. From those, Ewart constructed what was envisioned as the wall of a chapel. Added to the structure during the summer and fall were a few stones King managed to find from William Lyon Mackenzie's printing shop and more stones from London sent by Jack Patteson, including one engraved with the crest of the Speaker of the British House of Commons. The stones from his grandfather's shop were placed into the fireplace, since according to King, "that was where the fires of freedom for Canada were lighted." King was elated with the effect: "The moonlight last night on the stone was like white magic, ethereal—celestial. It was all most beautiful."

At the same time, the outlay of more money on this ever expanding project troubled his Presbyterian conscience, though this was quickly rationalized away. At stake, he insisted, was a far greater purpose. "I feel in some way I should not incur the expense," he wrote on September 1, 1937, "yet it seems to me there is something 'larger' which justifies it. I am a national figure, my example will have its effect on the Nation when I am gone—to foster the ideal—to teach the lesson of preservation and redemption in terms to which it can be referred so as to impress the youth . . . is part of my work. I am speaking to my own and other generations, expressing my soul—reconstructing the past, with the truly spiritual note, and making meanwhile a sanctuary for one's own soul and the soul of others."

That enduring objective was not entirely realized, since the ruins came to represent King's multitude of eccentricities rather than his love of history. Certainly that was what Lester Pearson thought. He was working for the external affairs department in London, serving in a high position at Canada House, and endured much of the German Blitz. A few days after Westminster Hall was bombed on May 10, 1941, his office received a "Secret and Most Immediate" telegram from Prime Minister King, with orders to collect stones and other discarded pieces from the Westminster salvage yard. Considering the seriousness of the hour, Pearson could hardly believe this strange request, yet he complied as any good civil servant would. Besides Pearson and Jack Patteson,

King had also asked Vincent Massey to scour the Westminster Hall yard for relics, but with the proviso that no more than £5 on the shipping charges be spent.

It took until 1947 before another crate from London arrived. Inside was a veritable treasure of ruins: "Railing from Buckingham Palace, pieces of sandstone and marble... from St. Paul's Cathedral, timber, lath and plaster from Westminster Abbey, a Tudor rose carving and a section of the noticeboard from the House of Commons as well as a lead finial from Westminster Hall." Curiously, King decided not to integrate these pieces into his ruins, nor did he leave any explanation of why not.

For the remainder of his life, the ruins continued to inspire and awe King's sense of idealism. Both the Pattesons' dog, Derry, who died in 1940, and King's beloved Pat 1, who succumbed a year later, were buried near the Abbey Ruins, making the ground even more hallowed in his imagination. In October 1949, King was thrilled when he was made the only honorary member of the Canadian Society of Landscape Artists and Town Planners. He told the gathering that if he had the option of doing his life over, he would have forsaken politics and become a professor of landscape architecture. That confession may not have been entirely honest, but few comments touched his soul as much as one by the chairman of the meeting, who referred to King as a "romanticist."

Kingsmere, as he also admitted to the landscapers that evening, kept him alive. The property and the ruins were his focal points and imbued him with a special connection to the "Great Beyond." Today, tourists visiting Kingsmere enjoy stopping at the ruins for photographs or to attempt to decipher the never-completed structure's hidden meanings.

MACKENZIE KING WAS only one of thousands in the United States, Canada, Britain and Continental Europe during the late nineteenth and early twentieth century to seek the answer to that most elusive of human conundrums: is there life after death? The idea that life merely ended with death was too painful to contemplate. "The ultimate agnostic's creed and his passing into the great darkness, as though life were but a lamp enkindled and going out, when its oil failed; with no replenishing, seemingly no hope of a hereafter; left on my own mind a shuddering sense of pain," opined Sir Daniel Wilson, a professor of history and English literature at the University of Toronto in a letter to his friend William Dawson of McGill University. "The separation from our loved one would be terrible indeed, if we could think that death meant such annihilation." For young and old, male and female, middle and upper class, neither traditional religion nor modern science offered an entirely satisfactory response to this enigmatic riddle. What these believers put their faith in, explains British historian Janet Oppenheim, was that science, so at odds with religion in the late nineteenth and early twentieth century, "might be harnessed to the search for philosophical or religious meaning in human existence."

Spiritual phenomena seemingly offered the proof of an afterlife and possible clues to its form and function. Soon mediums and soothsayers were popping up everywhere. As George Bernard Shaw described the whole spiritual phenomenon in his 1919 drama, *Heartbreak House,* in the years before the First World War, English society was "superstitious and addicted to table rapping, materialization, séances, clairvoyance, palmistry, crystal-gazing and the like to such an extent that it may be doubted whether ever before in the history of the world did soothsayers, astrologers and unregistered therapeutic specialists of all sorts flourish as they did during this half century of the drift to the abyss."

Beyond these fakers and frauds—and the movement attracted plenty of charlatans—such prominent intellectuals, inventors and writers as Susanna Moodie, Goldwin Smith, Dr. Maurice Bucke, who founded the University of Western Ontario Medical School, journalist Phillips Thompson, Reverend Salem Bland, Toronto writer and suffragist Flora MacDonald Denison, Alfred Russel Wallace, Alexander Graham Bell—the list is seemingly endless—probed, considered and debated the meaning of death and its mysterious world. Salem Bland, one of the staunch advocates of the social gospel in Canada and a talented teacher at Winnipeg's Wesley College in the early 1900s, had many doubts, but not about the hereafter. "The appearance or manifestation of the departed dead is to me a proven fact," he declared after attending a séance in Quebec City in 1894. "The only doubtful question, whether such manifestations could be, or should be sought at will." Similarly, Reverend Benjamin Austin, who was expelled in 1899 by the General Conference of the Methodist Church for his spiritualist beliefs, maintained that "spiritualism has removed forever the gloom and sorrow and torment which humanity suffered through the false church teachings concerning death." Finally, the respected Toronto physician and surgeon John King (no relation to Mackenzie King), who regarded himself as a man of "common sense [and] intelligence," wrote after attending a séance in 1894, "I am compelled to admit the existence of the phenomena of materialization; and many other apparently psychic manifestations. I cannot deny their existence." Meanwhile in Britain, the explosion of interest in the paranormal led to the establishment of the British Society for Psychical Research in 1882 (today, the Society for Psychical Research, or SPR), an organization that adopted an academic approach to investigating spirit sightings and other unexplained phenomenon. After 1920, one of its Canadian members was Mackenzie King.

Interest in spiritualism waned after 1904 or so, as more and more mediums and fortune tellers were exposed as frauds. Then, within a decade, the tragedy of the First World War and the deaths of millions of sons, husbands, fathers and brothers awakened a new, beleaguered generation to the possibilities that spiritualism could enable them to communicate with their departed loved ones. Like Mackenzie King, they read books about spiritualism published during or immediately after the war by such well-known writers and psychic researchers as Sir Arthur Conan Doyle of Sherlock Holmes fame. Asked once about his seemingly bizarre beliefs, Conan Doyle explained: "After weighing the evidence, I could no more doubt the existence of the [spiritualist] phenomena than I could doubt the existence of lions in Africa, though I have been to that continent and have never chanced to see one." Conan Doyle and many more like him believed that communication with those in the "Great Beyond" was quite possible. For the same reason and during the same era, as King lost his parents and two of his siblings within a five-year period, he was drawn to spiritualism, mediums and séances as the one way he could reconnect with those he cherished most. This was not a "crisis of faith" for him, but quite the opposite; he regarded his spiritualist activities as an evolution of his Christian beliefs and practices, not a rejection of them.

KING HAD ALWAYS found much-needed support from the spiritual world. In Geneva in September 1928 to speak at the League of Nations, he was more nervous than usual. He was calmed by a conversation he had with Dame Edith Lyttelton, a novelist and friend of Violet Markham, who told him of her otherworld experiences. She often felt the hand of her late husband, Alfred, the politician and footballer, upon hers and his presence everywhere. King was impressed. "Somehow or other," he wrote that day," I feel this moment is a fulfillment that all that has gone before. I cannot but believe dear Mother and father will inspire my thoughts and words for what I am to say... I can but trust in God and divine leading and inspiration."

This was a typical Mackenzie King diary entry, one of literally thousands in which he seeks the comfort of those in the "Great Beyond" to relieve his anxiety in the here and now. His multitude of debilitating insecurities, which became more acute during the Bennett years, propelled him into the world of the supernatural. "I continue to believe in evidences of survival and of personality and of the personal concern of some of the departed for our lives, and their power depending wholly upon their spiritual excellence, something not quickly or easily obtained," he wrote on July 6, 1934. "I believe in forces seeking to usurp other forces so long as they are near to earth—the desire of men for fame, etc. but I believe Truth and Justice are eternal and will work out for all in the end. In my own case I believe no spiritual power, of those who have departed, equals that of members of the family."

In his quest to find guidance and signs in absolutely everything that crossed his path, he became even more obsessed with messages allegedly imparted to him in his tea leaves and shaving cream lather and interpreted recurring numbers and the hands of the clock in ways few others did. On July 20, 1933, for instance, King was reading his little prayer book when his eyes fell first on the reading coming up for August 7. The meaning was clear. "[It was] the day I was chosen Leader of my party in 1919, the day I resigned as P.M. in 1926—it seemed to mark the place of new beginnings in 1933. Curious each 7 years apart as I have just counted—7 the mystical number." Or, on February 6, 1935, the anniversary of his mother's birthday—one of many sacred days on King's calendar—he woke at seven in the morning. "In reading the bible," he writes, "the chapters were revelations 8-9-10 all being part of the one vision of John. My eyes saw first the underlined words at the end of [chapter] 7 (the mystical number)... Something told me to keep on to the end of the vision which follows."

Tea leaves both amused and tantalized him. Dining in Toronto on April 12, 1934, at the Royal York Hotel, he was alone except for a cup of tea and his imagination:

> In the first cup there was very distinctively two birds soaring in opposite directions. The one a small bird having a staring course... the other a much larger bird, coming in an opposite direction and above the other, it seemed to have had to go through some obstacles... In the next teacup there again two figure prominent—the one very large which looked like a sort of hog or boar, occupied a good deal of space. Just behind him was a much smaller, lither and more active fellow... The third teacup was very distinct. There was the one small bird soaring alone...

And what did these various images mean? In King's imagination, the shapes of the leaves somehow pointed to a Liberal victory in a coming by-election, which did come to pass. The day after the Liberal candidate won by close to 1,500 votes, also predicted in the leaves, King added this bit of teacup wisdom: "At all events I have a feeling that these tea leaves were 'thought formations' sent as a further evidence of the truth that is being revealed in these many ways 'nothing too small' to be ordered by the Father of all, nothing that cannot be known by faith."

King frequently consulted with such "remarkable" psychics as Mrs. Quest Brown from England (who also read the palm of artist Marc Chagall) and was entranced by her insights into the connection with the past, future and reincarnation. "I know all this sounds foolish," King conceded in one of his more translucent moments, "but I believe in the unseen realities of spirit as the only realities." What he truly sought, however, was a more direct link with his departed loved ones. That overpowering need led him to attend his first séance on February 21, 1932, at a time when he was still dealing with the fallout from the Beauharnois Scandal.

The invitation had come from Mary Fulford, the widow of Senator George Fulford of Brockville, whom King had known for many years. The featured medium was to be Mrs. Etta Wriedt, a psychic of some note, who was visiting Mrs. Fulford from Detroit. That news excited King. "I felt this to be more important than anything else," he wrote upon reading Mrs. Fulford's letter. "I look forward to this event as one of the most remarkable in my life." His great anticipation was due to Mrs. Wriedt's reputation of possessing the gift to speak in the voices of the departed. "This is something too wonderful for words," King added. "It is all part of divine leading, I believe."

Long before Mackenzie King discovered her, Henrietta "Etta" Wriedt, probably in her early seventies in 1932, had already travelled far and wide conducting séances in London and across the United States. The British journalist William T. Stead (who died on the *Titanic*) and Vice-Admiral Usborne Moore, who profiled her in his 1913 book, *The Voices*, were two famous "sitters" at her séances in London. She was well known in the spiritual world as a direct voice medium; the spirits communicated through her. She often utilized a tin trumpet, a three-foot metal cone and the services of a spirit guide she called "Dr. Sharp." Mrs. Wriedt possessed the uncanny ability to speak through the trumpet in the voices of six or more "visitors" from the other world and in long dialogues in a variety of languages (German, Dutch and Arabic) that she did not know. In April 1898, she performed in Toronto at a three-hour séance at which Phillips Thompson participated. He described the wondrous experience in the *Globe*. "The most singular feature to a non-spiritualist or one not accustomed to the view of the cult," he wrote, "was the perfectly cool, calm, matter-of-fact demeanor of those who evidently fully believed that they were holding intercourse with the departed."

After his experiences in Brockville on February 21 and 22, 1932, King became one of those believers. He began keeping meticulous, lengthy and mind-boggling handwritten records of each séance he attended, which spirits communicated with him, and any other impressions he formed. He was so impressed with Mrs. Wriedt that he invited her and Mary Fulford to Ottawa the same week. Upon his return to Laurier House, almost the very first thing he did was to see Joan Patteson "to tell her of the conversations." Joan had always shared King's interest in spiritualism; now as he immersed himself even deeper, she became his willing partner in this endeavour. "She sees it as I see and feels as I feel on the great spiritual realities," he wrote sometime later. "Much would be lost if it could not be shared."

Despite dealing with Beauharnois, the snide comments of R.B. Bennett and a dinner at Government House, King, with Joan by his side, managed to convene three séances with Mrs. Wriedt in the "little dark room" on the third floor of Laurier House. This was where he kept his growing private collection of books and articles on spiritualism. Around the table sat Mrs. Wriedt, Mary Fulford, Joan and King. Mrs. Wriedt began all of her séances by reciting the Lord's Prayer. A high-pitched noise was heard, which King compared to the sound a radio made as it becomes "heated up." Then, through the power of the "telescopic trumpet" and "Dr. Sharp," the dear spirits, all of whom King and Joan wanted to hear from, made their appearances: King's mother, naturally, followed by the other departed members of his family. In subsequent séances, Bert Harper spoke to King, as did Sir Wilfrid Laurier and Peter Larkin. At that first séance, Isabel King reassured Joan that the child she had lost in infancy in 1906—the baby girl's name was Rose, but in the séance records she is referred to as "Nancy"—was playing with other children who were "taught about their parents and to love them." Joan later swore that she felt Nancy "brush" her cheek. The session was everything King had hoped it would be. "It was if anything," he raved, "even more remarkable than what took place at Brockville."

This séance lasted two hours, ending at five o'clock in the afternoon. Once the women had departed, King sat at his library desk and for the next three hours "wrote steadily," describing the "conversations" with the spirits. As soon as he was done, he had to attend a reception given by the wives of the Conservative cabinet ministers. But later that evening, he caught up with Joan, "who was full of wonder and mystery."

At the end of June, Mrs. Wriedt visited Kingsmere for another round of séances. (King paid her $200—by his estimation "a generous allowance"—for her travelling expenses and to compensate her for the books she left him.) "She seemed to bring a feeling of peace and rest with her," King observed the morning she arrived. Apart from Joan, he had also invited two trusted friends for this momentous occasion, his former private secretary Professor Norman Rogers of Queen's University and Dr. Thomas ("Tom") Eakin, whom King had known for many years and who was then principal of Knox College at the University of Toronto. A series of séances—"quite remarkable gatherings"—were convened during the next several days in a darkened room at Moorside, some with everyone present, some with just King alone. His mother was there again, along with the rest of the gang of dead family members and mentors.

"What impresses me greatly," King pointed out in his diary on June 27, "is not only what is said, but the apparent judgment and foresight with which subjects are approached or introduced. It would seem as if those we loved know not only our behaviour, but our spiritual needs, our thoughts, and were seeking mostly to minister to them." Three days later, he was even more enthusiastic, yet as pious as ever: "What I really should have done, having reference to this opportunity, was to have lived the week in prayer and writing out what was said immediately after the conversations. These can be no doubt whatever that the persons I have been talking with were the loved ones and others I had known and who have passed away. It was the spirits of the departed. There is no other way on earth of accounting for what we have all experienced this week. Just because it is so self-evident, it seems hard to believe. It is like those who had Christ with them in His day." Was King a gullible fool or a faithful follower? Perhaps he was both. His one regret during Mrs. Wriedt's week-long stay was that he missed having a wife (or maid), since he was forced to entertain his guests and was therefore unable to record the séance conversations in the timely fashion he desired.

Rogers and Eakin responded differently to Mrs. Wriedt and the journey into King's other world. Rogers was the more open of the two. "The unusual experiences of the last visit left a deep impression on my mind," he told King a few days later. "I will not say that there are not interrogations as I look back on it, but on the whole it has left me with a sense of peace and security about what lies beyond us." Yet he was also skeptical. "I should like to know more. On the other hand, we have our lives and responsibilities in the physical world."

Eakin was circumspect and moody for much of his stay, and King regretted inviting him in the depressed state he was in. The Presbyterian preacher and educator waited several weeks before assessing what he had witnessed. He did not entirely discount communication with the spirits and regarded Mrs. Wriedt as "ingenuous and honest," but he wondered about the authenticity of the voices he had heard, his dead father's among them. "I really don't know what to make of it," he admitted to King. "You seem to be convinced and I value your judgment but for myself I am mystified." Six months later, he attended another séance led by Mrs. Wriedt and told King that the experience had allowed him "to forget a good many of my worries and to make me determined to put them out of my mind." His interest and faith intensified after this. "He is beginning to believe that there is much in [spiritualism] and that it will give to the scriptures a certain verification which they seem to lack where these psychic experiences are related," wrote King in September 1933, after he and Joan dined with Eakin one evening.

For the next decade, until Etta Wriedt's death from a stroke in early July 1942 and even after King became prime minister again, she remained one of his key links with spiritualism. He visited her in Detroit on several occasions, staying in her house, and she returned to Ottawa, where he put her up at the Château Laurier. At each séance, he was comforted with words of encouragement and love from his parents, family and friends. (So real were these spiritual communications that in December 1933, for instance, King telephoned Mary Fulford in Brockville to relay a message to her from Laurier.) More importantly, he took solace in the fact that his family members were "all surviving after death, all happy, all doing helpful work." Mrs. Wriedt continued to amaze him. She "is a very remarkable person, one to be venerated," he noted after an encounter in late 1932.

During the Easter weekend in April 1933, Mrs. Wriedt held twelve sessions for him and Joan at which a small army of spirits appeared. King dutifully recorded each séance and "visitor" from the beyond and then indexed the notes for future reference. In bed after one of these sittings and trying to process everything he had heard, he was troubled by the comments about Jesus and the Resurrection, which he did not agree with, made by Mrs. Wriedt's guide, Dr. Sharp. "I thought to myself—it is again what I had to do when others were leaving Sir Wilfrid [in 1917] and I thought of his words about my loyalty and being true as any man." At that moment, he got out of bed and got down on his knees in the dark: "I felt I must declare my loyalty and I declared it absolutely to Jesus Christ... I felt he had 'died for me' as I had not felt it before."

Godfroy Patteson rarely, if ever, participated in séances. But on January 15, 1938, much to King and Joan's "amazement and delight," he joined them during one of Mrs. Wriedt's visits to Ottawa. Even more surprisingly, as King noted, during the sitting, "most of those who called were for him—3 bank friends, first his old teller in Toronto, Appelton who committed suicide, then McCullough of the Bank here and finally Edward Huston; all talked of matters of interest in Godfroy's world, and very affectionately with him—Godfory was quite overcome once or twice—Nancy came and played the violin... The power was strong—none of the voices very loud, but still fairly clear. Godfroy saw many stars, also clouds and shafts of light."

Mrs. Wriedt's death during the Second World War hit King hard. "She is another good friend whom I shall greatly miss," he wrote. Thereafter, she inhabited his dreams. The memory of those first séances at Laurier House and Kingsmere remained strong in his mind. As late as August 1948, he was still recalling verbatim conversations he had had with his parents—one in particular in which his departed father had broken down when he had referred to what a great son King had been—as related to him through Etta Wriedt's powers.

ONE EVENING IN mid-November 1933, King hosted an intimate dinner party at Laurier House for his close friend Dr. Arthur Doughty, the Dominion archivist (whom he had known since they were both deputy ministers during the Laurier years) and Madame Marika Pouliot, the wife of Quebec Liberal MP Jean-François Pouliot. The Pattesons were out of town. It was "an amazing evening," according to King, mainly because Doughty introduced him to table rapping. Sitting around a small table, King received the spiritualist version of Morse code. On that night, his mother, father, Max and Bella all communicated with him through a series of knocks on the table.

Ever since the young Fox sisters claimed in 1848 to have been spooked by spirits in their house in upstate New York—Margaret Fox later confessed that the girls had tied an apple to a string and swung it, which accounted for the strange noises—talking to the other world through rapping was as popular as the Ouija board with the Victorian upper and middle classes. Even the great French novelist Victor Hugo was for a time addicted to table-rapping séances.

As soon as Joan returned to Ottawa, she and King began experimenting with the "little table," as it came to be called. Thomas Eakin, visiting from Toronto, was also present. "Joan and I had started see if we could get results from placing our hands on the small table," as King described it. "It had spelt out the word 'Godfroy' and was under way with other words when we had to stop for Eakin, he joined us and the first word he spelled when he came in was 'Eak'—later we got quite a number of messages... It was our first experience at table wrapping [*sic*] and was amazingly successful. I was delighted to find Joan and I had the power."

A week later, it was Joan's sixty-fourth birthday, and King devoted much of his day to her. He had Joan and Godfroy to dinner and gave Joan "a petit point bag" that he had purchased while in New York—"the colouring is very pretty and flowers symbolic," King added. After dinner, it was time again for the little table "as the means of communication with those beyond." King wasn't disappointed: The sprits "came trooping in with their love and birthday greetings. Joan's family and mine, many friends, members of parliament." The actual method King and Joan used for table rapping was not described. But based on spiritualism guides, Charles Stacey suggested that "they placed their hands on the table and questioned the spirits, and the spirits replied by 'responsive raps.' Two raps meant Yes (we know this from a note by King) and one presumably No. Where detail was required the table 'spelled out' the answer in code; the method usually accepted seems to have been one rap meant A, two meant B, and so on through the alphabet."

In this way, King was able to carry on fairly long conversations with his departed family members and other notable personages. In this latter category were spirits whom King had respected or idolized such as Frederic Myers, the English poet, classics professor and founder of the Society for Psychical Research, who died at the age of fifty-seven in 1901. After church on Sunday, March 25, 1934, King recorded that he returned to Laurier House, had dinner with the Pattesons and then "a talk at the table—Had the great joy of a talk with Frederick [*sic*] Myers, a splendid evening, though I was very tired." Myers was celebrated in the psychic world for coming up with a scientific explanation for spiritualism and for originating the terms "telepathy" and "methetherial," or "beyond the ether, the transcendental world in which the spirits exist." King had studied Myers's work and read about his life (in 1918 Max had given him a book by William James with a significant section on Myers). Similarly, in late December 1936, he consulted with his grandfather about the abdication of King Edward VIII and was reassured that the fates of Canada and Great Britain were in his hands. Even Joan snickered a little at that message.

From King's perspective, table rapping was a convenient communication tool to the world beyond, because he did not have to depend on the presence or talents of a medium like Mrs. Wriedt. On New Year's

Eve 1933, after a tasty dinner of partridge and plum pudding with the Pattesons, Joan and King had a "long talk with some of the loved ones who were waiting to share the last of the old year with us. Father and mother and Max and Bell, both grandfathers, Mackenzie and King, and Joan's father and mother and grandfather; Marjorie Herridge and George Duncan [an old friend]." After a long and exciting day in London to participate in the ceremonies of King George vi's coronation on May 12, 1937, Prime Minister King returned by himself to the Ritz and tried the hotel room's table. "What I got was [grandfather] Mackenzie— also Mother—Love from all here," he wrote. He must have been the only dominion leader in attendance who could have made such a claim.

Occasionally, the little table delivered a misunderstood message or got it wrong entirely. At a session on March 9, 1934, King and Joan received words of wisdom from Kora Mura, an ancient "Hindoo" priest, who had visited King three months earlier. This time he declared, "Hail God who has ordained that you should love each other." At this, Joan became embarrassed and upset at the insinuation, and the table-rapping session soon stopped. That night, King described the experience as "grievously disappointing—but one of amazing enlightenment in some respects." By the next morning, he had changed his mind, likely feeling guilty about his muted love for Joan hinted at by the spirits. Now the uncomfortable moment between him, Joan and the "Hindoo" priest was a "hideous nightmare" and "the work of the devil."

Another more serious spiritual crisis unfolded a few months later. In 1919, the Conservative-led union government had halted the custom of proposing knighthood honours for Canadians, but R.B. Bennett, who revered the imperial connection, had reinstated it, much to King's annoyance. The list of Canadians who were to be honoured in 1934 was due out on June 3, King George's birthday. For some reason, King spent months fixated on who would receive a title; in the 153 days from January 1 to June 3, 1934, he commented on the issue in his diary at least 35 times or once every 4 days. On May 13, he was at Joan and Godfroy's house on Elgin Avenue for another "remarkable talk" at their little table during which his departed father provided him with a list of those to be knighted. It included Vincent Massey; Herbert Marler, a former Quebec Liberal MP who King had appointed as Canada's first envoy to Japan; and Henry Cody, the president of the University of Toronto. The morning of the announcement, King was certain that what he had been told by the spirits about the knighthoods "will be verified." However when he read the Canadian Press reports, the only two individuals knighted were Dr. Frederick Banting, one of the discoverers of insulin, and the agricultural scientist Charles Saunders, the inventor of Marquis Wheat.

King was stunned that the spirits had been wrong and suspected the manipulation of a demon, yet maintained that his faith in the guidance from the beyond was true. He was, if anything, imaginative. In his tea leaves he saw the letter *D* and the number one and reasoned that the knighthood announcements predicted by his father would be made by Bennett on Dominion Day, July 1. He received further confirmation of this twice more at sittings at which Laurier spoke to him. In fact, he was so positive that Massey, Marler and Cody would receive titles that he shared the news with Ernest Lapointe, but wisely did not tell him who his "inside informer" was.

On Dominion Day, both he and Joan waited and waited for the government to make a statement about conferring honours. There was nothing. By July 5, another session at the table revealed the only possible explanation—at least in the screwy reality in which they had constructed—"evil spirits" had gained the upper hand and purposely fed them falsehoods. The ordeal left King troubled and anxious. After reviewing all that had transpired at the little table, he concluded, "It comes back to the celestial and immortal voices of conscience—to the revelation that comes through the scriptures, to much told by those we love and who love us, to mother and father and Bell and Max, and especially grandfather, as continuing to shape my life. Sir Wilfrid also yes— but not to the degree of my own parents and grandparents, home, etc. I come back to the highest spirituality being the real power."

The following morning, King had come to one decision: he needed a break from the little table. "As dear mother said," he told himself, "let God send who He will, [it is] not for us to call those we want. I believe that [the table] brings both truth and error. I think the two are distinguishable—we have seen over and over again how an evil spirit has crept in where desires have been concerned." The break from the table sittings was only temporary. Upon further reflection, King drew up seven life lessons that he was determined to adhere to. They provide revealing insight into the inner workings of his complex imagination and oddball eccentricities a year before he was swept back into the prime minister's office:

1. Great care must be exercised in discerning truth and error, good and evil to be sure it is the spirit of good not the spirit of evil that is influencing the mind.

2. The two are subtly intertwined as joy and sorrow—but prayer and faith can separate the two and give the true vision.

3. Spiritual phenomena must be spiritually discerned—the distinction between the spiritual and material continually kept in mind.

4. We are not to be concerned with matters of position, events, etc. the times or seasons, it is the divine idea alone that is important.

5. It is advisable to give up looking for signs, etc. in *everything*. In some things where they impress themselves beyond mistake. It is well to recognize they are Heaven meant and Heaven sent but our business is not be looking for groundhogs but to be in search of the souls of men.

6. Hold fast to vision—as a method shown in the bible to be used by God in making His will and pleasure known to individuals.

7. It is safe to follow anything revealed in the bible—as to methods of knowing God—visions—significance of miracles, happenings, etc. but above all let us say 'Christ is the way the truth and the life' to follow Him is the surest guide of all.

HAD KING ACTUALLY followed this guide to survival, he might have been happier. Instead, given his obsessive-compulsive tendencies, his fascination with tea leaves and mystical numerology continued unabated; during 1934, the numbers zero, one and three, along with the always lucky seven, had a strange yet unexplained significance.

His interests and activities were diverse and non-stop. He collected volumes on spiritualism, corresponded with like-minded believers such as Lady Aberdeen, the Duchess of Hamilton, and Nandor Fodor, the research officer at the International Institute for Psychical Research in London. In 1933, he joined the American Psychical Institute under the name "M.K. Venice." These various letters were filed by his secretary Edouard Handy in boxes in the small room beside the library. He spent $7 for several back issues of the *Journal of Parapsychology* and ensured that he had time on an official trip to England in October 1936 to visit the London Spiritualist Alliance. And most notably, he continued to heed the maze of voices he heard at Mrs. Wriedt's séances. "I think she is a beautiful character," he wrote of his soothsayer in January 1938, after another of Mrs. Wriedt's weekends at Laurier House, "so simple, so childlike and yet very wise and keen insight. She told me that in all the séances she had never heard anything like father, speaking of kneeling and thanking God that I had been born, when he saw me at the head of the country, trying to guide its affairs aright."

Alone in his thoughts, King commented in his diary about every crazy dream he ever had, though rarely, if ever, did he offer any sophisticated interpretation. In the month of January 1935, for example, he made references to his "visions" nearly every day. Reading it more than seven decades later, it is impossible to comprehend how anyone, let alone the sixty-year-old leader of a political party, could devote so much of his valuable time to scribbling this nonsense. Yet scribble it he did. "Last night's dream was very indistinct," the entry for January 3, 1935 began.

> There were a series of persons in it I knew but just what their particular parts and significance I am not sure. [Conservative Cabinet Minister Hugh] Guthrie appeared and seemed to propose a toast to his own death. Chubby Power appeared and seemed to wish to say much to me, he had grown a heavy black beard and moustache and was difficult to recognize—he was speaking of elections. [*Ottawa Journal* writer] Grattan O'Leary appeared and seemed to wish to come close to me in an ingratiating sort of way. I kept clear of him... Most significant and clear of all was a table in Lord Oxford's London home set apparently for members of his family and ours... Mother and Bell seemed to be waiting to be shown their seats at one side, I was at another side... I felt much at home and congenial atmosphere... What Guthrie, Power and O'Leary signified I cannot say—they probably came into my thoughts [through] seeing Guthrie when calling and receiving New Year's day, Power from writing him (a beard is a sign of strength), O'Leary from communication prepared yesterday...

And so it went and went, with no rhyme or reason to any of it.

IN THE SUMMER of 1936, he had been back in office for about a year and was growing anxious. In Germany, Hitler was rearming the country's military, contrary to the Treaty of Versailles. Mussolini deliberately snubbed the League of Nations, as did Japan, which had designs on China. In such a tense period, King sought advice and support from the members of his cabinet, but he also found comfort in what the spirits revealed to him.

"Please help me father, I am beginning to feel unequal to the tasks—help me to realize that there is power from beyond and above." So began King's plea for assistance at a table-rapping sitting at Kingsmere on July 26, 1936. His father miraculously emanated from the Great Beyond to tell him to "eat less, take no wine and pray." With that sage counsel, Canada's leader steadied himself to deal with the most serious issues of his long career.

9

A United Canada Cry

*I spoke of the need of the Liberal Party stressing the unity
of Canada, and, at this time, in world affairs, the unity
of the Empire, and the further unity of countries holding
democratic form of Government.*—THE DIARY OF
WILLIAM LYON MACKENZIE KING, *January 20, 1937*

ONCE MACKENZIE KING was prime minister again, he
resumed his pre-1930 rigorous daily schedule. He was awakened around
eight o'clock in the morning by his trusty valet, Nicol, often described
as "omnipresent," "inscrutable," "sober" and "expressionless." He served
King loyally for many years, keeping all of his secrets and rarely, if ever,
angering him, an amazing feat in itself. Following breakfast in the sunroom, King perused the morning newspapers, the *Ottawa Journal* and
the *Montreal Gazette,* before spending up to an hour locked alone in his
library office. The secretaries on staff at Laurier House were not permitted to bother him during this period of meditation and study. Then, by
ten o'clock, his twelve- to fourteen-hour day of conducting the nation's
business began in earnest. No one who worked for him knew exactly
when each day was to end, since it was at the prime minister's whim.
King preferred to remain at Laurier House as long as possible, making it
inconvenient for his staff at the parliament buildings. Nicol was there to
act as gatekeeper.

Nicol's first name was John, but that was irrelevant in the self-
contained and peculiar world of Laurier House and the prime minister's office in the East Block. Like every other male who pressed King's
clothes, cooked his food, drove his car, wrote his speeches or answered

his correspondence, he was only and always called by his surname, as was the custom of the day. Other members of this select club during his long term in office from 1935 to 1948 were identified in his diary as: Handy, Henry, Pickering, Lay, Turnbull, Pickersgill and Heeney, among others who shared the unique and often exasperating experience of trying to figure out what precisely the prime minister wanted. The few women who were employed as stenographers were more formally referred to as "Miss," as in Miss Zavitske (or "Miss Z."), who was based in Laurier House and whom King sufficiently trusted that on occasion he dictated his diary to her—a task usually done by Edouard Handy after 1936. Gordon Robertson, who worked in King's office from 1945 to 1948 and became the clerk of the Privy Council in 1963, spoke to Miss Zavitske regularly, but he never knew until years later that her first name was Lucy.

King's PMO team was staffed by many accomplished and talented individuals, a fact he was aware of, yet usually took for granted. John Whitney "Jack" Pickersgill, who was with King for many years, rising from a seconded third secretary to head of the PMO, had been educated at Oxford University and taught history at Wesley College in Winnipeg. He was to go on to become clerk of the Privy Council in 1952 and then a cabinet minister in Louis St. Laurent's government. "His approach to government was so uniquely his own that his name found a place in Canada's political dictionary," according to Peter C. Newman. "The expression 'Pickersgillian' came to signify any partisan ploy that was too clever by half." But during the thirties, that was far in the future.

Pickersgill "promises to do well" was King's assessment of his young assistant in August 1938. "He is very keen and has a fine sense of duty." Pickersgill had the unenviable task of preparing King's speeches, or at least drafts of them, which King tinkered with right up to the moment he delivered them. As Pickersgill remembered, the prime minister insisted "that every word he uttered in public should be properly weighed in advance and he liked to have plenty of time to work out his phrases as he really wanted them. That was as true of relatively unimportant speeches as it was of the most momentous occasions." The real problem Pickersgill faced, as he explained, was that King's "style and vocabulary were frozen in the early years of [the twentieth] century." Nothing flamboyant was permitted, nor were, for reasons only King knew, the words "challenge," "sober" and "decent."

Gordon Robertson has described the near-impossible job requirements to survive in King's PMO as: "Flexibility, versatility, energy, and a capacity to interpret—and to anticipate—King's often obscure and

Byzantine reactions." Pickersgill, he adds, "had them to a degree not approached by anyone else... King was a constant problem; Jack never was." Pickersgill, confident and a bit brash, was respectful of King and tried mightily to stay on his good side, but he was not afraid of him.

Ross Martin, who joined the PMO after the Second World War, rubbed King the wrong way. In the spring of 1946, he had allegedly failed to properly say goodbye to King before a trip to Europe—he later claimed he had—and then had not showed up at the train station to bid him farewell, mainly because he was helping his wife with their newborn. Handy called Pickersgill from Halifax to tell him that King wanted Martin fired. Pickersgill refused to carry out this order, and King never challenged his decision—though he did allude to Martin's lack of respect in his diary. "I heard nothing more about firing him, however," Pickersgill recalled. "I suspect the prime minister had guessed there would be an unpleasant confrontation if I did."

Arnold Heeney was a young, bright Rhodes Scholar and up-and-coming civil lawyer in Montreal when he came to King's attention in July 1933. The connection of his first name with educator and historian Dr. Thomas Arnold and his son the poet Matthew Arnold, whom King revered, was no mere coincidence. After a lengthy courting period, King convinced Heeney to become his principal secretary in 1938 at an annual salary of $7,000 (by contrast, Jack Pickersgill as a third secretary received $2,280 a year).

Heeney envisioned his role to be like that of the British civil servant Sir Maurice Hankey, who had recently resigned his position as secretary of the cabinet and clerk of the Privy Council after more than a decade of distinguished public service in London. Heeney gradually made the job his own, bringing much-needed structure to the PMO. In those days, for example, King ran cabinet meetings without a written agenda, and no official minutes were kept. From time to time, there were disputes about precisely what his non-political job entailed, but King was generally pleased with Heeney's work. He told Violet Markham in May 1939 that Heeney's presence "has made a difference in my life." Within two years, Heeney became the Canadian version of Maurice Hankey and was named secretary to the cabinet and clerk of the Privy Council, a position he filled admirably until 1949. "He is an exceedingly lucky man to get this post," King wrote in his diary on the day of the announcement, "though I believe exceptionally well qualified for it, and I am glad it has all worked out as it has."

In his early sixties, King regularly complained about his excessive workload as prime minister; even considering his age, his obsession with his own exhaustion was constant. Between 1936 and 1940, the most overused word in his diary was "fatigued." (It appears no less than 326 times over a five-year period, or once every five days.) "I have seldom been in better condition physically," he confided to Violet Markham in the spring of 1937 as he was preparing to depart for that year's imperial conference, "but I have been more tired during the past session than words can begin to express. In this state of fatigue, the thought of having to make speeches in London has hung over my mind and spirit like a pall. Its effect has been paralyzing."

Many times during this period he contemplated "giving it all up," though probably not very seriously. King liked feeling sorry for himself, and in his self-centred bubble he rarely appreciated the diligence and effort of those around him. "Of course," he wrote to Markham in March 1937, "I know that my secretarial help, both at the Office and at Laurier House, is far from what it should be." On the way to London a month later, someone on his staff had misplaced a bag with an address book and stationary. "It is all part of the lack of an efficient organization," King wrote on board the ship. "I had to take up the answering of telegrams, letters, etc. as though I were dealing with it for the first time… This unfortunate incident has made a considerable difference in the happiness in the beginning of the voyage."

His saving grace was Edouard Handy. A French Canadian whom King had plucked from the Dominion Franchise Office (which prepared election lists and registered voters) on Skelton's recommendation, Handy served as King's personal secretary from 1936 to 1950. He was fiercely loyal and knew King probably as well as anyone. He also tended to the prime minister's every need. In 1945, while King was attending the San Francisco Conference to establish the United Nations, he and Handy went shopping for "shirts and undergarments." A few months later, after King caught a bad cold, Handy made him "some hot lemonade [with] a little Scotch." He was with King during the summer at Kingsmere and spent many evenings keeping him company. Beyond that personal contact, it must have taken a special and extremely patient person to listen to King dictate his diary for upwards of a decade. Perhaps that's the reason King often described Handy as "faithful."

In private, King was sporadically complimentary of his staff. To be fair, he was also capable of civility and generosity. He found time on a trip to Vancouver in July 1941 during a speaking tour about the war effort to visit with the parents of Norman Robertson, who had recently taken over as under-secretary of state for external affairs. When Robertson's parents visited Ottawa five years later, King held a dinner party for them, Norman and his wife, Jetty, Arnold and Peggy Heeney and Arnold's mother and father. "The details of menu and seating were planned as for a state occasion," recalled Arnold Heeney. "It was charming as, to do him justice, were most of King's parties. He was at his best, the considerate and entertaining host who quickly put our excited parents at ease. When we had finished the meal... King rose to his feet and embarked upon an extravagant and lengthy paean of praise for Norman and me." Still, a positive comment could just as easily be replaced by a negative one over a poorly written memo, a simple mistake of human error or any one of a thousand other reasons.

Howard Henry, his private secretary, and Edward Pickering, the assistant private secretary, were supposed to have arranged for tickets for the Pattesons to attend a Liberal Party election rally in October 1935, but the tickets were not delivered. "About 7:30 [PM] Joan phoned to tell me they were just leaving for the meeting but had not secured seats. I was terribly upset, fearing they might not get in, and greatly annoyed at my staff, Pickering and Henry for having failed, after being asked to get these seats," he recorded. "I spoke pretty sharply to Pickering about his and Henry's want of appreciation of the things I felt most about and which they should know about and attend to. I could not get down to work again in a quiet way. I tried to read but just tossed about."

On another occasion, Pickering was late with a memo and had assigned this imperative task to a stenographer. That too, "greatly annoyed" King. "Pickering and Henry between them, in their mechanical ways... put personal duties on to others... drive me almost to distraction." Pickering then had the audacity to question why he had to work at Laurier House rather than the East Block. "I am inclined to think it as well to drop [Pickering] altogether," King complained. "He is not happy about coming to work part of the time at Laurier House, and I do not want anyone around who is not happy in this environment and what it affords of association and opportunity. It is a bit embarrassing to have this additional problem just in the middle of the special season." Somehow, Pickering managed to survive in his position until 1938.

Everyone who toiled for King and later wrote about the ordeal is unanimous about one thing: he was an awful boss and a man who could never be satisfied. King could pass someone in the hallway of the parliament buildings who had been on his staff for more than a decade without a word or greeting. "I had learned that with King, the best one could hope for was a grudging, grumpy silence," writes Gordon Robertson. "Approval was impossible; praise was out of the question." According to

Jack Pickersgill, "Mackenzie King never admitted he was adequately served... Once or twice he even expressed satisfaction with my work though complaints were more frequent. I was anxious to give him no legitimate ground for dissatisfaction, no easy feat. There was no way to get firm direction from Mackenzie King or to be sure he would not change direction unpredictably at any moment."

King had absolutely no concept of family life and expected his staff to be on call twenty-four hours a day, seven days a week. Pickersgill's wife, Margaret, believed that King "had an uncanny knack" for knowing when the Pickersgills were hosting a dinner party, because, sure enough, as soon as they and their guests started eating, the telephone would ring and it would be the prime minister calling on some urgent matter. Woe it was to any of his secretaries or assistants who put their own personal lives before King's needs. James Gibson, who worked for the PMO during the Second World War, raised King's ire when on December 7, 1941, the day the Japanese attacked Pearl Harbor, he was away from Laurier House because he was seeing his mother. (If anyone should have understood obligations to one's mother, it was Mackenzie King.) "How little men realize what it is that causes one to succeed and the other to fail," King wrote angrily about Gibson's actions. "To be associated with the P.M. who himself is going without meals, recreation and anything—health, in trying to carry the country through the most critical of all moments, and not to be doing everything possible by way of assistance to that end, reveals a deplorable lack of appreciation in the light of opportunity and of responsibility."

Like an absolute monarch, King expected his staff to be there when he summoned them. On a Saturday morning in early October 1937, he was terribly upset because Henry, Pickering and Walter Turnbull did not arrive at Laurier House until after 9:30 AM. Or, on a Sunday in October 1944, he was out at Kingsmere and was in a controlled panic. "Pickersgill mentioned his desire to spend the day at the fishing club," he griped. "Turnbull is away on a 3 week holiday... No one to lend a hand—I took the situation calmly, realized I must do so."

Someone was required to be on call at the prime minister's parliamentary office until 11 o'clock each night, just in case King required something. One evening, it was Gordon Robertson's turn, and like clockwork King telephoned him at ten o'clock with what Robertson recalls was a minor issue about an inconsequential event happening the next day. "But King raised the question whether when X happened Y would follow: an issue about which there was, I thought, no possible doubt," Robertson writes. "I began to explain my clear understanding of the arrangements with an unfortunate choice of words: 'Well I assume Mr. King...' I got no farther. 'Robertson, as long as you are in this office you will *never* "assume" *anything*.'"

THE MEMBERS OF the parliamentary press gallery were happy to see King back as prime minister. Five years of R.B. Bennett's threats and bullying had been sufficient. When the *Ottawa Journal,* a newspaper supportive of the Conservative Party, published a mildly critical editorial about the Tory government, Bennett for a time refused to speak to Grattan O'Leary, its star journalist. He declared war on the *Financial Post,* or "The Fanciful Post," as he usually referred to it, and detested John Dafoe and the *Winnipeg Free Press* so much that he once halted a press conference in mid-sentence when he noticed Grant Dexter, the *Free Press*'s correspondent, among the throng of reporters in front of him.

Mackenzie King would never have acted in such a rude fashion, though he did have favourites, such as Dexter, Charlie Bishop of the *Ottawa Citizen* and Bruce Hutchison, who was the Ottawa reporter for the *Victoria Times* until he became an editorial writer and columnist for the *Vancouver Sun.* These were journalists who were generally loyal, or at least supportive, of the Liberals and whom King felt he could trust. Hutchison later recalled that he was "easily snared" by King's occasional attention and the access to the prime minister he was granted. In the case of Grant Dexter, King was quite conscious of the fact that anything he told the correspondent in private was sure to be relayed to John Dafoe back in Winnipeg, which was precisely what King desired. Dexter regarded King as "a queer sort of fish." He wrote about him in his celebrated private memos for Dafoe with a mixture of respect and sarcasm.

The line between independent journalist and partisan hack sometimes got blurred, however. In late August 1935, Vincent Massey and Norman Lambert of the National Liberal Federation arranged for Dexter to interview King for a profile in *Saturday Night* magazine. Dexter joined King out at Kingsmere for dinner one evening and was given a tour of the ruins-in-progress. Yet the discussion with the *Free Press* writer troubled King. He was taken aback by suggestions that he was a "delegator" of men rather than "a worker" and disliked having to clarify and explain his motives and actions. "One would think," he wrote in his diary after the visit, "that after 18 years of Leadership some one would know me well enough to spare that task." Once Dexter had written the extremely complimentary article, he sent it to the NLF office and King before it was published. That night, King phoned Dexter, who was, according to King, "quite pleased about my slight corrections of and appreciation of his article."

King was all for journalists such as Dexter doing his bidding; he was much less forthcoming at a press conference or when he was caught off guard by a reporter. One day, Tom Wayling of the ever-loyal *Toronto Star* grew impatient waiting for a few words from King. He marched into the prime minister's office and boldly demanded an interview, which was definitely not the right move. "I had him come in and spoke rather firmly to him, resenting his persistence in claiming the right to immediate approach," as King described the encounter with Wayling. "I also expressed my view that press interviews with the leader of a Party should be at proper times and places." Rarely was it ever the proper time and place for King to share much with journalists.

A Mackenzie King press conference was usually an utter waste of time. King had watched from a distance President Franklin Roosevelt charm newspapermen with witty repartee and frank answers to tough questions. In 1937, King decided that he would meet more regularly with the Ottawa gallery. Yet when the time came to actually answer a question or reveal a government secret, King clammed up. In the spring of 1937, he told the assembled reporters that he was off to Washington to see FDR. "What is the purpose of your visit?" a Quebec scribe asked. "To discuss the situation," King replied. "What situation?" the reporter inquired. "Matters of mutual interest," King answered. And that was the extent of it.

A decade later, nothing had changed. "Instead of spilling the main point of the announcement first and letting the details come out later in the form of anti-climax, Mr. King goes along in narrative style, filling in background with details and keeping all hands on the lookout for the payoff. Then, in his own good time, he comes across with the punch line," a Canadian Press story of late December 1946 observed. "When the correspondents go digging for more, the prime minister can demonstrate a knack, almost a fine art of giving a reply that answers the question, but tells just what the questioner knew before." The press gallery periodically complained about the lack of access to King—and even that the prime minister never entertained them—but King continued to ignore the reporters and used them to his own advantage in a way that would have made Prime Minister Stephen Harper, another "control freak," proud.

Any newspaper, whether it was friendly to the Liberals or not, that slipped up or misquoted him was the target of his wrath. "Nothing could be more damnable than the complete misrepresentation of one's position," he whined in his diary about a headline he didn't like in the *Globe and Mail* on April 1, 1939. (The *Globe* and the *Mail and Empire* merged in 1936 after the young entrepreneur George McCullagh purchased both newspapers.)

At the time, the situation in Europe was becoming worse by the minute, and King was naturally anxious. In a lengthy speech in Parliament, he asserted that Canada "is prepared to support whatever may be proposed by the Government at Westminster." The *Globe and Mail,* however, had it as: "Canada is prepared to support any action that may be decided upon by the Government at Westminster." King took a hissy fit. In the House, he called the *Globe and Mail* story "wilfully false" and "perverted" and recommended that the reporter, Harold Dingman, be "denied his privileges." The *Ottawa Journal* blamed King for the error, because he refused to give a "simple, clear-cut statement." King summoned Dingman and McCullagh, who happened to be in Ottawa, to his office. McCullagh appeased King by explaining that he had a right to be angry, but the fault was really the Canadian Press dispatch that had made the error in the first place. King was only partially mollified and proceeded to ream out Dingman for having done this "sort of thing before," to which the reporter took exception. "Dingman is a man without background or knowledge," King decided afterward, "and quite unfit for the grave responsibilities of reporting Parliament's proceedings at times like the present."

Control was indeed everything to King, so it should not have been surprising that he was unable to cope with a free-for-all event like the annual press gallery dinner—"one of the great agonies" of his life, as he called it. A kibitzer and one of the boys Mackenzie King was definitely not. He was far too insecure to laugh at himself; nor did he enjoy drinking or listening to racy stories. "At 8 I went to the dinner of the Press Gallery," he wrote on May 4, 1929. "I dislike it more than any event of the year. This year their publications and songs were very vulgar and I felt resentment at the cartooning and references to myself—the whole proceeding was most undignified and in some respects unworthy. I made a poor speech, nothing to it." The next morning, he was still seething about the "reference to myself and many of the others. The whole proceeding was wholly unworthy [to] thoughtful and serious minded men. As a matter of fact, it gives to press an idea of their power to destroy and to make reputations such as they should never be permitted to have."

As leader of the Opposition, he had survived through the dinners, even going so far as to be pleased with his 1935 speech, calling it the best he'd ever made. But once he was back in power, he was again at the centre of attention and detested every aching moment of it, with "its bar room cocktail approach." In the spring of 1936, Edward Pickering told him that what the journalists wanted from him was "humour," and advised against a speech with anything too serious in it. King, however, wanted to speak of what was "deepest in my heart." He suffered through the evening. "I felt different moods—uplifted at times out of the depression and the depths of it, joining the singing and at another completely at a loss for words or thought . . . The program was good in parts, hateful in others, ridiculing and perverting the truth and most sacred things—all this for something to laugh at—not quite so vulgar as usual, but very disconcerting and trying." He muddled his way through his speech, calling Robert Borden "Wilfrid Borden," and grumbled to the journalists about being tired. By the end of the dinner, he was, he claimed, "on the verge of a breakdown." Surrounded by "enemies" like John Stevenson, and even friends such as Charlie Bishop, King confessed that he "could never feel at ease." By 1947, he had finally figured out a solution to this painful occasion: have a cabinet minister fill in for him.

IF KING WAS brusque with his staff and secretive with the press, he was much more accommodating with the members of his cabinet—at least outwardly. Since his first days as prime minister in the early twenties, his preferred leadership style was to adhere to the philosophy of conciliation he had preached in *Industry and Humanity*. King was never shy about inviting accomplished, intelligent and independent-minded men into his cabinet and no more so than in 1935 and onward. Besides Ernest Lapointe, who maintained his role as Quebec chief and justice minister, the list included Charles Dunning in finance; James Ilsley in national revenue; Norman Rogers in labour; Ian Mackenzie in national defence; Thomas Crerar in mines and resources; Jimmy Gardiner, the former premier of Saskatchewan, in agriculture; Chubby Power in pensions and national health; William Euler in trade and commerce; and a newcomer who was to make his mark in Canadian politics, C.D. Howe, a successful civil engineer, who was given the revamped ministry of transport. These men represented the various regions of the country, and many would remain with King during the difficult years of the Second World War. The significant external affairs portfolio remained King's responsibility, even though Lapointe frequently lobbied him for it.

King was well known for operating a democratic cabinet where everyone was free to speak their minds (as long as they were not too frank) and consensus was reached through reasoned debate. During his terms as prime minister, the tone was as formal as King was; the only members of his cabinet he addressed by their first names were Lapointe and Rogers. "The opinion seems to be unanimous," Blair Fraser wrote in a 1946 profile, "that Mr. King is skillful, indeed, a brilliant chairman. His performance in cabinet, according to general report, is masterful."

So was his control of the Liberal caucus. He did not tolerate even the most innocuous questions about foreign affairs from his backbenchers, but as Paul Martin, Sr., who was a thirty-two-year-old rookie MP from Ontario in 1935, recalled, King permitted the caucus members to vent their various complaints and thereby preserved the unity of the party. "Every caucus was a political meeting which Mr. King would always end with a rousing speech which invariably provoked resounding cheers from the members. Reinvigorated, they would return to the backbenches and urge on Mr. King and the cabinet."

Likewise, in Parliament, King may not have been the most eloquent of speakers, but he knew how to handle himself. He had, according to Harry Ferns, who worked in the PMO during the forties, "a preternatural sense of what was up in the House of Commons and could anticipate a question before the opposition had even thought of it."

Behind this façade of unity and confidence frequently simmered a lot of resentment and tension. King detested that Chubby Power drank too much, and reluctantly took him into the cabinet after Ernest Lapointe convinced him to do so. Power thought that King was "needlessly polite." King was not crazy about Jimmy Gardiner, and James Ilsley felt the same way about King. And William Euler's stubbornness was a constant source of aggravation. In August 1938, after Euler refused to halt his vacation to chaperone an Australian dignitary, King castigated him in his diary as "the most selfish man I know."

Not valuing Thomas Crerar as he had in the twenties—and according to Norman Lambert, not liking him all that much—King left him hanging before making his ministerial appointment to mines and resources official. In fact, he had first asked *Free Press* editor John Dafoe to join the cabinet, but Dafoe, who had printer's ink coursing through his veins, turned him down. King then asked Dafoe whom he should choose from Manitoba. That was an easy question: "Crerar, of course," he replied. Despite questions raised about Crerar's ministerial abilities, primarily by Charles Dunning, who fancied himself King's chief western Canadian lieutenant, Crerar eventually received an official invitation. King felt it necessary to say to Crerar, "I should tell you there is much opposition to your coming in from Manitoba." To that jibe, Crerar, who later recalled that he had refused to "beg" for the position, politely suggested that if King could find someone who could serve the government more effectively, he should do so. That abruptly ended the conversation.

On most key issues, King was willing to let the cabinet have its say, and he was prepared to abide by its decisions, particularly on anything that might upset Quebec. Yet there were events that King considered so critical that he put his prime ministerial foot down. The abdication of King Edward VIII in 1936 was a good case in point. King was not initially caught up in the gossip swirling around Edward's torrid relationship with Mrs. Wallis Simpson, an American who was on her second marriage. (British prime minister Stanley Baldwin told King, who was in London at the end of October 1936, that he wished Edward "kept [Mrs. Simpson] at Belvedere [Edward's country retreat], or anywhere but not flaunt her on the public.")

After Mrs. Simpson divorced her husband, Ernest, and Edward proposed to her, the British government was thrown into a full-scale constitutional crisis. Mackenzie King was asked for his opinion, something he took very seriously. At first, he hesitated, since he did not want him or Canada blamed for making the situation worse. Oscar Skelton and Ernest Lapointe recommended that he not issue a public statement, but he insisted on doing so. This official declaration was "so cautiously worded, however, that it was almost incomprehensible," as Neatby rightly observed.

As the crisis unfolded and was ultimately resolved with Edward's sensational abdication on December 10, King devoted hours to contemplating each dispatch from London. He sought counsel from his ministers, even if most of them were not as troubled or caught up in it as he was. In this way, he covered himself, though there was no doubt about who was making the decisions. Ultimately, he quietly supported the British government's position that Edward had to abdicate if he married Mrs. Simpson, but he did so from a respectful distance.

In what was a strong statement of independence from King, he wrote in his diary on December 4: "I felt, regardless of judgment of all Cabinet combined, in some things I know absolutely what it is best and necessary to do." None of his ministers challenged him, in any event. More than a month later, after the installation of George VI, Mackenzie King patted himself on the back for a job well done and a catastrophe averted. "However exhausted I may be myself," he reflected on January 20, 1937, following the passage of legislation officially recognizing the abdication and transfer of royal authority, "I have the satisfaction of knowing that this, the most important historic incident in this country, has been handled, so far as the Canadian people and Parliament are concerned, practically to the satisfaction of all... That is something for which there is cause for great gratitude."

AS THRILLING AS it was playing a role, albeit a minor one, in the fall of one British monarch and the rise of another, Mackenzie King still had the Depression to solve and a long list of grumbling and unpredictable premiers to satisfy. It took a while for King (and he was by no means the only one in Canada and elsewhere) to get his head around the then-radical theories of British economist John Maynard Keynes, who advocated proactive government fiscal and monetary policies to stimulate growth. King was fairly rigid in his economic thinking and found it difficult to embrace deficit financing. Balanced budgets and less government regulation, not more, were his initial answers to Canada's woes. King did not want to dole out more money to the provinces, even if the premiers continually and legitimately complained about financial shortages in dealing with relief, nor to the members of his cabinet, who kept submitting increased department budgets. Many of his ministers, he believed, had "lost all sense of responsibility to the taxpayer, and are thinking only of making a showing in their particular fields, with public monies."

The minister of finance, Charles Dunning, felt exactly the same way, and it took a heated showdown in 1938 with Norman Rogers, the labour minister, who demanded $40 million for public works, for King to consent for the first time to a federal budget with a projected deficit (of $23 million). Once the deed was done, King was more reconciled to the dramatic shift in policy. "The world situation has headed the countries more and more in the direction of the extension of state authority and enterprise," he conceded, "and I am afraid Canada will not be able to resist the pressure of the tide."

Apart from planning a balanced budget, King's first inclination had been to open up trade with the Americans. In 1935, this tariff reduction initiative (which Bennett and the Conservatives had begun) had brought King into the White House and face to face with FDR. It was—in the immortal words of Humphrey Bogart in *Casablanca*—"the beginning of a beautiful friendship." FDR might have "intellectually dominated" King, as Roosevelt biographer Conrad Black asserts, but the president genuinely liked King, or "Mackenzie," as he called him, and was attentive to his overly sensitive nature and insecurities. King was FDR's elder by seven years but could rarely bring himself to call him "Franklin." To the more formal King, he was almost always "Mr. President." King was also too timid or polite to tell Roosevelt that his closest friends called him Rex, rather than Mackenzie. (During the war, Winston Churchill, following FDR's example, began to refer to King as "Mackenzie" as well.)

The two leaders were both graduates of Harvard, a fact that made King proud. They had never actually met as students, yet FDR still referred to King as being "an old personal friend." Journalist Bruce Hutchison, who witnessed the relationship up close, wrote that "King regarded Roosevelt with public adulation and privately with a mixture of amazement, skepticism, and some merriment." It was true that King had considered the New Deal as "political hokum," as Hutchison points out, but he still cherished his time with FDR. "Roosevelt had so few people around he could talk with who didn't have an axe to grind," according to Jack Pickersgill, "he liked to have King around to bounce his problems off him." The feelings were mutual. During the late thirties, King's relationship with the governor general, the novelist John Buchan, Lord Tweedsmuir, who was a bit imperialistic and haughty, was prickly. (Tweedsmuir insisted on calling himself "His Excellency" and his wife, Susan, "Her Excellency," which greatly irritated King). "I find the President of the United States a hundred times easier to talk to and get along with than the Governor General of my own country," King admitted to Violet Markham. "This ought not to be." (The strength of the King-FDR friendship was reinforced as well from the spirit world. At a session with the little table on November 13, 1938, the ghost of Laurier told King: "The President is very fond of you. He will treat you like a Prince.")

As they were apt to do then and later, the Americans wanted a lot in the trade negotiations and were not prepared to reciprocate quite as much as the Canadians desired. King arrived in Washington on November 7 and began negotiations with Secretary of State Cordell Hull, who greatly impressed him. Hull also gave him pointers on how to deal with the president. At the White House later that day—King's ego was boosted by the fact that the president had put his own car and driver at his disposal—he received the royal treatment. There was a formal dinner in his honour with Canadian and American diplomats and politicians in attendance.

Inevitably, the dinner conversation came around to trade. FDR said his piece, and then King, as he described in mind-numbing detail in his diary, presented the Canadian case, commodity by commodity— lumber, cream, seed potatoes and cattle. "I did not touch codfish at all," he recorded, "as I had been told it would only upset the President for me to mention cod." Before King retired at 11:30 PM, the bilateral trade agreement, which was officially signed a week later, was done. It was not full-scale reciprocity, nor did it automatically end the Depression. Yet as one American historian put it, by opening up the Canada-U.S. border again to a range of items, the 1935 agreement "formally marked the return to economic sanity by the two countries after several years of commercial folly." From King's perspective, the deal was a mini-triumph. The Canadian press praised him for doing something positive, and he had cemented his bond with FDR, which was to prove essential in Canadian-American relations in the difficult years ahead.

BACK IN OTTAWA, King now confronted an enduring headache: a ragtag group of provincial premiers, one more volatile than the next, and all demanding more money to offset the weaknesses of Canadian federalism. In King's ideal world, each level of government in a federal system was an independent entity that looked after its own constitutional responsibilities with its own financing. The Depression, however, exposed the frailties of such a structure and made a revamping of federalism essential. By 1937, Manitoba and Saskatchewan were bankrupt, and the federal government was forced to provide a temporary handout. King did not like this new economic reality, but he was also astute enough to know he could not ignore it. Steered in this matter by the creativeness of Norman Rogers, Oscar Skelton, Graham Towers, the first governor of the Bank of Canada, and others, he established the Royal Commission on Dominion-Provincial Relations in 1937, more commonly called the Rowell-Sirois Commission after its two chief commissioners, Newton Rowell (who served until 1938) and Joseph Sirois.

After three years of hearings, the commission's final report eventually served as the basis for a major overhaul of Canadian federalism, the collection of taxes, unemployment insurance and the allocation of grants to the provinces. Many of these changes occurred after King was gone, but he deserves some credit for advancing the discussion that had been set in motion by R.B. Bennett's New Deal legislation.

The immediate dire domestic situation was made worse by the independent authoritarians leading the biggest provinces: William Aberhart in Alberta, Mitchell Hepburn in Ontario and Maurice Duplessis in Quebec (and to a lesser extent, Duff Pattullo in British Columbia). "Some early and definite action by the Canadian Government to take control of the chaos that is developing throughout the Dominion seems vital," Skelton warned King in April 1937. "The disintegration of Canada is proceeding fast. Extreme assertions of provincial power, tendencies in several provinces of Governments to adopt an arbitrary and semi-Fascist attitude, [and] the increasing distrust of the East on the part of the Western provinces... make the situation in Canada today the most disturbing in my recollection." Skelton's doomsday prediction might have been exaggerated, but only slightly. As King later told Violet Markham, "the wretched spirit of fault-finding, blaming, and antagonizing in as many ways as possible has asserted itself to no small extent between different parts of the Dominion, and even between elements in our own party."

King's impression of William Aberhart, the Bible-thumping Social Credit radio preacher, had been favourable after he had spent a bit of time with him at a Dominion-Provincial Conference in December 1935. He went so far as to think that the Alberta premier might turn out to be a good friend and ally and even accepted the novel (and in retrospect, foolish) view of Louis Taschereau's wife, that Aberhart was "really a Liberal." He soon learned just how far off the mark such an assessment was; by 1937, the term "demagogue" was used to describe the Alberta leader in the pages of his diary.

Aberhart had won the provincial election of 1935 by promising to undo a corrupt system that had "poverty in the midst of plenty" and dangling before the dazed and susceptible voters of Alberta a Social Credit dividend of $25 a month and "just" prices. No other Canadian politician had so brilliantly meshed psychology, economics and religion (tinged with anti-Semitism) in a message effectively delivered via the wonder of radio. Yet as soon as Aberhart attempted to pass real Social Credit–style financial legislation, which clearly overstepped provincial jurisdiction on banking and credit, King and Lapointe had no choice but to do what was anathema to their political philosophy and ultimately disallow it.

The two dealt more delicately with the second demagogue premier they confronted: Maurice Duplessis, the indomitable force whose cult of personality shaped Quebec society in his own ultra-conservative image for two decades. If there ever was a politician in Canada akin to a much-beloved tyrant, it was Duplessis. As the leader of the provincial Conservatives, Duplessis had formed a short-lived alliance in 1935 with a group of left-leaning nationalist Liberals led by Paul Gouin. He soon seized control of their new offspring, the Union Nationale, and became premier of Quebec following the August 1936 provincial election. Once in power, Duplessis, who was backed by the unquestioning support of the Quebec Catholic Church, targeted socialists and communists as the enemies of the state. In the spring of 1937, his government pushed through the infamous Padlock Law, which gave the police wide and arbitrary discretion in closing, or literally "padlocking," any establishment that propagated, published or disseminated communism—which was conveniently left undefined. It was Quebec's version of the "Red Scare," with trade unionists, journalists, Jehovah's Witnesses (deemed "dangerous heretics" by the church) and anything or anyone remotely left of centre coming under attack.

King's liberal sensibilities were appalled by Duplessis's police-state actions, which the *Winnipeg Free Press* correctly labelled as "one of the most savage assaults upon freedom which Canada has ever seen." His best instincts repeatedly told him that the Padlock Law, which was "embarrassing and destroying Liberty" and clearly infringed on federal criminal law, should be disallowed as decisively as Aberhart's legislation. But King also convinced himself that there was a higher principle at stake and a dangerous risk in taking on Duplessis: he just might upset the seemingly tenuous strands of Canadian unity. And Lapointe, fearing that a federal intervention would be used by Duplessis to his own advantage, essentially vetoed any Liberal government action. In this matter, King abided by his Quebec lieutenant's dictate. Considering the tough treatment of Alberta, this was a hypocritical kowtowing to Quebec nationalists, and they both knew it. "In the circumstances, we were prepared to accept what really should not, in the name of Liberalism, be tolerated for one moment," King conceded in his diary on July 6, 1938. The greater national interest, however, took precedent. "I took the view," he added, trying hard to convince himself of the correctness of his position, "that in the last resort, the unity of Canada was the test by which we should meet all these things."

KING'S DECISION TO overhaul federalism put him on a collision course, as well, with the unpredictable premier of Ontario, Mitchell Hepburn. Mitch, as everyone called him, was a Liberal, but as he was later to declare, "not a Mackenzie King Liberal." Hepburn's biographer, Toronto historian John Saywell, lays much of the blame for the bitter Hepburn-King feud on King's paranoia and his fear that Hepburn could challenge him as leader of the federal Liberal Party. There is no dismissing King's overly suspicious nature, yet Saywell's assessment is far too one-sided. Hepburn may not have been the "cynical demagogue" he has often been depicted as—when he was Ontario premier from 1934 to 1942, he was responsible for improving health and education in the province, among many other accomplishments—but his erratic and often mean-spirited behaviour and actions were surely as much a factor in this Ottawa-Toronto clash as anything King did.

Hepburn was a populist, a fact even King had to admit. Mitch understood the common man in a way that King never did. "People believe he is honest; know he is fearless; and regard him as efficient in administration," King wrote begrudgingly after Hepburn had won a second majority government on October 6, 1937. "His manners, evidently, as well, catches the man on the street. It is the 'fellow' that counts and he is one of them in language and spirit." Hepburn was indeed King's polar opposite. He was impetuous, inconsistent, impulsive, vindictive, reckless and worst of all, from King's puritanical perspective, a sinner. Hepburn had a well-deserved and notorious reputation for being a drinker and a womanizer. Like someone who steals a glance at the lurid headlines of the *National Enquirer* in a grocery store, King savoured every morsel of gossip that was relayed to him about Hepburn. There was one story of Hepburn being caught with three young girls in his suite at the King Edward Hotel or another about the premier bedding one woman in a Kingston hotel and doing the same with another in Ottawa the next night. King had no doubt that Hepburn would "eventually destroy himself."

The feud simmered slowly. In October 1935, Hepburn demanded that Ontario Liberal MP Arthur Slaght, a man of some ability, yet who was also the premier's drinking buddy, be appointed to the federal cabinet. King said no and did so politely, but from Hepburn's standpoint the perceived snub implied that he should mind his own business. This was a slight, according to Saywell, that "Mitch never forgot, nor forgave." In the spring of 1937, Hepburn, the alleged man of the people, took a ruthless stance against strikers at the General Motors plant in Oshawa and treated the workers as if they were about to instigate another Bolshevik revolution. He was furious that Norman Rogers offered to mediate the dispute, wrongly considering it an attempt to embarrass him. Then he became madder still when King refused to put the Mounties at his disposal. Instead, he created his own force of special police—referred to derisively as "Hepburn's Hussars" and "Sons of Mitches"—to harass the strikers. Watching from a safe distance, King believed—and he was not the only one—that Hepburn had become a Fascist.

Hepburn had big plans to expand Ontario's hydroelectric power on the Niagara River, yet this met with problems from both the federal government and the Roosevelt administration, which had its own agenda. In this instance, King, in the role of conciliator, truly did try to broker a deal. When the various parties could not reach an agreement, Hepburn insisted he was the victim of an Ottawa-Washington conspiracy.

Upset as well by King's efforts to rewrite the rules of federal-provincial relations, Hepburn and Duplessis, who shared a disdain of communism and the federal Liberals, forged an "unholy alliance" with the stated purpose of driving Mackenzie King out of power. In Florida on a brief holiday when he first heard the news, King surprisingly took it in stride. "Just as well to have these two incipient dictators out in the open linked together," he recorded. "The public will soon discover who is protecting their interests and freedom... We will win on a 'United Canada' cry." A year later, there was talk that Hepburn had promised that in the next federal election he would deliver Ontario to Robert Manion, the new federal Conservative Party leader. King had no doubt the rumours were true and castigated Hepburn as a "traitor."

The battle with Hepburn and Duplessis was far from over. By 1938, however, King had far more serious concerns to consider than the recalcitrant premiers of Quebec and Ontario. Europe was on the verge of another bloody war, and if the tension could not be eased, Canada was going to be in the eye of the storm. Of that, King was certain.

MACKENZIE KING WAS the ultimate appeaser and no more so than during the final few years of peace in the late thirties. With all the best intentions in the world, he was juggling a handful of contradictory objectives. He desired to enhance his own status and friendship with FDR by playing matchmaker between the Americans and the British. As had been his guiding principle since the early twenties, he insisted on an autonomous made-in-Canada foreign policy. Yet in the deepest recess of his soul he understood that, as in 1914, if Britain was at war, Canada would be as well. As his principal secretary, Arnold Heeney, recalled of those years, "In the light of the record since revealed it is ironic that the impression over the war years [was] that Mackenzie King preferred Washington over London, the continental association to that with Britain and the Commonwealth. We who worked with him closely had reason almost daily to know the fact was quite otherwise. King was very wary of those he suspected of wanting to make of the British connection the fabric for a centralized political authority, but his attachment to Britain and things British were pervasive."

With memories of the divisive years of the First World War and conscription, King desperately sought to keep Ernest Lapointe, his other Quebec MPs and all of French Canada assured that Canada would not become embroiled in another foreign entanglement. Finally, like other statesmen of this era who did not yet truly grasp the malevolent machinations of the Nazi mentality, King believed and prayed regularly that a reasonable peace agreement with Adolf Hitler was a real possibility.

In trying to balance several conflicting goals, King was forced to be vague about Canada's position in a possible European conflict. "He had decided," asserts historian Norman Hillmer, "that ambiguity would offend the fewest." Offend the fewest, yes, but frustrate, exasperate, drive to distraction—that's another matter entirely. King's ambiguity and cautiousness led him to be unavoidably, albeit unfairly, portrayed as an isolationist and neutralist. As has now been established, he was never that. "King's backing and filling, his evasion and hesitation do not make inspiring reading," wrote historians Jack Granatstein and Robert Bothwell in a seminal article on the subject. "But his actions ... indicate his sure grasp of the public mood and his recognition that public opinion cannot be wished into existence simply because one course of action or another is 'right.' When King took a united Canada into the Second World War, he gave Canadians a policy that not only was right to him, but one that seemed right to them."

The ambiguity started with King's stand on the League of Nations. He was committed to peace and to Canada's participation in the beleaguered League, but was absolutely correct in his view that an international organization of collective security that did not include the United States, among other major powers, was an organization that was bound to fail. As he had made clear during the Abyssinia crisis and the question of confronting Mussolini, he was not about to permit Canada to be dragged into an untenable position—and certainly not into a war.

When Hitler ordered German troops into the demilitarized Rhineland, which was a clear violation of the Treaty of Versailles and the Locarno Pact, King became more anxious. "I pray that God may withhold the hand of the aggressor," he wrote in his diary on March 7, 1936. Pressed to explain the government's position by the CCF's J.S. Woodsworth two weeks after the first of Hitler's numerous belligerent actions, King replied with the haziness that has made him famous, "In a word the attitude of the government is to do nothing itself and if possible to prevent anything occurring which will precipitate one additional factor into the all important discussions which are now taking place in Europe." Two months later, he was more definite, reiterating his support of the League of Nations, but adding perceptively that "collective bluffing cannot bring collective security." (King's ego would have been pleased, though he might also have been astounded, that in January 2003, Donald Rumsfeld, President George W. Bush's embattled secretary of state, quoted King's snappy phrase in criticizing the United Nations Security Council's refusal to support a war against Saddam Hussein and Iraq.)

He was even more certain of his stand after he arrived in Geneva in September to attend the League assembly. There was an "absurdity," he felt, about "entrusting the affairs of one's country directly or indirectly to an aggregation of the kind which one sees in the Assembly Hall." Cheered on by Oscar Skelton and Norman Rogers—both of whom wanted to distance Canada from Europe's volatile problems more than King did—he told the delegates of Canada's support for the organization in one breath and then in the next declared that the League should "emphasize the task of mediation and conciliation rather than punishment." King had indeed gauged the mood of the country back home. The speech was praised by nearly everyone in the assembly (he made a point of listing each positive comment in his diary) and in newspapers in Quebec and the rest of Canada. Only in Winnipeg did John Dafoe predict that King had sunk the League, even though he would have conceded that it was likely sunk already.

King had more faith in the wisdom of British politicians to reach an acceptable compromise with Hitler. Before he headed home, he was reassured after discussions with Prime Minister Stanley Baldwin and Neville Chamberlain, then the Chancellor of the Exchequer. In particular, his conversation with Baldwin about dealing with Hitler, British royalty and other matters convinced him of his own and Canada's importance in the whole scheme of things. "What has been told to me today," he later reflected about "a most remarkable day," "has been a revelation as to how close one can be brought to the very summit of affairs and how much apparently in the way of authority and responsibility has come upon one's shoulders. There can be no doubt that the voice of the Prime Minister of Canada is very far reaching in the affairs of the British Empire." That probably would have been news to the British, who, despite Canada's gain in autonomy, more or less expected a response of "ready, aye, ready" from Canada and the other dominions if the situation with Germany got out of control.

This was the subtext at the imperial conference that brought King back to London in May 1937 for top-level meetings, in addition to attending the coronation of George VI. King had his own agenda, which included advocating for FDR's twin ideas for economic rather than strictly political appeasement of Germany and a "United Nations"–style world conference at which democracies and dictatorships could resolve their differences. The British pushed for a single empire voice, and King—with Skelton constantly peering over his shoulder to ensure that his boss's anglophile tendencies were kept in reasonable check—held fast, advancing the cause of "economic appeasement, Anglo-American harmony" and an independent Canadian foreign policy. Nothing much substantial was achieved at the conference and few of the delegates were pleased, except Mackenzie King. "It was a singular thing that at the end of 100 years," he wrote, "I was contending in No. 10 Downing Street, for a policy that would preserve the Empire while preserving national freedom to its parts, the very thing my grandfather was fighting for a hundred years ago."

If anything out of the ordinary transpired in London, it was King's private discussion with Joachim von Ribbentrop, who in 1937 was Hitler's ambassador to Britain. The future Nazi foreign minister, who was to be hanged for war crimes in 1946 at Nuremberg, was so impressed with King's enlightening attitude to the League and world events that he invited the prime minister to Berlin for a personal meeting with Hitler. This had been something King had been contemplating for more than a year.

King found von Ribbentrop, as he put it in his diary, "very prepossessing, very natural, a man I could get along with quite easily." (Maybe it was von Ribbentrop's Canadian connection. For a few years before the First World War, he lived in Montreal and then Ottawa, where he had established a small business importing German wine. If not for the outbreak of hostilities in Europe, von Ribbentrop might have remained in the country.) During their discussion, King thought it relevant to stress that he had been born in Berlin, Ontario, and that he had "knowledge of the German character at first hand." Then when von Rippentrop commented on Hitler's "sympathies with workers," King told the Nazi ambassador about *Industry and Humanity* and more: "I mentioned the struggles of '37 and the circumstances under which mother was born. I told him that all of these influences helped to give me, I thought, the sympathy with movements of the people, to appreciate motives, desires, etc." At that moment, King believed he might well be the one individual who could unite Britain, the U.S. and Germany in common cause and "friendship."

The auspicious encounter with Herr Hitler took place at the presidential palace in Berlin on June 29. It was a busy and exciting day. Accompanied by Pickering and assisted by a local translator, Mr. Schmidt, King first stopped to visit with Hermann Goering. The field marshal was polite, spoke intelligently of current issues and thanked King for arranging a shipment of Canadian bison to the Berlin Zoo. King made a point of letting Goering know that while Canada was independent, it was intrinsically linked to Britain. "I went on to say," King recalled, "that if any time we felt this freedom to be imperiled by any aggressive act toward Britain, our people would almost certainly respond immediately to protect our common freedom." Goering was intrigued and probed further about Canada's intentions.

Forty-five minutes after he bid farewell to Goering, King was at the palace. A guard led him upstairs, where he was greeted by Hitler's staff and shown into an adjoining room. The Führer, who was wearing formal dress accentuated by a white tie, was waiting for him. He shook King's hand warmly. They sat down on two large, comfortable chairs, with Schmidt the translator on Hitler's left. King had brought Norman Rogers's biography of him as a gift for Hitler. As he presented it, he opened the book to show Hitler photographs of his childhood and (as he had done in his conversations with von Ribbentrop and Goering) drew attention to the fact that he had grown up in Berlin, Ontario. Hitler smiled warmly and leafed through the volume. Before King departed, Hitler gave him in turn, as he described it, "a beautiful silver mounted picture of himself, personally inscribed." King was deeply touched by this gesture of "friendship."

The discussion was scheduled for thirty minutes, yet lasted an additional forty-five. Once or twice, Hitler's secretary reminded him that the allotted time had ended, but Hitler ignored him. The two men spoke about a wide range of topics, with the noticeable exception of the Nazi's anti-Jewish policies, which King would never have dared raise lest he insult his host. (King, however, did manage to let Hitler know that he had "the largest majority a Prime Minister had had in Canada.") The conversation mainly focused on the European situation, the injustice of the Treaty of Versailles, the need for German rearmament and the grave consequences of war. "My support comes from the people," Hitler told King, "[and] the people don't want war." King, in turn, reaffirmed what he had stressed to Goering: "That if the time ever came when any part of the [British] Empire felt that the freedom which we all enjoyed was being impaired through any act of aggression on the part of a foreign country, it would see that all would join together to protect the freedom which we were determined not to be imperiled."

In Paris a few days later, King repeated the same line for the press, which upset Skelton and the external affairs staff since it implied that Canada would defend the empire's interests and freedom no matter what the circumstances. That was too sweeping a statement for Skelton to stomach, and he gently told the prime minister so.

Like many others who came into contact with Hitler before the Second World War, King was plainly in awe of him, as is clear in the approximate 7,400-word moment-by-moment assessment of the meeting he dictated to Pickering for his diary. He praised Hitler more than once: "I spoke... of what I had seen of the constructive work of his regime, and said that I hoped that that would continue... That it was bound to be followed in other countries to the great advantage of mankind." To this, Hitler "spoke very modestly" of his accomplishments. He also felt compelled to tell Hitler "how all of our Ministers and I, myself, had been prejudiced against him on what we thought were narrow views and nationalistic and imperialistic policies, but that we had all come to feel quite differently and believed policies toward European countries would be wisely administered in his hand."

He left the palace feeling contented, satisfied and optimistic that war still might be averted. King's insight into Hitler is difficult to swallow today. Yet there is no denying the fact that in 1937 Hitler was hailed as a charismatic, if dangerous, visionary, somewhat as Mackenzie King portrayed him.

> To understand Hitler, one has to remember his limited opportunities in his early life, his imprisonment etc. It is truly marvelous what he has attained unto himself through his self education... He has much the same kind of composed exterior with a deep emotional nature within. His face is much more prepossessing than his pictures would give the impression of. It is not that of a fiery, over-strained nature, but of a calm, passive man, deeply and thoughtfully in earnest. His skin was smooth; his face did not present lines of fatigue or weariness; his eyes impressed me most of all. There was a liquid quality about them which indicate keen perception and profound sympathy. He looked most direct at me in our talks together... he then sat quite composed, and spoke straight ahead, not hesitating for a word... He has a very nice, sweet, and one could see, how particularly humble folk would come to have a profound love for the man... As I talked with him, I could not but think of Joan of Arc. He is distinctly mystic.

EVEN AS THE situation in Europe worsened in 1938, King held on to this numinous image of Hitler. Indeed, he convinced himself that the Führer was a fellow spiritualist, guided, as he was, by the spirit of his mother. "No one who does not understand this relationship—the worship of the highest purity in a mother can understand this power to be derived therefrom—or the guidance. I believe the world will yet come to see a very great man... in Hitler."

King and everyone else who did not grasp Hitler's intentions to keep on expanding German territory in the name of providing his people with *lebensraum,* or "living space," were in for a big disappointment. Nineteen thirty-eight was a year of constant worry. In one moment, war seemed imminent; in the next, peace and cooler heads prevailed. Such a dichotomy explains Mackenzie King's apparently inconsistent behaviour. On one hand, he put every ounce of his faith in Neville Chamberlain (who had succeeded Baldwin as British prime minister in May 1937) and his policy of appeasement to keep Hitler satisfied and stave off war. And on the other, he accepted the need for modestly increasing Canada's defence expenditures and speaking out about the country's obligation to the mother country—much to the astonishment and annoyance of Skelton, Lapointe and others.

In mid-September 1938, as the world held its collective breath and prayed that Chamberlain could make an acceptable deal with Hitler over his aggressive action in occupying the Sudetenland (the western area of Czechoslovakia with a large ethnic German population), King was clear in his own mind about the course Canada had to pursue. "We both agreed," wrote King after a discussion with Norman Rogers on September 13, "that it was a self-evident duty, if Britain entered the war, that Canada should regard herself as part of the British Empire."

As Chamberlain and Hitler jousted, King, who was battling a painful bout of sciatica, was overcome with a case of conviction. He wanted to take a more definite stand. From his bed, he sent a top-secret cable to Lapointe, who was in Geneva attending a League of Nations assembly. "The world might as well know that should the occasion arise, Canada will not stand idly by and see modern civilization ruthlessly destroyed." Here was King at his most decisive and courageous. Lapointe was "perturbed" and warned the prime minister of a Quebec fallout; Skelton was even more agitated, believing that his fourteen years fighting for an independent and level-headed Canadian foreign policy was about to be lost. With several other members of the cabinet pushing for a wait-and-see-what-happens strategy, King held back the declaration, even though he had not changed his mind. "Personally, I feel very strongly that the issue [of a Canadian commitment] is one of the great moral issues of the world," he wrote on September 24, "and that one cannot afford to be neutral on an issue of that kind."

The announcement of the last-minute Munich Agreement on September 29—in which Chamberlain and Édouard Daladier, the prime minister of France, sold out Czechoslovakia to appease Hitler and halt a war, if temporarily—naturally delighted King. "Great rejoicing Canada today," he cabled Lapointe who was on his way back to Canada. "Welcome home." (The *Winnipeg Free Press* was the lone Canadian newspaper that did not hail the news from Munich and London. "What's the Cheering For?" was the title of its lead editorial on September 30.)

King's relief, of course, was to be short-lived. Within a few months, the spectre of war had not faded. King continued to warn his ministers of the potentially dark road ahead. Then, on January 16, 1939, risking the anger of his Quebec colleagues, he declared the truth of the matter as he saw it. "If England is at war we are at war and liable to attack," he stated, echoing Laurier's oft-quoted dictum from 1910. "I do not say we will always be attacked; neither do I say that we would take part in all the wars of England. That is a matter that must be guided by circumstances upon which the Canadian parliament will have to pronounce and will have to decide in its own best judgment."

Lapointe was shocked by King's words, though his loyalty to the prime minister was unbroken. Despite loud objections from several ministers and the mandarins at external affairs and his own staff—one of his secretaries, probably Jack Pickersgill, had suggested to him that his speech was "appalling"—King was more certain of his stand than ever. The "only mistake" he had committed, he told Arnold Heeney, was that "[I] had not said in the past what was the real position about the reality side of things, thereby letting it be assumed that I was indifferent to this aspect, and was holding solely to the academic position—crisis or no crisis."

Hitler's occupation of the rest of Czechoslovakia on March 15, 1939, soon resolved the quandary. Publicly, King backtracked a bit. In a two-hour speech in the House on March 30, he deemed it wise to maintain a cautious approach, not wanting Neville Chamberlain to think for a minute he could take Canada for granted. The British prime minister did anyway. On March 31, without a word to King, Chamberlain issued Germany an ultimatum that Britain would defend the borders of Poland and Romania. King was resentful and annoyed.

Ernest Lapointe had the opposite reaction. The Nazis' latest act of aggression had finally convinced him that war was likely inevitable and that in such an event Canada had no choice but to support Britain. To allay any fears in Quebec, however, he swore that he would never be part of a government that instituted conscription.

King hailed Lapointe's speech, even if he was a bit jealous of all the positive press attention it received. Yet as he acknowledged in his diary, this was the turning point and a near guarantee that Canadian unity would be preserved should war break out. "Together, our speeches," King wrote on March 31, "constitute a sort of trestle sustaining a structure which would serve to unite divergent parts of Canada, thereby making for a united country." Now all there was left to do was wait and wonder.

FOUR MONTHS EARLIER and thousands of miles from Ottawa, seventeen-year-old Herschel Grynszpan, a Jewish student residing in Paris, was enraged by the Nazis' brutal treatment of his parents in Hanover, Germany, where his family had lived for twenty-seven years. On November 7, 1938, young Grynszpan walked into the German embassy, asked to see an official and was taken to the office of the diplomat Ernst vom Rath. Without a word, Gyrnszpan shot vom Rath, who died of his wounds two days later. That brazen act triggered a vicious pogrom against Jews throughout Germany and Austria on November 9 and 10, marking the true beginning of the Holocaust. On *Kristallnacht,* the night of broken glass, Jewish shops were destroyed and countless synagogues burned.

King was at heart a Christian humanitarian, and however he envisaged Hitler's talents, his conscience was deeply troubled by the news of this violence. "The sorrows which the Jews have to bear at this time are almost beyond comprehension... Something will have to be done by our country," he wrote. But what exactly was to be done? That was the question King and his cabinet had wrestled with since almost the day they returned to power in 1935. A month before King's victory, Hitler had instituted the anti-Jewish Nuremberg Laws, depriving German Jews of the rights of citizenship and equal treatment under the law. Escape and immigration from Germany was one possible solution, yet an almost impossible one for most German Jews. And for a simple reason: no country wanted them, including Canada.

Mackenzie King grew up in a decidedly un-multicultural and intolerant Canadian world, where the concept of race was everything. Black Canadians and Americans, known as Negroes back then, were more often referred to by King in his diary as "darkies,"—as in, "I breakfasted in the diner... greeting two darky porters on the way" (July 20, 1933); or "at the station this morning Nicol was talking to the darky porter" (March 3, 1940); or "the darky Willie was an outstanding waiter" (April 13, 1940); or when he wrote to Joan Patteson while holidaying in Jamaica in October 1938 that "one of the pleasantest moments of the whole journey was the service on the ship yesterday morning when a darky waiter who looked after Skelton and myself sang... "

As for "Jews and Jewesses," they were to be avoided and, as noted, King went to great lengths to ensure that no such "undesirables" purchased property anywhere near Kingsmere. He did, however, take a more enlightened view than most Protestant and Catholic Canadians. "There are good as well as bad Jews," he reasoned in an entry of August 1936, "and it is wrong to indict a nation or a race." At the same time, Mrs. Helen Lambert, a New York medium and friend of Etta Wriedt with whom King corresponded, sent him copies of various anti-Semitic books and propaganda, including the notorious *The Protocols of the Elders of Zion,* about a purported Jewish conspiracy to take over the world. A letter from Mrs. Lambert to King at the end of July 1934 told him, as he wrote, of her "belief in the influences at work by the Jewish element surrounding Roosevelt and of the control these people had come to have... there is much in Mrs. Lambert's letter that is very striking." (Three decades earlier, a younger King had concluded in his diary that "there is something in a Jew's nature which is detestable. The sucking of blood.")

Anti-Semitism and racism had certain respectability among most segments of Canadian society during the period from 1920 to 1950. Across the country, Jews found many professions and jobs closed to them, quotas and other discriminatory practices at universities, and clubs and beaches restricted. There is no official record of how many times the sign "NO JEWS OR DOGS ALLOWED" was posted in Canada during the thirties and forties. But historian Irving Abella asserts that it was spotted at "Halifax golf courses, outside hotels in the Laurentians and throughout the cottage areas of Ontario, the lake country of Manitoba and the vacation lands of British Columbia." A popular alternative, hardly less pointed, read "FOR CHRISTIANS ONLY."

Jews could not win, no matter what they did. They were blamed for being greedy capitalists and scheming bankers who destroyed the economy; they were blamed for being liberals and the creators of violent, immoral and vulgar films; and they were blamed for being evil communists intent on taking over the world and destroying Western civilization. Mackenzie King's Canada, it must be stressed, was not Nazi Germany. No one deprived Jews of their civil liberties. They had the right to vote and enjoyed the equal protection of the law. "Ours was a society with a well-defined pecking order of prejudice," explained author James Gray, who worked for the *Winnipeg Free Press* during the Depression. "On the top were the race-proud Anglo-Saxons, who were prejudiced against everybody else. On the bottom were the Jews, against whom everybody discriminated. In between were the Slavs and Germans." Or, put slightly differently by R.A. MacKay, dean of the Faculty of Arts at McGill University, whom King appointed to the Rowell-Sirois Commission: "The simple obvious truth is that the Jewish people are of no use to us in this country [and that] as a race of men their traditions and practices do not fit in with a high civilization in a very new country."

Few Quebeckers would have argued with Professor MacKay's assessment. Right, left, in the middle, it didn't matter. Everyone of any political ideology could agree on one thing: there were too many Jews in Montreal and elsewhere, and they definitely did not want any more, no matter how desperate the Jews' plight. "The French Canadians, though staunch defenders of 'minority rights,' are apt to define the term so as to exclude Jews," a lengthy anonymous article on "fascism in Quebec" in the prestigious journal *Foreign Affairs* astutely pointed out in 1938. That was putting it mildly. Adrian Arcand's Fascist party in Quebec boasted a membership of eighty thousand at the height of its popularity. In 1934, French-Canadian interns went on strike after the Notre Dame Hospital hired a Jewish doctor, Samuel Rabinovitch. French-Canadian Catholic intellectuals led by the still-revered Abbé Lionel Groulx sought to drive Jewish merchants out of the retail business by the "achat chez nous" ("buy from us") campaign. Groulx, to be fair, found anti-Semitism "a negative and silly solution," but he still endorsed the anti-Jewish boycott and wrote about race purity.

In the pages of the *L'Action Catholique* and other Catholic newspapers, Jews were denounced as parasites and Bolsheviks. According to Marcel Hamel, the editor of *La Nation,* communism was "a Jewish invention," a view shared by Quebec premier Maurice Duplessis and other members of the Union Nationale. In the nationalist organ *Le Devoir,* according to historian Esther Delisle, Jews in the 1930s were "aliens, circumcised, criminals, mentally ill, trash of nations, Tartars, infected with Semitism, malodorous—they smell of garlic, live in lice-ridden ghettos, have greasy hair and pot bellies, big crooked noses, and they are dirty." Such notions were shared to different degrees by a generation of such prominent Quebec leaders as journalist and activist André Laurendeau, who later became the editor of *Le Devoir;* Jean Drapeau, the future superstar mayor of Montreal; and even a young Pierre Trudeau, who in the late thirties wrote a "satirical comedy" entitled *Dupés* ("We've been had!") that was laced with anti-Semitic stereotypes. King's master of Quebec, Ernest Lapointe, was aware of all of this hostility and hatred and, as his biographer Lita-Rose Betcherman suggests, this "was doubtless the main reason for his implacable opposition to admitting Jewish refugees."

About the last issue Mackenzie King wanted to debate with his cabinet or publicly speak about in 1938 was whether or not Jewish refugees trying to escape from Nazi Germany should be permitted into Canada. After FDR had initiated the discussion among western countries that spring with talk of convening an international conference to discuss it further, King felt he could not ignore it any longer. Canada did not actually have a refugee policy at the time, so from a purely bureaucratic point of view any prospective Jewish immigrants, refugees or not, had to be considered under the restrictive immigration policy of the day, which was tricky. King also understood that any exceptions to these rules, especially for Jews, who were not regarded as "desirable" Canadians under any circumstances, would not be popular in Quebec and in many other areas of the country. "My own feeling is that nothing is to be gained by creating an internal problem in an effort to meet an international one," he wrote in his diary on March 29, 1938, echoing the popular eugenics sentiments of the day. "We must nevertheless seek to keep this part of the Continent free from unrest and from too great an intermixture of foreign strains of blood... I fear we would have riots if we agreed to a policy that admitted numbers of Jews."

All through this period, there was continual lobbying to open Canada's doors from the Jewish community as well as pressure from a tiny group of MPs, including the three most prominent Jewish ones in Parliament: Abraham Heaps of the CCF, and two Liberals, Sam Factor from Toronto (whom King considered appointing to the cabinet in July 1940 until that idea met with resistance from his Quebec colleagues) and his old friend Samuel Jacobs from Montreal. A succession of cabinet discussions revealed that only a few ministers were prepared to crack open the country's doors "on humanitarian grounds": primarily Thomas Crerar, who was responsible for immigration, Norman Rogers, William Euler, and to a lesser extent, James Ilsley. The majority, led by Lapointe, favoured keeping it tightly shut. Nor were Lapointe and his followers—and King was on side with them—swayed by the Jewish community's offer to accept complete economic responsibility for upwards of ten thousand refugees.

King frequently expressed his sympathy for the German Jews, yet he also maintained his belief, really nothing more than a convenient excuse, that allowing more Jews in would increase anti-Semitism in Canada. To this end, he instructed Canada's two delegates to the U.S.-backed conference on the refugees held at Évian-les-Bains in the south of France in early July 1938 "to make notes, say as little as possible and under no circumstances to make any promises or commitments." King needn't have worried. For nine days, there was much talk but little concrete action taken, a fact not lost on Hitler. Only the tiny Dominican Republic offered to admit a substantial number of Jewish refugees.

At stake for King in this matter were several sacred principles: cabinet solidarity, his eternal faith in solving difficult problems through consensus and, above all, preserving national unity. Not to mention political opportunism, with all of those Quebec votes to consider. King also permitted Frederick Blair, the deputy minister of immigration, whose anti-Semitic intransigence is now legendary, to wield enormous power over the fate of the Jewish refugees.

A few days after the violence of *Kristallnacht,* King attended the funeral of Abraham Heaps's wife, Bessie, who had died at the age of forty-nine from cancer. King was moved by the "deeply touching service," which he described at some length. Again, his conscience was triggered. "I feel Canada must do her part in admitting some Jewish refugees. It is going to be difficult politically, and I may not be able to get the Cabinet to consent, but will fight for it as right and just, and Christian."

That was in private; his public stance was much less conciliatory. Despite overwhelming newspaper support in the English-language press to accept some refugees in the wake of *Kristallnacht,* an appeal that grew stronger as the repression of Jews in Germany intensified— French-Canadian newspapers stuck firmly to their anti-Jewish immigration position during 1938 and 1939—and a series of demonstrations by Canadian Jews across the country, King and most members of his cabinet refused to bend. In another meeting with Jewish leaders on November 23, and another offer to cover all economic costs for ten thousand German-Jewish immigrants, King adopted a blunt tone, informing the delegation that "as long as there were unemployed, we had to consider them first." From his perspective, he was being both prudent and practical. Why make promises that he might not be able to keep?

At the next cabinet meeting, he did try again. And, in a prophetic moment, he urged his colleagues to "view the refugee problem from the way in which this nation will be judged in years to come, if we do not play our part, along with other democracies, in helping to meet one of humanity's... needs." He added, as well, his belief in the "fatherhood of God and the brotherhood of men," and that as a government, "we would have to perform acts that were expressive of what we believed to be the conscience of the nation, and not what might be, at the moment, politically most expedient." This plea fell on deaf ears, as it did a week later after Crerar declared that he was prepared to admit ten thousand refugees. (Australia had announced its intention to accept fifteen thousand German Jews.) "Quebec ministers strongly against admission," King recorded of that conversation. Various excuses were raised—the provinces had to be consulted and anti-Semitism might rise—but the bottom line was that most members of the government just did not want to let any Jews into the country. "We don't want to take too many Jews," astutely observed Norman Robertson, who worked under Oscar Skelton at the external affairs department, "but, in the present circumstances particularly, we don't want to say so." This was in line with recommendations being forwarded to the cabinet by Frederick Blair, which King strictly abided. At least once, however, during this tense period, the plight of the Jews took over his dreams and troubled his sleep.

On June 8, 1939, King was in Washington escorting King George VI and Queen Elizabeth on their royal tour and enjoying a fine lunch with President Roosevelt and his family at the White House. Over tea, the conversation eventually got around to the matter of the ss *St. Louis,* a ship that had left Hamburg with 907 German Jews on board bound for Cuba. Except when the ship reached Havana on May 30, the Cuban government refused to accept their entrance visas. More than a week later, no South or Central American country would provide the refugees with a safe haven. And the United States and Canada did not want them either.

King listened politely to FDR's views on the matter and then decided that this was not Canada's problem to solve. Other Canadians, among them historian George Wrong, B.K. Sandwell, the editor of *Saturday Night,* and Robert Falconer, the former president of the University of Toronto, beseeched King to show "true Christian charity of the people of this most fortunate and blessed country." All King was prepared to do was to seek the advice of Lapointe and Blair, and both, not surprisingly, were vehemently opposed to taking any action. In what was to be one of the definitive statements of this whole sorry ordeal, Blair told Skelton that the Jews on the *St. Louis* did not meet Canada's immigration regulations and that no country could "open its doors wide enough to take in the hundreds of thousands of Jewish people who want to leave Europe: the line must be drawn somewhere."

By the time Blair had written those sharp words, the *St. Louis* and its Jewish passengers were on their way back to Europe. Nearly 300 of them managed to get into Britain, and the rest found temporary refuge in France, Belgium and Holland. Of the 907 on the ship, 254 perished during the war.

Neither King, nor FDR, nor any other leader of a western country who refused Jewish immigrants could have predicted the tragic fate that awaited millions of European Jews at Auschwitz and the other Nazi death camps. At the same time, given the news reports of boycotts, violence against German Jews, synagogue burnings and forced fines, King could have shown true leadership on this issue if he had wanted to.

Yes, the Depression was not over and unemployment was still a factor in any discussion on immigration. Nevertheless, in one of his weakest moments as prime minister, he permitted political expediency to trump humanitarianism, when his own conscience told him to follow a more empathetic path. He conveniently heeded the rigid opposition from Lapointe and the other French-Canadian Liberals who, influenced by the anti-Semitism of the era, absolutely refused to budge or show an ounce of compassion towards the Jewish refugees.

As prime minister, King could have insisted on letting the German Jews in and perhaps even risked a cabinet revolt. He could have over-ruled his ministers. He could have had his way, as he did on many occasions during the war years. The key in those later debates, as Jack Granatstein observes, was King's "willingness to fight." Such a stance in this case, he believed, would have had detrimental consequences for Canadian unity—an objective far more important in his mind than accepting Jewish refugees and seemingly much more serious given the rise of Duplessis and the Union Nationale.

The final result of this inaction: between 1933 and 1945, Canada allowed fewer than 5,000 Jewish refugees into the country compared with 200,000 permitted into the United States, 70,000 into Britain and 25,000 into China. Even Bolivia and Chile made room for 14,000 each.

10

A Man on
a Tightrope

Before going down to my bed room, I knelt and prayed
that I might be able to renounce anything that would
stand in the way of a complete surrender of self to God
and my country, and that I might be given strength and
guidance to carry on.—THE DIARY OF WILLIAM LYON
MACKENZIE KING, *September 9, 1939*

EUROPE MAY HAVE BEEN moving towards a bloody abyss in
the spring of 1939, but Mackenzie King had far bigger issues to consider
and none more so than the flower scheme at Quebec City for the arrival
of King George VI and Queen Elizabeth. Following months of often
frustrating discussions back and forth between Ottawa and London,
King had prevailed with British officials, who he had caustically commented in his diary were "still living in the reign of the Stuarts" when
it came to seeing Canada as an autonomous Commonwealth nation
rather than a seventeenth-century colony. He was to be in charge of the
royal tour and be the "face" of Canada, rather than the governor general,
Lord Tweedsmuir. Understandably, King took this role very seriously.

The prime minister had devoted countless hours to working out the
protocol, dinner arrangements, even the vehicles to which the Queen's
ladies-in-waiting were to be assigned, in addition to a thousand other
minute details that had to be deliberated on for the tour. People noticed.
John Dafoe sarcastically remarked in a letter to *Free Press* managing editor George Ferguson that King likely had not been paying attention to
the recent grave reports from Europe because "his mind is running on

really important things such as—well let's say the kind of pillow the Queen will rest her head on as she travels through Canada." And, lo and behold, the prime minister had missed the fact that the flowers ordered for the King and Queen's first luncheon were to be Tory and Union Nationale blue. "This I know, will be resented by our Liberal friends, as it will be construed as though it was intended for the Government of Quebec," he wrote in his diary on April 25. At that moment, he stopped his dictation to Edouard Handy and made a phone call personally changing the colour of the flowers to more respectable Liberal red. Satisfied, he continued, "How true it is that in politics," he added, "it is what we prevent even more than what we accomplish that tells in the end."

King had reason to be nervous. This was, after all, the first time a reigning British monarch had set foot in Canada. On May 17, he and Ernest Lapointe, looking like two plump peacocks dressed in their Windsor uniforms and ostrich-plumed hats, greeted the royal couple as they disembarked. Sounding much like a knight at King Arthur's court, the prime minister declared, "Welcome, Sire, to Your Majesty's realm of Canada." Lord Tweedsmuir had boarded the ship before it docked and was more than happy to remain out of the way, as King wanted and protocol required. In typical Mackenzie King fashion, he made sure he told King George and Queen Elizabeth how wonderfully "helpful in every way" the governor general had been, even if Tweedsmuir occasionally annoyed him. Likewise, the governor general did not have a lot of patience for Mackenzie King's character flaws and quirkiness, but he was, if anything, a stickler for his constitutional responsibilities, and the tour proceeded with great flourish.

At the inaugural luncheon, King was impressed that King George VI was able to deliver his speech without stammering, an ailment that he had to cope with for much of his life. "I thought he did exceedingly well," King noted afterward, "and was immediately pleased to see how he had mastered his infirmity of speech." (In May of 1937, when King had attended King George VI's coronation, His Majesty had told him about his work with the Australian speech therapist Lionel Logue—depicted in the 2010 award-winning film *The King's Speech*.)

That day and for the rest of the month-long tour, the prime minister ensured that all went as smoothly as possible and that the king and queen were never out of his sight for any length of time. There was much excitement on May 20, when Their Majesties had lunch at Laurier House. Minutes before they arrived, King inspected the house and table setting one last time and fixed Pat's cute red bow. A crowd gathered outside on the street. King's sister Jennie had come in for the occasion and was by his side; Joan and Godfroy were not able to attend because Godfroy was ill. As soon as the King and Queen exited the car, Queen Elizabeth was immediately drawn to Pat, which pleased King immensely. Lunch was splendid; the cutlery included fish knives and forks that had belonged to William Lyon Mackenzie, which inevitably led to stories from the proud grandson about the 1837 rebellion. Little Pat curled up by the Queen's feet. After the first course had been savoured, the servants suddenly disappeared, much to King's embarrassment. Then he was informed that His Majesty preferred that the servants were not hovering during the meal.

Following lunch, King gave his royal guests the grand tour of the house. There was conversation about the portrait of Isabel King, and the dutiful son "drew attention to the fire light in my mother's face and a softer light on the back of her hair." By the time they reached Laurier's old library on the second floor, King was oozing with patriotic zeal. "I wished I could tell them what their lives and their example meant to the people of all parts of the Empire as in our country," he wrote. "That there was nothing I could say about the confidence they had given me, and the expression of their regard. That I was prepared to lay my life at their feet in helping to further great causes which they had at heart. What I was saying was quite unpremeditated and words came with very strong and sincere feelings."

In the ensuing weeks, as King accompanied Their Majesties on the train across the country and to Washington, D.C., for a historic meeting with FDR (it was also the first time a reigning British monarch had visited the U.S.), his emotions were at a constant high. To King's delight, FDR had requested that he, and not the British foreign secretary, act as the official "minister-in-attendance" during the side trip to the American capital. The president and King George ensured that Mackenzie King was included in one of their two private discussions, and FDR praised the prime minister and the positive nature of Canadian-U.S. relations at every opportunity. Hence, for King each day was more memorable than the previous one. He took solace that those "who were nearest and dearest" to him "happily" knew how he felt. The spontaneous applause and cheering from the crowds lining the tracks as the train passed through various towns especially pleased King. "Oddly enough," he recalled, "I heard some call me 'Billy King'; others 'Rex'; others 'Mackenzie' and 'Mackenzie King.' I begin to see how necessary it is that I should try to get out among people again." From the perspective of national unity, King could not have asked for anything more glorious.

Only a handful of problems nearly spoiled the party. The first occurred when Quebec Premier Maurice Duplessis failed to show up for the official lunch in Quebec City. He claimed that he had to attend to a personal matter, but did manage to show up for the evening event. King was not impressed. "He had nothing intelligent to say all day," King pointed out with exasperation. Then on May 22 in Toronto, King accompanied the royal couple to the race track for the running of the King's Plate, only to watch the detested *Globe and Mail* publisher George McCullagh's horse win and receive a prize from His Majesty. "I confess when I heard it was McCullagh's horse that won the King's Plate, I could not help feeling financial circumstances had accounted for it," he wrote. "Both Arnold [Heeney] and I had [a] feeling as though something had been done to insure McCullagh winning."

By the time King bid farewell to King George and Queen Elizabeth on June 15 in Halifax, he had elevated the royal tour to even loftier heights, convinced that it somehow served the "well-being of mankind." The final send-off was difficult, as King at his poetic best described it: "When I entered the room [on the ship the *Empress*] the Queen was standing to the left of the King. She made a gesture of a kind which suggested a woeful parting—making a motion like a curl of the sail of a ship when the breeze has blown it outward—a sort of bow with her hands on either side, and said something to the effect that we have come to a very sad moment. 'Alas, we have come to the moment of parting.'" King George presented the prime minister with a large volume, a book on gardening which contained a record of the royal oaks that had been planted at the time of the coronation. King was deeply touched. "God, I believe, has chosen you for a work which no other persons in the world can perform," he declared to them, "and which I believe you can. I truly believe that your example and influence what you may be able to do with the advice of the ministers around you should enable the peace of the world to be maintained." A simple goodbye was insufficient. "May God bless, keep and strengthen each of you," King added at the end with a touch of piety. (According to B.C. writer Louise Reynolds, so close did Mackenzie King stick to the royal couple during their visit that "an apocryphal story went the rounds that, on their return of Their Majesties to Buckingham Palace, the King looked behind the drapes on occasion to see if his Canadian Prime Minister was there.")

Ruining this golden moment was Lord Tweedsmuir's apparent deliberate decision not to follow King off the ship, but to linger in conversation, long enough to thoroughly aggravate the prime minister.

Nevertheless, King concluded that "no farewell could have been finer than that given [that] day [by] the people and the country. The happiest of relationships had been established in all directions."

KING'S RAPTUROUS JOY was short-lived. Throughout the tense summer of 1939, a resigned inevitability hung in the air. King planned for a fall election and prayed for a resolution in Europe. For a brief moment, he thought he might be the world's saviour. In January 1939, he had written to Hitler with a personal appeal that "regardless of what others may wish, or say, or do, you will, above all else, hold form to see the resolve not to let anything imperil or destroy what you have already accomplished."

He grew more apprehensive some weeks later, after Mrs. Quest Brown, the English clairvoyant, read King's palm and predicted the war would break out by June. King refused to believe her—he was skeptical as well of her prophecy that Hitler and possibly Mussolini would both die within the year—and his resolve to find a solution remained firm. He appreciated much more her declaration that the lines on his hand "indicated, above all, no diminishment of power or prestige but rather increased power [and] increased recognition in the coming years."

It took until July for Hitler to send him a reply to his diplomatic missive. The Führer wanted to invite a group of Canadians to Germany, and King's first inclination was that he would join them. The message from Hitler led King to conclude, curiously enough, that Hitler wanted to work through him to restore peace in the world. He wrote in a lengthy diary entry "that 'forces unseen'—loved ones in the beyond—were working out these plans, that there were no accidents, or chances in this but all part of a plan in which God was using man to effect his Will in answer to prayer, the Mediums being those in the beyond who were working for peace on earth, good will to men ... I recognized that great care would be needed in each step and the utmost secrecy preserved ... I felt I must communicate with Chamberlain ... " Encouraged by Joan Patteson, King sent a letter to Neville Chamberlain, who suggested he do whatever he could to ease the threat of hostilities. King planned his trip to Germany for November.

On August 22, King still thought that that there might be a way to stop the outbreak of war. "If Hitler wishes to gain world domination, this would appear to be his moment to strike. I have never believed this was his aim," he wrote naively. That afternoon, King learned of the Nazi-Soviet Non-Aggression Pact, which he felt "would occasion more caution on the part of the British." Three days later, he decided he had no choice but to send another cable to Hitler, as well as to Mussolini and the president of Poland, urging them to prevent the "impending catastrophe by having recourse to every possible peaceful means to effect a solution." Several members of the cabinet, in particular Ilsley, who was, according to King, "almost violently outspoken against sending a message," opposed the idea because they insisted it committed Canada automatically to war before Parliament had an opportunity to vote on such a resolution. King and, more importantly, Lapointe, did not agree, and the message was sent via diplomatic channels. Mussolini and the Polish government acknowledged the letters, but Herr Hitler did not.

On September 1, King was awakened by Nicol at 6:30 AM with news that the German *Wehrmacht* had crossed the border into Poland. The British and French declarations of war were issued on September 3. King, or rather Arnold Heeney, along with Lapointe, who was never far from the prime minister's side, sprang into action. King's cabinet was already used to the prime minister beginning any major discussion with the words, "I have already talked the matter over with Lapointe."

As he was shaving the next morning, he noticed that the lather resembled "a perfect swan with a figure like that of Siegfried [the main character in Richard Wagner's *The Siegfried Idyll*] rising out of the centre of it (as if it were a boat carrying him)." As he wrote, "It seemed to me to be a guide as to Hitler in some particulars ... I believe Hitler like Siegfried has gone out to court death—hoping for the Valhalla—an immortality to be joined by death—Wagner's emphasis on death to be aimed at." In the afternoon, he drove out to Kingsmere and had dinner with Joan Patteson. They sought reassurance from the little table about events in Europe. The spirit of John King told them that Hitler was dead, "shot by a Pole." When this turned out to be incorrect, King concluded two days later that he and Joan had been victimized by a "lying spirit" and questioned his faith in utilizing the little table to discern the truth.

A special session of Parliament was convened on September 7. On his feet in the Commons the next day, King was deliberate, decisive and, for a few minutes at least, uncharacteristically passionate: "We stand for the defence of Canada; we stand for the co-operation of this country at the side of Great Britain; and if this house will not support us in that policy, it will have to find some other government to assume the responsibilities of the present." The cabinet had gathered a night earlier. There was talk and fear of conscription by some ministers, which King quickly quelled. He later spelled out his position on this explosive issue in his diary: "There would be no conscription under the present Government, which means I would send the resignation of my colleagues and myself before allowing the measure of conscription of men for overseas to be introduced. It may conceivably come to conscription for our own defence; nothing has been said against that."

In the House, King suggested that a Nazi attack on Canada was likely since no other territory was as coveted. "There is no other portion of the earth's surface that contains such wealth as lies buried here," he said. "No, Mr. Speaker, the ambition of this dictator is not Poland... Where is he creeping to? Into those communities in the north, some of which today are going to remain neutral." When it was Lapointe's turn to speak, he said what he knew had to be stated for the record. "I hate war with all my heart and conscience," he declared, "but devotion to peace does not mean ignorance or blindness... I say to every member of this House that by doing nothing, by being neutral, we actually would be taking the side of Adolf Hitler." That did not mean conscription, he said, which he and other French-Canadian ministers would never accept, but it did mean sending a voluntary expeditionary force to Europe. He finished with a flourish that brought a rousing ovation from the government as well as opposition benches. "God bless Canada, God bless our Queen, God bless our King." Mackenzie King thought it a splendid speech, "a very noble utterance... very brave and truly patriotic." After Lapointe had resumed his seat, King "clasped his hand with the warmest friendship and agreement in all he had said."

With a near unanimous voice of support—only the CCF leader J.S. Woodsworth, a pacifist who resigned the leadership of his party over the issue, as well as a few Quebec Liberal MPs, spoke against participation—Canada declared war on Germany on September 10 and voted $100 million for the defence budget. The moment of the declaration signing late on September 9 in King's parliamentary office with the cabinet seated around his table—"the room seemed dark with only the one light in the ceiling above," as King described it—was, as might expected, not without mystical meaning. "As I was about to sign, I suddenly lifted my eyes off the paper and was surprised to see my grandfather's bust immediately opposite, looking directly at me—the eyes almost expressing a living light," King recalled. "He was the one person in my thoughts as I affixed my signature to the order-in-council." It was around two o'clock in the morning when he finally returned to Laurier House. He "roused little Pat from his slumbers" and prime minister and dog "shared a biscuit and some Ovaltine together." The entry continued, "I was particularly struck with his having given up his little basket and slept at the side of my bed tonight. He was first on the one side and later I found him sleeping on the other. He seems completely conscious of what is going on."

The War Measures Act had been proclaimed on September 1, and although few could have predicted it at the time, this meant government by order-in-council for the next five years, censorship of the press and radio and detention of enemy aliens. Other than *Le Devoir*, which insisted that Canada was headed towards adopting conscription, every other English- and French-language newspaper in the country stood behind the government. Still, as a sign of the problems to come, already there had been hostile nationalist protest meetings in Montreal and Quebec City at which King and Lapointe were denounced. Paul Bouchard, the separatist editor of *La Nation*, roused the large crowd at Montreal's Maisonneuve Market by declaring that "we do not want to see thousands of young Canadians die overseas to save international Jewry's finances." He was followed to the podium by lawyer René Chaloult of the Union Nationale, who proclaimed to loud cheers that "French Canadians would rather fight in the streets of Montreal than in Europe." More serious trouble was on the horizon.

In a drawn-out cabinet meeting on September 12, King maintained that despite the fact that he had been "persecuted" for his cautious approach to a declaration of war, such a strategy had in the end paid off. The country, he stated, "had come into war with a quietude and peace almost comparable to that of a vessel sailing over a smooth and sunlit lake." King innately understood that when the call came, Canada stood by Britain, as was her duty. And that was the only reason. No matter how the decision has been spun since, the declaration had nothing to do with any greater or more idealistic purpose. It was "not for democracy... not to stop Hitler... [and] not to save Poland," as Jack Granatstein and Desmond Morton argue. "Canada went to war only because Neville Chamberlain felt unable to break the pledges he had made to Poland in March 1939. Had he slipped free, as he tried to do, Canada would have sat by and watched the Reich devour Poland without feeling compelled to fight."

WITHIN WEEKS OF the declaration, there were political casualties before the real and more tragic ones on the European battlefield. A year earlier, the *Financial Post* and *Maclean's* magazine, both owned by the Maclean's Publishing Company, had begun an extensive investigation into alleged corruption of the department of national defence and the awarding of a contract to the Toronto manufacturer John Inglis to produce the Bren light machine gun. The articles, most notably ones researched and written by Arthur Irwin, exposed a pattern of wasteful spending, patronage, costly land flips and sheer incompetence—all of which occurred on the watch of King's weak defence minister, Ian Mackenzie. Irwin's investigation spurred on Ontario Conservative leader George Drew's demand for an official inquiry and public confirmation of many of the charges.

By early September 1939, King had decided that Mackenzie, whom he also castigated in his diary for having a drinking problem, had to be transferred out of national defence. He had summoned back First World War commander and lawyer Colonel James L. Ralston to Ottawa and reminded him of his promise to assist the government if war broke out. King wanted Ralston to replace Mackenzie, but Ralston requested the finance portfolio, which had become available after Charles Dunning decided to leave politics because of his poor health. This cabinet shuffle, the first of several that would take place during the war, was confirmed on September 6. Ralston was appointed the new minister of finance and eventually won a by-election on January 2, 1940. King and Ralston respected each other, and King had tremendous faith in Ralston's abilities. Nonetheless, the two men were headed towards a bitter breach over manpower and conscription that would play out like a poor version of a Shakespearean tragedy during the next four years.

King delayed dealing decisively with Ian Mackenzie for another two weeks. The truth was King was tired and, as he admitted in his diary, he did not want to be burdened with "details on defence matters." What he desperately needed was an independent and skilled minister in this most important of portfolios. The last straw with Mackenzie came on September 15. A discussion about the inadequacies of Canada's air force and lack of equipment angered King. "It was really pathetic," he wrote, "to see how helpless Mackenzie seemed in presenting any matters himself. He falls back on memoranda and is much more, as one of my colleagues said, the messenger of the heads of staff than their chief." How to move him out of defence became the next delicate task.

King knew that Mackenzie's "Highland blood and pride" meant that he had to handle this shuffle with great care. In tackling this personnel problem he was at his diffident best. Late on September 18, King had a private talk with Mackenzie, who eventually agreed to take over Chubby Power's portfolio of pensions and national health. Power then became postmaster general; Norman Rogers, whom King trusted as much as he did Lapointe, was made the new minister of national defence; and Norman McLarty, who had been postmaster general, took over from Rogers in the department of labour.

Power had wanted the defence job, but he accepted that given the volatility of the conscription issue it would have been unwise to have a Quebecker as defence minister. That was only half of it. The real issue was Power's excessive drinking. When Power was sober, he was a capable cabinet minister. Yet in the first month of the war, King was not about to replace one national defence minister who drank too much with another. Still, he could not ignore Power's talents; in July 1940, King took a leap of faith and appointed him to the newly created wartime portfolio of national defence for air. Reading over King's account in his diary of this shuffle, one cannot help but admire the deft manner in which he massaged the various egos and high-strung personalities involved. King permitted Mackenzie in his last act as defence minister to issue the press release announcing that a Canadian expeditionary force was headed to Europe.

Meanwhile in Quebec City, Maurice Duplessis was stewing and by all accounts drinking too much. The War Measures Act was a thorn in the premier's backside. Suddenly the federal government oversaw foreign exchange transactions and censored the airwaves. Duplessis could not tolerate that his radio speeches would now have to be vetted in advance by "some federal functionary." With Canada at war, the U.S. Neutrality Act prevented the Quebec government from borrowing much-needed funds from New York banks. The provincial debt was rising by the day, and there were rumours of gross mismanagement. The Bank of Canada even turned down Duplessis's request for a $40-million loan.

The Quebec premier had to do something decisive. On September 25, in the most ill-considered action of his brilliant and controversial political career, he called a provincial election, only three years into his first term. He brazenly made the war, the threat of conscription and the federal government's centralization of power the key issue of the election. Mackenzie King was dumbfounded when he heard the news. "It is a diabolical act on his part to have made the issue provincial autonomy versus Dominion Government," he wrote. Asked by King for his view, an angry Lapointe replied that it "was straight sabotage, the most unpatriotic thing he knew." Lapointe, along with the other two key Quebec ministers, Power and Arthur Cardin, charged that Duplessis had thrown down the gauntlet, and they were compelled to challenge him. In a bold move, they declared their intention to actively campaign in the election on behalf of the Quebec Liberal provincial leader, Adelard

Godbout. And they were prepared to publicly announce that should Duplessis win, they would interpret the results as lack of confidence in their leadership. In such an event, they would resign from the federal cabinet, leaving Quebeckers vulnerable to the possibility of conscription. King accepted the logic and courage of their convictions, but he argued then and later against them resigning over it. "The three deserve the support of all Canada in this contest," he added. "It will be a great contest between the highest patriotism and the lowest forms of disloyalty. It should make Lapointe a world hero."

The subsequent announcement by the Quebec trio of their position—"Canadian history's greatest turning of the political tables," as Conrad Black aptly calls it—caught Duplessis by surprise. With Lapointe, Power and Cardin on the radio and filling up halls, the nondescript Godbout was a politician possessed with renewed energy. He attacked Duplessis's "alleged administrative incompetence" and dangled the resignations of the Quebec ministers from the federal cabinet at every opportunity. No electorate was ever so blatantly manipulated as Quebec voters were in the fall of 1939. Although Duplessis was never an opponent to take lightly, the results on October 25 watched closely by the rest of the country were a complete vindication for the federal cause and the war effort. Godbout and the Liberals won seventy seats to Duplessis and the Union Nationale's fourteen, a crushing loss of forty-eight seats.

King was thrilled with the outcome and took great pleasure in the humbling of Duplessis, which he also regarded as an instructive lesson for his Ontario nemesis, Mitch Hepburn. In King's opinion, all three of his ministers had performed brilliantly, but it was truly Lapointe's day. "I venture to say that Lapointe's place today is as far in esteem as of Sir Wilfrid Laurier... He will have a place second to none in Canadian history, and well merited as a patriot. It is impossible to exaggerate the significance of today's victory."

The next morning, King led a contingent of happy Liberals to the Ottawa train station to welcome back two of the conquering heroes, Lapointe and Power (who, King noted with annoyance, smelled of liquor); Cardin returned by car. "There is always some flaw in the ointment," King bemoaned, and in this case it was Chubby Power's hurt ego.

Power felt that he had done all of the "dirty work" in the campaign and Lapointe had received all of the credit. He also harboured resentment against being passed over for the national defence portfolio despite having concurred with King's decision in mid-September to appoint Norman Rogers. Meeting privately with King, he "made a strong demand" to have a say in the appointments of Quebec's next lieutenant governor and a new senator. Power wanted the positions to go to two of his Quebec Liberal MP friends: Sir Eugéne Fiset as lieutenant governor and Fernand Fafard as senator. King explained "good naturedly" why he could not follow this advice at this particular time; however, Power was not placated and threatened to resign. King convinced him to attend the cabinet meeting, which he did. But there was a bad scene. Power became embroiled in an argument with Lapointe, and shouted that while Lapointe might have made the speeches, Power had actually won the Quebec election for the Liberals. King tried to calm him down. Power, according to King, was "obviously under the influence of liquor and excitement got [him] talking very querulously... [and] then [he] became more violent." He stormed out of the meeting room, marched to King's office and dictated his resignation to a stenographer. Within a few days, Power had forgotten the entire episode and was back pressing his case to make Fiset as lieutenant governor, an appointment that was finally made at the end of December. Likewise, King appointed Fafard a senator on January 29, 1940.

By no means was Power the only one to suffer from an inflated ego; it was almost a mandatory job requirement to serve in the Liberal cabinet. Although King had remained at a respectful distance during the Quebec election and was more than happy to bestow accolades upon Lapointe, Power and Cardin, by the end of the year he decided that he had been the true inspiration for what had transpired. He also forgot that he had opposed the trio's public threat to resign from the federal cabinet. The three Quebec ministers "will be the first to say that [the] result would never have been what it was had I not taken... the stand I did on conscription and had I not, over the years I have been Leader of the Party, gained the confidence of the people of Quebec," he wrote on New Year's Eve 1939, with some justification. "Above all I think I may claim to be responsible in no small measure for the unity of Canada at this time and, certainly for the united manner in which the country has entered the war at the side of Britain."

King's craving for recognition and his memory lapse still irritated Lapointe nearly a year later, when he raised the issue in an interview with Grant Dexter. "Of course, Dexter," Lapointe had told him, "after these events [the Quebec election] which have taken my strength, those who opposed me have quite happily shared in any credit that was going." Dexter added in a memo to John Dafoe that this was "the first time Lapointe has ever taken a dig at W.L.M.K. in talking to me. He repeated it two or three times, so I judge it rankles his mind."

MACKENZIE KING'S OTHER and arguably more dangerous adversary, the out-of-control premier of Ontario, Mitch Hepburn, had initially supported the federal Liberal government's early war efforts. It did not take long before Hepburn reversed course and began denouncing King for his inability to adequately prepare Canada and its soldiers for the fight in Europe. In this, he joined in a bizarre (though not, perhaps, for Hepburn) alliance with the Ontario opposition leader, George Drew, a lawyer and honorary colonel of the Royal Canadian Artillery, who detested King and the Liberals. After meeting with King and his cabinet in early October 1939, Drew described them as "a group of extremely tired and befuddled old men, with the exception of Ralston and Rogers. Ralston appeared anything but befuddled, although he did look extremely tired. Rogers looked young, befuddled, tired and insignificant."

King tried to make peace with Hepburn, but to no avail. The last straw was when King perused the transcript of an off-the-record interview Hepburn had given to a *Toronto Star* reporter in October 1939. "It would make you sick," the Ontario premier had said, "to see how confused and distraught everything and everyone is in Ottawa." King read the *Star* interview to his cabinet, later referring to it in his diary as a "thoroughly deceitful and lying report." He had no more use for Hepburn.

By mid-January 1940, both Hepburn and Drew were unreserved in their censure of the King government's war plans. The soldiers training in Toronto, Hepburn had complained to Wilson Southam, the publisher of the *Ottawa Citizen*, were lacking in proper uniforms, shoes, even underwear. Goaded by Drew, Hepburn stood in the Ontario assembly on January 18, 1940, and did something "very foolish" (as one of his cabinet ministers told him). He introduced a resolution "regretting that the Federal Government at Ottawa has made so little effort to prosecute Canada's duty in the war in the vigorous manner the people of Canada desire to see." Mitch also announced that if the resolution was defeated, he would resign. It passed easily with only ten Liberals defying their leader and voting against it.

King was getting ready for an evening at the movies with the Pattesons when he learned of Hepburn's latest act of treachery. He was stunned and irate that such a resolution could be passed, and with the votes of so many Liberals. "It shows how completely Hepburn has become a dictator," he wrote, "and how fearful men have become of not bowing to his will and word." Right then and there, King made up his mind to call an election and let the people of Canada judge whether they felt the same way as Hepburn, Drew and the rest of the Ontario legislature.

He went to the movies with Joan and Godfroy to see actress Greta Garbo in *Ninotchka,* the story of a Soviet agent in Paris—and returned to Laurier House intending to draft a response to the Hepburn attack. Except none of his staff were there, not even the always-reliable Handy. Uncertain how to proceed, he phoned Handy at home and then Heeney, but both were out. He then tried Pickersgill, who was supposed to be on duty at his House of Commons office; he was not at his post either. In an almost comical episode, he finally connected with young Deane Russell, a recently hired stenographer from Winnipeg, who arrived to assist the prime minister with the statement. King was definitely not amused at his staff's seemingly lax behaviour at what he judged a moment of enormous significance. "All of this illustrates so well what I have said that this is certain to happen unless rules are followed strictly," he noted. "This is the most important night since the war began. It is of such importance that on tonight's decision depends pretty much the whole future of the Government."

A brief statement was eventually issued after several phone calls to Ernest Lapointe and Norman Rogers. This was, however, merely the opening salvo. After war had been declared, King made a deal with Robert Manion, the Conservative Party leader, that no election would be called before a session of Parliament had run its course. Now he intended to break his promise. In a deft political move—or "a piece of unscrupulous trickery," as the opposition parties and some newspaper editors later called it—he decided to announce the dissolution of Parliament and set an election in his upcoming throne speech. "Hepburn has made an appeal absolutely necessary, and has made the issue that of the Government's conduct of the war," he wrote, adding that the Ontario premier, who had likely been drinking, was "filled with his own prejudice and hate and is entirely blind to the sentiment in other parts of Canada." Most distressing of all to King was the "shocking betrayal" of Ontario Liberals who had stood by Hepburn.

During the next week, he told only the members of the cabinet war committee about his election plan and kept the rest of his cabinet, caucus and even Senator Norman Lambert, the head of the NLF (who had been planning for an anticipated campaign for some months) in the dark until the last possible hour. Once the whole cabinet was let in on the secret strategy, there was a mixed reaction to identifying Hepburn's resolution as the main election issue. But King's mind was made up. Lambert reassured him that the party's finances were sound and the organization ready to go across the country, a fact King accepted with pleasure but never truly appreciated.

On January 23, two days before the House was to meet, he informed Lord Tweedsmuir of his plan; the governor general regarded the news as "quite dramatic." On the morning of January 25, King was understandably anxious. Then he recalled that twenty-three years ago to the day, he had been told by the family's doctor that his mother was too ill to survive. That memory reinforced his faith in what he was about to do. "Come what may," he wrote later that evening, "I know it is the right decision." He had also received the Lord's blessing as he prayed. "While these thoughts were in my mind, a sort of star in the far distance seemed suddenly to appear before my eyes," he recalled. "It was by itself, so clear. It made me think of the star that I have seen at times in the sky which always seems to me to speak of my mother. I felt it to be a guiding star."

Pat rode with him in the car to Parliament wearing his "little coat." King was so preoccupied that he forgot to say goodbye to the pooch, something he immediately regretted. The Speech from the Throne announcing the immediate dissolution of Parliament definitely had the desired effect; "the bomb has exploded," as Tweedsmuir remarked to King afterwards. At first, Manion was flabbergasted, not quite believing what he had heard. Then, rising in anger, he denounced King's "political trick," as did J.S. Woodsworth of the CCF. The leading Conservative newspapers—the *Montreal Gazette, Globe and Mail* and *Toronto Telegram*—soon to be branded by the *Winnipeg Free Press* as the "Mackenzie King Haters"—were less critical since they expected a Conservative victory followed by the formation of a Conservative-dominated union government. King was hailed by his caucus the next day, which filled him with "pride and satisfaction." And everyone in the room gladly accepted without reservation that as candidates during this election they were one and all "Mackenzie King Liberals." Hepburn's traitorous actions were not be condoned. King declared that he would never support a national coalition government, which he considered "the first step toward a dictatorship," and that Liberal Party would win this election. At that, he received a "great ovation."

The date of the vote was set for March 26, 1940. The campaign went exactly as King had anticipated, with the exception of the death of Lord Tweedsmuir. The governor general suffered a stroke on February 6 while he was in the bathtub and died five days later, on February 11. (The time of death, which King believed to be highly significant, was 7:13:30, exactly the moment, he made a point of noting in his diary, that he had glanced at his watch.) King had no idea that Tweedsmuir's death was to be only the first of several of those closest to him during the next year and a half.

King did the minimum amount of campaigning and travelling. He spoke on the radio on several occasions, working endlessly with Jack Pickersgill on every word of every broadcast he made. His message was simple and clear: only the Liberals, with him at the helm, could protect Canadian interests during the war. He faced little real opposition. Manion was no Bennett, or Meighen for that matter, and despite his best efforts to whip up enthusiasm for a national government—going so far as to change the name of the Conservatives to the "National Government"—the election was lost to the Liberals before the campaign had even begun. In 1940, voters, especially those outside of Ontario, would never have swallowed a union government, which was too closely linked with Drew, Hepburn, George McCullagh of the *Globe and Mail* and conscription. Behind the scenes for the Liberals, Norman Lambert, with skilled assistance from Toronto advertising agency Cockfield, Brown, ensured the wide distribution of campaign literature. In a display of political *chutzpah* at its most cynical, the same Liberal government that refused to permit German-Jewish refugees into the country now placed ads in Canadian Jewish newspapers promoting the party as the true fighters of Nazism.

On March 26, King received the kind of across-the-board endorsement every prime minister dreams of. Early that morning, Pat had accompanied King to the polls, and the dog "seemed fully conscious that a little event of real significance was taking place." The Liberals took 181 seats, a gain of 8 from 1935; the Conservatives (or National Government) Party won only 40 (and only one seat in Quebec), the same as in the previous election. Manion lost in his own riding and soon resigned as Conservative leader. He was replaced by an interim leader, Richard Hanson, from New Brunswick, who loathed King as much as any Tory, usually referring to the prime minister behind his back (according to Grant Dexter) as "the little son-of-a-bitch."

In all, King had a comfortable majority of 117 seats over all the other parties in the House. Even one of his most bitter critics and tormentors, John Stevenson, the *Times* of London's Ottawa correspondent, was forced to admit that he was "a shrewd political tactician [who] can regard the result as a great personal triumph." That, to be sure, King did.

Feeling vindicated, King wanted to lambaste the "Nazi mentality" that in his view had made its "evil presence felt" during the campaign (no doubt he had Hepburn in mind). He was convinced by Joan and his staff instead to take the high road. In his public remarks, he quoted passages from Abraham Lincoln's speeches about "how the victory of our party in time of war will always stand at the side of his own at the time of the Civil War, though his was an infinitely more difficult situation. No victory for democracy however could have been greater."

King was so swept up in the moment that he did not bother to telephone Lambert (who had half-expected an invitation to Laurier House) and thank him for his diligent efforts. Lambert took umbrage at this slight and made a point of complaining about it to King, who reacted at his defensive best. King could dish it out, but he rarely could take it. "I felt a little nettled at the unreasonableness [of Lambert and the NLF's] attitude. After all, it was their place to ring me up and congratulate me," he wrote with an air of justification. "It was scarcely mine, as Prime Minister, to be the first to congratulate them, though, naturally, they were in my thoughts for a word of very sincere thanks." Lambert had had enough of King and his unappreciated role as head of the NLF. The 1940 campaign was the last he ran, and the Liberals lost, in the words of one of the party's chroniclers, "the most talented and effective national organizer it had ever had."

MACKENZIE KING HAD little time to enjoy his electoral victory. Within a few months, most of Western Europe, including France, had been seized by the Nazis. At the end of May and beginning of June 1940, an estimated 330,000 British and Canadian soldiers had to be evacuated from Dunkirk, France, and another 30,000 were either killed or captured. The fall of Britain, originally thought to be impossible, now seemed very possible indeed to King, Roosevelt and many other political and military observers. One of the few who vehemently rejected such a tragic eventuality was Winston Churchill, as of May 10 the feisty, arrogant, intimidating, indefatigable and courageous British prime minister. On June 4, in one of his most celebrated speeches of the war, he declared:

> We shall fight on the beaches, we shall fight on the landing grounds, we shall fight in the fields and in the streets, we shall fight in the hills. We shall go on to the end... we shall never surrender, and even if, which I do not for a moment believe, this island or a large part of it were subjugated and starving, then our Empire beyond the seas, armed and guarded by the British Fleet, would carry on the struggle, until, in God's good time, the new world, with all its power and might, steps forth to the rescue and the liberation of the old.

King had never been much of a Churchill fan—as late as June 1939, he had regarded him as "one of the most dangerous men I have ever known"—but he considered Churchill's speech "the greatest feature of the day." Mind you, this was not so much because of Churchill's stirring words. King believed, most assuredly incorrectly, that a sketchy telegram he had sent the British prime minister a few days earlier—conveying FDR's promise to support the fleet should it have to do the unthinkable and retreat from Europe to North America and Australia—had inspired the last part of Churchill's remarks. "I am quite sure," King wrote, "that Churchill prepared that part of this speech which was the climax in the light of what I sent him and that I shall receive an appreciative word of thanks from him." There was no "word of thanks"; instead, the next day the British prime minister warned King "not to let Americans view too complacently prospects of a British collapse, out of which they would get the British fleet and the guardianship of the British Empire, minus Great Britain." No one knew, added Churchill for good measure, what would transpire if a "pro-German administration" took over in Britain.

Churchill recognized and respected King's political skills and genuinely regarded him as an "old friend," but he initially wondered about his resolve. And he wasn't the only one. Even before France fell, there were loud insinuations by Conservative politicians and in the Conservative press that Canada, like Britain, needed a true wartime prime minister. Replacing King with Colonel Ralston was one possibility bandied about. Ralston, however, was too loyal and honourable to have gone along with such a plot (and he likely concurred with Senator Norman Lambert's assessment that he "could not hold Quebec"). King himself dismissed such nasty talk as yet another Tory conspiracy aimed at destroying him and the government. Nevertheless, he was confronted by some impossible problems.

In the darkest years of the war, King was now a master juggler precariously balancing on a high wire. He had already sworn a thousand times to French Canada that there would never be conscription and keeping the country united was paramount. Yet he wanted to support Britain in any way possible—or he believed in his heart he did, at any rate. He did not view himself, as others did, as sacrificing Canada's British connection for closer ties with the United States. Nonetheless, at every opportunity he ingratiated himself with FDR; arguing or disagreeing with the U.S. president was never an option. The consequence of managing these often incompatible goals was years of criticism and aggravating stress.

It is equally impossible to ignore his adroit political manoeuvring and the implementation of policies that contributed positively to the defeat of Nazi Germany. "At no time during the war years was there the slightest doubt," Arnold Heeney argued, "that Mackenzie King was in fact and in law the head of the government and the master of his cabinet." That assessment rings true; however, more than once the pudgy master nearly plunged off the tightrope.

Coming to Britain's aid was his first priority, but in the classic Mackenzie King way it was arrived at finally with a truckload of qualifications. Negotiations between British and Commonwealth officials were underway for the ambitious British Commonwealth Air Training Plan, which was to prepare more than 130,000 British, Canadian, Australian and New Zealander pilots and their crews for active duty. King fussed about the money the Brits were demanding Canada contribute—$370 million, admittedly an enormous amount even if it was paid out over three and a half years. (The actual cost to Canada eventually exceeded $1.5 billion dollars.) And he argued for keeping the Royal Canadian Air Force separate, but "at the disposal" of the Royal Air Force and the British government. "Canadianization" of the air force was never properly sorted out until later.

King's endless contradictions exasperated the British, who did not truly understand or appreciate Canada's historical inferiority complex. Gerald Campbell, the British high commissioner in Ottawa, depicted King as "a very complex character. On the one hand he goes far beyond the average Canadian in his mystical and idealistic talk of a crusade or holy war against the enemies of civilization and democracy. On the other hand he is the narrowest of narrow Canadian nationalists. It is this twofold outlook which makes him one moment believed that he is serving humanity by dedicating Canada to the common cause, and at the next moment consider what the common cause can be made to help Canada."

King was certainly perplexing, and no more so than during the Second World War. Had he learned of Campbell's caustic comments, his retort would have been that Canadians elected him to safeguard their interests, no matter how ambiguously he defined them. For this reason, he wanted absolutely nothing to do with joining an imperial war cabinet. Nor did he have any desire to travel to England during the conflict or to leave the country alone for a minute. "As I see it," he later declared in June 1941, "my place, until the end of the war is over, is on this continent"—a sentiment Churchill said he understood, but repeatedly urged King to reconsider. In such a situation, he might have relied more on the Canadian high commissioner in London, Vincent Massey. That was out of the question, too, since he barely tolerated Massey and rarely entrusted him with handling anything too critical.

King's shunning of close imperial linkages made sense, except to several of the more practical Canadian mandarins who toiled under him. From their perspective, they wondered why their diffident leader would send young Canadians into battle and then purposely choose to opt out of a decision-making role over British war policy. "Personally, I dislike this role of unpaid Hessians," remarked Lester Pearson in April 1940 from his office at the high commission in London. "We can send a Division to England; refuse to send it to Norway; or recall it to Canada, in theory. In fact, however, we have no such powers and, so far as policy and planning in this war are concerned, our status is little better than that of a colony." Nevertheless, the imperial war cabinet was nixed.

Churchill next suggested a Commonwealth prime ministers' conference in London for the summer of 1941 to discuss war strategy. King initially refused to participate and leave Ottawa, a decision that young Charles Ritchie, who was also on Massey's staff in London, could not understand. "Mackenzie King has been putting on the most remarkable display of panic—was invited to come to get-together of Commonwealth Prime Ministers," wrote a frustrated Ritchie in his diary. "He has cabled the longest apologies to Churchill. 1. He cannot leave the country because of the problem of unity. 2. Labour difficulties. 3. Conscription. 4. External Affairs. 5. Possibility of the United States coming into the war. 6. Needed to campaign the country. 7. Knows nothing about strategy. I do not know why he does no add that he cannot leave because he is having the front parlour repapered and is needed to choose the design." Later, King expressed deep regret at not getting away from his desk more during the war—not necessarily travelling overseas, but certainly connecting more with the Canadian public outside of Ottawa. His excuse was "the burden of work" he constantly faced, but that merely masked his insecurity, even his irrational fear, of going out among the people. It was safer, personally and psychologically, to remain hidden away at Laurier House.

The fall of France and Mussolini's decision in mid-June 1940 to enter the war as Hitler's ally did lead King, with Ernest Lapointe's full support, to take decisive action on the home front. On June 18, the cabinet introduced what became the National Resources Mobilization Act (NRMA) to register men and women for possible military training and duty, but for the defence of Canada only. This was not, both King and Lapointe stressed as loudly and strongly as they could, conscription for overseas battle. Canada had to protect itself against the possibility of invasion; it was that simple. With French Canadians placated, the bill passed easily in a parliamentary vote.

The stress of the job and the responsibility of keeping Quebec content had finally taken its toll on the sixty-three-year-old Lapointe. On July 11, he arrived at Laurier House for a meeting with King. As soon as he sat down, Lapointe told King that he did not want to worry him unduly but that he was on the verge of a nervous breakdown. "Thereupon he began to cry like a baby," as King described the uncomfortable scene. "I went over and sat beside him. Said I was not at all surprised that he felt as he did. That I knew what that kind of strain was. He then said he could not sleep and moaned in a way: Is this not too bad this had to be. He said to me in his characteristically generous way: 'I hope I am not leaving too much to you and that you will be able to get along without me.' That was when I said to him there was only one thing for him to do. That was for him to go away at once and get a complete rest." Lapointe took that advice and soon regained his stamina—sufficiently to ensure that the outspoken mayor of Montreal, Camillien Houde, was arrested and interned for urging Quebeckers to defy the NRMA—but he was never quite the same again.

The defence of Canada also meant forging even closer ties with the United States. King did not support American neutrality in the early years of the war. Yet King's high regard for FDR sometimes subconsciously translated into acquiescence in the face of a greater power. It could not have been otherwise. In photographs of the two leaders together, King always appears to be enjoying himself immensely; just being in FDR's presence was uplifting. King did urge the president to supply much-needed training aircraft and destroyers to Britain, and into 1940 he continued to act as Churchill's chief go-between. For a time, the cables between "Mr. C., Mr. R. and Mr. K." were constant.

On August 16, 1940, FDR asked King to meet with him on his private railway car at Ogdensburg, New York, to discuss joint defence, a topic that King and Canadian diplomats had already raised with officials in Washington. Both the president and the prime minister were understandably concerned about protecting North America's coastlines. The discussion in Ogdensburg led to the establishment of the Permanent Joint Board on Defence (PJBD) and Canadian-American co-operation on defence matters. No matter how resolute King was—and to his credit he did put his foot down when FDR suggested establishing U.S. bases in Canada—the Ogdensburg Agreement, as it came to be known, was a deal between two unequal partners. Roosevelt assured King that he already had a military plan in place should Canada be invaded. In short, the U.S. president was prepared to do whatever was required to ensure the safety of American citizens, even if it meant stepping on Canada's toes.

Other than some staunch imperialists such as Arthur Meighen, who "lost his breakfast" when he read the account of the agreement, most Canadian politicians and journalists praised the defence arrangement without actually appreciating its future implications of a key and inevitable shift in Canadian foreign policy. Despite any shortfalls, the agreement made sense for Canada, politically and practically. "Had anyone said two years ago," Oscar Skelton remarked to his wife, "that [Canada] going to war with Britain would bring about a military alliance with the U.S. he never would have been listened to."

King himself later commented in the House about Canada "fulfilling a manifest destiny," a poor choice of words in view of the association of that phrase with the righteous American occupation of much of the continent. Given the circumstances and the real fear that Britain could be overwhelmed by the Nazi war machine, however, these sentiments were understandable. In his diary, King found an even deeper meaning. "I am convinced that that particular agreement made where it was, at Ogdensburg, is clearly the Hand of Destiny as anything in this world could possibly be," he wrote on August 22. "No matter of chance but a converging of the streams of influence over a hundred years ago as to place and time and of life purpose in the case of Roosevelt and myself—furthering good-will in industrial and international relations."

In London, Churchill, who anxiously understood that waning British power in the near future was to give rise to U.S. global hegemony, was more naturally circumspect. When he learned about the PJBD, he somewhat unfairly criticized King for giving away too much to the Americans, without acknowledging the importance of the agreement for Canadian security. After British high commissioner Gerald Campbell sent word back to London that Churchill's remarks had hurt King's feelings, another telegram from the British prime minister was dispatched, this time thanking King "for all you have done for the common cause and especially in promoting a harmony of sentiment throughout the New World."

King was out at Kingsmere when his staff phoned him to relay Churchill's message. He was elated, typically exaggerating its significance in the whole scheme of his life. "This cable gave me more pleasure than almost anything that has happened at any time," he recorded. "It made so clear my part in bringing together the English-speaking peoples and an appreciation by Churchill of my own efforts in connection with the war; also the significance of what I have striven to do on this continent in the preservation of democracy and the wresting of Europe from Nazism, as Churchill had expressed it." King tucked the telegram in his coat pocket and on cue displayed it proudly to political associates and reporters for some time.

The next spring, King's friendship with FDR—and, it must be said, his adept negotiating skills—again produced significant results for Canada and (to a lesser extent) the U.S. during a visit King made to the president's estate at Hyde Park, New York. The prime minister and his finance officials had been rightly concerned about the economic ramifications of the American Lend-Lease Act passed by Congress in March 1941. The act allowed the British to defer payments on their purchases of essential war supplies from the United States, but at the same time it threatened anticipated large British purchases in Canada. Supported by finance and external affairs bureaucrats, King made his case to Roosevelt and key members of his cabinet, and an acceptable agreement was quickly worked out.

By the terms of the Hyde Park Declaration, the U.S. pledged to increase its defence expenditures in Canada by between $200 to $300 million during the next twelve months, alleviating the growing trade deficit and thereby enabling Canada to buy what it needed for the war from the Americans. The deal also stipulated that component parts for equipment the British obtained from the Americans under Lend-Lease could be forwarded to Canadian munitions manufacturers to be used in the finished product.

King was exceptionally pleased with the agreement, considering it integral to the "economic defence of the western hemisphere," as he declared back in Ottawa. He also readily accepted the praise of his minister of munitions and supply, C.D. Howe, who called him the "greatest negotiator the country had."

This feeling of intense satisfaction was dampened when King was not invited to the secret meeting between FDR and Churchill and their respective staffs that took place on warships off the coast of Newfoundland in early August 1941 (and which ultimately led to the so-called Atlantic Charter and the two leaders' plans for a safer world after the war was won). When King had been at Hyde Park in April, FDR had intimated to him that such a conference with Churchill was likely to take place. Nevertheless, King's first reaction was disapproval at the thought of the U.S. president and British prime minister both leaving their countries during the war when communication by cables would have sufficed.

"To me, it is the apotheosis of the craze for publicity and show," he griped in his diary. "At the bottom, it is a matter of vanity." That was a rich comment from someone as consumed by vanity as King was.

His overly sensitive nature was in full overdrive the following day in a telephone conversation about the meeting with Malcolm MacDonald, who had replaced Gerald Campbell as the British high commissioner in Ottawa that spring. King, who was about to celebrate twenty years as leader of the Liberal Party, acted like the sulking school boy who had not been invited to the class party. While he had expected a personal visit between FDR and Churchill, he told Macdonald, "I had never thought of their bringing their representatives on foreign affairs... for conferences on war plans, having Canada completely to one side—simply saying that we would be told what had been done though having no voice in the arrangements." King added that he did not wish to make trouble, but he now understood that Canada was "a nation wholly on her own vis-à-vis both Britain and the United States." In essence, as Granatstein has suggested, Canada's exclusion from the meeting "led in a straight line... to Canada's assertion of responsible independence" in its relations with its two closest partners.

Gossip (probably exaggerated) was rampant in Ottawa about King's "unbounded" outrage at being left out of the Atlantic meeting, even though he later sent a congratulatory letter to FDR for his "vision and courage in bringing it about." What rightly troubled King was that the official communiqué about the charter, a document that was to impact on Canada, was released without the British or Americans letting the Canadians know first what it said.

Few observers of the prime minister were surprised when King finally relented and travelled to London to meet with Churchill at the end of August. To get there, he flew in an airplane for the first time, an experience that he found exhilarating and spiritual. In a briefing note before King's arrival, Malcolm MacDonald had told Lord Cranborne, the secretary of state for dominion affairs, of the absolute necessity of handling the Canadian prime minister with great delicacy. "He admires Churchill enormously, but between me and you he does not like him much," wrote MacDonald. "But I am sure that Mr. Churchill will handle him extremely well... Mackenzie King needs particularly careful handling."

That message must have been relayed to Churchill. Apart from King being reassured that Canadian conscription would not be necessary, not much of consequence in terms of war policy was decided during King's visit. Churchill, however, was on his best behaviour and accorded

King the respect he craved. He praised the Canadian prime minister in public, spent private time with him at his country estate (where, to the great amusement of everyone present, the two stocky sixty-six-year-old prime ministers danced together) and requested that King make a radio speech from London. "Mackenzie King did very well over here," Churchill later reported to the governor general, the earl of Athlone.

King was indeed so overwhelmed with gratitude that he easily accepted Churchill's reasonable explanation as to why he had been omitted from the Atlantic conference with FDR. Not only would it have made it difficult with the other Commonwealth leaders if only the Canadian prime minister had been in attendance, but it was also imperative, as Churchill told him, that he meet with FDR without the presence of a third party. For good measure, Churchill told King "of the President's great affection" for him. "Said in the presence of his colleagues," added King in his diary, "there was no one who knew the President as well as I did or had the same influence upon America."

That did it. King's attitude about Churchill was forever altered. "It is really a great delight to hear him converse," he wrote after his day at Chequers, Churchill's country home. "He is quite as eloquent in conversation and speech as in broadcasting. He ranges over such a field of knowledge and interest, always having something enlightening to say. What appeals to me most in him is his instinctive, innate love of truth and right and justice, and his tremendous courage in asserting their claims."

Before King left Britain, he twice visited with the Canadian troops under the command of General Andy McNaughton. He was compelled to say a few words to the men, which he resisted. On the first day at Aldershot, southwest of London, he was exhausted and feeling depressed. The men were divided into teams and engaged in sports activities. As King crossed the grandstand area, there was applause as well as booing. The reaction probably had more to do with sports than politics, yet King in his vulnerable state of mine found it "a little disconcerting... unfair and Tory tactics." (Churchill, too, commented to the governor general that the men were "'fed up' with having no fighting and nothing but drill and discipline instead.") That might have explained why three days later, when McNaughton reminded him again that he was to speak to another large contingent of Canadian soldiers, it almost made him ill. "I felt," he wrote, "what was like a dart pass through my bowels."

The insignificant event at the grandstand was blown out of proportion after a Canadian Press story mentioned that King had been booed by the soldiers, which was then commented upon negatively in several newspapers. This was nothing less, King insisted, than an attempt "to drive" him out of public life and return power to sixty-seven-year-old Arthur Meighen. The former Conservative leader had been convinced to resign his Senate seat, take over the helm of the Conservative party once more and promote conscription. "I have no doubt in my mind that his effort will fail and that I shall come through stronger than ever, if I can hold to my faith in the purpose made from Beyond," King wrote. "It is going to be a martyr's road, a difficult path and will take all the prayer that fasting and faith can do to fulfill the purpose in view." King still considered Meighen "the meanest and most contemptible of all political adversaries, bitter, unscrupulous [and] sarcastic."

LIKE MOSES WANDERING in the desert, Mackenzie King always believed that the Lord was testing his faith and steadfastness. This was no more so than in 1940 and 1941, when a series of deaths, each more devastating than the next, pushed him to the limit. Early on June 10, 1940, Norman Rogers, forty-six years of age, left Ottawa on a small plane for an Empire and Canadian Club speaking engagement in Toronto. About eighty kilometres east of his destination, near Newcastle, Ontario, the plane crashed, killing everyone on board. King heard the tragic news from Rogers's assistant, Grant MacLachlan.

He immediately thought of his last conversation with Rogers as well as a vision he had experienced a night earlier. He had been gazing out of his library window reflecting on the Pattesons' Irish terrier, Derry, who was ill and suffering from blindness, and on the likelihood of Italy entering the war on the side of Germany (which also occurred on June 10) when he saw a black crow. "It seemed to be an ominous sign," he recalled, "significance of death." Speaking later with Power and Ralston by telephone, it was decided that King should inform Rogers's wife, Frances, and her two young sons. (Initially King wanted MacLachlan to take on this difficult job because he had to speak in the House that day about Italy's entrance into the war and thought the "emotional strain" would be too much for him.) It was as difficult a moment as King had. "Surely, he is not gone," cried Frances upon hearing the devastating news. King replied, "No, not from your side; he is here with both of us now." If Frances grasped the message of spiritual comfort, she did not acknowledge it. Ralston arrived for further support, and King was able to depart.

As he drove to the House of Commons, all he could think about was the fact that he had advised Rogers not to fly and to keep free of engagements. "I could not help feeling badly that this advice had not been heeded," he later reflected. "However there is Destiny that shapes our ends... Rogers certainly put his last ounce of energy and breath into what he was doing." In the House, members of the Opposition naturally expressed their sympathy to King. "They all knew what close friends Rogers and I were. I felt, too, how deeply moved they were." It also finally sunk in that Rogers's death was a tremendous loss both personally and politically. "I told Arnold [Heeney] that Rogers had never failed me... Over the years of our association together, he was unfailing in his loyalty and devotion. He never thought of himself, but always of others—as beautiful a character as I have known... Rogers was the best man I had in the administration, bar none, for this period of the war. No loss could possibly be greater to the ministry." Within a week, some cabinet shuffling took place as a result of the tragedy. Among other changes, Ralston had become the new minister of defence, and Ilsley took over finance.

Seven months later, on January 28, 1941, Heeney interrupted King's work day with "very bad news." Oscar Skelton, the overworked mastermind of the external affairs department for more than fifteen years, had been driving in downtown Ottawa when he suffered a fatal heart attack. Skelton had had heart problems for a long time and ignored his doctor's orders to take it easy. King's first instinct was to rush to see Isabel Skelton, but she had already been notified about the accident and had gone to the hospital where Skelton had been taken. He saw her later at Laurier House. King tried his best to comfort her, sincerely pointing out to Isabel just how much he would "miss the Doctor."

Leaving Mrs. Skelton to rest, King, accompanied by Hugh Keenleyside, drove to the hospital. They were shown into the room where Skelton's body was lying covered on a stretcher. His eyes were opened slightly and his mouth was gaping. "Poor Skelton," King repeated. He moved around the examining table to study the corpse from every angle, unnerving Keenleyside, who finally left the room. King remained alone with Skelton for another twenty minutes. Keenleyside thought that King might be praying, but, in fact, he was contemplating the meaning of death. "I got the impression at once," King noted in his diary, "of a man who had gone to the very last gasp in pursuing a great purpose. I felt, however, that the man himself has vanished; that what I was witnessing was merely the shell that had contained his life." Back in the car, King, whom Keenleyside considered as self-centred a person as he ever knew, turned and said to him, "Well, Keenleyside, which do you think would be the better under-secretary, [Norman] Robertson or [John] Reid?"

Keenleyside's assessment of King's apparent lack of grief at Skelton's passing was somewhat unfair. King's diary is filled with honest, though

admittedly unconventional, reflection and remorse at the loss of someone as valuable as Skelton. In such events, he always looked for a deeper
meaning. "The fact that he had died at the wheel of his own car in one
of the busiest intersections of the city streets [the corner of Sparks and
O'Connor] seems to me to have been a most fitting close to his life and
his life's work. He was always independent, self reliant, liked to direct
things himself, and be in the thick of the tide, to share the life of the
active and busy world and of the men on the street."

When Heeney had first given him the news, King conceded that he
was "quite unmoved" and then quickly explained himself:

> I do not mean by that I did not realize instantly all that it meant in the
> way of loss both personally and in connection with my work and to the
> country at this time but rather that it seemed par to the inevitable...
> I felt a complete control of my feelings at the moment and throughout
> the day... There is no question, however, that so far as I am personally
> concerned, it is the most serious loss thus far sustained in my public
> life and work. However, there must be a purpose and, as I see it, it
> may be meant to cause me to rely more completely on my own judg-
> ment in making decisions.

In King's world, the business of government stopped for no one. Perhaps that was what the prime minister meant by his insensitive comment to Keenleyside about Skelton's replacement. Before the day was
finished, he had chosen thirty-six-year-old Norman Robertson, another
Rhodes Scholar who served the government, and by all accounts a brilliant foreign affairs strategist. Robertson, however, was never much of a
Mackenzie King fan.

Skelton's death was upsetting enough to King, but the worst was still
to come. No one, not Lapointe or even Joan, was as spiritually linked
to King as his Irish terrier, Pat. The bond between man and dog was
special, however you examine it (and however it makes readers shake
their heads in disbelief). Hardly a day went by that did not start with
King scratching Pat's tummy, an event duly recorded in his diary. But
that was merely skimming the surface of this intense human-canine
relationship.

One evening in mid-June 1931, King was kneeling in prayer before
his mother's painting in his library. "Little Pat came from the bedroom
and licked my feet," he recorded, "dear little soul, he is almost human.
I sometimes think he is a comforter dear mother has sent to me, he
is filled with her spirit of patience, and tenderness and love." Another

enough-to-make-you-cringe entry from August 21, 1937, found the prime minister out at Kingsmere. "Little Pat this morning looked the most adorable creature I have ever seen—a little angel dog if there were such—and there are no end of them—he lay in his little basket, his head on the green pillow, one leg under his head, his little paw turned out, sleeping on his right arm with palm of hand showing just as I have seen dear Mother sleep time and again... [His] little body so perfect in its shape—an expression of love in his nature. He made me think of dear Mother and its spirit seemed to me her spirit was speaking to me through him. I knelt down and kissed his little cheek and [put] his little paw around my neck, recalling the time I knelt at the bed side of dear Mother and put her arm around my neck."

The normal life expectancy for an Irish terrier is about fourteen years, and Pat reached that age in 1938. In early February of that year, the dog was in some pain, suffering from high blood pressure and requiring morphine to quiet him. A veterinarian tended to the dog, and much to King's relief, Pat after resting, ate a ginger snap and was back to normal. That close call shook King badly. "I confess, however, to experiencing great heaviness of heart throughout the day as I contemplated the possibility of development of trouble such as he had before," he added later in his diary. "I believe he would pull through as he did... I found as I was reading that I would occasionally see words suggestive of dog or bark which was an indication of how the mind may even see what it fears when it is fatigued."

Pat's health continued to deteriorate, as did his brother Derry, owned by the Pattesons. Derry went first, in a difficult but necessary decision made by Joan and Godfroy on August 23, 1940. King accompanied Joan to the Ottawa Humane Society to bid farewell to Derry. (Death was by electrocution in those days, as he pointed out.) He kissed him three times before he said goodbye. Wrapped in a coloured coat and placed in a small coffin, Derry was buried at Kingsmere near the Abbey Ruins.

Pat lasted for another year; he turned seventeen on January 3, 1941. At the end of June, King left on a three-week trip to Vancouver and western Canada, on a military recruiting campaign. He arrived back in Ottawa on July 12 and made his way out to Kingsmere only to find Pat dying. King "cried very hard," reluctantly accepting that "it would not be long before his little spirit, his brave, his noble spirit would [take] its flight." He remembered how he had come too late for his mother, but not for Pat. The dog's final and painful hours on July 14 and 15 were, like his mother's, recorded by King almost to the minute. He instructed Heeney to postpone a cabinet war committee meeting. He also berated himself for leaving the dog for so long and for enjoying himself with a few drinks of scotch at a dinner party given by Premier Pattullo in Vancouver. The veterinarian had prescribed a few drops of whiskey to relax the dog, which King soon regarded as sinful. "I felt it was all meant for me, a part of God's plan, to have me renounce for ever the use of alcohol"—which he did not.

In the middle of the night, Pat was restless. King held him in his arms and sang to him, "Safe in the arms of Jesus," looking at his mother's picture as he serenaded the distressed animal. Less than three hours later, Pat, at long last, died. "When I turned on the light," wrote an anguished King, "he lay there like a dog that had just taken a great leap—fore feet together outward, hind legs together stretched back as far they would go, his body in a straight line—a noble creature. He had bounded in one long leap across the chasm which men call death. My little friend, the truest friend I have had—or man ever had—had gone to be with Derry and the other loved ones. I had given him messages of love to take father, mother, Bell, Max, Sir Wilfrid and Lady Laurier, Mr. and Mrs. Larkin, and the grand-parents."

It is easy to dismiss the protracted and maudlin entry about Pat's passing as "a rather repulsive record" (to quote Charles Stacey) written by a prime minister in the midst of terrible war—as yet another example of King's "double life." But this misses the point of another highly significant event in the mystifying world of Mackenzie King. There is no denying that he did ignore the war for a day or two because of an animal. There is also no denying that he should have done what any other reasonable person would have done: take the poor dog to a veterinarian and have it put down. Yet anyone who studies King's diary cannot fail to appreciate just how much Pat meant to him. The Irish terrier was truly his child in almost every sense of the word. For this reason, of all the events and tragedies of 1941—the Nazi attack on the Soviet Union in June, the Japanese assault on Pearl Harbor in December and the deaths of Skelton and Lapointe—the death of Pat, King confessed on December 31, "touched me most deeply of all." He put the loss into perspective like this: "With Our years together, and particularly our months in the early spring and summer, have been a true spiritual pilgrimage. That little dog has taught me how to live, and also how to look forward, without concern, to the arms that will be around me when I, too, pass away. We shall all be together in the Beyond. Of that I am perfectly sure."

Later on July 15, he wrapped Pat in a white towel and left him outside. He had to leave Kingsmere and attend the delayed war committee meeting. "It was what Pat would wish me to do," he reasoned. He returned later in the evening and an intimate service was held. Pat was laid to rest close to where Derry was buried. Joan brought two boxes of flowers to place at Pat's head and feet. In attendance were Godfroy and members of King's staff who had cared for and loved the dog as much as King had. Before Pat was lowered into the ground, King, as he described it, "kissed his little face for a last time." The prime minister then read verses from the Bible. "Lift up your head, O ye gates, even lift them up ye everlasting doors..." He gazed upwards. "The sky was a little gray, with fleecy clouds, streaked with whole sections like veils, and with flashes of pink over parts of it—the sun setting quietly in the west beyond the trees." King finished the service with a reading of three special poems, including most notably "The Little Dog Angel" by Norah Holland, a Toronto journalist. Four years later, he was still dreaming of meeting Pat in heaven.

There is somewhat of a happy ending to this dog tale. By 1941, Jack Patteson and his wife, Molly, who were living in London, had sent their two youngest daughters, Ann and Mary, along with a governess and another Irish terrier, to live with Joan and Godfroy in Ottawa. This little dog, which King soon became fond of, was also called Pat.

When he returned from his trip to London that year, he noted that "the little dog Pat seems to have missed me greatly while I was away. The children and [Joan] and [Godfroy] all say he has been a different dog since I came back. It is very touching this affection of animals." King could not stop thinking of his own departed Pat; he saw his image in the clouds, in his shaving cream lather and his nightly visions. He stood by the dog's grave frequently and went on lonely walks along the same paths at Kingsmere that he and Pat used to journey on. At some point in October, the Pattesons and their granddaughters sent their Pat to live with King. On October 14, Pat II had "his first ride to the East Block." Back at Laurier House, the dog was "very frightened of the elevator but seems to be taking kindly to the house." The new terrier never quite replaced the original Pat, but as he began greeting King each morning with a wag of his tail, King grew fond of him. "Before going to bed, I had a little talk with Pat, in his basket," King wrote quite seriously on Christmas Eve in 1944. "We spoke together of the Christ-child and the animals in the crib." For the duration of the war, Pat II comforted King as much as any dog could.

ALL THROUGH THE fall of 1941, Ernest Lapointe continued to speak out against conscription, urging French and English Canadians to join up and support the war effort. His health was not good, however; his diabetes and a case of jaundice caused him pain. In early November, he was admitted into hospital in Montreal, where doctors discovered that he was suffering from pancreatic cancer. In 1941, there were no medical solutions or miracle drugs. King could not even contemplate losing Lapointe. "With Lapointe [in hospital] I have no one with whom to counsel as I need to at this time. No one to talk over Quebec situation which is very bad," he wrote on November 8. He felt so sorry for himself, so alone, strained and fatigued, that, as he wrote, "I felt I wanted to give up public life and avoid the break in my health which I feared might come." (He found some much-needed solace that evening with Joan at the "little table," at which they "had a good conversation.")

King finally phoned Lapointe in his hospital room on November 11. After a pleasant enough conversation, King realized just how serious Lapointe's health problems were. "He is the most loyal and truest of colleagues and friends. None has ever been more so to myself... Should he not recover strength enough to help me... shall be in a desperate plight." Three days later, King got word late at night that Lapointe's health had deteriorated. Alarmed that "no loss could be greater" to him, the country or the British Empire, he travelled to Ottawa the next afternoon. King was able to set Lapointe's mind at ease; at long last King said he would relinquish his position as president of the Privy Council and give it to Lapointe, moving him out of the far more stressful justice portfolio. The following day, King wrote what turned out to be a final letter to his lieutenant. "My chief joy in public life has been in sharing everything with you, just as my sense of security lies in having you at my side."

King was back in Montreal with Lapointe on November 19. Sitting by his bed, he told him with conviction that "no man ever had a truer friend. But for him I would never have been P.M. nor would I have been able to hold the office as I had held it all the years that we had had been associated together, in thought and work alike. That I was grateful to him from the bottom of my heart." King then leaned over and kissed him on his left cheek. Lapointe picked up his head and kissed King, and they kissed one more time before King departed.

A day later, King returned. In an attempt to make Lapointe peaceful, he spoke with him about the hereafter and his "exceptional experiences" in communicating with the departed ones in the Great Beyond. "Ernest," King said gently to him, "we will see each other again." Lapointe replied, "There is nothing truer than that." Seeing that Lapointe was overcome with emotion, King held his hand.

Ernest Lapointe died six days later. King arranged to send Lapointe off to the Great Beyond in the style he deserved with a state funeral in Quebec City and then burial in Rivière-du-Loup, close to the small farm where he had been born. The streets of Quebec City were lined with people on November 29. It was, thought King, "an amazing tribute," as he walked beside the horse-drawn hearse. "Truly Canada had become a nation in the fullest sense." He did not miss in Cardinal Villeneuve's eulogy a pointed reference to his friendship with Lapointe. Only John A. Macdonald and Wilfrid Laurier, King reckoned, had as grand a funeral as his French lieutenant had on that day.

Still, King felt lost and lonely. It finally hit him hard that he no longer had Lapointe to guide him through the murky waters of Quebec politics—and at the exact moment when the conscription issue was about to explode. Arnold Heeney travelled with King on the train to the cemetery in Rivière-du-Loup, and there was only one topic that he wished to speak of. Who would replace Lapointe? Premier Adelard Godbout was the obvious choice, but he had no desire to surrender the leadership of the province. Chubby Power had too many personal flaws, and no other French-Canadian ministers could fill Lapointe's large shoes. Only one name came up: Louis St. Laurent, a successful Quebec City corporate lawyer whom King barely knew. Heeney had met St. Laurent and assured King that he was a person "of cultivation and charm, and, most important of all character."

After receiving support from Power, Arthur Cardin and Senator Raoul Dandurand, King telephoned St. Laurent and requested he meet with him in Ottawa. That encounter took place in King's Laurier House library office two days before the disaster at Pearl Harbor. The dignified Quebec lawyer in a fine suit and neatly trimmed moustache immediately impressed him. King appealed to his sense of patriotic duty. The country needed a strong French Canadian to succeed the mantle of Laurier and Lapointe, and in the prime minister's opinion St. Laurent was that man. Would he join the cabinet as minister of justice? A House of Commons seat in a Quebec City constituency was waiting for him. St. Laurent said he realized that "in these times... every man should be prepared to do what he could to help the country." He was not certain, however, if he was the right choice. In the end, he told King he would need a few days to think about it further and consult with his clients, friends and Cardinal Villeneuve, whose political influence and support he considered essential.

On Tuesday, December 9, a day after King signed Canada's declaration of war on Japan, St. Laurent phoned the prime minister and told him he would be in Ottawa the next day to take his new office. King was delighted, as were members of the cabinet. Ottawa journalists, most of whom knew little about St. Laurent, were more skeptical. "If ever there was a case of a job seeking the man, this is it," a *Montreal Gazette* columnist opined.

From King's standpoint, he had not experienced a greater day in a long time than he did on February 9, 1942. On that day, he gained a new French lieutenant and rid himself of a detested enemy. Arthur Meighen's attempted comeback as Tory leader ended abruptly after he was stunningly defeated in a by-election in a largely working-class Toronto riding. His conqueror and only opponent was Joseph Noseworthy of the CCF, who had received $1,000 in campaign assistance from Norman Lambert. "I felt tonight," King wrote, "that public life in Canada had been cleansed."

Better still, Louis St. Laurent won his contest in Quebec easily over the nationalist candidate, Paul Bouchard. Of his new Quebec colleague, King wrote after watching him speak to the cabinet for the first time: "He is a great addition of strength to the Administration. I was immensely struck with his manner in getting in touch with different members. He gives a fine impression of strength and integrity." King had no idea just how much he was about to draw on that strength and integrity as he confronted the toughest dilemma of the war years: the ominous clash over conscription.

William Lyon Mackenzie King, age two. Library and Archives Canada (LAC) C-007332

LEFT Back row, left to right: The Mackenzie King siblings: Max (sitting), Jennie, William Lyon (age six) and Bella. LAC C-002854

ABOVE Left to right: Bella King, Mrs. Isabella King, William Lyon Mackenzie King, John King, 1880s. LAC C-007348

W.L.M.King. G6347
University College
Toronto.

September 6th 1893.

This diary is to contain a very brief sketch of the events, actions, feelings and thoughts of my daily life. It must above all be a true and faithful account. The chief object of my keeping this diary is that may be ashamed to let even one day have nothing worthy of its showing, and it is hoped that through its pages the reader may be able to trace how the author has sought to improve his time. Another object must here be mentioned and is this, The writer hopes that in future days — be they far or near — he may find great pleasure both for himself and friends in the remembrance of events recorded, surrounded, as they must be, by many an unwritten association. If either aim is reached this present diary will not have been in vain.
W.L.M.King.

TOP LEFT William Lyon Mackenzie King (right) as a student at the University of Toronto, 1890s. LAC C-055546

BOTTOM LEFT A page from King's diary, September 6, 1893. © Government of Canada. Reproduced with the permission of the Minister of Public Works and Government Services Canada (2010). Source: Library and Archives Canada.

ABOVE William Lyon Mackenzie King (left) with Max and John King, August 1899. LAC C-055520

ABOVE Jennie King, December 1906. LAC PA-139602

TOP RIGHT The woman King might have married: Mathilde Grossert Barchet, August 1907. LAC C-079189

BOTTOM RIGHT King's lifelong correspondent and friend, Violet Markham, 1912. LAC C-014179

TOP LEFT Left to right: Marjorie Herridge, Bert Harper, Isabel Grace
Mackenzie King and William Lyon Mackenzie King, Kingsmere, August 1901.
LAC C-014180

BOTTOM LEFT Clockwise from top left: William Lyon Mackenzie King,
Henry A. Burbidge, Norman Duncan and Wilfred Campbell, 1905. LAC C-002858

ABOVE William Lyon Mackenzie King, minister of labour, and John and
Isabel King, September 1911. LAC C-046521

TOP FAR LEFT William Lyon Mackenzie King (left) and Wilfrid Laurier,
August 1912. LAC C-018586

TOP NEAR LEFT Mackenzie King's friend Princess Julia Cantacuzène, 1924.
Library of Congress

BOTTOM LEFT Archie Dennison, William Lyon Mackenzie King (middle)
and John D. Rockefeller, Jr., Frederick, Colorado, 1915. LAC C-029350

ABOVE William Lyon Mackenzie King and John D. Rockefeller, Jr., Washington
D.C., 1915. LAC C-025281

ABOVE Lord and Lady Byng, 1922. LAC C-033995

TOP RIGHT Mackenzie King on the verandah at Kingswood Cottage, 1920s.
LAC PA-124436

BOTTOM RIGHT Left to right: Earnest Lapointe, Mackenzie King, Vincent
Massey and Peter Larkin, London, 1926. LAC C-001690

TOP FAR LEFT Giorgia de Cousandier in Montemurro, Italy, 1948. Centro di documentazione "Rocco Scotellaro e la Basilicata del secondo dopoguerra," Tricario, Italy

TOP NEAR LEFT Joan Patteson at Kingswood Cottage, 1930s. LAC PA-126153

BOTTOM LEFT Kingsmere Ruins, 1935. LAC C-086777

ABOVE Mackenzie King with his dog Pat 1, the spiritualist Etta Wriedt, Joan Patteson and Derry, Kingsmere, 1930s. LAC C-079191

ABOVE William Lyon Mackenzie King and Pat I at Laurier House, August 1939.
LAC C-087858

TOP RIGHT William Lyon Mackenzie King and Pat I, November 1940, Laurier House Library. LAC PA-165817

BOTTOM RIGHT William Lyon Mackenzie King and his nemesis, Ontario premier Mitch Hepburn, Toronto, 1934. LAC C-087863

TOP LEFT William Lyon Mackenzie King and Franklin D. Roosevelt, Kingston, Ontario, August 1938. LAC PA-052499

BOTTOM LEFT Left to right: Franklin D. Roosevelt, William Lyon Mackenzie King and Henry Stimson, 1940. Library of Congress

ABOVE William Lyon Mackenzie King and Ernest Lapointe greeting King George VI and Queen Elizabeth at the gangway of CPS *Empress of Australia*, Wolfe's Cove, Quebec City, May 1939. LAC C-035115

TOP LEFT Mackenzie King with members of the Liberal Party who accompanied him on a visit to Great Britain in August 1941. Front, left to right: Norman Robertson, Mackenzie King, Georges Vanier. Back, left to right: John Nicol, J.W. Pickersgill, Walter Turnbull, Edouard Handy. National Film Board of Canada. Photothèque/LAC PA -112769

BOTTOM LEFT Left to right: Rt. Hon. Mackenzie King, President Franklin D. Roosevelt and Rt. Hon. Winston Churchill at the Quebec Conference, August 1943. LAC C -014168

ABOVE William Lyon Mackenzie King and his cabinet, June 1945. Front, left to right: Louis St. Laurent, J.A. MacKinnon, C.D. Howe, Ian Mackenzie, Mackenzie King, J.L. Ilsley, J.G. Gardiner, C.W.G. Gibson, Humphrey Mitchell. Back, left to right: J.J. McCann, Paul Martin, Joseph Jean, J.A. Glen, Brooke Claxton, Alphonse Fournier, Ernest Bertrand, A.G.L. McNaughton, Lionel Chevrier, D.C. Abbott, D.L. MacLaren. LAC C-026988

ABOVE Mackenzie King sitting in front of the J.W.L. Forster painting of his mother, Isabel King, in the library at Laurier House, 1945. Gordon H. Coster/ LAC C-075053

RIGHT The Last Photograph, Kingsmere, July 18, 1950: Mackenzie King seated in a chair presented to him at Tyree, Scotland, in 1937. LAC PA-129854

ABOVE William Lyon Mackenzie King's funeral, July 1950. Queen's University
Archives, George Lilley fonds v25.5 13-5

11

Conscription if Necessary

*Speaking to my colleagues yesterday, I said I was prepared,
in order to reconcile views to assume myself a great trust;
that I thought the people had confidence in me, and that
I was prepared to ask them to trust me to see that conscription for overseas was not resorted to unless absolutely
necessary.*—THE DIARY OF WILLIAM LYON MACKENZIE
KING, *May 9, 1942*

DURING HIS TRIP out west in the early summer of 1941,
King was able to find a few minutes to meet with *Winnipeg Free Press*
editor John Dafoe and his associate George Ferguson. "The Chief had
a few minutes with Billy King," Ferguson reported back to Grant Dexter in Ottawa on July 12. "Told him bluntly that if K. were opposed to
conscription on principle, the FP would not follow him. K. assured the
Chief that this was not so, that if we needed conscription to meet the
needs, we'd tackle it." The prime minister also raised the possibility of
an election or referendum on the issue. Dafoe was not keen on either,
but he was not about to abandon the Liberals, because the alternative
was far worse.

For months, Dafoe had been increasingly angered by the spiteful personal attacks on King in the eastern Canadian Conservative press. "No
occasion passes on which the Canadian prime minister achieves some
outstanding success... that the little group of mean-minded and big-
oted men whose chief spiritual solace is to be found in the columns of
the *Toronto Telegram* and the *Montreal Gazette* begin to cry 'King must
go!,'" he wrote in a scathing editorial of May 1941 entitled "The Hymns
of Hate Once More." He soon began drawing comparisons between the

Tory journalists and Nazi propagandists and their use of the "big lie": "You say it long and loud enough and the public will buy it."

Dafoe had never given King unconditional support, but, like the prime minister himself, he came to the conclusion that in conscription lay the seeds of a national crisis. "No doubt [King] has made mistakes," Dafoe conceded in a letter of February 1942 to the decidedly anti-King journalist John Stevenson. "His methods of doing business may exasperate people and he may at times have confused national and party interests. In short, he is a miserable sinner. But aren't we all?"

The "miserable sinner" had much more juggling to do in the fall of 1941 and the first six months of 1942. The King government's management of the war economy (guided as it was by the autocratic minister of munitions and supply, C.D. Howe) was adequate. With the approval of labour and business, wage and price controls were instituted in October and rampant inflation was indeed checked. The mobilization of manpower, however, remained challenging. In Winnipeg in July 1941, the *Free Press*'s George Ferguson, who was active in military recruitment, told King that the appeal to volunteer for the armed services was "hollow." The prime minister "listened attentively," Ferguson related to Dexter, "but [he] did not or would not believe it... Said, if everyone were like me—out pushing for recruiting all would be well. Said there were no shortages in the navy and air force, that the dragginess in army was only because of inactivity in that line."

Ferguson might have questioned King's judgment, yet he was substantially right. In 1941 and into 1942, there was no military manpower shortage (and no casualties before the ill-fated raid on Dieppe in mid-August 1942, during which nearly two thousand Canadian soldiers were killed). This was confirmed when King met with General McNaughton, who was on leave in Ottawa in early February 1942. In a long conversation, the general expressed agreement with the prime minister about keeping the country united and, more importantly, that conscription was not necessary. At a dinner at Laurier House, McNaughton spelled it out like this, according to King's recollection: "As a matter of fact, if your Government had sent word to me overseas that they were contemplating conscription, I would have asked, for God's sake, not to do anything of the kind."

King was clearly relieved. Still, the general's assessment did not halt the conscription train wreck, which for months had been causing havoc at cabinet meetings, in news rooms and across the country. The conscription crisis of 1942 that never was, instead, as Jack Pickersgill put it, "was entirely political and psychological." To fend off the cries for immediate conscription, Mackenzie King had to juggle like he had never juggled before. He was attacked on all sides. There was his hated arch-enemy Arthur Meighen and the Conservative opposition and press; some English-speaking Canadians who believed that the Liberals had sold out Canadian soldiers on the verge of battle as well as England when it needed Canada's help the most; and above all, there was criticism from his minister of defence, James Ralston, and his main allies, finance minister James Ilsley and Angus Macdonald, the former premier of Nova Scotia and as of 1940, the minister of national defence for naval services.

Macdonald was proud to serve his country in wartime. He had arrived in Ottawa fully supportive of Mackenzie King and praising his leadership and stellar character. But in a matter of a few years that view was to change dramatically. Macdonald soon found King exasperating and he grew to detest him and the prime minister's penchant for equivocation. Throughout the conscription debate, he sided with Ralston, although never unreservedly.

Intellectually, Colonel Ralston understood that instituting conscription would pit English and French Canadians against each other and tear the country apart as it had in 1917. Yet he was a military man through and through rather than a politician. In the fall of 1941, with the Nazis in control of Western Europe and quickly advancing east towards Moscow, Ralston figured it would only be a matter of time before more Canadian troops would be required. He had little faith in volunteer recruitment—during May to July 1941 more than 34,000 men had signed up and passed the military's medical examination—and believed strongly that he owed his men his undying support. His plans also included expanding the number of divisions Canada maintained from four to five, a strategy that took on even greater urgency following the declaration of war against Japan and the loss of Hong Kong, where Canadians were stationed. Still, for a country of 11.3 million, that expansion was a huge undertaking.

King was not against conscription on principle. He simply approached it from a political and practical perspective. The army, air force and navy were not in serious need of more men, at least not yet. The labour requirements of the country's war industries needed to be met. And most importantly of all, King's government had made a sacred promise to Canadians, and Quebeckers specifically, that there would be no conscription for compulsory overseas duty. He knew full well that French Canadians—and that likely included most of his Quebec cabinet ministers—would never accept conscription for overseas service under any circumstances. How to manoeuvre through this dangerous minefield without losing Ralston and alienating Quebec became King's great quest. He did it by doing what he did best: taking it very slowly and occasionally threatening to resign, though never wholeheartedly. About one of many tense cabinet meetings in December, King stated that "we who were sitting at the table owed it to those from whom we got power, namely the people's representatives in Parliament, not to adopt any policy which involved conscription, until they, themselves, had approved of doing so." It was a logical approach, yet much to King's consternation, the cabinet remained divided. "Feels that some of our colleagues do not at all comprehend the Canadian situation, and that their action will destroy both the government and the party," King noted. "I am fearful of both consequences."

The solution, at least temporarily, was to hold a plebiscite to ask Canadians whether they would free the government from its no overseas conscription promise. (Or, as John Dafoe described the problem to Tom Crerar, "how to void the pledge with a minimum of heart-burning and bad feeling.") This had been gelling in King's brain for some time. He had no desire for Canadians to vote yes or no on conscription, but only to untie the government's hands. This was the reason he opted for a non-binding "plebiscite" instead of a more definitive "referendum." On December 17, King got some political help from the Manitoba legislature, which passed a pro-conscription resolution. He was not happy with the vote; however, it now gave him an excuse to pursue a plebiscite.

By no means was King advocating conscription; on the contrary, he told his cabinet that he saw no reason "why we should not hold to our policy of regarding conscription as unnecessary and inadvisable." Yet he also did not want "the government's hands tied" in case more drastic action was warranted. And he wanted to quash the charges being raised by Meighen and the Toronto and Montreal press that he was not doing enough for the war effort. Ralston and the other cabinet conscriptionists were prepared to go along with King's plebiscite, as were the Quebec ministers, with the exception of Arthur Cardin and Senator Dandurand, who feared the worst. King convinced them to wait until the new year before they did anything drastic.

The new year came more quickly than King would have liked. On January 5, 1942, the cabinet approved Ralston's demands for the fifth overseas division. Almost everyone supported this, provided it did not necessitate conscription. King rightly believed that Ralston would have resigned, followed by Macdonald, if the proposal had not passed. For the next few weeks, King wrestled with himself over announcing the plebiscite and framing the question that would be asked. He finally decided that he would address the issue in the Speech from the Throne. By then, he had won over Cardin, who had no desire to create an untenable situation that might allow the Tories to seize power.

King confirmed the plebiscite on January 22; the government would ask the people to "release" it from "any obligation arising out of any past commitments restricting methods of raising men for military service." Once the House met, King set out his case in favour of the plebiscite, but again declared that he personally did not favour conscription for service beyond Canada's borders. "What I was aiming at above all else," he wrote after he had drafted his speech, "was to keep faith with the people." The date of the plebiscite was set for April 27.

The turmoil inside the cabinet meeting room never really stopped. Ralston and Macdonald continued to press King for an explanation on how the plebiscite vote was to be interpreted. Did a vote in favour of releasing the government from the pledge mean that conscription could be automatically instituted? Or, did it mean that a parliamentary debate and vote was required? These questions were to cause yet another cabinet crisis in the near future. For the time being, King was evasive in his reply.

The "no" side immediately coalesced. Quebec nationalists led by André Laurendeau, Paul Gouin, *Le Devoir* editor Georges Pelletier and Liberal MP Maxime Raymond denounced the government's action and established la Ligue pour la Défense du Canada, which urged a no vote. Besides rallies and speeches, the group's most clever and effective tactic was obtaining tapes of Lapointe's no conscription promises and replaying them endlessly on the radio. Raymond also hit the nail on the head when he pointed out in his address in the House that Lapointe's promises in 1939 really had been made to Quebec, so why was the government asking the rest of the country to release it from its pledge essentially made to French Canadians? Raymond deserved a response, but he never got one. King merely dismissed him and the few other Quebec Liberal-nationalists who voted against the throne speech as "very narrow" in outlook.

The "yes" campaign was much less passionate and more disorganized. King was furious with the lack of effort put forth by Norman McLarty, the cabinet minister responsible for the yes campaign publicity. "All we had was evidence of what the opponents were doing," King complained after a fruitless meeting with his cabinet on April 1. "McLarty, I thought, on the wrong lines, in talking of establishing a national committee, instead of getting men to organize non-partisan committees and taking efficient steps toward that end. He seems incapable of getting men around him to work in an efficient way... I felt it useless to try and take on a load of the kind myself and that the matter would just have to go as best it could."

That it did. A series of Gallup polls, the first in Canada—the nascent American pollster George Gallup had set up his Canadian operation, the Canadian Institute of Public Opinion, in November 1941 with support from the *Toronto Star* and other newspapers—predicted King's worst nightmare: English Canada would vote overwhelmingly in favour of releasing the government from its pledge; French Canada would not. On April 27, the polling results proved all too true. Nationally, 63 per cent of Canadians voted yes, and 37 per cent voted no (Gallup's figures had been 68 to 32 in favour). But where it mattered, in Quebec, the figures were reversed: only 27.5 per cent voted yes, and the vast majority, 72.5 per cent, voted no.

Those figures depressed King, but he tried to put a positive spin on it. The government, he insisted, had been given permission to repeal Section 3 of the National Resources Mobilization Act, making it possible—but almost certainly not necessary—to conscript men for overseas service. Ralston interpreted the result of the vote quite differently, believing that conscription must be implemented; so did St. Laurent. Angus Macdonald, curiously enough, saw the situation the same as King: that the people had given the government the "moral right" to impose conscription, yet it did not have to do so immediately.

THE NEXT FEW months, as Bill 80—as the repeal of Section 3 of the NRMA was called—was debated in the House, was as trying a time for King as there was throughout his political career. He knew that surviving the public and private discussion was going to require a "careful strategy." That was an understatement. The critical issue now became whether conscription could be instituted by order-in-council or whether Parliament ultimately had to decide. King believed strongly in the latter option and insinuated that he might resign if he was forced to adopt a contrary policy; Ralston and Macdonald argued fanatically for the former. From the other side, Cardin's principles could take no more. On May 9, he decided that he could not remain in the cabinet if Section 3 was repealed and submitted his resignation to King. It was accepted two days later. This was, King wrote glumly, "the greatest crisis that faced not only the party but the country... It will now be said that I am responsible for the whole situation."

Cabinet meetings during the rest of May and into June were as unpleasant for King as ever. Following a speech to the caucus on May 12, Angus Macdonald complained to Grant Dexter that King was incapable of making a "direct statement." According to Macdonald, "weasel words popped out of every sentence." Be that as it may, King stood his ground on referring overseas conscription to a parliamentary vote. On June 10, he spoke in the House on Bill 80 for more than two hours; all that is remembered from that oratorical homily is King's celebrated dictum of alleged uncertainty. His policy, he declared, was "not necessarily conscription, but conscription if necessary." King had borrowed the line right from a *Toronto Star* editorial of April 28, which is generally ignored.

More importantly, he deliberately left open to interpretation whether he would send conscription back to the House for a vote before he imposed it. At a cabinet meeting on June 12, King "wobbled very badly," in Dexter's words, and he heard it from Crerar, Gardiner, Power, Ralston and Ilsley, among the many ministers who were regularly feeding him information. "He began with Grandpop William Lyon [Mackenzie] and his great struggle for the supremacy of parliament ... This shook the boys down plenty." As soon as the meeting concluded, Ralston told King that he was going to resign. King was not surprised, but convinced Ralston to think further about taking such a drastic step. Macdonald, who was more miserable than Ralston, told Dexter that King was nothing but "a twister and a wobbler who does not mean ever to have conscription."

Another two and a half weeks passed before Bill 80 was finally voted on. King had declared to his cabinet that he considered it a vote of confidence, and if the bill did not pass he was prepared to ask the governor general for a dissolution or suggest that he ask someone else to form a government. Assuming the bill did pass, his policy on conscription would be to permit a debate in Parliament for a maximum of two days. That evening, before King was to speak in the House and wind up discussion on Bill 80, Ralston personally delivered his letter of resignation. He wrote to King that in light of the clear vote in the plebiscite, at least as he saw it, he could not support further discussion and another potential vote on conscription in Parliament.

Since the Conservatives had determined they could not vote against a bill that advanced the cause of conscription, Bill 80 passed easily, with almost every Liberal, except a few disgruntled Quebec MPs, supporting the government. The next morning, after conferring with St. Laurent, King met with Ralston and informed him that he could not accept his resignation, that the country's entire war effort hinged on unity and such a split between the prime minister and the minister of defence could have devastating repercussions. Somehow over the next few days, King's logic, mixed with threats of his own resignation and its calamitous consequences, convinced Ralston to stay in the cabinet. King left his future options open, as he spelled out to Ralston in a "pretty stiff" letter (as Chubby Power called it). More than one minister remarked, as King did, how strained and pale Ralston looked. In Power's opinion, on July 9 Ralston "appeared to be in a state of great perturbation and more incoherent that I have ever seen him before." King did not lose much sleep during this ordeal, according to his regular diary entries about his nocturnal activities. For the time being, Ralston's signed resignation letter remained in his desk drawer.

Grant Dexter, who was working overtime during the early summer of 1942, could only marvel at King's skilful handling of Ralston and the entire so-called conscription crisis of 1942. He might as well have been a member of the cabinet, given the breadth of his inside knowledge. "T.A. [Crerar] believes that King has never had but one point in his mind," Dexter wrote to Dafoe on July 17. "He is determined to bring in conscription without losing Quebec—without forfeiting national unity. He thinks this has become the great ambition of King's life." Four days later, Dexter sat down with St. Laurent. "He could see no alternative to King. If King went out, Quebec would be lost irretrievably. And provided that the need for conscription could be reasonably proved, Quebec would accept it under King." Mackenzie King had become the linchpin holding Canada together.

IN THE SAME moment as he was doing his damnedest to keep the country united, King made another fateful decision. Compared with the amount of time he spent deliberating over conscription, this one was made with seemingly little reflection or discussion. Its consequences, however, would impact negatively on his legacy and still be hotly debated decades later. The conscription decision was based on diligent politicking and an intelligent assessment of the situation in Quebec, while this one was a reaction to a paranoid hysteria steeped in persistent and intransigent Edwardian-era attitudes about race.

The King government's abysmal treatment of Japanese Canadians occurred during the height of the Second World War, when the possibility of a Japanese attack on the west coast seemed imminent. Nevertheless, the forced evacuation from British Columbia of more than twenty thousand Japanese, nearly 75 per cent of whom were either naturalized or Canadian citizens, is a shameful episode in the country's history, and one that rests squarely on the slouched shoulders of Mackenzie King. Moreover, it was unnecessary.

On Sunday, December 7, 1941, "a date which will live in infamy," as President Roosevelt declared, King had been notified that day of rumours about a possible Japanese attack on the United States. He spent the afternoon at Kingsmere walking the trails and resting. At four-thirty, Norman Robertson telephoned to tell him the rumours were true: Japan had brazenly bombed Pearl Harbor in Hawaii and Manila in the Philippines. King told Robertson to assemble the cabinet for a meeting three hours later. "Hands of the clock were at 5:25 as we started out from Kingsmere," he later thought it relevant to include in his diary, "and the hall clock was straight 6:00 pm, all but a minute, when I looked at it on reaching [Laurier House]."

Given Japan's declaration of war on the United States and the British Empire, Canada was also at war with Japan, and the opposition leaders agreed with King that no additional parliamentary vote on whether to enter the war was required. By the time the cabinet convened, Norman Robertson, who was cool and collected throughout this episode, had the order-in-council with Canada's war declaration ready for King to sign, thereby making it official. The mood in the cabinet room was understandably sombre, and concern was raised about the Japanese living in B.C., though there was more apprehension about the possibility of anti-Japanese demonstrations than about Japanese-Canadian loyalty. Nonetheless, 1,200 Japanese fishing boats were impounded by the Royal Canadian Navy, an act that marked the gradual purging of the civil rights of Japanese Canadians. King went on the radio the next day to explain the situation—what he regarded privately as the "most crucial moment in all world history"—and call for calm.

Little had changed in British Columbia or in King's attitudes since he had, as the young deputy minister of labour, investigated the anti-Asiatic riot in Vancouver in 1907. B.C.'s white population was more alarmed than ever about the "peaceful penetration" of an increasing number of Japanese immigrants, newcomers who, according the white majority, would never assimilate and become "true" Canadians. The perceived unfair economic competition from Japanese fishermen was also deeply resented. (In fact, only 16 per cent of the Japanese were fishermen.) Plus the sheer concentration of the Japanese on the west coast—96 per cent of the Japanese living in Canada in 1941 resided in B.C.—fed the fears of invasion and the "yellow peril." Finally, the rise of Japan itself as a militaristic empire during the twenties and thirties, confirmed forever by its vicious sneak attack on the Americans, hardened prejudices and caused panic.

By the racism standards of the day, King might be fairly called "moderate" in his opinions about Asians, as well as Jews and Blacks. He was never hysterical or crude in the way that the key B.C. spokesman in the cabinet, Ian Mackenzie, the hard-drinking minister of pensions and health, was. Mackenzie, who was first elected to the House of Commons in 1930 and had served in the B.C. legislature before that, had, with one exception, "endorsed every anti-Asian proposal raised in the Legislative Assembly, in Parliament and in Cabinet," lawyer and historian Ann Sunahara points out. (In 1929, Mackenzie finally supported giving the right to vote to Asian veterans of the First World War.) He freely distributed what could only be labelled as hate literature and in 1944 declared at his nomination meeting: "Let our slogan be for British Columbia: 'No Japs from the Rockies to the seas.'"

King was different. At least that was the impression of Thomas Shoyama, the editor of the *New Canadian*, a Japanese-Canadian newspaper in Vancouver. "The Prime Minister," Shoyama wrote a month after the Pearl Harbor attack, "is not the type of man to be swayed by prejudice or irrational emotion." Still, King shared in the widely accepted vision of Canada as a white man's country. On August 6, 1945, when he learned that the U.S. had dropped an atomic bomb on Hiroshima, he wrote in his diary, "It is fortunate that the use of the bomb should have been upon the Japanese rather than upon the white races of Europe." That comment might not have been entirely representative of his prejudices, but it was probably more common a sentiment among non-Japanese Canadians than we would care to admit today.

On the specific question of the treatment of the Japanese in British Columbia in 1942, King was governed by three main considerations, each open to question and criticism. First, he wanted to guarantee Liberal Party support in the next election from the vast majority of British Columbians, who were decidedly not Japanese. Second was his strong desire, both sincere and practical, to ensure that there were no pogroms on the coast, and in early 1942 that threat seemed very real. And third, in light of accurate reports that the Japanese navy were in the vicinity of the Aleutian Islands in the northern Pacific (part of which the Japanese occupied by June 1942), King needed to counter any possible espionage in support of the Japanese enemy.

On the last point he wavered. One week in February 1942, he gave a grim warning to the cabinet about an impending Japanese assault on the west coast, and the next he corrected journalist Bruce Hutchison, who made the very same argument. "Hutchison seemed to think the Japanese invasion would come via the Aleutian Islands, Alaska and Canada," King wrote. "I told him that was my own view, if the Japanese succeeded in getting the whole of the Far East, but I thought that the immediate developments would be in India."

This much is clear: no matter how many reasonable recommendations were given to him by such external affairs advisors as Norman Robertson and Hugh Keenleyside—Keenleyside particularly adopted a liberal and humanitarian attitude to this issue—the chiefs of staff and the RCMP, all of whom counselled against a massive evacuation of Japanese Canadians, King listened primarily to the reactionary voice and threats of Ian Mackenzie and B.C.'s hysterical and racist politicians and press. "Public prejudice is so strong in B.C. that it is going to be difficult to control the situation," he admitted. Action was required, King believed, and action was taken. Almost as soon as the U.S. decided to evacuate its Japanese citizens from California, Canada naturally followed suit.

During the week of February 24, approximately 23,500 Japanese Canadians lost their civil rights and were ordered to immediately evacuate to areas east and away from the supposedly vulnerable coastline. They left behind their homes, property and businesses. As Japanese-Canadian writer Ken Adachi, who was twelve years old in 1941, put it: "born in Canada, brought up on big-band jazz, Fred Astaire and the novels of Rider Haggard, I had perceived myself to be as Canadian as the beaver. I hated rice. I had committed no crime. I was never charged, tried or convicted of anything. Yet I was fingerprinted and interned." That was not technically correct, even if it felt like it. Japanese Canadians were never "formally imprisoned," according to historian Patricia Roy, but the majority wound up in isolated federal government camps, or as workers on Prairie farms. And all faced humiliation, prejudice and discrimination.

Hundreds of German and Italian Canadians, who were considered security threats for one reason or another, were interned during the war, but these two larger communities of the main European enemies were never similarly targeted for mass evacuation. It is difficult not to agree with Keenleyside's conclusion that the Japanese evacuation was "a cheap and needless capitulation to popular prejudice fanned by political bigotry or ambition or both." Indeed, in several polls conducted by Gallup on this question during 1942 to 1945, a majority of Canadians—many of whom who were undoubtedly angered by tales of cruelty perpetuated on Canadian POWs in Japanese camps—supported the government. There would have been little or no protest if all the Japanese, Canadian born or not, had left Canada for good.

This ordeal was not quite finished. In January 1943, King and his government permitted all confiscated Japanese property to be liquidated at prices that may not have been at fire-sale levels, but were hardly fair. Any way you look at it, the Japanese were cheated, which was why in September 1988 the government of Brian Mulroney formally apologized to the Japanese-Canadian community and offered proper compensation.

Finally, there was the issue of what to do with Japanese Canadians once the war was over. At external affairs, Norman Robertson maintained that the federal policy "has not been unduly harsh and can be defended as reasonable in the circumstances." However, he also pointed out to King that "the people involved are in the great majority of cases British subjects and Canadian nationals who personally are guilty of no offence other than of having Japanese ancestry. I think it would be very desirable if we could remove as far as possible the aspects of racial discrimination that are involved." That was reasonable enough. Except King again worried about Liberal fortunes in the coming election and paid too much attention to the voices of prejudice in B.C. and around his own cabinet table. All too typical was a speech given by Ian Mackenzie to the Canadian Legion in Vancouver in early June 1944, in which he declared that he would "not remain 24 hours as a member of any Government or a supporter of any Party—that ever allows them [the Japanese Canadians] back again to these British Columbia shores."

Before the end of the summer, King announced in the House of Commons (following Robertson's suggestion) that although no sabotage had been carried out by Japanese Canadians during the war, any disloyalty discovered would result in automatic deportation for non-British subjects, and "voluntary repatriation" was encouraged. Restrictions about future settlement, even for Japanese who were born in Canada, were to be imposed as well. This announcement was slightly sugar-coated, but there was never any doubt about the government's intentions. A *Vancouver Sun* headline later summed it up: "All Japanese ordered out of B.C.: East of the Rockies or Back to Japan." No one in Parliament, not even the CCF, spoke against the resolution.

Of the 23,500 Japanese in Canada in 1945 after the war ended, about 3,900, a third of whom were Canadian, returned to Japan. Most left voluntarily. The terrible treatment of Japanese Canadians did not start with the attack on Pearl Harbor, as Sunahara argues. "Rather, the uprooting, confinement, dispossession, dispersal and attempted deportation of Japanese Canadians were the culmination of a long history of discrimination resulting from Canadian social norms that cast Asians in the role of second-class citizens." King likely would have disputed that harsh judgment. If he had lived to see the day on which the Japanese were accorded a proper apology and formally compensated, he likely would have agreed with his secretary Jack Pickersgill's more moderate assessment. "I did not like the action, which those with responsibility felt was necessary as an act of war, but it was certainly less painful than most acts of war. I felt regret but no sense of guilt at the time and later."

BEYOND THE JAPANESE question, the entrance of the United States into the war had for King far more serious ramifications. By the summer of 1943, he was under no illusion of who was running the war, and it wasn't him or the Canadian military. "The Canadian government had accepted the position that the higher strategic direction of the war was exercised by the British Prime Minister and the President of the United States, with the Combined Chiefs of Staff," read the nondescript yet highly telling Canadian war cabinet meeting minutes of a discussion with British representatives on August 11, 1943. King had recognized this reality of global politics reluctantly, but also with a sense of relief that such an enormous responsibility was not on his shoulders. What Mackenzie King demanded was respect. Respect for the fact that Canada's contribution of men, arms and financial aid was unprecedented. And most of all, respect for his position as an elder statesman. Try as he might to be assertive on both counts, King was not quite as successful as he wished. In the midst of a brutal battle with Nazi Germany, Canada's perennial inferiority complex was a nuisance to most British and American military types and barely tolerated by their more open-minded political leaders and diplomats. King persisted, nevertheless.

As much as he deferred to Churchill and revered FDR, King was far too insecure to tolerate anything that remotely demeaned his standing in the eyes of his colleagues, the Conservative opposition, the press and the public. Misunderstandings turned into personal slights. As the major assault on Sicily got under way on July 10, 1943, involving Canadians in their first real taste of action, the British government sent King a draft of the official announcement. It mentioned only the participation of British and American soldiers. Canadian officials requested that it be corrected to reflect the Canadian contribution as well, which the British agreed to. In the communication confusion, King was led to believe that only the Americans had complied and not the British. Before obtaining confirmation, he caused a hullabaloo by making a statement in the House about this alleged slap in the face. Malcolm MacDonald, who had a good relationship with King, reported the situation to London, concluding his memo with the warning, "I fear the evil effects of this deplorable incident will last for many years." After reading MacDonald's report, Churchill scribbled "rubbish" near the last section. Well aware of King's multi-faceted anxieties, Churchill took the high road and did his best to clear up this mix-up. In the same cable, he also asked King if he and FDR could meet in Quebec in four weeks' time.

Two months earlier, in mid-May, Churchill and Roosevelt had invited King to join them in Washington (after he had proposed the idea) for a high-level meeting at the White House. There, King had been briefed on the impending attack on Italy and learned of "atomic energy experiments." Reading King's drawn-out account of his interactions with the British prime minister and U.S. president, excerpts that are enough to make anyone squirm, it is difficult to dispute an astute observation made by Lester Pearson in his memoirs. Pearson was the second-highest ranking diplomat in Washington from mid-1942 to the end of 1944, at which time he replaced Leighton McCarthy as Canadian ambassador. "The leaders of the British and American governments," Pearson wrote, "seemed to feel that all would be well if they played up to Mr. King's egoism ... by posing with him for pictures or other occasions sending him fulsome letters. His reaction to this treatment was all they could desire in terms of friendly understanding. As long as it *seemed* to Canadians that he was one of a 'Big Three,' neglect to consult and inform him did not appear to matter quite so much."

Buttering up Mackenzie King was inexplicably at the top of the agenda. "Is it not a fact," Churchill pondered at one meeting, "that we three men who are this table, now, have had more experience in government than any other men in the world today?" To this King agreed, praising both the prime minister and president for saving the free world. The conversation, King added, then "returned to my years in public life and the part I had taken." In other discussions, FDR spoke about the role Canada might play at the future United Nations as "moderator," while Churchill requested that King review the draft of his speech to the U.S. Congress. King suggested that Churchill ensure he included a reference to the military participation of the dominions rather than implying the war was only being waged by the British and the Americans. Churchill said he would do so.

King listened to Churchill's address sitting in the executive box close to the Duke of Windsor, the former King Edward VIII, who was also one of FDR's guests. Churchill "brought in a very kind reference to myself," he commented afterwards, "but I did not see that he brought in the Dominions in relation to the war effort in quite the way that I had hoped he would. It is still the combination between Britain and the United States." He did not forget that perceived snub.

He was hesitant to agree to the Churchill–FDR meeting at Quebec in the middle of August, without some prior agreement that he was "more than in the position merely of host" to the two leaders. He was supported wholeheartedly in this decision by Norman Robertson as well as Malcolm MacDonald, who, oddly enough, was advising the Canadian prime minister to take a tough stand with his own prime minister. Churchill was more amenable than FDR, who had no desire to involve a veritable United Nations in these private talks at Quebec. The president feared that if Canada played a major role, other dominions as well as China, Mexico and Brazil would have demanded an invitation.

In the end, it was agreed that King and Canadian military officials could attend the plenary sessions, but that the important decisions would be made in private meetings between the British and the Americans. That King could live with. "My own feeling," he recorded on July 20, "is that Churchill and Roosevelt being at Quebec, and myself acting as host, will be quite sufficient to make clear that all three are in conference together and will not only satisfy but will please the Canadian feeling, and really be very helpful to me personally." It was not as if Canada, a middle power, was about to play a major strategic role in planning.

The Quebec gathering opened on August 17 and lasted for a week. While Churchill and FDR strategized the D-Day landing, King ensured that proper protocols were followed and appearances kept up. Even the flagpole was an issue. King was angered when he found the unofficial Canadian flag with the red ensign flying below Britain's Union Jack. Before the day was out the Canadian flag was flying alongside the British and American ones. No one was the wiser that Canada's main task at this gathering, as Charles Stacey caustically commented, "was to provide the whiskey and soda."

King proudly accompanied Churchill to meet Roosevelt and was included in other excursions, official photographs, and press conferences. At the opening dinner, both leaders extolled King's virtues. Churchill did likewise in his radio broadcast from Quebec, as did FDR in his speech to the members of the House of Commons, following the conference and during the first official visit of a U.S. president to Ottawa. King duly recorded these events, in addition to each and every other compliment and accolade, in his diary. Even Clementine

Churchill boosted King's ego. During tea one afternoon prior to Roosevelt's arrival, Mrs. Churchill, as King wrote, "spoke about how much Churchill relied upon me and spoke of how glad he was to be with me and to share days together."

Late one evening, King found Churchill, FDR and Anthony Eden, a key member of the British cabinet, gathered in an intimate conversation. All three men were wearing smoking jackets and slippers. King was about to leave, when Churchill insisted that he join them. A wideranging discussion on the war and the future of the world ensued, which raised King's sense of himself and his position.

"You have made Mackenzie King a very happy man," the governor general, the earl of Athlone, wrote to Churchill a few days after the conference had concluded, "and I must say he deserves far more confidence than his fellow Canadians place in him but it was ever so in Politics as you yourself know! He has done extraordinarily well in this war." With King so appeased and the conference work fruitful, Churchill soon started planning for a second Quebec conference.

THREE MONTHS BEFORE hosting Churchill and FDR in Quebec City, King had attended a joint meeting of the Canadian Federation of Mayors and Municipalities and the Canadian Club of Ottawa. The guest speaker was the noted British economist Sir William Beveridge, the author of the recent study *Social Insurance and Allied Services,* more popularly known as the Beveridge Report. In his speech that day, he outlined his eye-opening, reform-minded blueprint for post-war Keynesian Britain to combat poverty and disease with unemployment insurance, a national income minimum, health care and other social legislation that ultimately laid the foundation for England's welfare state. King was most impressed with Beveridge's remarks, not the least because he had envisioned a nearly identical program in his book *Industry and Humanity.*

He was at one with Beveridge and, when asked to do so, had no difficulty saying a few words to the large gathering. "I was delighted to find that I was able to speak with exceptional fluency," he later recorded. "Thoughts came rapidly to my mind and shaped themselves into utterances that in part were really of an inspirational character. I was given, if anything, a greater round of applause than Sir William himself."

The loud applause that day notwithstanding, truth for King, as we have seen time and again, was often twisted and contorted beyond all recognition to fit his own curious version of reality. Life was easier that way, and it certainly allowed him to sleep soundly at night. Despite only the most minimum of social welfare accomplishments, King in 1943 still imagined himself as the compassionate forty-year-old social reformer who had worked for the Rockefellers. In fact, he had rejected a far-reaching unemployment insurance program in 1937 put forward by the National Employment Commission and its chairman, businessman Arthur Purvis, as being too "Tory" in its indifference to the provinces' constitutional rights. Three years later, convinced by the Rowell-Sirois report's recommendations for the creation of a Canadian welfare state with Ottawa at the helm, King's government instituted the 1940 Unemployment Insurance Act. This legislation was not perfect; as James Struthers argues, the 1940 act "represented more of an attempt to limit federal responsibility for the jobless than it did the recognition of a new social right of citizenship."

Even if this analysis had been pointed out to King, he would not have accepted it. For him, Arnold Toynbee's "gospel of duty" was as integral to his life's work as it had been when he was a young student. "This is really a great achievement for the Liberal Party," he wrote on August 1, 1940, the day the Senate passed the Unemployment Insurance Bill. "We inaugurated social legislation to the old age pensions measure... It will be law in a few days. For all time to come, that will remain to the credit of the Liberal Party under my leadership."

King failed to mention that he had first proposed unemployment insurance more than two decades earlier. Talk about the slow wheels of government! Inside the cabinet room and on the floor of the House of Commons, King's pragmatism, his fear of upsetting his supporters in the business community and his intense aversion to spending public money tempered his left-wing reform impulses. Weighing heavily on him was the powerful memory of the economic turmoil and labour strife which followed the First World War. So, too, was the inexplicable rise in popularity of the CCF and the rather radical decision of the Tories to invite John Bracken, the long-time Liberal-Progressive, soft-spoken and decidedly non-partisan premier of Manitoba (and far too nice of a guy for down-and-dirty federal politics), to become their new leader. The Tories even accepted Bracken's terms to change the name of the party to "Progressive Conservative." In the midst of a war in which the Allies' fortunes in 1943 were only gradually improving, King was shrewd enough to realize that if the Liberals did not start contemplating Canada of the future they could conceivably be replaced. And the next federal election was fast approaching.

Canadians' expectations were not going to diminish when peace was finally won. The majority were not about to accept a return to the massive unemployment problems of the Depression. The federal government was going to have to take charge and show leadership, whether or not King and some of his conservative cabinet ministers were reluctant to do so. Such drastic change during a war was difficult, of course; the government had to concern itself first and foremost with defeating the enemy and providing for its soldiers. But future priorities could not be ignored.

If, in his increasing old age, Mackenzie King was distracted by the minutiae of day-to-day governing, he was surrounded by enough bright, creative and committed politicians, staff and bureaucrats to keep him moving in the right direction—in this case, left of centre on the political spectrum. Among this cadre of progressive thinkers who influenced King were Ian Mackenzie, who in the prime minister's eyes had redeemed himself as minister of pensions and health, despite his racist attitudes and actions on the Japanese question; Brooke Claxton, a skilled Montreal lawyer who had been elected as a Liberal in the 1940 federal election and became King's parliamentary assistant in the spring of 1943; and Jack Pickersgill in the PMO, whose political antennae were unmatched. ("I've fixed it with Jack" soon became tantamount to prime ministerial approval among members of the Liberal cabinet.)

Behind the scenes, there was the McGill University economist Leonard Marsh, who had trained with Beveridge at the London School of Economics. In 1943, as a consultant to the federal Committee on Post-War Reconstruction, he produced Canada's version of his mentor's chart for the future with the dull but accurate title "Report on Social Security for Canada." With his idea for a national health care plan, Marsh was decidedly ahead of his time. Further constitutional changes were required to institute his plan, and the matter had yet to be negotiated with several of the stubborn provinces, which resisted federal control of social welfare. Nevertheless, King was in the thick of this pivotal transformation—which soon affirmed that Canadians now considered it a fundamental right that the state had a duty to look after them—and not in his customary cautious role, but as an active advocate, albeit one who just needed a bit of a push.

Back in early January 1943, King had paid attention to Ian Mackenzie's advice about what it would take to lead the party to another victory in a general election: a post-war platform that incorporated "a complete programme of social security": old-age pensions, improved unemployment insurance, even health insurance. That, King immediately concluded, "would be a natural rounding out of my life-work." In the months ahead, when he was not completely preoccupied by the war, he heeded the various social reform recommendations advanced by Claxton and Pickersgill and some practical policies began to take shape.

There was cabinet resistance from the likes of James Ilsley, C.D. Howe, James Ralston, Thomas Crerar and others, who were worried about the seemingly never-ending costs of the war and who, in Crerar's case, were philosophically opposed to the idea that the government was there to guarantee incomes by such measures as family allowances. On the "baby bonus," which Marsh had recommended and King eventually accepted as a worthy addition to his social security legacy, Crerar also pointed out that it would favour French Canadians, who tended to have many children. From King's point of view, half of his cabinet were either too old or rigid, or like Crerar, "losing their grip." Meanwhile, witnessing the sixty-eight-year-old King's frequent mood swings, several of his ministers wondered if their elder leader's "hardening of the arteries" was causing "fits of depression."

Hard political realities propelled King forward. In the Ontario provincial election on August 4, George Drew and his newly revamped Progressive Conservatives won thirty-eight seats and formed a minority government, barely defeating the CCF, which had its greatest electoral victory to date—capturing thirty-four seats from having none previously, and an impressive 31.7 per cent of the popular vote. The formerly reigning Liberals, now led by Harry Nixon, a King loyalist who had replaced an ailing Mitch Hepburn, were reduced to a humiliating fifteen seats and third-party status. King blamed the dramatic upset on Hepburn's erratic actions. Then on August 9, in four by-elections (two in Quebec and the other two in Manitoba and Saskatchewan) all four Liberal incumbents lost. King had half-expected these poor results. He was determined to reconnect with labour and to show the entire country that he held "a place which no other man in Canada holds in the way of a key position to help the peace and post-war policies." He had, he believed, one more great victory in him.

Some months earlier, he had declared to the Liberal caucus that he would "never allow an appeal to the people on social security measures at a time of war with a view to bribing them to support the [government] because of what it would pay out of the public treasury." That was then. Now, circumstances dictated a different approach. Social security legislation, along with an all-out effort to support Canadian soldiers in the field, he believed, would translate into success at the polls. This was the message—largely framed by Jack Pickersgill and Brooke Claxton—King delivered at key meetings with his cabinet, caucus and the almost defunct National Liberal Foundation in September. But he demurred slightly.

As he reflected in his speech to the caucus on September 25, "Canada had yet her greatest ordeal through which to pass—her Gethsemane. That I felt until [the war] was over, we had better not give all our time to questions which related to restrictions, etc., asking the taxpayers to carry a still heavier burden to meet expenditures on social questions. That, by all means, to have everything in readiness for the moment the war was over." His words of encouragement to the NLF, however, were more upbeat and progressive. "The task of Liberalism," he announced, "will not be finished when the war is won. That great moment will but mark a new place of new beginning. The future of the Liberal party will not be found in defending the privileges of the few or in arousing the prejudices of the many... In meeting the problems of the post-war period, the task of Liberalism will remain the preservation and extension of freedom."

On September 27, 1943, just as he was preparing for the NLF meeting, King learned that his nephew Lyon, one of Max's twin boys, who was a surgeon, navy lieutenant and father of two young children, had been lost and was presumed dead after the destroyer he was on sunk. There is no doubting King's love for Lyon (his full name was William Lyon Mackenzie King) or his brother, Arthur. Yet he opted not to phone either his sister-in-law May or Lyon's wife because he first had to finish his speech for the NLF gathering and he felt "he could not trust himself." That task was completed at 6:30 PM, but there was no communication with the family until the next morning. Lyon's death did hit King hard, though he classically interpreted it from his own unique perspective. "In many ways, I cannot but feel that this sacrifice of life is meant in part to make plain to the people of Canada that I myself have been among those that have suffered in an immediate way, as Lyon was almost as much as a son could be," he wrote in a diary entry of September 28. "I had been his foster father. Last night, I had the feeling that I could wish Lyon had been my own son, that I might have had this grief to bear alone, rather than others, and the honour which one feels in having one's own to do what is heroic and right."

The problems in planning for post-war Canada were complex, but not insurmountable. By the second week in January 1944, the cabinet had approved the creation of three new federal departments in time known as: reconstruction, which was to be the new purview of C.D. Howe, soon to be regaled as "the Minister of All the Talents" and the "Great Pooh-Bah"; national health and welfare, which was given to Brooke Claxton as a reward for his industriousness; and veterans affairs, headed by Ian Mackenzie as recognition for his forward-looking understanding of the country's affairs. King prepared to unveil the first major step in his social reform package in the upcoming Throne Speech. Caught up in the historical connections of the moment, as he was apt to do, he noted in his diary a few days before the speech that the planned social legislation "rounds out what I have worked for through my life." He perceived this as the natural evolution of his thinking dating back to *Industry and Humanity*. (Chubby Power, however, later argued that King was merely being politically opportunistic.) When his cabinet read the final draft of the proposed legislation, there were still objections, but King, in yet another example of his willingness to play the dictator if he needed to, quickly silenced the naysayers.

Universal health insurance would have to wait for further negotiations with the provinces, but there was no backing down on family allowances. "While the post-war objective of our external policy is world security and general prosperity," the governor general stated on January 27, "the post-war objective of our domestic policy is social security and human welfare... [The] establishment of a national minimum of social security and human welfare should be advanced as rapidly as possible." That meant guarantees of "useful employment for all those willing to work," family allowances and improved old-age pensions. There was another seven months of debate and wrangling until the family allowance bill was passed. King himself, although committed in principle, worried obsessively about the cost of the program and the perception of doling out money to Canadians in an election year, lest he be accused of trying to bribe voters. Many Conservatives condemned the measure as wasteful, dangerous and an unfair remuneration to French Canadians with large families.

One day, as King battled incessant fatigue, he asked Jack Pickersgill why he was so in favour of family allowances. Pickersgill told him that if not for allowances for the children of pensioned widows (his father Frank, a First World War veteran, had died in 1920), he would not have been able to obtain an education and would never have been working in the PMO. King "made no comment to me," as Pickersgill recalled, "but he subsequently told Handy, who repeated it to me later, that the account of our family experience had convinced him." At the end of July 1944, the family allowance bill passed unanimously; no Tory wanted to be singled out as being against such a measure.

King felt a tremendous sense of accomplishment when the bill made it through the House, and rightly so. Politically expedient or not, it did signal a monumental shift in Canadian society whose ramifications are still with us. "Though very tired," he wrote on July 29, "I feel a great peace at heart and an assurance that I am carrying out the purpose of my life in the social reforms in which we have entered . . . It will be part of a memorable period in the life of the Dominion. How far reaching it may be, God alone knows."

THAT FEELING OF peace did not last long. The war again intruded, precipitating one of the great crises of King's prime ministerial career. In terms of sheer drama, only his legendary clash with Lord Byng rivals the personality conflicts and uproar of the second battle over conscription in the fall of 1944.

Besides the success of the social security legislation, the year had been positive. By D-Day on June 6, the end of the war in Europe was on the horizon. King had spent nearly a month in London at the prime ministers' conference during late April and much of May. There was much socializing and luncheons with the king and queen and others, but not many concrete policies were decided upon. The meetings did permit King to extol the virtues of the Commonwealth and his own role in shaping its current form. He had also enjoyed his private time with Churchill, who in an unguarded moment had suggested that the assault on France was to be at the end of the month rather than the beginning. Hence, on June 6, King was surprised and somewhat annoyed when he heard the news and did not have a proper statement prepared.

Six months earlier, at the end of January, Lord Halifax, the British ambassador to Washington, was in Toronto to deliver an innocuous speech before the city's board of trade. Or so he thought. Halfway through the talk he pointed out how Canada and the other dominions had mutual interests and how it would be wise for the Commonwealth countries to work in unison in the future. His remarks were hardly radical; nevertheless, when King was given a copy of the speech, he went into what Vincent Massey accurately called a "paranoiac fury." King was "dumbfounded" by what he perceived to be Halifax's support for a centralized Commonwealth foreign policy. "It seemed such a complete bolt out of the blue," he expounded in his diary, "like a conspiracy on the part of imperialists to win their own victory in the middle of the war." Not surprisingly, Lord Halifax was bewildered by King's anger and Malcolm MacDonald was forced yet again to smooth things over. When Churchill, who had a lot of patience for Mackenzie King, was informed of the kerfuffle, he muttered glibly: "It seems to be suitably vague and recalls the boneless wonder." This minor tempest in a teapot was indicative of King's not altogether unfounded suspicions that the British had yet to accept the new Commonwealth order.

Before King left London on May 21, Churchill had asked him if he would permit a second British-U.S. conference at Quebec City. King was more than happy to approve such a meeting and was touched to be consulted on it. By the time Churchill and FDR arrived again at Quebec City in mid-September, King had celebrated twenty-five years as Liberal leader. This was an enormously memorable and gratifying occasion for him. On August 7, he had been accorded a "great ovation" in the House of Commons and feted at a dinner at the Château Laurier with a thousand people in attendance—including his sister Jennie and Joan Patteson, who was escorted for the evening by Senator Duncan Mac-Tavish, a King loyalist. Apart from an aggravating argument with Ilsley about wartime economic policy in Quebec, the day and evening were perfect. At the dinner, King had been asked to speak for fifteen minutes and instead went on for an hour with stories about his parents and grandfather.

During the second Quebec Conference, King once more was cast in the role of head waiter, though he did not have much choice in the matter. The Canadian prime minister was invited to ride in the car, attend the joint press conference (as an observer, mind you), and get briefed over lunch with his two powerful guests; Churchill even gave him a tour of the map room set up in the Quebec Citadel, which King thoroughly enjoyed as much as a young kid in a candy store. As Malcolm MacDonald told Lord Cranborne:

> Mr. Churchill has been at great pains to keep Mr. Mackenzie King and his colleagues informed. Between you and me, his attitude with the Canadians is far better now than it was up to eighteen months ago. He is friendly and generally approachable, gives them lots of confidential information, and is more or less ready to discuss patiently and reasonably any question which they wish to raise. They used to admire him enormously, but from a distance and with a certain sense that they were naughty children whom the headmaster saw occasionally

with no particular pleasure. They now admire him at least as much as they ever did, not only as a great but forbidding leader, but also as a friend and good companion.

The substantive issues of defeating Germany and Japan and the reconstruction of the world, however, were deliberated behind closed doors, where no Canadians were allowed. FDR, who was about to win an unprecedented fourth term as president, was ailing badly, and as soon as King saw him on September 11, he noticed that he looked gaunt, "older and worn." Churchill, on the other hand, was animated and, according to King, "looked as fresh as a baby."

Speaking to the large contingent of journalists present at a press conference, the British prime minister again was effusive in his praise for King and Canadian sacrifices made during the conflict. His impromptu speech received sustained applause from the newspapermen. A farewell intimate dinner with the Churchills—at which Winston, emotional and teary-eyed, expressed his genuine appreciation of everything Mackenzie King had done for him and Britain—touched the Canadian prime minister deeply. Churchill presented King with a gift: small silver models of Mulberry harbours, the huge barges akin to a temporary harbour that were used to transport equipment to Normandy. The next day, he wrote to Churchill, a man he once detested, but for whom he now had heartfelt affection:

> My dear Winston,
> No remembrance could have begun to express in like manner the intimacy of the friendship and the completeness of the confidence you have accorded me throughout these years of war as do the little silver models, so beautiful in themselves, and which symbolize, so perfectly, the secret of the success of the invasion. To have received this gift from your own hand, here at [Quebec City], where so many of the plans have been worked out ... lends to your gift a preciousness quite beyond words.
> Your devoted friend,
> Mackenzie.

WHILE KING WAS at Quebec City, he was invited to speak to the local reform club at the Château Frontenac. Most members of his cabinet, except James Ralston, were present, along with a group of Quebec Liberals. If Mackenzie King's life were a shlocky Hollywood movie, the ominous background music track would have been heard as he marched up to the podium. For in the course of his off-the-record remarks, King raised the sticky issue of conscription. He announced again that there would be no conscription, "unless it were absolutely necessary," adding that he never believed it was or would be necessary, especially with the war in Europe nearly won. The volunteer system had worked well, he maintained. In the audience, several of his pro-conscription ministers, most notably Angus Macdonald and Thomas Crerar (who was already mad at King about family allowances), were livid. "Trust Willie," Grant Dexter wrote to George Ferguson after both Macdonald and Crerar had briefed him. "The cheering, I am told was right from a thousand tummies—shook the Chateau like the earthquake shook Ottawa."

Like the first of a row of toppling Cold War dominoes, that rousing single speech by King sparked a series of actions and reactions, political jousting of the first order, which led to yet another and even more melodramatic debacle over conscription. Within hours, Macdonald like an excitable schoolgirl had run to find Ralston to blurt the news about King's wholly honest anti-conscription indiscretion. The colonel immediately confronted the prime minister, who restated that in view of the positive military situation in Europe, he would never be part of a government that imposed conscription for the war against Japan. There matters uneasily stood, but Ralston and Macdonald were on edge again. "So nothing much happened," Dexter quipped in a follow-up memo about Ralston's tense meeting with King, "except that the Col. hates his guts more than ever."

The Tories in Ottawa, as well as Ontario Premier George Drew, now yelled even louder about forcing the National Resources Mobilization Act recruits, sarcastically known as the Zombies, to serve overseas rather than remaining safe and secure protecting Canada's shores as they had been promised by the government. The vast majority of the sixty thousand home defence troops had little interest in venturing into the European battlefields at this juncture in the war. Besides, Canadian generals kept on reassuring King and Ralston that the volunteer system was adequate to sustain the country's five divisions and two armoured brigades.

The real trouble started after Ralston returned from a trip to Italy and Belgium in mid-October. He reported to King that while reinforcement numbers remained at a barely acceptable level, he anticipated problems as the war dragged on to an ultimate conclusion. A more recent memo from the Canadian chief of staff in London, General Kenneth Stuart, who two months earlier had essentially guaranteed to King that the volunteer system was working well enough, now more or less supported Ralston's position that dispatching a contingent of Zombies to Europe might be required.

Here was the crux of the final collision between the two stubborn politicians. Ralston, ever the military man, could not accept a tenuous status quo that might put the Canadians already overseas in harm's way, leave the army undermanned and impinge on his reputation. When he read the military reports, he could see only that the general and NRMA volunteer rates could not keep up with the death and casualty estimates. But "estimates" was all they were, and Mackenzie King was not about to backtrack on the biggest promise of his career because of a hunch, educated or otherwise. Given that Nazi Germany was in retreat, "the people of Canada would hardly understand why we should resort to conscription at this time," King told Ralston at one of their tense face-to-face meetings. "To do so would create confusion, which would undo much of the good which our war effort up to the present had effected."

Ralston's sincerity and moral code were above reproach. But he had blinders on with respect to any political fallout or national unity issues, and as he had done during the first conscription crisis of 1942, he threatened King with resignation at each step of the negotiations. From the end of October to the end of November 1944, King's cabinet was embroiled in a bitter stalemate over conscription. By most accounts, the prime minister managed this crisis exceptionally well like the politico he was. Through it all he had the full support of Louis St. Laurent, who agreed with him that only if victory in Europe was in jeopardy should overseas conscription of the NRMA soldiers be instituted. In such an unlikely eventuality, St. Laurent said he would lead the fight to convince French Canadians to back the government.

As for his defence minister, by October 20, King had pretty well decided that Ralston had to go. He had not forgotten that Ralston's 1942 resignation letter remained in his desk drawer and had never been withdrawn. King's first and only choice for a new minister of defence was General Andy McNaughton, who had recently returned from Europe. Due to stress and disputes with Ralston and the British and Canadian military, McNaughton had been relieved of his command of the Canadian army at the end of 1943, although the public had been told that he was ill.

Following six months of much deserved and needed rest, the General toyed with the thought of running for the Conservatives in the next election. Then King, who had always liked and respected McNaughton, approached him about becoming the first Canadian governor general. McNaughton loved the idea and ended his talks with the Tories. Before anything was finalized, the conscription issue interceded and King requested that the General become the minister of defence. McNaughton assured King that he would figure out a way to raise enough volunteers to satisfy any potential manpower shortages.

They were of the same mind about Ralston. "I feel more and more, as McNaughton does... that there is something inhumanely determined about [Ralston] getting his own way, regardless of what the effects may be on all others," King noted in his diary after a contentious cabinet session. He also warned Ralston and his other ministers that on the conscription issue, the future of the Liberal Party hung in the balance. With his conviction that conspiratorial bogey men were after him, King soon came to a spurious conclusion. He believed that Ralston and the conscriptionists—chiefly Macdonald and Ilsley, though King regarded Howe and Crerar, who had different objectives, as part of the group—who had opposed the family allowance bill were involved in a plot to move the Liberals to the right instead of the left, as he was doing. That was a bit of a leap in logic. Yet there was no doubt that several of those cabinet ministers had lost faith in and patience with King's wavering and were contemplating resignation.

The King-Ralston brouhaha soon reached its not-so-glorious climax. For a moment or two, King had thought of resigning and letting Ralston or Ilsley form a government. When challenged, however, neither of the disgruntled ministers wanted to act so decisively, though Ilsley did say that he'd think about it. After further reflection, King decided he would not step down before he had a chance to face Parliament. Late on October 31, Ralston composed another resignation letter. The next morning, King advised the governor general of what was about to transpire and spoke again with McNaughton.

At the critical cabinet meeting on November 1, King took charge. Ralston, who probably guessed at the prime minister's strategy, played it safe and did not present his resignation letter. Perhaps, as Macdonald's biographer, T. Stephen Henderson, suggests, Ralston was "following Macdonald's advice and trying to resolve the crisis." King interpreted his actions differently.

The preliminary discussions lasted more than two hours. Then, when it was evident that Ralston was not about to change his mind, King did what he had to do. He took out Ralston's original resignation letter from the inside pocket in his suit jacket and stated that he was now accepting it and asking McNaughton to join the cabinet. Other versions of the meeting suggest that King asked for and got Ralston's resignation a second time. (King later insisted in a conversation with Grant

Dexter that Ralston "quit" rather than being fired.) Ralston went around the room shaking hands with his colleagues before he left.

Macdonald, who wrote scathing commentary about King in his own diary almost daily, was as "thunder-struck" (to quote Dexter) as everyone else around the cabinet table. He called the dismissal "the most cold-blooded thing I have ever seen." That was hyperbole at its finest, considering his and Ralston's continual complaining about King's personal and professional flaws. Crerar believed, too, that Ralston, whose military policies he had not agreed with, had been treated terribly by King and the Liberal Party and "deserved much better from its leader."

If anyone, however, was the architect of his own fate in this wartime song and dance, it was James Ralston. He had pushed King as far as he could be pushed and received more understanding from the prime minister than likely any cabinet member in Canadian history. Mackenzie King had demonstrated the true extent of his political proficiency and toughness, yet with the civility and politeness he had adhered to his entire life. Moreover, while none of Ralston's allies resigned, they were tempted to do so. This was partially because of an overarching sense of duty to the country during a war and more specifically because of assurances from McNaughton that conscription was still an option, but also surely because they enjoyed the power and prestige that came with their positions—in other words, for the very same reasons that King, the man they all so detested, did not quit either.

Now it was Andy McNaughton's turn to prove Ralston wrong. He was sworn into the cabinet on November 2 and went to work on a recruitment drive. He made an emotional appeal to the NRMA Zombies to volunteer for overseas duty. But within three weeks, he finally realized his volunteer targets could not be met. Most of the NRMA men refused to be converted, and while there were many volunteers for general service there were not enough to keep the flow of men to Europe at the rate the military deemed necessary.

King had a big headache to deal with. He was depressed, and his initial instinct was to resign. "More and more," he recorded after McNaughton had delivered one of several gloomy reports, "I come back to my firm convictions that the thoughts I had when the matter first came up were the right ones and that I should stand firmly against agreeing to conscription and not following the volunteer enlistment, because I doubt if it would help to get the men, and secondly the great possibility of making the situation worse for the present and for all time to come in Canada." After more discussion and reflection, King finally told his cabinet that he would give the volunteer appeal another three weeks. If after that time the numbers were still insufficient, he would stand aside to permit another minister to take over the government and an order-in-council could then be passed to send NRMA conscripts overseas.

Once King finished speaking, the meeting room was silent for what seemed like an eternity. "When they began to speak, one after the other spoke of standing with me," as King recalled. This act of loyalty pleased him. In truth, almost every minister was not prepared to serve in an administration without him. Only Macdonald, Ilsley and Crerar thought it best that they resign; while Chubby Power, alone among the Quebec ministers, declared that he would never be part of an administration that instituted conscription.

Quick confirmation from McNaughton that the numbers of volunteers and NRMA conversions would not improve tested King's resolve further. Under the impression (as questionable as it was) that men were indeed required on the European front lines, he now decided that his duty necessitated that he pass the order-in-council to force about sixteen thousand NRMAs to head overseas, but not before there was a vote of confidence on the resolution in the House. No prime minister enjoys being caught in an about-face, but there are some infamous cases, including King's in late 1944. He had five days to sell this to his divided cabinet, three-quarters of whom were about to resign for contrary reasons: the Quebec ministers and Jimmy Gardiner from Saskatchewan because of conscription, and the ones from English-speaking parts of the country because of King's insistence on delaying instituting conscription as long as he could.

Louis St. Laurent, ever the loyalist, was not happy about this decision, yet he agreed to remain at King's side (possibly because he believed King's dire warning of a "revolt of the generals"). Chubby Power, in contrast, was adamant that he would have to resign, a position that Ernest Lapointe, whose animosity to conscription was deep, most likely would have followed as well. Meanwhile, unknown to King, half of his cabinet were meeting privately in Crerar's office; almost everyone had drafted letters of resignation. "They decided to act as a body and either force conscription or resign," as Dexter later described it. When that conclave broke up, they were informed of King's new strategy, and no resignations—to King they were all "Judases"—were forthcoming.

A morning caucus meeting followed on November 23, "a real corker," as Dexter put it. "Gardiner was the headliner with a strong speech against conscription. He sure made a hit with Quebec and most of the lads took the excessively realistic view that he probably clinched the leadership in succession to King." The order-in-council limiting the conscripts to sixteen thousand NRMA troops was drafted, and Power officially resigned. As Jack Granatstein wryly observes, "Ralston had gone because he wanted conscription, Power because he opposed it."

The nearly two-week ordeal of resolving the conscription issue in the House of Commons began on November 27. As it turned out, it was Joan Patteson's seventy-fifth birthday. As a gift, King gave her a bottle of vitamins "in a small white casket, carrying reliefs of Dante and Beatrice." Joan loved the present and "spoke of vitamins as nova vitae and of the symbolism." Much of the day was spent preparing his address to the House. He had Pickersgill gather for him statements made by Laurier and Lapointe, which were integrated into his speech. Before he left for Parliament, he prayed at the side of his bed that he might be given "strength and guidance" to carry on.

King felt later that he "spoke well" without having to refer much to his notes. *Maclean's* Ottawa correspondent Blair Fraser called his performance "magnificent," adding that "for a man of 70 years old it was a remarkable feat of energy and endurance." Jack Pickersgill also remembered it as "certainly the most momentous speech of his life." In a three-hour oration, with relevant and moving quotations from John A. Macdonald and Wilfrid Laurier, King set out his case for conscription of sixteen thousand NRMA men with reasoned passion. "At times," Pickersgill stated, "he turned his back on the Opposition to appeal directly to the Liberal members." When he finally finished he was "given a real ovation," but was taken aback because no one approached him to shake his hand.

General McNaughton, too, was compelled to champion this policy switch. Jack Pickersgill and Brooke Claxton wrote a speech out for him justifying the change and handed it to him. He barely glanced at it. "General, aren't you going to read it?" Pickersgill inquired. "No, I'm in such a state I can't," said McNaughton. And he did not, but he did deliver it word for word. In a 1986 interview, Pickersgill recalled it as "a gruesome business, one of the most odd experiences of my life."

As soon as King returned to Laurier House, he took a long nap. He was back at the House after dinner. During the long evening session, he spoke somewhat angrily to Crerar, who he felt had betrayed him. "I said to him . . . as he sat beside me, drawing little squares and pictures in blue and red like a child in a nursery, that I thought he and others supporting conscription should never have put me in the position I was in unless they were prepared to take over the government themselves. He coloured a bit but said nothing. Nor had he anything further to say." (Three years later, on May 21, 1948, Ralston died at the age of sixty-six. At the funeral, King watched the procession of the pallbearers, which included Angus Macdonald, Ilsley and Crerar, and all he could think about was "how these men had joined together to try and secure conscription and also to get Ralston to head the Ministry.")

King finally made it back home at 2:40 AM to find little Pat II waiting for him. Before he retired, he prayed again and "thanked God for the help given me through the day." In his tired and emotional state, he felt compelled to fondle a locket of his mother's hair, which he kept in a small box. "I kissed her white hair which shone with a beautiful luster in the light of the lamp," he noted. He and Pat then had a little biscuit together and he fell asleep content that he had once again saved the country.

The vote of confidence was taken in the wee hours of December 8, a little after 12:30 AM, when "the hands of the clock would be in a perfectly straight line" (King's obsession with the clock was especially evident on this anxious day and was commented upon numerous times in the entries for December 7 and 8, 1944.) He received what he considered the "amazing" support of 143 members to 73 against. To his utter joy, 19 French-speaking Quebec Liberals stuck with the government, whereas the Conservatives "found themselves in the oddest of all situations," as King explained with relish, since they were forced to vote against a conscription resolution along with the Quebec nationalists.

King was overwhelmed by this endorsement in his leadership, his conscription policy, and his administration. He accepted the accolades with as much humility as he was capable of showing. Jack Patteson and his wife, Molly, were in Ottawa and watching from the gallery. They congratulated him, as did McNaughton, who was as giddy as a "school boy." This was more than good fortune shining on him, he knew in his heart. The spirits were in play. "Really this has all been more amazing than any experience I have known in life," he dictated to Edouard Handy later. "What I seemed to realize most clearly tonight were the forces that were working from Beyond to bring the vote to the high figure at which it is and to completely confound political forces that were hoping to make it just the opposite . . . I felt most deeply in my heart that if I ever believed in God and in the power of prayer it was tonight." He even thought that the speech he had delivered in the House earlier in the evening might well be his last. "If there are others that will rest with Higher Powers," he added, "if they still wish to use as their instrument in God's service . . . There are terrific problems ahead, I feel the solution of them should come to rest on younger shoulders."

In fact, it was not so-called "Higher Powers" that were to decide the next step in King's political career, but rather the Canadian people.

12

The Sense of Triumph

*I am a great believer in the power of truth and, having
the truth, I feel certain that we shall win, and that now
we may find the real path of peace which will be letting
the masses of the people know how they are controlled by
the few and bringing about a real brotherhood among the
common people of the earth.*—THE DIARY OF WILLIAM
LYON MACKENZIE KING, *February 16, 1946*

IN THE EARLY HOURS of December 17, 1944, the day of his
seventieth birthday, Mackenzie King was restless. He awoke at regular
intervals and, as he recalled, "seemed to be experiencing in a subjective
way the struggle of winning through the recent serious crisis... The
sensation seemed to be that of an upper and lower stone; the lower one
sliding past and getting through all right on its onward way. The mills of
God grind slowly but they grind exceedingly small." He prayed that his
life "might be consecrated to the fulfillment of God's will." When the
sun came up, he peered outside to gaze at a beautiful winter morning. A
heavy snowfall had blanketed the streets.

He telephoned Joan Patteson, "to have a word with her before anyone else." The call woke her, but she was happy to be the first to wish
him birthday greetings. Amazingly, at least from King's point of view,
when he picked up his watch it had stopped around 7 AM, the hour he
had been born. Suddenly, as he moved some articles on his desk, a coat
check ticket with the number eighty on it "sprang up." This signified the
number of years he was to live, or so he believed.

The rest of the day was a whirlwind of birthday wishes from his staff,
sister Jennie, nephew Arthur and members of his cabinet and caucus,

including James Ralston. He even received a congratulatory cable from the king and queen. There was more prayer and reflection, a bit of work and finally an intimate dinner with Joan and Godfroy. After dining, Joan read out loud from the book she had given King as a present, *The Life of Charles Lamb* by Edward Lucas. King was enthralled by the story of the popular nineteenth-century writer. There was no sitting at the little table—communion with the spirits had stopped during the war years—yet it was for King a perfect birthday and brief reprieve from his hectic pace.

The new year soon brought more challenges. At the top of the list was ensuring that General McNaughton had a seat beside him in the House of Commons. A Prairie boy, McNaughton would have preferred to run in Saskatchewan, but a supposedly safe seat was found for him in the rural Ontario riding of Grey North, in the vicinity of Collingwood and Owen Sound. The sitting Liberal, William Telford, who had just turned seventy-seven, was convinced to retire in favour of the General. It seemed like a foolproof plan, except it backfired badly.

In normal times, the opposition parties might have permitted a new cabinet minister to win a by-election by acclamation. Yet McNaughton was no ordinary minister, and there was nothing normal about early 1945. Not only had the manner in which King handled the conscription issue angered the Conservatives and the CCF, but partisanship that should have receded during a war instead intensified in Ottawa. A national coalition government probably would have stemmed those intense party loyalties. King, however, had consistently eschewed such an idea. Even the Progressive Conservative leader, the mild-mannered John Bracken (who inexplicably did not run for a seat until the federal election of 1945), was fighting mad. In an appearance in Grey North at the end of January, Bracken attacked McNaughton's competency and implied that despite the new Liberal conscription policy, the General had deserted the Canadian men fighting in Europe.

McNaughton fought as best as he could; yet he was a better soldier than politician. The local press described his public speaking style as "cold, precise, and even pedantic." He was regarded as an interloper in the constituency, as opposed to his main rival, Garfield Case, the popular mayor of Owen Sound, who was running for the Tories. (Mayor Case's catchy and effective slogan during the campaign was "Send a Grey North man to Ottawa, not an Ottawa man to Grey North.") The CCF candidate was the highly respected and decorated retired Air Vice Marshal Albert Earl Godfrey. A final headache for the Liberals was religion. The general was an Anglican, but his wife, Mabel, was Catholic. In the uneasy climate of the day and among voters in rural and very Protestant Ontario, that was an issue.

A month before the by-election, King remained confident in a McNaughton victory. "He was sure [the General] would win," reported Grant Dexter following a wide-ranging interview with the prime minister. "But that wasn't the point," King had told him. "The point was that the Tories had not only challenged the vital war policy of the government but were now doing their utmost to prevent the minister of national defence from entering Parliament. Success would force an election. They want an election."

King sent out several pro-McNaughton circulars with ominous warnings of doom if the General was kept out of the House of Commons. Nothing worked, and King's decision not to make a personal appearance during the campaign did not help. On February 5, a bitter cold winter day, Mayor Case defeated General McNaughton by 1,236 votes (7,333 to 6,097). Many former Liberal supporters undoubtedly had opted for Godfrey and the CCF, who received more than 3,000 crucial votes. McNaughton was so enraged by the results that he hurled chairs around his committee room.

At Laurier House, King had received hourly election updates throughout the evening. By around nine o'clock, with the news disappointing, he was forced to concede that McNaughton had lost. He released a statement to the press indicating that a general election was now all but a certainty. King feared the real possibility that in future three-way contests the CCF would continue to siphon off Liberal votes. He went to bed angry that night, with images of Ralston in his thoughts for his "precipitation of the conscription issue" and causing the predicament he found himself in.

He soon met with McNaughton in Ottawa, who had calmed down yet was still angry. The General denounced statements made by John Bracken as "reckless," deplored how the Tories had distributed liquor in the riding to buy votes, and how his public meetings had been interrupted by paid hecklers in uniform. King counselled him to refrain from attacking Bracken any further and making him into a martyr. McNaughton reassured King he would not abandon him or the government and that he would run in the next federal election in Saskatchewan. Victory assuredly would come soon enough.

MEANWHILE, KING'S OLD antagonist Mitch Hepburn, the former erratic premier of Ontario, was back. Actually, he hadn't really left.

Hepburn had resigned as premier but had remained in the provincial legislature. He won his seat in the 1943 Ontario election as an independent (that is, a not–Mackenzie King) Liberal and made life difficult for his successor, Harry Nixon. By December 1944, Hepburn had been reinstated as Liberal House leader, and Nixon had graciously stepped aside. A party convention was scheduled for the spring with Hepburn the only real candidate. The federal Ontario Liberals, however, were not certain Hepburn could be trusted.

At the beginning of March 1945, speaking to the Canadian Club of Toronto, Hepburn stated that he had "probably erred in attacking Mackenzie King in a personal way." King felt vindicated after he read a story on the speech in the newspaper the next morning. "I had waited long for this confession," he noted in his diary, "but he has found he has had to make it in the hope of getting back into a strong position in the party. I have never asked for anything from him." Within a month, the CCF and Liberals had teamed up to defeat Premier George Drew on a confidence motion. A week after that, Hepburn's contrition about King, as insincere as it likely was, propelled him back as Ontario provincial Liberal leader. Drew set June 11 as the date for an election.

King and his advisers had wanted the federal election on June 25. Hepburn's return aside, most everyone in Ottawa expected Drew to be back as premier, and no one wanted a potential Conservative victory in Ontario to upset the federal Liberal Party's chances. It was C.D. Howe who came up with a solution at a cabinet meeting on April 12: hold the federal election on June 11 as well. King liked the idea, and the following day he announced the date for the general election to be held together with Ontario's on June 11.

Later that afternoon, while a physician was giving him a massage, King learned that President Roosevelt had died. Given FDR's poor health, it was not unexpected. (He likely had cancer, according to new research). King was, he wrote, "too exhausted and fatigued to feel any strong emotion." He did phone Joan Patteson, who had heard the news and provided King with more information. King then tried to put a few personal thoughts on paper, but was unable to do so. By then, Jack Pickersgill had arrived at Laurier House and was working on the official statement. As that was being prepared, King worried incessantly about his exhaustion. Within him rose deep feelings and "a great gratitude for having been privileged to know the President so well."

King travelled to Hyde Park, New York, three days later for the funeral. Despite the sombre occasion, he was feeling at one with nature. "The fields were green. The little leaves were coming out on the trees, birds were singing," he recorded. "Many shrubs were already in bloom. Some blossoms on the trees. The sun shining brightly and the air fresh and balmy. It really filled one's soul with a feeling of delight." He immediately thought of the lines of a "darky" folk song, "Kitty Wells."

As he walked onto the grounds, he discovered to his horror that no one except him was wearing a silk hat. He had argued with John Nicol about it, but his valet had insisted he doff it. He was able to borrow a more appropriate black homburg from Walter Turnbull, who went bare headed. At what he perceived to be the right moment, he placed flowers on FDR's grave, and stood quietly for a few seconds. Then, at the behest of Secretary of State Edward Stettinius and a movie camera man shooting a film of the funeral, he did it again for posterity. Upon returning to Laurier House, he "knelt down and thanked God for the great friend he had given me in the President ... [and] asked for strength to carry on his work."

Election preparations continued. Mitch Hepburn was thrilled at the thought of holding the federal and Ontario elections on the same day in June and sent a telegram to King: "Liberals everywhere can now join forces for the defeat of Toryism and get the whole job done at once." King genuinely appreciated the gesture and telephoned Hepburn to thank him for the message. But George Drew wanted nothing to do with what the *Globe and Mail* denounced as King's "desperate scheming" and "web of chicanery." "Confusion is the primary purpose of his plan," the editorial concluded. On April 15, while King was at Hyde Park, Drew announced that the Ontario election would now be a week earlier, on June 4. When King found out, he was livid, probably more because he had been outmanoeuvred, though he claimed in his diary it was all about democracy. "I cannot understand Drew's action ... It is a form of highway robbery to seek to get into power by destroying right constitutional procedure." Upon further reflection, he decided that it would all turn out well under any circumstance. "I can see that our forces will be working hard with the Ontario Liberals," he added. "All will be welded into one and even if our men do get the worst of it provincially, they will be perhaps keen to get even federally."

Ontario was likely the key to another Liberal majority. Winning most of Quebec's sixty-five seats was equally essential, as it had been in every election King had contested. But there were problems, a residual of the two conscription crises. Through the winter, King and St. Laurent had been lobbying Chubby Power to return to the cabinet. Power stubbornly refused. In late February, he had told King and St. Laurent that he and other disaffected Liberals intended to run in the election as independents, as "authentic Liberals, inheritors of the Laurier-Lapointe tradition" who had stood up against conscription. (Power was telling anyone who would listen to him that his group's slogan was to be: "Not Necessarily Mackenzie King, but Mackenzie King if Necessary.") King took great exception to Power's insinuation that he and St. Laurent were conscriptionists, and dismissed his carping as his eternal desire for control of the Quebec wing of the party and his foul mood that day as yet another hangover.

Power could not be appeased. In the weeks ahead, he spoke with another unhappy former Liberal, Arthur Cardin, who formed his own independent Liberal faction—Cardin opposed King's social welfare legislation and remained bitter about conscription—called le Front National. Power remained skeptical of Cardin as well. "I got the impression of a great deal of sincerity," he recalled, "and a vast amount of bluff about their power and prestige." That assessment was accurate. Two weeks after the formation of le Front National, Cardin deserted his new cause and ran (and won) as an independent. Power, on the other hand, bided his time in April and May, slowly inching his way back to King, the Liberal Party, and another cabinet position.

King's election planning was interrupted for a few weeks by the United Nations Conference on International Organization in San Francisco, which began on April 25. After a long train journey, King arrived exactly (and apparently highly significantly) at 10:10 AM on April 22, as he noted in a letter to Joan Patteson. (The hardest part of leaving Ottawa was not foregoing election preparation, but leaving Pat II behind. "The little chap knew as we neared the station that I was leaving him. He took a last upper glance at me and then lay flat pointing his little nose in an opposite direction.") With news that the war was almost at an end, King arrived in California with a sense of optimism. "It is wonderful to think that these days at San Francisco may mark the period of transition between an old order that has been full of tyranny, strife and injustice and the new order which may help to maintain peace and to further good-will," he added in his note to Joan.

King was accompanied by a large delegation that included St. Laurent, representatives of the opposition parties, Gordon Graydon for the Conservatives and M.J. Coldwell of the CCF; Norman Robertson, Humphrey Hume Wrong and Lester Pearson, among other advisers from external affairs; and his diligent staffers, Jack Pickersgill, James Gibson and a few others. Gordon Robertson, then with external affairs, served as secretary of the Canadian delegation. While King attended the opening sessions of the conference, he was nervous, as he had been about the defunct League of Nations, about committing Canada to a global organization that might compel the country into military action. Nor was he interested in competing with Herbert Evatt, the Australian minister of external affairs (whom he later grew to dislike), "as the vociferous champion of the middle-sized nations against the great power."

Mostly King was preoccupied, to the point of obsession, with polishing his speech, which he was to broadcast on the radio as soon as victory in Europe was declared. By the end of April 1945, American and Soviet forces had met by the Elbe River near Torgau, about 150 kilometres south of Berlin; Mussolini had been executed by Italian partisans; and on the last day of the month, Hitler committed suicide in his Berlin bunker. News of the Führer's death took King back to his meeting with Hitler in 1937. "My thoughts were much of what I urged him to consider most," he reflected in his diary, "his chance to be of service to the world by championing the cause of the poor and seeking to improve the standards of the many but avoiding at all costs the temptation of war."

Jack Pickersgill's assignment was to assist King with his anticipated radio talk, and as he recalled, the prime minister "was apt to want to work on the speech with me at almost any hour." King insisted on including a passage that thanked God "for his Mercy as revealed in the deliverance which has come at last from the evil forces of Nazi Germany." What exactly did he mean by "evil forces"?

Confirmation of Nazi atrocities at Auschwitz, Buchenwald, Bergen-Belsen and the other concentration and death camps was already well known in the West, as well as by the Canadian external affairs department, and had been since at least May 1944, if not earlier. Certainly an assortment of unsettling reports and news clippings sent to Ottawa by Lester Pearson and others would have crossed King's desk. No matter. Despite the best efforts of the America section of the Jewish Agency, the horrific treatment of European Jews or the future of a Jewish state in Palestine was not raised at San Francisco. The subject barely dented King's consciousness then or later, and there are only a handful of references to Jews as "victims of Hitler's persecution" in his diary.

A year earlier, in June 1944, the German journalist Emil Ludwig—who had recently completed a short biography of King—was visiting Kingsmere and raised the issue of permitting Jewish refugees into Canada after the war. King was sympathetic. He promised Ludwig that "Canada would have to open her doors and fill many of her large waste spaces with population once our own men had returned from the front and that we would have to be generous and humanitarian in our attitude." He said he would meet with the Jewish community to discuss it further. But he also had to explain to Ludwig "the nature of the political problem [and] the difficulty of a leader of a government bringing this question on the eve of an election." This cautious attitude was to shape Canada's rigid post-war refugee policy, especially when it came to survivors of the Holocaust.

Early in the morning of May 7, Nicol woke King to tell him that Norman Robertson had sent a message: the war in Europe was over. The first thing King did was to utter a "prayer of thanksgiving and rededication to the service of my fellow-men." He spent the rest of the day reviewing his speech and fussed excessively, even for him, about whether the Canadian Red Ensign should be hoisted to the mast of the flagpole at Parliament Hill and remain there for the day, which it finally was.

Jack Pickersgill was caught up in King's Red Ensign dilemma. Asked for his opinion, he replied somewhat sardonically, "Once you put it up, you will never be able to take it down." King was not pleased with that impudent remark, as Pickersgill later discovered when he read King's diary entry about it. "I also felt a little indignant and hurt at the way in which Pickersgill shot at me that I would not think of taking down the flag once it had been up," King wrote. "Really a piece of impertinence for any member of a staff to speak in that way to a Prime Minister."

The broadcast on CBC was aired on May 8 without any glitches. To his credit, King touched the right chord, with moving references of "gratitude for the deliverance from the forces of evil of Nazi Germany" and "humble and reverent thanksgiving to God." King was delighted when he was informed that the broadcast "had the largest coverage of any the [CBC] has thus far given to anyone. It included the entire Canadian network, U.S. national network and short-wave to Britain, also to sailors and soldiers. It was supposed to have gone to all the battlefronts."

There were luncheons and meetings to attend in San Francisco, and King did push for and win wording in the UN charter that protected the independence of countries such as Canada to participate in any military action when called upon to do so by the Security Council. The text of this article, number 44, was "a tribute to the tenacity of the Prime Minister of Canada," according to political scientist James Eayrs. King prided himself that his "part has been done quietly, unobtrusively but effectively behind the scenes." Still, the upcoming election took precedence over the future of the United Nations.

KING LEFT SAN FRANCISCO on May 14 fatigued as ever, but arrived in Vancouver on May 16 with a renewed strength to fight for his job. At a gathering at the Hotel Vancouver, he gave his first real speech of the campaign. The Liberals had adopted as their slogan, "Keep Building a New Social Order for Canada," which was mainly aimed at winning over any left-leaning supporters who toyed with the idea of voting for the CCF. For much of the campaign, King, following Pickersgill's advice, stressed his own experience as giving him the ability to guide Canada in the post-war world. Why, he asked repeatedly, would the citizens of the country want to entrust domestic and international issues to "unknown and untried hands?"

That strategy worked much better in Quebec than in the rest of the country, where doubts about King's judgment and leadership persisted, including the fact that at seventy, he was well past his prime. With either the departure to the Senate or the retirement of several cabinet ministers—among them Crerar, who was appointed to the Senate, and Angus Macdonald, who had had enough of Ottawa and Mackenzie King (the feelings were mutual) and returned to Halifax to become the premier of Nova Scotia again—King had elevated the next generation of Liberals. New significant additions to the cabinet included Lionel Chevrier, Paul Martin, Sr., David McLaren and Douglas Abbott.

He returned to Ottawa from Vancouver by train, stopping at each major city to promote national unity, social security, full employment, and himself. He even suggested that it was time for Canada to have its own flag, a good idea but twenty years too early. King was still as formal a campaigner as he had ever been. He greatly enjoyed visiting different parts of the country and being feted by local dignitaries; nothing delighted him more than a boisterous send-off at a train station. At various election events, he wanted nothing more than to connect with his audiences, which he felt he frequently did. But he was also his toughest critic. He usually blamed exhaustion. "I must avoid being jammed up against so much and without adequate rest," he wrote during a stop at Prince Albert, Saskatchewan, on May 19. "I should have made a very memorable and appealing speech tonight, but disappointed myself, and I am sure fell far short of the impression that should have been created on this occasion."

The most notable radio broadcast of the campaign was done from London, Ontario, in which King highlighted family allowances and the Liberal social welfare vision for the future. It was not state socialism of the variety the CCF alluded to, but it was bold enough to counter anything John Bracken and the Conservatives could offer. Besides, the Tories foolishly harped on conscription for the Pacific conflict, a somewhat meaningless strategy since the war in Europe had been won. This tactic merely alienated Quebec voters, as the party had been incessantly doing for decades. The Conservatives' inability to be more flexible and alter their policies sufficiently to steal seats from King and the Liberals in Quebec must rank as the worst and most ill-advised politicking in Canadian history. (The party stalwarts in Toronto were so disenchanted with Bracken's performance that they had George McCullagh, the outspoken owner of the *Globe and Mail,* make the campaign's final radio broadcast.)

Near the end of the campaign, things came together in Quebec for King. He had been a bit uneasy about his appearance, yet by the time he arrived in the province on June 6, Chubby Power had decided to jump back on the Liberal bandwagon. In Quebec City, King was about to take a nap in his suite at the Château Frontenac and Pickersgill was polishing that evening's speech, when there was a loud knock on the door. It was Power, "half-sober," as Pickersgill benevolently put it, or "under the weather," to use King's words. Pickersgill told Power that the prime minister was sleeping. A moment later, the bedroom door opened and King appeared in his long underwear. Normally, King would have glowered with contempt at Power if he was intoxicated; the uncertainty of the election being what it was, however, he embraced his former minister, and the two chatted warmly for the next hour or so. That evening at the Palais Montcalm, Power, who according to Pickersgill, "could barely stand on the platform," wholeheartedly endorsed King and the Liberal Party as French Canadians' only option.

By the time King spoke, it was midnight. He launched into an oration about Laurier and Lapointe and his actions during the war. He did not apologize, as he recalled, "but came out openly and said I had not sought the leadership in the war. I took it because it was my duty." He described the new Liberal world of full employment and social security, and to his great satisfaction he received a "thunderous" applause.

For King's final broadcast two days before the vote, Pickersgill wanted to spice up his remarks by referring to the fact that Bracken "had been advertised like a new breakfast food or soap." King found the phrasing "much too flamboyant." Pickersgill spent much of the evening writing a new draft. In the interim, King, without letting his overworked assistant know, had decided to rework the first version. Pickersgill then had to spend another six hours reviewing each word with King prior to the broadcast. Despite the all-too-typical Mackenzie King–inspired commotion, the speech flowed well and even included the "flamboyant" marketing reference about Bracken.

"The choice on Monday next is not between a Liberal majority and a majority for some other party. No party but the Liberal party has the remotest chance of securing a majority in the next House of Commons," King declared. "The choice, therefore, is between a Liberal majority, and no majority of any party." He beseeched Canadians to "take the only course which will ensure a continuance of strong and stable government."

George Drew's massive victory on June 4—the Ontario Conservatives won 66 of 90 seats—initially upset King. Yet, as was his custom, he looked on the bright side. The routing of the CCF, he felt, boded well for the federal election, and "the cry Drew in [Ontario] and Mackenzie King in the Federal field may still hold true." His optimism was not entirely misplaced. On June 11, King and the Liberals won another, albeit slimmer, majority with 125 seats in a House of 245. Again, the key was the margin of victory in Quebec, where the Liberals won 53 seats; French Canadians had balked at supporting the nationalists and forgiven King for his conscription policy. That was all the more crucial since in Ontario, the Liberals captured only 34 seats to the Conservatives' 48 (the CCF were shut out).

The news was not all good, however. General McNaughton lost his second bid for a seat, this time in his native province of Saskatchewan; he resigned—and was encouraged strongly but gently to do so by King—as minister of defence two months later. More of a "shock" was King's own defeat in Prince Albert, which was confirmed on June 14 after the soldiers' votes from overseas were counted. (For three days he thought he had won.) He lost to the CCF candidate by 159 votes. It was "cruel it should be my fate, at the end of the war, in which I have never failed the men overseas once," he lamented, "that I should be beaten by their vote." In August, the seventy-year-old Liberal backbencher Dr. William McDiarmid resigned his seat in Glengarry, near Ottawa, which King easily won.

Grant Dexter of the *Winnipeg Free Press,* among other commentators, did not find "much elation" in the election results. The country remained divided: the Conservatives were strong only in Ontario, the CCF won all of its twenty-eight seats in the west and the Liberals were being propped up by Quebec. Discontented Liberals like Ilsley felt that it was time for King to retire. And there was no doubting his weariness. Yet whereas Winston Churchill and the Conservatives suffered a stunning defeat to the Labour party in the British general election of July 1945, King had figured out a way to embrace the social changes demanded by Canadians and responded accordingly.

Whatever his many detractors said behind his back, the overall victory, the final one of his life, was his to savour. He rested on June 12 at Kingsmere. Reflecting on the election, he wrote:

I sat quietly for a short time in the sunroom, in my heart thanking and praising God for his goodness. The relief of mind that I experienced is indescribable. It brings with it peace of heart as well. Little Pat and I took a walk together. I felt... as if I had had a bath after a dusty and dirty journey, with the storm of lies, misrepresentations, insinuations and what not in which I have had to pass during the past few weeks— I might even say over most of the years of the war... I felt a real vindication in the verdict of the people and the sense of triumph therefrom.

THE WAR IN Europe had been won; the surrender of Japan was soon to follow and with it the horrifying consequences of the atomic bomb. "I feel that we are approaching a moment of terror to mankind, for it means that, under the stress of war, men have at last not only found but created the Frankenstein which conceivably could destroy the human race," King wrote in his diary a week before the first atomic bomb was dropped by the Americans on Hiroshima.

As he waited anxiously with the rest of the Western world that summer, he occupied himself with the more mundane task of dealing with the premiers at the first Dominion-Provincial Conference in four years. This one, which was intended to be about post-war reconstruction, was in reality the initial step in implementing the centralizing fiscal controls recommended in the Rowell-Sirois report. As King expected, Maurice Duplessis, newly installed again as the premier of Quebec, and Ontario Tory premier George Drew, among several other provincial leaders, did not want to relinquish their taxing powers in return for federal payments. Not much was decided when this conference ended, but the federal-led changes would be instituted soon enough. King was not troubled in the least by the lack of progress in the negotiations and instead congratulated himself on delivering such a fine closing speech. "The kind of speaking I did tonight," he wrote as if he were a novice politician, "is the only thing that is really worthy of the name."

As fate would have it, the first day of the conference was August 6, the day Hiroshima was bombed. (It was also the day King swept to victory in his by-election in Glengarry, which the premiers good-naturedly ribbed him about.) King broke the news of the A-bomb to the premiers in the appropriate solemn fashion. He and Jack Pickersgill immediately got to work on another radio speech for VJ day. The taping was done on Sunday, August 12, at the CBC's studio in the basement of the Château Laurier. He returned to Kingsmere later that day. Joan was visiting in the evening when they both heard the announcement on the radio that Japan had surrendered. As planned, the tape of King's speech was played. Both King and Joan were elated, and King got down on his knees and thanked God for the dawn of peace again.

Except the CBC news broadcast was mistaken; the Japanese did not officially surrender until August 14, five days after the dropping of the second bomb on Nagasaki on August 9. When the full realization of this gaffe hit King, he was mortified. He quite correctly rationalized that it had been the CBC's fault that his speech had aired on the wrong day. With his insecurities at full tilt, this seemingly innocent human error was blown hugely out of all proportion and was followed by several days of mental flagellation, as only King could inflict upon himself. He read in anguish the *Ottawa Journal*'s August 13 headline, "No word from Japs on terms. Erroneous broadcast recording by P.M. sets off V.J. celebrations in Canada," among other similar headlines in newspapers across the country and abroad (the *Times* of London had a note about it on August 13, for example). And he could not help but notice how the members of his cabinet silently enjoyed his discomfort when he reviewed what had occurred. His only relief, as he wrote on August 15, was the "many evidences of the spiritual presence of those who wish to bring me comfort at this time; comfort is what one needs at this moment in the light of the appalling blow which came on Sunday night with the knowledge that in every part of the world, in the minds of the people of all classes and stations of life, there would be adverse and bitter comment on my action in having made any announcement as to the end of the war. The world will never understand that the mistake was that of the [CBC]."

THE SECOND WORLD WAR was over, and the Cold War was about to begin with the first shot, so to speak, fired in Ottawa of all places. On the evening of September 5, 1945, Igor Gouzenko, a nervous twenty-six-year-old Russian cipher clerk, left his post at the Soviet embassy with a large pile of classified documents stuffed under his shirt. He was like many on the staff at the embassy, an agent of the Glavnoye Razvedyvatel'noye Upravleniye, or GRU, the foreign military intelligence directorate of the Red Army General Staff. Risking his life and the lives of his wife, Svetlana (also known as Anna), who was about six months pregnant, his two-year-old son, Andrei, and his family back home, Gouzenko had decided to defect instead of going back to the Soviet Union in a month's time as ordered.

The documents he absconded with that night, and others in his possession he had likely removed during the previous few weeks, contained the sensational revelation that the Soviets had been operating an espionage ring in Ottawa. The GRU had recruited a collection of about fifteen or so willing Canadian-based civil servants, secretaries, research scientists and the only communist MP in the House of Commons, Fred Rose, each of whom had either allegedly coordinated the espionage or passed on various bits of military, technical and privileged information. The aim of this clandestine operation was to satisfy Stalin's urgent need for any and all details about the American atomic bomb. For the most part, the actual information passed on was nothing extraordinary and was easily obtainable, "especially in a country like Canada or the United States," as the *New York Times* later argued. But that was beside the point. There was no escaping the fact that a supposed ally of Canada was spying on it.

The absurd Keystone Kops–like story of Gouzenko's defection is well known. Over a terribly tense two days, he bounced from the offices of the *Ottawa Journal* to the Department of Justice to the Ottawa Police and back again, until he finally found someone in authority who could understand his broken English and assist him. Sounding very much like Ensign Pavel Chekov of *Star Trek* fame, Gouzenko told Elizabeth Fraser, one of several *Ottawa Journal* reporters he encountered, that "it's 'des' if you can't help me." Given the danger he had placed himself in, Gouzenko was understandably stressed and largely incoherent for much of this ordeal. Once the police verified that Gouzenko's apartment had been broken into by Soviet NKVD (the Narodnyi Komissariat Vnutrennikh Del, or the People's Commissariat of Internal Affairs) security agents, he was handed over to the RCMP and placed in protective custody, in which he more or less remained for decades.

There may not have been anyone as fatigued as Mackenzie King in September 1945; "over-fatigued and bedraggled" was how he put it. A new session of Parliament was set to begin on September 9, which King dreaded more than usual. When he arrived at his East Block office on the morning of September 6, about the last thing he expected was to find himself hip deep in a Soviet spy ring. Waiting impatiently for him were Norman Robertson and Hume Wrong of the external affairs department, and they looked grim. Robertson told him, as King later recorded, that "a most terrible thing had happened. It was like a bomb on top of everything and one could not say how serious it might be or to what it might lead." (King kept the story of the Gouzenko affair in a separate secret and confidential diary from September 6 to November 9, 1945. A second volume covering November 10 to December 31, 1945, vanished when King's papers were being organized. This is the only major section of his diary from 1893 to 1950 that was lost or destroyed.)

King's first instinct was reflective: don't do anything rash or take any action that might upset Canada's relations with the Russians, "a trusted ally." Robertson warned him that if the information from Gouzenko— soon code-named "Corby"—was accurate, then there were serious consequences for the United States and Britain as well. Soviet spies may have even infiltrated the White House. He also suggested that Gouzenko was suicidal. On that first day, King probably wished the Soviet defector had taken that route so the whole affair could be quietly disposed of. After all the details were uncovered, however, he changed his mind. Gouzenko, he decided, was a "true world patriot," whose "soul and conscience" had been "roused" by the example of Canadian democracy. When King finally met Gouzenko in mid-July 1946, he found him "youthful in appearance," "clean cut," with a "keen intellect." He told the young Russian that he "was very pleased at the way in which he had conducted himself throughout the period of great anxiety... I appreciated his manliness, his courage and standing for right."

Today, it is easy to mock King's naïveté about Stalin's ruthlessness, the Soviet paranoia about the West and the resulting Red Scare paranoia in the United States (and to a lesser extent in Canada and Britain). But in 1946, a third and more devastating world war seemed likely, and certainly most politicians and most everyone else believed it was only a matter of time until another atomic bomb was dropped from the sky. King rightly feared that if the Soviets and Americans went to war it would be a real war, not a cold one, and that Canada could potentially become the "battlefield."

In the hectic days following Gouzenko's defection, King and Robertson consulted with William Stephenson, the Winnipeg-born British Security Coordination chief (otherwise known as "Intrepid"), then based in New York, who stressed the gravity of the situation. He dissuaded King from playing the role of conciliator and speaking directly to the Russians about what had transpired. Instead, Canadian authorities pretended to help the Soviets search for Gouzenko—whom embassy officials sharply accused of theft—and arranged for a secret order-in-council to re-enact certain sections of the War Measures Act (which had technically expired at the end of 1945), allowing the police to detain anyone suspected of espionage against Canada. For the time being, apart from King, Robertson and Arnold Heeney, only a handful of cabinet ministers knew about the order, which was kept in a vault and not officially passed until October 6. The urgency for the order was due to that fact that Gouzenko's files had revealed that Professor Alan Nunn May, a British physicist and nuclear expert who was working at the National Research Council in Montreal, was in league with the GRU. May was about to return to Britain, and King wanted the power to arrest him if it became necessary. (May was eventually arrested in London in March 1946. He confessed to espionage and was sentenced to ten years' hard labour; he was released in 1952.) King also decided that he had to personally brief both U.S. president Harry Truman and British prime minister Clement Attlee about everything he knew.

As King continued to learn more about the extent of the espionage, he was inevitably drawn into the affair as if he were one of the main characters in a John le Carré spy novel. There is naturally a serious tone to the extensive diary entries King wrote about the affair, but you also can sense the sheer excitement he felt by being at the centre of it all. He ruminated endlessly on the ideological conflict at the heart of the Gouzenko situation, blamed the CCF for "lending themselves, unconsciously perhaps, to the spread of this old kind of communist influence" and expressed sadness at the degree to which he now distrusted the Russians and shock at the actions of MP Fred Rose, who it was eventually confirmed functioned as a recruiter for the GRU.

In time, King came to believe that the hand of destiny was once more directing his path. As he wrote in February 1946:

> I cannot interpret these events and how all of this has come direct to my own doorstep in any way other than in the play of world forces and unseen forces beyond that I have been somehow singled out as an instrument on the part of unseen forces to bring about the exposure that has now taken place. There has never been anything in the world's history more complete than what we will reveal of the Russian method to control the continent as a result of Corby [Gouzenko] fleeing to the Department of Justice and the course then taken, which has been taken under direction of my office.

After much fussing over his travel arrangements, the likes of which only Mackenzie King could engender, he and Norman Robertson made it to Washington and London to speak with Truman, Attlee and their respective intelligence officials. King had wanted to travel in his private rail car to Washington and then go on to New York, where he and Robertson would travel on the *Queen Mary* ocean liner to Britain. When King was told, however, that it cost $300 to transfer his rail car from Grand Central Station to Penn Station in New York City, he quickly vetoed that plan. Against his initial wishes, he was forced to fly to Washington, where he was met by Lester Pearson, who often found

King's various demands and quirks somewhat hard to take and wrote to Hume Wrong with his usual wit: "We went to particular pains to see that the Prime Minister's visit was a pleasant one and for that purpose kept off a storm until ten minutes after their arrival; arranged to have the temperature drop from 92 to 68 within three hours of their arrival, and had the clock put back that night one hour so that Mr. King would be able to get some additional sleep. I don't really see how hospitality could go further!"

The Brits and the Americans—including J. Edgar Hoover, the dictatorial director of the FBI and a fierce opponent of communism—at first agreed with the Canadians that more evidence had to be gathered before any arrests were made. By the beginning of 1946 and following the defection in the U.S. of another spy working for the Soviets, American Elizabeth Bentley—who implicated about 150 people, most of whom were innocent—Hoover had changed his mind. Now he urged King to order the RCMP to act decisively, something Stuart Wood, the RCMP chief commissioner, was eager to do. Yet King hesitated, hoping that diplomacy might still produce some positive results; in particular, he was not keen to arrest any spies before the British and Americans did so as well. To Hoover, this Canadian waffling was "spineless."

Hoover finally forced King's hand by almost certainly leaking the Gouzenko story to Drew Pearson, the hard-hitting syndicated columnist and radio personality, who publicized the case on his NBC radio show on February 3. Two days later, King revealed most of the details of the affair to his cabinet and authorized through an order-in-council for the appointment of a special Royal Commission.

The secret investigation of Gouzenko and his evidence were to be undertaken by two Supreme Court of Canada justices: Robert Taschereau, a former professor of law at Laval University, and Roy Kellock, an Ontario lawyer and judge before his appointment to the high court in 1944. The two justices soon recommended that those Canadians alleged in the Soviet documents to have participated in espionage should be detained and questioned.

From the start, King was torn. He understood that he had to take the proper steps, but feared that his actions "might help to set aflame a controversy of extensive and bitter proportions, throughout this country and the U.S. and also to further suspicion and unrest in other countries." Moreover, unlike with the relocation of Japanese Canadians during the war, King worried more this time about the infringement of civil liberties. "I can see where a great cry will be raised, having had a Commission sit in secret, and men and women arrested and detained under an order in council passed really under War Measures powers," he wrote in his diary on February 13. "I will be held up to the world as the very opposite of a democrat. It is part of the inevitable."

Under the power of the order-in-council of October 6, 1945, the first police arrests were made on February 15 and several more followed the next day. The police had planned to swoop down on the suspected spies at three o'clock in the morning, but King ordered the time changed to the more civilized hour of seven o'clock. "We are not going to behave like the Soviets," he declared. In all, twelve men and one woman, a secretary from the British High Commission, Kathleen Willisher, were arrested and incarcerated at the RCMP's Rockcliffe barracks. The suspects were kept in small rooms with the windows nailed shut. Like those individuals rounded up during the FLQ crisis almost twenty-five years later, they were not permitted to call a lawyer or their families and were interrogated for hours on end. Their guards were not allowed to speak to them, and they were not allowed to speak to each other. Without divulging too many specifics—including that the Soviet Union was involved, a moot point since Drew Pearson had already identified the Russians as the culprits on the radio and in his newspaper column—King released a public statement about the commission and the arrests.

For a while, the Gouzenko affair caused a frenzy of newspaper coverage. Journalists converged on Ottawa, and sensational stories, many of which were inaccurate, were published daily during the spring of 1946. Ignoring that he had earlier wanted to keep the Gouzenko affair hushed up, King now reveled in the attention. "It can be honestly said that few more courageous acts have ever been performed by leaders of the government than my own in the Russian intrigue against the Christian world and the manner in which I have fearlessly taken up and have begun to expose the whole of it," he noted on February 17.

King's vision of himself as a Christian crusader owed to the fact that several of the suspects detained or alleged to have ties to the Soviets (most notably Fred Rose) were Jewish. And King subscribed to the propaganda that Bolshevism and Jews went hand in hand. He was troubled further by James Byrnes, Truman's secretary of state, who for the moment opted to play it cautiously and publicly denied that there was any similar spy ring operating in the United States. To King, this was proof of the impact of Jews in Washington.

"I am coming to feel," he reflected, "that the democratic party have allowed themselves to be too greatly controlled by the Jews and the Jewish influence and that Russia has sympathizers in high and influential places to a much greater number than has been believed." He recalled what his old mentor Goldwin Smith had said to him once about Jews, "that they were poison in the veins of a community." But then, echoing what he had written in his diary in August 1936, he decided that such a sweeping judgment was a mistake. "I myself have never allowed that thought to be entertained for a moment or to have any feeling which would permit prejudice to develop, but I must say that the evidence is strong, not against all Jews, which is quite wrong, as one cannot indict a race any more than one can a nation, but that in a large percentage of the race there are tendencies and trends which are dangerous indeed."

A month later, after several more Jewish suspects were arrested, he added, "It is a rather extraordinary thing that most of those caught in this present net are Jews, or have Jewish wives or of Jewish descent." In fact, he was wrong. Among those individuals who were arrested and eventually charged with violating the Official Secrets Act (and in the case of Carr, with conspiracy to obtain a false passport) only four were Jewish—David Shugar, Israel Halperin, Fred Rose and Sam Carr—and Shugar and Halperin were acquitted. One of those found guilty, Gordan Lunan, had a Jewish wife.

The more he read the commission's reports and Gouzenko's testimony, the more King's suspicions were heightened. He had what could be called a Red Scare of the mind. He questioned the communist sympathies of John Grierson, the founder of the National Film Commission (later the National Film Board), who lost his job in 1945 because of his alleged communist leanings. Oscar Skelton's son Sandy, a Rhodes Scholar, who worked for the Bank of Canada (and became the deputy minister of trade and commerce in 1948), was mentioned by Gouzenko (in the most cursory fashion, which King ignored) as being "an intermediary" (he was not). This got him thinking about the late Oscar himself, his most trusted adviser. "Indeed there was a strong feeling I know among many that Dr. Skelton was much too sympathetic with the Communist outlook," King wrote.

The external affairs department was thought by "some people," at least according to King, to be "too closely associated with ultra-radical positions." The Red menace had even supposedly infiltrated Laurier House. In a fantastical leap of the imagination, King was suddenly uncertain about the patriotic loyalties of his long-serving chauffer and valet, Robert Lay. "I have never been too sure of Lay's feelings and attitudes. He has openly confessed his sympathy with the Reds and has even discussed it with me." Nonetheless, King lived long enough to see Lay appointed Chief of the Senate's stationery division.

The Royal Commission produced three interim reports during March and submitted its final report at the end of June. The two Supreme Court justices recommended that charges for violating the Official Secrets Act be brought against almost everyone who had been picked up and detained by the police. Fred Rose was arrested in March 1946, and Sam Carr, the Labour-Progressive (Communist) Party's organizer, who had left the country, was not apprehended until 1949. Of the sixteen individuals in the Gouzenko affair who faced trials between 1946 and 1949, seven were acquitted for lack of evidence.

As King predicted, the civil liberties of the accused spies quickly became a major issue. M.J. Coldwell of the CCF observed that "the war was fought to destroy states which made such police activities a general practice. To say that it is necessary to resort to authoritarian methods in order to secure evidence is no valid excuse for abrogating the elementary principles of Canadian justice." Similarly, the *Globe and Mail* condemned the government for permitting the commission to employ a "totalitarian procedure." The security of the state was paramount, the newspaper's editors accepted—in an editorial that could have been written in 1970 during the FLQ crisis or in the post-9-11 age—"but there is nothing in the acceptance of this which licenses the Government to suspend all the judicial safeguards in order to facilitate police work or make easier the conduct of an official inquiry."

A self-proclaimed, if selective, civil libertarian, the prime minister agreed with this criticism. As he had told Norman Robertson at the end of February, "I said I thought it was wrong that those who are suspected should be detained indefinitely and that some way should be found to shorten the enquiry and give them full rights of protection which the law allows them." It also troubled him that "the commissioners should let the Liberal party get into the position where it may take a long time for the party itself to be freed of the charge of having acted in a very arbitrary manner with respect to civil liberties." At the same time, the fear of the Soviet Union and communism was strong; according to a poll taken in May 1946, a majority of the Canadians asked believed that the government had acted appropriately.

On March 18, King, despite his personal feelings, defended his and the RCMP's actions in the House of Commons and recounted much of the Gouzenko tale. He even included his misguided faith in "Mr. Stalin" and the hope that Canada's good relations with the Soviet Union would be maintained. Naturally, he was weary when he rose to speak. "I could feel the whole weight of my body from my neck down and also the drawing of my throat from fatigue which made it difficult for me to raise my voice and speak out clearly," he recalled. Somehow, he managed to give an account of the espionage saga in a riveting fashion, speaking for nearly ninety minutes without notes. The galleries applauded when he was finished. "Altogether the scene was quite an impressive one," he added. Still, he thought he could have done better; in his view, it "might have made a very great speech" instead of just a satisfactory one. The *Winnipeg Tribune,* for one, disagreed. "Calm in gesture—he might have been talking about change in the tariff—Mr. King last night told the most dramatic story ever unfolded in the House of Commons," the paper suggested.

The fallout from Gouzenko's defection reverberated for years. In the immediate aftermath, the exposure of the espionage ring led to the recall of several officials at the Soviet embassy in Ottawa who were identified as spies in the commission's final report. Notwithstanding King's conciliatory gestures to the Russians—after all was said and done by the commission, he sent a personal message of friendship to Stalin, which was not answered—Canada's relations with the Soviets suffered, as the communist regime's relations with most western countries did during the tense decades of the Cold War. Even before King retired in 1948, talks were underway for Canada to be part of the North Atlantic Treaty Organization (NATO), a mutual defence pact and a deterrent against a Soviet attack. The information provided by Gouzenko (as well as Elizabeth Bentley) led to new security protocols in Canada and to the hysteria of McCarthyism in the United States, characterized by a rash of false accusations against innocent individuals whose lives were ruined after they were suspected of being communist subversives.

Gouzenko, who lived in Ontario under an assumed name for the rest of his life (he died in 1982) wrote his memoirs and sold his story to Hollywood, which was turned into the 1948 movie *Iron Curtain,* starring Dana Andrews in the lead role. For that deal, he received about $135,000. His daughter Evelyn, who was born in December 1945 at Camp X, the intelligence training centre outside Toronto, only discovered who her father really was when she was sixteen.

As for Mackenzie King, while there would be other European and American meetings and conferences in the final two years of his political career, the Gouzenko affair marked the last time he was truly at the centre of a major international incident.

IN THE IMMORTAL words of American real estate tycoon Donald Trump, "wisdom is in knowing when to stay and when to go." From that perspective, Mackenzie King in 1946 was akin to the scarecrow in *The Wizard of Oz*: he did not have the brains, nor could he take a hint. Mind you, according to Jack Pickersgill, no one, with the possible exception of C.D. Howe, could drum up the courage to tell him that he had fulfilled his public duty. Believing that he was indispensable to the country and the Liberal Party, King stuck around for another two very long years, when it was clear to everyone around him that he should have retired after the Gouzenko affair had been dealt with. He was exhausted nearly all of the time and more irritable than usual with his staff and his colleagues. As Gordon Robertson, who worked closely with King from 1945 to 1948, recalled, "his interest in both the House of Commons and the cabinet flagged. More than once he left his ministers arguing over some point in a cabinet meeting while he went around the corner to his office in the East Block to have tea." Deep down, King probably did know that his time as prime minister was up, yet with no wife or children to spend his remaining years with, the idea of truly being alone must have frightened him. The spotlight had shined on him for so many years that it was next to impossible to fade to black.

His last years were spent in large part trekking back and forth between Canada and Europe, allegedly to help rebuild Europe and reshape the Western world. Ironically, King, as he had throughout his career, still feared any notion of a centralized Commonwealth foreign policy. "Imperial defence," he told Vincent Massey, his underappreciated and underutilized high commissioner in London, "is a phrase that we should like to see dropped from the current vocabulary as it leads to unnecessary misunderstandings and irritations and has little value in relation to the strategic realities of to-day [*sic*]." In truth, rebuilding Europe held little real interest for him. After a private heart-to-heart discussion with Ernest Bevin, the new British Labour Party foreign secretary in Paris, in August 1946, King proclaimed in his diary, "what folly it would be for any of our people to go poking their fingers into the European pie too deeply. There is a mass of intrigue, grievance which none but those who are parties to them will ever be able to know or understand."

Adamant that Prime Minister Attlee was surreptitiously trying to turn the clock back on Commonwealth policy, King did not rush to attend the prime ministers' conference in London in May 1946. When he finally did arrive, all he stressed—with the Gouzenko affair still raging—was that the Soviets should be handled delicately. When Attlee and Bevin raised the matter of sharing the empire's defence costs in view of Britain's heavy war debt, King wanted nothing to do with it. John Holmes, who worked with Massey at Canada House, described

King as acting as he frequently did, "like a teetotaler at a party into which alcohol has been introduced." Nothing made King happier than when the prime ministers' meeting adjourned and no significant decisions or commitments had been made.

Much more notable during this visit to London was a séance he attended at the Queensbury Club. "It was," he asserted, "one of the most remarkable experiences of my life." The medium of the hour was Mrs. Helen Hughes, with whom King had been acquainted since at least 1936. Almost as soon as the group was seated, Mrs. Hughes, as King later recorded, "immediately spoke of those who were around us and named... my father and mother, Max and Bella and my grandmother Christina calling her by name and other members of the family." There was more. "My mother putting her hand on my forehead; my brother standing by me and stroking me on the back. Through the interview, she brought special messages from each in turn. Lent her voice to them to speak to me. There was no mistake in anything she said. Everything as natural and true as could possibly be."

Once the excitement of these visits had died down, there was one more mundane matter to deal with before he departed for Canada. Vincent Massey had informed King in February that he was resigning his post in London at the end of May 1946. Before Massey left London, King George wanted to award him with a Companion of Honour as gratitude for his good service during the past decade and especially throughout the war. According to Canadian government policy, however, such an honour required cabinet approval, and King, who generally loathed Massey, did not rush to grant such approval. It was all very embarrassing for everyone except Mackenzie King, who conjured up various excuses to procrastinate and torment Massey. "He would rather get a recognition from the British Government than from the Canadian government," King told Norman Robertson about Massey, with disdain in his voice. "He had become an Imperialist, and would be such from now on." In King's tainted mind, it was all about "vain glory" and Massey's craving for "Royal recognition; preferring another country." (In his memoirs, Massey recalled King's nasty and humiliating treatment of him with much more politeness than the prime minister deserved.)

As Buckingham Palace waited for the Canadians to do the right thing, King finally cabled St. Laurent and the necessary approval arrived just in time for His Majesty to present Massey with his honour. Quite unbelievably, Mackenzie King, who earlier in the day had admitted he "despised" Massey, now congratulated himself on a job well done. "I confess it made me feel very happy to have felt that I had submerged all my own feelings and to see the complete change in Vincent's mind and nature as a consequence." In this change of heart, King felt that he had "received some direction from Beyond." This incident showed the extent of King's pettiness and a mean streak that was most unbecoming of a prime minister.

Who would replace Massey now became the million-dollar question. Following much negotiation and with due regard to King's perception of each man's talents and personality, Norman Robertson, who needed a break from the hectic pace at external affairs in Ottawa and from Mackenzie King, went to London to serve in the less demanding high commissioner's job. Lester Pearson returned from Washington to become under-secretary of state for external affairs. And Hume Wrong, who had been second in command under Robertson, but who did not get along with King quite as well as Pearson did, took over as ambassador in Washington. Pearson figured that he had his work cut out for him. "I will be in a very tough spot," he told Norman Robertson in the summer of 1946. "After all you [and] Hume have taken a beating, but there have at least been the two of you. I would have no one like either of you to help. I would be new and I would have the P.M. in his declining difficult months."

Pearson's apprehension about King was not without reason. His behaviour continued to be of great concern to those he worked with. He had his staff compute the exact day he was to surpass John A. Macdonald's 6,937 days as prime minister. For King the magic date was June 7, 1946—that it was the seventh day of the month was highly significant—while he was still in London. After some deliberation, Norman Robertson purchased on behalf of everyone in the PMO Frederick S. Oliver's *The Endless Adventure,* the three-volume biography of Sir Robert Walpole, the first and longest-serving British prime minister. King was "deeply touched" by the gift and enjoyed the day immensely, including the bevy of congratulatory telegrams sent to him from Canada marking the occasion. King George also acknowledged the achievement.

King arrived home at the end of June and was back in Europe on July 26, first in London for a brief stay and then in Paris for the peace conference. The Paris conference, which ran from July 29 to October 15, 1946, "was not much of a conference," as C.P. Stacey observes. "Essentially, it consisted of representatives of the great Allied powers revealing to their smaller partners the arrangements they had made for making peace with some of the enemy countries, and giving them an opportunity to offer comments." Nevertheless, King decided to attend, even though he could have sent St. Laurent in his place. A question from Conservative leader John Bracken had convinced him. "I thought Bracken was correct when he said the country and Parliament would expect me to represent Canada." King added, with little humility, "among the public men of the world today, I was one of the few that had led his country through the war and was in the position now of the one who had held office longer than any living man. I thought Canada would expect me to go."

As in London in May, King's interest in the conference waned as soon as the proceedings began. He delivered one speech stressing his long-held views about conciliation, justice and equity. "There seemed to be an astonishing silence while I was speaking and my voice sounded to me much stronger than I expected it would be," he noted later in his diary, somewhat oblivious to his own lack of connection with the audience. For the most part, however, he found the conference "wearisome," strengthening his firm conviction that Canada must not "get unnecessarily involved in international entanglements." The best part about being in Paris was the artichoke at the Hôtel de Crillon's dining room.

Before he returned to Canada on August 25, leaving Brooke Claxton in charge of the Canadian delegation, a tour of the Normandy battlefields was arranged. His guide was none other than historian Colonel Charles Stacey, then stationed in London, who years later was to expose the many secrets of King's diary and the oddities of his character.

Stacey had been instructed by Claxton and Arnold Heeney to keep the tour low-key since the prime minister was so worn out. Stacey flew to Paris to meet King for the first time and found him "most affable." As the tour proceeded through Normandy, small ceremonies were held at which King and Canada were honoured for their part in liberating France from the Nazis. At the first gathering at Caen, King was officially welcomed. Then, as Stacey tells it with great relish, "King drew from his pocket a half-sheet of notepaper on which somebody had written for him three short paragraphs in French. These he read in an accent so excruciatingly bad that I suspect most of his hearers thought he was speaking English." At every subsequent stop, he insisted on reading the same message, as Stacey and the other members of the Canadian party cringed. While walking along the beaches of the D-Day landing, Stacey did his best to educate the prime minister, but in his opinion, King "was not very much interested in the battlefields." In general, Stacey was not impressed by King, a feeling that definitely influenced his many writings about the Liberal leader. "The Prime Minister's conversation," he recalled, "was just what you might have heard from any old gentleman in the back of a Toronto streetcar."

King was fairly miserable on the ocean voyage back to Canada. He lashed out at his loyal staff, essentially for desiring to do their jobs. King rarely appreciated the true talent of those who served and catered to him. His only concern as he saw it was that no one had offered to assist him with his voluminous correspondence. This was not true, although that was beside the point. A tense (and quite unbelievable) private conversation with Heeney on the deck of *Queen Mary* went as follows, according to King: "I then said that I had never felt so let down in all my life as I had been by some of those around me in the last couple of weeks. That, except, where there was something that they wanted for themselves, there had not been even a question as to whether there was anything they could do in furthering my interests or wishes ... I again repeated I had never felt so let down in my life nor had I suffered as much in mind and heart as I had in the last while." The tongue lashing went on and on and all poor Heeney could do was apologize for his and the staff's perceived inadequacies.

King's retirement was a long, drawn-out process. The first step in passing the torch to St. Laurent, whom King had anointed as his successor, occurred as soon as he had returned from Paris. After much contemplation, he decided that he would relinquish his position as minister of external affairs and give the portfolio to St. Laurent so that he could concentrate on his prime ministerial duties.

Apart from the great sigh of relief at external affairs, the change necessitated a significant cabinet shuffle. James Ilsley was moved to justice, Douglas Abbott became the finance minister, Brooke Claxton took over defence and Paul Martin, Sr., was promoted to minister of health and welfare. With Gordon Robertson's assistance, Martin in his former capacity as secretary of state had already worked diligently to have the Canadian Citizenship Act passed in June. The legislation came into effect on January 1, 1947, and at a special ceremony a few days later with the judges of the Supreme Court presiding, Mackenzie King received certificate number one. Nothing gave him more pleasure, as he recalled, than to address the audience "as a citizen of Canada." After finishing his remarks, which included references to the unity of the country and the bonds between the English and the French, the applause "was so long" that he had "to rise and acknowledge it with a bow to the audience and the Chief Justice."

IN THE SPRING and summer of 1947, the Liberal government toyed with the idea of a new reciprocity trade agreement with the United

States—mainly as a way to ease Canada's post-war shortage of American dollars—until King got cold feet and killed it in 1948. Harking back to Laurier's reciprocity treaty with the U.S. in 1911, he feared the deal was really "commercial union" in disguise and the first step towards the American conquest of Canada. No matter how long and hard the mandarins, diplomats and ministers, such as Douglas Abbott and C.D. Howe, urged him to reconsider, King refused to budge. Reinforced by a message from the Beyond, he stood as he always had for an autonomous Canada within the British Commonwealth. The United States was Canada's great friend, ally and trading partner, but there was a limit to how close the two countries would become under King's watch.

Negotiations that were to eventually bring Newfoundland into Confederation two years later were also underway. In mid-August 1947 those meetings were postponed when Frank Bridges, the young minister of fisheries from New Brunswick, died after a brief illness. King accepted Bridges's passing with a certain degree of stoicism. He was far more emotional when the veterinarian told him that his dog, Pat 11, had a malignant tumour on his leg.

This time, unlike the dreadful experience with Pat 1, King allowed the terrier mercifully to be put to sleep. The last day of Pat 11's life was on August 11. This was the occasion for another incredibly lengthy minute-to-minute diary entry. It is at once painful to read—you can truly feel King's agony as he said farewell to the distressed animal—but it is also impossible to review without wincing at his over-the-top raw emotionalism.

King and Pat began the day at the farm house at Kingsmere, which had been extensively renovated during the past several years (at a cost exceeding $15,000) and had replaced Moorside as King's principal summer residence. "The little fellow was sleeping most of the morning in the hall," King wrote, "[and] been remarkably quiet, dear little soul." King had to return to Ottawa for Bridges's funeral and to attend to a few duties before returning later in the afternoon. The diary entry continues: "On the way kept thinking of little Pat, of these being my last few hours with him—only a very few and to get back as soon as I could to my dear little friend. I found it hard to control my emotions on the way as I went over hymns that spoke of God's love and tender care—Safe in the arms of Jesus ... I thanked God I still had the heart of a child."

Before dinner, Joan and Godfroy arrived to say their sad goodbyes and comfort King. The three of them ate the last meal together, and Pat received "a little bit of sausage." When the time came to take the dog to the veterinarian, King kissed Pat three times "on his little head."

As they reached Ottawa, "the sun was setting, a great ball of fire in the west, a glorious sight through the trees." There was another sorrowful parting at the veterinarian's office. King kissed the dog three more times and pronounced a benediction before leaving him.

> He lay horizontally at my feet for a time, then turned his face towards the lighted part of the room, the shadow fell across the lower half of my body, the sun light across his head… I noticed a beautiful white spot, like a star of white hair between his ears, and again a silvery whiteness… just beyond down his nose, he kept his head erect and his paws stretched out before him. It was a beautiful picture, one I shall never forget, his flesh subdued, the mind and spirit exalted. I could ask for no lovelier memory.

Soon, back at Kingsmere, the expected telephone call from the vet came confirming that Pat had been put to rest. He and Joan commiserated for a while. "We spoke of the place he played as holy ground."

His driver, Lay, brought Pat's body to the country the next day, and the dog was laid to rest at a serene spot close to the farm house, near King's sacred Bethel Stone. Joan placed a large white rose on the grave. Alone at the Farm, King felt little Pat's spirit everywhere. "I knelt and prayed to God to help me to be like my little friend in strength and spirit." Now both Pats were "with the angels" and he was certain they would watch over him.

THE DECISION TO retire was near. On the first of October 1947, King met with representatives of Macmillan Publishers to discuss the publication of his memoirs, a mammoth project he planned to undertake in the years ahead. He and Pickersgill had already arranged for the Dominion Archives to hire a young historian, Fred Gibson (who later taught at Queen's University for more than three decades), to organize King's massive collection of papers, which was to form the basis of his book. Gibson already had been working on John Dafoe's papers, and after interviewing him, King felt he would be the right fit for the project. Most importantly, as King stressed, Gibson "himself had always been sympathetic to our administration and he himself an admirer of myself."

That same day in October, King also had a private talk with St. Laurent, who had more or less made up his mind to accept the Liberal leadership and succeed King as prime minister if that was the will of the party and the people. St. Laurent was concerned about stirring up "racial and religious" divisions as only the second French Canadian to lead the country since Laurier. But King eased his mind, and the conversation that day with St. Laurent brought him much "relief."

There is a difference, of course, in thinking of relinquishing power and actually doing so. Earlier, in May, King had strenuously objected to St. Laurent leading a discussion on Canada's response to India being declared an autonomous dominion within the Commonwealth. According to Bruce Hutchison, King actually "snatched" a letter from British Prime Minister Attlee out of St. Laurent's hands, insisting that he was still the prime minister and the letter must have been addressed to him. St. Laurent "was astonished, as were all present."

At the end of November, King, Pickersgill and a small entourage journeyed to Britain so that the prime minister could attend the marriage of Princess Elizabeth and Lieutenant Philip Mountbatten at Westminster Abbey. In London, he found the time to sneak away from any official duties to participate in two spiritualist sittings on the same day: one in the morning with a well-known British medium, Mrs. Sharplin, whom he had first encountered on an earlier trip to London in 1945, and in the afternoon with Geraldine Cummins, a fifty-seven-year-old Irish Protestant novelist and playwright and one of the founders of the Society for Psychical Research. Miss Cummins used "automatic writing" to convey messages from the other world.

At Mrs. Sharplin's, King received supportive messages from FDR and his parents. When the session was over, he wrote that he "felt a great peace of mind and clearness of perceptions." Later that day, FDR, who was clearly in King's thoughts, reappeared at Miss Cummins's along with Laurier. Both urged King not to retire. "I beg of you at whatever cost to continue in public life," the apparition of the late U.S. president told King, as he recorded in the transcript of the session. "It is wiser from the point of view of health to retire but I feel it is your duty not merely to your country but to the world to stay on." After King returned to Canada, Miss Cummins heard from FDR's spirit again, with an opposite message that she relayed: the president now advised that because of King's deteriorating health, he had to retire, which was precisely what he did.

There were also more pedestrian visits to Belgium, France and Holland. In Belgium, King received an honorary degree from the University of Brussels. As Pickersgill recounted in his memoirs, that event turned into a Marx Brothers comedy after King's luggage went missing on the train ride from Paris to Brussels. Panic ensued, but his trusty secretary, the ever-efficient Handy, saved the day by arranging for a tailored formal suit to be delivered to King at his hotel. King made it on time to the university ceremony, and as Pickersgill remembered, "was taken into the robing room and encased in a heavy gown, which covered him completely from neck to toes. He could just as well have been wearing pyjamas."

While King was away from Ottawa, St. Laurent was in charge and made a minor decision—or, at any rate, what should have been regarded as minor decision. Under the auspices of the United Nations, the Americans had set up a temporary commission to supervise developing issues in Korea caused by the end of the Japanese occupation. The U.S. requested that Canada appoint a representative to the nine-member board. Ilsley, who led the Canadian delegation at the UN, reluctantly consented, and St. Laurent approved the decision. That should have been the end of it, and with any other prime minister it probably would have been. But definitely not Mackenzie King, who saw in this issue the undermining of his authority and a potentially dangerous international commitment. As soon as he learned about it, he took great umbrage at his cabinet's actions and refused to sign the order-in-council approving Canada's participation. "We knew nothing about the situation," he told his colleagues, "and should keep out of it." St. Laurent (and Ilsley) respectfully disagreed, and there matters hung for the next few weeks.

King was determined to have his way. He wrote to President Truman explaining that Canada did not want to serve on the commission. Truman asked King to reconsider, and the prime minister drafted a reply affirming his position. Before that second letter was sent, King allowed St. Laurent to read it. The new external affairs minister was not happy. As a matter of principle, St. Laurent could not accept King's decision. He told Pickersgill that he was prepared to resign from the cabinet, thereby throwing King's succession plan into disarray.

On the evening of January 7, 1948, King and St. Laurent dined at Laurier House and, according to King's account, "had a very pleasant talk together." The two men adjourned to the library, where St. Laurent stated that if Canada was not represented on the Korea commission, he and Ilsley would have to resign. King did not care much about Ilsley, but he was dumbfounded that St. Laurent was taking such a stand. He said that he would resign first, even though he wished to surpass Sir Robert Walpole's record as longest-serving prime minister of a western parliamentary-style government. St. Laurent replied that "it would never do for [King] to go. That would mean the complete break-up of the government." They were, it seemed, going to "out polite" each other.

St. Laurent did not really want to quit, since he had already psychologically prepared himself to succeed King. An acceptable compromise

was agreed to when St. Laurent explained that Soviet acceptance of
the commission was required for the supervision of elections in North
Korea and it was almost a certainty that such approval would not be
forthcoming (which was exactly what happened). King said as long as
that was made clear by St. Laurent in Parliament, he would not object
to Canadian participation on a commission that was not going to function. To everyone's relief, this unnecessary disagreement was resolved.

Three weeks later, at the annual National Liberal Federation dinner
held at the Château Laurier, Mackenzie King was piped in and received
"a great ovation" as he took his seat at the head table. "To tell the truth,"
he later claimed, "being fatigued, I hardly noticed it." When it was King's
turn to address the full ballroom, he spoke "vigorously" for about an
hour. He was later "amazed at the power and strength" that he possessed.
Amidst the "intense silence" he presented his state-of-the-union review
of national and international issues before announcing that he was retir-
ing as leader of the Liberal Party after twenty-nine years and that a leadership convention was to be held in the summer. He received another
standing ovation with sustained applause and cheering. He soaked up
this glorious personal moment of triumph and "historic occasion." On
the way out, King stood near the doors and shook the hands of his well-
wishers, many, he recalled, "with tears in their eyes." (He also dutifully
recorded the reaction and body language of each member of his cabinet.)

Back at Laurier House, he telephoned Joan to recount the evening
and had a sandwich and Ovaltine, along with a glass of port to ease his
heart strain. He was a man content. "I must confess that I felt a sense
of relief and of joy in having at last taken the first step in retirement,"
he wrote, especially with the Liberal Party "at the very zenith of its
strength." He fell asleep with all of his loved ones in his thoughts. A little later he wrote to Violet Markham with the news, adding that "I shall
be happy to be free from what is fast becoming an intolerable strain."

King at last surpassed Walpole's prime ministerial record on April
20, 1948. It was another day of telegrams, flowers, and lots of reflection
on the meaning of life. He received another standing ovation, this time
in the House of Commons, but noticeably not from the members of the
opposition parties. He believed that their unkind refusal to acknowledge him was a negative reaction to a Liberal Party press release
announcing his remarkable achievement. "It was clear to me that they
were sick and tired of those indifferent recognitions," he conceded, "and
heaven knows, I felt that way very much myself."

Thomas Crerar, then in the Senate, and admittedly at odds with
King, later dismissed the prime minister's career scorekeeping as "silly

talk." Still, that feat, as well as his long career in general, received the requisite positive press attention. "The Canada of the last quarter of a century... has been the Canada of Mackenzie King and will be so remembered in history," proclaimed a *Winnipeg Free Press* editorial when King first announced his intention to retire. King himself was feeling philosophical when he agreed to speak to journalist Harold Dingman of the Toronto magazine *New Liberty* for a feature article published in the July 1948 issue. "I am so near the end of the road now that I do not believe that what any one man can do will greatly alter things," King told Dingman. "There is a power beyond... a man—one man—can do little things, do the things we must do... but there is a greater power in control." When King read the piece, he enjoyed it but thought that Dingman had exaggerated, or at least misunderstood, his ardent religious beliefs.

At least Dingman, unlike in an article he had penned a year earlier, and similar to so many other profiles of King, had not harped on his apparent loneliness. King was not, he had insisted in his diary with annoyance, "lonely... solitary, aloof and monastic." And he said so publicly as well, but no one in the press gallery seemed to be paying attention. "I see that I am sometimes referred to as a lonely old man," he had commented to the Canadian press in August 1944, on the occasion of his twenty-fifth year as Liberal leader. "I doubt if there is a less lonely man in Canada. I don't recall a lonely moment in the many years since my parents died." That was hardly an honest assessment—the pages of his diary, especially in the post-war years, reveal a somewhat bitter and lonely soul—yet King had no qualms about suggesting otherwise.

The Liberal convention was set for August 5 to 7, the exact dates it had taken place in 1919. Obsessed as he was with anniversaries and numbers, this was a plan that King wholeheartedly supported. Challenging St. Laurent were Jimmy Gardiner and Chubby Power. King himself declared his neutrality, although he fooled no one; the worst-kept secret in Ottawa in the summer of 1948 was that Mackenzie King had deemed St. Laurent as his successor. King was sufficiently worried about the outcome that he convinced C.D. Howe, Douglas Abbott, Brooke Claxton, Paul Martin and a few others all to be nominated and then immediately withdraw, declaring their support for St. Laurent. This "demeaning and unnecessary" charade (in Pickersgill's view) was carried out to great effect on August 6. The next day, and even after he gave the worst speech of the convention, St. Laurent won the Liberal Party leadership on the first ballot. As King watched St. Laurent receive accolades from the crowd, he turned and said in jest to St. Laurent's wife, Jeanne, that "I was right when I said at Quebec that the nigger was drafted for the duration. She looked remarkably happy." And so was he.

That evening, King made another decision. Three weeks earlier, Handy had presented the prime minister with a retirement gift: a four-month-old Irish terrier. King was touched but uncertain whether he wanted another dog. On the closing night of the Liberal convention, he decided to keep the dog and, not surprisingly, named it Pat. After dining with Handy and Pat III, he was driven out to Kingsmere at eleven in the evening. "I can truly say as I leave that I do so with a much lighter heart tonight than I had 29 years ago. A happier heart and a more peaceful mind. I have no regret—only thankfulness to God that all has gone so well, and that this chapter of my life has closed at the beginning of my 30th year of leadership of the party which came to me through my loyalty to Sir Wilfrid and which has been handed on to St. Laurent today through his loyalty to me."

Everyone in Ottawa, and the rest of the country, assumed that Mackenzie King would quickly hand the reins of power over to St. Laurent. There was, however, one more piece of unfinished business. While it was not known at the time, St. Laurent and King had agreed back in May that assuming St. Laurent won the leadership, King was to attend the UN General Assembly in Paris as well as the Commonwealth prime ministers' conference in London in the fall. (Even Pickersgill did not know of this arrangement and only learned of it when he obtained access to King's diary a few years later.)

King was in Paris from September 20 to October 5, where the UN discussion focused on the Soviet control of Berlin. King made his final speech to the assembly, which was well received, though was not satisfied with it. (Worse, he discovered once he had finished that he had skipped over page thirteen, significant in itself). "I have no one around me who is worth his weight in mud," he had declared to his staff as they had worked tirelessly on the speech.

King began feeling ill, and by the time he reached London he had come down with a serious bout of influenza and heart problems. He was so sick and stressed that he wound up being unable to leave his suite at the Dorchester Hotel and summoned St. Laurent from Ottawa to take his place at the conference. A parade of visitors made the pilgrimage up to his room to wish him well. Among the more prominent were: King George, who gossiped about Attlee and the other prime ministers; Winston Churchill, who grudgingly obeyed the doctor's orders and butted out his cigar before he saw King; and Princess Alice, the duchess of Gloucester, who shared with King the "inside story" of how some Jews were allegedly monopolizing the building trade, hoarding materials needed for post-war reconstruction and reaping huge "rake-offs" in commissions. The prime minister of India, Jawaharlal (or Pandit) Nehru, also showed up, and King did his part to help resolve India's desire to remain in the Commonwealth but become a republic. "It is an exaggeration to imply that as its senior statesman, [King] dominated the conference from his suite at the Dorchester Hotel," argues James Eayrs. "But these bed-side conversations [with Nehru and other leaders] were not without influence on the outcome."

Making their way more discreetly to the Dorchester were King's favourite British mediums, Mrs. Sharplin, Geraldine Cummins and Edith Thomson from Glasgow. One day, however, they were not discreet enough. On the afternoon of October 23, Cummins and her associate Miss Beatrice Gibbs arrived at the hotel to find the lobby crowded with reporters expecting King George to make another appearance. The two women were allowed up to Mackenzie King's suite, which the reporters found odd. Many of them wondered who the two "strange (and very ordinary looking) nameless ladies were." The journalists never found out, and the two women left the hotel by a side door. During the séance Cummins conducted by his bedside, King was reassured again by his loved ones. His mother told her "dearest Boy" to pay attention to his health. The ever-hovering Franklin Roosevelt warned King about dangers in the Far East and encouraged him to write his memoirs.

Finally well enough to travel, King arrived back in Ottawa on November 7. As he stepped into Laurier House, he noted at once that "the hands of the clock were in a straight line a little after half past twelve." He looked at his mother's picture and also "got a glimpse of the Parliament Buildings." His hair was thinner and greyer, and his midriff was much portlier than it had been so many years ago when as young Rex he endeavoured to make a name for himself at the University of Toronto.

On Monday, November 15, 1948, he resigned as prime minister. The age of Mackenzie King ended with heartfelt farewells and a huge sigh of relief from just about everyone in Ottawa. Pat III accompanied him in the car to see Governor General Field Marshal Viscount Alexander so that he could make it official. He was glad to have the spunky pooch by his side. The hands of the clock were together at five to eleven when he entered Government House. "Snow was falling quite freely," he wrote. "It was a beautiful sight."

13

With the Loved Ones at Last

I seemed to be conscious of dear mother's spirit being
very near to me. It was like an angel's presence, not
visible and yet not invisible, but both. Her lovely face as
if floating by and leaving a consciousness of its presence.
—THE DIARY OF WILLIAM LYON MACKENZIE KING,
February 20, 1949

THERE ARE GENERALLY two kinds of retirees: those who take great pleasure in their new-found freedom by spending more time with their families and indulging in their hobbies; and those who never learn to adjust to a slower pace of life, who always look behind at what could have been, rather than enjoying the precious moments they have remaining. Mackenzie King convinced himself and told his friends and associates that he was the former type, when it was clear that he was a textbook case of the latter. Two months after he had relinquished control of the country to Louis St. Laurent, he updated Violet Markham on his state of mind. "I doubt if I ever experienced a sense of relief equal to that I did on the afternoon of that day," he wrote. But he also added, "the business of disassociating oneself from government after so many years of office is, if anything, I think, more difficult and perplexing than what is involved in the taking on of new duties."

King found that disassociation extremely difficult, mentally and physically. His blood pressure was high and his heart not as strong as it should have been. During this period, he was understandably obsessed with his health—reporting almost daily, as well, on his irregular bowel movements. For such an insecure person, this loss of control over his own body was intolerable. He became morose and felt pangs of loneliness, whether awake or in the *Twilight Zone* world of his dreams. Handy and Nicol and a few other members of his household staff were the people he spent the most time with, but those one-sided relationships, as good as they were, did not satisfy him. At one particularly low moment in early February 1949, he confessed, "I felt life was not worth living." A visit from his sister Jennie in the summer was pleasant, yet it did little to improve his disposition. She was optimistic, but he found it hard to talk to her. "Her subjects of conversation are very limited and she is so positive in all her statements and talks and talks," he griped.

Much of King's anxiety in the winter and spring of 1949 owed to the fact that he had yet to give up his House of Commons seat. He attended parliamentary sessions infrequently after November 1948 and when he did, he rarely enjoyed it. "I find my seat most uncongenial," he wrote, "persons to [the] right and back, shallow and small and unpleasant in many remarks."

When St. Laurent obtained a dissolution at the end of April 1949 and called an election, King took the final difficult step and ended his thirty-year political career by resigning as a member of Parliament. His final speech before the Liberal caucus was on April 27, and he was treated warmly. "I was given a great, a real old time ovation," he later recalled. "Members standing and applauding for some time." He spoke of his gratitude to all who had served with him, declaring, "If I had my life to live over again, I would choose to take part in public life. [I] spoke of the opportunity of help to fellow men politics offered if pursued as a calling [and] not as a trade." He fittingly chose to end his comments with a quote from Morley's biography of his hero, William Gladstone: "Be inspired by the belief that life is a great and noble calling. Not a mean and slovenly thing that we are to struggle through as one can, but an elevated and lofty destiny."

Three days later, he arrived on Parliament Hill for the farewell. "When we next meet, I will be a freer man," he remarked to Robert Lay, his chauffer, as he exited the car. Remembering his last moment in the Commons, he was at his poetic best.

> As I looked beyond the Press Gallery to the Public Gallery above, I was impressed by the beauty of the golden light which was coming through the windows. They were the only windows in the room though which other than ordinary daylight seemed to be streaming. These windows however were a beautiful golden glow... It seemed to

me that I was seeing and feeling what may be experienced from now on in increasing measures, the glow that comes at eventide the westward being bright horizon of the closing years and days of light, and that I was looking on the larger freedom... I had no feelings of regret at leaving a past that has carried many burdens for many years—a past that has brought with it many burdens over many years but which are lessening more and more.

AFTER THE SESSION adjourned, a throng of MPs congratulated him, and he felt again like a prime minister. Even the new Conservative leader and his old enemy from Ontario, George Drew, shook his hand, which made a positive impression. Likewise, the former Tory leader John Bracken wished him well, except in all the excitement King could not remember his name.

"When the doors were opened I was almost startled at the moment to see the light in the lobby—a light bluish in colour, almost like the moonlight shining down on the plane on the larger Capital of Canada," his diary entry for April 30 continued. "It was almost like a vision that one was being given of the future. It made me very happy. I thought how strange it was that these two lights should above all else be moving my soul and my very being at this time..." He retreated by himself to his office and shut the door. His thoughts were of his first day in the House back in 1908. He remembered the sight of his beaming parents in the gallery as he was escorted in by Laurier. And then there was the symbol of his grandfather. "When I saw the bust of my grandfather," he writes with emotion, "I could not refrain from reverentially and quietly kissing its lips, realizing what his life had meant to mine and what I trust mine has meant to him."

Lonely or not, King was lucky to have had such a loyal and caring friend as the white-haired Joan Patteson, who turned eighty in 1949. Even with Godfroy not well (he did not die until 1957), she continued to dote on King as much as her time allowed. Near the end, he remarked that Joan had "filled the place of my mother in my heart over the years." This was a revealing comment, although Joan had been much more than a maternal figure to him over so many decades. She was a true companion, avid listener, fellow spiritualist and Christian and his connection with normalcy. Without her anchoring his life—and her husband, Godfroy, was there too to do his part—King would have been truly alone and more of a lost soul than he was.

On December 17, 1948, King's seventy-fourth birthday, Governor

General Viscount Alexander (whose term had begun in April 1946; he would be the last British dignitary to occupy the viceregal position) hosted a large party for King at Government House. Out of courtesy to King, Joan and Godfroy had been invited, but only Joan could attend. King had not expected to speak but was "coerced" into doing so. It was a perfect evening for him. "Tonight seems to give its true significance to the meaning of it all," he wrote. "The reward of patient waiting and great humility." Neither were characteristics that King possessed an abundance of, although he surely would have disputed such a contention.

That night, he confessed that he had never seen Joan happier. "She really looked lovely in the black gown she wore," he recorded, "and the sweet expression on her face... I think I derived my greatest happiness through her pleasure. She has been self-[ef]facing all her life." In February, they had a quiet dinner. "I can see she feels real concern about my loneliness and depression for such it is becoming [through] no one to share my life and work and thoughts with—we had the little table for a short time."

There were honours bestowed upon him, too, which temporarily cheered him up. France, Belgium, Luxembourg and the Netherlands presented King with medals of honour for the courage he and Canada had showed during the war. In particular, the Grand Cross of the Order of the Netherlands Lion, which was decreed on February 6, 1949, his mother's birthday, and presented to him on March 11, the day before his grandfather's, resonated with spiritual implications. A year later, in May 1950, plans were announced for the construction of a new bridge in Ottawa across the Rideau Canal from Albert and Slater Streets on the west to Nicholas Street on the east side. The bridge was to be named after Colonel John By, the founder of the city. Solon Low, a Social Credit MP, however, had another proposal. He recommended that King receive that accolade instead, a fine idea that the former prime minister quickly embraced. Jack Pickersgill was discreetly made aware of his wishes, which were passed on to St. Laurent and in due course the "Mackenzie King Bridge" became a reality. It was, as Charles Stacey somewhat cynically put it, "a monument both to King's interest in Ottawa and to his considerable ego." Still, the *Ottawa Journal,* which had never been a great King supporter, approved of the honour.

King had two main literary projects to occupy his time and weigh on his mind: a new biography of his grandfather, which he supported, and the all-consuming writing of his memoir. After several months of negotiations, he had come to terms with Macmillan for a study of William

Lyon Mackenzie, a study that was to portray his grandfather's human side and the "great injustice" done to him. For King, this was to be the fulfillment of "one of the great purposes" of his life.

He insisted that the author of the book be Catherine Macdonald Maclean, a Scottish writer whom he held in high regard. Among other works, she had published a well-received biography of the English poet Dorothy Wordsworth (the sister of William Wordsworth) in 1932. He and Joan had read that book in the fall of 1933 and it had made a tremendous impact on them both. "I could see from the first chapters that I was going to enjoy it exceedingly," King had noted at the time. "It is written with an understanding of the forces that influence conduct and character." When he finally met Miss Maclean in London in May 1944, it was one of those Mackenzie King magical moments in which he was led to her by "some great power." He committed $4,000 over two years as an advance to Miss Maclean and agreed to cover her trans-Atlantic travel expenses. Alas, both King and Maclean died before the book was completed.

His autobiography was also never written. He did oversee the start of the organization of his vast public and private documents, which Frederick Gibson and one of his former students, Jacqueline Côté, who later married King biographer Blair Neatby, were undertaking. Yet just thinking about starting the memoir wore him out. That was understandable. "The amount of material I saw in Ottawa at the Record Office and Laurier House made me gasp!" Violet Markham later told Fred McGregor after King's death. "It can't possibly all be used in a [biography of King]."

In her survey of the history of the King collection, archivist Jean Dryden adds, "the King Papers were stored in an astonishing array of containers in rooms all over the house. Binders, boxes, folders, tubs, cabinets, a cedar chest and a 'large flat leather trunk'... had all been used, and could be found in the Dark Room, the Sun Room, the Cold Room, the Warm Room, the Valet Room, and the Office." His old friend John D. Rockefeller, Jr., tried to make his life easier. On King's seventy-fourth birthday, in December 1948, Rockefeller had gifted him $100,000 to ease any financial concerns. Then, a few months later, the Rockefeller Foundation provided King with another $100,000, no strings attached, for his memoir project.

This generous funding permitted King to recruit his former secretary Fred McGregor—"If I only had McGregor," he repeated to Violet Markham—who had been working for many years as a commissioner of the Combine Investigation Act. At first, no doubt remembering how difficult a boss the former prime minister was, McGregor, sixty-one, demurred. King soon convinced him to come on board, even if he was disappointed that McGregor smartly held out for an annual salary of $12,000. Nevertheless, he was immediately delighted to have him involved. "[McGregor] does not yet begin to realize the task in its true perspective," King wrote in early 1950. "However it is a godsend feeling he is there to take responsibility for seeing that things get done."

ON BOXING DAY 1949, King was miserable. "A terrible night," he lamented, "up but little or no urine passed at any time during night, bowels more or less tight." He prayed that he would recover his strength. It was not to be. On December 27, King suffered a mild heart attack. Joan was by his side and helped him arrange for nurses to visit Laurier House. His physician prescribed him lots of rest. "I really was alarmed," his diary entry read. "Felt I might die during the night without Will made or any disposition of property, possessions etc... It is a frightful situation to face. Have never felt more up against complete possibility of collapse than last night." For the first time since he arrived in Ottawa as a young man in 1900, he could not attend the governor general's levee.

By the spring, he seemed to have recovered and ensured that his will and estate were in order. At the end of June, he moved out to the farm at Kingsmere for what would be the final time. Joan was there to dine and read aloud to him, for which he was most grateful. "She has been amazingly kind," he confessed. "Has come every evening to read or talk since the beginning of the year. So helpful and cheerful. Godfroy equally so. I can never repay all their kindness." He spent some of his time, trying to get his "papers and letters up to date. Those destroyed that should be destroyed," he noted. Who knows what he tossed?

By the first week of July, he was unwell again and able to walk only small distances. One of the doctors caring for him advised him to go into the hospital, but he stubbornly refused. "I said that my own feeling was that my last hope lay in the summer at Kingsmere. That if I had to lose part of the summer to which I had been looking forward since last autumn to go into the hospital for observation, it would be a serious matter. I said if now my condition was such that the summer at Kingsmere would not improve matters, I would rather end my days in the country."

The next week, his condition was up and down. On July 15, he marked the ninth anniversary of the death of his beloved Pat 1 and arranged for roses to be placed on the grave. But he was anxious again about his bodily functions. On July 19, he dictated the following to his recently hired secretary, Rolland Lafleur: "I really should have gone into the sun, but was very tired, and feeling weary, went to bed instead." After fifty-seven years of chronicling the significant moments and minutiae of his daily life, this was Mackenzie King's final diary entry.

The following day, late in the afternoon, he was relaxing in the sun on the verandah, when at 5:20 PM he had a massive heart attack. Joan rushed over from Shady Hill; the doctor took about an hour to arrive. Joan recounted the dreadful episode to Violet Markham a week later: "I did not think he could endure the pain much longer and the terrible effort to get a breath—nor could he have, the Doctor said. So the morphine gave him relief and he became unconscious and never spoke again. While we were enduring those last conscious hours, he was so sweet and kept praying that God would let him live to finish his work, thanked me for what I had done, but I don't think he realized that it was the very final attack." On July 21, several of his nephews came to say their farewells, as did Godfroy Patteson, who was immediately in tears.

William Lyon Mackenzie King—the greatest and most peculiar prime minister Canada has ever had and likely will ever have—died peacefully at 9:42 PM on July 22, 1950. "It was a lovely evening," Joan recalled in her letter to Violet, "but at the moment he died thunder and lightning and torrents of rain came without warning. So many have remarked it—for the rain fell only at Kingsmere—not Ottawa even... The country took charge and he was out of our care."

A month later, Markham reflected further on King's passing. "The penalty of long life (and I am 77) is that inevitably one outlives many friends," she wrote to Fred McGregor. "But no loss could be more painful to me than that of Rex." On King's friendship with Joan, she added, "her gracious personality is the expression of a beautiful character and I cannot say how thankful I am that Rex, who was a lonely man, had such loyal friendship and affection at his side. The blank he must leave in the lives of Mr. and Mrs. Patteson is painful to contemplate." She was right. "I am in a strange sort of restiveness," Joan wrote to Violet in the spring of 1951." I miss him. I miss the thoughts we shared, the bigness of his life and the delight of companionship."

Prime Minister St. Laurent decreed that King was deserving of a state funeral and fittingly left all the arrangements to Jack Pickersgill, who had tended to King for so many years. His body lay for a time at the Parliament Buildings, and long lines of Canadians, many who had undoubtedly cast aspersions on him when he was alive, paid their final respects. The funeral was at nearby St. Andrew's Presbyterian Church, where he had been a regular congregant for five decades and where more praise of his accomplishments and magnitude echoed through the sanctuary. If anyone ever should have attended their own wake, it was Mackenzie King. Then, again, perhaps he was there hovering like an angel and enjoying every minute of it.

From St. Andrew's, the casket was taken to the train station, and King made his final journey to Toronto, where he was laid to rest beside his beloved parents at the family plot in Mount Pleasant Cemetery. "His death... aroused deep feeling everywhere in Canada," remembered Pickersgill. "It was as though people suddenly realized, at the same moment, that the country had lost one of its great citizens, and that a period of history had ended. If there was not an outpouring of grief, there was a solemn feeling of awe."

Editorialists and journalists tried to outdo themselves finding just the right adjectives to describe King's life and career. He was a "tremendous statesman," "a lover of liberty" and the "Great Conciliator." He had "a genius of bringing men of opposite views together and composing their differences," Blair Fraser opined in *Maclean's*. He added in a CBC Radio commentary: "In public he was cold and remote; a little inhuman. But face to face, he could make you feel like you were the one person in Canada he really wanted to see." Even the *Globe and Mail*, which had hounded King for years, now hailed his political savvy, albeit with one last dig. "Such a feat," the newspaper's editors commented in reference to King's political success, "indicates not only unusual qualities of mind, but an understanding of human nature and a mastery of political forces possessed by few. In comparison with world leaders like President Roosevelt or Mr. Churchill, Mr. King notably lacked the capacity to inspire multitudes by personal magnetism, or sway them with his oratory. But in political sagacity he was unrivalled in his generation."

Alex Hume of the more supportive *Ottawa Citizen* wrote the best howler. King, he suggested, "was never rude to a political opponent, nor harsh, nor bitter." Either Hume was merely being polite or he had never heard King lambaste Arthur Meighen and R.B. Bennett, among other Tory enemies. The *Winnipeg Free Press* provided the most astute analysis. "Many years must pass, the secret documents will be studied and a large literature will be written before history pronounces its final verdict on the career of William Lyon Mackenzie King," its editors wrote. "But in one respect history's verdict cannot differ from that already given by the Canadian people who knew him—Mr. King was one of the greatest products of this nation's life." The newspaper grouped him with Robert Baldwin, John A. Macdonald and Wilfrid Laurier.

How he had pulled it off was, and remains, a trickier question to answer. The retrospectives pointed to his aloofness, isolation, detachment and loneliness, all factors that he would have taken great exception to, because they hit too close to the bone. Yet that was only part of the explanation.

KING'S BIZARRE SECRETS gradually trickled out. As soon as his will was probated, his enormous wealth was revealed. In property and investments, he was worth approximately $750,000, a notable sum in 1950, with the equivalent purchasing power today of at least $7 million. Besides Laurier House and Kingsmere, much of this had derived from the generous gifts he had received from the Peter Larkin fund, John D. Rockefeller, Jr., and Sir William Mulock (who had left King $50,000 when he died in 1944). Nevertheless, considering how neurotically frugal King was throughout his life, even his closest friends were shocked.

The year before, King had anticipated such reactions. He wrote on March 31, 1949:

> To my amazement I found that my investments had been so carefully husbanded that they were really about double of what I had believed them to be. I had not realized that my investments mostly in Dominion Bonds and a few private [company] shares had reached proportions that would not be understood by the public. With Mr. [Rockefeller's] additional gift I have now a new problem to face. I can honestly say I would almost prefer to be without more than a mere subsistence, could I be sure of this to the end of my life. I would not like my life's record of service to be overshadowed by anything savouring of wealth, and that is what I have now come to possess, believing the fact that I was none too well off for one being out of office, and a calling of any kind.

"I was staggered," Violet Markham wrote to Fred McGregor, "to read the amount of money in the Will! I always thought Rex a poor man and when he retired I was worried as to how he was going to afford a motor car and a secretary and keep a country house! ... I was puzzled at Ottawa by the obvious affluence of the household and was relieved to see there was no ground to worry about his personal circumstances. But where on earth has all this money come from?"

King and Joan Patteson had shared a most intimate friendship and discussed every subject imaginable—except, it seemed, money. "We always [thought] that he had very little to look forward to when he retired," Joan confessed to Violet. "I think one reason may have been that he did not want any severance from the ordinary men and women—which capitalism would have made, and the dread of long illness and needs at the end, doubtless preyed on him . . . Dear soul, it only increases my tenderest affection and the wish that he had confided in us."

For someone who, according to the press, was "friendless," King left a long list of bequests. His family, friends, staff and the Canadian people were all considered. Jennie, his surviving sibling, received a $5,000 annuity for the rest of her life (she died in 1962), and Max's wife, May, his three nephews, two nieces and any children they had at the time also benefited from King's generosity. Joan and Godfroy were left a $2,500 annuity to continue until they both died. (Joan outlived her husband by three years, passing away in 1960 at the age of ninety-one.) Members of his staff, including Robert Lay, Lucy Zavitske, John Nicol and Edouard Handy, received amounts ranging from $500 to $10,000 (in the case of Handy). Fred McGregor was named one of the executors of King's will and was also left a gift of $10,000.

King deemed that money from Rockefeller's $100,000 gift support university graduate travel scholarships. And most significantly, Laurier House—along with a bequest of $225,000 for its future upkeep from the Larkin fund—and his five-hundred-acre Kingsmere estate were left to the government "in trust for the citizens of Canada." King had hoped that part of Kingsmere might become a wildlife sanctuary. No doubt he would be pleased today that both properties are thriving and operate as popular historic sites and tourist attractions. The Farm, where King spent his last days, became the official residence of the Speaker of the House of Commons.

THE REVELATION THAT he was a spiritualist was the next eye-opener. In May 1937, while King was in Britain, he had visited with the duchess of Hamilton, an avid believer in the world beyond (and an early animal rights activist), with whom he corresponded. At some point during his visit, King encountered J.J. MacIndoe, a Scotsman and member, like the duchess, of the London Spiritualist Alliance. A few months after King died, MacIndoe wrote a letter to the *Psychic News* revealing that King had been a believer. The *News* dispatched a writer to interview the duchess, who confirmed MacIndoe's story. Back in Canada, the *Ottawa Citizen* found out about this and reprinted the *Psychic News* article about King's fascination with spiritualism, setting off a journalistic investigation.

The ever-resourceful Blair Fraser of *Maclean's* started digging for the truth and with confirmation from Handy eventually tracked down and interviewed several of King's favourite mediums, including Geraldine Cummins in London. A few months later the magazine published Fraser's sensational piece, "The Secret Life of Mackenzie King, Spiritualist," which revealed that King received messages from the "other side"—from Laurier, FDR, his parents and even Pat I, who spoke to him through his sister Bella. There were intimate details about the "direct-voice" medium from Detroit, Etta Wriedt. Fraser stressed more than once, as if he felt compelled to protect King's reputation, that the late prime minister was not seeking advice on how to run the country from the dead, "but simply [wanted] to talk to them."

King had not guarded his secret as closely as he thought he had. Many people in Ottawa knew about his extra-curricular activities: his closest friends, relatives, PMO and household staff, Liberal Party associates, several cabinet ministers and senators, and likely some members of the opposition parties. According to Percy Philip, the *New York Times* correspondent in Ottawa, King had blabbed to the governor general, the earl of Athlone, at a Christmas party in 1945 that he had recently chatted with Franklin Roosevelt. "President Truman, you mean," Athlone said, correcting him given that Roosevelt had died seven months earlier. "Oh, no, I mean the late President Roosevelt," King responded nonchalantly.

The prime minister also engaged Philip himself in an intense conversation about life after death during a voyage across the Atlantic. "It is inevitable that one must believe that the spirits of those who have gone take an interest in the people and places they loved during their lives on earth," King told Philip, who later claimed on CBC Radio that he had had a conversation with Mackenzie King's ghost while he was sitting on a bench on the grounds of Kingsmere. "It is the matter of communication that is difficult. For myself I have found that the method of solitary, direct, communion is best."

Similarly, one day in 1945 King had a long talk with Colonel Charles Lindsey, the son of his late cousin George. As Lindsey later told Fred McGregor, "he spoke casually of a talk he had had the week before with my father... who died several years before. He said my father was very much interested in and pleased with what I was doing. He referred to the conversation as if it had been the other day with a living person."

For one reason or another, none of those who knew, not even apparently the opposition MPs or Philip, had anything to gain by telling the world that the country's esteemed leader was conferring with the spirits. In those days, a prime minister's private life truly remained private. Gossip floated around the capital about King's supernatural interests, but, according to Grattan O'Leary, it was "shoved under the rug," where it remained. However, journalists such as Bruce Hutchison and Alex Hume who spent considerable time with King and who believed they knew him well, were stunned when they read Fraser's article and learned the truth. Cabinet minister Paul Martin, Sr., was told about King's beliefs by Philip, the day after King's funeral. Philip opted not to reveal anything before Fraser's piece had been published. Clearly, King had been very lucky that the story was not exposed while he was still alive.

As sensational as the news about King was, the stories did not yet derail his reputation or legacy. Subsequent press stories reaffirmed that King was not gazing at crystal balls while guiding the country's affairs. His spiritualism, it was suggested, was of a more harmless variety. "Those of his Canadian associates who were aware of his odd belief," wrote the *Globe and Mail* columnist J.V. McAree, "probably regarded them as an amiable weakness or hobby like stamp collecting or pigeon breeding." The idea, however, that he was communicating with his terrier, Pat, was refuted strongly by unnamed friends, who told the *Ottawa Journal* that much as King loved the dog, he "would not have claimed it had a soul like mankind."

The entire spiritualism issue would have been a minor footnote in King's career, something for historians to joke about at academic conferences over coffee and stale doughnuts. But his diary, or rather the decision by his literary executors to make the diary public, changed everything, altering perceptions and interpretations about Mackenzie King forever.

King would have been shocked and appalled by this perceived damage to his reputation. On the occasions that he contemplated his own legacy, he envisioned himself as a humanitarian champion of progress, a defender of liberal ideals and the guardian of his family's heritage. One Sunday in April 1935, some months before he regained the prime minister's office in the federal election of that year, he spent a pleasant evening with Joan Patteson. "I told Joan," he later wrote in his diary, "I should like to have 10 years of office—write grandfather's life—so live as Prime Minister as to give grandfather his place in history, then to do like honour to father and mother—leave this as a legacy to young people who may follow—the truth that what we are good, is what is passed on from sacrifices parents and forebears make for us. Would like to write my own life from that point of view—if done worthily." Suffice it to say that the inclusion of his most private secrets would not have been a remote possibility.

Conclusion: Canada's Greatest Prime Minister

As I read grandfather's life I feel more and more what in truth is my place and call in the history of the country. I begin to feel as in earlier days the hand of Destiny guiding me to a Future that has been planned.—THE DIARY OF WILLIAM LYON MACKENZIE KING, *March 3, 1904*

ON THE DAY THAT King died, Norman Robertson, then back from England and serving as clerk of the Privy Council and secretary to the cabinet, was visiting with fellow Canadian diplomat and professor Douglas LePan at LePan's cottage on Georgian Bay. "I turned to Norman and made the kind of anodyne remark about the late Prime Minister that the circumstances seemed to require," LePan later recalled. "I will never forget his reply. He said quite simply, 'I never saw a touch of greatness in him.'" Neither did a lot of people who knew or worked closely with him.

After spending those few trying days guiding King around Normandy in the summer of 1946, Charles Stacey jotted down his impressions: "There was no sense of a forceful penetrating mind at work; and one was left wondering just how the Prime Minister concealed the qualities which he must have possessed in order to rise to the highest office in an important state and retain that office for so long a period." And in his memoir, Jack Pickersgill, who worked side by side with King for more than a decade, had no qualms about admitting that "he felt no real sorrow at Mackenzie King's death."

As harsh they were, there is a certain degree of honesty in these critical assessments. William Lyon Mackenzie King lacked many of the characteristics we tend to visualize in a "great leader." He did not have the zest and vitality of John A. Macdonald, the charm and skill of Wilfrid Laurier or the charisma and intellect of Pierre Trudeau. Nor was he in the same league as the monumental British Liberal Party giant William Gladstone, whom he worshipped, or Franklin Roosevelt and Winston Churchill, whom he came to revere. King was petty, autocratic, selfish and, from our twenty-first-century perspective, an awful bigot. His attitude and treatment of Blacks, Jews and Asians was prejudiced and discriminatory, even if such racist beliefs were acceptable in pre-1950 Canada.

As this biography has demonstrated time and again, King's negative qualities were related to his deep-rooted insecurity, which affected and frequently tainted his relationships with almost everyone he encountered. Classically, he perceived himself exactly the opposite way. Suggestions in newspaper and magazine profiles that portrayed him as a dictator frustrated and angered him. "All these writers seem to assume that men in public life were actuated by ambition alone, desire for power for power's sake [and] motivated by vanity," he wrote in his diary in early March 1947. "As a matter of fact in appointments and everything else, I have sought only what I thought was the highest good for the public. In the same way with regard to measures before the House, etc. I have not tried at any time to assert myself against what I believe to be the best interest of the country and the party. It would be ludicrous to think that one could have carried on for twenty years with the kind of motives which so many of these reporters who don't really know one at all keep attributing to one."

Was he merely fooling himself? Or, was there a truth here missed by everyone else who came into contact with him? King's insecurity did not prevent him from appointing accomplished, principled and opinionated men to his cabinets, from Sir Lomer Gouin to James Ralston and C.D. Howe, who could and did challenge his authority as prime minister. Utilizing the conciliation skills he had acquired as a young deputy minister of labour and mastered during his stint with the Rockefeller Foundation, King sought compromise in almost everything he did. To outsiders, and even to members of his cabinets, this cautious go-oh-so-slow approach seemingly lacked conviction and tenacity. But few could dispute King's unprecedented political success.

That's the main reason why scholars of late have consistently ranked King as one of Canada's greatest prime ministers. "The King diaries are not the record of a public success and a private failure... Carefully read, his public and private records portray an extraordinarily gifted and sensitive man, the product of a certain moment in cultural history, who dedicated his life to public service and succeeded beyond even his own ambitious dreams," argues historian Michael Bliss in his 1994 book, *Right Honourable Men.* "Willie King did a good job both as a politician and as a human being."

Harry Ferns, ironically enough, was of the same mind. A historian who worked briefly with King in the forties, Ferns co-authored (with Bernard Ostry) the book *The Age of Mackenzie King,* published in 1955, which skewered the late prime minister and his alleged insatiable ambition. "Mackenzie King was not a good man or a normal man or a straightforward man or a likeable man," Ferns observed. "He was simply a great man at his trade, in the way that Michelangelo was a great artist or Marlborough a great soldier."

THE ONE RULE of politics, which King understood better than anyone else of his generation, is that you can be as brilliant and committed to your principles as you wish, but if you cannot get elected to office and maintain power for any length of time, what exactly is the point? This is the ongoing dilemma faced by today's bewildered Liberals, but that was definitely not the case in King's day.

So how did he do it? How and why did someone seemingly so average, and whom so many voters either disdained or frowned upon, maintain power for so many years? Persistence was one reason. Without a wife and children, King devoted every ounce of his time and energy (as wavering as that usually was) to strengthening the Liberal Party and pursuing an agenda that appealed to, or at least did not offend, the greatest number of voters. He believed in his heart that he carried Laurier's mantle and kept national unity always at the forefront of everything he did—no matter how long it took him to do.

His leadership during the Second World War and his management of the conscription crisis may not have been pretty, yet it kept his government and the country from tearing itself apart, which happened in 1917. King was never a great orator, but he possessed the political smarts, skilful parliamentary tools and uncanny ability to do what had to be done in order to stay in power. In 1926, his actions during the crisis with Governor General Byng, for example, were cunning and perhaps slightly deceitful, but they certainly were effective.

To his credit, he refocused—with a lot of help from Oscar Skelton and Norman Robertson and others—Canada's relations with Britain, shaping the Commonwealth in his own image, but without sacrificing the country's historic link to the mother country. Any suggestion that

King somehow sold out Canada and our British heritage and steered the country firmly into the American orbit is pure nonsense. After 1900, only a foolish Canadian prime minister would have ignored the United States, and King was hardly that. Closer economic, military and social and cultural ties with the U.S. were inevitable, and King managed relations with the Americans efficiently and effectively.

His claim to fame as one of the country's pioneer Christian social reformers is equally legitimate, even if it took him decades to go beyond old-age pensions, which, to be fair, labourites like J.S. Woodsworth and Abraham Heaps had promoted much more strongly than he had in the mid-twenties. King did not agree with the activist government theories of economist John Maynard Keynes, yet he was intelligent enough to listen to the advice he was getting from his ministers, advisers and bureaucracy to restructure Canadian federalism and in the process create, for better or worse, the fairly centralized welfare state we have today.

The dictum that Walter Lippmann penned after FDR died in 1945 is applicable: "The final test of a leader is that he leaves behind him in others the conviction and will to carry on." That Mackenzie King did, and his vision of a united and compassionate Canada was fostered by his Liberal successors, Louis St. Laurent, Lester Pearson and even Pierre Trudeau to a certain extent.

His astounding electoral success was based even more on his ideological flexibility, or opportunism, depending on your perspective; his unrepentant partisanship, which stocked the Senate with generally loyal and rich supporters who helped establish the Liberals for a long time as the "natural "governing party"; his grasp of the brittle regional realities that defined Canada and his appointment of semi-autonomous provincial czars like Charles Dunning, James Gardiner, C.D. Howe and others to ensure local concerns were properly addressed; and above all, on his innate understanding that if he could not hold Quebec, the Liberals were doomed.

King's treatment and concern for Quebec was especially critical. Three related factors fused in his favour. He might not have truly understood Quebec (as Gordon Robertson argues), but he accepted its distinctiveness and trusted the generally sound judgment of Ernest Lapointe and later Louis St. Laurent to keep the province happy. Next, he was fortunate to govern in an era when the nationalist bloc had not yet manifested itself, as it so dramatically did between 1993 and 2011. During King's reign of power, a majority of French Quebeckers, federally at any rate, were not prepared to cast their ballots for a separate nationalist party, and they had no other choice but to vote for the

Liberals (unlike in the election of May 2011 in which Jack Layton and the NDP swept the province, the CCF were not a viable option in King's day). Even in the failed election of 1930, King still won forty of the province's sixty-five seats.

This brings us to the last and most significant reason to explain his longevity. No matter what policy he proposed or implemented, he could consistently count on the rival Conservatives adopting something far less palatable. For what seemed like an eternity, his Tory opponents continued to find new ways to alienate Quebec, as they had since Louis Riel was hanged in 1885. And they did the same thing to western Canadians with their rigid promotion of economic protectionism and a high tariff. The only person who benefited from these politically wrong-headed policies was Mackenzie King. He may have ardently believed that his life was guided "by the hand of destiny," but in truth King's many political achievements were the result of his diligence, superior political skills and imprudent opposition.

King might not have evoked the adoration from Canadians that other prime ministers have; however, that is beside the point. "King so dominated Canadian politics for almost three decades that it is impossible to think of Canada... without his special contribution," asserts intellectual historian Paul Roazen. "In fact he played such an immense role in Canadian public affairs that I am inclined to think that it is a nonsensical question to ask whether what he did was good or bad for the country; it would be like questioning whether an extraordinarily long-standing marriage was successful or not. Canada and King are inconceivable without each other."

AND WHAT, FINALLY, of the inane perspective of his life gleaned in the pages of his diary and his communing with the spirits? Was Canada actually run by, to borrow Harry Ferns's phrase, "a superstitious lunatic"? There can be no denying that King was a strange individual with enough idiosyncrasies to keep an army of psychiatrists busy full-time for years. He was crippled with insecurity, seemingly powerless to cope with the death of his parents and siblings, and unable to find a woman to marry him and carry on the King line—which, as Reginald Whitaker pointed out many years ago, he should have desperately wanted to do, considering his "obsessive concern with his family tree." Add to this the ticklish fact that, except for maybe a half-dozen occasions in close to six decades, King lacked any kind of sexual life. Is it any wonder that he found something comforting and quite sacrosanct in his contacts with the "loved ones" in the other world?

His friend and fellow believer the duchess of Hamilton offered the best explanation of the fascination that she shared with King and thousands of others in an article she wrote in 1937:

> Spiritualism, I think we would all agree, denotes primarily that we are aware that we are spirits here and now, that our deeper selves are rooted in a spiritual universe, and that we act through our material bodies in matter, which can be and is moulded by spirit into diverse forms and shapes. In other words—that spirit is supreme in the Universe and that Spiritualism is the exact opposite of Materialism, which plumbed the depths of degradation in the last century, when scientists held that matter evolved itself and that brain secreted thought.

The final word should be his. On the warm August night in 1948, the day before he officially relinquished the leadership of the Liberal Party, he arrived back at Laurier House with his heart and mind bursting with sentimentality. As he gazed at the painting of his father in the outer hall, it seemed to come alive.

"I was impressed by what seemed to be a sort of subdued feeling of intense emotion as though he could hardly restrain the feeling he had," he recorded in his diary. "It made me think of the experience I had with Mrs. Fulford when my father came and spoke to her and [the medium] Mrs. Wriedt who was acting as an interpret[er]. He broke down when he referred to what a good son I had been to him and to my mother. I shall never forget the expression on that face as I looked at the painting. I then looked at the one of mother. Thought much of it, particularly in relation to the painting of my grand-father... I thought of how Lady Laurier had presented me with Laurier House; how Sir Wilfrid had taken me into his Cabinet as a very young man, then of the painting of Gladstone; what I owed to him as inspiration in my political life; then of the painting of dear Mr. Larkin and all that he had done for me. It was as though those that I had loved the most and had meant the most in my life had all gathered around me at that time. Sir William Mulock came much in my thoughts with his generosity and real affection for me. I thought too of Dandurand, Lapointe, and Skelton, Norman Rogers and one or two others whose lives had meant very much to me... In the quiet of that hour, in the peace that was in my mind, my soul roused as it was after speaking, particularly, at this moment of great significance in my life. I felt as though we were all together. Certainly we were in spirit. No one will ever make me believe that they were not all with me at the time."

And no one ever will.

Acknowledgements

AS WITH ALL of my books during the past decade, the idea for a biography of William Lyon Mackenzie King started with a "what do you think about this?" e-mail to my proficient literary agent, Hilary McMahon of Westwood Creative Artists. I thank her once again for her encouragement, friendship and advice and for always looking after my best interests.

Special thanks as well to everyone at Douglas & McIntyre, who were enthusiastic about King's biography from the moment we started working on it. In particular, I owe a debt of gratitude to Scott McIntyre for his keen interest in the project and to my enthusiastic and supportive editor, Trena White, who skilfully improved the manuscript and was a pleasure to work with. Thanks, too, to Emiko Morita, Jessica Sullivan, Peter Cocking, Lara Kordic, Corina Eberle, Alison Cairns and proofreader Peter Norman.

For taking the time to comment on parts of the manuscript and offer sage advice on Mackenzie King's life and career, I thank historians and writers Lita-Rose Betcherman, Jack Granatstein, Charlotte Gray, Amy Knight, Peter Waite and especially Norman Hillmer of Carleton University, a friend and a font of wisdom, particularly about Canada's foreign affairs during the twenties and thirties. He also kindly permitted me to research and quote from his forthcoming biography of O.D. Skelton. At Library and Archives Canada in Ottawa, Maureen Hoogenraad, whom I first met more than twenty years ago, again guided me through King's voluminous papers and answered my many queries in a timely fashion. And a heartfelt thanks, as well, to Peter C. Newman for searching through his files for several relevant articles about King, which I put to good use, for his warm friendship during the past three decades and for always being an inspiration on how to write well.

I greatly appreciate the ongoing support of my uncle and aunt, Mel and Eve Kliman, who spent a pleasant Sunday in Kitchener with me touring and photographing Mackenzie King's childhood home of Woodside. I also extend my gratitude to Stan Carbone at the Jewish Heritage Centre of Western Canada for drafting a letter in Italian, which helped me acquire a photograph from an archive in Italy; Dinah Jansen for assisting me with journalist Grant Dexter's papers at Queen's University; and John McCallum, the member of Parliament, for putting me in touch with his mother and aunts, the granddaughters of Joan Patteson, Mackenzie King's closest friend. Research funds for this project were kindly provided by the Winnipeg Arts Council and the Manitoba Arts Council.

My wife, Angie, and our two children, Alexander and Mia (along, of course, with Maggie, our beagle, whom I know Mackenzie King would have loved), offered their usual support, for which I am as always grateful beyond words.

It goes without saying that all omissions, misinterpretations and errors of fact and judgment are solely my own.

A.L.
Winnipeg, March 2011

Timeline

1837   William Lyon Mackenzie leads a failed rebellion in Upper Canada.

1843   John King is born in Toronto to Christina King, a widow, in Toronto. Isabel Grace Mackenzie, the thirteenth child of William Lyon Mackenzie and Isabel Baxter Mackenzie, is born in New York.

1849   The Mackenzie family returns to Upper Canada from New York State.

1861   William Lyon Mackenzie dies in Toronto.

1872   John King and Isabel Grace Mackenzie marry on December 12.

1873   Isabel Grace King gives birth to Isabel "Bella" Christina Grace King on November 15.

1874   William Lyon Mackenzie King is born at Berlin (Kitchener), Ontario, on December 17.

1876   Janet "Jennie" Lindsey King is born on August 27.

1878   Dougall Macdougall "Max" King is born on November 11.

1886   The King family moves into Woodside in Berlin (Kitchener), Ontario.

1891   William Lyon Mackenzie King (WLMK) enrolls at the University of Toronto.

1893   The King family relocates from Berlin to Toronto. WLMK begins his diary on September 6.

1895 WLMK receives his bachelor of arts in political economy from the University of Toronto and works for the Toronto-based *Globe* newspaper.

1896 Receives his bachelor of laws from the University of Toronto and in the fall enters the University of Chicago's graduate program as a fellow in political economy.

1897 Receives his master of arts from the University of Toronto and works for the Toronto *Mail and Empire*. He attends Harvard University in the fall on a scholarship.

1898 Receives his master of arts from Harvard University.

1899 Spends the summer at Newport, Rhode Island, and later in the year studies at the London School of Economics.

1900 Travels throughout Western Europe. At the end of June, he receives an offer to edit the *Labour Gazette* and returns to Canada to begin his new position in Ottawa. In September, he is appointed the deputy minister of labour.

1901 His best friend Bert Harper dies on December 6 in a drowning accident. Purchases property in Gatineau, Quebec, which eventually becomes part of his Kingsmere estate.

1906 His first book, *The Secret of Heroism,* a testimony to Bert Harper, is published.

1908 In September, resigns as deputy minister of labour to enter politics and is elected to Parliament on October 26 as the member from North Waterloo, Ontario.

1909 Receives his doctor of philosophy from Harvard University. In June, he is appointed Canada's first minister of labour.

1911 The Liberals, along with WLMK, are defeated in the general election on September 21.

1914 Is hired as labour consultant by the Rockefeller Foundation.

1915 His sister Bella dies on April 4.

1916 His father, John, dies on August 30.

1917 Is defeated in North York in the general election on December 17. His mother, Isabel, dies on December 18.

1918 As part of his work with the Rockefeller Foundation, he publishes his second book, *Industry and Humanity.* Becomes friends with Joan and Godfroy Patteson.

1919 At the Liberal convention in Ottawa on August 19, he is elected leader of the Liberal Party, a position he shall hold until 1948. On October 20, he is elected as a member of Parliament in a Prince Edward Island riding.

1921 On December 6, the Liberals win the general election, and WLMK, who wins a seat in North York, becomes the prime minister of Canada and secretary of state of external affairs.

1922 His brother, Max, dies on March 18.

1923 Early in January, he moves into Laurier House, given to him as a bequest by Lady Laurier. Attends Imperial Conference in London in the fall, where he argues for Canadian autonomy within the British Empire.

1924 The Pattesons give WLMK an Irish terrier he calls Pat.

1925 On October 29, he is defeated in North York in the general election, but does not relinquish power. He governs with support of the Progressive Party.

1926 Is elected to Parliament on February 15 in a by-election in Prince Albert, Saskatchewan. On June 28, he asks Lord Byng, the governor general, for a dissolution of Parliament, but is refused. He resigns as prime minister and Arthur Meighen, the leader of the Conservative Party, becomes prime minister. The Meighen government is defeated on July 2. On September 25, WLMK becomes prime minister again following the general election.

1927 On July 1, hosts the Diamond Jubilee of Confederation. In November, his collection of speeches, *The Message of the Carillon,* is published.

1928 Establishes "Moorside" as his main summer residence on his Kingsmere estate.

1930 Is defeated in the general election on July 28. Conservative leader R.B. Bennett becomes the prime minister.

1931 The Beauharnois Scandal tarnishes the Liberal Party. WLMK delves further into spiritualism. The Statute of Westminster is passed by the British Parliament, confirming Canada as a self-governing dominion.

1935 On October 14, the Liberals win a majority government, and WLMK becomes prime minister again.

1937 Meets with Adolf Hitler in Berlin on June 29.

1939 Hosts King George VI and Queen Elizabeth during Royal Tour. The Second World War begins on September 3. Canada declares war on Germany on September 10.

1940 On March 26, the Liberals and WLMK win another majority in a general election. On August 17, meets with President Franklin Roosevelt and signs Ogdensburg Agreement.

1941 His key advisor, Oscar Skelton, the under-secretary of external affairs, dies on January 28. His dog, Pat I, dies on July 15 and he soon is presented with another Irish terrier, Pat II. His friend and chief Quebec cabinet minister Ernest Lapointe dies on November 26. Japan attacks Pearl Harbor, Hawaii, on December 7, and the following day Canada declares war on Japan.

1942 Successfully manages first conscription crisis. A national plebiscite is held asking Canadians to release the federal government on its pledge not to impose conscription.

1943 Hosts first Quebec Conference with President Franklin Roosevelt and British prime minister Winston Churchill.

1944 Canadian troops participate in D-Day Landing on June 6. Family Allowance Bill is passed in Parliament in September. Hosts second Quebec Conference in mid-September. Successfully contains second conscription crisis in November.

1945 The Second World War ends in Europe on May 8. Attends United Nations Conference in San Francisco. On June 11, the Liberals win another majority in a general election, but WLMK is defeated in Prince Albert. Wins by-election on August 6 in the riding of Glengarry, Ontario. War against Japan ends on August 15. Igor Gouzenko, a cipher clerk at the Soviet embassy in Ottawa, defects, setting off the first crisis of the Cold War.

1946 Attends Paris Peace Conference.

1947 Becomes first Canadian citizen under new Citizenship Act. His dog Pat II dies on August 11.

1948 Is given his third Irish terrier, Pat III. Resigns as leader of the Liberal Party on August 7. Louis St. Laurent becomes the new leader and, on November 15, the prime minister. WLMK sets the record for the longest-serving prime minister in parliamentary democracy.

1949 On April 30, Parliament is dissolved, and WLMK relinquishes his seat in the House of Commons, ending his political career.

1950 Dies at Kingsmere July 22.

Notes

INTRODUCTION: By the Hand of Destiny

1 *"Like most days"*: Library and Archives Canada (LAC) William Lyon Mackenzie King (WLMK) Diaries, R10383-0-6-E, MG26 J13, January 15, 1929 (hereafter King Diary). The diary can be found at www.collectionscanada.gc.ca/databases/king/001059-100.01-e.php.

1 *Mrs. Bleaney soon*: LAC, WLMK Papers, R10383-15-8-E, MG26 J9, Spiritualism Series, Vol. 3, File 26, H-3037, King to Rachel Bleaney January 15, 1929; Bleaney to King January 1929 (no day).

2 *"He wasn't a crackpot"*: *Globe and Mail*, January 7, 1975, 3.

3 *"You are obviously torn"*: Spiritualism Series, Vol. 7, File 21, H-3042, "Interview with Mrs. Quest Brown February 7, 1939."

3 *The bear "seemed to be"*: King Diary, January 20, 1948. See also September 2, 1939, March 6, 1943, February 6, 1944 (the lather looked like Pat), June 8 1946, November 24, 1946, and December 27, 1946.

3 *At his magical Kingsmere estate*: Cited in Clyde Sanger, *Malcolm MacDonald: Bringing an End to Empire* (Montreal: McGill-Queen's University Press, 1995), 219.

4 *"Life is contingent"*: Michael Shermer, *Why People Believe Weird Things: Pseudoscience, Superstition, and Other Confusions of Our Time* (New York: A.W.H. Freeman/Owl Book, 2002), 5, 274, 281–83.

4 *"I may be prepared"*: King Diary, March 15, 1908. See also, May 8, 1898; July 25, 1907.

5 *"Personally, I do not"*: Ibid., October 22, 1942.

5 *"He succeeded with hardly"*: Frank Underhill, "Concerning Mr. King," *Canadian Forum* 30 (September 1950), 122–23.

5 *The academics pointed*: Norman Hillmer and J.L. Granatstein, "Historians Rank the Best and Worst Canadian Prime Ministers," *Maclean's*, April 21, 1997, 34–39. See also J.L. Granatstein and Norman Hillmer, *Prime Ministers: Ranking Canada's Leaders* (Toronto: HarperCollins, 1999), 83–101.

6 *In the most recent:* Norman Hillmer and Stephen Azzi, "Canada's Best Prime Ministers," *Maclean's,* June 20, 2011, 20–24.

6 *Short, squat, stocky:* John English and J.O. Stubbs, eds., *Mackenzie King: Widening the Debate* (Toronto: Macmillan of Canada, 1977), 5.

6 *"We will frankly confess":* London Free Press, January 25, 1943.

6 *Conservative Party guru:* Dalton Camp, *Gentlemen, Players and Politicians* (Toronto: McClelland and Stewart, 1970), 1.

6 *"How shall we speak of Canada":* "W.L.M.K." in F. R. Scott, *The Eye of the Needle: Satire, Sorties, Sundries* (Montreal: Contact Press, 1957) at Canadian Poetry Online, www.library.utoronto.ca/canpoetry/scott_fr/poem5.htm.

6 *Twenty years later:* Dennis Lee, *Alligator Pie* (Toronto: Macmilllan of Canada, 1974), 28.

6 *Novelist Robertson Davies took:* Robertson Davies, *The Manticore* (Toronto: Macmillan of Canada, 1972), 98–99.

7 *Similarly, in his witty:* Mordecai Richler, *Joshua Then and Now* (Toronto: McClelland & Stewart, 1980), 175–81. In classic Richler fashion, the focus of his vignette on King is the "William Lyon Mackenzie King Memorial Society," whose members are all Jewish.

7 *More disparaging:* Heather Robertson, *Willie: A Romance* (Toronto: Lorimer, 1983); see in particular, 226–28, 308–17. See also, Charlotte Gray, "Crazy Like a Fox," *Saturday Night* 112:8 (October, 1997), 42–43.

7 *Likewise, in Donald Brittain's:* Val Ross, "Restoration of a Tarnished King," *Globe and Mail,* May 31, 1997, C1.

7 *And in Allan Stratton's:* See, www.allanstratton.com/strattonplays.html.

7 *After speaking with:* Paul Roazen, *Canada's King: An Essay in Political Psychology* (Oakville, Ontario: Mosaic Press, 1998), xliii.

7 *His staff, however:* Hugh Keenleyside, *Memoirs: Hammer the Golden Day,* Vol. 1 (Toronto: McClelland and Stewart, 1981), 442.

7 *"It is amusing to hear":* Cited in Frederick W. Gibson and Barbara Robertson, eds. *Ottawa at War: The Grant Dexter Memoranda, 1939–1945* (Winnipeg: The Manitoba Record Society, 1994), 158.

8 *A month earlier:* John English, *Shadow of Heaven: The Life of Lester Pearson,* Vol. 1: *1897–1948* (Toronto: Lester & Orpen Dennys, 1989), 234.

8 *"He really is excessively friendly":* Cited in J.L. Granatstein and Norman Hillmer. *For Better or Worse: Canada and the United States to the 1990s* (Toronto: Copp Clark Pitman, 1991), 79.

8 *Leonard Brockington, the witty:* The story is recounted in H.S. Ferns, *Reading from Left to Right: One Man's Political History* (Toronto: University of Toronto Press, 1983), 151.

8 *His tactics to:* Paul Martin, *A Very Public Life: Far From Home,* Vol. 1 (Ottawa: Deneau Publishers, 1983), 194; Interview with Paul Martin, Jr., Montreal, August 18, 2010.

9 *He had a:* F.A. McGregor, *The Fall and Rise of Mackenzie King* (Toronto: Macmillan of Canada, 1962), 9.

9 *Late one night:* Blair Fraser, "Mackenzie King: A Tribute," *Macleans,* September 1, 1950, in John Fraser and Graham Fraser, eds., *"Blair Fraser Reports": Selections, 1944–1968* (Toronto: Macmillan of Canada, 1969), 39.

9 *The renowned Ottawa-based:* Yousuf Karsh, *Karsh Canadians* (Toronto: University of Toronto Press, 1978), 90; Maria Tippett, *Portraits in Light and Shadow: The Life of Yousuf Karsh* (Toronto: House of Anansi Press, 2007), 134; King Diary, August 21, 1940.

9 *"You will never"*: King to Karsh, January 10, 1948, cited in Tippett, ibid., 134.

10 *Pencils in his office*: H. Blair Neatby, *William Lyon Mackenzie King: The Lonely Heights, 1924–1932* (Toronto: University of Toronto Press, 1963), 303 (hereafter Neatby ii); James A. Gibson, "At First Hand: Recollections of a Prime Minister," *Queen's Quarterly* 61:1 (Spring 1954), 16.

10 *"I soon learned"*: Gordon Robertson, *Memoirs of a Very Civil Servant: Mackenzie King to Pierre Trudeau* (Toronto: University of Toronto Press, 2000), 52.

10 *Of the eleven behaviours*: See, Scott Wetzler, *Living with the Passive-Aggressive Man* (New York: Simon & Schuster, 1992), 34–37.

10 *"Problems arise with the"*: Ibid., 22.

10 *King was "a literary"*: Peter C. Newman, "The Nobility of Sailor Jack's Last Career," *Globe and Mail,* January 17, 1976.

10 *The collection of King's papers:* Jean Dryden, "The Mackenzie King Papers: An Odyssey," *Archivaria* 6 (Summer 1978), 40.

11 *In 1904, when Prime Minister:* F.A. McGregor, *The Fall and Rise of Mackenzie K ing* (Toronto: Macmillan of Canada, 1962), 21–22.

11 *In December 1926:* King Diary, December 27, 1926.

11 *He was in an anxious:* Ibid., June 6, 1935.

11 *"I think this is one"*: Ibid., October 23, 1943.

11 *"It shows how completely"*: Violet Markham, *Friendship's Harvest* (London: Max Reinhardt, 1956), 161.

12 *"What does one"*: C.P. Stacey, *A Date with History: Memoirs of a Canadian Historian* (Ottawa: Deneau Publishers, 1985), 192.

12 *In his diary:* King Diary, August 16, 1945.

12 *Similarly, after King had died:* *Ottawa Citizen,* July 24, 1950.

12 *"This diary is to contain"*: King Diary, September 6, 1893.

12 *King's diary, journalist and critic:* Cited in LAC, *A Real Companion and Friend: The Diary of William Lyon Mackenzie King,* "Introduction," www.collectionscanada.gc.ca/ king/index-e.html.

13 *Until about 1935:* Ibid., "A Private Record: The Diary in King's Lifetime Part 1: The Handwritten Diary, 1893–1935" and "Part 2: The Dictated Diary, 1935–1950" www.collectionscanada.gc.ca/king/023011-1020-e.html.

13 *Over the course:* Dryden, "The Mackenzie King Papers," 55. In this book, when I refer to diary entries after 1935 as having been written or recorded by King, the reader can assume that these entries were dictated to his secretary, unless it is indicated that he personally wrote it himself by hand, which he also did while he was out at Kingsmere. As well, throughout the diaries, King used various abbreviations, including the "&" sign. For clarity's sake I have substituted the word "and' for that symbol; other editing corrections are made in square brackets.

13 *He occasionally recognized:* King Diary, April 10, 1900.

13 *"It was quite curious"*: Ibid., November 5, 1941.

14 *Ever since Charles Stacey:* C.P. Stacey, *A Very Double Life: The Private World of Mackenzie King* (Toronto: Macmillan of Canada, 1976). A notable exception was University of Toronto political scientist Joy Esberey's intriguing *Knight of the Holy Spirit: A Study of William Lyon Mackenzie King* (1980). She used a Freudian perspective and examined King as a whole person, exploring his various neuroses. As useful as this book is, it was also written in stuffy academic language excessively

burdened with psychoanalytical jargon. See Joy E. Esberey, *Knight of the Holy Spirit: A Study of William Lyon Mackenzie King* (Toronto: University of Toronto Press, 1980).

14 *"Much of what Stacey wrote":* Michael Bliss, *Right Honourable Men: The Descent of Canadian Politics from Macdonald to Mulroney* (Toronto: HarperCollins, 1994), 129.

14 *Civil servant par excellence:* Robertson, *Memoirs of a Very Civil Servant,* 70.

14 *So, too, did Grattan:* Grattan O'Leary, *Recollections of People, Press and Politics* (Toronto: Macmillan of Canada, 1977), 81.

14 *Stacey recalled that after:* Stacey, *A Date with History,* 264.

14 *Michael Bliss was:* Bliss, *Right Honourable Men,* 129. See also, Robert H. Keyserling, "Mackenzie King's Spiritualism and His View of Hitler in 1939," *Journal of Canadian Studies* 20: 4 (Winter 1985–86), 26–27. Keyserling argued for a similar fusion of King's private and public lives. In a conversation with Bert Harper in early September 1895, King addressed the "saint" and "devil" side of his personality. King Diary, September 7, 1895.

15 *Over the centuries:* On Puritan diaries, see, Gerald Garth Johnson, *Puritan Children in Exile* (Bowie, MD: Heritage Books, 2002), 113; Perry Miller, Thomas H. Johnson, eds., *The Puritans: A Sourcebook of their Writings* (Mineola, NY: Dover Publications, 2001), 461.

15 *Keeping a diary is also supposed:* Ian Sample, "Keeping a Diary Makes You Happier," *Guardian* (London), February 15, 2009.

15 *As might be expected:* John R. Graham, "William Lyon Mackenzie King, Elizabeth Harvie, and Edna: A Prostitute Rescuing Initiative in Late Victorian Toronto," *The Canadian Journal of Human Sexuality* 8:1 (Spring 1999), 52.

15 *In its self-reflection:* See, H.C.G. Matthew, *Gladstone 1809–1898* (London: Oxford University Press, 1997), 90–94.

15 *"Oh God I want to be like that man":* King Diary, August 12, 1900. In her fascinating Queen's University doctoral dissertation (2008) examining the marginalia in King's voluminous library, Margaret Bedore discovered that one of King's favourite books, which he returned to on many occasions, was John Morley's *The Life of William Ewart Gladstone* (1903). She found that King "dated the first page of the chapter 'Prime Minister' from Volume 2 eleven different times between 1921 and 1950." See, Margaret E. Bedore, "The Reading of Mackenzie King." PhD Thesis, Queen's University, 2008, 88.

15 *It is akin to a journey:* Reginald Whitaker, "Mackenzie King in the Dominion of the Dead," *Canadian Forum* 55 (February 1976), 7.

15 *Take as only one:* King Diary, September 15, 1939.

16 *He detested card games:* McGregor *The Fall and Rise,* 10–11.

17 *However, some years ago:* Roazen, *Canada's King,* 137.

17 *When King's physician:* King Diary, November 10, 1916. According to Douglas How, who in 1945 was a young reporter in Ottawa, one story circulating about King went as follows: The wife of a Liberal MP told King that she would sleep with him if he agreed to appoint her husband to the cabinet. Flushed, King said that such a request required a telephone call to the governor general. He left the woman alone "and fled." Douglas How, "One Man's Mackenzie King," *The Beaver* 78:5 (October/November, 1998), 34.

17 *At the age of nineteen:* Ibid., January 1, 1894.

18 *He wanted the diary:* LAC, WLMK Papers, R10383-23-7-E, MG 26 J 17, Literary executors of the King Estate, Copy of the Last Will and Testament of the Right Honourable W.L. Mackenzie King, clauses 6-11; LAC, Violet Markham Carruthers Papers, R5859-0-X-E MG 32 F6, Vol. 1, File: F.A. McGregor, McGregor to Markham, November 2, 1950.

18 *Did he truly want:* See also, Stacey, *A Very Double Life,* 11.

18 *There was a strong indication:* King Diary, March 4, 1950. See also, J.W. Pickersgill's comments in *Globe and Mail,* January 7, 1981, 8.

19 *"In the first place":* King Diary, March 4, 1950.

19 *That was, at any rate:* LAC, F. A. McGregor Papers (Part of WLMK Papers) R10383-0-6-E, MG 26 J18, Vol. 1, File 1, Varcoe to Pickersgill, December 14, 1950 (hereafter McGregor Papers).

19 *As Colonel Charles B. Lindsey:* Ibid., File 3, Lindsey to McGregor, January 26, 1952.

20 *"Frankness is the":* Ottawa Journal, March 20, 1952. See also, Dryden, "The Mackenzie King Papers," 52; *Montreal Gazette,* July 15, 1955.

20 *"Mr. McGregor and I":* Violet Markham Carruthers Papers, File: Joan Patteson, Joan Patteson to Markham, March 3, 1951; Stacey, *A Very Double Life,* 13.

20 *He did not even want:* Violet Markham Carruthers Papers, File: R.M. Dawson., Dawson to Markham, September 13, 1953; March 18, 1954; J.L. Granatstein, *A Man of Influence: Norman A. Robertson and Canadian Statecraft 1929–1968* (Toronto: Deneau Publishers, 1981), 203.

20 *At first, Fred McGregor:* Violet Markham Carruthers Papers, File: F.A. McGregor, McGregor to Markham, December 27, 1950, February 22, 1951.

20 *Dawson finally asked Violet Markham:* Violet Markham Carruthers Papers, Vol. 1, Dawson file, Dawson to Markham, September 13, 1953; March 18, 1954; Granatstein, *A Man of Influence* 203.

20 *"I regard the decision":* Newman, "Nobility of Sailor Jack's Last Career."

21 *Nevertheless, these publications:* Dryden, ibid., 62.

21 *That action was initially:* Dryden, ibid., 58–66; Granatstein, *A Man of Influence* 203.

21 *Gordon Robertson tells that story:* Robertson, *Memoirs of a Very Civil Servant,* 55. Dryden, "The Mackenzie King Papers," 66.

21 *"I found little on external policy":* Stacey, *A Very Double Life,* 14; Stacey, *A Date with History,* 261–64.

22 *He was "the loner":* Nate Hendley, *William Lyon Mackenzie King: The Loner Who Kept Canada Together,* illustrated by Jordan Klapman (Toronto: JackFruit Press, 2006), 6; Bedore, "The Reading of Mackenzie King," 61.

22 *"No figure in contemporary history":* Violet Markham, *Friendship's Harvest,* 144.

CHAPTER 1: His Mantle Has Fallen on Me

23 *"I was delighted":* King Diary, September 9, 1947.

24 *Some years earlier:* Murray W. Nicholson, *Woodside and the Victorian Family of John King* (Ottawa: National Historic Parks and Sites Branch, 1984), 77–95.

24 *The winding stone walkway:* Charlotte Gray, *Mrs. King: The Life and Times of Isabel Mackenzie King* (Toronto: Viking, 1997), 79–80; Nicholson, ibid., 30–31.

24 *As Charlotte Gray has noted:* Gray, ibid., 79.

24 *That's just the way King:* Cited in Nicholson, *Woodside,* 10.

24 *When the area around:* John English and Kenneth McLaughlin, *Kitchener: An Illustrated History* (Waterloo, Ontario: Wilfrid Laurier University Press, 1983), 1, 6–10.

Berlin was easy to miss. "On April 16, 1833 young James Potter had gone toward Berlin," the *Galt Reporter* recounted fifty years later. "In hunting for the 'Town' he walked through bush and swamp... and found his way to Bishop Eby's when he was told that he had missed the 'Town' which consisted then of a few straggling houses around the corner of King and Queen Streets." *Galt Reporter,* June 7, 1883.

25 *Or, as the* Toronto Mail: Cited in English and McLaughlin, *Kitchener,* 217 n.6.

25 *Yet making your way:* Ibid., 48; Gray, *Mrs. King,* 64.

25 *Tramps and vagabonds:* English and McLaughlin, ibid., 49.

25 *He conceded that:* Ibid., 32–34.

26 *Without a husband:* Nicholson, *Woodside,* 14.

26 *In an era:* Gray, *Mrs. King,* 54.

26 *In a beautiful example:* Ibid., 34.

26 *Presbyterian dictates were:* Ibid., 5–7.

27 *When Isabel, or Bell:* Ibid., 70.

27 *Historians generally have:* See, Joy E. Esberey, *Knight of the Holy Spirit: A Study of William Lyon Mackenzie King* (Toronto: University of Toronto Press, 1980), 12–13; Stacey, *A Very Double Life,* 68; Nicholson, *Woodside,* 68; Gray, ibid., xi-xiii.

28 *John, who was attending:* Reginald H. Hardy, *Mackenzie King of Canada* (London: Oxford University Press, 1949), 3; Bruce Hutchison, *The Incredible Canadian* (Toronto: Longmans, Green and Company, 1953), 12.

28 *At the time of Jennie's birth:* In 1927, the Benton Street house was torn down and a Pentecostal church was built on the property. The church gave way in 1968 to an art gallery, which was transformed again in 1980 into apartment block called the Mackenzie King Manor, honouring the fact that it was the spot of King's birth. However, the owner of the apartment incorrectly capitalized King's name as "MacKenzie King," which it remains to this day. See, Chris Masterman, "Future Leader Born in Kitchener House," *Kitchener-Waterloo Record,* December 8, 2007, W11.

28 *Woodside's large grounds:* Nicholson, *Woodside,* 13–14, 30.

28 *At the same time:* Gray, *Mrs. King,* 54; R. MacGregor Dawson, *William Lyon Mackenzie King: A Political Biography 1874–1923* (Toronto: University of Toronto Press, 1958), 34, 53; LAC, WLMK Papers, R10383-0-6-E, MG 26 J7, Family Papers Series, Vol. 1, File 3, Max King to WLMK, May 8, 1891.

28 *Somewhat isolated:* Dawson, ibid., 10–11.

28 *He enjoyed debating:* Dawson, ibid., 12; Hardy, *Mackenzie King of Canada,* 18.

29 *He was not a tall boy:* King Diary, August 26, 1899.

29 *"It seems to me you":* Cited in Esberey, *Knight of the Holy Spirit,* 14.

29 *Jennie's challenge:* Ibid., 15.

29 *"As for Max":* WLMK Family Papers, WLMK to Jennie King, Vol. 1, File 23, February 15, 1898.

29 *Willie was not above:* King Diary, August 8, 1899.

29 *She insisted that:* Dawson, *William Lyon Mackenzie King,* 11–12.

30 *When Ontario newspaper:* Gray, *Mrs. King,* 97; Dawson, ibid., 20–21.

30 *Like other businessmen:* Nicholson, *Woodside,* 24.

30 *She accepted without:* Gray, *Mrs. King,* 101.

31 *Isabel and her daughters:* Dawson, *William Lyon Mackenzie King,* 19.

31 *In a very good year:* Nicholson, *Woodside,* 54.

31 *His solution to:* LAC, WLMK Papers, R10383-20-1-E, MG 26 J14, King Family Series, Vol. 3, "John King Financial," May 31, 1905.

31 *"I came into this world"*: WLMK Family Papers, Vol. 2, File 13, John King to WLMK, December 27, 1899; Dawson, *William Lyon Mackenzie King*, 22.

31 *"I am constantly reminded"*: Cited in Gray, *Mrs. King*, 211.

31 *"I have awakened"*: King Diary, June 18, 1895.

31 *From January to December 1901*: LAC, WLMK Papers, R10383-0-6-E, MG 26 J11, Finances Series, Vol. 6, Personal Account Book, 1900–1905.

32 *In November 1900*: Gray, *Mrs. King*, 208.

32 *"Now Billie"*: Cited in ibid., 292–94.

32 *"The last word"*: King Diary, September 22, 1898.

33 *"You children could"*: Cited in Gray, *Mrs. King*, 206.

33 *During one of Mackenzie King's*: McGregor Papers, Vol. 1, Norman Robertson to McGregor, October 19, 1954.

33 *At Laurier House*: Hutchison, *Incredible Canadian*, 82; King Diary, March 16, 1901; Gray, *Mrs. King*, 216–17; Janet Adam Smith, *John Buchan: A Biography* (London: Rupert Hart-Davis, 1965), 439.

33 *"No portrait of her"*: LAC, WLMK Papers, R10383-0-6-E, MG 26 J8, Personal Papers Series, Vol. 15, WLMK to Mathilde (Grossert) Barchet, January 29, 1918.

34 *"Men are more likely to confess"*: William Sutcliffe, "Men and Mothers: What is it all about?" *The Sunday Times* (London), April 27, 2009.

34 *In Victorian Canada*: See, Bret E. Carroll, *American Masculinities: A Historical Encyclopedia* (Thousand Oaks, CA: Sage Publications, 2003), 319–20; Richard D. Altick, *Victorian People and Ideas* (New York: W.W. Norton and Company, 1973), 50–54.

34 *Novels of the period*: Barbara Z. Thaden, *The Maternal Voice in Victorian Fiction: Rewriting the Patriarchal Family* (New York: Garland Publishing, 1997), 107; A.N. Wilson, *The Victorians* (London: Hutchinson, 2002), 312; Gray, *Mrs. King*, 210–11.

34 *Boys were encouraged*: Carroll, *American Masculinities*, 320.

34 *The mother-son relationship*: Ibid., 320.

34 *"She is like a little girl"*: King Diary, January 2, 1899.

35 *Some months later*: Ibid., July 29 1899.

35 *While he was travelling*: Cited in Dawson, *William Lyon Mackenzie King*, 83.

35 *"She is, I think"*: King Diary, September 4, 1900.

35 *"Billie likes all my clothes"*: Cited in Gray, *Mrs. King*, 248.

35 *In any age*: Ibid., 248.

35 *"It would be most"*: Markham, *Friendship's Harvest*, 151.

36 *In 1891, a large sign*: Martin L. Friedland, *The University of Toronto: A History* (Toronto: University of Toronto Press, 2002), 158.

36 *One day he wanted*: See, King Diary, November 7, December 31, 1893; January 15, 1894; Dawson, *William Lyon Mackenzie King*, 42.

36 *King, who had a genial nature*: Dawson, ibid., 39–41.

36 *Somehow she endured*: King Diary, August 20–27, 1895; Stacey, *A Very Double Life*, 37–38.

36 *Kitty's feelings were*: King Diary, August 30, 1895.

37 *"I cried very hard"*: Ibid., December 7, 1895.

37 *The nickname stuck*: Dawson, *William Lyon Mackenzie King*, 30.

37 *"There were several"*: Cited in ibid., 49.

37 *"Who in the hell"*: Cited in Friedland, *The University of Toronto*, 168–69.

37 *To protest the university's firing*: William Dale got in serious trouble after a letter he sent to the *Globe* was published in February 1895 in which he brazenly chastised

the university administration and the provincial government's consent of the appointment of George Wrong as a full-time professor in the history department. Wrong, who would go on to become a pioneer in the field of Canadian history, was the son-in-law of former federal Liberal leader Edward Blake, then the chancellor of the University of Toronto. Dale, a lecturer, believed he had been robbed of the appointment due to nepotism. Martin Friedland, the author of a comprehensive history of University of Toronto, could not determine 100 per cent whether or not Blake interfered in Wrong's appointment, but he argues that there is suggestive evidence that the family connection did play a role. King had taken courses from Wrong and had been angered by poor marks he had received on a history essay and exam. See, ibid., 162–63. On King's relationship with George Wrong, see, King Diary, June 13, July 8, October 1, 1894; Robert H. Blackburn, "Mackenzie King, William Mulock, James Mavor, and the University of Toronto Students' Revolt of 1895," *Canadian Historical Review* 69:4 (December 1988), 492–93.

38 *At a boisterous rally:* Hector Charlesworth, *More Candid Chronicles* (Toronto: Macmillan of Canada, 1928), 78.

38 *In his diary:* King Diary, February 15, 1895.

38 *King's commitment:* Ibid., February 20, 21, 22, 1895; Lynn McIntyre and Joel Jeffries, "The King of Clubs: A Psychobiography of William Lyon Mackenzie King, 1893–1900," Unpublished essay, n.d. in University of Manitoba Archives and Special Collections, Heather Robertson Papers, MS 77, Box 1, Folder 2. Dr. McIntyre, now a professor of the Department of Community Health Sciences, Faculty of Medicine at the University of Calgary, wrote this draft with the plan of turning it into book. See also, Lynn McIntyre, "William Lyon Mackenzie King: The Defence Mechanisms of a Social Reformer," *University of Toronto Medical Journal* 56 (1979), 136–40. I thank Dr. McIntyre for sharing this information with me.

38 *Recounting this episode:* Cited in Henry Ferns and Bernard Ostry, *The Age of Mackenzie King* (Toronto: James Lorimer & Company, 1976), 21, n.2.

38 *The commission's final report:* Friedland, *The University of Toronto,* 168.

38 *In words:* McIntyre and Jeffries, "The King of Clubs," 15; Toronto *Globe,* April 18, 1895 (hereafter *Globe*).

39 *There might have been:* King Diary, January 20, February 5, March 3, 6, 23, 1895; Blackburn, "Mackenzie King, William Mulock, James Mavor, and the University of Toronto Students' Revolt of 1895," 494.

39 *Several key influences:* Graham, "William Lyon Mackenzie King," 51; Sara Z. Burke, *Seeking the Highest Good: Social Service and Gender at the University of Toronto, 1888–1937* (Toronto: University of Toronto Press, 1996), 4–5.

39 *The Mackenzie factor:* Stacey, *A Very Double Life,* 68.

39 *In a 1953 review:* Thomas A. Crerar, "The Incredible Canadian: An Evaluation," *International Journal* 8:3 (Summer 1953), 155.

39 *He also memorized:* Hutchison, *Incredible Canadian,* 15.

40 *As soon as he glanced:* King Diary, June 17, 1895.

40 *A few days after:* Ibid., June 22, 1895.

40 *Again, three years later:* Ibid., February 26, 1898.

40 *Woe to anyone:* In 1908, the civil servant, journalist and historian William Dawson LeSueur, who had been commissioned to write a biography for George Morang's "Makers of Canada" biographical collection, produced a critical volume essentially

disputing the "Whig" thinking that had elevated Mackenzie to hero status. Mackenzie King was by then a member of Parliament. He urged his first cousin George Lindsey, Charles's son, to launch a protracted legal battle, which ultimately lasted five years, to prevent LeSueur's manuscript from being published. This was based on the fact that the family had given LeSueur permission to use Mackenzie's private papers as well as Charles Lindsey's notes on the understanding that his biography would be fair, at least by the family's standards. George Lindsey accused LeSueur of using "false pretenses" to gain access to the papers. This nasty legal battle ultimately meant that this critical biography was not available until it was finally released in the Carleton Library series in 1979.

From the start, King had strenuously objected to Morang's selection of LeSueur as his grandfather's biographer. He regarded LeSueur as a secularist, a fair assessment, and also, more oddly, as "a Tory of Tories." Given King's penchant to put a personal twist on most confrontations, he suggested more than a decade later that LeSueur's prime motivation was his animosity against King because in 1900 King had been promoted to deputy minister of labour and LeSueur had to remain as a lowly secretary in the post office department.

On the LeSueur case see, William Dawson LeSueur, *William Lyon Mackenzie: A Reinterpretation*. Edited and with an Introduction by A.B. McKillop (Toronto: Macmillan of Canada, 1979), vi–xxx; King Family Series, Vol. 4, Lindsay v. LeSueur;" LeSueur to Morang, May 11, 1908; Personal Papers Series, WLMK to Barchet, January 29, 1918.

40 *"Underneath the seemingly"*: Cited in Marianne Valverde, *The Age of Light, Soap, and Water: Moral Reform in English Canada, 1885–1925* (Toronto: McClelland & Stewart, 1991), 132; Allan Levine, *The Devil in Babylon: Fear of Progress and the Birth of Modern Life* (Toronto: McClelland & Stewart, 2005), 14.

41 *"I hope that my life"*: King Diary, October 15, 1893.

41 *"I have learned"*: Personal Papers Series, Vol. 24, WLMK to Markham, August 28, 1912.

41 *His predilection to*: Esberey, *Knight of the Holy Spirit*, 109.

41 *He was certain*: King Diary, October 13, 1898; Ramsay Cook, *The Regenerators: Social Criticism in Late Victorian Canada* (Toronto: University of Toronto Press, 1985), 198–201.

41 *He was, in the words*: Cook, ibid., 196.

41 *Toynbee had been deeply*: Burke, *Seeking the Highest Good*, 4–14.

41 *"I was simply enraptured"*: King Diary, July 11, 1894. Margaret Bedore found that Toynbee's *Lectures on the Industrial Revolution* was the "most annotated book in the King collection." He specifically highlighted Toynbee's thesis on page 25: "The historical method has revolutionised Political Economy not by showing its laws to be false, but by proving that they are relative for the most part to a particular stage of civilisation. This destroys their character as eternal laws and strips them all of their force and all of their sanctity." Bedore, "The Reading of Mackenzie King," 70.

42 *"The reference he made"*: King Diary, December 7, 1895.

42 *King became somewhat*: Burke, *Seeking the Highest Good*, 20; Cook, *The Regenerators*, 201.

42 *"Today I had a curious"*: King Diary, December 2, 1939; Dawson, *William Lyon Mackenzie King*, 46.

42 *"Houses of ill fame"*: C.S. Clark, *Of Toronto the Good* (Montreal: Toronto Publishers, 1898), 106; Carolyn Strange, *Toronto's Girl Problem: The Perils and Pleasures of the*

City, 1880–1930 (Toronto: University of Toronto Press, 1995), 93–96.

42 *He felt the police court:* King Diary, November 14, 1895; March 13, 1896.

42 *Despite biographer Charles Stacey's:* Stacey, *A Very Double Life,* 41–48.

43 *He prayed with:* King Diary, February 6-15, March 9, June 5, 1894.

43 *They talked until:* Ibid., October 2, 1894.

43 *"Poor little girls":* Ibid., October 5, 1894.

43 *Edna told King:* Ibid., October 11, 19, 20, 1894; Graham, "William Lyon Mackenzie King," 50.

43 *During the next few:* King Diary, November 22, 1894; July 24, August 25, 1895; April 12, 1896; Graham, ibid., 51.

43 *He later turned:* King Diary, December 12, 16, 31, 1894; January 15, February 3, 10, 1895.

43 *Did Mackenzie King:* Stacey, *A Very Double Life,* 41–48; King Diary, February 1–2, 1894.

43 *"Tried to avoid temptation":* King Diary, April 25, 1896. See also, the entries of May 23, December 6, 1896; May 4, October 15, 1897; McIntyre and Jeffries, "The King of Clubs," 34.

44 *As Charlotte Gray suggests:* Gray, *Mrs. King,* 136.

44 *Willie King and his male:* Graham, "William Lyon Mackenzie King," 52.

44 *As for masturbation:* See, Levine, *Devil in Babylon,* 280–81.

44 *At the end of each lecture:* Angus McLaren, *Our Own Master Race: Eugenics in Canada 1885–1945* (Toronto: McClelland & Stewart, 1990), 70–71; Levine, ibid., 281.

44 *"I committed a sin":* King Diary, October 26, 1893.

44 *Some months later:* Ibid., February 2, 1894.

44 *His sense that he:* Ibid., February 13, 1898.

45 *"What man could have":* Ibid., January 4, 1898; Stacey, *A Very Double Life,* 49–50.

45 *Remarkably, King was:* King Diary, October 27, 1916; September 15, 1939.

45 *King, who had genuinely:* King Diary, April 2, September 30, 1895; June 18, 1896; Blackburn, "Mackenzie King, William Mulock, James Mavor, and the University of Toronto Students' Revolt of 1895," 495.

46 *He immediately went:* King Diary, July 13, 1897; Cook, *The Regenerators,* 203.

46 *As he wrote to:* Cited in Dawson, *William Lyon Mackenzie King,* 55.

46 *That, plus the fact:* Family Papers Series, Vol. 1 File 13, WLMK to John King October 31, 1896; Dawson, ibid., 56–57.

46 *Veblen opened King's:* Cook, *The Regenerators,* 203–4.

46 *Later, he concluded:* King Diary, July 5, 1899; Cook, ibid., 206.

47 *Veblen also accepted:* W.L. Mackenzie King, "Trade Union Organization in the United States," *Journal of Political Economy* (March 1897), 201–15; "The International Typographical Union" *Journal of Political Economy* (September 1897), 458–84; Dawson, *William Lyon Mackenzie King,* 63.

47 *She had, King wrote:* King Diary, March 18, 1898.

47 *Yet a few weeks later:* Ibid., April 7, 1898; Stacey, *A Very Double Life,* 50–1.

47 *Mathilde had come to Chicago:* Louise Reynolds, *Mackenzie King Friends & Lovers* (Victoria: Trafford Publishing, 2005), 8.

47 *He was instantly:* King Diary, May 23, 1897.

47 *One example, a letter:* The letter from WLMK to Grossert of August 30, 1897, is included in the King Diary entry for December 31, 1897.

48 *According to his diary:* King Diary, June 23, 1897.

48 *By January 1898:* Ibid., January 5, February 24, 27, March 11, 1898.
48 *One after the other:* Ibid., April 6, 7, 1898.
48 *Isabel's letter arrived:* Family Papers Series, Vol. 1, File 25, Isabel King to WLMK, April 6, 1898; Stacey, *A Very Double Life,* 55.
48 *"Why did Mother":* Ibid., Jennie King to WLMK, April 6, 1898.
49 *This was followed:* Ibid., John King to WLMK, April 9, 1898.
49 *He noted in his diary:* King Diary, April 9, 10, 1898. See also, March 6, 14, 23, 31; April 1, 1898.
49 *Had King married:* Gray, *Mrs. King,* 144–45.
49 *Near the end:* King Diary, April 19–20, 1898.
49 *By the time:* Ibid., August 18, 1898.
49 *Thereafter, King remained:* See Personal Papers Series, Vol. 15; Stacey, *A Very Double Life,* 66.
49 *"It gave me a great":* Personal Papers Series, ibid., WLMK to Barchet, May 6, 1940.
50 *Professor James Mavor:* Friedland, *The University of Toronto,* 176.
50 *"Divine Providence":* King Diary, July 12, 1897.
50 *The fact was:* Bryan D. Palmer, *Working-Class Experience: Rethinking the History of Canadian Labour, 1800–1991* (Toronto: McClelland & Stewart, 1992), 135.
51 *"What a day":* King Diary, September 18, 1897.
51 *Within days:* Ibid., September 19, 1897.
51 *He told his parents:* Family Papers Series, Vol. 1, File 20, WLMK to John and Isabel King, November 28, 1897; Dawson, *William Lyon Mackenzie King,* 69.
52 *At Harvard, King:* Dawson, ibid., 73–76.
52 *Much to King's delight:* King Diary, May 20, June 23, 1898; Cook, *The Regenerators,* 204.
52 *After much hand-wringing:* Gray, *Mrs. King,* 157–59, 163–64.
52 *"I go away not":* Family Papers Series, Vol. 2, File 10, WLMK to John and Isabel King, September 24, 1899; Dawson, *William Lyon Mackenzie King,* 69.
53 *After receiving a tour:* King Diary, February 24, 1900; Burke, *Seeking the Highest Good,* 38.
53 *He wrote in his diary:* King Diary, November 16, 1899.
53 *"Two of the women":* Ibid., January 3, 1900; Dawson, *William Lyon Mackenzie King,* 86–87.
53 *"This has been an important":* King Diary, July 9, 1900.
54 *"I am a Canadian":* Family Papers Series, Vol. 2, File 20, WLMK to John King, July 10, 1900.

CHAPTER 2: The Peacemaker
55 *All four men:* Globe, March 18, 2005.
55 *A year later:* Dawson, *William Lyon Mackenzie King,* 173.
56 *"His rise to prominence":* Globe, July 29, 1906.
56 *"Gloomy" was his:* King Diary, July 24, 1900.
56 *Liberal prime minister:* Sandra Gwyn, *The Private Capital: Ambition and Love in the Age of Macdonald and Laurier* (Toronto: McClelland and Stewart, 1984), 228.
56 *A more fitting nickname:* John H. Taylor, *Ottawa: An Illustrated History* (Toronto: James Lorimer & Company, 1986), 119.
56 *One large enterprise:* Gwyn, *Private Capital,* 228.
56 *The entire civil service:* Taylor, *Ottawa,* 120.

56 *More than fifteen thousand:* Gwyn, *Private Capital,* 228.

56 *The countess of Aberdeen:* Cited in ibid., 228.

56 *Sparks Street already:* Taylor, *Ottawa,* 94; Ibid., 229.

57 *"I have a nobler":* Family Papers Series, Vol. 2, File 28, WLMK to John and Isabel King, March 9, 1901.

57 *"What matter the credit":* King Diary, August 3, 1900.

57 *"I think I can take":* Ibid., February 24, 1904.

57 *Historian Paul Craven's:* Paul Craven, *'An Impartial Umpire': Industrial Relations and the Canadian State 1900–1911* (Toronto: University of Toronto Press, 1980), 209.

57 *"If I can save":* King Diary, February 10, 1902.

58 *Any story or statistic:* Ibid., August 5, 1900; Craven, *'An Impartial Umpire,'* 210.

58 *"I was scarcely":* King Diary, August 5, 1900.

58 *In September 1900:* Dawson, *William Lyon Mackenzie King,* 104.

58 *"He is not yet":* Ottawa Journal, December 17, 1904.

59 *In a letter to his:* Family Papers Series, Vol. 2, File 24, WLMK to John and Isabel King, November 4, 1900; Dawson, *William Lyon Mackenzie King,* 107–9.

59 *Some months later:* Canada. Parliament. *House of Commons. Debates* (hereafter House of Commons Debates) April 3, 1901, 2590; Dawson, ibid., 111–12.

59 *Harper, who also:* King Diary, August 12, 1900.

59 *"No man who desires":* Harper to WLMK, December 3, 1900, in William Lyon Mackenzie King, *The Secret of Heroism* (New York: Fleming H. Revell, 1906), 97–98.

59 *A favourite book:* King Diary, January 5–6, 1901; Stacey, *A Very Double Life,* 80.

59 *Harper began his:* King, *The Secret of Heroism,* 155. See also, 152, 159.

59 *"I miss you, Rex":* Ibid., 159.

60 *King did not like:* King Diary, March 30, 1901.

60 *"The deeds of":* Ibid.

60 *"The Information":* Diary of Bert Harper, March 31, 1901, cited in Dawson, *William Lyon Mackenzie King,* 119–20.

60 *Some months earlier:* Gray, *Mrs. King,* 247.

60 *"The trees were":* Cited in ibid., 247.

60 *On the summit:* Family Papers Series, Vol. 2, File 23, WLMK to John King, October 20, 1900; George F. Henderson, "Mackenzie King's first visit to Kingsmere," *Up the Gatineau* 19, http://outaouais.quebecheritageweb.com/article_details.aspx?articleId=140.

60 *The day:* WLMK to John King, October 20, 1900; King Diary, May 8, 1900; Dawson, *William Lyon Mackenzie King,* 116–17.

61 *King and Mrs. Herridge:* King Diary, September 7, 8, 1901.

61 *He soon purchased:* Stacey, *A Very Double Life,* 88.

61 *Late that same afternoon:* LAC, WLMK Papers, R10383-12-2-E, MG 26 J6, Pamphlets and Clippings, Vol. 16, File 131, Amaryllis, "Notes From the Capital," *Saturday Night,* December 14, 1901.

61 *Harper's group was:* King, *The Secret of Heroism,* 11; *Manitoba Free Press* (hereafter MFP), December 7, 1901; Gwyn, *Private Capital,* 323–24.

61 *"He has gone":* King Diary, January 2, 1902; *Ottawa Citizen,* December 10 and 11, 1901.

62 *"The quality of a man's":* King, *The Secret of Heroism,* 21.

62 *As Stacey caustically:* Stacey, *A Very Double Life,* 84; King, ibid., 151–52.

62 *Ethel Chadwick:* Gwyn, *Private Capital,* 326. See also, *Globe,* April 21, 1906,
Saturday Magazine section, 4; *Saturday Night,* March 3, 1904, 14; *Times Literary
Supplement,* March 9, 1906, 79; *Toronto Star,* March 3, 1906.

62 *On the eighth anniversary:* King Diary, December 6, 1909.

63 *As King's friendship with her:* Personal Papers Series, Vol. 20, Marjorie Herridge to
WLMK, July 8, n.d., May 5, May 18, June 13, June 18, June 29, 1906.

63 *He found the quiet:* King Diary, October 12, 1901.

63 *Yet the more:* Ibid., February 10, 1901; February 26, 1902.

63 *On the night:* Ibid., February 26, 1902.

64 *"I long for rest":* Ibid., February 28, March 1, 1902.

64 *"The story of my":* Ibid. September 21, 1902.

64 *"There would have been":* Gray, *Mrs. King,* 253; Stacey, *A Very Double Life,* 93;
Reynolds, *Mackenzie King Friends & Lovers,* 56.

65 *King adored the:* King Diary, September 26, 1904.

65 *As fate would:* Dawson, *William Lyon Mackenzie King,* 117.

65 *One evening:* King Diary, September 16, 1904.

65 *That spring:* Reynolds, *Mackenzie King Friends & Lovers,* 61–62.

66 *She made the:* King Diary, December 17, 1914.

66 *Five years later:* Reynolds, *Mackenzie King Friends & Lovers,* 66.

66 *"The past ten":* King Diary, March 22, 1924; Stacey, *A Very Double Life,* 103–4.

66 *In the following:* King Diary, July 1, 1934; Reynolds, *Mackenzie King Friends &
Lovers,* 68.

66 *Whatever small scale:* Craven, *'An Impartial Umpire,'* 209–10. King wanted Mulock
to have stationary printed with the heading "Minister of Labour" emboldened on it.
As a wealthy man, Mulock was reluctant to do so. "They will make fun of me if
I come out as the champion of labour," he told King. King eventually convinced him
otherwise. See, King Diary, August 17, 1900; Dawson, *William Lyon Mackenzie
King,* 103–4.

67 *This was an age:* See, Palmer, *Working-Class Experience,* 164–65.

67 *"I feel at times":* King Diary, August 4, 1907.

67 *Leonard Barrett:* "The Winnipeg General Strike," www.timelinks.merlin.mb.

68 *"They are driven to it":* See, Nicholson, *Woodside,* 74.

68 *Only the CMA:* Craven, *'An Impartial Umpire,'* 211–12; King Diary, August 18, 1900;
February 24, 1904.

68 *The investigating committee's:* Craven, ibid., 85.

68 *Still, between 1900 and 1907:* Palmer, *Working-Class Experience,* 205; Gregory
S. Kealey, *Workers and Canadian History* (Toronto: University of Toronto Press,
1995), 427.

69 *He was indeed:* See John Haynes, *The Fundamentals of Family Mediation* (Albany,
NY: State University of New York Press, 1994), 1–8.

69 *"There is a relation":* Cited in Dawson, *William Lyon Mackenzie King,* 144. King
could lose patience with the stubborn attitudes of management and labour. On
June 3, 1902, he was trying to solve a strike at Port Burwell on Lake Erie and was
not impressed by what he found. As he noted that day in his diary, "The whole thing
was caused by a stupid blunder of a stupid man dealing with stupid men, was over
a desired increase of wages and if those in authority had acted shrewdly would not
have happened." See, King Diary, June 3, 1902.

69 *King regarded Dunsmuir*: King Diary, May 10, 1903; Craven, 'An Impartial Umpire,' 248.

69 *He was not blind*: Craven, ibid., 233–35, 246.

69 *In 1904*: Ibid., 254–61.

70 *By November*: Ibid., 264–65.

70 *"I was desirous"*: LAC, WLMK Papers, R10383-0-6-E, MG 26 J4, Memoranda and Notes Series, Vol. 13, File 80, "Confidential Memorandum Re Lethbridge Strike," 1906; Ibid., 267.

70 *In the end*: Palmer, *Working-Class Experience*, 205; William Baker, "The Miners and the Mediator: The 1906 Lethbridge Strike and Mackenzie King," *Labour/Le Travillieur* 11 (Spring 1983), 116–17.

70 *King's detractors*: Ferns and Ostry, *The Age of Mackenzie King*, 69; Dawson, *William Lyon Mackenzie King*, 140.

70 *King, too, was greatly*: Craven, 'An Impartial Umpire,' 43–44, 76–77.

70 *"The measure is my own"*: King Diary, January 3, 1907.

71 *After the prime minister*: Ibid., January 10, 1907; Craven, 'An Impartial Umpire,' 281.

71 *Another Ontario newspaper*: Cited in Gray, *Mrs. King*, 283.

71 *Yet between 1907*: Palmer, *Working-Class Experience*, 163.

71 *In certain instances*: Ibid., 206.

71 *From 1895 to 1914*: Ibid., 163.

72 *"Whatever King's impenetrable"*: Ian McKay, "Strikes in the Maritimes, 1901–1914," *Acadiensis* 13 (1983), 43.

72 *The governor general*: King Diary, December 12, 1900.

72 *She was a handsome*: "Miss Violet Markham," *Times of London*, February 3, 1959; "James Carruthers," June 29, 1936; *A Real Companion and Friend*, "Violet Markham."

72 *Ever ambitious*: Dawson, *William Lyon Mackenzie King*, 144.

72 *Grey introduced King*: Markham, *Friendship's Harvest*, 143, 152; Violet Markham, *Return Passage: The Autobiography of Violet R. Markham* (London: Oxford University Press, 1953), 82.

73 *For a brief moment*: Dawson, *William Lyon Mackenzie King*, 127; Gwyn, *Private Capital*, 293–97.

73 *In the back*: King Diary, November 4, 1905; January 2, February 3, 1906; Dawson, ibid., 180–83.

73 *Martha was a great*: Gray, *Mrs. King*, 299.

73 *"Unless she can surpass"*: King Diary, August 28, October 8, 1907.

73 *In 1909, Martha*: "Senator George T. Fulford, Ontario Heritage Foundation, www.heritagetrust.on.ca/getattachment/Programs/Commemoration/Provincial-Plaque-Program/Plaque-of-the-Month/Archives/Senator-Fulford-ENG.pdf.aspx.

74 *He gushed*: King Diary, February 8, 1901.

74 *He held her tightly*: Ibid., September 2, 1901.

74 *That style played*: See Pamphlets and Clippings, Vol. 32, File 244, "W.L.M. King Scrapbook of Clippings, 1906–08."

74 *Popular writers*: Levine, *Devil in Babylon*, 34.

75 *At the top of*: Ibid., 39; Peter W. Ward, *White Canada Forever* (Montreal: McGill-Queen's University Press, 1978), 59–60.

75 *"I have very"*: Cited in Ward, ibid., 58–59.

75 *Mackenzie King and*: See, Ibid., 83; King Diary, January 11, 1909; Stephanie Bangarth, "Mackenzie King and Japanese Canadians," in John English, John Kenneth

McLaughlin and P. Whitney Lackenbauer, eds., *Mackenzie King: Citizenship and Community* (Toronto: Robin Brass Studio, 2002), 105–7, 111.

75 *Slightly more enlightened*: King Diary, January 11, 1909; King, *Industry and Humanity* (Boston: Houghton Mifflin, 1918), 323 (hereafter *Industry and Humanity*, 1918).

75 *Yet it was*: "The Oriental Influx Problem," *Globe*, May 7, 1908. See also, "Report by W.L. Mackenzie King, C.M.G., Deputy Minister of Labour, on Mission to England to Confer with British Authorities on the Subject of Immigration to Canada from the Orient and Immigration from India in Particular," Canada, *Sessional Papers* (Ottawa, 1908), No. 36A; John Price, "'Orienting' the Empire: Mackenzie King and the Aftermath of the 1907 Race Riots," BC Studies 156 (Winter/Spring 2007/08), 70–71.

75 *But the same assessment*: In his 1909 book, *Strangers Within Our Gates* Woodsworth ranked different nationalities and ethnic groups according to how easily he believed they would assimilate and adapt to "Canadian values." Naturally, British, Scandinavians and Germans were high on his list; Slavs, "Hebrews" and Italians placed slightly lower, while "Negroes" and "Orientals" were deemed unwelcome and unwanted. See, Levine, *Devil in Babylon*, 57–58.

76 *Vancouver Liberal* MP: Ward, *White Canada Forever*, 66.

76 *The fear and hatred*: *Vancouver Province*, September 9, 1907; Ward, ibid., 68–69.

76 *At first, he thought*: Ward, ibid., 70-4; Memoranda and Notes Series, Vol. 42, File 221, "Royal Commission to investigate into losses sustained by the Japanese population of Vancouver, B.C., 1907;" *Industry and Humanity*, 1918, 75–6.

76 *Several Chinese opium*: W.L, Mackenzie King, "Report on the Need for the Suppression of the Opium Traffic in Canada," (Ottawa: King's Printer, 1908); House of Commons Debates, January 26, 1911, 2518–53; Bangarth, "Mackenzie King and Japanese Canadians," 109–10.

76 *For many years*: Bangarth, ibid., 110.

76 *Further inquiries King*: Kirk Niergarth, "William Lyon Mackenzie King's 1908 Adventure in Diplomacy: Coming Back a 'Public Man,'" Unpublished paper presented at the Canadian Historical Association Meetings, (May 2009), 6–7; Bangarth, ibid., 110.

77 *Almost overnight*: Bangarth, ibid., 110.

77 *King's stringent stand*: See, Niergarth, "William Lyon Mackenzie King's 1908 Adventure in Diplomacy," 4–21.

78 *"There is in England"*: King Diary, April 3, 1908; Dawson, *William Lyon Mackenzie King*, 168.

78 *"Sir Wilfrid told me"*: Family Papers Series, Vol. 4, File 18, WLMK to John and Isabel King, April 23, 1906; Dawson, ibid., 172–73.

78 *That sent King*: King Diary, June 29–July 3, July 25, 1907.

79 The Globe, *among*: *Globe*, September 22, 1908.

79 *"The brilliant young"*: Ibid., September 25, October 21, 1908; See also, *London Advertiser*, September 25, 1908; *Berlin Daily Telegraph*, September 25, 1908; Rych Mills, "On the Hill Over Yonder," in English, et al., *Mackenzie King: Citizenship and Community*, 43–7.

79 *"It was a most"*: *London Advertiser*, October 21, 1908.

79 *The* Montreal Herald *claimed*: *Montreal Herald*, October 30, 1908. King was not the only grandson of a former rebel leader to be elected in 1908. Louis-Joseph Papineau, the grandson of Louis-Joseph Papineau the 1830s leader of the Lower Canadian

rebellion, was also elected for the Liberals in Quebec. He later crossed the floor and won as a Conservative in the 1911 federal election, serving until 1917 when he rejoined Laurier and Liberals in the election over conscription during the First World War. He remained in Parliament until 1925.

79   *In the days leading:* The story of the feud with the governor general's office and King's entry into the House of Commons is in the King Diary, November 9, 10–26, 1909. See also, Gwyn, *Private Capital,* 233.

CHAPTER 3: Industrial Prophet

82   *"Congratulations, with a ring":* LAC, WLMK Papers, R10383-0-6-E, MG 26 J1, Primary Series Correspondence, Vol. 15, C-1913, 14013, Murray to King, August 3, 1910.

83   *King's thirty-eight:* Cited in Craven, 'An Impartial Umpire,' 329–41.

83   *On such contentious:* G.R. Stevens, *Canadian National Railways,* Vol. 2 (Toronto: Clarke, Irwin, 1962), 251; Craven, ibid., 349–50.

83   *"Railroading in the United States":* Craven, ibid., 347–48; Memoranda and Notes, Vol. 13, File 81, "Industrial Disputes: Grand Trunk Railway, 1910." Charles Hays did not have long to relish his victory. In 1912, he travelled to Britain to discuss with the GTR's board of directors the company's situation and a plan to possibly sell the enterprise to the Canadian government. He made the fatal mistake of booking his return passage on board the *Titanic.* He died with his son-in-law (the husband of his daughter Orian), Thornton Davidson, the son of Sir Charles Peers Davidson, chief justice of the Quebec Supreme Court. See, *Toronto World,* April 17, 1912; www.encyclopedia-titanica.org/c-m-hays-career.html; www.encyclopedia-titanica.org/titanic-victim/thornton-davidson.html.

83   *If there were flaws:* Dawson, *William Lyon Mackenzie King,* 211; Craven, 'An Impartial Umpire,' 352.

83   *The Conservatives:* Ulrich Frisse, "The Missing Link: Mackenzie King and Canada's 'German Capital,'" in English, et al., *Mackenzie King: Citizenship and Community,* 28–29.

84   *The 1911 federal:* See, John English, *The Decline of Politics: The Conservatives and the Party System 1901–20* (Toronto: University of Toronto Press, 1977), 53–69; André Pratte, *Wilfrid Laurier* (Toronto: Penguin Canada, 2011), 179–84.

84   *He did learn:* J.W. Pickersgill, "Mackenzie King's Political Attitudes and Public Policies: A Personal Impression," in John English and J.O. Stubbs, eds., *Mackenzie King: Widening the Debate* (Toronto: Macmillan of Canada, 1977), 18.

84   *In a long letter:* Cited in Charles W. Humphries, "Mackenzie King Looks at Two 1911 Elections," *Ontario History* 56:3 (1964), 205.

84   *Three days after:* King Diary, September 24, 1911.

85   *When Jennie:* Gray, *Mrs. King,* 271–72.

85   *Laurier remarked to:* McGregor, *The Fall and Rise of Mackenzie King,* 64.

85   *As the wonder boy:* See, Augustus Bridle, *The Masques of Ottawa* (Toronto: Macmillan of Canada, 1921), 52.

85   *"One feels one":* King Diary, December 14, 1911.

86   *"Somehow I believe":* Ibid., October 31, 1911.

86   *For the moment:* McGregor, *The Fall and Rise of Mackenzie King,* 80.

86   *She sent him:* Cited in Dawson, *William Lyon Mackenzie King,* 224.

86   *At first, he was:* King Diary, September 24, 1911.

87 *"To go into politics"*: Ibid., January 1–14, 1919.

87 *After spending an:* Ibid., January 25, October 31, December 31, 1911; January 17, 1912; Stacey, *A Very Double Life,* 107–8.

87 *"It was a shock"*: King Diary, May 7, 1914; Stacey, ibid., 110–12.

88 *So, too, did Rockefeller's wealth:* Accounting for inflation, the value of the dollar and wealth as a fraction of GDP, *Fortune Magazine* in 2007 ranked him as the richest American in history, ahead of Cornelius Vanderbilt, John Jacob Astor, and Bill Gates, among others. "The Richest Americans," *Fortune,* http://money.cnn.com/galleries/2007/fortune/0702/gallery.richestamericans.fortune/index.html.

88 *"God gave me"*: Cited in John Thomas Flynn, *God's Gold: The Story of Rockefeller and His Times* (Chautauqua, NY: Chautauqua Press, 1932), 401.

88 *Rockefeller, Sr., and:* Howard M. Gitelman, *Legacy of the Ludlow Massacre: A Chapter in American Industrial Relations* (Philadelphia: University of Pennsylvania Press, 1988), 7.

88 *"I am so glad"*: Cited in Ron Chernow, *Titan: The Life of John D. Rockefeller, Sr.* (New York: Random House, 1998), 188.

88 *Following a liberal arts:* Ibid., 355.

88 *He did not drink:* Raymond B. Fosdick, *John D. Rockefeller, Jr.: A Portrait* (New York: Harper and Brothers, 1956), 43; Chernow, ibid., 350-55; Jules Abels, *The Rockefeller Billions* (New York: The Macmillan Company, 1965), 304.

89 *In turn, CFI's two key:* At the end of October 1913, President Woodrow Wilson appealed to Bowers and Welborn to negotiate in good faith with the miners and UMWA. Bowers's response must have shocked Wilson, (as Rockefeller biographer Ron Chernow notes). "We shall never consent," Bowers told the president, "if every mine is closed, the equipment destroyed, and the investment made worthless." See, Chernow, ibid., 576.

89 *The miners, many:* Fosdick, *John D. Rockefeller,* 146; Gitelman, *Legacy of the Ludlow,* 11.

89 *Back in New York:* Fosdick, ibid., 146; Chernow, *Titan,* 574.

89 *The bloodiest moment:* See, Gitelman, *Legacy of the Ludlow,* 18–20.

90 *He was denounced:* Chernow, *Titan,* 579.

90 *In public statements:* Ibid., 585.

90 *"Those who listened"*: Walter Lippmann, *Early Writings* (New York: Liveright, 1970), 264–65 (from *New Republic,* January 30, 1915).

91 *While Lee is credited:* Gitelman, *Legacy of the Ludlow,* 35; Ray Heibert, *Courtier to the Crowd: The Story of Ivy Lee and the Development of Public Relations* (Ames, Iowa: Iowa State University Press, 1966), 101.

91 *"How terribly broken"*: King Diary, June–December 1914; McGregor, *The Fall and Rise of Mackenzie King,* 96.

91 *His brother, Max:* Family Papers Series, Vol. 6, File 27, Max King to WLMK, July 21, 1914.

91 *"It all depends"*: Cited in McGregor, *The Fall and Rise of Mackenzie King,* 101.

92 *King was so paranoid:* Gitelman, *Legacy of the Ludlow,* 115–16.

92 *The "unseen hand"*: Ibid., 45.

92 *Reflecting on their:* Fosdick, *John D. Rockefeller,* 154, 421.

92 *Likewise, King raved:* Cited in Dawson, *William Lyon Mackenzie King,* 232.

92 *At a meeting:* Cited in Stephen J. Scheinberg, "Rockefeller and King: The Capitalist and the Reformer," in English, et al., *Mackenzie King: Widening the Debate,* 92.

93 *Quite soon, the:* Dawson, *William Lyon Mackenzie King,* 233.

93 *"The truth is":* King Diary, March 4, 1915; Ibid., 234.

93 *It was, he wrote:* King Diary, August 4, 1914.

93 *He urged for unity:* Ibid., August 1, 4, 5, 1914.

93 *Yet apart from raising:* See, ibid., August 4, 1914; Dawson, *William Lyon Mackenzie King,* 256–57.

93 *"I talked with":* Ibid., November 25, 1914; McGregor, *The Fall and Rise of Mackenzie King,* 112.

93 *King reminded Rockefeller:* King to Rockefeller, February 9, 1915, cited in Gitelman, *Legacy of the Ludlow,* 90.

94 *During his testimony:* Ibid., 73–74.

94 *"True unionism is":* Testimony of William Lyon Mackenzie King before the U.S. Commission on Industrial Relations, May 25, 1915, 108–9; Dawson, *William Lyon Mackenzie King,* 245.

94 *"You make a mistake":* "William Lyon Mackenzie King: Capital and Labor Specialist," *Everybody's Magazine* 31 (December 1914), 763.

94 *As Fred McGregor:* Dawson, *William Lyon Mackenzie King,* 240-41. Of all of the facts that King learned, the one that troubled him the most was that John Osgoode, the hard-nosed director of the Victor-American Fuel Company and the recognized leader of the local coal mine executives opposed to UMWA, had had an affair with his wife some years before marrying her and while both of them were married to other people. See, Gitelman, *Legacy of the Ludlow,* 119.

94 *King spoke and socialized:* Gitelman, ibid., 136.

95 *In the privacy:* King Diary, March 27, 28, April 25, 1915.

95 *"One could not help":* Ibid., March 27, 1915.

95 *But he also:* Gitelman, *Legacy of the Ludlow,* 143.

95 *His goal, he maintained:* United States Commission on Industrial Relations (Washington, 1916), Vol. 9, 8792–93.

95 *The scheme King:* Gitelman, *Legacy of the Ludlow,* 190–91.

96 *Shrewdly, he set up:* "Labor Wins Over Mr. Rockefeller," *New York Times* (NYT), January 28, 1915, 1.

96 *A gaggle of reporters:* See NYT, September 21, 22, 24 1915.

96 *King later admitted:* WLMK to Abby Rockefeller, October 6, 1915, cited in Gitelman, *Legacy of the Ludlow,* 184-86.

96 *"I was merely":* Fosdick, *John D. Rockefeller,* 161.

97 *"I have never felt":* Cited in ibid., 337.

98 *"The truth is":* Ibid., January 9, April 26, 1917.

98 *Out at Kingsmere:* Ibid., June 22, 1916.

98 *The doctor told:* Roazen, *Canada's King,* 64.

99 *"My mind was":* King Diary, November 7, 1916.

99 *"I came back":* Ibid., October 26, 1916.

99 *"I felt this man":* Ibid., November 1, 14, 1916. Dr. Meyer was especially keen on self-help books, all the rage since at least the mid-nineteenth century. King later favoured such titles as *How to Live on Twenty-Four Hours a Day,* by Arnold Bennett (1910), *Why Worry?* by George Lincoln Walton (1916) and *The Efficient Life,* by Luther Halsey Gulick (1910). The latter book, which King finished reading in the spring of 1917, argued that "the health of the thinker, of the financier, of the executive genius, demands a momentary alertness of all the faculties, an ability to grasp, to originate,

to carry out, a trained perception and an intelligent discrimination. He must be the master of a delicate, high-grade machine calculated to carry on high-grade work." See, Luther Halsey Gulick, The Efficient Life (1910), 9, http://hearth.library.cornell. edu/cgi/t/text/text-idx?c=hearth;idno=4287414; King Diary, October 27, 1916.

99 *"My mind is greatly"*: King Diary, October 27, 1916.

99 *As he reported*: Roazen, *Canada's King*, 80–81.

100 *"Some days I have"*: Cited in Gray, *Mrs. King,* 360.

100 *Max, who at the time*: King Diary, January 7, 1922; Stacey, *A Very Double Life,* 156.

100 *In a moment*: King Diary, January 24, 25, 1917.

101 *"More and more"*: Ibid., February 1, 1917.

101 *"He is prepared"*: Ibid., March 7, 1917.

102 *In Winnipeg*: Allan Levine, *Scrum Wars: The Prime Ministers and the Media* (Toronto: Dundurn Press, 1993), 96–98; "Sir Robert Borden," Dictionary of Canadian Biography (DCB), Vol. 16, www.biographi.ca/index-e.html.

102 *In his imagination*: King Diary, October 31, 1917.

102 *"My whole argument"*: Personal Papers Series, Vol. 24, WLMK to Markham, January 2, 1918.

102 *"Poor soul"*: Gray, *Mrs. King,* 353–54.

103 *"This will be"*: WLMK to Markham, January 2, 1918.

103 *Thereafter, Isabel King*: Gray, *Mrs. King,* 363.

103 *"Her spirit is everywhere"*: Cited in ibid., 357.

103 *In King's vivid*: King Diary, February 15, 1918.

103 *These spiritual encounters*: Gray, *Mrs. King,* 363.

103 *Nonetheless, Jennie was*: Family Papers Series, Vol. 9, File 2, WLMK to Max King January 23, 1918; Vol. 9, File 4, Max King to WLMK, April 4, 1918; Vol. 11, File 25, WLMK to Jennie King, December 3, 1918; Nicholson, *Woodside,* 75–76.

104 *To his surprise*: King Diary, September 16, 1918.

104 *What precise meaning King*: Dawson, *William Lyon Mackenzie King,* 252.

104 *It is difficult to argue*: McGregor, *The Fall and Rise of Mackenzie King,* 243.

104 *"I think I get"*: Cited in Dawson, *William Lyon Mackenzie King,* 249.

104 *"Who do you expect"*: King Diary, February 17, 1918.

104 *His thesis*: Memoranda and Notes Series, Vol. 33, File 190, WLMK to Rockefeller Foundation, May 10. 1916, 13.

104 *The trick*: William Lyon Mackenzie King, *Industry and Humanity: A Study in the Principles Underlying Industrial Reconstruction* (Toronto: University of Toronto Press), 97-104 (hereafter *Industry and Humanity,* 1973).

104 *As he maintained*: Bliss, *Right Honourable Men,* 136.

105 *Industry and Humanity was*: King Diary, November 4, 11, 1918; January 1–14, 1919; Bliss, ibid., 136.

105 *Co-operation and democracy*: See, M. Donald Hancock, John Logue, Brent Schiller, *Managing Modern Capitalism: Industrial Renewal and Workplace Democracy in the United States and Western Europe* (Westport, CT: Greenwood Press, 1991) and D. D'Art and Thomas Turner, "Independent Collective Representation: Providing Effectiveness, Fairness, and Democracy in the Employment Relationship," *Employee Responsibilities and Rights Journal* 15:4 (December, 2003), 169–81.

105 *Survival of the fittest*: Kirk Hallahan, "W.L. Mackenzie King: Rockefeller's 'Other' Public Relations Counselor in Colorado," *Public Relations Review* 29:4 (November 2003), 409.

105 *So, too, does his:* Ibid., 409–10.

106 *Sounding very much:* Scheinberg, "Rockefeller and King," 100.

106 *"It is highly gratifying":* Personal Papers Series, Vol. 33, Rockefeller to WLMK, January 23, 1924.

106 *"Industry's basic function":* Nathaniel Benson, "The Community Must Come First: An Interview with Mackenzie King," *Forbes Magazine of Business* 49:11 (June 1, 1947), 14.

106 *His income for:* King Diary, December 17, 1918; McGregor, *The Fall and Rise of Mackenzie King,* 271–72.

107 *"I see in her":* King Diary, February 21, 1918.

107 *"It made me":* Ibid., February 24, 1918.

107 *"There is a case":* Ibid., March 13, 1949; Stacey, *A Very Double Life,* 115.

108 *"I confess at times":* King Diary, September 17, 1918.

108 *King was told:* Ibid., November 23, 1918.

108 *"All the Liberal":* Ibid., February 18-22, 1919; McGregor, *The Fall and Rise of Mackenzie King,* 322–23.

108 *Still, King was unsettled:* King Diary, August 5–9, 1919; McGregor, ibid., 322, 341.

109 *"Unmarried," he confided:* Family Papers Series, Vol. 9, File 8, WLMK to Max King, April 13, 1919; McGregor, ibid., 329.

109 *As a precaution:* McGregor, ibid., 337.

109 *King would not:* As Ferns and Ostry rightly pointed out, King took a fairly middle-of-the-road approach to labour issues largely out of the pages of *Industry and Humanity.* As might be expected, there were no concrete statements of support for unions or collective bargaining and only the most perfunctory comments of solving the high cost of living, all factors which had led to the Winnipeg General Strike in May 1919. Ferns and Ostry, *The Age of Mackenzie King,* 318.

109 *The same criticisms:* MFP, August 8, 1919.

109 *On August 7, the day:* The story of the convention is from King Diary, August 5–9, 1919.

CHAPTER 4: A Not Unacceptable Party Leader

111 *"King is by no means":* LAC, W.K. Lamb Papers MG 31 D8, Vol. 9 File: "King, WLM Biography and Access to Papers" Excerpt from Borden Papers J.A. Calder to Borden Oct 28, 1919; Diary of Sir George Foster Nov 7, 1919, cited in Dawson, *William Lyon Mackenzie King,* 330.

111 *Right from the start:* Dawson, ibid., 318.

112 *Nonetheless, his opponent:* Ibid., 318.

112 *King "is as full":* LAC, John W. Dafoe Papers, R1831-0-5-E, MG 30 D45, Vol. 2, Stevenson to Dafoe, January 23, 1920; Levine, *Scrum Wars,* 129.

112 *That only embittered:* Dafoe to Sifton, March 5, 1923, in Ramsay Cook, ed., *The Dafoe-Sifton Correspondence 1919–1927* (Winnipeg: Manitoba Record Society, 1966), 157.

112 *The great Winnipeg:* Pierre Berton, *The Promised Land: Settling the West 1896–1914* (Toronto: McClelland and Stewart, 1984), 34.

112 *At the posh:* J.E. Rea, *T.A. Crerar: A Political Life* (Montreal: McGill-Queen's University Press, 1997), 32.

112 *Privately Dafoe did:* Robert A. Wardhaugh, *Mackenzie King and the Prairie West* (Toronto: University of Toronto Press, 2000), 83; Bliss, *Right Honourable Men,* 135–36.

113 *King regarded:* On King and the *Globe,* see Levine, *Scrum Wars,* 143–50.

113 *Journalist Melville Rossi:* Primary Series Correspondence, Vol. 211, C-3684, 181693-93, Rossi to WLMK, January 5, 1935; Levine, ibid., 143.

113 *He did have:* Dawson, *William Lyon Mackenzie King,* 319.

113 *Back in the House:* Ibid., 318.

114 *"Their stand on":* Gerald Friesen, *The Canadian Prairies: A History* (Toronto: University of Toronto Press, 1984), 370.

115 *King recognized the allure:* King Diary, October 27–November 1, 1919; Dawson, *William Lyon Mackenzie King,* 316.

115 *He was at one with:* Personal Papers Series, Vol. 24, WLMK to Markham, December 2, 1925.

115 *"What we want":* House of Commons Debates, June 1, 1920, 2982-94; Wardhaugh, *Mackenzie King and the Prairie West,* 49; King Diary, January 18, 1920.

116 *"If the three":* King Diary, December 22, 1919; Lita-Rose Betcherman, *Ernest Lapointe: Mackenzie King's Great Quebec Lieutenant* (Toronto: University of Toronto Press, 2002), 34.

116 *That also went for:* Wardhaugh, *Mackenzie King and the Prairie West,* 52; Dafoe Papers, M-73, Dafoe to Sifton, November 10, 1920.

116 *This meant placating:* Conflict of interest rules for Canadian politicians were much different than they are today. Gouin worked as a lawyer while he was premier and was a director of many major corporations, including the Bank of Montreal, the Royal Trust Company, the Crédit Foncier Franco-Canadien, the Shawinigan Water and Power Company and the Mutual Life Assurance Company of Canada. See, DCB, "Sir Lomer Gouin," Vol. 15, www.biographi.ca/index-e.html.

116 *An ardent protectionist:* Betcherman, *Ernest Lapointe,* 41; Dafoe to Sifton, February 14, 1921, in Cook, *Dafoe-Sifton Correspondence,* 54–55.

116 *During the Liberal:* Betcherman, ibid., 23.

116 *"It was too good":* King Diary, July 8, 1920.

116 *"There was never":* LAC, Alex Hume Papers, R5833-0-X-E, MG 32 G6, Vol. 1, "CBC Interview transcript," December 4, 1972.

116 *Adds Meighen's sympathetic:* Roger Graham, *Arthur Meighen: And Fortune Fled,* Vol. 2 (Toronto: Clarke, Irwin and Company, 1963), 10.

117 *He was, King:* King Diary, July 8, 1920, cited in Bliss, *Right Honourable Men,* 98.

117 *Ernest Lapointe, never:* Norman Ward, ed., *The Memoirs of Chubby Power* (Toronto: Macmillan of Canada, 1966), 77.

117 *"In a country like ours":* Primary Series Correspondence, Vol. 184, C-2324, 156671–73, WLMK to Ames, January 29, 1931; Neatby II, 355.

117 *"Meighen was incapable":* Roger Graham, *Arthur Meighen: The Door of Opportunity,* Vol. 2 (Toronto: Clarke Irwin, 1960), 295–96. Many immigrants living in thePrairies also never forgot that it was Meighen who had drafted the unpopular and draconian Wartime Elections Act in 1917, which disfranchised thousands of Canadians born in enemy countries. Meanwhile, in Montreal during the election of 1921, a popular poster announcing an upcoming visit by the Conservative leader said it all: "Meighen is coming. Bring eggs." See, Wardhaugh, *Mackenzie King and the Prairie West,* 48; Levine, *Scrum Wars,* 112.

117 *It was not:* Dawson, *William Lyon Mackenzie King,* 342.

117 *"The issue, so far":* House of Commons Debates, May 19, 1921, 3613.

118 *"Those words are"*: *Halifax Morning Chronicle*, October 10, 1921; Dawson, *William Lyon Mackenzie King*, 351.

118 *On election day*: King Diary, December 6, 1921.

119 *"It was quite"*: Ibid., October 2, 1918.

119 *Joan MacWhirter and Godfroy Patteson*: Author's interview with Joan McCallum, October 17, 2009; Stacey, *A Very Double Life*, 120.

119 *Joan Patteson's granddaughter*: Interview with Joan McCallum, ibid., and interview with Mary Fox-Linton, October 20, 2009. See also, Personal Papers, Vol. 30, Mary Patteson to WLMK, December 30, 1943; Ann Patteson to WLMK, January 2, 1944; Robert K. Barney, Malcolm Scott, and Rachel Moore, "'Old Boys' at Work and Play: The International Olympic Committee and Canadian Co-option, 1928–1946," *Olympika: The International Journal of Olympic Studies*, Vol. 8 (1999), 90-7. In 1947, Joan married Alexander McCallum. They had a son, John, born in Montreal in 1950. He became an economist and rose to be senior vice-president and chief economist of the Royal Bank of Canada. In 2000, John McCallum, the great-grandson of Godfroy and Joan Patteson, was elected to Parliament as a Liberal. In the governments of Jean Chrétien and Paul Martin, McCallum held several Cabinet portfolios, including Minister of National Defence, Minister of Veterans' Affairs, Minister of Natural Resources and Minister of National Revenue. He was re-elected in the Ontario riding of Markham-Unionville in the federal election of 2011.

120 *Had the two of them*: King Diary, May 21, 1922; Stacey, *A Very Double Life*, 122.

120 *She was discreet*: King reciprocated this trust. In early January 1949, he read through a draft of Reginald Hardy's biography of his life, *Mackenzie King of Canada*, published by Oxford University Press later that year. He did not approve of Hardy's references to his interest in spiritualism, but the most "objectionable reference," as he put it, was "to [his] friendship with Joan, not that it is not nicely written, but to single her out by name to exclusion of all others, is most unchivalrous and will cause her with a sensitive nature to be greatly embarrassed." King raised the issue with Hardy and all references to Joan Patteson were removed from the book. See, King Diary, January 7, 1949.

120 *"It is so delightful"*: Ibid., May 8, 1920; Stacey, *A Very Double Life*, 120.

120 *When Walter Turnbull*: Cited in Reynolds, *Mackenzie King Friends & Lovers*, 170.

120 *"She has a nimble"*: September 6, 1920; Stacey, *A Very Double Life*, 121.

120 *"It seems so hard"*: Stacey, ibid., 121.

121 *"Now I must bend"*: King Diary, September 14, 1920; Stacey, ibid., 122.

121 *As Charles Stacey suggests*: King Diary, September 16, 1920; Stacey, ibid., 122.

121 *"I never saw"*: King Diary, August 4, 1923; Stacey, ibid., 126.

121 *It was reminiscent*: Reynolds, *Mackenzie King Friends & Lovers*, 179–81.

122 *"I hope Jane"*: Personal Papers Series, Vol. 25, WLMK to Markham, May 1, 1939.

122 *On September 29, 1924*: On this episode with Pat's disappearance, see, King Diary, September 29, 30, October 1, 1924.

122 *His athletic stature*: Granatstein and Hillmer, *For Better or Worse*, 79.

122 *Canada did not*: Laurier L. LaPierre, *Sir Wilfrid Laurier and the Romance of Canada* (Toronto: Stoddart, 1996), 237.

123 *The furnace, plumbing*: King Diary, January 19, 1922; January 16, 1922.

123 *King's white knight*: Roy MacLaren, *Commissions High: Canada in London, 1870–1971* (Montreal: McGill-Queen's University Press, 2006), 228.

123 *According to former Canadian*: Ibid., 228.

123 *As a first step*: King Diary, January 23, 1922; Finances Series, Vol. 5, WLMK to Larkin, April 22, August 22, 1922; Larkin to WLMK, September 2, 1922.

124 *"Its imposing and"*: Finances Series, Vol. 5, WLMK to Larkin, November 23, 1922.

124 *"Am writing for"*: King Diary, January 12, 1923.

124 *The house is still*: Markham, *Friendship's Harvest*, 163.

125 *"It is really rather"*: MacLaren, *Commissions High*, 227 note 3.

125 *As he had done*: Finances Series, Vol. 5, "Memorandum on Laurier House," January 10, 1951; King Diary, January 14, 1929; August 24, 1930; Betcherman, *Ernest Lapointe*, 106.

125 *Such party funds*: Betcherman, ibid., 106.

125 *"Larkin has been"*: King Diary, December 31, 1927; MacLaren, *Commissions High*, 227.

125 *Max was one*: Dawson, *William Lyon Mackenzie King*, 382.

125 *"Let me impress"*: Family Papers Series, Vol. 9, File 17, Max King to WLMK, September 22, 1921; Dawson, ibid., 382.

126 *He questioned God's*: King Diary, January 8, 1922.

126 *"I was glad"*: Ibid.

126 *"He loved you"*: Family Papers Series, Vol. 9, File 23, May King to WLMK, March 22, 1922; King Diary, March 18, 1922.

127 *"King was at"*: Dawson, *William Lyon Mackenzie King*, 383.

127 *One cutting political*: Grant Dexter, "This Man Mackenzie King," *Saturday Night*, September 21, 1935, 5, 8.

127 *"To my great"*: King Diary, March 13, 1922.

127 *As historians Jack Granatstein*: Granatstein and Hillmer, *For Better or Worse*, 79-80.

127 *"You keep the disease"*: "Canada: The Dominion: Preventive Medicine," *Time* 47, January 7, 1946, www.time.com/time/magazine/article/0,9171,797731,00.html. Thomas Crerar had tougher assessment. "I am afraid that one of King's difficulties is that he is not able to distinguish between good advice and poor advice," he wrote to one associate, "and that he has scarcely anyone whom he can take into his confidence, and whose advice he seeks. He has a lot of faith in his own star, but keeps his eye so intently on it that he may miss seeing the pitfalls at his own feet." Cited in Wardhaugh, *Mackenzie King and the Prairie West*, 94.

128 *Falling grain prices*: Rea, *T.A. Crerar*, 93.

128 *Upon hearing the*: King Diary, November 12, 1922.

128 *"But for you"*: Ibid., November 19, 1941; Betcherman, *Ernest Lapointe*, ix.

129 *Born in rural Quebec*: On Lapointe's early years, see, Betcherman, ibid., 4–7, 24.

129 *"I told him"*: King Diary, December 10, 1921.

129 *Three days later*: Betcherman, *Ernest Lapointe*, 45.

130 *The prime minister was*: Dafoe to Sifton, July 11, 1922, in Cook, *Dafoe-Sifton Correspondence*, 118–20. Thomas Crerar concurred with his friends. "To my mind," he wrote in June 1922, "Gouin is the boss of the administration. He gives the impression of having great reserve power... he sits in his seat, with his square head and determined jaw, alert and keen, he impresses you as a man of strong purpose and determined will." See, Crerar to H.B. Mitchell, June 10, 1922, cited in Wardhaugh, *Mackenzie King and the Prairie West*, 70.

130 *"I seem to have"*: King Diary, February 5, 1922.

130 *Westerners shook their*: Wardhaugh, *Mackenzie King and the Prairie West*, 73,

130 *"We took up"*: King Diary, April 28, 1922. See the entry for April 10, 1922, for a more sympathetic comment on Gouin.

130 *"I pointed out their"*: Ibid., November 15, 16, 1922.

130 *There were further*: Wardhaugh, *Mackenzie King and the Prairie West*, 73.

131 *"I feel a great"*: King Diary, January 6, 1924; January 3, 1924; Wardhaugh, ibid., 91–98.

131 *On the afternoon of Saturday*: King Diary, September 16, 1922.

131 *"I confess it"*: Ibid., October 4, 1922.

131 *On a visit*: Ibid., March 5, 1932.

132 *"Never again will"*: Primary Series Correspondence, Vol. 82, C-2249, 69059–60, Skelton to WLMK, October 13, 1922.

132 *King was already*: Norman Hillmer, Unpublished Skelton Biography, Chapter 1, 12–13.

132 *In words that*: Cited in Graham, *Arthur Meighen*, Vol. 2, 209–10.

132 *Meighen's predecessor*: Bliss, *Right Honourable Men*, 143–44.

133 *As he clarified*: Ibid., 145.

133 *Granatstein and Hillmer explain*: Granatstein and Hillmer. *For Better or Worse*, 81.

133 *In September 1923*: King Diary, September 11, 1923.

133 *His friends and foes*: Norman Hillmer, "Foreign Policy and the National Interest: Why Skelton Matters," Fourteenth O.D. Skelton Memorial Lecture, December 17, 2008, 3.

134 *Skelton, he says*: Ibid., 3–5.

134 *As Frank Underhill once*: Frank H. Underhill, "J.W. Dafoe," *Canadian Forum* 13 (1932), 23.

134 *"You could undoubtedly"*: Cited in Ramsay Cook, "J.W. Dafoe at the Imperial Conference, 1923," *Canadian Historical Review* 41:1 (March 1960), 19.

134 *"I have very"*: Dafoe Papers, M-74, Dafoe to Sifton September 12, 1923.

135 *Once in London*: Hillmer, Unpublished Skelton Biography, Chapter 2, 9.

135 *There was lunch*: Dawson, *William Lyon Mackenzie King*, 457; Bedore, "The Reading of Mackenzie King," 36 note 123; King Diary, October 10, 1923.

135 *At the conference*: King Diary, November 5, 1923; Dawson, ibid., 474.

135 *"Our attitude is"*: Cited in MacLaren, *Commissions High*, 250.

135 *Or, put another*: Leopold S. Amery, *My Political Life*, Vol. 2 (London: Hutchinson, 1953), 273.

135 *"Mackenzie King, you"*: King Diary, November 7, 1923; Dawson, *William Lyon Mackenzie King*, 476.

135 *"His efforts to make"*: Cited in MacLaren, *Commissions High*, 251.

136 *He angrily dismissed*: Grace E.T. Curzon, *Reminiscences* (London: Hutchinson, 1955), 181–82.

136 *Yet as several*: Philip G. Wigley, *Canada and the Transition to Commonwealth: British-Canadian Relations, 1917–1926* (New York: Cambridge University Press, 1977), 3.

136 *Even a curmudgeon*: Dafoe Papers, M-74, Dafoe to W.A. Buchanan, December 6, 1923.

137 *"If we could only get"*: Primary Correspondence Series, Vol. 103, C-2266, 87349-53, Larkin to WLMK, August 29, 1924; WLMK to Larkin, September 6, 1924; Levine, *Scrum Wars*, 145.

137 *Lewis soon resigned*: King Diary, August 15, 1925; Mark Moher, "The 'Biography' in Politics: Mackenzie King in 1935," *Canadian Historical Review* 55:2 (June 1974), 239–40.

138 *Hector Mackinnon*: Levine, *Scrum Wars*, 146–47.

CHAPTER 5: Vindication and Victory

139 *On the first day:* See, King Diary, March 1, 1925; Spiritualism Series, Vol. 3, File 26, H-3037, File 41, WLMK to Bleaney, November 26, 1925. Eight months later, King learned from the military that George Bleaney had been discharged in 1919 with no disabilities, though he had suffered a gunshot wound. He had syphilis, which he had contracted after leaving the service, and was experiencing nervousness and headaches. The military felt no obligation to provide George with a pension. King wrote Mrs. Bleaney with the news that he might be able to help her son find work.

140 *"I feel ... in a personal":* King Diary, August 12, 1925.

141 *After the meeting:* Ibid., August 17, 1925; Neatby II, 62.

141 *Massey himself:* Vincent Massey, *What's Past Is Prologue* (Toronto: Macmillan of Canada, 1963), 97–99. Arthur Meighen wanted to embarrass Massey during the campaign by publishing private correspondence between the two of them from the previous year. Then, Massey had argued in favour of a high tariff. Massey, however, would not consent, arguing that he had written to Meighen in his capacity as the president of Massey-Harris. He told Meighen that he was now an individual running for a seat in Parliament and eventually conceded publicly that he had changed his mind about the protection issue. This tactic nearly worked, although he still lost the election in his Ontario constituency to the Conservative candidate by one thousand votes.

141 *The activity:* Allan Levine, *The Devil in Babylon: Fear of Progress and the Birth of Modern Life* (Toronto: McClelland & Stewart, 2005), 213, 219–20; John Thompson with Allen Seager, *Canada 1922–1939: Decades of Discord* (Toronto: McClelland and Stewart, 1985), 121.

141 *Corruption was rife:* Ralph Allen, *Ordeal by Fire: Canada, 1910–1945* (Toronto: Doubleday Canada, 1961), 4; Lita-Rose Betcherman, "The Customs Scandal of 1926," *The Beaver* 81: 2 (April/May 2001), 14–15.

141 *Asked in the:* House of Commons Debates, February 2, 1926, 689; Betcherman, *Ernest Lapointe,* 87.

142 *Bisaillon was actually:* Thompson, *Canada 1922–1939,* 121; King Diary, September 1, 1925; Betcherman, ibid., 88–89.

142 *King's political instincts:* Neatby II, 69–71.

142 *The 1925 campaign:* See, Levine, *Devil in Babylon,* 247.

143 *Today it is almost:* Levine, *Scrum Wars,* 169; *Globe,* June 19, 1930; *Ottawa Journal,* August 9, 1935.

143 *"I confess":* Personal Papers Series, Vol. 24, WLMK to Markham, October 15, 1925.

143 *In a telling:* Levine, *Scrum Wars,* 135.

143 *The encounter was:* King Diary, October 20, 1925.

144 *A day before:* Ibid., October 28, 1925.

144 *With his faith:* Ibid., October 29, 1925; Thompson, *Canada 1922–1939,* 117–18; Primary Correspondence Series, Vol. 117, C-2278, 89919-27, WLMK to Larkin, November 26, 1925.

145 *The Globe, on:* *Globe,* October 30, 1925; Thompson, ibid., 118; WLMK to Larkin, ibid. William Jaffray had absolutely no regrets about his actions. "We feel that country comes before party and we could not conscientiously support the Liberal party in the last election," he told Senator Allen Aylesworth in mid-November, 1925. "What claim has the official Liberal party upon the support of the Globe, may we ask? The Globe has given a very great deal of support, during all the years of its history, to the

Liberal party and debt, it seems to me, is all on the other side. The Liberals should not expect the Globe to be subservient." See, Primary Correspondence Series, Vol. 3, c-2273, 94722-25, Jaffray to Aylesworth, November 18, 1925.

145 *The following day:* King Diary, October 30, 1925.

146 *Field Marshal Julian:* Jeffrey Williams, *Byng of Vimy* (London: Secker and Warburg, 1983), 5.

146 *In retrospect:* Ibid., 283; Reynolds, *Mackenzie King Friends & Lovers,* 110.

146 *"I cannot tell":* Personal Papers Series, Vol. 24, WLMK to Markham, September 29, 1921; King Diary, August 12, 1921.

146 *In one letter:* Spiritualism Series, Vol. 1, Lady Byng to WLMK, May 31, 1923.

147 *One of King's:* King Diary, September 2, 1921.

147 *"She congratulated me":* Ibid., December 7, 1921.

147 *"Well, I can't tell":* This first discussion with Lord Byng in the aftermath of the federal election can be found in the King Diary, October 30–November 4, 1925.

149 *This was rather:* Neaby II, 82-3.

150 *Byng also told:* Williams, *Byng of Vimy,* 305; Graham, *Arthur Meighen,* Vol. 2, 353–54; Neatby, ibid., 84–85; Peter B. Waite, "Mr. King and Lady Byng," *The Beaver* 77:2 (April/May 1997), 28.

150 *"I feel I am":* King Diary, November 8, 1925.

150 *"To cling to":* Saskatoon Phoenix, November 6, 1925; Graham, *Arthur Meighen,* Vol. 2, 354.

150 *Meighen did not:* Levine, Scrum Wars, 120-21; *Winnipeg Tribune,* December 3, 1925; LAC, J.S. Willison Papers, R10010-0-X-E, MG 30 D29, Vol. 15, Folder 116, Arthur Ford to Willison, December 2, 1925; LAC, P.D.Ross Papers, R2236-0-7-E, MG 30 D98, Vol. 2, Ross to Meighen, December 2; Meighen to Ross, December 4, 1925; O'Leary, *Recollections,* 63.

151 *Yet as his biographer:* Graham, *Arthur Meighen,* Vol. 2, 367.

151 *Mackenzie King had:* King Diary, November 18, 1925; WLMK to Mulock, November 10, 1925, cited in Neatby II, 86.

151 *There were loud:* Wardhaugh, *Mackenzie King and the Prairie West,* 110-11; Peter S. Regenstreif, "A Threat to Leadership: C.A. Dunning and Mackenzie King," *Dalhousie Review* 44 (April 1964), 272–89.

151 *During this period:* See, Neatby II, 77–78.

151 *His leadership difficulties:* King Diary, November 19, 1925.

152 *He was seated:* Williams, *Byng of Vimy,* 308; Reynolds, *Mackenzie King Friends & Lovers,* 114–15.

152 *Without a seat:* King Diary, January 7, 1925.

152 *He did, however, finally:* There was no love lost between Dunning and Jimmy Gardiner, who became the new premier of Saskatchewan. Gardiner also moved into the federal Liberal cabinet, although not until 1935. Decades later, when Gardiner was months away from death in January 1962, he commented to an interviewer that in 1926 he had told King to please "take Dunning to Ottawa rather than leave him in Saskatchewan because in the larger political arena he would do less 'harm.'" See, Wardhaugh, *Mackenzie King and the Prairie West,* 113–14.

152 *Suddenly the Liberals:* Thompson, *Canada 1922–1939,* 119–20.

152 *Once Woodsworth's:* King Diary, January 8, 18, 1926; Harry Gutkin and Mildred Gutkin, *Profiles in Dissent: The Shaping of Radical Thought in the Canadian West* (Edmonton: NeWest Publishers, 1997), 281.

153 *The bill passed the:* King Diary, January 26, 1926; Neatby II, 110, 126.

153 *Premier Jimmy Gardiner:* King Diary, February 16–26, 1926; Personal Papers Series, Vol. 24, WLMK to Markham, February 26, 1926; Wardhaugh, *Mackenzie King and the Prairie West,* 117–18; Thompson, *Canada 1922–1939,* 120.

154 *King had convinced:* King Diary, February 17, 1926. In early March, King was also angry when the Byngs's schedule prevented them from hosting a gathering for a visiting British politician, Oswald Mosley, and his wife, Lady Cynthia, the daughter of Lord Curzon. At the time, Mosley was a member of the Labour Party, but within six years, he was to become the leader of the British Union of Fascists. "I like both Mosley and Lady C immensely," King noted. "I shall be surprised if some day he is not Prime Minister of England." See King Diary, March 6, 1926; Williams, *Byng of Vimy,* 311.

154 *According to Vancouver:* Allen, *Ordeal by Fire,* 268-72; Neatby II, 133–34.

154 *Sensing the worst:* King Diary, June 18, 1926; Betcherman, *Ernest Lapointe,* 103; House of Commons, Special Committee Investigating the Administration of the Department of Customs and Excise, Ottawa: King's Printer, 1926. www.archive. org/stream/CCcustomsexcise1926proco1uoft/CCcustomsexcise1926proco1uoft_ djvu.txt.

154 *"It did not seem to":* King Diary, June 19, 1926; Allen, *Ordeal by Fire,* 273.

154 *On June 22:* House of Commons Debates, June 22, 1926, 4832.

155 *"What is the penalty":* Ibid., June 23, 1926, 4923.

155 *"Our men were":* King Diary, June 24, 1926. Watching from the gallery for nearly all of this debate was Eugene Forsey, then a thirty-year-old political science lecturer at Carleton and McGill Universities and a friend and adviser to Arthur Meighen. Forsey later recollected that his most vivid memory of the debates was Meighen's speech on the customs scandal, which also lasted for several hours "and was as usual replete with names, dates, precise quotations, all delivered without a note of any kind." At one point, he recalls, George Boivin interrupted to request Meighen's permission to ask a question of him. Meighen consented and as he sat down, Forsey writes, he mumbled, "You'll be sorry." The remark was not recorded in Hansard, but Forsey and Meighen recalled it later. After Boivin had made his query, adding few comments in his defence, "Meighen immediately made a devastating reply," notes Forsey, "demolishing every one of Boivin's points, word by word, comma by comma." See, Eugene Forsey, *A Life on the Fringe: The Memoirs of Eugene Forsey* (Toronto: Oxford University Press, 1990), 102.

155 *King was back:* King Diary, June 26, 1926.

155 *King's thinking on:* Ibid.

156 *King believed to:* Ibid., June 9, 1935 written at the time of Byng's death.

156 *"I had not":* Forsey, *A Life on the Fringe,* 103.

156 *King had "played":* J.L. Granatstein, *Mackenzie King: His Life and World* (Toronto: McGraw-Hill Ryerson), 1977, 46. See also, Christopher Moore, "That King-Byng Thing," *The Beaver* 88:6 (December 2008/January 2009), 57–58.

157 *Byng, as King:* King Diary, June 26, 1926.

157 *"If dissolution can":* Williams, *Byng of Vimy,* 323; LAC, Lord Byng Papers, R4376-0-9, MG 27 III A2, Byng to Amery, June 30, 1926.

157 *In a last-ditch:* Byng Papers, Amery to Byng, July 3, 1926; Neatby II, 149–50.

157 *Another tense meeting:* King Diary, June 28, 1926.

158 *On Monday, with:* Ibid.; Byng Papers, WLMK to Byng, June 28, 1926.

158 *"If Lord Byng":* Crerar, *The Incredible Canadian,* 156.

158 *"Mr. King, whose":* Byng Papers, Byng to Amery, June 30, 1926.

159 *In a letter of:* Cited in Williams, *Byng of Vimy,* 322.

159 *"No decision he":* Graham, *Arthur Meighen,* Vol. 2, 420.

160 *Some months later:* MacRae to Dafoe, November 21, 1926, cited in Ramsay Cook, "A Canadian Account of the 1926 Imperial Conference," *Journal of Commonwealth Political Studies* 3:1 (March 1965), 61.

160 *Meighen, however, in a:* Memorandum by Lord Bessborourgh, March 6, 1932, cited in David Dilks, *The Great Dominion: Winston Churchill in Canada 1900–1954* (Toronto: Thomas Allen Publishers, 2005), 130.

160 *Recounting the events:* Personal Papers Series, Vol. 24, WLMK to Markham, July 12, 1926.

160 *Whether the Canadian:* Granatstein, *Mackenzie King: His Life and World,* 48; *Toronto Star,* July 4, 1926; King Diary, July 2, 1926.

160 *In the west:* Ramsay Cook, *The Politics of John W. Dafoe and the Free Press* (Toronto: University of Toronto Press, 1963), 162–67, 179–80; Dafoe Papers, M-74, Dafoe to Grant Dexter, September 17, 1926; Dafoe to Sifton, September 27, 1926. Asked once by Ottawa journalist and Conservative Party insider Grattan O'Leary why he was so dedicated to the Liberals, Dafoe replied: "I simply think of all the sons of bitches in the Tory party, then I think of all the sons of bitches in the Liberal party, and I can't help coming to the conclusion that are more sons of bitches in the Tory party." See, Levine, *Scrum Wars,* 67.

161 *One episode says:* John G. Diefenbaker, *One Canada: The Crusading Years 1895–1956* (Toronto: Macmillan of Canada, 1975), 148–49, 152–53.

161 *"Mr. King's luck":* Cited in Granatstein, *Mackenzie King: His Life and World,* 48.

161 *In words with more:* Dafoe Papers, M-74, Dafoe to Willison, September 17, 1926.

162 *Relaxing at Kingsmere:* King Diary, July 4, 1926.

162 *By ten in the evening:* Ibid., September 14, 1926.

162 *It was left:* Ottawa Citizen, September 26, 1926. Most recently, in September 2008, Governor General Michaëlle Jean accepted Prime Minister Stephen Harper's advice and dissolved Parliament for a federal election of October 14, 2008. Because of Canada's new fixed-date election law (passed in 2006), this was regarded as an incorrect decision by a cadre of academics led by Michael Behiels of the University of Ottawa. Some political scientists, however, disagreed and insisted the prime minister still had the right to seek a dissolution, just as King had tried to do. See, Moore, "That King-Byng Thing," 57-8.

162 *On arrival:* King Diary, September 23, 1926.

163 *"This seemed pretty":* Ibid., September 27, 1926.

163 *King was invited:* Ibid., May 12, 13, 1932; Reynolds, *Mackenzie King Friends & Lovers,* 116.

164 *Byng, ever the gentleman:* Waite, "Mr. King and Lady Byng," 29.

164 *King hesitated to:* King Diary, June 6, 1935.

164 *"Throughout the service":* Ibid., June 9, 1935.

165 *Late that evening:* Reynolds, *Mackenzie King Friends & Lovers,* 116.

165 *"I don't suppose that":* Cited in Waite, "Mr. King and Lady Byng," 24.

165 *There was one:* King Diary, December 9, 1940.

CHAPTER 6: A Place in History

166 *"I do not like"*: Ibid., July 22, 1927.

166 *"I tire a little"*: Finances Series, Vol. 6, WLMK to Larkin, January 20, 1928.

166 *When he couldn't quite*: See King Diary, February 1, 1927; January 23, October 17, 1929; Pierre Berton, *The Great Depression 1929–1939* (Toronto: McClelland and Stewart, 1990), 54.

167 *During the last week*: McDougald was married to Mary Hannan, whose family owned the lucrative Ogdensburg Coal and Towing Company in Ogdensburg, New York. McDougald became an executive in the firm and expanded its importing business to Canada, making a fortune for himself. King appointed him to the Senate in June 1926 as gratitude for his support of the Liberals. See, T.D. Regehr, *The Beauharnois Scandal* (Toronto: University of Toronto Press, 1990), 16–17.

167 *Normally, he insisted*: LAC, WLMK Papers, R10383-18-3-E, MG 26 J12

167 Personal Miscellaneous, Vol. 1, WLMK to Barker, February 27, 1927.

167 *He dutifully recorded*: King Diary, February 25, 26, 27, 1927.

167 *"This led me"*: Ibid., February 28, 1927.

167 *He liked to sleep*: This section on King's routine is based on the description in Neatby II, 200.

167 *It was "intended"*: Mary Tileston, *Daily Strength for Daily Needs* (Boston: Roberts Brothers, 1889), i.

167 *During his second term*: Howard Measures was with King for about five years and then became R.B. Bennett's personal secretary before becoming the chief protocol officer at the department of external affairs. His last day of work for King was in mid-September 1930, about a month after the decisive summer election. In all too typical fashion, King sulked about Measures's decision to work for Bennett. "Measures leaves on Monday," he wrote, "and in many ways I am glad to be free of him, he is more and more uncertain, good meaning but with but little judgment and liable at any moment to become more or less of a charge. He has a kindly and agreeable side but his readiness to turn the moment he saw his opportunity was disappointing." See, King Diary, September 13, 1930.

168 *In January 1928*: Ibid., January 19, 1928.

168 *On another occasion*: LAC, WLMK Papers, R10383-16-X-E, MG 26 J10, Laurier House and Kingsmere Series, Vol. 160, WLMK to C.B. Garland, Managing Director, Red Line Limited Sightseeing Tour Company, July 27, 1929; Garland to WLMK, July 30, 1929.

168 *Then, early in 1927*: Edwina Von Baeyer, *Garden of Dreams: Kingsmere and Mackenzie King* (Toronto: Dundurn Press, 1990), 94.

168 *"I believe the property"*: King Diary, March 27, 1927.

169 *He had to*: Von Baeyer, *Garden of Dreams*, 94–95.

169 *King reasoned*: King Diary, April 3, 11, May 15, August 14, 1927; Laurier House and Kingsmere Series, Vol. 28, Harris to WLMK, June 25, 1925.

169 *The* Ottawa Journal *was*: *Ottawa Journal*, September 30, 1927.

169 *Even Grant Dexter*: Levine, *Scrum Wars*, 134.

169 *And even if*: Ibid., 134.

170 *"This is calculated"*: King Diary, April 26, 1927.

170 *A few weeks later*: Laurier House and Kingsmere Series, Vol. 23, WLMK to Cranston, February 21, 1928; "Kingsmere," *Toronto Star Weekly Magazine*, February 4, 1928.

170 *At first, it:* King Diary, April 3, 29, May 14, 1927; April, 27, June 3, 16, July 24, 1928; Von Baeyer, *Garden of Dreams,* 183. The first time King had a lamb killed for dinner was equally a traumatic moment. "I feel genuinely pained at the thought and doubt if I can eat it," he wrote after calling McBurnie to order the deed done. "I love to see the little creatures on the hill side. It is a reposeful picture in one's mind and they are as children." There is no follow up diary comment on the lamb dinner, but presumably King enjoyed it. King Diary, May 9, 1927.

170 *As he noted:* Ibid., June 3, 1928.

170 *"The moonlight through":* Ibid., August 3, 1928.

171 *"I can see wherein":* Ibid., July 6, 1929.

171 *Accompanying the prime minister:* Betcherman, *Ernest Lapointe,* 120, 123; MacLaren, *Commissions High,* 262; Massey, *What's Past Is Prologue,* 111.

171 *When he felt better:* Hillmer, Unpublished Skelton Biography, Chapter 3, 29–30.

171 *"King is in the":* MacRae to Dafoe, October 29, 1926, in Cook, "A Canadian Account of the 1926 Imperial Conference," 54. In light of the *Manitoba Free Press's* support of King in the recent federal election, Dafoe decided that it was best that he not accompany King again to London and instead sent a younger editor, D.B. MacRae, who later became the editor of the *Regina Leader-Post.* As Dafoe explained to Clifford Sifton in a letter of September 27, 1926, "It is, I think, highly necessary that we [should] do nothing now to strengthen any impression that may have been created by our fight that the Free Press has [been] turned into a government organ. For me to travel to London in King's entourage and there be thrown into close contact with him and his associates... [would] have precisely that effect." See, Dafoe Papers, M-74, Dafoe to Sifton, September 27, 1926.

172 *The* London Daily News: Cited in Neatby II, 196.

172 *As he wrote to:* Terence De Vere White, *Kevin O'Higgins* (London, Methuen, 1948), 221. O'Higgins, only thirty-five, was assassinated eight months after the conference, an early victim of the IRA. The killing shocked King, who noted in his diary on the day he heard the news, "What a strange race, a certain type of Irish." See, King Diary, July 11, 1927.

172 *One day, forty:* MacRae to Dafoe, November 4, 1926 in Cook, "A Canadian Account of the 1926 Imperial Conference," 58.

172 *Always thinking of:* King Diary, October 24, 1926; MacRae to Dafoe, November 4, 1926.

172 *A believer in:* "Oliver Joseph Lodge," www.spiritwritings.com/oliverjosephlodge.html.

172 *"When I had concluded":* King Diary, November 12, 1926.

173 *He remained a:* Oswald Pirow, *James Barry Munnik Hertzog* (Cape Town: Howard Timmins, 1958), 11.

173 *"'Autonomy' was the popular":* Hillmer, "Foreign Policy and the National Interest: Why Skelton Matters," 6.

173 *Still, King was as:* Wigley, *Canada and the Transition to Commonwealth,* 266.

173 *Back in June 1925: Documents on Canadian External Relations* (DCER), Vol. 3 (Ottawa: Department of External Affairs, 1967), Larkin to WLMK, June 15, 1925, 379–80.

174 *King "at once":* Amery, *My Political Life,* Vol. 2, 377; King Diary, April 1, 1925.

174 *King reassured the:* C.P. Stacey, *Canada and the Age of Conflict 1921–1948,* Vol. 2 (Toronto: University of Toronto Press, 1981), 81.

174 *All the British requested:* Massey, *What's Past Is Prologue,* 111.

174 *The time had:* King Diary, October 22, 1926; Neatby II, 182–88.

175 *"Little as the general public":* Harold Begbie, *Mirrors of Downing Street: Some Political Reflections by a Gentleman with a Duster* (Port Washington, NY, Kennikat Press, 1970), 65.

175 *Or, put more bluntly:* Ibid., 65.

175 *At long last:* Sir Peter Marshall, "The Balfour Formula and the Evolution of the Commonwealth," *The Round Table* 90: 361 (September 2001), 541, 544; Amery, *My Political Life,* Vol. 2, 390.

175 *"Poor old Balfour":* Cited in MacLaren, *Commissions High,* 266; Cook, "A Canadian Account of the 1926 Imperial Conference," 59–60; Massey, *What's Past Is Prologue,* 112.

175 *Later, Amery and:* Amery, *My Political Life,* Vol. 2, 392.

175 *Sir Maurice Hankey:* Dafoe Papers, M-74, W.L. Grant to Dafoe, December 29, 1927.

175 *Oscar Skelton was:* Cited in Betcherman, *Ernest Lapointe,* 127.

175 *Most succinct was:* Cited in Hillmer, Unpublished Skelton Biography, Chapter 3, 33.

176 *"It is evident from":* Dafoe Papers, M-74, Dafoe to Sifton, November 20, 1926; Sifton to Dafoe, November 23, 1926.

176 *He privately oozed:* John MacFarlane, *Ernest Lapointe and Quebec's Influence on Canadian Foreign Policy* (Toronto: University of Toronto Press, 1999), 73.

176 *As he confided:* King Diary, December 29, 1926. John Dafoe certainly understood the inner workings of King's mind. "The inclination of Mr. King, I think will be to bask in the glory of the 1926 achievement and do nothing for, let us say, the life of this parliament," he wrote perceptively to Sifton on November 29, 1926. "It is quite evident that if King had had his way at London nothing would have come out of the conference except perhaps the modernization of the Governor-General's office, with one or two other minor adjustments. We undoubtedly owe the declaration to General Hertzog, and the *Free Press* has said so, in effect, in an editorial which, no doubt, you have noted." See, Dafoe Papers, M-74, Dafoe to Sifton, November 29, 1926.

177 *"I felt immensely relieved":* King Diary, February 3, 1927.

177 *The new reality:* Betcherman, *Ernest Lapointe,* 129.

177 *He met President Calvin Coolidge:* The asking price for the mansion was $500,000 with furniture, which Skelton objected to as extravagant. King liked the house, however. After some bartering and the furniture being removed from the deal, the price dropped to $275,000. This residence was Canada's headquarters until 1946. It is currently the embassy of Uzbekistan. See, Massey, *What's Past Is Prologue,* 150; King Diary, March 21, 1927.

177 *"It is but the doctrine":* House of Commons Debates, April 13, 1927, 2472; Stacey, *Canada and the Age of Conflict,* 91.

177 *At this stage:* C.P. Stacey, *Mackenzie King and the Atlantic Triangle* (Toronto: Macmillan of Canada, 1976), 112; MacLaren, *Commissions High,* 267. In 1928, Massey and Skelton also occupied themselves with the "critical" issue of whether Canada should have its own diplomatic uniform. During the legation's first year, Massey and his staff wore their British diplomatic dress, "but it soon seemed inappropriate," recalled Massey. Skelton favoured adopting a "distinctively" Canadian uniform with maple leaf embroidery on the collar and the cuff and the use of special Canadian buttons. The Canadians were so outfitted. See, Massey, *What's Past Is Prologue,* 139.

177 *"We would have":* Hillmer, Unpublished Skelton Biography, Chapter 4, 19.

177 *He regarded the professor:* King Diary, March 1, 1928.

178 *But it should not:* Hillmer, Unpublished Skelton Biography, Chapter 4, 19–20.

178 *To King's chagrin:* Ibid., Chapter 3, 28.

178 *He cabled King:* Primary Correspondence Series, Vol. 144, C-2297, Lapointe to WLMK, August 12, 1927; King Diary, June 8, 1927; MacFarlane, *Ernest Lapointe,* 71–72.

178 *In the heat of the moment:* MacFarlane, ibid., 72.

179 *"I think it is a mistake":* King Diary, September 4, 1927.

179 *The British, too:* Betcherman, *Ernest Lapointe,* 133–34.

179 *Reviewing the modest:* King Diary, May 29, 1928.

179 *"[I] began to sing":* Ibid., August 22, 1928.

180 *As he wrote his:* Ibid., August 27, 1928.

180 *The election:* Ibid., September 3, 4, 7, 1928; Hillmer, Unpublished Skelton Biography, Chapter 4, 29–33

180 *King spent another:* King Diary, ibid., September 25, 26, 1928.

181 *There was, as:* Allen, *Ordeal by Fire,* 293–94.

181 *King was certain:* King Diary, June 20, July 1, 1927. The aura of the Diamond Jubilee had worn off by mid-August, when the governor general, the Marquess of Willingdon, (who had replaced Lord Byng) wrote to King asking for a $5,000 grant to cover entertainment expenses for the royal princes and Prime Minister Baldwin. King wasn't happy about it. "I am surprised at the request which may occasion comment, but I do not see how we can do other than to seek to meet it. It will be necessary to explain to the Governor, however, that it may bring his name into public discussion in an unfavourable light... I am beginning to be 'fed up' with this 'English invasion' and all the Diamond Jubilee celebrations and ceremonies." See King Diary, August 13, 1927.

182 *"What is the greatest":* NYT, May 22, 1927; Levine, *Devil in Babylon,* 262.

183 *Mackenzie King was:* King Diary, July 3, 1927.

184 *Between 1908:* Levine, *Devil in Babylon,* 233–36.

185 *"We got into a good":* King Diary, October 20, 1927.

185 *Good government demanded:* Neatby II, 233.

185 *"The old age pension":* King Diary, March 31, 1927; Neatby II, 219; Bernard L. Vigod, *Quebec Before Duplessis: The Political Career of Louis-Alexandre Taschereau* (Montreal: McGill-Queen's University Press, 1986), 149–50.

185 *The kids called:* Peter G. Oliver, *Howard Ferguson: Ontario Tory* (Toronto: University of Toronto Press, 1977), 145.

186 *Like almost every:* Vigod, *Quebec Before Duplessis,* 78, 119, 124.

186 *King was prepared:* King Diary, May 17, 1930.

186 *The court's decision:* Ibid., February 5, 1929; Oliver, *Howard Ferguson,* 354–55.

186 *"The conference broke up":* King Diary, November 10, 1927; Oliver, ibid., 305.

187 *"May I hasten to":* Primary Correspondence Series, Vol. 158, C-2307, 134854–74, WLMK to Taschereau, February 20, 1928.

187 *Or that Canada's export market:* Berton, *The Great Depression,* 23–27.

187 *Instead, on January 10, 1929:* Spiritualism Series, Vol. 3, File 46, H-3037, Memorandum on Bleaney Visit, January 10, 1929; King Diary, January 10, 1929.

188 *Bennett, said journalist:* O'Leary, *Recollections,* 69; John Boyko. *Bennett: The Rebel Who Challenged and Changed A Nation* (Toronto: Key Porter, 2010), 20.

188 *"[Bennett] will have difficulties":* Dafoe Papers, M-74, Dafoe to Sifton, October 15, 1927; M-75, Dafoe to Stevenson, November 15, 1928; Cook, *The Politics of John W. Dafoe,* 188.

188 *In words more:* Ibid., M-75, Dafoe to Stevenson, November 15, 1928.

188 *"I should have preferred":* King Diary, October 12, 1927.

189 *"It is not a red-blooded":* House of Commons Debates, April 9, 1929, 1404; King Diary, June 17, 1929.

189 *In January of 1929:* LAC, W.L. McDougald Papers, R4664-0-3-E MG 27 III C24, Vol. 1, File: "Correspondence Mackenzie King 1927–1931," Bank deposit slips, December 29, 1927, for $10,000; and October 1, 1928, for $15,000. See also, Neatby II, 383; J.L. Granatstein, "Was King Really Bribed? Diaries cast doubt on it," *Globe and Mail,* January 18, 1977.

189 *"I felt the embarrassment":* King Diary, January 14, 1929.

189 *King was always:* Neatby II, 373.

190 *This deal was contingent: Journals of the Senate of Canada,* Second Session of Seventeenth Parliament, Vol. 69, Special Committee on Beauharnois Power Project August 1, 1931, 326. Regehr, *The Beauharnois Scandal,* 54–55.

190 *Thereafter, the Ebbs-Haydon:* Regehr, ibid., 54.

190 *And finally, for good:* Primary Correspondence Series, Vol. 234, C-2329, 162912-18, "Memorandum Presented by the Honourable Andrew Haydon Respecting the Beauharnois Power Corporation Enquiry; Special Committee on Beauharnois Power Project, 332; King Diary, April 23, 1932; Neatby II, 371–72.

190 *He had not wanted:* King Diary, January 21, February 1, 1928.

190 *There is no reason:* Regehr, *The Beauharnois Scandal,* 109; Neatby II, 374–75.

191 *In a classic case:* King Diary, November 1, 1930; Neatby II, 296.

191 *A few days before:* King Diary, November 26, 1929.

191 *Asked by the press:* Berton, *The Great Depression,* 37–38.

191 *"I thanked God":* King Diary, December 31, 1929.

192 *King kept all:* Neatby II, 302; Betcherman, *Ernest Lapointe,* 170; Berton, *The Great Depression,* 37.

192 *Then he instituted pro-labour:* Lee E. Ohanian, "What—or Who—Started the Great Depression?" Journal of Economic Theory 144:6 (November 2009), 2310–35; "Hoover's Pro-Labour Stance Spurred Great Depression," *University of California Newsletter,* August 29, 2009, www.universityofcalifornia.edu/news/article/21795. Thanks to Prof. Ohanian for sending me a copy of his paper before publication.

192 *Many of Canada's: Globe,* January 3, 1930; Boyko, *Bennett,* 185.

193 *First, King's view of:* Neatby II, 310-12; House of Commons Debates, April 24, 1922, 1073. See also, Doug Owram, "Economic Thought in the 1930s: The Prelude to the Keynesianism," Canadian Historical Review 66:3 (September 1985), 360–65. Owram discusses how challenging it was for Canadian economists to comprehend the full extent of the Depression and to accept the necessity of adopting more innovative solutions.

193 *When a Prairie delegation:* King Diary, February 26, 1927; Thompson, *Canada 1922–1939,* 197–98.

193 *His view had not:* House of Commons Debates, March 31, 1930, 1119–27; April 1, 1930, 1158; April 2, 1930, 1210, April 3, 1930, 1231–57.

194 *Alone in his study:* King Diary, April 1, 2, 3, 1930.

194 *He began by reiterating:* House of Commons Debates, April 3, 1930, 1225–28; Neatby II, 318.

194 *Isaac MacDougall:* Ibid., April 3, 1930, 1256.

194 *King realized later:* King Diary, April 3, 1930

195 *According to John Diefenbaker:* Diefenbaker, *One Canada*, 171–72; Oliver, *Howard Ferguson*, 365–66.

195 *Within days, the Conservatives: Globe*, April 4, 5, 1930; *Toronto Star*, April 3–8, 1930; King Diary, April 8, 1930.

195 *King did not:* See Neatby II, 314–26.

195 *As fate was to:* King Diary, July 29, 1930; Neatby II, 330–31; Reginald Whitaker, *The Government Party: Organizing and Financing the Liberal Party of Canada 1930–1958* (Toronto: University of Toronto Press, 1977), 10–12. King eventually discovered that Sweezy had given the Liberals a whopping $700,000 for the 1930 election campaign. According to King's calculations, $300,000 of these funds had remained in the hands of Quebec Liberals under the control of Senator Donat Raymond, who was a close associate of Quebec premier Taschereau. The company's largesse was easily under-standable: it was simply protecting its interests and the future anticipated wealth which the project was to produce—and Beauharnois wanted the Liberals to win. As a safeguard, Robert Sweezy, Beauharnois's engineer and chief promoter, had also offered the Conservatives $200,000, funds that the Tory campaign manager Major-General Alexander McRae never collected. The Conservatives instead took money from the president of the Royal Bank, Sir Henry Holt, who was connected to Beau-harnois's chief rival, Montreal Light, Heat and Power Consolidated. See King Diary, July 28, 1931; Transcripts of Hearings of the Special Committee on Beauharnois Power, published as part of Appendix 5, *Journals of the House of Commons*, Session 1931, 833–34; Regehr, *The Beauharnois Scandal*, 113–14.

196 *Bennett and Hugh Graham:* Levine, *Scrum Wars*, 156.

196 *It proved to him:* King Diary, April 2, 1930.

196 *From King's perspective:* Ibid., April 9, 1930.

196 *John Dafoe urged King:* Cook, *The Politics of John W. Dafoe*, 187.

196 *She told him that June:* Spiritualism Series, Vol. 3, File 47, H-3037, "Reading by Mrs. Bleaney February 8, 1930."

197 *His family:* The death of Peter Larkin, despite the fact that he was seventy-four years old, was naturally a blow to King. "The more I think of him, the finer his character appears to me to be," he wrote in his diary, "a truly noble-minded, generous, kind good man. He has gone as he wished to go—The High Commissioner for Canada in office, with his life lived and before his faculties failed. He reminded me much of father and his death must be almost at the same age. No man ever had a truer friend that I had with him. He was indeed to me 'a Sheltering Tree.'" See King Diary, February 3, 1930.

197 *King enjoyed the:* Ibid., April 13, 15, 18, 19, 1930; McDougald Papers, Vol. 1, File: "Correspondence 1931–1938: Affidavits," Affidavit July 16, 1931 of W.L. McDougald.

198 *Back in Ottawa, he:* King Diary, April 27, 1930.

198 *The two-month campaign:* Ibid., June 24, July 3, 21, 27, 1930.

198 *In early June, Major-General:* Thompson, *Canada 1922–1939*, 202.

198 *In Winnipeg, where:* Levine, *Scrum Wars*, 152, 157; King Diary, June 9, 1930; Dafoe Papers, M77, Dexter to Dafoe, January 4, 1935.

199 *King didn't get it:* On the night of June 17, 1930, King had another vivid dream. In it, he was boarding a ship heading for England. He was confronted by two naked men, Bennett and Conservative MP, Robert Manion. "I offered to help them a bit," King recorded in his diary, "and did lend them some dress, instead of getting on the vessel they went in another direction and turned into a sort of club, one man [Bennett] was

particularly nasty towards myself, both were clearly enemies." Interpreting his own dream, King decided that the ship was the campaign and Dunning's budget, given the connection to England. And the allegation was that the Liberals had stolen the Conservatives' clothes. "The fact that the dream came just on starting out on the campaign was I thought very significant," he concluded. See, King Diary, June 17, 1930.

199   *Asked in Edmonton:* Cited in Betcherman, *Ernest Lapointe,* 176.

199   *In Ontario, Premier Howard:* Oliver, *Howard Ferguson,* 365.

199   *Sensitive as ever:* King Diary, May 17, July 21, 24, 1930.

199   *A week before:* Ibid., July 21,1930.

199   *Suffering from a bad:* Ibid., July 27, 1930.

200   *"The authority and prestige":* Ward, *The Memoirs of Chubby Power,* 115.

200   *King, too, was:* King Diary, July 29, 1930.

200   *Because he was given:* Neatby II, 342.

200   *"I feel I must":* King Diary, November 2, 1930.

CHAPTER 7: Scandal and Resurrection

201   *It did not take:* Regehr, *The Beauharnois Scandal,* 133.

201   *"I confess I am":* King Diary, July 10, 1931.

201   *Raymond had little:* In later testimony before a Senate committee, Haydon claimed that a $200,000 payment had been made to the then premier of Ontario, Howard Ferguson, by businessman and Ontario Conservative Party fundraiser John Aird, Jr. This was purportedly in exchange for Ferguson's (and Ontario Hydro's) support of Beauharnois. Ferguson, who in 1931 was in London as Canada's new high commissioner, returned to Ottawa to vehemently deny these allegations. It was confirmed by Robert Sweezy that Aird had, in fact, received $125,000 as a lobby fee and that this money did not go to Ontario Hydro or Ferguson. No other concrete evidence against Ferguson ever turned up, though, as his biographer, Peter Oliver notes, the gossip about his possible impropriety continued for years. See, Oliver, *Howard Ferguson,* 398–99.

202   *"It would seem that":* King Diary, July 10, 1931.

202   *Hidden in the mass:* McDougald Papers, Vol. 1, File: "Correspondence 1931–1938: Affadavits," Affidavit July 16, 1931; Regehr, *The Beauharnois Scandal,* 135.

202   *When Ian Mackenzie:* King Diary, July 13, 1931.

203   *When the story:* House of Commons Debates, July 22, 1931, 4028–29.

203   *The Kingston Whig-Standard:* Reprinted in the *London Advertiser,* July 27, 1931.

203   *King's interpretation of it:* King Diary, July 15, 1931.

203   *The "sensation" and "suspicion":* Ibid., July 17–20, 1931.

204   *The official report:* Report of the Special Committee on Beauharnois Power, *Journals of the House of Commons,* Session 1931, Appendix 5, 631–45; Regehr, *The Beauharnois Scandal,* 140–42.

205   *As the* Financial Times: *Financial Times* (Toronto), July 31, 1931; Regehr, ibid., 143.

205   *"Individual members of":* House of Commons Debates, July 30, 1931, 4387–88.

205   *At home later:* King Diary, July 30, 1931.

205   *The real trouble:* Regehr, *The Beauharnois Scandal,* 152.

206   *At a tense meeting:* King Diary, October 23, 1931.

206 *A few weeks later:* Ibid., November 16, 1931; Primary Correspondence Series, Vol. 189, C-2328, 163904-05, WLMK to Charles Stewart, August 26, 1931; Whitaker, *The Government Party,* 20.

206 *Instead, he wrote:* King Diary, March 2, 1932; Regehr, *The Beauharnois Scandal,* 156–57.

207 *When it was over:* Canada. Parliament. Senate. *Special Committee on Beauharnois Power Project* (Ottawa: King's Printer, 1932), xvii–xix.

207 *"Two Liberal Senators":* "Three-hour Condemnation of McDougald and Haydon Called the Most Violent in Senate's History," *Globe,* April 29, 1932, 1.

207 *King saw the report:* King Diary, February 12, April 23, 1932.

207 *He was more upset:* The few Liberals on the Senate committee were also sympathetic to Haydon yet felt they had no choice but to sign the report in which he was harshly criticized. According to Senator Charles Murphy, a former Liberal MP, who was not fond of King, even though King had appointed him to the Senate in 1925: "It broke the hearts of the Liberals on the [Senate] Committee to sign such report, but that they had to sacrifice Haydon in order to save King! In other words, they knew that Haydon was innocent; but, with amazing stupidity and lack of logic, as well as lack of honesty, that characterizes so many people in their attitude towards a party and towards a Party Leader, they deliberately did a grievous wrong to an honourable and innocent man in order to shield one who did not deserve to be shielded." See, LAC, R.B. Bennett Papers, MG 26 K, Vol. 682, M-1339, 419141-45, Charles Murphy to George Lynch-Staunton, December 30, 1931.

207 *That barely satisfied Meighen:* Meighen Papers, Vol. 149, C-3552, 91387-90, R.S. Stevens to John P. Ebbs, April 30, 1932; J. Fenton Argue to John P. Ebbs, April 30, 1932.

207 *Grant Dexter of:* Grant Dexter, "Senator Andrew Haydon," *Winnipeg Free Press* (WFP), November 12, 1932.

207 *He berated himself:* King Diary, November 10, 12, 1932.

208 *Following the 1935:* MacDougald Papers, Vol. 1, File: "Correspondence Mackenzie King," MacDougald to WLMK, October 31, 1935.

208 *At one meeting:* King Diary, May 20, 1936.

208 *McDougald tried again:* MacDougald Papers, Vol. 1, File: "Correspondence 1931–1938," MacDougald to WLMK, June 27, 1938.

208 *"I cannot but feel":* King Diary, June 19, 1942.

208 *Moreover, the Beauharnois:* Regehr, *The Beauharnois Scandal,* 179.

209 *In assessing King's culpability:* Ibid., 190.

209 *"I have grown nearly":* Primary Series Correspondence, Vol. 184, C-2325, 157568-73, WLMK to Crerar, November 7, 1931.

209 *In the immediate aftermath:* Whitaker, *The Government Party,* 21.

209 *"I hope you realize":* Primary Series Correspondence, Vol. 184, C-2324, 156728-29, WLMK to Béique, March 7, 1931.

209 *King himself, not:* King Diary, November 26, 1931; Whitaker, *The Government Party,* 19.

210 *King made his:* King Diary, August 14, 1931.

211 *He "felt a mental":* Ibid., August 24, 1931.

211 *"I have loved the cause":* Massey, *What's Past Is Prologue,* 209.

211 *"I can see it all":* Ibid., 210.

211 *In words that:* J.A. Spender, *Sir Robert Hudson: A Memoir* (London: Cassell, 1930), v; Personal Papers Series, Vol. 24, WLMK to Markham January 28, 1933; Primary Series Correspondence, Vol. 189, C-2327, 160812-15, WLMK to Spender, November 4, 1931; Bedore, "The Reading of Mackenzie King," 94–5.

211 *To that end:* King Diary, November 23–25, 1931.

212 *King found the entire:* Ibid., March 9, 1932; Whitaker, *The Government Party*, 26.

212 *"He has a fine":* King Diary, July 20, 1933; Whitaker, ibid., 43–44.

212 *He was sure:* Ibid., November 27, 1932.

213 *More significantly, Lambert's:* Personal Papers Series, Vol. 24, WLMK to Markham, December 11, 1933; Larry Glassford, *Reaction and Reform: The Politics of the Conservative Party under R.B. Bennett 1927–1938* (Toronto: University of Toronto Press, 1992), 177.

213 *Provincial Conservative governments:* Quebec was the exception for the NLF. The organization and fundraising there was under the tight control of Ernest Lapointe, Chubby Power, and other top Quebec Liberals. There was no sister provincial association for the NLF to work with, as in the other provinces, and there was not to be until the 1950s. When Lambert assumed the presidency of the NLF in 1936, he was told by King "to leave the province of Quebec alone." Memoranda and Notes Series, Vol. 192, File 1757, C135210-13, "Re: Meeting of Executive Committee of National Liberal Federation," March 16, 1937; Whitaker, *The Government Party*, 27; 432 note 67.

213 *King's own insecurities:* Personal Papers Series, Vol. 24, WLMK to Markham, January 28, 1933.

213 *In September 1933:* King Diary, September 8, 1933.

213 *One Catholic journal:* Thompson, *Canada 1922–1939*, 234–35.

213 *King distinguished between:* King Diary, March 25, 1934; H. Blair Neatby, *William Lyon Mackenzie King: The Prism of Unity, 1932–1939* (Toronto: University of Toronto Press, 1976), 29 (hereafter Neatby III).

213 *He entitled his:* Massey, *What's Past Is Prologue*, 214–15.

214 *When he read the:* King Diary, March 31, 1933.

214 *Massey pitched the:* Ibid., September 5–9, 1933; Massey, *What's Past Is Prologue*, 212; Whitaker, *The Government Party*, 41.

214 *"Mackenzie King himself":* Massey, ibid., 211.

214 *In truth, King was:* King Diary, May 11, 1933.

215 *"There was nothing":* Ibid., July 7, 1933.

215 *Speaking at a public:* Massey, *What's Past Is Prologue*, 215-16; Whitaker, *The Government Party*, 40.

215 *At dawn on:* King Diary, September 3, 1933.

215 *From his perspective:* Massey, *What's Past Is Prologue*, 212.

215 *Given King's negative:* King Diary, September 9, 1933.

216 *Judging from his:* Ibid., September 5, 6, 1933.

216 *One day, he:* Ibid., September 5, 1933.

216 *Massey understood that:* Massey, *What's Past Is Prologue*, 212–13; King Diary, April 10, 1934.

216 *That didn't mean:* Personal Papers Series, Vol. 24, WLMK to Markham, December 11, 1933.

216 *King told him:* King Diary, April 10, 1934.

217 *After that, King:* Ibid., August 5, 1934; Whitaker, *The Government Party*, 42.

217 *Personally, he regarded:* See, for example, King Diary, July 10, August 16, 1931; February 23, August 10, 15, 1932; May 9, 1933.

217 *On Remembrance Day:* Ibid., November 11, 1932. For the record, the press felt the same way about Bennett as King did. Bennett bullied reporters and threatened to throw editors in jail for publishing the truth. "It must be galling for a man who dramatized himself as the Saviour of his country to find that all his plans are going badly and he became the target for continual and well-founded criticism," journalist John Stevenson wrote to John Dafoe. "Of course, [Bennett] is afflicted with Mussolinian megalomania and wants in his blind rage to trample upon everybody who criticizes or thwarts him." In an editorial of February 12, 1934, entitled "Why Bennett Hates Newspapers," the *Vancouver Sun* opined, "The reason that Mr. Bennett finds fault with reporters and editors is that our Prime Minister is an orator rather than a thinker. Meaningless words flow from his agile tongue like water over a millrace while his brain is as arid of thoughts as an African river in the dry season." See, O'Leary, *Recollections,* 74; Dafoe Papers, M-75, Norman Smith to Dafoe, September 25, 1930; John Stevenson to Dafoe, undated letter; *Vancouver Sun,* February 12, 1934; Levine, *Scrum Wars,* 126.

217 *During a heated exchange:* Ward, *The Memoirs of Chubby Power,* 290.

217 *King rarely felt:* King Diary, September 5, 1930; WLMK to Markham, January 28, 1933.

218 *The conference was:* Boyko, *Bennett,* 249–53; Thompson, *Canada 1922–1939,* 220; Levine, *Scrum Wars,* 166. Glassford is more positive about the conference: Glassford, *Reaction and Reform,* 116.

218 *British officials regarded:* Neville Chamberlain, then a minister in the new British national government of Ramsay MacDonald, attended the conference and detested Bennett more than King did. Chamberlain's diary entry for the last day of the conference summarizes Bennett's atrocious behaviour as follows: "He alternately blustered, sobbed, bullied, prevaricated, delayed and obstructed to the very last moment." Chamberlain, exhausted, wrote to his wife, "it will take me a week or two to recover from the last week. I never want to see Canada again!" To be fair, however, as John Boyko points out, Canadian trade with Britain did increase during the next two years, partly owing to the negotiations made at the conference. See, Iain McLeod, *Neville Chamberlain* (London: F. Muller, 1961), 161; Neatby III, 21; Robert C. Self, *Neville Chamberlain: A Biography* (London: Ashgate Publishing Company, 2006), 172; Boyko, ibid., 253.

218 *King was certain:* Primary Series Correspondence, Vol. 184, C-2324, 156671-73, WLMK to Ames, January 29, 1931; Neatby II, 355.

218 *Charles Dunning, who had:* Neatby III, 15.

219 *After Beauharnois, Cromie:* Levine, *Scrum Wars,* 139.

219 *"Opposition," he concluded:* King Diary, July 20, 1932.

219 *The imperial economic conference:* WLMK to Markham, January 28, 1933.

219 *"I have never seen":* King Diary, November 25, 1932.

219 *Within a year:* Glassford, *Reaction and Reform,* 101.

219 *There were no opinion:* Bliss, *Right Honourable Men,* 113.

219 *As listed by:* Larry A. Glassford, "Retrenchment—R.B. Bennett Style: The Conservative Record Before the New Deal," *American Review of Canadian Studies* 19:2 (Summer 1989), 142, 153; Glassford, *Reaction and Reform,* 135–36. See also Boyko, *Bennett,* 17–18.

220 *He was up at:* King Diary, December 17, 1934.

221 *He took the doctor's:* Personal Miscellaneous, Vol. 1, WLMK to Barker, December 4, 1933; Barker to WLMK, December 7, 1933; WLMK to Barker, February 14, 1934; WLMK to Barker, February 16, 1934; Barker to WLMK, November 5, 1938.

221 *The Liberal political:* Whitaker, *The Government Party*, 72.

222 *In what can:* For a revisionist view of Bennett's New Deal see, Boyko, *Bennett*, 364–65.

222 *On the evening of:* Levine, *Scrum Wars*, 170-71; Boyko, ibid., 364–69.

222 *He could not believe:* Neatby III, 88; Thompson, *Canada 1922–1939*, 261.

222 *Most Tory newspapers:* Neatby, ibid., 88–89; Dafoe Papers, M-77, Dafoe to Dexter, January 10, 1935; Thompson, ibid., 264; Boyko, *Bennett*, 370–77.

222 *"In its egotist style":* King Diary, January 2, 9, 1935.

223 *Bennett's program:* Thompson, *Canada 1922–1939*, 265. "Was there a Bennett New Deal?" asks Larry Glassford. His answer: "Only in the public-relations sense, if by new New Deal is meant a distinctive policy of radical reform, originating in the New Year's broadcasts of 1935, and translated into legislation over the next six months. Bennett's public image was indeed, radically transformed from the corpulent, prosperous tycoon with friends in high places to the fire-breathing prophet of a new economic and social order, with justice for all. Neither caricature of the prime minister was particularly accurate, though there were elements of truth in each." See, Glassford, *Reaction and Reform*, 173.

223 *Fearful of being:* King Diary, July 2, 1935; Bill Waiser, "King and Chaos: The Liberals and the 1935 Regina Riot," in English, et al. *Mackenzie King: Citizenship and Community*, 74–75.

223 *"I should have":* King Diary, July 5, 1935.

223 *The list included:* See, Allan Levine, "Desperate Canadians Turned to Political Messiahs," *Compass* (March/April 1993), 21–23; H. Blair Neatby, *The Politics of Chaos: Canada in the Thirties* (Toronto: Macmillan of Canada), 1972.

224 *King spent hours:* Moher, "The 'Biography' in Politics," 242–44.

224 *This volume was:* Primary Series Correspondence, Vol. 209, C-3682 180230-35, WLMK to Morang, February 22, 1935; 180, 286–87, Morang to WLMK, March 18, 1935.

224 *Despite enormous pressure:* William J. McAndrew, "Mackenzie King, Roosevelt, and the New Deal: The Ambivalence of Reform" in English, et al., *Mackenzie King: Widening the Debate*, 138; William Lyon Mackenzie King, "The Issues As I See Them," *Maclean's*, September 15, 1935, 11, 31–32.

225 *"I did not intend":* King Diary, June 26, 1935.

225 *As Escott Reid:* Escott Reid, "The Canadian Election of 1935—And After," *American Political Science Review* 30 (February 1936), 116.

225 *It needs to be said:* Stacey, *A Very Double Life*, 183–84.

225 *King was never:* There is a 1936 memo in King's papers on "how to talk on the radio," which recommended that you "speak to the listener as you would in the privacy of your own home." Jack Pickersgill, who joined King's staff in 1937, recalled telling King repeatedly that it was sound rather than content that truly mattered in a radio broadcast. "Mr. King," Pickersgill said in a 1990 interview, "paid no attention to this advice." See, Memoranda and Notes Series, Vol. 204, File 1943, "Memo: Radio," 1936; author's interview with J.W. Pickersgill, 1990; Levine, *Scrum Wars*, 169.

225 *"What this country":* Hutchison, *Incredible Canadian*, 198.

225 *After the first broadcast:* Ottawa Journal, October 1, 1935; *Mail and Empire*, August 1, 1935.

226 *Paul Martin, Sr., too:* Paul Martin, "King: The View from the Backbench and the Cabinet Table in English, et al., *Mackenzie King: Widening the Debate,* 31–32.

226 *For these editors:* J.M. Beck, *Pendulum of Power: Canada's Federal Elections* (Scarborough: Prentice Hall of Canada, 1968), 213; Glassford, *Reaction and Reform,* 181.

226 *The Tories had more:* Memoranda and Notes Series, Vol. 204, File 1943, "Mr. Sage Broadcasts September-October, 1935"; Levine, *Scrum Wars,* 173–74.

226 *He was in a particularly:* Massey, *What's Past Is Prologue,* 222.

226 *King's diary entry:* King Diary, October 5, 1935.

227 *Speaking before a cheering:* Ibid., October 8, 1935; *Toronto Star,* October 9, 1935.

227 *Historians and journalists harped:* Beck, *Pendulum of Power,* 217–19; Whitaker, *The Government Party,* 84; Alvin Finkel, "Origins of the Welfare State in Canada," in J.M. Bumsted, *Interpreting Canada's Past,* Vol. 2 (Toronto: Oxford University Press, 1993), 541–42; Boyko, *Bennett,* 409–14.

228 *With the victory secure:* King Diary, October 14, 1935.

CHAPTER 8: A Romantic Among the Spirits

229 *He was constantly:* Thomas Carlyle, *Heroes, Hero-Worship and the Heroic in History,* in the *Collected Works,* Vol. 12 (London: Chapman and Hall, 1869), 80; King Diary, October 9, 23, 1900; October 10, 1929; January 6, 1947; March 6, 1949. Bedore, "The Reading of Mackenzie King," 65–66, 302.

230 *"I wondered what":* King Diary, August 31, 1930.

230 *Fifteen years later:* Personal Papers Series, Vol. 29, WLMK to Joan Patteson, April 29, 1945; King Diary, April 29, 1945; October 15, 1948; June 12, 1949.

230 *There were other:* King Diary, April 20, 1938.

230 *"There will not be":* Ibid., May 26, 1929.

231 *One night, she:* Ibid., October 16, 1927.

231 *"She has come to see":* Ibid., January 23, 1926.

231 *"I wish I were":* Ibid., September 8, 1928.

231 *He described what he:* Ibid., September 20, 1931.

232 *When Beatrix Henderson Robb:* NYT, April 21, 1957.

232 *For about a month:* King Diary, August 28, 31, September 1, 11, 28, October 1, 2, 1928.

232 *During his years:* The tutoring paid off. Bob Gerry (1877–1957) became a successful businessman and owner of a stable of winning thoroughbred racehorses. Peter Gerry (1879–1957) became a U.S. congressman and then a senator from Rhode Island. The brothers died within a day of each other. They were also the great-grandsons of Elbridge Gerry (1744–1814), the vice-president in 1813 to 1814 under President James Madison, and famous for being associated with the term "gerrymandering" for realigning the electoral map to favour the ruling party. See, "Elbridge Gerry" and "Peter Gerry," Biographical Dictionary of the United States Congress, http://bioguide.congress.gov/scripts/biodisplay.pl?index=G000139 http://bioguide.congress.gov/scripts/biodisplay.pl?index=G000141.

233 *He found she had:* King Diary, September 30, 1928.

233 *He wrote to thank:* Neatby II, 404-05; King Diary, May 5, 1932.

233 *"I fear that I have":* King Diary, May 8, 10, 1932.

234 *On June 3:* Ibid., June 3,4, 21,1932; Neatby II, 405.

234 *They continued to write:* Ibid., October 21, 1932; Neatby, ibid., 406.

234 *Julia later told:* Neatby, ibid., 409.

235 *He had fallen asleep:* King Diary, November 6, 7, 1932.

235 *"To me it is quite":* Ibid., November 11, 1932.

235 *In private, they:* Ibid., July 17, 1939.

236 *Julia Grant may:* Fred McGregor told historian Blair Neatby that after King's funeral in 1950, he invited six women back to his house for tea. All were married and each one believed she had had a close friendship with King. As Neatby relates the story, several of the women did not know each other and eyed each other suspiciously throughout the intimate gathering. Thanks to Blair Neatby for sharing this story with me.

236 *Little of it made:* See, ibid., October 29, 30, November 4, 1934.

236 *"I prayed earnestly":* Ibid., October 31, 1934.

237 *At the studio:* Ibid., November 5, 1934.

237 *What made this:* Ibid., November 8, 1934.

237 *Years later, and despite:* King's letters to Giorgia were found by historian Peter Waite in the voluminous papers of Lord Beaverbrook, who had purchased her correspondence in 1952 from the Cuban Ambassador in Rome, E. de Blanck. The letters can be found at the Parliamentary Archives in London. They were first referenced by Professor Waite in his article "Mackenzie King and the Italian Lady," *The Beaver* 75:6 (December 1995/January, 1996): 4–10. I thank Professor Waite for sharing information on this correspondence with me.

237 *"She is one of":* King Diary, November 8, 11, 1934.

238 *Other English-speaking ministers:* Ibid., October 25, 29, 1935; Neatby III, 138–40.

238 *He stubbornly maintained:* House of Commons Debates, February 8, 1932, 29–30; Neatby, ibid., 138–40; King Diary, May 25, 1934.

238 *In Geneva, Walter Riddell:* Neatby, ibid., 140–42; King Diary, November 4, 28, 29, 1935; Robert Bothwell, Ian Drummond and John English, *Canada 1900–1945* (Toronto: University of Toronto Press, 1987), 307–8; English, *Shadow of Heaven,* 178–83.

239 *"It has broken my heart":* Parliamentary Archives, London, Lord Beaverbrook Papers, BBK/A/243, WLMK Correspondence, WLMK to Cousiander, November 28, 1935; King Diary, November 28, 1935.

239 *"When your letter came":* Beaverbrook Papers, ibid., King to Cousiander, April 23, 1937; November 22, 1937.

239 *His last letter:* Ibid., King to Cousiander, January 13, 1940; Waite, "Mackenzie King and the Italian Lady," 8.

239 *Giorgia divorced her:* Beaverbrook Papers, ibid., King to Cousiander, February 12, 1945; August 16, 1946; Waite, ibid., 8. In late August 1943, Ludwig interviewed King at Kingsmere and was given the grand tour. Ludwig was most impressed and, as King noted, kept saying in German, "Wondershem" (he meant *Wunderschön*). In his book, Ludwig described Moorside as "simple and dignified, the home of a man who is deeply connected with European culture, and who, in old British manner, prefers privacy to brilliancy." See, Emil Ludwig, *Mackenzie King: A Portrait Sketch* (Toronto: Macmillan of Canada, 1944), 57; King Diary, August 26, 1943; Von Baeyer, *Garden of Dreams,* 105.

239 *When he was in London:* Beaverbrook Papers, King to Cousiander, October 29, 1945.

239 *"It is deeply touching to see":* King Diary, August 23, 1946; Ibid., King to Cousiander, August 26, 1946.

240 *"He lavished such tenderness"*: Waite, "Mackenzie King and the Italian Lady," 9.

240 *"[He] asked me in all"*: Ibid., 9.

240 *When they parted*: Beaverbrook Papers, King to Cousiander, August 28, October 16, November 7, 1947; January 31, March 20, May 3, November 5, 1948; March 20, October 14, 1949; January 21, 1950; Waite, ibid., 6.

240 *"Our friendship is"*: Beaverbrook Papers, King to Cousiander, July 19, 1950. Giorgia, known as the "blonde baroness," continued to write journalism and poetry and during the 1950s she became the companion and then the husband of the Italian poet Leonardo Sinisgalli. See, "Leonardo Sinisgalli," http://utenti.lycos.it/BiagioRusso1/biografia-r.html.

241 *"Bits of statuary"*: King Diary, November 6, 1934.

241 *There was a sundial*: Ibid., April 3, 1934; Von Baeyer, *Garden of Dreams*, 151; "The Community Must Come First: An Interview with Mackenzie King," 28.

241 *He believed the grounds*: Von Baeyer, *Garden of Dreams*, 12.

241 *While he was in Paris*: King Diary, October 31, 1934; May 16, 1935; August 23, 1937.

242 *Initially, he placed*: Ibid., August 23, September 2, 1937; Von Baeyer, *Garden of Dreams*, 153.

242 *"As I stood beside"*: King Diary, August 21, 1937.

242 *King was stirred*: Laurier House and Kingsmere Series, Vol. 25, File 6, "Kingsmere in Chronology," King Diary, September 3, 1932; June 21, 1937.

242 *In this, he was hardly*: Rudy J. Favretti and Joy P. Favretti, *For Every House a Garden: A Guide for Reproducing Period Gardens* (Lebanon, NH: University Press of New England, 1990), 30; Glenn Hooper, *Landscape and Empire 1720–2000* (Aldershot, England: Ashgate Publishing, 2005), 79; John Dixon Hunt, *Gardens and the Picturesque: Studies in the History of Landscape Architecture* (Cambridge, MA: Massachusetts Institute of Technology, 1992), 179.

242 *In June 1937*: King Diary, June 21, 1937.

243 *Still standing was*: Ibid., May 8, 1935; May 9–19, 1935; Von Baeyer, *Garden of Dreams*, 161–63.

243 *That difficult decision*: King Diary, May 15, 1935.

243 *In quick order*: Ibid., May 24,25, 1935.

244 *"I am anxious"*: Ibid., May 26, June 11, 12, 1935; Von Baeyer, *Garden of Dreams*, 164–65.

244 *He and Joan walked*: King Diary, July 5, 1935.

244 *In March 1936*: Ibid., July 4, 1936.

245 *Other stone pieces*: Ibid., June 27, 1936; June 21, August 8, 30, September 4, 10, October 17, 1937; Von Baeyer, *Garden of Dreams*, 169–71.

245 *"I feel in some"*: King Diary, September 1, 1937.

245 *He was working*: Lester B. Pearson, *Mike: The Memoirs of the Rt. Hon. L.B. Pearson*, Vol. 1 (Toronto: University of Toronto Press, 1972), 187–88; Primary Correspondence Series, Vol. 223, C-3691, 191787, WLMK to Massey, May 9, 1936.

246 *Inside was a*: Von Baeyer, *Garden of Dreams*, 172.

246 *Both the Pattesons' dog*: King Diary, April 19, 1942; Von Baeyer, ibid., 172.

246 *In October 1949*: King Diary, October 15, 1949.

246 *"The ultimate agnostic's"*: Wilson to Dawson, February 9, 1888, cited in Carl Berger, *Science, God, and Nature in Victorian Canada* (Toronto: University of Toronto Press, 1983), 68.

247 *What these believers:* Janet Oppenheim, *The Other World: Spiritualism and Psychical Research in England, 1850–1914* (Cambridge, England: Cambridge University Press, 1988), 200.

247 *As George Bernard Shaw:* George Bernard Shaw, *Heartbreak House: A Fantasia in the Russian Manner on English Themes* (New York: Penguin Books, 2000), 13–14.

247 *Beyond these fakers:* Ramsay Cook, "Spiritualism, Science of the Earthly Paradise," *Canadian Historical Review* 65:1 (March 1984), 4–5; Stan McMullen, *Anatomy of a Séance: A History of Spirit Communication in Central Canada* (Montreal: McGill-Queen's University Press, 2004), 25–41.

247 *"The appearance or manifestation":* Salem Bland, "My Experience of Spiritualism," cited in Cook, ibid., 5.

247 *Similarly, Reverend Benjamin:* B.F. Austin, *The Mission of Spiritualism and Original Poems* (Toronto: Austin Publishing Company, 1902), 19; Cook, ibid., 16.

247 *Finally, the respected Toronto:* McMullen, *Anatomy of a Séance*, 88–91.

248 *Asked once about:* Cited in Michael W. Homer, "Arthur Conan Doyle's Adventures in Winnipeg," *Manitoba History* 25 (Spring 1993), 9.

248 *This was not a "crisis of faith":* Allison C. Bullock, "William Lyon Mackenzie King: A Very Double Life?" MA Thesis, Queen's University, 2009, 76–79. King had been grappling with these complex issues surrounding the meaning of life for many years. In late November 1899, while he was in London, he attended a lecture by the British socialist, Herbert Burrows. Entitled "Materialism and Modern Science," the wide-ranging cerebral discussion prompted King to consider the existence of an afterlife. "I was called upon to say a word or two," he later wrote, "and briefly suggested that as the material and mental were different, the spiritual also might be, and granting the German psychologist his materialistic conception of the origin of the mind in matter the spiritual might still stand beyond either as the real force giving life to both and a greater life to the extent to which it was operative … I believe that everything acts according to law and nothing to chance, but the laws of the spiritual world are different in kind from those of the natural, though the latter may be part but not all of the former and I believe we have yet to discover to what a great extent the so-called spiritual laws, e.g., faith and result of, are acting in our lives regarding daily material experiences." See, King Diary, November 29, 1899.

248 *"Somehow or other":* King Diary, September 5, 1928.

249 *"I continue to believe":* Ibid., July 6, 1934.

249 *In his quest:* Ibid., July 20, 1933; February 6, 1935. See also entries for April 20, 1933, May 17, 1935, March 4, 1937, May 5, 1938, July 21, August 7, 1939.

249 *Dining in Toronto:* Ibid., April 12, 1934.

250 *"At all events":* Ibid., April 12, 1934. See also, April 19, May 17, June 11, 1934; January 27, February 6, May 17, September 28, 1935; March 21, 1937.

250 *"I know all this sounds":* Ibid., February 18, 1931; Jonathan Wilson, *Marc Chagall* (New York: Schocken, 2007), 154.

250 *"I felt this to":* King Diary, February 13, 1932.

250 *She was well known:* McMullen, *Anatomy of a Séance*, 66; Patricia Fanthorpe, *Death: The Final Mystery* (Toronto: Hounslow Press, 2000), 136; Arthur Conan Doyle, *The History of Spiritualism*, Vol. 2, (San Diego: Book Tree, 2007), 158–59 (originally published in 1926). According to the website "Skeptiseum," there were many trumpet tricks: "If the medium's hands were not controlled (a practice intended to

prevent trickery) he or she could simply move the trumpet about, a rubber tube being attached through which the medium spoke. Sometimes, a removable luminous band was employed and moved about at the end of a telescoping rod. In these instances, the whispered voices did not actually emanate from the trumpet; the illusion that they did worked on the ventriloquism principle: it is not easy to locate the source of a sound, especially if misdirection takes place. If controlled, the medium had clever techniques of getting one hand free or could use a secret assistant dressed all in black." See, www.skeptiseum.org/index.php?id=189&cat=ghosts.

251 *"The most singular feature"*: Phillips Thompson, "Trumpet Seances," *Globe,* April 20, 1898, 4.

251 *He began keeping*: Spiritualism Series, Vol. 4, File 2, H-3037, Records of séances of Mrs. Wriedt's visit to Brockville-Ottawa, February, 1932.

251 *Upon his return to Laurier House*: King Diary, February 4, 1932.

251 *Around the table*: Spiritualism Series, Vol. 4, File 2, H-3037, Séance Records, February 24–26, 1932; King Diary, February 24, 25, 1932.

252 *"She seemed to bring"*: Spiritualism Series, Vol. 7, File 2, H-3041, Séance Record, June 26, 1932.

252 *"What impresses me greatly"*: King Diary, June 27, June 30, July 1, 1932.

252 *"The unusual experiences"*: Spiritualism Series, Vol. 4, File 3, H-3037, Rogers to WLMK, July 1, 1932.

253 *He did not entirely*: Ibid., Eakin to WLMK, August 22, 1932; Vol. 3, File 15, H-3037, Eakin to WLMK, January 12, 1933.

253 *"He is beginning"*: King Diary, September 25, 1933.

253 *At each séance*: Ibid., December 14, April 17, 1933. See also, January 6–7, October 15, December 13–14, 1933; January 15, 1938; January 9, 1939.

253 *During the Easter weekend*: Spiritualism Series, Vol. 7, File 4, H-3042, Séance Records, April 14, 15, 1933; King Diary, April 17, 1933.

254 *Godfroy Patteson rarely*: King Diary, January 15, 1938.

254 *"She is another good friend"*: Ibid., December 10, 1942; August 6, 1948.

254 *It was "an amazing evening"*: King Diary, November 13, 1933; Stacey, *A Very Double Life,* 171.

254 *Ever since the young*: Harry Houdini, *A Magician Among the Spirits* (New York: Arno Press, 1972), 5–6 (original printing in 1924); Carl Murchison, *The Case For and Against Psychical Belief* (Manchester, NH: Ayer Company Publishers, 1975), 322; Lewis Spence, *Encyclopedia of Occultism* (London: G. Routledge and Company, 1920), 767; Victor Hugo and Table Rapping," NYT, August 3, 1906; Oppenheim, *The Other World,* 28.

254 *As soon as Joan*: King Diary, November 18, 1933.

255 *A week later*: Ibid., November 27, 1933.

255 *But based on*: Stacey, *A Very Double Life,* 172.

255 *After church on Sunday*: King Diary, March 25, 1934; Leonard Zusne and Warren H. Jones, *Anomalistic Psychology: A Study of Magical Thinking* (Mahwah, NJ: Lawrence Erlbaum, 1989), 215–16.

255 *King had studied Myers's*: King Diary, March 18, 1926; Bedore, "The Reading of Mackenzie King," 157–58.

255 *Similarly, in late December*: Stacey, *A Very Double Life,* 173.

255 *On New Year's Eve*: King Diary, December 31, 1933.

256 *After a long and exciting:* Ibid., May 12, 1937.

256 *At a session on:* Ibid., January 27, March 9, 10, 1934; Stacey, *A Very Double Life,* 183; Neatby III, 74–75.

256 *Another more serious:* King Diary, June 3, 1934; Stacey, ibid., 178.

257 *King was stunned:* King Diary, June 12, 24, 1934; Neatby III, 75.

257 *On Dominion Day:* King Diary, July 5, 1934.

257 *"As dear mother said":* Ibid., July 6, 1934.

258 *Had King actually:* Here is a diary entry about "mystical numbers" from July 25, 1936: "A further amazing thing today, was the presence in my papers along with the letter form England of statistics re: my tenure of office as Prime Minister to July 20, 1936—July (seventh month) 20—mystical number The table reveals in first column years—7—months 20, days 70: a total of 9 years on June 10, 1936 (another 10)."

258 *His interests and activities:* Spiritualism Series, Vol. 1, File 1, H-3034, Lady Aberdeen to WLMK, October 13, 23, 1934; File 9, Marie Carrington to WLMK, November 11, 1933; File 17, Duchess of Hamilton to WLMK, October 4, 1936; WLMK to the Duchess, May 17, 1937; Vol. 2, File 14, H-3036, WLMK to Nandor Fodor, March 22, 1938. For a comprehensive study of King's spiritualist collection and reading habits see, Bedore, "The Reading of Mackenzie King," 216–80.

258 *"I think she":* King Diary, January 16, 1938.

258 *Alone in his thoughts:* See, Neatby III, 76–79.

259 *"Last night's dream":* King Diary, January 3, 1935. See also, Stacey, *A Very Double Life,* 185–86.

259 *"Please help me":* Cited in Stacey, *A Very Double Life,* 185.

CHAPTER 9: A United Canada Cry

260 *He was awakened:* Keenleyside, *Memoirs,* 216, 442; Blair Fraser, "The Prime Minister," *Maclean's,* June 1, 1946, 76–7.

261 *Gordon Robertson, who:* Robertson, *Memoirs of a Very Civil Servant,* 50.

261 *"His approach to":* Peter C. Newman, The Distemper Of Our Times (Toronto: McClelland & Stewart, 1968), 231.

261 *Pickersgill "promises to":* King Diary, August 2, 1938.

261 *As Pickersgill remembered:* J.W. Pickersgill, "Mackenzie King's Speeches," *Queen's Quarterly* 57 (Autumn 1950), 304–7.

261 *Gordon Robertson has described:* Robertson, *Memoirs of a Very Civil Servant,* 48.

262 *Handy called Pickersgill:* King Diary, June 15, 1946; J.W. Pickersgill, *Seeing Canada Whole: A Memoir* (Markham, Ontario: Fitzhenry and Whiteside, 1994), 288.

262 *The connection of his first:* King Diary, July 9, 1933.

262 *After a lengthy courting period:* Personal Miscellaneous, Vol. 6, WLMK to Heeney, July 13, 1939; Heeney to WLMK, August 24, September 4, 1938; Arnold Heeney, *The Things That are Caesar's: Memoirs of a Canadian Public Servant* (Toronto: University of Toronto Press, 1972), 38–42.

262 *Heeney envisioned his role:* Heeney, ibid., 37–49; King Diary, August 24, 1938; Arnold Heeney, "Cabinet Government in Canada: Some Recent Developments in the Machinery of the Central Executive," *Canadian Journal of Economics and Political Science* 12 (August 1946), 285; J.L. Granatstein, *The Ottawa Men: The Civil Service Mandarins 1935–1957* (Toronto: Oxford University Press, 1982), 196–200.

262 *He told Violet Markham:* Personal Papers Series, Vol. 25, WLMK to Markham, May 1, 1939.

262 *"He is an exceedingly lucky"*: King Diary, March 22, 1940.

263 *"I have seldom been"*: Personal Papers Series, Vol. 25, WLMK to Markham, April 23, 1937.

263 *Many times during*: Keenleyside, *Memoirs,* 447.

263 *"Of course," he wrote*: Personal Papers Series, Vol. 25, WLMK to Markham, March 17, 1937.

263 *"It is all part of"*: King Diary, April 24, 1937. In late August 1937, King discovered that his butler, McLeod, had been drinking heavily, together with a member of his RCMP detail, and had been nearly arrested for public drunkenness. King had been away in England at the time and consequently his dog, Pat, had been left alone during the spree. "It was a shocking story," wrote King when Joan Patteson told him all the sordid details. King did not fire McLeod but was determined to save him and his wife, who also admitted to drinking, from "ruin." He delivered a stern lecture about sobriety, honesty and the rise of the Mackenzie family out of poverty, followed by a Bible reading. See, King Diary, August 30, September 1, 1937.

263 *A French Canadian*: A Real Companion and Friend: "A King's Who's Who Biographies: Edouard Handy," www.collectionscanada.gc.ca/king/023011-1050.53-e.html.

263 *In 1945, while King*: King Diary, April 28, October 9, 1945; Pickersgill, *Seeing Canada Whole,* 151.

263 *He found time*: Granatstein, *A Man of Influence,* 201–2.

264 *"The details of menu"*: Heeney, *The Things That are Caesar's,* 91.

264 *Howard Henry, his private*: King Diary, October 12, 1935.

264 *On another occasion*: Ibid., October 8, 1937; September 15, 1930.

264 *King could pass*: Keenleyside, *Memoirs,* 443.

264 *"I had learned that with"*: Robertson, *Memoirs of a Very Civil Servant,* 62. See also, Neatby III, 263; Heeney, *The Things That are Caesar's,* 91–92; Ferns, *Reading from Left to Right,* 139–40; Granatstein, *A Man of Influence,* 202.

264 *According to Jack Pickersgill*: Pickersgill, *Seeing Canada Whole,* 271.

265 *Pickersgill's wife, Margaret*: Ibid., 272.

265 *"How little men"*: King Diary, December 7, 1941; Ferns, *Reading from Left to Right,* 144.

265 *On a Saturday morning*: King Diary, October 9, 1937; October 1, 1944.

265 *"But King raised"*: Robertson, *Memoirs of a Very Civil Servant,* 49.

266 *When the* Ottawa Journal: O'Leary, *Recollections,* 74

266 *He declared war*: Levine, *Scrum Wars,* 162, 166–67; O'Leary, ibid., 75.

266 *Hutchison later recalled*: Bruce Hutchison, *The Far Side of the Street* (Toronto: Macmillan, 1976), 69; author's interview with Bruce Hutchison, June 1990.

266 *In the case of Grant Dexter*: Levine, *Scrum Wars,* 132–33.

266 *"One would think"*: King Diary, August 28, 1935.

267 *One day, Tom Wayling*: Ibid., September 21, 1926.

267 *In 1937, King decided*: "Mr. King's Press Conferences," *London Free Press,* October 21, 1937; Bruce Hutchison, "Press Conference in Room Sixteen," *Victoria Times,* March 19, 1937.

267 *"Instead of spilling"*: "PM Parries Questions from Reporters," *St. John's Telegraph-Journal,* December 16, 1946.

267 *The press gallery periodically*: Allan Levine, "In Defence of Political Control Freaks," *Globe and Mail,* October 15, 2007, A13.

267 *"Nothing could be more"*: King Diary, April 4, 1939; House of Commons Debates, March 5, 1939, 2605.

268 *Control was indeed:* King Diary, May 28, 1936.

268 *"At 8 I went":* Ibid., May 4, 5, 1929.

268 *As leader of the Opposition:* Ibid., March 2, 1935; May 28, 1936.

269 *By 1947, he had:* King did show up at the 1948 gallery dinner to share plans for his retirement. According to Blair Fraser of *Maclean's* magazine, he was funny and witty that night: "For the first 15 or 20 minutes he gave a burlesque of himself—first an ambiguous sentence then a demonstration of all the fantastic meanings editorial writers would read into it. We all laughed until our sides ached." See, Fraser, "Mackenzie King," 40.

269 *Since his first days:* Neatby III, 127–32.

269 *King was well known:* Paul Martin, *A Very Public Life,* 382.

270 *During his terms:* Heeney, *The Things That are Caesar's,* 58.

270 *"The opinion seems":* Fraser, "The Prime Minister," 80.

270 *He did not tolerate:* Paul Martin, Sr. "King: The View from the Backbench and Cabinet Table," 33–34. See also, Martin, *A Very Public Life,* 172.

270 *He had, according to:* Ferns, *Reading from Left to Right,* 164.

270 *King detested that:* King Diary, October 17 1935; April 20, 1937.

270 *Power thought that:* Ward, *The Memoirs of Chubby Power,* 73.

270 *King was not crazy:* Robertson, *Memoirs of a Very Civil Servant,* 70–71; King Diary, August 4, 1938

270 *Not valuing Thomas Crerar:* King Diary, October 18, 1935; Rea, *T.A. Crerar,* 171–72.

271 *The abdication of King Edward VIII:* See, King Diary, October 23, 1936; Neatby III, 184–85. The message King sent on December 8, 1936, which must have made Edward scratch his head, was as follows: "There is no doubt in our minds that a recognition by Your Majesty of what as King is owing by you to the Throne and to Your Majesty's subjects in all parts of the British Commonwealth should, regardless of whatever the personal sacrifice may be, be permitted to outweigh all other considerations." See, House of Commons Debates, January 18, 1937, 39.

271 *As the crisis unfolded:* King Diary, December 4, 1936; January, 20, 1937; Gordon Beadle, "Canada and the Abdication of Edward VIII," *Journal of Canadian Studies* 4:3 (August 1969), 37–38, 41–42.

272 *King was fairly rigid:* Ibid., October 31, 1935; February 23, 1937.

272 *The minister of finance:* Thompson, *Canada 1922–1939,* 300–301.

272 *"The world situation":* King Diary, April 1, 1938.

272 *FDR might have:* Conrad Black, *Franklin Delano Roosevelt: Champion of Freedom* (New York: Public Affairs, 2003), 789.

272 *King was FDR's:* King Diary, August 17, 1941; Elizabeth R.B. Elliott-Meisel, "A Grand and Glorious Thing: The Team of Mackenzie and Roosevelt," in Thomas C. Howard and William D. Pederson, eds., *Franklin D. Roosevelt and the Formation of the Modern World* (Armonk, NY: M.E. Sharpe, 2003), 138.

273 *Journalist Bruce Hutchison:* Hutchison, *Incredible Canadian,* 218.

273 *"Roosevelt had so":* Lawrence Martin, *The Presidents and the Prime Ministers: Washington and Ottawa Face to Face* (Toronto: Doubleday Canada, 1982), 123.

273 *During the late thirties:* J.L. Granatstein, *Canada's War: The Politics of the Mackenzie King Government 1939–1945* (Toronto: Oxford University Press, 1975), 87; Personal Papers Series, Vol. 25, WLMK to Markham, January 5, 1939. See also, King Diary, December 17, 1935.

273 *The strength of the King-FDR:* Notes of a Séance, November 13, 1938, cited in Stacey, *A Very Double Life,* 189.

273 *King arrived in Washington:* King Diary, November 8, 1935.

273 *At the White House:* King was given the room in the White House in which Abraham Lincoln had signed the Emancipation Proclamation in 1863. The symbolism of slavery was not lost on him. "I knelt and prayed by the side of the bed," he wrote unable to sleep, "that God would bless our two countries, that he would use me as an instrument to free numbers of men and women of some of their bonds—of poverty, of unemployment—that he would make the visit a blessing not only to this continent but to the world . . . I was very tired, but so happy to be there in that room, and to have dear mother's picture on the dresser before me." Ibid., November 8, 1935.

274 *Yet as one American:* Richard N. Kottman, "The Canadian-American Trade Agreement of 1935," *The Journal of American History* 52: 2 (September 1965), 275.

274 *In King's ideal world:* Neatby III, 150; King Diary, January 8, 1937; Granatstein, *The Ottawa Men,* 59–60.

274 *Steered in this matter:* One of the chief voices against calling the royal commission was Ernest Lapointe, who feared that Quebec premier Maurice Duplessis would regard it as an unwarranted intrusion into provincial affairs. King eventually changed Lapointe's mind, one more example that King did not always heed Lapointe's view on every matter. See, King Diary, January 8, 1937; Betcherman, *Ernest Lapointe,* 235–36.

274 *Many of these changes:* See, Bliss, *Right Honourable Men,* 163; Bennett, 418–25.

274 *"Some early and definite action":* Primary Series Correspondence, Vol. 82, C-2249, 69059–60, Skelton to WLMK, April 20, 1937.

275 *As King later told:* WLMK to Markham, May 1, 1939.

275 *He went so:* King Diary, December 11, 1935.

275 *In the spring of 1937:* Herbert F. Quinn, *The Union Nationale: A Study in Quebec Nationalism* (Toronto: University of Toronto Press, 1963), 126. As John Thompson points out, "publications seized under the [Padlock] law included the *Canadian Forum,* the *Labour World,* and copies of the Liberal organ *Le Canada.*" The movie *The Life of Emile Zola* was banned. Quite absurdly, the police also seized a *Time* magazine from one home because it had a photo of Leon Trotsky on the cover. See, Thompson, *Canada 1922–1939,* 285.

276 *King's liberal sensibilities:* WFP, March 27, 1937; King Diary, August 6, 1937; July 6 1938.

276 *In this matter:* Betcherman, *Ernest Lapointe,* 239–40.

276 *"In the circumstances":* King Diary, July 6, 1938.

276 *Hepburn's biographer:* John T. Saywell, *'Just Call Me Mitch': The Life of Mitchell F. Hepburn* (Toronto: Univesity of Toronto Press, 1991), 532–33.

276 *"People believe he":* King Diary, October 7, 1937.

277 *Hepburn had a well-deserved:* Ibid., January 5, 1937; November 24, 1934; June 4, 1937; Saywell, *'Just Call Me Mitch,'* 287, 192.

277 *King said no:* Saywell, ibid., 239.

277 *He was furious:* King Diary, April 13, 1937; Thompson, *Canada 1922–1939,* 289.

277 *When the various:* King Diary, December 17, 1937; Neatby III, 200–204.

278 *"Just as well":* Ibid., December 10, 1937.

278 *A year later:* Saywell, *'Just Call Me Mitch,'* 396–402; King Diary, December 8, 1938.

278 *As his principal secretary:* Heeney, *The Things That are Caesar's,* 56.

278 *"He had decided":* Norman Hillmer, "The Pursuit of Peace: Mackenzie King and the 1937 Imperial Conference," in English, et al., *Mackenzie King: Widening the Debate,* 166.

279 *"King's backing and filling":* J.L. Granatstein and Robert Bothwell, "'A Self-Evident National Duty': Canadian Foreign Policy, 1935–1939," *Journal of Imperial and Commonwealth History* 3:2 (January 1975), 222.

279 *"I pray that God":* King Diary, March 7, 13, 1936.

279 *Pressed to explain:* House of Commons Debates, March 23, 1936, 1332–33; June 18, 1936, 3866; Neatby III, 172–74.

279 *King's ego would:* Richard W. Stevenson and David E. Sanger, "Threats and Responses, NYT, January 16, 2003.

279 *There was an "absurdity":* King Diary, September 21, 1936.

280 *Cheered on by:* R.A. MacKay and E. B. Rogers, *Canada Looks Abroad* (New York: Oxford University Press, 1938), 367; Stacey, *Canada and the Age of Conflict,* 186.

280 *Only in Winnipeg:* King Diary, September 29, 1936; James Eayrs, *In Defence of Canada: Appeasement and Rearmament* (Toronto: University of Toronto Press, 1965), 39–40; WFP, October 1, 1936.

280 *"What has been told":* King Diary, October 23, 1936. Grant Dexter of the *Winnipeg Free Press,* who spent a lot of time in London during the thirties, saw the situation much differently. As he told his editor George Ferguson, whenever he met Englishmen who then learned that he was a Canadian, they would "pontificate, high hat, talk you down to rat-hole dimensions and, finally, urinate on you and go their way." Cited in Patrick H. Brennan, *Reporting the Nation's Business: Press-Government Relations during the Liberal Years, 1935–1957* (Toronto: University of Toronto Press, 1994), 21.

280 *This was the subtext:* Hillmer, "The Pursuit of Peace," 150–52.

280 *The British pushed:* Ibid., 158; King Diary, June 7 1937.

281 *King found von Ribbentrop:* King Diary, May 26, 1937; Michael Bloch, *Ribbentrop* (London: Bantam, 1992), 4–7. Shortly before King left London, Violet Markham gave him some sage advice about meeting Hitler: "Apparently he as a very attractive personality and makes a considerable impression on nearly everyone who sees him. All the same don't let him hypnotize you! . . . He is head of a detestable system of force and persecution and real horrors go on in Germany today for which he is responsible." See, Personal Papers, Vol. 25, Markham to WLMK, June 15, 1937; Eayrs, *In Defence of Canada: Appeasement and Rearmament,* 46.

281 *The auspicious encounter:* See, King Diary, June 29, 1937, and King's memo on his meeting with Hitler, Eayrs, ibid., 226–31.

282 *The two men spoke:* The next day, King met with Hitler's foreign minister, Baron Konstantin von Neurath, who was more outspoken on the Jewish issue. King described part of their talk as follows: "He said to me that I would have loathed living in Berlin [before 1933] with the Jews, and the way in which they had increased their numbers in the city, and were taking possession of its more important part. He said there was no pleasure in going to a theatre which was filled with them. Many of them were very coarse and vulgar and assertive. They were getting control of all the business, the finance, and had really taken advantage of the necessity of the people. It was necessary to get them out to have the German people really control their own city and affairs." To this, King had no comment or observation. See, King Diary, June 30, 1937.

282 *That was too sweeping*: Granatstein and Bothwell, "'A Self-Evident National Duty,'" 219.

282 *He praised Hitler*: King Diary, June 29, 1937.

283 *"No one who does not"*: Ibid., March 27, 1938. See also, May 10, July 22, 1938; September 2, 1939; C.P. Stacey, "The Divine Mission: Mackenzie King and Hitler" *Canadian Historical Review* 61:4 (1980), 507–9.

284 *On one hand, he put*: King Diary, September 12, 13, 1938; Granatstein and Bothwell, "'A Self-Evident National Duty,'" 221–22.

284 *"We both agreed"*: King Diary, September 13, 1938.

284 *"The world might"*: Primary Series Correspondence, Vol. 262, C-3740, 223110–20, WLMK to Hume Wrong, September 23, 1938; Lapointe to WLMK, September 24, 1938; Betcherman, *Ernest Lapointe*, 250.

284 *With several other members*: King Diary, September 24, 1938; Granatstein and Bothwell, "'A Self-Evident National Duty,'" 222–23.

284 *"Great rejoicing Canada"*: Primary Series Correspondence, Vol. 252, C-3735, 215287, WLMK to Lapointe, September 30, 1938; Betcherman, *Ernest Lapointe*, 251; WFP, September 30, 1938.

285 *"If England is at war"*: House of Commons Debates, January 16, 1939, 52; King Diary, January 16, 1939.

285 *Lapointe was shocked*: Betcherman, *Ernest Lapointe*, 259.

285 *Despite loud objections*: King Diary, January 27, 1939; Betcherman, ibid., 260.

285 *In a two-hour speech*: Granatstein and Bothwell, "'A Self-Evident National Duty,'" 226–27; King Diary, March 20, 29, 1939; House of Commons Debates, March 30, 1939, 2409–28.

285 *On March 31*: King Diary, March 31, 1939; Granatstein and Bothwell, ibid., 227–29.

285 *Ernest Lapointe had the*: House of Commons Debates, March 31, 1939, 2468; Betcherman, *Ernest Lapointe*, 265–66.

285 *King hailed Lapointe's*: King Diary, March 31, 1939.

286 *On November 7, 1938*: Martin Gilbert, *The Holocaust: The Jewish Tragedy* (London: Fontana, 1987), 67–74.

286 *"The sorrows which"*: King Diary, November 12, 1938.

286 *Mackenzie King grew up*: The U.S. philosopher Horace Kallen coined the term "cultural diversity" in 1924, in which there was "unity in diversity." But apart from the Chicago social worker Jane Addams, few Americans or Canadians embraced the idea until well after the Second World War. See, Levine, *Devil in Babylon*, 37, 48.

286 *Blacks*: See also, King Diary, October 26, 1938; April 26, May 12, 1940; July 21, 1941; April 15, August 16,1945; Personal Papers, Vol. 29, WLMK to Joan Patteson, October 24, 1938.

286 *"There are good"*: King Diary, August 9, 1936.

286 *At the same time*: Ibid., August 2, 1934; Spiritualism Series, Vol. 1, File 19, H-3034, Lambert to WLMK, July 30, 1934; Bedore, "The Reading of Mackenzie King," 228–29. See also, King Diary, February 3, 1900.

287 *Across the country*: In the early thirties, both Clifford Clark, the deputy minister of finance, and Oscar Skelton had wanted to recruit Louis Rasminsky, a brilliant economist who in 1961 was to become the third governor of the Bank of Canada and the first Jew to hold such a high position. But at the time, Rasminsky's background was an impediment. As Skelton pointed out in a letter to a friend in 1934, Canadian anti-Semitism was an obstacle for "such men." Rasminsky was not interested in the job

Clark had him in mind for, yet "even if he had been willing," added Skelton, "there would probably have been difficulties because of the prejudice in question." See, Granatstein, *The Ottawa Men, 138.*

287 *There is no official record:* See, Allan Levine, *Coming of Age: A History of the Jewish People of Manitoba* (Winnipeg: Jewish Heritage Society of Western Canada/Heartland Associates, 2009), 251–52; Gerald Tulchinsky, *Branching Out: The Transformation of the Canadian Jewish Community* (Toronto: Stoddart, 1998), 186; Irving Abella, "Anti-Semitism," *Canadian Encyclopedia,* http://thecanadianencyclopedia.com.

287 *"Ours was a society":* James Gray, *The Winter Years,* (Toronto: Macmillan of Canada, 1966), 13.

287 *Or, put slightly differently:* Cited in Paul Axelrod, *Making a Middle Class: Student Life in English Canada During the Thirties* (Montreal and Kingston: McGill-Queen's University Press, 1990), 33.

288 *"The French Canadians":* "Embryo Fascism in Quebec," *Foreign Affairs* 16:3 (April 1938), 456.

288 *Adrian Arcand's fascist:* Tulchinsky, *Branching Out,* 185

288 *French-Canadian Catholic:* Ibid., 173–75, 185.

288 *According to Marcel Hamel:* Cited in Michael Oliver, *The Passionate Debate: The Social and Political Ideas of Quebec Nationalism 1920–1945* (Montreal: Véhicule Press, 1991), 185.

288 *In the nationalist organ:* Esther Delisle, *The Traitor and the Jew: Anti-Semitism and Extremist Right-wing Nationalism in Quebec from 1929 to 1939* (Montreal: R. Davies Publishing, 1993), 43. See also, Max Beer, "The Montreal Jewish Community and the Holocaust," *Current Psychology* (2007) 26, 193–94.

288 *Such notions were shared:* Tulchinsky, *Branching Out,* 186–88; Max and Monique Nemni, *Young Trudeau: Son of Quebec, Father of Canada, 1919–1944* (Toronto: McClelland & Stewart/Douglas Gibson Books, 2006), 58–59, 187–89, 194–95.

288 *King's master of Quebec:* Betcherman, *Ernest Lapointe,* 241.

289 *Canada did not:* Claude Bélanger, "Why did Canada Refuse to Admit Jewish Refugees in the 1930s?" *L'Encyclopédie de l'histoire du Québec / The Quebec History Encyclopedia,* http://faculty.marianopolis.edu/c.belanger/quebecHistory/readings/CanadaandJewishRefugeesinthe1930s.html; Tulchinsky, *Branching Out,* 59.

289 *"My own feeling":* King Diary, March 29, 1938.

289 *All through this period:* Ibid., May 17, 1938; Primary Series Correspondence, Vol. 249, c-3733, 213016–20, Heaps, Factor and Jacobs to WLMK, March 11, 15, 1938; Vol. 251, c-3734, 214191, Heaps to WLMK, May 16, 1938; Vol. 252, c-3734, 21432–36; Heaps, Factor and Jacobs to WLMK, April 24, May 3, 1938; Irving Abella and Harold Troper, *None Is Too Many: Canada and the Jews of Europe, 1933–1948* (Toronto: Lester & Orpen Dennys, 1982), 24–37. On appointing Sam Factor to the cabinet, see King Diary, July 2, 1940.

289 *A succession of cabinet:* King Diary, March 29, 1938.

289 *To this end:* Abella and Troper, *None Is Too Many,* 28–32.

290 *King was moved:* King Diary, November 13, 1938.

290 *Despite overwhelming newspaper:* Abella and Troper, *None Is Too Many,* 41, 59.

290 *In another meeting:* King Diary, November 23, 1938.

290 *And in a prophetic:* Ibid., November 24, December 1, 13, 1938.

291 *"We don't want to":* Abella and Troper, *None Is Too Many,* 46–50.

291 *This was in line:* Following the Munich Agreement of September 1938, Canada, at

Britain's request, accepted several thousand anti-Nazi non-Jewish refugees. As the high commissioner in London, Vincent Massey, who brokered the arrangement, told King, "If we could take a substantial number of them it would put us in a much stronger position in relation to later appeals from and on behalf of non-Aryans." The Sudeten Germans, Massey added, included, "many persons who would be much more desirable as Canadian settlers and much more likely to succeed in our country than certain other types of refugees." See, ibid., 48–89.

291 *At least once:* King Diary, April 10, 1939.

291 *King listened politely:* King Diary, June 8, 1939

291 *Other Canadians:* Betcherman, *Ernest Lapointe,* 41; Abella and Troper, *None is Too Many,* 64.

291 *In what was to be one:* Abella and Troper, ibid., 64; See also Primary Series Correspondence, Vol. 280, c-3750, 237087–89, Skelton to WLMK, June 8, 1939; 237095–96, Skelton to WLMK, June 9, 1939.

291 *Of the 907:* "The Voyage of the St. Louis," United States Holocaust Memorial Museum, www.ushmm.org/wlc/article.php?ModuleId=10005267.

292 *The key in those:* J.L. Granatstein, "King and his Cabinet: The War Years," in English, et al., *Mackenzie King: Widening the Debate,* 184–87.

292 *The final result of this:* Abella and Troper, *None is Too Many,* x.

CHAPTER 10: A Man on a Tightrope

293 *Following months:* King Diary, February 21, 28, 1939; Primary Series Correspondence, Vol. 264, c-3741, 224759–65, WLMK to Neville Chamberlain, March 14, 1939. See also, David Reynolds, "FDR's Foreign Policy and British Royal Visit to the U.S.A., 1939," *The Historian* 45:4 (August 1983), 461–66.

293 *John Dafoe sarcastically:* Neatby III, 313; Heeney, *The Things That are Caesar's,* 50–51.

294 *"This I know":* King Diary, April 25, 1939. See also, Fraser, "The Prime Minister," 76–77.

294 *On May 17:* Berton, *The Great Depression,* 491.

294 *Sounding much like:* King Diary, May 17, 1939.

294 *Lord Tweedsmuir:* Reynolds, *Mackenzie King Friends & Lovers,* 129–30; Smith, *John Buchan,* 395–98, 421, 437–41, 455. At the same time, a week after Buchan's death in February 1940, King wrote to Lord Amery about the late governor general, "Nothing could have been happier than our official relations. We never had a real difference of view on anything."

294 *At the inaugural luncheon:* King Diary, May 17, 1939.

294 *There was much excitement:* Ibid., May 20, 1939.

295 *To King's delight:* David Reynolds, "FDR's Foreign Policy and British Royal Visit to the U.S.A., 1939," 464–66; Transcript of King George VI's Handwritten Notes for a Memorandum on his Conversations with FDR on June 10 and 11, 1939, Franklin D. Roosevelt Library and Museum, http://docs.fdrlibrary.marist.edu:8000/memorand.html; King Diary, June 10, 11, 1939

295 *He took solace:* Ibid., May 22, 1939.

296 *"He had nothing":* Ibid., May 17, 22, 1939.

296 *The final send-off:* Ibid., June 15, 1939.

296 *According to B.C. writer:* Louise Reynolds, *Mackenzie King Friends & Lovers,* 130.

297 *In January 1939:* DCER, Vol. 4, (Ottawa: Department of External Affairs, 1967), 1221–22; Stacey, "The Divine Mission," 507–9.

297 *He grew more apprehensive:* Spiritualism Series, Vol. 7, File 21, H-3042, "Interview with Mrs. Quest Brown," February 7, 1939.

297 *He wrote in:* King Diary, July 21, 1939.

297 *"If Hitler wishes":* Ibid., August 22, 1939.

298 *Three days later:* DCER, Vol.6, (Ottawa: Department of External Affairs, 1972), 1246–47, 1259, 1266–67; Eayrs, *In Defence of Canada: Appeasement and Rearmament,* 79; Stacey, "The Divine Mission," 511.

298 *On September 1:* King Diary, September 1, 1939; Heeney, *The Things That are Caesar's,* 57–58; Betcherman, *Ernest Lapointe,* 275.

298 *As he was shaving:* King Diary, September 2, 1939; Stacey, *A Very Double Life,* 190–92.

298 *On his feet in the Commons:* House of Commons Debates, September 8, 1939, 21–22; King Diary, September 7, 1939.

299 *"There is no other portion":* House of Commons Debates, September 8, 1939, 22; September 9, 1939, 61–69; King Diary, September 9, 1939.

299 *The moment of:* King Diary, September 9, 1939.

299 *He "roused little Pat":* Conversations with Pat were not out of the ordinary. Some months earlier he had written to Violet Markham, noting, "I hope [Violet's dog] Jane still remembers me. I think of her often. Only last night I was telling my old dog Pat about the chats we had together, as he was lying at my feet while I was preparing material for the Royal visit. Poor Pat is now fifteen years of age and the years are telling upon him. No man ever had a truer or more faithful friend." See, Personal Papers Series, Vol. 25, WLMK to Markham, May 1, 1939.

300 *Other than* Le Devoir: Primary Series Correspondence, Vol. 270, C-3744, 229115–33, Heeney to Lapointe, October 4, 1939; Memoranda and Notes Series, Vol. 155, File 1345, "Memo on Press Information," November 15, 1939.

300 *Paul Bouchard, the separatist editor:* Betcherman, *Ernest Lapointe,* 275; Conrad Black, *Render Unto Caesar: The Life and Legacy of Maurice Duplessis* (Toronto: Key Porter Books, 1998), 167.

300 *In a drawn-out:* King Diary, September 12, 1939.

300 *It was "not for democracy":* J.L. Granatstein and Desmond Morton, A Nation Forged in Fire: Canadians and the Second World War, 1939–1945 (Toronto: Lester & Orpen Dennys, 1989), 11.

301 *The articles:* Levine, *Scrum Wars,* 178–79; David Mackenzie, "The Bren Gun Scandal and the Maclean's Publishing Company's Investigation of Canadian Defence Contracts, 1938–1940," *Journal of Canadian Studies* 26:3 (Fall 1991), 141–52; Brennan, *Reporting the Nation's Business,* 35–40.

301 *King wanted Ralston:* King Diary, September 5, 1939.

301 *King and Ralston:* Granatstein, "King and his Cabinet," 187.

301 *King delayed dealing:* King Diary, September 18, 19, 1939.

302 *Yet in the first:* Drinking also did in Norman McLarty. In November 1941, King received a report from Leighton McCarthy, who was then head of the Canadian legation in Washington, D.C., that "McLarty had more or less humiliated and disgraced them all at the Labour Conference in New York, in drinking. It as been obvious to me that he has been doing this of late. I am afraid I shall have to get rid of him altogether." Less than a month later, McLarty became the innocuous secretary of state. See, ibid., November 4, 1941; Granatstein, "King and his Cabinet," 184.

302 *Still, he could not ignore:* Granatstein, *Canada's War,* 82.

302 *King permitted Mackenzie:* King Diary, September 19, 1939; Granatstein, "King and his Cabinet," 180.

302 *Duplessis could not:* Black, *Render Unto Caesar,* 168–69.

302 *"It is a diabolical":* King Diary, September 25, 1939. While Chubby Power always respected Ernest Lapointe, he was envious of his high standing in the cabinet and in King's eyes. According to Power's version of these events, he had had to convince both Lapointe and Cardin to accept this bold course of action, including the threat to resign, should Duplessis be victorious. See, Ward, *The Memoirs of Chubby Power,* 346–47.

303 *The subsequent announcement:* Black, *Render Unto Caesar,* 172.

303 *"I venture to say":* King Diary, October 25, 1939.

303 *The next morning:* Ibid., October 26, 1939.

304 *The three Quebec:* Ibid., December 31, 1939.

304 *"Of course, Dexter":* University of Manitoba Archives and Special Collections, John W. Dafoe Papers, Box 8, Folder 2, Grant Dexter Memo, October 25, 1940.

305 *After meeting with King:* Cited in Saywell, *'Just Call Me Mitch,'* 433.

305 *"It would make":* King Diary, October 10, 1939.

305 *The soldiers training:* Granatstein, *Canada's War,* 76.

305 *Goaded by Drew:* Saywell, *'Just Call Me Mitch,'* 437.

305 *"It shows how":* King Diary, January 18, 1940.

306 *Now he intended:* Beck, *Pendulum of Power,* 224.

306 *"Hepburn has made":* King Diary, January 18, 1940.

306 *During the next week:* Whitaker, *The Government Party,* 118–19. The cabinet war committee was established by an order-in-council on December 5, 1939. It was the executive group that led the country's war effort. Besides the prime minister, it also included the ministers of justice, finance, national defence, mines and resources and the leader of the Senate. Other ministers were later added. See, Granatstein, *Canada's War,* 10.

306 *Lambert reassured:* Whitaker, *The Government Party,* 96.

307 *On January 23:* King Diary, January 23, 25, 1940. Ian Mackenzie told Chubby Power that a few days before dissolution, King had allegedly "consulted one of his spiritualistic soothsayers about the line he proposed to follow, and the chances of the party in the election." As far as can be determined, there is no record of such a séance or session with the little table. In mid-January 1940, he was, however, reading the book *My Psychic Life* (1931), by Marianne Bayley-Worthington. His friend the duchess of Hamilton had written the introduction. The book had been sent to him by his acquaintance Miss Lind-af-Hageby, the president of the London Spiritualist Alliance. In thanking her, King opined, "Matter is destructible; the spirit indestructible. We may be called upon to witness a condition as near to Armageddon as anything this world had known before inheriting the higher plane of thought and understanding which surely will be the outcome of the travail of our times." See, Ward, *The Memoirs of Chubby Power,* 353; Spiritualism Series, Vol. 2, File 13, H-3036, WLMK to Lind-af-Hageby, January 13, 1940.

307 *The Speech from the Throne:* King Diary, January 25, 1940; House of Commons Debates, January 25, 1940, 10–20; Granatstein, *Canada's War,* 79–80; Levine, *Scrum Wars,* 178.

307 *And everyone in the*: King Diary, January 26, 1940. In February 1941, King similarly rejected the suggestion of CPR president Sir Edward Beatty that he invite former Conservative prime minister R.B. Bennett, who had retired to England, into the cabinet as finance minister. King listened politely to Beatty, but as he wrote, "I certainly knew [Bennett] well enough to know that he could not be in the Cabinet a day without having some disagreement and wishing to assert himself above everyone else." See, King Diary, February 19, 1941; Boyko, *Bennett*, 452–53.

308 *In 1940, voters*: Granatstein, *Canada's War*, 84; Saywell, *'Just Call Me Mitch,'* 441–42.

308 *Early that morning*: King Diary, March 26, 1940.

308 *He was replaced*: Dexter Memo, August 20, 1940, Gibson and Robertson, *Ottawa at War*, 73.

308 *Even one of his most*: John Stevenson, "The General Election in Canada," *Fortnightly* 147 (May 1940), 513; Beck, *Pendulum of Power*, 237. Stevenson's overall harsh impression of King, however, remained unchanged. "King [is] my beau ideal of the successful political charlatan," he wrote to his editor in London September 1941. "He has a masterly skill in the arts of political manipulation and evasion of issues ... rarely has there been such a blend in human form of incompetence, chicanery, and moral cowardice. He lives on other people's brains ... he is temperamentally unfit to be a war leader ... his primary concern is not winning the war but the fortunes of the Liberal Party." See, Stevenson to Geoffrey Dawson, September 8, 1941, cited in MacLaren, *Commissions High*, 354.

308 *Feeling vindicated*: King Diary, March 26, 29, 1940; Queen's University Archives, Grant Dexter Papers, Box 2, Folder 17, Dexter Memo, April 15, 1940.

309 *The 1940 campaign*: Whitaker, *The Government Party*, 131. Walter Herbert, the overworked secretary of the NLF, had also had it with King. According to Grant Dexter, Herbert had worked "day and night chiefly on matters affecting only the prime minister ... [He] emerged from the campaign with so strong an antipathy to King that he simply could not bear to see him ... King telephoned him personally and asked him for lunch. Walter replied that he did not feel like going. King was very nice, saying that whenever he felt up to it to phone Handy and ask Handy to tell King. Walter did not do this. In fact, Walter had about made up his mind that he would tell King that he had no desire to know him socially, or to have any direct contact with him. This would have been madness and we all argued with Walter—Norm, Joe Clark of Toronto and others—until we got him to work out a line of approach which would not offend King." Herbert soon quit his position at the NLF. See, Dexter Memo, April 15, 1940, Gibson and Robertson, *Ottawa at War*, 52.

309 *One of the few*: "My object is to preserve the maximum initiative energy," Churchill confessed in an unguarded moment to one of his private secretaries in August 1940. "Every night I try myself by court-martial to see if I have done anything effective during the day. I don't just mean pawing the ground—anyone can go through the motions—but something really effective." Cited in Dilks, *The Great Dominion*, 141.

309 *On June 4*: See, www.theyworkforyou.com/debates/?id=1940-06-04a.787.0.

309 *King had never*: King Diary, June 10, 1939; June 4, 1940.

310 *There was no "word of thanks"*: Cited in Martin Gilbert, *Churchill and America* (Toronto: McClelland and Stewart, 2005), 189; Granatstein, *Canada's War*, 123.

310 *Ralston, however, was*: Grant Dexter Papers, Box 2, Folder 17, Dexter Memo, May 31, June 7, 1940.

311 *"At no time during"*: Heeney, *The Things That are Caesar's*, 58.

311 *Negotiations between British and Commonwealth*: Granatstein, *Canada's War*, 43–59; MacLaren, *Commissions High*, 342.

311 *Gerald Campbell*: Cited in Joe Garner, *The Commonwealth Office, 1925–68* (London: Heinemann, 1978), 225.

311 *"As I see it"*: Cited in MacLaren, *Commissions High*, 348. See also, King Diary, October 18, November 3, 1940; Churchill to King, June 11 and July 25, 1941 in Martin Gilbert, ed., *The Churchill War Papers: The Ever-Widening War, 1941*, Vol. 3 (London: Heinemann, 2001), 795, 982; and English, *Shadow of Heaven*, 214–15, for King's attitude toward the Canadian High Commission in London.

312 *"Personally, I dislike this"*: Cited in C.P. Stacey, *Arms, Men and Government: The War Politics of Canada, 1939–1945* (Ottawa: Information Canada, 1974), 141–42.

312 *"Mackenzie King has been putting"*: Charles Ritchie, *The Siren Years: A Canadian Diplomat Abroad 1937–1945* (Toronto: Macmillan of Canada, 1974), 110–11.

312 *Later, King expressed*: Clyde Sanger, *Malcolm MacDonald: Bringing an End to Empire* (Montreal: McGill-Queen's University Press, 1995), 220.

313 *"Thereupon he began"*: King Diary, July 11, 1940.

313 *King did urge*: Granatstein, *Canada's War*, 119–24.

313 *On August 16, 1940*: Granatstein and Hillmer. *For Better or Worse*, 13–42; J.W. Pickersgill, *The Mackenzie King Record 1939–1944*, Vol. 1. (Toronto: University of Toronto Press, 1960), 130–39 (hereafter Pickersgill 1).

314 *"Had anyone said two"*: Cited in Terence A. Crowley, *Marriage of Minds: Isabel and Oscar Skelton Reinventing Canada* (Toronto: University of Toronto Press, 2003), 252–53.

314 *King himself later commented*: House of Commons Debates, November 12, 1940, 57.

314 *"I am convinced"*: King Diary, August 22, 1940.

314 *In London, Churchill*: English, *Shadow of Heaven*, 240–41; J.L. Granatstein, "The Man Who Wasn't There: Mackenzie King, Canada, and the Atlantic Charter," in Douglas Brinkley and David R. Facey-Crowther, eds., *The Atlantic Charter* (New York: St. Martin's Press, 1994), 117.

314 *After British high commissioner*: Primary Series Correspondence, Vol. 286, C-4568, 241597–98, Churchill to WLMK, September 12, 1940; Granatstein, *Canada's War*, 131.

314 *"This cable gave"*: King Diary, September 13, 1940; Granatstein, ibid., 131.

315 *The next spring*: J.L. Granatstein and R.D. Cuff, "The Hyde Park Declaration 1941: Origins and Significance," *Canadian Historical Review* 55:1 (March 1974), 72–79; Dexter Memo, April 21–22, 1941, Gibson and Robertson, *Ottawa at War*, 152–54.

315 *The act allowed*: R. Douglas Francis, Richard Jones and Donald B. Smith, *Journeys: A History of Canada* (Toronto: Nelson Education, 2010), 442; Granatstein and Cuff, ibid., 59–60; 79–80.

315 *Supported by finance*: King Diary, April 20, 1941; Dexter Memo, April 21–22, 1941; Granatstein and Cuff, ibid., 73–74.

315 *By the terms*: Francis, et al., *Journeys: A History of Canada*, 442; Granatstein and Cuff, ibid., 79–80.

315 *King was exceptionally*: House of Commons Debates, April 28, 1941, 2289; King Diary, April 21, 1941; Granatstein and Cuff, ibid., 59, 74.

316 *"To me, it is the"*: King Diary, August 6, 1941.

316 *While he had expected:* Ibid., August 7, 1941.

316 *In essence:* Granatstein, "The Man Who Wasn't There," 125.

316 *Gossip (probably exaggerated):* Dexter Memo, September 16, 1941, Gibson and Robertson, *Ottawa at War,* 193–94; King to Roosevelt, August 15, 1941, cited in Granatstein, ibid., 123–25.

316 *What rightly troubled:* DCER, Vol. 7, Part 1 (Ottawa: Department of External Affairs, 1974), 237; Granatstein, ibid., 123.

316 *To get there:* He described his inaugural flight in his diary as follows: "I thoroughly enjoyed the assent [*sic*] itself and from the moment we began to fly on the level, enjoyed the whole sensation of floating through and above the clouds, getting glimpses of the country below. I was impressed with how plainly everything was visible ... The words that kept coming to my mind were: terrestrial and celestial; seeing a new heaven and a new earth ... It gave one a feeling of greater reverence for God and a greater regard for man and his achievements." See, King Diary, August 19, 1941.

316 *"He admires Churchill":* Cited in Dilks, *The Great Dominion,* 152.

316 *Churchill, however, was on:* Pickersgill 1, 238–50. See also, Churchill's speech of September 4, 1941, and Churchill to Athlone, September 12, 1941, in Gilbert, *The Ever-Widening War, 1941,* 1155–56, 1572.

317 *King was indeed so:* King Diary, August 23, 1941.

317 *On the first day:* Ibid., August 23, 1941. See also, Churchill to Athlone, September 12, 1941.

317 *"I felt," he wrote:* King Diary, August 26, 1941.

318 *"I have no doubt":* Ibid., November 17, 1941; November 8, 1941.

318 *He immediately thought:* Ibid., June 10, 1940.

319 *Seven months later:* Ibid., January 28, 1941.

319 *Skelton had had heart:* Crowley, *Marriage of Minds,* 248.

319 *King's first instinct:* King Diary, January 28, 1941.

319 *He moved around:* Ibid., January 28, 1941; Keenleyside, *Memoirs,* 446.

320 *In such events:* King Diary, January 28, 1941. The death of Skelton stimulated King's spiritualist antennae. He saw the image of a camel in his tea cup, only to hear on the radio news that night that the Free French Camel Corps operating in Chad in north-central Africa had started an offensive against the Italians in Libya. "The significance of this to me," King wrote, "is that it is evidence that some mind has been seeking to communicate with me in advance, thereby giving an assurance of the presence of others who are with me helping and guiding. In other words, a testimony of the existence of the spiritual world as I believe it to be."

320 *Robertson, however, was:* See, Granatstein, *A Man of Influence,* 149–51; 202.

320 *"Little Pat came from":* King Diary, June 16, 1931.

321 *"Little Pat this morning":* Ibid., August 21, 1937.

321 *A veterinarian tended:* Ibid., February 8, 1938.

321 *Derry went first:* Ibid., August 26, 1940. The Pattesons were heartbroken when the dog died. "Little Derry's death has made a great blank in the life of [Joan] and [Godfroy]," wrote King. "[Joan] said the day following the death that something very vital had gone out of the house. Yesterday she spoke of how, for over 16 years, Derry had been continuously in the house with them both. They are two very sorrowful people. [Godfroy] is particularly lonely and I can see that it has saddened [Joan] very much. If they had lost a child, they could not feel more deeply than they do." See, King Diary, August 28 1940.

321 *He arrived back*: Ibid., July 14, 15, 1941; Stacey, *A Very Double Life*, 141–42.

322 *It is easy*: Stacey, ibid., 142–43.

322 *For this reason*: King Diary, December 31, 1941.

322 *Later, on July 15*: Ibid., July 15, 1941.

323 *This little dog*: Ibid., August 18,1941; author's interview with Joan McCallum, October 17, 2009.

323 *When he returned*: Ibid., September 9, 1941.

323 *At some point*: Ibid., October 12, 1941.

323 *"Before going to bed"*: King Diary, December 24, 1944.

324 *He was admitted*: Betcherman, *Ernest Lapointe*, 340–42.

324 *"With Lapointe [in hospital]"*: King Diary, November 8, 1941.

324 *"He is the most loyal"*: Ibid., November 11, 1941.

324 *Alarmed that "no loss"*: Ibid., November 14, 1941.

324 *King was able*: Ibid., November 15, 1941.

324 *"My chief joy in"*: Primary Series Correspondence, Vol. 307, c-4864, 259726–27, WLMK to Lapointe, November 16, 194; Betcherman, *Ernest Lapointe*, 343.

324 *Sitting by his bed*: King Diary, November 19, 1941.

324 *In an attempt*: Ibid., November 20, 1941.

324 *Ernest Lapointe died*: In the coming months, there were two more deaths of note. Senator Raoul Dandurand, eighty years old, died on March 11, 1942. The senator, a member of King's first cabinet in 1921, had been a loyal but not uncritical supporter. His death, as King pointed out in his diary, "leaves me as the only member of the original [1921] Cabinet." Ten days later, J.S. Woodsworth of the CCF, who had succumbed to a stroke, also passed away in Vancouver. King had rarely agreed with Woodsworth's politics, but he had respected him as a man of principles. "His life… was clearly one of fine Christian public service," he wrote, "and he has left real impression on the country." King Diary, March 11, 23, 1942.

324 *King arranged to*: Ibid., November 26, 29, 1941; Betcherman, *Ernest Lapointe*, 342–43.

325 *Arnold Heeney travelled with*: Heeney, *The Things That are Caesar's*, 59.

325 *The dignified Quebec lawyer*: King Diary, December 5, 1941.

325 *In the end*: Conflict of interest rules were definitely different in that era. St. Laurent asked King if becoming a cabinet minister meant giving up his position as a director of several corporations, including the Bank of Montreal and the Gatineau Power Company. Today, no cabinet minister could serve in such positions. King, however, told St. Laurent that he "never exacted conditions of the kind of Ministers, but left that to their own sense of duty; they naturally would be more independent if perfectly free." See, ibid., December 5, 1941.

326 *"If ever there was"*: Montreal Gazette, December 11, 1941; King Diary, December 9, 10, 1941.

326 *On that day*: Dexter Memo, January 27, 1942, Gibson and Robertson, *Ottawa at War*, 268; Granatstein, *Canada's War*, 220–21. There was talk in Ottawa that the Liberals had contributed more funds to Noseworthy's campaign, but according to Granatstein, Lambert had acted alone with his $1,000 contribution. See, J.L. Granatstein, *The Politics of Survival: The Conservative Part of Canada 1939–1945* (Toronto: University of Toronto Press, 1967), 109.

326 *"I felt tonight"*: King Diary, February 9, 1942.

326 *Of his new Quebec*: Ibid., February 11, 1942.

CHAPTER 11: Conscription if Necessary

327 *"The Chief had a"*: Grant Dexter Papers, Box 2, Folder 20, Ferguson to Dexter, July 12, 1941; King Diary, July 10, 1941.

327 *"No occasion passes"*: WFP, May 10. 1941.

328 *"No doubt [King] has"*: Dafoe Papers, M-79, Dafoe to Stevenson, February 12, 1942.

328 *The mobilization of*: For a more critical view see, Michael D. Stevenson, *Canada's Greatest Wartime Muddle* (Montreal: McGill-Queen's University Press, 2001), 3, 14.

328 *In Winnipeg in July*: Ferguson to Dexter, July 12, 1941.

328 *This was confirmed*: King Diary, February 5 1942.

328 *The conscription crisis*: Pickersgill 1, 333.

329 *Macdonald was proud*: T. Stephen Henderson, *Angus L. Macdonald: A Provincial Liberal* (Toronto: University of Toronto Press, 2007), 88, 92–93, 102–3.

329 *Still, for a country*: A manpower study in November 1941 revealed that there were only 600,000 "fighting men" available in Canada, and 209,000 were in Quebec. The air force and navy required 200,000, but all had to be English-speaking. In other words, as Grant Dexter informed Dafoe, "the army must take Quebeckers and therefore Ralston must realize that his capital of manpower is 400,000 of which 209,000 is in Quebec. Victor [Sifton, the publisher of the *Free Press* who was serving as Master-General of the Ordnance] pointed out to him, that he must draw French Canadians man for man with English speaking boys." See, University of Manitoba Archives and Special Collections, John W. Dafoe Papers, Box 8, Folder 2, Grant Dexter Memo, November 20, 1941.

330 *He did it*: See one episode when King told Ralston and Macdonald he wanted to resign, described in Angus Macdonald's diary, June 10, 1941, cited in Henderson, *Angus L. Macdonald*, 102–3. King's version of this discussion recorded in his diary makes it clear, however, he was merely trying to show both ministers how valuable he was as head of the government. See, King Diary, June 10, 1941.

330 *About one of many tense*: King Diary, December 10, 1941; Granatstein, *Canada's War*, 210–12.

330 *Feels that some*: Ibid., December 11, 1941.

330 *Or, as John Dafoe*: Dafoe Papers, M-79, Dafoe to Crerar, January 12, 1942.

330 *This had been gelling*: King Diary, December 16, 1941; Pickersgill 1, 333.

330 *On December 17*: Ibid., December 17, 1941; Granatstein, *Canada's War*, 214–15.

330 *By no means*: Ibid., December 18, 23, 1941.

330 *King rightly believed*: Ibid., January 5, 1942; Henderson, *Angus L. Macdonald*, 106–7.

330 *For the next few weeks*: Memoranda and Notes Series, Vol. 354, File 3807, Memo on Conscription, January 8, 1942; King Diary, January 15, 1942.

331 *"What I was aiming"*: King Diary, January 19, 1942.

331 *These questions were*: Ibid., January 31, 1942; Henderson, *Angus L. Macdonald*, 107.

331 *Quebec nationalists led*: Betcherman, *Ernest Lapointe*, 345; Primary Series Correspondence, Vol. 330, C-6810, 282099–100, W.E.G. Murray to WLMK, April 28, 1942; Vol. 335, C-6814, 288010–12, J.T. Thorson to Walter Turnbull, April 29, 1942.

331 *Raymond also hit*: Granatstein, *Canada's War*, 222–23; King Diary, February 19, 1942. The day of Raymond's speech was also the day King met with McNaughton to discuss military manpower needs. But the real issue that day was an accident at Laurier House. Another strange dream or "vision" had awoken King that morning, containing images of a crooked stick and a dark pencil. He had written it out when the clock

was at ten past seven. A few hours later, he learned from his butler, MacLeod, that a loose bracket on a shelf in the dining room had given way and seven china plates painted by his mother had shattered. "That it should be the china associated with my mother, her own handwork, gave an added significance to the whole vision," King wrote. Moreover, the accident according to MacLeod had occurred at seven past ten, "just the reversal of the figures of 10 past 7" when his vision had been recorded. See, King Diary, February 5, 1942.

331 *"All we had was"*: Ibid., April 1, 1942.

332 *A series of Gallup polls*: Daniel J. Robinson, *The Measure of Democracy: Polling, Market Research, and Public Life, 1930–1945* (Toronto: University of Toronto Press, 1999), 66–74.

332 *The government, he insisted*: King Diary, April 27, 1942; Granatstein, *Canada's War*, 229.

332 *He knew that surviving*: King Diary, May 1, 1942.

332 *On May 9, he decided*: Ibid., May 9–11, 1942.

333 *Following a speech*: Dexter Memo, May 13, 1942, Gibson and Robertson, *Ottawa at War*, 322–23; Henderson, *Angus L. Macdonald*, 109.

333 *On June 10, he spoke*: House of Commons Debates, June 10, 1942, 3236; Granatstein, *Canada's War*, 234; author's interview with J.W. Pickersgill, 1990. King's line has been a staple of political punditry ever since. Most recently, in June 2009 *Globe and Mail* columnist Jeffrey Simpson, commenting about current Liberal leader Michael Ignatieff's threat to force an election, wrote that Ignatieff "bowed … to one of his predecessors, Mackenzie King, and essentially said: An election this summer if necessary, but not necessarily an election." See, Jeffrey Simpson, "Election if necessary, but not necessarily an election," *Globe and Mail*, June 18, 2009, A21.

333 *King "wobbled very badly"*: Grant Dexter Papers, Box 3, Folder 22, Dexter Memo, June 16, 1942.

333 *King had declared*: King Diary, July 7, 1942.

333 *The next morning*: Ibid., July 8, 1942.

334 *Somehow, over the next*: Ibid., July 10, 1942; Granatstein, *Canada's War*, 240–41; See, the correspondence between Ralston and King in Primary Series Correspondence, Vol. 332, C-6811, 283340–91, July 7–15, 1942.

334 *"T.A. [Crerar] believes"*: University of Manitoba Archives and Special Collections, John W. Dafoe Papers, Box 8, Folder 2, Grant Dexter Memo, July 17, 1942.

335 *At four-thirty, Norman Robertson*: King Diary, December 7, 1941.

335 *Nonetheless, 1,200 Japanese*: Ann Gomer Sunahara, *The Politics of Racism: The Uprooting of Japanese Canadians During the Second World War* (Toronto: James Lorimer & Company, 1981), 29; Patricia Roy, J.L. Granatstein, Masako Lino and Hiroko Takamura, *Mutual Hostages: Canadians and Japanese During the Second World War* (Toronto: University of Toronto Press, 1990), 76–77.

335 *B.C.'s white population*: Ward, *White Canada Forever*, 107.

336 *Mackenzie, who was*: Sunahara, *The Politics of Racism*, 16–17; Canadian Race Relations Foundation, "From Racism to Redress: The Japanese Canadian Experience," www.crr.ca/divers-files/en/pub/faSh/ePubFaShRacRedJap.pdf.

336 *"The Prime Minister"*: New Canadian, January 10, 1941, cited in Sunahara, *The Politics of Racism*, 14.

336 *On August 6, 1945*: King Diary, August 6, 1945.

336 *And third, in light of:* King was not the only one who feared an attack or invasion of B.C. Soon after Pearl Harbor, Victoria mayor Andrew McGavin publicly declared that, "we expect [the Japanese] here at any time. The situation is very grave." In the House of Commons in February 1942, Conservative MP Howard Green, from Vancouver, announced that "sooner or later British Columbia would be bombed." The following month, General Kenneth Stuart, in an off-the-cuff remark that found its way into the press and heightened the panic, remarked casually to a group of journalists, "the Japs could take Alaska with two divisions." See, House of Commons Debates, January 29, 1942, 152.

337 *"Hutchison seemed to":* King Diary, February 27, 1942.

337 *This much is clear:* Ibid., February 7, 1942; Roy, et al., *Mutual Hostages,* 83; Sunahara, *The Politics of Racism,* 34, 48; Granatstein, *A Man of Influence,* 159–60.

337 *"Public prejudice is":* King Diary, February 19, 1942.

337 *As Japanese-Canadian: Toronto Star,* September 24, 1988, cited at "From Racism to Redress: The Japanese Canadian Experience," www.crr.ca/divers-files/en/pub/faSh/ePubFaShRacRedJap.pdf. On the evacuation and internment policy, see, Patricia Roy, "Internment," *Canadian Encyclopedia* www.thecanadianencyclopedia.com/index.cfm?PgNm=TCE&Params=A1ARTA0004039.

337 *It is difficult not to:* H.L. Keenleyside, "The Canada–United States Permanent Joint Board of Defence, 1940–1945," *International Journal* 16:1 (1960–61), 63.

337 *Indeed, in several polls: Public Opinion Quarterly* 8:1 (Spring 1944), 160; 9:1 (Spring 1945), 107.

338 *Any way you look:* After a three-year investigation, Justice Henry Bird in his 1950 report recommended compensation of $1.2 million, which was only a measly $52 a person. A Price-Waterhouse study in 1987 "estimated real property loss at $50 million [and] total economic loss at $443 million." See, "From Racism to Redress: The Japanese Canadian Experience."

338 *At external affairs:* Granatstein, *A Man of Influence,* 165–66; Roy, et al., *Mutual Hostages,* 157.

338 *All too typical:* Cited in Roy, ibid., 157.

338 *Before the end of the summer: Vancouver Sun,* March 16, 1945; House of Commons Debates, August 4, 1944, 5915–19, 5924–31.

338 *Of the 23,500 Japanese:* Roy, et al., *Mutual Hostages,* 162.

338 *The terrible treatment:* Sunahara, *The Politics of Racism,* 161.

339 *"I did not like the action":* Pickersgill, *Seeing Canada Whole,* 261–62.

339 *"The Canadian government":* Cited in Stacey, *Canada and the Age of Conflict,* 337.

339 *What Mackenzie King demanded:* Granatstein, *Canada's War,* 294. In 1943, Canada had a population of 11.8 million, compared with the United States with 136.7 million. However, Canada was contributing 25 per cent of the country's expenditure to the war effort in Britain, while under the Lend-Lease Act, the U.S. was sending funds overseas equal to less than twelve and half per cent of its total budget expenditure. See, Dilks, *The Great Dominion,* 237.

340 *Malcolm MacDonald:* Dilks, ibid., 239–40.

340 *Reading King's drawn-out:* King Diary, May 18, 1943; Pickersgill 1, 502–14.

340 *"The leaders of the British":* Pearson, *Mike,* 215.

340 *"Is it not a fact":* King Diary, May 20, 1943.

340 *In other discussions:* Ibid., May 19, 26, 1943.

341 *He was supported:* Ibid., July 19, 1943; Stacey, *Mackenzie King and the Atlantic Triangle,* 56–58.

341 *"My own feeling":* King Diary, July 20, 1943.

341 *Before the day:* Ibid., August 12, 1943; Stacey, *Mackenzie King and the Atlantic Triangle,* 58.

341 *King duly recorded:* Pickersgill 1, 548–50; Dilks, *The Great Dominion,* 293; Martin, *The Presidents and the Prime Ministers,* 142–43.

341 *Even Clementine Churchill:* King Diary, August 12, 1943.

342 *Late one evening:* Ibid., August 21, 1943; earl of Athlone to Churchill, August 29, 1943, cited in Dilks, *The Great Dominion,* 288.

342 *King was most:* King Diary, May 25, 1943. See also, Alfred F. Havighurst, *Britain in Transition: The Twentieth Century* (Chicago: University of Chicago Press, 1985), 330–32.

342 *"I was delighted":* King Diary, May 25, 1943.

343 *Despite only the:* Ibid., January 19, February 12, 17, March 5, 8, May 16, October 2, 1943.

343 *In fact, he had rejected:* Ibid., April 4, 1938; Finkel, "Origins of the Welfare State, 544–45.

343 *This legislation was not:* James Struthers, "Unequal Citizenship: The Residualist Legacy in the Canadian Welfare State"; English, et al., *Mackenzie King: Citizenship and Community,* 171–72.

343 *"This is really":* King Diary, August 1, 1940.

343 *King failed to mention:* Granatstein, *Canada's War,* 254. See also, Finkel, "Origins of the Welfare State," 548–49.

344 *Nevertheless, King was:* Bliss, *Right Honourable Men,* 163.

344 *Back in early January:* King Diary, January 7, 12, 1943; Granatstein, *Canada's War,* 250–51, 267–68.

345 *On the "baby bonus":* King Diary, January 12, 1943; Rea, *T.A. Crerar,* 222.

345 *From King's point:* King Diary, September 2, 1943; Rea, ibid., 220.

345 *King blamed the:* King Diary, August 4, 1943; Saywell, *'Just Call Me Mitch,'* 508–9. One of the decisive factors in Harry Nixon's demise was probably King's wartime policy to restrict beer production, which in the hot summer of 1943 enraged workers in Toronto and elsewhere. When this was pointed out to King, he dismissed it as "ridiculous." But he also could not explain why the police were required to stand on guard in front of Ontario beer stores and pubs. See, King Diary, September 7, 1943; Saywell, ibid., 509.

345 *King had half-expected:* King Diary, August 9, 1943.

345 *Some months earlier:* Ibid., March 24, 1943.

346 *As he reflected:* Ibid., September 25, 1943.

346 *"The task of Liberalism":* William Lyon Mackenzie King, Canada and the Fight for Free-dom* (Toronto: Macmillan, 1944), 298 (reprinted Freeport, NY: Books for Libraries Press, 1972). Grant Dexter was not impressed by King or his rhetoric for a Liberal future. "The more I watch things here the more it seems to me that the government is comprised of burnt out men," he wrote to Dafoe on November 1, 1943. "They are finished, except that they can run the war and, unless King chooses to dissolve suddenly, will do so until the victory is won. After the recent caucus and meeting of the [National Liberal Federation] there was a very encouraging flurry of optimism but

this has all seeped away. They are back where they were and I can see no disposition to stand up to the [CCF] to contest the field." See, Grant Dexter Papers, Box 3, Folder 25, Dexter Memo, November 1, 1943.

346 *There is no doubting:* King Diary, September 27, 28, 1943.

347 *Caught up in:* Ibid. January 22, 24, 25, 1944; Power to Bruce Hutchison, March 26, 1952, cited in Henderson, *Angus L. Macdonald,* 125.

347 *"[The] establishment of":* House of Commons Debates, January 27, 1944, 1–2.

347 *Pickersgill told him:* Pickersgill, *Seeing Canada Whole,* 233. The first family allowance payments began in July 1945. The money was tax-exempt, and the amount received was determined by the age of the children. A child under the age of five years was worth $5 per month; six- to nine-year-olds $6; ten- to twelve-year-olds $7; and thirteen- to fifteen-year-olds $8. See, J.W. Pickersgill and D.F. Forester, *The Mackenzie King Record 1939–1944* Vol. 2 (Toronto: University of Toronto Press, 1968), 37 (hereafter Pickersgill II); "Family Allowances," *Canadian Encyclopedia,* www.thecanadianencyclopedia.com.

348 *"Though very tired":* King Diary, July 29, 1944.

348 *Hence, on June 6:* Ibid., June 6, 1944; Stacey, *Mackenzie King and the Atlantic Triangle,* 55.

348 *His remarks were hardly:* Massey, *What's Past Is Prologue,* 393; Pickersgill I, 636–40; King Diary, January 25 1944; Dilks, *The Great Dominion,* 304–5.

349 *On August 7:* August 7, 1944, was the day before the decisive Quebec provincial election that saw Duplessis and the Union Nationale defeat Godbout and the Liberals and reclaim power. King was determined to do everything he could, including helping Quebec taxi drivers who had been adversely affected by the federal wartime prices and trade policy. King, supported by most of the cabinet, wanted to ease the restrictions as a favour to Godbout. But Ilsley, the finance minister, was his "obstinate" old self and he "incensed" King. See, King Diary, August 7, 1944.

349 *As Malcolm MacDonald told:* Cited in Dilks, *The Great Dominion,* 348.

350 *FDR, who was:* King Diary, September 11, 1944.

350 *"My dear Winston":* Primary Correspondence Series, Vol. 356, C-7049, 309478, WLMK to Churchill, September 17, 1944; Dilks, *The Great Dominion,* 346–47; King Diary, September 16, 1944.

351 *For in the course:* King Diary, September 14, 1944.

351 *"Trust Willie":* Grant Dexter Papers, Box 3, Folder 26, Dexter to Ferguson, September 22, 1944.

351 *The colonel immediately:* King Diary, September 15, 1944; Dexter Memo, September 25, 1944, Gibson and Robertson, *Ottawa at War,* 484.

352 *But "estimates" was:* King Diary, October 13–19, 1944; Granatstein, *Canada's War,* 340–43.

352 *Given that Nazi Germany:* King Diary, October 18, 1944. During the height of the cabinet battle, journalists Grant Dexter and Bruce Hutchison met one evening with Angus Macdonald and asked him a good question for which he did not have an answer: How "is it that in this war the Army, with some 40,000 casualties out of a total force of 400,000, is now so short of reinforcements that it has to resort to conscription whereas in the last war we had some 200,000 casualties out of, say 500,000, and did not invoke conscription until the last year of the war?" See, Granatstein, *Canada's War,* 353.

352 *Through it all:* King Diary, October 19, 20, 1944; Granatstein, ibid., 344–45.

352 *Then King, who:* Ibid., September 23, October 13, 1944.

353 *"I feel more":* Ibid., October 21; October 24, 1944.

353 *With his conviction:* Ibid., October 30, 1944. For a slightly different interpretation of these events and motivations of those invovled, see, J.E. Rea, "The Conscription Crisis: What Really Happened?" *The Beaver* 74:2 (April/May 1994), 18–19.

353 *The King-Ralston brouhaha:* Ibid., October 31, November 1, 1944.

353 *Perhaps, as Macdonald's:* Ibid., 135.

353 *King later insisted:* Dexter Memo, January 9–10, 1945, Gibson and Robertson, *Ottawa at War,* 492; Fraser, "Mackenzie King: A Tribute," 37; Granatstein, *Canada's War,* 356–57.

354 *Macdonald, who wrote:* Grant Dexter Papers, Box 3, Folder 26, Dexter to Ferguson, November 6, 1944; Ward, *The Memoirs of Chubby Power,* 156; Granatstein, *Canada's War,* 357; Henderson, *Angus L. Macdonald,* 135–36.

354 *Crerar believed, too:* Rea, *T.A. Crerar,* 227; Rea, "The Conscription Crisis," 19; Crerar, "The Incredible Canadian: An Evaluation," 155.

354 *Moreover, while none:* Henderson, who is a sympathetic biographer, attributes Macdonald's actions to noble factors. He details the great lengths to which Macdonald rationalized staying in the cabinet, mainly to "avoid a governmental collapse" and for "the sake of infantry and the country." See, *Angus L. Macdonald,* 136–42; 145–46.

354 *Now it was Andy McNaughton's:* Ontario premier George Drew, who envisioned McNaughton as a prospective Conservative, was disgusted by his Liberal metamorphosis. "Your latest move destroys the last vestiges of respect I had for you," Drew wrote to him in mid-November. "Now I would not insult a yellow dog by calling you one." Cited in J.L. Granatstein, *The Generals: The Canadian Army's Senior Commanders in the Second World War* (Toronto: Stoddart, 1993), 81.

354 *"More and more":* King Diary, November 20, 1944.

355 *"When they began":* Ibid., November 21, 1944.

355 *Under the impression:* Ibid., November 22, 1944.

355 *No prime minister:* Two of the most celebrated are Pierre Trudeau instituting wage and price controls in 1975, when he had ridiculed the Opposition for proposing it some months earlier; and Jean Chrétien promising to abolish the GST during the 1993 election and then deciding not to do so because the government's books were far worse than he anticipated.

355 *Louis St. Laurent, ever the:* Dale C. Thomson, *Louis St. Laurent: Canadian* (Toronto: Macmillan, 1967), 150–51; Ward, *The Memoirs of Chubby Power,* 163–67; Henderson, *Angus L. Macdonald,* 145.

355 *"They decided to act":* Grant Dexter Papers, Box 3, Folder 26, Dexter Memo, November 23, 1944; Dexter Memo, January 9, 1945; Henderson, *Angus L. Macdonald,* 143–45.

356 *As Jack Granatstein wryly:* Granatstein, *Canada's War,* 371.

356 *As a gift:* King Diary, November 27, 1944.

356 *King felt later:* Granatstein, *Canada's War,* 371; Pickersgill, *Seeing Canada Whole,* 248.

356 *"General, aren't you":* Sanger, *Malcolm MacDonald,* 227–28.

356 *During the long:* King Diary, November 27, 1944.

356 *Three years later:* Ibid., May 25, 1948.

357 *To his utter joy:* Ibid., December 7, 1944. Rounding up the NRMA conscripts was easier said than done. Many refused to report and there was a violent anti-draft riot in Drummondville, Quebec, at the end of February 1945. After all the political upheaval, however, only 2,463 NRMA conscripts saw any battle action. Of those, 69 were killed, 232 were wounded and 13 captured. "The estimates of need that had provoked the reinforcements in the first place," Jack Granatstein writes, "proved, as estimates often do, wrong, as the Canadian troops fortunately suffered fewer casualties than had been expected." See, Granatstein, *Canada's War*, 373; Beck, *Pendulum of Power*, 245.

357 *King was overwhelmed:* King Diary, December 7, 1944.

CHAPTER 12: The Sense of Triumph

358 *In the early hours:* King Diary, December 17, 1944.

359 *In an appearance:* Globe and Mail, February 1, 1945; John Kendle, *John Bracken: a Political Biography* (Toronto: University of Toronto Press, 1979), 218–20.

359 *The local press:* Andrew Armitage, "Like Today, Local Riding Shunned Parachute Candidate in the 1940s," *The Sun Times*, October 2, 2009.

360 *In the uneasy:* Ibid.

360 *"He was sure":* Dexter Memo, January 9, 1945.

360 *McNaughton was so:* Armitage, "Like Today, Local Riding Shunned Parachute Candidate in the 1940s."

360 *He went to bed:* King Diary, February 5, 1944.

360 *He soon met:* Ibid., February 7, 1945.

361 *He won his:* Saywell, *'Just Call Me Mitch,'* 515–16.

361 *At the beginning:* Ibid., 519.

361 *"I had waited long":* King Diary, March 2, 1945.

361 *King was:* Ibid., April 12, 1945.

361 *"The fields were":* Ibid., April 15, 1945; Martin, *The Presidents and the Prime Ministers*, 145–46.

362 *Mitch Hepburn was:* Primary Correspondence Series, Vol. 383, C-9874, 343316, Hepburn to WLMK, April 13, 1945.

362 *But George Drew:* "Mr. King's Desperate Scheming," *Globe and Mail*, April 14, 1945.

362 *"I cannot understand":* King Diary, April 16, 1945.

362 *In late February:* Ibid., February 20, 1945; Pickersgill, *Seeing Canada Whole*, 267.

363 *"I got the impression":* Ward, *The Memoirs of Chubby Power*, 175.

363 *"The little chap":* King Diary, April 22, 1945.

363 *King was accompanied:* Robertson, *Memoirs of a Very Civil Servant*, 43–44. Conservative MP (and future prime minister) John Diefenbaker was also in San Francisco as Gordon Graydon's adviser. Diefenbaker enjoyed haranguing King in the House of Commons, and King regarded him as a Tory troublemaker. "Why Diefenbaker is here I cannot think," King noted with exasperation in his diary on April 29. See, King Diary, April 29, 1944.

364 *Nor was he interested:* Ibid., July 29, 1946; Pickersgill II, 375.

364 *"My thoughts were":* King Diary, May 2, 1945.

364 *Jack Pickersgill's assignment:* Pickersgill, *Seeing Canada Whole*, 259; King Diary, April 29, 1945.

364 *Confirmation of Nazi atrocities:* Abella and Troper, *None Is Too Many*, 185–87.

364 *The subject barely*: See, King Diary, February 1, 13, June 11, 1944. One non-Jewish
Liberal in his caucus who did fight anti-Semitism and advocate civil liberties was
Toronto MP Arthur Roebuck, who had also served as attorney-general of Ontario in
Mitch Hepburn's cabinet from 1934 to 1937. King respected Roebuck and generally
paid attention to his statements on human rights.

364 *He promised Ludwig*: Ibid., June 11, 1944.

365 *Early in the morning*: Ibid., May 7, 1945.

365 *Jack Pickersgill was*: Ibid., May 7, 1945; Pickersgill 11, 381.

365 *To his credit*: Ibid., May 8, 1945; "Mackenzie King Addresses the Nation," Broadcast Date: May 8, 1945, http://archives.cbc.ca/war_conflict/second_world_war/
clips/11573/.

365 *The text of this*: James Eayrs, *In Defence of Canada: Peacemaking and Deterrence*
(Toronto: University of Toronto Press, 1972), 159.

366 *The Liberals had adopted*: According to CCF leader M.J. Coldwell, an unnamed Liberal Party emissary, possibly Grant Dexter, was sent by King to ascertain if Coldwell
was interested in joining a Liberal-CCF coalition cabinet as deputy prime minister
following the election. King later denied this story and there is nothing of it in his
diary. Dexter reported that King did tell him in mid-1944 that "it was most regretta-
ble that Coldwell was not in the Liberal party." Coldwell insisted in his unpublished
memoirs that a genuine coalition offer had been made. See, Walter Stewart, *M.J.:
The Life and Times of M.J. Coldwell* (Toronto: Stoddart, 2000), 148–49; Grant Dexter
Papers, Box 3, Folder 26, Dexter to Ferguson, November 6, 1944.

366 *Why, he asked repeatedly*: Granatstein, *Canada's War*, 403.

366 *With either the departure*: King Diary, April 11, 1945; Henderson, *Angus L.
Macdonald*, 148–49. As T. Stephen Henderson points out, Macdonald, Crerar, Power
and others kept a sharp eye on biographies of King to ensure that the correct version
of the conscription crises, as they saw it, was recorded. Following the publication
of H. Reginald Hardy's laudatory 1949 biography (critiqued by King before it was
released), Macdonald responded to a positive article about the book in *Saturday
Night* defending Ralston's honour. Macdonald and Crerar also did not agree with
Bruce Hutchison's conclusion in his 1953 book, *The Incredible Canadian,* that King
"as a public figure was greater than [John A.] Macdonald or Laurier." See, Henderson,
ibid., 203–4. On Hardy's biography, see, King Diary, January 5, 6, 7, October 14, 21,
28, 1949.

366 *King was still*: King Diary, May 17–22, 1945.

367 *In Quebec City*: Ibid., June 6, 1945; Pickersgill, *Seeing Canada Whole*, 267–68.

367 *For King's final broadcast*: Pickersgill, ibid., 268.

367 *"The choice on"*: *Toronto Star,* June 9, 1945.

368 *George Drew's massive*: King Diary, June 4, 1945; Spiritualism Series, Vol. 7, File 14,
H-3042, Séance Records, June 10, 1945.

368 *More of a "shock"*: King Diary, June 14, 1945.

368 *Grant Dexter*: Cited in Gibson and Robertson, *Ottawa at War,* 500–01.

368 *Yet whereas Winston Churchill*: Like many others, King was surprised by the results
of the election in Britain. "I am personally sorry for Churchill," he wrote on July
26. "I would like to have seen him continue his coalition until the Japanese war was
over and then drop out altogether. I think he has made a mistake running again...
His ambition has over-reached itself and he has fallen from a terrible height. It will

be an awful blow to him ... I doubt very much if the mass of the people care for the ostentatious cigar, drinking, etc., which was inseparable from one side of Churchill's nature." See, King Diary, July 26, 1945.

369 *"I sat quietly"*: Ibid., June 12, 1945.

369 *"I feel that we are"*: Ibid., July 27, 1945.

369 *"The kind of speaking"*: Ibid., August 8, 1945.

370 *His only relief:* Ibid., August 15, 1945.

370 *The documents he absconded:* The best and most thorough account of the Gouzenko affair is Amy Knight, *How the Cold War Began: The Gouzenko Affair and the Hunt for Soviet Spies* (Toronto: McClelland and Stewart, 2005), 14–33. See also, NYT, March 5, 1946, 5.

371 *Sounding very much:* John Sawatsky, *Gouzenko: The Untold Story* (Toronto: Macmillan, 1984), 22. Gouzenko was eventually denounced as a traitor by the Soviet embassy and tried *in absentia* in the Soviet Union. While some high-ranking officials of the GRU and the NKVD wanted to dispose of him, Stalin forbade it because he wanted to maintain whatever goodwill remained from the war alliance. But, as historian Amy Knight points out, there were still punishments meted out by Laverntii Beria, the brutal head of the NKVD, and GRU chief Fedor Kuznetsov. Gouzenko's boss in Ottawa, Colonel Nikolai Zabotin, wound up in a Siberian labour camp and was not released until after Stalin died in 1953. Gouzenko's mother was arrested and died in Lubianka prison. Members of his wife Anna's family were also harassed, and several were sent to Siberian labour camps as well. See, Knight, *How the Cold War Began,* 100–101.

371 *When he arrived:* King Diary, September 6, 1945.

372 *On that first day:* Ibid., February 20, July 16, 1946.

372 *King rightly feared:* Ibid., February 16, 1946.

372 *In the hectic days:* J.W. Pickersgill and D.F. Forester, The *Mackenzie King Record 1945–1946* Vol. 3 (Toronto: University of Toronto Press, 1970), 28 (hereafter Pickersgill III); Knight, *How the Cold War Began,* 69–70.

372 *The urgency for:* Pickersgill III, 18–19; Knight, ibid., 29–30; "Alan Nunn May, 91, Pioneer In Atomic Spying for Soviets," NYT, January 25, 2003.

373 *King also decided:* None of the three Western leaders knew that Kim Philby, then head of counter-intelligence at MI6 (the British Secret Intelligence Service) in London and an infamous double-agent, was secretly supplying information to the Soviets about the Gouzenko investigation—including the impending arrest of Alan Nunn May. According to Keith Jeffery in his recent book, *The Secret History of MI6: 1909–1949* (2010), as soon as Philby learned of Gouzenko's defection, he warned other Soviet spies and did everything in his power to influence and impede the investigation in the Soviet's favour. See, Keith Jeffery, *The Secret History of MI6: 1909–1949* (London: Penguin Press, 2010), 656–58. See also, Randy Boswell, "Notorious Turncoat Philby Ran Interference in Gouzenko Spy Sensation, Author Finds," *National Post,* October 1, 2010.

373 *He ruminated endlessly:* King Diary, September 11, 24, 1945; Pickersgill III, 19, 29–30.

373 *As he wrote in:* King Diary, February 16, 1946. See also, March 4, 1946.

373 *Against his initial:* Ibid., September 28, 1945.

374 *"We went to particular":* Pearson to Wrong, October 1, 1945, cited in Knight, *How the Cold War Began,* 72. Before they departed for London, King and Robertson spent

one night in New York at the Harvard Club, where King always stayed. Gordon Robertson, who was working in the PMO, had joined them for a meeting with Jacques Gréber, the French architect, who had been assigned the task of beautifying Ottawa. Once the meeting had concluded, King, the teetotaler, suggested it was time for drinks. At the bar, Gordon Robertson recalls, "King drummed with his pudgy fingers on the wooden surface, a gesture I learned, often occurred when he was not entirely comfortable in some situation. He looked at his guests, bar-order form in hand, and said, 'Well gentlemen, we will all have lemonade?' No one had the courage to suggest a preference for anything else." See, Robertson, *Memoirs of a Very Civil Servant*, 51.

374 *By the beginning:* Kathryn S. Olmsted, *Red Spy Queen: A Biography of Elizabeth Bentley* (Chapel Hill, NC: The University of North Carolina Press, 2002), 92–112; Athan Theoharis, *Chasing Spies: How the FBI Failed in Counter-Intelligence but Promoted the Politics of McCarthyism in the Cold War Years* (Chicago: Ivan R. Dee, 2002), 42–43.

374 *To Hoover:* Knight, *How the Cold War Began*, 93.

374 *The secret investigation:* Robert Bothwell and J.L. Granatstein, eds, *The Gouzenko Transcripts* (Ottawa: Deneau Publishers, 1982), 11.

374 *He understood that:* King Diary, February 5, 13, 1946.

375 *In all, twelve men:* Ibid., February 15, 1946; Bothwell and Granatstein, *The Gouzenko Transcripts*, 11; Pickersgill III, 137–42.

375 *For a while, the Gouzenko:* *Globe and Mail*, February 18, 1946; WFP, February 21, 1946; Ibid., 15; King Diary, February 17, 1945.

375 *"I am coming to feel":* King Diary, February 20, 1946.

376 *"It is a rather extraordinary":* Ibid., March 21, 1946.

376 *He questioned the:* Ibid., February 20, 1946.

377 *The two Supreme Court justices:* "Spy Trials: Sentences," *Canada's Rights Movement: A History*, www.historyofrights.com/sentences.html.

377 *M.J. Coldwell:* *Toronto Star*, March 8, 1946, 1. See also, Coldwell's personal reflections on the affair in which he reiterates on the "serious blots on the administration of justice," in Stewart, *M.J.*, 162–64.

377 *Similarly, the* Globe and Mail: "Totalitarian Procedure," *Globe and Mail*, March 6, 1946; "Still on Trial," March 22, 1946; WFP, April 11, 1946.

377 *As he had told Norman:* King Diary, February 27, 1946. By the middle of March, King was growing "indignant" at the amount of time the commissioners were taking. He and St. Laurent, he noted in his diary, "were both astonished that Kellock was going to adjourn sittings for some days to keep some engagement with a Y.M.C.A. meeting... It seems to me that everybody is going crazy these days." Ibid., March 12, 1946.

377 *At the same time:* *Public Opinion Quarterly* 10:2 (Summer 1946), 265.

377 *On March 18:* House of Commons Debates, March 18, 1946, 46–54; King Diary, March 18, 1946.

378 *The Winnipeg Tribune, for:* *Winnipeg Tribune*, March 19, 1946; Knight, *How the Cold War Began*, 167.

378 *The information provided:* Granatstein, *A Man of Influence*, 181–82; NYT, November 22, 1998.

378 *Gouzenko, who lived in:* Sawatsky, *Gouzenko*, 99–107; Daniel J. Leab, "The Iron Curtain" (1948): Hollywood's First Cold War Movie," *Historical Journal of Film, Radio, and Television* 8:2 (1988), 153–88; "Gouzenko's Daughter Speaks," CBC Radio,

October 14, 2002, Interview with Evelyn Wilson, http://archives.cbc.ca/war_conflict/cold_war/clips/14833/.

379 *Mind you, according:* Pickersgill, *Seeing Canada Whole,* 292.

379 *As Gordon Robertson:* Robertson, *Memoirs of a Very Civil Servant,* 62.

379 *"Imperial defence":* WLMK to Massey, March 23, 1946, DCER, Vol. 12, www.international.gc.ca/department/history-histoire/dcer/details-en.asp?intRefid=11872.

379 *After a private heart-to-heart:* King Diary, August 14, 1946.

379 *John Holmes, who worked:* John W. Holmes, *The Shaping of Peace: Canada and the Search for World Order, 1943–1957,* Vol. 1 (Toronto: University of Toronto Press, 1979), 151; Pickersgill III, 229.

380 *Much more notable:* King Diary, May 30, 1946.

380 *Before Massey left:* Massey was well aware of how King felt about him, and it only got worse while he was in London. "In my correspondence with Mackenzie King I misjudged how some of my suggestions would be received. I did not realize at first the amount of suspicion and the capacity for misinterpretation of one's ideas that I had to face. I was deeply hurt by Mackenzie King's use of the word 'self-aggrandizement' in one of our more difficult conversations in London about the conduct of my post." See, Massey, *What's Past Is Prologue,* 448.

380 *"He would rather":* May 22, 1946. About a year later, King was speaking to Gordon Robertson about his desire to appoint a Canadian governor general. Robertson made the mistake of suggesting that Vincent Massey would be the ideal candidate. King glared at him. "Robertson, I will not have this country run by Alice Massey!" The Masseys got in the last word and laugh, of course. In 1952, two years after King's death, Louis St. Laurent had Vincent Massey appointed as the first Canadian-born governor general. See, Robertson, *Memoirs of a Very Civil Servant,* 62.

380 *"I confess it":* King Diary, May 22, 1946.

381 *Pearson figured that:* cited in Granatstein, *A Man of Influence,* 200–201.

381 *The Paris conference:* Stacey, *A Date with History,* 181.

382 *"I thought Bracken":* King Diary, July 5, 1946.

382 *"There seemed to be":* Ibid., August 2, 5, 1946; Maurice A. Pope, *Soldiers and Politicians* (Toronto: University of Toronto Press, 1962), 320.

382 *Then, as Stacey tells it:* Stacey, *A Date with History,* 183–84, 187.

383 *A tense:* King Diary, August 30, 1946.

383 *Apart from the great:* Pickersgill III, 336; Pearson, *Mike,* 279–80; English, *Shadow of Heaven,* 315.

383 *Nothing gave him:* Ibid., January 3, 1947.

384 *King and Pat:* Ibid., August 11, 1947.

385 *Most importantly:* Ibid., September 10, 1946; Dryden, "The Mackenzie King Papers," 41–42.

385 *St. Laurent was concerned:* King Diary, October 1, 1947.

386 *According to Bruce:* Hutchison, *Incredible Canadian,* 424; King Diary, May 28, 1947.

386 *In London, he found:* Spiritualism Series, Vol. 7, File 14, H-3042, "Marked Secret," October 25, 26 and 31, 1945; File 15, "Séance with Mrs. Sharplin at Queensbury Club," May 27, 1946; File 16, "Sitting with Mrs. Sharplin, November 22, 1947 at 25 Jubilee Place." See also, Ged Martin, "Mackenzie King, The Medium and the Messages," *British Journal of Canadian Studies* 4 (1989), 109–14. In her book about her spiritualist experiences, *Mind in Life and Death,* published in 1956, Geraldine Cummins wrote of King that he "saw no wrong in seeking to obtain communications

from the spirits of the dead. For he was following the example of Christ, who, on the mountain, spoke with the spirits of Moses and Elias, and they had been dead for many years." See, Geraldine Cummins, *Mind in Life and Death: Review of Recent Evidence of the Survival of Franklin Roosevelt and Others* (London: Aquarian Press, 1956), 110–11; Bedore, "The Reading of Mackenzie King," 291.

386 *At Mrs. Sharplin's:* "Sitting with Mrs. Sharplin, November 22, 1947 at 25 Jubilee Place."

386 *"I beg of you":* Spiritualism Series, Vol. 6, File 6, H-3041, "Sitting, November 22, 1947." Three days later, when King saw Churchill, he told him of his "conversation" with FDR. Churchill asked to see the transcript of the séance, which King gladly gave him. As King recalled, "he spoke as one who was familiar with the whole phenomenon of mediumship. He seemed interested in what related to the Hereafter… He was quite reverent in his attitude. Seemed to think it quite natural that I should have brought up the subject with him." After studying the document, Churchill gave King back his papers. "Thank you so much for letting me see these interesting documents." See, ibid., Vol. 7, File 16, H-3042, Churchill to WLMK, November 25, 1947.

386 *After King returned:* Blair Fraser, "The Secret Life of Mackenzie King, Spiritualist," *Maclean's,* December 15, 1951, 9.

386 *King made it:* Pickersgill, *Seeing Canada Whole,* 294–97.

387 *As soon as he:* King Diary, December 18, 1947; Stacey, *Canada and the Age of Conflict,* 414–15.

387 *As a matter of principle:* King Diary, January 7, 1948; Pickersgill, *Seeing Canada Whole,* 300.

387 *On the evening:* Ibid., January 7, 1948; J.W. Pickersgill, *My Years with Louis St. Laurent: A Political Memoir* (Toronto: University of Toronto Press, 1975), 40–45.

388 *Three weeks later:* King Diary, January 20, 1948.

388 *A little later:* Personal Papers Series, Vol. 25, WLMK to Markham, March 7, 1948.

388 *"It was clear":* King Diary, April 20, 1948.

388 *Thomas Crerar:* Crerar, "The Incredible Canadian," 155; WFP, April 20, 1948.

389 *"I am so near":* Harold Dingman, "Mackenzie King Sums Up," New Liberty 25:5 (July 1948), 7–10; King Diary, June 23, 1948.

389 *He was not:* King Diary, March 6, 1947.

389 *"I see that":* Vancouver Province, August 7, 1944.

389 *The Liberal convention:* King Diary, August 6, 7, 1948; Pickersgill, *Seeing Canada Whole,* 307–8; Thomson, *Louis St. Laurent,* 235–40.

390 *"I can truly say":* King Diary, August 7, 1948.

390 *Even Pickersgill did:* Pickersgill, *Seeing Canada Whole,* 309.

390 *King was in Paris:* J.W. Pickersgill and D.F. Forester, *The Mackenzie King Record 1947–1948,* Vol. 4 (Toronto: University of Toronto Press, 1970), 391–93 (hereafter Pickersgill IV); Robertson, *Memoirs of a Very Civil Servant,* 63–64.

390 *A parade of visitors:* Pickersgill IV, 410–27; King Diary, October 22, 1948.

391 *"It is an exaggeration":* Eayrs, *In Defence of Canada: Peacemaking and Deterrence,* 238–39.

391 *Making their way:* King Diary, October 22, 1948; Fraser, "The Secret Life of Mackenzie King, Spiritualist," 7; Martin, "Mackenzie King, The Medium and the Messages," 125–27; Spiritualism Series, Vol. 7, File 27, H-3042, "Account for Services," October 27, 1948; Bedore, "The Reading of Mackenzie King," 293.

391 *As he stepped:* King Diary, November 7, 1948.

391 *The hands of the clock:* Ibid., November 15, 1948.

CHAPTER 13: With the Loved Ones at Last

392 *"I doubt if"*: Personal Papers Series, Vol. 25, WLMK to Markham, January 5, 1949.

393 *He became morose*: Personal Miscellaneous, Vol. 1, WLMK to Dr. G.R. Brow, May 5, 1949; King Diary, March 7, April 24, July 24, 25, September 19, 21, 1949.

393 *At one particularly*: King Diary, February 1, 1949.

393 *A visit from his sister*: Ibid., July 9, 1949.

393 *"I find my seat"*: Ibid., February 1, 1949.

393 *"I was given"*: Ibid., April 27, 1949.

393 *"When we next meet"*: Ibid., April 30, 1949.

394 *Near the end*: Ibid., June 25, 1950; Esberey, *Knight of the Holy Spirit*, 99–100.

395 *"Tonight seems to"*: King Diary, December 17, 1948.

395 *"I can see she"*: Ibid., February 12, 1949.

395 *There were honours*: Ibid., March 11, 1950; Stacey, *A Very Double Life*, 218.

395 *Solon Low, a Social Credit*: Ibid., June 20, 1950 (*Ottawa Citizen* clipping); *Ottawa Journal*, June 12, 1950; Stacey, ibid., 219.

396 *For King, this was*: King Diary, March 12, 1949.

396 *"I could see from"*: Ibid., September 25, 1933.

396 *When he finally*: Ibid., May 13, 1944.

396 *He committed $4,000*: Ibid., March 12, 1949; May 4, 1950; Stacey, *A Very Double Life*, 217.

396 *"The amount of material"*: Personal Papers Series, Vol. 25, Markham to McGregor, August 16, 1950.

396 *In her survey*: Dryden, "The Mackenzie King Papers," 43.

396 *This generous funding*: Markham to McGregor, August 16, 1950.

397 *"[McGregor] does not yet"*: King Diary, April 8, November 1, 4, 1949; January 5, 1950.

397 *"A terrible night"*: Ibid., December 26, 1949.

397 *"I really was alarmed"*: Ibid., December 27, 1949; January 1, 1950.

397 *"She has been amazingly"*: Ibid., June 25, 1950; Dryden, "The Mackenzie King Papers," 51.

397 *"I said that my"*: King Diary, July 7, 1950.

397 *On July 15*: Ibid., July 15, 1950.

397 *On July 19*: Ibid., July 19, 1950.

398 *"I did not think"*: Violet Markham Carruthers Papers, File: "Joan Patteson," Joan Patteson to Markham, July 28, 1950.

398 *"The penalty of long life"*: Markham to McGregor, August 16, 1950.

398 *"I am in a strange sort"*: Violet Markham Carruthers Papers, File: "Joan Patteson," Joan Patteson to Markham, March 3, 1951.

398 *Prime Minister St. Laurent*: Pickersgill, *Seeing Canada Whole*, 353.

399 *"His death"*: Ibid., 353.

399 *Editorialists and journalists*: Fraser, "Mackenzie King: A Tribute," 37–38; CBC Digital Archives, "King Dies at 75," July 25, 1950; http://archives.cbc.ca/politics/prime_ministers/clips/7243/; *Globe and Mail*, July 24, 1950, *Ottawa Journal*, July 24, 1940.

399 *Alex Hume*: *Ottawa Citizen*, July 24, 1950. A few weeks after Meighen had lost the February 1942 by-election, King had a private meeting with Grant Dexter and Bruce Hutchison. He had told the two journalists how thrilled he was that Meighen had lost, adding, "I hate him. I hate him so much." See, Grant Dexter Papers, Box 3, Folder 21, Dexter Memo, February 28, 1942.

399 *"Many years must"*: WFP, July 24, 1950.

400 *As soon as his will*: LAC, WLMK Papers, R10383-25-0-E, MG 26 J19, Royal Trust Estate Papers, Vol. 1, Memo, September 7, 1950; *Toronto Star*, August 9, 1950; Violet Markham Carruthers Papers, File: "F.A. McGregor," McGregor to Markham, September 14, 1950.

400 *"To my amazement"*: King Diary, March 31, 1949.

400 *"I was staggered"*: Markham to McGregor, August 16, 1950.

400 *"We always [thought] that"*: Violet Markham Carruthers Papers, File: "Joan Patteson," Joan Patteson to Markham, December 17, 1950.

401 *His family, friends, staff*: Royal Trust Estate Papers, WLMK Estate Correspondence, Part III, H.C. Buxton to beneficiaries, August 11, 1950.

401 *King deemed that*: *Toronto Star*, August 9, 1950; *A Real Companion and Friend*: "Mackenzie King Slept Here."

401 *In May 1937, while King*: Levine, *Scrum Wars*, 135; Fraser, "The Secret Life of Mackenzie King, Spiritualist," 61. In 1950, Geraldine Cummins was preparing to publish her book, *Unseen Adventures* (London: Rider and Company, 1951). After the *Psychic News* story about King appeared, she added a section in the book's appendix about his attending séances she conducted in 1947 and 1948. She had sent the proofs to a friend in Ottawa, who in turn showed it to Senator Duncan MacTavish, one of the executors of King's estate. In what was a futile effort, he and Leonard Brockington visited Miss Cummins in London and asked her not to expose King. She reluctantly agreed to their request and instead identified him as a "British Commonwealth Statesman" and "Mr. S." when referring to him in her book. Six months after the book was published, Fred McGregor was satisfied that journalists in Canada had not uncovered the book's secret, although the *Maclean's* correspondent in London eventually spoke to Cummins, who confirmed the truth and the identity of the "British Commonwealth Statesman." See, Fraser, ibid., 61; Violet Markham Carruthers Papers, File: "F.A. McGregor," McGregor to Markham, February 22, June 29, 1951; Geraldine Cummins, *Unseen Adventures* (London: Rider and Company, 1951), 177–83.

401 *The ever-resourceful*: Thanks to Graham Fraser, the Commissioner of Official Languages and Blair Fraser's son for sharing this information with me. See: "Notes for an Address at the International Council for Canadian Studies, International Conference, Canada Exposed," May 27, 2008, www.ocol-clo.gc.ca/html/speech_discours_27052008_e.php; Violet Markham Carruthers Papers, McGregor to Markham, October 15, 1951.

401 *A few months*: Fraser, "The Secret Life of Mackenzie King, Spiritualist," 7–9, 60–61.

402 *According to Percy Philip*: Spiritualism Series, Vol. 7, File 20, H-3042; Percy J. Philip, "Fantasio," CBC Broadcast Transcript, September 24, 1954. See also, Percy J. Philip, "My Conversation with Mackenzie King's Ghost," *Liberty* 31:11 (January 1955), 17, 58–59.

402 *As Lindsey later told*: Ibid., Vol. 7, File 20, H-3042, "Memo by F.A. McGregor," December 30, 1952.

402 *Gossip floated around*: O'Leary, *Recollections*, 87; author's interview with Bruce Hutchison, June 1990; Hume Papers, Vol. 1, Hume to Neatby, November 19, 1963.

403 *Cabinet minister Paul Martin, Sr.*: Roazen, *Canada's King*, 7.

403 *"Those of his Canadian"*: J.V. McAree, "Mr. King and the Spirits," *Globe and Mail*, September 14, 1953.

403 *The idea, however*: James McCook, "Spiritualist Mackenzie King Tried 'Communication with the Dead,'" *Ottawa Journal*, December 10, 1951.

403 *"I told Joan"*: King Diary, April 26, 1935.

CONCLUSION: Canada's Greatest Prime Minister

404 *"I turned to Norman"*: D.V. LePan, "The Spare Deputy: A Portrait of Norman Robertson," *International Perspectives* (July/August 1978), 4; Granatstein, *A Man of Influence*, 202.

404 *After spending those*: Stacey, *A Date with History*, 191.

404 *And in his memoir*: Pickersgill, *Seeing Canada Whole*, 353.

405 *"All these writers"*: King Diary, March 6, 1947.

405 *"The King diaries"*: Bliss, *Right Honourable Men*, 129.

406 *"Mackenzie King was"*: Ferns, *Reading from Left to Right*, 308.

407 *Closer economic, military*: See, J.L. Granatstein, *How Britain's Weakness Forced Canada into the Arms of the United States* (Toronto: University of Toronto Press, 1989), 24–26.

407 *He might not have*: Robertson, *Memoirs of a Very Civil Servant*, 69.

408 *"King so dominated"*: Roazen, *Canada's King*, 148.

409 *His friend and fellow believer*: Spiritualism Series, Vol. 6, File 3, H-3041, Duchess of Hamilton, "The Letter Killeth but the Spirit Giveth Life," *Light: A Journal of Spiritualism, Psychical, Occult and Mystical Research* 62 (June 1937), 369; Bedore, "The Reading of Mackenzie King," 249.

409 *"I was impressed"*: King Diary, August 6, 1948.

Bibliography

PRIMARY

William Lyon Mackenzie King's voluminous papers, including his diary (which is also available online), are kept at Library and Archives Canada in Ottawa. Other archival sources consulted include the papers of Wilfrid Laurier, Arthur Meighen, R.B. Bennett, W.L. McDougald, Violet Markham Carruthers, Arnold Heeney, John W. Dafoe, Oscar Skelton and Norman Lambert.

SECONDARY

Abella, Irving, and Harold Troper. *None Is Too Many: Canada and the Jews of Europe, 1933–1948*. Toronto: Lester and Orpen Dennys, 1982.

Abels, Jules. *The Rockefeller Billions*. New York: The Macmillan Company, 1965.

Addams, Jane. *Twenty Years at Hull-House*. New York: The Macmillan Company, 1910.

Allen, Ralph. *Ordeal by Fire: Canada, 1910–1945*. Toronto: Doubleday Canada, 1961.

Altick, Richard D. *Victorian People and Ideas*. New York: W.W. Norton and Company, 1973.

Baker, William. "The Miners and the Mediator: The 1906 Lethbridge Strike and Mackenzie King." *Labour/Le Travillieur* 11 (Spring 1983): 89–118.

———. "The Personal Touch: Mackenzie King, Harriet Reid, and the Springhill Strike, 1909–1911." *Labour/Le Travillieur* 13 (Spring 1984): 159–76.

———. "A Case Study of Anti-Americanism in English-Speaking Canada: The Election Campaign of 1911." *Canadian Historical Review* 51:4 (December 1970): 426–49.

Bangarth, Stephanie. *Voices Raised in Protest: Defending North American Citizens of Japanese Ancestry, 1942–48*. Vancouver: University of British Columbia Press, 2008.

Beck, J.M. *Pendulum of Power: Canada's Federal Elections*. Scarborough: Prentice Hall of Canada, 1968.

Bedore, Margaret. "The Reading of Mackenzie King." PhD Dissertation, Queen's University, 2008.

Berton, Pierre. *The Promised Land: Settling the West 1896–1914*. Toronto: McClelland and Stewart, 1984.

———. *The Great Depression 1929–1939*. Toronto: McClelland and Stewart, 1990.

Betcherman, Lita-Rose. *Ernest Lapointe: Mackenzie King's Great Quebec Lieutenant*. Toronto: University of Toronto Press, 2002.

———. "The Customs Scandal of 1926," *The Beaver* 81:2 (April/May 2001): 14–19.

Black, Conrad. *Render Unto Caesar: The Life and Legacy of Maurice Duplessis*. Toronto: Key Porter Books, 1998.

Blackburn, Robert H. "Mackenzie King, William Mulock, James Mavor, and the University of Toronto Students' Revolt of 1895." *Canadian Historical Review* 69:4 (December 1988): 490–503.

Bliss, Michael. *Right Honourable Men: The Descent of Canadian Politics from Macdonald to Mulroney*. Toronto: HarperCollins, 1994.

Borden, Robert Laird: *His Memoirs*. Vol. 2. Toronto: McClelland and Stewart, 1969.

Bothwell, Robert, Ian Drummond and John English. *Canada 1900–1945*. Toronto: University of Toronto Press, 1987.

———. *Canada Since 1945: Power, Politics and Provincialism*. Toronto: University of Toronto Press, 1989.

Bothwell, Robert, and J.L. Granatstein, eds. *The Gouzenko Transcripts*. Ottawa: Deneau Publishers, 1982.

Boyko, John. *Bennett: The Rebel Who Challenged and Changed A Nation*. Toronto: Key Porter, 2010.

Brennan, Patrick H. *Reporting the Nation's Business: Press-Government Relations during the Liberal Years, 1935–1957*. Toronto: University of Toronto Press, 1994.

Brinkley, Douglas, and David R. Facey-Crowther, eds. *The Atlantic Charter*. New York: St. Martin's Press, 1994.

Brown, R. Craig and Ramsay Cook. *Canada, 1896–1921: A Nation Transformed*. Toronto: McClelland and Stewart, 1974.

Bullock, Allison C. "William Lyon Mackenzie King: A Very Double Life?" MA Thesis, Queen's University, 2009.

Burke, Sara Z. *Seeking the Highest Good: Social Service and Gender at the University of Toronto, 1888–1937*. Toronto: University of Toronto Press, 1996.

Byng, Evelyn. *Up the Stream of Time*. Toronto: The Macmillan Company, 1946.

Careless, J.M.S. *Brown of the "Globe": The Voice of Upper Canada*. Vol. 1. Toronto: Macmillan, 1959

———. *Brown of the "Globe": Statesman of Confederation*. Vol. 2. Toronto: Macmillan, 1963.

Chernow, Ron. *Titan: The Life of John D. Rockefeller, Sr.* New York: Random House, 1998.

Clarkson, Stephen. *The Big Red Machine: How the Liberal Party Dominates Canadian Politics*. Vancouver: University of British Columbia Press, 2005.

Cook, Ramsay. "J.W. Dafoe at the Imperial Conference, 1923." *Canadian Historical Review* 41:1 (March 1960): 19–40.

——. *The Politics of John W. Dafoe and the Free Press*. Toronto: University of Toronto Press, 1963.

——. "A Canadian Account of the 1926 Imperial Conference." *Journal of Commonwealth Political Studies* 3:1 (March 1965): 50–63.

——, ed. *The Dafoe-Sifton Correspondence 1919–1927*. Winnipeg: Manitoba Record Society, 1966.

——. "Spiritualism, Science of the Earthly Paradise." *Canadian Historical Review* 65:1 (March 1984): 4–27.

——. *The Regenerators: Social Criticism in Late Victorian Canada*. Toronto: University of Toronto Press, 1985.

Craven, Paul. *'An Impartial Umpire': Industrial Relations and the Canadian State 1900–1911*. Toronto: University of Toronto Press, 1980.

Crowley, Terence A. *Marriage of Minds: Isabel and Oscar Skelton Reinventing Canada*. Toronto: University of Toronto Press, 2003.

Dawson, R. MacGregor. *William Lyon Mackenzie King: A Political Biography 1874–1923*. Toronto: University of Toronto Press, 1958.

Dilks, David. *The Great Dominion: Winston Churchill in Canada 1900–1954*. Toronto: Thomas Allen Publishers, 2005.

Documents on Canadian External Relations. Ottawa: Department of External Affairs, 1967. Volumes 3 to 15.

Dryden, Jean. "The Mackenzie King Papers: An Odyssey." *Archivaria* 6 (Summer 1978): 40–69.

Duffy, Dennis. "Love Among the Ruins: The King of Kingsmere." *American Review of Canadian Studies* 37:3 (Autumn 2007): 355–96.

Eayrs, James. *In Defence of Canada: Appeasement and Rearmament*. Toronto: University of Toronto Press, 1965.

——. *In Defence of Canada: Peacemaking and Deterrence*. Toronto: University of Toronto Press, 1972.

Ellis, Lewis E. *Reciprocity 1911: A Study in Canadian-American Relations*. New York: Greenwood Press, 1968.

English, John. *Shadow of Heaven: The Life of Lester Pearson*. Vol. 1: 1897–1948. Toronto: Lester and Orpen Dennys, 1989.

—— and J.O. Stubbs, eds. *Mackenzie King: Widening the Debate*. Toronto: Macmillan of Canada, 1977.

——, Kenneth McLaughlin and P. Whitney Lackenbauer, eds. *Mackenzie King: Citizenship and Community*. Toronto: Robin Brass Studio, 2002.

—— and Kenneth McLaughlin. *Kitchener: An Illustrated History*. Waterloo, Ontario: Wilfrid Laurier University Press, 1983.

Esberey, Joy E. *Knight of the Holy Spirit: A Study of William Lyon Mackenzie King*. Toronto: University of Toronto Press, 1980.

Ferns, Henry, and Bernard Ostry. *The Age of Mackenzie King*. Toronto: James Lorimer and Company, 1976 (first published 1955).

Forsey, Eugene. *A Life on the Fringe: The Memoirs of Eugene Forsey*. Toronto: Oxford University Press, 1990.

——. *The Royal Power of Dissolution of Parliament in the British Commonwealth*. Toronto: Oxford University Press, 1943.

Fosdick, Raymond B. *John D. Rockefeller, Jr.: A Portrait*. New York: Harper and Brothers, 1956.

Friedland, Martin L. *The University of Toronto: A History.* Toronto: University of Toronto Press, 2002.

Friesen, Gerald. *The Canadian Prairies: A History.* Toronto: University of Toronto Press, 1984.

Gibson, Frederick W., and Barbara Robertson, eds. *Ottawa at War: The Grant Dexter Memoranda, 1939–1945.* Winnipeg: The Manitoba Record Society, 1994.

Gilbert, Martin, ed. *The Churchill War Papers: Never Surrender May 1940–December 1940.* Vol. 2. New York: W.W. Norton, 1995.

———. *The Churchill War Papers: The Ever Widening War, 1941.* Vol. 3. London: Heinemann, 2001.

———. *Churchill and America.* Toronto: McClelland and Stewart, 2005.

Gitelman, Howard M. *Legacy of the Ludlow Massacre: A Chapter in American Industrial Relations.* Philadelphia: University of Pennsylvania Press, 1988.

Glassford, Larry A. "Retrenchment—R.B. Bennett Style: The Conservative Record Before the New Deal." *American Review of Canadian Studies* 19:2 (Summer 1989): 141–57.

———. *Reaction and Reform: The Politics of the Conservative Party under R.B. Bennett 1927–1938.* Toronto: University of Toronto Press, 1992.

Graham, John R. "William Lyon Mackenzie King, Elizabeth Harvie, and Edna: A Prostitute Rescuing Initiative in Late Victorian Toronto." *The Canadian Journal of Human Sexuality* 8:1 (Spring 1999): 47–60.

Graham, Roger. *Arthur Meighen: And Fortune Fled* Vol. 2. Toronto: Clarke, Irwin and Company, 1963.

Granatstein, J.L. *Canada's War: The Politics of the Mackenzie King Government 1939–1945.* Toronto: Oxford University Press, 1975.

——— and Robert Bothwell. "'A Self-Evident National Duty': Canadian Foreign Policy, 1935–1939." *Journal of Imperial and Commonwealth History* 3:2 (January 1975): 212–33.

——— and J.M. Hitsman. *Broken Promises: A History of Conscription in Canada.* Toronto: Oxford University Press, 1977.

———. *Mackenzie King: His Life and World.* Toronto: McGraw-Hill Ryerson, 1977.

———. *A Man of Influence: Norman A. Robertson and Canadian Statecraft 1929–1968.* Toronto: Deneau Publishers, 1981.

———. *The Ottawa Men: The Civil Service Mandarins 1935–1957.* Toronto: Oxford University Press, 1982.

——— and Norman Hillmer. *For Better or Worse: Canada and the United States to the 1990s.* Toronto: Copp Clark Pitman, 1991.

———. *The Generals: The Canadian Army's Senior Commanders in the Second World War.* Toronto: Stoddart, 1993.

———. *Canada's Army: Waging War and Keeping the Peace.* Toronto: University of Toronto Press, 2002.

Gray, Charlotte. *Mrs. King: The Life and Times of Isabel Mackenzie King.* Toronto: Viking, 1997.

———. "Crazy Like a Fox." *Saturday Night* 112:8 (October 1997): 42–46, 48, 50 and 94.

Gwyn, Sandra. *The Private Capital: Ambition and Love in the Age of Macdonald and Laurier.* Toronto: McClelland and Stewart, 1984.

Hallahan, Kirk. "W.L. Mackenzie King: Rockefeller's 'Other' Public Relations Counselor in Colorado." *Public Relations Review* 29:4 (November 2003): 401–14.

Hardy, H. Reginald. *Mackenzie King of Canada.* London: Oxford University Press, 1949.

Harney, Robert F. and Harold Troper. *Immigrants: A Portrait of the Urban Experience, 1890–1930.* Toronto: Van Nostrand Reinhold, 1975.

Heeney, Arnold. *The Things That Are Caesar's: Memoirs of a Canadian Public Servant.* Toronto: University of Toronto Press, 1972.

Henderson, George F. *W.L. Mackenzie King: A Bibliography and Research Guide.* Toronto: University of Toronto Press, 1998.

Henderson, T. Stephen. *Angus L. Macdonald: A Provincial Liberal.* Toronto: University of Toronto Press, 2007.

Hillmer, Norman. "The Foreign Policy that Never Was, 1900–1950." Canadian External Affairs.

———. "O.D. Skelton and the North American Mind." *International Journal* (Winter 2004–2005): 93–110.

Holmes, John W. *The Shaping of Peace: Canada and the Search for World Order 1943–1957.* Toronto: University of Toronto Press, 1979.

Hoogenraad, Maureen. "Mackenzie King in Berlin." *Archivist* 20:3 (1994): 19–21.

How, Douglas. "One Man's Mackenzie King." *The Beaver* 78:5 (October/November 1998): 31–37.

Howe, Daniel Walker. "American Victorianism as a Culture." *American Quarterly* 27 (December 1975): 507–532.

Humphries, Charles W. "Mackenzie King Looks at Two 1911 Elections." *Ontario History* 56:3 (1964): 203–6.

Hutchison, Bruce. *The Incredible Canadian.* Toronto: Longmans, Green and Company, 1953.

Kealey, Gregory S. *Workers and Canadian History.* Toronto: University of Toronto Press, 1995.

Keenleyside, Hugh L. *Memoirs: Hammer the Golden Day.* Vol. 1. Toronto: McClelland and Stewart, 1981.

Keyserling, Robert H. "Mackenzie King's Spiritualism and His View of Hitler in 1939." *Journal of Canadian Studies* 20: 4 (Winter 1985–86): 26–44.

———. "'Agents within the Gates': The Search for Nazi Subversives in Canada during World War II." *Canadian Historical Review* 66: 2 (June 1985): 211–239.

Kilbourn, William. *The Firebrand: William Lyon Mackenzie and the Rebellion in Upper Canada.* Toronto: Clarke, Irwin and Company, 1956.

King, William Lyon Mackenzie. *The Secret of Heroism.* New York: Fleming H. Revell, 1906.

———. *Industry and Humanity: A Study in the Principles Underlying Industrial Reconstruction.* Toronto: University of Toronto Press, 1973. (First edition: Boston: Houghton Mifflin, 1918.)

Knight, Amy. *How the Cold War Began: The Gouzenko Affair and the Hunt for Soviet Spies.* Toronto: McClelland and Stewart, 2005.

Kottman, Richard N. "The Canadian-American Trade Agreement of 1935." *The Journal of American History* 52: 2 (September 1965): 275–96.

LaPierre, Laurier L. *Sir Wilfrid Laurier and the Romance of Canada.* Toronto: Stoddart, 1996.

Lederle, John W. "The Liberal Convention of 1919 and the Selection of Mackenzie King." *Dalhousie Review* (April 1947), 85–92.

LeSueur, William Dawson. *William Lyon Mackenzie: A Reinterpretation.* Edited and with an introduction by A.B. McKillop. Toronto: Macmillan of Canada, 1979.

Levine, Allan. *Scrum Wars: The Prime Ministers and the Media.* Toronto: Dundurn Press, 1993.

———. *The Devil in Babylon: Fear of Progress and the Birth of Modern Life.* Toronto: McClelland and Stewart, 2005.

Library and Archives Canada. "A Real Companion and Friend: The Diary of William Lyon Mackenzie King." www.collectionscanada.gc.ca/king/index-e.html.

MacFarlane, John. *Ernest Lapointe and Quebec's Influence on Canadian Foreign Policy.* Toronto: University of Toronto Press, 1999.

MacLaren, Roy. *Commissions High: Canada in London, 1870–1971.* Montreal: McGill-Queen's University Press, 2006.

Mallory, J. R. "Mackenzie King and the Origins of the Cabinet Secretariat." *Canadian Public Administration* 19:2 (1976): 254–266.

Markham, Violet. *Return Passage: The Autobiography of Violet R. Markham.* London: Oxford University Press, 1953.

———. *Friendship's Harvest.* London: Max Reinhardt, 1956.

Marshall, Peter. "The Balfour Formula and the Evolution of the Commonwealth." *Round Table* 90: 361 (September 2001): 541–53.

Martin, Ged. "Mackenzie King, The Medium and the Messages." *British Journal of Canadian Studies* 4 (1989): 109–35.

Martin, Joe. "William Lyon Mackenzie King: Canada's First Management Consultant?" *Business Quarterly* 56:1 (Summer 1991): 31–35.

Martin, Lawrence. *The Presidents and the Prime Ministers: Washington and Ottawa Face to Face.* Toronto: Doubleday Canada, 1982.

Massey, Vincent. *What's Past Is Prologue.* Toronto: The Macmillan Company of Canada, 1963.

McGregor, F.A. *The Fall and Rise of Mackenzie King.* Toronto: Macmillan of Canada, 1962.

McMullen, Stan. *Anatomy of a Séance: A History of Spirit Communication in Central Canada.* Montreal: McGill-Queen's University Press, 2004.

McNaught, Kenneth. *A Prophet in Politics: A Biography of J.S. Woodsworth.* Toronto: University of Toronto Press, 1959.

Moher, Mark. "The 'Biography' in Politics: Mackenzie King in 1935." *Canadian Historical Review* 55:2 (June 1974): 239–48.

Morton, W.L. *The Progressive Party in Canada.* Toronto: University of Toronto Press, 1950.

Neatby, H. Blair. *William Lyon Mackenzie King: The Lonely Heights, 1924–1932.* Toronto: University of Toronto Press, 1963.

———. *The Politics of Chaos: Canada in the Thirties.* Toronto: Macmillan of Canada, 1972.

———. *William Lyon Mackenzie King: The Prism of Unity, 1932–1939.* Toronto: University of Toronto Press, 1976.

Nicholson, Murray W. *Woodside and the Victorian Family of John King.* Ottawa: National Historic Parks and Sites Branch, 1984.

Niergarth, Kirk. "William Lyon Mackenzie King's 1908 Adventure in Diplomacy: Coming Back a 'Public Man.'" Unpublished paper presented at the Canadian Historical Association Meetings (May 2009).

Nolan, Brian. *King's War: Mackenzie King and the Politics of War 1939–1945*. Toronto: Random House, 1988.

Oliver, Peter. *G. Howard Ferguson: Ontario Tory*. Toronto: University of Toronto Press, 1977.

Palmer, Bryan D. *Working-Class Experience: Rethinking the History of Canadian Labour, 1800–1991*. Toronto: McClelland & Stewart, 1992.

Perlin, George C. *The Tory Syndrome: Leadership Politics in the Progressive Conservative Party*. Montreal: McGill-Queen's University Press, 1980.

Pickersgill, J.W. "Mackenzie King's Speeches." *Queen's Quarterly* 57 (Autumn 1950): 304–11.

———. *The Mackenzie King Record 1939–1944*. Vol. 1. Toronto: University of Toronto Press, 1960.

——— and D.F. Forester. *The Mackenzie King Record 1939–1944*. Vol. 2. Toronto: University of Toronto Press, 1968.

———. *The Mackenzie King Record 1945–1946* Vol. 3. Toronto: University of Toronto Press, 1970.

———. *The Mackenzie King Record 1947–1948*. Vol. 4. Toronto: University of Toronto Press, 1970.

———. *My Years with Louis St. Laurent: A Political Memoir*. Toronto: University of Toronto Press, 1975.

———. *Seeing Canada Whole: A Memoir*. Markham, Ontario: Fitzhenry and Whiteside, 1994.

Pirow, Oswald. *James Barry Munnik Hertzog*. Cape Town: Howard Timmins, 1958.

Pope, Maurice A. *Soldiers and Politicians*. Toronto: University of Toronto Press, 1962

Prang, Margaret. *N.W. Rowell: Ontario Nationalist*. Toronto: University of Toronto Press, 1975.

Quinn, Herbert F. *The Union Nationale: A Study in Quebec Nationalism*. Toronto: University of Toronto Press, 1963.

Rasporich, Anthony W., ed. *William Lyon Mackenzie*. Toronto: Holt, Rinehart and Winston of Canada, 1972.

Rea, J.E. "The Conscription Crisis: What Really Happened?" *The Beaver* 74:2 (April/May 1994): 10–19.

———. *T.A. Crerar: A Political Life*. Montreal: McGill-Queen's University Press, 1997.

Regehr, T.D. *The Beauharnois Scandal*. Toronto: University of Toronto Press, 1990.

Reynolds, Louise. *Mackenzie King Friends & Lovers*. Victoria: Trafford Publishing, 2005.

Ritchie, Charles. *The Siren Years*. London: Macmillan, 1974.

Roazen, Paul. *Canada's King: An Essay in Political Psychology*. Oakville, Ontario: Mosaic Press, 1998.

Robertson, Gordon. *Memoirs of a Very Civil Servant: Mackenzie King to Pierre Trudeau*. Toronto: University of Toronto Press, 2000.

Rogers, Norman McLeod. *Mackenzie King*. Toronto: George N. Morang and Thomas Nelson & Sons, 1935. (A revised and extended edition of a biographical sketch by John Lewis, published in 1925).

Roy, Patricia, J.L. Granatstein, Masako Lino and Hiroko Takamura. *Mutual Hostages: Canadians and Japanese During the Second World War*. Toronto: University of Toronto Press, 1990.

Russell, Bob. *Back to Work? : Labour, State, and Industrial Relations in Canada.* Toronto: Nelson Canada, 1990.

Sanger, Clyde. *Malcolm MacDonald: Bringing an End to Empire.* Montreal: McGill-Queen's University Press, 1995.

Sawatsky, John. *Gouzenko: The Untold Story.* Toronto: Macmillan, 1984.

Saywell, John T. *'Just Call Me Mitch': The Life of Mitchell F. Hepburn.* Toronto: Univesity of Toronto Press, 1991.

Schenkel, Albert F. *The Rich Man and the Kingdom: John D. Rockefeller, Jr. and the Protestant Establishment.* Minneapolis: Fortress Press, 1995.

Schieman, Scott. "Socioeconomic Status and Beliefs about God's Influence in Everyday Life." *Sociology of Religion* 71:1 (Spring 2010): 25–51.

Shermer, Michael. *Why People Believe Weird Things.* New York: Henry Holt, 2002.

Spaulding, William B. "Why Rockefeller Supported Medical Education in Canada: The William Lyon Mackenzie King Connection." *Canadian Bulletin of Medical History* 10 (1993): 67–76.

Stacey, C.P. *Mackenzie King and the Atlantic Triangle.* Toronto: Macmillan of Canada, 1976.

———. *A Very Double Life: The Private World of Mackenzie King.* Toronto: Macmillan of Canada, 1976.

———. "The Divine Mission: Mackenzie King and Hitler." *Canadian Historical Review* 61:4 (1980): 502–12.

———. *Canada and the Age of Conflict 1921–1948.* Vol. 2. Toronto: University of Toronto Press, 1981.

———. *A Date with History: Memoirs of a Canadian Historian.* Ottawa: Deneau, Publishers, 1985.

Stevens, Paul. *The 1911 General Election: A Study in Canadian Politics.* Toronto: Copp Clark Pittman, 1970.

Stevenson, Michael D. *Canada's Greatest Wartime Muddle.* Montreal: McGill-Queen's University Press, 2001.

Strange, Carolyn. *Toronto's Girl Problem: The Perils and Pleasures of the City, 1880–1930.* Toronto: University of Toronto Press, 1995.

Sunahara, Ann Gomer. *The Politics of Racism: The Uprooting of Japanese Canadians During the Second World War.* Toronto: James Lorimer & Company, 1981.

Swatsky, John. *Gouzenko: The Untold Story.* Toronto: Macmillan of Canada, 1984.

Taylor, John H. *Ottawa: An Illustrated History.* Toronto: James Lorimer & Company, 1986.

Thompson, John with Allen Seager. *Canada 1922–1939: Decades of Discord.* Toronto: McClelland and Stewart, 1985.

Thomson, Dale C. *Louis St. Laurent: Canadian.* Toronto: Macmillan, 1967.

Tulchinsky, Gerald. *Branching Out: The Transformation of the Canadian Jewish Community.* Toronto: Stoddart, 1998.

Underhill, Frank. "Concerning Mr. King." *Canadian Forum* 30 (September 1950): 121–27.

Vigod, Bernard L. *Quebec before Duplessis: The Political Career of Louis-Alexandre Taschereau.* Montreal: McGill-Queen's University Press, 1986.

Von Baeyer, Edwina. *Garden of Dreams: Kingsmere and Mackenzie King.* Toronto: Dundurn Press, 1990.

Waite, Peter B. "Mackenzie King and the Italian Lady." *The Beaver* 75:6 (December 1995/January, 1996): 4–10.

———. "Mr. King and Lady Byng." *The Beaver* 77:2 (April/May 1997): 24–30.

Ward, Norman, ed. *The Memoirs of Chubby Power*. Toronto: Macmillan of Canada, 1966.

Ward, W. Peter. *White Canada Forever*. Montreal: McGill-Queen's University Press, 1978.

Wardhaugh, Robert A. *Mackenzie King and the Prairie West*. Toronto: University of Toronto Press, 2000.

Wetzler, Scott. *Living with the Passive Aggressive Man*. New York: Simon & Schuster, 1992.

Whitaker, Reginald. "Mackenzie King in the Dominion of the Dead." *Canadian Forum* 55 (February 1976), 6–11.

———. *The Government Party: Organizing and Financing the Liberal Party of Canada 1930–1958*. Toronto: University of Toronto Press, 1977.

Wigley, Philip G. *Canada and the Transition to Commonwealth: British-Canadian Relations, 1917–1926*. New York: Cambridge University Press, 1977.

Williams, Jeffrey. *Byng of Vimy*. London: Secker and Warburg, 1983.

Wilson, A.N. *The Victorians*. London: Hutchinson, 2002.

Copyright Permissions

Index

MIA LEVINE

ALLAN LEVINE is the author of ten books, including *The Devil in Babylon, Scattered Among the Peoples, Fugitives of the Forest, Scrum Wars* and *Coming of Age: A History of the Jewish People of Manitoba,* which won the McNally Robinson Book of the Year in 2010. His op-ed pieces have been published in the *Globe and Mail,* the *Winnipeg Free Press* and the *National Post*—among many other publications. He lives in Winnipeg with his wife, Angie, and their two children. www.allanlevinebooks.com